8th International Conference on Manufacturing Science and Education (MSE 2017)

Trends in New Industrial Revolution

MATEC Web of Conferences Volume 121 (2017)

Sibiu, Romania
7 - 9 June 2017

Part 1 of 2

Editors:

Ioan Bondrea
Carmen Simion
Marinela Inta

ISBN: 978-1-5108-4705-7

Printed from e-media with permission by:

Curran Associates, Inc.
57 Morehouse Lane
Red Hook, NY 12571

Some format issues inherent in the e-media version may also appear in this print version.

Printed by Curran Associates, Inc. (2017)

For additional information, please contact EDP Sciences – Web of Conferences
at the address below.

EDP Sciences – Web of Conferences
17, Avenue du Hoggar
Parc d'Activité de Courtabœuf
BP 112
F-91944 Les Ulis Cedex A
France

Phone: +33 (0) 1 69 18 75 75
Fax: +33 (0) 1 69 28 84 91

contact-edps@webofconferences.org

Additional copies of this publication are available from:

Curran Associates, Inc.
57 Morehouse Lane
Red Hook, NY 12571 USA
Phone: 845-758-0400
Fax: 845-758-2633
Email: curran@proceedings.com
Web: www.proceedings.com

TABLE OF CONTENTS

PART 1

PART 2

PREFACE

The 8th International Conference on Manufacturing Science and Education – **MSE 2017** "Trends in New Industrial Revolution" was organized by the Industrial Engineering and Management Department and Industrial Machines and Equipment Department from Faculty of Engineering – Lucian Blaga University of Sibiu, Romania during 7 – 9 June, 2017.

The conference website is located at http://conferences.ulbsibiu.ro/mse/2017/.

The goal of MSE 2017 Conference was to bring together specialists from universities, institutions and companies to provide an opportunity for the exchange of knowledge and ideas regarding trends in new industrial revolution, to promote modern and innovating manufacturing processes and systems as well as modern educational methods used in technical universities.

In the context of the European integration of the technical higher education, MSE Conference has become over the years a tradition of the Faculty of Engineering from Sibiu, Romania and it is our belief that the edition of this year has represented an emblematic scientific event.

Some raw numbers are as follows: 222 papers were received and 161 of them have been accepted to be presented at the conference. All the submitted papers were subjected to a rigorous peer review process, each paper being reviewed by minimum two members of the Reviewing Committee. From the accepted papers, 154 works have been presented at MSE 2017 Conference and people from 16 countries have been involved as authors and co-authors. Extended abstracts of all accepted papers were published in the Conference Book of Abstracts and all presented papers at MSE 2017 Conference are published online in this volume of MATEC Web of Conferences.

Finally, we would like to thank all people who have been supportive to MSE 2017 Conference: the expert refeers whose enormous work for assessing the papers is gratefully appreciated, the invited keynote speakers and those who serve as chairmen for their invaluable contribution, the authors themselves for contributing research papers without whose expert input there would have been no conference, the sponsor organizations for their financial sponsorship, the members of the International and Organizing Committee that have been extremely helpful in successfully organizing of the conference and last but not least, we would like to express our gratitude to MATEC team for its support in online publishing of MSE 2017 Proceedings.

We consider that the 8th International Conference on Manufacturing Science and Education – **MSE 2017** has been a stimulating and memorable conference that has generated numerous fruitful discussions and layed the foundations of new future collaborations.

Chairman of MSE 2017 Conference
Prof. Eng. Ioan BONDREA, Sc. D.,
Rector of Lucian Blaga University of Sibiu, Romania

DARABONT Doru Costin, Romania
DOBRESCU Tiberiu, Romania
DOICIN Cristian, Romania
DRĂGHICI Anca, Romania
DRĂGOI Mircea, Romania
DUMITRAŞ Cătălin Gabriel, Romania
DUMITRAŞCU Dănuţ, Romania
DUŞE Carmen Sonia, Romania
FETECĂU Cătălin, Romania
FRANZ Charles, USA
GAVRUS Adinel, France
GĂMAN George Artur, Romania
GONET Andrzej, Poland
HEINISCH Martin, Germany
HORODINCĂ Mihăiţă, Romania
HUANG Regina, China
HUSI Géza, Hungary
IONESCU Florin, Germany
KOLLÁTH Ľudovít, Slovakia
KOŠŤÁL Peter, Slovakia
KRÁLIKOVÁ Ružena, Slovakia
KROLCZYK Grzegorz, Poland
RADA Elena Cristina, Italy
RADOVANOVIĆ Miroslav, Serbia
RAGAZZI Marco, Italy
RATH Sethik, Cambodia
RYBICKI Czeslaw, Poland
ROTHMUND Johannes, Germany
ŠAJN Viktor, Slovenia
SAVU Tom, Romania
SCHUPP Peter, Germany
SIEMEK Jakub, Poland
SIMION Carmen Mihaela, Romania
SŁOMKA Tadeusz, Poland
STAN Simona, USA
ŞTEFĂNESCU Dan Paul, Romania
TOMOVIĆ Radoslav, Montenegro
TORRETTA Vincenzo, Italy
TROTMAN Dave, United Kingdom
ULIASZ-MISIAK Barbara, Poland
VELISEK Karol, Slovakia
WISNIOWSKI Rafae, Poland
ZELJKOVIĆ Milan, Serbia
ZUHLKE Detlef, Germany

ORGANIZING COMMITTEE

CHAIRPERSON

SIMION Carmen Mihaela

MEMBERS

BEJU Livia Dana
BIRIŞ Cristina Maria
BONDREA Ioan
CIOCA Lucian-Ionel
CIOCA Marius
CHICEA Anca
COFARU Nicolae
CRENGANIŞ Mihai
DEAC Cristian
GLIGOR Alina
GÎRJOB Claudia Emilia
GRECU Valentin
INŢĂ Marinela
LOBONŢ Lucian
MIRICESCU Dan
MORARU Gina
PETRUSE Radu
POPA Daniela
POPESCU Liliana
PURCAR Carmen
OLEKSIK Mihaela
ROŞCA Liviu
ROTARU Ionela
ROTARU Mihaela
TERA Melania
ZERBEŞ Mihai

REVIEWING COMMITTEE

INVITED KEYNOTE SPEAKERS

Dorel BANABIC
Romanian Academy

INDUSTRY 4.0-APPLICATIONS IN METAL FORMING

George DRĂGHICI
Politehnica University of Timişoara, Romania

PROJECT QUALITY MANAGEMENT IN AUTOMOTIVE INDUSTRY

Johannes ROTHMUND
Institute for Machine Tools at University of Stuttgart, Germany

CURRENT RESEARCH IN MACHINE TOOL ENGINEERING

Marco RAGAZZI
University of Trento, Italy

UNCONVENTIONAL MONITORING TO IDENTIFY THE ROLE OF SINGLE SOURCES OF AIR POLLUTION

Nicolae BÂLC
Technical University of Cluj-Napoca, Romania

DESIGN FOR ADDITIVE MANUFACTURING AND AM APPLICATIONS DEVELOPED AT TUC-N

PARALLEL SESSIONS

SESSION 1 – Design, Manufacturing and Management of Technological Equipment and Systems

SESSION 2 – Management, Modelling and Monitoring of Manufacturing Processes

SESSION 3 – Advanced Manufacturing Technologies

SESSION 4 – Virtual Engineering

SESSION 5 – Quality Engineering and Management

SESSION 6 – Transport Engineering and Road Vehicles, Traffic Management

SESSION 7 – Management and Economic Engineering

SESSION 8 – Mechatornics and Robotics

SESSION 9 – Engineering in Oil and Gas Field

SESSION 10 – Environmental Engineering

SESSION 11 – Safety and Health at Work

SESSION 12 – Academic Education and Management

WORKSHOPS

INDUSTRY – EDUCATION: INDUSTRY REPRESENTATIVES IN DIALOG WITH THE DEANS OF MECHANICAL AND INDUSTRIAL ENGINEERING SCHOOLS

Chairmen: Cristian Doicin, Liviu Roşca

SAFETY AND HEALTH AT WORK: PRESENT AND PERSPECTIVE

Chairmen: Lucian-Ionel Cioca, Roland-Iosif Moraru, Artur Găman, Doru Darabont

The 8th INTERNATIONAL
CONFERENCE
ON
MANUFACTURING
SCIENCE
AND
EDUCATION
June 7-9, 2017
Faculty of Engineering, Lucian Blaga University of Sibiu
http://conferences.ulbsibiu.ro/mse/2017/
swoboda hartmann
Continental
bertrandt
wengler
CBS
EXIT

ULBS

Content
1. Introduction
2. The proposed methodology
3. Case study - A practical application of t
intervention usability evaluation using
methodology
4. Conclusions and future work

fast growing
innovative global
ambitious
solution driven
reliable
responsible
KENDRION
WE MAGNETISE THE WORLD
solution driven
ambitious
flexible
innovative
global
responsible
reliable
fast growing
Achim MUNTEAN
Eugen PĂMINTAȘ

Carmen PURCAR

competence
high quality
SWITCHES
JOB SECURITY
SATISFACTION
SPECIALIST
manufacturing
competence future oriented

Statement of Peer review

In submitting conference proceedings to MATEC Web of Conferences, I certify to the Publisher that I adhere to the **Policy on Publishing Integrity** of the journal in order to safeguard good scientific practice in publishing.

1. All articles have been subjected to peer review administered by the proceedings editors.
2. Reviews have been conducted by expert referees, who have been requested to provide unbiased and constructive comments aimed, whenever possible, at improving the work.
3. Proceedings editors have taken all reasonable steps to ensure the quality of the materials they publish and their decision to accept or reject a paper for publication has been based only on the merits of the work and the relevance to the journal.

Title, date and place of the conference

8th International Conference on Manufacturing Science and Education – **MSE 2017** "Trends in New Industrial Revolution"
Sibiu, Romania, June 7 – 9, 2017

Proceedings editor(s):

Ioan BONDREA, Lucian Blaga University of Sibiu, Romania
Carmen SIMION, Lucian Blaga University of Sibiu, Romania
Marinela INŢĂ, Lucian Blaga University of Sibiu, Romania

Date and editor's signature

July 12, 2017

Fundamental and practical aspects concerning the characterization of smart textiles

Gianina Broasca[1], *Gabriela* Borcia[2], *Nicoleta* Dumitrascu[2], *Marius* Cioca[3], *Diana* Coman[3], *Nouredine* Ourfelli[4], and *Narcisa* Vrînceanu[3,*]

[1] "Gheorghe Asachi" Technical University of Iaşi, 53 Mangeron Street, Romania
[2] "Alexandru Ioan Cuza" University of Iaşi, 21 Carol I Street, Romania
[3] Lucian Blaga University of Sibiu, 4 Emil Cioran St, Romania
[4] Université de Tunis El Manar, Laboratoire de Biophysique et Technologies Médicales; Abdulrahman Alfaisal University, Department of Chemistry, College of Science, Saudi Arabia

Abstract. The research aims towards a novel methodology of ZnO impregnation onto polyamide supports, in order to attain a special inorganic-organic hybrid polymer material, with increased UV-barrier attributes and high hydrophobicity. The experimental approach employed ZnO micro particles powder. Solutions with different concentrations of ZnO powder dispersed in methanol are prepared as anti-UV finishing agent and applied onto polyamide fabric. The responsive behaviour of this complex polymer network to UV irradiation, the photo protective performance and its time stability, also the hydrophobic character are assessed by different characterization techniques. The proposed method has advantages, like: easy standardization, at lower production cost, resulting a high homogeneity and dispersion of ZnO micro particles into the textile polymeric network. The covered polymeric supports show better photo protective response, higher hydrophobicity and control on the surface charge induced by irradiation. This method granted enhanced quality, comfort and long term stability of the intelligent/smart garments.

1 Introduction

The textile area intents to design fabrics accommodated to strict surroundings conditions, as to promote the human physiological comfort, by facile and reduced cost finishing procedures. Commonly, the fabrics sector needs relevant procedures from the processing of raw materials to the distinct finished items [1-3].

The attributes of garments are functions of the processing technique. In this regard, the nanotechnologies polarized much significance in the textiles area, granting to realize multifunctional materials with attractive properties, controlled in the 0.1 - 100 nm range. Recently, the embodiment of metal particles (e.g., zinc, silver, gold, platinum, etc.) was analysed, in order to attain novel textiles [4]. For instance, the silver nanoparticles have been embodied within different polymeric fibers belonging to various fabrics, primarily to provide added value features, like: stain-repellence, bactericidal properties, or blocking of

* Corresponding author: vrinceanu.narcisai@ulbsibiu.ro

smelly and odorous vapours [4]. Nevertheless, these metals occur in the form of nanoparticles, meaning that these can be destructive to health if absorbed by the human body [4]. Such issues need extended research efforts for back-up covering/incorporating polymeric supports.

The barrier against the UV radiation is a practical demand for the safety of garments designed for particular applications, as extended sun exposure during outdoor activities. Due to their unique properties, like: electrical, photocatalytic, optical, dermatological and antibacterial, zinc oxide (ZnO) powders are widely used semiconductor materials, having numerous structures. Furthermore, for medical applications ZnO is bio-safe and biocompatible. ZnO is experienced as a UV-blocking system, mainly in the UV-A region. In order to apply ZnO powder as UV-absorbing system, it has to be assimilated into different matrices, by means of diverse approaches developed to improve the stability of ZnO in different dispersions media and to block particles clustering.

Considering all these, a novel methodology based on ZnO impregnation of polyamide supports in order to engineer an inorganic-organic hybrid polymer network, with augmented UV-barrier attributes and high hydrophobicity was nominated. The experimental component consists of employing ZnO microparticles powder. As anti-UV finishing agent different concentrations of solutions with of ZnO powder dispersed in methanol were prepared and applied onto polyamide fabric. Different characterization methods were used to evaluate the responsive behaviour of this complex polymer matrix to UV irradiation, such as: the photoprotective work, long term stability, the hydrophobic character. As advantages of this solution we can mention:

- Easy standardization,
- Lower production cost,
- High homogeneity and dispersion of ZnO microparticles into the textile polymeric network.

2 Experimental

2.1 Materials and preparation

Woven polyamide fabrics (83.5 g/m^2) were identified and prepared for immersion in the form of 3 cm - 7 cm strips. The micrometer-sized ZnO powder was supplied by Ensait Laboratory, France. Laser diffraction was employed to perform the particle-size distribution, showing that 75% of the ZnO powder consists of 1.7 μm size particles. The solutions are based of ZnO powder dispersed in methanol (99.8%), at different concentrations: 1%, 3%, 5% and 7%. The magnetic stirring was used for 15 minutes, in order to obtain the solutions, adding 4 drops of Vitexol (BASF) to avert foaming and 65 g/l Apretan (Clariant) to provide the fixation of ZnO to fibers.

The studied samples were immersed into ZnO solutions by padding, on a Wemer Mathis AG Laboratory machine, followed by wet-picking, pressing the textile between hydroextraction cylinders, in order to eliminate a part of liquid/paste from the polymeric surface. The samples are weighted before and after impregnation. The immersed specimens were exposed in a Vetter machine for drying, for 3 minutes, at 110 °C. The mixture is then fastened on the fabric, heated at 150 °C for 3 minutes, by thermal induced crosslinking [5].

2.2 Characterization methods

Scanning electron microscope (SEM) images of the samples are obtained with a Quanta 200 3D Dual Beam type microscope (FEI Holland), coupled to an energy-dispersive X-ray spectroscopy (EDS) analysis system (EDAX - AMETEK Holland) equipped with a SDD (silicon drift detector) type detector. Taking into account the sample type, the analyses are performed using Low Vacuum working mode, allowing the probes tested in their initial state, without a previous metallization. Both for the acquisition of secondary electrons images and EDS elemental chemical analysis, LFD (Large Field Detector) detector is used, running at 60 Pa pressure and 30 kV voltage.

The wicking of the polymeric support is assessed by contact angle measurement. The sessile drop technique was used in order to obtain the contact angles, under controlled conditions of room temperature and humidity. An automated system stored the drop images, via a Canon A85 camera, with PC-based control, acquisition and data processing.

Ten measures on the imaged sessile liquid drop profile, for 1 μl drop size were performed. By averaging these values, the values of the static contact angle were obtained. As test liquid distilled water was employed. The adhesion activity of water on the surface, meaning wicking, is calculated as follows:

$$W_a = \gamma_{lv}(1+\cos\theta) \qquad (1)$$

Where W_a is the contact angle and γ_{lv} is the surface tension.

3 Results and discussion

Figure 1 displays SEM microphotographs of polyamide supports covered with ZnO. The surface belonging to the reference polyamide fabric is smooth (Figure 1a), exposing only tiny detached particles, like impurities.

Different concentrations of ZnO solutions lead to different layer covering the polyamide samples, having different morphology based on the increasing ZnO concentration of the finishing agent.

For the sample treated with 1% ZnO, a thin film, with lamellar morphology and good uniformity, was obtained (Figure 1b). The thickness of this layer increments with an increased ZnO concentration, displaying a uniform coating of the supports for 3-5% ZnO (Figure 1c).

In case of higher concentration in zinc oxide, a lower dispersion of the microparticles was noticed, along with the visible clusters leading to non-uniform covering of the supports (Figure 1d). Consequently, an uniform covering of fibers is achieved due to the best dispersion of the particles, correlated to ZnO concentration lower than 5%.

The covering of the supports is sustained by EDS measurement, where the Zn quantity on the specimens displays correlation to the ZnO concentration of the solutions. It was demonstrated that the content is the highest for the 7% ZnO sample.

The wicking of the polymeric support surface evaluates the material behaviour in presence of water or humid environment. The values are gathered in Table 1. Moreover, rerun measurement demonstrates that the contact angle is not altered by macroscopic non-uniformity of the woven materials, since all liquid drops deposited onto the surface show regular, symmetric profile.

Fig. 1. SEM photos of polyamide supports covered with ZnO: a) reference, b) 1% ZnO, c) 5% ZnO, d) 7% ZnO.

Generally speaking, the polyamide fabric is hydrophobic, having a water contact angle higher than 90°. In this case the surface is not wet. In addition, the polymeric support does not absorb the water, because the drops deposited onto the surface vanish after long intervals, by evaporation.

The padding with ZnO pointed out, in a visible manner, a higher hydrophobicity, as the augmentation of contact angle. This aspect can be regarded as a diminution of the adhesion activity, with an average 50%. This fact certifies that the polymeric supports are covered with an oxide layer, which is more hydrophobic than the polymer itself.

Another interesting fact is the lowest wicking noticed for 3% ZnO impregnated polymeric support, while incremented ZnO concentration conducts to higher wicking, i.e. reversal of the surface attributes. Between the 3% and 7% ZnO samples a difference of 68% has been calculated, which is relevant.

This response can be regarded as a function of a better dispersion of ZnO particles in the 3% specimen, which is more diluted, granting more homogenous covering of textile supports, and more hydrophobic attribute due to oxide film. In case of samples coated with higher oxide concentration, the cluster of ZnO particles leads to a low levelness, consequently the fibers are irregularly covered with films of different thickness.

Table 1. Contact angle and adhesion responsive attitude of water on polyamide supports covered with ZnO, before and after UV irradiation.

Samples	**before UV irradiation**		**after UV irradiation**	
	θ (°)	W_a **(mJ/m²)**	θ (°)	W_a **(mJ/m²)**
polyamide	92°	67	88°	75
polyamide + 1% ZnO	120°	32	115°	30
polyamide + 3% ZnO	131°	20	135°	27
polyamide + 5% ZnO	129°	26	124°	32
polyamide + 7% ZnO	113°	33	116°	48

The UV irradiation of the specimens leads, as expected, to the decreasing of the contact angle. This decreasing is noticeable for the polyamide fabric, in terms of the surface chemistry modification of polymers under exposure to energy sources. The modification is important less for the supports covered with ZnO. In this manner the UV-barrier function of the oxide layer was stressed. It is noteworthy to mention that all the studied ZnO covered probes show a relevant hydrophobicity after UV irradiation. The 7% ZnO sample showed the highest level of modification. The synergic effect provided by the better dispersion of the microparticles (3-5% ZnO solutions) and the good levelness of the surface layer, showed an improved UV-protection.

4 Conclusion

A novel methodology is oriented toward the impregnation of ZnO powder in a polymeric matrix, applicable, in order to obtain textiles with a multifunctional performance: UV barrier, high hydrophobicity.

The method provides a homogenous dispersion of ZnO particles through the polymeric support. New attributes like higher photoprotection, higher hydrophobicity and assessment on the irradiated surface.

This experimental approach maintains the mechanical features of the material unmodified, providing an improved quality, comfort and long term stability of the clothing.

References

1. Y.W.H. Wong, C.W.M. Yuen, M.Y.S. Leung, S.K.A. Ku, H.L.I. Lam, Autex Research Journal, **6** (2006)
2. K.L. Hatch, U. Osterwalder, Dermatologic Clinics, **24** (2006)
3. J.M. Menter, K.L. Hatch, in: P. Elsner, K. Hatch, W. Wigger (Eds.), Textiles and the Skin, Current Problems in Dermatology, 31, Karger, Basel (2003)
4. C.H. Xue, W. Yin, S.T. Jia, J.Z. Ma, Nanotechnology, **22** (2011)
5. G. Broasca, G. Borcia, G. Dumitrascu, N. Vrinceanu, Applied Suface Science, **279** (2013)

Study regarding the physical-mechanical properties of knits for garments – pilling performance

Alina Mihaela Coldea[1,*], and *Dorin* Vlad[1]

[1]Lucian Blaga University of Sibiu, Department of Industrial Machinery and Equipment, 10 Victoriei Bd., 550024 Sibiu, Romania

Abstract. The effect of the pilling degradation results is an important decreasing in garment quality and a negative influence on the user's comfort to wear. There are a lot factors, such as the yarn spinning system, fabric geometry and finishing operation, which affect the pilling performance. This research paper presents the investigation of fabric surface concerning pilling formation, especially knits, influenced by the fibrous composition. As samples were used knitted fabric panels, obtained from carded yarns, with fibers composition like cotton, wool or mixed one. Regarding fabric geometry 3 types were used – single jersey, 2x2 rib and moss stitch. In order to establish which of the yarns are most appropriate for knitting garments from the point of view of pilling resistance were realized tests using the standard SR EN ISO 12945-2.The tests were performed with the Nu-Martindale – Abrasion and Pilling Tester, Model 864.For assessment were used two methods – the first one is a comparison with visual standards, tested fabrics or photographs of fabrics, to find the degree of pilling on a scale from 5 (no pilling) to 1 (very severe pilling) and the second one by determination of mass loss percent.

1 Introduction

The process of pilling appearance consists of three steps [1-2]: because of mechanical impact to the surface of garment, several fibers tips are pulled out creating a peeled surface (figure 1). Then the broken fibers come to separate tips and forms, gradually increasing pills.

The fibers that retain these balls are still subjected to mechanical stresses, due to other factors (wear, washing, etc.), resulting in the friction wear effect that leads to their fall. [3-4].

The resistance to pilling of the fabric depends on the fabric density, i.e. when the length of loop structure decrease and the surface density increases, the pilling resistance grows [5-7]. The correct form of yarn, fabric and fabric formation can improve the quality of knit fabric [8-9]. Washing intensity, cleaning process [10] that varies depending on chemical softeners used during washing [11-12], however, their influence is not fully investigated.

* Corresponding author: dorin.vlad@ulbsibiu.ro

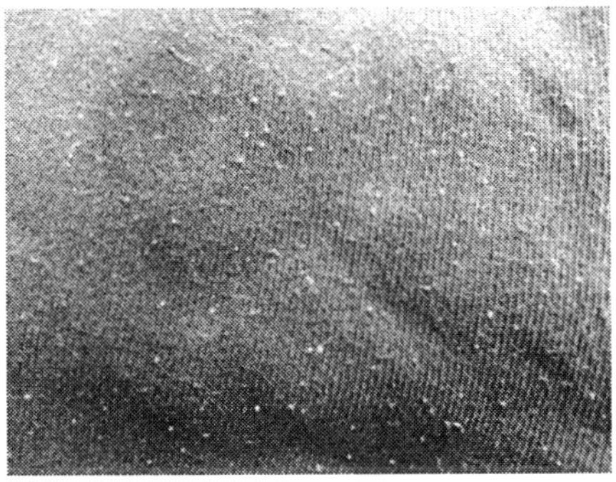

Fig. 1. Pilling effect appearance.

2 Background

There are many factors, like the yarn spinning mode, fabric geometry and finishing operations, which affect pilling performance [9, 15]. Özdil et all compared knitted fabrics from compact spun yarns with classic ring spun yarns and find that knitted fabric from compact yarns have better pilling performance [16]. Candan & Önal test the pilling effect of weft knitted fabric made of OE and RS yarns. They say that 100% cotton samples knitted from RS yarns goes to have lower pilling rates than those made from 100% cotton OE spun yarns [17]. Beltran et al. [18] compared piling effect between fabrics made from conventional worsted spun yarns, solospun ring spun yarn and jetwind modified yarns, and report that compared to conventional ring spun yarn fabric, the pilling performance of solospun yarn fabric and jetwind modified yarn fabric showed a half grade and full grade improvement.

As well as an expensive textile fabric, cashmere fiber exhibits a tiny diameter, short length and very smooth surface. The degree of pilling effect of cashmere knitted fabric has required the attention of consumers, manufacturers and researchers. Li et al. [19] found that the relationship between the pilling degree of cashmere knitted fabric and yarn properties was obtained using optimal scaling regression analysis. For the same spinning method, the dye color of cashmere fiber is of primary importance by influencing fabric pilling rates, due by the interaction of the actual yarn twisting and its CV and the yarn tensile strength.

Moreover, Liu and Wang [20] compared the cashmere yarn structure of ring, mule and rotor spinning by means of SEM observations. They showed that the structure of cashmere mule yarn was compacted and the hairiness of cashmere mule yarn was low than that of cashmere ring yarn and rotor yarn.

3 Method and apparatus

In order to establish which of the yarns are most appropriate for knitting garments regarding the pilling performance were realized tests according to the standard EN ISO 12945-2:2000 Textiles – Determination of Fabric Propensity to Surface Fuzzing and to Pilling – Part 2: Modified Martindale method [13, 23].

The tests were performed with the Nu-Martindale – abrasion and pilling tester, model 864. Were prepared 150 mm diameter specimens of the cloth to be tested. A circular sample, mounted in a specimen holder and subjected to a specific load, is rubbed in the face of an abrasive standard fabric in a translational movement tracing a Lissajous figure, also the specimen holder is additionally freely rotatable around its own axis, perpendicular on the plane of the specimen. [24].

One type of the evaluation for the pilling appearance of the knitted fabric is determined from the inspection interval to degradation of the sample. In according to the standard mentioned above, the samples were inspected every 2000 cycles with a magnifying glass and compared with standard photographic samples to establish the degree of degradation [2]. The other type of evaluation is the calculation of the weight loss percent of the fabrics, after 4 000 rubs.

4 Samples

Like samples were used knitted fabric panels, catted from garments, obtained on a flat knitting machine Cotton with different gauge (table 1), using carded yarns from fibers like cotton, wool or mixed one e.g. Wool 40% + viscose 30% + polyamide 20% + Cashmere 10%, wool 70% + nylon 30%. Whole samples were conditioned [14].

Knits geometry are shown in the figure 2 and their characteristic are described in the table 1.

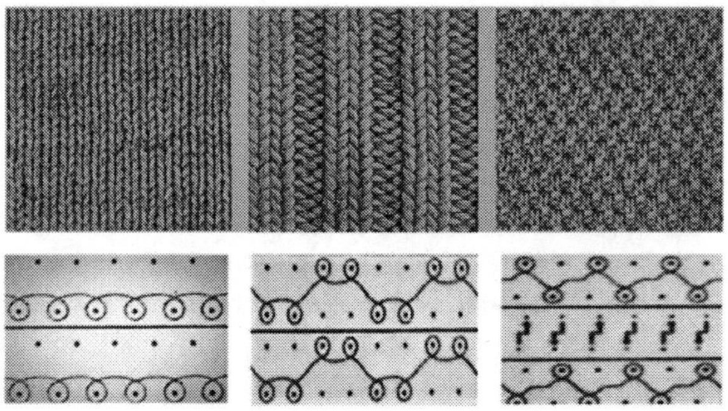

a. b. c.

Fig. 2. Knits geometry: a – single jersey, b – 2x2 rib, c – moss stitch.

Table 1. Knitted fabric characteristic.

Code	Fabric geometry	Raw material	Yarn fineness [Nm]	No. of ply yarn	Fabric Density		Machine gauge
					Wales /50mm	Rows /50mm	
S1	Single jersey	Cotton	70/2	2	43	50	12 gg
S2	Single jersey	Cotton	50/2	2	33	45	21gg
S3	Single jersey	Cotton	50/2	3	28	35	15 gg
S4	Single jersey	Wool	30/2	1	38	45	12 gg
S5	Single jersey	Wool 70% + Nylon 30%	15/2	1	25	35	7 gg
S6	Rib 4x2	Wool 40%+ Viscose 30%+ Polyamide 20%+ Cashmere 10%	15/1	2	38	45	12 gg
S7	Moss Stitch	Cotton	70/2	2	33	30	12 gg

Cotton knitted fabric is breathable and transmits moisture, liquid and vapor, far away from the body and is absorbent and removes liquid from the skin, like a towel, according to Cotton Incorporated [26].

Cotton as a natural cellulose fiber, has a lot of characteristics such as: comfortable, soft hand, good absorbency, color retention, prints well, machine-washable, dry-cleanable, good strength, drapes well, easy to handle and sew [21].

Wool fiber is composed of protein substance called as keratin. The wool fibers have crimps or curls, which create air pockets and give the wool a spongy feel and create insulation for the wearer.

The characteristics of wool fiber are as follows: excellent absorbency, moisture regain is good, tend to be warmer than others, have poor resistance to alkalis but good resistance to acids, have good elasticity and resiliency [22].

5 Assessment

An objective evaluation of the pilling tendency of woven cloth and knitted fabric is a broad requirement of the textile industry [25].

To evaluate the pilling tendency in accordance with SN EN ISO 12945-2 a photo standard can be used, see figure 3. Each photo standard consists of a series of 4 comparable photographs. Each comparable photograph shows the boundary between 2 marks (mark 4-5, 3-4, 2-3, 1-2).

The surface condition of each test sample is judged against a photo standard, it can then be decided if the test sample is better, equal, or worse than the photo standard, resulting in a pilling mark [25].

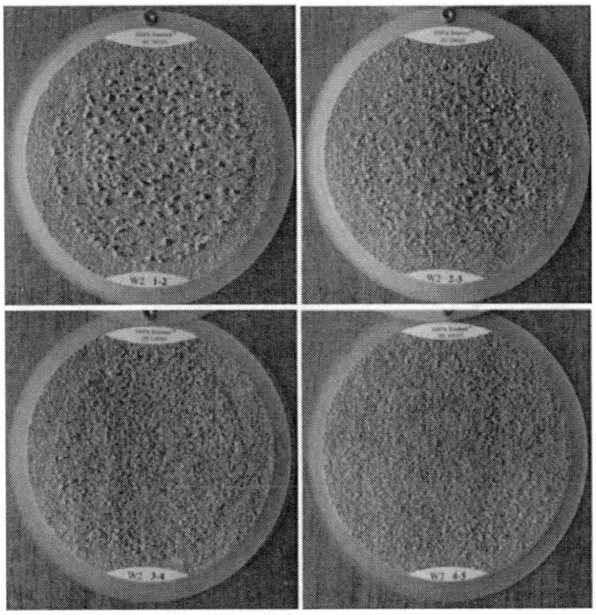

Fig. 3. Marking system [25]. Mark 5 = no or very little pilling, Mark 4 = light pilling formation, Mark 3 = moderate pilling formation, Mark 2 = distinct pilling formation, Mark 1 = strong pilling formation, (Half marks are allowed).

DOI: 10.1051/matecconf/201712101002

6 Results

In the table 2 are presented the samples appearance after 2000 and 4000 rubs and the table 3 show the values of mass loss percent obtained for all knitted fabric geometry and raw material from this study.

Table 2. Sample appearance after 2000 and 4000 rubs.

Code	Sample view after 2000 rubs	Sample view after 4000 rubs	EMPA Degree after 4000 rubs
S1			4
S2			3 - 4
S3			4
S4			2
S5			1-2

S6			2
S7			3-4

Table 3. The percent of mass loss.

Code	Knitted fabric geometry	Raw material	Initial Weight [g]	Final Weight [g]	Loose Weight [%]
S1	Single jersey	Cotton	3.928	3.911	0.43
S2	Single jersey	Cotton	4.583	4.555	0.61
S3	Single jersey	Cotton	5.921	5.910	0.19
S4	Single jersey	Wool	4.236	4.172	1.51
S5	Single jersey	Wool 70% + Nylon 30%	6.505	6.452	0.81
S6	Rib 4x2	Wool 40%+ Viscose 30%+ Polyamide 20%+ Cashmere 10%	9.501	9.450	0.54
S7	Moss Stitch	Cotton	3.417	3.402	0.44

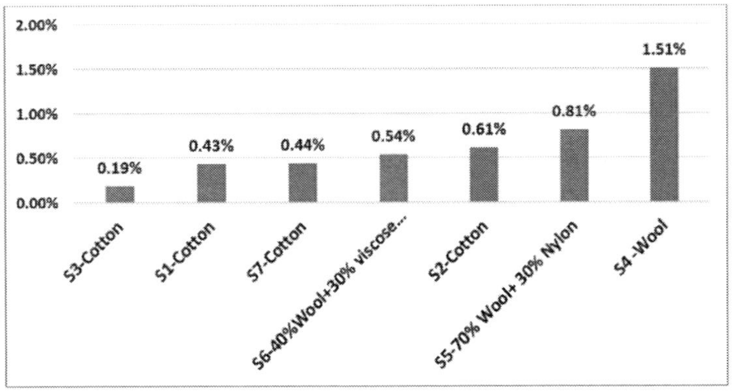

Fig. 4. Histogram of weight loss, according to the raw material and fabric geometry.

7 Discussions and conclusions

In this study, the pilling tendency on a textile's surface for three type of weft knitted fabric geometry - single jersey, 2x2 rib and moss stitch - produced from ring yarns and raw material like cotton, wool and mixed yarns were evaluated. According to the raw material, percent of mass loss is higher for knitted samples from wool yarn than cotton samples.

If we increase the yarn fineness (S5 and S6), the mass loss will increase, see figure 4 and table 3.

For the same raw material (cotton 70/2 Nm), but different geometry (S7 and S1), percent of mass loss is higher in the case of S7.

For the same raw and pattern (S2 and S3) if we add one more ply yarn (e.g. S3) mass loss percent decrease by 68,85%, see figure 4 and table 3.

Concerning the influence of fabric geometry on the pilling tendency:

- If we analyze the appearance of the sample surface after 4000 rubs can observe that, single jersey has a good resistance than rib and moss stitch structure, table 3 show.
- It can be said that the knit geometry and raw material have an effect on the pilling performance of knit fabrics. To be able to produce knitted fabrics for garments whose abrasion resistance and pilling performance are high, compact yarns should be used.
- Try to avoid fabrics that are made from fiber blends. Knitted or woven fabrics that combine different types of threads - especially those that combine natural and synthetic fibers - are more likely to pilling degradation.
- We can choose woven fabrics than knits. Woven fabrics goes to pilling, less than knits. Of course we love our knits, so choose one that is tightly knit over a looser knit.
- For clothes that you suspect will pill, use the washer's gentle cycle. The slower agitation and shorter wash cycle will protect your clothes. Or, choose hand washing which is even more gentle.

References

1. http://internet.ktu.lt/lt/mokslas/zurnalai/medz/pdf/medz0-106/15%20Textile...(pp.297-301).pdf
2. I. Gykytė, E. Strazdienė, R. Titas, V. Urbelis, Pilling of Knitted Materials Science, Medžiagotyra **8**, 316 (2002)
3. https://www.researchgate.net/publication/267772861_The_Influence_of_Structure_Parameters_of_Weft_Knitted_Fabrics_on_Propensity_to_Pilling
4. C. Candan, L. Önal, Textile Research Journal, **72**, 164 (2002)
5. https://www.researchgate.net/publication/228544267_Structure_Properties_of_Knits_from_Natural_Yarns_and_their_Combination_with_Elastane_and_Polyamide_Threads
6. N. Emirhanova, Y. Kavusturan, Fibers & Textiles in Eastern Europe, **16**, 69 (2008)
7. N. Uçar, S. Ertuğrul, Prediction of Fuzz Fibers on Fabric by Using Neural Network and Regression Analysis Fibers & Textiles in Eastern Europe, **2**, 58 (2007)
8. F. Ceken, Pilling of Flat Knitted Fabrics Knitting Technology, **2**, 16 (2000)
9. M. Akaydin, Y. Can, Fibres & Textiles in Eastern Europe, **18**, 51 (2010)
10. http://www.fibtex.lodz.pl/article166.html
11. N. Çelik, Z. Değirmenci, H.K. Kaynak, Tekstil ve Konfeksiyon, **1**, 41 (2010)
12. http://internet.ktu.lt/lt/mokslas/zurnalai/medz/pdf/medz0-106/15%20Textile...(pp.297-301).pdf
13. ISO 12945-2:2000 Textiles – Determination of Fabric Propensity to Surface Fuzzing and to Pilling – Part 2: Modified Martindale method
14. ISO 139:2005 Textiles – Standard Atmospheres for Conditioning and Testing.
15. Long Li, Meijing Zhu, Xiaojuan Wei, Fibres & Textiles in Eastern Europe, **22**, **1**, 74 (2014)
16. http://www.fibtex.lodz.pl/50_11_39.pdf.
17. C. Candan, L. Önal. Textile Research Journal, **72**, 164 (2002)

18. https://researchbank.rmit.edu.au/view/rmit:5858
19. L. Li, G. Jia, W. Zhou, Fibers & Textiles in Eastern Europe, **17**, 76 (2009)
20. J. Liu, S. Wang, Shanghai Textile Science & Technology, **26**, 19 (1998)
21. http://www.swicofil.com/products/001cotton.html, Access date: March 2017
22. http://textilelearner.blogspot.ro/2011/08/wool-fiber-properties-of-wool-fiber_5920.html, Access date: March 2017
23. A.M. Coldea, D. Vlad, C. Budulan, Annals of the Oradea University. Fascicle of Textile-Leatherwork, June 1st-2nd, **XIII**, 63 (2012)
24. Technical book of Nu-Martindale – abrasion and pilling tester, model 864
25. http://www.testfabrics.com/download.php?id=uploadfiles/TmpJPQ==.pdf&filename=Empa
26. https://www.thespruce.com/how-to-prevent-clothes-from-pilling-2146667. Access date: March 2017

MATEC Web of Conferences **121**, 01003 (2017) DOI: 10.1051/matecconf/201712101003
MSE 2017

Superficial hardened layer of cut surface by turning

Sorin Mihai Croitoru[1,*], *Ionelia* Voiculescu[2] , and *Roxana Doina* Trușcă[3]

[1]Politehnica University of Bucharest, Faculty of Engineering and Management of Technological Systems, Machines and Production Systems Dept., 313 Spl. Independentei, 6 Sector , 060042 Bucharest, Romania
[2] Politehnica University of Bucharest, Faculty of Engineering and Management of Technological Systems, Materials Technology and Welding Dept., 313 Spl. Independentei, 6 Sector , 060042 Bucharest, Romania
[3] Politehnica University of Bucharest, Faculty of Engineering in Foreign Languages, 313 Spl. Independentei, 6 Sector, 060042 Bucharest, Romania

Abstract. One of research methods in metal cutting process is to measure hardness in the contact zone between cutting tool and workpiece. The objective of the performed research was to determine thickness and hardness of the superficial layer of cut surface due to cutting process, both orthogonal and complex cutting. The most important finding was that thickness of the superficial hardened layer is very thin under considered conditions, less than 0.01 … 0.02 mm. This research should be continued.

1 Introduction

This paper is part of a study related to the hardness of the cut surface in case of orthogonal cutting [1] and complex cutting [2] of C45 steel. It presents a review of the mentioned papers and some new SEM micrographs. These lead to some new conclusions.

There are several classic methods to research the chip formation process in the workpiece material [3, 4]: method of deformed net, method of metallographic examination, method of measuring hardness in different points of the contact zone between workpiece, chip and cutting tool. Until present (see references [5-6]), the last method did not detail characteristics of the deformed and hardened superficial layer of the cut surface.

Research on chip formation was performed since a half century ago [7], but new technologies related to computers development, both hardware and software, allow today greater possibilities to perform this research [5-6, 8].

Present paper is part of the studies on depth and hardness of the cut surface performed in Politehnica University of Bucharest. Cutting experiments were performed in the Cutting Process and Cutting Tools Laboratory, Department of Machines and Manufacturing Systems. Hardness measurements and optical microscopy were performed in the Metallographic Tests Laboratory, Department of Materials Technology and Welding. SEM microscopy was performed in the National Centre for Scientific Research on Food Safety.

* Corresponding author: sorin.croitoru@gmail.com

DOI: 10.1051/matecconf/201712101003

2 Orthogonal cutting by turning

Orthogonal cutting is a theoretical case of cutting process. It is obtained in special cases of plane and radial turning, when width of the workpiece is less than of the cutter.

Design of experiments is presented in the following.

Cutter was made of HSS for easiness of sharpening different geometrical parameters of the cutting edge. Time of effective cutting process is small, so the wear was not important.

Clearance angle was fixed at α=10°. Rake angle and tool edge radius were considered to have the following values: γ= –10°, 0°, 10°, respectively ρ=0.1 mm, 0.2 mm, 0.4 mm.

Workpiece was made of C45 steel so maximum cutting speed was v_c=60 m/min.

In case of orthogonal turning cutting depth is the radial feed. Its values were chosen a_p=f_r= 0.1 mm; 0.2 mm and 0.4 mm.

Experimental plan is not presented in the paper. However, there were performed 40 tests, both theoretical and experimental, for different combinations of the mentioned cutting process parameters. For these experiments we present only samples which obtained higher hardness and, as well, qualitative aspects and trends observed in all tests.

A Romanian lathe, SN400, was used for the orthogonal cutting experiments because on this lathe the dynamometer's table could be assembled.

2.1 Theoretical research and preliminary presumptions

DEFORM™ (2D and 3D) [9] is a software package used to simulate plastic deformation processes of metallic parts using Finite Element Method for plastic or elastic-plastic deformation processes, in motion. Tool and workpiece are defined as meshes inside a closed contour (2D) or a closed surface (3D). Active part of the tool is simulated incrementally in its relative movement to the workpiece and, in case the deformation of finite elements mesh of the workpiece lead to errors, a new mesh is defined automatically and the data processing is continued.

In case of orthogonal cutting process DEFORM 2D™ software was used and in case of complex cutting process (see next paragraph 3.1) DEFORM 3D™ software was used.

Presumptions for the simulations were the following:

- Cutting tool edge is considered to be perfectly rigid, made of metal carbide (24% Cobalt).
- Cutting movement is translation, 20 mm stroke.
- Mesh generation and ambient conditions were chosen automatically by the software;
- Cutting speed, cutting depth, rake angle and clearance angle were set as in the experimental plan.
- Only the interaction zone between workpiece and cutter's edge was modeled.

The main output data were the following:

- Deformed shape of the workpiece, (see Fig. 1,*a* – test No. 28, step 182).

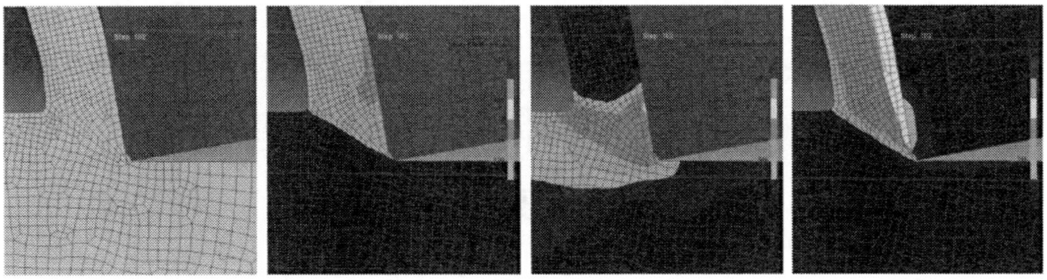

a) b) c) d)

Fig. 1. Output simulation data for test No. 40 (v_c=60 m/min; a_p=0.4 mm, α=10°, γ=-10°, ρ=0.4 mm). a) Deformed shape; b) Strain Effective; c) Stress Effective; d) Temperature.

- Strain – effective (see Fig. 1, b– test No. 28, step 182).
- Stress – effective (see Fig. 1, c– test No. 28, step 182).
- Temperature distribution during the cutting process (see Fig. 1, d– test No. 28, step 182).

Because of the presumption that hardening of the superficial layer of cut surface is caused by the crushing of the workpiece material, the most important output data of the simulations were considered the effective stress (possibly above the yield strength of the material) and vertical component of the cutting force.

After simulations the following qualitative conclusions were noticed:

- Stress distribution varies, along the tool point trajectory.
- The most important input variables which influence hardening of the cut surface in orthogonal cutting are cutting depth and tool edge radius. Stresses over yield strength of the workpiece's material seem to affect (crush) this superficial layer on a depth of (1.5...2)a_p. Stresses of almost the same size are revealed in the surface layer even in case the tool edge radius ρ is null.

2.2 Experimental research and results

Experimental stand for orthogonal cutting research by turning is presented in Fig. 2.

Fig. 2. Experimental stand - orthogonal cutting.

In the universal chuck *1* of the lathe an axle with a disc *2* is fixed on which the workpiece *3* is fixed. This fixture of the workpiece was chosen in order to obtain greater cutting speed at a lower number of spindle's rpm, for a better control of the experiments. On the transversal slide *8* of the lathe the dynamometric table *5* of a KISTLER dynamometer is fixed on cutter's slide fixture *7*.

The cutter 4 is fixed onto the dynamometric table by means of a clamping device. Measuring data is sent to the dynamometer's amplifier by means of a data cable 6.

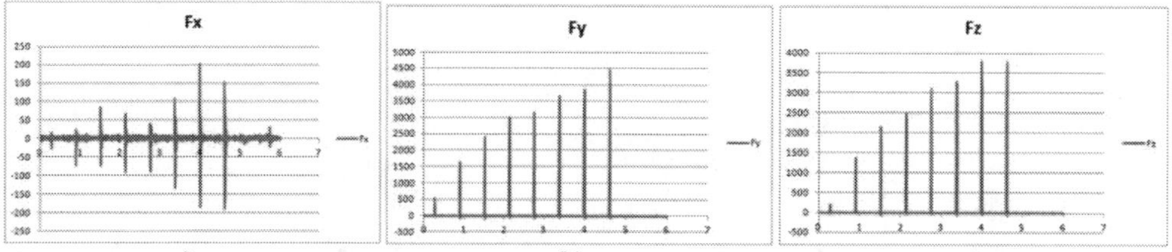

a) b) c)

Fig. 3. Measured cutting forces in test No. 28.

Cutting forces were measured using the experimental stand presented in figure 2 (KISTLER dynamometer), see figure 3. Cutting forces were as following: F_x – axial force, F_y – radial force and F_z – tangential force. F_x variation – which should have been null –

showed the lathe used for cutting experiments was not stiff enough. This lathe should be changed for the next experiments.

Micrographs, both optical and SEM (see Figs. 5 and 6) showed superficial deformation (crushing) on cut surface, which was expected.

Microhardness HV 0.1 was measured. After all tests were done, the samples were polished on lateral surface and microhardness HV 0.1 was measured in maximum 6 points from the cut/manufactured surface, at distances of 0.05 mm, up to 0.3 mm (see Fig. 4).

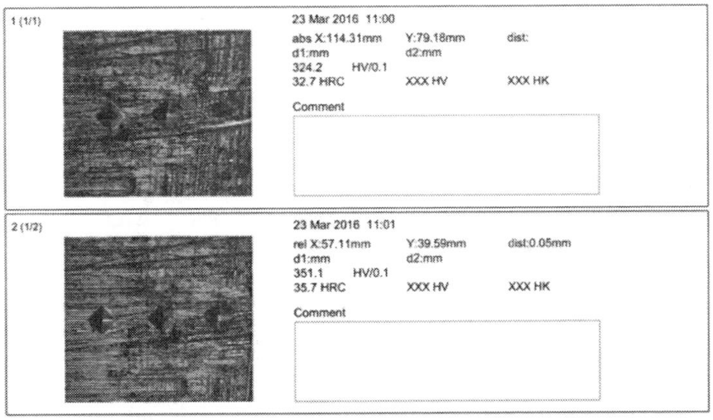

Fig. 4. HV 0.1 measurement in points 1 and 2 for Test No. 28 sample.

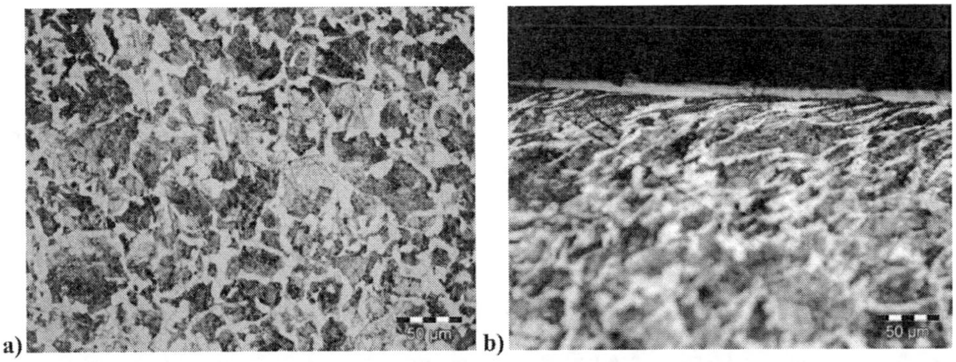

Fig. 5. Optical microscopy – Test No. 28. *a)* base material, *b)* deformed superficial layer.

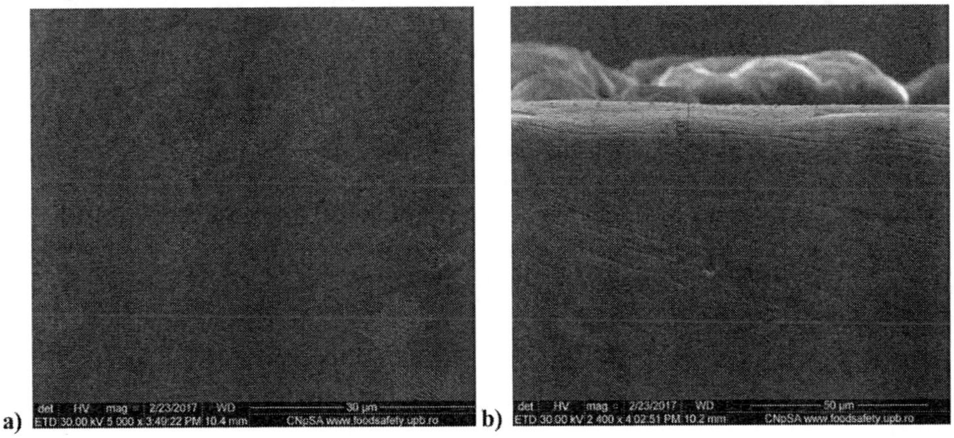

Fig. 6. SEM microscopy – Test No. 28. a) base material, b) deformed superficial layer.

In all tests minimum hardness was HV 0.1 = 253.34, maximum hardness was HV 0.1 = 338.28, average hardness was HV 0.1 = 300.66. This could mean that C45 steel is not homogenous. However, all hardness measurements had the diminishing trend of HV 0.1 hardness as the distance from the cut surface increases.

The only problem was the crushed and hardened layer was very thin, thinner than expected (0.01…0.02 μm). This implies the measurement of superficial hardness is not exactly credible, because the pushing prism of the hardness measuring machine punches this layer, so the measured hardness may vary significantly.

Furthermore, C45 steel was composed of ferrite and pearlite, phases which have different hardness [10]: 60…100 HV for ferrite and 240…400 HV for pearlite.

3 Complex cutting by longitudinal turning

Complex cutting is the real general case of cutting process. The most common cutting operation is longitudinal turning, which was chosen for the study.

Design of experiments is presented in the following.

Cutter was made of HSS for easiness of sharpening different geometrical parameters of the cutting edge. Time of effective cutting process is small, so the wear was not important. Effectively, after each test the tool was resharpened.

Clearance angle value was fixed at α=10°. Rake angle and tool edge radius were considered to have values: γ= –5°…15°, respectively ρ=0.1…0.2 mm.

Workpiece was made of C45 steel.

Cutting regime parameters were as follows: cutting speed v_c=20 … 60 m/min, cutting depth a_p= 0.5 … 2.5 mm, longitudinal feed f= 0.05 … 0.2 mm.

Experimental plan is not presented in the paper. However, there were performed 17 experiments for different combinations of the mentioned cutting process parameters. Some of them failed. From these experiments we present only sample of test No. 1, which failed due to the hardest cutting conditions (highest cutting speed) and obtained higher hardness and, as well, we present qualitative aspects and trends observed in all complex cutting tests.

3.1 Theoretical research and preliminary presumptions

Presumptions for the 3D simulations were the following:
- Cutting tool edge is considered to be perfectly rigid, made of metal carbide (24%Cobalt);
- Cutting movement is translation, 10 mm stroke.
- Mesh generation and ambient conditions were chosen automatically by the software.
- Cutting speed, cutting depth, rake angle and clearance angle were set as in the experimental plan (not presented).
- Only the interaction zone between workpiece and cutter's edge was modeled (see Fig. 7).
- Workpiece is fixed into a clamping device (see Fig. 7, a).

The main output data was considered to be the effective stress (possibly above the yield strength of the material) (see Fig. 7,b– test No. 1, step 65) because of the presumption that hardening of the superficial layer of cut surface is caused by the deformation/crushing of the workpiece material by the cutting tool.

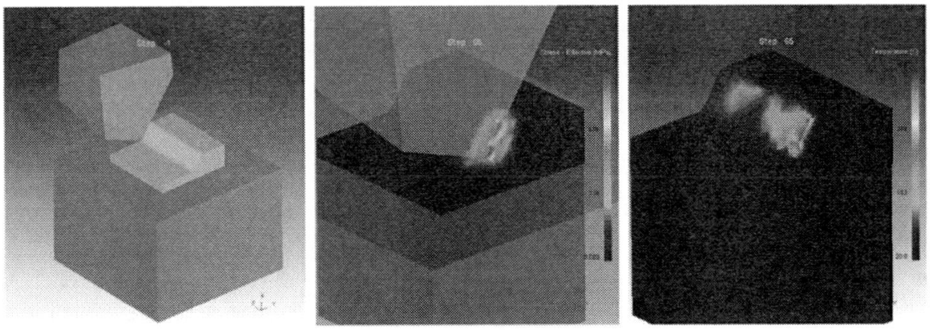

a) b) c)

Fig. 7. Output simulation data for test No. 1 (v_c=60 m/min; a_p=1.5 mm, f=0.1 mm/cicle, α=10°, γ=10°, ρ=0 mm). a) Initial set; b) Stress Effective (step 65); c) Temperature (step 65).

After simulations some interpretations and conclusions issued, as follows.

1. As in orthogonal cutting process, in complex cutting process occur stresses above yield stress of the workpiece material (σ_rC45=630...700 MPa).
2. Effective stress distribution in the surface layer is different in case of intermediary cut surface connecting the initial workpiece surface and the final manufactured surface than in case of the very final manufactured surface. Values of the effective stress are higher in first case. This confirms the higher hardness of the chip than the hardness of the manufactured surface. As well, higher hardness occur in the intermediary cut surface connecting the initial workpiece surface and final cut surface than in the very final cut surface.
3. Cutting temperature is not so high to imply a hardening heat treatment. Consequently, hardening occurs due to the plastic deformation of the workpiece material. In case of using coolant this is obvious.
4. Probable crushed/hardened surface layer thickness is higher under the intermediary cut surface connecting the initial workpiece surface and final cut surface, than under the very final cut surface. This means there are important differences between orthogonal and complex cutting processes related to deformation of superficial layer of the cut surface.

3.2 Experimental research and results

Experimental stand of complex cutting by longitudinal turning is presented in Fig. 8. A MAZAK Quick Turn 6T lathe was used for the longitudinal turning of the samples. Cutter 1 was fixed into the turret of the lathe and setting its position was made using the positioning device 2. Workpiece 3, having tracks for the different experiments, was fixed into the chuck of the lathe.

Some experiments failed due to the following reasons: too high cutting speed (above 60 m/min), too high feed (operator's error), collision (operator's error).

Measurements of hardness were performed both onto the external cylindrical surface and onto the frontal surface, along the radius of the sample.

DOI: 10.1051/matecconf/201712101003

Fig. 8. Experimental stand for complex cutting (detail).

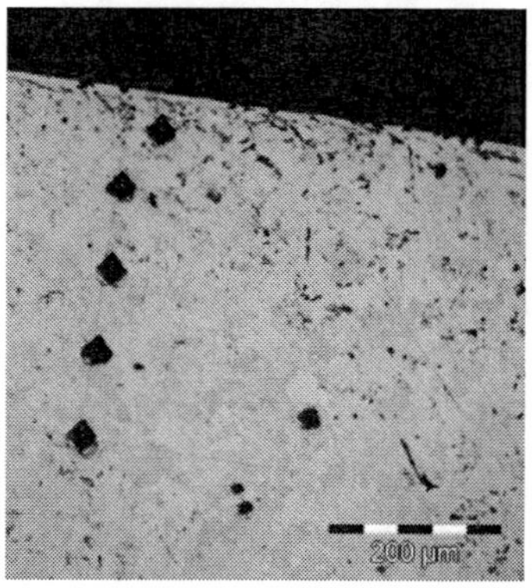

Fig. 9. Optical microscopy Test No. 1.

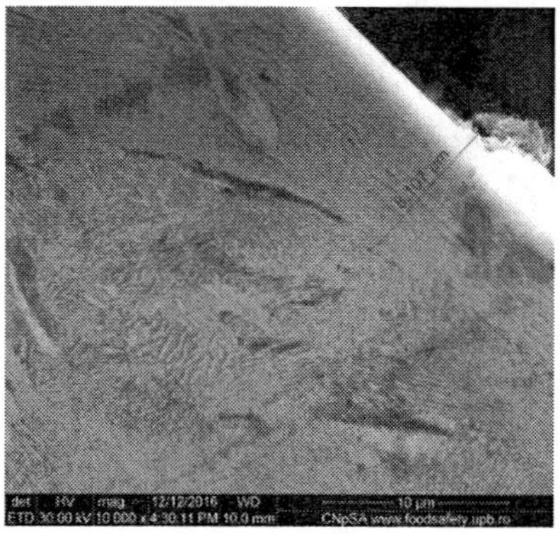

Fig. 10. SEM microscopyTest No. 1.

Values of the HV hardness of the external cylindrical surface varied between 185-240 HV and values of microhardness HV 0.2 varied between 230-300 HV for workpiece and between 300- 400 HV for the detached chip.

It may be stated that hardness of the cut surface is not constant from point to point, even if they are close to each other. It is the same in case of orthogonal cutting. The explanation is C45 steel is not homogeneous, being composed of ferrite and pearlite, which have different hardness. Furthermore, depth of the deformed/crushed superficial layer of the cut surface is very thin and not constant, as we should see in the following.

Micrographs, both optical and SEM (see Figs. 9 and 10) showed superficial deformation (crushing) on cut surface in complex cutting case, which was expected. The deformed crushed superficial layer seems to be even thinner in case of complex cutting than in case of orthogonal cutting. Its thickness was not constant and it was rarely above 10…15 μm.

This shows that the deformed/crushed/hardened superficial layer of cut surface is thicker in case of orthogonal cutting than in case of complex cutting. However, its thickness is very small, possibly not significant.

4 Conclusions

Research presented in this paper is not final. This research objective was to determine thickness and hardness of superficial layer of cut surface due to cutting process, both orthogonal and complex cutting. The most important finding is thickness of the superficial hardened layer is very thin under studied conditions, less than 0.01…0.02 mm. This research should be continued.

Performed research had good aspects: revealing the hardened cut surface layer both by means of theoretical simulations and by real experiments, and not so good aspects: the low stiffness of the used lathe in case of orthogonal cutting and the not homogenous workpiece material of the samples.

Theoretical research was performed by simulations using DEFORM 2D™ v.9.0 and DEFORM 3D™ v.6.1 software package [9]. These simulations showed that stress in the just cut surface (just under the tool point) is not constant. It has values above yield strength of the workpiece material. Such stress values occur down to a depth of approximately (1.5...2)a_p below the cut surface. This leads to the idea of deformed/crushed/hardened surface layer occurrence. Both stress in the workpiece and probable thickness of the crushed/hardened layer of the cut surface depend more on the cutting depth in case of orthogonal cutting, respectively on the feed in case of complex cutting, they depend more on tool edge radius and less on the cutting speed. Superficial layer hardness increases when cutting speed decreases. This reminds the idea of hardening by broaching.

Experimental research was performed for C45 steel by orthogonal turning on a SN 400 lathe and by complex cutting on a MAZAK Quick Turn 6T lathe.

Cutting forces were measured online and HV 0.1 microhardness was measured off-line, after orthogonal turning. Each orthogonal cutting test meant to detach several chips from a disc sample fixed onto a flange, which was fixed into the lathe's spindle. The samples were laterally polished in order to measure hardness under the cut surface. Some of the samples were super-polished for optical and SEM microscopy.

In case of complex cutting measurements of hardness HV, microhardness HV 0.2 and optical and SEM microscopy were performed. Complex cutting experiments revealed the hardened cut surface layer, which imply greater hardness of the workpiece on the cut surface layer. However, hardness increase is not so important on the cut surface layer than in the detached chip. Greater hardness of the chip is one important reason of tool crater wear occurrence.

Important differences between orthogonal and complex cutting were revealed. This means the hardened cut surface layer should be studied for each and every cutting case.

Cutting parameters which influence hardness increase are the following: cutting speed decrease, cutting depth decrease, feed increase, null to negative rake angle and greater tool edge radius.

As consequences of the presented research some questions occurred:
- Does the coolant influence the hardened cut surface layer related to its microstructure, deformed superficial layer thickness and hardness?
- Other cutting operations than turning lead to a greater hardness of the cut surface layer?

Hardness is relatively difficult to be determined, having dispersion due to the inhomogeneous workpiece material and cutting process parameters variation.

Future research related to the deformed/crushed/hardened superficial layer of cut surface should be done, at least, in the following ways:
- Redo the experimental research in better conditions: lathe with greater stiffness, homogenous C45 steel, setting its initial hardness by heat treatment.
- Possibly change the studied workpiece material with another homogeneous one.
- Change the values range of the input parameters, in order to set the case of hardening by broaching, grinding or the case of burnishing [11].
- Research other complex cutting operations than turning or even plastic deformation operations (because cutting process is, essentially, a plastic deformation process: hardening broaches suggested by [1] or ball burnishing suggested by [1, 11].
- Research the influence of all cutting process conditions, including coolant use.

References

1. S.M. Croitoru, V. Majstorovic, *10th CIRP Conference on Intelligent Computation in Manufacturing Engineering - CIRP ICME '16, Ischia, Gulf of Naples* (2016)
2. S.M. Croitoru, V. Majstorovic, *16th CIRP Conference on Modelling of Machining Operations, Cluny, France* (2017)
3. C. Minciu, Şt . Velicu, S.M. Croitoru, *Bazele aşchierii şi generării suprafeţelor (Basics of cutting process and generating surfaces),* (Litografia UPB, Politehnica University Publishing House, Manual for students, Bucharest, Romania, 1998)
4. A. Oprean et. All., *Bazele aşchierii şi generării suprafeţelor (Basics of cutting processes and generating surfaces)*, (Editura Didactică şi Pedagogică, Romanian Education Ministry Publishing House, Bucharest, Romania, 1981)
5. R.W. Maruda, G.M. Krolczyk, M. Michalski, P. Nieslony, and S. Wojciechowski, J Mater Eng Perform, **26**, 431 (2017)
6. G.M. Krolczyk, P. Nieslony, R.W. Maruda, S. Wojciechowski, J Clean Prod, **142/4**, 3343 (2017)
7. R. Piispanen, Annals of the CIRP, **XVII**, 561 (1969)
8. S.N. Melkote, R. Liu, P. Fernandez-Zelaia, T. Marusich, Cirp Ann-Manuf Techn, **64**, 65(2015)
9. *** DEFORM 2D™ v.9.0 and DEFORM 3D™ v.6.1. perpetual license, POLITEHNICA University of Bucharest, National Research Center for Performances of Technological Systems OPTIMUM, Romania (2008)
10. ***, *Structure/Property Relationships in Irons and Steels, Metals Handbook Desk Edition*, Second Edition, ASM International®, Metals Handbook Desk Edition, Second Edition, J.R. Davis, Editor, 153 (1998)
11. V. Chomienne, F. Valiorgue, J. Rech, C. Verdu, CIRP Journal of Manufacturing Science and Technology, **13**, 90 (2016)

Regarding the influence of the particle size of crumb rubber from waste rubber on the physical and mechanical characteristics of reclaimed rubber

Dan Dobrotă[1,*], and *Gigel* Paraschiv[2]

[1]Lucian Blaga University of Sibiu, Bd-ul Victoriei 10, Sibiu, Romania
[2] University Politehnica of Bucharest, 313 Splaiul Indepentei, Romania

Abstract. Thus, at present, a large amount of rubber waste cannot be higher harnessed due to recovery technologies used. Also, the requirements for new products of rubber worldwide are greater than the production capacity and the availability of raw materials. To offset this balance is necessary to improve technologies for recycling waste rubber so to can get regenerated rubber with the best mechanical physical characteristics. The main technology for obtaining the regenerated rubber is based on the rubber waste shredding and making the powder, so that it can be carried out de-vulcanization in autoclave. All the time it was monitored as the de-vulcanization process to be as economical and less polluting, and this requires that the sizes of the crumb rubber to be optimal. Thus, in the conducted researches was aimed to establish the particle size of crumb rubber from waste rubber so as to obtain an economical de-vulcanization, but also a reclaimed rubber with superior physical and mechanical characteristics. The organization of researches was made so that crumb of rubber waste was divided into 5 groups, bowing from crumb rubber with particle's size of 1000 μm (RPS1), and reaching up to crumb rubber with particle's size of 200 μm (RPS 5).

1 Introduction

Regarding the concerns about the reintroduction in the processes for obtaining various rubber products, rubber waste they are as old as the rubber processing industry. Over time it was manifested a distinct interest from the recovery and regeneration of waste rubber which suffered constant changes depending on the quantities required by the market and its price development [1-2]. Thus, it was found that there is a constant competition between products made from reclaimed rubber and the resources obtained from the original resources. The problem of valorification through the rubber processing from waste is an increasing topic in recent years, as a result of the general crisis of raw materials and energy, and because large quantities of rubber waste are withdrawn from economic circuit [3-5].

* Corresponding author: dobre97@yahoo.com

The presence of an intense waste rubber causes environmental pollution and the removal from economic circuit of significant material resources.

For an efficient utilization of rubber waste there can be taken into account the variant in which it is converted into a material that can be returned to the manufacturing process from which they come. Such a variant is possible if the rubber waste undergoes a technological process (Fig. 1) so that by grinding and devulcanization is obtained the regenerated rubber, which can be used in the composition of various rubber products [6].

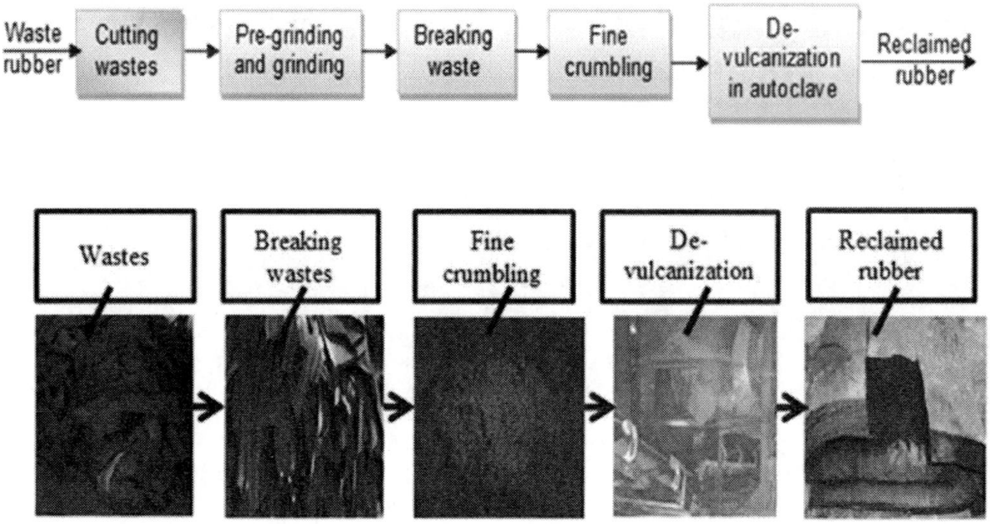

Fig. 1. Classic flow process diagram for obtaining reclaimed rubber.

Through the regeneration of waste rubber, there is obtained the reclaimed rubber, which can have a whole series of economic and technical advantages, namely [7, 8]:

- the stimulation and stability of the price are determined by broad based of raw material. It should be taken into account that at the regeneration of waste there is obtained the regenerated rubber and other components with a high economic value (textile inserts or metal);
- short mixing cycles are characteristic for compositions that include reclaimed rubber because it contains all the necessary ingredients and the swelling agents and/or the regeneration favors the mixing process;
- low power consumption at mixing is due to specific structure of reclaimed rubber. Realizing charging of and plasticizing at the reclaimed rubber proportions, mixing rubber NR type requires 20% more energy, and rubber SBR 1712 with 43% more;
- calendering is made easier due to the fact that the development of heat in the mixture bench is lower, it can work in a wide range of temperature. Even if the mixture contains regenerated rubber, it has a shorter pre-vulcanization time (Mooney) and it may be safer in processing in calender;
- calendering and rapid and uniform extrusion of the rubber is sustained by the introduction of various types of regenerated rubber in the compositions;
- the penetration of the compositions in fabric at the operations of adding rubber solution is facilitated by the presence of reclaimed rubber.

For the evaluation of the quality of reclaimed rubber, over time there have been proposed several indicators. About these indicators it can be said that some of them are antagonistic, others have become anachronistic etc., but it is always difficult to make an overall follow-up after several indicators simultaneously.

Therefore, it has been proposed [9] a generalized indicator (QIG) defined by the equation:

$$\mathrm{QIG} = \frac{100 - C}{100} \cdot M \cdot R \cdot \frac{A}{100} \tag{1}$$

where:
C – ash content, %;
M – Mooney viscosity;
R – traction resistant, N/mm^2;
A – tensile elongation limit, %.

In the relation of the QIG indicator is observed that are taken into account only a small number of physical-mechanical characteristics and thus in practice, to assess the quality of a product of rubber there should be more considered the physical and mechanical characteristics.

The uses of reclaimed rubber are limited by the fact that it can reduce the abrasion resistance. Breakage strength may also be adversely affected, although there are known cases (the use of regenerate from integral tire) [10] when the influence is positive. Moisture absorption of the reclaimed rubber is relatively large in comparison with the compositions of new tires and therefore its use is not recommended when water absorption is an essential criterion [10-13].

According to data presented in [4], there can be taken as guidance the following distribution of reclaimed rubber in fields of use, Table 1.

Table 1. Distribution of reclaimed rubber by the fields of use.

Fields of use	Quantity used, %
Tires	55
Pressed technical articles	19
Conveyor belts, belts, hoses	7
Ebonite products	6
Gaskets for installation and construction, household products, dispersions, adhesives, footwear etc.	13

Assortment standardization and quality of reclaimed rubber is very approximate. Selecting the type and proportions used in the manufacture of certain products is done on experimental basis depending on the demands made on the finished product [14]. It is indicated that for each case to consult the supplier specifications on qualitative indicators of the regenerated product which is offered. The obtained results with the regenerated product provided by a particular manufacturer can not expand unconditionally and the switching of the provider requires reverification of the use potential [15, 16].

2 Materials and methods

In the technological process of obtaining reclaimed rubber, two operations present an important role in achieving rubber products with superior mechanical characteristics, namely the fine grinding operation and the devulcanization operation [17, 18]. By fine grinding of waste rubber there is obtained an increase in specific surface of the particles produced which allows an improvement in devulcanization conditions in the autoclave and a possible change of the physical and mechanical characteristics of the regenerated rubber.

In the research there was considered a technological process of regeneration of waste rubber from conveyor belt reinforced with metal inserts. It was considered a waste rubber because the rubber from these products is one with superior physical and mechanical characteristics. Thus, in the research there were determined physico-mechanical properties of the obtained rubber taking into account the equipment and standards outlined in Table 2.

Table 2. Equipment and standards used to determine physical and mechanical characteristics.

Physical and mechanical characteristic	Equipment	Standard
Hardness, ^{0}Sh	Durometer BAREISS HPE tip Shore A	SR ISO 27619-1/ 2011
Elasticity, %	Zwick 5109	ISO 4662 / 2009
Viscosity Mooney ML(1'+ 4')100 °Cb	Viscometer Mooney Model: TO-230	ASTM D1646
Grade of rubber vulcanization	Rheometer Model: TO-220	ASTM D 1646
Tensile strength, N/mm^2	Machine of Traction Lloyd instruments a trademark of AMETEK tip LR10K PLUS 01/3052	SR ISO 37:2010
Elongation at break, %	Machine of Traction Lloyd instruments a trademark of AMETEK tip LR10K PLUS 01/3052	SR ISO 37:2010
Residual elongation, %	Machine of Traction Lloyd instruments a trademark of AMETEK tip LR10K PLUS 01/3052	SR ISO 37:2010
Tear strength, N/mm	Machine of Traction Lloyd instruments a trademark of AMETEK tip LR10K PLUS 01/3052	D 41 1507/C
Density, g/cm^3	Balance SARTORIUS 2462	SR ISO 2781+C1/ 2001
Wear, mm^3	Abrazimeter Model C5000	ISO 11998

3 Results and discussions

The process of obtaining reclaimed rubber is quite complex and includes multiple operations, but two of them are of particular importance, namely the grinding operation respective the de-vulcanization operation.

Both operations greatly influence the quality of reclaimed rubber. Thus, by a very fine grinding it is obtained an increase in the specific surface of the rubber particles and this may result in an improvement of the process of de-vulcanization in autoclave by better breaking of the links of the rubber.

In these circumstances, the researches aimed at establishing the influence of particle size obtained in the grinding operation on the physic-mechanical properties of reclaimed rubber.

Thus, waste rubber from conveyor belt which contains rubber with superior physical-mechanical characteristics of the type TRDZ were subjected to the regeneration. Also, it was considered a type of waste because the rubber quantity from such waste is very large and has a considerable economic value.

There were 5 groups of particles, Figure 2, obtained by grinding of tire, namely:
- RSP 1 crumb rubber with a particle size of 1000 µm;
- RSP 2 crumb rubber with a particle size of 800 µm;
- RSP 3 crumb rubber with a particle size of 600 µm;
- RSP 4 crumb rubber with a particle size of 400 µm;
- RSP 5 crumb rubber with a particle size of 200 µm.

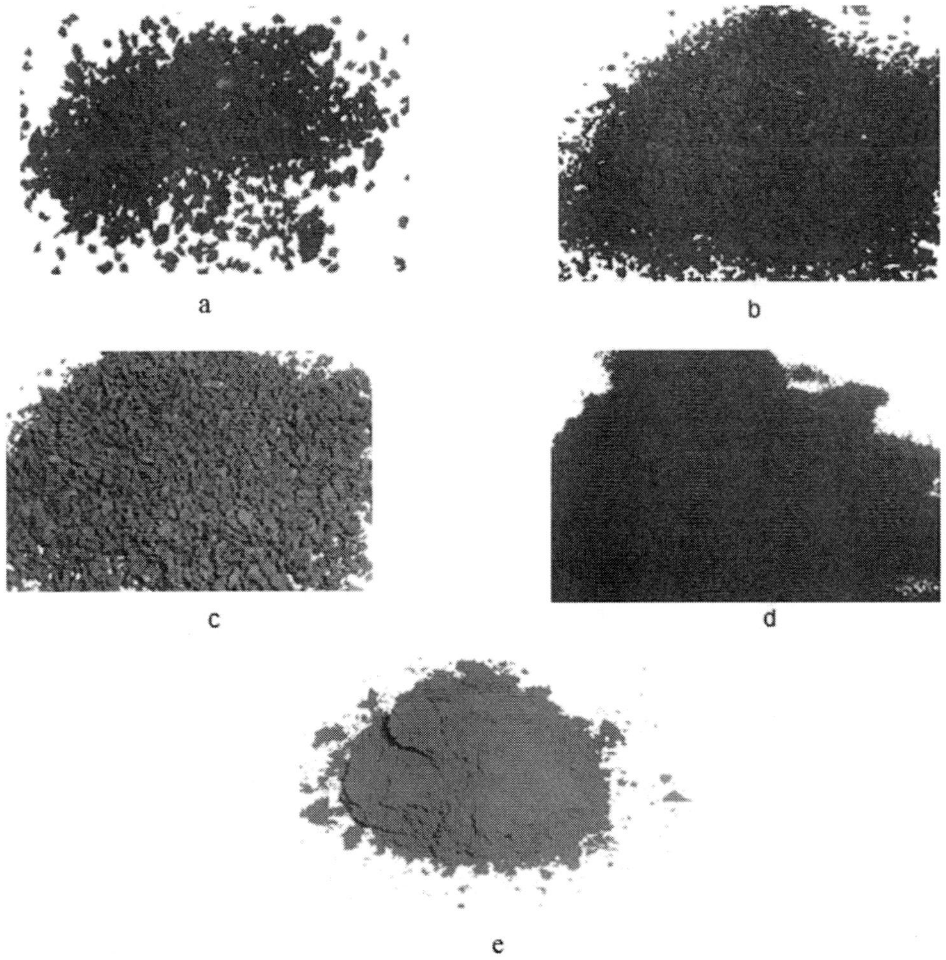

Fig. 2. The crumb rubber resulting from waste crumbling: a - crumb rubber with a particle size of 1000 μm; b - crumb with a particle size of 800 μm; c - crumb with a particle size of 600 μm; d - crumb with a particle size of 400 μm; e - crumb with a particle size of 200 μm.

After fine grinding operation the five groups of particles from the waste rubber have been subjected to the operation of de-vulcanization compliance with the same technological parameters, respectively temperature, time, pressure. Following the de-vulcanization and refining there were obtained 5 types of reclaimed rubber which may enter into the composition of various rubber recipes.

The waste which was left from the conveyor belts have been monitored as the 5 kinds of reclaimed rubber to enter into the same type of rubber that is the TRDZ. This type of rubber must have good physical and mechanical characteristics because the conveyor belts are subjected to high mechanical stress and work under very heavy conditions.

To obtain a TRDZ rubber type with best characteristics it must have in composition different types of material in addition to reclaimed rubber. Thus, the composition of rubber type TRDZ, Table 3, was realized so that a batch of 320 kg of rubber to get into 100 kg of reclaimed rubber obtained from the five groups of particles (RSP 1, RSP 2, RSP 3, RSP 4, RSP 5).

Due to the fact that there were obtained five kinds of the regenerated rubber based on the particle size it was possible to obtain five kinds of rubber type TRDZ (TRDZ 1 when using reclaimed rubber from particles SRP 1, TRDZ 2 when using reclaimed rubber from particles SRP 2, TRDZ 3 when using reclaimed rubber from particles SRP 3, TRDZ 4 when using reclaimed rubber from particles SRP 4, TRDZ 5 when using reclaimed rubber from particles SRP 5).

Table 3. The TRDZ rubber composition using reclaimed rubber.

Materials	**Mass, Kg**
Regenerated rubber (RSP 1, RSP 2, RSP 3, RSP 4, RSP 5)	100
Natural rubber SIR-20	20
Styrene butadiene synthetic rubber SBR	50
Poly-butadiene synthetic rubber SKD ND	30
Naphtha	30
Carbon black HAF	50
Zinc oxide ZnO	10
Antioxidant 4010NA/LG(IPPD)	5
Antioxidant HS/LG (TMQ)	5
Stearin	5
Bitumen	20
Inhibitor of pre-vulcanization CTP	5
Vulcanization accelerator CZ (CBS)	5
Vulcanization accelerator D	5
Sulfur	10
Total	350

For all the five kinds of TRDZ rubber, there were determined different physical and mechanical characteristics, and the obtained values are presented in Table 4.

Table 4. Physico-mechanical characteristics for TRDZ rubber.

Physico-mechanical characteristic	**TRDZ 1**	**TRDZ 2**	**TRDZ 3**	**TRDZ 4**	**TRDZ 5**
Hardness ^{0}Sh	72	70	68	66	64
Elasticity %	14	15	17	19	21
Viscosity Mooney $ML_{(1+4)}$ 100^{0}C	76	77	79	80	83
Grade of rubber vulcanization	10.8	11.5	12.3	12.7	12.9
Tensile strength N/mm^2	79	81	85	91	97
Elongation at break %	361	367	378	386	392
Residual elongation %	21	19	17	15	14
Tear strength N/mm	57	55	54	51	43
Density g/cm^3	1.29	1.27	1.25	1.24	1.22
Wear mm^3	147	145	143	141	139

The results obtained in experimental research and presented in Table 4 demonstrate the following:
- the hardness of the rubber is reduced in the case where for the realization of the rubber type TRDZ is used regenerated rubber made of crumb rubber with lower dimensions;
- the influence of particle size from crumb manifests on the elasticity of rubber in the sense that for smaller particles there is achieved a greater elasticity which is a great advantage if it is sought to obtain rubber products which will be used in high temperature conditions and high mechanical stress;
- when the rubber particle size is smaller there was observed an increase in viscosity that allows the production of rubber products with superior technical features;
- rubber vulcanization degree increases with decreasing of particle size of the rubber which shows what connections occur during the vulcanization of rubber are much stronger;

- tensile strength greatly increases with the size reduction of the rubber, and this allows the obtaining of rubber products such as tires and conveyor belts with high performance in operation;
- elongation at rupture is much higher if there is used reclaimed rubber obtained from the crumb with very small particles size, and this allows to obtain different types of products which gives up very hard if they are subjected to requests for stretching;
- the use of crumb rubber with small sizes allows to produce products with high wear resistance and in these conditions, in practice there should be used a lot of reclaimed rubber from waste rubber which are undergo to intense grinding.

4 Conclusions

Recycling and the regeneration of waste rubber constituted and constitute still one of the activities which must be given special attention. The research results obtained demonstrate that the transformation of waste rubber in reclaimed rubber and its use for obtaining different types of products is an effective technical solution which can have economic effects and it reduce the environmental pollution.

Through an efficient waste rubber grinding there are considerably improved the de-vulcanization conditions in autoclave, and this allows to obtain reclaimed rubber mixtures with superior physical and mechanical characteristics. It also requires the optimization of grinding and de-vulcanization process so there is a correlation between their technological parameters to obtain low cost, but also in order to achieve superior regeneration of various types of waste rubber.

References

1. D. Dobrotă, Journal of Adhesion Science and Technology, **27**, 125 (2013)
2. J.Y. Xu, M. Shen, X.Q. Wang, C.H. Chen, Z.X. Xin, Journal of Macromolecular Science, Part B, **53**, 1182 (2014)
3. J. Karger-Kocsis, L. Mészáros, Journal of Materials Science, **48**, 1 (2013)
4. V. Torretta, E.C. Rada, M. Ragazzi, E. Trulli, I.A. Istrate, L.I. Cioca, L.I., Waste Management, **45**, 152 (2015)
5. S. Saiwari, E. Lohyi, C. Nakason, Advanced Materials Research, **844**, 437 (2014)
6. X. Sun, A.I. Isayev, Journal of Materials Science, **42**, 7520 (2007)
7. D. Dobrotă, Materiale Plastice, **43**, 65 (2006)
8. P. Krawczak, Express Polymer Letters, **5**, 838 (2011)
9. D. Dobrotă, G. Dobrotă, Journal of Cleaner Production, An innovative method in the regeneration of waste rubber and the sustainable development http://doi.org/10.1016/j.jclepro.2017.03.022 (2017)
10. S. Ramarad, C. Ratnam, M. Khalid , A. L. Chuah, S. Hanson, Radiation Physics and Chemistry, **130**, 362 (2017)
11. S. Ramarad, C.T. Ratnam, M. Khalid, A. L. Chuah, J. Appl. Polym. Sci., **132**, 416 (2014)
12. P. Sae-oui, C. Sirisinha, K. Hatthapanit, Express Polymer Letters, **1**, 8 (2007)
13. M. Magioli, A. Sirqueira, B. Soares, Polymer Testing, **29**, 840 (2010)
14. C. Searcy, D. McCartney, S. Karapetrovic, Corporate Social Responsibility and Environmental Management, **14**, 135 (2007)
15. H. Su, J. Yang, T.-C. Ling, G.S. Ghataora, S. Dirar, Journal of Cleaner Production, **91**, 288 (2015)
16. B.S. Thomas, R.C. Gupta, V.J. Panicher, Journal of Cleaner Production, **112**, 504 (2016)
17. A. Tripathy, D. Williams, R. Farris, Polymer Engineering & Science, **44**, 1338 (2004)
18. F. Hashim, M. Almaamori, F. HamoodInternational Journal of ChemTech Research CODEN (USA), **48**, 325 (2016)

Tool machinery vibrations frames comparison concerning welded or moulded manufacturing structures

Dan Ovidiu Glavan[1,*], and *Theoharis* Babanatsas[1]

[1] University "Aurel Vlaicu" of Arad, Faculty of Engineering, 310130 Arad, Romania

Abstract. Precision of manufactured object is the main goal for all engineers in modern times. The times of massive structures with low displacement have passed. We are looking forward now to develop new type of structures with controlled deformations easy to predict and therefore to correct in order to obtain the „ideal" product. The paper is analysing tool machinery frames vibrations in welded and moulded manufacturing structures ,the vibrations being studied in both horizontal and vertical planes and the loads are simulated as manufacturing loading pulses generated by special excitation electromagnetic devices, in order to determine advantages and disadvantages of using them for frames in projecting, producing and exploitation of tool machinery.

1 Theoretical considerations

In welded and moulded structures in any machinery bed frame structure appears vibration in the moment of work.

This vibration are unmaintained free dynamic balance equation (1), vibrate movement is of the form:

$$m\ddot{x} + c\dot{z} + kx = 0 \tag{1}$$

where: m - mass; c - damping coefficient; k - spring constant; x – acceleration; $\ddot{x}$-displacement; $\dot{z}$-speed [1-3].

King m and noting (2):

$$\frac{c}{m} = \lambda \; and \; \frac{k}{m} = p^2 \tag{2}$$

we get (3):

$$\ddot{x} + \lambda\dot{z} + p^2x = 0 \tag{3}$$

where p - pulsation of the system (4):

* Corresponding author: glavan@fortuna.com.ro

$$p = 2\pi f_m \frac{2\sqrt{c}}{T} \tag{4}$$

where: f - frequency oscillation; T - period oscillation.

Now results the differential equation which solving this equation yields with solutions that provide information about the movement of the structure, from characteristic points [3, 7-8].

We calculate the amplification baseline vibration (5):

$$x_1 = x_0 e^{-\lambda_t} \tag{5}$$

The vibration amplitude until the touch (6) result:

$$x_2 = x_0 e^{-\lambda_T} \tag{6}$$

The intensity of depreciation results (7) are:

$$\frac{x_1}{x_2} = e^{\lambda_T} \tag{7}$$

The logarithmic decrement damping (8) is:

$$\delta = \frac{ln x_1}{x_2} = \lambda_T \tag{8}$$

And with this we can calculate the logarithmic decrement damping factor λ (9):

$$\lambda = \frac{\delta}{T} \tag{9}$$

2 Experimental conditions

Bed frame structures can be manufactured with welded frame or with casted frame under the same conditions of settlement in the foundation. That means that a rubber hammer can be applied in order to generate a signal pulse on the bed frame structure, in the headstock and in the movable doll [6].

This pulses applied in both directions (vertically and horizontally). This impulse give a feedback value from frameworks in the form of a value that can be interpreted and recorded. For each point and place where was applicated this impulse taking both values of impulse (vertically and horizontally). In this way was registered the values that was obtained from vibration amplitude, oscillation period and amortization period.

With these parameters, we obtained the intensity depreciation δ, λ and the pulsation damping response p. Means vibrometer measuring and transducer [1-5].

To the framework poured appear higher frequency response (own bed frame and no settlement construction) which do not appear welded to the framework. Comparing the damping factors λ there is better the frame welded [9-11].

The values of vibration to tailstock are in table 1.

Table 1. Vibration to tailstock.

Parameters	Vertical impulse Vertical reaction (Figure 1)		Horizontal impulse Vertical reaction (Figure 2)		Vertical impulse Horizontal reaction (Figure 3)		Horizontal impulse Horizontal reaction (Figure 4)	
	B.S.	**B.T.**	**B.S.**	**B.T.**	**B.S.**	**B.T.**	**B.S.**	**B.T.**
T_1 (ms)	150	175	35	120	35	110	56	80
T_2 (ms)	25	50	18	50	25	50	18	50
f_1 (Hz)	7	6	29	8	29	9	20	12
f_2 (Hz)	40	20	56	40	40	20	56	20
p_1 (rad/s)	44	38	182	50	182	57	126	75
p_2(rad/s)	251	126	352	251	251	126	352	126
$\overline{x}_1$	2	2,2	1,0	1,2	1,1	1,2	1,4	1,2
x_i+1	1,6	1,5	0,7	0,9	0,5	1,0	1,2	0,5
δ	0,2	0,3	0,4	0,3	0,8	0,3	1,5	0,9
$\lambda*10^{-3}$	1,4	1,7	11,4	2,5	22,8	2,8	26,7	11,2

The graphic value of vibration to tailstock are in figures 1, 2, 3 and 4.

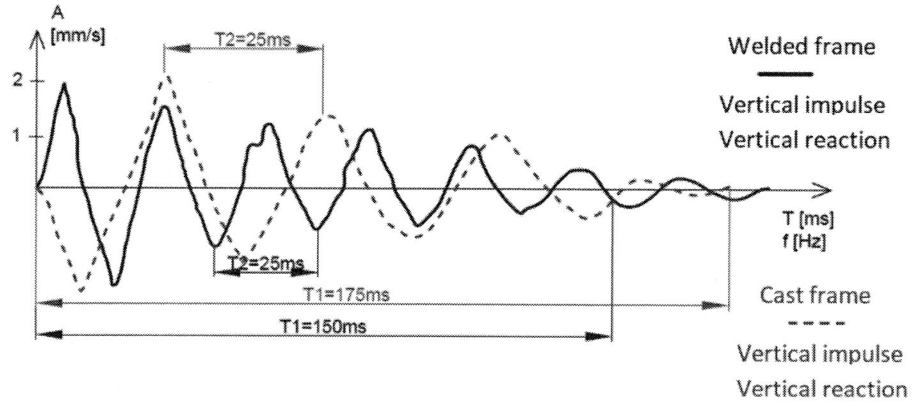

Fig. 1. Vertical impulse and vertical reaction.

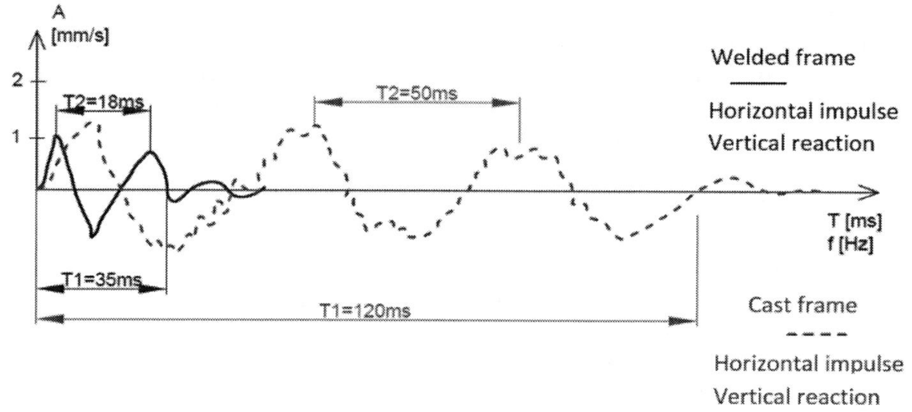

Fig. 2. Horizontal impulse and vertical reaction.

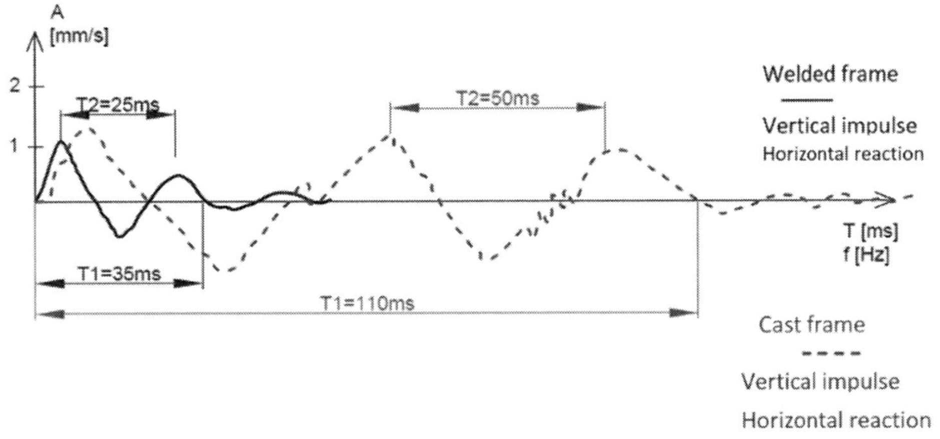

Fig. 3. Vertical impulse and horizontal reaction.

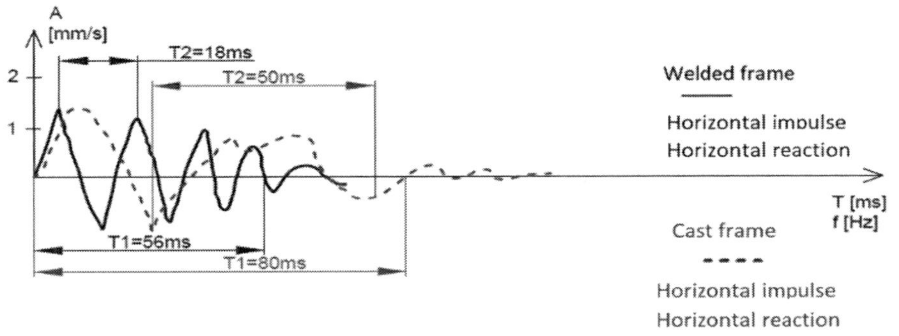

Fig. 4. Horizontal impulse and horizontal reaction.

The value of vibration to headstock are in table 2.

Table 2. Vibration to headstock

Parameters	Vertical impulse Vertical reaction (Figure 5)		Horizontal impulse Vertical reaction (Figure 6)		Vertical impulse Horizontal reaction (Figure 7)		Horizontal impulse Horizontal reaction (Figure 8)	
	B.S.	B.T.	B.S.	B.T.	B.S.	B.T.	B.S.	B.T.
T_1 (ms)	150	130	35	50	35	120	65	80
T_2 (ms)	25	50	18	20	25	50	18	20
f_1 (Hz)	7	8	29	20	20	8	15	20
f_2 (Hz)	40	20	56	50	40	20	56	50
p_1 (rad/s)	44	50	182	126	182	50	94	75
p_2(rad/s)	251	126	352	314	251	126	352	314
$\overline{x}_1$	1,6	1,3	1,2	0,9	1,1	1,1	1,1	1,1
x_i+1	1,1	1,1	0,8	0,6	0,7	0,6	0,8	0,6
δ	0,4	0,2	0,4	0,4	0,5	0,6	0,3	0,3
λ*10^{-3}	2,6	1,5	11,4	8	14,3	5	4,6	3,7

The graphic value of vibration to headstock are in figures 5, 6, 7 and 8.

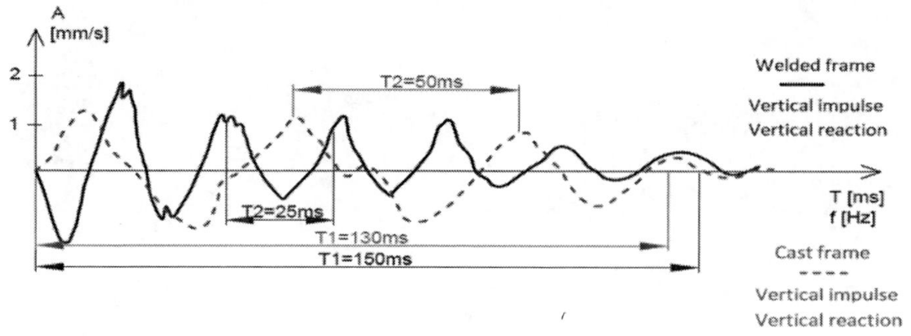

Fig. 5. Vertical impulse and vertical reaction.

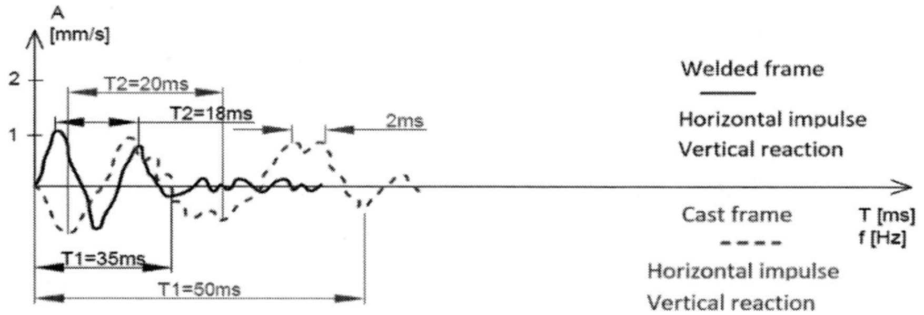

Fig. 6. Horizontal impulse and vertical reaction.

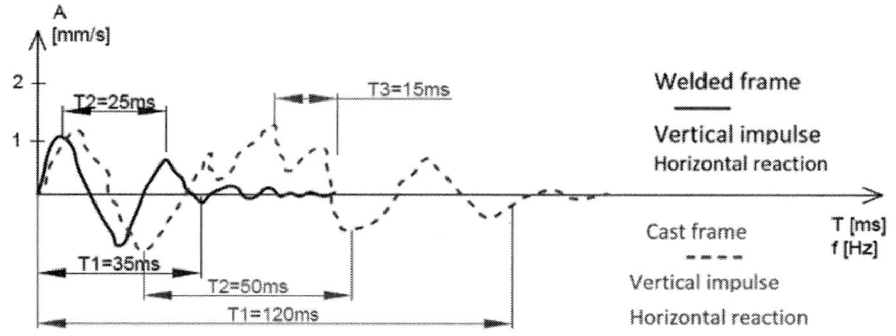

Fig. 7. Vertical impulse and horizontal reaction.

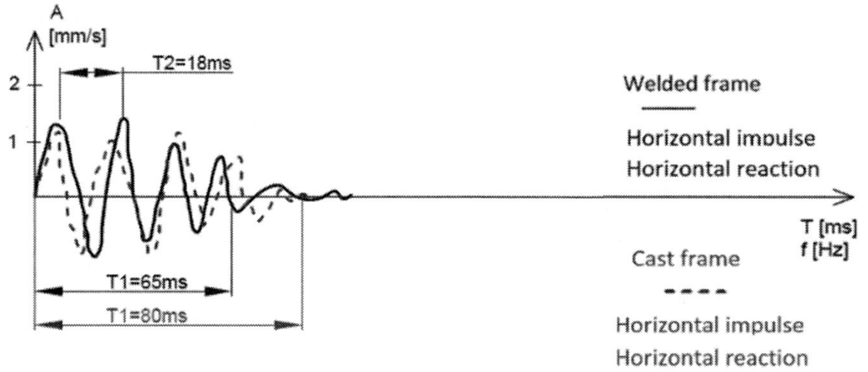

Fig. 8. Horizontal impulse and horizontal reaction.

3 Results and conclusions

Besides headstock arousal and response. Welded to the frame of the vertical frequency of the impulse response is 40 Hz, the period of 150 ms damping being greater in vertical direction than in the horizontal direction 35 ms. To the framework poured vertical impulse frequency response is 20 Hz, the amortization period is approximately the same in horizontal and vertical direction. Comparing welded frameworks and poured observed that the cast machine frame has a damping factor better than λ welded machine frame.

The frequency response of frameworks with excitement next of the tailstock have the same values as the next of the headstock excitement under the same excitement and response.

The momentum in the horizontal direction on the bedframe environment shows that it's response takes a place on other frequencies and payback lower than the vertical impulse.

From factors we can observed that the framework coefficient λ is better than the molded that welded structures.

References

1. D. Glăvan, I. Radu, *Vibrațiile mecanice în tehnică* (Editura Universității „Aurel Vlaicu", Arad, 2001)
2. D. Glăvan, I. Radu, *Elemente vibrații mecanice* (Editura Universității „Aurel Vlaicu", Arad, 2001)
3. D. Glăvan, T. Babanatsas, *Mașini unelte și aggregate* (Editura Universității „Aurel Vlaicu", Arad, 2002)
4. H. Meyer, F. Fuchs, and K. Thiel, *Manufacturing Execution Systems: Optimal Design, Planning, and Deployment.* (McGraw Hill, 2009)
5. L. Cîndea , C. Hatiegan, G.R. Gillich, C. Marta, M. Nedeloni, Scientific Bulletin of the "Politehnica" University of Timisoara, Transaction on Mathematics & Fhysics, **57,** 68 (2012)
6. Y. Rodichev, O. Soroka, V. Kovalov, Y. Vasilchenko, V. Maiboroda, Journal of Research and Development in Mechanical Industry, **7**, 1 (2015)
7. V.D. Kovalev, Y.V. Vasilchenko, and P. Dašiü, Journal of Mechanics Engineering and Automation, **4**, 269 (2014)
8. L. Hakansson, *Adaptive Active Control of Machine-Tool Vibration in a Lathe – Analysis and experiments* (Lund Univ., Sweden, 1999)
9. I. Claesson, L. Håkansson, Int. J. Acoust. Vib., **3**, 4 (1998)
10. N. Deshpande, M.S. Fofana, Comput. Integr. Manuf., **17**, 1 (2001)
11. W. de Silva Clarence, *Vibration Signal Analysis, in: Mechanical Engineering Series, Frank Krith, Vibration and Shock Handbook* (Taylor and Francis Group, LLC, 2005)

Design and construction of an inspection robot for the sewage pipes

Jana Knedlová[1,*], *Ondřej* Bílek[1] , *David* Sámek[1] , and *Petr* Chalupa[2]

[1]Tomas Bata University in Zlín, Faculty of Technology, 275 Vavrečkova Zlín, 76001 Czech Republic
[2]Tomas Bata University in Zlín, Faculty of Applied Informatics, 4511 Nad Stráněmi Zlín, 76005 Czech Republic

Abstract. Robotic systems are used wherever human limits are achieved. The inspection robots are used for securing the access to the manholes, pipes and sewage systems, for monitoring faults and for the eventual minor repairs. The paper presents the design, construction and assembly of a four-wheeled inspection robotic rover used in sewage pipes with diameters larger than 200 mm. We managed to design the robust, waterproof and dirt-resistant washable rover solution, which has been proved to be very functional and easy to produce.

1 Introduction

Robotics is one of the fastest developing branches of industry and robots are presently used wherever working conditions are difficult, unsafe, often recurring or otherwise limited. Attention is focused on the robotic systems capable of controlled movement and the inspection of inaccessible pipe systems. It is the only way how to inspect cleanness, leakage, mechanical damage or the corrosion level and also to identify recovery spots [1-5] directly beneath the surface.

It is currently usual that piping systems are inspected by full autonomous robots or by cable-tethered robots with a remote control by human operator being on the ground [6-7]. Fully autonomous inspection robots are able to inspect vast area of sewage pipes. Moreover, fully autonomous robots depend on some form of intelligence [8-10] allowing them to move in various piping systems, where the only limitation is the battery capacity. In this case, a robot without a cable connection is difficult to be picked up from a pipeline. On the other hand, cable-tethered robot can be pulled by the cable out of the pipeline in case of failure. The length of the cable is the limiting factor for the distance to be inspected. The operator controls the robot from the outside of the piping system and monitors the area by an inspection camera placed on the robot.

In-pipe robots have long history and several mechanical concepts [11] were developed until now. The geometry and environment are the determining factors to which the robots are able to adapt. A wide-spread type is a wheeled robot with differential driving [12]. The caterpillar type is preferably used on uneven surfaces to overcome obstacles [13].

* Corresponding author: knedlova@ft.utb.cz

Legged or snake robots are used for branched pipe systems [14], both are characteristic by their mobility. Further, the inchworm type of robot is able to move in curved pipes. The screw type robot has rather simple structure, nevertheless the control reacts slowly [15].

Recently, a number of developers are dedicated to the construction of more complex robots with hybrid locomotion [16-18].

In this paper, we focus on the design, development and construction of the in-pipe inspection robot intended to be used in the sewage pipes with diameters larger than 200 mm. The wheeled moving mechanism was considered in order to realize the inexpensive and robust inspection robot, equipped with an inspection camera at the front, and operated by the human operator on the ground surface. Following chapters describe the main construction units of the robot that were designed and assembled in Solid Edge software. Unless otherwise specified below, non-standard parts were fabricated on a CNC machine tool from duralumin work material.

2 Requirements

Following criteria that the inspection robot must meet were determined. The robot should consist of a small size inspection camera with a minimal viewing angle of 180°. The robot must have regulated movement back and forth at a constant speed. Cabling must include not only a power wire and wire for Ethernet communications, but also a string for towing in case of failure.

Wheels should be appropriate for the environment of sewage pipes both for tire tread design and material, and must ensure their interchangeability. An important condition is the ability to upgrade the entire control system, as well as individual elements and the CCD camera also for the reason of repair.

Water resistance is the next requirement for washing up after pulling out of sewage pipe, together with overpressure in the body of the robot, so that the external environment will not pervade inside, furthermore there is requirement for stiffness, LED lights for the camera system and easy portability.

3 Vision system

Structural arrangement of the whole camera (Fig. 1) has been designed to meet the demands of working in sewage polluted environments, and it has been done with the requirement to be leak-proof.

For the sensing of the area inside of the drain, the pipe camera Rasberry Pi has been used, which capable is of transferring the pixel data at extremely high speed. Lens OFISH-125MIR XtendLan with a range of 185 degrees have been used. A lens are attached to the camera holder, which we designed using CAD software and manufactured on a CNC milling machine.

The three LEDs (each with an output of 1 W) are attached to the surface of the camera holder. Camera housing was made of (spare parts for industrial camera) cover hemisphere and the ring of PMMA produced by CO2 laser machining equipment. The two housing parts have been joined after assembly into a single unit sealed with silicon. The clamping ring keeps attached the cover hemisphere to the camera holder. Sealing rings have provided protection against intrusion of dirt.

DOI: 10.1051/matecconf/201712101006

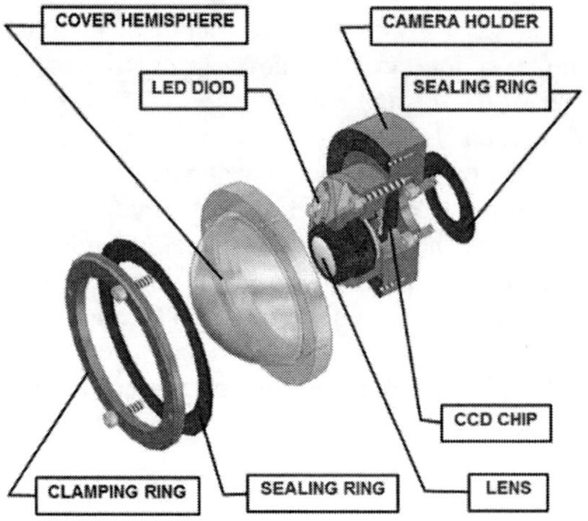

Fig. 1. Camera module.

4 Motion unit

The drive unit (Fig. 2) was designed to operate at speed of 0.2 ms^{-1}. It was made by the company Maxon (power rating 90W, speed 7270 rpm stall torque 966 m). The transfer of the drive torque to the front axle of the robot has been achieved through the transmission (Haberkorn 2 T20 / 20), the transfer ratio has been 1:1.

For more efficient movement of the rear axle gear we used a sprocket (6 x 2.8 T23) and chain. Wheels have been compiled from three parts of the hub, plates and tires. The hub and plates were manufactured using CNC milling. Similarly, a mold for vulcanizing the tires was made.

The modular design allows more solutions. We have chosen interchangeable wheels for a larger diameter waste pipe. Tires can be exchanged according to the surface of the pipe walls.

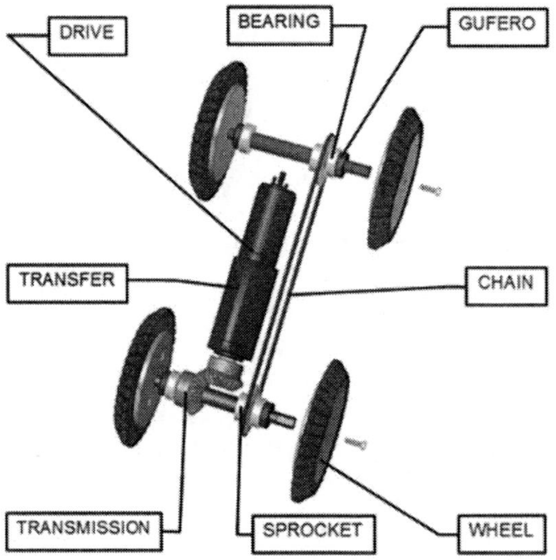

Fig. 2. Drive unit.

5 Electronics unit

Electronics unit (Fig. 3) includes a microcomputer (Raspberry PI), H-Bridge and Convertor. These were assembled by using the robot mounting board. The mounting board was modelled in 3D with CAD and CNC machining milling using polymer PA. Their interconnection is shown schematically in Fig. 4.

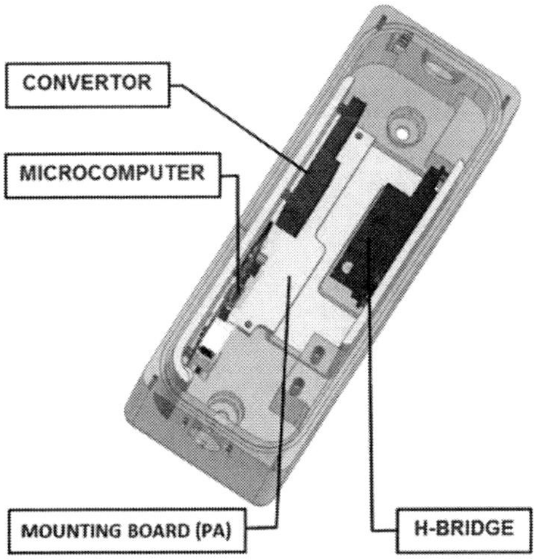

Fig. 3. Electronics arranged on plastic mounting board.

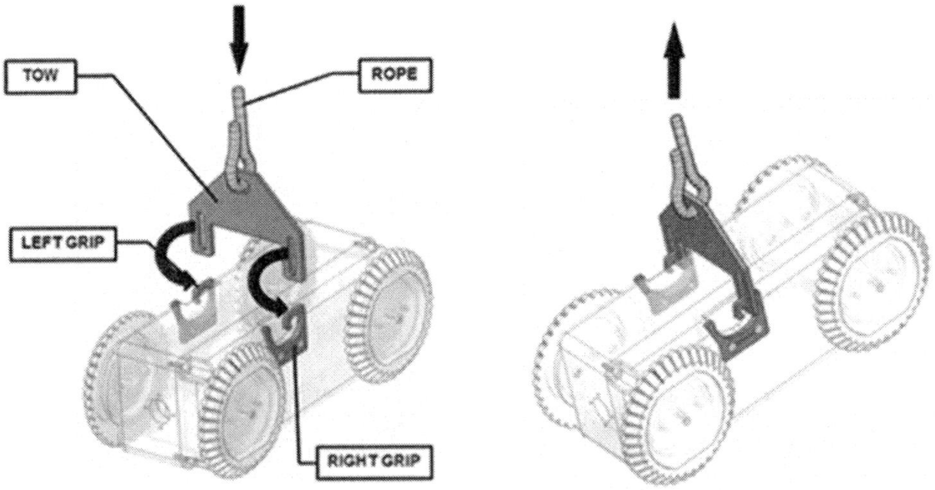

Fig. 4. Diagram of the electronic part of the robot.

6 Handling

Handling parts (Fig. 5) consist of the left and right grips for the location of the hinge for the rope, by which the robot is inserted into the waste pipe. Before the action the tow together with a rope are pulled out of the pipe. After the inspection the pipeline robot is removed by the same tow with a rope. Handling elements (left and right grip, tow) were created using CAD and produced using a CNC milling machine.

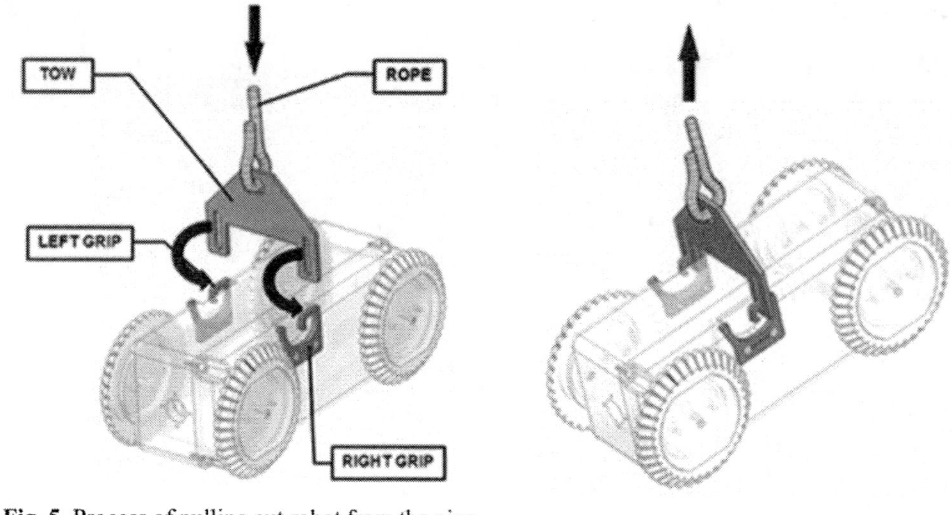

Fig. 5. Process of pulling out robot from the pipe.

7 Outer body

The outer body (Fig. 6) which we proposed is divided into the left and right section for easy assembly and disassembly. The design of the inside of the covers has been adapted to meet the requirements for placement of Motion unit (into mounting protrusion) and Electronics unit (pocket created for electronics). External surfaces of the body are manufactured so that the grips can be flush fixtured. The seal groove has been milled in the left cover. The seal is intended to prevent the intrusion of impurities inside the body. A hole has been created for insertion of a cycling valve in the left cover. The outer body was modelled using CAD and produced on CNC milling machine. Both parts of the outer body were manufactured from a single blank of aluminium alloy.

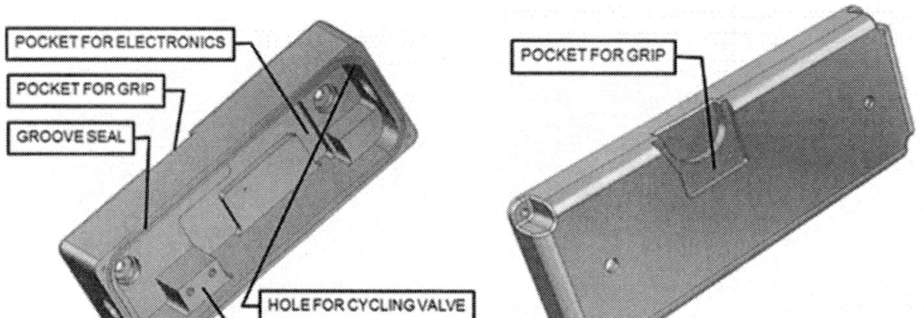

Fig. 6. Left and right cover of outer robot body.

8 Final robot assembly

The completed robot is shown in the figure in the CAD version (left) and the real rover (right) in Fig. 7. After assembling all the mechanical components of the robot, the sealing has been carried out, followed by pressurizing the air through the valve. Gauge air pressure ensures protection against intrusion of polluted wastewater into the robot body. Next, a leak test was carried out by immersion in water (the extreme environment to which the robot should not be exposed). After the sealing test, testing of electronics and robot motion has been carried out. Moreover, software that controls the movement of the device there and back a controls lighting has been designed.

Fig. 7. Final CAD assembly and manufactured prototype.

9 Conclusions

A prototype of final design was created. The operating test was completed successfully; it demonstrated that:

- the robot moves forward and backward without slipping,
- the robot unit can be upgraded, malfunctioning parts can be replaced,
- the material and robot cover allow easy maintenance (washable) and sealing of the gas pressurized inside,
- the robot can be pulled out of sewage pipeline by the cable or by the tow on the rope.

References

1. R.R. Marangoni, R. Baron, The 58th Ilmenau Scientific Colloquium, 8 - 12 September 2014 at the Technische Universität Ilmenau (2014)
2. A. Nayak, S.K. Pradhan, Procedia Eng., **97,** 2081 (2014)
3. J. Novak-Marcincin, M. Janak, D. Takac, Manuf. Technol., **14** (2014)
4. P. Li, S. Ma, Y. Li, *Proc.* – IEEE Int. Conf. Robot. Autom. Pasadena, California (2008)
5. H. Lim, T. Ohki, ICROS-SICE International Joint Conference Fukuoka, Japan (2009)
6. F. Nassiraei, Y. Kawamura, A. Ahrary, Y. Mikuriya, K. Ishii, Proc. - IEEE Int. Conf. Robot., Angelicum University Roma, Italy (2007)
7. M. Bachratý, P. Hubinský, A. Vitko, Proceedings of the 14th International Conference on Climbing and Walking Robots and the Support Technologies for Mobile Machines, CLAWAR (2012)
8. C. Zhu, Ind. Robot Int. J., **34** (2000)
9. S. Hirose, H. Ohno, T. Mitsui, K. Suyama, IEEE Int. Conf. Robot. Autom. Detroit, MI, USA (1999)
10. Kawaguchi, I. Yoshida, H. Kurumatani, T. Kikuta, Y. Yamada, Proc. 1995 IEEE Int. Conf. Robot. Autom., Nagoya, Aichi, Japan, ***1*** (1995)
11. V. Kepplin, K.U. Scholl, and K. Berns, IEEE/ASME Int. Conf. Adv. Intell. Mechatronics September, GA, USA (1999)
12. S.G. Roh, H.R. Choi, IEEE Trans. Robot., **21**(1), 1 (2005)
13. S.G. Roh, D.W. Kim, J.S. Lee, H. Moon, H.R. Choi, Int. J. Control. Autom. Syst., **7**, 105 (2009)

14. O. Tătar, D. Mândru, I. Ardelean, Mechanika, **68**(6), 60 (2007)
15. N.A.B.H. Yahya, N. Ashrafi, A.H. Humod, IOSR Journal of Mechanical and Civil Engineering, **11**(4), 1 (2014)
16. N.S. Roslin, A. Anuar, M.F.A. Jalal, K.S.M. Sahari, Procedia Eng., **41,** 1456 (2012)
17. M.R.A. M. Zin, K.S.M. Sahari, J.M. Saad, A. Anuar, A.T. Zulkarnain, Procedia Eng., **41,** 1469 (2012)
18. J. Slamka, M. Jedinák, M. Tolnay, M. Bachratý, O. Staš, Appl. Mech. Mater., **332**, 423 (2013)

Volume optimization of gear trains with spur gears using genetic algorithm

Nenad Marjanović[1] , *Nenad* Kostić[1,*], *Nenad* Petrović[1], *Mirko* Blagojević[1] , and *Miloš* Matejić[1]

[1]University of Kragujevac, Faculty of Engineering, 6 Sestre Janjić, 34000 Kragujevac, Serbia

Abstract. Gear train volume optimization presents a complex problem tied to practical application in gear train manufacturing. This paper is oriented on the analysis of the problem of gear train volume minimization from a shaft axes positioning aspect. An original mathematical model has been developed where the objective function gives a minimum volume with changed shaft (spur gear) axes positions, while at the same time complying with all physical constraints. An original optimization software has also been developed using RCGA (Real Coded Genetic Algorithm) optimization methods. The general mathematical model was applied to three real conceptions of gear train as well as a comparative analysis of initial and optimal values. The results show a decrease of volume being directly linked to a decrease of not only space but material used to make the housing, costs, documentation formulation rate, etc.

1 Introduction

Gear train presents a complex technical system due to the complexity of its mathematical formulation and large number of variable values. From a research and practical aspect there is a need for constant improvement theoretical and practical segments in order to justify gear train application. Achieving better operational gear train characteristics, and a lower mass, volume and cost are basic objectives when designing a gearbox. Optimization is the most effective way of achieving desired gear train characteristics, it allows for advances in this field as a current and alternative method of improvement. This paper only considers parallel shaft axes gear train.

The optimization method which has proven itself as significant for solving practical problems is the heuristic method of genetic algorithm, which is why it is interesting for this field of research. In this specific case the RCGA (Real Coded Genetic Algorithm) was developed, dedicated to solving general engineering problems.

Aside from knowing the construction of gear train, the process, and method of optimization, the development of a mathematical model is necessary which can be representative of the problem and improve gear train performance. This paper contains relevant examples which can present results of performance improvements with gear trains.

* Corresponding author: nkostic@kg.ac.rs

Chonget al. [1] presented a general methodology for optimizing gear ratios, sizes and housing volume for multi-stage gear trains in preliminary design phases. Marjanovic et al. [2] developed a practical approach to optimizing gear trains with spur gears based on selection matrix, selection of optimal materials, optimal gear ratios and shaft axes positions. Golabiet. Al [3] presented gear train volume/weight minimization optimizing single and multistage gear trains' gear ratios. Mendi et al. [4] aimed to optimize gear train component dimensions to achieve minimal volume comparing GA results to analytic method parameter volume. Savsani et al. [5] described gear train weight optimization comparing various optimization methods to genetic algorithm (GA) result values. Gologlu and Zeyveli [6] performed preliminary design automation through optimization of gear parameters and properties using a GA based approach. Tand and Tseng [7] presented a new mutation operator to adaptive direct mutation that focuses on simplicity, robustness, and efficiency within the context of RCGAs. Shopova et al. [8] presented an adjustable genetic algorithm deigned to deal with engineering optimization problems with both real and integer problems. Mushavati and Hamouda [9] implemented cyclic crossover operation to modify GA for multiple part manufacturing line optimization. For the manufacturing process optimization it is necessary to employ theoretical models as presented in [11]. Such a process is practical [12], economical [13], and relatively easy for use [14].

There is a real need to explain and mathematically formulate the gear train volume optimization problem, to find influential factors, and to find approaches and methods through which optimal gear train characteristics are possible.

Motivation is based on gear train performance; achieving a minimum volume, consequently a smaller mass, smaller embedded measurements, less material used, decreased expenses and other benefits.

The results obtained by the here presented method are compared to common concepts through specific examples.

2 Problem definition

Gear trains with spur gears are most frequently made so their shaft axes are in the same plane, which is most frequently the horizontal plane.

Gears placed in this way take up a large volume which can be reduced through optimizing the transmission. An optimal solution defines a new position of gear train shaft axes which results in a decreased overall volume of the gear train [15-16].

Gear train volume can be defined as a result of multiplying length, height, and width based on which a mathematical model can be made. Since width considered as a constant due to calculated values of spur gears and other components' constructive dimensions, the optimization problem becomes planar, based on optimizing the area defined by multiplying length (L), and width (H) of the transmission.

Fig. 1 shows the positions of the gears relative to the x and y axes. The position of the input shaft is constrained in the origin, while the positions of other shafts is defined by the angle, θ_i which is also shown in Fig.1. Gearbox volume is directly dependent on the change of these angles. A constant, c_1, is left on all sides as a technical clearance and is included in both the initial and optimal models, which is added to form the overall length and height.

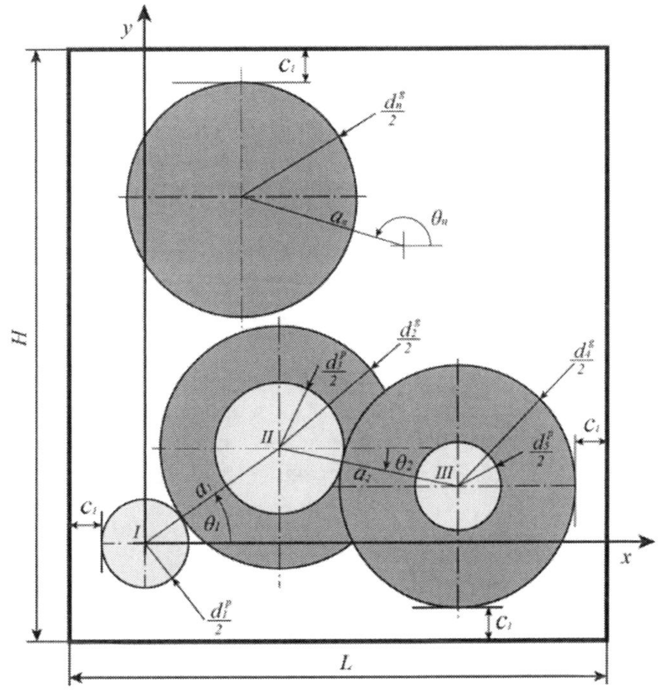

Fig. 1. Formulation of optimal gear train shaft axes.

Before optimization, in the initial design phase the following condition must be met for paired gears u and v in cases of gear trains with three or more stages:

$$a_{uv} > \frac{d_{u-1}^{p}}{2} + \frac{d_{v-1}^{g}}{2} \tag{1}$$

This ensures there is no clash between non paired elements placed in the same plane. If more than one stage can be placed in the same plane, this will initially give a decrease of volume even before optimization (Fig. 2a).

3 Mathematical model

Once all initial design conditions are met, and gear dimensions calculated, the typically positioned gear train can be optimized. Gear train volume changes when shaft position angles are changed. Minimum volume is achieved through achieving a minimal area of the profile plane. A difference of the rightmost coordinate of the gear train from the origin placed in the first pinion, and the leftmost coordinate from the origin gives the length of the gear train for the observed angles. In the same way the topmost, and bottom most coordinates subtracted give the height of the gear train. The function to determine the area which the gear train occupies is defined as follows:

$$f = \left[\max\left(\frac{d_{ai}^{(p)}}{2}, \sum_{i=1}^{n}(a_i \cdot cos\theta_i) + \frac{d_{ai}^{(g)}}{2}\right) + c_1 - \min\left(-\frac{d_{ai}^{(p)}}{2}, \sum_{i=1}^{n}(a_i \cdot cos\theta_i) - \frac{d_{ai}^{(g)}}{2}\right) + c_1\right] \cdot$$

$$\cdot \left[\max\left(\frac{d_{ai}^{(p)}}{2}, \sum_{i=1}^{n}(a_i \cdot sin\theta_i) + \frac{d_{ai}^{(g)}}{2}\right) + c_1 - \min\left(-\frac{d_{ai}^{(p)}}{2}, \sum_{i=1}^{n}(a_i \cdot sin\theta_i) - \frac{d_{ai}^{(g)}}{2}\right) + c_1\right] \tag{2}$$

This research presents a gear train optimization approach based on changing positions of gear shaft axes in relation to the horizontal plane. The angle between shaft angle positions can vary in the range of: $-180° \leq \theta_i \leq 180°$. The previous equation represents a general mathematical model. The equation includes the addition of 15mm on all sides of the gear train area as technical clearance for a housing. This clearance is also included as a clearance on all sides in the horizontal plane, as well as being a minimal distance between two gears on a single shaft.

In order to properly constrain the complex mathematical model, so that the result of the optimization gives a realistic model, many design factors need to be considered. Multilevel gear trains with spur gears can be set up in different ways in the initial design phase depending on spur gear dimensions. In case two sets of gears can be placed in the same plane and maintain a minimal clearance between non-paired gears, the volume can be decreased initially by reducing the overall width of the gear train.

Optimization constraints, similarly to the initial condition (Fig. 2a), in this case are maintaining a minimal clearance between non-paired gears of these stages when the shaft axes angles change (Fig.2b). For all stages in separate planes there is an obligatory constraint for avoiding clash of shaft and gear, and limiting their distance to a minimal possible value of 15 mm, where this clash is possible (Fig.2c).

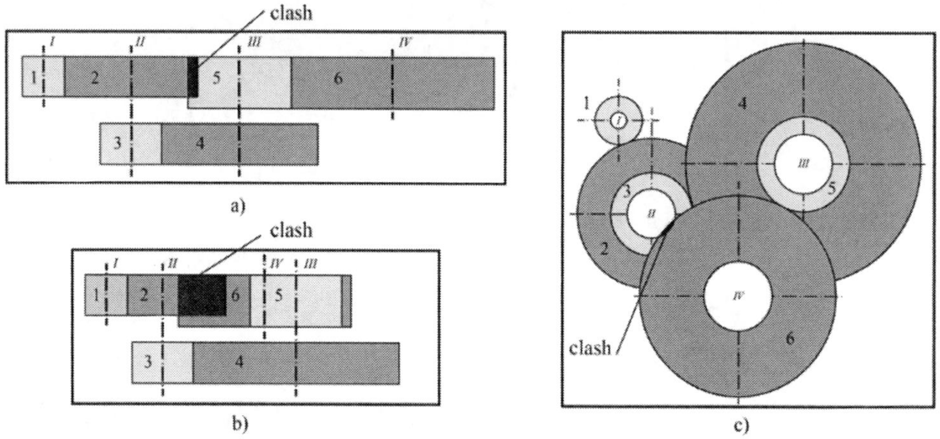

Fig. 2. Possible clashes between a) non paired gears in initial design phase, b) non paired gears in optimized model, c) shaft and gear in optimized model.

The objective function for all concepts has an addition of a minimum distance between gears and housing on all sides. The minimal volume which is achieved through optimization includes clearances not only in the observed plane, but also on shaft lengths. These technical clearances have constant values.

Constraints which ensure against clash of any non-paired elements *m* and *n* are given as a minimal center distance between those elements which must be greater than or equal to the sum of their respective radiuses and a constant technical clearance. The general equation for this constraint is:

$$a_{mn} = \sqrt{x_{mn}^2 + y_{mn}^2} \geq \frac{d_m}{2} + \frac{d_n}{2} + c_2 \, ; \qquad c_2 = 15mm$$

$$x_{mn} = \sum_{j=m}^{n} a_j \cdot cos\theta_j \, ; \qquad y_{mn} = \sum_{j=m}^{n} a_j \cdot sin\theta_j \, . \qquad (3)$$

4 Optimization method and software

Gear train volume optimization is a complex problem and as such is not possible without the use of a computer. In order to optimize, it is necessary to use an optimization method flexible for complex problems with numerous constraints. In this case, genetic algorithm method was chosen due to its favorable characteristics, and an original software has been developed for optimization based on this method. A real coded genetic algorithm (RCGA) was used, the algorithm is presented in Fig.3. The software used was developed in C++.

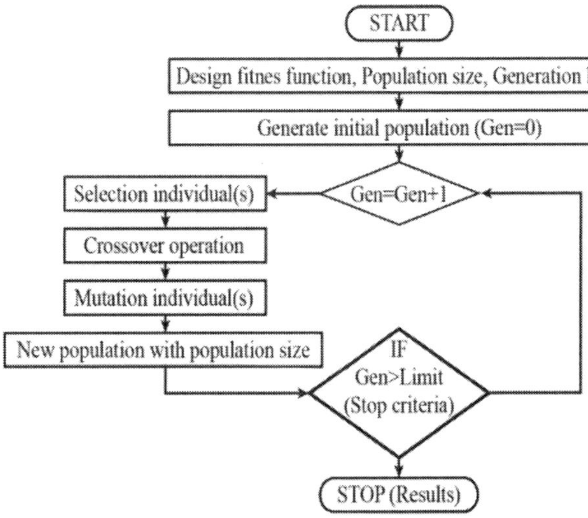

Fig. 3. Genetic algorithm for real coding (RCGA).

Genetic algorithm (GA) is a heuristic method for optimizing whose operation is based on mimicking natural/evolutionary processes [8]. The algorithm contains three basic operators: selection, crossover, and mutation. Selection is the process of transferring genetic information through generations. Crossover represents the process/operations between two parents, where an exchange of genetic information and new generations are made. The mutation operator creates a random change in the genetic structure of some individuals for overcoming early convergence. Algorithm operation is based on survival of the fittest individuals through evolution which exchange genetic material. Selection ranks individuals in the population using values from the fitness function, which defines the ability/quality of the individual.

Genetic algorithm, due to its convergence characteristics has a widespread application. Researchers are inspired to use this algorithm for scientific purposes, industrial application, business applications and to further increase its use.

Before use the software was tested according to available literature and its operation meets all necessary criteria. The original software is intended for all problems formulated by a mathematical model.

5 Test examples

For the purpose of verifying the mathematical model, in this paper three examples of various multistage gear trains were put through the optimization process. As the model can be applied to any gear train with two or more stages the examples used are of a two stage (Example 1) [2], three stage (Example 2) [2], and four stage (Example 3) [1] gear trains. Selected gear trains are representative of real practical examples used in the industry and are comparable to results from literature. The input values for the mathematical model for these examples are shown in Table 1.

Table 1 also gives initial gross dimensions of each example with included technical clearances. As the mathematical model aims to find the minimal volume by changing the area normal to shaft axes optimized examples are hypothesized to change in some measure correlating with the change of shaft position angles. Volumes of test examples in the initial concept arrangement are 9 925 631.1mm^3 for Example 1, 72 341 883mm^3 for Example2, and 39 697 456.5mm^3 for Example 3. The concept of Example 2 does not have the 1st and 3rd gear pairs in the same axes due to the initial design conditions not being met. Example 3 has the 4th gear pair face aligned with the same side face of the 1st gear pair which also places it behind the 2nd gear pair due to the width of the 4th gear pair. The initial concept of Example 3 therefore has an initially decreased volume than if the gear pairs were all in their own respective planes.

Table 1. Input values for creating a mathematical model for all examples.

	Example 1(2 stage)			**Example 2 (3 stage)**			**Example 3 (4 stage)**			
Stage	1	2		1	2	3	1	2	3	4
Module (mm)	2	3		2	3	4.5	1.5	2	3	4
Number of teeth in pinion	20	23		24	24	23	14	18	20	25
Number of teeth in gear	83	67		110	84	56	77	79	74	82
Pitch diameter of pinion (mm)	40	69		48	72	103.5	21	36	60	100
Pitch diameter of gear (mm)	166	201		220	252	252	115.5	158	222	328
Face width (mm)	103	135		134	162	177.75	68.25	97	141	214
Net length, height, width	388.5	231	110.6	653.75	282	392.4	724.75	358	153	

6 Results

The mathematical model is formed based on geometric characteristics of multistage gear trains. Optimization of initial concepts was conducted in the aforementioned software using RCGA. Input values were inserted into the mathematical model from Table 1.

As the width of the gear train is unchanged, the mathematical model gives a minimal area. Convergence of the area and shaft axes angles are shown in Figures 4, 5, and 6 for a smaller number of iteration steps to illustrate rate of convergence. Optima were achieved in 1500 iteration steps for Example 1, 3000 for Example 2, and 6000 for Example 3.

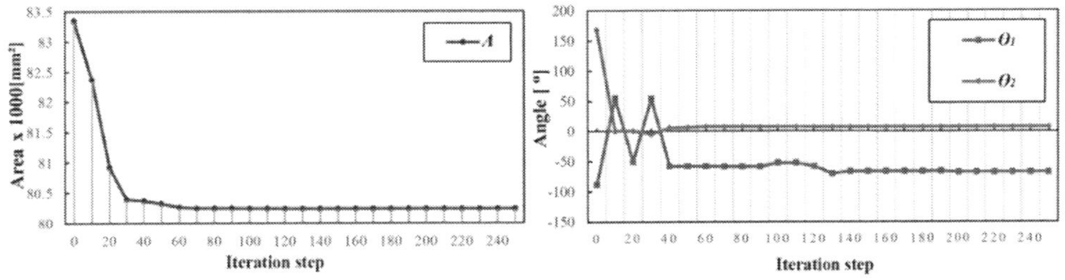

a) b)

Fig. 4. Example1 convergence of a) minimal area, b) angle values, at iteration step.

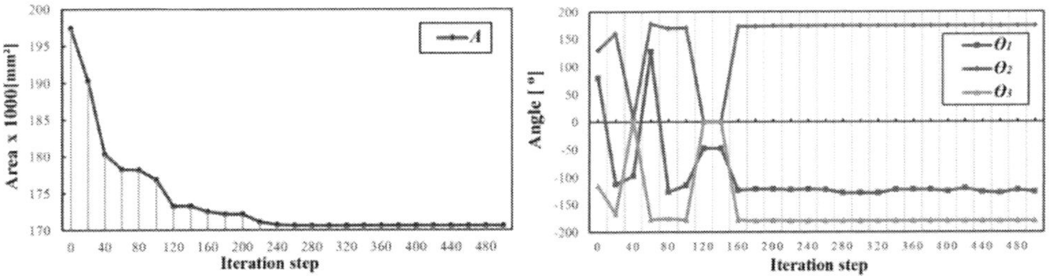

a) b)

Fig. 5. Example 2 convergence of a) minimal area, b) angle values, at iteration step.

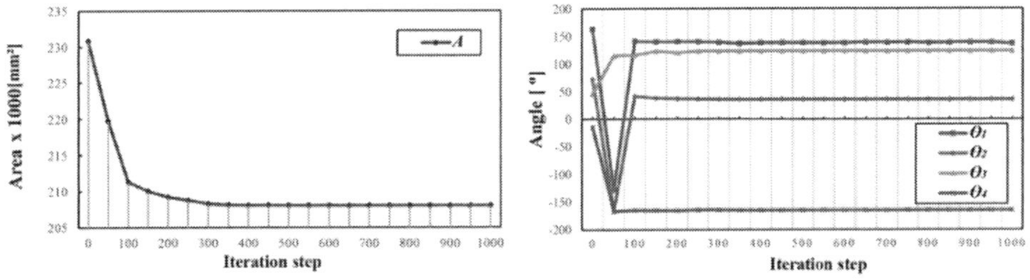

a) b)

Fig. 6. Example 3 convergence of a) minimal area, b) angle values, at iteration step.

The optimization results shown in Table 2 show an overall decrease in volume for all examples. Examples 1 and 2 show an unchanged height and decreased length, while Example 3 has an increased height and significantly decreased height.

Table 2. Optimal values of shaft position angles, overall dimensions, and volume for examples.

	Example 1	**Example 2**	**Example 3**
θ_1 [°]	-55.0359	-61.3595	-154.2624
θ_2 [°]	7.4482	5.6681	-144.9345
θ_3[°]	-	0	-57.7319
θ_4[°]	-	-	14.3393
L [mm]	347.361	604.958	555.61
H [mm]	231	282	373.702
B [mm]	110.6	392.4	153
V [mm^3]	8874585.592	66942709.8	31767763.44

Initial angles have changed significantly from original concepts with all shaft axes in the same plane resulting in optimal concepts with a minimal volume (Fig. 7).

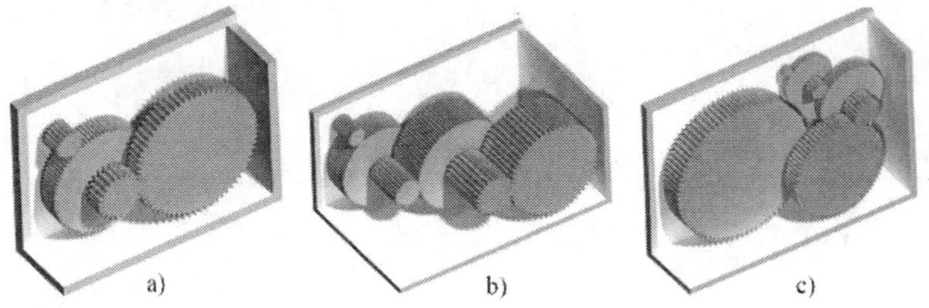

Fig. 7. Optimal concept positions for a) Example 1, b) Example 2, and c) Example 3.

Examples in figure 7 are given as a section view of a simplified housing in order to better show the optimal interior volumes and concepts of the gear trains. Example 3 in figure 6 is shown with the housing rotated by 90° for better visualization.

Unlike in [1] constraints for Example 3, as well as for all other examples, do not allow for an overlap of shafts and gears. All shafts have an intended clearance to allow for shafts to be designed in any desired length without forcing the design of a more complex housing to compensate for shaft embedding.

Example 1 gives an optimal area (volume) for the range of $-52.4912° \leq \theta_1 \leq -72.0083°$. All concepts also give optimal volumes for completely mirrored or completely normal values of all angle optima.

7 Conclusions

Optimization of gear trains with spur gears is a complex process. One such practical approach is presented in this paper. Presented in the paper is a universal mathematical model which can be applied to gear trains with two or more stages. Three examples are shown optimized using the presented mathematical model calculated in an original software based on a GA method. Real coded genetic algorithm (RCGA) is used for optimization. A large number of iterations is necessary to find optima due to the complexity of the mathematical model.

The examples initial volumes are calculated for the common case where all shaft axes are in the same plane. Comparison of initial volumes and optimization results for concepts of Example 1 show a decrease in volume of 10.589%, for Example 2 a decrease of 7.463%, and for Example 3 a decrease of 19.975%.

Gear trains with 2, 3 and 4 stages are used as test examples to show the influence of the number of stages on the percentage of volume decrease. Example 2 is used to show that the number of stages does not directly influence volume minimization. In other words the two-stage gear train (Example 1) has a greater relative decrease in volume from its original conception than the three-stage gear train (Example 2) due to a small difference in pinion sizes relative to their corresponding gears. If pinion to gear ratios are greater the area, and in turn volume is expected to have a greater difference in volume for optimal shaft positions.

Initial conditions and optimization constraints are incorporated to ensure the optimal concepts are practically applicable. The achieved minimization of volume in turn results in lower costs and weights of the housing for such gear trains.

References

1. T.H. Chong, I. Bae, G.J. Park, MMT, **37**, 295 (2002)
2. N. Marjanovic, B. Isailovic, V. Marjanovic, Z. Milojevic, M. Blagojevic, M. Bojic, MMT, **53**, 1 (2012)

3. S.I. Golabi, J.J. Fesharaki, M. Yazdipoor, Mechanism and Machine Theory, **73,** 197 (2014)
4. F. Mendi, T. Başkal, K. Boran, F.E. Boran, ESA, **37,** 8058 (2010)
5. V. Savsani, R.V. Rao, D.P. Vakharia, MMT, **45,** 531 (2010)
6. C. Gologlu, M. Zeyveli, Computers & Industrial Engineering, **57,** 1043 (2009)
7. P.H. Tang, M.H. Tseng, Applied Soft Computing, **13,** 600 (2013)
8. E.G. Shopova, N.G. Vaklieva-Bancheva, Computers & Chemical Engineering, **30,** 1293 (2006)
9. F. Musharavati, A.S.M. Hamouda, ESA, **38,** 10770 (2011)
10. S. Wojciechowski, R. Maruda, P. Nieslony et al., International Journal of Mechanical Sciences, **119,** 360 (2016)
11. R. Maruda, E. Feldshtein, et al., AJSE, **41,** 661 (2016)
12. J.B. Krolczyk, G.M. Krolczyk, S. Legutko, et al, Technical Gazette, **22,** 1447 (2015)
13. G.M. Krolczyk, S. Legutko, MMS, **21,** 759 (2014)
14. G.R.S. X. Li, G. Cockerham, MMT, **31,** 717 (1996)
15. S.P.N. Mushtaq M.A. Kader, G.K. Grover, MMT, **33,** 839 (1998)

General aspects regarding the kinematic and dynamic analysis of the fabric take up/ cloth roller

Cristian Matran[1,*], and *Nicolae* Bercan[1]

[1]Lucian Blaga University of Sibiu, Faculty of Engineering, Machines and Industrial Equipments Department, 550012 Sibiu, Romania

Abstract. The aim of this paper is to presents different aspects regarding the kinematic and dynamic behaviour of the fabric winding at the weaving loom. In this paper shall be presented the kinematic diagram for operating the cloth roll on the weaving machine, making the study of kinematic and dynamic cloth behaviour, setting up the expression of fabric tension. Finally, where drawn the charts of the variation of the fabric load depending by kinematic and constructive parameters.

1 Introduction

In present days, the request for the high quality garments imposes certain trends in the weaving loom development/construction. The modern weaving loom uses positive take-up mechanisms, acted by individual c.c. motors and clutches, protecting the fabric for over loading.

The paper presents a kinematical and a dynamical analysis of the take-up and cloth roller mechanism. The command schema and the kinematic chain that acting the take up cylinder contain the computer, gears, chain and frictional clutch as is shown in Figure 1 [1-2], where: 1 – loading cell, 2 – computer, 3 – d. c. motor, 4 - z_i gears, 5 – cloth take up roller, 6 – chain and chain wheels, 7 – clutch, 8 – cloth roller.

2 Theoretical kinematical and dynamical aspects of the fabric loading

In the fabric take up and rolling process, the fabric element is extracted with a constant velocity. The characteristics of the cloth (from the paper's point of view) are:
- specific weight "ρ";
- fabric thickness "δ";
- the length of the cloth roller "l_0";
- the diameter of the fabric roller "d_0";
- the axial moment of inertia "J_0".

* Corresponding author: cristian.matran@ulbsibiu.ro

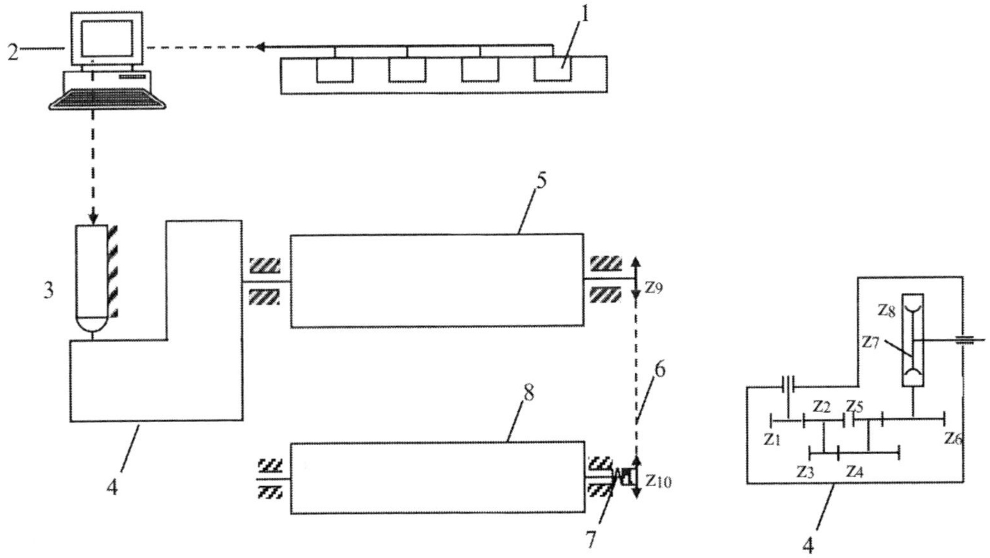

Fig. 1. The command schema of the take up and the fabric rolling mechanism (1- loading cell, 2- computer, 3- d. c. motor, 4- z_i gears, 5- cloth take up roller, 6- chain and chain wheels, 7- clutch, 8- cloth roller).

In the process of the rolling cloth, its axial momentum of inertia (J_Δ) represents a sum of the both empty fabric roller and fabrics' axial momentum of inertia (1) [3-4].

$$J_\Delta = J_0 + J_\Delta^{tes} \quad (1)$$

The axial moment of inertia of the empty fabric roller (J_0) could be calculated using:

$$J_{cil} = J_{cil}^{plin} - J_{cil}^{gol} = \frac{m_{cil}^{plin} \cdot r_0^2}{2} - \frac{m_{cil}^{gol} \cdot r_{int}^2}{2} \quad (2)$$

Where: m_{cil}^{plin}, m_{cil}^{gol} – the mass of the filled/empty fabric roller, [kg], r_0, r_{int} – external/internal radius of the fabric roller, [m].

Using the constructive particularities of the beam and the density of its material (aluminium), the equation (2) became:

$$J_{cil} = \frac{\pi \cdot \rho \cdot h}{2}\left(r_0^4 - r_{int}^4\right) \quad (3)$$

Where: ρ – Aluminium density [2700 kg/m^3], h – Length of the fabric beam [m], r_0, r_{int} – external/internal radius of the fabric beam, [m].

The fabrics' axial momentum of inertia , J_Δ^{tes} depends by time (indirectly) and could be calculate using the relation:

$$J_\Delta^{tes} = \int r^2 dm \quad (4)$$

Where: r_0 – the radius of the empty fabric beam, [m], R – maximal radius of the fabric roller, [m], r – the radius of the fabric roller at a specific time, [m], dm – elementary mass, [kg].

 DOI: 10.1051/matecconf/201712101008

And

$$dm = \rho \cdot dV \tag{5}$$

dV – elementary volume, Figure 2.
It results:

$$dm = \rho \cdot dV = \rho \cdot dl \cdot dA = \rho \cdot dl \cdot dl_1 \cdot dr = \rho \cdot dl \cdot r \cdot d\theta \cdot dr \tag{6}$$

for $dl \in [0, l_0]; d\theta \in [0, 2\pi]; dr \in [r_0, r]$
Where *dl* – elementary length of cloth roller.

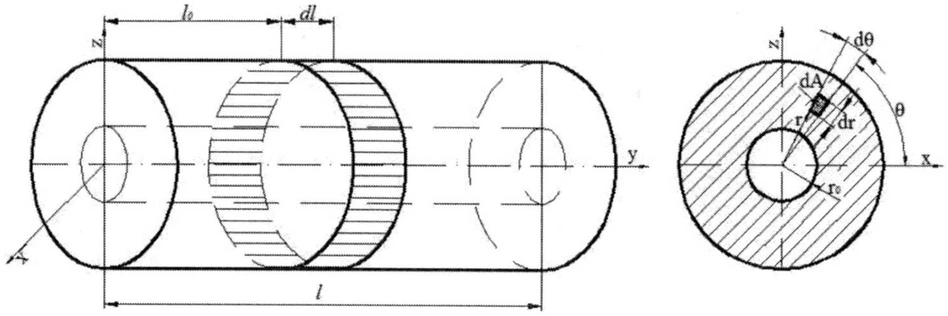

Fig. 2. The elementary volume calculus, dV.

dA –surface element;
dθ –; the elementary angle;
$dl_1 = r \cdot d\theta$
dl_1 – elementary length of the circles' arch.
Replacing (6) in (4), it results:

$$J_{\Delta}^{tes} = \rho \cdot \int_{r_0}^{R} r^3 dr \cdot \int_{0}^{l} dl \int_{0}^{2\pi} d\theta \tag{7}$$

Where

$$r = r_0 + n \cdot \delta$$
$$n = \frac{\theta}{2\pi} \tag{8}$$

n –the number of cloth layers at a specific moment.
Finally, will obtain:

$$J_{\Delta}^{tes} = \frac{\pi \rho l}{2}\left[\left(r_0 + \delta \cdot \frac{\theta}{2\pi}\right)^4 - r_0^4\right] \tag{8}$$

θ – the warping angle, [rad];
δ – fabrics' thickness, [m].
The elements of the fabric winding are show in the figure 3.
The theorem of the kinetic moment variation, calculated in relation with the rotation axis of the fabric beam, is given by:

$$\dot{K}_{\Delta} = M_{\Delta} \tag{9}$$

Where

$$K_{\Delta} = J_{\Delta} \cdot \omega \tag{10}$$

$$\dot{K}_{\Delta} = \dot{J}_{\Delta} \cdot \omega + J_{\Delta} \cdot \dot{\omega} \tag{11}$$

$$M_{\Delta} = M_a - T \cdot r \tag{12}$$

It results:

$$\dot{J}_{\Delta} \cdot \omega + J_{\Delta} \cdot \varepsilon = -T \cdot r + M_a \tag{13}$$

Where M_a –drive momentum, [N·m],

$$M_a = 9.55 \cdot 10^3 \frac{P}{n_s} \tag{14}$$

P –the power of the driving motor [W];
n_s –speed of the fabric roller [rpm].

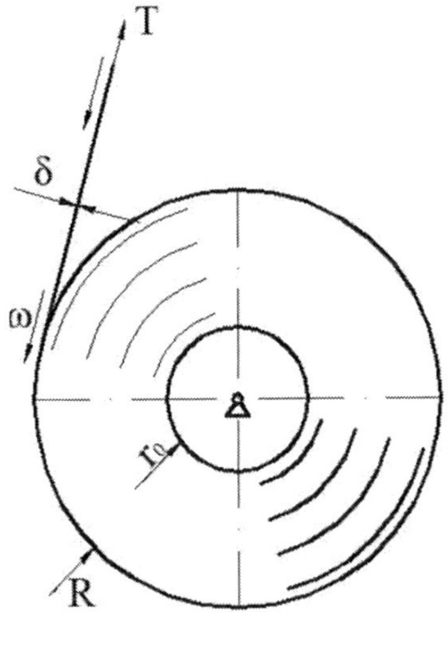

Fig. 3. Elements of the fabric winding [5].

From the equation (13), we can determine the fabric tension:

$$T = \frac{1}{r} \cdot \left(M_a - \dot{J}_{\Delta} \cdot \omega - J_{\Delta} \cdot \varepsilon \right) \tag{15}$$

Where: ε –angular acceleration, [rad/s^2].
Deriving the equation (1), we get:

$$\dot{J}_{\Delta} = \dot{J}_0 + \dot{J}_{\Delta}^{tes} \tag{16}$$

And

$$\dot{j}_0 = 0 \tag{17}$$

Deriving and successive processing the equation (8), we get:

$$j_{\Delta}^{tes} = \frac{4\pi\rho l}{2}\left(r_0 + \frac{\delta\cdot\theta}{2\pi}\right)^3 \cdot \frac{\delta\cdot\dot{\theta}}{2\pi} = \rho\cdot l\cdot\delta\cdot r^2\cdot v_T \tag{18}$$

From figure 3 results the angular acceleration deriving the (9) in relation to time:

$$\varepsilon = \dot{\omega} = \frac{d\omega}{dt} \tag{19}$$

Realizing a change of variable in (19), we'll obtain:

$$\varepsilon = \frac{d\omega}{dr}\cdot\frac{dr}{dt} \tag{20}$$

Where:

$$\frac{d\omega}{dr} = \frac{d}{dr}\left(\frac{v_t}{r}\right) = -\frac{v_t}{r^2} \tag{21}$$

$$\frac{dr}{dt} = 0 + \delta\cdot\dot{n} = \delta\cdot\left(\frac{\dot{\theta}}{2\pi}\right) = \frac{\delta}{2\pi}\cdot\dot{\theta} = \frac{\delta}{2\pi}\cdot(\dot{\omega}\cdot t) = \frac{\delta\cdot\omega}{2\pi} = \frac{\delta\cdot v_t}{2\pi\cdot r} \tag{22}$$

Replacing in (20), it became:

$$\varepsilon = -\frac{v_t}{r^2}\cdot\frac{\delta\cdot v_t}{2\pi\cdot r} = -\frac{\delta\cdot v_t^2}{2\pi\cdot r^3} \tag{23}$$

Replacing the calculated sized in relation (15), the final relation will be:

$$T = \frac{1}{r}\left[M_a - \rho\cdot l\cdot a\cdot r^2\frac{v_t^2}{r} + \frac{\delta\cdot v_t^2}{2\pi\cdot r^3}\cdot\frac{\pi\cdot\rho\cdot l}{2}\left(r^4 - r_{\text{int}}^4\right)\right] \tag{24}$$

3 Experimental aspects to calculate fabric loading

The theoretical values were applied to the Somet Thema weaving loom, with the main parameters shown in table 1.

The working parameters of the loom being optimized by the loom's computer, the transmission of the movement to the fabric roller presents different technological capabilities:

- z_1=48/**24**/17, z_2=80/**96**/103, z_3=16, z_4=48, z_5=16, z_6=48, z_7=2, z_8=90, z_9=45, z_{10}=31;
- main motor speed, max. 2350 rpm.;
- take up roller speed, 100 – 1400 rpm;
- pick density, 5 - 150 picks/cm.

Table 1. Combination of gears to calculate maximal fabric density.

Type	Z	yarns/cm	yarns/inch	Comment
B	$z_1 = 40$ $z_2 = 80$	5 – 40	12,7 – 101,6	At request
C	$z_1 = 24$ $z_2 = 96$	10 – 80	25,4 – 203,2	Standard
D	$z_1 = 17$ $z_2 = 103$	15 - 150	38,1 - 381	At request

Appling the specific parameters of the loom in equation (27) it results the values of the fabric loading as shown in table 2.

Table 2. Fabric loading.

n_{et} [rpm]	Ma [Nm]	δ [m]	J_0 [kg·m²]	ρ_s [kg/m²]	l_0 [m]	r_0 [m]	r [m]	T [N]
470	355.5851	0.00429	0.18881	0.1414	1.9	0.1	0.1000	3555.8509
470	355.5851	0.00429	0.18881	0.1414	1.9	0.1	0.1125	3160.7564
470	355.5851	0.00429	0.18881	0.1414	1.9	0.1	0.1250	2844.6808
470	355.5851	0.00429	0.18881	0.1414	1.9	0.1	0.1375	2586.0735
470	355.5851	0.00429	0.18881	0.1414	1.9	0.1	0.1500	2370.5673
470	355.5851	0.00429	0.18881	0.1414	1.9	0.1	0.1625	2188.2160
470	355.5851	0.00429	0.18881	0.1414	1.9	0.1	0.1750	2031.9149
470	355.5851	0.00429	0.18881	0.1414	1.9	0.1	0.1875	1896.4539
470	355.5851	0.00429	0.18881	0.1414	1.9	0.1	0.2000	1777.9255
470	355.5851	0.00429	0.18881	0.1414	1.9	0.1	0.2125	1673.3417
470	355.5851	0.00429	0.18881	0.1414	1.9	0.1	0.2250	1580.3782
470	355.5851	0.00429	0.18881	0.1414	1.9	0.1	0.2375	1497.2004
470	355.5851	0.00429	0.18881	0.1414	1.9	0.1	0.2500	1422.3404
470	355.5851	0.00429	0.18881	0.1414	1.9	0.1	0.2625	1354.6099
470	355.5851	0.00429	0.18881	0.1414	1.9	0.1	0.2750	1293.0367

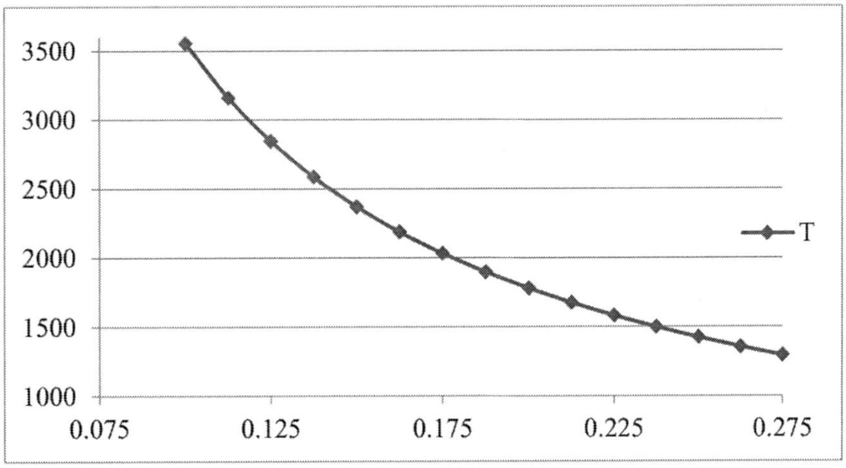

Fig. 4. The variation of the fabric loading depending by winding radius.

4 The Clutch's influence

As is seen in the weaving process, the fabric load is falling down according with the rising of the radius, as shown in table 4 and figure 4.

On the other hand, using the gear ratio between take up roller and the cloth beam, the peripheral speed of the fabric roller is rising, overloading the fabric (the gear ratio is greater than 1). This overload of fabric could be lowered using a frictional clutch. Because the same element of fabric extracted by the take up roller must be wrapped on the cloth roller, and keeping in mind that the both peripheral speeds of the rollers are different, we could calculate the clutch's slippage.

Starting from the equation of the rollers' peripheral velocity depending by its speed, we obtained:

$$l_t = l_s \tag{25}$$

Where:
$l_{t,s}$ – the length of a fabric element, [m];

On the other hand, the expression of the cloth roller's speed, based on the take up roller's speed becomes:

$$n_s = n_t \cdot \frac{z_9}{z_{10}} \cdot \frac{100-a}{100} \tag{28}$$

Where:
z_9, z_{10} – sprockets' size;
a – clutch slippage, [%].

$$n_s = \frac{v_t}{2 \cdot \pi \cdot r_x} = n_t \cdot \frac{z_9}{z_{10}} \cdot \frac{100-a}{100} \tag{29}$$

Where:
v_t – peripheral speed of the take up roller, [m/s];
r_x – the radius of the cloth roller at a specific time, [m].
According with (29), we can calculate the radius of cloth roller:

$$r_x = \frac{100 \cdot v_t}{2 \cdot \pi \cdot (100-a)} \cdot \frac{z_{10}}{z_9} \tag{30}$$

The slippage of the clutch will result according with (30):

$$a = 100 - \frac{100 \cdot v_t}{2 \cdot \pi \cdot n_t \cdot r_x} \cdot \frac{z_{10}}{z_9} \tag{31}$$

Considering that a=0% (the working moment of clutch), the cloth roller's radius is:

$$r_x = \frac{v_t}{2 \cdot \pi \cdot n_t} \cdot \frac{z_{10}}{z_9} \tag{32}$$

At limit, when $r_x \to \infty$,

$$a = 100 \tag{33}$$

Table 3. Clutch's slippage.

n_{et} [rpm]	n_t [rpm]	v_t [m/s]	r [m]	a [%]
470	0.29	0.255	0.1000	3.5923
470	0.29	0.255	0.1125	14.3043
470	0.29	0.255	0.1250	22.8738
470	0.29	0.255	0.1375	29.8853
470	0.29	0.255	0.1500	35.7282
470	0.29	0.255	0.1625	40.6722
470	0.29	0.255	0.1750	44.9099
470	0.29	0.255	0.1875	48.5825
470	0.29	0.255	0.2000	51.7961
470	0.29	0.255	0.2125	54.6316
470	0.29	0.255	0.2250	57.1521
470	0.29	0.255	0.2375	59.4073
470	0.29	0.255	0.2500	61.4369
470	0.29	0.255	0.2625	63.2732
470	0.29	0.255	0.2750	64.9426

Based on the Table 3, the graphical variation of the slippage is presented in figure 5.

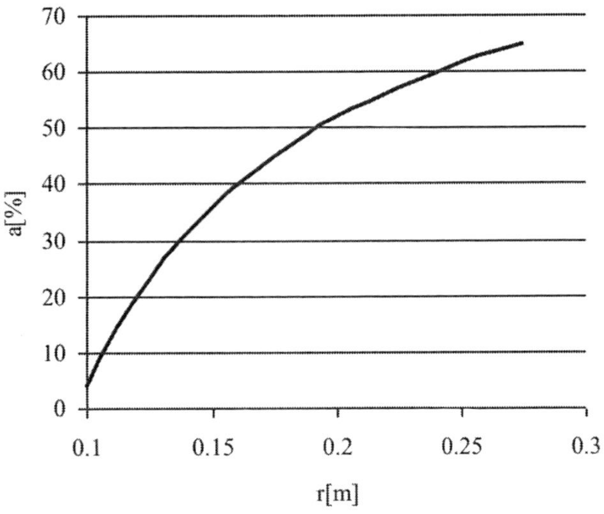

Fig. 5. The clutch's slippage.

5 Conclusions

Starting from kinematical and dynamical analysis, it results a calculus formula of the fabric loading in the wrapping process, depending by the geometrical dimensions of the cloth roller.

Also, we obtained the mathematical equation and diagrams of speed variation and clutch's slip, depending by the radius of the cloth roller, too.

As is seen in the weaving process, the fabric load is falling down according with the rising of the radius, as shown in table 2 and figure 4.

On the other hand, using the gear ratio between take up roller and the cloth beam, the peripheral speed of the fabric roller is rising, overloading the fabric. This overload of fabric could be eliminated by the worker by using a clutch (raising the slippage of the clutch's discs).

The worker's action in the fabric rolling is more often with the radius raising and it decrease the number of weaving loom surveyed by a worker.

The state-of-the-art weaving looms use an electronic controlled take up and individual motors [6-7].

References

1. *** *Somet Thema 11E User's Manual*
2. S. Maity, K. Singha, M. Singha, AJSS, **1**, 7 (2012)
3. I. Gheorghe, *Statica si dinamica structurilor* (ULBS, Sibiu, 2009)
4. I. Gheorghe, N. Bercan, V. Oleksik, *Culegere de probleme de mecanică* – Dinamica (ULBS, Sibiu, 2013)
5. D. Liuțe, D.D. Liuțe, *Bazele prelucrării firelor textile* ("Gh. Asachi" Pub. House, Iași, 2002)
6. *** ITEMA R9000 Rapier Weaving Machine (http://www.itemagroup.com/en/products/rapier/r9000/, 2015)
7. H.A. Mannan, *Automation in Weaving Shed, Let-off, Take-up and Stop Motion*, (http://textilelearner.blogspot.ro/2014/02/automation-in-weaving-shed-electronic.html)

Load capacity of worm gear transmission from aspect of maximal use of available resources

Aleksandar Miltenović[1,*], *Milan* Banić[1], and *Đorđe* Miltenović[2]

[1]University of Niš, Faculty of Mechanical Engineering, 14 A. Medvedeva, 18000 Niš, Serbia
[2]The College of Textile, V. Pušmana, 16000 Leskovac, Serbia

Abstract. Worm gear transmission has number of advantages over the other types of transmission (high transmission ratio, compact design, high load capacity, vibration damping, reliability, etc.) and thus is widely used as drives in application with short-term load and for the transmission of movement. Previous studies in this area show that in their work can occur a larger number of damage forms that are limiting the load capacity of this drives and that can be identified based on the condition of working surfaces. The occurrence of any limit state represents the termination of work function, no matter that resources for other limit states are not reached. Theoretically, the best use of transmission would be if the load capacity for all limit states has uniform use. The paper is focusing a detailed analysis of the load carrying capacity of worm gear transmission from aspect of the available resources.

1 Introduction

Worm gears have a wide range of applications, particularly in the short-term drive because of their significant advantages. Applications of worm gear drives such as machine tools, transport equipment and vehicles are primarily linked to the transfer of power, as well as fine and precision devices for the transfer of movement. The most important advantage is their large transmission ratio, which in one stage can be achieved in the range of $i = 5$ to $i = 70$. In this way, it is possible to achieve a very large area of output speeds.

Contact between tooth flans is a line and depending on gear parameters contact can be on number of teeth (usually 2 ... 4). Because of the contact line between flanks occurs relatively large sliding. Meshing is realized with oil layer between the flanks which decreases the internal dynamic forces, with vibration damping [1-2]. This benefit primarily relates to worm gear sets with greater efficiency [3].

High slip between flanks of worm and worm gear leads to high local surface pressure, which requires the use of materials worm pairs that are resistant to scuffing damage form [4]. A possible solution for this is a combination of dissimilar soft and hard material of worm gear pair. Therefore, in modern structural solutions usually worm gear made of tin bronze a worm tempered and brushed steel. The disadvantage of this solution is low strength and high price of the production of worm gear [5].

* Corresponding author: amiltenovic@yahoo.com

This paper shows that the small changes in parameters of worm gear drives can lead to the significant savings which is not the case in other drives such as drives with spur, helical or bevel gears. Therefore, this is one more advantage of worm gear drives which product development engineers are often not aware.

2 Load Capacity and Available Resources of Worm Gear Transmission

Criteria for load capacity of worm gear drive are torque on worm gear or the output torque, which is directly coupled to the driven machine. The values of the output torque are limiting boundary conditions that can occur during operation of the transmission. The most common limit states are:

- destruction of tooth flanks due to pitting,
- wear of tooth flanks,
- fracture of the worm gear tooth at the root,
- heating and heat distortion stability of transmission,
- scuffing.

For the analysis of load capacity of worm gear drive is considered a family of transmission $a = 63$, $a = 100$ and $a = 250$, whose characteristics are given in Table 1. Calculation of load capacity of the worm gear transmission according to DIN 3996 [6] regarding to wear, pitting, teeth fracture and the thermal stability. The critical layer thickness is estimated by thickness of wear layer on the tip of tooth.

Table 1. Transmission data for a = 63 mm; a = 100 mm; a = 250 mm.

	Transmission ratio	u		**9.75**	**19.5**	**29**	**39**
a = 63 mm	Number of teeth of worm	z_1		4	2	1	1
	Number of teeth of worm gear	z_2		39	39	29	39
	Axial module	m_x	mm	2.5	2.5	3.15	2.5
	Operative diameter of worm	d_{m1}	mm	26.5	26.5	33.5	26.5
	Root diameter of worm	d_{f1}	mm	20.5	20.5	25.94	20.5
	Root diameter of worm gear	d_{f2}	mm	93.5	93.5	84.94	93.5
a = 100 mm	Transmission ratio	u		10.33	20.5	29	40
	Number of teeth of worm	z_1		3	2	1	1
	Number of teeth of worm gear	z_2		31	41	29	40
	Axial module	m_x	mm	5	4	5	4
	Operative diameter of worm	d_{m1}	mm	45	36	53.175	40
	Root diameter of worm	d_{f1}	mm	33	26.4	41.175	30.4
	Root diameter of worm gear	d_{f2}	mm	143	154.4	134.82	150.4
a = 250mm	Transmission ratio	u		10	20	31	40
	Number of teeth of worm	z_1		4	2	1	1
	Number of teeth of worm gear	z_2		40	40	31	40
	Axial module	m_x	mm	10	10	12.5	10
	Operative diameter of worm	d_{m1}	mm	95	95	112	95
	Root diameter of worm	d_{f1}	mm	71	71	82	71
	Root diameter of worm gear	d_{f2}	mm	381	381	358	381

For comprehensive review of load capacity of worm gear drive according to different criteria are varied the following parameters (Table 1):

- Axial distance (a = 63 mm; a = 100 mm; a = 250 mm),
- Transmission ratio (i = 10; i = 20; i = 30; i = 40),

- Worm speed (n_1 = 200 min^{-1}; n_1 = 500 min^{-1}; n_1 = 1000 min^{-1}; n_1 = 1500 min^{-1}; n_1 = 2000 min^{-1}),
- Material of worm gear (CuSn12-C-GZ; CuAl10Fe5Ni5-C-GZ; EN-GJS-400-15)
- Oil (mineral, synthetic),
- Lubrication (immersed, injection),
- Lifetime L_h = 10000h.

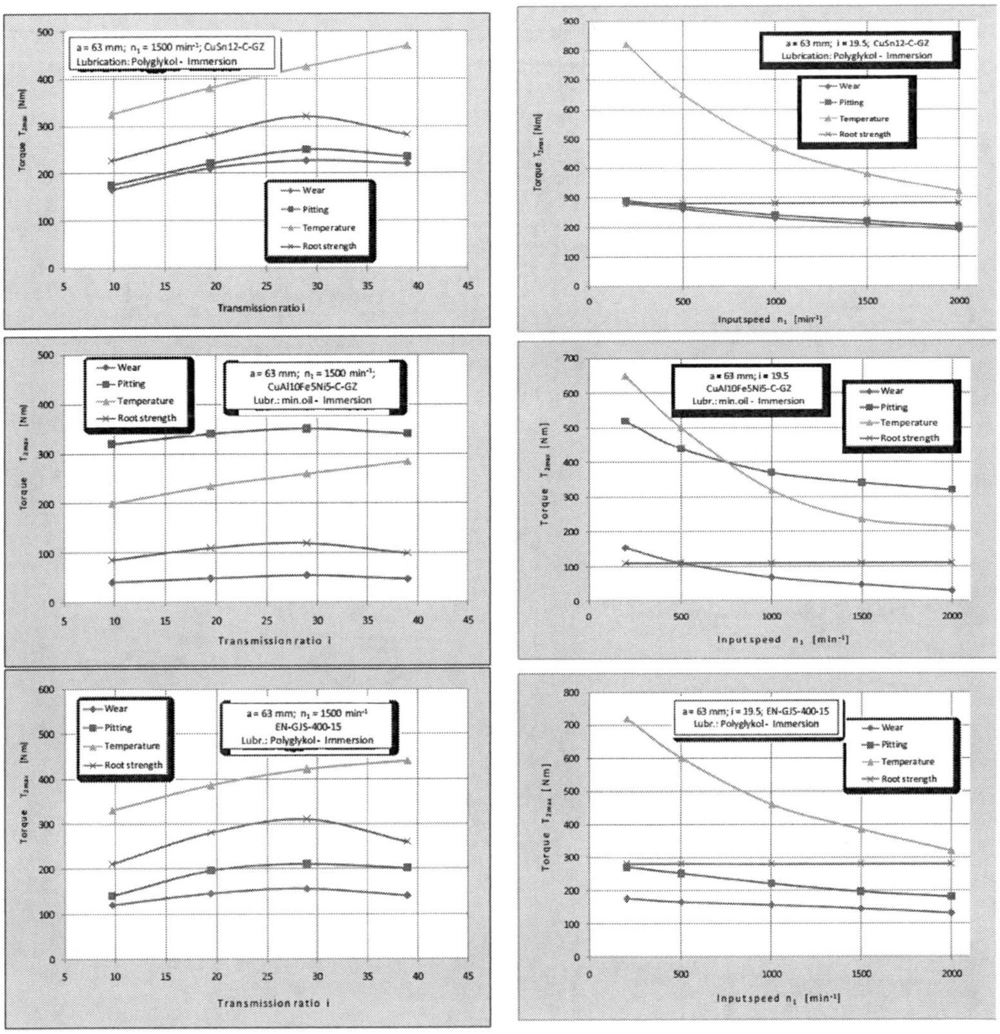

Fig. 1. Load capacity of drive with a = 63 mm for different transmission ratio, speeds and materials of worm gears.

2.1 Transmission a-63

The calculation results of the load capacity of the drive a-63 (a = 63 mm) are shown in figure1.

For drive a-63 with worm made of alloy steel and worm gear from tin bronze CuSn12-C-GZ with the varying transmission ratio for constant speeds n_1 = 1500 min^{-1} optimum load capacity is obtained for the area of transmission ratio i = 20-30. For the transmission ratio

$i = 19.5$ and vary speeds respectively sliding speed, nearly all criteria obtained uniform load capacity, which is good from the aspect of transmission resources.

Transmission a-63 with worm made of alloy steel and worm gear of Al-bronze CuAl10Fe5Ni5 C-GZ with varying transmission ratio and at varying speeds according to all criteria, the calculation results not uniform load capacity.

Transmission a-63 with worm made of alloy steel and worm gear of ductile iron EN-GJS 400-15 at varying transmission ratio for constant speeds $n_1 = 1500$ min^{-1} optimum load capacity is obtained for the area of transmission ratio $i = 20 - 30$. Even with varying transmission ratio and at varying speeds on all the criteria of the calculation results in good uniformity load capacity.

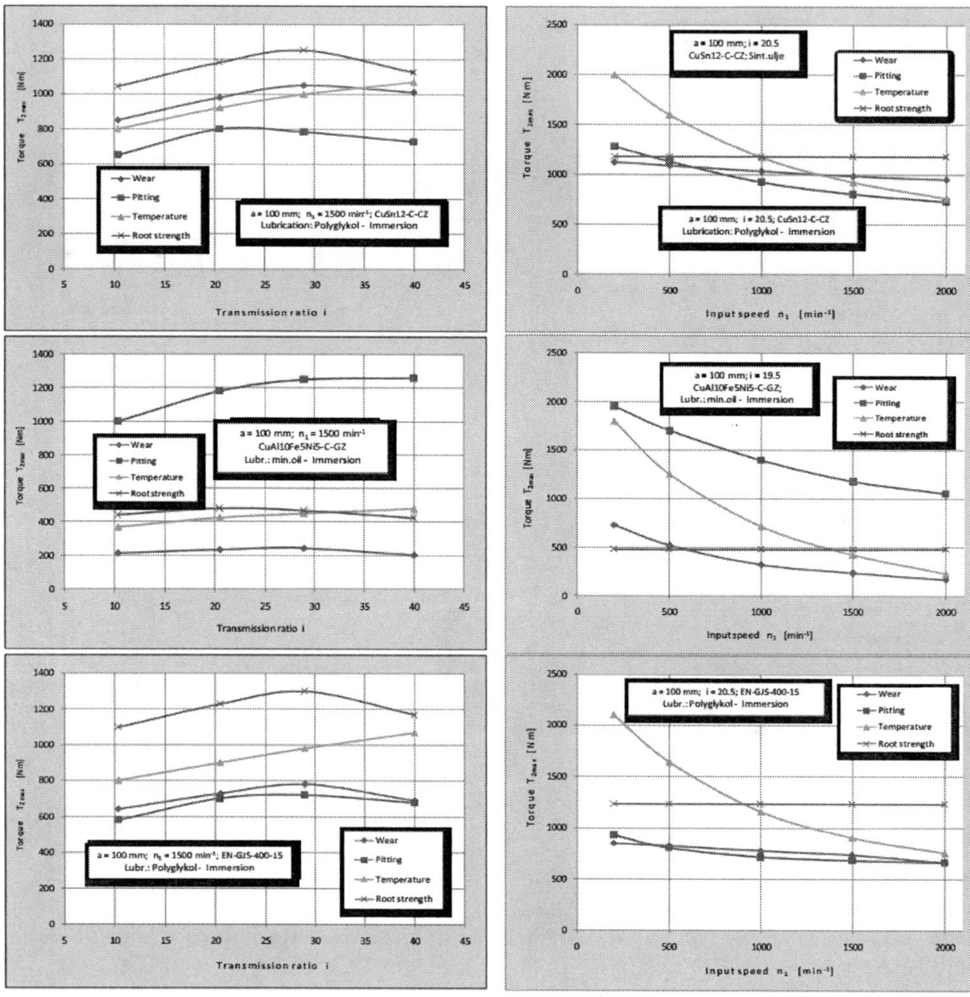

Fig. 2. Load capacity of drive with $a = 100$ mm for different transmission ratio, speeds and materials of worm gears.

2.2 Transmission a-100

Calculation of load capacity of drive a-100 ($a = 100$ mm) was performed in a similar manner as the calculation of load capacity of the drive a-63. The calculation results are given in the load capacity at figure 2.

 DOI: 10.1051/matecconf/201712101009

Drive a-100 with worm made of alloy steel and worm gear from tin bronze CuSn12-C-GZ can be seen that with the varying transmission ratio for constant speeds $n_1 = 1500$ min^{-1} optimum load capacity is obtained for the area of transmission ratio $i = 20$-30. For the gear ratio $i = 20.5$ and vary speeds regarding sliding speed nearly all criteria, the calculation results with balanced load capacity, which is not the case with varying transmission ratio.

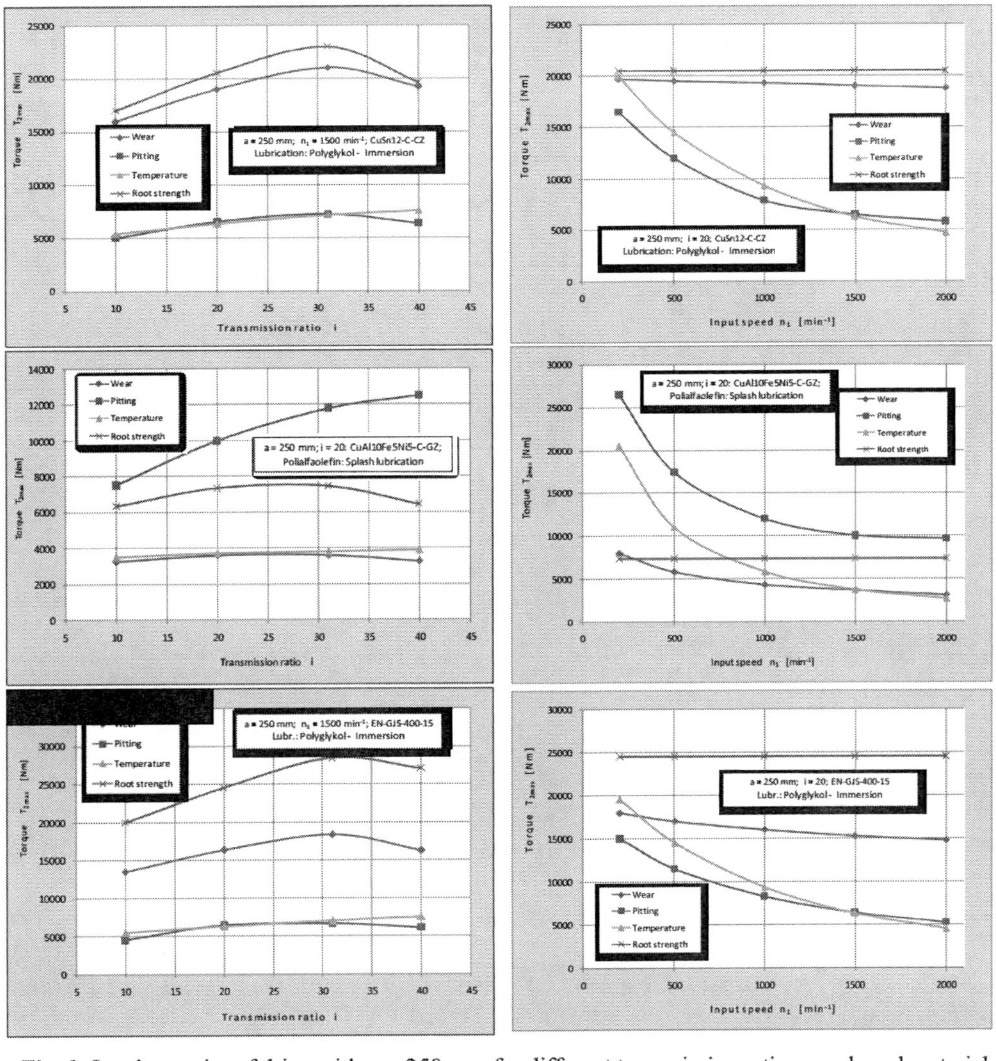

Fig. 3. Load capacity of drive with $a = 250$ mm for different transmission ratio, speeds and materials of worm gears.

When drive a-100 with a worm made of alloy steel and worm gear of Al-bronze CuAl10Fe5Ni5 C-GZ can be seen that with varying transmission ratio and at varying speeds according to all criteria, the calculation results are showing the not uniform load capacity. Consequently, the resources available of transmission are only partially used.

For worm drive a-100 with a worm made of alloy steel and worm gear made of ductile iron EN-GJS-400-15 can be seen that with the varying transmission ratio for constant speeds $n_1 = 1500$ min^{-1} optimum load capacity is obtained for the area of transmission ratio $i = 20$ - 30.

Table 2. Limitation of load capacities of transmission.

<table>
<tr><th>No</th><th>Worm gear material</th><th>Transmission ratio/ Speed</th><th>a = 63 mm</th><th>a = 100 mm</th><th>a = 250 mm</th></tr>
<tr><td>1</td><td rowspan="9">Tin bronze</td><td>$i = 10$</td><td rowspan="4">H (B); UN</td><td rowspan="4">B; UN</td><td rowspan="3">T, B, NUN</td></tr>
<tr><td>2</td><td>$i = 20$</td></tr>
<tr><td>3</td><td>$i = 30$</td></tr>
<tr><td>4</td><td>$i = 40$</td><td>B, NUN</td></tr>
<tr><td>5</td><td>$n_1 = 200$ min^{-1}</td><td rowspan="5">H, B, UN</td><td rowspan="2">H; UN</td><td rowspan="3">B, NUN</td></tr>
<tr><td>6</td><td>$n_1 = 500$ min^{-1}</td></tr>
<tr><td>7</td><td>$n_1 = 1000$ min^{-1}</td><td rowspan="3">B; UN</td></tr>
<tr><td>8</td><td>$n_1 = 1500$ min^{-1}</td><td rowspan="2">T, NUN</td></tr>
<tr><td>9</td><td>$n_1 = 2000$ min^{-1}</td></tr>
<tr><td>10</td><td rowspan="9">AL bronze</td><td>$i = 10$</td><td rowspan="4">H; NUN</td><td rowspan="4">H; NUN</td><td rowspan="4">T, H, NUN</td></tr>
<tr><td>11</td><td>$i = 20$</td></tr>
<tr><td>12</td><td>$i = 30$</td></tr>
<tr><td>13</td><td>$i = 40$</td></tr>
<tr><td>14</td><td>$n_1 = 200$ min^{-1}</td><td rowspan="2">P, NUN</td><td rowspan="2">P, NUN</td><td>P, NUN</td></tr>
<tr><td>15</td><td>$n_1 = 500$ min^{-1}</td><td rowspan="2">H, NUN</td></tr>
<tr><td>16</td><td>$n_1 = 1000$ min^{-1}</td><td rowspan="3">H, NUN</td><td rowspan="3">H, NUN</td></tr>
<tr><td>17</td><td>$n_1 = 1500$ min^{-1}</td><td>H, T, NUN</td></tr>
<tr><td>18</td><td>$n_1 = 2000$ min^{-1}</td><td>T, NUN</td></tr>
<tr><td>19</td><td rowspan="9">Ductile iron</td><td>$i = 10$</td><td rowspan="4">H; ≈UN</td><td rowspan="4">B, H</td><td rowspan="4">T, B, NUN</td></tr>
<tr><td>20</td><td>$i = 20$</td></tr>
<tr><td>21</td><td>$i = 30$</td></tr>
<tr><td>22</td><td>$i = 40$</td></tr>
<tr><td>23</td><td>$n_1 = 200$ min^{-1}</td><td rowspan="5">H; UN</td><td rowspan="5">B, H, ≈UN</td><td rowspan="3">B, NUN</td></tr>
<tr><td>24</td><td>$n_1 = 500$ min^{-1}</td></tr>
<tr><td>25</td><td>$n_1 = 1000$ min^{-1}</td></tr>
<tr><td>26</td><td>$n_1 = 1500$ min^{-1}</td><td>B, T, NUN</td></tr>
<tr><td>27</td><td>$n_1 = 2000$ min^{-1}</td><td>T, NUN</td></tr>
<tr><td colspan="6">H – wear (45); B – pitting (36); P – tooth-root strength (5); T – thermal stability (18); UN – uniform load capacity; NUN – not uniform load capacity</td></tr>
</table>

2.3 Transmission a-250

The calculation results of load capacity of drive a-250 (a = 250 mm) are given in the diagram figure 3.

For drive a-250 with worm made of alloy steel and worm gear from tin bronze CuSn12-C-GZ can be seen that, on average, have about the same on one side load capacity for pitting compared with thermal stability ($T_2 = 6.400$ Nm) and on other side load capacity for wear comparing to the tooth root safety ($T_2 = 19.400$ Nm).

For a-250 with a worm made of alloy steel and worm gear of Al-bronze CuAl10Fe5Ni5 C-GZ can be seen that with varying transmission ratio and varying speeds according to all criteria calculation results in not uniform load capacity.

For drive a-250 with a worm made of alloy steel and worm gear made of ductile iron EN-GJS-400-15 can be seen that for varying the transmission ratio the load capacity of drive is about the same compared the safety factor temperature of the drive with pitting safety ($T_2 \approx 6.240$ Nm).

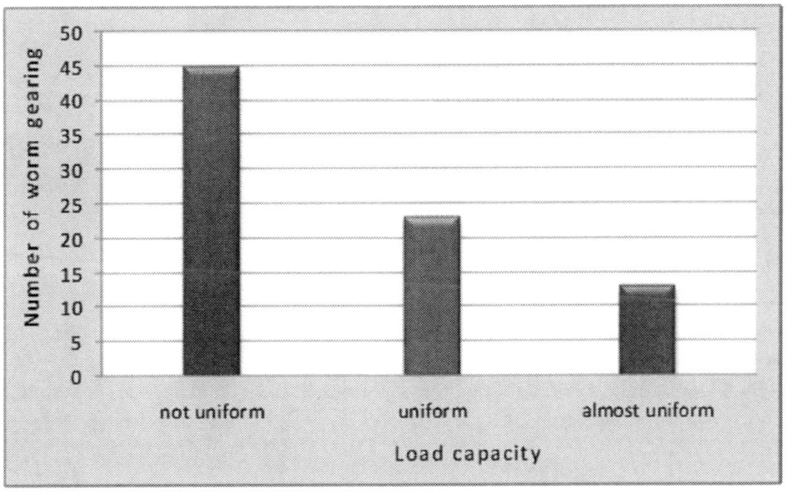

Fig. 4. Load capacity of transmission.

Calculation is performed for 81 cases which take into account the variation of the transmission ratio, speeds, material of worm gear and the axial distance. Above analysis (Table 2) shows the load capacity for different load capacity limitations. It may be noted that the largest load capacity of worm gear drives is calculated for wear, pitting and thermal stability.

From the aspect of use of available resources worm drives is very important that the drive has a similar load capacity for all limit states. This is indicated in the analysis as uniform (UN) or the not uniform load capacity (NUN).

From 81 calculations, only at 23 drives have uniform load capacity, 13 approximately uniform, while 45 drives have a not uniform load capacity (Fig. 4). The best use of resources of drives was obtained in the calculation with worm gear made of tin bronze. This is the reasons why majority of producers of worm gear drives use combination of material – worm made of tempered (hardened) steel with ground teeth and worm gear of centrifugally cast tin bronze.

3 Conclusion

In this paper is presented calculation of load capacity for wear, pitting, fracture at the tooth root and thermal stability for selected family of worm gear drives with axial distance $a = 63$ mm, $a = 100$ mm and $a = 250$ mm. In calculation are varied the following parameters: axial distance, transmission ratio, speed, worm gear material, type and method of lubrication oil. Based on the calculation results can be conclude following:

- Results of the calculation (Table 2) show that the largest load capacity of worm gear drive for wear, pitting and thermal stability.
- From the aspect of use of available resources of worm gear drives is very important that the drive has a similar load capacity for all limitations.
- The results of analysis are shown in Table 2 and in Fig. 4 show that of the 81 analysed drives only 23 drives have uniform load capacity, in 13 is approximately uniform, while 45 have not uniform load capacity. Therefore, the maximum use of available resources was achieved in 23 drives (28% of all analysed drives).

This approach shows that the variation of parameters have decisive importance for load capacity of worm gears as well as the extension of working life. Small changes in the parameters can lead to significant savings which is good approach for product development engineers.

References

1. B-R. Höhn, K. Michaelis, C. Weisel, International Conference on Gears - Gears 2010. VDI Berichte 2108.2 (2010)
2. A. Miltenović, M. Milovančević, Đ. Miltenović, Proceedings of the 5th KOD, Novi Sad, Serbia (2008)
3. Đ. Miltenović, M. Tica, A. Miltenović, M. Banić, Proceedings of the12th DEMI, Banja Luka, BiH *(2015)*
4. Đ. Miltenović, *Istraživanje termičke stabilnosti i habanja pužnih prenosnika* (PhD, Banja Luka, 2017)
5. V. Miltenović, M. Banić, A. Miltenović, Proceedings of the 7th KOD 2012, Balatonfured, Hungary (2012)
6. DIN 3996: Tragfähigkeitsberechnung von Zylinder-Schneckengetrieben mit sich rechtwinklig kreuzenden Achsen (2012)

MATEC Web of Conferences **121**, 01010 (2017) DOI: 10.1051/matecconf/201712101010
MSE 2017

About the best value of rake angle for cutting tools

Eugen Pămîntaş[1,*], and *Felicia Veronica* Banciu[1]

[1]Politehnica University Timişoara, Mechanical Faculty, Materials and Manufacturing Engineering Department, 300222 Mihai Viteazu Av., Timişoara

Abstract. Regardless of whether the case of processing through methods of cutting metal, wood, plastics, stone, composite materials etc., the rake angle (γ) is of major importance for energy efficiency and dynamic behaviour of cutting process, dimensional accuracy, work surface quality and pieces made cost. The paper examines the possibility of establishing the best value for the rake angle supposing two idealizations: one regarding the conditions in which cutting process takes place, and the second related to the simplifications aiming to facilitate a first stage of theoretical development of mathematical description for the chip forming process. Both idealizations are intended to individualize the influence that has the value of rake angle of the cutting tools on the process of chip forming in order to perceive the meaning of the link between the geometry of the cutting tool with other parameters of chips and by default with the work parameters set to machine-tool. While both validity and applicability of the mathematical relations obtained are negatively marked by the simplifying assumption which was appealed in achieving the approach, the highlighted findings may be good starting points for further research on the kinematics cutting depth and in the structure of the material being processed.

1 Introduction

When treating the problem of angle values for active part of cutting tools, the scientific literature abounds with recommendations, motivations and logical explanation of the variety of situations and practical working conditions.

All assertions summarized above represent in fact a very brief overview of these situations, to introduce an old question: it might determine the optimum value of rake angle at cutting from conditions which take into account the energy efficiency at metal cutting?

Attempts to answering this question were highly much in the last hundred years. The results are very good and briefly presented below: „At present, tool forces, tool temperatures, tool wear, component surface finish, and ease of chip disposal are all impossible to predict a priori with any certainty and/or without a large number of quasi-empirically obtained constants. ''Machining is an unconstrained metal working operation'' [1]. Indeed, even more, because of the abundance of methods in an attempt to respond to the challenges caused by new materials for workpieces [2], by new materials [3] and new

*Corresponding author: eugen.pamintas@upt.ro

types of tools [4], by the performance of new machine tools and more stringent requirements of customers in quality and cost, was said: „...such an approach can hardly be called a science" [1].

The purpose of this paper is to examine the possibility of establishing the best value for rake angle (γ).

2 Problems of the main cutting angles

In any chip removing process the sizes of main angles: clearance angle (α) and rake angle (γ) are widely known to be crucial for the process [5-7].

The size of the clearance angle (α) must necessarily be positive for the cutting can be held and its maximum value is limited, particularly in terms of resistance of the active part of the tool and by the size of wear (h_α) on the major flank (A_α) of the tool (usually not exceeding 12 ° and often coming within 5 ° ÷10 °).

The value of rake angle (γ) is however of great importance to the mechanical work of plastic deformation and splitting chip, for:
- The temperature dissipated in tool.
- Chip and piece.
- The form of chip detached.
- Wear and tool life.
- The dynamic behaviour.
- The overall energy efficiency of the cutting process.
- The dimensional accuracy, surface quality, and for the final cost of the processed part.

If we were to limit ourselves only to the effects listed above, the best value for departure angle would be that which could provide:
- Minimum cutting efforts.
- The most powerful chip fragmented.
- The better surface smoothness of the processed part.
- The highest durability for the tool edge.
- The smallest amplitudes of vibration intensities of machine tools.
- The least harmful effects on others geometrical parameters of tool cutting as a whole.
- Lowest cost of operating the tool etc.

The rake angle being of such importance, it is not surprising that even industry standards normalized no less than three types (6 variants) of sharpening the rake face of cutting tool (A_γ), e.g. STAS350-82 and STAS R6781-83 [8], as opposed to one way of sharpening for clearing face of the cutting tool.

However, for choosing the value of the rake angle there are numerous recommendations in university courses, handbooks, design guides and professional literature, normative and standards of companies that manufacture cutting tools (e.g. Sandvik, Walter etc.).

Unlike the clearance angle, the rake angle can have values both negative and positive:
- Negative rake angle have the disadvantage of very big stresses on the part, high power consumption to the machine tool, poor quality of part surface manufactured and greater resistance of cutting tool peak and greater durability thereof;

All these reasons make that negative values for rake angle must only be used to roughing operations in the case of variant of sharpening with negative facet and their value does not exceed about -8°.
- Positive rake angles are limited primarily by higher flexural strength of the cutting tool material and only subsequently by the coefficient of friction chip-tool.

For these reasons:
-it is usual recommended for rake angle values below 10° (very rarely in steel finishing);
- and disc or prism shaped knives is reached and the value of 25°.

3 About Chip Formation

3.1 Work hypothesis

The approach is based on two idealisation hypothesis:
- One concerning the conditions in which the cutting process takes place.
- The other, in connection with simplifications designed to ease, in the first stage development of theoretical mathematical description regarding the process of chips forming.

Both idealizations are intended to individualize the influence of rake angle cutting tools value has in the process of chip forming in order to perceive the meaning of the link between the geometry of the cutting tool with other chips parameters and default working parameters set to machine-tool.To facilitate understanding this paper's aim and the extent of the validity of its conclusions, it will bring some clarification on the meaning of the word "ideal" as used in paragraph beginning.

Ideal in the sense of this approach refers to conditions in which the cutting process, on the one hand:
- The contact between the workpiece and tool edge sector is made in a direction perpendicular to the main resultant cutting speed vector (Fig. 1).
- The length of the tool major cutting edge is at least equal to the length of depth of cut projection on that edge.
- The thickness (b) and the width (a) of the material layer to be removed from the workpiece (undetached chip) are constant throughout the machining phase;
- Undetached chip length is at least twice the depth of cut (a_p),
- Kinematic parameters of the cutting process (v, f, a_p) are constant during machining.

On the other hand, refers to introduce simplifications to ease - in this first phase of development theory- the mathematical description of the process of detachment of the chip:
- Cutting without using coolant.
- The tool major cutting edge is perfectly sharpened.
- The workpiece material is homogenous and isotropic.
- There is no friction between the tool-piece-chip and does not form build-up edge and other similar phenomena.

In these circumstances, the ideal longitudinal turning will be characterized by parameters defined in the tool-in-hand system, as presented, with the notation according to ISO 3002 / I-1977 [8] in Figure 1.

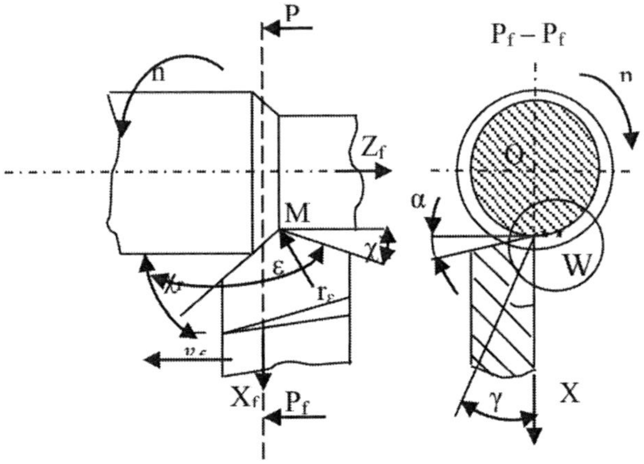

Fig. 1. The ideal longitudinal turning scheme.

For example and support of theoretical future development in Figure 1 is shown schematically the case of longitudinal turning, that captured a Δt time in a turning phase of a cylindrical steel workpiecewith mechanical and structurally homogeneous and isotropic known,using constant cutting conditions.

3.2 Ideal chip model

As result of cutting movements, the tool tensions a small material volume placed ahead it (Fig.2). Packages of plastic and elastic deformed crystals begin to slide against each other, orienting after sliding plane B-B (Fig.2) inclined with Φ_2 angle upon the cutting direction. When the value of strengths exceeds the tensile strength limit of the workpiece material, chip detaching occurs after shearing plane A-A, inclined with angle Φ_1 to cutting direction.

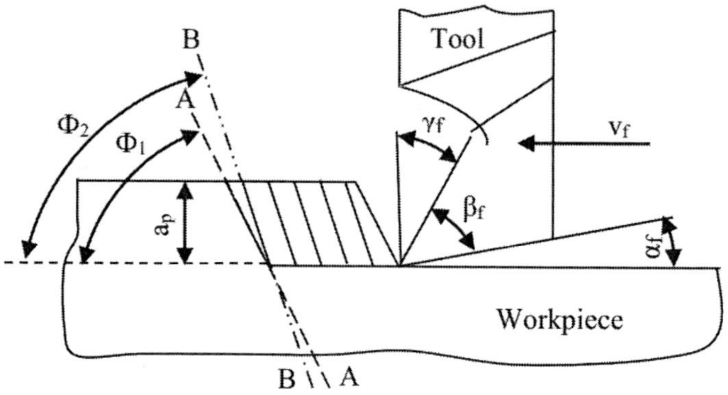

Fig. 2. Chip detaching mechanism.

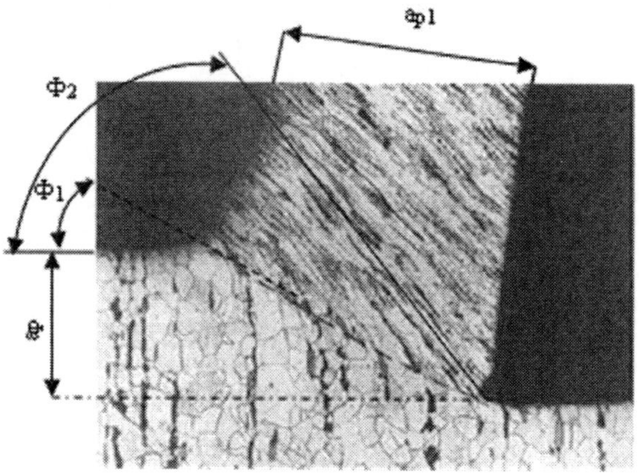

Fig. 3. Practical example of undetached chip material [9].

This theoretical mechanism of chip formation has often been highlighted by numerous experiments and metallographic analyses ofundetached chip material [9, 1], as shown in Figure 3.

From the same experiments among conclusions was find that angle values Φ_1 and Φ_2 depend mainly on chip elements (a- thickness, b-width), on frictional forces between the chip and tool rake face (A_γ), rake angle γ etc. and that the difference between their sizes is about: $\Phi_2 - \Phi_1 = 20° \text{ to } 30°$ [10].

Also, detached chip shape is severely affected by: the physical properties of the material to be cut, the chip area, the angle γ, and the cutting speed (v_c).

 DOI: 10.1051/matecconf/201712101010

Because the friction phenomena, the mechanical and thermal stresses, detached chip suffer deformation thus occurring chip shortening and thickening: $K_m = a_{pl}/ a_p$ (a_{pl} - detached chip thickness a_p - unbroken chip thickness; experimental observation of the compression ratio of the thickness of the chip Km values in the range 1.5 ÷ 4) [11].

In the ideal case of an isotropic and homogeneous workpiece material, in the absence of friction between tool-chip-workpiece, for orthogonal turning with constant working parameters, it can be assumed that the twinning angle and the shear have the same value and more this value is one of the characteristic peculiar to type of material.

The chip resulting in these assumptions would be one idealized, undeformed and cold for which the angles $\Phi_2= \Phi_1= \Phi$, undetached chip has the same dimensions with the detached one (see Fig. 4) and so would have $K_m=1$ ($a=a_{pl}= a_p$), $b= b_1$ width, length $l = l_0$ etc.

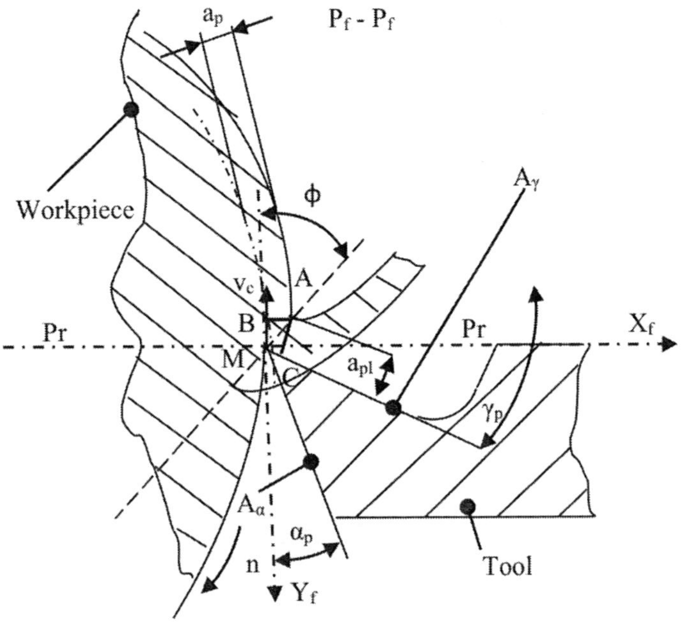

Fig. 4. Chip formation mechanism scheme for calculus.

4 Ideal rake angle

4.1 Establishing of calculus relation

In the above mentioned assumptions are searched in the following an expression of a relationship of dependency between the cutting tool rake angle γ and the workpiece material natural shear angle Φ. Thus, from MAB and MAC triangles in Figure 4, $\overline{MA}$ segment is developed:

$$\overline{MA} = \frac{a_p}{\sin\Phi} = \frac{a_{pl}}{\cos(\gamma_p - \Phi)}. \quad (1)$$

Using the notation $K_m=a_{pl}/a_p$from equation (1) results:

$$K_m = \cot\Phi\, \cos\gamma_p - \sin\gamma_p \quad (2)$$

In equation (2) $\sin\gamma_p$andcos $\gamma_p$will be expressed using $\tan{}^{\gamma_p}/_2$, designating $\tan{}^{\gamma_p}/_2 = t$. Next, developing the equation 2 will be obtained a quadric with t variable.

Solving this eq. it is obtained for $\gamma_p$the dependency relation in form of equation (3).

$$\gamma_{p1,2} = 2\tan^{-1}\frac{\tan \Phi \pm \sqrt{(1-K_m^2)\, tg^2\, \Phi + 1}}{K_m \tan \Phi + 1} \tag{3}$$

4.2 Testing the calculus relation

In the specialized literature are presented ranges of values among which - depending on working conditions - are usually coefficients, reports, cutting angles, etc., among which are found also those of interest in this case: Km, Φ and γ.

Table 1. Some solutions of equation (3) solving.

Φ / K_m	50°		60°		70°		80°	
	γ_p		γ_p		γ_p		γ_p	
	1	2	1	2	1	2	1	2
1	90°	10°	90°	30°	90°	50°	90°	70°
0.67	109°07'	-09°7'	114°33'	05°28'	120°59'	19°1'	128°38'	31°14'
0.50	117°30'	-17°29'	124°21'	-04°21'	131°57'	08°1'	140°32'	19°31'
0.4	122°08'	-22°09'	129°45'	-09°44'	137°53'	02°6'	146°48'	13°12'
0.33	125°22'	-25°22'	133°23'	-13°23'	159°29'	-01°56'	151°	08°58'
0.29	127°09'	-27°9'	135°29'	-15°28'	144°13'	-04°11'	153°24'	06°36'
0.25	128°56'	-28°57'	137°31'	-17°30'	146°25'	-06°25'	155°46'	04°15'
0.22	130°16'	-30°17'	139°	-18°59'	148°03'	-08°4'	157°30'	02°39'

By most times the extent of the ranges where can take values the cutting sizes covers all the working conditions and cases. However, most often, the practice has demonstrated that measurements that are at limits of ranges, and those near the middle thereof, are rarely attained. For this reason, in Table 1 was tested relationship (3) using the recommended values for K_m (0.22 ÷ 0.67) and Φ (45° to 80°) and the obtained results for rake angle were compared with those recommended by theorists and researchers in the cutting field and with those used by manufacturers and industrial users of turning tools [12-14].

Table 1 presents the full results of calculations (including ideally $K_m = 1$, regarded as the simplifying assumptions used in reference to the relationship deduction for calculating the angle γ) for some random values of the parameters K_m and Φ. Both solutions are obtained from equation (3) for γ_{p1} and γ_{p2} were tabulated in order to observe both values and as such the tendency of their variation.

Analysing the results in Table 1 it can be retained as relevant for the purposes of this paper the following basis:

- γ_{p1} is a solution without practical correspondent and should not be considered only as a marker for checking the keeping of upward trend with increasing the value of shear angle Φ and increasing the chip deformation (decrease of K_m);
- γ_{p2} solution is credible but other factors involved in cutting (failure conditions for chip detachment, friction angle between the material and tool, the prohibitive cutting forces, tool cutting edge strength, machined surface quality, etc.) make that only the values marked in grey in table 1 to be relevant for practice;
- Cutting conditions considered in case of ideal assumptions are obtained for γ_p angle values in range: $\gamma_p \in$ (-8°÷ 13°), when the chip contraction coefficient $K_m \in$ (0.22 ÷ 0.5) and shear angle of workpiece material is in the range of Φ ∈ (70°÷ 85°).

5 Conclusions

Although both the validity and applicability of mathematical relations obtained are marked negative by simplifying assumptions (which was called in realizing the presented approach), highlighted findings may be good starting points for further research into the kinematics of cutting and structure of the material being processed thoroughgoing study:
- Experimental testing meaning to evaluate the validity of these theoretically findings will be restricted to areas $K_m \in (0.25 \div 0.4)$ (low degree of chip shrinkage is very rare in practice and high degrees of shrinkage are prohibitive from an energetic point of view) and $\Phi \in$ (75° to 80°).
- γ_{p2} solutions will be considered only in relation (3) - with the "minus" between factors from fraction denominators.

Who would not like that knowing the shear angle of workpiece material Φ - provided by the manufacturer of raw materials – and requiring a certain value for the desired degree of chip contraction - for reasons of energy, dynamic machine-tool behaviour, chip removing process etc. – choosing the tool rake angle γ, the most favourable or optimal?

The idea of looking for a link between the shear angle Φ of each type of material and the rake angle γ that would provide the smallest chip breaking effort to the cutting operations, more than the validity of the deduced relationship, should be considered - more chosen due to the many simplifications in the hypothesis - as just a "scaffold", necessarily to lift any subsequent "construction". It is known that operating parameters, cutting conditions, material properties and cutting tool etc. influences the values of the cutting angles optimal to the intended purpose. However, the authors believe it would be desirable to know the value of the rake angle that would provide the lowest energy consumption. This is the point around which they revolve choice of values to optimize the criteria chosen by the client for each case. It would therefore be desirable to develop future research, not wasting precious time mastering scaffolding, confusing "platform" with "construction".

It should be established multidisciplinary research teams with specialists in cutting, materials physics, metallographic analyse that would work after such an experimental program that tries to solve the question of naturally shear angle of the material to be cut.

It could be another step in the sum of which would provide one of the most desired technical problems: full control of the cutting process

This research acknowledges to Walter Tools SRL, Timişoara Romania and Universitary Association of Manufacturing Engineering in providing information on the practice of metal cutting and support financial funding for this paper. The valuable comments from anonymous reviewers are greatly acknowledged to help improve the paper's quality.

References

1. P.K. Wright, D.A. Bourne, *Manufacturing Intelligence* (Addison-Wesley Publishing Company, USA, 1988)
2. M.P .Groover, *Fundamentals of Modern Manufacturing Materials*, (Processes and Systems, 4th ed., John Wiley & Sons Inc., New York, 2010)
3. S.R. Maity, P. Chatterjee, S. Chakraborty, Materials & Design, **36**, 372 (2012)
4. ***http://www.ttonline.ro/tag/walter-tools, accessed on 13.03.2017
5. Y. Altintas, *Manufacturing automation: metal cutting mechanics, machine tool vibrations, and CNC design* (Cambridge University Press, 2012)
6. L. Juneja, G.S. Sekhon Nitin Seth, *Fundamentals of metal cutting and machine tools* (New Age International Publishers, New Delhi, 2005)
7. M.E. Trent, K. Paul, K.P. Wright, *Metal Cutting* (Elsevier Inc., 2000)

8. *** *Cutting tools for metal manufacturing. Romanian standard*, **II,** (Technical Ed., Romania, 1987)
9. J.S. Colton, *Manufacturing processes and Systems. Machining overview & basics of chip formation mechanics,* (Georgia Institute of Technology, GIT, 2009)
10. I. Pop, *Cutting tools design* (Politehnica Publishing House, Timişoara, Roamania, 1992)
11. I. Dumitrescu, *Cutting tools design. Lab. Guide* (UniversitasPublisching House, Petroşani, Romania, 2016)
12. ***http://engineeringhut.blogspot.ro/2010/11/types-of-chips.html, accessed: 13.03.2017
13. ***https://www.slideshare.net/ilieureche/bazele-aschieriiaranjata, accessed: 26.02.2017
14. R. Kreheľ, D. Krchova, M. Kočiško, Key Engineering Materials, **669**, 382 (2016)

MATEC Web of Conferences **121**, 01011 (2017) DOI: 10.1051/matecconf/201712101011
MSE 2017

The effect analysis of the morphing concept on the small swept flying wings

Vasile Prisacariu[1,*], *Mircea* Boşcoianu[1], and *Ionică* Cîrciu[1]

[1] "Henri Coandă" Air Force Academy, Braşov, Romania

Abstract: The UAV manoeuvrability when classical control surfaces are used is limited. At present, the research focuses on the biological implications regarding total, partial or local morphing changes which might enable better trajectory control. The current paper shows an application of the flying wings with elements of twist morphing at plane edges meant to reveal a global optimum area of turning features, by use of theoretical and numerical methods.

1 Introduction

According to [1-2], UAVs are aerial vehicles with no pilot aboard which use aerodynamic forces for movement along 3D non ballistic trajectories, either controlled or independently, and which carry loads important to the areas of operational interest. The UAVs are largely used both in civilian and operational theatre missions. The use of this type of aerial system is cost-effective in terms of design, construction, and exploitation as compared to other manned aerial vehicles, [20].

The classical "aileron" [7] placed on the leading wing edge of the wing tips for ensuring the rolling motion was patented by the British Mattew Piers Watt Boulton in 1868. Later on, in 1906, in the USA, the Wright brothers were awarded an extended patent, (US 821393). Therefore, the geometry of the flying wing UAV is advantageous in terms of mass balance versus overall size, in the case of the same missions performed by conventional design UAVs (see Figure 1), [19].

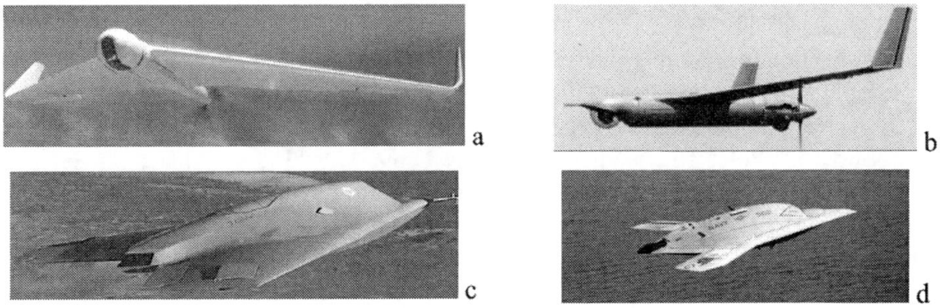

Fig. 1. Flying wing UAV (a. Orbiter-Israel, b.EagleScan-USA, c. Taranis-France. d. X46-USA), [22-25].

* Corresponding author: aerosavelli73@yahoo.com

Wing flying compared with classical configuration offers the following advantages: lower charging that involves a higher payload; a lower total resistance to progress due to absence tail and fuselage (with their interference) witch implies an increase aerodynamic characteristics (C_z / C_x, V_{min}); portability and modularity feature improved.

The classical control (incidence, skidding, and side slope) by turning the corresponding control surfaces (elevators, rudder, and ailerons) generates pitching movements, yaw and roll. Through turning these control surfaces can be done after the controls all six degrees of freedom of movement. The ability to change the angular velocity and flight speed (manoeuvrability) when classical control surfaces are used is limited. At present, the research focuses on the biological implications regarding total, partial or local morphing changes which might enable better control along the trajectory (see Figure 2).

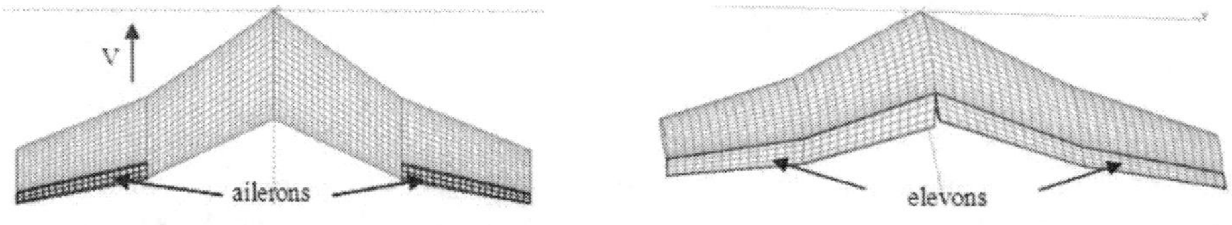

Fig. 2. Ailerons / elevons –flying wing.

Recent approaches to manoeuvrability aimed at applying the innovative morphing concepts [3-6, 12-13, 21] through replacement of classical ailerons with flexible geometries, be it local or on the surface. Manoeuvre control on the trajectory has required research, according to a series of aerial factors, of the concept of morphing which is classified as follows: 2D morphing (lift surface deformation) by means of variable geometry profiles (bending and thickness), the telescopic wing (simultaneously and alternatively), the variable-sweep wing; 3D morphing with variable-twist wing, morphing winglets (morphlets); morphing combined with the blow of the boundary layer with Coanda effect; total morphing or a combination of these (see Figure 3) [8]. The most important aspects when designing elevons are: effectiveness at low speeds and small deviations, effectiveness at high speeds and small deviations, reasonable action forces in the entire flight envelope, for reduced energy consumption.

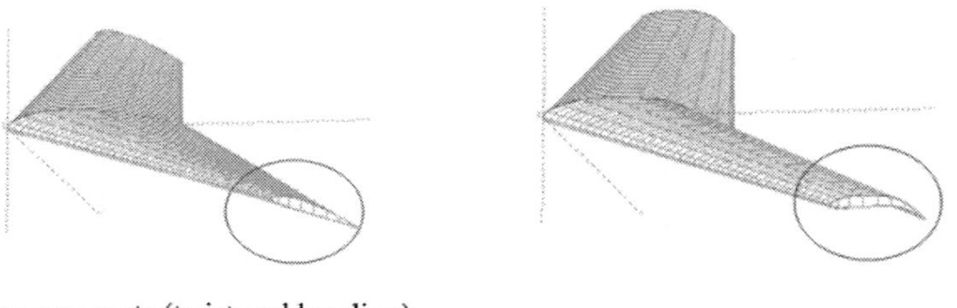

Fig. 3. Morphing concepts (twist and bending).

The application proposed is related to the flying wings (2m wing span) with elements of twist morphing at plane edges for revealing a global optimum area of turning features, by use of theoretical, numerical, and experimental methods.

In contrast by Peter Ifju studies at Florida University [14, 15] made on the flexible wings of small and medium aspect ratio we used to study the medium aspect ratio of the semi-rigid wings. For achieve higher performance ($C_{z\alpha}$ and C_{Xi}) so a better drag-lift ratio in a incidence range of interest. This wing falling within miniUAV class that offers advantages of portability, scalability (by extending bearing surfaces extreme easy to install depending on the specific mission). The introduction of morphing considers an increase of manoeuvrability in the average aspect ratio simultaneously with a reduction of energy use on the actuators in the use of innovative materials. The morphing advantages solution exists

DOI: 10.1051/matecconf/201712101011

in the border areas spoiler/aileron-wing (the shear zones), see Figure 4 where this class of UAVs significant losses occur.

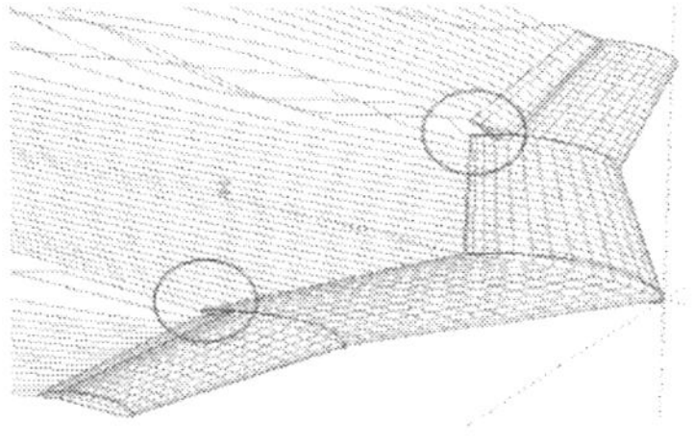

Fig. 4. The border areas spoiler/aileron-wing.

2 Simplified analytical approach

Proposes a simplified approach provides an analytical overview of the correlation aerodynamic elements, analytical approach that is intended as a filter so that possible solutions in the following section present an analysis of aerodynamic 3D.

The simplified analytical approach offers better understanding of the influence of basic geometric parameters of the wing on the aerodynamic characteristics. The general conclusions are extremely relevant to the applications in the field of the morphing wing concept (significant twist and variable rigidity of the wing tip structures). Dimensioning of the control surfaces of a flying wing according to the morphing concept consists in considering a morphing surface of elevon type capable of generating rotation around the centre of gravity in order to perform 3D flight. For this we take a trapezoidal morphing surface as in the figure bellow.

$$C_y = C_0 + y\cdot(tg\beta_F - tg\beta_A) \quad (1)$$

-we assume the linear variation $\alpha_y = \alpha_y(C_y)$, see Figure 6.

$$\alpha_y = \alpha_e \cdot \frac{C_0 - C_y}{C_0 - C_e} \quad (3)$$

where α_e - the morphing control angle. For determining lifting (see Figure 6), we have:

$$dF_z^{\alpha} = \frac{\rho\cdot V_\infty^2}{2}\cdot c_z(\alpha)\cdot dS \quad (4)$$

$$dS = C(y)\cdot dy \quad (5)$$

$$dF_z(\alpha_y) = \frac{\rho\cdot V_\infty^2}{2}\cdot C(y)\cdot d(y)\cdot c_z(\alpha_y) \quad (6)$$

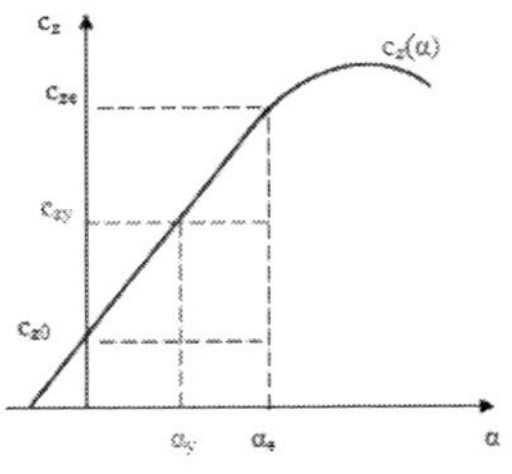

Fig. 5. Variation C_z-α.

$$F_{z_M} = \int_0^{b_M} \frac{\rho \cdot V_\infty^2}{2} \cdot C(y) \cdot c_z(y) \cdot d(y) \tag{7}$$

-estimating the coefficient $c_z(y)$, from Figure 5 we have:

$$\frac{\alpha_y}{\alpha_e} = \frac{c_z(y) - c_{z0}}{c_{ze} - c_{z0}} \tag{8}$$

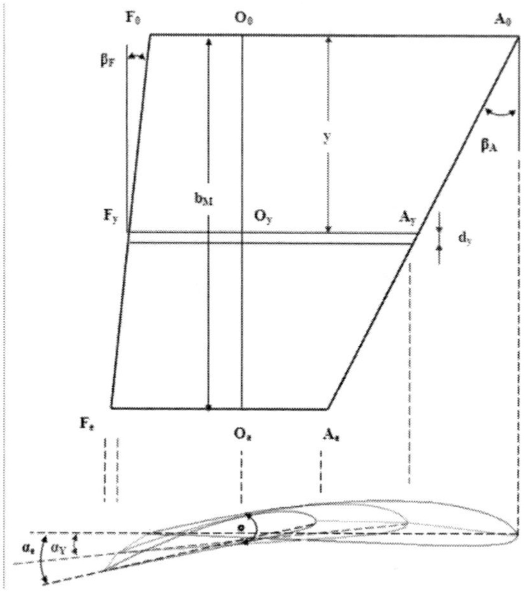

Fig. 6. Calculation of morphing parameters.

From where we obtain:

$$c_z(y) = c_{z0} + \frac{\alpha(y)}{\alpha_e}(c_{ze} - c_{z0}) \tag{9}$$

For calculating $c_z(y)$, we replace equation 1 in equation in 3, and we have:

$$\alpha_y = \alpha_e \cdot \frac{C_0 - C_0 - y \cdot (tg\beta_F - tg\beta_A)}{C_0 - C_e} \tag{10}$$

$$\alpha(y) = \alpha_e \cdot y \cdot \frac{tg\beta_A - tg\beta_F}{C_0 - C_e}$$

-we put equation 8 in equation 10 and we get:

$$c_z(y) = c_{z0} + y \cdot \frac{(tg\beta_A - tg\beta_F)}{C_0 - C_e} \cdot (c_{ze} - c_{z0}) \tag{11}$$

-for calculating the integral on the morphing surface, we have:

$$F_{zM} = \int_0^{bM} \frac{\rho \cdot V_\infty^2}{2} \cdot C(y) \cdot c_z(y) \cdot d(y) \tag{12}$$

-and after integration:

$$F_{zM} = \frac{\rho \cdot V_\infty^2}{2} \cdot \{C_0 \cdot c_{z0} \cdot b_M + k_a \cdot \left[c_{z0} - \frac{C_0 \cdot k_0}{k_c} \right] \cdot \frac{b_M^2}{2} - \frac{k_a^2 \cdot b_M^3}{3}\} \quad (13)$$

Where:

$k_a = tg\beta_F - tg\beta_A$

$k_c = C_0 - C_e$

$k_0 = c_{ze} - c_{z0}$

The control force on the lift surface is:

$$\Delta F_z = F_{zM} - F_{z0}, \quad F_{z0} = \frac{\rho \cdot V_\infty^2}{2} \cdot S \cdot c_{z0} \quad (14)$$

So the equation 13 the morphing surface lift force depends on directly proportional to the surface the morphing (C_0*b_M) and inversely proportional to the difference (k_c) chord the morphing surface (C_0-C_e).

3 CFD-3D analysis of the aerodynamic performance

We propose for analysis the flying wing in figure 4 having the characteristics in table 1.

The simulation was based on the use of a quantization network (about 6 million cells) with maximum density, as well as the effective use of the calculation resources available (CPU 2,4 GHz, RAM 24 Mb). The objective of the CFD analysis is to reveal the aerodynamic behaviour of the lifting surface of the flying wing type (see Figure 7 and the characteristics in table 1) in response to a set of angles of incidence.

Table1. Geometrical characteristics.

Span	**2000 mm**	**Area**	**0,66 m^2**
Profile	Clark Y	Aspect ratio	6,67
Angle of sweepback for c/4	25^0		

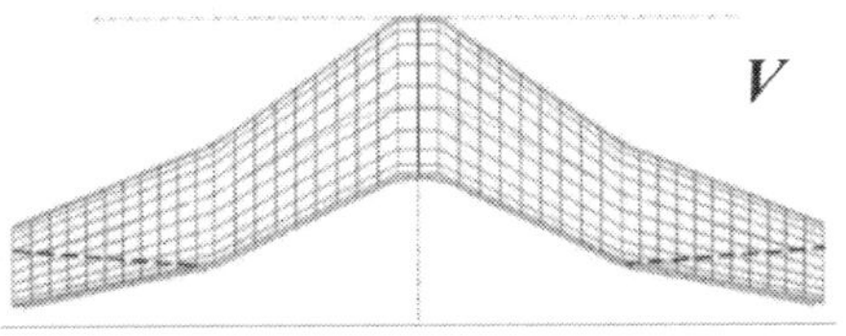

Fig. 7 Flying wing subject to CFD analysis.

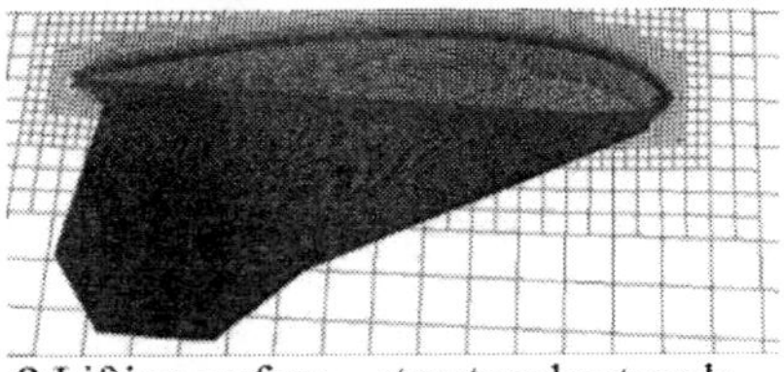

Fig. 8 Lifting surface – structured network.

The 3D numerical simulations have been achieved with the help of the Fluent solver of ANSYS 14 package by means of the method of finite elements, for investigating the aerodynamic behaviour of the lifting surface. The geometry used within CFD simulation was created so that it would be similar to models on scale 1:1. For the study of the flow, an unstructured grid has been used. The grid was in accordance with the parameters in table 2 (see Figure 8), [9, 10].

Table 2. Input parameters.

Speed	**10 m/s**	**Model**	**Turbulent SST K-ω**
Version/method	3D double precision	Turbulence	4,99 %
Iterations / elements	1000 / 6,12 million		

The aerodynamic behaviour of the chosen geometry (based on the simplified approach) can be observed with the help of the distribution of the pressure coefficient depending on the chord (C_p-c), along five reference sections (see figure 9).

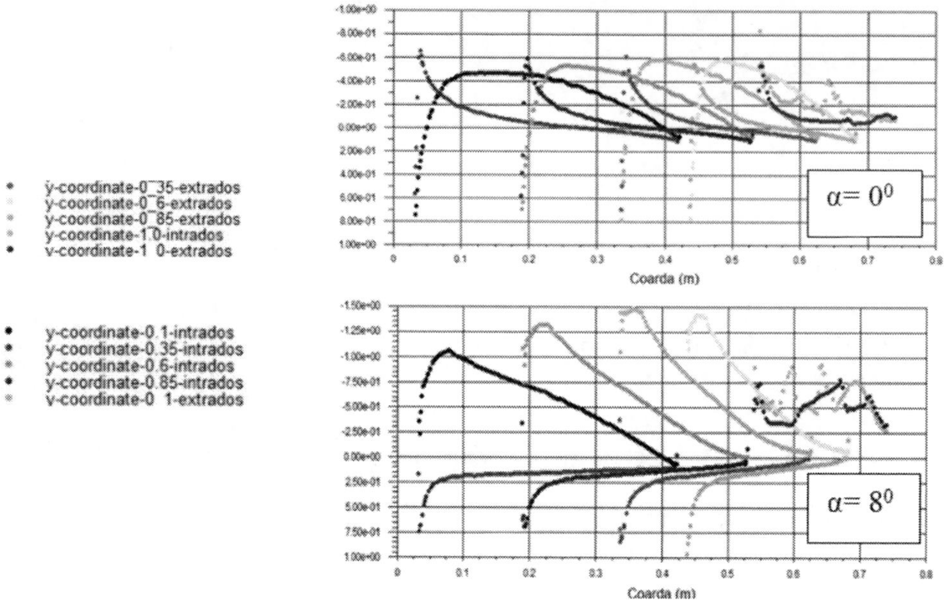

Fig. 9. Diagrams of the pressure coefficient C_P.

Table 3. Aerodynamic characteristics of flying wing.

AoA	C_z	C_x	C_z/C_x
-40	-0.046	0.020	-
00	0.243	0.023	13.500
40	0.536	0.029	18.483
80	0.778	0.050	15.560
120	0.867	0.113	7.673
160	0.688	0.119	5.782

The diagrams of C_p (Figure 10) show the increased turbulent behaviour regarding the flow at the tip of the wing along with the increase of the angle of incidence (see Figure 10 at $\alpha=4^0$ and $\alpha=4^0$). A maximum of C_z can be realized when the angle of incidence is of around 12^0, (Fig. 10).

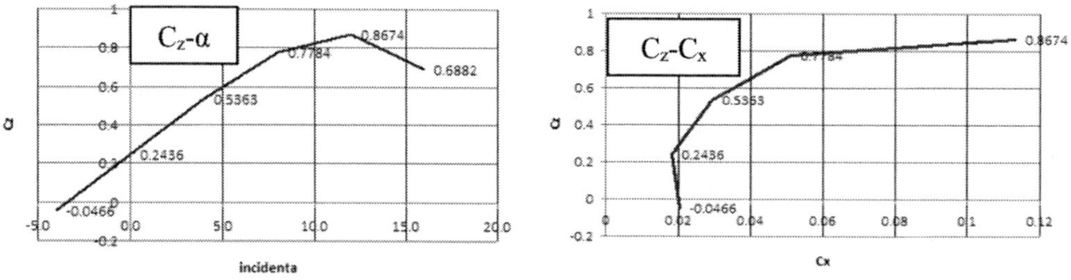

Fig. 10. Polar diagram of the flying wing.

The main aerodynamic characteristics are highlighted in Figure 10 and table 3. CFD analysis at an incidence of 12^0 reveals the tendency of detachment of the boundary layers on the leading wing edge and on the wing tips when performing at an incidence of over 8^0 (see Figure 10 and 11) and having the coefficients in table 3. It can be noticed that at 16^0 the aerodynamic performance does not deteriorate significantly. In order to compare the results of the CFD analysis, we have used some free software: XFLR5 6.09 [11].

We introduced geometrical parameters identical to the input ones in table 4; for similar incidences, we have obtained the values of the graphs below (see Figure 11).

Table 4. Input parameters.

Speed	**10 m/s**	**Iterations / elements**	**500 / 2400**
Model	VLM 1 (horseshoe vortex)	Limit conditions	Dirichlet

There are differences between the values of aerodynamic coefficients especially when the angles of incidence are big, where there is detachment of the boundary layer and when the VLM (XFLR5 6.06) method is not effective anymore.

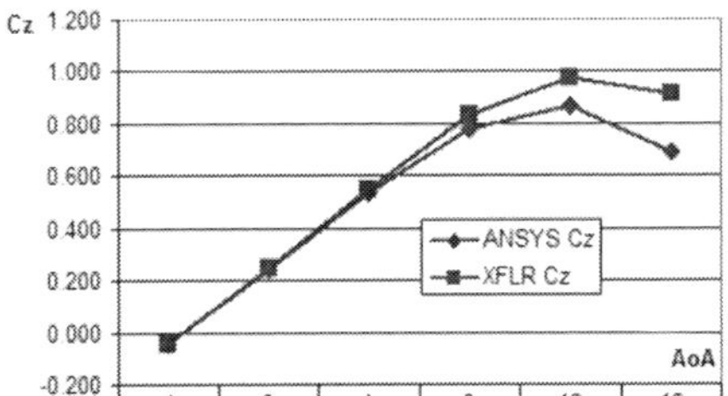

AoA	-4	0	4	8	12	16
ANSYS Cz	-0.047	0.244	0.536	0.778	0.867	0.688
XFLR Cz	-0.047	0.252	0.547	0.833	0.979	0.915

a. C_z-AoA

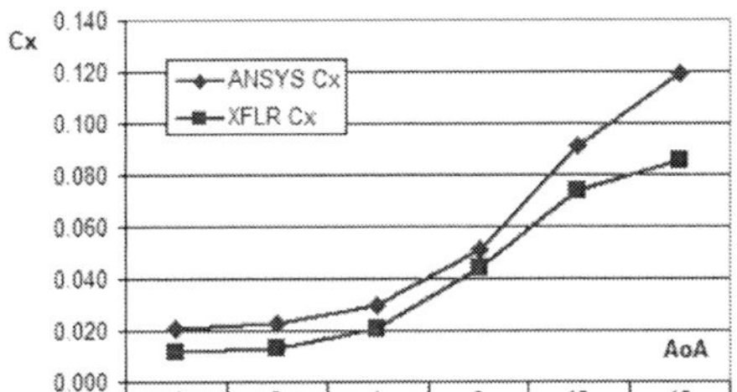

AoA	-4	0	4	8	12	16
ANSYS Cx	0.020	0.023	0.029	0.051	0.091	0.119
XFLR Cx	0.012	0.013	0.021	0.044	0.074	0.086

b. C_x-AoA

Fig. 11. Comparative diagrams of the performances with focus on flight at AoA near –stall.

The graph in Figure 11a does not show significant decreases in the values C_z critical angle which shows a post-stall behaviour acceptable. The graph in Figure 11b provides an acceptable increase C_x values which largely justifies the geometrical characteristics.

4 Conclusions

In conclusion, the results of the simulation of the flow are mainly influenced by the way network meshing is done as well as by the quality of the turbulent flow model used.

Comparing these results to the experimental data based on speed vector distribution and on Reynolds tensor distribution is also helpful. Although it is obvious that the use of the thick/light model influences the final results, we can use both types in future CFD analysis. CFD analyses by means of the VLM (vortex lattice method) provide a satisfactory degree of resolution and the results can be used in the phase preceding the project, more precisely in establishing the aerodynamic architecture, because they offer a quick image of the factors of influence. For detailed design, 3D CFD analysis, with high degrees of resolution, are recommended, especially for the study of the flow in the area of the boundary layer.

CFD design and analysis play an important role; they contribute to shortening the periods of time awarded for design, production, and testing. CFD analysis enables an early analysis with multiple possible interventions at the expense of minimum resources, which makes the difference between an expensive model which does not rise to the expectations and an optimized one.

The authors wish to thank "Henri Coandă" Air Force Academy of Braşov and INCAS Bucureşti for supporting the research necessary for writing this article. This work is supported by the Executive Agency for Higher Education, Research, Development and Innovation Funding (UEFISCDI) under MASIM project (PN-II-PT-PCCA-2013-4-1349).

References

1. OSD UAV Roadmap 2002-2027, Office of the Secretary of Defence Acquisition, Technology, & Logistics, Air Warfare, December (2002)
2. *Unmanned Aircraft System ROADMAP 2005-2030*, US DoD, Washington DC, 213 (2005)
3. S. Barbarino, O. Bilgen, R.M. Ajaj, M.I. Friswell, D.J. Inman, Journal of intelligent material system and structures, **22**, 823 (2011)
4. B. Beguin, C. Breitsamter, N. Adams, *AIAA Journal*, **50**, 2588 (2012)
5. M.D. Skillen, and W.A. Crossley, Modeling and Optimization for Morphing Wing Concept Generation, (2007), Langley Research Center NASA/CR-2007-214860, http://www.sti.nasa.gov, accessed on 01.10.2017
6. R.W. Wlezein, G.C Horner, A.R. McGowan, S.L. Padula, M.A. Scott, R.J. Silox, and J.O. Simpson, The aircraft morphing program, AIAA-98-1927 (1998)
7. ***http://www.etymonline.com/index.php?term=aileron, accessed on 22.02.2017
8. J. Valasek, *Morphing aerospace vehicles and structures* (A John Wiley & Sons, Ltd., Publication, 2012)
9. S. Dănăilă, C. Berbente, *Metode numerice în dinamica fluidelor* (Editura Academiei Române, 362, 2003)
10. ANSYS FLUENT User's Guide, **14**, 2948 (2011)
11. M. Drela, H. Yungren, Guidelines for XFLR5 v6.03, (2011), http://sourceforge.net /projects/xflr5/files, accessed: 22.02.2017
12. V. Prisacariu, M. Boşcoianu, I. Cîrciu, C-G. Rău, Advanced Materials Research, **463**, 1564 (2012)
13. M. Boşcoianu, V. Prisacariu, I. Cîrciu, A. Luchian, C. Ciufudean, *Latest Trends in Applied and Theoretical Mechanics WSEAS Conference, Salerno, Italy, 9* (2014)
14. P.G. Ifju, D.A. Jenkins, and others, Flexible wing based micro air vehicles, American Institute of Aeronautics and Astronautics (AIAA), 13 (2002)
15. W. Shyy, P. Ifju, D. Viieru, *Applied Mechanics Reviews*, **58**, 283 (2005)
16. D.P. Raymer, *Aircraft design: A conceptual approach (4th edition), AIAA Education series, California-USA, 840* (2006)
17. J.D. Anderson jr., *Aircraft performance and design* (McGraw-Hill Book-Co Singapore, 580, 1999)
18. N.V. Constantinescu, S. Găletuşe, *Mecanica fluidelor şi elemente de aerodinamică* (Editura Didactică şi Pedagogică, Bucureşti, 506, 1983)
19. UAS Yearbook, *Unmanned aircraft systems, The Global Perspective 2011/2012*, Blyenburg & Co, Paris, 216 (2011) available www.uvs-info.com
20. R. Austin, *Unmanned Aircraft Systems – UAVs design, development and deployment* (Aerospace series, Wiley and Sons Ltd publication, 365, 2010)
21. V. Prisacariu, M. Boşcoianu, I. Cîrciu, Lile Ramona, Applied Mechanics and Materials, **811**, 157 (2015)
22. ***http://www.israeli-weapons.com/weapons/aircraft/uav/orbiter/orbiter_8.jpg, accessed: 22.04.2017
23. ***http://www.boeing.co.in/resources/en-in/media/BoeingIndia/Products-and-Services/Defense-Space-and-Security/BDS-in-India/scaneagle_headera878.png, accessed: 22.04.2017
24. ***http://www.defense-aerospace.com/dae/articles/features/Taranis_BAE.jpg, accessed: 22.04.2017
25. ***http://www.naval-technology.com/features/featureuclass-kicks-off-4647440/featureuclass-kicks-off-4647440-1.html, accessed: 22.04.2017

Analysis of the small flying wings performances in the morphing concept

Vasile Prisacariu[1,*], *Ionică* Cîrciu[1], *Doru* Luculescu[1], and *Laurian* Gherman[2]

[1]"Henri Coandă" Air Force Academy, Department of Aviation, Braşov, Romania
[2]"Henri Coandă" Air Force Academy, Department of Air Defence, Braşov, Romania

Abstract. Using of the flying wing (the UAVs category) in various fields (civilian and military) determine interests sustained of the aerodynamic and trajectory optimizations. The article presents analysis of wing flying performance (in morphing concept) by studying optimizing the maneuvers control and wind tunnel tests.

1 Theoretical aspects

The geometry of the wing (dimensions, shape within the plane, profile, span thickness evolution, dihedral angle) is critical to the flight performance and characteristics of the aerial vehicle.

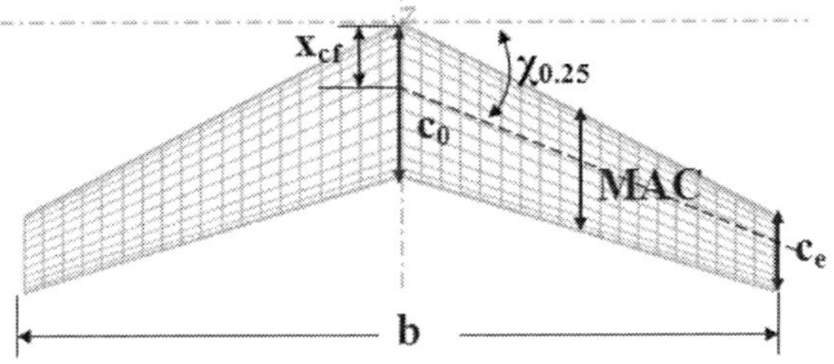

Fig. 1. Geometrical parameters of the swept wing.

If the line of the focal points does not coincide with axis *Oy*, then the sweepback angle χ is revealed. If the Sweepback angle is reduced $\chi<15\div20^0$, the sweepback effect is reduced as well and the wing can be considered to be straight. If $\chi>30\div40^0$ the effect is significant and the wing is considered to be swept, [13]. The sweepback angle allows the adjustment of the centering of the aerial vector.

The positive wing sweep has a stabilizing dihedral effect. So, we have a wing with a sweepback angle χ - measured on the line uniting ¼ of the chords, in the speed current *V* and the positive angle of deflection β. Only the regular components create lift forces on the two semi-wings. These components are $V\cdot cos(\chi-\beta)$, on the right semi-wing and $V\cdot cos(\chi+\beta)$, on the left semi-wing. According to these speeds, a difference in lift appears (ΔF_z):

* Corresponding author: aerosavelli73@yahoo.com

$$\Delta F_z = \frac{\rho \cdot S}{2} \cdot C_z \cdot V^2 \cdot \left[\cos^2(\chi - \beta) - \cos^2(\chi + \beta)\right] \quad (1)$$

Or, because β is small, we approximate $sin\beta = \beta$, $cos\beta = 1$, therefore:

$$\Delta F_z = S \cdot \beta \cdot C_z \cdot \frac{\rho}{2} \cdot V^2 \cdot \sin 2\chi \quad (2)$$

-the rolling moment is:

$$L = -k \cdot \beta \cdot C_z \cdot \sin 2\chi \quad (3)$$

The neutral point coordinate is:

$$x_F = \frac{c_0}{4} + \frac{2b}{3\pi} \cdot tg\varphi_{0,25} \quad (4)$$

If the ratio of trapezoids is >0,375 and

$$x_F = \frac{c_0}{4} + \frac{b \cdot (1 + 2\lambda)}{6 \cdot (1 + \lambda)} \cdot tg\varphi_{0,25} \quad (5)$$

If the taper ratio is < 0,375.

The aspect ratio influences the aerodynamic characteristics of the wing, the inertia moments and the weight of the structure is according to [11, 12, 14, 15].

$$C_z = \frac{2\pi \cdot \alpha \cdot \lambda}{\lambda + 2} \quad (6)$$

But for approximate calculation the following formula can be used:

$$C_z^{\alpha} = \frac{5{,}73 \cdot \pi \cdot \lambda}{5{,}73 + \pi \cdot \lambda} \quad (7)$$

The global drag coefficient is expressed by:

$$C_x = C_{x0} + \frac{C_z^2}{\pi \cdot \lambda} \quad (8)$$

And the induced angle:

$$\alpha_i = \frac{C_z}{\pi \cdot e \cdot \lambda}(1 + \delta) \quad (9)$$

The coefficients for the sweepback angle of the wing are:

$$C_z = C_{zn} \cdot \cos^2 \chi \, , \, C_x = C_{xn} \cdot \cos^3 \chi \quad (10)$$

2 Optimizing the maneuvers control

The level of autonomous behavior of UAV increased recently, autonomy depending directly on the level of involvement human or resolution sensors. According to scientific references sense and avoid capability to UAVs can be optimized by making the leading

DOI: 10.1051/matecconf/201712101012

characteristics and which contribute to adapt quickly to dynamic situations. Research has focused on numerical trajectory planning methods, such as direct (direct shooting, direct collocation, pseudospectral methods) and indirect methods, [5-7].

The maneuvers control along the trajectory can be optimized through a linear system, invariant to duration, and is associated through its reaction to stimulus. The reaction to stimulus is, almost always, dimensionally infinite. Since in the case of the stable systems the stimulus exponentially tends to 0, we can assume that it can be simplified. It can be neglected for a period of quantization, in comparison to the constants of the system, which cannot. A significant number of parameters are necessary for indicating the simplified signal. Therefore, we have an example of the finite response to stimulus [1, 10].

This example can be shown by means of the following relations:

$$y(p)=x_1\cdot n\cdot(t-1)+x\ \cdot n\cdot(t-2)+\ldots+x_p\cdot n\cdot(t-p) \quad (1)$$

$$y(p)=\varphi^T\cdot(p-1)\cdot\phi \quad (2)$$

Where:

$$\phi=[x_1,x_2\ldots x_p]^T$$

With

$$\varphi^T\cdot(p-1)=[n\cdot(t-1)\cdot n\cdot(t-2)\cdot\ldots\cdot n\cdot(t-p)]^T$$

The structure can be viewed as a system with entries n and y and exit ϕ. Double correction can be noticed: the first correction through the proper correction loop itself and the second one by a supplementary comparison to a preset program (see fig. 2).

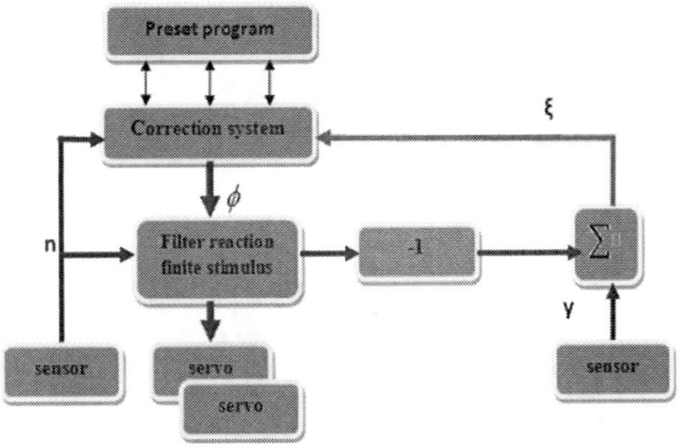

Fig. 2. Simplified block-diagram of a recursive system.

In the correction loop we can introduce constant initial values of the flight altitude (altitude sensor) and/or of the turning speed for the angle of inclination while in the preset program we can set these values before the mission according to the mass of the UAV. The proposed system highlights the way a flight stabilizer works and can influence the maneuverability of the UAV along the trajectory depending on the type of mission.

Figure 3 show the two situations, with and without longitudinal stabilization (2D) during a 10 second turning.

For example: the rolling is set on 25^0 and the measured rolling is 15^0; an error will immediately be displayed at the rolling control and the ailerons will respond until the value of 25^0 with 0 error will be reported on the rolling channel. Throughout the rolling, the nose of the UAV will be descending while the pitch controller will control the pitch until 0 errors is present.

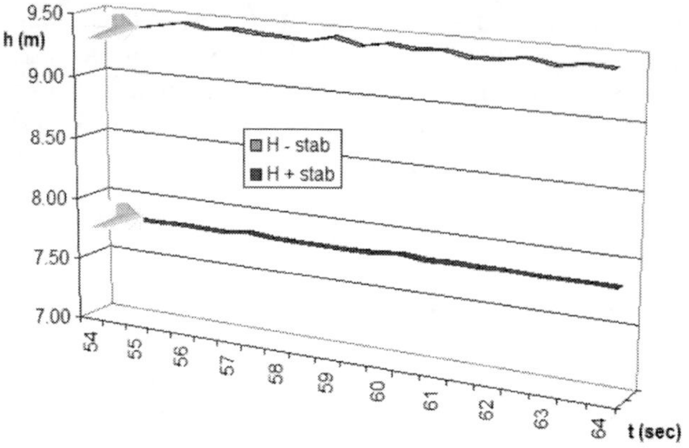

Fig. 3. Longitudinal stability.

3 Experimental research with tunnel tests

The approach used in this section aims at a comparative analysis of the results taken from tests in the wind tunnel by the two configurations (classical and the morphing) with results obtained from CFD numerical analysis.

3.1 Characteristics of the aerial vector proposed for the tunnel tests

The tests in the aerodynamic tunnel were performed in the subsonic tunnel of INCAS, Bucharest, in order to provide data for investigating the benefits of the morphing concept [2]. The model used for the experimental testing had been made of materials available on the market. The lifting surface is of expanded polystyrene and covered with a thermo retractable foil. The configuration is double trapezoidal (see figure 4) having the geometry in Figure 4.

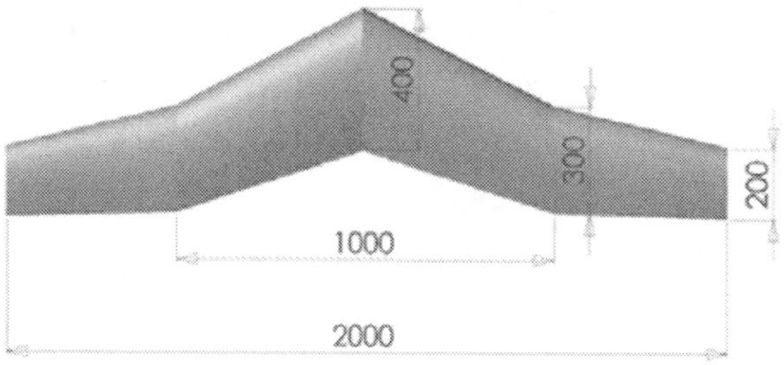

Fig. 4. Experimental model.

The wings subject to tunnel tests have a Clark Y airfoil and do not display any constructive torsion at plane edges but there are pairs of winglets mounted on them for stabilization. The changes in angles of torsion of the wings from the basic configurations are made by use of two standard Futaba S3003 servomechanisms (see table 2).

DOI: 10.1051/matecconf/201712101012

Table 2. Futaba S 3003 features.

Signal	analogue	Drive limit	3.2 Kg·cm at 4.8 V
Execution speed	0.23 sec/60° at 4.8V	Weight	38 g

They are activated through the command and control system [3], both for the longitudinal channel and lateral channel. The basic configurations of the flying wings have been fitted with identical pairs of winglets, the classical wing being equipped with articulated elevons activated by servomechanisms while the morphing wing is activated by a torsion bar type system the tips ribs being activated by servomechanisms (see figure 5).

Classical flying wing

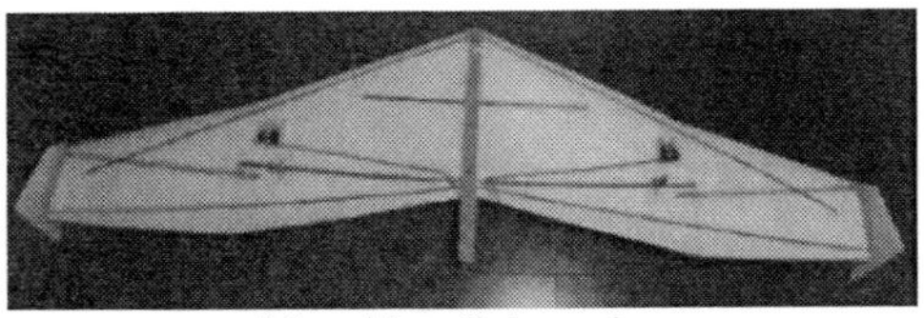

Morphing flying wing

Fig. 5. Experimental models.

3.2 Description of the subsonic aerodynamic tunnel where the tests were performed

The aerodynamic tunnel used is closed-circuit, subsonic with continuous flow generated by a direct current electric motor (3.5 m per diameter) and a 20KW electro-ventilator for separate cooling. It is equipped with an external measurement balance (six components – forces and moments) with precision of 1/25000 and a data acquisition system with LABVIEW software.

The acquisition system is controlled by the computer and based on a National Instruments base with eight PXI 1042 slots [4] with synchronization at 10 MHz for the existing models. Table 3 shows the main characteristics of the tunnel [2].

Table 3. Characteristics of the subsonic wind tunnel.

Testing room dimensions	4 x 2 x 2.5 m	K contraction ratio	10
Speed	7÷110 m/s	Angle of turning	$-140^0 \div 216^0$
Reynolds	$1x10^6$	Angle of incidence	$-45^0 \div 45^0$
Turbulence factor	1,11	Fan motor	1200 kw

The experimental models were mounted on three points, leaving the rear assembly mobile for making possible the modification of the angle of incidence during the tests (see figure 6).

Classical flying wing

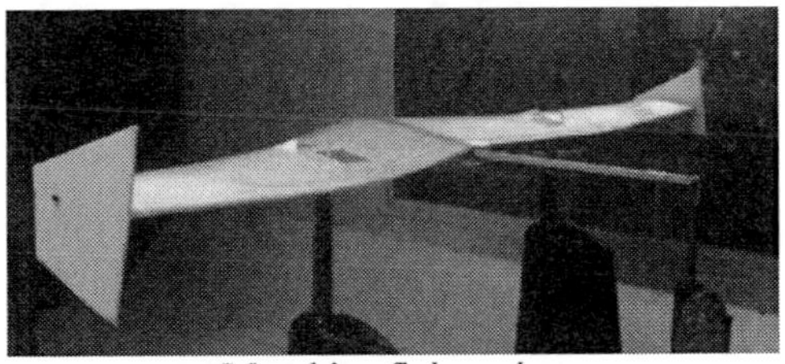

Morphing flying wing

Fig. 6. Testing room of the subsonic tunnel.

External balance calibration and the estimations were made before the wind tests. The tests covered a range of -4^0 to 10^0 for a speed of 10 m/s, the experimental models were controlled by means of a command and control system on 2.4 GHz and two channels [3] and the calibration of the control surfaces (classical/morphing elevons) was performed using a digital inclinometer with an error of 0.1^0.

The experimental models (classical and morphing) were tested under the same conditions, without any activation of the control surfaces, for the comparative analysis of the physical measurement data collected, [9].

3.3 Results and discussions

The following conventions are applied when presenting results: the moments are shown in axes system (axis Ox opposite to the flight direction, axis O_z upwards, axis O_y to the right) (see figure 7); the mean aerodynamic chord (MAC) was the reference length for the diving moment. The moments that have been measured are referenced to the centre of gravity (CG).

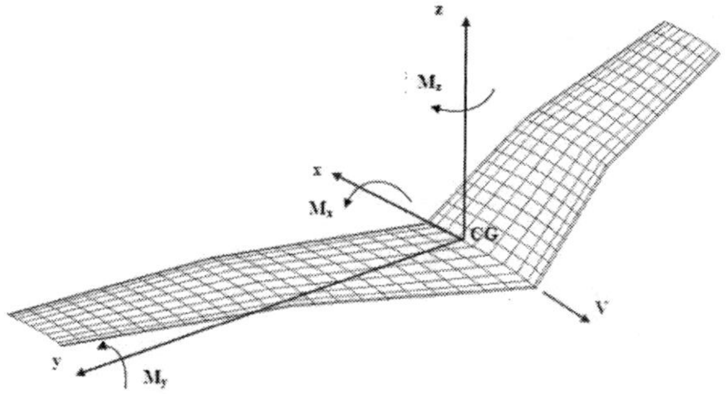

Fig. 7. Reference system.

The data collected during the tunnel tests provided values for the following: the lift force (F_z), the force of resistance to movement (F_x), the lateral force (F_y), the pitch moment (M_y), the rolling moment (M_x) and the gyration moment (M_z). The data collection system provided unprocessed data which needs to be corrected according to instructions for use of the aerodynamic balance. After data processing we obtain the aerodynamic parameters in figure 8.

The variation of the aerodynamic coefficients according to CFD analysis with ANSYS provides a numerical reference for the comparative analysis of the tunnel tests (see figure 8). Compared to the classical wing, in the case of the morphing wing, an increase of coefficients C_z, C_x and C_m on a range of 0^0-8^0 of the incidence angle can be noticed.

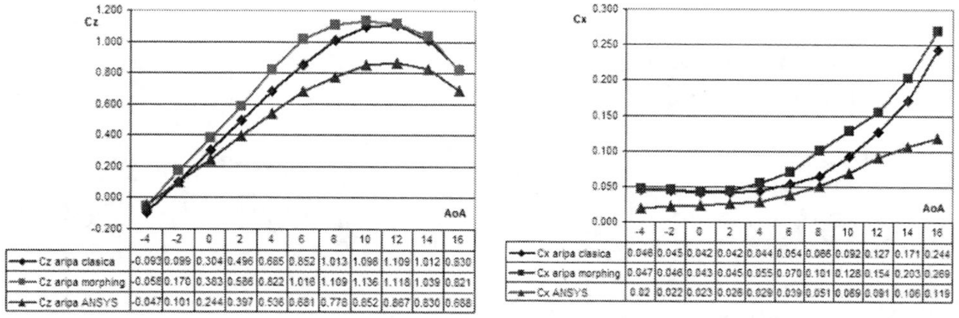

	-4	-2	0	2	4	6	8	10	12	14	16
Cz aripa clasica	-0.093	0.099	0.304	0.496	0.685	0.852	1.013	1.098	1.109	1.012	0.830
Cz aripa morphing	-0.058	0.170	0.383	0.586	0.822	1.018	1.109	1.136	1.118	1.039	0.821
Cz aripa ANSYS	-0.047	0.101	0.244	0.397	0.536	0.681	0.778	0.852	0.867	0.830	0.688

	-4	-2	0	2	4	6	8	10	12	14	16
Cx aripa clasica	0.046	0.045	0.042	0.042	0.044	0.054	0.066	0.092	0.127	0.171	0.244
Cx aripa morphing	0.047	0.046	0.043	0.045	0.055	0.070	0.101	0.128	0.154	0.203	0.269
Cx ANSYS	0.02	0.022	0.023	0.026	0.029	0.039	0.051	0.069	0.091	0.106	0.119

Fig. 8. Comparative curves of the polar diagrams (C_z-AoA, C_x- AoA).

Figure 8 shows a good behavior (for V=10 m/s and Re=2.15×10^5) in the angles of incidence stall and post-stall aerodynamic performance is not significantly affected.

4 Conclusions and future research considerations

The analysis of both aerodynamic and flight mechanics aspects of the morphing concept flying wing aims at providing an overall idea regarding the aerodynamic performance of these small aerial vehicles in order to better understand the compromise between aerodynamic efficiency and superior maneuvers characteristics in the context of increased power efficiency.

The main advantages of the flying wing are increased aerodynamic performances (reduced weight, increased maneuverability). The literature in the field rarely mentions the swept flying wing of the UAS type in terms of aerodynamic performance depending on maneuvers characteristics.

The step by step analysis as part of a simplified theoretical approach, the CFD-3D simulations and the aerodynamic tunnel tests offer the opportunity for improving existing options regarding the projection of the aerial vector.

The comparison of CFD-3D simulation results to the aerodynamic tunnel test ones ensures the accuracy of the aerodynamic performances as well as the formulation of conclusions related to important aspects of flight mechanics.

The applications proposed are relevant to the field of miniature UAVs research in what regards the two maneuvers described (turning and spiral), which are typical for this type of aerial vehicles.

Our contribution is a new, simple, and intuitive way of connecting aerodynamic performances to flight mechanics aspects regarding the two maneuvers. It offers good accuracy results. The examples for the turning and the spiral as simple maneuvers are also relevant to the development of new tactical capabilities of small UAVs through improving the performance itself. Moreover, the applications could be helpful in the development of new tactical capabilities of UAVs related to the global performance indicator in a synergic way.

The authors wish to thank "Transylvania" University and "Henri Coandă" Air Force Academy of Braşov and INCAS Bucureşti for supporting the research necessary for writing this article. This work is supported by the Executive Agency for Higher Education, Research, Development and Innovation Funding (UEFISCDI) under MASIM project (PN-II-PT-PCCA-2013-4-1349).

References

1. K. Nonami, M. Kartidjo, and others, *Autonomus control systems and vehicles. Inteligent Unmanned Systems,* (Springer Series, **65**, 2013)
2. M.V. Pricop, C. Stoica, C. Nae şi alţii, INCAS Bulletin, **4**, 67 (2012)
3. Futaba Corporation, Instruction manual for Futaba 6EXAP 6-channel, PCM/PPM (FM) selectable, Radio control system for aircraft, 2005, 24p, available at http://www.manualslib.com/manual/378276 /Futaba-6exap.html, accessed on 19.09.2014
4. National Instruments, User Manual 6023E/6024E/6025E, Multifunction I/O Devices for PCI, PXI, CompactPCI, and PCMCIA Bus Computers, Part Number 322072C-01, Austin Texas
5. Betts, J.T., Journal of Guidance, Control, and Dynamics, **121**, 193 (1998)
6. Yakimenko, O.A., Y. Xu, and G. Basset, Collection of Technical Papers - AIAA Guidance, 2008, Navigation, and Control Conference, **1**, 1 (2008)
7. G.T. Huntington, D.A. Benson, and A.V. Rao, Collection of Technical Papers - AIAA Guidance, Navigation, and Control Conference, **1**, 840 (2007)

8. B. Geiger, Unmanned Aerial Vehicle trajectory planning with direct methods, The Pennsylvania State University, available at https://etda.libraries.psu.edu /files/final_submissions/767
9. V. Prisacariu, *Managementul integrării soluțiilor tehnice inovative la sisteme aeriene robotizate, teză de doctorat,* (Universitatea Transilvania Braşov, 2014, available at webbut.unitbv.ro/teze/rezumate/2014/rom/PrisacariuVasile.pdf)
10. I. Cîrciu, V. Prisacariu, *Review of the Air Force Academy, Braşov, Romania,* ***1,*** 13 (2013)
11. D.P. Raymer, *Aircraft design: A conceptual approach (4th edition), AIAA Education series, California-USA,840* (2006)
12. J.D. Anderson jr., *Aircraft performance and design* (McGraw-Hill Book-Co Singapore, 580, 1999)
13. N.V. Constantinescu, S. Găletuşe, *Mecanica fluidelor şi elemente de aerodinamică* (Editura Didactică şi Pedagogică, Bucureşti, 1983)
14. V. Prisacariu, M. Boşcoianu, I. Cîrciu, C-G. Rău, Advanced Materials Research, **463-464**, 1564 (2012)
15. V. Prisacariu, M. Boşcoianu, I. Cîrciu, Applied Mechanics And Materials, **332,** 44 (2013)

Analysis of justification for applying two gear sets within the universal gear reducers

Milan Rackov[1,*], *Siniša* Kuzmanović[1], *Ivan* Knežević[1], *Maja* Čavić[1], and *Marko* Penčić[1]

[1]University of Novi Sad, Faculty of Technical Sciences, 6 Trg Dositeja Obradovića, 21000 Novi Sad, Serbia

Abstract. In the area of universal gear reducers, standard does not prescribe relations between load capacity, gear ratios and axis heights. Therefore, larger load capacity of some gear reducers manufacturers has advantage for lower gear ratio and, vice versa, smaller load capacity of the gearbox achieves higher values of gear ratio. However, there are some manufacturers who produce gear reducers with smaller or larger load capacity for higher or lower gear ratio, within the same axis height. This paper deals with analysis of justification of applying two gear sets, especially in the case when so-called intermediate sizes of gear reducer are not used.

1 Introduction

Axis heights of universal gear reducers are prescribed according to the standard row R20. Since this row is very dense, for a while manufacturers of universal gear reducer produced reducers with axis height from a standard row R10 (these numbers belong to the standard row R20). But soon they increased axis heights in order to increase gear ratio, and now reducers are produced with axis height in a standard row R20/2. In the area of the most used gearbox sizes, some manufacturers produce so-called intermediate sizes with the axis heights in a row R20.

There is no standard which defines particular load capacities within different sizes (axis heights), so manufacturers have open field for defining load capacities [1].

Load capacity values of leading gear reducer manufacturers are followed, so that particular interchangeability could be provided. In order to achieve some advantage at the market, smaller manufacturers of gear reducers produce reducers with smaller torque capacity, but with higher gear ratios. In this way they are able to provide cheaper lower stage gearbox in particular segment of gear ratios.

However, leading gearbox manufacturers recognized this occurrence, and started to offer two sets of gear pair within the same size, and also even in the same of gearbox housing. With this two gear sets, they provide particular gear ratio for larger and for smaller load capacity, so that they annulled advantage of smaller gear reducer manufacturers [2].

* Corresponding author: racmil@uns.ac.rs

2 Problem description

Depending of the gear ratio value, there are single-, two-, three- and multi-stage gear reducers. One-stage gear reducers are always assembled in the housing for single-stage gear reducer. However, due to lower demand for these reducers, there are manufacturers which do not produce single-stage gearbox at all, and their gear ratio is replaced by lower gear ratio of more expensive two-stage gear reducer [2-4].

Two-stage gear units can be installed in separate housing for two-stage gear reducer [5], or in the universal housing for two- and three-stage gear reducers [6]. Of course, there are manufacturers which produce both type of housing, two-stage and three-stage gear reducer housing [7]. It is understandable that producing of two-stage gear reducers in special housing for only two-stage gear pairs has slightly lower production costs than using universal housing for two- and three stages of gear pairs.

Three-stage gear reducer can be assembled in universal housing for two- and three-stage gearbox, but also combined of two-stage and single-stage gear unit [1, 5]. Production costs of three-stage gear reducer assembled in universal gear unit housing are lower than producing the same reducer combined of two-stage and single-stage gear unit. However, there are some manufacturers that rather produce three-stage gear units in this way, combining two-stage and single-stage gear unit, but some manufacturers produce them in both ways [7, 8].

Four-stage gear units are usually manufactured combining two-stage gear units. Five-stage gear units are combined of three- and two-stage gear units, etc.

Which concept of gear reducer will be adopted depends on marketing orientation of their manufacturers. The largest gear reducer demand is within the speeds between 20 and 50 rpm. For driving with four-pole asynchronous motor, that roughly corresponds to the gear ratio between 30 and 70, which covers higher gear ratio values of two-stage gear units and lower gear ratio values of three-stage gear units [9, 10]. Until recently it was usual to produce single-stage gearbox with highest gear ratio of about 6.3, two-stage gearbox with highest gear ratio of about 20 and three-stage gearbox with highest gear ratio of about 100. Today, values of gear ratio are increased, so highest gear ratio for one-stage gearbox is up to 12.5, for two-stage up to 50 and for three-stage gearbox up to 300 [5, 8]. This increasing of gear ratio has caused the using of simpler (lower-stage) and thus cheaper gearbox. However, it should be noted that gear reducer with the highest gear ratio is always more expensive than gear reducer with some lower gear ratio. It is because of larger axis distance and thus larger diameter of driven gears. Nevertheless, producing of lower-stage gear reducer makes certain savings which compensate increased production costs. Additionally, with increasing gear ratio, some manufacturers succeed to increase maximal load capacity of their gear reducers and in this way to offer cheaper gear units. These interventions of leading gear reducer manufacturers improved their position at the market, but very soon smaller manufacturers have done the same thing. However, there is some reduction of competition at the market since small manufacturers were not be able to financially follow the interventions of major manufacturers, mainly due to the high cost of producing special casting tools [11, 12].

3 Descriptions the ways of solving problems

Gear ratio of single-stage gear reducer is performed in a standard row R10; only for the reason row R20 needs more gears, but especially driving gears. Since motor gear reducer is usually used, positioning of driving gear is done on shaft of gear unit motor. Each size of electric motor will need different gear because of its different shaft diameter and therefore row R20 is not good for single-stage gear units. Gear ratio of two-, three- and multi-stage gear reducer is performed in standard row R20. In that case, two possible gear ratios are added on the first gear pair. Their combination gives two gear ratios and combination of other first gear pairs gives the values of standard row R20 for output gear ratio of the gearbox (Fig. 1) [2, 9, 10].

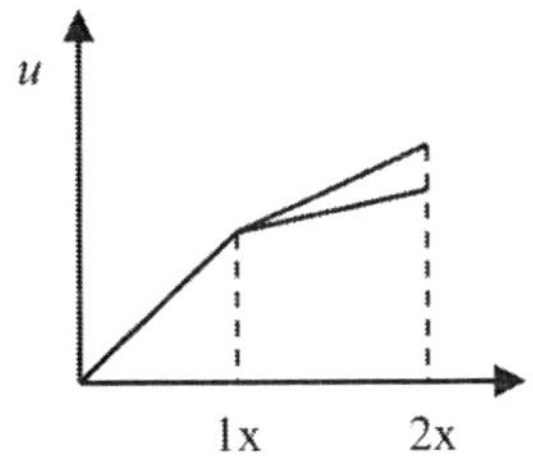

Fig. 1. The way of performing gear ratio for two-stage gear reducer.

Gear ratio of three-stage gear reducer is also offered in standard row R20. These gear ratios can be obtained in two ways. If the output gear pair with higher load capacity is added on two-stage gear reducer, it can be obtained gear ratio in the row R20 (Figure 2-a). The other way is to use the output gear pairs from two-stage solution with inserting a gear pair with lower load capacity as the second and using the first gear pair from a smaller size of gearbox (Figure 2-b). Of course, this is always possible except for the smallest size of gear reducer. Using the first gear pair for smallest size of three-stage gear reducer is not economical since the gears have higher load capacity than it is required [1, 2, 9, 10].

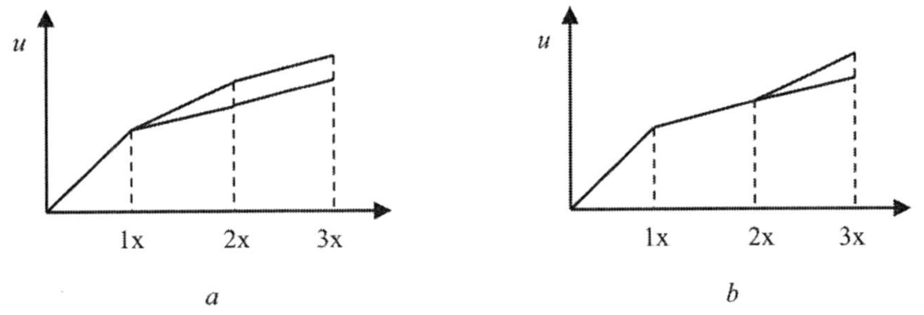

Fig. 2. The way of performing gear ratio for three-stage gear reducer, (a) with additional output gear pair, (b) with additional lower second gear pair and using the first gear pair from a smaller size of gearbox.

If the first solution is used (Fig. 2.a) with additional third gear pair, the same uniform housing can not be used for two- and three-stage gear reducer. It is because of big difference between load capacities of two- and three-stage load capacity which will increase production costs. Therefore, the other solution with common output gear pair for two- and three-stage gear reducer will be used (Fig. 2.b). This solution provides assembling two and three gear pairs in the same housing, thereby significantly reducing production costs. This approach achieves the same load capacity of two- and three-stage gear reducer (Table 1, Table 2). Therefore it is required to install different sizes of electric motor on the same housing in order to use all available advantages of the gear unit.

Values of load capacities are adopted in such a way that growth factor of torque is calculated as follows [1, 2]:

$$q_T = q_L^3 = 1{,}24^3 = 2 \quad (1)$$

which approximately corresponds to the maximum value of service factor f_B (where q_T – growth factor of torque, q_L – growth factor of linear dimensions, shaft height in this case).

Considering the application area of universal gear unit (Figure 3), it is evident the gear unit is mostly used in the area ($T_N - T_N/f_B$).

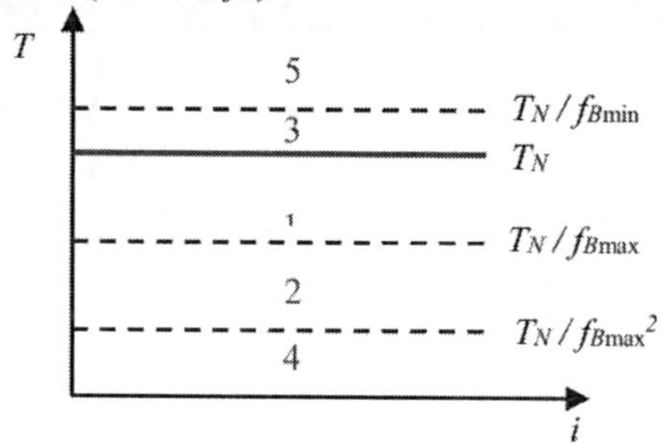

Fig. 3. Schematic review of application field of universal gear reducer according to torque values: 1 - the main area, 2 – the additional area, 3 - critical area (used only for short drives), 4 - the area where the reducer is oversized (not economically used) and 5 - critical area where breakdown of the reducer happens immediately.

Defining the values of output torque within the family of universal gear reducers in such a way, for the approximately same gear ratio interval, provides full coverage of market demands for output torque. In this way the main application area of gear unit is covered by the family of gear reducers (Fig. 4).

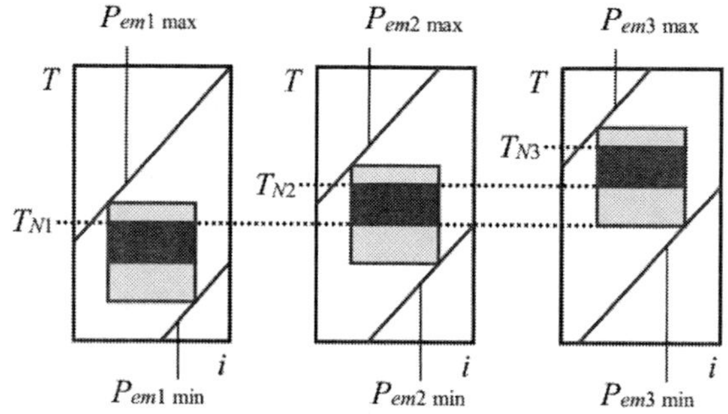

Fig. 4. Position of the main application areas of universal gear reducers within three neighbour sizes of gear unit.

However, the main application area of only one size of gear unit should be considered with using two sets of gear pairs. One set of gear pair has large load capacity and lower gear ratio (Figure 5, area 1), and the other set has smaller load capacity with higher gear ratio (Figure 5, area 2). It is clear that the gear unit with smaller load capacity will be cheaper since the gears have smaller module and smaller production costs. Considering higher gear ratio, for slightly lower capacity, three-stage gear reducer will be much cheaper than four-stage gear reducer with higher load capacity (shaded area, Fig. 5) [9, 10].

If it is required smaller load capacity and higher gear ratio, universal three-stage gear reducer can be much improved. For example three-stage gearbox with highest gear ratio of 100, can be modified and produced with some smaller load capacity, but with much higher gear ratio, for instance 112, 125, 140, 160, 180, 200, 224, 250, or even 280. In such way, it is offered cheaper solution than four-stage gear ratio, with the same values of gear ratio. This approach increases the offer of cheaper solutions, but also it should be emphasized that in the area of smaller capacities and lower gear ratio than 100, it can be also offered lower cost solution. This is the space where smaller manufacturers of gear units can find their place for existence on the market.

Table 1. Examples of nominal torque capacities and highest gear ratio for two-stage gear reducer of different manufacturers.

Axis height *h*, **mm**	**Designation of gear unit**	**Nominal output torque,** T_N, **Nm**	**Max gear ratio** i_{max}
Manufacturer: SEW			
90	R 27	130	28,37
90	R 37	200	28,32
115	R 47	240	33,79
115	R 57	450	26,31
Manufacturer: Siemens-Flender			
90	Z28	145	51,35
90	Z38	210	44,12
Manufacturer: Nord - NORDBLOC			
90	SK 272	145	64,75
90	SK 372	181	51,23
115	SK 472	334	65,56
115	SK 572	431	51,7
Manufacturer: Rossi - ES07			
90	MR 2I 2	156	28,9
90	MR 2I 3	195	31,1
115	MR 2I 4	286	31,4
115	MR 2I 5	398	23,4
Manufacturer: Rossi - E04			
90	MR 2I 40	75	22,1
90	MR 2I 41	74	13,3
106	MR 2I 50	143	22,3
106	MR 2I 51	200	18,3
132	MR 2I 63	302	24,3
132	MR 2I 64	401	19
160	MR 2I 80	650	24,5
160	MR 2I 81	730	20,1
195	MR 2I 100	1100	23,4
195	MR 2I 101	1230	19,3
236	MR 2I 125	2280	24,3
236	MR 2I 126	2560	19

Table 2. Examples of nominal torque capacities and highest gear ratio for three-stage gear reducer of different manufacturers.

Axis height *h*, **mm**	**Designation of gear unit**	**Nominal output torque,** T_N**, Nm**	**Max gear ratio** i_{max}
Manufacturer: SEW			
90	R 27	130	135,09
90	R 37	200	134,82
115	R 47	300	176,88
115	R 57	450	186,89
Manufacturer: Siemens-Flender			
90	D28	142	207,96
90	D38	224	191,75
Manufacturer: Nord - NORDBLOC			
90	SK 273	137	330,9
90	SK 373	214	238,68
115	SK 473	336	309,58
115	SK 573	465	309,58
Manufacturer: Rossi - ES07			
90	MR 3I 2	155	126
90	MR 3I 3	230	136
115	MR 3I 4	350	181
115	MR 3I 5	500	178
Manufacturer: Rossi - E04			
90	MR 3I 40	88	92,8
90	MR 3I 41	98	74,4
106	MR 3I 50	160	130
106	MR 3I 51	224	107
132	MR 3I 63	335	142
132	MR 3I 64	450	111
160	MR 3I 80	670	154
160	MR 3I 81	900	126
195	MR 3I 100	1320	144
195	MR 3I 101	1800	118
236	MR 3I 125	2650	150
236	MR 3I 126	3550	117

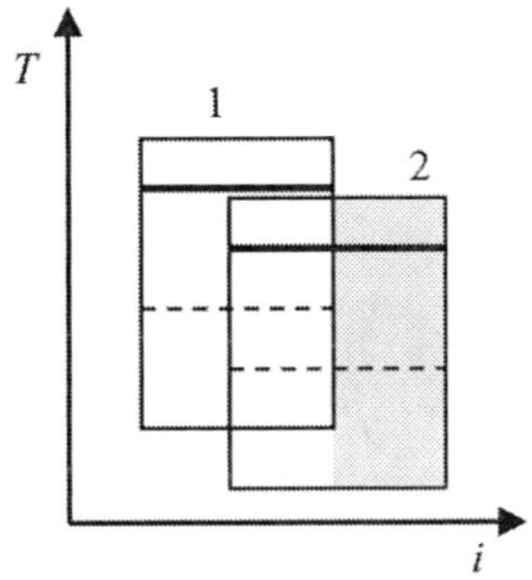

Fig. 5. Characteristic application areas of three-stage gear unit with larger torque capacity (1) and with smaller torque capacity, but higher gear ratio (2). Shaded area represents less expensive solution.

Some large manufacturers are trying to create so-called intermediate size of gear units and thus offer a cheaper solution (Figure 6, Table 3). With this additional size, they increase the number of gear unit components they must produce (housings, gears, shafts, bearings, etc.), so this procedure can not be considered as favourable as the application of housing with two sets of gear pairs.

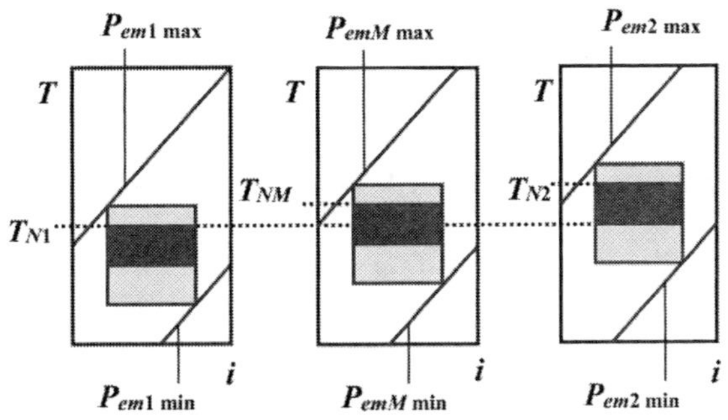

Fig. 6. Schematic review of torque and gear ratio areas that are covered by certain sizes of gear reducers by introducing additional size.

Table 3. Examples of providing additional size of two-stage gear reducer of different manufacturers.

Axis height h, mm	**Designation of gear unit**	**Nominal output torque, T_N, Nm**	**Max gear ratio i_{max}**
Manufacturer: SEW			
115	R 57	450	26,31
130	**R 67**	**580**	**28,13**
140	R 77	705	23,37
Manufacturer: Nord - NORDBLOC			
115	SK 572	431	51,7
130	**SK 672**	**534**	**64,18**
140	SK 772	858	62,32

Using this approach, almost the same load capacity can be obtained as with second gear pair of gear reducer with two sets of gear pairs. Production costs will be a little lower, but production will be increased because of introducing additional size and keeping in stock the components of all sizes of gear reducer. Large manufactures do not have this problem since they have large production and large stocks. However, if some small manufacturers want to be competitive, they have to produce every special dimension of the additional gears, but sometimes they are not able to provide absolute interchangeability.

4 Conclusions

The using of two gear sets of gear pairs within the same gear reducer housing is very justified procedure. Production is a little bit complicated and it is required larger stock for all necessary gears, but only in this way it is always possible to offer cheaper solution and in that way remove the possible advantage of competitive solution. In order to increase the offer of their reducer, large manufacturers often introduce additional size of gear unit. Small manufacturers try to compete with them and stay at the market by offering cheaper solution with high gear ratio and smaller torque capacity, so they offer gear units with one stage less.

This paper is part of a research on project "Research and Development of a New Generation of Wind Generators of High Energy Efficiency" TR 35005, supported by the Ministry of Education and Science, Republic of Serbia.

References

1. S. Kuzmanović, *Universal Gear Reducers with Cylindrical Gears* (in Serbian), (University of Novi Sad, Faculty of Technical Sciences, Novi Sad, 2009)
2. S. Kuzmanović, M. Rackov, *Proc. of the Int. Conf. on Power Transmission – BAPT, Kallithea, GR, 145* (2009)
3. Rossi, *E04 – Coaxial Gear Reducers and Gearmotors*, 4002BRO.ECO-en 1210 HQM (2015)
4. Rossi, *ES07 – Coaxial Gearmotors*, STANDARDFIT, 4063BRO.EST-en 1210 HQM (2015)
5. ***https://www.lenze.com/fileadmin/lenze/documents/de/catalogue/CAT_GKR_GKS_GSS_MF_15593792_de_DE.pdf, Accesed 01.03.2017
6. ***http://download.sew-eurodrive.com/download/pdf/16997611_G01.pdf, Accesed 10.03.2017
7. NORD Drivesystems, *Standard Helical Gearboxes*, G (2012)
8. ***https://www.automation.siemens.com/md-static/D87-1_en_2011.pdf, Accesed 01.02.2017
9. S. Kuzmanović, M. Vereš, M. Rackov, T. FAMENA **38**(1), 77 (2014)
10. M. Rackov, *Conceptions of Development of Universal Gear Reducers* (Ph.D. thesis in Serbian), (University of Novi Sad, Faculty of Technical Sciences, 2013)
11. S. Kuzmanović, M. Rackov, I. Knežević, M. Vereš, *Proc. of the Int. Symp. on Machine and Industrial Design in Mechanical Engineering – KOD, Balatonfured, HU, 73* (2016)
12. M. Rackov, I. Knežević, S. Kuzmanović, M. Čavić, M. Penčić, S. Ianici, *Proc. of the Int. Conf. on Mechanical Engineering Technologies and Applications – COMETa, East Sarajevo, BA, 301* (2016)

The simulation of operating of a positioning system actuated by electrical stepping motor

Dana Rizescu[1], and *Ciprian Ion* Rizescu[1,*]

[1]University POLITEHNICA of Bucharest, Mechatronics and Precision Engineering Department, 313 Splaiul Independentei, 060042, Bucharest, Romania

Abstract. The work presents the current situation concerning positioning systems manufactured by well-known companies in the industry: Festo, SMC, etc, or patented solutions for positioning system. The authors suggest, following a thorough analysis, as a original solution, a positioning system driven by an electrical stepping motor. There are presented some particular elements of design, modeling, execution, testing and optimization of the proposed system. For validation, the mathematical model, authors have developed, in Matlab environment, a graphic user interface in which one can simulate the operation of several solutions with stepping motors, different positioning strokes.

1 Introduction

There is known that among the benefits of the pneumatic systems may be: rapid operation, easy maintenance, flexibility in use, good weight /power ratio, are clean, properties of selfcooling and the relatively low level costs. In comparison with hydraulic actuators, in the case of pneumatic actuators there is no risk of contamination of the driven load with leaking liquids and are not necessary additional mechanisms that transforms the rotation in linear movement such as the cases in which use electric rotational motors. Another important advantage is the availability of compressed air in most industrial units. All of these features have been immeditately tappet in a large number of applications in the fields of robotics, machine tools, machines, the processing industry, food processing systems, in the health-care and rehabilitation systems.

There are known solutions of some companies in the field the precise adjustment of the position such as: Festo, SMC but we have to mention here the solutions of patents. The authors have taken into account more solutions. One of them is to be known from the U.S. Patent 6,802,242 B1, and shall relate to the control of the flow of fluid and in particular a control system for an air circuit with improvement in the sensitivity of the cylinder carried out through the use of selective control valve for the components of the sealing O ring at each end of the cylinder [1]. In such applications, it may be desirable a quick operation and in a repeated positioning of the piston with the precision of a thousandths of an inch.

* Corresponding author: ciprian.rizescu@upb.ro

Another solution is that of the US 7,021,191 B2 patent, where is described a position control system that is used to control a pneumatic cylinder having a room, defined by a piston, located in a housing for moving between two proportional valves for the flow control, electrically actuated and connected to each union of the cylinder, to control the flow of fluid selectively and proportional in one of the chambers [2]. A pressure sensor is provided for measuring the pressure of the fluid corresponding to each rooms.

2 Experimental Setup

In figure 1 is presented the principle sketch of the solution proposed, in which there were noted: M - guided mass which must be stopped with high precision, PLM - pneumatic linear motor, UPO - stop unit with high precision, S1, S2 - sense valves, DP – pneumatic distributors, EM - electromagnets, O – stopper, SP – position sensor.

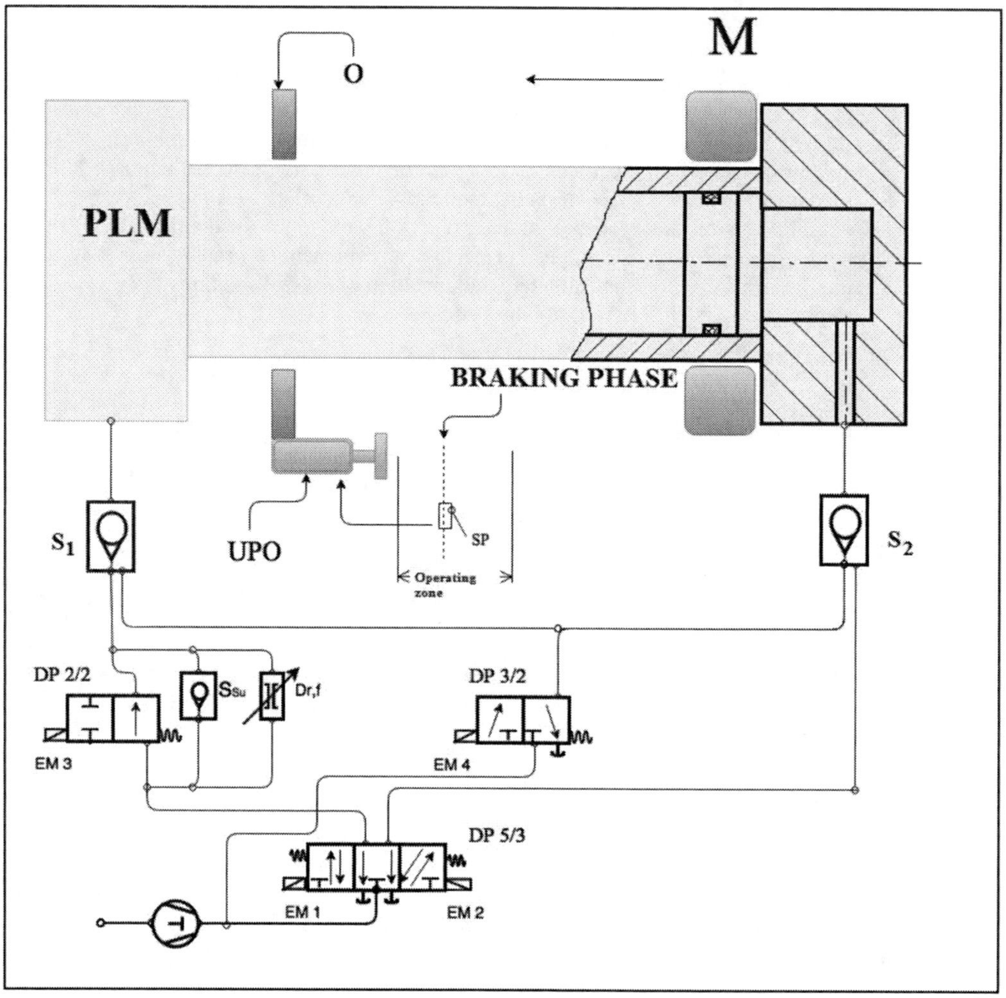

Fig. 1. Principle sketch of the experimental setup.

The operation mode of the equipment. The incremental electric motor (stepping motor) operate for positioning the stopper O in the desired position and locks the stopper to prevent the displacement when the contact with the mobile assembly is obtained [3]. The stepping motor actuates a nut-screw mechanism for a high precision positioning. Then switch the pneumatic distributor DP1 toward the right, feeding with air the room from the right-hand side of cylinder. The cylinder piston moves to the left. When the mobile assembly shall enter into range of the position sensor, SP, this command the pneumatic distributor DP2 which closes. At this time starts the braking process of piston (mobile assembly) as follows: air is purged from the camera from the left hand side of the cylinder through restrictor Dr, achieving an air cushion (therefore a back-pressure) in left hand side of the cylinder. The cylinder piston continue driving on at low speed. Selective unlockable sense valves S1 and S2 air locks cylinder rooms, this being stopped firmly on position.

The authors have developed an graphic user interface in the Matlab environment for simulating the stop unit operation, considering the various electric stepping motors having different step angles. As a result, one can study different actuation solutions for the stepper motors with the different step angles.

3 Experimental Setup Simulation

There is known that there are three commonly used excitation modes for step motors; these are full step, half step and microstepping.First mode: in full step operation, the motor moves through its basic step angle, i.e., a 1.8° step motor takes 200 steps per motor revolution. There are two types of full step excitation modes. In single phase mode, also known as "one-phase on, full step" excitation, the motor is operated with only one phase (group of windings) energized at a time [4]. This mode requires the least amount of power from the driver of any of the excitation modes. In dual phase mode, also known as "two-phase on, full step" excitation, the motor is operated with both phases energized at the same time. This mode provides improved torque and speed performance. The second mode: half step excitation is alternating single and dual phase operation resulting in steps that are half the basic step angle. Due to the smaller step angle, this mode provides twice the resolution and smoother operation. Half stepping produces roughly near 15% less torque than dual phase full stepping. The third mode: microstepping is a technique that increases motor resolution by controlling both the direction and amplitude of current flow in each winding. Current is proportioned in the windings according to sine and cosine functions. Microstepping can divide a motor's basic step up to 256 times. Microstepping improves low speed smoothness and minimizes low speed resonance effects. Microstepping produces roughly 30% less torque than dual phase full stepping.

The graphic user interface – main panel is presented in figure 2. There are some fields that must be filled in or some pop-up selection have to be made. The cylinder stroke range runs from 100 mm to 350 mm [5-6]. This stroke represents the displacement of guided mass M (figure1). For example, the cylinder stroke was considered 225 mm, then there was selected motor 1 from a database developed by authors. There were considered some representative stepping motors with different number of windings, step angles and excitation modes [7] . In figure 2 was selected motor 1 with Torque diagram shown in a dedicated window. When this selection is completed a motor characteristic is displayed [8]. The motor excitation mode can be chosen : stepping or microstepping. The authors used for this exeprimental setup full step and half step excitation modes.The stepping motor stroke could be chosen up to 20 mm. This stroke represents the displacement of the UPO unit, actuated with the stepping motor. If the operator "forget" to fill in one field or to choose one selelction a error message will be displayed on the screen. At this moment the SIMULATION button has to be pressed.

In the first step, the stepping motor stroke is set, considering the actuation mode, the desired stroke. This simulation is shown in figure 3. The number of steps/10 versus stepping motor stroke is presented also in the figure 3. After the motor stroke is set the system is ready to actuate the slider mass M. There is another window which presents the impact moment between the slider mass M and the mobile nut from the stop unit.

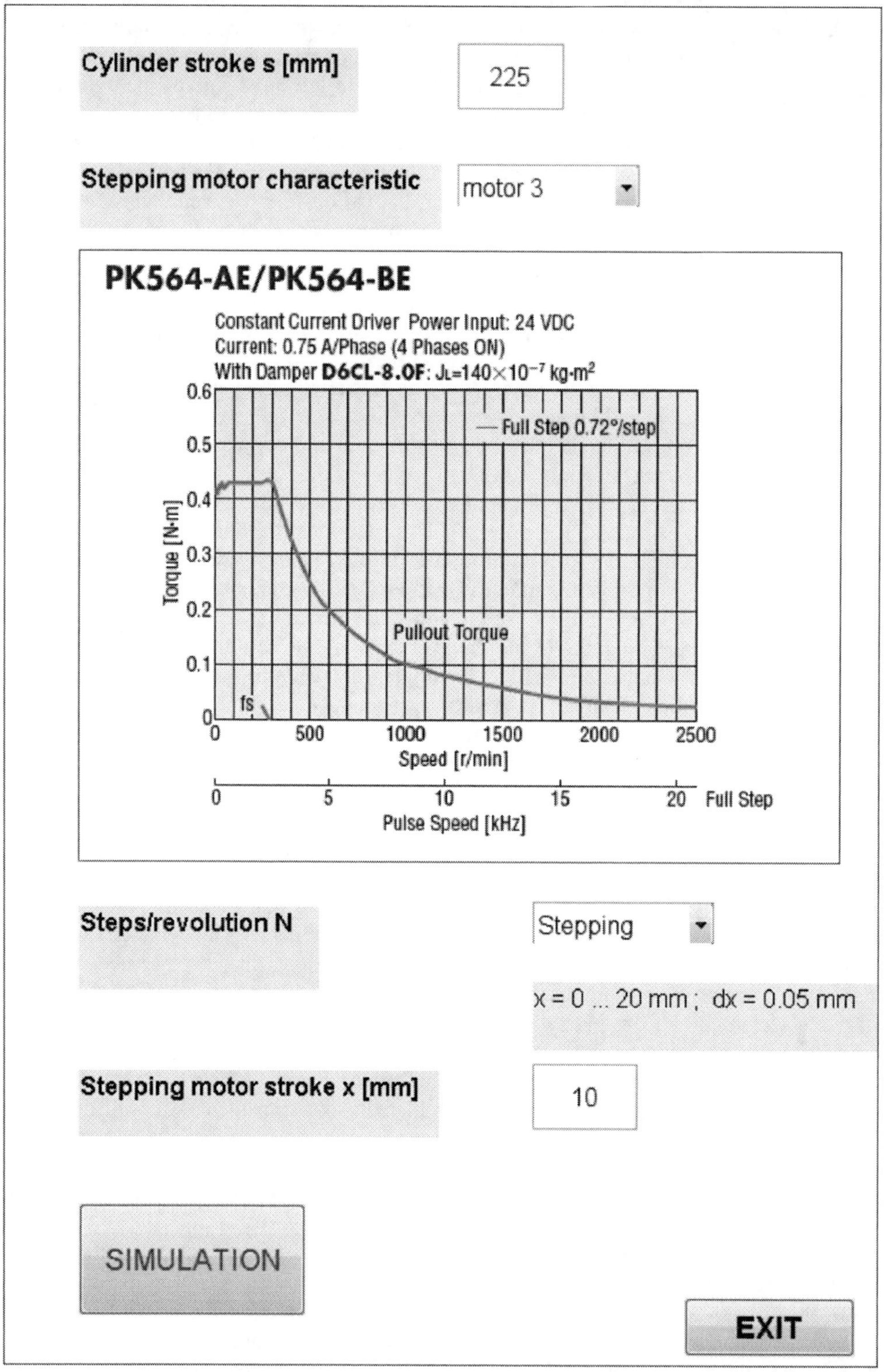

Fig. 2. Graphic user interface – main panel.

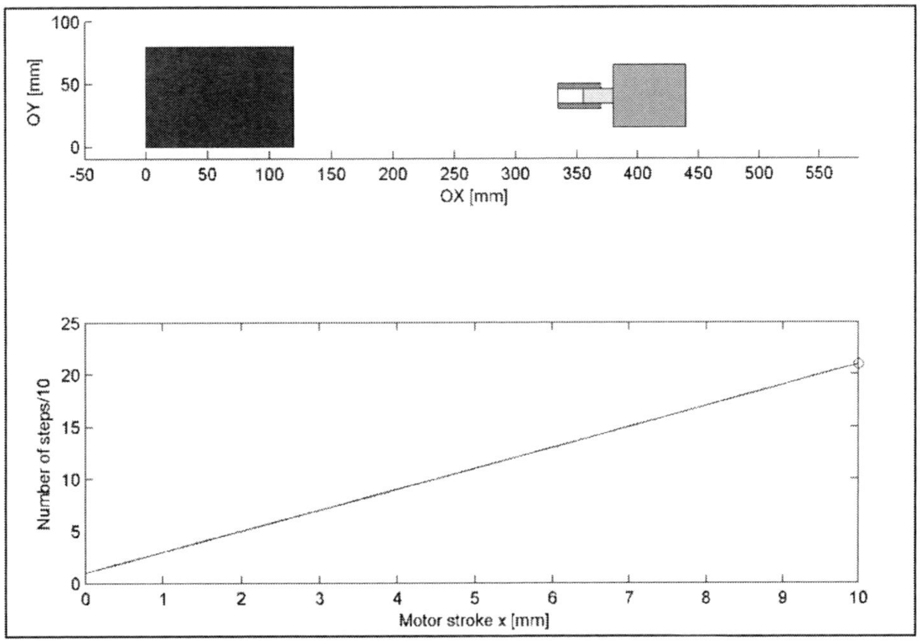

Fig. 3. Setting the stopper position with the stepping motor.

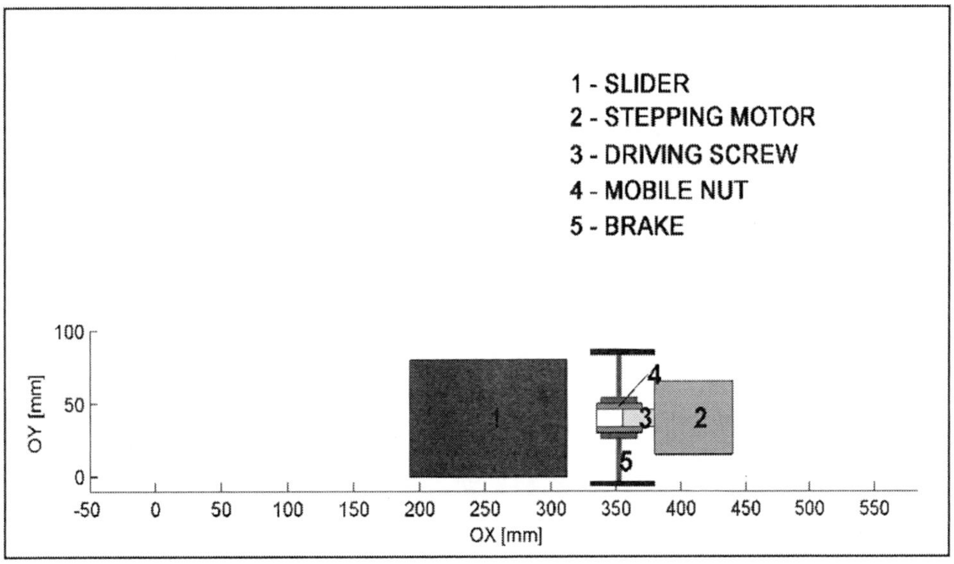

Fig. 4. The impact moment between the slider and the stopper unit.

The impact is shown in figure 4. Here the slider mass moves with the cylinder stroke. The simulation allows to compute stepping motors actuation times.

4 Conclusions

The simulation allows to compare different actuation solutions considering various stepping motors and actuation modes. A simulation allows for investigation of phenomena not directly measurable with the experimental apparatus. This research demonstrates the capabilities and limitations of advanced control strategies with stepping motors. Based on this simulation the authors developed an experimental setup.

References

1. Patent: US 6,802,242 B1, Pneumatic Circuit Control System (2004)
2. Patent US 7,021,191 B2, Accurate Fluid Operated Cylinder Positioning System (2006)
3. Gh. I. Gheorghe, I. Ilie, Optics & Mechatronics, **46**, 40 (2014)
4. Gh. Prisacaru, V. Carlescu, D. Olaru, *Actuatori neconventionali în mecatronică* (Ed. Tehnopres, 2013)
5. G. Belforte, 6th trien. Int. Symp. on fluid control measurement and visualization: FLUCOME (2000)
6. C. Ferraresi, G. Quaglia, *Proceedings of the Forth JHPS International Symposium on Fluid Power,* Tokyo, 207 (1999)
7. X. Luo, H. Sun, J. Wang, American Control Conference on O'Farrell Street, San Francisco, CA, USA (2011)
8. Parker Hannifin Corporation Parker - Origa Glendale Heights, Illinois www.parker.com/pneu/rodless ORIENTAL MOTOR (EUROPA) GmbH www.orientalmotor.de

OOA composite structures applicable in railway industry

Soňa Rusnáková[1*], *Milan Žaludek*[1], *Milena Kubišová*[1], and *Vladimír Rusnák*[2]

[1]Tomas Bata University in Zlín, Department of Production Engineering, Vavrečkova 275, 760 01 Zlín, Czech Republic
[2] VŠB – Technical University of Ostrava, Faculty of Metallurgy and Materials Engineering, Department of Material Engineering, 17.listopadu 15, 708 33 Ostrava – Poruba

Abstract. Composite sandwich structures offers several advantages over conventional structural materials such as lightweight, high bending and torsional stiffness, superior thermal insulation and excellent acoustic damping. In the aerospace industry, sandwich composites are commonly manufactured using the autoclave process which is associated with high operating cost. Out-of-autoclave (OOA) manufacturing has been shown to be capable of producing low cost and high performance composites. In this paper we present results of experimental testing of various sandwich materials according various standards and actual requirements in transport industry. We compared the different types of surface and paint systems, because these layers are the most important in contact with the surrounding environment and load conditions. In the experimental measurements were used various materials. For the core of the sandwich structure were selected aluminium honeycomb, aramid honeycomb and PET (Polyethylene terephthalate) foam core. Support layers were chosen two kinds of predimpregnated materials. The conditions of measurements were requirements for strength and rigidity, safety - flame resistance and reflectivity resistance. The samples were tested at the 3 - point bending test according to standard EN ISO 178, by modified test to determine the force required to rapture threaded insert, by test of reflectivity according to UIC CODE 844-4 R and according to standard EN 45545-2 fire protection of railway vehicles.

1 Out-of autoclave technology

The majority of high performance composite structures are manufactured in autoclaves. Autoclave processing of composites uses primarily pre-impregnated unidirectional fibers or fabrics (prepregs) that are cured at 180°C and consolidated with 6 atm of pressure, usually. A large database of design allowable for autoclave prepregs is available and accessible to product designers. This well-established process requires high infrastructure investment for

* Corresponding author: rusnakova@ft.utb.cz

autoclaves as well as high recurring tooling costs. The maximum component size is limited by the size of available autoclaves. Alternative processing methods need to be investigated in order to reduce production costs and broaden the supply base of composite structures. This presentation addresses the application of a relatively new class of composite materials using out-of-autoclave (OOA) technologies. The advantages of OOA processes include substantially lower infrastructure, tooling, and overall manufacturing costs. [1-2]
For OOA technologies using predimpregated materials (pre-preg) belong to curing of prepregs in oven and PCM (Prepreg Compression Moulding).

2 Fire resistance

In case of fire, the composite is protected by foam formation, which is isolated from oxygen. To be absolutely halogen-free, smoke toxicity and opacity is dramatically reduced. Fire resistance in the railway industry defines standard EN 45545-2 Fire protection of railway vehicles. The standard specifies reaction to fire-intensive materials and products used in rail vehicles. It is based on existing fire safety regulations for rail vehicles in the International Union of Railways and various European countries. Each state has its own testing methods that are adapted to local conditions and customs. Individual test methods cannot be compared together, slow down the adjustment process. [3-4]

For each product is determined by the specific requirements set by standards - R:

- Interior (panels, baggage, window frames, flat bar),
- Exterior (seat cushions, covers, air section),
- Furniture (seat cushions, covers, mattresses, bedding),
- Electronic equipment (cables, housings, insulation material, resistors),
- Mechanic equipment (flexible steel / rubber parts).

The test materials are classified according to this standard (EN 45545) into the category R1, which is valid for the entire European Union since 2013 and unites the most important requirements for non-flammability, toxicity and emissions of flue gas and heat. [5] For this category of material are required to test these basic tests:
Determining Hazard Level (HL)

Fire resistance levels (HL1,2,3) is measured using the following five tests:

- CFE (kW / m^2): Critical Flux at extinction or lateral spread of flame
- MAHRE (kW / m^2): The maximum rate of heat release
- Ds (4): Specific optical density of smoke
- VOF4: Cumulative value of specific optical densities in the first 4 minutes of the test
- CITG 8min Conventional Toxicity Index (C02, CO, HF, HCl, HBr, HCN, NOx, SO2)

Table 1. Construction categories [6]

Operation Category	**N** Standard vehicles	**A** Automatic train without trained personal	**D** Double decked vehicles	**S** Sleeping and couchette vehicles
1 On infrastructure	HL1	HL1	HL1	HL2
2 Underground Tunel <5km With side evacuation	HL2	HL2	HL2	HL2
3 Underground Tunel >5km	HL2	HL2	HL2	HL3
4 Underground Without side evacuation	HL3	HL3	HL3	HL3

3 Experimental part

3.1 Tested samples

The samples were made in the form of flat plates by hand-laying pre-preg technology and subsequent curing in the oven. The technology is more suitable for piece production. Used materials for the production of sandwich panels:

- Pre-preg 1 - Fiber-reinforced, thermosetting pre-impregnated material with excellent FST and mechanical properties. It is halogen free modified phenolic system with 42% phenolic resin. Fabric of E-glass Yarn, 300g/m2, 8H satin weave.
- Pre-preg 2 - Fibrous-reinforced thermosetting pre-impregnated material with excellent FST and mechanical properties. It is a halogen-free phenolic system with 47% phenolic resins. Fabric of E-glass Yarn, 300g/m2, 8H satin weave. It is highly adhesive to aluminium.
- Foam - Closed thermoplastic pores and recyclable polymer foam with excellent smoke, toxic and fire properties. It is the ideal sandwich core material applied where high fire resistance is required.
- Aramid honeycomb - It is the main component of sandwich in industrial sectors with high dimensional stability in heat and humidity. The honeycomb used per square meter has a density of 32 cells, a cell size of 4.8 mm, a compressive strength of 0.9 MPa, a shear strength of 0.55 MPa, a shear modulus of 18 MPa.
- Aluminium honeycomb - core of aluminium honeycomb. The honeycomb used has a density of 77 kg / m^3, a cell size of 4.8 mm, a compressive strength of 3.94 N / mm^2, a shear strength of 2.29 N / mm^2.

3.2 Description of samples

Sample 2A
Consisting of Aramid honeycomb and 2 layers of prepreg on each side - referred to as 2A

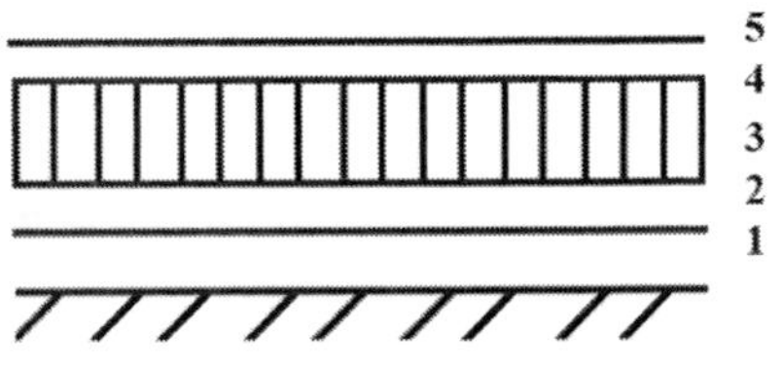

Fig.1. Composition of sample 2A - 1,2,4,5 pre-pregs 1; 3 Aramid honeycomb.

Sample 3A
Consisting of Aramid honeycomb and 3 layers of pre-preg on each side – referred to as 3A

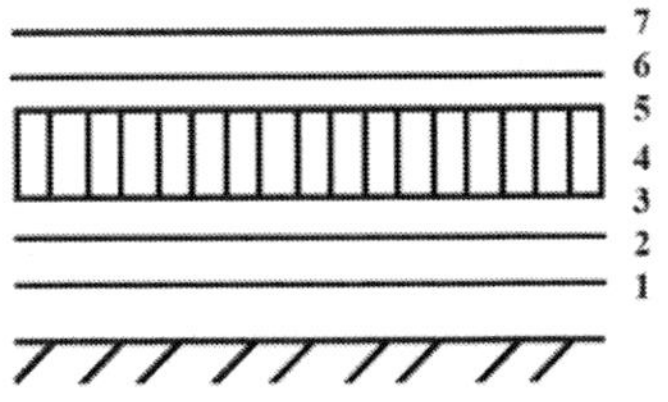

Fig.2. Composition of sample 3A - 1,2,3,5,6,7 pre-pregs 1; 4 Aramid honeycomb.

DOI: 10.1051/matecconf/201712101015

Sample 4A
Consisting of Aramide honeycomb and 4 layers of pre-preg on each side referred to as 4A

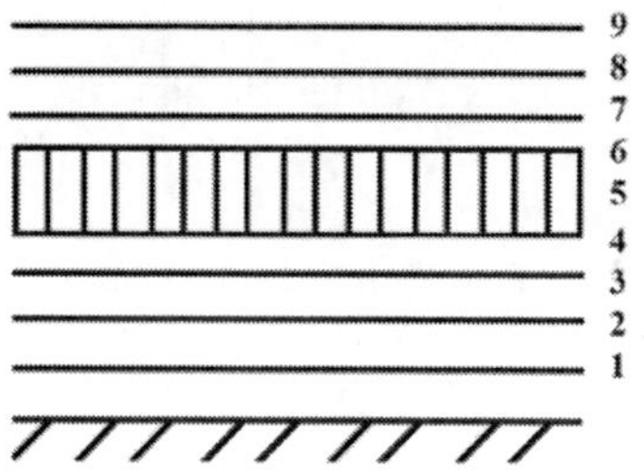

Fig. 3. Composition of sample 4A - 1,2,3,4,6,7,8,9 pre-pregs; 5 Aramid honeycomb.

The samples referred to as 2B, 3B and 4B are the same composition like samples referred 2A, 3A and 4A, but instead of Aramid honeycomb is like a core material used polymer foam.

The samples referred to as 2C, 3C and 4C are the same composition like samples referred 2A, 3A and 4 A, but instead of Aramid honeycomb is like a core materials used aluminium honeycomb.

4 Experimental testing of mechanical and FST properties

4.1 Three-point bending test 5

The test was carried out on Zwick 1456 according to ČSN EN ISO 178. The test samples were tested at room temperature 23 ° C. Sample sizes were determined at 100x20 mm.

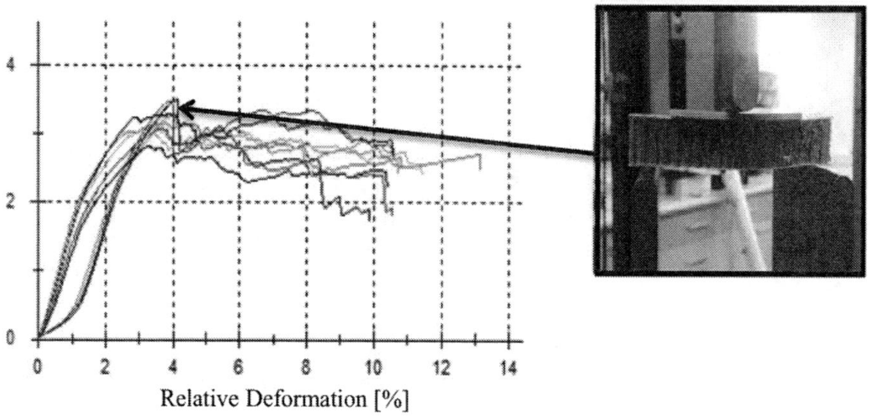

Fig. 4. Results for sample 2A, with view on arrangement during 3-point bending test.

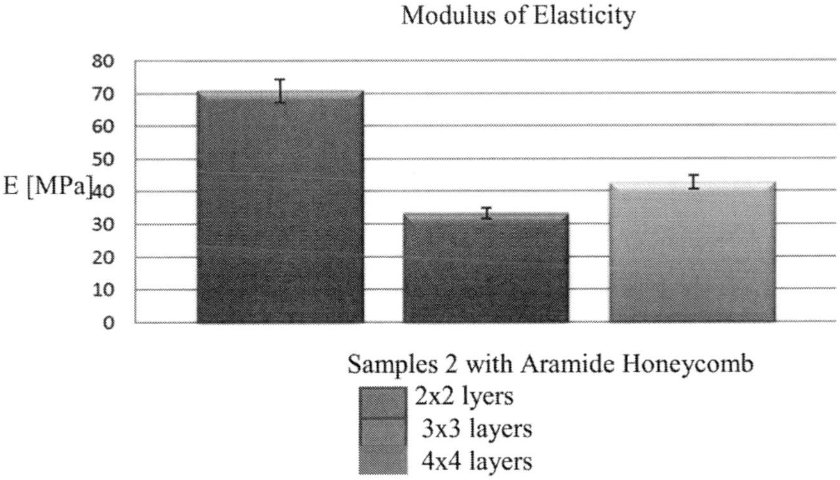

Fig. 5. Evaluation of Modulus of Elasticity in 3-point Bend Testing for Samples A.

The best fit in the bending test a Type C samples with an aluminium honeycomb core. The largest modulus of elasticity was measured 1860 MPa for the C sample with 3x3 pre-preg layers. On the contrary, the most severe modulus of elasticity resembled Type B samples with a foam core. Throughout the course of the test, the malfunctions were also recorded, which in most cases were of a generic nature and most often caused a violation of the cohesiveness between the core and the outer layer.

The highest value of the strength limit was 29.7 MPa for C type samples. Again, it turned out that the best type of core is the aluminum honeycomb, which shows the best results of the strength of all the samples tested. Worst results were obtained with the Aramid honeycomb sample A, which reached the minimum strength limits, the lowest strength was measured in a type A sample with 2x2 layers of pre-preg, namely 3.21 MPa. The most suitable materials are C-samples with aluminum honeycomb. In both cases, as with the strength limit or the modulus of elasticity, it comes out with the highest values.

4.2 Cone calorimeter

Samples of sandwich composite structures with two different painting systems. The size of the samples used for the conical calorimeter test were in accordance with the requirements of the technical standards ISO 5660-1 and ISO 5660-2. Samples were conditioned prior to the test according to ISO 5660-1.

4.2.1 Painting systems

Painting system 1 - consists of a filler and a top coat

1. Filler - It is a two-component, high-particle polyurethane based filling material. It has high chemical and mechanical resistance.
2. Top layer - This is a two-component polyurethane-based structural varnish with anti-graffiti properties. It is possible to create single-layer smooth or scarred (Nar-ben) surfaces and two-layer pearl structural effects. Cured film is abrasion-resistant, scratch-free, repellent, and easy to clean.

Painting system 2 - contains filler and top coat.

1. Filler - is a two-component polyurethane spray anti-corrosion filler for the creation of a high quality substrate under the top layer. Depending on the degree of dilution, either thick layers can be created to fill rough surface roughness, or thin, highly rigid and even layers.

2. Top layer - Two-component varnish based on polyurethane resin for semi-gloss. It is possible to create smooth thin layers. These treatments have high film strength, good chemical resistance and weather resistance.
Samples were tested at a heat flow density of 50 kWm^{-2}. The orientation of the samples during the test was horizontally.

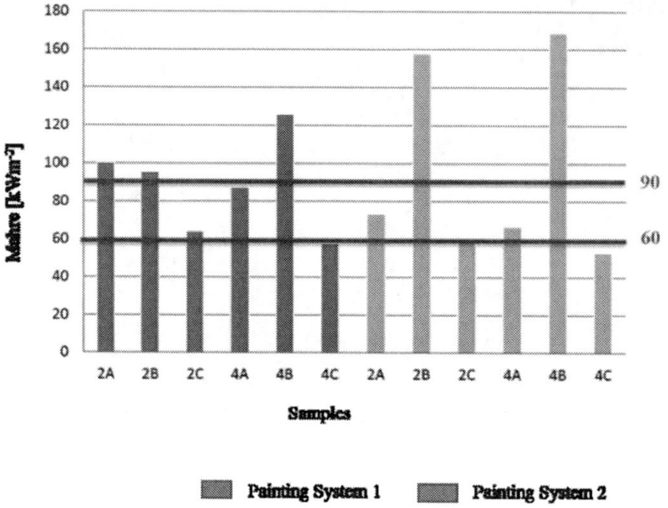

Fig. 6. Measured values with marked limits for the category HL2.

Conditions during the exam:

- Ambient temperature: 20 to 21°C
- Relative humidity: 52 to 55%
- Atmospheric pressure: 98.8 to 99.5 kPa
- Density of the heat flow reaching the surface of the tested samples: 50 kWm^{-2}
- Air flow through the hood: 0,024 m^3/s

The maximum average heat release rate is a basic feature that highlights the fire risk of the material. Initiation time and Maximum Average Rate of Heat Emission (Marhe).

All samples tested showed a maximum average heat release rate of less than 40 kW m^{-2} within 20 s of the test. During the entire test duration, the Marhe value was less than 90 kW m^{-2}, only the samples 2C, 4A, 4C for the coating system 1, and for the samples were 2A, 2C, 4A and 4C. The data on the maximum average heat release rate, the time-dependent rate of heat release, the total amount of released heat, and the specific CO release rate depend on the atmospheric pressure at which they were determined, with the increasing altitude decreasing both the maximum and average heat release rates. The nature of the samples also results in the high sensitivity of the data obtained from the way they are stored in the holder. In the case of insufficient mineral wool support in the holder, the tilt can occur during the test and subsequent very rapid burning of the filling part of the samples examined. Significant impact on the data obtained also has the side of the sample from which it is subjected to thermal radiation during the test. It is possible to see how the individual test results differ significantly, due to the different material composition and non-homogeneity of the materials in the samples. All samples were tested according to EN 45545-2. It was mainly about finding out whether the tested materials met the HL2 category and had to move between 60-90 [kWm^{-2}]. In order to be used in most design categories.

We evaluated two coating systems for samples 2A, 2B, 2C and 4A, 4B, 4C. We focused on the maximum average heat release rate called Marhe [kWm^{-2}]. Samples with paper honeycomb 2A and 4A at both coating systems reached values below 90 [kWm-2] up to sample 2A with a painting system 1 at 100 [kWm^{-2}] and therefore not meeting the HL2 category. Poor results were obtained with the foam core 2B and 4B in both paint systems, when the outer layers burnt rapidly, and the foam core was rapidly burnt. These samples have reached values of 95-169 [kWm^{-2}] and do not meet the requirements of category HL 2. On the other hand, the best results were obtained with samples of aluminum 2C and 4C of both paint systems. Samples 4C showed the best values of all 52 and 57 [kWm^{-2}] and thus even meet the higher HL3 category. Better results were obtained with the sample system 2, which had lower values than system 1.

5 Results

Research paper serves as a basis for the development of a new structure of the sandwich structure, which could start to be applied in the railway industry produced by OOA technology.

Various types of surface and coating systems were compared. Different materials were used for experimental measurements. For the core of the sandwich structure, aluminium honeycombs, Aramid honeycombs, and a foam polymeric PET core were selected, and on the outer layers 2 types of pre-pregs.

Based on requirements for strength and stiffness, safety self-extinguishing, wear resistance, samples were tested for 3-point bending according to EN ISO 178 standard, modified test to determine the force required to pull the thread insert, UIC CODE 844 - 4 R and according to the standard ČSN EN 45545 - 2 fire protection of railway vehicles, the cone calorimeter test according to ISO 5660-1 was selected.

In all tests, the samples with the foam core have the lowest values and have not met either standard requirement.

The solution is to use a different type of foam with better mechanical properties or to adjust the number of individual pre-preg layers that can be replaced by other types of pre-pregs. Sandwich structures with a foam core are very easily mouldable and can also produce very complex products. [7]

For samples with Aramid honeycomb, variable results were obtained for all tests. In the bending test, they reached the lowest values of all tested materials. Self-extinguishing samples matched, except for one that had only two layers of pre-preg. For wear resistance the best result was achieved by sample 4A, but it didn't meet the requirements of the standard. The next proposal is to change the honeycomb, to another with a higher cell density, thereby increasing strength and stiffness.

The highest and best values were obtained with aluminium honeycomb samples. In the bending test, they reached up to several times the strength and modulus of flexural strength compared to the paper and foam core samples. They achieved a conical calorimeter evaluation all samples of very good results and meet the HL2 category. Quadruple samples also met the HL3 category.

Appropriate combinations of materials that would already meet all requirements for the railway industry must be further developed and tested. The resulting product must have sufficient strength and stiffness, safety, functionality, wear resistance, but a very important condition is also the low weight.

This work and the project is realized with the financial support of the internal grant of TBU in Zlín No. IGA/FT/2017/002 funded from the resources of specific university research.

References

1. NIU, M. C. Y. *Composites Airframe Structures; Practical Design Information and Data.* 1st ed. (Hong Kong: Conmilit Press Ltd., 664 s., 1992)
2. VINSON, J. R., *The Behavior of Sandwich Structures of Isotropic and Composite Materials*. 1st (Ed. New York: CRC Press, 378 s., 1999)
3. BITZER, T. N. Honeycomb Technology: Materials, *Design, Manufacturing, Applications and Testing.* 1st ed. (London: Chapman & Hall, 233 s., 1997)
4. ZENKERT, D. Nordic Industrial Fund, *The Handbook of Sandwich Construction* (Worcestershire: EMAS Publishing, 447 s., 1997)
5. LEHMHUS D., BUSSE M., HERRMANN A., KAYVANTASH K. *Structural Materials and Processes in Transportation*, (1st ed. New York: Wiley-VCH Verlag GmbH, 598s., 2013)
6. DOSTAL C. A. *Engineered Materials Handbook: Composites*, Volume **1**, (Ohio: ASM International, 983 s., 1987)
7. GUPTA K. M. *Engineering Materials: Research, Applications and Advances*, (1st ed. Boca Raton: CRC Press, 636 s., 2014)

DOI: 10.1051/matecconf/201712101016

Study on the reliability of the underground conveyor belt system installed at Vulcan Mine, the Jiu Valley Basin in Romania

Ovidiu-Bogdan Tomuş[1,*], *Andrei* Andraş[1], and *Mihai* Magyari[2]

[1] University of Petroşani. Dept. of Mechanical, Industrial and Transports Engineering, Romania
[2] INCD Insemex Petroşani, Romania

Abstract. The paper deals with a comprehensive reliability analysis of the conveyor belts belonging to an underground coal mine in the Jiu Valley, Romania. As resulted from the mine management reports, the transportation system is responsible for many downtimes and is a real bottleneck in the constant and adequate production, and the suspicion is the weak state of belt conveyers, which are the spinal column of the extraction process. For this reason, a comprehensive reliability analysis has been decided, in order to deliver a maintenance-upgrading plan.

1 Introduction

Vulcan mine plant is one of the oldest underground hard coal mines in the Jiu Valley coalfield. Jiu Valley hard coal is mined out since 1850, actually in 4 plants, other 3 being in downsizing and closure process. Vulcan is one of active mines, belonging to the Hunedoara Energetic Complex, with an output of about 2 M. tons/year, with a foreseen lifespan until the year 2022. The location of Vulcan mining field in the layout of Jiu Valley coalfield can be seen in figure 1 [1].

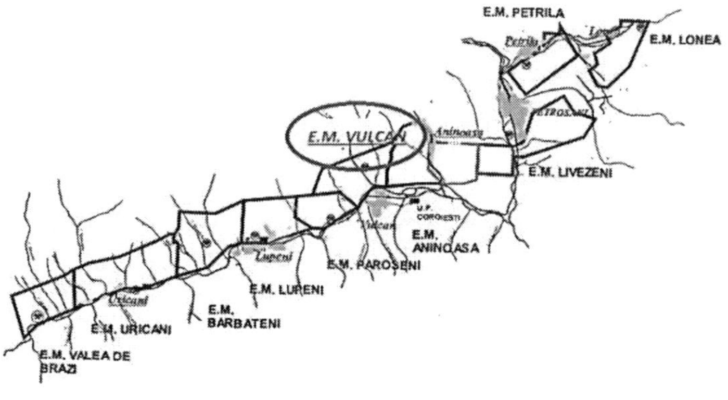

Fig. 1. The location of Vulcan mining field in the layout of Jiu Valley coalfield.

* Corresponding author: tobogdan2002@yahoo.com

The mine plant, due to its age and the tectonics of the coal seams, has been developed on horizontal and vertical directions in a way which led to a very large extension. In plus, the used technology, specific to thick inclined seams in horizontal slices, with top coal caving, in the past 20 years, has as result a very complex and irregular structure of the conveying system, as can be seen in figure 2 [2].

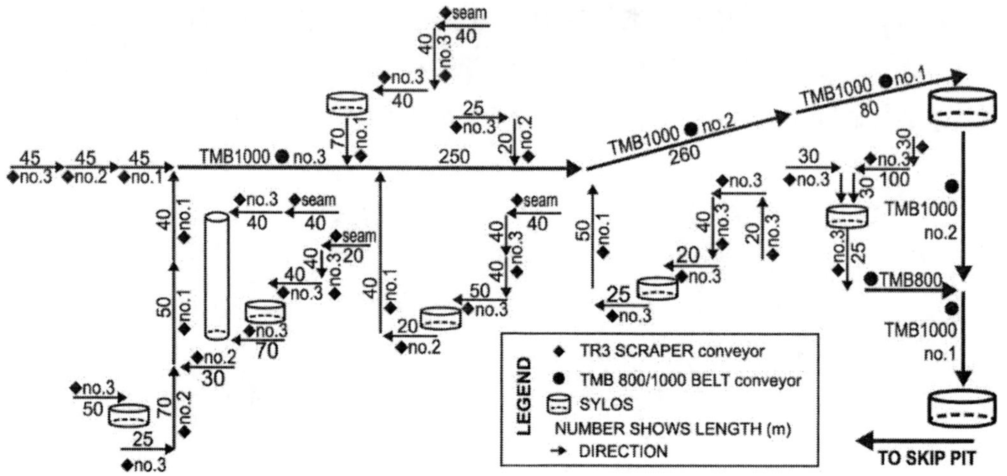

Fig. 2. Structure of the conveying system.

The main conveying system consist in belt conveyors TMB 800 and TMB 1000 (the figure indicating the width of the belt) on main gates at the levels 250, 315 and 360 (elevation related to the sea level) until the intermediary silos and finally to the main hoisting shaft [5].

As resulted from the mine management reports, the transportation system is responsible for many downtimes and is a real bottleneck in the constant and adequate production, and the suspicion is the weak state of belt conveyers, which are the spinal column of the extraction process [1].

For this reason, a comprehensive reliability analysis has been decided, in order to deliver a maintenance-upgrading plan.

2 Reliability analysis of main belt conveyors

The reliability study of the main belt conveyors at Vulcan mine plant, has been performed using the data from the conveyor system monitoring books taking into accounts the records in the period June 2012 – March 2013.

The structure and occurrence frequency of faults of the conveyors in the period are presented in table 1 and in figures 3 and 4.

Table 1. Structure of faults occurred at belt conveyors.

No.	Subassembly failed	Absolute frequency of faults	
		No. of faults	%
1	Belt - staples	119	46.48
2	Upper and lower idlers	76	29.69
3	Belt tensioning system	37	14.45
4	Belt – requiring replacement	11	4.30
5	Mechanical drive system	7	2.74
6	Line supports	6	2.34
Total		256	100.00

The table above shows that the most frequent is the belt sections coupling staples , at about 47%.

This kind of fault is more relevant because they are 9 conveyers, each with one, two or three joints, remediation of a failure paralyzing for about two-three hours all the haulage system, which dramatically influence the operation cycle in all working faces [4].

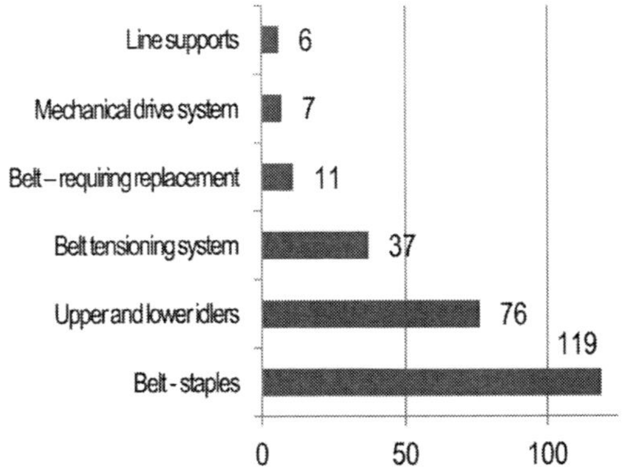

Fig. 3. Absolute frequency of faults.

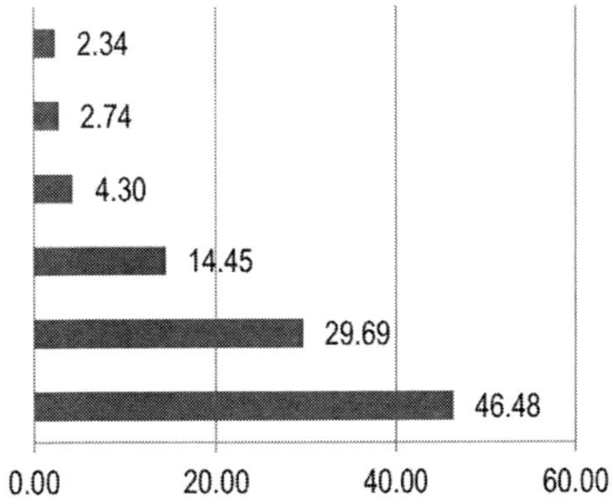

Fig. 4. Relative frequency of faults in %.

The idlers fault and replacement has a less impact on overall mining plant availability, despite their large amount, and its large share in total of faults - about 30% because of reduced time consumption for their replacement. The large share of staple joint faults in one year of operation was the main concern leading to the necessity of the present reliability analysis [2].

Analysing the occurrence of stapled joints failures, we selected from the overall 9 conveyors 3 ones which are similar length, working conditions and width of belt.

A factor which is more difficult to be quantified is the degree of ageing of belt's material, the three conveyors having, at the beginning of the study different operating hours.

Table 2. Empirical repartition function $\hat{F}(t_i)$ calculation.

i	t_i, hours	Absolute frequency, n_i	Relative frequency, f_i	Empirical repartition function, $\hat{F}(t_i)$
1	50	9	0.097826	0.097826
2	58	8	0.086957	0.184783
3	60	21	0.228261	0.413043
4	100	4	0.043478	0.456522
5	110	3	0.032609	0.489130
6	135	2	0.021739	0.510870
7	147	6	0.065217	0.576087
8	150	4	0.043478	0.619565
9	154	3	0.032609	0.652174
10	170	2	0.021739	0.673913
11	200	8	0.086957	0.760870
12	210	5	0.054348	0.815217
13	220	7	0.076087	0.891304
14	230	1	0.010870	0.902174
15	250	1	0.010870	0.913043
16	340	1	0.010870	0.923913
17	400	1	0.010870	0.934783

i	t_i, hours	Absolute frequency, n_i	Relative frequency, f_i	Empirical repartition function, $\hat{F}(t_i)$
18	420	1	0.010870	0.945652
19	460	1	0.010870	0.956522
20	620	1	0.010870	0.967391
21	760	2	0.021739	0.989130
22	1100	1	0.010870	1,000000
		$\sum_{i=1}^{22} n_i = 92$	$\sum_{i=1}^{22} f_i = 1{,}000000$	

From the available data, we inferred the running hours until the failure, obtaining the time series as follows: 50; 50; 50; 50; 50; 50; 50; 50; 50; 58; 58; 58; 58; 58; 58; 58; 58; 60; 60; 60; 60; 60; 60; 60; 60; 60; 60; 60; 60; 60; 60; 60; 60; 60; 60; 60; 60; 88; 88; 88; 88; 88; 88; 88; 88; 100; 100; 100; 100; 110; 110; 110; 135; 135; 147; 147; 147; 147; 147; 147; 150; 150; 150; 150; 154; 154; 154; 170; 170; 200; 200; 200; 200; 200; 200; 200; 200; 210; 210; 210; 210; 210; 220; 220; 220; 220; 220; 220; 220; 230; 250; 340; 400; 420; 460; 620; 760; 760; 1100

This statistical series have 92 values and it is a statistical series of type S2 with redundant values. For this series of running hours between two faults, the empirical repartition function, $\hat{F}(t_i)$ the calculi are presented in table 2.

The values of the Empirical repartition function are calculated with the formula:

$$\hat{F}(t_i) = \sum_{j=1}^{i-1} f_j, \quad \text{for } i = 1,2,\ldots,22 \tag{1}$$

Considering the nature of the subassembly in the study, it is assumed that the times of failure, considered between two consecutive failures are distributed following a Weibull distribution law, as this will be confirmed or infirmed using concordance tests [4].

The probability density of failures for tri-parametric Weibull distribution is expressed by the relation:

$$f(t;\eta,\beta,\gamma) = \frac{\beta}{\eta}\left(\frac{t-\gamma}{\eta}\right)^{\beta-1} \exp\left[-\left(\frac{t-\gamma}{\eta}\right)^{\beta}\right] \tag{2}$$

where β is the shape parameter, η is real scale parameter and γ is the initializing parameter.

The parameters of a tri-parametric Weibull distribution can be calculated using the method of moments. The shape parameter β is obtained by solving the equation:

$$CV = \frac{\sqrt{\Gamma\left(\frac{2}{\beta}+1\right) - \left[\Gamma\left(\frac{1}{\beta}+1\right)\right]^2}}{\Gamma\left(\frac{1}{\beta}+1\right)} \tag{3}$$

where CV is the coefficient of variation, which is obtained using the relation

$$CV = \frac{s}{m} \tag{4}$$

where s is the standard deviation and m is the mean value of the string.

The scale parameter η is calculated with

$$\eta = s / C_{\beta} \tag{5}$$

and initializing parameter γ with the relation

$$\gamma = m - \eta\, K_{\beta} \tag{6}$$

In these relations K_β and C_β are coefficients dependent on the shape parameter β, which are calculated from the relations:

$$K_{\beta} = \Gamma\left(\frac{1}{\beta} + 1\right) \tag{7}$$

$$C_{\beta} = \sqrt{\Gamma\left(\frac{2}{\beta} + 1\right) - \left[\Gamma\left(\frac{1}{\beta} + 1\right)\right]^{2}} \tag{8}$$

For the data above, with the mean value m = 288.363636, standard deviation s = 258.257447 and CV = 0.895596 the parameters of the Weibull distribution are obtained are: β = 1,118460; η = 300.455874 ore; $\gamma = -9{,}0421\times10^{-5}$; K_β = 0.959754; C_β = 0.859552.

By calculating the elements needed to define the distribution and verification of the Weibullian character of the analysed product behaviour using the Kolmogorov-Smirnov concordance test, we obtain the maximum distance, D_{max} = 0.367631 ≈ $D_{\alpha,\, n} = D_{99{,}5,\, 22}$ = 0.357818, $D_{99{,}5,\, 22}$ 0.085254 < $D_{\alpha,\, n} = D_{80.\, 71}$ = 0.124985, $D_{80.\, 71}$ being the Kolmogorov-Smirnov test feature for a confidence level of 99,5% and n = 22 values, so that the Weibullian character of failure times distribution is validated.

The tri-parametric Weibull distribution parameters characteristics are calculated with the equations:

- Reliability function:

$$R(t;\eta,\beta,\gamma) = \exp\left[-\left(\frac{t-\gamma}{\eta}\right)^{\beta}\right], \% \tag{9}$$

- The non-reliability function:

$$F(t;\eta,\beta,\gamma) = 1 - \exp\left[-\left(\frac{t-\gamma}{\eta}\right)^{\beta}\right] \tag{10}$$

- Intensity or rate of failure:

$$z(t;\eta,\beta,\gamma) = \frac{\beta}{\eta}\left(\frac{t-\gamma}{\eta}\right)^{\beta-1} \tag{11}$$

- Average uptime:

$$m = \gamma + \eta \, \Gamma\left(1 + \frac{1}{\beta}\right) \tag{12}$$

- Median of uptimes:

$$t_{med} = \gamma + \eta(-\ln 0{,}5)^{\frac{1}{\beta}} \tag{13}$$

In figures 5, 6, 7, 8 these reliability parameters variation are presented.

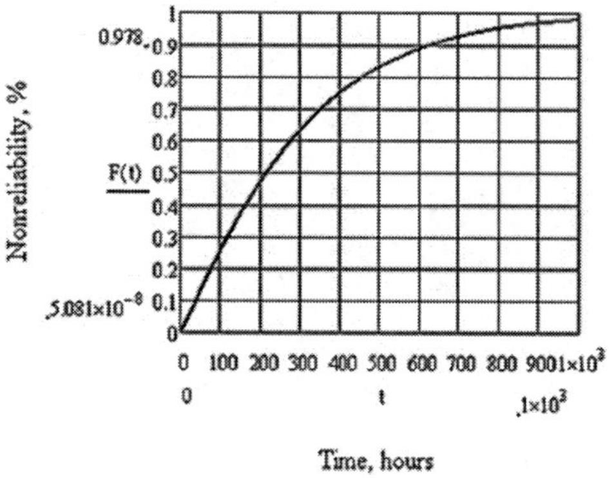

Fig. 5. Variation of the non-reliability function.

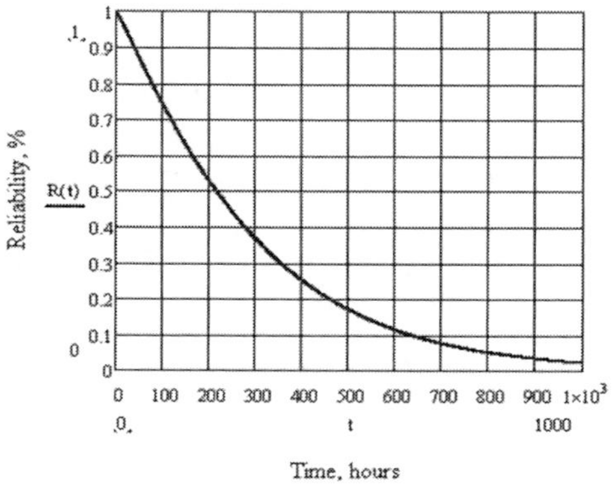

Fig. 6. Variation of the reliability function.

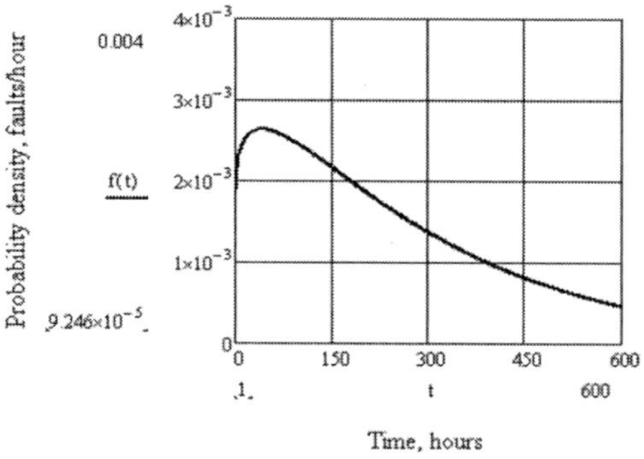

Fig. 6. Probability density of failure occurrence.

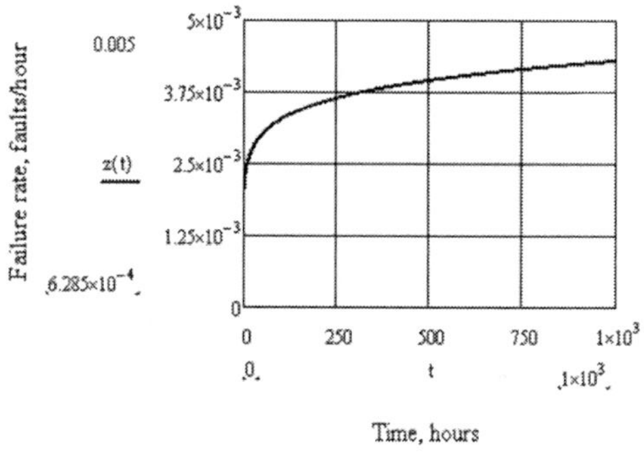

Fig. 7. Failure rate.

3 Conclusion

The analysis performed shows that:

The mean time to failure of belt joint is 288 hours, and the median is 216 hours. Considering that the running time is about 20 hours per day, it is expected to have a fault of joint each 14 days. This low reliability is shown by the diagrams above, the probability to not fail after 100 hours, i.e. one week is about 75%. In order to reduce downtimes, the quality of belt joining must be improved, by replacing stapled joints with vulcanized joints.

References

1. Ş. Covaci, *Exploatări miniere subterane* (Ed. Didactică şi Pedagogică, Bucureşti,1983)
2. F.D. Popescu, *Proceedings of the 9th WSEAS International Conference Automation And Information (Icai'08),* Bucharest, Romania, 120 (2008)
3. G.D. Mihai, I. Kovacs, M.S. Nan, N. Iliaş, I. Andras, S.M. Radu, *Proceedings of the 2nd WSEAS International Conference on Engineering Mechanics, Structures and Engineering Geology EMESEG 09,* Rodos, Greece (2009)
4. I. Kovacs, S.M. Nan, I. Andraş, G.B. Urdea, *Proceedings of the Microcad International Conference* (2013)
5. I. Marian, *Utilaje de încărcare şi transport minier* (Ed. Didactică şi Pedagogică, Bucureşti, 1984)

Development of elbow rehabilitation equipment using pneumatic muscles

Georgiana Vetrice[1,*], and *Andrea* Deaconescu[1]

[1]Transilvania University of Braşov, Department of Industrial Engineering and Management, 5 Mihai Viteazul Street, Braşov, Romania

Abstract. Applying continuous passive movements (CPM) to the injured joints represents a viable solution for the rehabilitation of patients with posttraumatic disabilities of the elbow. In order to effectively implement CPM-based rehabilitation procedures, the development of new solutions for high performance equipment serves the interest of both medical facilities and patients. This paper deals with a study of the kinematics, construction and actuation of such new rehabilitation equipment using continuous passive motion. The actuation of this equipment is achieved by two pairs of linear pneumatic muscles.

1 Introduction

The elbow is one of the most regularly used joints in the body [1]. It is a synovial hinge joint, formed by the lower part of the humerus and the upper part of the radius and ulna. Situated between the shoulder and the hand joint, the mobility of this joint is subordinated to the function of the shoulder and contributes to the mobilization of the hand. Its functional importance derives from the fact that after having sustained elbow injuries, regardless of the cause, both daily activities and professional tasks are compromised.

2 Biomechanics of the elbow

Designing such rehabilitation equipment requires knowledge about the anatomy and biomechanics of the upper limb, and especially of the elbow.

The elbow joint possesses two degrees of freedom. It is responsible for the movements known as flexion-extension and pronation-supination [2]. As shown in Fig. 1, flexion consists in rotating the forearm towards the arm, by a maximum angle of 150°. The functional range of motion for elbow is, however, of 30° to 130°. Extension consists in rotating the arm in the opposite direction, by a maximum angle of 10° (hyperextension). Fig. 2 shows the pronation-supination movement, which consists in rotating the forearm by its longitudinal axis, by a maximum angle of 90°; the thumb is moving closer to the body in the case of pronation and is turned away from the body in the case of supination. The functional motion for elbow pronation-supination ranges from 50° to -50°.

[*] Corresponding author: vetrice.georgiana@unitbv.ro

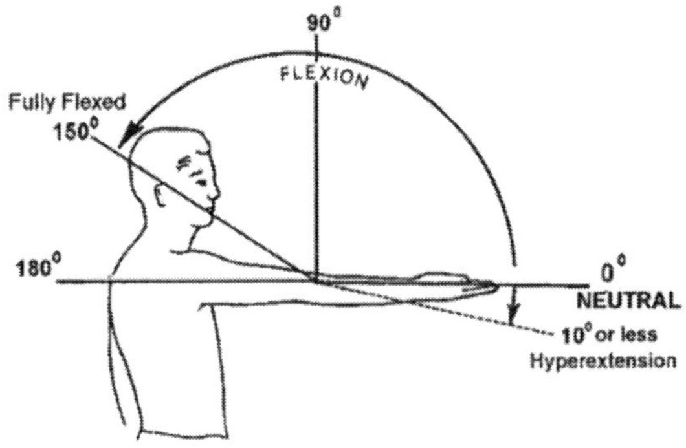

Fig. 1. Elbow flexion-extension [3].

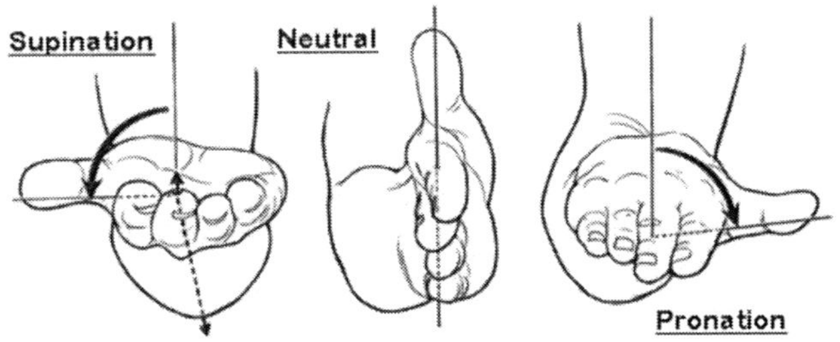

Fig. 2. Elbow pronation-supination [4].

Diminution or loss of elbow functions may be caused by a variety of diseases and medical conditions. Some of these are tendinitis, bursitis, fractures, dislocation, arthritis, etc. Posttraumatic pathologies are the most severe. The goal of posttraumatic rehabilitation is to achieve an optimum level of independent functioning for the patient. The main objective of posttraumatic recovery is to maintain the normal range of motion of the healthy segments, to rehabilitate the parts injured by trauma and to regain the original functions of joint.

The most important method used in posttraumatic recovery is physical therapy. Continuous passive motion, as part of the so-called passive physical therapy, is designed for patients suffering from posttraumatic affections of the elbow who cannot move autonomously. This method consists in applying constant motion to the joint without requiring any effort from the patient.

3 Design of the elbow rehabilitation equipment

The elbow movements that need to be recovered are exactly the movements that the rehabilitation equipment should provide. Although rehabilitation equipment is nowadays available on the market, it is driven by electric motors entailing less than favourable dynamic characteristics.

Pneumatic artificial muscles represent an alternative driving modality of these devices. Compared to the most frequently encountered drives at present, their price is lower, they benefit from a better weight-to-power ratio, are adjustable during the rehabilitation process and due to their specific properties the interaction with patients is safe [5].

Among the advantages of using such equipment is the faster recovery of the patients, safe and easy use, increased patient comfort, better rehabilitation results and the possibility of deployment in both medical units and in patients' homes.

Performing a kinematic analysis of the elbow is very important for designing the rehabilitation equipment, because the kinematics of the rehabilitation equipment must reflect the kinematics of the joint it is intended for. Fig. 3 shows the modified kinematic model of the elbow.

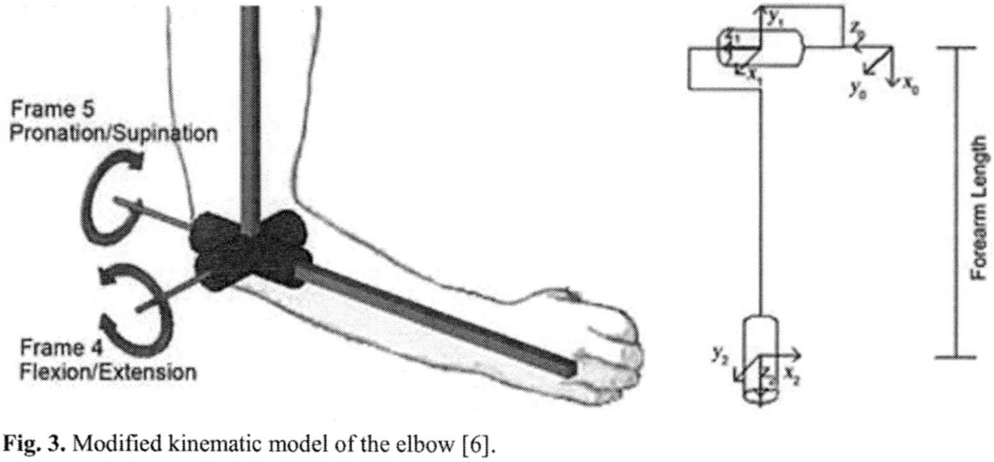

Fig. 3. Modified kinematic model of the elbow [6].

The rehabilitation equipment must have the following characteristics [7]:

- kinematics that allow the replication of human motions.
- ergonomic design, for a comfortable human-machine interaction.
- safety and compliance.

Taking into account the requirements of the rehabilitation process and the kinematic analysis of the elbow, a variant of elbow rehabilitation equipment was designed, the novelty of which consists in the utilization of linear pneumatic muscles and in allowing movement in two planes (Fig. 4).

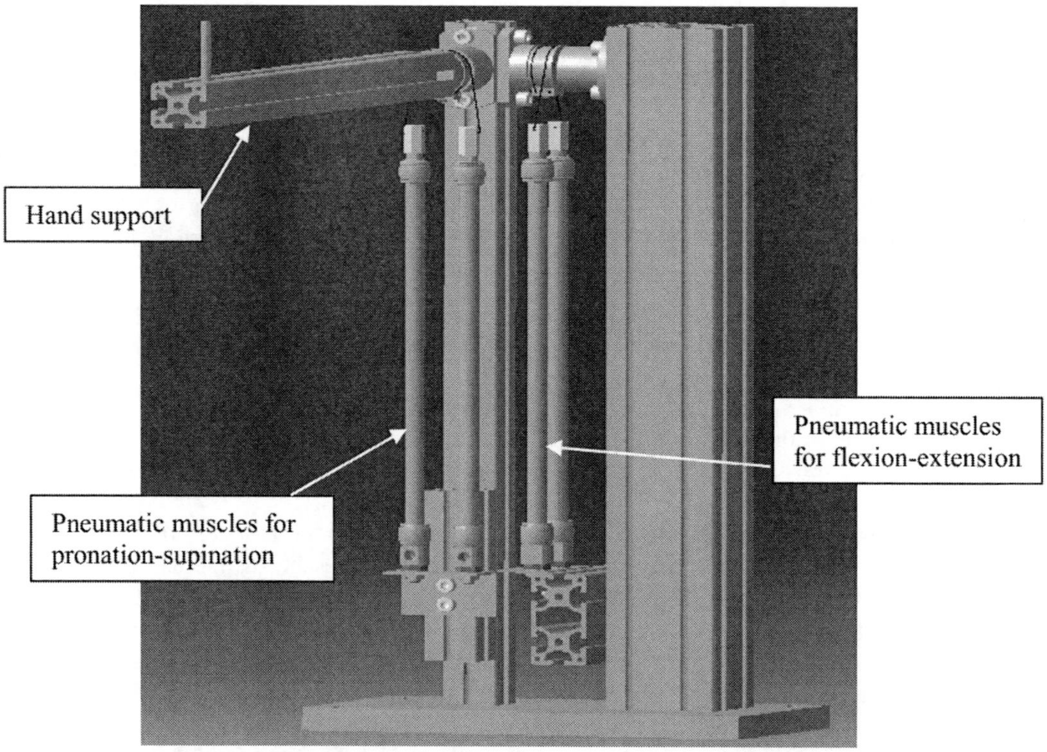

Fig. 4. Elbow rehabilitation equipment.

As stated above, the linkage of the equipment has to be compatible with linkage of the human limb. The axes of the joint of the equipment have to be identical to the human joint axes, in order to ensure an ergonomic human-machine interaction. During the rehabilitation process, the equipment has to ensure maximum comfort to the patient and must not compromise or limit for any reason the patient's ability to perform natural motions.

Fig. 5 shows the construction of the rotation modules for the rehabilitation equipment. The hand support is adjustable in length for different sizes of the upper limb. Patients place their forearm on the hand support and start the rehabilitation exercise by holding the bar. The upper ends of the pneumatic muscles are connected by a steel cable, drawn over a pulley and fixed on it. These cables ensure the transmission of motion between actuators and joints. The pneumatic muscles work one against the other, causing the rotation of the shaft and also the movement of the hand support. It is important to mention that the movements performed by the equipment can be achieved only separately and not simultaneously.

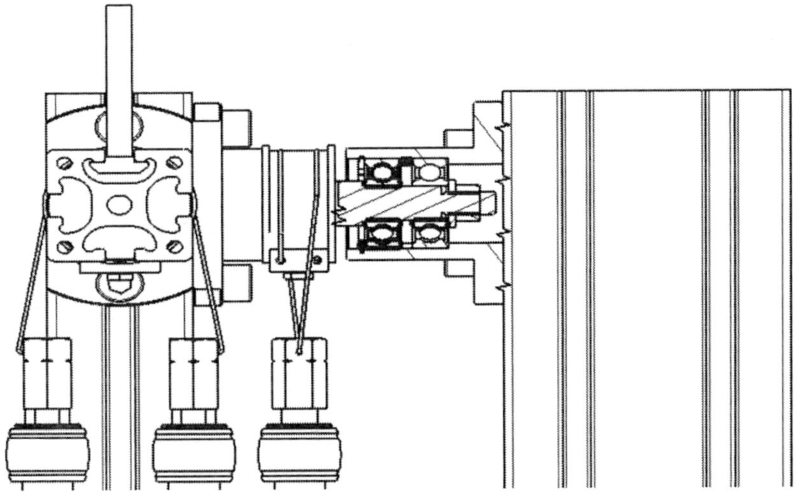

Fig. 5. Breakout view of the rotation module of the rehabilitation equipment.

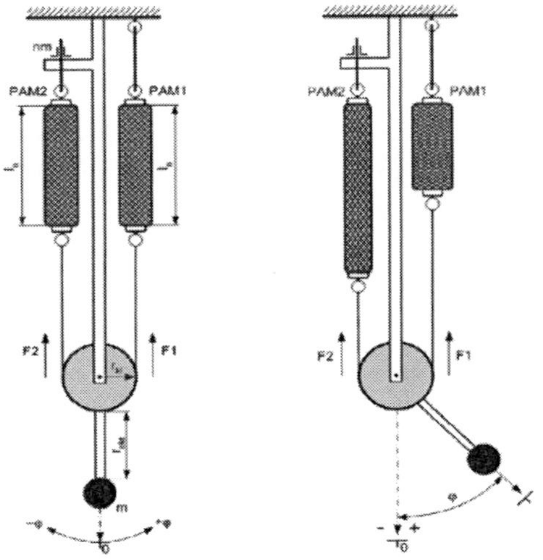

Fig. 6. Pneumatic artificial muscles in antagonistic connection [9].

The equipment has two rotational degrees of freedom, placed perpendicularly. In this case two pairs of muscles were used, one for each movement performed by the equipment.

Thus, each rotation module of the equipment is driven by a couple of pneumatic muscles arranged in antagonistic connection, as shown in Fig. 6. For achieving the neutral position both muscles are fed the same pressure, and the feeding of compressed air causes the movement to take place.

The four pneumatic muscles used in the construction of the equipment are of 10 mm interior diameter and 300 mm initial length. The maximum possible stroke that can be performed is of approximately 20% of the relaxed muscle length, in this case 60 mm.

3 Conclusions

The elbow is a joint that easily develops stiffness, slowing down the rehabilitation process. Regaining the mobility of this joint is indispensable for the mobility of the upper limb; else, a non-rehabilitated elbow can have a major impact on personal and professional life. This desideratum underlies the need for conceiving high performance and cost effective rehabilitation equipment for patients suffering from posttraumatic affections of the elbow joint. The ones developed so far are electrically driven and have a rigid structure.

The research discussed in this paper is aimed at developing a mono-articular device based on continuous passive motion, with two degrees of freedom, which ensures a complete recovery (flexion-extension and pronation-supination, one at a time, not simultaneously). This equipment has also the advantage of being deployable in both medical units and in patients' homes.

This paper presents and discusses the kinematics, construction and actuation system for such rehabilitation equipment, intended for patients lacking the necessary biological resources for achieving the movement. Its design combines knowledge from anatomy, biomechanics and kinematics. The specific requirements for this equipment are due to the fact that it comes into direct contact with patients and also due to its non-conventional drive - pneumatic muscles. Two pairs of linear pneumatic muscles were used in this case, one for each movement. This type of drive is suitable for medical recovery due to its low weight, high performance and low cost.

Currently the rehabilitation equipment is undergoing a test phase, in order to verify the compliant behaviour of these actuators.

References

1. ***http://www.healthline.com/human-body-maps/elbow#seoBlock, accessed: 11.02.2017
2. B.F. Morrey, J. Sanchez-Sotelo, *The elbow and its disorders* (The Mayo Clinic Foundation, 2009)
3. ***http://www.titaniumgeek.com/medical-journal/fracture-cast/, accessed: 11.02.2017
4. ***https://tt.tennis-warehouse.com/index.php?threads/pronation-and-isr-clarification-question-for-chas_tennis.546324/, accessed: 11.02.2017
5. B. Jobbagy, D. Simsik, J. Marek, J. Karchnak, D. Onofrejova, AJME, **7**, 299 (2014)
6. H.A. Abdullah, C. Tarry, R. Datta, G.S. Mittal, M. Abderrahim, JRRD, **44**, 43 (2007)
7. O. Liska, M. More, D. Janacova, H. Charvatova, MMOTE, 151 (2013)
8. T. Deaconescu, A. Deaconescu, *Applications of pneumatic muscles developed within the Festo National Fluid Actuation and Automation Training Centre in Brașov* (available at http://www.recentonline.ro/050/Deaconescu-R50.pdf, 2016)
9. J. Pitel, M. Balara, AT&P, **4** (2009)

Comparison between two models of elbow rehabilitation equipment

Georgiana Vetrice[1,*], and *Andrea* Deaconescu[1]

[1]Transilvania University of Braşov, Department of Industrial Engineering and Management, 29 Eroilor Bd., 500036, Braşov, Romania

Abstract. International studies have revealed that using continuous passive motion (CPM) equipment as part of a rehabilitation program improves the recovery period and also diminishes the rehabilitation costs by about 50%. This explains the need for conceiving elbow joint rehabilitation equipment to help persons suffering from posttraumatic disabilities. In order to create the best model to meet the patients' requirements this paper presents two constructive solutions of such equipment and a comparison between their designs. Both achieve the desired movements of the elbow joint: flexion-extension and pronation-supination and are actuated by pneumatic muscles, due to the multiple advantages of pneumatic actuation: low cost, compliance and favorable response to commands.

1 Introduction

The elbow joint consists of bone, cartilage, ligaments and fluid. The movement of the elbow also requires muscle and tendon input. If any of these parts is injured, the function of the elbow joint is affected. Situated between the shoulder and the hand joint, the elbow plays an important role in the functionality of the upper limb. Therefore, it is bound to be overworked and since the joint is not wrapped in layers of muscle it is prone to injuries. These traumas can be treated with or without surgery, depending on the severity of the injury. Either way, the elbow joint needs posttraumatic recovery, in order to restore its original function.

The original function of the elbow entails a movement known as flexion/extension (rotation of the forearm toward the arm and rotation of the forearm away from the arm, respectively) and another movement known as pronation/supination (rotation of the forearm by its longitudinal axis, the thumb moving closer to the body and the thumb being turned away from the body, respectively), as shown in figure 1.

In designing a posttraumatic rehabilitation program several aspects related to trauma, location and severity must be taken into account, along with other specific features of each type of trauma. Medical recovery addresses patients of all ages, suffering from various diseases and it is essential to reduce the consequences of the motor deficit on the patient's

* Corresponding author: vetrice.georgiana@unitbv.ro

life. The most important method used to restore the diminished functions and increase joint mobility is physical therapy, which includes active and passive movements.

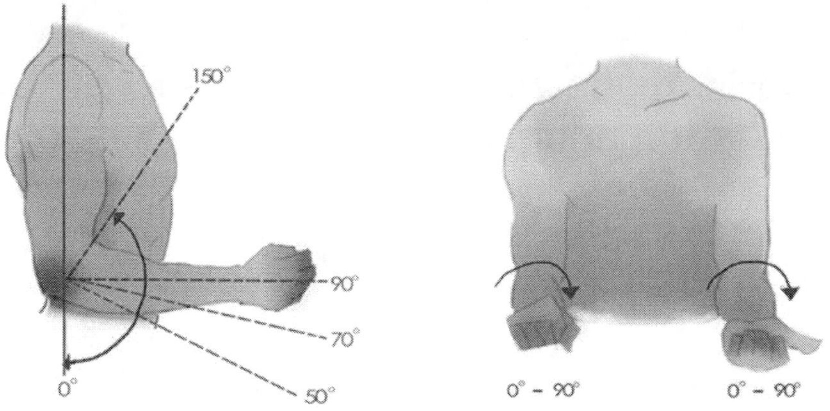

a. flexion/extension **b.** pronation/supination

Fig. 1. Elbow movements [1].

In case of patients that cannot achieve these movements by themselves, the recommended solution is continuous passive motion. This therapeutic procedure provides regular movement to the joint, without the tensing of the muscles.

Research has proved that rehabilitation equipment using this method is useful in the case of patients with posttraumatic disorders, improving the recovery period and reducing the quantity of pain medication.

Therefore, the rehabilitation costs are significantly reduced [2]. Although rehabilitation devices are available nowadays on the market, most of them are electrically driven and have rigid structure. This explains the need for developing new elbow rehabilitation equipment.

2 Proposed elbow rehabilitation equipment

2.1 Kinematics of the rehabilitation equipment

The equipment is modelled based on the kinematics of the human elbow joint (Fig. 2.). Therefore, the equipment will have two rotational degrees of freedom (one for flexion/extension – joint 1 and one for pronation/supination – joint 2).

Considering a secure recovery, it is very important for the rehabilitation equipment to allow for the replication of human movements, to have an ergonomic design that will minimize the discomfort of human – machine interaction, and to satisfy increased demands for safety and compliance [3].

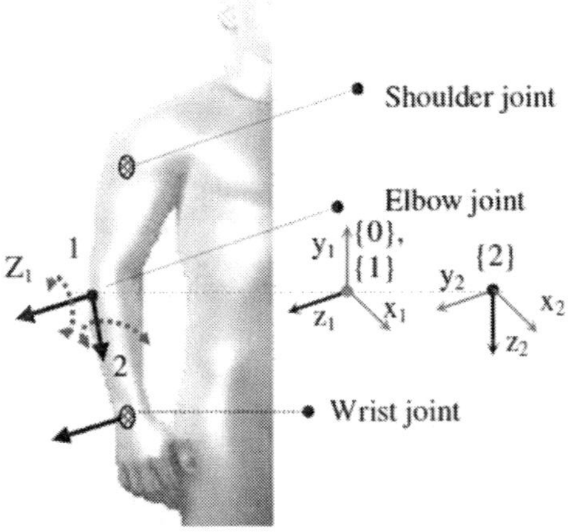

Fig. 2. Kinematics of the elbow joint [4].

2.2 Actuating system

Due to the specific characteristics and constructive advantages of pneumatic actuating systems, such as safe deployment, reduced overall size, easy generation, storage and transport of compressed air, the research was oriented towards the development of rehabilitation equipment actuated with pneumatic muscles.

The pneumatic artificial muscle (PAMs) is an elastic component, which increases its diameter and reduces its length under air pressure, thus carrying out a stroke (Fig. 3.a.). Pneumatic muscles are usually grouped in pairs: one agonist and one antagonist. Thus, by feeding compressed air, the pair of muscles causes the movement.

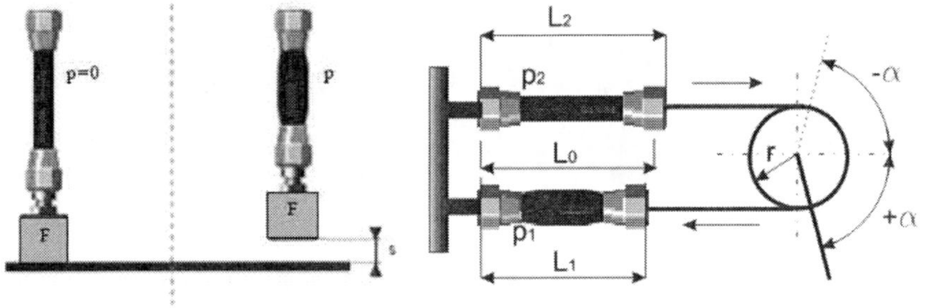

a. The stroke carried out by a PAM **b.** Kinematics of the rotation module

Fig. 3. Working principle of the pneumatic artificial muscle [5, 6].

The proposed equipment has two rotational degrees of freedom. Each rotation of the joint is generated by means of a pair of antagonistic pneumatic muscles. In Fig 3.b. the kinematics of such as module is shown.

2.3 Models of elbow rehabilitation equipment

The proposed models are shown in figure 4 and figure 5. The linkage of the equipment is similar for both rehabilitation devices. They have a hand support adjustable in length, for different sizes of the upper limb.

 DOI: 10.1051/matecconf/201712101018

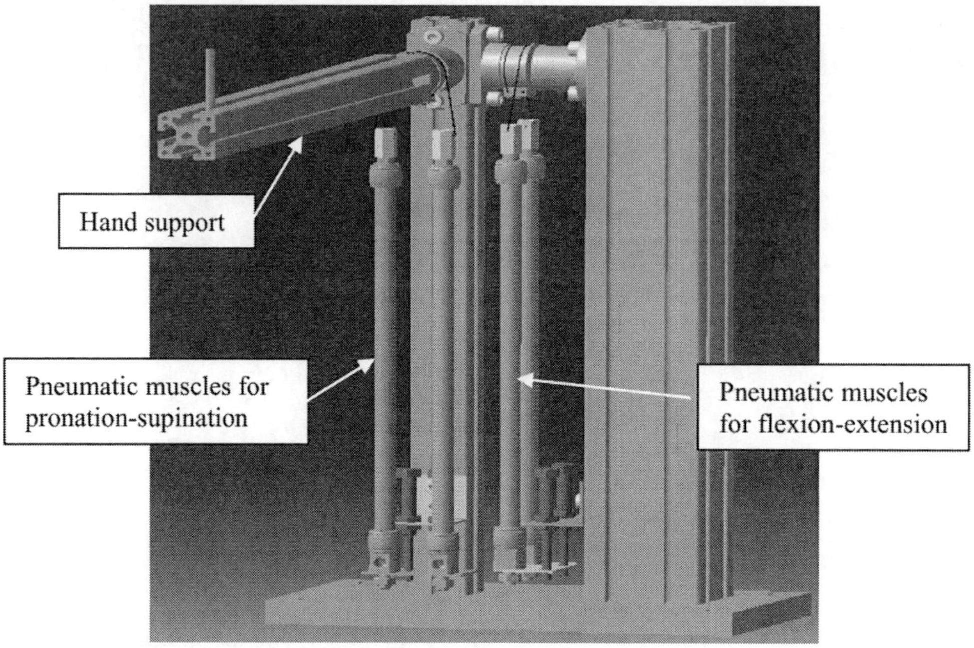

Fig. 4. First constructive variant of the elbow rehabilitation equipment.

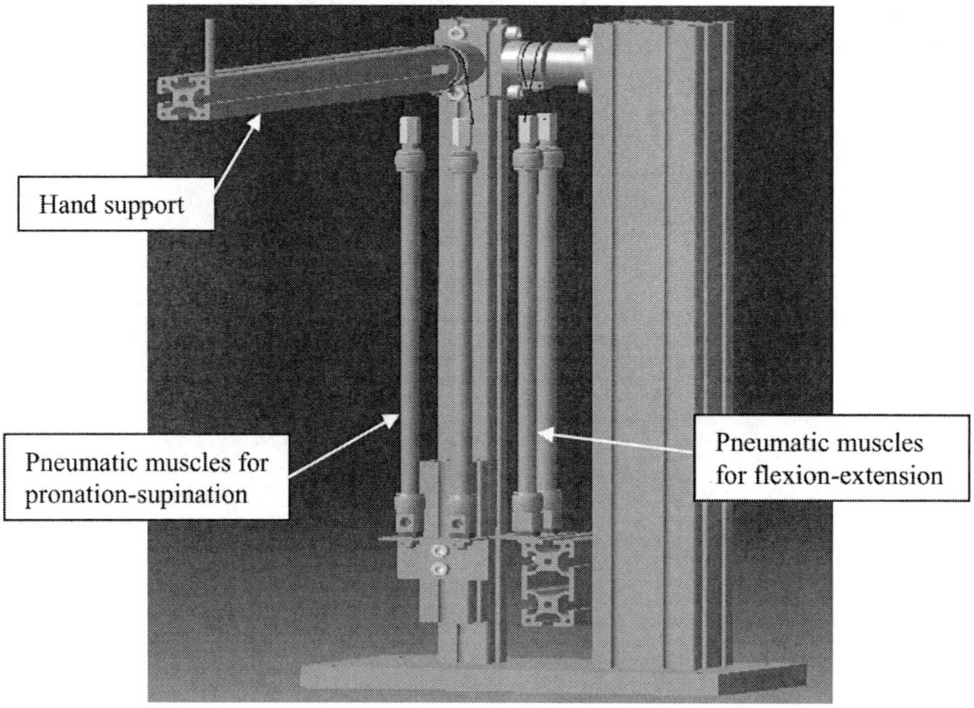

Fig. 5. Second constructive variant of elbow rehabilitation equipment.

Before rehabilitation starts, the patient places the forearm on the hand support and holds the bar. For these rehabilitation devices two pairs of pneumatic muscles were used, placed perpendicularly, one for each movement necessary to be recovered.

For the construction of the equipment four pneumatic muscles were used, of 10 mm interior diameter and 300 mm initial length. The maximum stroke that can be achieved by these is of approximately 20% of the relaxed muscle length, which means 60 mm.

The lower ends of the pneumatic muscles have to be fixed. For this purpose, in the first devised variant the muscles were fasten by two sheets: one tied to the metallic profile of the holding structure, the other one affixed to the free ends of the pneumatic muscles, both held together by a screw.

The second devised variant suggests fixing the lower ends of the muscles to a sheet and the sheet to the metallic profile by means of a short aluminium profile, such as to enhance stability.

In both cases, the upper ends of the pneumatic muscles are connected by a steel wire, drawn over a pulley and fixed in a specially designed structure. These wires ensure the transmission of motion between the pneumatic muscles and the joints. They are working one against the other, causing the rotation of the shaft and implicitly the rotation of the hand support. Fig. 6 presents the model of the rotational assembly.

Taking into consideration these variants, the first one was discarded and the second one maintained, namely the variant where the fixing involves a sheet and an aluminium profile, giving the equipment stability and safety in performing the movements.

Fig. 6. The rotational assembly.

3 Conclusions

The upper limb is one of the most important parts of the body. Regaining the mobility of the elbow (flexion/extension and pronation/supination) is crucial for the individual's physical autonomy. Otherwise, such lack of movement results in barriers in achieving daily tasks. In the case of patients suffering from posttraumatic affections of the elbow joint rehabilitation equipment using continuous passive motion is highly eligible, as it helps reducing the recovery period and improves the efficiency of the rehabilitation process.

Although there are rehabilitation devices available on the market, these are less human-friendly because of their electrical drives and inherent rigidity. This explains the purpose of developing new high performance and cost effective equipment for patients suffering from posttraumatic afflictions of the elbow joint.

Two variants of rehabilitation equipment of the human elbow have been developed, such as to achieve both elbow movements. The kinematics and geometric models were presented in this paper.

Both constructive variants are actuated by pneumatic muscles, due to the numerous advantages of pneumatic actuation: low weight, high power-to-weight ratio and low cost. For the movement to take place, each pneumatic muscle must have a free end and a fixed one. The free ends are connected to the rotational assembly by a steel wire, which ensures the transmission of the motion between pneumatic muscles and joints. The fixed ends can be fastened by two sheets or with a sheet and a short aluminium profile.

Since this equipment comes into direct contact with patients the second variant was selected, with a sheet and a profile, such as to ensure patient safety.

References

1. ***http://www.militarydisabilitymadeeasy.com/elbowandforearm.html, Accessed: 21.02.2017
2. I. Petre, T. Deaconescu, ICEEMS, IN Recent, **10**, 373 (2009)
3. O. Liska, M. More, D. Janacova, H. Charvatova, MMOTE, **151** (2013)
4. M.H. Rahman, M. Saad, J.P. Kenné, P.S. Archambault, MCCA, 1567 (2010)
5. A. Deaconescu, T. Deaconescu, DAAAM, **1**, 365 (2008)
6. A. Deaconescu, T. Deaconescu, IMCECS (2009)

MATEC Web of Conferences **121**, 02001 (2017) DOI: 10.1051/matecconf/201712102001
MSE 2017

Highlighting product issues during design phase using functional analysis

Felicia Veronica Banciu[1,*], and *Eugen* Pămîntaş[1]

[1]Politehnica University of Timişoara, Mechanical Faculty, IMF Department, 300222 Mihai Viteazu Bv. Timişoara, Romania

Abstract. The major costs of a product take place in product design stage and essential elements like i. e. ease of assembly, components accessibility in an assembly, disassembly, number of parts have to be taken into account during product design stage. The paper aim is to gain understanding and improvement of overall engineering skills and thinking of engineering students with regard to the design/redesign process for a product from a functional perspective with the aim of extracting information for design and manufacturing associated design. The model is based on functional modelling and analysis of product using Function Block Diagram. Functional analysis helps to identify components' functions and interactions between products parts, they describing also, the functioning. Analysing these functions and interactions between product's parts offer an overview on their number, their interactions useful for example in assembly. Starting from a functional modelling and analysis, creativity techniques could be used in order to generate alternative solutions with regard to products design, manufacturing and assembly.

1 Introduction

Designers must be able to see the effects that their decisions produce on the manufacture, assembly, maintenance and recycling processes. They must be familiar with the general principles of design, manufacturing and specialized knowledge to solve complex design tasks like assembly-disassembly-reassembly. Assembly and a majority of the problems attached are determined by design.

The issue of expressing functional decomposition, the taxonomy of this area was and still is a current theme. In literature, there are many strategies on functional decomposition and functional models [1-3]. Functional decomposition is an extensive subject in engineering, its development and study reasons are mainly finding a common language for designers and the need for sharing engineering knowledge, understood by everyone involved in design process. This is important from the perspective of using a common language between designers and reuse knowledge attached functional models [1].

A better sedimentation of knowledge, completing the design tasks with the overall picture of the importance that functional products decomposition occupies in analysing similar products is considered in this paper. In order to emphasise the functional blocs that

*Corresponding author: felicia.banciu@upt.ro

accomplish the same task but using different structural components offering in the same time a concise and visual link between product parts and their links to each other will be used a functional approach based on FBD.

Examining existing solutions and new solutions involves both logic and creation and it is about a complex phenomenon that must train in order to develop the creative thinking, the following: divergent thinking: association, extrapolating, generalizing combining imagination, the ability to view static and dynamic [4].

Thus, the paper proposes an approach to the design/redesign process for a product from a functional perspective aiming to extract information related to functional structure, number of product part's associated to functions (needed to achieve them) from conceptual design phase.

2 The functional approach

Fundamental to existing approaches is the notion of function and functions decomposition, possibly on hierarchical levels and then highlighting behavioural links between these functions. Functional structures are attached to constructive structures and follows the logic of what we want and how we get, logic meet practical both in systematic as well in the axiomatic approaches on product design [3, 5]. A highlight of functions and information extraction related to assembly and what implies induce an alternative thinking, extended possibly to other solutions or oriented to solve the existing ones and different methods of creativity could be used. A representation of functions by functional blocks, the functions being arranged hierarchically - FBD Model- is presented in [2]. Structurally, the functions are divided into basic functions and secondary functions that help achieve basic functions. The function is a pair verb-noun the verb expressing an action and the noun is measurable. Between the various elements of functions exists links like and / or temporal link, the causal links that cause so-called side effects- effects of the implementation of certain functions. A hierarchy of functions leads to a hierarchy of parts and components. FBD elements, functions, does not specify uniquely a physical form but the sequencing logic functions can stimulate creativity. The product is basically a sum of functions that are attached to components and subassemblies.

Thus, the starting point for a design consists of the main function, the role that products have, the reason of their existence. This primary function is then decomposed into levels of sub-functions leading to sub-assemblies and components of the product.

3 Applied approach

As a starting point for achieving the purpose of the work was chosen FBD approach - in American literature presented by [2] and in the French by the APTE society [6].

For example, consider two types of devices, mandrels used for fixing and centring parts on parts inner cylindrical surfaces, one with wedges and plungers in figure 1 and one with balls in figure 2. To analyse the reference solutions it is used FBD (Functional Block Diagram). This tool makes easier to emphasize the aggregate of elementary functions and their assignment to basic and design functions. More information about developing a FBD can be found in [6-7].

To realize a FBD we have to build a graphic and dispose all the composing elements of the product or subassemblies in contact relation, to identify external and internal contacts and also the open streams and the closed streams that are relations between different product's elements because of its designing. They can be external opened streams, which integrate the relations between some of the products components with one or more external environments, respectively internal closed streams that take into account just the product's components. Closed streams are sources for potential optimization as they could be achieved through different solutions. Using an internal functional analysis decomposition, it can be developed a functional structure that in its turn conducts to the solution dependent functions. Each technical solution has associated good functioning criterions or conditions that allow their performance level definition.

The principle functional scheme of mandrel with wedges and plungers used for centring and fixing work pieces on their inner cylindrical surfaces [8] is presented in figure 1. Its components are: 2 plungers, 3 wedges, 4 rolls, 5 stem with double slope area, 6 elastic elements, 7 springs, 8 mandrel body, 9 locating pin, 10 cover and 1 work piece. The stems is pulled to left, the balls "mount" on stem's sloped area and next transmit movement to multiple wedges 3 that transmit movement and force to plungers 2 that will move synchronous on radial direction. In this way, first it is consumed the clearance up to work piece and next is done the clamping.

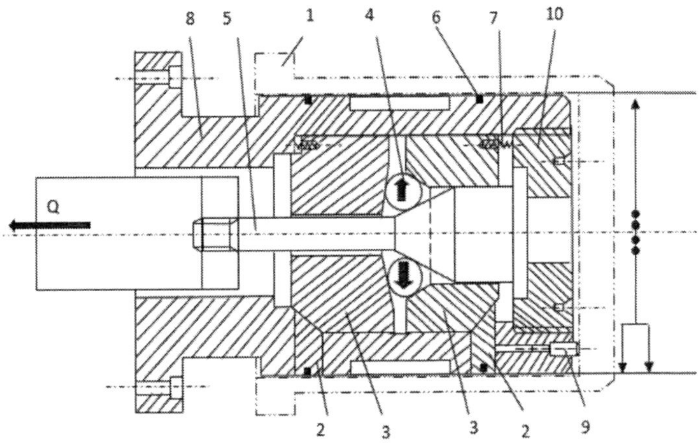

Fig. 1. Scheme of mandrel with wedges and plungers [8]

The functional principle scheme of mandrel with balls used for centring and fixing work pieces on their inner cylindrical surfaces is presented in figure 2 [9]. Its components are 1 the mandrel body with sloped area, 2 stem with threaded end, 3 balls with different diameters, 4 balls cover, 5 nut, 6 washer.

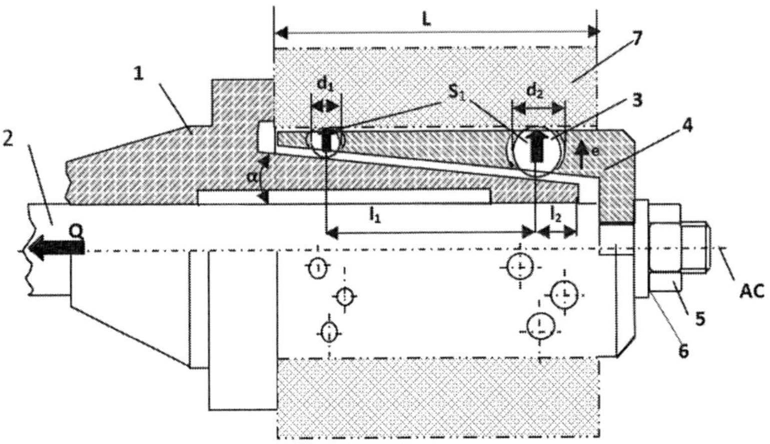

Fig. 2. Scheme of mandrel with balls [9]

The balls cover moves together with stem. It is mounted on the stem using a washer 6 and a nut 5. The stems is pulled to left by the actuating mechanism, the balls cover 4 (jointed with stem through nut and washer) is moved to left and the balls 3, that are in balls cover, "mount" on CD' sloped area and next consume the clearance up to work piece 7 and (their movement being stopped by the work piece) is realised its clamping.

Table 1 presents the main functions and their corresponding technical functions and they are the same for analysed examples.

Table 1. The main functions and their corresponding technical functions

The main functions (noted FP)	Corresponding technical functions for main functions
FP1 centring the work piece on inner cylindrical surface and prevent the movement of the piece	FT1 –use up clearance between work piece and elements that move radial, their contact with the work piece and work piece clamping
FP 2 Hold the piece against locator	FT2 – Block work piece axial movement

In figures 3 and 4 are presented the FBD diagram and the functions identified following a FBD are next depicted and table 2.

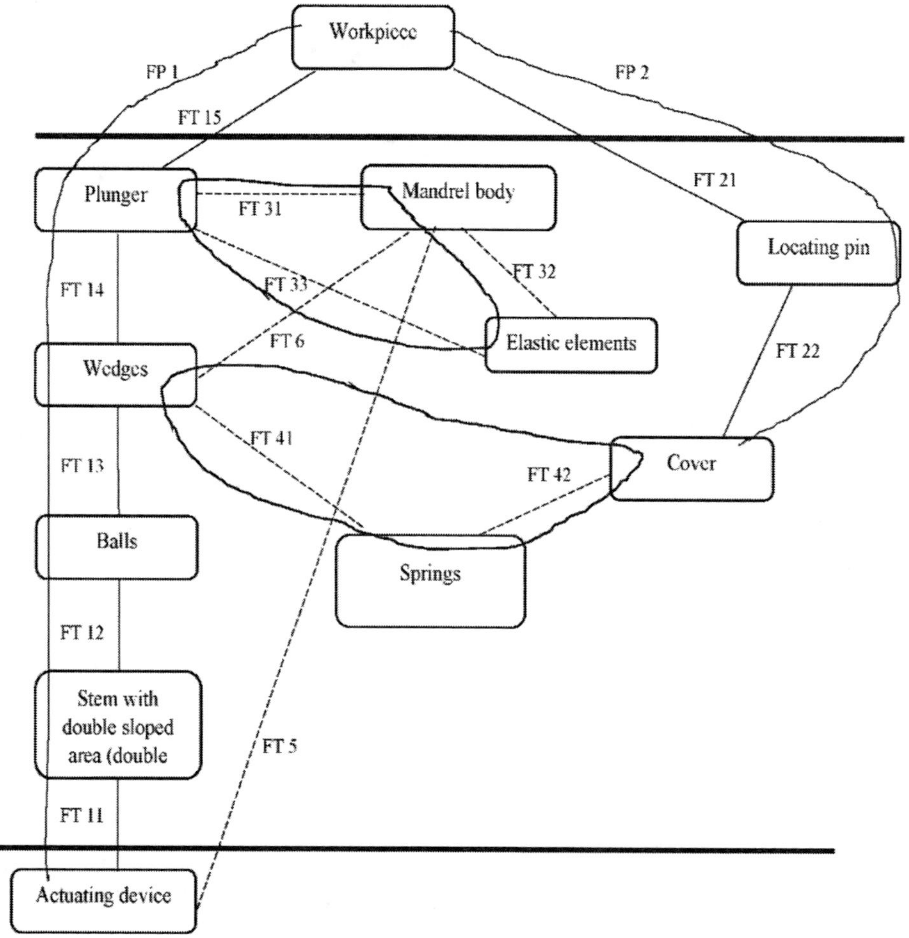

Fig. 3. FBD for mandrel with wedges and plungers - centering and fixing on inner cylindrical surfaces

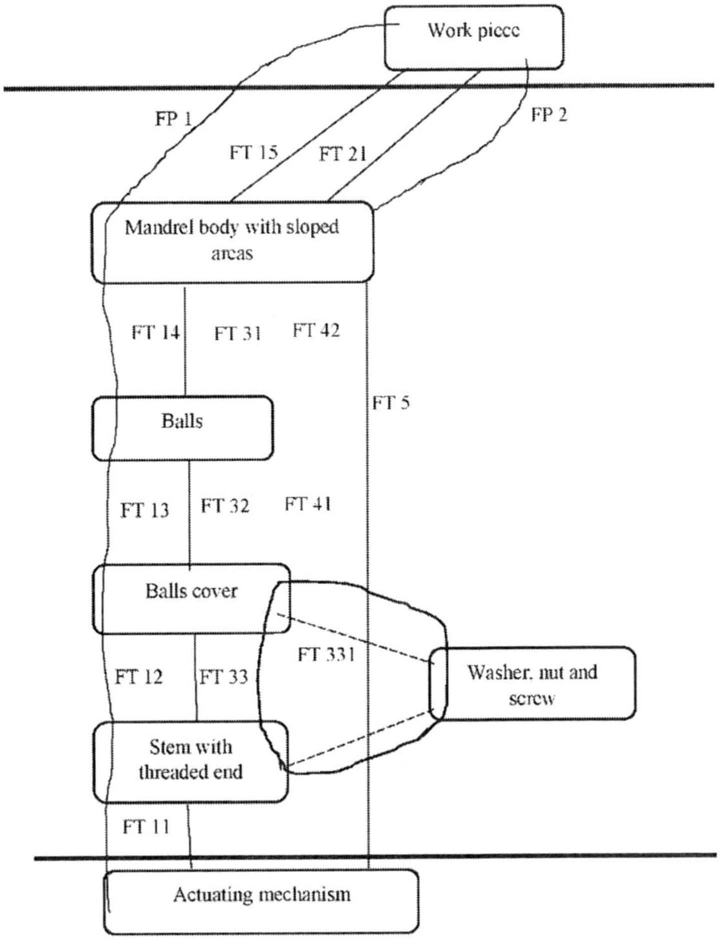

Fig. 4. FBD for mandrel with balls – centring and fixing on inner cylindrical surfaces

The way that these functions are accomplished differ, namely the way in which technical functions are realised is different. The functional analyse FBD for these mandrel emphasise the following:

An opened stream is that between the actuating device and the work piece through the mandrel's elements that transmit movement and force in order to achieve the main functions. Resuming, the main function FP1 uses functions FT11 - FT15. In both cases, the route is the same, only the components involved are different. In figure 1 and 2 the main functional chains needed to achieve the main functions FP1 and FP2 are drawn with free lines and link work piece with actuating device. The ways that these functions are accomplished differ, namely the way in which technical functions are realised is different. The functional analyse FBD for these mandrel emphasise the following:

For first example the second opened stream is between the cover, locating pin and work piece and the main function FP2 is achieved through FT21 and FT22.This second opened stream, in figure 4 is obtained through work piece and a locating surface – the mandrel body. In this case, this solution eliminates the need for a third piece, and obviously the necessity to assemble them for example. In figure 3 the closed streams are that given by contacts between elements in order to hold them in place and to assure the reversal movement for wedges (see figure 3): realise technical function FT3 - hold the plungers in place against the mandrel body - between the plungers-mandrel body and the elastic elements and the second –realise the technical function FT4 to assure the reverse movement for wedges between the wedge-spring- cover. For the mandrel with balls, we could observe that technical functions, realising FT3 and FT4 is done using the same functional chain as that for FP1 that implies no other components. In these cases, it remains to compare the rest of eventual assembly process that could say if one or another.

In the same time, each function could be analysed also from a useful/harmful point of view. In this way a complete function structure can be obtained, a more clear view and maybe opportunities to improve functions that lead to other solutions. In this regard friction forces has both positive and negative aspects (they are reduced by using balls, also they are needed to assure the contact and force transmitting between parts), the materials for pieces –e.g. it is necessary that balls do not leave marks on sloped surfaces.

Taking into account that this mandrels are with moving elements, their accuracy relatively to another types of mandrels proper to deliver the main functions can be discussed.

Table 2. The technical functions (FT) and their decomposition (left column – right column)

	Mandrel 1	**Mandrel 2**
FT1 needed for achieving the primary useful function FP1 is achieved following the next functional chain between actuating system and piece	FT11- transmit movement from actuating mechanism to the stem with double sloping area (double wedge zone) FT12-transmit the movement to rolls FT13- transmit the movement to wedges FT14 - transmit the movement synchronous to plungers FT15 – due to their movement use up clearance between work piece and mandrel body and next transmit clamping force to the work piece	FT11 transmit movement from actuating mechanism to the stem with threaded end FT12 transmit movement from stem to balls cover FT13 transmit movement to balls FT14 balls movement on mandrel's body sloped areas FT 15 due to their movement use up clearance between work piece and mandrel body and next transmit clamping force to the work piece
FP2 Hold the piece against locator is done through technical function FT2 that implies:	FT21 - a contact between cover-locating pin FT22 - contact between piece and locating pin	FT21 contact between work piece and mandrel body
FT3-hold the plungers in place against the mandrel body – auxiliary function FT3 Hold the balls in place against mandrel body	FT31 contact between plungers and mandrel body FT32 contact between mandrel body and elastic elements FT33 contact between plungers and elastic elements	FT31 contact between balls and sloped part of mandrel body FT32 contact between balls and balls cover FT33 contact between balls cover and stem FT331 screw on stem's threaded end in contact with balls cover
FT4 - assure the reverse movement for wedges - auxiliary function FT4 - assure the reverse movement for balls - auxiliary function	FT41-contact between wedge and spring FT42-Contact between spring and cover	FT41 contact between balls and balls cover FT 42 contact between balls and mandrel body

FT5 contact between actuating device and mandrel body - auxiliary function	X	x
FT6 – contact between wedge and mandrel body - auxiliary function	x	-

4 Conclusions

By decomposing a product functionally then identifying its components according to functional decomposition it is achieved a better understanding of implications, number of parts, their contacts and links in order to accomplish the main function. Each identified function, its link between two parts means a process of joining, and adding parts together – assembly. Possibilities of achieving a closed stream generates a series of alternatives and the selection of a variant or the other can be done by analysing these solutions in terms of assembly.

For the work described in this paper, the approach proposed conduct to the following achievements: 1) inventory of functions and ways to achieve them (to carry them out); 2) identify the open and closed streams; 3) analyse the base solution and alternatives (the parts number as a way of facilitating the extraction of alternatives and choosing the best from the assembly process point of view). Information attached to functional models are numerous and as the design progresses they are multiplied so even in the case of simple products with a reduced number of components, inventory all aspects of the design and what involves it is an overwhelming work, that is why it must be carried out by a team, not just by one person.

This paper presented an approach from a functional perspective of design process that helps to build an overall picture of product itself, its internal links between its components, which leads to improvement of overall engineering skills, and thinking of a design team members.

This research acknowledges to Walter Tools SRL, Timişoara Romania and Universitary Association of Manufacturing Engineering in supporting literature research, information gathering and support financial funding for this paper. The valuable comments from anonymous reviewers are greatly acknowledged to help improve the paper's quality.

References

1. C.F. Kirschman, G.M. Fadel, C.C. Jara-Almonte, J. Mech. Des. **120(3)**, 475 (1998)
2. Robert H. Sturges, *Artificial Intelligence in Design '92,* (Springer Netherlands, 1992)
3. G. Pahl, et al., *Engineering Design - A Systematic Approach*, (Springer, 2007)
4. V. Belous, *Technical Innovation in Machine Building*, (Ed. Junimea Iasi, 1986)
5. Dingmar van Eck, *Functional Decomposition: On Rationality and Incommensurability in Engineering* (Available at: https://repository.tudelft.nl/islandora/object/uuid:b2423a0a-6d5b-4149-9e1e-95f0396f6f74?collection=research)
6. *** http://www.wikiwand.com/fr/M%C3%A9thode_APTE, Accesed 05.01.2017
7. C. Petitdemange, *Design Control through Value Analysis*, (Editions Lavoisier Tec & Doc, Paris, 1992)
8. S. V. Roşculeţ, N. Gojineţchi, C. Andronic et al., *Proiectarea Dispozitivelor*, (E.D.P., Bucureşti, 1982)
9. I. Grozav, *Dispozitive în constructi de maşini*, (Politehnica Publishing House Timişoara, (2008)

Researches regarding cutting tool condition monitoring

Marinela Inţă[1,*], *Achim* Muntean[1], and *Sorin-Mihai* Croitoru[2]

[1]Lucian Blaga University of Sibiu, Industrial Engineering and Management Department, Victoriei, 10, 550025, Sibiu, Romania
[2] Politehnica University of Bucharest, IMST Faculty, Spl. Independentei, 313, Bucharest, Romania

Abstract. The paper main purpose is monitoring of tool wear in metal cutting using neural networks due to their ability of learning and adapting their self, based on experiments. Monitoring the cutting process is difficult to perform on-line because of the complexity of tool wear process, which is the most important parameter that defines the tool state at a certain moment. Most of the researches appraise the tool wear by indirect factors such as forces, consumed power, vibrations or the surface quality. In this case, it is important to combine many factors for increasing the accuracy of tool wear prediction and establish the admissible size of wear. For this, paper both the theoretical data obtained from FEM analyze and experimental ones are used and compared in order to appreciate the reliability of the results.

1 Introduction

The most updated techniques for intelligent tool condition monitoring which are based on information from different sensors integrated with a neural network [1]. The artificial neural networks (NN) consist of many massive interconnections of rather simple neurons simulating the biological nervous system. These networks are also referred to as parallel distributed processing. Based on the modification of weights, which came from the experience, the network can learn from its past experience, so as to simulate the human brain. So, we can say these type of networks is intelligent, [2].

In any machining process, the survey of the tool state plays a central role in the process diagnosis system. During this process, the tool also loses the capacity of cutting, which is referred as tool wear. The size of the tool wear is the common way to evaluate the tool state.

The monitoring of gradual wear requires the development of sensitive, accurate, and reliable devices. On-line monitoring and compensation of the tool wear would be of a great help to avoid the increase in cutting force [3] loss of accuracy, deterioration in surface finish, increase in cutting temperature and increase in vibration due to tool wear.

* Corresponding author: marinela.inta@ulbsibiu.ro

The main types of monitoring the tool wear are direct and indirect methods. Although, both methods use different sensors to monitor the tool wear, but they differ in some aspects.

In the direct method [4], sensors directly measures the tool wear, such as optical scanning technique, electrical resistance, radioactive technique, measurement of tool geometry, change in work piece size, and analysis of tool wear particles in the chips. The difficulties of direct methods lead to the indirect measuring techniques to measure accessible process variables (machine tool vibrations, acoustic emission, temperature, noise etc.), which are related to tool wear and correlating the changes in these parameters to the change in the tool wear. The static methods use some static characteristics of the monitoring signal such as the mean the RMS [5]. Unfortunately, static methods are often too sensitive to the variation of the cutting conditions and not suitable for on-line monitoring. Therefore, the dynamic methods are developed, which use the dynamic characteristics of the monitoring signal to identify and predict the tool condition.

The majority of the techniques discussed [6] above for tool wear sensing are limited due to influence of external disturbances unrelated to tool wear, and inability of instrumentation to operate reliably in the immediate vicinity of the cutting process. Investigations have been carried out by several researchers on various tool wear sensors based on radioactive isotopes. The input settings of the process are selected by evaluating, on-line, a set of feasible alternatives with respect to several criteria. Relevant performance measures such as process cost and production rate can be directly influenced in this approach by establishing appropriate definitions for the decision criteria.

2 Monitoring of cutting tool state

Monitoring of the machining process has an important role in avoiding to long idle time of the machine or preventing undesired events like loss of accuracy, excessive tool wear or even the failure of the cutting tool.

The researches looking the conception of any tool state monitoring systems is continuing for a while and even if the technology of sensors and computers has evolved, the use in industry is poor, [7]. As [1] "suggested", despite more a decade of intensive scientific research, the development of the monitoring systems of the tool wear is still an attempt in course of development", even if a big variety of monitoring techniques have been reported and many publications presented satisfactory performances. Thus, the success of a monitoring system of the tool state depends in a great measure upon the capacity of the system to identify any anomaly and to answer on-line with a corresponding action.

Measuring and analyze on purpose of diagnosis and monitoring assumes the conception of certain hierarchic structures in a modular construction so that each module has to solve the necessary input data for the next modules based on the data from previous ones. This could be achieve based on information for each module and by inclusion in it of a specialized monitoring function depending on the possible defect to appear.

For realizing an on-line tool condition monitoring system, we assume that:

- Normal function of the process depends directly on the tool condition; bad function means an altered condition of the tool, that could be detected by the system;
- After detecting of an aberrant signal, the system must analyze and decide in accordance with imposed conditions. The answer is send to the machine tool system for execution of the received order;
- Uses of artificial intelligence by means of neural networks/ fuzzy logic increase the accuracy and reliability of the system.

From the earlier researches, one can notice that the main parameters used to monitor the tool wear are the cutting forces, vibrations, acoustic emission and the power consumed [8].

Because all these parameters are indirect indicators of tool wear the use of single or only two parameters causes a bad accuracy and reliability of the system. In a few cases, also the temperature of the tool is considered, but due to difficulties encountered at the practical measure of that, it is not practically useful.

This paper analyzes the possibility of using temperature of the cutting tool to be one of the three parameters to indicate the state of wear of the cutting tool: main cutting force, temperature and acoustic emission.

The researchers are focused on introducing the temperature as the third monitored parameter of the wear, because all analytical models for the tool wear contain the temperature as main parameters and everybody assumes that the main causes of toll wear are mechanical and thermal loads. Also, the prediction of tool wear has the form of temperature field in FEM analyze.

3 Tool wear simulation with FEM

Concerning the topic of the present paper the introduction of the FEM in the field of deformation in elastoplastic media made possible the use of this method in the metal cutting process. For establishing a theoretical model for the cutting temperature and forces, Deform 2D Machining software is used, for modeling and simulating of the cutting process, using the assumptions of orthogonal cutting.

The software can simulate the cutting process using different cutting parameters like cutting speed, feed and depth of cut for computing the cutting forces, temperature of cutting zone, stress and strain state in chip and work piece, appreciation of chip form and estimation of the tool wear and tool life.

The data referring to the force, temperature and tool wear prediction (based on the Usui model) obtained from the simulation with FEM are used for preliminary training of the NN. In this way, we can obtain a theoretical function of the tool state, which links the tool wear with the three parameters monitored.

Based on Usui Model (eq. 1) a program for solving the tool wear was developed and the theoretical values of wear were calculated:

$$\frac{dw}{dt} = C_1 \cdot \sigma \cdot v_s \cdot e^{\left(\frac{-C_2}{T}\right)} \tag{1}$$

where $\frac{dw}{dt}$ – wear rate;

C1, C2 – coefficients depending on tool and piece materials

v_s - chip shear velocity, [m/min],

T - cutting temperature

σ - normal pressure

In order to obtain some input data for the algorithm of calculus of tool wear, referring at load, forces, temperature a series of simulations were performed in DEFORM 2D software.

For the process of simulation there was used a tool made of hard metal (WC), covered with a layer of 5 microns of titanium carbo-nitrid (TiCN), and for the work piece material a C45 steel annealed and a 50 mm in diameter specimen. Cutting parameters are presented in table 1 together with working conditions that are introduced in pre - processor.

Simulation with FEM is made for determine the size of forces in order to obtain a model of forces for the cutting process, taking into consideration the tool and process parameters and also to determine the temperature of the tool. In case of a new one and considering different degrees of the tool wear, making the assumption that the temperature increase with the degree of wear due to the change in tool geometry and increase of the friction forces.

Table 1. Input data for simulation

Nr.	Tool geometry	Feed, [mm/rot]	Depth of cut, [mm]	Cutting speed, [m/min]
1.	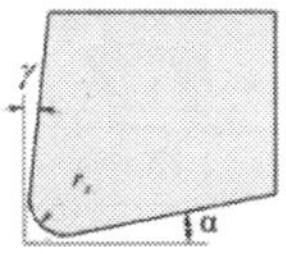 $\alpha = 6^0$; $\gamma = 5^0$; $r_\varepsilon = 0{,}4$	0,07	0,5	52
2.		0,14	1,0	100
3.		0,28	1,5	200
4.	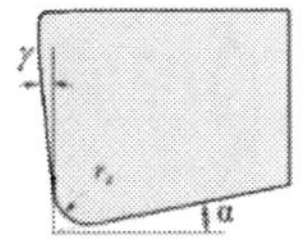 $\alpha = 6^0$; $\gamma = -5^0$; $r_\varepsilon = 0{,}4$	0,07	0,5	52
5.		0,14	1,0	100
6.		0,28	1,5	200
7.	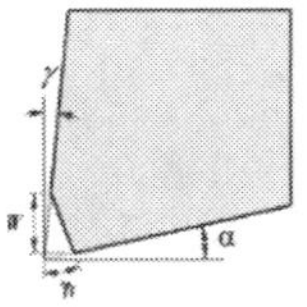 $\alpha = 6^0$; $\gamma = 5^0; w = 0{,}2$; $\gamma_f = -10^0$	0,07	0,5	52
8.		0,14	1,0	100
9.		0,28	1,5	200

Consequently, to the run of the program of finite elements used for the simulation of the turning process the following values for output parameters were obtained:

- the temperature distribution field in the cutting zone (figure 1);
- the isotherms of temperature field in the cutting tool, very useful for the estimation of the tool wear;
- plane stress state in work piece and chip with effective stress;
- the load on X and Y for the chip elements in N.

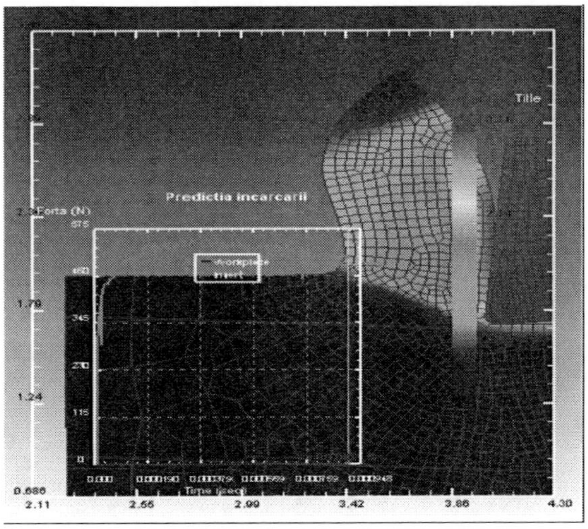

Fig. 1. Evolution of temperature in the turning process (no.3 from table 1)

With data from FEM simulation, a software product named "calcul_uzura.m" in Matlab software. After running, the program and considering a real cutting time data were obtained and presented in table 2.

Table 2. Calculated data (theoretical values)

Temperature [0C]	Force F_y [N]	Wear [mm]	Temperature [0C]	Force F_y [N]	Wear [mm]
20.00	0.00	0.00	298.00	93.82	0.50
111.00	9.79	0.05	302.00	109.22	0.55
153.00	10.70	0.10	306.00	122.42	0.60
202.00	36.82	0.15	307.00	135.62	0.65
227.00	56.47	0.20	315.00	148.82	0.70
259.00	56.57	0.25	326.00	162.02	0.75
265.00	70.45	0.30	338.00	175.22	0.80
278.00	81.91	0.35	349.00	186.20	0.85
283.00	88.56	0.40	353.00	232.50	0.90
290.00	92.10	0.45	357.00	298.76	0.95
298.00	93.82	0.50	359.00	338.56	1.00
302.00	109.22	0.55	361.00	387.95	1.05
306.00	122.42	0.60	363.00	432.99	1.10
307.00	135.62	0.65	367.00	547.32	1.15

The calculated wear after a certain time cycle could be represented and used in the monitoring system, as the neural network learning curve.

4 Experimental test

In order to estimate tool wear by measuring noise experimental tests were carried out on the same machine tool and under the same cutting conditions as in the case of attempts to force, temperature and noise.

Starting from the parameters of data acquisition board National Instruments USB6008, three signals were acquisitioned (F, T and Z) with the sampling frequency imposed by hardware, these being the instant values for input in the NN. The three acquisitioned signals are performed using:

- workpiece made of C 45 steel with length 500 mm having diameter 30 mm with the following characteristics: R_{p02} = 380 daN/mm^2; R_m = 640 daN/mm^2; HB 218;
- measure of the temperature, T variation in cutting area, performed by a natural thermocouple tool-work-piece method;
- cutting forces, F it has been measured by dynamometer in turning;
- measurement of noise, Z was made using Sonometer Quest Technologies 2100.

4.1 Tool temperature measurement using natural thermocouple

For measurement and acquisition on-line of cutting tool, temperature there was realized an experimental stand and virtual instrument developed in LabView 8.5 software. It is well known that the main problem encountered in tool temperature using the natural thermocouple is its calibration.

That problem is caused because of the different materials used for the cutting tool and for the work pieces that are normally different in each operation and product [9]. Using the stand for different natural thermocouple calibration presented in [10] which is table to calibrate simultaneous three types of natural thermocouples with a reference K type thermocouple we were able to calibrate the used couple of materials. The following calibration equation for the couple P30 – AISI 1045 steel was obtained:

$$y = 9.78 + 0.382 \cdot x \tag{2}$$

where:
y is temperature in [°C] corresponding to the measured voltage x in μV.

For obtaining the mathematical model of temperature was used a factorial experiment type 3^2, and the following mathematical model was derived:

$$T = 36.4105 + 2.0035 \cdot v - 1.00776 \cdot f - 0.0155 \cdot v \cdot f + 1.8452E - 003 \cdot v^2 + \\ + 0.010331 \cdot f^2 \tag{3}$$

Using the ANOVA analysis of the temperature model obtained, we can find influences of each parameter and its concordance. Temperature dependence of the cutting parameters is illustrated by the graphic below (fig.2).

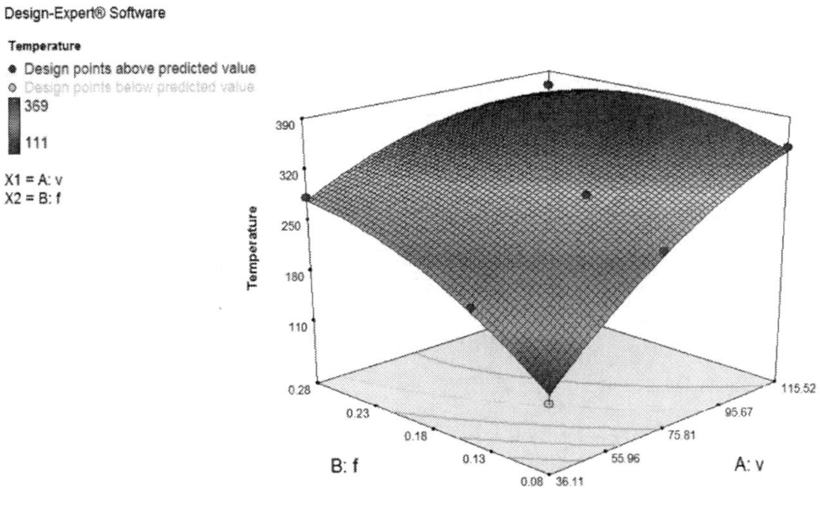

Fig. 2. Interacting factors

We can observe that, the temperature increase more at small feeds that for the greater feeds, where the chip thickness is bigger.

4.2 Cutting forces measurement

To determine the influence of the cutting forces on tool wear an inductive dynamometer for measuring cutting forces in turning was used.

To find out a model of the cutting force a full factorial experiment type 3^3 for the three factors and three levels of variation was used, which results in 27 experiments. After data processing with Design of Experiment software the following mathematical model was obtained for the main cutting force:

$$F_y = 386.76 - 1908.07 \cdot f - 153.33 \cdot a_p - 7.26 \cdot v + 1119.86 \cdot f \cdot a_p + 23.28 \cdot f \cdot v + \\ + 4.82 \cdot a_p \cdot v \quad (4)$$

The model obtained was analyzed using statistical tests ANOVA and Fisher, which give us the influence of each factor, reliability and diagnostics for each factor. In this case, the degree of confidence used is 95%.

Fisher coefficient calculated for testing the suitability of the proposed model shows that the main cutting force model is completely having factors interactions not considered in the model and influence in a significant way values resulting from measurements.

Table 3. Significance testing factors

Source of variation	Sum of squares	DF	Fcalc	Prob>F
A factor - feed	83891.70	1	4.35	0.0574
B factor -depth of cut	4.90 e^{005}	1	25.40	0.0002 ⇒The most significant
C factor - cutting speed	59155.32	1	3.06	0.1036
AB factor	46563.01	1	2.41	0.1444
AC factor	1.177 e^{005}	1	6.10	0.0282 ⇒ significant
BC factor	1.145 e^{005}	1	5.93	0.0300 ⇒ significant

Comparing coefficients calculated by Fisher's test for F_y component, it appears that, B factor (depth of cut) is the most significant parameter.

To be able to achieve a prediction of wear (VB) in terms of cutting force, F_y we assume that the forces measured during the experiments are distributed equally over the entire area of contact between tool and workpiece. Also in this case it were made a series of measurements on the same cutting conditions, keeping constant cutting conditions and using different wears of inserts. To obtain the mathematical model a factorial experiment of type 3^3 was developed, and after analysis with software Design-Expert the following mathematical model:

$$F_y = 415.80 - 420.44 \cdot a_p - 93.24 \cdot f - 1204.58 \cdot VB + 196.51 \cdot a_p \cdot f + \\ 1164.97 \cdot a_p \cdot VB + 3295.42 \cdot f \cdot VB \quad (5)$$

Analyzing the obtained results, the following conclusions can be drawn:

- analysis of cutting force signal (sudden increase or decrease) one can detect a collision, breaking or other undesirable random events;
- for achieving a correlation wear-force a mathematical model was derived both for a new tool and a worn one having a wear rate VB = 0.25-1.2 mm;
- in the initial experiment the parameter cutting speed was suppressed due to the low influence and replaced with parameter wear, VB that allows to determine a correlation force-wear;
- in the new model the order of influence in the output is VB, a_p, f that indicating that force could be used as monitored parameter of the wear process;

4.3 Noise measurement

The results of the measurements were processed using a virtual instrument in LabView 8.5 software presented in figure 3. To obtain a clearer signal and for eliminating other factors to intervene in the cutting process a measurement chain of noise band-pass filter was used, which will completely attenuate any signal outside the desired range.

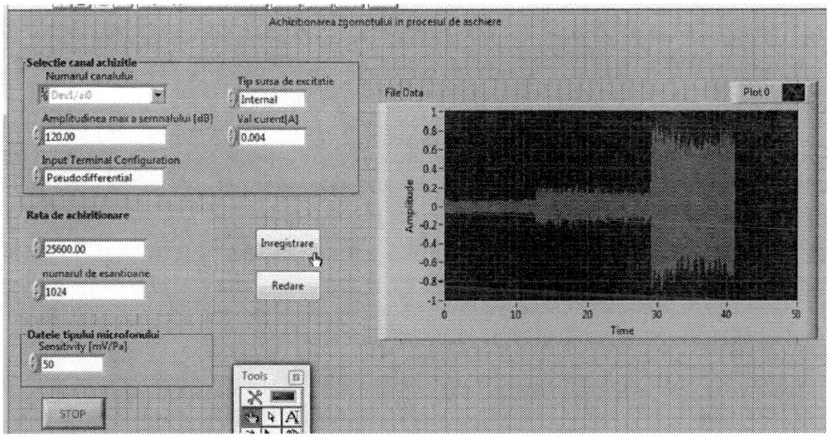

Fig. 3. Virtual instrument for noise analyze

Analyze and interpretation of measurement results for the cases mentioned wear were performed using the software SigView.

The following features were obtained and analyzed:

- noise amplitude (Figure 4),
- analyze FFT (Fast Fourier Transform);
- highest 5 peaks;
- determining the fundamental harmonic signals, etc.

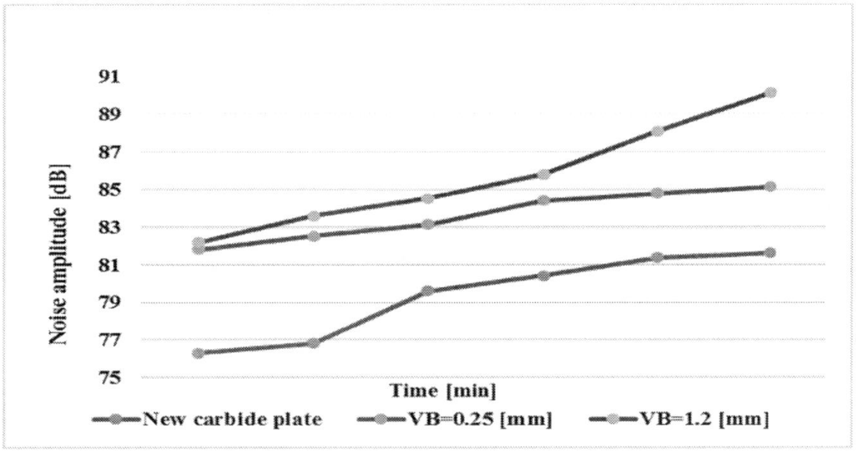

Fig.4. Variation of noise amplitude in time

In order to use the theoretical instead of the experimental data for training the neural network those were compared to the experimental values obtained for forces and temperature with the ones obtained in FEM:

- The temperature difference between the maximum temperature measured and that obtained by finite element simulation is greater (about 30%), because in experiment we measured the average temperature of tool- chip contact and in FEM it was the maximum tool temperature on a discrete element.
- If we consider the average chip-tool temperature, model accuracy is much better, with less than 3.5 % error.

5 The model of neural network (NN) in monitoring

For better on-line monitoring of machining processes, it is necessary to have the actual process of tool wear. Tool wear can be estimated from the signals of sensors.

The use of NN for monitoring the tool wear involves a series of measures linked with data acquisition, processing and input of the measured values in the NN. The main steps are:

- data acquisition of the signals for forces, temperature (theoretical values) and noise (real values);
- normalization of the data;
- transmission of the data in the input layer of NN.

The Neural network is composed by 3 layers using the inner product DOTPROD of bias functions, the net input function NETSUM and the transfer function TANSIG. The designed network for wear prediction is composed as follows: 3 neurons on the input layer (for force, temperature and noise), 3 neurons in the hidden layer and a neuron for the output (normalized value of the wear). Due to the training back propagation algorithm needs that activation functions must be limited and differentiable a virtual instrument has been realized for the activation function as sigmoid type both for the neurons in the hidden layer and for the output level.

For data normalization, a program in Microsoft Visual Basic was developed. It takes the acquired data in an time interval from Excel Cells in 3 vectors as described in the program code presented below. The result of this processing is that it keeps a constant activity level at the level of a neuron layer. Because of the heterogeneous 3 values of the input data their scales are completely different and that must be eliminated. The normalized data for NN training having the values between 0 and 1 are presented in figure 6 and the values were used as example for the network training.

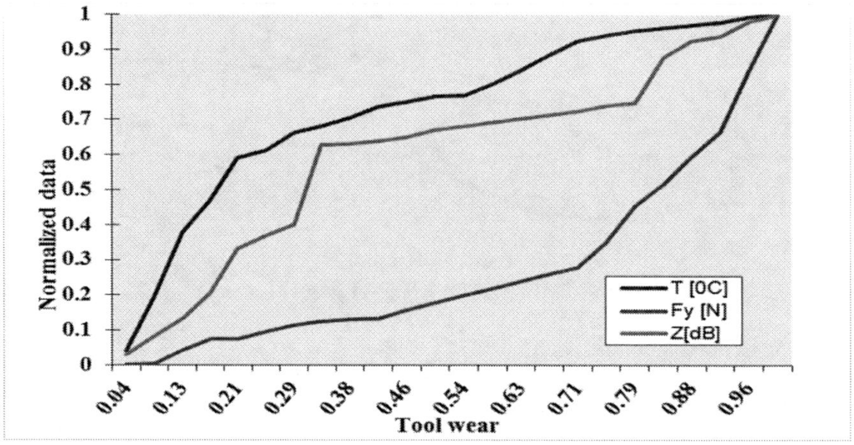

Fig. 6. Template learning using input data

With proposed neural network, model and theoretical data obtained from FEM a NN training was done using theoretical wear function as output and adding the third parameter as measured noise considering that as an improvement of accuracy.

6 Conclusions

The main goal of the paper was to show that introducing tool temperature as input factor in neural network training besides forces and noise or vibrations improve accuracy and stability of the results. Important researches were performed for improving the method of natural thermocouple for measuring the variation of tool temperature.

Considering both theoretical results and experimental ones, we can obtain an ideal set of input data for neural network training for different sets of tool and workpiece materials. The complex function obtained can be tuned in function of real conditions for each cutting operation, taking into consideration the measured parameters from on line sensors.
The difference between the measured and computed temperature is inside a reasonable limit (approx. 3% for average temperature) and this can be explained by the fact that the natural thermocouple measures the average temperature of tool-piece interface. Achieved small difference between theoretical and experimental values justify the use of the calculated values instead of experimental and can thus generate databases for preliminary training neural networks

The present paper aims to contribute to the integration of neural network capabilities in the Lab View environment for improving computer-aided control and monitoring of machining processes.

Further researches are proposed for improving the accuracy, reliability and communication between the tool state monitoring and the NC of the machine tools.

References

1. B. Stick. Mechanical Systems an Signal Processing, **16(4),** 487 (2002)
2. R. J. Schalkoff. *Artificial neural networks*, (Publisher McGraw- Hill Inc., New York, 1997)
3. A. Salimi, M. Zadshakouyan. Australian Journal of Basic and Applied Sciences, **3(1),** 283 (2009)
4. T. Özel, Y. Karpat International Journal of Machine Tools and Manufacture, **45,** 467 (2005)
5. R. Allen, H. Shi. International Journal of COMADEM, **4(1),** 26 (2001)
6. C. Byrne, D. Dornfeld, I.I.G. Ketteler, W. Konig, R. Teti. CIRP J. Man. Sci. Tec **44,** 541 (1995)
7. L. Huajie, J. Dong. Journal of Intelligent Manufacturing, **7 (3),** 243 (1996)
8. C. Scheffer, H.,Katz, P.S. Heyns, F. Klocke, Int J Mach Tool Manu, **43,** 973 (2003)
9. U. Zuperl, F. Cus, B. Mursec, T.A. Ploj. Journal of Materials Processing Technology, **157,** 82 (2004)
10. M. Inţă, A. Muntean, Advanced Materials Research, **1036,** 274 (2014)

Reliability estimation of the milling machines teeth obtained by welding deposition process

Daniela Maria Iovanas[1,*], and *Adela-Eliza* Dumitrascu[2]

[1]Transilvania University of Brasov, Department of Materials Engineering and Welding, 1 Colinei Street, 500084 Brasov, Romania
[2]Transilvania University of Brasov, Department of Manufacturing Engineering, 5 M. Viteazul Street, 500174 Brasov, Romania

Abstract. Currently, governments and authorities worldwide allocate annually significant amounts of money in an attempt to maintain the existing road infrastructure functional and safe. In our country too there is this concern to rehabilitate a significant number of damaged road structures and to create new modern ones. The research presented in this paper focused on applying modern manufacturing technologies, loading by welding the milling teeth for stripped asphalt road structures in order to increase their reliability. It was thus estimated the reliability of classical teeth made from 41Cr4 and teeth manufactured through loading by welding with special filler materials in the active surfaces, which grants them self-protection systems from wear and locking in rotation and which equipped a Wirtgen 50 milling machine in operation on one of the national roads.

1 Introduction

One of the most important types of transport which recorded a significant development both countries wide and worldwide is road transport.

The special role played by the road in the development of road transport is represented by the diversification of the network, which allows the performance of transport between the loading and unloading points, as well as by the effects of the state of roads on vehicles.

A great importance in the accomplishment of the above is played by the earthmoving equipment, respectively, milling machines for topsoil stripping.

In the literature are presented numerous examples of milling cutters made of various materials, with different geometries, with carbide or diamond inserts, with planar or mobile active surfaces, working at variable pressures and having different wear [1-6]. The disadvantage of the advanced solutions consists in the complexity of the manufacturing processes, plus high costs and low reliability.

The research presented in this paper focused on the implementation of some modern manufacturing technologies, namely, loading by welding the milling teeth for topsoil stripping in order to increase their durability and reliability.

*Corresponding author: daniela.iovanas@yahoo.com

2 Milling teeth

Worldwide, there is a large number of companies which produce asphalt milling machines in various constructive versions (Fig. 1, a, b, c) which differ not only in terms of shape and sizes of the active parts, but also in terms of the nature of the milled material, the used milling technology, the milling depth and the characteristics of the milling equipment [7].

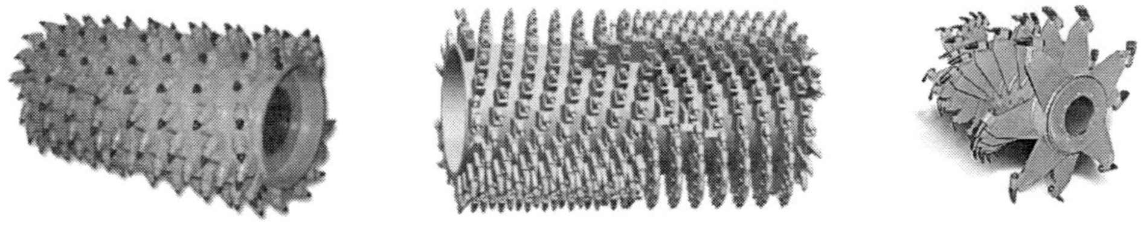

(a) (b) (c)

Fig. 1. Constructive versions of mills: (a) standard mill, (b) mill for micro-profiles and small depths, (c) recycling mill.

The main part of the mills is the milling drum (Fig. 2), made of a metal cylinder (1), on which the teeth are attached (2) through a special base (3).

Fig. 2. Road structure milling drum: 1 - drum, 2 - milling tooth, 3 - mounting base.

The important active part of the mills used for stripping asphalt road structures is the milling tooth (Fig. 3), also named cutting tool. The main elements which make up the tooth itself are: the tooth body (1), the tooth tip (2) made of wolfram carbides and fixed in the body through abrasion and the elastic mounting element (3).

In practice, several versions are used for the manufacture of teeth, which differ not only in terms of the shape and size of the active parts, but also in terms of the shape and sizes of the wolfram carbide tip, according to the manufacturing company and the destination [7].

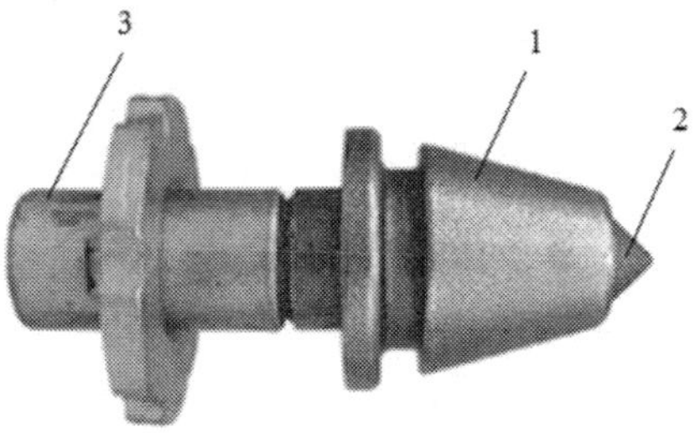

Fig. 3. Milling tooth: 1 - body, 2 - wolfram carbide point, 3 - fixing element.

During the milling process, the tip of the sintered wolfram carbide tooth undergoes intense usage through high pressure abrasion. On the other hand, the tooth body also undergoes wear due to the rough contact with the particles of the asphalt mixture. All of these lead, eventually, to the self-locking of the milling teeth in the rotation around their own axis and, consequently, to their early disuse [7, 8].

To that end, in order to increase wear resistance and to fight back these wear phenomena, research focused on obtaining, through loading by welding, intelligent wear and locking protection systems of the various types of milling teeth for the stripping of asphalt road structures, in order to improve their reliability [7-9].

Fig. 4 illustrates a milling tooth with the two active areas [8], respectively, the wolfram carbide tip (1) and the tapered part of the tooth body (4), whose wear resistance is meant to be improved by placing, through fusion welding, several layers (2) of proper width and thickness, with high resistance to wear and to locking during the rotation around its own axis.

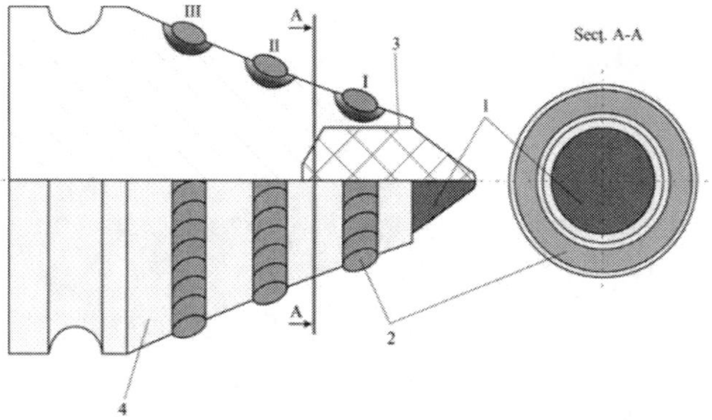

Fig. 4. Loaded milling tooth: 1 - wolfram carbide tip, 2 - circular bands obtained through loading by welding, 3 - abrasion area of the wolfram carbide tip in the frontal orifice, 4 - tapered part of the milling tooth.

The fusion welding procedures used in the research for the achievement of the loading by welding of the tapered part of the milling tooth, as well as the filler materials used in experiments are illustrated in Table 1, [10].

Table 1. Procedures and filler materials used in the experiments.

No.	Welding procedure	Welding procedure symbol (EN ISO 4063)	Filler material
1.	WIG	141	Tubular wire FILEUR DUR 606 B
2.	WIG	141	Tubular rod VTCr2.5TiD
3.	MIG / MAG - Cold metal	135 / 136	Tubular rod FILEUR DUR 606 B

3 Estimating the reliability of the milling teeth manufactured through loading by welding

The study consists in estimating the reliability of the milling teeth manufactured through various loading by welding procedures to be used in road rehabilitation structures. The data required for appreciating reliability were gathered directly from the exploitation of a

Wirtgen 50 milling machine used for the topsoil stripping of a road in Arad County, fitted with 50 teeth which worked for an 8-hour cycle and then were entirely replaced [8]. Given the rough working conditions, the milling machine is stopped at regular intervals (approximately 2 hours), for cooling and inspection. The periodic inspection, in this case, consisted in the thorough analysis of each tooth (after flushing it with water), estimating the wear degree of each and the manner in which it was produced. Even if the degree of wear in some milling teeth was high, the teeth were kept until the end of the working cycle (8 hours).

Practically, we managed to fit the milling machine with the tested milling teeth, per tooth batches, as follows:

- with classical teeth made of 41Cr4, type B,- used in this type of mill, (version I);
- with teeth obtained through loading by means of the MIG/MAG (CMT) welding procedure; filler material: tubular wire type FILEUR DUR 606B, (version II);
- with teeth obtained through loading by means of the WIG welding procedure; filler material: tubular wire type FILEUR DUR606B, (version III);
- with teeth obtained through loading by means of the WIG welding procedure; filler material: tubular rod VTCr2,5TiD, (version IV).

For analysing reliability, first we identified the statistical model most adequate for such an analysis [11-15]. The experimental data were validated using Anderson-Darling adequacy test in order to determine whether a sample comes from a population with a specific distribution. After applying the test, we adopted normal distribution.

The estimation of the reliability of the milling teeth for the four procedure types was made using the Minitab software. Based on the experimental data, we estimated the main reliability indices of the milling teeth (PDF-probability density function, reliability function and hazard rate) using least squares method (Fig. 5 to Fig. 8).

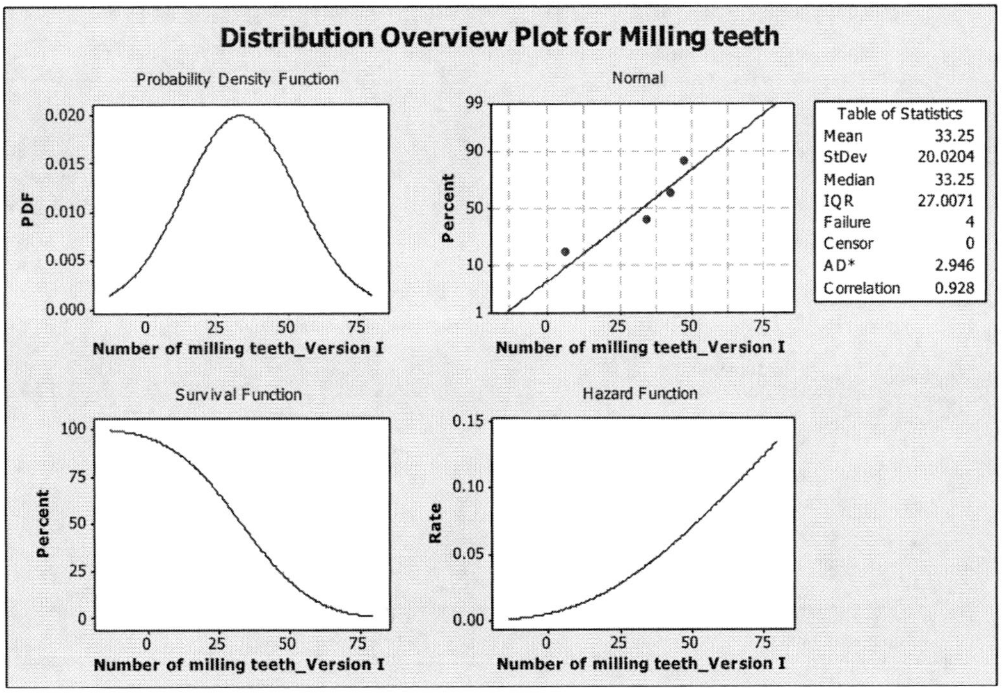

Fig. 5. Reliability indices estimation of milling teeth – version I.

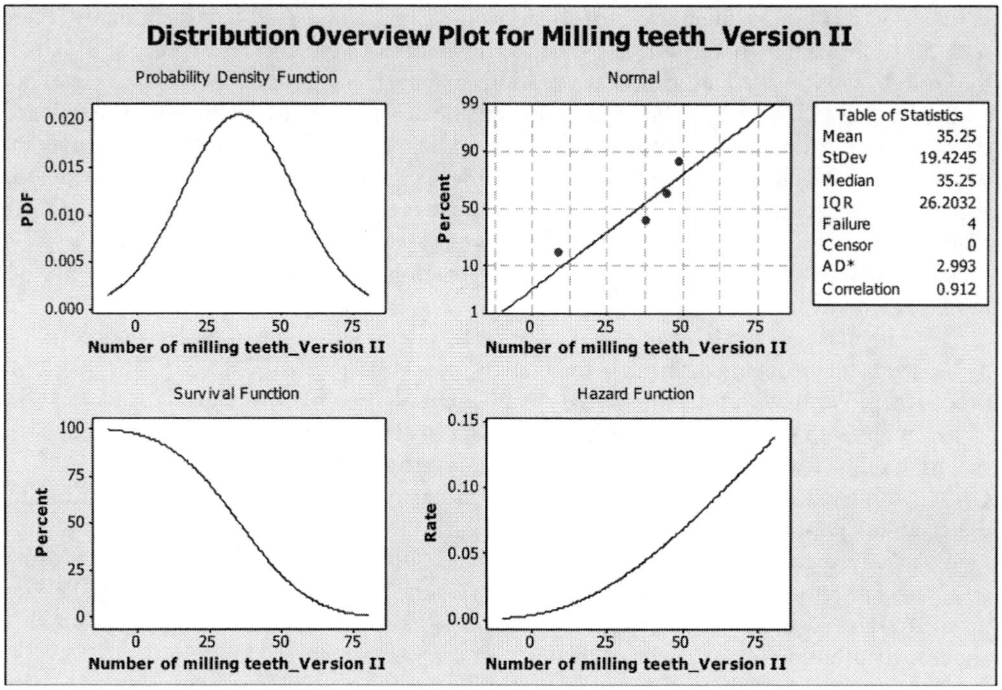

Fig. 6. Reliability indices estimation of milling teeth – version II.

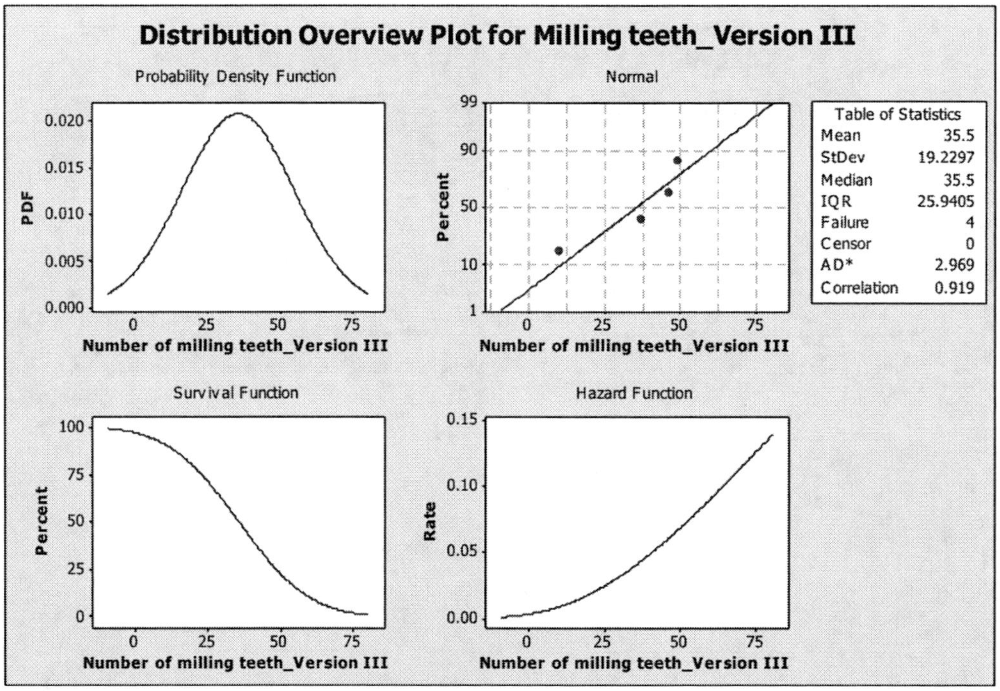

Fig. 7. Reliability indices estimation of milling teeth – version III.

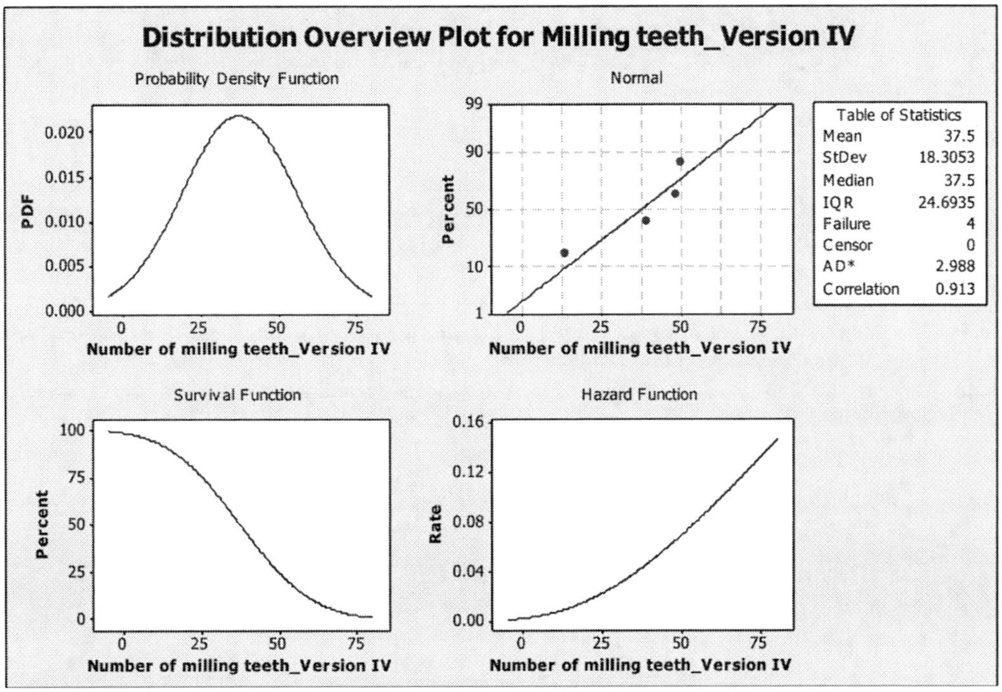

Fig. 8. Reliability indices estimation of milling teeth – version IV.

4 Conclusions

Based on the analyses made on the milling teeth, we found that the evolution of wear is exacerbated after 4 hours of operation, most of it being caused by self-locking during rotation, thus resulting a preferential wear (on only one side). This was seen especially in the milling teeth tested in versions I, II and III. In version IV, the wear was fairly symmetrical (sharpening effect), which means that the self-locking effect was significantly reduced.

In terms of reliability, after 6 hours of operation, we notice a significant decrease of reliability in all the analysed types. Analysing the reliability of the milling teeth obtained through the different loading by welding procedures, the milling teeth obtained by loading using the WIG welding procedure, with tubular rod type VTCr2,5TiD have a better reliability as compared to the other studied versions.

References

1. E.J. Oles, L.E. Thomas, US Patent 61619090 (2000)
2. D. R. Hall, R. B. Crockett, J. Jepson, J. Fox, M. Barnhill, US Patent 7469972 (2008)
3. D. R. Hall, J. Bailey, R. B, Crockett, US Patent 7669938 (2010)
4. S.G. Patel, M. George, R.J. McClellan, P.E. Pastusek, I.R. Rugashoborola, N.J. Lyons, US Patent 7814998 (2010)
5. D.R. Hall, R.B. Crockett, US Patent 8028774 (2011)
6. A. Binchiciu, I. Voiculescu, V. Geantă, H. Binchiciu. R. Ştefănoiu, D. Iovănaş, E. Binchiciu, R. M. Negriu, RO Patent 125760 (2011)
7. R.F. Iovănaş, Theoretical and experimental research on increasing the durability and reliability of equipments, manufactured by weld cladding, visible on earth drilling equipments, PhD thesis, Transilvania University of Brasov (2012)

8. *** http://www.matfrez.ro/, accesed 10.02.2017
9. E. Binchiciu, V. Geanta, I. Voiculescu, R. Stefanoiu, R. F. Iovanas, A. Binchiciu, H. Binchiciu, RO Patent A/00375 (2012)
10. D. M. Iovanas, *Welding load of the used parts to the earthmoving equipments (in Romanian: Încărcarea prin sudare a pieselor uzate la utilajele terasiere)*, (Lux Libris Publishing House, 2013)
11. M. Comandini, G. Olmi, A. Freddi, Transactions of Famena **31**, 2 (2007)
12. J. Stodola, P. Stodola, Transactions of Famena **33**, 1 (2009)
13. A.-E. Dumitraşcu, B. Lepadatescu, D.-I. Dumitraşcu, A. Nedelcu, D.V. Ciobanu, The Scientific World Journal, Article ID 296762, **2015** (2015)
14. D.C. Montgomery, G.C. Runger, *Applied statistics probability engineers. 5th Edition* (John Wiley & Sons, Inc., New-York, 2011)
15. Anderson-Darling Test Engineering Statistics Handbook, (2013), available at: www.itl.nist.gov/div898/handbook/eda/section3/eda35e.htm; accesed 17.03.2017

Comparison of problem solving tools in lean organizations

Maria Virginia Iuga[1+2,*], and *Liviu Ion* Rosca[2]

[1]Marquardt Schaltsysteme S.C.S., 2 München st., 550018 Sibiu, Romania
[2]Lucian Blaga University of Sibiu, Faculty of Engineering, 10 Victoriei Bvd., 550024 Sibiu, Romania

Abstract. As global market competition is getting fiercer, and companies are looking at ways to stay on top, more and more organizations are looking at Lean Manufacturing and lean tools to support them in achieving their goals. Especially within the automotive industry, lean practices are very well received. The speed at which the automotive industry is evolving, especially but not only, in countries like Romania, leads to the need to carefully analyze lean manufacturing concepts, examine them against local production conditions, and to develop and standardize them. One of the most important things to take into consideration here is the application of an adequate problem solving technique to avoid waste. The objective of this research paper lies in analyzing and comparing the problem-solving methods recommended by the Toyota Production System, and to propose their appropriate application at shop floor, in relation to the specific problem type.

1 Introduction

The word „problem" is widely used in daily, organizational and business language. According to the etymology dictionary, the origins of the term „problem" is the Latin word „problema", the Greek „πρόβλημα" or the Old French „problème" meaning „a difficulty, a barrier" [1]. Generally, the term is defined as a question involving doubt, uncertainty, or difficulty [2]. In business terms, a problem is defined to be the gap perceived between an existing status and the desired one, or a deviation from a standard [3].

All organizations deal with large pools of problems and challenges. Often, problems are swept under the rug due to either lacking capabilities in resolving them, lack of resources or processes, or due to the fact that they are not being credited with a high importance factor. Frequently organizations will face problems when they are either threatening existing processes and the success of the organization, or when they are so called "quick fixes". Currently not many organizations have solid problem solving techniques and processes in place and even less organizations look at problems as being opportunities rather than threats. The problem solving philosophy from lean manufacturing comes from the main idea of reducing waste, specific the waste of resources generated by an ineffective approach to organizational problems.

* Corresponding author: virginiaiuga@gmail.com

2 Problem solving philosophy in Lean Manufacturing

The starting point of this paper is Toyota's view of a problem. The problem is viewed as being difficult, a matter of dispute and the antonym to the solution. A problem is therefore unique, time limited, contains a certain risk and complexity while having unknown solution. If it is possible to solve the „problem" with the available knowledge or rules, then it must be regarded as a task and is no longer a problem. On the other hand, Toyota has a very mature and efficient way of looking at organizational challenges as being opportunities [4]. Within his research Puvanasvaran [5] states that any problem should be appreciated as an opportunity to improve process and develop people.

There are many ways companies, which identified problems as being organizational opportunities, are dealing with these. Some organizations even assign specific problem solving teams to detect and solve organizational challenges. Some of the more famous methods used by organizations are techniques like audits, brainstorming, checklists for risk identification or benchmarking.

The lean manufacturing problem solving philosophy is an accessible and easy to understand notion. Problem solving in lean manufacturing is a main part of entire Toyota Production System (TPS) philosophy.

Toyota divides problems into three categories and utilizes the resources appropriately for all three levels (Fig.1). Larger issues are generally addressed through management-detected and management-controlled activities. The way this company manages to increase performance is through leverage and focus. Toyota enforces basic problem-solving to all employees. This way every employee can become a problem solver making it possible for the company to effectively leverage its people resources. For more complex problems which require a high degree of problem solving skills, members of management are trained via Management Kaizen events. According to the Toyota philosophy, the needed information to describe a problem are: the current situation description and the historical trends, the target including the difference between the target and current status, and the detailed description or problem [4].

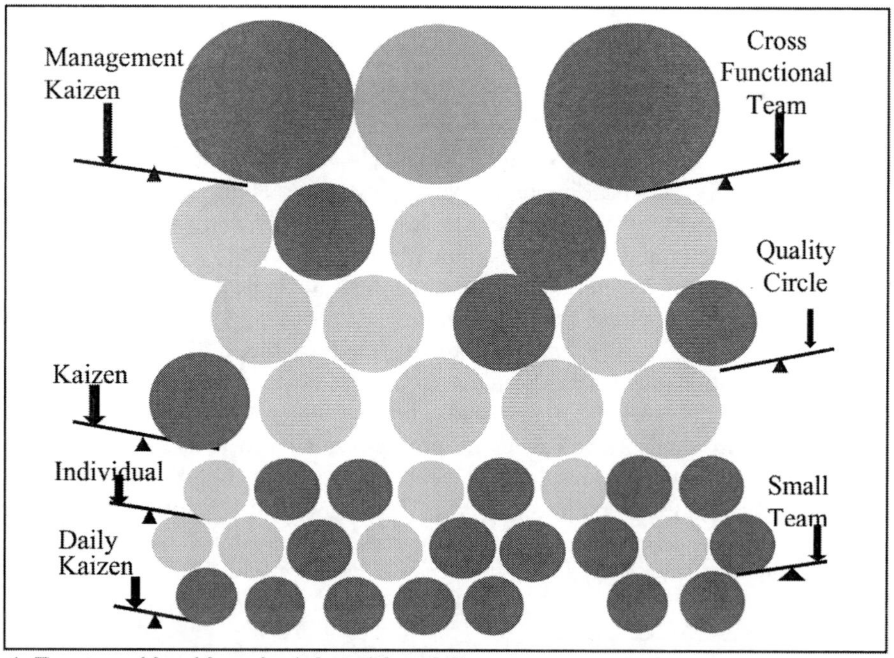

Fig. 1. Toyota problem hierarchy (adapted from [4])

3 Problem solving methods in Lean Manufacturing

3.1 Brainstorming

Increasingly, brainstorming has become an integral part of developing products and processes. Companies utilize brainstorming as a method to reach conclusions about an existing issue or as a means of developing new directions and solutions. It is a method that encourages creative thinking and new ideas. Alex Faickney Osborn coined the term and method back in 1957 in his book, "Applied Imagination". He named the method after the idea, namely "using the brain to storm a problem" [6].

Typically, in the beginning stages of working with teams, producing new ideas begins by attempting to solve existing organizational issues. During the initial idea-generation stage of the decision-making process, brainstorming is used to try and solve issues that can have major or minor consequences on the organization. [7]. A group will function together and consistently generate new ideas and solutions freely without deep assessment or criticism. Ideas are documented visually for members of the group to help develop any possible new concepts.

Various guidelines exist to create a successful brainstorming session amongst organizational members. Originally Alex Faickney Osborn created his own special means to cause an optimal brainstorming session that fosters creativity. Brainstorming directions are as followed: (1) No explanations about contributed ideas; (2) when a lull is happening reengage the group by restating the issue at hand; (3) support or incite non-participating members to add ideas; (4) recommend participants revisit categories when not creating new ideas [8]. The approach developed by Faickney allows each member to fully participate in the brainstorming session. Each member is held accountable for active participation and idea generation to further develop the session. Moreover, by simply focusing on generating ideas, and not dwelling on implications or difficulties involved, members are able move to new ideas and the session can advance with more engagement leading to more solutions. To create many ideas for a given subject, brainstorming is expended as a technique with the focus of deferring assessment until new ideas have been proposed [9].

For many organizations, it is a problem-solving technique that usually requires approximately 20 minutes. Required resources involve flipcharts, pin boards, presentation cards and pens.

Conversely, it is possible for social influences, such as evaluation apprehension, to have an impact on the performance of group members within the organizational brainstorming session and lead to productivity losses. For example, if some group members do not consider themselves to be of the same intellectual or cognitive abilities as others, it may impede their ability to willingly contribute with ideas. A study by Thompson [10], found that if an individual notices that their performance is moderately similar to that of others, they may identify their performance in a more positive way. Furthermore, research has shown that people in groups exert less effort into answering an issue than those who are attempting to solve a problem on their own; this is referred to as free-riding or social loafing [11].

Brainstorming, while still an effective means of collective intelligence, has progressed with technology. Electronic brainstorming is a practice that organizations are using to further expand on group collaboration. Stemming from traditional brainstorming, electronic brainstorming is where members of a group use computers to generate ideas and alternative solutions [12]. Typically, a mediator is present which details the issue at hand as well as the specific targets of the organization. Electronic brainstorming has shown to help participants be more creative without the help of other participants influencing idea generation or inspiration. Results have shown that when groups are lacking in creativity and generating new ideas, electronic brainstorming is effective [13].

While the future of brainstorming and idea generation is unknown, it remains a useful tactic for organizations. Discovering new ways to generate solutions is useful to solve existing or potential organizational issues. Managers can use traditional and electronic brainstorming to engage creativity and find resourceful means of sustainability.

3.2 Ishikawa

The Ishikawa technique (Cause and Effect Analysis) provides a useful way to explore all the things that could potentially cause a problem. Ishikawa diagrams were popularized by Kaoru Ishikawa in the 1960s and the tool is considered one of the seven basic quality tools. The diagrams are known as Ishikawa Diagrams or Fishbone Diagrams because a completed diagram can look like the skeleton of a fish. Mazda Motors famously used an Ishikawa diagram in the development of the Miata sports car [14].

The Ishikawa diagram shows the relationship between an effect or problem and the potential causes. It allows the categorization of the causes. The seven main factors/ categories of causes are all beginning with "M": Material, Machine, Method, Man, Management, Milieu (Mother Nature, Environment), Measurement. Depending on the problem statement, several „M"s could be omitted or added.

Technically seen, and concerning the required materials, the tool is simple, powerful (particularly in the context of teamwork), easy to learn, and requires little effort.
The cause-effect diagram normally produces good results in a relatively short time while the visual presentation is easy to understand.

3.3 5 Why

The 5Why questioning technique serves to find the real causes ("root causes") of a problem. Determining the root cause of the problem is a technique that implies looking at the problem and asking "Why?" and "What caused this problem". The general rule is to ask five times "why", until the true cause is determined.

Usually the answer to the first "Why" question leads to another "Why" question, therefore, it is called the 5Why technique. The technique was first used by S. Toyoda within the Toyota Company, and is a main component of problem-solving training of the Toyota Production System. Taiichi Ohno (1988) describes this method as being "the basis of Toyota's scientific approach, by repeating "why" five times the nature of the problem as well as its solution becomes clear" [15].

Tracking the causes to the true cause prevents from working on symptoms and helps to find sustainable solutions and to solve a problem on a long term.

The technique is easy to perform and offers a deep root cause analysis. Nevertheless, there is a risk to stop too early with the question of technology, and thus the actual cause is potentially not found or, on the other hand, there are answers that can lead to a dead end, e.g. "The customer wants it that way."

3.4 Therefore technique

The "therefore technique" is a technique dedicated to problem identification and has its roots in TPS. While the 5Why technique is used mostly to identify the root cause of a problem to create an approach to solve it, the "therefore technique" helps to find the true problem that occurs in a process. This technique focuses on the problem identification and can be considered as a former step to the application of problem solving tools (PST) such as

5 Why, Ishikawa, etc. This step assures the discovery of the true problem and then applies the root cause analysis (e.g. 5 why) to the real problem. Also, it will assure that the most effective measures will be applied. Applying the "therefore technique" is mandatory within the Toyota culture in order to make the difference between problems and their symptoms and to correctly detect the root causes and solutions to real problems, and not symptoms.

The general tendency is to find solutions for the symptoms. Analyzing the symptoms can lead to a "dead end", especially when the problems are technical ones. E.g. the 5 Why technique applied to a technical issue can lead to a root cause such as "machine design mistake". The solution to this "cause" can be expensive and complicated. Applying the "therefore technique" leads to an understanding of the real problem. This approach leads to alternative, easy and less expensive solutions, in order to reduce interruptions [4]. The technique will maximize results with minimal efforts and finally will resolve the true problem by correcting causes in addition to the ones initially identified.

The main advantage of the technique is ensuring that the most important opportunities are taken into consideration.

3.5 Questioning technique (5W&2H technique)

The "Questioning" technique concentrated on looking at a problem from all possible angles. It is a method to identify the causes of problems [16]. Its name comes from the letter "w" or "h" that is contained in every of the 7 questions that are: What?, Who?, Where?, When?, Why?, How?, How often / many / much?.

The method can be used in different areas every time when situations have to be analyzed and when it comes to clarifying problems. The main objective consists in a fast and comprehensive analysis of the actual state.

3.6 Pareto diagram (ABC Analysis)

The "Pareto Diagram" is a data analysis tool that is used for the selection of a limited number of causes that produce a significant overall effect. In terms of quality improvement, a large majority of problems (80%) are produced by a few key causes (20%).

A Pareto diagram is a simple bar chart that ranks related measures in decreasing order of occurrence. The principle was developed by Vilfredo Pareto, an Italian economist and sociologist who conducted a study in Europe in the early 1900s on wealth and poverty. He found that wealth was concentrated in the hands of the few and poverty in the hands of the many [17]. The application of the Pareto analysis in risk management allows management to focus on those risks that have the most impact on the project. The method allows to make complex, large amounts of data manageable by sorting and clustering, to visualize a rough picture of the actual work situation, to separate the "essential" from the "inessential" and to set priorities (80:20 rule).

The method requires reasonable effort, is easy to use and allows a very clear graphical representation of results.

3.7 Seven alternatives (7 ways)

The Seven Alternatives (7 ways) technique is rooted from new product/ process development, formally known as Production Process Preparation (3P). The method is used to encourage creativity with the purpose to generate multiple solutions. Initially used as a part of the 3P process, the method`s objectives are the visualization of relationships between construction and the use of components and assemblies.

This technique is applicable in all manufacturing processes. The steps to perform the method are as followed:

1. Showing and describing seven alternatives to solve the problem. Usually it is difficult to find seven possible alternative solutions for a problem. Nevertheless, the number of seven must be reached. This forces the participants to seek creative and unusual ideas. For this reason, the collection of ideas is done in the spirit of brainstorming. The focus lies in the invention of new alternatives.
2. Selection of evaluation criteria. These are chosen in relation to the purpose e.g.: Cost, Benefit, Efficiency, Feasibility, Acceptance, Effort for Approval, Duration of Decision Making Processes
3. Evaluation of the alternatives through scoring each solution for all established criteria
4. Selection of the "best" solution – best score

3.8 Force field analysis

Analyzing the forces "For" and "Against" change is a common instrument used for systematically studying complex problems. It helps with the visualization of all the factors for and against a plan, with the prioritization of these based on importance of these factors, and with the decision of whether a plan is worthy of being implemented or not. Force Field Analysis was developed by Kurt Lewin [18]. This system was originally meant to be used as a tool in social psychology. Nevertheless, Force Field Analysis is nowadays used in business for taking and communicating go/ no-go decisions. The Force Field Analysis is often used in the initial phase of a problem-solving process.

The method follows the next steps:

- writing the topic, the problem, the target state as a heading on the pin board
- labeling columns ("+"driving forces and "-"restraining forces)
- determining the forces that support the desired goal and the forces that impede the desired state
- identifying the positive and negative forces with priorities, such as "length of an arrow" or visualized with adhesive point
- analyzing and editing the results further. After finding a solution in a problem-solving process, e.g. a counter-check of whether all the positive factors are in place and all the blocking factors are eliminated or weakened.

The method is fast, easy, requires short time (about 10-20 min), takes into consideration all the aspects determined by a team, delivers approaches to positive changes, and visualizes the balance of the effective factors.

Nevertheless, the application of this method alone will lead to solutions and the findings and results from the force field analysis must be further processed.

3.9 A3 Problem solving story

The A3 Problem Solving Story is the most specific and complete TPS problem solving technique. This tool was developed by Toyota and represents their approach in problem solving.

The method uses a simple way to collect data and information, followed by a standardized approach for team based problem solving. It is also the easiest way to visually communicate information and ideas – it tells the story and is perfect for facilitating team based problem solving.

The method was part of Toyota's Quality Circle problem solving efforts in the 1960s and it allowed teams to gather the most important information on one sheet of paper (easy to read, understand and base decisions upon). The philosophy starts from the idea of concision: if a problem story can't be described in one page, it means the result is not concise enough (the problem solving process is similar to a story, hence the term "Problem Solving Story"). There is an introduction (steps 1-3: problem statement, information, actual situation through facts and figures, objectives) followed by a middle part (step 4: cause and analysis), a main part (steps 5-6: counteractive measures/ alternative solutions; rollout scenario), and finally a conclusion (steps 7-8: documentation/ further actions).

The originality of this method lies in the fact that in order to find a solution it does not need a computer, but only pen, paper, common sense, knowledge, creativity, and sustainability. The method uses the power of other lean hard tools to perform a good analysis. If there is a solution it is important to know, why the solution works. According to Toyota, the method can be used by every employee independent of his/ her experience and education. The method allows stake holders which are closest to the problem to make an impact, while any interested parties can review and draw conclusions. The main disadvantage is that the collected and analyzed data cannot be stored on an IT-system.

The "A3 Problem Solving Story" tool can be applied in a wide range of situations: from half hour problem solving meetings up to several day's problem solving workshops or within large Kaizen projects [4].

4 METHODS COMPARISON

Lean Manufacturing problem solving tools can be used depending on problem complexity and their impact by themselves or combined. TPS proposes very different methods: quick and creative ones, analytical, methods that can be applied in small teams or complex methods which require deep analyses and follow-ups of the iterative four-step management method Plan-Do-Check-Act Cycles (PDCA), or methods which help validate solutions, such as A3- Problem Solving.

In a fierce competition market, the decision maker should find rapidly solutions. Time usually do not allow the use of all the above tools. So, what method should be used and when? In order to asses this issue we suggest to compare these methods using six different criteria. These criteria represent, in our opinion and literature, the defining features of these methods (Table 1). Literature provides many scales, scores, utilities, items and so on to be used to differentiate methods or alternatives. For exemplification purposes each method was granted a score between 0 and 3. In our case, 0 means that the criterion does not describe the method at all, while 3 means that the criterion perfectly describes the method.
So, we suggest:

1. CO - Creativity orientated
2. RC- Root cause analyze orientated
3. CM - Complexity of method
4. IS - Immediate solution oriented
5. AS - Solutions analyze orientated
6. CPX- Appropriate for complex problems

Table 1. Overview of PST

Method/Criteria		CO	RC	CM	IS	AS	CPX
A	Brainstorming	3	0	1	3	1	1
B	Pareto	0	2	1	0	0	2
C	7 Alternatives	3	3	1	2	3	3
D	Force field analyze	2	0	1	0	3	2
E	5 Why	2	2	2	1	0	1

	Method/Criteria	CO	RC	CM	IS	AS	CPX
F	Therefore Technique	0	3	2	1	0	2
G	Ishikawa	1	3	3	0	0	3
H	5W&2H	0	2	3	0	0	1
I	Problem Solving Story	3	3	3	3	2	3

To obtain a better visualization for choosing the most appropriate method different scatter diagrams can be design as in (Fig. 1-2).

Depending on the problem, needs and possibilities, this kind of visualization will facilitate the decision of which method or methods should be used and when.

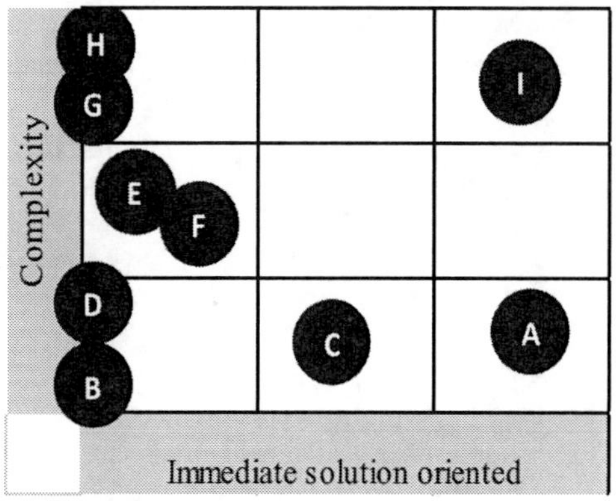

Fig. 1. Scatter diagram: PST complexity vs. immediate solution

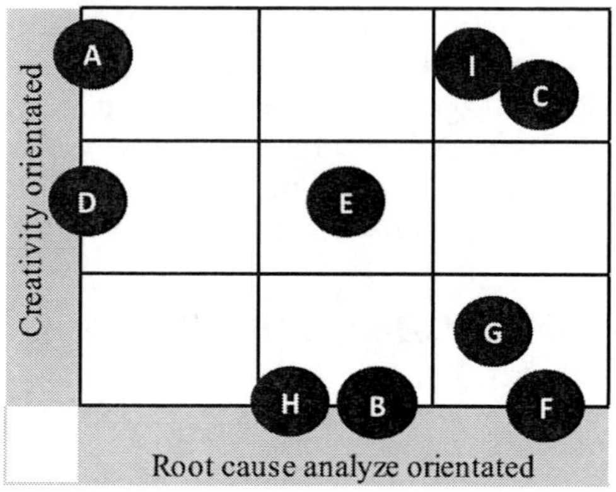

Fig. 2. Scatter diagram: PST creativity vs root cause analyze orientation

5 CONCLUSIONS

In conclusion, the concept of problem solving can be assessed from different angles as a set of tools or their applications combined with the organizational know-how culture. In fact, it is all about efficiency, not through complex solutions but through waste elimination and through the selection of the most efficient way to solve organizational problems. Further on, for a better foundation of the method, a discussion about and analysis of the appropriate criteria and scores should be addressed.

References

1. ***http://www.etymonline.com/index.php?term=problem, Accesed 5.01.2017
2. ***http://www.dictionary.com/browse/problem?s=t, Accesed 5.01.2017
3. ***http://www.businessdictionary.com/definition/problem.html, Accesed 5.01.2017
4. J. Liker, D. Meier, *The Toyota way fieldbook: A practical guide for implementing Toyota's 4Ps* (New York: McGraw-Hill, 2006)
5. P. Puvanasvaran, H. Megat., T. S Hong, and M. M Razali, Journal of industrial engineering and management, **2**(1), 128-152 (2009)
6. A. F. Osborn, *Applied Imagination*, (Charles Scribner's Sons, New York, 1957)
7. R. Griffin and G. Moorhead, *Organizational behavior* (Nelson Education, 2011)
8. P. B Paulus, B. A Nijstad, *Group creativity: Innovation through collaboration* (New York: Oxford University Press, 2003)
9. D. L Nelson, J. C Quick, *Organizational behavior: Science, the real world, and you* (Mason, OH, USA: South-Western Cengage Learning, 2013)

10. L. L Thompson, *The social psychology of organizational behavior: Key readings* (New York: Psychology Press, 2003)
11. L. K. Stroh, G. B Northcraft, M. A Neale, *Organizational Behavior: A Management Challenge* (Psychology Press, 2003)
12. C. Williams, *Principles of management* (Mason, OH: South-Western Cengage Learning, 2009)
13. N. Kock, *Emerging E-collaboration concepts and applications* (IGI Global, 2006)
14. R.L. Goodden, *Lawsuit!: Reducing the Risk of Product Liability for Manufacturers*, (John Wiley & Sons, 2009)
15. T.Ohno, *Toyota production system: beyond large-scale production,* (Diamond. Inc., Tokyo, 1988)
16. C.V. Kifor, C. Oprean, *Ingineria calității: îmbunătățirea 6 sigma* (Editura Universității " Lucian Blaga", 2006)
17. V. Pareto, *The mind and society* (Рипол Классик, 1935)
18. K. Lewin, Force field analysis, The 1973 Annual Handbook for Group Facilitators, 111-13 (1946)

Visual communication in lean organizations

Maria Virginia Iuga[1+2,*]

[1]Marquardt Schaltsysteme S.C.S., 2, München st., 550018 Sibiu, Romania
[2]Lucian Blaga University of Sibiu, Faculty of Engineering, 10 Victoriei Bvd., 550024 Sibiu, Romania

Abstract. Many researchers have studied the role and importance of communication and communication types for organizations and their work force. The topic of organizational communication is a constantly changing one. The thing that remains stable throughout this very wide topic is its purpose: communication is vital for the effectiveness and success of organizations of all sizes, types and cultures. In terms of type, one could split up topic by type as followed: verbal, visual and written. This research paper focuses on the visual component at shop floor within automotive organizations. It aims at giving the reader a structured overview on the most appropriate but also lean visual communication practices. In order to establish these methods at shop floor, a survey was conducted among nine big automotive companies in Romania. It was then followed up by a second survey dedicated to check the effectiveness of implementation of these methods in a big automotive organization. The paper presents the analyses, results and the conclusions of this surveys.

1 Introduction

In the very competitive automotive industry, but not only there, the lean concept and philosophy gain credibility more and more, due to the really improvements of organizations' results. In the actual global market economy, the focus on efficiency increased dramatically and lean manufacturing implementation helped organizations to get the best results through very simple approaches, primarily orientated to avoiding wastes. Organizational communication is one of the most important components of lean philosophy and the savings potentials due to a lean communication must be taken into consideration.

Current, at the shop floor there are three arts of communication which coexist: visual, verbal and written (typed information through email). In order to remove the waste (Jap: muda) created by misunderstandings, written communication must be as visual as possible. In lean manufacturing, it is essential to use the appropriate visual languages in general and visualization in particular to maximize human visual perceptual power for rapid and effective communication in management [1]. According to current trends, the visualization in shop floor management can be defined as the slogan "five minutes shop floor instead of fifty management minute of presentation" [2]. An advanced visual management means "the ability to understand the status of a production area in five minutes or less, by simple

* Corresponding author: virginiaiuga@gmail.com

observation, without using a computer and without talking to anyone [3]. The idea of lean communication refers to a visually transparent management system, leading to the identification of wastes in Gemba and therefore to the necessary action to eliminate these.

The research targeted an analysis of the actual stage of shop floor visual communication in order to define the most appropriate and efficient methods and tools. The method used within this research paper consisted of two questionnaires. The first step used a survey, conducted in Romania among eight automotive companies. A ranking of most appropriate visualization methods was obtained. The second step performed an analysis of changes after the implementation of shop floor management in a big automotive company, having over 2500 employees in area Sibiu, Romania. This analysis was performed using the second survey and it targeted on one hand, the perception of employees regarding the effectiveness of new implemented lean visualizations methods and, on the other hand, the improvements regarding time management due to improved shop floor communication.

2 Survey design and procedures

In order to project the ideal model of visual management at shop floor, a survey named "Effectiveness and Visualization at Shop Floor" was designed (2013). The questioned persons were asked to assess the efficiency of nine different visualization methods at shop floor. The question was used in combination with a five-point scale (1-"very efficient" to 5 – "inefficient"). Depending on the area of application, the visualization methods were clustered in three categories: central visualizations – representative for the entire production area, zonal ones – representative for groups of machines or production lines and local visualization – representative for one line/machine. The proposal of other methods from the respondents remains open.

The questionnaire was handed out using a mailing action within automotive companies from Romania but was also posted within various specialty groups regarding lean manufacturing using the professional social media platform "LinkedIn". The survey was distributed using two different hyperlinks which served the purpose of segmenting two different types of respondents. The hyperlink to the survey was then distributed via online and social media channels. One link was distributed via the social media channel LinkedIn. The second hyperlink was distributed via an emailing action throughout production companies within nine companies from the automotive sector. The completion of the surveys was completely anonymous. For the completion of the questionnaire, no incentive or reward was offered to the participants. The time frame for which the questionnaire was available for completion was three weeks in March 2013.

Within the data analysis, the descriptive statistics along with the frequency distributions were analyzed. The measures of central tendency have also been taken into consideration. While the mode indicated the score that appeared most frequently in the data set, the median helped indicate the center of a distribution and locate the middle score when scores are ranked in order of magnitude. The mean is the measure of central tendency that shows the average score of the data set [4]. All procedures reported here utilized SPSS 20.0.

3 Data collection and sample profile

3.1 Data collection

The situation of collected responses was centralized Table 1. Within the data collection process, 88 surveys were filled out without any missing values from totally of 157 participants. After a thorough examination of the responses, one of these surveys was filtered out even if it did not contain any missing values. The reason for the exclusion of this survey was the indication of a response pattern. Therefore, 87 surveys went into the data analysis process. Table 1 illustrates the situation of collected responses:

Table 1. Collected responses

	Total responses	Valid responses
LinkedIn	37	24
Romanian automotive organizations	120	63

3.2 Sample profile

In order to identify the provenience of the answers respondents were asked to indicate:

- Which area of work they activate in.
- What company type they work at.
- What hierarchical level they are working on.
- Size of the company the respondents currently work at.
- Whether they are leading shop floor employees or not.
- Whether they are directly leading personnel or not.

The answers are presented in figures 1 to 4. The data sets proved that the selected sample was very representative for a general population of employees which work in industry (96%), especially in automotive industries (80.5%) and in big companies with more than 1000 employees (77%). Furthermore, the sample represents in its majority persons working within production departments, proportionally for each hierarchical level.

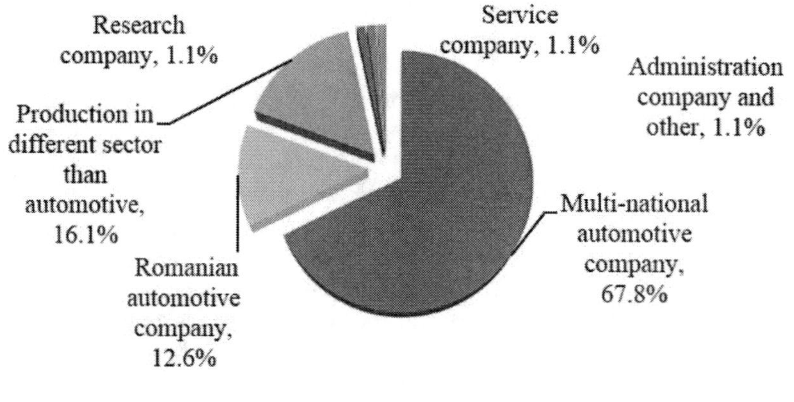

Fig. 1. Sample company type

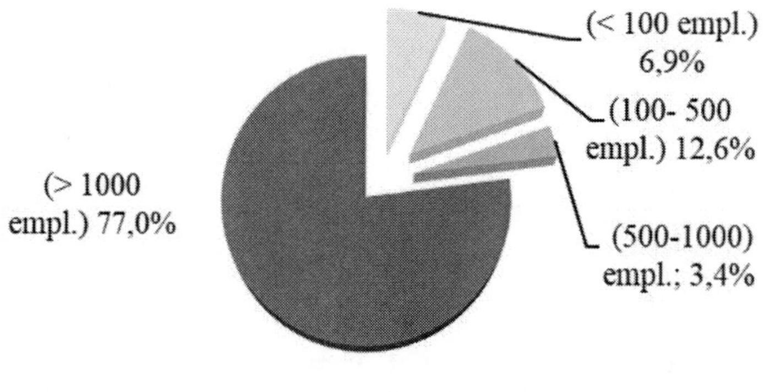

Fig. 2. Sample company size

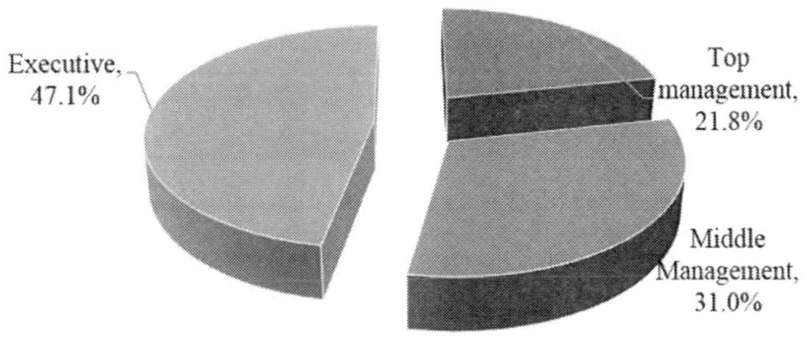

Fig. 3. Sample hierarchy

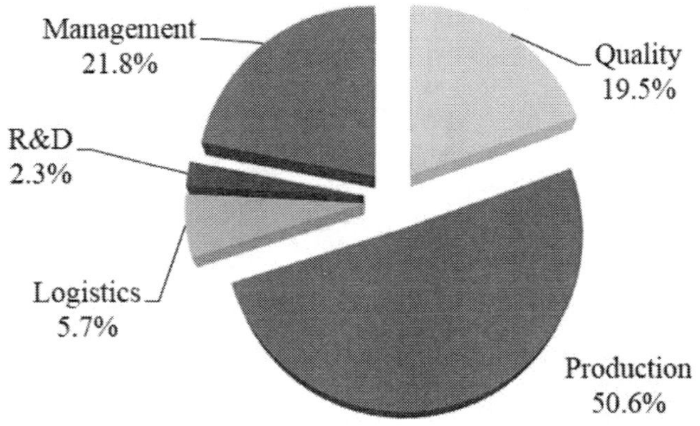

Fig. 4. Respondents sample area of work

4 Survey results

In order to assess the shop floor visualization methods to the previous mentioned questionnaire the participants were asked to evaluate the efficiency of nine most usual forms of visualization at shop floor. This question was used in combination with a five-point scale (1-"very efficient" to 5 -"inefficient"). The respondents were asked to indicate how efficient the nine shop floor visualization methods are in their opinions and to propose other methods which were not asked. Table 2 gives a structured overview of the nine chosen visualization methods. The respondents had to evaluate five shop floor central and zonal visualization methods and four visualization methods used only local, per machinery/production line.

Table 2. Communication through visualization at shop floor

Item No.	**Visualization method**	**Abbreviation**
1	Central/ zonal: panel (electronic) for a production area showing the operating status of all machines in the area (ANDON CENTRAL)	CZELPA
2	Central/ zonal: displays for general information (statistics, events, days off, etc.) with data derived from a central computer	CZDIS
3	Central/ zonal: information panels on production areas containing statistical data listed on paper and updated at least once a month	CZINPA
4	Zonal: Paper holders (flip Figures) with interactive data updated by employees	ZPAHO
5	Zonal: White board with interactive data updated daily by employees	ZWHBO
6	Per production line: electronic screen at every line (machinery) based	PPLELS

Item No.	Visualization method	Abbreviation
	on electronic data updates of production data (produced quantities, defects, etc.) and real time visualization	
7	Per production line: board at the end of the line (machinery) based on interactive updates of the shop floor employees	PPLBILB
8	Per production line: stoplight indicator, operated automatically, indicating the operating status of the production line (ANDON)	PPLSTA
9	Per production line: stoplight indicator, operated manually, indicating the operating status of the production line (ANDON)	PPLSTM

The order in which the respondents assessed these visualization methods to be efficient was randomly selected as follows: PPLELS, PPLSTA, ZWHBO, CZELPA, PPLBILB, PPLSTM, ZPAHO, CZDIS and CZINPA.

The visualization method regarding an electronic screen at every line based on electronic data updated and real time visualization (PPLELS) was assessed as being very efficient by 66.7% of the sample. Only 2.3% of the sample (two respondents) considered this visualization method to be inefficient (M=1.60, SD=1.017).

The stoplight indicator operated automatically (ANDON) (PPLSTA) was considered to be the next most efficient visualization method, 59.8% of the sample considered this visualization method to be very efficient. Only three respondents indicated that in their opinion this visualization method is very inefficient (M=1.71, SD=1.056). White board with interactive data updated daily by employees (ZWHBO) was assessed as being the third most efficient visualization method with a valid percent of 55.2% indicating that the method is very efficient (M=1.76, SD=1.078).

The fourth most efficient visualization method was CZELPA, the panel (electronic) for a production area showing the operating status of all machines in the area (ANDON CENTRAL), 56.3% of the sample assessed this method as being very efficient and only two respondents as being inefficient (M=1.77, SD=1.064). Table 3 and Figure 5 offer a preview of all frequency distributions for the nine visualization methods.

Table 3. Visualization methods ranking

		CZELPA	CZDIS	CZINPA	ZPAHO	ZWHBO	PPLELS	PPLBILB	PPLSTA
N	Valid	87	87	87	87	87	87	87	87
	Missing	0	0	0	0	0	0	0	0
Mean (Ti)		1.77	2.66	2.82	2.38	1.76	1.60	1.80	1.71
Std. Error of Mean		.114	.163	.161	.140	.116	.109	.118	.113
Median		1.00	2.00	3.00	2.00	1.00	1.00	1.00	1.00
Mode		1	1	1	1	1	1	1	1
Std. Deviation		1.064	1.516	1.498	1.305	1.078	1.017	1.098	1.056
Variance		1.133	2.298	2.245	1.703	1.162	1.034	1.206	1.114

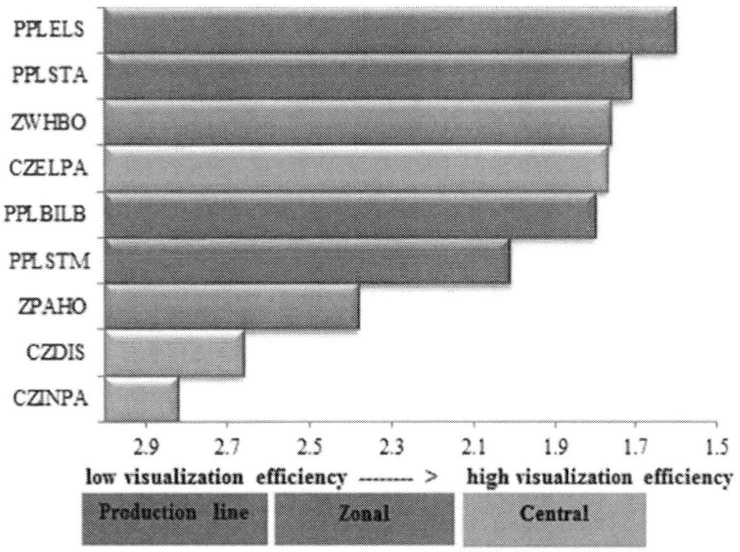

Fig. 5. Range of visualization methods

5 Case study results

Considering the efficiency of the visualization methods, the following arts of visualization are the most efficient: for the production line: PPLELS, PPLSTA, PPLBILB, and PPLSTM and for larger production areas ZWHBO and CZELPA. Illustrations of these visualizations are presented below in Figure 6.

According to the results of survey regarding the visualization methods the ones which should be avoided, are CZDIS, CZINPA, and ZPAHO.

Effectiveness analyse for implementation of these visualization methods was performed in a big automotive company (over 2500 employees) in Romania using a questionnaire. The analyse was performed about one year after implementation of shop floor management which included the recommended visualisation methods. The following results have been recorded:

- Regarding employees perception, the respondents were asked to indicate whether the implemented visualization methods have a higher, lower or the same utility as other visualization methods used before introducing the lean shop floor management standards. Most of the respondents (68.2%) indicated that the new visualization methods have a much higher utility than older visualization methods used in the company. 22.4% indicated that there is no difference regarding utility and only less than 10% (9.4%) indicated that the new visualization methods have a lower utility than older ones.
- Regarding time management, due to implementation of lean communication at shop floor, improvements were found at the top management level, where a saving of 5.5-7.6 hours/ month was recorded, followed by savings consisting on an average of three hours a day for middle management [5].

 DOI: 10.1051/matecconf/201712102005

PPLELS/CZELPA: Production unit /Central/ zonal electronic panel showing the operating status in real time

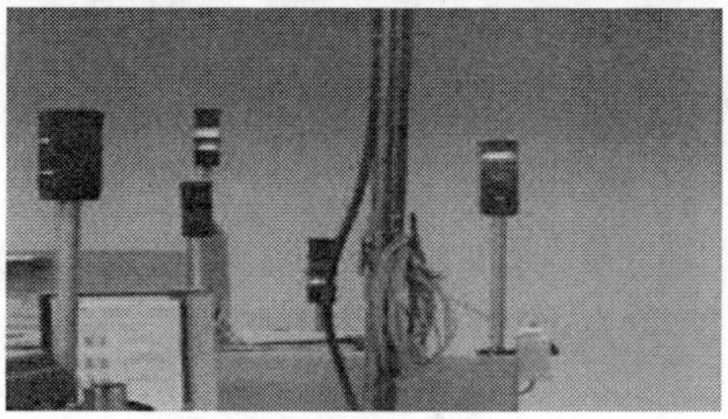

PPLSTA Stoplight indicator, operated automatically, indicating the operating status of production unit (line/ machinery)

PPLSTM: Stoplight indicator, operated manually, indicating the operating status of the production line (ANDON)

PPLBILB Board at the end of the line (machinery) based on interactive updates of the shop floor employees

ZWHBO Zonal: White board with interactive data updated daily by employees

Fig. 6. Types of visual communication at shop floor

6 Conclusions

The conclusions of the research can help manufacturing organizations select the most efficient art of visualization at shop floor according to the lean manufacturing philosophy, in order to avoid the waste created through inappropriate visual communication. Adopting one or several combined recommended solutions can lead to a better shop floor information flow with a minimum of resources consumption. The visualization of relevant information directly at shop floor should assure a high level of transparency for everybody.

The best ranked lean solutions for production units (machinery/line) are to apply real time visualization assured through electric boards at each production unit and /or stoplight indicators, operated automatically. At the same time for zonal/central visual management white boards with interactive data updated daily by employees are recommended to be standardized at shop floor. These boards should contain KPIs, which can be tracked with standardized work sheets in handwritten visualizations.

The implementation of this kind of interactive visualization lead to an intensely deal with the processes and problems from employees and managers. The information boards on shop floor serve on one hand as a communication tool throughout the employees and on the other hand to spread direct shop floor information to all management levels.

The implementation of most appropriate visualization methods has led to measurable improvements and savings.

Limitations: beside the LinkedIn area of research, the questionnaire regarding visual management was directly conducted in only nine Romanian organizations. Further research in this area should concentrate on iterating these analyses in other automotive organizations in Eastern Europe where this industry is younger and growing. The aim of this entire action is to design lean bases visualization in shop floor and transparent visual management system, so that the administration can recognize the need to act in a timely manner.

References

1. K. Zhang, J. Visual Languages & Computing, **23** (6), 340 (2012)
2. http://www.staufen.ag/fileadmin/Brochures/brochure-staufen-shopfloor-management-en.pdf, Accesed 18.03.2017
3. M. Imai, *Gemba Kaizen: A Commonsense, Low-Cost Approach to Management* (McGraw Hill Professional, 1997)
4. A. Field, *Discovering statistics using IBM SPSS statistics* (Sage 2013)
5. M.V. Iuga, C.V. Kifor, I. Bondrea, Proceedings of 18th QMOD-ICQSS Conference, Seoul, 한국품질경영학회 추계학술발표논문집, 458 (2015)

Value Stream Mapping as a tool for the optimization of production – case study

Jolanta B. Krolczyk[1,*], *Stanislaw* Legutko[2], and *Anna* Szczepańska[2]

[1]Opole University of Technology, Faculty of Mechanical Engineering, 76 Proszkowska Street, 45-758 Opole, Poland
[2]Poznan University of Technology, Faculty of Mechanical Engineering and Management, 3 Piotrowo Street, 60-965 Poznan, Poland

Abstract. The aim of the paper is to present the possibility of introducing improvements in processes concerning production logistics basing on the use of value stream mapping. An analysis of the current situation in order to identify waste and problems in the stream flow has been carried out. Subsequently, a future shape map has been drafted with a proposal concerning the manner in which future state ought to be achieved. The thesis is concluded with a general summary as well as conclusions regarding the studied subject and the proposals for further work.

1 Introduction

Material flow and Value Stream Mapping (VSM) has been recognized for its effectiveness in improving productivity [1], product quality [2-4] and on-time delivery. Many manufacturers try to collect information connected with product quality such as defect rates [5-7] or error rates [8]. VSM is a method which is applied to the analysis of the production system. It is also a tool which gives the opportunities to the improvement of the material and information flow processes' consistency in the Lean Manufacturing. This tool helps to understand the flow of information and materials during the route which is spend by the product in the stream of values. One should trace the production route of the product from the customer to the supplier, with the accurate graphical representation of the each of processes participating in the material and information flow. The mapping can be successfully transferred into the all divisions and areas, from the production sphere to administrative and office processes. The common feature of the all business processes is a time, which elapses from the instant of initiation process to the finish. The waste in the business process manifests always in a form of an excessively long time. The time reduction is a primary objective of the advancement, according to the Lean Management, independently on the discussed area, e.g. production, administration, service or health service. Mapping allows to clearly illustrate the ineffectiveness of action, measures the

* Corresponding author: j.krolczyk@po.opole.pl

waste level and supports to design the appropriate changes, independently on economic activity's character [9].

The aim of the paper is to present the possibility of introducing improvements in processes concerning production logistics of aircraft parts basing on the use of value stream mapping.

2 Stages of Value Stream Mapping

Value stream mapping allows the identification of activities which constitute the waste. The wastes in the production can be found in the easiest way after the total presentation of process flow in the graphical form and indication of material and information flow in the process. After the map development, it is easy to identify the locations in which the materials, semi-finished products or final products are halted in the particular place for the longer time, instead of flowing to the customer [10]. In the value stream mapping, the specified graphical symbols are being applied, which enables the clear and transparent process presentation, as well as the identification of places intended to the advancement.

Value stream mapping consists of a few following stages [11]:

1) The selection of products' family. The family of products can be defined as products, which pass through the same working places and technological operations, however it is unnecessary to pass any type of product through the all working places.
2) The selection of mapping stream manager. His task is maintenance of the value stream flow and reaction to the appearing problems. Manager should have responsibilities and rights for making decisions correlated with the work of the all organizational cells affected by the flow of the particular products' stream.
3) Acquiring the information about the process. In order to create the map of the current state, one should cross the route, which is physically passed by the products „from door to door", measure how many time they spend on the particular working places, where and how long they await for the next operation and how they get to the client. The following information are primarily gathered: suppliers' requirements regarding the ordered batch size, the type of products or the method of supply; requirements of our company concerning the deadline of orders by customers; requirements of our suppliers to order materials for the production of the analyzed product family; types of manufacturing operations; durations of operations; changeover times; availability of jobs; shift-work organization; the number of operators of machines and equipment used in the process; way of production planning and allocation of production tasks; planning shipments to the customer, inventories of materials, products in the production and finished products; ways of communication process between the customer and our organization, the organization and the supplier, the division of production planning and production departments.
4) Development of maps of the current state. Graphical presentation of the activities using specific symbols.
5) Analysis of maps: the search for losses, identification of problems, proposing solutions.
6) Development of a future state map. Graphical presentation of proposals for the elimination of waste.
7) Development of a plan for implementation of the future state. Proposing changes according to a fixed plan with specific objectives, tasks, deadlines and responsibilities.
8) Implementation.
9) Evaluation of the results.

3 The specification of the selected production process

The analyzed manufacturing company employs 1200 employees who work in 10 locations around the world. Branch in Poland, specializes in precision machining of difficult-to-cut components of aircraft engines, and currently employs approx. 130 employees. The company is working on 3 shifts. Main products are turbine steering parts - for the aerospace and turbine blades - for the energy industry. Typical types of manufactured parts are shown in Fig. 1.

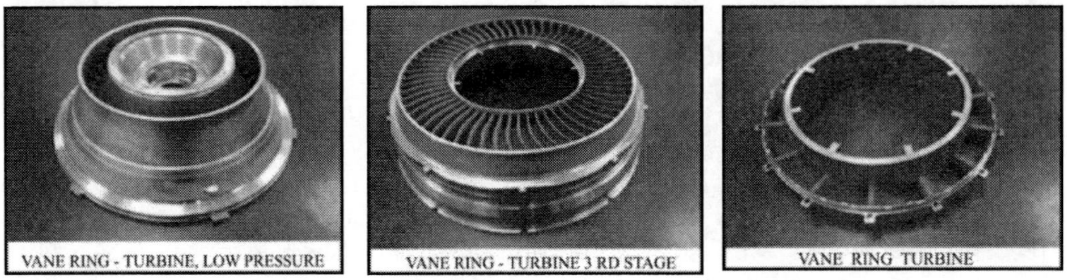

Fig. 1. Typical types of manufactured parts.

The selected optimization process is the production process of the "Turbine Control Apparatus" (Fig. 1 middle). Parts of this type are included in the aircraft engine. This part is responsible for directing the combustion exhaust gases. These are the non-rotational, complex-shaped components in the engine. The starting material for producing these parts is the precise mold made of aircraft material with high nickel content. According to the literature [12-14], the machining of parts with complex shapes requires the selection of the appropriate cutting conditions enabling the reduction of forces and vibrations occurring during the machining process. Structurally this part consists of two outer and inner rings which are connected by vanes. This part is being manufactured in different configurations. The company produces 32 types of this part. Each part is produced in a different monthly amount, according to customer demand for a specific type and class. The annual production is in the range of 9 000 parts. Due to the similarity of all parts, the technology can be divided into families and thus it has been formed:

- Family I - parts machined roughly and in finishing conditions with heat treatment.
- Family II - parts machined roughly and in finishing conditions with hard brazing of honeycomb type rings.
- Family III - parts machined roughly and in finishing conditions with a plasma protective coating imposed on selected surfaces.
- Family IV - parts machined rough and in finishing conditions and additional assembly (riveting of lids on the inner ring).

All mechanical processes and assembly (riveting) are performed on the spot in the company. Special processes such as heat treating, brazing and plasma coating of protective layers are performed in cooperation approved by the customer source. In an article the optimization of the selected part of the family I is presented. The Family I consist of 6 part numbers and represents approximately 25% of the company's production. Because of the very great technological similarity of the parts from the family I, it was assumed that the parts can be described in one standard technological process (Fig. 2), which is shown below:

✓ Turning – preparation the bases of machining,
✓ Turning II - roughing,
✓ Turning III - roughing,
✓ Grinding I - roughing,
✓ Bench work roughing,

- ✓ Cooperation - heat treatment,
- ✓ Grinding II – improvement of the bases after heat treatment,
- ✓ Turning IV - finishing,
- ✓ Grinding III - finishing,
- ✓ Grinding IV - finishing,
- ✓ EDM sink, WEDM, milling - finishing,
- ✓ Bench work finishing,
- ✓ Control FPI, flow control, control CMM.

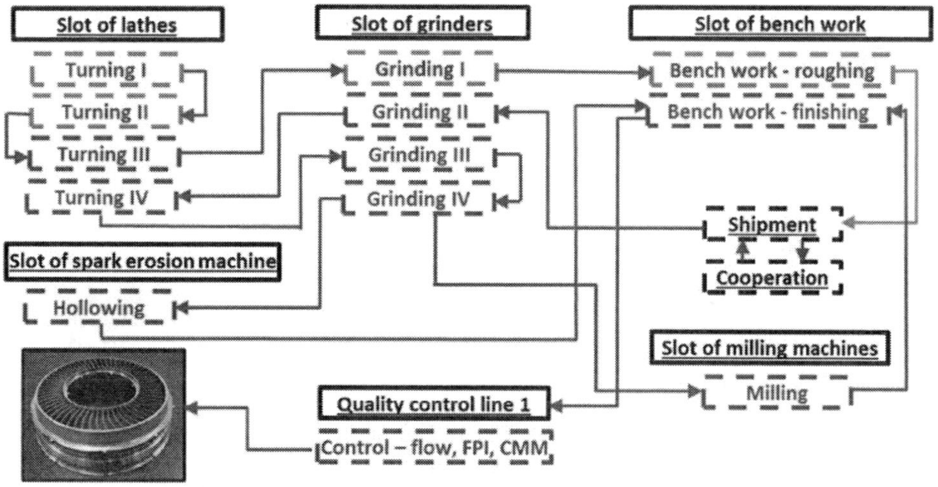

Fig. 2. Graphical representation of the technological process for the Family I.

Due to the similarity of parts from the family I, one can adopt the following assumption for the analysis needs, and treat the part of the family as one part representing 25% of total production in quantities of 2356 units/year which gives 196 units/month and 50 units a week. The production process ends with the transfer of part to a warehouse for finished products. Parts are sent to the customer normally once a week on Tuesday. In exceptional cases, parts can be shipped on Friday as additional shipping. In the analysis of the current situation, only process of the part flow through the production unit was taken into account. It did not take into account the technological process, i.e. the technical details of manufacturing. The analysis starts with determining: what is the cycle time for the selected part of the family. Calculation of the cycle time begins from the determination of the time available per one change in the production company which is 465 minutes (8 hours decreased by the breakfast break – that is 15 minutes). The standard unit of time to determine the tact time is a second. However in this work, due to the long cycle times and long operation times of changeovers, the unit of a minute has been applied. Then, one should divide the obtained value by the customer's requirements, which are equaled to 3 parts, based on the change in production. Subsequently, the value of tact time of 155 minutes is obtained. The resulting cycle time means that to meet customer demands within the available time, the company should exempt the customer "Turbine Control Apparatus" every 155 minutes. The resulting value does not include the time of machine failures, the times required for retooling and lost time for the production of poor quality products, popularly called deficiencies. These wastes are things that a company must deal with, because they reduce the time available for production. Another element of map for the analysis is the technological process, used resources, work stations (machines) and their arrangement on the manufacturing floor. Browsing the technological route, it can be seen that the process is divided into three main stages:

✓ rough treatment - includes: turning I, II, turning, turning III, IV turning, grinding and rough bench works,
✓ The heat treatment - carried out in cooperation,
✓ Finishing treatment - includes: grinding II, V, turning, grinding III, V, grinding, milling, bench work in finishing conditions and control.

In contrast, the current layout for machines on the manufacturing system is a nest. In this system, we have such a slot:

✓ Lathes – 9 fully convertible machines in a 3 slot system per 3 machines,
✓ Grinders - 9 machines in a 1 x 3 machines plus 3 slot per 2 machines,
✓ EDMs - 6 machines in a 3 slot system per 2 machines,
✓ Milling machines - 2 machines in a 1 slot system, 2 machines,
✓ Bench work – 3 stands in one slot.

Fig. 3 shows the map of the current state of the production process.

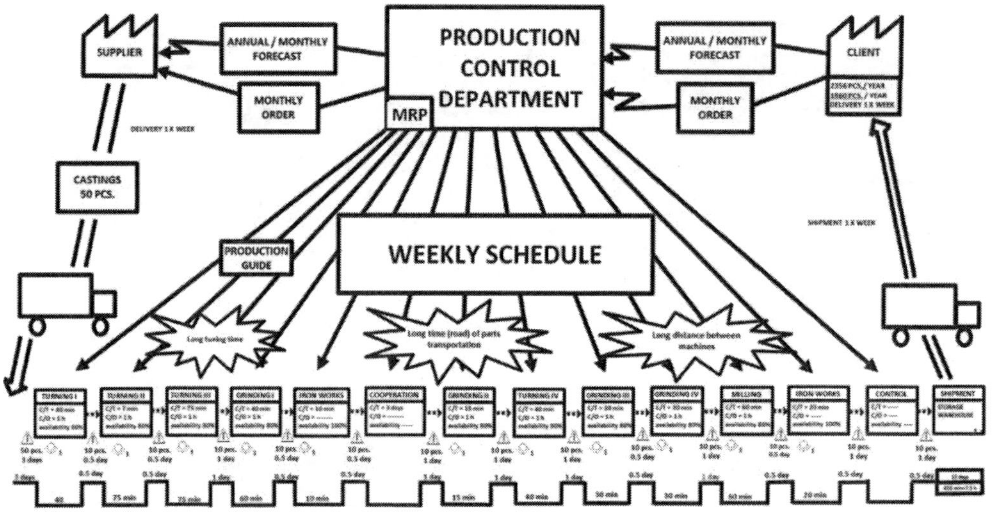

Fig. 3. The map of the current state

The analysis of the current situation maps and the collected information allowed the identification of value streams, i.e. all activities, both adding and not adding the values. Summing up the analysis of current state, the following elements have been specified to improve:

✓ Reduction of failure and increase the availability of machines,
✓ Shorter setup times,
✓ Reorganization of the machine settings,
✓ The establishment of the zone dedicated to roughing and finishing,
✓ Introduction of the FIFO method for all machines where possible.

4 Optimization of production – the development of the future state map

All parts manufactured in the company are grouped in families according to the similarity technology, which also shows that for every part the division to the aforementioned stages (roughing, finishing and cooperation) occurs. Therefore, a certain conclusion can be formulated. In order to improve the flow of parts through the production department, one should distinguish the **two production zones,** namely: roughing and finishing (Fig. 4).

Between them an additional area for shipping parts to the cooperation can be created. According to this assumption, the time associated with the transport of parts between operations should be reduced, which at large distances between the machines and high frequency, should give tangible effect of shortening the time wasted on these activities.

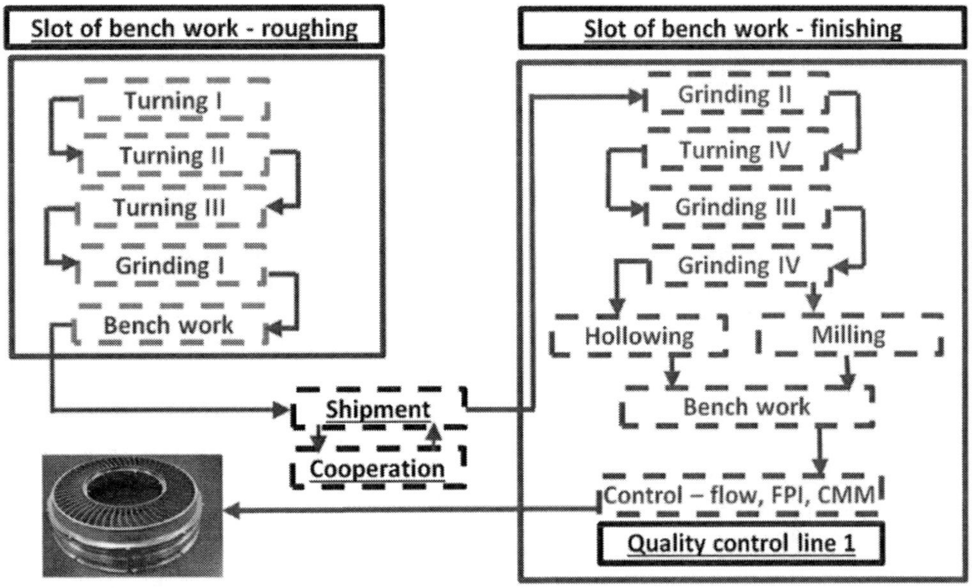

Fig. 4. Future state – graphical presentation of the technological process for the Family 1

The reduction of machine downtime and high availability of machine time can be achieved through a well-executed assumptions of TPM (Total Productive Maintenance). TPM goals are realized in two areas: human and machine. In the first area, the task of TPM is to increase the efficiency of employees by expanding their skills and knowledge - which is synonymous with an increase in their responsibilities. From the perspective of machine, operation workers should focus on the maintenance of machinery and equipment in a state of high availability in such a way that the maintenance department will receive from the operators the information regarding the state of the whole machine park, in order to plan actions. Maintenance services move from reaction to the predictive maintenance of machines (analyzing the occurring defects to predict what might happen in the future), thus increasing machine availability, and reliability. To shorten setup times, the SMED methodology is applicable [15]. According to this methodology, the information regarding retooling for operations on the parts from the family I, as well as for roughing and finishing have been collected. Tooling hire place is located at a distance from the particular machines therefore the transport of holders takes the longest time. It can be assumed that it is on average about 60% of the total changeover time. It will not be possible to eliminate this time completely, but it is possible to significantly shorten its value. The solution proposal is to put the necessary handles on the shelves within the close distance to the machine tools in the field of crane located above them. Optimization per month, only for a part from the family I, gives 81 extra hours to produce. One of the general principles of lean manufacturing is the introduction of continuous flow wherever possible. Continuous flow is a term referring to the production of one part transmitted between neighboring operations without downtime and waiting and thus without the generation of other types of waste. In the presented case, the introduction of continuous flow is not possible because the heat treatment is carried out at the external provider. The solution in this case may be the use of so-called FIFO queue, i.e. the type of queue: "first in, first out". The establishment of such a

FIFO queue in front of the machine can order the parts and operations in terms of the sequence of their execution and do not require the creation of a daily or shift production schedule for individual machines. The actions implementing the FIFO principle will require the least procedures. These will include the selection of places by the machines on a newly designed machines' setting plan. Already at the design stage of a new set of machines, the introduction of a FIFO queue should be taken into account. According to the future state map (Fig. 5) FIFO should be included on all machines intended for roughing and finishing. Queues should be marked on the floor as squares forming a trolley's projection with a suitable FIFO designation.

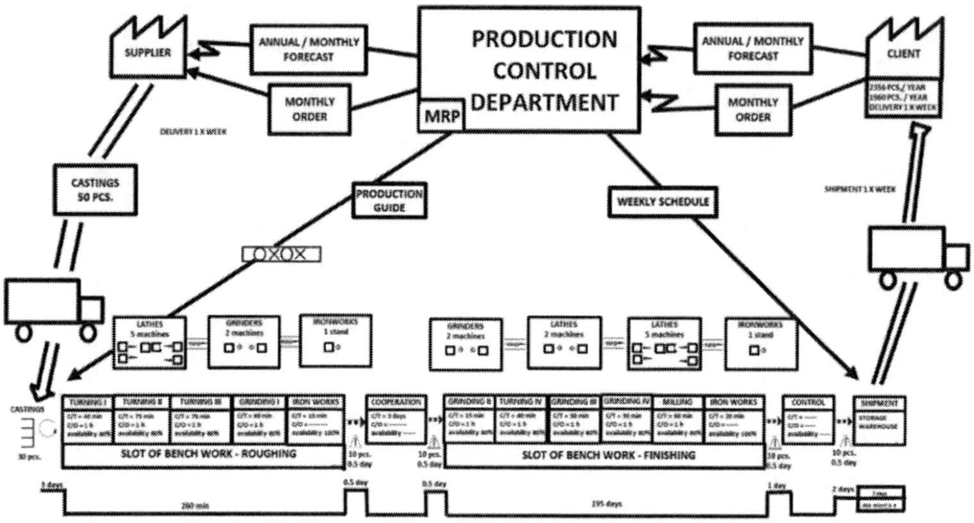

Fig. 5. Future State Map

5 Conclusions

1. Value stream mapping is a method by which one can get closer to achieving continuous flow from the beginning to the end of the stream values in the company. The consequence of actions and their implementation in order to achieve a future state becomes a very important element. Multiple mapping of the same value stream can discover another source of waste to eliminate. Therefore, a continuous improvement of the stream value that can be successfully accomplished by utilizing the method of Value Stream Mapping and Lean Manufacturing tools is essential.
2. Value Stream Mapping has helped to achieve this goal by limiting the production cycle from 12 days to 7 days. It should be noted that 5 days should be add to this time in order to complete the co-operation (which is not shown on the map, and is governed by a trade agreement between the company and cooperator).
3. The elimination of the identified waste is proposed to implement with the available tools of Lean Manufacturing. The main improvements are: shorter setup times, reducing downtime and increasing machine availability, reorganization of the settings of the machine, creating a zone for roughing and finishing, the introduction of the FIFO principle for all machines.
4. Achieving a future state for the selected group of the parts from the family 1, according to the Map of Future State should not complete the improvements throughout the whole stream value. Further work should focus on the execution of the same activities for families

in the field of II - IV. During the improvement of machines' settings for the Family I and the organization of roughing and finishing sockets, all parts manufactured in the company were taken into account. The calculations for the amount of the necessary machines in the slots were made for the entire production.

5. The result of each subsequent mapping should include the saved activities to be performed in order to achieve the future state. According to the lean philosophy, one should not try to solve all the problems at one time. The quest for the perfect flow by making small steps can give a more tangible effect than a creation of a comprehensive schedule that will be difficult to implement.

References

1. J.B. Krolczyk, G.M. Krolczyk, S. Legutko, J. Napiorkowski, S. Hloch, J. Foltys, E. Tama, Teh. Vjesn., **22** (6), 1447-1456 (2015)
2. S. Wojciechowski, P. Twardowski, M. Pelic, P.W. Maruda, S. Barrans, S., G.M. Krolczyk, Precis. Eng., **46**, 158-165 (2016)
3. D. Lehocka, J. Klich, J. Foldyna, S. Hloch, J.B. Krolczyk, J. Carach, G.M. Krolczyk, Measurement, **82**, 375–383 (2016)
4. A. Glowacz, Meas. Sci. Rev., **15**, 167–75 (2015)
5. C. Zhang, Z. Li, C. Hu, S. Chen, J. Wang, X. Zhang, Meas. Sci. Technol., **28**, 35102 (2017)
6. J.B. Krolczyk, B. Gapinski, G.M. Krolczyk, I. Samardzic, R.W. Maruda, K. Soucek, S. Legutko, P. Nieslony, Y. Javadi, L. Stas, Teh. Vjesn., **23** (1), 301-306 (2016)
7. P. Hreha, A. Radvanska, L. Knapcikova, G. M. Królczyk, S. Legutko, J.B. Królczyk, S. Hloch, P. Monka, Metrol. Meas. Syst., **22** (2), 315-326 (2015)
8. J.V. Saraph, P.G. Benson, R.G. Schroeder. Decision sciences, **20** (4), 810-829 (1989)
9. M. Rother, J. Shook, *Naucz się widzieć* (LEI Polska, 2009)
10. M. Rother, R. Harris, *Tworzenie ciągłego przepływu* (Wrocławskie Centrum Transferu Technologii Politechniki Wrocławskiej, Wrocław, 2001)
11. M. Rother, J. Shook, *Naucz się widzieć. Eliminacja marnotrawstwa przez Mapowanie Strumienia Wartości* (Wrocławskie Centrum Transferu Technologii Politechniki Wrocławskiej, Wrocław, 2002)
12. S. Wojciechowski, Int. J. Mach. Tool. Manu., **89**, 110-123 (2015)
13. S. Wojciechowski, P. Twardowski, M. Pelic, *Procedia CIRP*, **14**, 113-118 (2014)
14. S. Wojciechowski, R.W. Maruda, P. Nieslony, G.M. Krolczyk, Int. J. Mech. Sci., **119**, 360-369 (2016)
15. T. Kafel, M. Walczak, *Koncepcje zarządzania systemami wytwórczymi* (ed.: M. Fertsch, S. Trzcieliński, Wydawnictwo Politechniki Poznańskiej, Poznań, 2005)

The modernization of the production system of the selected type of synchronizer ring

Jolanta B. Krolczyk[1,*], *Stanislaw* Legutko[2], and *Piotr* Bialek[2]

[1]Opole University of Technology, Faculty of Mechanical Engineering, 76 Proszkowska Street, 45-758 Opole, Poland
[2]Poznan University of Technology, Faculty of Mechanical Engineering and Management, 3 Piotrowo Street, 60-965 Poznan, Poland

Abstract. The aim of the work focuses on the modernization concept of the production system of the selected type of synchronizer ring. The synchronizer rings are applied in the gearboxes of the vehicles in order to synchronize the gears. The range of this work includes the analysis of the previous production system and identification of its specified elements affecting the limitations of productivity or additional costs. As a result of analysis, the set of innovative solutions, enabling the better exploitation of company resources, have been presented.

1 Introduction

Contemporaries manufacturers are making efforts to obtain or maintain their position in a competitive environment [1]. More efficient production by reducing production errors, increases the total companies' productivity [2]. The main problem companies are too high cost of the final product caused by defects [3-4]. Several methods and techniques successfully have been used to reduce the production defects and optimization of product [5-9].

One of the primary objectives of a company is manufacturing of products with high quality, which can satisfy the consumer's needs. The quality of products is strongly affected by the selected technological input parameters, condition and properties of a machine, as well as the process dynamics [2, 10]. The modern company should produce in the short time and in the low level of costs. Fulfilling all of the above mentioned conditions allows the competetiveness of the company. Nowadays, selling the products with high quality is insufficient. The introduction of the product to the market in the shorter time than the competitors is thus essential.

The application of Cax computer techniques enables the partial automation and robotization of manufacturing processes in order to improve the productivity, stablility, and the elimination of shortages, as well as safety improvement of the employees. Robots and manipulators substitute the human in the realization of processes which require the large physical effort or include the hazard, like in example: welding, casting, assembly [11]. The application of Cax techniques is in favor of popular JIT (Just-in-time) strategy,

* Corresponding author: j.krolczyk@po.opole.pl

implemented in the companies [12-14]. The selection of the appropriate managament strategy is a difficult task and requires comprehensive analysis of the actual company's management system.

The aim of the work focuses on the modernization concept of the production system of the selected type of synchronizer ring. The synchronizer rings are applied in the gearboxes of the vehicles in order to synchronize the gears. The range of this work includes the analysis of the previous production system and identification of its specified elements affecting the limitations of productivity or additional costs. As a result of analysis, the set of innovative solutions, enabling the better exploitation of company resources, have been presented.

2 The description of the technological process

The analyzed manufacturing process of the synchronizer ring has been carried out in the company, which is specialized in the production of forged elements and forgings made of non-ferrous alloys and steel. Above 40% of products is intended to the export to the European countries and USA. The company is medium, according to the size. The manufactured forgings are formed by the hot drop forging with the application of modern apparatus and tools. The mass of the produced forgings (after the removal of fash) is in the range of 0,025 to 3,0 kg and the maximal diameter is of 220 mm.

The synchronizer's ring is the axis-symmetrical product with the length to diameter ratio $L/D \leq 0,5$. Therefore, it can be classified as bush or disc. In this kind of parts, the external and internal cylindrical surfaces are being machined [15]. The important issue is also the limitation of roundness errors occurring during the manufacturing process. The manufactured rings are formed as forgings in the range of diameters from 90 mm to 250 mm. The rings made of preforms are forged in the specially formed tools. Forgings are subjected to the heat treatment and preliminary machining without the finishing. Nowadays, the company manufactures the 100 types of rings to the following companies: Mercedes Benz, Scania, MAN and IVECO. The exemplary synchronizer rings are depicted in Fig. 1. The synchronizer ring is made of 16MnCr5 steel, intended to carbonization, which is characterized by the high ductility and hardness of the outer surface.

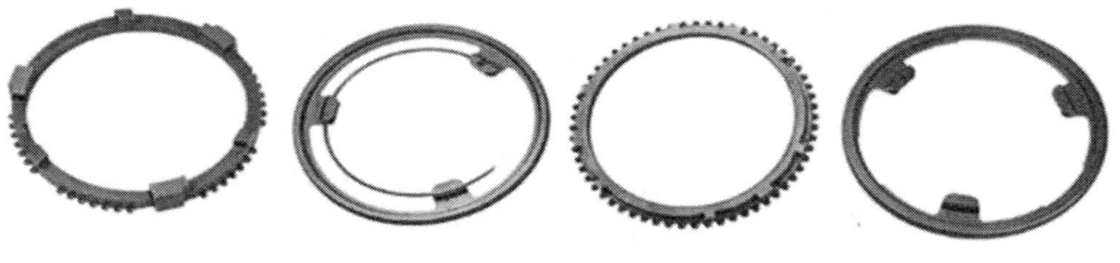

Fig. 1. The exemplary synchronizer's rings of the gearbox.

The blank is a preform made of steel supplied by the outer supplier. The first operation involves the preform turning on the semi-automatic turret lathe. During this operation, the following surfaces are being cut: face, outer diameter and chamfer. One in ten of semi-finished products is being measured by the calliper. Figs. 2 and 3 present the preforms, namely: before and after the turning. The semi-finished product is then drop forged on the LKM 1000 press. The preform is being heated in the heater for the 75 seconds, and then is formed in the die under the press' impact (Fig. 4). The dimensional control with the calliper is carried out for the one in ten of semi-finished products.

The next operation in the ring's manufacturing process is the fash removal, which is created during drop forging. This operation is conducted on the stand equipped with CNC

TZC-32 lathe. After the fash removal, the rings are transported to the OWT – 400 rotor-tape cleaner. This process is carried out with the cleaning agent's stream.

The rings after the cleaning are normalized. This process is applied in order to obtain the homogenous fine-grained structure, the improvement of product's mechanical properties and unify the structure and properties of the steel products in the mass production. Heat refining lasts for the 6 hours and 20 minutes. The rings are being heated for the 3 hours to the 910° C then kept in the constant temperature for a 1 hour and finally cooled down for the 2 hours and 20 minutes to the room temperature.

Fig. 2. Preform before turning. **Fig. 3.** Preform after turning. **Fig. 4.** Forging.

The next operation is ring calibration on the calibrating press PH-750. This process creates the oval shape of the ring and forms the external teeth, which are not being machined in the subsequent production process stages. This operation is necessary, because of customer's requirements regarding the shape errors of product. The control measurement of teeth with the use of calliper is carried out by the employee. The calibrated rings are further subjected to turning on the TZC-32 stand. During this operation, the teeth external diameter and face are machined in the finishing conditions. However, the internal diameter is bored in roughing conditions. The additional treatment during this operation is the burr removal from the teeth external diameter by the brush. Figure 5 presents the ring before and after the turning.

Fig. 5. The ring before machining (left) and after machining (right).

Because of the longest machining time of singular product and a large number of dimensions to control, this operation is particularly important during the whole manufacturing process. Therefore, the work on this stand requires high experience enabling the fulfilling of production norm. The roundness control, carried out on special device, includes the each of manufactured parts. Moreover, the employee controls the one in ten of products, in accordance to specially designed instruction.

The next operation is finish turning of chamfer and conic on the WEMAS DZ 270 stand. Because of the short time of the machining, which amounts to about 12 seconds, this is the one of the more difficult processes for the operator. The ring is clamped in the specially designed chuck enabling the precise machining. The construction of the chuck allows the stable clamping of the element, without the influence of the adverse circumferential forces causing the deflection. The operator during the boring process controls the roundness of the each ring on the specially designed measurement device. The one in ten of products is being checked according to the special instruction – similarly to the previous operation.

The machined rings are transported to the stand on which the maintenance (by the drowning in the fluid) is carried out. The parched rings are packed up and transported to the warehouse. The quality control is carried out on the specialized CMM machine (Fig. 6) and profile meter (Fig. 7).

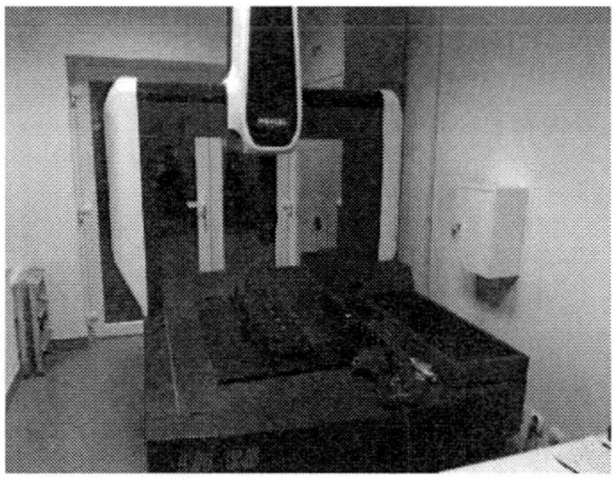

Fig. 6. Coordinate measurement machine WENZEL.

Fig. 7. Profile meter.

3 Modernization concepts

3.1 The identification of system's elements limiting the productivity and generating costs

Modernization of production system is based on its sustainable improvement by the upgrade. It can involve the changes regarding the applied technology or the work organization. The primary objective of modernization is the improvement of productivity and products' quality.

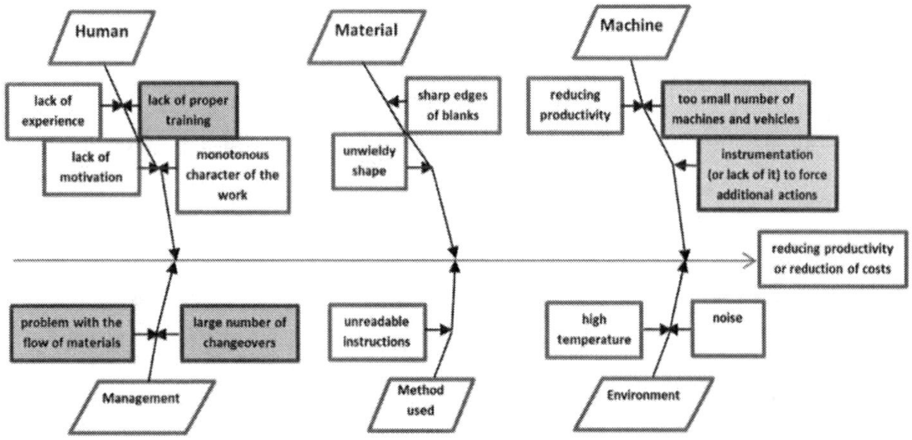

Fig. 8. Ishikawa diagram depicting significant factors limiting the productivity.

An example of the tool enabling the identification of problems occurring in the specified system is the Ishikawa diagram. The Ishikawa diagram depicting significant factors limiting the productivity of the analyzed synchronizer ring production system is presented in Fig. 8.

3.2 Modernization concepts

The first modernization proposition is related to the upgrade of LZK 1000T press' induction heater. Preform is placed in the induction heater in order to prepare it to the forging. The heating process lasts for 75 seconds.

The stand is equipped with two heaters, which allows the simultaneous preparation of two preforms, while forging the third one. The solution which enables the process time shortening and therefore the productivity improvement is the exchange of the heater and application of the newer one with the shorter heating time. The application of the specialized inductor's construction which simultaneously heats the external and internal preform diameter would allow the reduction of a heating time by a 50% in comparison to the actual method. The another solution can be the purchase of an additional heater with a higher power.

The second proposal is upgrading the equipment of lathe TZC-32 with chip conveyors. The lathe is used to remove burrs and roughing processes. Operator working on this machine has to periodically remove chips into a special container. This process takes about one minute. It is repeated twice within an hour on the burr removal stand - but during roughing operations - once an hour. Typically, an operator working in a particular slot controls both machines at the same time. The use of chips cartridgebox can save the time associated with chip removal on a scale of 16 -30 minutes per shift.

Another proposed amendment concerns additional holes in the WEMAS DZ 270 lathe's chuck. During the machining of the ring, the phenomenon of accumulating the coolant and chips inside the chuck and the insert's grooves has been noticed. The effect of this phenomenon was a problem with the preparation of the machine for the cutting of the each subsequent part. Inaccurate cleaning of the chuck caused the deflection of ring's axis toward the tool's axis and as a result the generation of a lack resulting from exceeding the permissible value of a roundness error and burr formation.

The solution to the above problem is to drill three holes in the chuck for removing the cooling fluid by the action of centrifugal force during machining of the ring. It would allow the removal of chips using compressed air, which significantly shortens the preparation time of machine. Processing of the product on the WEMAS stand takes about 12 seconds, while preparing the machine takes about 35 seconds and includes cleansing the chuck and clamping the work piece.

This means that the machine capabilities are used only in about 25%. Based on the analysis of a preparation time of a similar ring with the application of cleaning with the compressed air, it was estimated that the implementation of the above solution would shorten the cycle up to 40%.

During the analysis of the production system a problem has been noticed with the flow of resources between stands. During the each shift, the production department is supported by only one forklift. The operator of the trolley supports a variety of processes which affects the production delays. The use of an additional vehicle would avoid the losses resulting from delays in the delivery of the necessary semi-finished products for all stands of the production department.

The need for an additional training of employees to control and reduce the number of changeovers has been also observed. An important issue concerning the permanent improvement of the production system is to minimize the number of defects generated by the company.

The analysis conducted by the Quality Control Department shows that the most common deficiencies are resulting from the machining errors. They can be the result of the inappropriate selection of technological parameters or improper inspection by an operator at a given stand. The solution might be to carry out more detailed training in the use of measurement tools.

Another type of failure are the defects and cracks in the forging process. They arise particularly when starting a new order, when the optimal process parameters are not identified. The solution to the problems associated with forging and adjustment of machine tools is to reduce the number of changeovers. Where possible, the production should be organized in such a way that the batch of products was as long as possible. The above concepts of modernization are aimed at increasing the efficiency of the production system and minimizing the number of deficiencies. Due to the lack of detailed data on the number of units produced and the number of deficiencies, as well as individual times in operations, the authors have estimated effects on the basis of research conducted in the company.

The heater used currently in the process of forging heats the preform at the internal diameter side. The innovative solution of simultaneous heating of the inner and outer parts of the preform would shorten the process by the 50%.

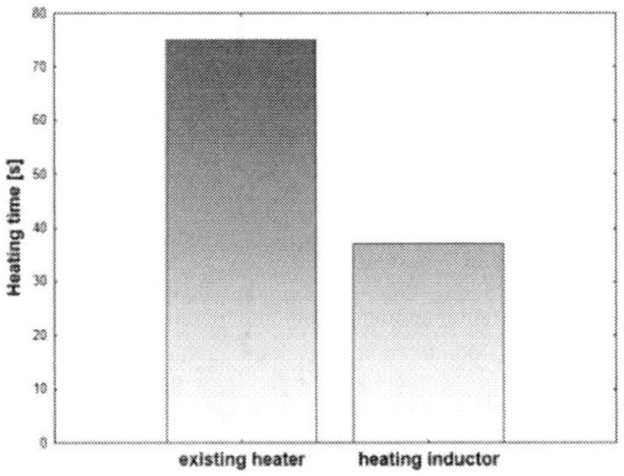

Fig. 9. Shortening heating time.

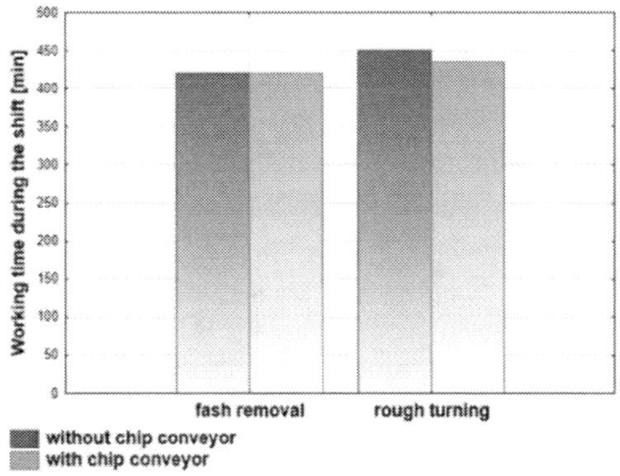

Fig. 10. Estimated extension of time in positions removal of fash and rough turning.

Figure 9 presents the estimated benefits of using this solution.

The existing equipment of TZC – 32 lathe's stand forces the employee to chip removal from the machine, which induces the downtimes of the machine ranging from 16 to 30 minutes per shift. The use of chip conveyor enables to save this time, thereby increasing the efficiency of the stand. The estimated benefits are shown in Figure 10.

The current use of the WEMAS DZ 270 machine's capabilities is approx. 25%. Implementation of the solution described above, involving the execution of additional holes in the chuck of WEMAS DZ 270 lathe would shorten the time to prepare the chuck and increase the efficiency of stands by almost 40% (Fig. 11).

DOI: 10.1051/matecconf/201712102007

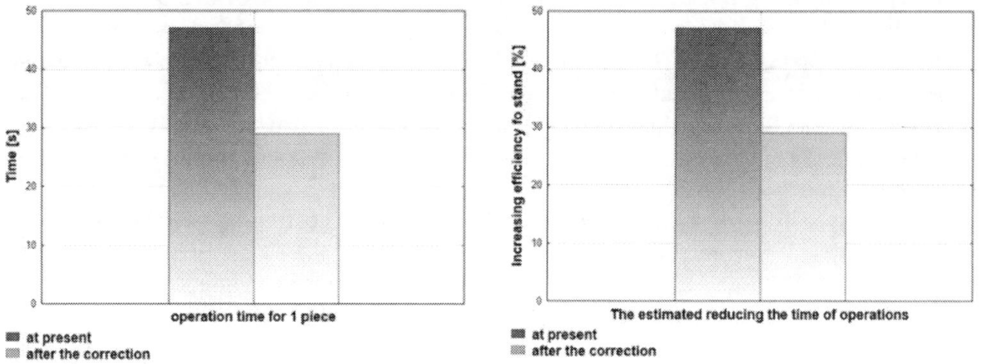

Fig. 11. The estimated benefits of using the modernized design of the handle.

4 Conclusions

1. Efficiency of the LZK 1000T stand, where the forging operation of preforms is performed, is limited by the generator of heaters. Because of the possibility of failure, the heating time is prolonged. The solution is to buy new heaters, which allows faster preparation of preforms for processing.
2. From the observation of TZC-32 stands it can be concluded that the employee spent 16 to 30 minutes per shift on the process of removing chips from the machine. The use of chip conveyor in these machines would save time and increase the production quantity of rings.
3. On the basis of research conducted by the authors it can be concluded that the introduction of changes into the design of the WEMAS lathe's chuck would shorten the manufacturing time of the ring up to 40%. The removal of cooling fluid through the additional holes, during the machining of ring would enable faster chuck's cleaning and better exploitation of machine's capabilities.
4. Modernization of the production system may also include organizational issues. The authors proposed the execution of more detailed training in the use of measuring tools and reduce the number of changeovers. The purpose of these changes would be to reduce the number of defects and improving the quality of manufactured products.
5. An important issue that covers the entire production system of the enterprise, is the flow of material between the stands. The decision to purchase an additional vehicle to support the production department must be preceded by detailed studies on the viability of such an investment.
6. During the fash removal operation on the TZC-32 stand, the frequent cutting tool's cracking at the irregular intervals has been observed. This demonstrates the high instability of the cutting process. One should carry out a study to compare whether more is preferable to apply the high cutting parameters, or increase the tool life.

References

1. J.B. Krolczyk, G.M. Krolczyk, S. Legutko, J. Napiorkowski, S. Hloch, J. Foltys, E. Tama, Teh. Vjesn., **22**, (6), 1447 (2015)
2. S. Wojciechowski, P. Twardowski, M. Pelic, R.W. Maruda, S. Barrans, G. Krolczyk, Precis. Eng., **46**, 158 (2016)

3. J.B. Krolczyk, B. Gapinski, G.M. Krolczyk, I. Samardzic, R.W. Maruda, K. Soucek, S. Legutko, P. Nieslony, Y. Javadi, L. Stas, Teh. Vjesn., **23** (1), 301 (2016)
4. A. Glowacz, Meas. Sci. Rev., **15**, 167 (2015)
5. D. Lehocka, J. Klich, J. Foldyna, S. Hloch, J.B. Krolczyk, J. Carach, G.M. Krolczyk, Measurement, **82**, 375 (2016)
6. P. Hreha, A. Radvanska, L. Knapcikova, G. M. Królczyk, S. Legutko, J.B. Królczyk, S. Hloch, P. Monka, Metrol. Meas. Syst., **22** (2), 315 (2015)
7. G.M. Krolczyk, R.W. Maruda, P. Nieslony, M. Wieczorowski, Measurement, **94**, 464–70
8. J. B. Królczyk, Int. Agrophys., **30**, 193 (2016)
9. C. Zhang, Z. Li, C. Hu, S. Chen, J. Wang, X. Zhang, Meas. Sci. Technol., **28**, 35102 (2017).
10. S. Wojciechowski, P. Twardowski, M. Pelic, Procedia CIRP, **14**, 113 (2014)
11. J. Żurek, *Podstawy robotyzacji. Laboratorium* (Wydawnictwo Politechniki Poznańskiej, Poznań 2006).
12. N. T. Thomopoulos, *Elements of Manufacturing, Distribution and Logistics* (Springer International Publishing, 2016).
13. P. Hakala, *Direct Modeling in Global CAD Environment,Phd thesis Tampere University of Technology, Finland* (2015).
14. J. Liu, X. Liu, Z. Ni, H. Zhou, J. Intell. Manuf., 1 (2016).
15. M. Feld, *Podstawy projektowania procesów technologicznych typowych części maszyn* (Wydawnictwa Naukowo-Techniczne, 2007).

Application of IoT concept on predictive maintenance of industrial equipment

Radu Constantin Parpala[1,*], and *Robert* Iacob[1]

[1]University Politehnica of Bucharest, Machine and Manufacturing Systems Department, Bucharest, Romania – 060042

Abstract. The Internet of Things (IoT) concept describes an intelligent connectivity of smart devices using the internet network. Nowadays, companies try different approaches for predictive maintenance as a solution to reduce costs and the frequency of maintenance activities. The IoT platforms provide a good support for predictive maintenance as it can integrate information from different machines and manufacturing systems. The main drawback in integrating production system with IoT dedicated platforms is the communication framework, knowing that the main industrial communication protocols are incompatible with modern communication protocols implemented on IoT platforms. In this context, the present paper proposes a new and simple method for on-line monitoring and predictive maintenance of industrial equipment. This method has two features of connected manufacturing. One of these is process monitoring for constant quality assurance, the other one is condition monitoring in order to prevent unplanned downtimes. A case study is presented to demonstrate the feasibility of the proposed method.

1 Introduction

The Internet of Things was described as the intelligent connectivity of smart devices by which objects can sense one another and communicate, thus changing how, where and by whom decisions about our physical world are made [1]. At first, the IoT concept was implemented on some commercial products: refrigerators, washing machines, control devices for smart homes, allowing them to directly connect to the internet using mobile or private networks. As the concept evolved, other types of products where equipped with the necessary technology and today many companies are trying to implement the IoT concept into modern manufacturing systems. Collecting usage data from different production equipment using an IoT platform is the main step in building a general, cloud based, predictive maintenance system, thus simplifying the factory upkeep. The main drawback of using the IoT technology in manufacturing systems is the communication (connection) between standard industrial devices and the web platforms. To solve this issue, the current paper proposes a new method to translate industrial equipment language to IoT ready devices like Arduino or Raspberry-Pi microcontrollers.

*Corresponding author: radu.parpala@gmail.com

2 IoT platforms

In the early 2000s, before the widespread of the IoT concept, different techniques were deployed in industry for machinery diagnostics and prognostics. A well-known method is Condition-Based Maintenance (CBM) which consists of three main steps: data acquisition, data processing and maintenance decision-making. It is a maintenance program that recommends maintenance decisions based on the information collected through condition monitoring [2].

Knowing that a single product is no longer sufficient to cope with the dynamic market environment, more recently, firms are trying to provide the integrated offering of products and services. In this regard, a generic structure of product-service integrated roadmap was developed [3].

Nowadays, the manufacturing paradigm changes towards predictive manufacturing, the role of maintenance function within manufacturing needs to be refined as a value creation function for achieving more sustainable operations. In this context, new techniques for maintenance are proposed [4]. Knowing that fault prediction represents a key technology to ensure sustainable operations, currently different fault prediction systems, based on IoT, are created [5].

The purpose of IoT devices is to connect with other devices or applications, this task being achieved by IoT platforms. Generally, an IoT platform is used as front end or a dashboard to present and process machines data. For simple IoT applications, there are several commercial or open source platforms that are already configured and ready to use. Most of the IoT platforms come with programming libraries for common developer boards like Arduino, Linux based devices (Raspberry Pi) or ARM based devices. For complex industrial application [6-7], developers can install an IoT platform like Easy-IoT, Kaa, Microsoft Azure IoT suite, PTC ThingWorx etc. These platforms come with a server installation and SDK libraries for different OS platforms.

Interoperability, which is dependent upon communication protocols and standardization level, is the ability of different IT systems and software to communicate, exchange, and use data. It's key to crafting a successful and resilient solution, regardless of whether an organization builds its IoT system in-house or engages a vendor. The communication between devices and IoT platforms is based on web protocols like HTTP GET and HTTP POST or Java Script data-interchange format (JSON). Given that most of those communication protocols are not implemented on industrial devices, one rapid solution is to use Arduino as a translator between industrial machines and IoT platforms (Figure 1).

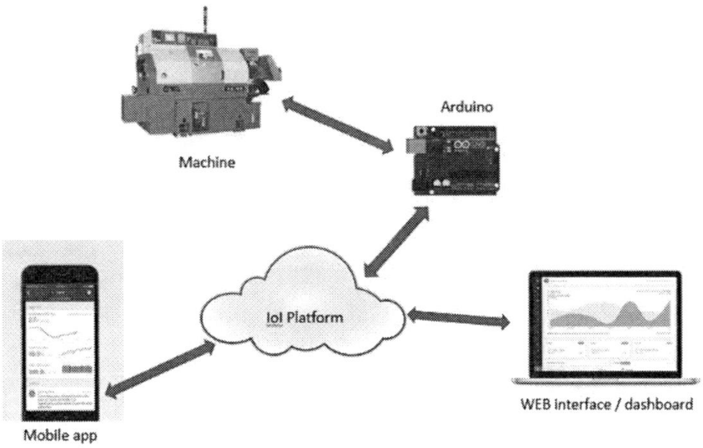

Fig. 1. Simplified scheme of an IoT platform that uses an Arduino board

Table 1. Serial communication standards

Serial Standard	Operation mode	Total nr. of devices	Cable length	Speed	Wires
RS-232	Single Ended	1 Sender / 1 Receiver	15 m	20 Kbits/s	min. 3
RS-422	Differential	1 Sender / 10 Receiver	1200 m	10 Mbit/s	4
RS-485	Differential	32 Sender / 32 Receiver	1200 m	10 Mbit/s	2

3 Industrial communication protocol

Industrial devices use several serial protocols to communicate one with another. A serial communication protocol can only send one bit at a time. This is slower than parallel communication standards, but it can be used over longer distances. When using serial communication, two aspects should be considered: the hardware specification and the software format of the data.

3.1 Hardware standards used in industrial communications

The main difference between different communication standards, as can be seen in the Table 1, consists in the number of devices that can communicate in the same time, the speed and the maximum cable length.

Because of specific hardware limitation on industrial machines, RS-232 standard is usually used to connect to a computer or to some types of HMI devices.

3.1.1 Standard RS-232

The RS-232 protocol is the most common serial interface mainly because almost every computer has at least one RS-232 serial port. The main disadvantages of this standard are: limited cable length, possibility of noise occurrence and limited number of devices that can be connected at the same time.

Usually, the RS-232 standard connects only two devices. Thus, in order to be able to capture usage information that is sent over the serial port for the IoT platform, a new device could be connected as an echo device – if it has two serial ports (Figure 2.a.), or it can be connected only to a send line of a master device (Figure 2.b.) – in this case it can only receive information from a single device.

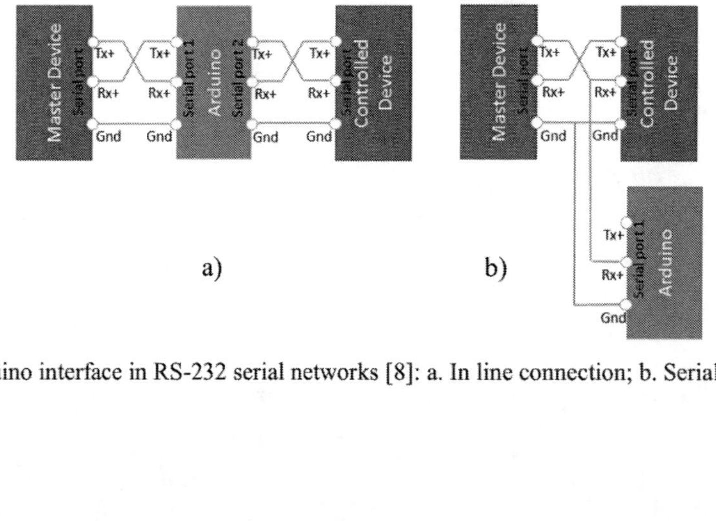

Fig. 2. Arduino interface in RS-232 serial networks [8]: a. In line connection; b. Serial connection

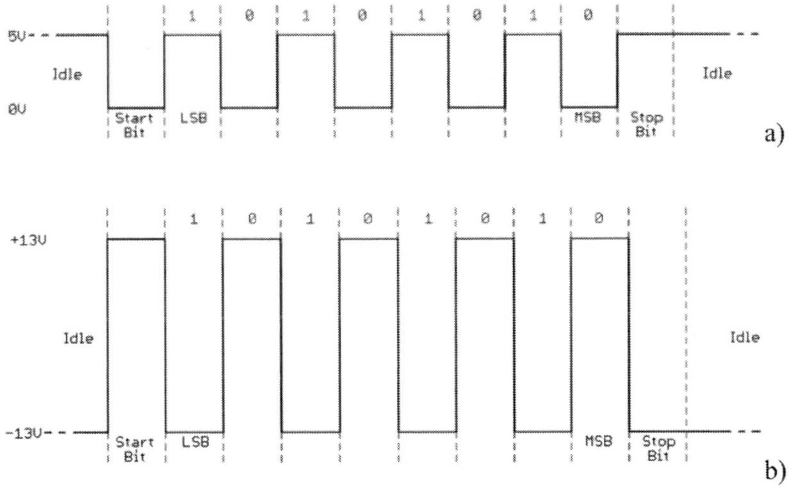

Fig. 3. Serial port voltages [9]: a. TTL Serial; b. Standard Serial

Another drawback of interconnecting industrial serial and Arduino consists in the difference between voltages used by communication ports. Arduino uses TTL serial and industrial equipment uses standard serial. The Arduino serial port uses a voltage of 3.3v or 5v (Figure 3.a.), while all industrial equipment use a voltage between -13v and 13v (Figure 3.b.), so, obviously, connecting the Arduino to a machine's serial port will result in burning the Arduino port. In this case a convertor is required to transform Standard serial to TTL serial.

3.1.2 Standard RS-422

The RS-422 protocol was initially implemented on Macintosh as the serial standard interface. It became fast an industrial standard because it uses differential electrical signal, thus permitting more devices to use the same bus, and it also has greater noise immunity, thus allowing longer cable lengths. With RS-422 every device has a unique ID, data is sent from the master device to all the devices that are connected in the same network, but only the destination station, with the required ID, will process the request.

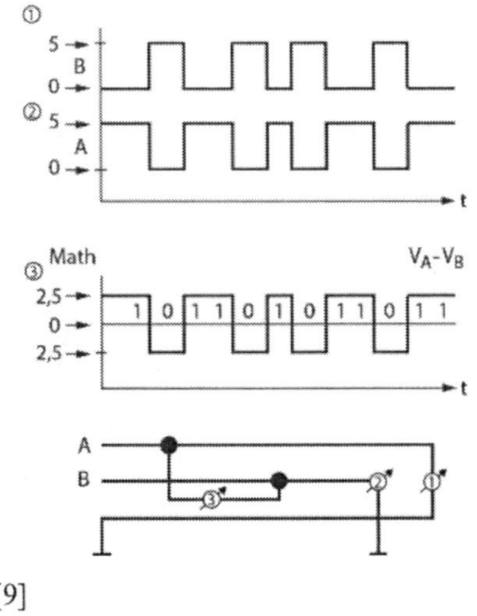

Fig. 4. RS-485 voltage levels [9]

DOI: 10.1051/matecconf/201712102008

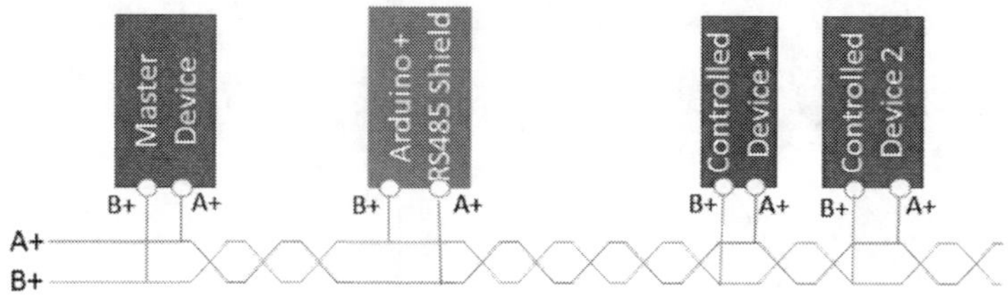

Fig. 5. Arduino interface in a RS-485 network [8]

3.1.3 Standard RS-485

The RS-485 protocol is an improvement of the RS-422 protocol, as it can handle more devices and it accepts more than one master device as long as only one master device will send data at one time. RS-485 is a superset of RS-422, so any device that supports RS-485 will be able to communicate with a device that only supports RS-422. As the voltage level between the two differential wires usually has a value lower than 5v (Figure 4) it is possible to connect an Arduino board directly to an RS-485 line (Figure 5).

3.2 Software format of the serial standards

The RS-485 data is sent using the ModBus protocol. ModBus was first developed for Programmable Logic Controllers (PLCs) and it became a standard communication protocol for many industrial devices. ModBus enable data communication between several devices connected in the same network. A typical ModBus data packet consists of: a destination station address, function code, data area and cyclic redundancy check (CRC). There are two protocol versions used over RS-485 serial: ModBus RTU and ModBus ASCII.

ModBus RTU is the most common implementation used in serial communication. It uses a compact binary representation of data. In ModBus RTU data is sent without spaces, the messages being marked at the beginning and at the end with a silence period of 31/2 characters. In ModBus RTU each byte is sent as a string of eight binary characters enclosed by a start bit and a stop bit.

ModBus ASCII marks the start of each message with a colon character and the end of the message with a carriage return and line feed characters, thus allowing spaces inside data messages. In ModBus ASCII there are only 7 data bits, but there is one extra parity bit before the ending bit.

4 Case study

To demonstrate the viability of the proposed method, the operating parameters of an automatic machine were monitored (Figure 6). The equipment used for testing is an automatic polishing and sanding machine for high gloss lacquered furniture components (painted MDF – medium-density fibber). The automatic machine offers the possibility to obtain quality products in a short time with limited human intervention. The equipment is accessible to small firms, has a reduced size: 1500×1100×1200 [mm] ideal for workshops with a reduced floor space, allows rapid reconfiguration – depending on the size of the plate, and simple operation. The machine has three types of moves: the motion of the conveyor table, with the fixed wood panel; the motion of the brush head in a transversal direction to the wood panel; continuous rotation of the brush.

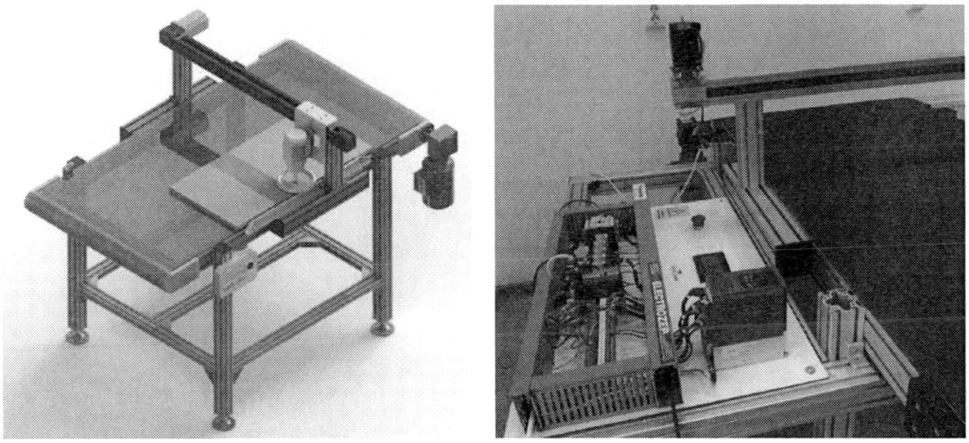

Fig. 6. Automatic polishing and sanding machine: a. Virtual prototype; b. Equipment

The machine's PLC uses RS-485 / ModBus RTU communication standard to control the 3 motor drivers. For preventive maintenance purposes, the speed and the functioning time of all motors are recorded and exported to an IoT platform.

For this case study, an open source IoT platform named Carriots (Figure 7) was chosen. It offers some predefined functions that are indispensable for the proposed predictive maintenance system. In the first step, only 7 parameters where included in the developed monitoring system: working status of the motors, speed of the motors and the temperature measured near the main (first) motor.

The hardware required to connect the polishing and sanding machine to the IoT platform include an Arduino Uno board, equipped with an Ethernet shield, and one serial RS-485 shield. The Arduino waits continuously for the RS-485 serial commands and, if a Start or a Stop command is received, based on the station number, it then sends the data to the IoT platform (Figure 8).

For this proof-of-concept application, the Arduino device was used because it packs considerable power on a very small board and it provides many opportunities for automation, networking and data collection. It should be noted that we are aware of the existing drawbacks that generally limit the use of Arduino based devices in industrial application: low level voltage, robustness and safety. However, nowadays, more and more research systems are based on Arduino and similar devices as the main disadvantages are overcomes: underwater sound recorders for marine research, robots for assembling gene-editing tools and even a system developed for an outreach project on the International Space Station [10].

Fig. 7. Carriots Platform

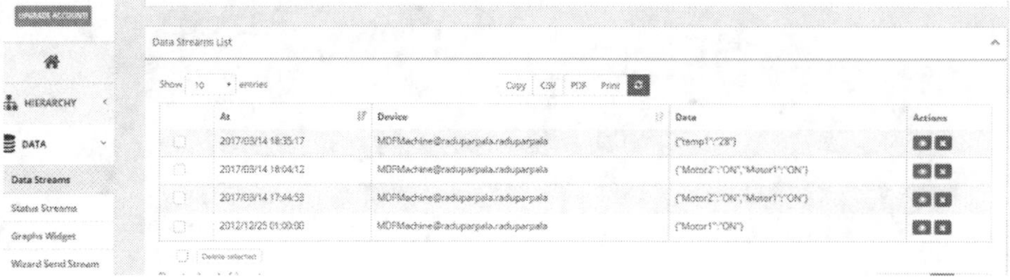

Fig. 8. IoT Stream data

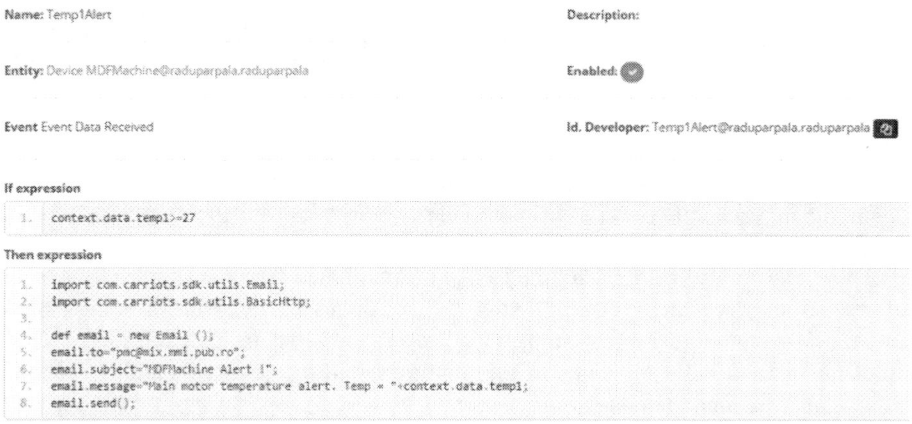

Fig. 9. E-mail rule definition

Fig. 10. E-mail alert

The main function of the developed predictive maintenance system is to send usage alert via e-mail or other messaging services. For testing purposes, a new alert was defined for a temperature of 27 °C (Figure 9). Using this function, each time the defined threshold is reached, an e-mail alert is send to the user defined (Figure 10). In a similar way, other parameters can be added and different alerts can be defined. Thus, using the proposed method, an on-line monitoring and predictive maintenance system can be easily developed for any type of industrial equipment.

5 Conclusions

Currently, IoT is proving to be a game-changer for many companies as a variety of industries begin employing IoT-enabled architectures and experimenting with how IoT solutions can bring new benefits. From self-driving cars to smart homes filled with voice activated devices, the innovation of IoT's connectivity never fails to impress. However, many of the machines used in skilled trades or manufacturing are still not connected to IoT platforms because they lack sensors, software and connections to IT systems [11].

In this context, the paper presented a new and simple method for on-line monitoring and predictive maintenance of industrial equipment. This method is based on an integration of devices with software and it contains a procedure to translate different industrial equipment language into web protocols, thus creating a way to efficiently implement predictive maintenance in industrial systems.

In order to demonstrate the feasibility of the developed method, a case study is presented. The operating parameters of an automatic polishing and sanding machine for high gloss lacquered furniture components were monitored and a threshold alert was defined. In a similar way, using the presented method, an on-line monitoring and predictive maintenance system can be easily developed for any type of industrial equipment. It should be noted that for complex industrial applications a more robust device is envisaged – a similar system based on a FPGA process computer.

This work has been funded by University Politehnica of Bucharest, through the "Excellence Research Grants" Program (UPB-GEX). Identifier: UPB-EXCELENTA-2016. Research project title: "Monitorizarea inteligenta integrata a sistemelor de fabricatie pentru asigurarea mentenantei preventive a acestora". Contract number: 34 / 26.09.2016.

References

1. M. Albert, Modern Machine Shop Magazine (2015)
2. A.K.S. Jardine, D. Lin, D. Banjevic, Mechanical Systems and Signal Processing, **20** (2006)
3. Y. Geum, S. Lee, D. Kang, Y. Park, Journal of Engineering and Technology Management, **28** (2011)
4. J. Lee, M. Holgado, H.A. Kao, M. Macchi, IFAC Proceedings, **47** (2014)
5. X. Xu, T. Chen, M. Minami, Computers & Mathematics with Applications, **64** (2012)
6. Q. Wang, J. Gao, Journal of Loss Prevention in the Process Industries, **25** (2012)
7. L. Dong, R. Mingyue, M. Guoying, Procedia Engineering, **174** (2017)
8. *** https://www.arduino.cc/en/Main/Products, Accesed 10.02.2017
9. *** http://www.omega.co.uk/techref/das/rs-232-422-485.html, Accesed 15.02.2017
10. D. Cressey, Nature journal, **544** (2017)
11. ***http://www.bosch-presse.de/pressportal/de/en/69632.html, Accesed 18.03.2017

Comparative study between different methods for the efficiency of the production activity in the wood industry

Radu Dan Paltan[1], *Cristina* Biriş[2,*], and *Loredana Anne-Marie* Rădulescu[1]

[1]Transilvania University of Braşov, B-dul Eroilor nr.29, 500036 Brasov, Romania
[2] Lucian Blaga University of Sibiu, Department of Industrial Machines and Equipments, 10. B-dul Victoriei, Sibiu, Romania

Abstract. Generally speaking,the quality and cost of goods and in particular of the wood products is a decisive condition in their sales on the market, attracting buyers to satisfy their wishes and requirements. Starting from the general idea, by which it is considered that the manufacturing activity is effective if production is obtained at maximum quality with a low (minimum) cost, or when the revenue from the sale of products exceeds the market expenditure which is necessary to achieve it, this research article aims to study ways to make production more efficient by methods which could also be applied in the wood industry, by presenting a comparative study on production optimization methods in this industry.

1 Introduction

Starting from the general idea that a manufacturing activity is effective if production is obtained at maximum quality with low (minimum) cost, or when the proceeds from the sale of products exceed the market expenditures that were necessary to achieve them, any optimization of machining processes must consider primarily an optimization on economic criteria.

These criteria once applied to the studied process will inevitably lead to finding technical solutions to meet the requirements of the market. Optimizing a process can be done by applying various methods of assessing the performance of the studied process, such as the Target Costing method, the Taguchi Method, the Six Sigma method and so on.

By applying them, all listed methods bring more or less important improvements, but the starting questions are: which of the applied assessment methods are most effective? and which of them would lead to those improvements that converge towards a real optimization? These questions, once clarified facilitate the process management, positively influencing both the decrease of losses (in material or energy) and especially the growth of the company profit.

* Corresponding author: cristina.biris@ulbsibiu.ro

As a result of the unfolded researches, the authors of this paper have found the answer to the two questions, an answer that will be detailed and fundamented in the presentation below.

2 Means and methods of work

The goal of any company and therefore of those in the wood processing industry is to obtain maximum profit. For this, special attention needs to be granted to the management that must find the right method for optimizing the production flow, increasing the quality and the degree of polishing for the finished products and be the easiest to implement and also give the fastest results and the most significant results in terms of quantity and quality [1].

Of the many techniques and methods of economic analysis that are used to optimize production costs, following a meeting of the company management of SC Famos SA, active in the wood processing industry, with the engineers directly involved in the production, there were selected and surveyed three methods for economic efficientization:

The Target Costing method: is part of a strategic management approach according to which each cost carrier is analyzed throughout their life cycle. The target or objective cost is a cost management concept used and developed in Japanese companies, especially in the automotive industry since the 70s. Underlying this target cost concept there was the need to produce smaller series of products that can better adapt to the market needs, the introduction of the new production organization methods (Just In Time operating system) and the introduction of automation-based technologies (CIM - Computer Integrated Manufacturing - systems).

The Taguchi method imposed itself as being more effective than other methods of experiment planning. It derives from the method of factorial experiments and proposes an alternative method for calculating the average effects of factors and interactions, thus making mathematical modeling much easier. Its efficiency is due to reducing the number of tests given by the split plan method, which allows modeling with much less experiments than the full plan method. The Taguchi method is one of these split plan methods and has also the advantage that it is easy to apply in practice. When determining the number of experiments involving the study of a phenomenon by this method, stricter conditions must be applied.

The Six Sigma Method represents the standard deviation in statistics and it is a management methodology aimed at increasing product quality by determining and removing the continuous causes of defects and process variability (potential or detected) in order to ensure customer satisfaction, based on the methods FMEA (Failure Mode and effect and Analysis) and QFD (Quality Function Deployment) and modern management methods applied to the joint teams made up of manufacturer, supplier, client, professional, research centers etc [2].

Each of the three methods aims to achieve the economic efficiency of the company by taking into account certain parameters that significantly or less significantly influence the process optimization. Among the projected (common and / or different) parameters for each method, there were studied only the parameters that the three methods have in common, so that a managerial decision can be taken on the method to be applied within the company.

The optimal decisions for the company management, since there are several alternatives (methods), can be obtained by applying a multi-criteria analysis method (AMC). The goal of this method is to conduct a comparative evaluation of the proposed options (methods). For the AMC, all common parameters of the selected methods were taken into account simultaneously in a complex situation. The method is designed to assist decision makers by integrating different options, reflecting the views of the actors involved in a prospective or

DOI: 10.1051/matecconf/201712102009

retrospective framework. The analyzed parameters reflect how objectives are achieved. The best option will be the one that will be closest to achieving most objectives and that will obtain the highest scoring.

The comparison was made between the efficiency models that could be selected taking into account the specific characteristics of the wood processing industry.

The analysis consists of the calculation of value hierarchizing coefficients, for the performance of the objects under comparison.

The result of the multi-criteria analysis, properly applied, provides scientific and effective results for optimal solutions in the technological process which will be reflected in the economic cost/price of the final product.

3 Research methodology

The multi-criteria analysis (MCA) describes any structured approach to be used in determining the general preferences of several alternative options, options that lead to achieving a number of objectives. In the current case, our main goal is to optimize production by reducing costs at SC Famos SA. The steps that were followed are:

3.1 Establishing methods and evaluation criteria in the decision-making context

For identifying the optimization method with the greatest applicability in the wood processing industry, the research was based on the use of brainstorming techniques and multi-criteria analysis.

Within this step there were identified the alternative criteria to be taken into account. In our case the options considered are: alternative 1 the Target Costing method; alternative 2 the Taguchi method. Alternative 3 the 6 Sigma method.

Through the multi-criteria analysis (MCA) there race, eliminating the subjectivity, since the order of the criteria is determined by comparing each criterion relative to other selected criteria.

The brainstorming consisted in organizing a work meeting with a group consisting of the company management and 10 engineers from SC Famos SA Odorheiul Secuiesc. During the brainstorming session, held in the company's meeting hall, there were presented several ways to optimize production from which there were chosen the 3 mentioned methods and the participants generated a number of ideas regarding the applicability of these methods in the technological process.

After continuing the brainstorming session, there were generated following evaluation criteria that must be followed in the process of multi-criteria analysis:

1. Cost of implementing the methods (C);
2. The simplicity of using the method (S);
3. The effect of applying the method reflected in a reduction of the energy consumption (E);
4. Acceleration of the technological process - eliminating some steps in the process => reducing the time and costs that are reflected in the price (A);
5. Ease of implementing the methods (U);
6. The quality of the product (O);
7. Number of failures / waste (N).

3.2 Determining the share of criteria common for the three methods selected for relative quantifying

At this stage, after identifying the 7 common criteria which are relevant for solving the problem in selecting the decisional problem for the efficientisation method, they have been ranked. This ranking took into account the major categories of costs and benefits resulting from the options considered. The scoring was done according to the principle – a criterion is more important than the other (= 1), as important (= 1/2), less important (= 0) Table 1 presents the calculation of points, the level and the weighting coefficient Y_i [3].

Table 1. Point calculation, the level and the weighting coefficient Yi.

Criteria	C(1)	S(2)	E(3)	A(4)	U(5)	O(6)	N(7)	points	level	Yi
C(1)	0.5	0	0.5	0	0	0.5	0	1.5	7	0.27
S(2)	1	0.5	0.5	0.5	1	1	0	4.5	2	2.89
E(3)	0.5	0.5	0.5	0	0.5	1	0	3	5	1.17
A(4)	1	0.5	1	0.5	1	0	0	4	3	2.2
U(5)	1	0	0.5	0	0.5	0	0.5	2.5	6	0.77
O(6)	0.5	0	0	1	1	0.5	0.5	3.5	4	1.64
N(7)	1	1	1	1	0.5	0.5	0.5	5.5	1	4.57

3.3 Hierarchizing the options

As a result of the inter-criteria comparison process, there resulted a matrix whose sum by lines (sum of each criterion in part) (column 9) resulted in a hierarchy (column 10) of the criteria's significance as seen in table 1.

The next step in the MCA procedure was calculation of the weighting coefficient (Y_i) column11, which is calculated using the FRISCO formula (1)

$$\gamma_i = \frac{p + \Delta p + m + 0.5}{-\Delta p' + \frac{N_{crt}}{2}} \tag{1}$$

Where:
p – is the sum of points obtained on the line by the considered item;
Δp – the difference between the score of the considered item and the score of the item at the last level.
m – represents the number of surpassed criteria, i.e. the number of criteria with scores below the item;
Ncrt – number of criteria taken into account;
Δp` - the difference between the score of the considered element and the score of the element in the first rank (a negative result).

3.4 The standardization of score

The standardization of score for each criterion was made in a common scale ranging from 1 to 10 and taking into account the effect that it will have on the company's activity. The result of the analysis is presented in Table 2.

Table 2. Criteria and their value range.

Criteria	Indicator	Value ranges	Effect
C1(C)	Cost of implementing the methods (C);	1-10	-
C2(S)	The simplicity of using the method (S);	1-10	+
C3(E)	Effect of applying the method reflected in a reduction of the energy consumption (E);	1-10	+
C4(A);	Acceleration of the technological process (A);	1-10	+
C5(U)	Ease of implementing the methods (U);	1-10	+
C6(O);	The quality of the product (A);	1-10	+
C7(N).	Number of failures / waste (N).	1-10	-

3.5 The creation of the performance matrix

The creation of the performance matrix describes the expected performances of each option according to the chosen criteria. The information on the dimension of each criterion can be expressed in units. Table 3 presents the performance matrix and the assigned values were determined by the personnel of SC Famos SA within the organized brainstorming session.

Table 3. Performance matrix.

	TC Method	T Method	6S Method
C(1)	8	9	10
S(2)	9	8	10
E(3)	9	10	9
A(4)	10	8	9
U(5)	10	9	10
O(6)	10	10	10
N(7)	9	10	9

3.6 Examining of results

Considering the performance matrix and the weighting coefficient there was calculated the product between the Yi coefficient and the performance degree Ni, then the products were added by criterion to achieve the final result. The data obtained are presented in Table 4 and can be compared.

Table 4. Calculation of the products between the grades N and the weighting coefficients

		Method TC		**Method T**		**Method 6S**	
Criteria	**Yi**	Ni	Ni*Yi	Ni	Ni*Yi	Ni	Ni*Yi
C(1)	0.27	8	2.16	9	2.43	10	2.7
S(2)	2.89	9	26.01	8	23.12	10	28.9
E(3)	1.17	9	10.53	10	11.7	9	10.53
A(4)	2.2	10	22	8	17.6	9	19.8
U(5)	0.77	9	6.93	9	6.93	10	7.7
O(6)	1.64	10	16.4	10	16.4	10	16.4
N(7)	4.57	9	41.13	10	45.7	9	41.13
Final Score			**125.16**		**123.88**		**127.16**

The final analysis was done taking into account the sums of all criteria that influence each method. It can be noticed that 6S is the most effective method, totaling the highest value. In MCA procedure, there is also a sensitivity analysis stage that was not required to be unfolded here, given that the sensitivity of the criteria in relation to changing the method is very close.

4 Conclusions

Following the multi-criteria analysis applied at the wood processing company SC Famos SA in order to choose the assessment method for optimizing the technological process, there resulted that the most efficient method that the company management should choose in order to apply it in production is the 6Sigma method that earned the highest score - 127.16, compared to the Target Costing method that obtained 125.16 and the Taguchi method that scored only 123.88.

These results confirmed the discussions and opinions of the engineers from SC Famos SA expressed during the brainstorming session that the Taguchi method is obsolete, due to the managerial principles it takes into account, while the Target Costing method is limited in its application to certain industrial branches, being incomplete compared to the multitude of managerial aspects and decisions that the management must make in the wood processing industry.

After applying the 6Sigma method for optimizing the milling process that is found in the manufacturing technology at SC Famos SA and that significantly affects production, one of the criteria analyzed was energy consumption. The most telling parameter of energy consumption in wood milling is the specific strength (K), that characterizes the timber processed and that can be calculated for milling function of the absorbed chipping power, using the relation [4]:

$$K = \frac{P \cdot 6 \cdot 10^3}{b \cdot h \cdot u} \ [\text{N/mm}^2] \quad (2)$$

Where:
P = cutting power measured in kW;
b = cutting width, measured in mm;
h = cutting depth, measured in mm;
u = feed ratthat ie, measured in m / min

It was found that a significant difference in energy consumption was recorded in processing on the transversal direction of the wood, compared to that on the longitudinal direction, the power consumption being much higher in the first case. Also, during the transversal processing, apart from the higher energy consumption, it also requires an additional stage in processing, in order to finish the products at the same quality as the products obtained through processing on the longitudinal direction [5]. These steps are generating additional costs and are also responsible for extending the production time. Also, during the finishing in the additional phases there is generated more waste and also the risk of generating more failures can increase (requiring, for more complicated models the applying of manual finishing) [6].

Despite the slightly higher costs to put in practice the management decision using 6 S noticed during the implementation, these will be eventually annihilated and will produce the effect of optimizing costs by minimizing the amounts of failures and waste produced.

There was a unanimous opinion on the ease with which this method can be applied at all stages of production, especially since it implies the operating personnel's involvement in the continuous measurement, assessment and improvement of parameters.

The quality and cost of wood products and of other finished products prove decisive in their offering on the market, in attracting buyers and satisfying their wishes and requirements. The sellable product quality and value depends heavily on the equipment performance, on the level of the technical endowment, on the organization of the manufacturing and processing processes [7].

A higher quality level can be achieved by eliminating as much as possible of the elements that negatively affect yields, consumption of raw materials, energy, some of them leading to the need to retool production lines and in the end the product quality from a technical point of view.

All the listed items as well as others that have resulted from the analysis have as effect the increase of production costs, reflected in the cost of the finished product and therefore in reducing the production yield by decreasing the company's profitability [8].

References

1. L. Lobont Mihai, V Zerbes, *Metode tehnici si instrumente pentru imbunatatirea calitatii – lucrari practice*, (Editura Universităţii „Lucian Blaga" din Sibiu, 2014)
2. C.V Kifor, C Oprean, *Ingineria calităţii. Îmbunătăţirea 6 Sigma, (*Editura Universităţii „Lucian Blaga" din Sibiu, 2006)
3. Ş. Bobancu*, Creativitate şi Inventică (C&I) Suport Curs,* (Universitatea „Transilvania", Braşov, 2015)
4. ***http://old.fonduri-ue.ro/res/filepicker_users/cd25a597fd62/Documente_Suport/Studii/0_Studii_Instrumente_Structurale/Pag.3_ACB/19_Analiza_Multicriteriala.pdf, Accesed 12.04.2017
5. M. Marthy, I. Cismaru, ProLigno, Journal, **3** (2009)
6. E. Salca, A. Fotin, I Cismaru, Pro Ligno, Journal, **4**(2) 57 (2008)
7. L. Duguleana, C. Duguleana, Economia contemporana. Prezent si perspective, **4**, 417, (Editura AGIR, Romania, Pitesti, 2004)
8. C. Duguleană, L. Duguleană, Proceedings of International Conference -Small and Medium Enterprises in European Economies", Babes-Bolyai University Cluj-Napoca, Faculty of Business, Cluj-Napoca, Romania, **11** (2003)

Preliminary study on occurrence of composite material delamination processed by abrasive water jet cutting

Ioan Alexandru Popan[1,*], *Nicolae* Bâlc[1], and *Alina Ioana* Popan[1]

[1]Technical University of Cluj-Napoca, B-dul. Muncii, No. 103-105, Cluj-Napoca, Romania

Abstract. The paper presents a preliminary experimental study on processing composite materials (CFRP) using abrasive water jet cutting (AWJC) process, analysing the possibilities of occurrence of material delamination. AWJC is a proper solution for cutting CFRP because of reduced interface temperature, high flexibility, low mechanical loading and reduced cutting forces. Cutting CFRP using AWJC involves several challenges like material delamination due to the high velocity impact of the jet. To understand the delamination, three experimental tests were made: in the first test the cutting and the drilling was made with high water pressure (350 MPa), in the second test the cutting was made with high water pressure and for drilling was used low water pressure (200 MPa) and in last test a pre-drilled hole was used. Within those experiments was observed the CFRP delamination appears just during the drilling, not during the cutting. By decreasing the water jet pressure, the jet energy is decreased and in this way the delamination decrease.

1 Introduction

A wide range of composite materials (Carbon Fibre Reinforced Plastics - CFRP) are used for many industrial applications. The main characteristics of CFRPs are: high strength to weight ratio, high modulus, high fracture toughness, corrosion and thermal resistance [1]. These materials have wide applications in aerospace, automotive, marine goods, etc. [2].

AWJ cutting is a good solution for processing CFRP because of reduced interface temperature, low mechanical loading, better surface integrity, low tool wear, and small cutting forces [3-5]. The AWJC uses pressurized water, which acts either by itself at a very high pressure, or in connection with the erosion of the abrasive particles, added into the water jet [6-7].

Figure 1 illustrates the main components of the AWJ head, including the mixing tube, where the pressurized water jet is mixed together with abrasive particles, which are needed for metals cutting.

* Corresponding author: ioan.popan@tcm.utcluj.ro

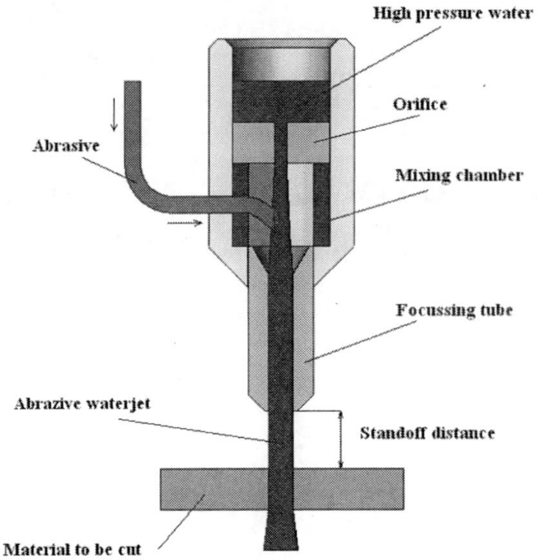

Fig. 1. AWJ head [6]

The AWJ working principle is based on a very high pressure water jet, which takes the abrasive particles into the mixing chamber, in order to create a cutting device [6-7].

Cutting composite materials using AWJC involves several challenges. The most important problem is material delamination due to the high velocity impact of the jet [4-5]. In figure 2 is illustrated the delamination in case of CFRP [8].

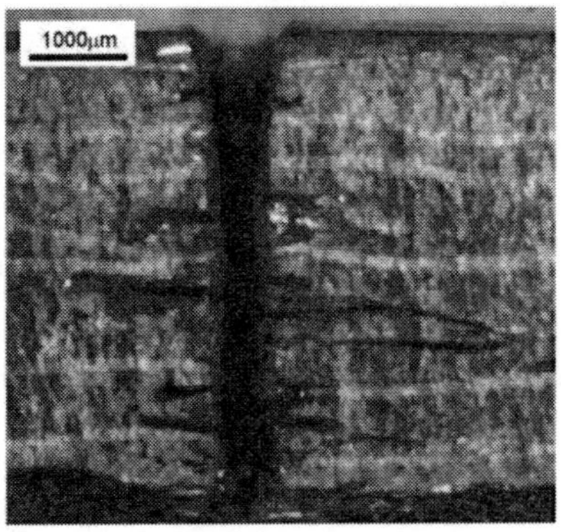

Fig. 2. Delamination effect [8]

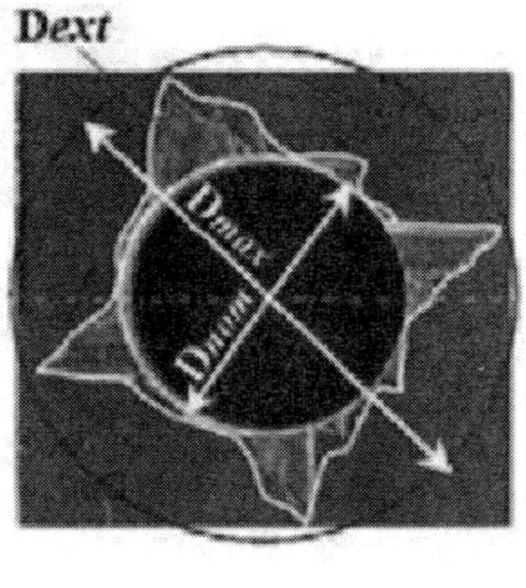

Fig. 3. Typical image for the visualization of the damage [10]

In figure 3 is presented a sketch for visualization of the damage evaluated by DeFu et al. (2012). The delamination extent (D_{ext}) is defined as a difference between the maximum diameter (D_{max}) of the damaged zone and the nominal exit hole diameter (D_{min}) [9-10].

To minimise the composite material delamination, a starter hole can be made, but this introduce time consuming and extra costs [11].

This paper presents an experimental investigation on delamination of composite materials during the AWJ cutting process.

2 Experimental procedure

The material used in this study is a multi-layer composite material consisting of carbon fiber reinforced plastics (CFRP): a biaxial balanced carbon fibber material, of 200 g/m^2, plain-weave 5 layers in epoxy resin, type Epiphen RE4020 for the matrix produced by Bostic. In the case of the material, 50% of filaments are orientated along the longitudinal direction, while the remaining of 50% is oriented along the transverse direction [7].

The experimental trials have been conducted on a 3-axis, Omax 2626 abrasive waterjet system (Figure 5). It is equipped with an OMAX high pressure pump with an output pressure of maximum 400 MPa. The CNC Omax controller can move the cutting head in 3 axis simultaneous with maximum 4500 mm/min.

The constant parameters used on this study are: orifice diameter (diamond) 0.35 mm, nozzle tube diameter 0.76 mm, nozzle length 101.65 mm, abrasive type Australian Garnet mesh size 80, impingement angle 90° and standoff distance 2 mm.

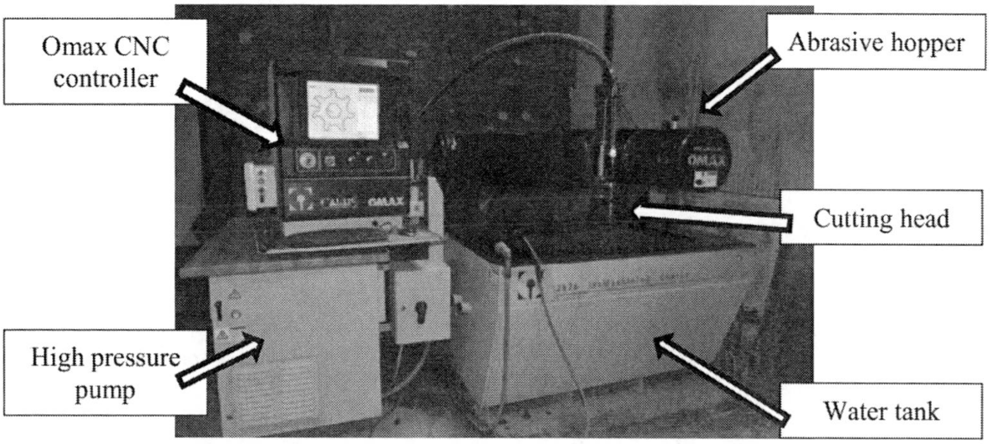

Fig. 5. AWJ equipment, Omax 2626.

The 2D cutting sketch was designed in the *Omax Layout* software. The part origin was selected in the centre of the part. Another step on process setup was choosing the lead In/Out. The lead In/Out was choose automatically within this application and was selected just the length of lead In of 15 mm, to avoid the material delamination on drilling point (figure 6).

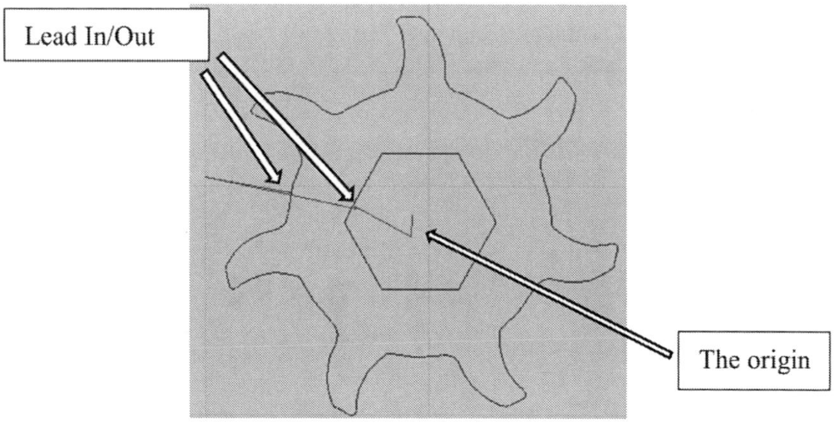

Fig. 6. The origin and the lead In/Out points.

In this experimental study was analysed the possibility to occur the delamination during the processing composite materials in three cases:

i: In the first case for the cutting and drilling processes was used with high water pressure.

For processing the part were selected the following process parameters: cutting speed V= 300 mm/min; water pressure: P= 350 MPa and abrasive flow rate: M_a=0.45 Kg/min.

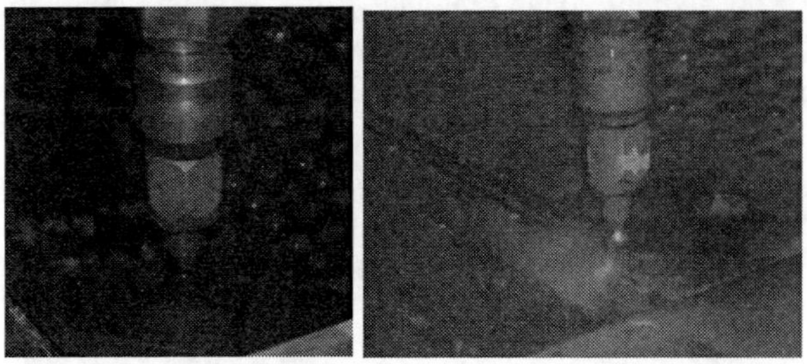

Fig. 7. The part made using pre-drilled hole.

ii: In second case for the cutting process was used a high water pressure and for drilling a low water pressure .

The selected process parameters are: cutting speed V= 300 mm/min; water pressure for cutting: P= 350 MPa; water pressure for drilling: P= 200 MPa, abrasive flow rate: Ma=0.45 Kg/min.

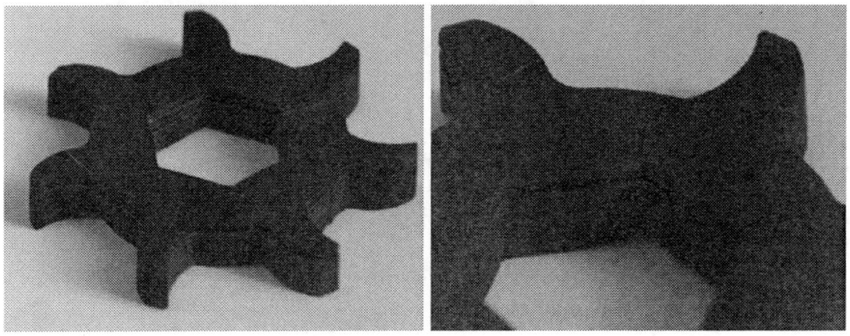

Fig. 8. The part made only with AWJC.

iii: In third case a pre-drilled hole was used.

To avoid the composite material delamination in the drilling point, a 6 mm hole was made. This hole was made by using a carbide drill. In figure 9 is illustrated the drilling process and the hole. The process parameters are similar to those used in the second experimental case.

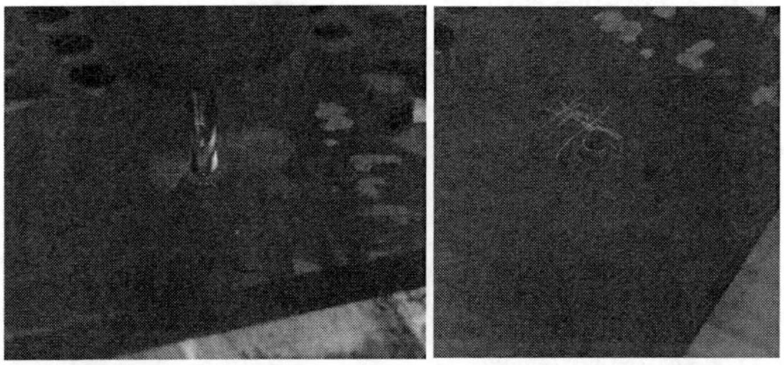

Fig. 9. The pre-drilled hole.

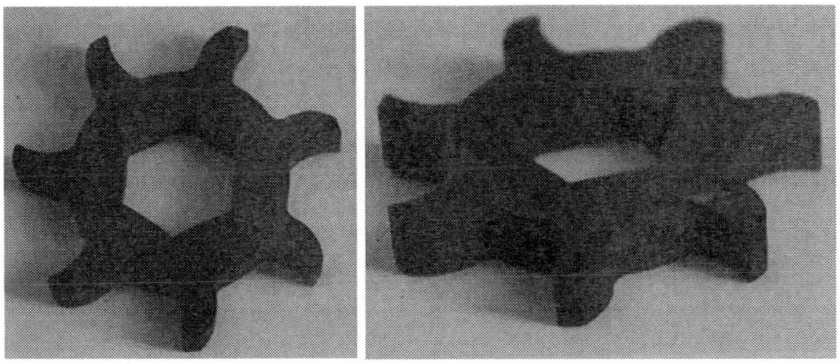

Fig. 10. The part made using pre-drilled hole.

3 Results and discussions

The maximum extent of delamination up to 60 mm, was obtain during the drilling of CFRP using abrasive water jet at high pressure 350 MPa. This powerful delamination effect results in splitting the part into two pieces, destroying the part (figure 11).

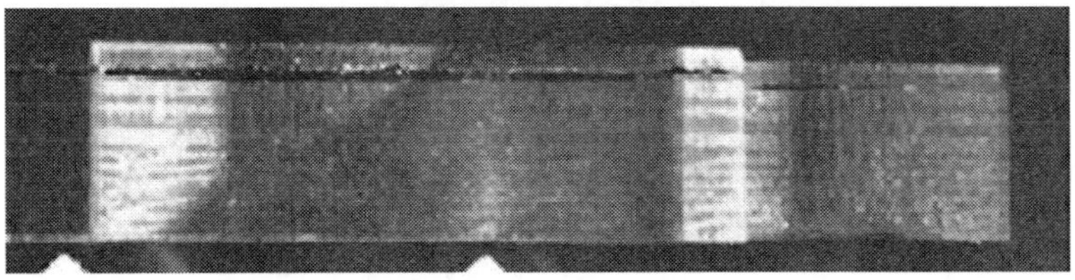

Fig. 11. The part drilled using abrasive water jet at high pressure 350 MPa.

In figure 12 is illustrated an image of the drilled part using abrasive water jet at low pressure 200 MPa and cut with a high pressure 380 MPa.

On this case study the maximum extent of delamination was 5 mm and abrasive embedment do not appear.

It can be observed by decreasing the water pressure from 380 MPa up to 200 MPa the extent of delamination decreasing from 60mm up to 6mm, during the drilling process.
This proves that by decreasing the water jet pressure, the jet energy is decreasing and in this way the delamination can be avoided.

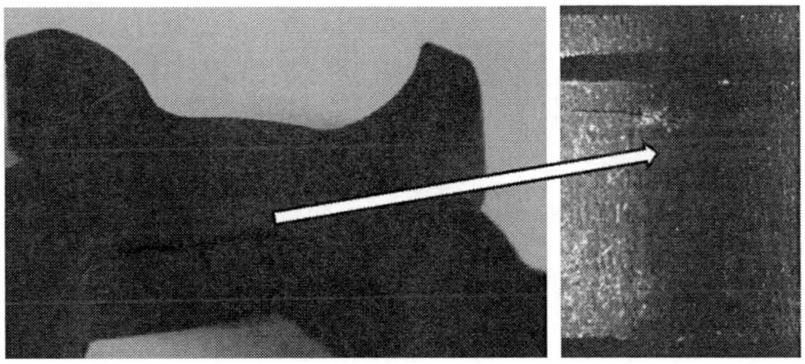

Fig. 12. The part drilled using abrasive water jet at low pressure 200MPa.

In figure 13 is illustrated a microscopic image of the part manufactured onto the third experiment.

In this case the material delamination and abrasive embedment do not appear. In this experiment the abrasive jet doesn't hit the material during the drilling process, because of start point a starter hole was made. This proves that due to the high velocity impact of the jet during the drilling the delamination appears.

In conclusion the delamination does not appear during the cutting process only during the drilling process.

In the case of using a starting hole the material delamination can be avoided. An extra cost is the main disadvantage to use a pre-drilled hole during the cutting complex parts from composite materials. Ramulu and Kraja (2002) obtained the same results in their studies in minimization the delamination by using a pre-drilled hole.

Fig. 13. The part made using pre-drilled hole.

Measuring the parts dimensions a good dimensional accuracy of ±0.05 mm was obtained.

Analysing the part contours was observed a shape deviation in the cutting starting and ending points. This shape error is illustrated in figure 14. The dimension of this over cut was 0.2x1 mm. To obtain precise parts from CFRP manufacture using AWJ cutting this effect must be analysed. This type of defects can lead to a damaged part. [12]

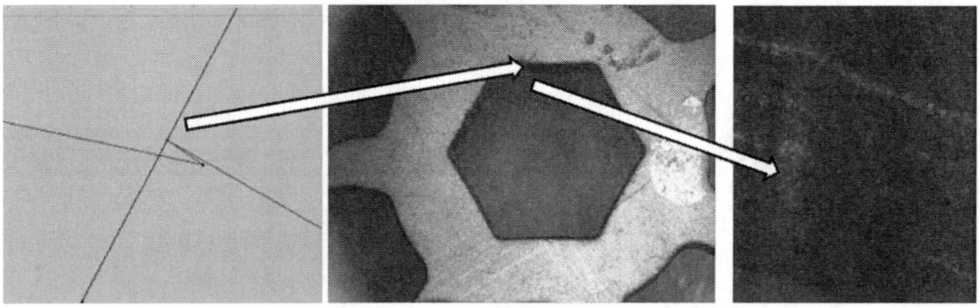

Fig. 14. The cutting start and end error.

4 Conclusions

This paper presents a preliminary study regarding the possibilities of delamination occurrence in case of CFRP materials manufacture using abrasive water jet cutting.

The specific conclusions from this experimental study are:

- The maximum extent of delamination, 60 mm, was obtain during the drilling of CFRP using abrasive water jet at high pressure of 350 MPa.
- Decreasing the water jet pressure during the drilling, up to 200 MPa, the maximum extent of delamination was 5 mm.

- Decreasing the water jet pressure, the jet energy is decreased and in this way the delamination is decreasing. Using the optimal process parameters the delamination can be avoided.
- Using a starter hole the material delamination can be 100% eliminated. The main disadvantage of this solution is that an extra operation is introduced, this means extra costs.
- The CFRP delamination appears just during the drilling, not during the cutting process.
- In the cutting "start" and "end" points a shape deviation (an over cut) was observed. This shape error has the dimensions of 0.2x1 mm.

The research was supported within the PreMCo project (Bridge Grant No. 99BG/2016), PN-3-P2 – UEFISCDI.and the H2020 AMaTUC project (GA 691787).

References

1. A. Popescu, L. Hancu, P. Bere, Applied Mechanics and Materials **371**, 394 (2013)
2. E. Lemma, F. L. Chen, Compos Struct; **57**, 297 (2002)
3. P. Hreha, A. Radvanská, Metalurgija **53** (4), 533 (2014)
4. A. Filip, M.Vasiloni, MATEC Web of Conferences **94**, 03003 (2016)
5. J. Wang, Int J Adv Manuf Technol, **15**, 757 (1999)
6. A. Popan, N. Balc, Academic journal of manufacturing engineering, **7**(1), 17 (2009)
7. A. Miron, A. Popan, *Studies on water jet cutting of 2D parts made from carbon fiber composite materials*, IMANE (2013)
8. D.K. Shanmugam, Composites: Part A **39 (6)**, 923 (2008)
9. D. Liu, Y. Tang and WL. Cong, Composite Structures **94,** 1265 (2012)
10. M. Ramulu, S.Kraja, *33rd International SAMPE Proceedings*, **33**, 1327 (2002)
11. K. Phapale, R. Singh, Procedia Manufacturing **5,** 521 (2016)
12. M. Harničárová, J. Valíček, R. Čep, J. Petrů, L. Čepová, *Surface integrity analysis in laser beam cutting process,* Metal, Brno, Czech Republic, EU (2013)

Reduction of dimensional variability of pieces, resulted in AWJ process, using robust design method

Bogdan Barabas[1,*], *Tudor* Deaconescu[1], and *Sorin* Barabas[1]

[1]Transilvania University of Brasov, Faculty of Technological Engineering and Industrial Management, Department of Engineering and Industrial Management, Brasov, Romania

Abstract. Dimensional variability arises in AWJ system from uncontrolled factors intervention or due to the variability of input signals considered fixed or due to interactions of the input signals with unexpected effects, leading to deterioration of unpredictable outcome. Number of noise factors, negatively influence the results registered, dispersion of the result values is directly proportional to the number and their effect. Stochastic influence and his unpredictable nature determine a thorough research of the hydroabrasive process. The research refers to the study of dimensional variability for different configurations of hydroabrasive liquid pressure, abrasive type and traverse velocity. This is achieved by performing the experiments statistically processed by combining of control parameters, which allow favorable conclusions on reducing variability results and on optimizing the entire process of hydroabrasive machining.

1 Introduction

Abrasive water jet (AWJ) erosion is an extremely complex system wherein the plurality of parameters that determine the phenomena occurring at extremely high pressures, stochastic responses resulted, are leading to the need of using probability and statistical calculation for the study of this processing methods. Robust design method of systems that rely on the use of instruments mentioned above may contribute decisively to the mathematical modelling of processing technology by hydro-abrasive jet.

Using robust design in optimization of hydro-abrasive jet processing consists in choosing of an optimal combination of parameters that influence the process, so that it becomes insensitive to sources of variability, and the result obtained is maximal in terms of quality [1]. Dispersion values reduction is aimed and their focusing on a target value, tending to constant values by judiciously combining controlled parameters that define the system.

Advantages of a robust approach of hydro-abrasive jet processing optimization consist in the possibility of studying the dynamic of the whole factors ensemble, being in interdependence. Applying robust design on the technological process of hydro-abrasive jet

* Corresponding author: barabas.bogdan.florin@unitbv.ro

processing is achieved through determining noise factors z and preventing the propagation of their stochastic variation effects on a system y [2].

$$y = f(x, z) \tag{1}$$

where x represents values of controlled parameters of technological system. It is noted that the values of system responses, y, are also random. Developing the relationship (1) as a Taylor series, lead to expression:

$$f(x, z) = f(x, \mu z) + f'(x, \mu z)(z - \mu z) + \frac{f''(x, \mu z)}{2!}(z - \mu z)^2 \tag{2}$$

where μ is average of stochastic variables z. Equation (2) shows the influence of variable z on system y, respectively on transfer function f that represent AWJ system. Average of variance of answers given by system y is calculated using the equation (3):

$$Var(y) = \frac{1}{n}\sum_{i=1}^{n}\left(x_i - R_y\right)^2 \tag{3}$$

where n represents the number of random values system responses y; x_i represents values of controlled factors of system and R_y represents the average of those values.

$$R(y) = R[f(x, z)] \approx f(x, \mu_z) + f'(x, \mu_z)R(z - \mu_z) = f(x, \mu_z) \tag{4}$$

Of (3) and (4) equations, resulting Gauss approximation formula:

$$Var(y) = Var[f(x, z)] \approx Var[f'(x, \mu_z)(z - \mu_z)] = [f'(x, \mu_z)]^2\sigma_z^2 \tag{5}$$

where μ_z represent standard deviation for values of answers of system y for values variation of stochastics factors z.

For a hydro-abrasive jet processing system, z is the noise factor (variable which can´t be controlled completely), and x is the factor on which there is total control. The system responses y can be, where appropriate, cut surface roughness, processing cost, processing time, or other response. They may be controlled, their dispersion can be minimized, leading to a target value, by using the Gauss´s approximation [3]:

$$\mu_y \approx f(\mu_z) \tag{6}$$

$$\sigma_y^2 \approx [f'(\mu_z)]^2\sigma_z^2 \tag{7}$$

where μ_z şi σ_y represent average, respectively standard deviation for values of measured answers.

Optimization of technological process of hydro-abrasive jet processing is achieved by determining the parameters, their type and the variation interval of values [4].

2 Optimizing hydroabrasive jet processing by reducing the variability of responses

The variation of responses by hydro-abrasive erosion processing system is characterized by a measurable error that occurs from one piece to another, even if it retains control the same controlled parameters. This variation may be due to: unforeseen interactions of control factors; unexpected changes in any controlled parameter; environment (temperature, humidity fluctuations); user.

The difference between the obtained response and the expected response is the response error, which is a nonlinear function related to entries into the system, controlled parameters and uncontrollable factors (noises) and interactions occurring [5-6]

Errors that occur from one piece to another, although they are executed under the same conditions can be mitigated and managed to a minimum, thus optimizing the entire technological process.

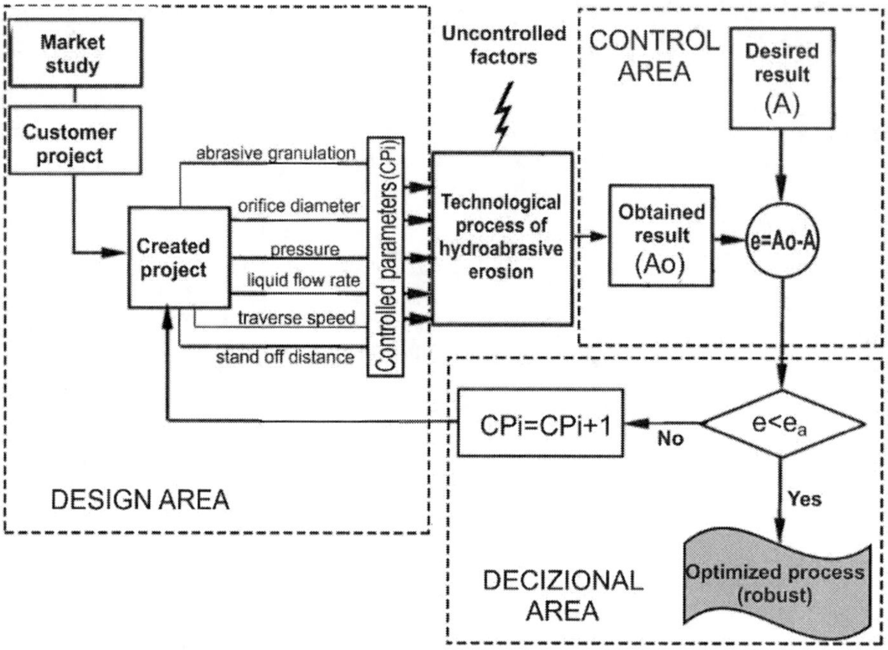

Fig. 1. Algorithm for optimization of erosion hydroabrasive process

Arrangement of controlled parameters so that whatever action of uncontrollable factors is, the difference between the desired result and the obtained result ($e = A_o - A$) is less than a permissible variation e_a conducting to robust results. If $e > e_a$, then move to a new combination of controlled factors until the algorithm will generate a controlled parameters solution that insensitize the system to uncontrolled actions and influences.

Considering, for example, a tolerated dimension the objective of optimizing the hydro-abrasive process, algorithm from Fig. 1, allow the following assertion:

$$Optim = Minim|A_0(CP_i) - A(CP_i)|, A_0(CP_i) \in [a, b] \quad (8)$$

where: A_0 is obtained dimension [mm]; A is desired dimension [mm]; CP_i represents controlled parameters setup; [a, b] represents tolerated dimensional interval accepted for nominal dimension A.

Optimisation is realized by minimizing the error and by insensitivity of results to any other influence outside that resulting to controlled parameters. Average error e_{med}, caused by noise, becomes:

$$e_{med} = \frac{A_{01} + A_{02} + \cdots + A_{0i}}{i} - A = \mu_{A0i} - A \quad (9)$$

By applying the relationships (8) and (9):

$$Optim = Minim|\mu_{A0i} - A|, A_0(CP_i) \in [a, b] \quad (10)$$

3 Determination of experiments plans

Reducing the dimensional variation of the pieces resulting from processing with hydroabrasive jet is quantified by minimizing the difference between the obtained dimension A_0 and nominal dimension, A (10) [7]. The formula used for minimization of dimensional variation is calculated using minimum effects caused by chosen parameters

$$e = e_{med} + \sum_{i=1}^{n} minE_i \tag{11}$$

where: - e is difference between effective obtained size and nominal size[mm]; e_{med} is arithmetic average of the measured e in all experiments [mm]; E_i is the effect of each combination of parameters.

For verification was used Taguchi method [8, 9]. The optimal combination of parameters is made using criterion of minimization of signal-noise ratio:

$$\frac{S}{N} = -10\log(s^2 + \bar{y}^2) \tag{12}$$

where: - S/N is ratios depending of parameters values (performance indicator); s is standard deviation of data; y is arithmetic average of data.

3.1 Configuration of parameters that influence the hydroabrasive processing

Setting of processing with hydroabrasive jet was applied on a machine Maxiem Jetcutting center 1530. Were used high pressure with following values: p_1 = 345 MPa and p_2 = 180 MPa. Were used two types of garnet abrasive: Mesh 80 and Mesh 100, with following characteristics (Table 1).

Table 1. Properties of used abrasives

No.	Type of abrasive	Grain size [mm]	Density [g/cm³]	Hardness [Mohs]	Purity
1.	Garnet Mesh 80	0.18	2.31	>8	>80%
2.	Garnet Mesh 100	0.15	2.40	>8	>80%

The used material is alloyed austenitic stainless steel X5CrNi 18-10 with following properties: Hardness Vickers, HV = 190; machinability, M = 80.8; modulus of elasticity, E = 200 GPa. Samples consisted in execution of rectangles with thickness H=10 mm, the measured nominal dimension, A=20 mm and B=50 mm. Five samples were achieved for each combination of factors, overall executed 40 samples.

The used software for AWJ setting and cutting was Intelli-Max Make, version 23.0. In Table 2 is shown the two levels of variance for input parameters.

Table 2. Parameters values used in experiments

Parameters	Traverse speed [mm/min]	Pressure [MPa]	Stand-off distance [mm]	Type of garnet [Mesh]	Abrasive mass flow [Kg/min]
	A	B	C	D	E
Level 1	60	180	1	80	0.05
Level 2	90	345	2	100	0.08

3.2 Experimental results

The plan of experiments is presented in Table 3. Obtained dimension A_0 is measured and in table is written the difference to the desired dimension A, considered theoretical dimension, by calibrating the software. Combinations of configurations parameters are influenced differently by uncontrollable noise factors [10-11].

Table 3. Matrix of experiments and their results

	A	B	C	D	E	Measurement $\|A_0-A\|$ [mm]					Arithmetic mean (y)	Standard deviation(s)	*S/N*
						1	2	3	4	5			
1	1	1	1	1	1	0.02	0.03	0.02	0.04	0.03	0.028	0.0083666	30.685
2	1	1	1	2	2	0.01	0.02	0.01	0.01	0.02	0.014	0.0054772	36.458
3	1	2	2	1	1	0.02	0.05	0.05	0.04	0.03	0.038	0.0130384	27.921
4	1	2	2	2	2	0.03	0.02	0.02	0.04	0.03	0.028	0.0083666	30.685
5	2	1	2	1	2	0.06	0.04	0.04	0.05	0.06	0.05	0.01	25.850
6	2	1	2	2	1	0.05	0.04	0.04	0.06	0.05	0.048	0.0083666	26.245
7	2	2	1	1	2	0.03	0.02	0.01	0.02	0.01	0.018	0.0083666	34.045
8	2	2	1	2	1	0,02	0.03	0.01	0.03	0.00	0.0175	0.015	32.747

In classical method is calculated the average response for each factor and for each level. The calculated responses is shown in Table 4, where in row 1 is average response and in row 2 is shown the average effect.

Table 4. The average response and calculated effect of combined parameters

	A1	A2	B1	B2	C1	C2	D1	D2	E1	E2
1	0.027	0.03337	0.035	0.02537	0.0193	0.041	0.0335	0.0269	0.03287	0.0275
2	-0.003	0.003	0.005	-0.005	-0.011	0.011	0.003	-0.003	0.0027	-0.0027

Using equation (11) is obtained $e_{min}= |A_0-A| = 0.006$ [mm] and parameters value combination: 12122 (Fig.2).

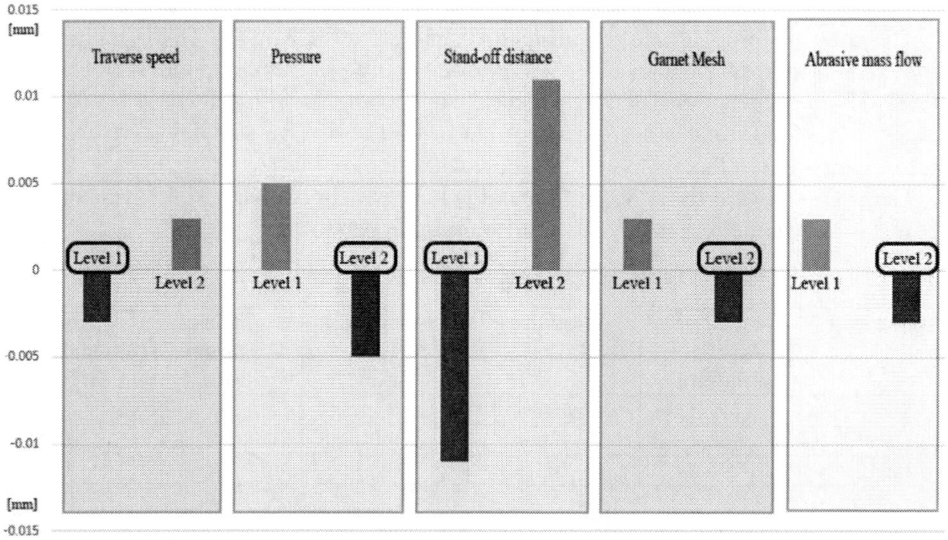

Fig. 2. Effect of different levels of parameters on dimensional variability in AWJ process

With Taguchi method is calculated the response for every ratio *S/N* and the resulted effect shown in Table 5, where in row 1 is average S/N response and in row 2 is shown the average effect.

Table 5. The average response of S/N and calculated effect of combined parameters

	A1	A2	B1	B2	C1	C2	D1	D2	E1	E2
1	31.437	29.721	29.809	31.349	33.484	27.675	29.625	31.534	29.399	31.759
2	0.857	-0.857	-0.769	0.769	2.904	-2.904	-0.954	0.954	-1.180	1.180

With Taguchi method, the optimal combination of input parameters is 12122.

4 Conclusions

Were executed 40 samples which consisted in cutting rectangles from steel plate X5CrNi 18-10, with thickness H = 10 mm. Were chosen 5 parameters with notable influence on dimensional precision, receiving 2 values, each parameter. To the 5 parameters with 2 levels, was attributed matrix of plans of experiments from tab. 3 and for each combination of parameters were performed 5 measurements, introducing in tab.3, the arithmetic mean and standard deviation. The signal-noise report was calculated with relation (12) for minimization of dimensional variability.

The arithmetic average of all measurements taken, representing the difference in absolute value, compared to the nominal size was e_{med} = 0.03 mm, equivalent to a tolerance field, equal with 0.06 mm. For certain combinations, the tolerance field was situated at 0.1 mm (21212). In Fig. 2 it is observed that certain levels of parameters, influences dimensional variability in the sense of magnifications of tolerance field, and other levels reduces tolerance field. Choosing those levels which minimizes the tolerance field, implicitly dimensional variability is obtained according to (11) e_{min} = 0.006 mm, equivalent with a tolerance field equal with 0.012 mm. Using plans of experiments, allows choosing a combination of parameters, that, even if was not used in performing tests, it is the optimal combination resulting in reduced variability and increased dimensional accuracy of execution.

References

1. V. Sharma, S. Chattopadhyaya, S. Hloch, Int J Adv Manuf Technol **56**, 1019–1025 (2011)
2. M.Arner, *Statistical robust design,* (Wiley, 2014)
3. J. Kechagias, G. Petropoulos, N. Vaxevanidis, Int J Adv Manuf Technol **62**, 635–643 (2012)
4. C. W. Kang, P. H. Kvam, *Basic Statistical Tools for Improving Quality*, (Wiley 19-25, 2011)
5. D. K. Shanmugam, S. H. Masood, J. Mater. Process Technol. **209,** 3887-3893 (2009)
6. A. Perec, Proc.Eng., **149,** 366-373 (2016)
7. D. Trainer, Int., J. Multi. Flow, **81,** 77-87 (2016)
8. A. Radvanska, T. Ergic, Z. Ivandic, S. Hloch, J. Valicek, J. Mullerova, Strojarstvo **51,** 347-354 (2009)
9. G. Taguchi, S. Chowdhury, Y. Wu, *Taguchi's Quality Engineering Handbook,* (Wiley, 239-263, 2005)
10. R. Shukla, D. Singh, Swarm and Ev. Comp. (to be published)
11. Jagadish, S. Bhowmik, A. Ray, Int J Adv Manuf Technol **87**, 1359–1370 (2016)

Researches regarding influence of traverse speed and stand-off distance to the roughness in AWJ process

Bogdan Barabas[1,*], and *Tudor* Deaconescu[1]

[1]Transilvania University of Brasov, Faculty of Technological Engineering and Industrial Management, Department of Engineering and Industrial Management

Abstract. This paper aims to establish a relationship between surface roughness results from hydroabrasive processing and parameters with controlled variability, traverse speed and stand-off distance. Measuring the influence of these parameters is done using statistical and experimental tools. Establishing the relationship between the variation of input signals and the variation of result, is performed by introducing coefficients whose values determine the interdependence function between the input signal and response. Have been obtained diagrams of behavior of hydroabrasive process which allowed conclusions about the variability of the results obtained. Thus it can be concluded that the traverse speed and stand-off distance are input signals and they depends on the initial configuration of the system and its adjustment possibilities by variability of controlled parameters. Roughness is a result of this application, clearly defined which can be optimized by setting controlled parameters.

1 Introduction

Hydro-abrasive jet processing system is an open system with inputs set according to technical possibilities and outputs consisting of finished pieces with clear odds and technological conditions, with parameters of influence that can be controlled with variable parameters acting without control possibilities. All these elements are in a state of interdependence, changing of an element having predictable actions or not.

By changing the value of an element of influence we will get many possible outcomes that do not have a linear evolution. This is due to the effects that changing of the influence element value has on others, depending to it. Thus, for example, a finer granulation of abrasive does not necessarily achieve a best roughness (granulation abrasive influences mass flow, which can cause an increase in roughness) [1-2] or a higher pressure does not necessarily mean dimensional accuracy increased (increased pressure leads to accelerated aging, so a possible decrease in accuracy, etc). In Fig. 1 is shown the jet hydro-abrasive processing. Its optimization by using the method of robust design is achieved by applying criteria such as *Optimum = m (smaller, better)*, for roughness, and *Optimum = T (target achievement)* for dimensional accuracy.

* Corresponding author: barabas.bogdan.florin@unitbv.ro

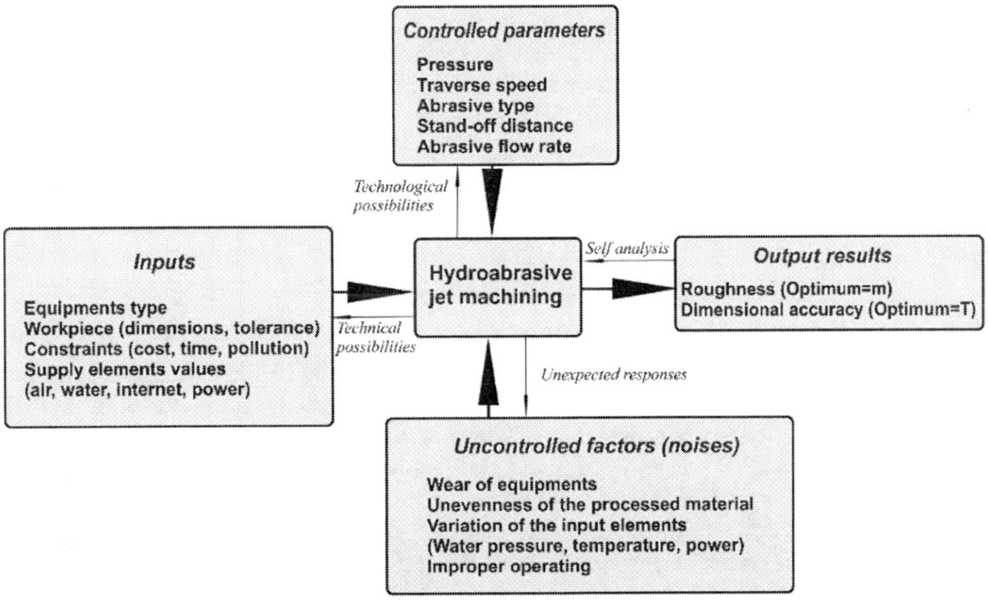

Fig. 1. Hydroabrasive jet processing – parametric diagram.

From the analysis of the presented diagram the dependence of the results, both for the system inputs and for the controlled and disruptive factors is observed. The controlled parameters image acts on the system as a whole. The elements of the respective matrix are strongly interrelated (quality feature does not increases or decreases linearly with each parameter), and the effects of random factors and inputs to the system are added to quantify the results [3, 4]

The factors that characterize jet hydro-abrasive processing are: external factors (abrasive material quality, hardness of the work piece, ambient temperature, environmental humidity, vibrations occurred during processing, variations in electric current, etc.), internal factors (wear of the nozzle, of the concentration tube, of the pump, etc.), processing factors (water pressure, processing speed, tube diameter, processing distance, cutting forces, etc.), human factors (operator experience, operator fatigue, etc.).

All these factors can be controlled, partially controlled or uncontrollable. Their management through a robust approach leads to optimization of technologic process by increasing its insensitivity to the effect of uncontrollable factors. Constance of results allows directing them to the intended target without damaging variability of target. Given that input elements have fixed values and noise factors can't be controlled, it results that the choice of optimal configurations of controlled parameter values will decide the hydro-abrasive processing results.

The optimization criterion chosen (to *minimize* – in the case of roughness, or processing cost, to *maximize* if productivity aim is chosen, or to *achieve the target* if dimensional accuracy aim is chosen) is satisfied by choosing the optimal configuration of controlled factors values [5].

In the case of jet hydro-abrasive technological processing system, as main requirement appears the part's drawing. This is accompanied by part's quotas and tolerances (thickness, roughness, accuracy, etc.), the material's properties (hardness, workability, material type, chemical structure, etc.). The part's drawing is correlated with the parameters of jet processing facilities parameters, for the part be performed according to the requirements and possibilities of the machine.

Determining the execution possibility of the piece is followed by conditioning (other than those pursued as optimized result): production cost, execution time. Other input signals are constituted by elements of external supply of the hydro-abrasive processing system: pressure, temperature and water quality, values of tension and intensity of electricity supply, air supply pressure value.

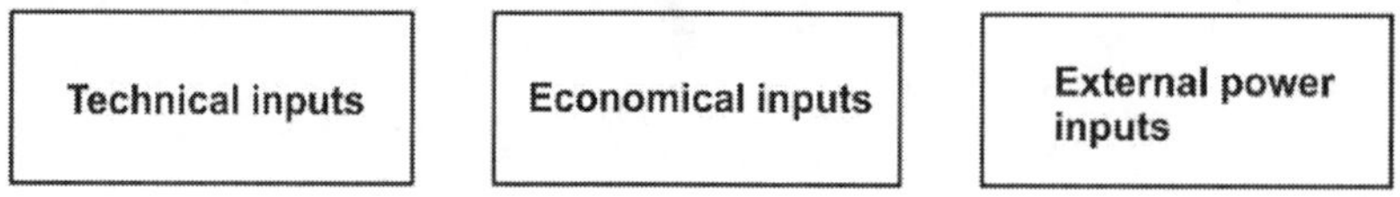

Fig. 2. Input signals in hydroabrasive jet system.

To the left of the scheme is shown the target to be reached (if the part can be executed) economic type inputs are positioned in the middle of the scheme (part execution must be economically justified) and to the right of the scheme are highlighted input values to the system related to system powering from external networks. Signals sent towards the hydro-abrasive jet technological processing system are of "technological requirement", "economic conditionality" and "feed values from the external network" type. The outputs from the system (work piece) must correspond to conditionings. A non-compliant output is an error (scrap). The multitude of not-compliant outputs is directly proportional to the variation of responses offered by the system. The objective of processing is to obtain a compliant result to initial conditionings.

2 Theoretical considerations

Optimization of hydro-abrasive jet process is to achieve the target proposed in accordance with the initial objective [target criterion if it is about a nominal quota proposed by the design or minimizing / maximizing criterion for the case of roughness, processing time, processing cost (minimizing) or profit (maximizing)] (Fig. 3).

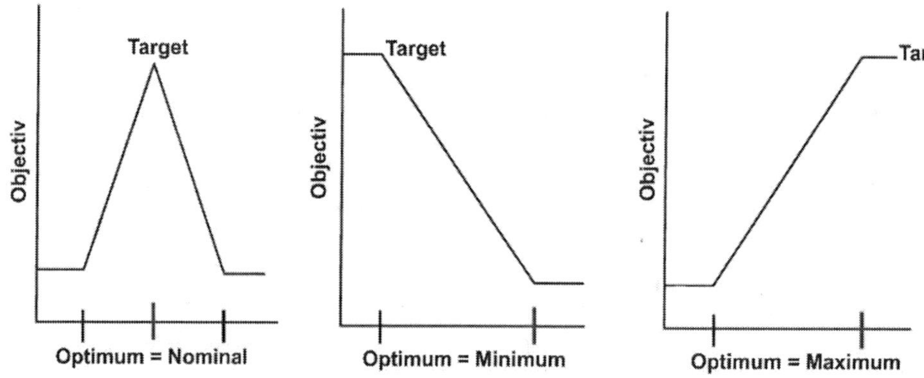

Fig. 3. Optimization of an open system based on defined objectives.

Reducing variation in responses of hydro-abrasive jet processing system is done by choosing operating parameters at which the system to respond consistently, regardless of randomly acting uncontrolled factors [6-7].

Technical and economical-type input values are conditionality in defining qualitative compliance being quantified within assessment process of responses, becoming optimized targets. System response to input values K, to control factors Pc, noise factors Z and interactions occurring I is noted by Y and has the form [8-9]:

$$Y = f\ (K,\ P_c,\ Z,\ I) \quad (1)$$

In the case of a technological process, Taguchi [10] conditions the acquisition of a response according to customer requirements by browsing three steps: selecting the system required for receipting the response; defining controlled parameters and choice of optimal level of thereof; identifying sources of noise and interactions that affect response (response tolerance).

Considering response Y as a desired result of the introduction in the system of input values and controlled factors, its variability is an undesirable outcome, caused by uncontrolled factors (noise) and unforeseen interactions occurred within the system between controlled factors. Relation (1) become:

$$Y = \alpha K + \beta P_c + [f(K, P_C, Z, I) - (\alpha K + \beta P_c)] \quad (2)$$

where: $(\alpha K + \beta P_c)$ is the signal received by the system to release the desired response; $[f(K, P_c, Z, I) - (\alpha K + \beta P_c)]$ is noise induced in the system by uncontrolled factors and interactions with unpredictable effects, occurring between controlled factors.

The existence of a large number of input signals leads to an enormous volume of calculations in assessing the final result. To reduce the volume of calculations the following rules are introduced:

1. Input signals from of fixed type (material properties of the piece, values of power supply elements, certain parameters of the machine – axles number, machining precision) are considered to have zero influence on the outcome variability. Instead, the variability of these signals act on the result as an element of noise.
2. Variable input signals, with the possibility of control, whose influence on the results is smaller than the louder noise influence, are considered negligible, being appreciated at their turn as noise [11].

Measuring the influence of each input signal is done by using statistical tools (regression analysis) and experimental tools. Establishing a relationship between the input signal variation and result variation represents how to measure the influence of that signal. An example of application relations cause - effect in the case of hydro-abrasive jet technological processing system can be seen in Fig. 4.

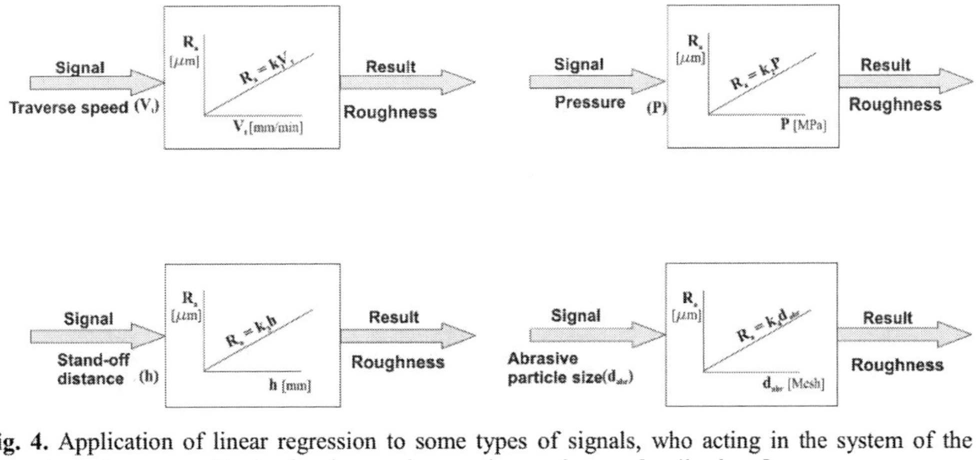

Fig. 4. Application of linear regression to some types of signals, who acting in the system of the technological process of hydroabrasive erosion, on the roughness of realized surfaces.

Applying equations (1) and (2) to the diagram shown in Fig. 4, wherein the response Y is exemplified as value of surface roughness R_a, and the configuration parameters are the speed of traverse V_t, the pressure P, the distance from focus tube h, and the abrasive particle diameter d_{abr} there were obtained ideal dependence functions of the response R_a and each configuration parameter. Thus, for the relationship between roughness R_a and traverse speed V_t the ideal function is:

$$R_a = k_1 V_t \quad (3)$$

where k_1 is a coefficient that takes account of the value of traverse speed.

Also, the respone R_a is influenced by other factors [12]:

$$R_a = f(V_t, P, h, d_{abr}) \quad (4)$$

If there are more parameters x, relation (4) can be written:

$$R_a = f(V_t, x_1, x_2, \ldots, x_n) \quad (5)$$

$$R_a = \gamma V_t + [f(V_t, x_1, x_2, \ldots, x_n) - \gamma V_t] \quad (6)$$

It follows that the first term γV_t it is the signal that lead to the answer R_a, and the second term, f (V_t, x_1,x_2,...x_n)-γV_t, is the noise who causes variability of response R_a.

3 Experimental results

Determination of the influence of the traverse speed on the roughness was carried out on samples of austenitic stainless steel X5CrNi 18-10 with following properties: Hardness Vickers, HV = 190; machinability, M = 80.8; modulus of elasticity, E = 200 GPa. Depth of cut was chosen H_1 = 10 mm, H_2 = 20 mm, H_3 = 30 mm. The input configuration is: pressure P = 345 MPa, type of garnet d_{abr} = 80 Mesh, orifice diameter d = 0.28 mm, stand-off distance h = 1 mm, length of cut L = 100 mm. Traverse speed, V_{t1} = 20 mm/min, V_{t2} = 30 mm/min, V_{t3} = 40 mm/min, was determined based on the time and distance, modifying material thickness in processing setup in Maxiem Intelli-Max software version 23.0, from Maxiem 1530 equipment. The measurements of roughness was made at 2 mm of bottom of cut material and are shown in Table 1.

In classical method is calculated the average response for each factor and for each level. The calculated responses is shown in Table 1.

Table 1. Roughness values for different traverse speed and material thickness

No.	Traverse speed [mm/min]	Material thickness [mm]	Roughness [µm]				
			1	2	3	Arithmetic mean	
1.	20	10	2.105	2.213	2.078	2.132	2.888
2.		20	2.964	3.251	3.187	3.134	
3.		30	3.340	3.642	3.219	3.400	
4.	30	10	3.641	3.447	3.592	3.560	4.328
5.		20	4.155	4.409	4.273	4.279	
6.		30	5.261	4.903	5.271	5.145	
7.	40	10	4.111	4.453	4.066	4.213	5.782
8.		20	5.827	6.081	5.767	5.891	
9.		30	7.402	7.045	7.281	7.242	

Applying equation (3) on the values obtained by experiment, are obtained for coefficient k_1, 3 values, respectively, k_{11} = 0.144; k_{12} = 0.1442; k_{13}= 0.1445

To determine the influence of stand-off distance on roughness were chosen the same basic configuration value, with a constant traverse speed for all samples, V_t = 30 mm/min.

Stand-off distance has following values $h_1 = 1$ mm; $h_2 = 2$ mm; $h_3 = 3$ mm. The results are shown in Table 2.

Table 2. Roughness values for different stand-off distance and material thickness

No.	Stand-off distance [mm]	Material thickness [mm]	Roughness [μm]				
			1	2	3	Arithmetic mean	
1.	1	10	3.641	3.447	3.592	3.560	4.328
2.		20	4.155	4.409	4.273	4.279	
3.		30	5.261	4.903	5.271	5.145	
4.	1.5	10	4.264	4.189	4.190	4.214	5.582
5.		20	5.910	5.812	5.673	5.798	
6.		30	6.705	6.589	6.911	6.735	
7.	2	10	5.395	5.510	5.481	5.462	6.862
8.		20	6.895	7.012	7.141	7.016	
9.		30	8.262	8.093	7.969	8.108	

Applying equation (3) on the values obtained by experiment, according Fig.4, are obtained for coefficient k_3, 3 values, respectively $k_{31} = 4.328$; $k_{32} = 3.721$; $k_{33} = 3.431$.

4 Conclusions

The experiments shown in table 1 leads to the conclusion that traverse speed influences the roughness of processed surfaces, with increasing of traverse speed, increasing the roughness. For different thickness of material processed, are obtained different graphics of influence. (Fig.5). Coefficient k_1 still remains nearly constant. Based on experiments, optimal traverse speed, $V_{topt} = R_{aopt}/k_1$, can be fixed on the machine, using thickness variation to yield a required surface roughness.

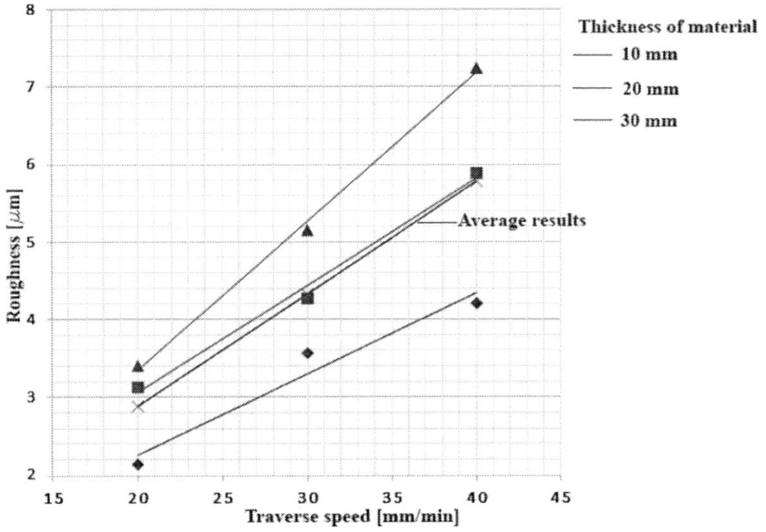

Fig. 5. Influence of traverse speed V_t, on roughness R_a for different thickness of material.

Increased stand-off distance leads to increased roughness (Table 2). With the growth of stand-off distance, the hydro-abrasive jet loses the cutting energy, due to scattering of abrasive particles. The coefficient k_3 is not a constant coefficient, decreasing with growth of thicness of material In Fig. 6 are shown graphs of influence of stand-off distance on roughness.

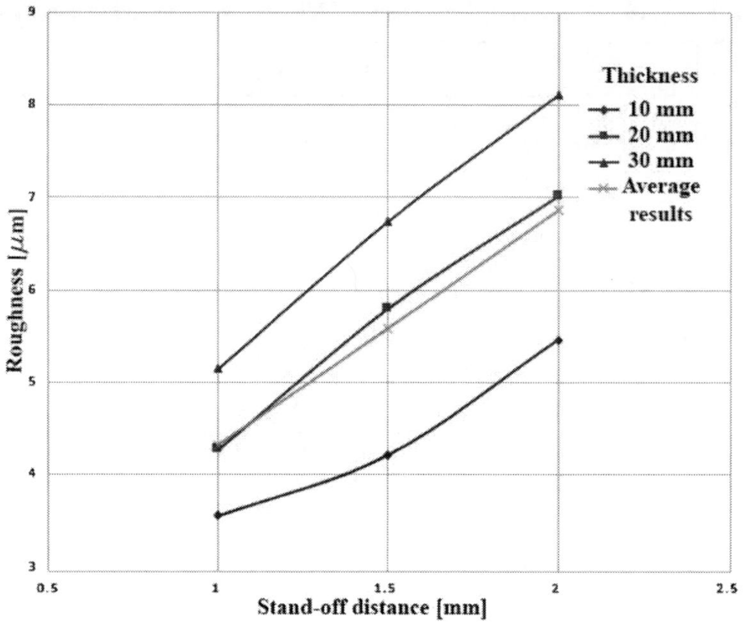

Fig. 6. Influence of stand-off distance h on roughness R_a for different thickness of material.

This research opens the door to extensive research in this area, in order to optimize cutting abrasive water jet for different types of materials with variable thickness.

References

1. P. Loschner, K. Jarosz, P. Niesłony, Proc. Eng. **149** 276 – 282 (2016)
2. M.Arner, *Statistical robust design,* (Wiley, 2014)
3. M. Bagci, Trib. Int. **94,** 336–345 (2016)
4. D. Begic-Hajdarevic, A. Cekic, M. Mehmedovic, A. Djelmic, Proc. Eng. **100**, 394 – 399 (2015)
5. J. Kechagias, G. Petropoulos, N. Vaxevanidis, Int J Adv Manuf Technol. **62**, 635–643 (2012)
6. M. Santhanakumar, R. Adalarasan, M. Rajmohan J. Mec. Sci.Tech. **30** 1371-379 (2016)
7. S. A. Barabas, A. Florescu, G. Calefariu, F. Sarbu, Adv. Mater. Res., **1128,** 293-303 (2015)
8. K. S. Kumar, S. Karthick, App. Mech. Mat, **592-594,** 106-111 (2014)
9. R. Jafar, J.K.Spelt, M.Papini, Wear, **303,** 138–145 (2013)
10. G. Taguchi, S. Chowdhury, Y. Wu, *Taguchi's Quality Engineering Handbook,* Wiley, 239-263 (2005)
11. D.V. Srikanth, M. S. Rao, Mat. Today: Proc. **2** 3308 – 3317 (2015).
12. P. Lozano Torrubia, D. A. Axinte, J.Billingham, Int J Machine Tools Man. **95** 39–51 (2015)

Study of hydrodynamic behaviour of large bearings depending on the viscosity of the lubricant

Sorin Barabas[1,*], *Adriana* Florescu[1] , and *Flavius* Sarbu[1]

[1]Transilvania University of Brasov, Faculty of Technological Engineering and Industrial Management, Department of Engineering and Industrial Management

Abstract. The research refers to study of hydrodynamic behavior of large bearings equipped with automatic lubrication system, where the liquid used may have a viscosity with variable values depending on the temperature or depending on the additive used. The introduction of additives with nanoparticles causes modification of viscosity and a significant reduction of friction coefficient. This study establishes a direct link between the friction, viscosity and tensions that occur in the contact area. Experimental studies confirm improving tribological properties of the contact area between the bearing rollers and raceways, through attracting nanoparticles on surfaces in contact and reducing friction coefficient. Accomplished finite element analysis showed decrease of the contact stresses, therefore decrease of the wear and increase the life of the bearing. Research has concluded that influencing viscosity and friction coefficient of lubricant can result in reducing wear of the bearing components, as well as increase the lifetime of bearing.

1 Introduction

Large bearings, used in wind energy assemblies, are mobile components that support heavy loads, in variable conditions. The rolling elements are designed to withstand at working conditions, counting on Hertz contact theory [1], lifetime theories [2], type of movements, dynamic loads and other phenomena who acts in bearings use. A decisive role in increasing bearings lifetime is owned by lubrication.

The quality of lubricants [3] and computerized methods of lubrication [4] protect the rolling elements again wear. The presence of additives with nanoparticles, or other thickener, determine the lubrication process, in terms of increasing the carrying capacity of fluid, reducing the roughness of surfaces in contact, decreasing the working temperature and modifying the viscosity of lubricant. Changes occurred at nanometric level and interactions that occur between influenced factors, have an effect that is not fully explained by elastohydrodynamic theory. Knowledge of lubrication system depends by necessity for analysis of the phenomenon, and conducting experiments that highlights the link between the viscosity of lubricant, dynamic load and contact stress appeared in rolling elements.

* Corresponding author: sorin.barabas@unitbv.ro

The influence chart of nanoparticles insertion on the hydrodynamic behaviour of bearings is presented in Fig.1.

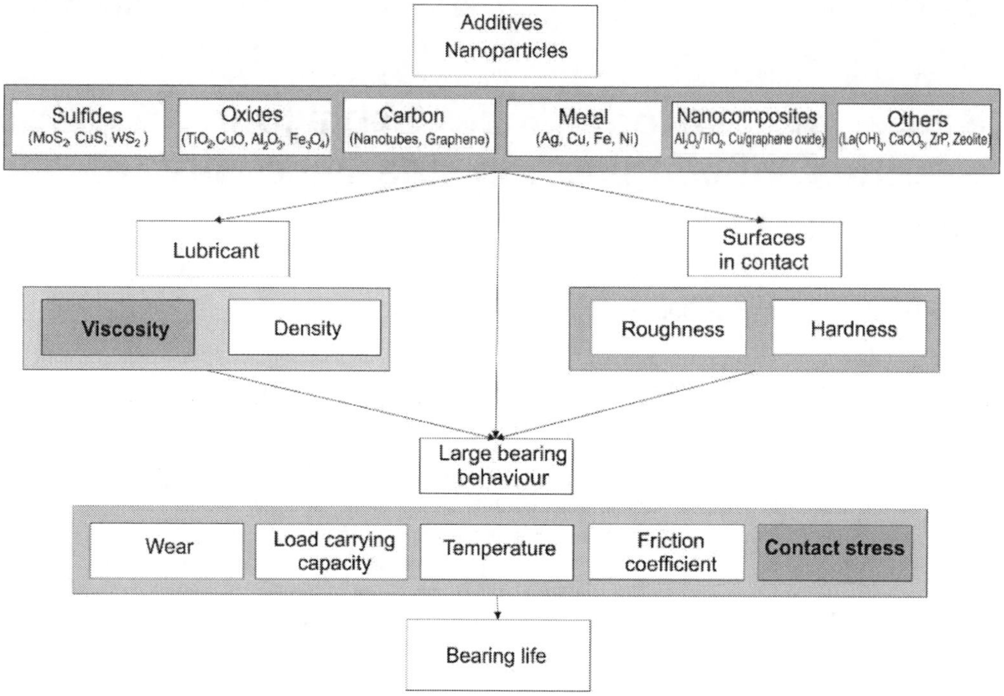

Fig. 1. The influence of nanoparticles insertion on the hydrodynamic behavior of bearings

The insertion of nanoparticles change the lubrication mechanism, improving tribological performance and decreasing the friction coefficient. Several nanoparticles like sulfides, oxides, carbon-based nanoparticles, nanocomposites (MoS_2, TiO_2, SiO_2, carbon nanotube, graphene, Al_2O_3/TiO_2), mixed in the lubricant, have multiple effects: nanoparticles penetrate the surface in contact, decreasing roughness and thus, the friction coefficient, also, grow carrying capacity of the lubricant film with raising of viscosity and decrease of temperature. Depending on the type and size of nanoparticles, friction reduction is 25% to 50% and wear reduction is aprox. 25% [5]. For metal nanoparticles (Ag, Fe, Ni, Cu, Sn) is recorded an effect of forming a protective film, raising the sliding process and the sintering effect, due to chemical absorption of nanoparticles in surfaces in contact and form a thin layer at nanometric range, having hydrodinamic properties. The multiple actions of nanoparticles insertion in lubricant on the bearing behavior, make this important subject, a matter difficult to modeling because of interdependence of considered effects.

Because the resulted viscosity depend of type, quantity and consistency of thickener, of type and viscosity of base lubricant, temperature of process, sliding speed, size of surfaces, this paper aims to establish only a relation between viscosity of lubricant (modify by nanoadditives) and contact stress occurred in contact zone with use of FEM analysis and some constraints placed to rolling elements.

Large bearings with cylindrical rollers, utilized in wind power energy, have a moderate rotational speed but the presence of large dynamic load forces require a good lubrication with monitoring the thickness of lubricant film [6]. Are usually used a grease with a mineral oil base and lithium thickener. For maintaining a lubricant film at medium speed and heavy loads, a good results was obtained with MoS_2 and carbon-based nanoparticles additives. [7].

2 Mathematical modeling

Computation of the contact stress occurred in contact zone, between roller and raceway is based on hypothesis that the surfaces in contact are separated by a lubrication film whose thickness is changing with viscosity of grease. Some constraints are made: the lubricant is incompressible, on temperature interval of bearing function, the variability of viscosity of lubricant has negligible influences on chemical composition, the curvature of thickness of lubricant film is ignored and effect of turbulence is considered null because the speed of elements in contact is reduced [8].

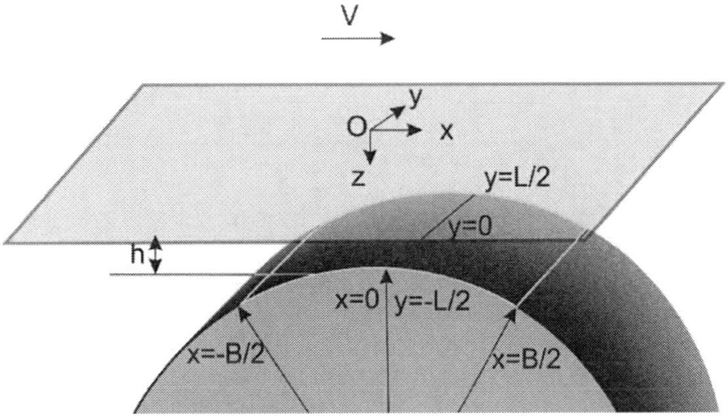

Fig. 2. The geometry of formation of lubricant film

The geometry of lubricant film is shown in Fig.2. Film formation occurs to area 2πRxL and Reynolds equation is:

$$\frac{\partial}{\partial X}\left[\frac{h^3}{\mu}\frac{(\partial p)}{(\partial X)}\right] + \frac{\partial}{\partial Y}\left[\frac{h^3}{\mu}\frac{(\partial p)}{(\partial Y)}\right] = 6\mu V\frac{\partial h}{\partial X} \tag{1}$$

where: h is film thickness [mm]; μ is lubricant viscosity [Pa.s]; p is pressure of lubricant [MPa]; V is linear velocity [mm/s]; X is distance on direction parallel to sliding X=2πR [mm]; y is distance on direction normal to sliding $Y=y/L$ [mm], $\partial p/\partial X$ and $\partial p/\partial Y$ are the pressure gradients in the bearing.

Is noted integrals of thickness lubricant [9] I_1 and I_2 [mm/Pa.s]:

$$I_{1(x,y,t)} = \int_0^h \frac{1}{\mu(x,y,t,\xi)} d\xi \tag{2}$$

$$I_{2(x,y,t)} = \int_0^h \frac{\xi}{\mu(x,y,t,\xi)} d\xi \tag{3}$$

where: ξ is parameter of thickness lubricant.

By integration is obtained [5]:

$$Q = \int_0^L \int_0^{2\pi R} p(x,y)dxdy \tag{4}$$

where: Q is load capacity [daN]

The viscosity model is changed for very little particles. It was studied by Brinkman in 1952 [11], who define the relative viscosity ($\bar{\mu}$) the ratio between viscosity of normal lubricant (μ_{nl}) and viscosity of lubricant with nanoadditives (μ_{na}). In 1959 was developed Krieger-Daugherty equation [12]:

$$\bar{\mu} = \frac{\mu_{nl}}{\mu_{na}} = \left[1 - \frac{\phi}{\phi_m}\right]^{\mu_i \phi_m} \quad (4)$$

where: $\bar{\mu}$ is non dimensional relative viscosity; μ_{nl} is viscosity of normal lubricant [Pa.s]; μ_{na} is viscosity of lubricant with nanoadditives [Pa.s]; Φ is nanoparticle volume fraction; Φ_m is maxim particle volume fraction

The equation of viscosity (4) was modified by Chen in 2007, for lubricant with nanoparticle aggregate structure [13]:

$$\phi_a = \left[1 - \frac{\phi_a}{\phi_m}\right]^{-2.5\phi_m} \quad (5)$$

where:

$$\phi_a = \phi \left(\frac{a_a}{a}\right)^{3-D} \quad (6)$$

where: a is radius of primary nanoparticles [nm]; a_a is radius of aggregate of nanoparticles [nm]; D is a fractal index.

The relation [14] between shear stress given by viscous behaviour and viscosity is:

$$\overline{\tau_{ij}} = \bar{\mu}\left(\frac{\partial u_i}{\partial x_j} + \frac{\partial u_j}{\partial x_i} - \frac{2}{3}\delta_{ij}\nabla\vec{V}\right) \quad (7)$$

where: τ_{ij} is shear stress given by viscous lubricant; $\vec{V}$ is velocity vector in nanolubricant; u_i and u_j are components of velocity on the x, y directions; δ_{ij} is function Kronecker delta.

Deformation h_0, occurred between roller and inner ring surfaces [15] can be written as:

$$h_0 = \frac{2Q}{\pi LE}\left[ln\left(\frac{2R_1}{b}\right) + 0.407\right] + \left[ln\left(\frac{2R_2}{b}\right) + 0.407\right] \quad (8)$$

where: h_o is deformation occurred to the contact between roller and inner ring [mm]; Q is load capacity [daN]; E is Young's elastic modulus [daN/mm^2]; R_1 is roller radius [mm]; R_2 is inner ring radius [mm]; B is length of solid contact between surfaces [mm].

3 FEM analysis for roller bearings with lubricant provided with nanoadditives

Using FEM analysis for study the bearing behaviour in EHD lubrication is made through discretization of equilibrium equations. Bearing chosen for finite element analysis is a large bearing with D=1600 mm, d=1100 mm, and B=440 mm. The roller have following dimensions: D=90 mm, L=180 mm. Material for rollers was SAE 3310. Other properties of used material: Rockwell Hardness – 63 HRC, E=200 kN/mm^2. For simulation, are considered 4 types of lubricant with different kinematic viscosity (ISO VG 100, ISO VG 220, ISO VG 460, ISO VG 1000)

The influence of viscosity on the contact stress and deformations are shown in (1), (7) and (8) equations. The researches done using finite element analysis are executed for 4 different load capacity: 100 kN, 150 kN, 200 kN and 250 kN.

The results of FEM analysis are show in Table 1 and Table 2.

In Fig. 3 and Fig.4 are shown graphics of Von Mises stress and deformations occurs in the roller bearing, depending on viscosity for different values of load capacity.

DOI: 10.1051/matecconf/201712103003

Table 1. Values of stress and deformation for different type of lubricant

Type of lubrifiant	Von Mises stress (N/mm^2)	Deformation (mm)	Type of lubrifiant	Von Mises stress (N/mm^2)	Deformation (mm)
Q=100 kN			Q=150 kN		
1	1028	0.0151	1	1033	0.0154
2	1016.8	0.0142	2	1026.8	0.0148
3	1005	0.0129	3	1011.7	0.0133
4	997	0.0115	4	1006.4	0.0121

1 - ISO VG 100 (μ=100 cSt) ; 2 - ISO VG 220 (μ=220 cSt) ; 3 - ISO VG 460 (μ=460 cSt) ; 4 - ISO VG 1000 (μ=1000 cSt).

Table 2. Values of stress and deformation for different type of lubricant

Type of lubrifiant	Von Mises stress (N/mm^2)	Deformation (mm)	Type of lubrifiant	Von Mises stress (N/mm^2)	Deformation (mm)
Q=200 kN			Q=250 kN		
1	1037	0.0163	1	1041	0.0175
2	1031.4	0.0154	2	1036.8	0.0170
3	1023.3	0.0143	3	1030.5	0.0163
4	1016.4	0.0130	4	1026.1	0.0148

1 - ISO VG 100 (μ=100 cSt) ; 2 - ISO VG 220 (μ=220 cSt) ; 3 - ISO VG 460 (μ=460 cSt) ; 4 - ISO VG 1000 (μ=1000 cSt).

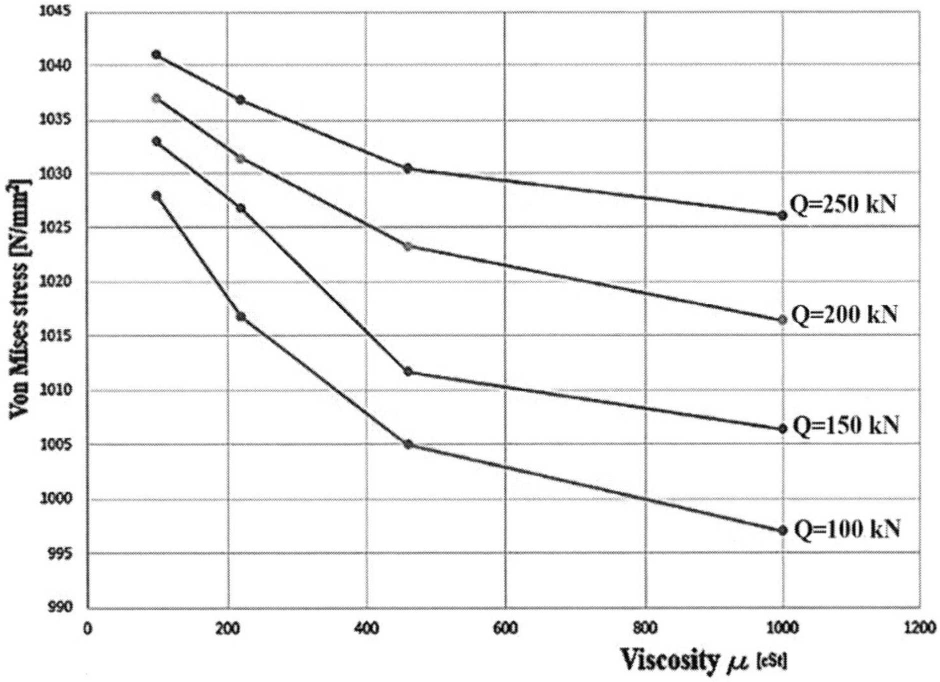

Fig. 3. The graphics of Von Mises stress occurs in the roller bearing, depending on viscosity for different values of load capacity

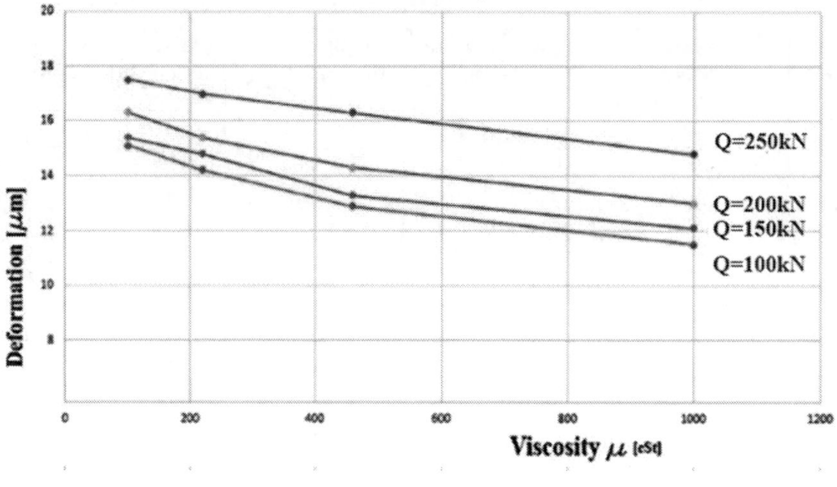

Fig. 4. The graphics of deformations occurs in the roller bearing, depending on viscosity for different values of load capacity

4 Conclusions

In the two graphics is evident influence on viscosity on stress occurs in rolling elements and on deformations of rollers. In Fig.4, adding of nanoadditives increase viscosity of lubricant and decrease Von Mises stress, with significant value. In Fig. 4 was observed a slight decrease of deformations with increasing the viscosity of lubricant.

Use of nanoadditives gives to the lubricant much better properties who confirm improving tribological properties of the contact area between the bearing rollers and raceways, through attracting nanoparticles on surfaces in contact and reducing friction coefficient.

Finite element analysis performed with Nastran program revealed a decrease of stress and deformations.

References

1. S. Timoshenko, J. N. Goodier, *Theory of Elasticity* McGraw-Hill, 466- 490 (1970)
2. E. Laniado-Jacome, J. Meneses-Alonso, D. Lopez, Trib. Int. **43**, 2175-2182 (2009)
3. G. Xie, D. Guo, J. Luo, Trib. Int. **84,** 22–35 (2015)
4. S. Barabas, A. Florescu, Matec Conf. COSME 16, **94** (2017)
5. W. Dai, B. Kheireddin, H. Gao, H. Liang, Trib. Int. **102,** 88–98 (2016)
6. X. Liu, N. Xu, W. Li, M. Zhang, L. Chen, W. Lou, X. Wang Trib. Int. **109,** 467–472 (2017)
7. A. Kornaev, L. Savin, E. Kornaeva, A. Fetisov Trib. Int. **101,** 131–140 (2016)
8. D. Bonneau, A. Fatu, D.Souchet, *Thermo-hydrodynamicLubrication in Hydrodynamic Bearings,* (Wiley, USA, 2014)
9. X. Meng, M.M. Khonsari Tribology International **107,** 116–124 (2017)
10. H.C. Brinkman, J. Ch. Phys. **20,** 571 (1952)
11. K.G. Binu, B.S. Shenoy, D.S. Rao, R. Pai, Proc. Mat. Sci. **6,** 1051 – 1067 (2014)
12. H. Chen, Y. Ding, C. Tan, New J. Phys. **9,** 13-19 (2007)
13. H. Shahmohamadi, R. Rahmani, H. Rahnejat, C.P. Garner, N. Balodimos, Trib. Int. (to be published).
14. F. J. Ebert, Chin. J. Aer. **23**, 123-136 (2010)

The influence of the weld toe grinding and wig remelting weld toe rehabilitation techniques, on variable stresses, in case of cross fillet welds, reinforced with additional welding rows

Claudiu Babis[1,*], *Andrei* Dimitrescu[1], *Gabriel* Iacobescu[1], *Gheorghe* Solomon[1], and *Oana* Chivu[1]

[1]Politehnica University of Bucharest, Materials and Welding Technology Department, 313Splaiul Independenţei, 060042 Bucharest, Romania

Abstract. Variable stresses where the load value varies between a maximum and a minimum value, or varies the position in time, cause after accumulating a large number of load cycles in those structures, the emergence of drug fatigue. Fatigue is characterized by failure on values of the applied stress from the load cycles, below the material flow, values which in case of static stress would not have caused problems. Knowing that the variable stressed structures are sensitive to stress concentrators, the paper aims to highlight the influence of two techniques to reduce stress concentrator weld toe grinding and WIG remelting weld toe, on the behavior of variable tensile test of cross corner welded specimens, reinforced with additional welding rows.

1 Introduction

There are many types of welded structures like welded bridges, ship structures; aircraft structures; bodies of the vehicles that are stressed variable and in time, they are affected by the phenomenon of fatigue which causes failure to values below the flow limit [1-2].

Fatigue failure are specific to welded structures which have fillet welds, which shows the stress concentrators at the intersection between the base and the filler material. Reducing these stress concentrators is accomplished by applying two reconditioning techniques: weld toe grinding and WIG remelting weld toe on cross welded samples, from which the specimens are taken for the variable tests [3-4]. Both reconditioning techniques determine either by polishing or by WIG remelting the reduction of stress concentrators at the junction base material / filler, causing an increase fatigue lifetime of the structure on which it is applied.

We considered telling this research to be done on the cruciform fillet welded joints as a welded structure, this type of connection showing the biggest failure by fatigue susceptibility.

* Corresponding author: claudiubbs@gmail.com

Research is extracted from a large study in which 5 cross welded samples were used denoted by A; B; C; D and E depending on the welding process used.

On all samples were applied the reconditioning techniques weld toe grinding and WIG remelting weld toe, then extracted samples for static and variable tests. [5-6].

The cruciform welded sample which is the subject of the sample with sign C, was welded by manual welding process with coated electrode in horizontal position on vertical wall that was reinforced with additional welding lines.

2 Experimental procedure

The shape and size of welded sample C on which were applied two reconditioning techniques weld toe grinding and WIG remelting weld toe in order to reduce the tensile concentrator and implicitly to enlarge the fatigue lifetime are shown in figure 1.

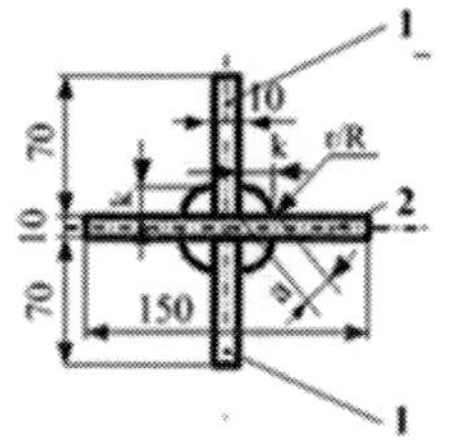

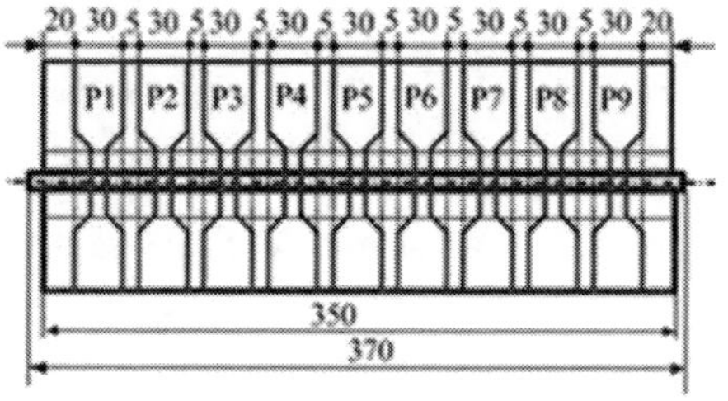

Fig. 1. Shape and size of welding sample C

It is seen from Figure 1 that the welding sample C consists of two vertical components 1 and one horizontal component 2.

Specimens C1; C4 and C7 will be subjected to static tensile testing static and are not subject to this paper, and specimens C2; C3; C5; C6; C8 and C9 will be subject to variable loads, the reconditioning techniques are applied before the test as follows: C2 and C3 without reconditioning; C5 and C6 with weld toe grinding and C8 and C9 with WIG remelting weld toe.

Fatigue testing specimens have been obtained by cutting the cross welded sample C yielding strips with width of 30 mm as shown Figure 1.c.

After mechanical cutting, vertical parts of the specimen attached to the horizontal plate will be milled to reduce the width from 30 mm to 10 mm in welding seams area, as shown in Figure 1.c. and Figure 2 a.

This milling is done in order to, during the variable loads to be submitted specimens, the section being reduced and the tensile increased on its surface, conduct the breaking in the seam or in the thermos-mechanical influenced area of the seam.

It is noticed that section is not reduced sharply from 30 to 20 mm but milling is done with a radius R = 10 mm, in order to not introduce additional stress concentrations.

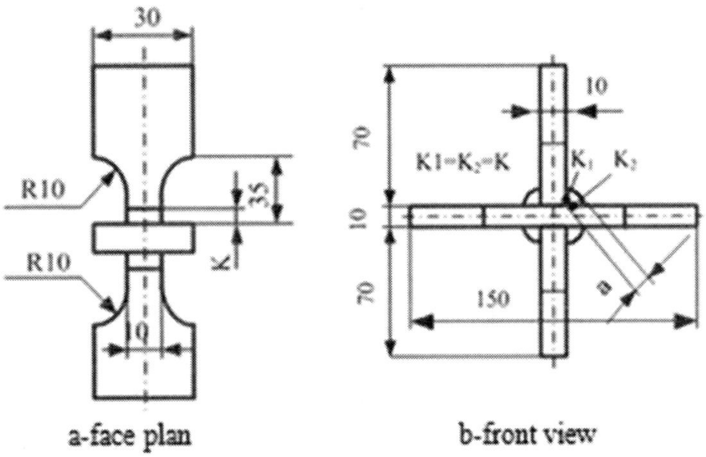

Fig. 2. Specimen shape after milling

The specimens are made of steel type S 235 JR.
Welding regime parameters used to obtain sample C are shown in Table 1

Table 1. Welding regime parameters for sample C

Row	Is [A]	Ua [V]	Ts [s]	Lc [cm]	vs [cm/s]	Vas [m/min]	El [kJ/cm]
1-3	220 228	19,9 20,1	85	35	0.41	5,3	7,67

Breaking sections are presented below.
In Figure 3 are shown breaking sections for specimens C2 and C3 without reconditioning.

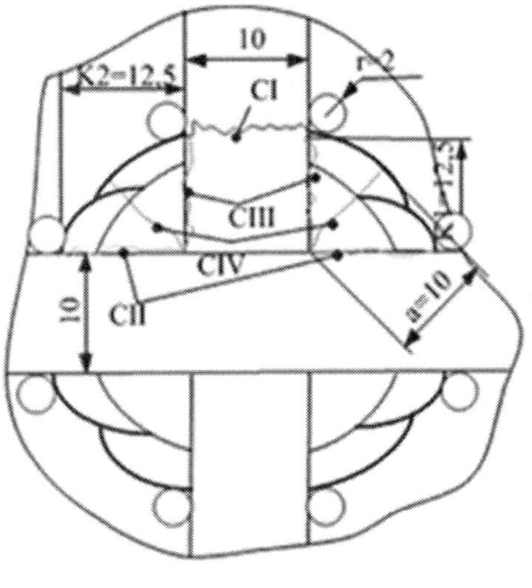

Fig. 3. Breaking sections for C2 and C3

In Figure 4 are shown schematically possible breaking sections in case of applied techniques weld toe grinding – Figure 4.a and WIG remelting weld toe - Figure 4.b.

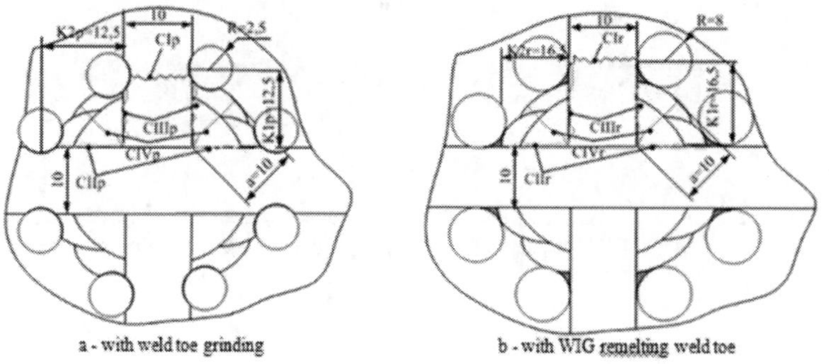

Fig.4. Breaking sections in case of applying the reconditioning

Geometric elements; K1; K2 and r of the welding seams of the specimens from sample C are shown in Table 2.

Table 2. Geometric elements of the welding seams – specimens sample C

Geometric element	Without reconditioning C2 si C3	weld toe grinding C5 si C6	WIG remelting weld toe C8 si C9
a	10	10	10
K1=K2	12.5	12.5	16.5
r	2	2.5	8

where: a is the thickness of the welding seam; K1 and K2 are the seam legs and r is the radius between seam and base material.

Fatigue tests were carried out with LVF 1000-HM unit. All attempts were made at 10 Hz frequency, the loadd cycle being an symmetrical alternating one, with asymmetry coefficient R = -1, the application being tensile-compression. For each set of specimens, with reconditioning techniques applied or not, they were raised durability curves that they have been compared to each other. For raising the durability curves were applied the the following loads variations, as follows: for C2; C5 and C8 ±14 kN and for C3; C6 and C9 ±7,5 kN. The fatigue test samples are shown in figure 5.

Fig.5. The fatigue test samples

A few frames during fatigue tests for C2 and C3 samples without reconditioning, are presented in figure 6.

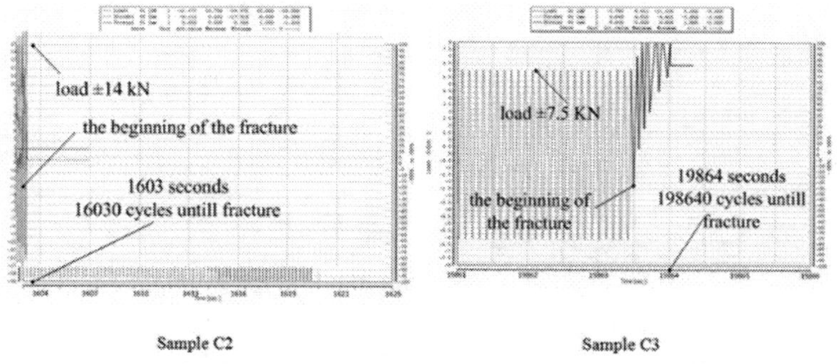

Fig.6. Frames for fatigue tests for C2 and C3 samples

We observe that, with decreasing the solicitation applied force for samples C3 and C2, increase the number of cycles untill the fracture: 198640 cycles for C3 and 16030 cycles for C2.

Another two frames for fatigue tests in case of C5 and C6 samples with weld toe grinding applied technique are shown in figure 7.

We also aobserve that, with decreasing the solicitation applied force for samples C6 and C5, increase the number of cycles untill the fracture: 287992 cycles for C6 and 22400 cycles for C5.

The last two frames for fatigue tests for C8 and C9 samples with WIG remelting weld toe applied technique, are shown in figure 8.

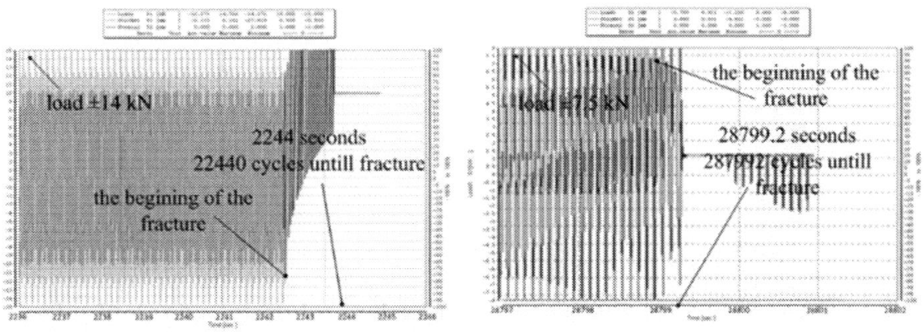

Sample C5 Sample C6

Fig.7. Frames for fatigue tests for C5 and C6 samples

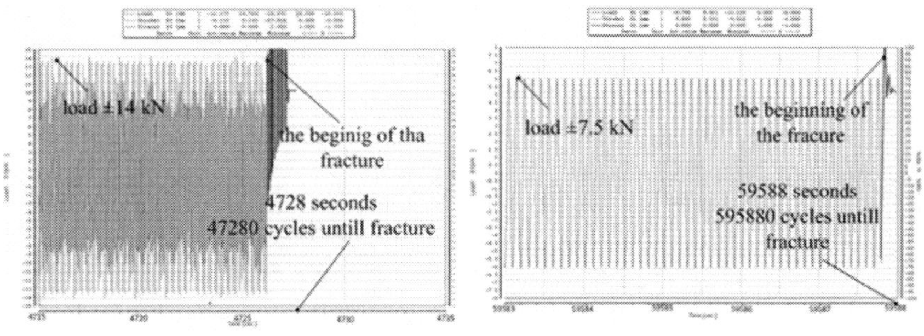

Sample C8 Sample C9

Fig.8. Frames for fatigue tests for C8 and C9 samples

3 Results and discussions

Forces applied to the specimens and results obtained from fatigue tests are presented centralized in table 3

Tabel 3. Results obtained from fatigue tests

Denotation	Reconditioning	F[Hz]	+/Fi[KN]	t[s]	N
C2			±14	1603	16030
C3	Without		±7,5	19864	198640
C5	Weld toe	10	±14	2244	22440
C6	grinding		±7,5	28799.2	287992
C8	WIG remelting		±14	4728	47280
C9			±7,5	59588	595880

We notice that, with decreasing tensile stress on the samples C2 and C3, increases the number of cycles up to fracture. We mention that on specimens without reconditioning we have the highest value of stress concentrator at the junction welding seam / base material, because of the smallest radius connecting seam to the base material, namely r = 2mm.

The sample C5 on which we apply ± 14kN, was broken after 22440 cycles and C6 on which we have applied ± 7.5 kKN broke after 287 992 load cycles. We found an increase in the number of cycles compared to homologous specimens C2 and C3. This is explained by the reduction of the stress concentrator at the top of the seam connection by increasing the radius of 2 to 2.5 mm as shown in table 2

We also observed that the sample C8 on which we apply ± 14kN, broke after 47 280 cycles and the sample C9 on which we have applied ± 7.5 KN, broke after 595 880 load cycles. We determinate an increase in the number of cycles compared to the homologous C5 and C6 specimens. This is also explained by the stress concentrators reduction at the top seam according to table 2, by increasing the radius from 2.5 to 8 mm.

Radii measurements were made with a specialized software by entering the cross-sections pictures of the specimens, scale 1:1.

4 Conclusions

Conclusions can be drawn after raising the durability curves from figure 9, with Mathcad calculation program.

With Mathcad calculation program, for samples C2 and C3, without technical reconditioning, we found values p = 2 and r = 6.7, for which the chart function $\sigma 1c(n)$, approaching most of our points c1 = (16030; 40027; 198 649) and f representing force vector f = (14;9;6). The shape of this curve is show in figure 9 with dashed point line.

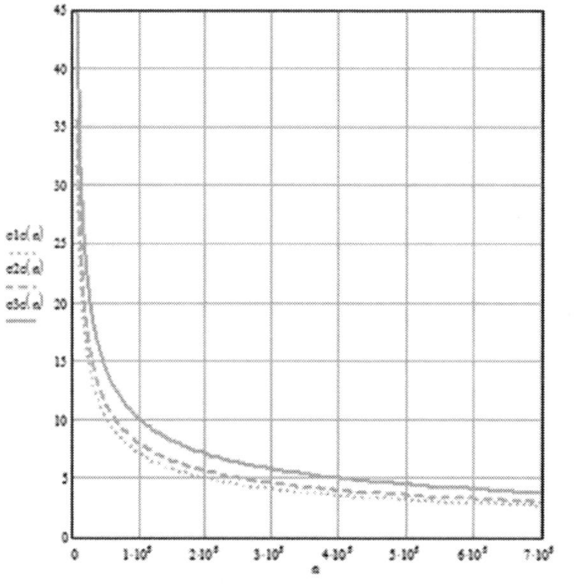

Fig.9. Raising the durability curves with Mathcad program

For the second set of samples C5 and C6, with weld toe grinding rehabilitation technique applied, with Mathcad software, were found the values p = 2 and r = 6.8 for the function chart, approached most of our points c2 = (22440; 56038; 287 991) and force vector f = (14;9;6). The shape of this curve is shown also in figure 9 with all dashed line.

For the third set of samples C8 and C9, on which we have applied WIG remelting weld toe, with the Mathcad software, were found the values p = 2 and r = 6.5 where the function chart is closest to our points c3 = (47281; 118080; 595888) and the force vector f = (14;9;6). The shape of this curve is shown in figure 9 with continuous line.

References

1. H.-K. Juna, J.-W. Seoa, I.-S. Jeonb, S.-H. Leec, Y.-S. Chang, Engineering Failure Analysis, **59**, 478-492 (2016)
2. J. Barle[a,], V. Grubisic[a, b], F. Vlak[a]3. Engineering Failure Analysis, **18**, 3, 1076–1084 (2011)
3. R. Baptistaa, V. Infanteb, C.M. Brancob. International Journal of Fatigue (2008)
4. H. C. Yıldırım. International Journal of Fatigue (2015)
5. D. T. Ngoula, H.Th. Beier, M. Vormwald. International Journal of Fatigue (2016)
6. T. Skriko, M. Ghafouri, T. Björk. International Journal of Fatigue (2017)

Technological particularities of laser manufacturing

Daniel Besnea[1,*], *Alina Spanu*[1] , *Victor Constantin*[1], and *Dontu Octavian*[1]

[1] Politehnica University of Bucharest, Mechatronics and Precision Engineering Department, Splaiul Independenţei, nr. 313, Bucharest, Romania

Abstract: The paper presents some investigations about the influence of the Nd:YAG laser welding parameters on the penetration and metal evaporation of single and dual pass weld in the case of thin sections of stainless steel sample. The metal loss during welding process was measured in order to establish the optimal values of welding parameters. The geometric size of the welded zone was measured using an SEM microscope in order to establish the correlation between the penetration and with at different values of welding parameters.

1 Introduction

Due to the advantages that laser provides, such as high temperature and strictly controlled time for warming, the minimum area of the narrow boundary along the welding line with thermal influence, the high speed for welding without supplementary material, the opportunity of entire automation, it is used widely in industry especially in the fields where there are stainless steel and aluminum. Some examples are from chemical, food, aeronautical and pharmaceutical industries.

We have to choose the laser provider device correctly in order to obtain the right laser welding. Meantime it is needed to adapt the type of joining and the used materials according to the strength of the process, which is using the optimum length of wave providing a great absorption of radiations.

The temperature increasing T_m that is developed inside the welding area should be greater than the melting temperature $T_{melting}$ but lower than the temperature of vaporization, so that the parameters of laser radiations should be controlled as function of material type and part positions.

$$T_{melting} < T_m < T_{vaporisation} \tag{1}$$

The working time of laser radiation is chosen according with maximum melting depth that should be done when the material of outside area has the boiling value, because of the melting front penetration. The penetration should start before the vaporization of shallow layer.

* Corresponding author: d_bes@yahoo.com

The rapid conduction of heat inside both materials along the welding area is depending on the material thermal diffusion *d* as well as on the type of connection. Before the welding process, the parts should be cleaned from oxides and some other impurities, so the connection is an ideal one meaning there are not temperature discontinuities.

It is possible to evaluate the melting material depth using the laser welding, because we suppose there are no changes of specific heat coefficient of the material when they pass through melting point and the spatial laser profile is a Gaussian one.

The mathematical equation of material temperature as function of absorbed flow intensity I_0 is given below

$$T_{(r,z,t)} = \frac{I_0 a^2}{k} \sqrt{\frac{d}{\pi}} \int_0^{\tau_i} \frac{e^{\frac{z^2}{4dt}} e^{r^2(4dt+a^2)}}{\sqrt{t}(4dt+a^2)} . \quad (2)$$

As we infer from the equation above, it is depending on: area unit which is constant for period shorter than the period of laser pulse τ_i; radial distance r from the center of laser wave; the material penetration depth z; time t

We may compute I_0 as function of laser ray period taking into account the condition that the ray center ($r = 0$) on the material surface ($z = 0$) the temperature has its maximum value but lower than the vaporization temperature.

$$T(r,z,t)\Big|_{\substack{r=o \\ z=0}} = T(0,0,t) \leq T_{vaporisation} \quad (3)$$

The depth for material melting inside for specific laser working period is limited by the way the thermal field is spread inside the part. On its turn, it is depending on the material properties, the thermal characteristics before the laser ray action and the dimension of the part. We may define the thermal time constant C_{iT} as the period when the temperature of melting material is about the same level of value with the temperature of surface warmed by the laser ray. For a extruded part with the thermal diffusion *d* and the thickness *l*

$$C_{iT} = \frac{l^2}{4d} \quad (4)$$

and it is a rough result for the time needed for the thermal flow propagation through a material layer during its melting. When the working period of laser ray is small than the thermal time constant of the material used for welding, it is not possible to obtain a homogeneous melting for the metalograph components of this material.

The focal distance of the optical system used for focusing the rays and the position of welding parts in directly relationship with the focal plane are other controlled parameter, which may influence the thermal distribution of temperature inside the material as well as the area melting by the laser.

Another important characteristic which may influence the quality of laser welding frame is given by the chemical reaction for oxydation or nitration with the gases inside the atmosphere by using high temperature level. It may causes some problems in case the oxides or some other resulted chemical elements have the temperature of dissociation higher than the melting point of the steel. Consequently the resulted welding area will be spongy and for avoiding it we should use an actionless gas [1-4].

In the following table 1 are shown some results of laser welding in the case of materials often used in mechatronics systems.

Among all the other conditions specified above, we have to impose a high accuracy positioning of the part in front of laser ray for manufacturing the better quality laser welding taking into account the presence of the supplementary material needed.

This accurate positioning demands the part preparation very carefully as well as tolerance system dimension by comparison with other welding type. The positional errors along the movement axis should not exceed the focal ray radius. Consequently, the tolerance value of 0.1 mm is recommended, but for smaller dimensions it has to be reduced [5-6].

Table 1.

Material	Comments
Aluminum 1100	Welds well; no cracking problem or transformation
Aluminum 2219	No cracks; no filler metal required
Cu-Zn Brasses	Out-gassing of Zn prevents good welds
Carbon Steel	Good welds with carbon content under 0.25%; for greater carbon content, may be brittle and crack
Galvanized Steel	Severe Zn boil-off causes porosity
Steel, 300 Stainless	Welds well, except 3030 and 303SE, which crack
Steel, 400 Stainless	Generally welds somewhat brittle; may require pre- and post-weld heat treating
Steel, 17-4PH Stainless	Needs post-weld heat treating to strengthen
Tantalum	Ductile welds; special precautions against oxidation required
Titanium	Ductile welds; special precautions against oxidation required
Tungsten	Brittle welds; requires high energy
Zirconium	Ductile welds; special precautions against oxidation required.

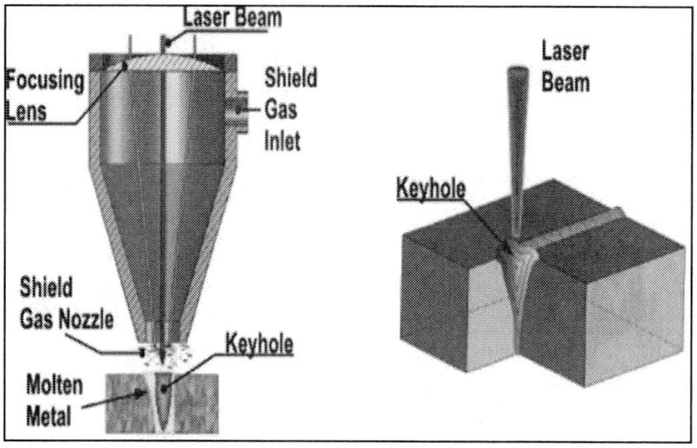

Fig. 1. The positioning of focal ray in the welding plane of part [5].

The positioning along the Z axis of laser ray is the easiest way due to the depth of the material if it is placed in the XY plane. For the smaller dimensions the ray is placed on the part surface and inside it for bigger ones.

Usually, the following values are used in practice:
– for thickness between 0.2 and 0.5 mm – it is accepted the join J;
– for thickness between 1 and 2 mm – it is accepted the join;
– for thickness between 3 and 10 mm – it is accepted the join J.

In table 2 there are presented the most important assembly types that could be manufactured with laser welding.

Table 2. Assembly types

No.	Welding Type	Exemplary
1	**Butt welding**	
	Weld correctly  	**Weld positioning** 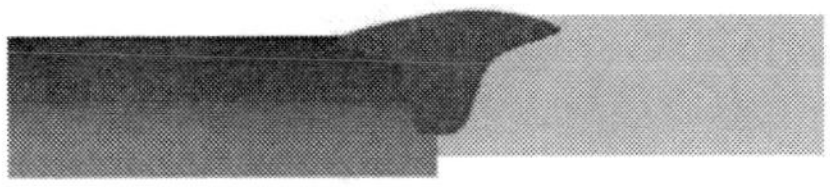 
2	**Over edge joint** 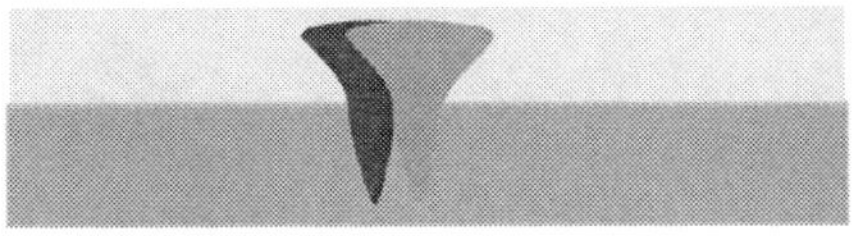	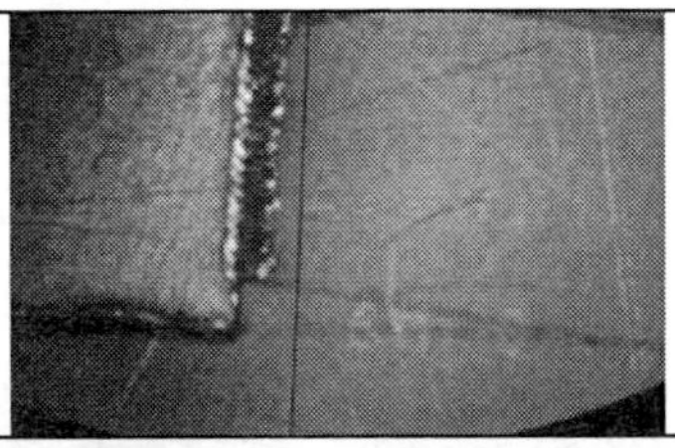
3	**Corner joint** 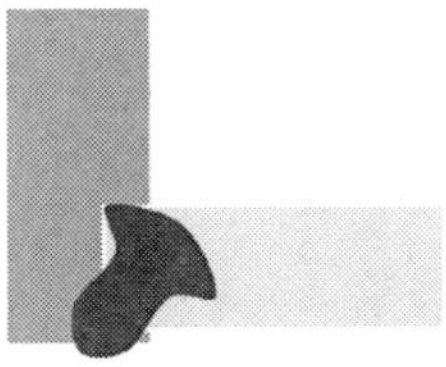	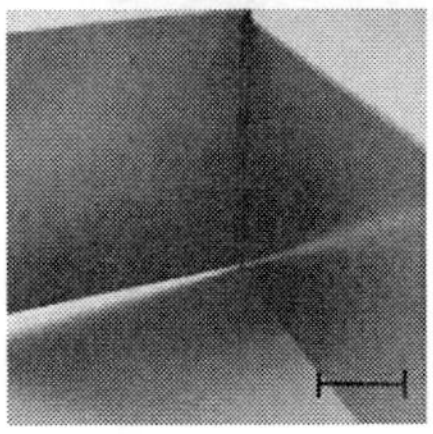 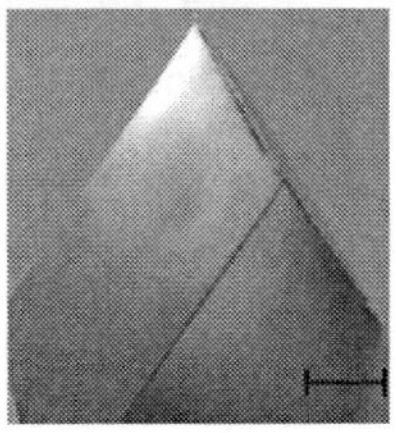

2 Experimental researches

The laser welding for high alloy stainless steel is widely used for accurate mechanical systems and industrial products. We present an example for Nd:YAG laser using for automatic welding of radiation source tightness made of stainless steel .

The capsule for radiation sources should be welded following high security conditions inside controlled atmosphere conditions. The experimental researches have been done using the laser system with solid activ environment Nd:YAG ROFIN Starweld SWP 6002 [9], equipped with numerical control, fig. 2. By choosing correctly the welding parameters, especially the energy values and the impulse period, there were possible to manufacture some welding capsules with high thermal module without a heavy load of additional material. The body and the head parts used as capsule assembly components are manufactured with circular welding. The capsules are made of 304 type stainless steel 10TiNiCr180 with a balanced austenitic structure (chemical formula Fe < 0.08% C, 17.5-20% Cr, 8-11% Ni, < 2% Mn, <1%Si, < 0.045% P < 0.03 % S).The enclosed capsules for nuclear radiation used in non destructive testing or measuring application require a good sealing during the working period. The capsule assembly contains two pieces, the capsule body and the cover, which needs to be joined together by circular welding made on the

DOI: 10.1051/matecconf/201712103005

capsule's top. In order to obtain a good quality sealing, laser welding process was applied to join cover to the capsul's body without filler metal. The dimensions of a capsule, figure 2 are: 5 mm in diameter and 8 mm in length [2, 5-6].

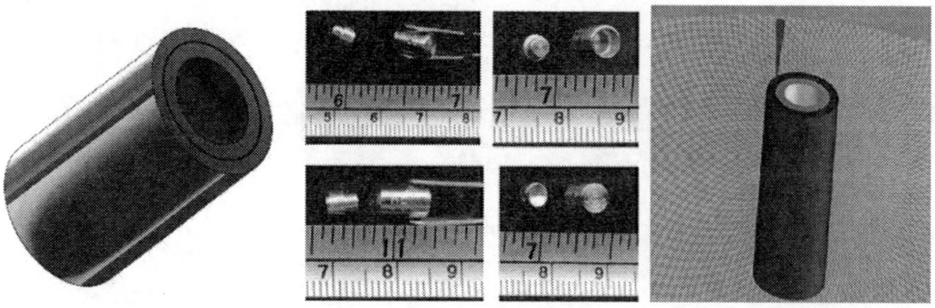

Fig. 2. The components and dimensions of the capsule.

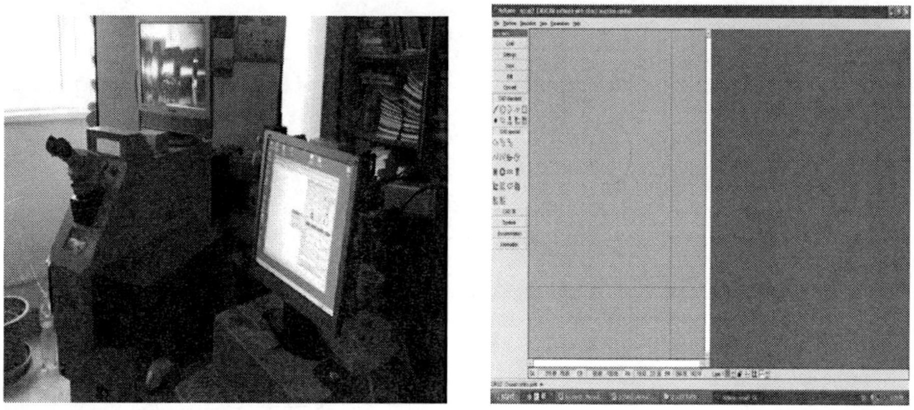

Fig. 3. The Nd:YAG Starweld SWP 6002 Rofin-CNC laser device for welding and the menu for the 2D design [9].

The software nccad_6 affords the opening of new file for designing or for changing the file. In the CNC menu we have the text editor for editing and for running. For a new model we use Fichier CAD/CAM – *Nouveau dessin* figure 3. If we want to work in CNC or CAD/CAM we may use the menu for function CAM manufacturing, the edit group that allows the changes of a model, the CAD group that allows the design using the XY axis fig. 3. The nccad_6 software allows working by following the rules imposed by ISO 66025. At first the files are written inside a text editor, the graphical simulation could be done by working with Menu Simulation and finally the manufacturing is done by working with Menu Machine. Mainly, the G functions are used affording the movement and the machine functions are based on M functions . Nearby all the working facilities specified before, we may use some complementary functions that may be recognized by their name with some blank spaces. As we may infer from fig. 4 the CNC package allows the transfer inside the working area based on the coordinate system [9].

3 Experimental results and conclusions

The micro-welding laser that works with impulses, produced by ROFIN company provides a better solution for rapid welding with high accuracy for all the steel materials and their used alloys inside a solid activ environment Nd:YAG with wave length 1. 06 μm; the maximum energy of a pulse is 30...110 J.

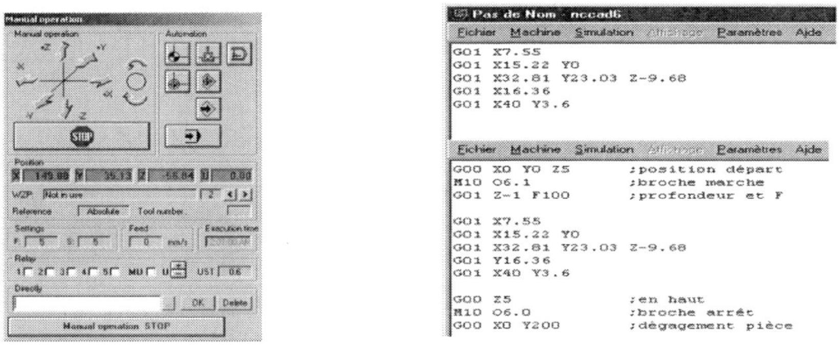

Fig. 4. The software written in nccad_6 language is transferred in the working environment [9].

The working power is 35...60 W the maximum power of a pulse is 4.5...10.6 kW the pulse frequency is 1...20 Hz; the pulse period is from 0.5 to 50 ms; the working class is 4.

The maximum length: $80 \times 80 \times 140$, maximum speed: 10 mm/s, accuracy: < 20μ, resolution: 2.5 μm, maximum load: 20 kg, maximum distance allowed for welding $400 \times 200 \times 100$ mm, the software for control nccad_6, the file type imported: 3D CAD (2D DXF, HPGL, 3D STL), the direct language programming NC ISO 66025. The control is made by using the microcontroller with external programming through RS 232 [9].

Finally we have tested the hardness by using a Shimadzu HMV 2 T and electronic microscopy with SEM INSPECT S microscope. The mass losses during the welding process was measured with an electronic balance ABJ 220 – 4M with 1 mg accuracy. Before the welding the parts were cleaned in an ultrasonic system. In the Table 3 there are presented the working parameters for three tested parts [10-11].

A condition imposed by SR ISO 9978 standard stipulates that inside the capsule that contains radioactive material, the temperature should not exceed 95°C. To determine the most suitable welding parameters to meet these requirement, a mathematical model of heat propagation inside the material was constructed [7-8].

This model was implemented in ANSYS software and a simulation was generated.

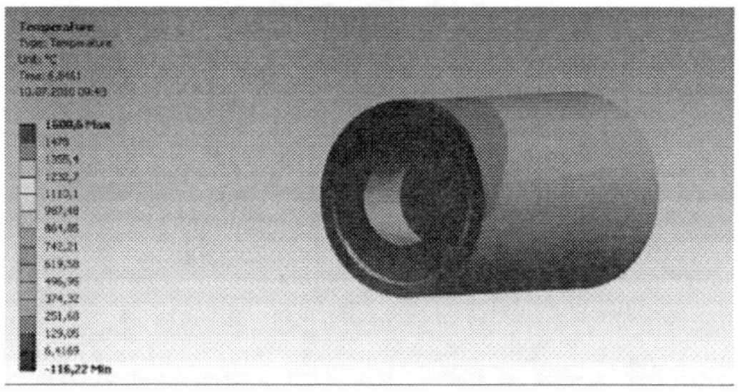

Fig. 5. Heat propagation simulation.

The simulation took into account an initial temperature of 20°C, a perfect contact between the top of the capsule and the capsules body and a heat exchange with the environment by convection.

The simulation figure 5, showed a fast rise in temperature and also a fast cooling at the welding spot and a slow rise in temperature at the bottom of the capsule. This increase in temperature is not generated directly by the laser beam but by heat dissipation in material after applying the welding spot.

For the radioactive capsule having the working parameters P2. For the highest value of laser power the steel losses due to vaporization and penetration are maxim. The diminution of laser pulse period has produced an increase of width of welding stitch, breakings and porous surfaces (Fig. 7). Instead modifying the working parameters for welding sample P1 was executed correctly.

Table 3. Working parameters for three tested parts

Sample	Voltage [V]	Pulse time [ms]	Beam energy, [J/cm^2]	Pulse frequency [Hz]	Average penetration [μm]	Metal loss [%]	Spot [mm]
P1	233	2.3	1275	2.6	337.54	0.36	0.775
P2	245	3.1	1895	2.3	577.72	1.28	0.775
P3	230	3.0	1249	2.0	413.22	0.39	0.95

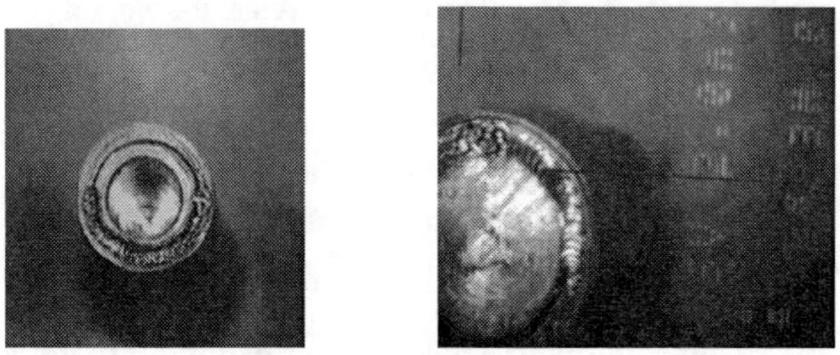

Fig. 6. The manufactured welding [9].

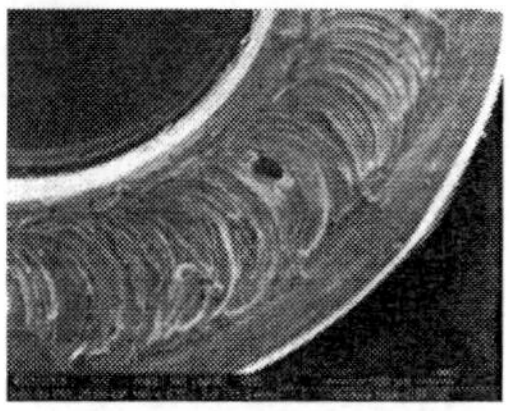

Fig. 7. The macrostructure for working parameters P2 [9].

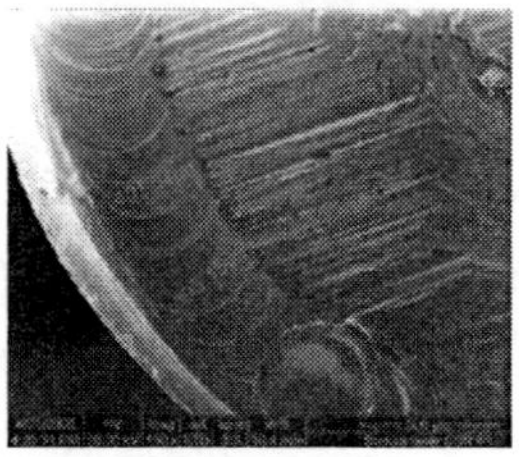

Fig. 8. The macrostructure for working parameters P1 [9].

The laser using for inscription, engraving and putting a mark on are very recent time applications.

These activities could be done with good efficiency on the laser systems characterized by continuous emissions (CO_2 + He+ Ne) as well as on the laser systems with discreet emissions, the later one with some disadvantages regarding the efficiency and the continuity of shapes.

The inscriptions, engraving and putting the marks on by laser using could be applied for steel, crystals (diamond, sapphire, ruby), glass and some other.

The part movement is done in front of laser ray controlled strictly by software and microcontrollers for laser provider and for mechanical part. In such case the symbol for engraving – putting the mark on is usually a numeric one, meaning a fully process automation (Fig. 11).

In order to design a new model, we begin with a CAD /CAM file named *Nouveau dessin* (Fig. 9). The working space has the table dimensions Y left – right and X axis for the machine axis.

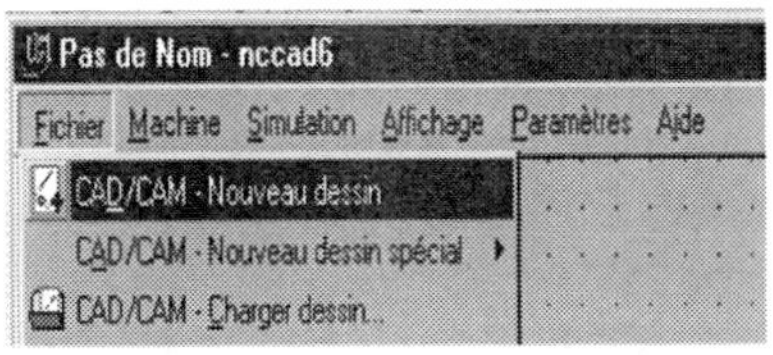

Fig. 9. The opening of CAD /CAM working environment in a new file and the icon CAD standard.

For CAD file manufacturing we use the Technologies icon from group CAM. We continue with the placement of working point over the rectangle and we put the mark on. Inside the Technologies window we approve the values (Fig. 10).

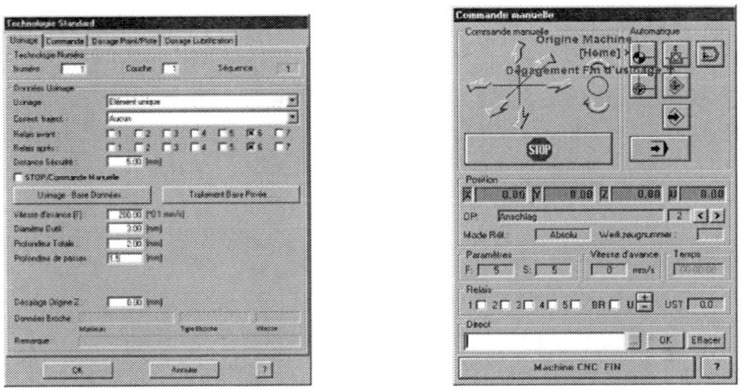

Fig. 10. The approval of the values and the command for setting the part origin.

From the window named *Comande manualle* belonging to Machine/Machine CNC menu we push the button *Degagement fin d'usinage,* figure 10. We continue with the operation for establishing the part origin, so that we start with *Lancement du programe* finally [9].

Welding Program G00 Z-95 Y35 X35 M05; G91; M10 O2.1; Gas Power M30 P18; Pause M10 O3.1 Laser Power G02 X0 Y0 I2 J2 F15; M10 O3.0; Laser stop M10 O2.0; Gas stop M05; G90	Gas Power M30 P18;Pause M10 O3.1 ; Laser Power M30 P25;Pause M10 O3.0; Laser stop G00 X35.24 Y33.02 M10 O2.1; M10 O2.0; Gas stop G00 Y26.10 M10 O2.1; Gas Power M30 P18 ;Pause M10 O3.1 ; Laser Power M30 P25 ;Pause M10 O3.0 ; Laser stop M10 O2.0 ; Gas stop	M10 O2.1 ; Gas Power M30 P18 ;Pause M10 O3.1 ; Laser Power M30 P25 ;Pause G00 X39.21 M10 O3.0; Laser stop M10 O2.1 ; Gas Power M30 P18 ;Pause M10 O3.1 ; Laser Power M30 P25 ;Pause M10 O3.0 ; Laser stop M10 O2.0 ; Gas stop M05 ; G00 X37.17 Y34.1	Gas Power M30 P18 ;Pause M10 O3.1 ; Laser Power G01 Y-5 F15 ; G01 Y-3 X2 ; M10 G91 ; M10 O2.1 O3.0 ; Laser stop M10 O2.0 ; Gas stop .... Gas Power G02 X0 Y0 I0.2 J0.2 F15 ; M10 O3.0 ; Laser stop M10 O2.0 ; Gas stop G90 ; T

Fig. 11. The engraving sequence.

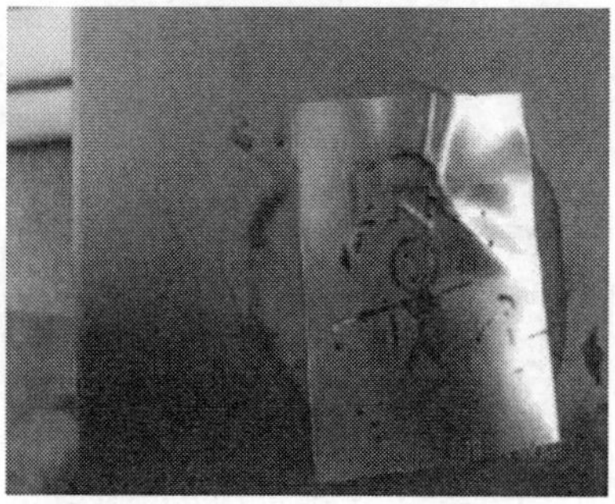

Fig. 12. Example of engraving with numerical control and the software sequence [9].

4 Conclusions

The originality of the paper is that it approaches a field of unconventional technology - laser-welding with solid active medium YAG:Nd in CNC mode for obtaining radioactive capsules with applications in medical and nuclear fields, according to ISO2919 standards and ISO9978. Since the rules on nuclear radiation and radiation protection sources are very restrictive, the experimental research presented in the paper offers a solution to the realization of a viable welding cord and meeting these criteria of sealing. The metallographic analysis carried out on our team's laser micro-welding of radioactive capsules revealed the advantages of the method and the fulfillment of current ISO standards.

We have pointed out the opportunity of using the laser device Nd:YAG for welding the parts with small dimensions and width, as we have explained with the examples presented above. The way the ray is sending forth allows a good control of energy in the working area, especially for thermal influences around the welding point and minim deformation for the welding materials.

The main advantages of laser using in the manufacturing technologies are:

– the possibility of manufacturing any steel or dielectric material, without taking into account its hardness, but we have to pay attention to its absorption power of radiations;
– the shortening of manufacturing time because of instantaneous way of process task;
– from the kind of the contact point of view, we may say it is not any deformation of the part, because of contactless characteristics;
– we may assess the manufacturing of parts with very small dimensions or with complex shapes;
– the process could be automatized entirely and could be done by applying the numerical control;
– we may place very accurately the part in front of laser ray with precise focal, so the accuracy is the highest one.

References

1. D. Besnea, O. Dontu, M. Avram, A. Spanu, C. Rizescu, T. Pascu, The 7th International Conference on Advanced Concepts in Mechanical Engineering, ACME, Iasi, Romania (2016)
2. O. Dontu, J.L. Ocana Moreno, R. Ciobanu, M. Branzei, D. Besnea, , J. of Optoelectronics and Adv. Mat., **7**, nr. 9-10 (2015)
3. R. Ciobanu, Gh.I. Gheorghe, O. Dontu, D. Besnea, D. Cioboata, The Romanian Review Precision Mechanics, Optics&Mechatronics, **48** (2015)
4. R. Ciobanu, O. Dontu, D. Besnea, A. Spanu, I. Avarvarei, AMM **811** (2015)

5. D. Besnea, O. Dontu, Gh.I. Gheorghe, R. Ciobanu, A. Cuta, *Laser microwelding of mechatronics components*, Romanian Review Precision Mechanics, Optics&Mechatronics, **46** (2014)
6. R. Ciobanu, O. Dontu, D. Besnea, V. Constantin, D. Cioboata, , 5th Internat. Conf. on Materials Science and Technologies, 15-17 Octombrie, Bucharest, Romania (2014)
7. ISO 2919 Surse inchise de radiatii nucleare
8. ISO 9978/1992 Radioprotectie
9. ROFIN - Laser micro welding SW 6002
10. Microscope electronic Inspect S, FEI, Netherlands– Technical book
11. Hardness tester SHIMADZU HMV – 2TE, Japan – Technical book

The influence of laser heat treatment on the geometric structure of the surface and condition of the surface layer and selected properties of Waspaloy

Aneta Bartkowska[1,*], *Mateusz* Kukliński[2], and *Piotr* Kieruj[2]

[1] Poznan University of Technology, Institute of Material Science and Engineering, Jana Pawła II 24, 60-965, Poznan, Poland
[2] Poznan University of Technology, Institute of Mechanical Technology, Piotrowo 3, 60-965, Poznan, Poland

Abstract. In the aviation industry the nickel-based superalloys such as Waspaloy are very often used. Conventional machining of this alloys is difficult and expensive. Therefore a Waspaloy requires the new techniques for machining like e.g. Laser Assisted Machining (LAM). New development directions of mechanical engineering are focusing on this type of hybrid machining, where materials are heated and cut in the single process. LAM enables increasing the material machinability through the increase of its temperature in cutting zone. It is assumed that additional increased temperature in the contact zone between surface of material and cutting tool causes the increase of machinability. This paper presents the results of laser heat treatment process on Waspaloy material. The influence of laser heat treatment parameters on the surface condition and selected properties of Waspaloy were analyzed. Laser heat treatment was carried out using diode laser with nominal power equaled to 3.0 kW. The laser beam power density and its scanning laser beam velocity were analyzed. In the first step the single laser tracks were prepared and analyzed. In the second step the multiple laser tracks were prepared. The special attention was directed on change in microstructure. It was found that the laser heat treatment have significant influence on microstructure. The dendritic microstructure of Waspaloy which is characterized by a lower microhardness, causes better machinability of this material.

1 Introduction

One of the new techniques which improves the machinability of material is a Laser Assisted Machining (LAM). This technique focuses on increasing machinability of material through the increase of its temperature in cutting zone, as a result of laser beam effect [1-7]. However, before beginning such treatment, it is necessary to know what happens to the material only after the heating [8]. In many papers are analyzed problems of laser heating

* Corresponding author: aneta.bartkowska@put.poznan.pl

of material (iron and non-ferrous alloy) and occurring changes in it [9-12]. In the last time a lot of attention is directed to non-ferrous alloys such as nickel alloy. Also, the emphasis is put on the machinability improvement of difficult-to-cut materials [13-14]. The group of difficult-to-cut materials undoubtedly include nickel alloys such as Waspaloy [9-10] or alloys reinforced with SiC [1, 3-5], WC [7, 15-16] particles. There is a tendency to create products of higher quality in a shorter time and at a lower cost. To achieve some of the stated objectives, there is a need to use hard and resistant materials which usually have low machinability.

Therefore for the aim of this study was to determine the appropriate laser heat treatment parameters on dimensions of laser tracks and their microstructure. This step is a very important during the Laser Assisted Machining of material.

2 Research method

Studies were carried out on the specimens made on nickel-based superalloys such as Waspaloy. The samples of Waspaloy alloy with the following dimensions 20 mm x 30 mm x 12 mm were subjected to laser heat treatment. Chemical composition of this material is shown in Table 1.

Table 1. Chemical composition of Waspaloy [%wt.]

C	Cr	Co	Mo	Ti	Al	B	Fe	Ni
0.06	19.5	13.5	4.25	3.1	1.4	0.007	1.0	balance

The influence of laser heat treatment parameters on the surface condition, dimensions of laser tracks, microstructure and microhardness of Waspaloy were analyzed. In the first step the single laser tracks were prepared and analyzed. Then, in the second step the multiple laser tracks were prepared. The aim of this step was to obtain a uniform remelted area dependent on distance of laser tracks between each other. Laser heat treatment by using the TRUDIODE 3006 diode laser with a nominal power of 3 kW was conducted. The diode laser was integrated with the robot arm KUKA. The parameters of the laser heat treatment are presented in tables 1 and 2. LHT parameters for the single laser tracks are presented in Table 1.

Table 2. The parameters LHT for the single laser tracks

P [W]	q [kW/cm^2]	v [m/min]
800	102	5
1400	178	5
2000	255	5
2600	331	5
1400	178	25
1400	178	50
1400	178	75
1400	178	100

LHT parameters for the multiple laser tracks are presented in Table 2. Table 1 and 2 present: laser beam power P [W], laser beam power density q [kW/cm^2], scanning laser beam velocity v [m/min] and laser tracks overlap O [%], which depends on distance between the laser tracks f [mm].

Overlap of laser tracks is defined by the formula:

DOI: 10.1051/matecconf/201712103006

$$O = [(b_l - f)/b_l] \cdot 100\% \quad (1)$$

where:
b_l - diameter of laser beam [mm]
f - distance between the laser tracks [mm].
In this study the diameter of laser beam was of 1mm.

Table 3. The parameters LHT for the multiple laser tracks

P [W]	q [kW/cm^2]	v [m/min]	f [mm]	O [%]
1400	178	75.0	1.00	0
			0.50	50
			0.25	75
			0.125	87.5

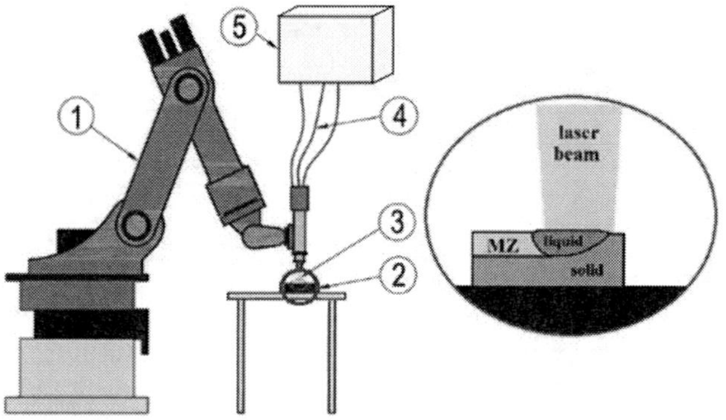

Fig. 1. Scheme of laser heat treatment process: 1) KUKA robot, 2) specimen, 3) laser beam, 4) optic fibers, 5) optic resonator

The view of the test stand of diode laser with the KUKA robot and diagram of remelting of surface layer by using the laser beam are shown in Figure 1. The dimensions of laser tracks were measured on the cross-sections of laser tracks, using Axio Vision software. Width (W) and depth (d) were measured for single laser tracks, while for multiple laser tracks the effective layer thickness (t_{ef}) was measured. The designation of measurement of dimensions on laser tracks cross-sections sample are shown in Figure 2.

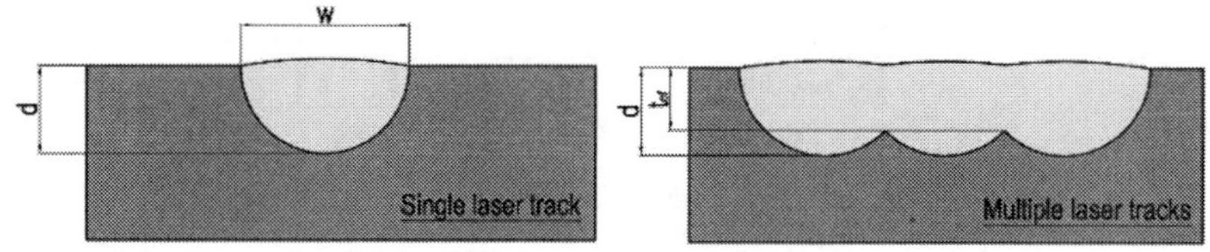

Fig. 2. Laser tracks dimensions locations: W – width of laser track, d – depth of laser track, t_{ef} – effective layer thickness

Macroscopic photos were made using Zeiss Stereo Discovery V20 camera. 3D topography images were made by profile meter Hommel tester H8000.

Microstructure observations were carried out on cross-sections of specimens ground with abrasive papers of grit ranged from 120 to 2000 and subsequently polished using aluminum oxide suspension after etching in Marble's reagent. For this purpose the Carl Zeiss Metaval microscope equipped with Moticam 2300 3.0 MP Live camera was applied. To determine microhardness profiles, Zwick 3212B Vickers hardness tester was used. Indentation load of 100 G and loading time 15 seconds were applied in these studies.

3 Results and discussion

Figure 3 presents the macroscopic images of Waspaloy surface after laser heat treatment for single laser tracks. Figures 3a to 3d presents the influence of laser beam power density at the constant scanning laser beam velocity on surface of Waspaloy. The movements direction of liquid material is clearly visible in Figure 3a.

With increasing the laser beam power density in the central area of laser tracks the solidified drops of metal appear on the sample surface. Those drops are clearly visible in Figures 3c and 3d. Figure 3e - 3h presents the macroscopic images when the laser beam power density is constant, and scanning laser beam velocity is variable. Here is visible similar dependency, which shows that with increasing of scanning laser beam velocity the dimensions of laser tracks are decreased. Increasing the scanning laser beam velocity affects the rapid cooling of material which can be also observed as solidified drops of metal on Waspaloy sample's surface (Fig. 3g and Fig. 3h).

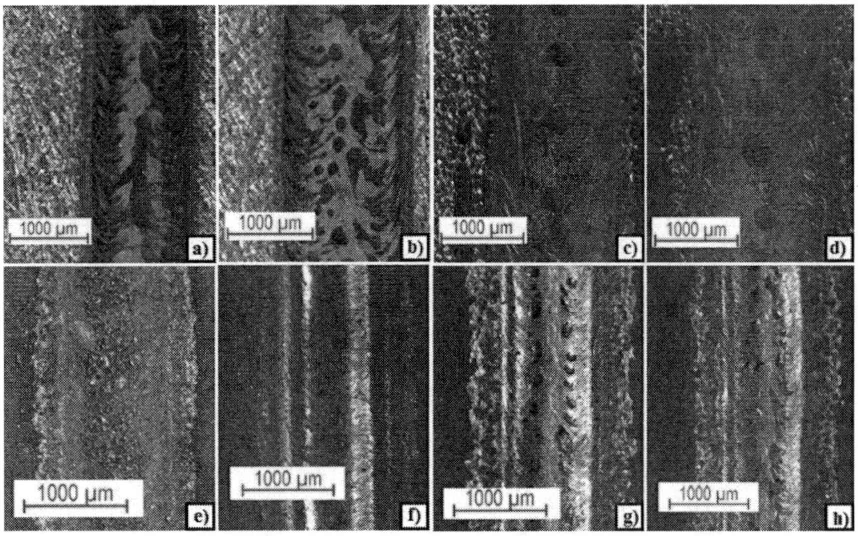

Fig. 3. Macroscopic image of Waspaloy surface after laser heat treatment – single laser tracks: a) LHT: q = 102 kW/cm^2, v = 5 m/min; b) LHT: q = 178 kW/cm^2, v = 5 m/min; c) LHT: q = 255 kW/cm^2, v = 5 m/min; d)LHT: q = 331 kW/cm^2, v = 5 m/min; e) LHT: q = 178 kW/cm^2, v = 25 m/min; f) LHT: q = 178 kW/cm^2, v = 50 m/min; g) LHT: q = 178 kW/cm^2, v = 75 m/min; h) LHT: q = 178 kW/cm^2, v = 100 m/min.

Figure 4 presents the macroscopic images of surface Waspaloy after laser heat treatment for multiple laser tracks. Figures 4a-4d present surface of laser tracks where is visible the entire width of laser track, whereas the Figures 4e-4h present the magnified areas of Figures 4a-4d. It can be seen the influence of distance between the laser tracks on surface of material. In Figure 4a the distance between the laser tracks was of 1 mm. With decreasing distance between adjacent laser tracks, the overlap of laser tracks is increasing (Figs. 4b-d). Figure 5 presents 3D images of roughness of the Waspaloy surface after laser heat treatment for multiple laser tracks. With increasing the overlap of laser tracks, the roughness of the Waspaloy surface after laser heat treatment decreases. The results of S_a

and S_z roughness parameters are presented in Table 4. The S_a parameter is an average roughness deviation and S_z parameter is a ten-point surface irregularities height. It was found that surface roughness decreases when laser tracks are closer to each other.

Table 4. The roughness parameters after LHT for the multiple laser tracks

P [W]	q [kW/cm²]	v [m/min]	f [mm]	O [%]	S_a [µm]	S_z [µm]
1400	178	75.0	1.00	0	10.0	86.6
			0.50	50	4.6	41.8
			0.25	75	5.6	50.1
			0.125	87.5	1.9	20.2

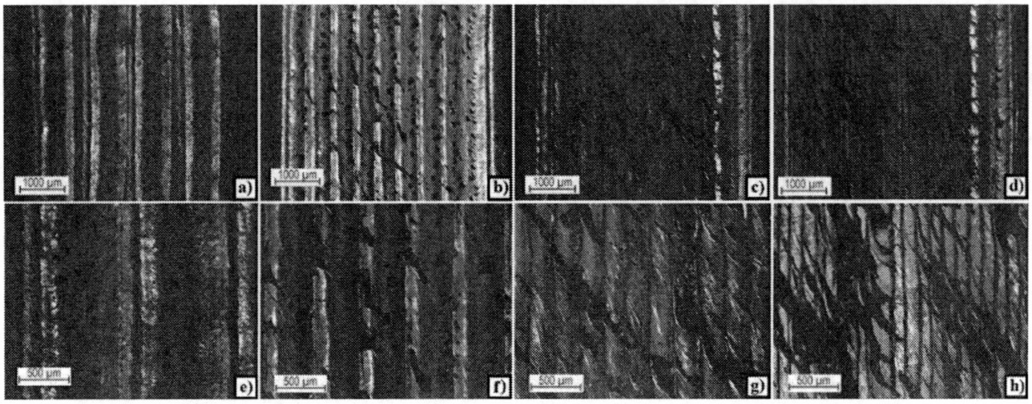

Fig. 4. Macroscopic image of Waspaloy surface after laser heat treatment – multiple laser tracks: a, e - LHT: q = 178 kW/cm², v = 75 m/min, O = 0; b, f - LHT: q = 178 kW/cm², v = 75 m/min, O = 50; c, g - LHT: q = 178 kW/cm², v = 75 m/min, O = 75; d, h - LHT: q = 178 kW/cm², v = 75 m/min, O = 87.5

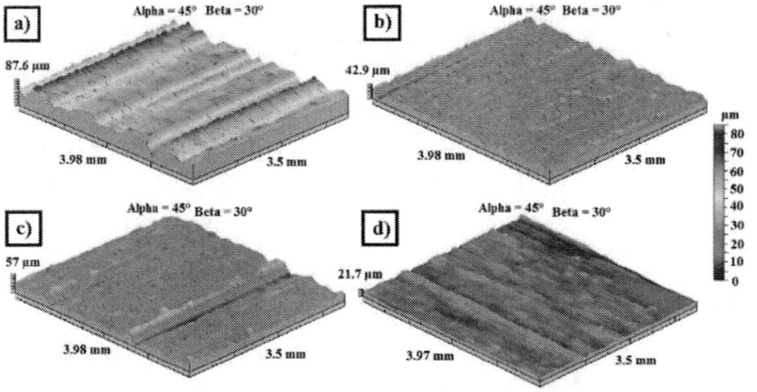

Fig. 5. 3D images of roughness of the Waspaloy surface after laser heat treatment:
a) LHT: q = 178 kW/cm², v = 75 m/min, O = 0; b) LHT: q = 178 kW/cm², v = 75 m/min, O = 50;
c) LHT: q = 178 kW/cm², v = 75 m/min, O = 75; d) LHT: q = 178 kW/cm², v = 75 m/min, O = 87.5

Figures 6a-6c show a single laser tracks obtained at a constant scanning laser beam velocity v = 5.1 m/min and a variable laser beam power density. It can be seen that the with the increasing the laser beam power density the dimensions of the laser tracks are increasing.

In the plane perpendicular to the direction of melting, a dendritic and column crystals oriented in the direction of heat dissipation can be observed. In the remelted zone, where it followed a change of direction of heat dissipation, the crystals grow almost parallel to the direction of heat, so the microstructure is equated. When is a sufficiently large supercooling, the character of crystals changes, and second order branches characteristic to a dendritic growth appear. This dependency is clearly visible in Figures 6b and 6c. The columnar and dendritic crystals in the microstructure are bigger when the higher laser beam power density is used.

Figures 6d-6f show a single laser tracks produced at constant laser beam power density q = 178 kW/cm^2 and variable scanning laser beam velocity. In this study the speed of laser heating corresponds to the most commonly used cutting speeds of applied materials. There is a certain dependence that with increasing scanning laser beam velocity, the dimensions of laser tracks are decreasing.

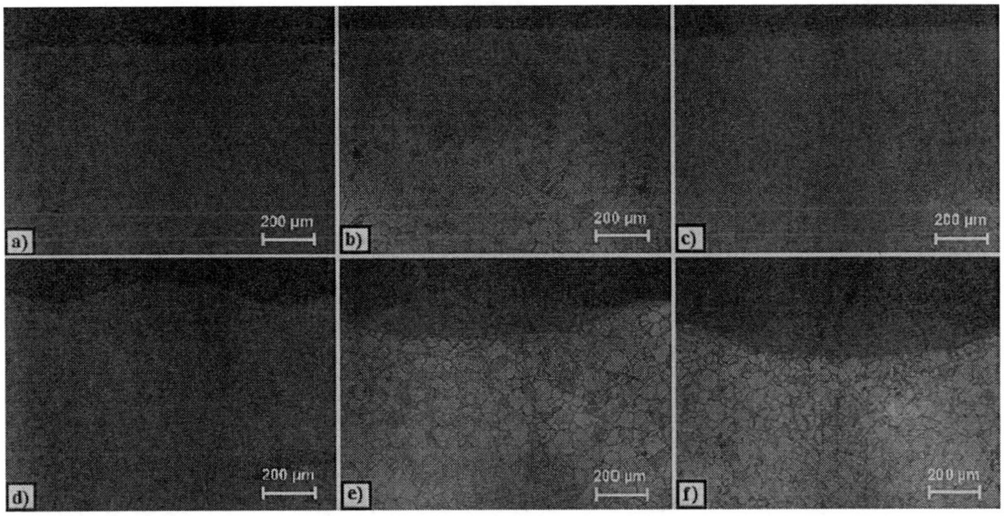

Fig. 6. Macrostructure of Waspaloy surface after laser heat treatment – single laser tracks: a) LHT: q = 102 kW/cm^2, v = 5 m/min; b) LHT: q = 178 kW/cm^2, v = 5 m/min; c) LHT: q = 255 kW/cm^2, v = 5 m/min; d) LHT: q = 178 kW/cm^2, v = 50 m/min; e) LHT: q = 178 kW/cm^2, v = 75 m/min; f) LHT: q = 178 kW/cm^2, v = 100 m/min

Figure 7 presents the microstructure of multiple laser tracks. Figures 7e-7h presents the magnified area from Figures 7a-7d. In the microstructure of the laser track cracks can be observed which are likely related to the high scanning speed of the laser beam, as well as heating and cooling of the material using a small distances between adjacent tracks. Probably the cracks may arise during the contraction of material after rapid cooling and solidification to the ambient temperature, and then re-heating it.

For the scanning laser beam velocity v = 75.0 m/min and the distance between the laser tracks f = 0.25 mm (Fig. 7c) and f = 0.125 mm (Fig. 7d), the numerous cracks were observed in the microstructure of surface layer. Selection of parameters for the laser treatment is important with regard to obtaining a continuous surface layer with uniform thickness. It can be seen that the smaller the distance f, the thicker effective layer.

When the value of the parameter f is higher (distance between laser track is are large) the depth in the axis and in the borders of laser tracks are various. A lower value of the parameter f affects the total fulfillment of the work surface. In this case the best parameters can be considered as: q = 178 kW/cm^2, v = 75 m/min, f = 0.5 mm. Too large distance between laser tracks is not conducive to completing the entire surface workpiece, which can affect the worse machinability in areas where the material was not heated. On the other

DOI: 10.1051/matecconf/201712103006

hand, too small distance between laser tracks can cause cracks in the material after laser heating, which may result in propagation of cracks into the based material.

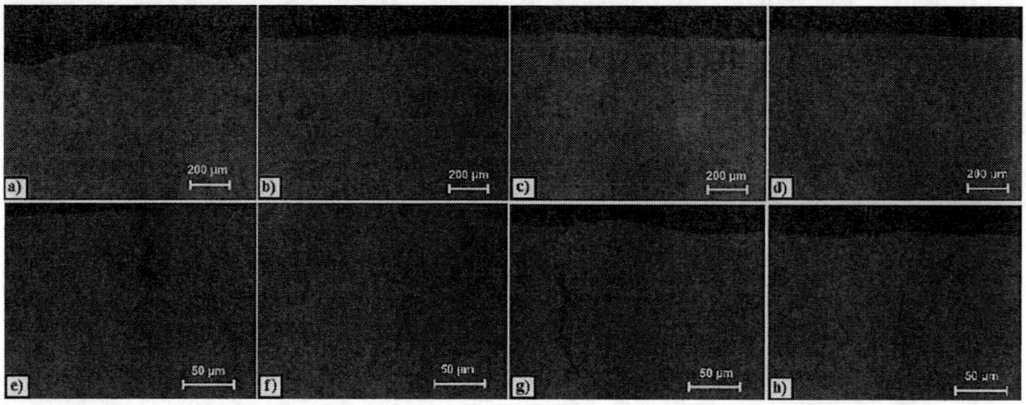

Fig. 7. Macrostructure of Waspaloy surface after laser heat treatment – multiple laser tracks: a, e - LHT: q = 178 kW/cm^2, v = 75 m/min, O = 0; b, f - LHT: q = 178 kW/cm^2, v = 75 m/min, O = 50; c, g - LHT: q = 178 kW/cm^2, v = 75 m/min, O = 75; d, h - LHT: q = 178 kW/cm^2, v = 75 m/min, O = 87.5

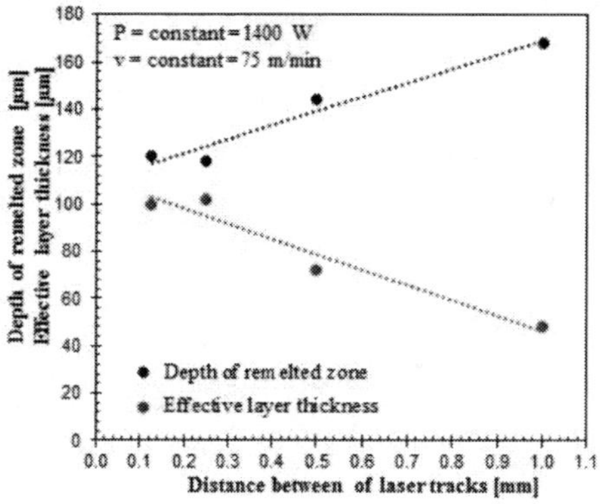

Fig. 8. The influence of distance between of laser tracks on a depth of remelted zone and effective thickness; P = constant = 1400 W, q = constant = 178 kW/cm^2 and v = constant = 75 m/min

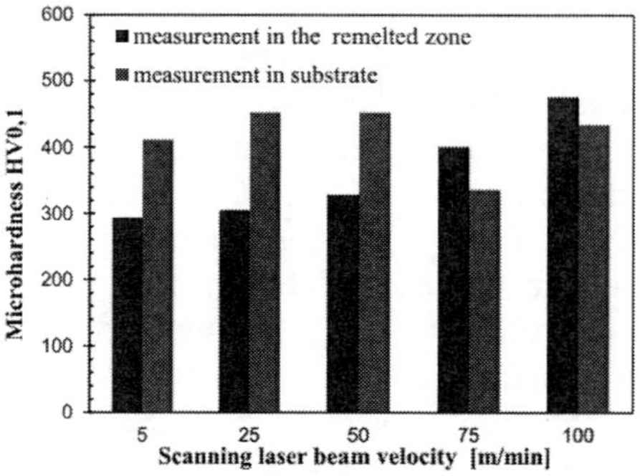

Fig. 9. The influence of scanning laser beam velocity on microhardness in remelted zone and the substrate for a constant of laser beam power density q = 178 kW/cm^2

Figure 8 presents the influence of distance between the laser tracks on a depth of remelted zone and effective thickness for a q = constant = 178 kW/cm^2 and v = constant = 75 m/min. Reducing the distance between laser tracks influences the increase of the effective thickness of layer. For the largest completing the working surface, the depth of remelted zone and the effective thickness layer have a similar value. However, such high scanning laser beam velocity and high laser beam power density is not recommended, because in the surface layer the numerous cracks can appear. Figure 9 shows the influence of scanning laser beam velocity on the microhardness of the layer and the substrate at a constant laser beam power density q = 178 kW/cm^2. It can be seen that with the increase of the scanning laser beam velocity, the microhardness decreases. Its value was equaled to about 500 HV0,1 in a substrate, and about 300 HV0,1 in surface layer. This phenomenon is advantageous because it allows easier machining of the material.

4 Conclusions

Based on the results of laser heating of Waspaloy it was found that laser heat treatment parameters have a significant impact on the microstructure and the microhardness of Waspaloy. By adjusting the laser heat treatment parameters such as the beam power density, scanning laser beam velocity and the distance between the laser tracks, the appropriate thickness of the melted zone can be obtained. As the most advantageous parameters of laser heating, the q = 178 kW/cm^2, v = 75 m/min, f = 0.5 mm can be selected. These parameters allow the reduction of hardness in cutting zone, which can cause the tool life improvement. Proposed parameters of laser heating for the Laser Assisted Machining (LAM) method are most preferred because they provide a continuous layer, free of cracks.

Acknowledgement: The presented research results, executed under the domestic project PBS of No 244445, were funded with grants for education allocated by the National Centre for Research and Development

References

1. M. Kawalec, D. Przestacki, K. Bartkowiak, M. Jankowiak. ICALEO 27th International Congress on Applications of Lasers and Electro-Optics, Congress Proceedings, 895-900D (2008)
2. H. Ding, Y. C. Shin. Int J Adv Manuf Tech, **64**, 475-486 (2013)
3. D. Przestacki, M. Jankowiak. Journal of Physics: Conference Series, **483**, 012019, 1-7, (2014)
4. D. Przestacki. Procedia CIRP **14**, 229-233 (2014)
5. D. Przestacki, P. Szymanski, S. Wojciechowski. Composites: Part A, **91**, 370-379 (2016)
6. S. A. Tadavani, R. S. Razavin, R. Vafaei. OPT Laser Technol, **87**, 72-78 (2017)
7. D. Przestacki, T. Chwalczuk, S. Wojciechowski. Int J Adv Manuf Technol, 1-12 (2017)
8. Steen M. W., Mazumder J. *Laser Material Processing*, (Springer, 2010)
9. K. A. Mumtaz, P. Erasenthiran, N. Hopkinson. J Mater Process Tech, **195**, 77-87 (2008)
10. R. S. Razavi. Opt Laser Technol, **82**, 113-120 (2016)
11. Y.-J. Liang, J. Li, A. Li, X.-T. Pang, H.-M. Wang. Scripta Mater, **127**, 58-62 (2017)
12. Y. Li, L. Zhou, W. He, G. He, X. Wang, X. Nie, B. Wang, S. Luo, Y. Li, Sci. Technol. Adv. Mater. **14**, 055010, 1-9 (2013)
13. S. Wojciechowski et al. Precis Eng, **46**, 158-165 (2016)
14. S. Wojciechowski et al. Int J Mech Sci, **119**, 360–369 (2016)
15. D. Bartkowski, G. Kinal. Int J Refract Met H, **58**, 157-164 (2015)
16. D. Bartkowski, A. Młynarczak, A. Piasecki, B. Dudziak, M. Gościański, A. Bartkowska, Opt Laser Technol, **68**, 191-201 (2015)

Technological aspects regarding machining the titanium alloys by means of incremental forming

Octavian Bologa[1,*], and *Carmen* Cotigă[1]

[1]Lucian Blaga University of Sibiu, Department of Industrial Machines and Equipment, 550025 Victoriei 10, Sibiu, Romania

Abstract. Titanium alloys are materials with reduced formability, due to their low plasticity. However, today there are high demands regarding their use in the automotive industry and in bio-medical industry, for prosthetic devices. This paper presents some technological aspects regarding the machinability of titanium alloys by means of incremental forming. The research presented in this paper aimed to demonstrate that the parts made from these materials could be machined at room temperature, in certain technological conditions.

1 Introduction

Lightweight materials are demanded nowadays mostly in the automotive industry. However, titanium alloys, due to their biocompatibility are also suitable for use in bio-medical industry, for manufacturing prosthetic devices [1-2]. Because prosthetic devices are prototypes, being tailored for each patient, incremental forming is one of the machining processes which could be used for manufacturing them. However, titanium alloys are well known for their low-plasticity and processing them by plastic deformation could be a problem. Also, parts machined by incremental forming are subject to dimensional errors.

To overcome these aspects, in [3] is presented an approach which uses high speed forming reducing the geometrical errors of the parts machined by incremental forming. Heating the parts by different methods is also a method reported in the literature as a way for increasing the plasticity of the titanium alloys during machining them by plastic deformation. In [4], the heating is achieved combining local heating with the heat generated by friction due high tool speed. The research presented in [5] is based upon heating the parts using direct current. Paper [6] also presents some promising results regarding the incremental forming of parts heated by means of electrical methods. However, the process of heating the parts has also some significant drawbacks: the temperature is very difficult to be controlled and heating could lead to oxidation, which can dramatically affect the biocompatibility of the titanium alloys.

The approach presented in this paper is to machine the parts made form Ti6Al4V titanium alloy, by means of incremental forming at room temperature, in order not to affect their biocompatibility.

* Corresponding author: octavian.bologa@ulbsibiu.ro

2 The experimental layout and the processed parts

Incremental forming involves the coordinated control of the processed sheet workpiece in XY plane, while the tool (punch) is moving incrementally on Z axes. Consequently, CNC milling machine are suited to be used as technological equipment for this process.

For processing parts by means of incremental forming, a CNC milling machine Haas MiniMill 1 was used. A CAD/CAM approach was used for generating the NC code for driving the punch on the processing toolpaths, due to the closeness between a milling process using a spherical mill and an incremental forming process using a punch with a hemispherical head.

The experimental layout is presented in figure 1. This experimental layout was successfully used before by the authors for some other incremental forming machining processes [7-8].

Fig. 1. Experimental layout: machine, fixture and processed part

The approach targeted the process of manufacturing prosthetic devices, mainly cranial prostheses. The slope angles for these kinds of parts are limited, so the initial assumption made by the authors was that the low plasticity of Ti6Al4V titanium alloy will not hinder the machining process at room temperature.

As test shape, the cone-frustum was selected to test if incremental forming could be performed on the parts, at room temperature without the occurrence of cracks. The starting slope angle of the cone-frustum parts was chosen 30° and it was gradually increased.

Figure 2a shows the imposed geometry of a processed parts (side view), while figure 2b shows the 3D model which was used for generating the CNC code for machining the part.

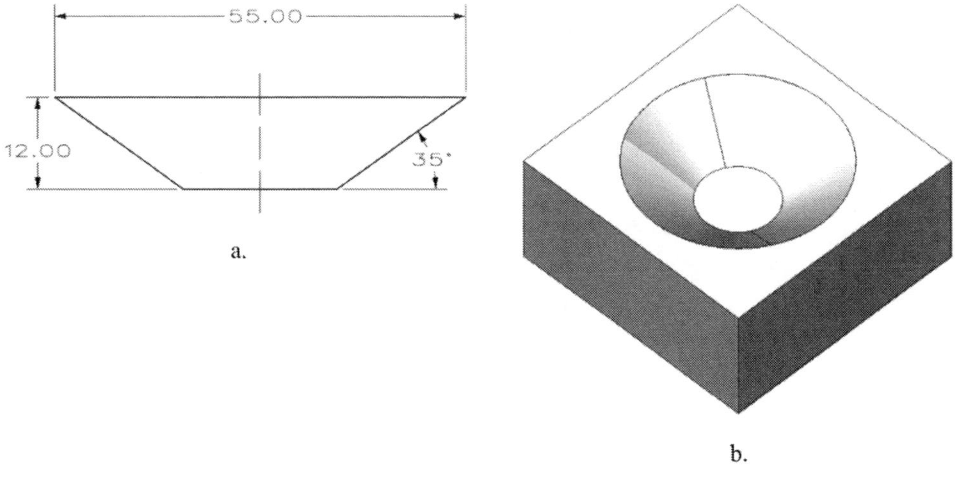

Fig. 2. Example of processed part, side view (a) and 3D model (b)

3 The processing strategies and toolpaths

To process the parts by means of incremental forming at room temperature, several technological conditions were selected. Preliminary researches have demonstrated that while the slope angle exceeds 30°, the vertical increment should be chosen in the range of 0.2-0.6 millimeters. Bigger vertical increments leaded to crack occurrence to more than 50% of machined parts.

Three types of toolpaths were considered: circular toolpaths, the tool was drive along one simple circular trajectory on every XY plane, each plane being situated at one vertical increment form each other (figure 3a). The second approach used Archimedes' spiral in each plane situated at one vertical increment from each other (fig. 3b). Finally, the third approach used a continuous spatial spiral toolpath for machining the whole part (fig. 3c)

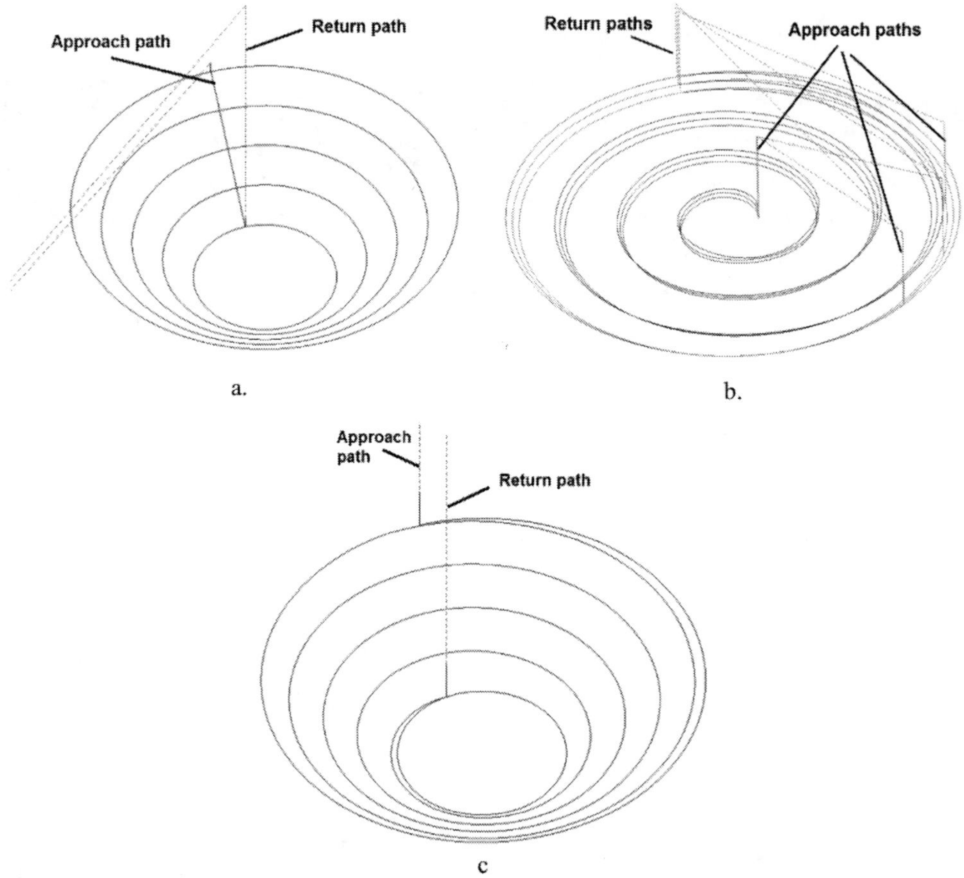

Fig. 3. Processing toolpaths, circles (a), Archimedes spiral (b), spatial spiral (c)

Table 1 presents a synthesis with regard of the results. It can be noticed the fact that the best results have been obtained by using the spatial spiral toolpath. This fact could be explained by observing that for this toolpath, there is only one entry point and one exit point, while between them the toolpath is continuous. For the other two approaches, there are multiple entry and exit points Entry points are considered stress concentrators for the incremental forming process and aligning them on a line as in figure 3a favors the apparition of cracking lines. Archimedes spiral toolpaths have shown slightly better results as simple circles, however the complexity of generating them is not compensated by this results. Archimedes spirals are not standardized toolpaths in most of the CAM programs, so the toolpaths had to be generated in a semi-automatic way.

From table 1 it can also be seen that the cracks started to occur at the angle of 37° for circular toolpaths and a vertical increment of 0.6 millimeters and at 38° for vertical increments of 0.4 millimeters. At a slope angle of 38° and 0.6 vertical increment, the parts cracked every time, while for same conditions but Archimedes spirals the crack only occurred from some of the parts.

Finally, at slope angles bigger than 38°, the cracks occurred for every toolpaths and increment step. It was considered that this slope angle is the physical limit of machining parts from this material (Ti6Al4V titanium alloy), at room temperature, without heating the part by means of incremental forming.

In figure 4a a processed part is presented (with a slope angle of 35°, while in figure 4b a processed part with cracks is shown (with a slope angle of 38°).

At this stage of the research the rotational speed of the tool was chosen 150 rev/min and the feedrate 150 mm/min, for all processed parts.

a. b.

Fig. 4. Processed part (a), processed part with cracks (b)

Table 1. Comparison

Slope angle	**Increment step**	**Trajectory**		
35°	0.2	Circles	Archimedes spiral	Spatial spiral
	0.4	Circles	Archimedes spiral	Spatial spiral
	0.6	Circles	Archimedes spiral	Spatial spiral
36°	0.2	Circles	Archimedes spiral	Spatial spiral
	0.4	Circles	Archimedes spiral	Spatial spiral
	0.6	Circles	Archimedes spiral	Spatial spiral
37°	0.2	Circles	Archimedes spiral	Spatial spiral
	0.4	Circles	Archimedes spiral	Spatial spiral
	0.6	Circles **CRACKS OCCURENCE**	Archimedes spiral	Spatial spiral
38°	0.2	Circles	Archimedes spiral	Spatial spiral
	0.4	**Circles CRACKS OCCURENCE**	Archimedes spiral **CRACKS OCCURENCE**	Spatial spiral
	0.6	Circles **CRACKS**	Archimedes spiral **CRACKS OCCURENCE**	Spatial spiral

4 Conclusion

The main goal of the research presented in this paper was to demonstrate that, if the technological conditions are selected properly, parts made from Ti6Al4V titanium alloy can be processed by incremental forming at room temperature. Heating the parts was not considered as an option, in order not the affect the biocompatibility of the parts.

The experimental tests have demonstrated that a slope angle of 38° is the upper limit for cone-frustum parts machined at room temperature. While this value could be considered small, it is here noticeable that the parts targeted by this research were cranial prostheses, were the slope angles are also small. For these kind of parts, 38° could be considered as totally acceptable.

Three types of toolpaths were taken into consideration. The best results have been found for the spatial spiral toolpath. The assumption made by the authors was that this toolpath is a continuous one, with only one entry and exit point, a fact which does not favor stress concentrators and consequently does not favor cracks occurrence. Further researches, based by a joint FEM analysis and experimental test process will be performed in order to validate this assumption.

References

1. A.L. Chicea, R.E. Breaz, 9th International Technology, Education and Development Conference INTED 2015, Madrid, Spain, 3568-3574 (2015)
2. Muresan, G., Morar, L., Breaz, R.E., International Technology, Education and Development Conference Valencia, Spain, 4115-4122 (2014)
3. G. Ambrogio, F. Gagliardi, S. Bruschi, L. Filice, CIRP Annals - Manufacturing Technology, **(62/1)**, 243–246 (2013)
4. G. Palumbo, M. Brandizzi, Materials & Design **(40)**, 43–51 (2012)
5. G. Ambrogio, L. Filice, F. Gagliardi, Materials & Design **(34)**, pp. 501–508 (2012)
6. G. Fan, L. Gao, G. Hussain, Z. Wu, International Journal of Machine Tools and Manufacture, **(48/15)**, 1688–1692 (2008)
7. R. Breaz, O. Bologa, M. Tera, G. Racz, The 14th Conference on Metal Forming, Krakow - Poland, 427-430 (2012)
8. M. Tera, R.E. Breaz, O. Bologa, G. Racz, *Proceedings in Manufacturing Systems*, **(9/2)**, 99-104 (2014)

Study of the Formability of Laminated Lightweight Metallic Materials

Claudia Girjob[1,*], and *Gabriel* Racz[1]

[1]Lucian Blaga Univerity of Sibiu, 10 B-dul Victoriei, Sibiu, Romania

Abstract. The main objective of this work was to test the formability of laminated materials. Laminated materials are considered a good choice when parts with reduced weight are considered. Thus, a laminated material, aluminum - polypropylene - aluminum (Al-PP-Al), as sheet 1.2 mm and 1.4 mm thickness was used. Before processing the material by means of unconventional plastic deformation, its formability was determined by running the Nakajima test. After obtaining the forming limit curves, the material was machined by means of incremental forming.

1 Introduction

At present days, there is a continued interest in using composite materials made by means of laminated layers for building structural elements for the automotive industry. These materials are lightweight metallic materials [1]. A widespread use as composites is reported for aluminum-plastic-aluminum, polymer-steel plastic and steel-polymer-steel.

These materials possess some significant advantages, because they can replace more expensive ones, while preserving the qualities of the final product, or even increasing the overall performance and lifespan of it.

Laminated workpieces made of metal-polymer-metal layers show a good machinability by means of plastic deformation, which is considered similar with the one of workpieces made of usual metallic materials. The work presented in [2] has experimentally proven that the workpieces made of steel-polypropylene-steel layers shows a better formability compared with AA5128 aluminum layer. Some specific characteristics for plastic deformation processes, such as the limiting drawing ratio were found to be better or at least equal for some composite materials than the ones of conventional materials used for manufacturing body parts within the automotive industry [3]. The work presented in [4] has proven that the forming limit for laminated materials composed of aluminum - polymer-aluminum layers is heavily influenced by the size of the aluminum layers.

To assess the machinability by means of plastic deformation processes, together with their mechanical proprieties, a series of tests are unfolded, such as uniaxial tensile test, bending test and shear tests [5].

Laminated materials were targeted lately for manufacturing parts with reduced of small weights. Aside of automotive industry, prosthetic devices can also be produced using these materials as workpieces [6].

[*] Corresponding author: claudia.girjob@ulbsibiu.ro

In order to machine parts made of composite materials, unconventional technologies are often the best choice [7]. Incremental forming, due to its high degree of flexibility, has been proven itself as machining process, mainly when producing parts in small batches or even prototypes [8-9]. One of the requirements when machining by incremental forming is that the formability of the workpiece must be known, consequently some specific tests, such as Nakajima test must be unfolded, to determine the forming limit curves of the processed material [10].

Incremental forming may be performed using either CNC multi-axis milling machines, or serial industrial robots as technological equipment. Preventive measures should be taken, because technological forces cannot be accurately predicted and could exceed the allowable forces on the CNC machines or robots. There are some research works reported in the literature with regards of how to calculate the size of the technological forces which occur during the incremental forming process [11-13].

A literature survey reports in [14] a research work which considered composite material mild steel-polypropylene-mild steel aluminum-polypropylene-aluminum respectively as workpiece for manufacturing parts by incremental forming. The experimental tests had shown good results with regards of formability, allowing to be machined without the occurrence of cracks.

2 Materials used for researches

For the approach presented here, test samples from aluminum - polypropylene - aluminum (Al-PP-Al) laminated material were used. The workpieces were chosen as sheets of 1.2 mm and 1.4 mm thickness. The workpiece consists of two layers of aluminum (AA5182) of 0.2 mm respective 0.24 mm thickness, which embed between them a layer of polypropylene (PP). This material shows remarkable proprieties such as flexural stiffness and very high lightness, which make it suitable for aeronautic and automotive industry [15].

The proprieties of the test samples made of Al-PP-Al (1.2 mm thickness) laminated material are presented in Table 1.

Table 1. Hylite properties [14]

Weight	1.8 kg/m^2
Maximum stretch	22 %
Plain strain stretch	18 %
Peel strength	4 N/mm
Flexural stiffness	7.1 kNmm
Aluminum yield point	140 MPa
Aluminum tensile strength	280 MPa
Expansion coefficient	28*10^{-6}/K

3 Determination of mechanical characteristics and formability of the test samples

3.1 Uniaxial tensile test

In this stage of the research, the strain hardening curve and other parameter of the material were determined by running the uniaxial tensile tests. As experimental layout, an Instron 5587machine and an ARAMIS optical measurement system were used.

DOI: 10.1051/matecconf/201712103008

Figure 1. shows the experimentally determined hardening curve.

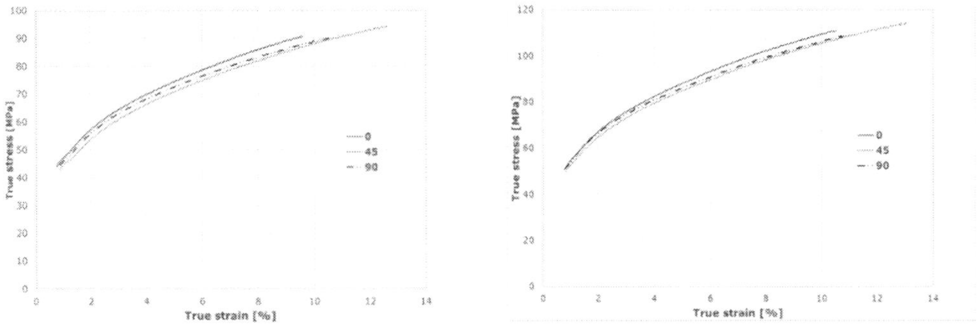

a. 1.2 mm laminated material b. 1.4 mm laminated material

Fig. 1. The hardening curve

The Also, the plane anisotropy coefficients were determined, per SR ISO 10113:1996 standard, table 2.

Table 2. Plastic anisotropy coefficients

Material	Cutting angle with respect to the rolling direction [0]	Plastic anisotropy coefficients
1.2 mm composite	0	1.06
	45	1.28
	90	1.09
1.4 mm composite	0	1.11
	45	1.32
	90	1.14

3.2 Determination of the forming limit curves

Nakajima test is widely used to obtain the forming limit curves for a given material. The test was performed on an experimental layout designed and built by the authors, using a hemispherical shape of the punch and a circular shape for the die. The Hasek samples proposed in 1978 were used. The state of stresses for this test is regarded as a combination between uniaxial tensile and echi-biaxial stretch. The data were gathered and processed by means of a data acquisition system. The ARAMIS measurement system was also included in the experimental chain. After running the test, the forming limit curves were determined. The obtained curves, for both thicknesses of the test samples used are shown in figure 2.

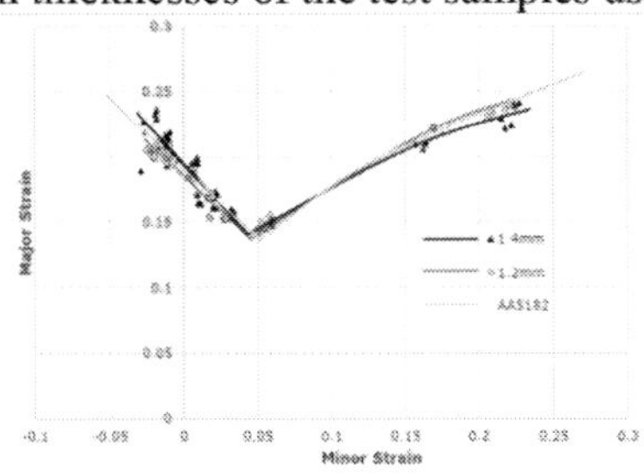

Fig. 2. Forming limit curves

DOI: 10.1051/matecconf/201712103008

3.3 Processing the parts by means of incremental forming

As stated above, CNC milling machines are widely used for machining by means of incremental forming. For this research, a 3-axis vertical machining center was used. The considered tool was a hemispherical punch with a radius of 5 mm and the test samples were 120 mm squares. The incremental forming process was unfolded by driving the punch on the machining trajectories with 150 mm/min (feed), while the rotational speed around its own axis was 150 rev/min.

An off-line measuring process was used for strains and thickness reduction, by means of the ARGUS optical measurement system.

A cone-frustum shape was imposed for the part. The geometrical parameters for the cone-frustum were the following: diameter of the bases 45 mm and 4 mm, slope angle 35°. d. The processing trajectories were generated by intersecting the cone-frustum with parallel planes (all of them being also parallel with XOY plane, figure 3). The vertical distance between these planes, seen as the vertical increment was 1 mm.

Fig. 3. Circular trajectory

The behavior of the part processed by single point incremental forming can be evaluated by measuring the distribution of the values of major and minor strains (Fig. 4, 5), the thickness reduction of the sample (Fig. 6) for the two sheets of workpiece.

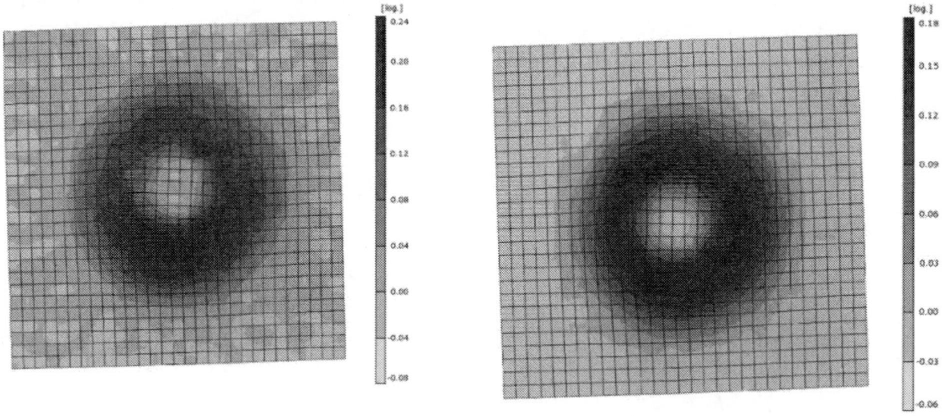

a. 1.2 mm laminated material   b. 1.4 mm laminated material

Fig. 4. Major strain distribution

For the major strain, there is a homogenous distribution throughout the circumference of the pieces and shows a major strain at the start point of the every circle, in each plane. The maximum values of major principal strains are: 0.2377 for 1.2 mm laminated material respectively 0.1776 for 1.4 mm laminated material.

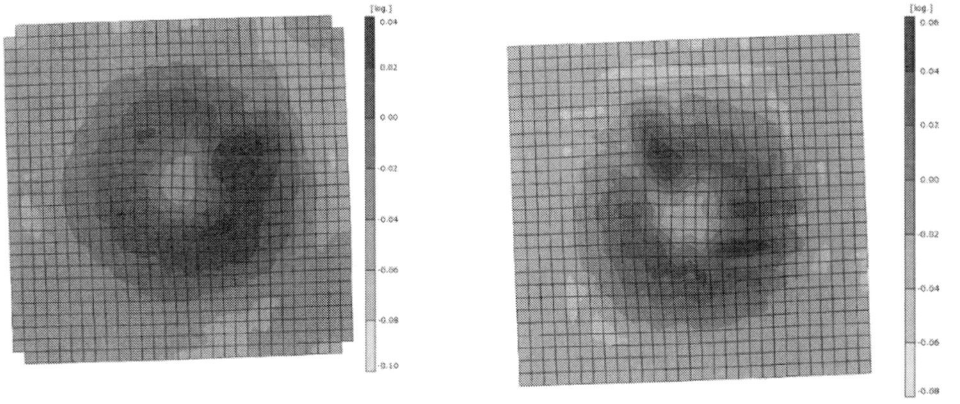

a. 1.2 mm laminated material b. 1.4 mm laminated material

Fig. 5. Minor strain distribution

The minor strain distribution is similar to the major strain distribution. The values of minor strains are: 0.0421 for 1.2 mm laminated material respectively 0.0707 for 1.4 mm laminated material.

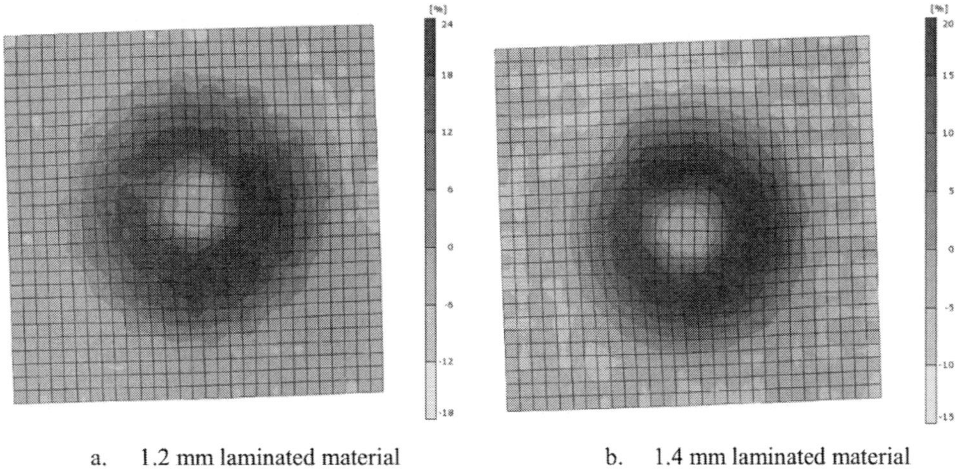

a. 1.2 mm laminated material b. 1.4 mm laminated material

Fig. 6. Thickness of the specimen

For 1.2 mm laminated material reduction is 18.5% respectively 22.8% for 1.4 mm laminated material in the descent point of the last circular punch trajectory. In this case the work piece thickness reduction is not uniform distributed throughout its circumference.

4 Conclusion

The results of the researches have shown that laminated lightweight metallic materials have a good formability.

Single point incremental forming process of laminated material sheets 1.2 mm respectively 1.4 mm material were unfolded and the preliminary results have demonstrated that a proper selection of technological parameters (diameter of the punch, feed, speed) and processing trajectories can lead to parts with good structural and dimensional characteristics.

Further experimental tests will be performed in order to evaluate the influence of the geometry of the punches and the cutting parameters upon the surface roughness of the machined parts.

References

1. C. Girjob, O. Bologa, G. Racz, C. Biris, Applied Mechanics and Materials (**760**), 391-396 (2015)
2. J.A. DiCello, SAE, (**800078**), 1 (1980)
3. L.W. McKenna, M.H. Wohl, J.C. Woodbrey, (**800079**), SAE, 506 (1981)
4. T.S.V. Somoyajulu, *Vibration and formability characteristics of aluminum-polymer sandwich materials*, PhD Thesis, University of Michigan (2004)
5. B. Engel, J. Buhl, *Metal Forming - Process, Tools, Design.* (Edited by Mohsen Kazeminezhad, ISBN 978-953-51-0804-7, 2012)
6. A.L. Chicea, R.E. Breaz, 9th International Technology, Education and Development Conference INTED 2015Madrid, Spain, 3568-3574, (2015)
7. C. Biriş, Academic Journal of Manufacturing Engineering, (**14**), 2, pp. 33-39 (2016)
8. M. Tera, R.E. Breaz, O. Bologa, G. Racz, *Proceedings in Manufacturing Systems*, (**9/2**), ISSN 2067-9238, pp. 99-104 (2014)
9. R. Breaz, O. Bologa, M. Tera, G. Racz, The 14th Conference on Metal Forming, Krakow - Poland, 427-430 2012
10. H. Chalal, S.G. Racz, T. Balan, International Journal of Mechanical Sciences, (**59**), 1, 104-114 (2012)
11. J. R. Duflou, A. Szekeres, and P. Vanherck, Advanced Materials Research, (**6-8**), 441-448 (2005)
12. R. Aerens, P. Eyckens, A. Van Bael, J. R. Duflou, International Journal of Advanced Manufacturing Technology, (**46**), Issue 9-12, 969-982 (2010)
13. M. Tera, R.E. Breaz, O. Bologa, S.G. Racz, Key Engineering Materials, (**651**), 1115-1121 (2015)
14. K.P. Jackson, J.M. Allwood, M. Landert, Journal of Materials Processing Technology (**204**), 290 (2008)
15. EAA (European Aluminum Association), *The Aluminum Automotive Manual*, http://www.eaa.net/aam (2002)

Innovative Industrial Technologies for Preventive Anti-Graffiti Coating

Martin Heinisch[1], and *Dan* Miricescu[2,*]

[1]MHI Ingenieurgesellschaft mbH, B+H Solutions GmbH in 73630 Remshalden, Germany; also visiting professor Engineering Faculty, Lucian Blaga University of Sibiu, Romania
[2] Lucian Blaga University of Sibiu, Engineering Faculty, Sibiu, Romania

Abstract. Graffiti is the main issue in terms of acts of vandalism with public transport. Hardly any suburban train, commuter train or goods train, hardly any engineering building, station or noise barrier wall escape graffiti spraying. This means enormous expenses every year which one would rather use to the benefit of its customers. Furthermore, the temporary and permanent anti-graffiti protection systems used up to now do not ensure adequate graffiti protection. The costs for the application of graffiti protection and/or graffiti removal to a large part depend on the circumstances of the individual case in question, i.e. on the size of the surface, the type and size of the ground, the accessibility of the location and of the fact whether the graffiti need to be removed from unprotected or protected surfaces. A unique innovation based on nanotechnology can provide a durable graffiti protection on a wide range of surfaces at economical prices. The protection makes the application of lettering, images or signatures significantly more difficult, yet facilitates their removal without any harm to the ground surface.

1 Damage due to graffiti in Germany and with Deutsche Bahn AG

In the year 2014 alone, more than 35,000 acts of vandalism and graffiti were recorded throughout Germany. The damage caused by unauthorized graffiti on buildings and public transport vehicles is estimated to range between 200 million Euros and 500 million Euros per year. The annual expenses, including also prevention and taking out insurance, are even estimated to be a billion Euros.

In accordance with Czycholl, Harald (2011) [1], the economic damage resulting from acts of vandalism is 250 million Euros per year, as stated by the Deutscher Städtetag [German Cities and Communities Council], public transport accounting for half of this amount, private real estate for one third and public buildings for one fifth.

In accordance with the DB information service issue 03/2015, [2] Deutsche Bahn AG annually incurs damage of about 30 million Euros due to about 14,000 graffiti damage

* Corresponding author: dan.miricescu@ulbsibiu.ro

cases; a new paint coat for a complete train costing up to 15,000 Euros and taking about seven days.

2 Currently used systems - graffiti protection

Although Deutsche Bahn AG uses protective paint coats and anti-graffiti film to protect their buildings and trains, the effort and environmental pollution for graffiti removal is enormous and cost-intensive. Despite the use of lacquer coats and so-called "sacrificial layers", from which graffiti can be removed more easily, DB incurs the high amount of damage as stated so that new technological innovations in terms of preventive graffiti protection are urgently required.

Basically with the presently used systems a distinction must be made whether these serve the removal of graffiti and cleaning or whether they provide protection in the form of a temporary or permanent coating which is to enable relatively simple cleaning.

Normally graffiti paint has a high percentage of solvents which can be absorbed particularly by absorbent grounds to a depth of several millimeters. Part of the non-absorbent grounds is attacked by the solvents to such a degree that the surface is damaged when removing them. Simultaneously, in addition to graffiti paint, marker pens and wax crayons as well as coal-tar colors are used on smooth surfaces.

Spray can paint on the basis of acrylates and alkyd resins, which are available in almost all hues, are frequently used. Furthermore, products with binding agents such as polyalkene, polyvinyl acetate or copolymers or even two-component resins and oil paints are used.

Fig. 1. Graffiti train close to Frankfurt/Main Central Station, Germany.

A particular challenge for anti-graffiti systems are hardly soluble silver, bronze and gold paints or felt-tipped pens which have a very high penetration into the ground. Increasingly car under seal in spray cans is used which can be removed - if at all - only with a great deal of effort and at great expenses.

There are different mechanical or chemical methods for removing undesired graffiti, e.g.:

- manual removal
- mechanical removal using the low-pressure wet sand blasting method
- chemically supported removal by hot water high-pressure jet
- removal by hot water high-pressure jet on grounds with a temporary or permanent coating
- laser technology

The use of temporary or permanent coatings depends heavily on the surface to be coated: mineral or organic, porous and optical.

Accordingly, different protective systems are feasible:

DOI: 10.1051/matecconf/201712103009

- protective sealing with barrier layers
- temporary protective layers using sacrificial systems
- permanent protective layers using firmly adhering layers
- graffiti protection coats
- anti-graffiti film

Temporary sacrificial layers are cost-efficient protective layers which can be easily removed. However, such layers are high-maintenance as they have poor adherence qualities and their effect is reduced by mechanical damage.

In contrast, regular renewal of the layer is not necessary for permanent protective layers, which, however, have a low degree of water vapor permeability and in part tend to develop surface yellowing. In the case of a fire, poisonous carbon and nitrogen compounds can develop.

Graffiti protection coatings are normally applied to defaced surfaces; first a primary coat is applied, then it is coated with a fresh layer of paint. Such paint is commercially available in different hues and transparent.

Anti-graffiti film is used on glass and is as good as invisible. It is used on outdoor as well as indoor glass panes and are pulled off and replaced by new film when they were subjected to a spraying attack.

What all systems have in common is the fact that they do not (initially) prevent the spraying of graffiti, but (are meant to) facilitate graffiti removal.

3 Nanotechnology

What is desirable is an anti-graffiti system which either prevents the application of lettering, images ("pieces") or signatures ("tags") with spray cans or felt-tipped pens on the surfaces from the defaced surfaces from the outset and after cleaning of the surface, which should be as simple as possible, looks as it did before.

B+H Solutions GmbH is a nanotechnology company focusing on the research, development, production and marketing of products containing nanoscaled materials – against the backdrop that almost all chemical properties of any material result from its surface, not from its mass. For example, steel does not oxidize in the component core, but on the components outside on the surface. Almost all chemical reactions and compounds happen on the surface of the materials in question. This is an effect which nanotechnology uses, thus being able to generate new properties for a large number of materials. This generates new potential for new solutions in the event of low material resources.

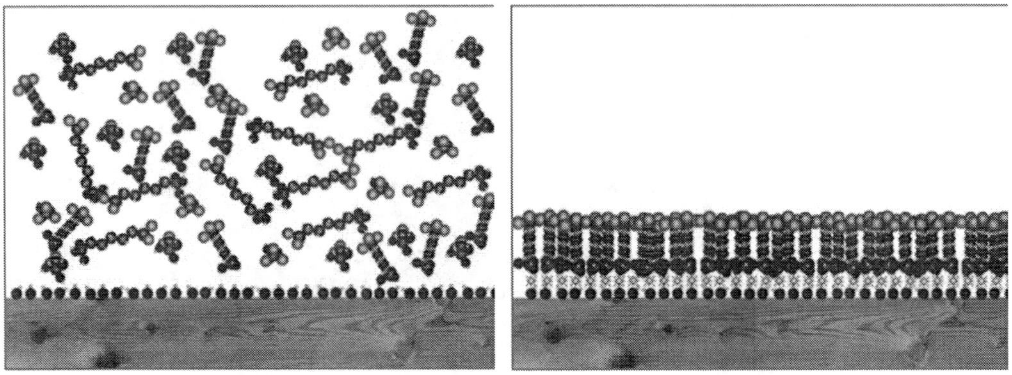

a b

Fig. 2a. Orientation of the molecules when applying the coating (short moment), **2b.** Molecules after application and drying of the coating; perfectly aligned = newly defined surface with new properties

Significant for nanoparticles is the development of other properties, i.e. for example a larger surface relative to the mass, higher reactivity and another color (silver turns to yellow and gold turns to red).

Nanomolecules are self-orienting and contrary to microparticles which are spread on the surface in a nondirectional manner, are aligned. Such alignment ensures a consistent surface.

Nanotechnology enables the creation of new structures, e.g. a lotus effect structure which is water, oil and dirt repellent. Nanotechnology is extremely resource-friendly, as little mass can have great effect.

One of the products developed by B+H Solutions GmbH is AGraffit®. No paint (and no dirt) will stick to surfaces treated with AGraffit® so that the sprayer cannot spray a clear image or pattern there, which will result in the sprayer quickly losing their motivation.

In addition to the product AGraffit®, there are many more technological innovations, and the company continues developing and formulating new products. Another mentionable product is ClearView intended for use on glass and ceramics. After its application water will simply roll off. Dirt cannot adhere any more. This considerably reduces the maintenance costs.

Compared to the products of other manufacturers, the product ClearView® stands out by it long durability, reducing the contamination of glass and ceramics and by the active combination of hydrophobisising (repellence of water) and oleophobisising (repellence of oil). This results in a self-cleaning effect in the event of torrential rain and/or enables cleaning of the surfaces with just water. When applied to photovoltaic systems, a clearly increased performance is stated due to the lower level of contamination of the element surfaces.

Furthermore, surfaces and coating layers can be coated in such a way that chewing gum will not stick to them anymore.

4 Engineering innovation for preventive graffiti protection by B+H Solutions GmbH

Thanks to the use of nanotechnology, the product AGraffit is doused with hydrophobic and oleophobic properties, which enable all grounds to be manufactured as repelling contamination, paints and solvents.

Fig. 3. Lotus effect you find in nature: Rolling-off of water

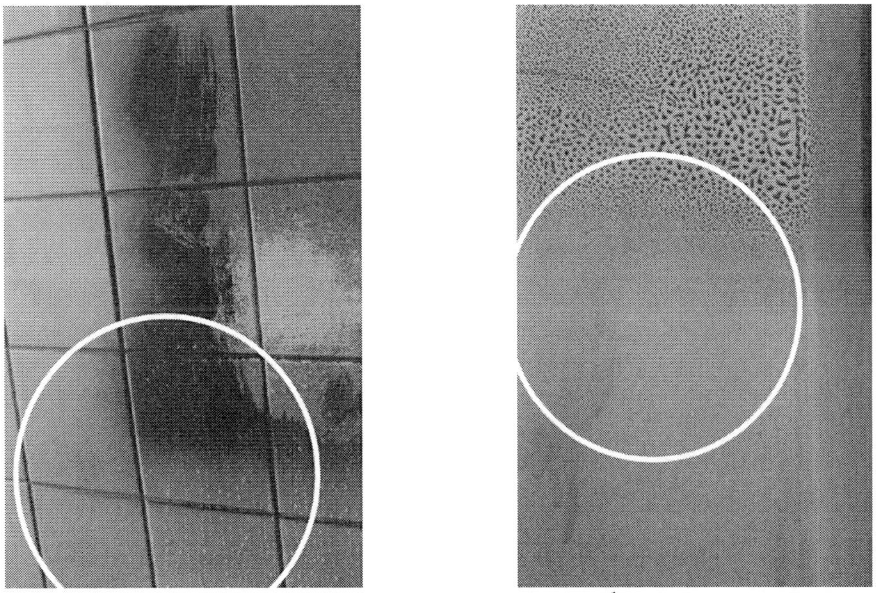

a b

Fig. 4a. Tiles were pre-treated with AGraffit® in the test area (light-coloured area). It is clearly visible that the paint spray hardly adheres and can be removed without problem (white area). The rolling-off effect is clearly visible; **4b.** Application of AGraffit® on sheet metal

AGraffit® is a quickly drying one-component coat with excellent adherence on steel, aluminum, nonferrous metal, glass-fiber reinforced plastic, plastics and a large number of paints. It generates a surface which is neither shiny nor dull, yet highly weatherproof and resistant to ultraviolet light.

a) b)

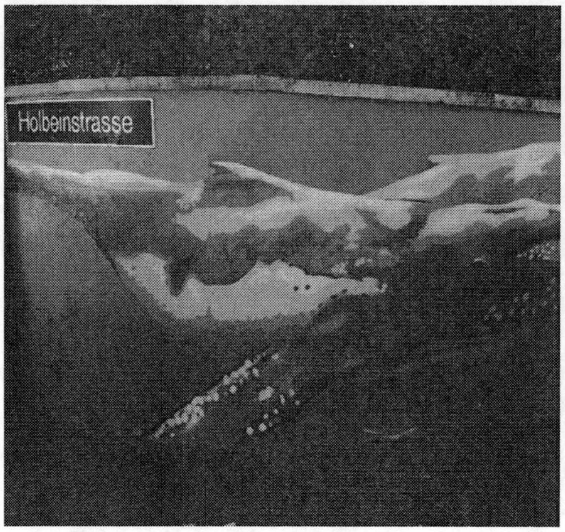

c)

d)

Fig. 5a. Original position: Soiled surface of mural (concrete); **5b.** During cleaning process; **5c.** Result after cleaning; **5d.** Prevention measures to protect genius art work

Sprayers cannot generate a precise image on the surface, the color rolls off. It is virtually impossible to spray lettering. Furthermore it is clearly visible that the paint can be removed without any problem.

Cleaning of surfaces treated with AGraffit® is simple and fast and in case of contamination requires nothing but water. If graffiti spray is dried, the manufacturer recommends the use of solvent (e.g. general-purpose thinner). The nano proofing will not suffer from cleaning; it is not a sacrificial layer. The effect of AGraffit® will remain unchanged.

5 Application

The product is applied on a dust-free ground by wiping it on with a cloth, rolling with paint roller or spraying with a layer thickness of 5 to 10 μm. On smooth and non-absorbent grounds, the manufacturer recommends applying the product with a cloth, on porous and absorbent grounds in an airless method, a thinner can be added, and the product should be applied in a cross layered pattern over the entire surface.

Depending on the degree of thinning and the type of thinner used, AGraffit® is dry after half an hour at room temperature and hard-dried after four hours and fully cured and chemically stable after seven days.

Forced drying and baking at 40 to 80 °C is possible in order to achieve faster resistance to chemicals. It can be removed easily by using 120 °C hot water vapors.

These products are tested in practice by the DB AG and already in use at train and suburban railway stations in and around Frankfurt/Main, Germany.

6 Cost effectiveness

The costs for the application of graffiti protection and/or graffiti removal to a large part depend on the circumstances of the individual case in question, i.e. on the size of the surface, the type and size of the ground, the accessibility of the location and of the fact whether the graffiti need to be removed from unprotected or protected surfaces. This means that the target prices cannot be quoted more accurately than within a wide scope. Depending on the type of ground, whether absorbent or not, one liter of AGraffit® protects a surface from 20 to 50 square meters.

7 Summary

This innovation can provide a durable graffiti protection on a wide range of surfaces at economical prices. The protection makes the application of lettering, images or signatures significantly more difficult, yet facilitates their removal without any harm to the ground surface.

References

1. H. Czycholl. *Nur teure Zusatz-Police helfen gegen Graffiti - Only expensive additional policies help against graffiti*, (Berliner Morgenpost, 2011)
2. J.-O. Voss, *Kunst oder Straftat? Graffitischäden bei der DB - Art or criminal act? Graffiti damage with DB*, DB Themendienst, **03** (2015)

MATEC Web of Conferences **121**, 03010 (2017) DOI: 10.1051/matecconf/201712103010
MSE 2017

Factors that influence the quality constant of the manufacturing process for asphalt milling knifes

Daniela Maria Iovanas[1,*], *Horia* Binchiciu[2], *Ionelia* Voiculescu[3] and *Emilia Florina* Binchiciu[2]

[1]Transilvania University of Brasov, Department of Materials Engineering and Welding, 1 Colinei Street, 500084 Brasov, Romania
[2]SC Sudotim AS SRL Timisoara, 30 Mihai Viteazu Street, 300222 Timisoara, Romania
[3]University POLITEHNICA of Buchares, Department of Materials Technology and Welding, 313 Splaiul Independentei Street, 060042, Bucharest, Romania

Abstract. The quality constant for mill knifes used to strip asphalt is significantly influenced by the quality of the reinforcement which, in its turn, is influenced by the thermic brazing process and by manufacturing the protection system at blockage through welding when it spins around its axis. It's also influenced by the quality of the intelligent wear and blocking self-protection systems that in their turn are influenced by oxidation and diffusion processes of W and C that make simmered carbides from the reinforcement and brazed joints. Overheating during welding and brazing of the knife reinforcement and/or blockage self-protection reinforcement favours the oxidation of the W carbides leading to a fast degradation of the affected zones, even in exploitation. Exceeding optimum temperature during brazing of the reinforcement in the low chromium alloyed steel support leads to Zn evaporation in certain areas from the brazing material and lowers the brazed joint resistance to wear this causes the knife reinforcement to detach from the support. Taking into consideration the above mentioned facts it is recommended that the production stages of the mill knifes are done mechanized and/or automatic constantly monitoring the execution parameters.

1 Basic aspect

The above mentioned method is an experiment developed at Sudotim and it is a worldwide novelty.

The research objective, that of obtaining better asphalt cutting mill knifes and their manufacturing process, was reached by equipping the knife support with an intelligent protection system to self-locking when the knife is spinning [1-5], figure 1, around its own axis and that of implementing in the simmered tungsten carbide reinforcement the wear self-protection system, made from rhomb grids by melting with laser bean [1-3, 6].

* Corresponding authors: daniela.iovanas@yahoo.com

The technical-scientific difficulties are those generated to assure a quality constant:

- Composite deposition from the self-locking system when spinning;
- Deep joint brazing with technical-mechanical constraint between the simmered tungsten carbide reinforcement and the low alloyed with chromium steel support;
- The laser bean melting technology needs to be without flaw at oxidation and/or vaporizing the rhomb grid that consolidates the reinforcement on the tip of the cutting knife.

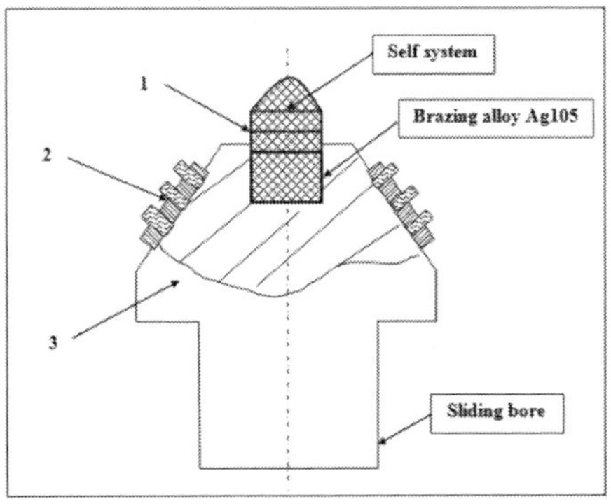

a)

b)

Fig. 1. Mill asphalt cutting knife.

a) constitutive parts [2], b) assemble; 1-reinforcement-cutting body with self-protection at wear, 2-knife support made from low allowed steel with chromium, 3-self-locking when spinning system.

2 Experiments to develop the protection system when self-locking at spinning

Knife supports, developed in two ways, by casting or by forging, are characterized by:

– Chemical composition, Table 1, [3];

Table 1. The supports chemical composition.

Mark		Constitutive elements [in mass %]							
		C	Mn	Si	Cr	Ni	S	P	Base
Forged	Prescribed	0.95 ÷ 1.10	0.90 ÷ 1.20	0.40 ÷ 0.65	1.30 ÷ 1.65	Max. 0.30	0.020	Max. 0.027	Fe
	Determined	1.03	0.95	0.45	1.42	0.21	0.020	0.018	Fe
Cast	Prescribed	0.85 ÷ 0.95	1.30 ÷ 1.60	--	1.80 ÷ 2.20	1.80 ÷ 2.20	0.020 ÷ 0.040	Max. 0.35	Fe
	Determined	0.92	1.4	0.62	1.8	2.12	0.033	0.029	Fe

– Structural and sclerometric characteristics of the material. The supports hardness is changed during their manufacturing process, determined by softening thermal treatments and by enhancing mechanical processing at final odds;

– Precise geometric configuration.

The self-protection to spinning system is characterised by:

- The welding behaviour of the addition material at welding;
- Matrix structure and reinforcement morphology;
- Configuration and thickness of the deposited layers.

Depositing the self-locking to spinning system was done using MIG [3] process, Fig. 2.

Fig. 2. MIG deposit using tubular wire.

Additional material (AD) is flux cored wire with composition: ST Fe-25%W-4%Cr used for welding with MIG process deposits thin layers of alloy type Fe-25%W-4%Cr-Ti, rich in complex carbides, needle like [3]. The deposited layers have a harsh aspect and a high resistance to abrasion combined with corrosion and thermo-mechanical wear.

Chemical composition (Table 2), micro-structure (Fig. 3), and micro-hardness measured values (Base Material - BM, Heat Affected Zone - HAZ, Deposited Material – DM), (Table 3), of the deposited material using MIG procedure on a carbon steel plate fit into standard SF-SDT 14/2016, [7]. The tungsten deviation is explained by the fact that the carbides inside the reinforcement contain most of the tungsten.

Table 2. Chemical composition of the deposited metal (DM).

Prescribed	**0.7-1.2%C**	**~0.7%Mn**	**max.0.7%Si**	**3.5-4.5%Cr**	**~25%W**	**~0.3%Ti**
Spectral determined	0.9	0.65	0.35	4.2	12.8	0.35

Fig. 3. Deposited metal micro-structure, nitric attack 2% - 25sec, X500, dendritic structure, uneven.

Table 3. Micro-hardness measured values.

Measured points values HV $_{0.2}$		
BM	**HAZ**	**DM**
402; 410; 415	365; 372; 353	535; 544; 565 – on matric 1200; 1135; 1208 – on carbides

3 Experiments to qualify the brazing process

The brazed joint, support-tungsten carbide (WC) reinforcement, is characterized by the chemical-physical proprieties of the addition material and by the manufacturing process of the joints, prescribed by the execution procedure in order to assure constant quality [8-9].

The thermal dilution coefficient of the support, made from the above presented steels, is available with temperatures between the interval 12, 19...19.4x106 [1/grd.], at temperatures from 0°C to 1200°C.

The reinforcement used in both situations is from the hard alloys category, made by powder metallurgy, type VC-TiC-Co. For testing we used tips from the category P30, characteristics are presented in Table 4, in a non-affected thermal state. We used the above mentioned taking into account resistance to bending and cost.

Table 4. Alloy proprieties VC-TiC-Co.

Group	Chemical composition [%]			Medium values of the chemical- physical characteristics			
	WC	TiC	Co	Hardness HV [daN/mm²]	Bending resistance [N/mm²]	Compression resistance [N/mm²]	Thermal dilution coefficient x10⁶[1/grd]
P30	82	8	10	1500	1700	5000	5.5

Technical characteristics provided by material manufacturers and those determined by experiments highlight the framing in prescriptions.

Physical-chemical and structural characteristics of the addition material, mark VIAg40SnR, [3, 7], are shown in Table 5, 6 and figure 4.

Table 5. Chemical composition of the constituents used to make VIAg40SnR.

Code	Composition [mass %]				
	Ag	Cu	Zn	Sn	Si
MDVIAg40SnR	40.3	30.2	26.1	2.4	0.05

Table 6. Hardness valued determined on the deposited material using VIAg40SnR.

Hardness value HB/DM	Micro-hardness VICKERS-HV0.1/DM	Average value
185; 190; 183; 188; 189	208; 206; 212; 215; 210 Measuring uncertainty: 2.14%	187 HB 210.2 HV0.1

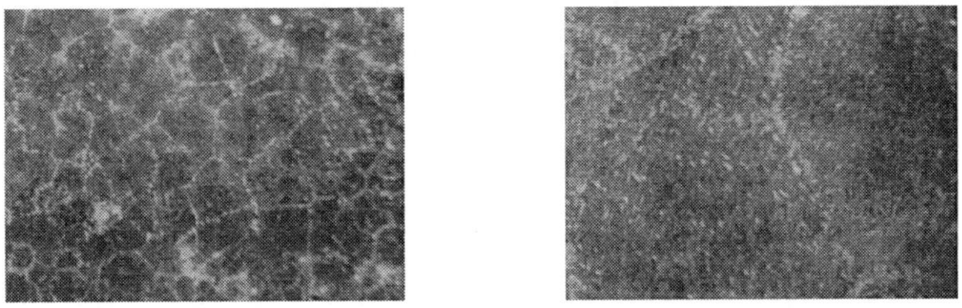

a) b)

Fig. 4. Structural characteristics of the deposited material using VIAg40SnR, a) 100X, b) 500X.

Sclerometric results confirm detected structures and chemical composition on the deposited material.

Joint design, fig. 1, was determined on the principle of developing capillary joint, which have a high resistance to wear by abrasion under high pressure and reduced costs. To this end we choose a joint of 0.1÷0.15 mm, made by mechanical processing. Parts positioning, in order to start brazing, were done after examining the brazing surface and that of the remains left after processing. Checking the geometric constant of the joint was done by calibrating, thickness 0.1 mm. In order to heat and deposit the addition material into the joint, we choose a nozzle and adjust the oxy-acetylene flame character. To avoid losing carbides and reinforcement oxidation when heating, we used a carbide flame and a protection for the reinforced tip made from a mixture of graphite and alumina and sodium silicon, dried and burnt.

Deposition temperature for the addition material in deep joint measured with pyrometer with laser bean was 625÷780°C. Thus obtained joint was cleaned by brushing and activated by washing with water and activating solution.

We extracted samples from the accomplished joint in order to characterise the product, by destructive and non-destructive testes, according to SR EN 13134:2002. Results were subjected to visual examination in two stages, unfinished state and end product (according to SR EN 12799:2002).

Metallographic results of the joints at the passing zone of the addition material and knife support, influenced by the brazing process, are highlighted below, Fig. 5.

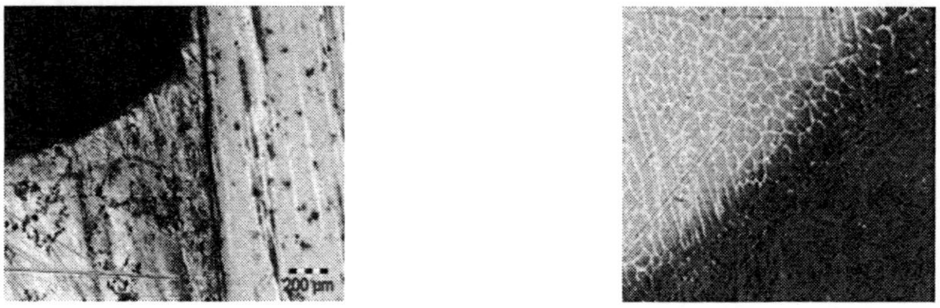

a) b)

Fig. 5. Macro-micrographic analysis of the joining zone in cross section, a) macroscopic aspect of knife support-carbide (20X); b) microscopic aspect addition material-carbide (100X).

We can observe a good diffusion of the addition material into the base material (BM), also changes like reduced grain size of the recurrence structure of the base material in the adjacent area of the addition material. These phenomena can be explained by overlapping the thermal effect produced at brazing with the one from the recurrence one of the support.

Metallographic analysis was accompanied by hardness determinations, using Vickers method, (Table 7).

Table 7. Hardness test results (HV0.1).

Test area	**MB support**	**HAZ support**	**AM**	**WC**	**Matric**
Micro-hardness Vickers HV0.1	680	550	160	1480	352
	700	515	160	1560	339
	710	535	181	1410	351

Hardness structures confirm detected structures. Results obtain on tungsten carbide constituents show a decrease of 1%. Hardness decrease is insignificant in concrete industrial situations.

Conducted tests validate the brazing process and justifies it's qualification proposal, according to SR EN 13134:2002.

4 Experiments to determine the melted grid on the tip the reinforcement-cutting knife, using laser bean

The self-protection system to wear is characterised thus:

- By the reinforcement quality;
- By the constant quality of the melting process of the grid;
- By geometrical parameters of the grid.

Work parameters and their values are presented in Table 8, [1].

Table 8. Work parameters.

Work parameters	Work parameters value		
	1	**2**	**3**
Focalize [mm]	5	5	5
Movement speed [mm/s]	1	1.5	2
Pulse frequency [Hz]	6633.100	6633.100	6633.100
Distance between impulses [mm]	0.05	0.05	0.05
Welding time [min]	24	20	16

To analyse micro-hardness characteristics we made measuring on the external surfaces of the reinforcement-cutting tip, using Vickers method.

Micro-hardness values reinforcement tip-cutting body before laser processing are shown in Table 9, and the ones measured on the laser processed surfaces in Table 10, with the remark that values that are below average can be explained by the fact that the diamond tip had excess binder in some areas.

Table 9. Micro-hardness values determined on the tip of the reinforcement-cutting body made from tungsten carbide before laser processing.

Measured values, $HV_{0.2}$	Average value, $HV_{0.2}$
873; 1102; 910; 1124; 1128; 1191; 1151; 1456; 890; 1143	1092

Table 10. Micro-hardness values HV0.2 determined on reinforcement tip – cutting body made from tungsten carbide after laser processing.

Measuring area	Measured values, $HV_{0.2}$	Average value, $HV_{0.2}$
1	1273; 1357; 1426; 1134; 1088	1256
2	1221; 1220; 1284; 1305; 1189	1244
3	1381; 1244; 1052; 1039; 1399	1223

Micro-structural analysis was made by electronic microscopy SEM, they can be observed in Fig. 6. The image presents micro-structural aspects of the tungsten carbides for areas 1, 2 and 3. After laser processing we can clearly observe melted areas under the lasers action.

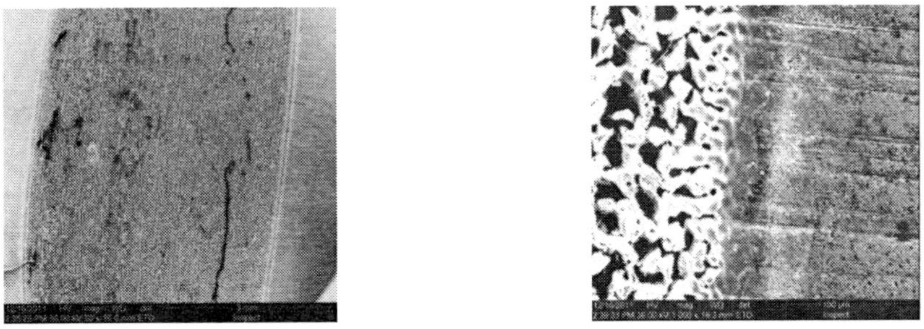

a) b)

Fig. 6. SEM analysis images of the specific areas, a) crack due to a major energetic impact on the melted grid, b) passing area.

In figure 7, on area 3 we can notice micro-cracks developed from the melted area and the forming of a melted crust with length of approximately 45 μm, which in exploitation determines the reinforcements degradation.

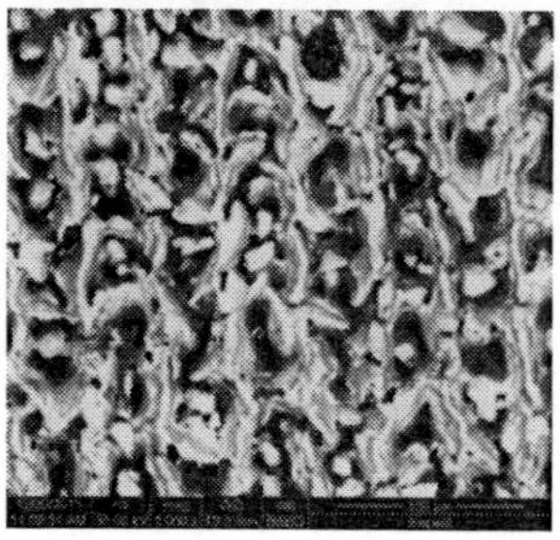

Fig. 7. Micro-crack area, growth 1000X.

5 Conclusions

The solution for implementing in production consists of:

- Developing the reinforcement tip, by simmering tungsten carbides, and the tooth support, made of low alloyed steel with chromium, by casting or precision forging;
- Loading by welding, with harsh deposits, of the knife support in its conical area;
- Brazing the reinforcement tip in holes in the body of the knife support.

Applying the solution imposes solving the following problems:

- discharging or reducing the loss of carbides and/or that of corrosion by oxidation of the reinforcement under the action of heat when brazing;
- Assuring compatibility at brazing, with oxy-acetylene flame, of the base materials with the addition ones, in intense conditions of the cycle efforts of dilution and contraction of the tooth support, of the reinforcement at brazing and when depositing the joining layers;
- Assuring the joint removal temperature, superior to that of the brazed ensemble, when developing the joint.

References

1. R.F. Iovanas, *Theoretical and experimental research on increasing the durability and reliability of equipments, manufactured by weld cladding, visible on earth drilling equipments*, (PhD thesis, Universitatea Transilvania din Brasov, Brasov, Romania, 2012)
2. E.F. Binchiciu, V. Geanta, I. Voiculescu, R. Stefanoiu, R. F. Iovanas, A. Binchiciu, H. Binchiciu, *Cutit de freza pentru decopertat asfalt cu sistem de autoprotectie la uzare si autoblocare la rotire si procedeu de fabricatie*, (RO Patent 129863, 2014)
3. ***, http://www.matfrez.ro/, accessed 1.02.2017
4. T. D. Tanasache, C. Florea, V. Geanta, E. Binchiciu, AMR, **1029**, 78 (2014)
5. C. Buzatu, A.-E Dumitraşcu., D.I., Dumitraşcu, I. Orzan I., AMM, Eng. Sol. and Techno. in Manuf., **657**, 48 (2014)
6. R. F. Iovanas, D. M. Iovanas, I. Voiculescu, N. I. Trif, Metalurgia, J, **4**, 20 (2012)
7. ***, www.sudotim.ro, accessed 11.02.2017
8. E. Binchiciu,T. Fleser, I. Voiculescu, AMR, **1029**, 72 (2014)
9. E.F. Binchiciu, *Integrated brazing technologies with advance precursors (*PhD thesis, Universitatea Politehnica Timisoara, Romania, 2016)

DOI: 10.1051/matecconf/201712103011

Tool life of diamond inserts after laser assisted turning of cemented carbides

Piotr Kieruj[1,*], and *Mateusz* Kuklinski[1]

[1]Poznan University of Technology, Institute of Mechanical Technology, Piotrowo 3, 60-965, Poznan, Poland

Abstract: The paper presents results and discussion on parameters influence on tool life when turning laser cladded cemented carbides under laser assisted machining (LAM) conditions. The layers are known as Technolase 30S. This material is widely recognized as hard-to-cut. The experimental investigation was carried out with diamond inserts. The laser power density, feed rate and laser scanning speed were analysed. Flank wear (VBc) indicator was measured and cutting tool life was calculated. It was proved that Technolase 30S can be machined with LAM technology. The laser enhanced turning is more efficient than conventional turning and grinding.

1 Introduction

For many years a trend in manufacturing technologies is machining of hard-to-cut materials. Its accompanying themes are hard cutting [1-2], machining with reduced cooling, hybrid machining and machining assisted with various energy sources. Hard-to-cut materials are materials with specific strength properties which make their machining difficult, and mostly with high hardness. Among them titanium- and nickel-based alloys, ceramics, metal-matrix composites [3-4] and cemented carbides [5] can be identified. The last group of materials was used so far as tool materials only but becomes increasingly important as a construction materials and materials used for regeneration of machines and devices parts [6]. The regeneration with metal carbides is realized by laser cladding. Moreover, they are used for improving properties of surface layers of machine parts, e.g. for corrosion resistance or wear resistance improvement [7-8]. During laser cladding, metal carbides are fed as powders and sintered by the influence of laser beam [9]. Unfortunately, coatings produced this way are characterized by high roughness, waviness and shape tolerance [10-11]. If specific surface quality is necessary, it is required to carry out machining for obtaining surface properties which comply prescribed requirements [6]. According to metal carbides' producers, these materials are unmachinable [12] but there are some experimental researches describing this process. It is therefore necessary to mention publications confirming this status, in particular earlier works of our research team [5,11,13]. Intensive researches make this method capable of industrial application.

* Corresponding author: piotr.a.kieruj@doctorate.put.poznan.pl

The most commonly cladded material which improves wear resistance is the mixture of powders including tungsten carbides WC and metallic matrix, e.g. Ni, Co, NiCr or Stellite. The analysis of literature showed that WC with Ni matrix is the most commonly used in research [9,14,15]. Tungsten carbide particles increase wear resistance and hardness of cladded layer. Moreover, tungsten carbide is characterized by the highest ductility among the metal carbides used for cutting tools technologies.

The machining of composites containing WC requires using the hardest cutting materials: polycrystalline diamond (PCD) and cubic boron nitride (CBN) [6,11,13,16]. Unfortunately, hard WC particles significantly impede machining of materials which contain them and increase tool wear. The porosity of cladded carbide layers has also a negative influence on condition of cutting edge [8]. These factors combined affect the possibility of cutting materials catastrophic wear because PCD and CBN are vulnerable to chipping, crumbling and breaking out. It is significant to set cutting parameters which guarantee the highest durability and the largest volume of cut material simultaneously to minimalize costs of cutting. Finishing of very hard materials is good known aspect in industrial application but its costs are significant part of final product price [13,17]. Finishing is characterized by numerous disadvantages including:

- low volume efficiency of the process,
- tensile stress in surface layer which causing cracks,
- usage of coolants which is ecological burden,
- intensive tool wear. [17]

At the Poznan University of Technology, studies on turning of cladded cemented carbides layers are being conducted. Attempts are successfully implemented both in classic turning and laser assisted machining. Recently, the research allowed to select a cutting edge with the best machinability. Moreover, the influence of machining parameters and laser beam treatment on cutting edge durability was identified. Laser assisting has positive impact on machining, by stabilizing the process and improving tool durability.

2 Research method

2.1 Tool and workpiece materials

The research was carried out on Technogenia Technolase 30S layer (from now on, 30S) which was laser cladded on a steel roller (Fig. 1). The material – 30S – contains hard particles of tungsten carbide WC as Spherotene® with nickel matrix [12].

Fig. 1. The view of the test sample prepared to research.

The matrix contains also Cr, Si, Fe and B. Matrix hardness is 30 HRC and mass content of WC particles is about 60% which is equal to approximately 40-45% of cladded layer volume. Size of Spherotene® particles is 40-160 μm and their hardness is ranged between 2500 and 3300 HV [8]. Cladded layer thickness was 3 mm and shaft diameter after cladding was equal to 56 mm. Sample after cladding was trimmed to provide machining continuity. Workpiece was painted with the special coating in each case to improve an absorption of laser beam radiation. In order to limit the variability of subsequent tests of laser assisting, painted layer has constant thickness. A method has been developed for manual cladding of coverage layer with control of its thickness using thickness meter.

Cutting material carried out in the research was multi-modal polycrystalline diamond KD1425 of grain size between 2 and 30 μm. Cutting inserts had a geometrical symbol TPGN110304F single-tip type according to ISO. The cutting insert was characterized by: corner radius r_ε = 0,4 mm, neutral rake angle γ_o = 0° and main cutting edge angle Kr = 90°. The producer guarantees sharp cutting edge with r_n = 0 μm.

2.2 Experimental conditions and procedure

The research was based on longitudinal turning of WC/Ni sample assisted with laser treatment. The test stand consisted: numerically controlled lathe DMG Mori Seiki CTX 310, industrial robot KUKA KR16-2 and diode laser Trumpf TruDiode 3006. Laser head was mounted to the wrist of industrial robot, which allowed to insert it to the operating space of lathe. The view of the test stand is shown in Figure 2.

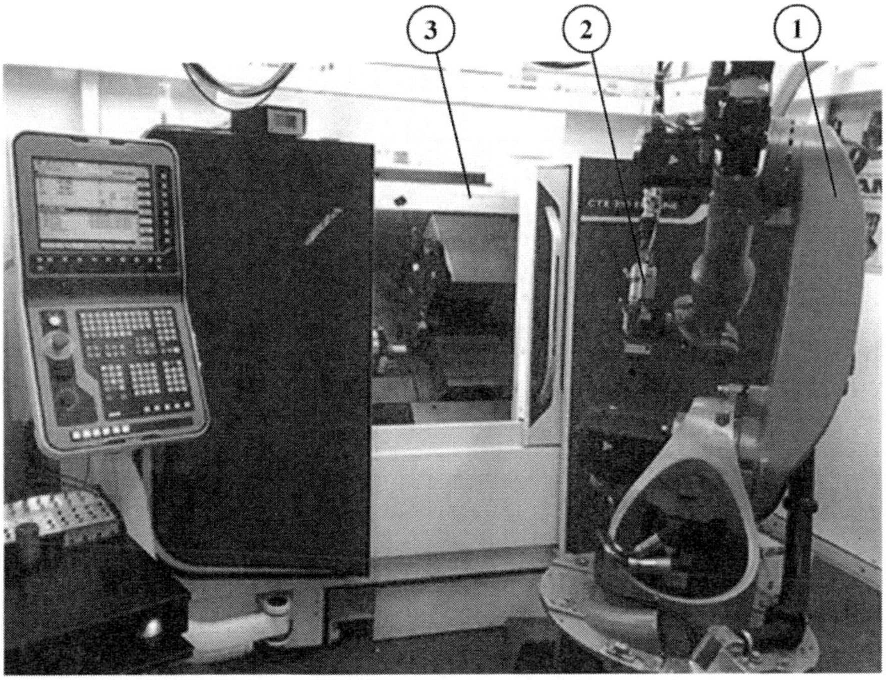

Fig. 2. The view of the test stand: 1) industrial robot, 2) laser head, 3) CNC lathe.

Industrial robot provided longitudinal movement with velocity accordant with feed movement of cutting tool and with constant distance from the workpiece. Constant angle position of laser beam incidence was set on 35° before cutting tool edge (Fig. 3). During the research a tool life was identified. In the process of longitudinal turning a variable cutting speed v_c,, feed rate f and laser beam power P were carried out.

The following constant parameters for processing were set: laser dot diameter on the workpiece surface d_l = 1,2 mm and depth of cut a_p = 0,1 mm. The values of turning parameters and the research program are given in Table 1.

 DOI: 10.1051/matecconf/201712103011

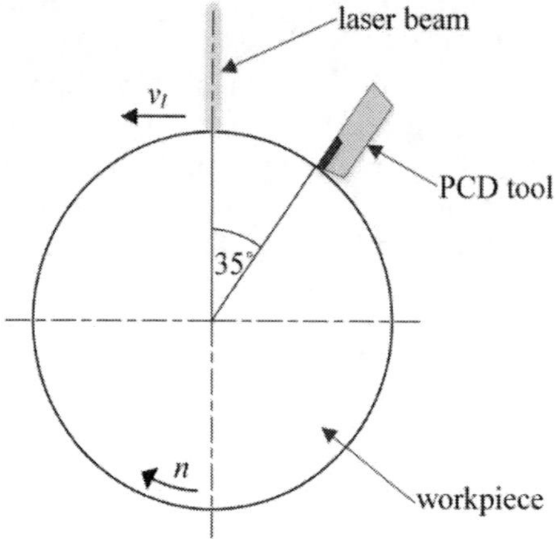

Fig. 3. The view of the schematic tool angular position from the laser beam.

Table 1. Research program.

		v_c [m/min]	*f* [mm/rev]	*P* [W]
Following number of edge/test	1.	35	0,12	1500
	2.	50	0,12	1500
	3.	35	0,18	1500
	4.	35	0,12	2000

During the research the wear ratio value was measured on tool flank wear *VBc* with accuracy of 0,01 mm. Measurements were carried out at set time intervals. The research was proceeded until *VBc* reached 0,3 mm value. In order to describe investigated functions, mathematical model in a form of regression function was applied:

$$y = ax^3 + bx^2 + cx + d \tag{1}$$

To assignment relations between variable values t_{ci} and investigated parameter VBc_i squared correlation factor R^2 was determined.

3 Results and discussion

Obtained results were shown as diagrams. The influence of machining times t_c on the value of tool flank wear *VBc* is shown in figures 4 – 6. Each diagram illustrates different results summary for various parameters v_c, *f* and *P*.

In figure 4 the wear progress of edges KD1425 with various machining speeds is shown. With first test parameters (Table 1) time of tool life was less than 12 minutes. An increase of cutting speed (v_c = 50 m/min) led to tool life decrease to about 4 minutes. Both progresses of tool wear are compatible with wear curve in the form of 3rd order polynomial (as shown in figure 4).

Figure 5 shows wear curves for cutting edges working with different feed rates. The progress for f = 0,12 mm/rev is corresponding to diagram shown in figure 4. An increase of feed rate to f = 0,18 mm/rev led to decrease of tool life to 8 minutes. Wear curves also form a polynomial of 3rd order, characteristic for hard-to-cut materials.

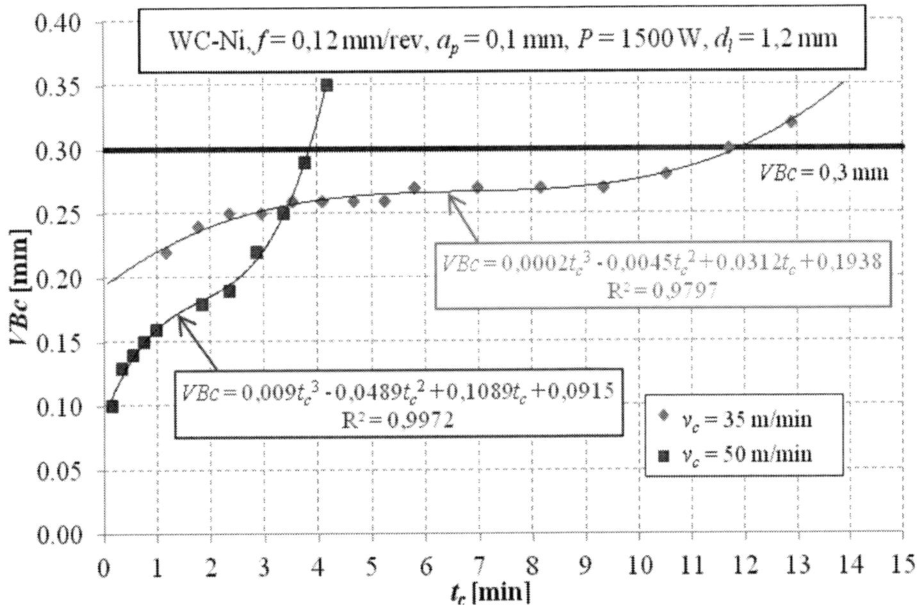

Fig. 4. Tool flank wear *VBc* as a function of cutting time t_c for a various cutting speed v_c.

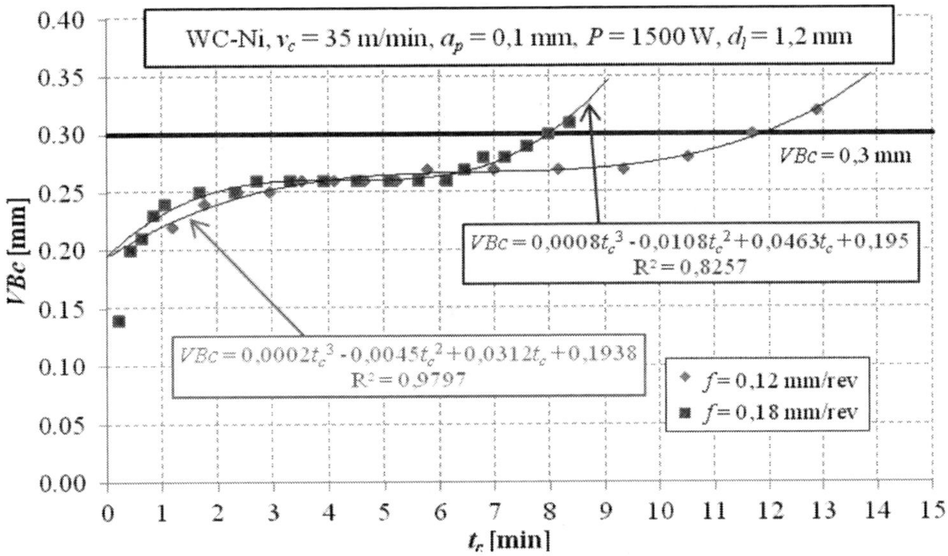

Fig. 5. Tool flank wear *VBc* as a function of cutting time t_c for a various feed rate *f*.

Another diagram (Fig. 6) illustrates an influence of power of laser beam which assists machining until the cutting edge will exceed the wear criterion VBc = 0,3 mm. Increase of laser beam power, with other parameters kept constant, caused tool life time insignificantly increase. However, the difference is sight. For laser beam power P = 1500 W tool life time was T = 12 min and for P = 2000 W about 1 minute more, T = 13 min.

In figure 7 the summary of two results are shown: removed material volume V until exceeding criterion Vbc = 0,3 mm, and machining volumetric efficiency Q. For both feed rate and laser beam power values, the similar results of removed material volume were obtained, ranged from od 5,11 cm^3 to 5,43 cm^3.

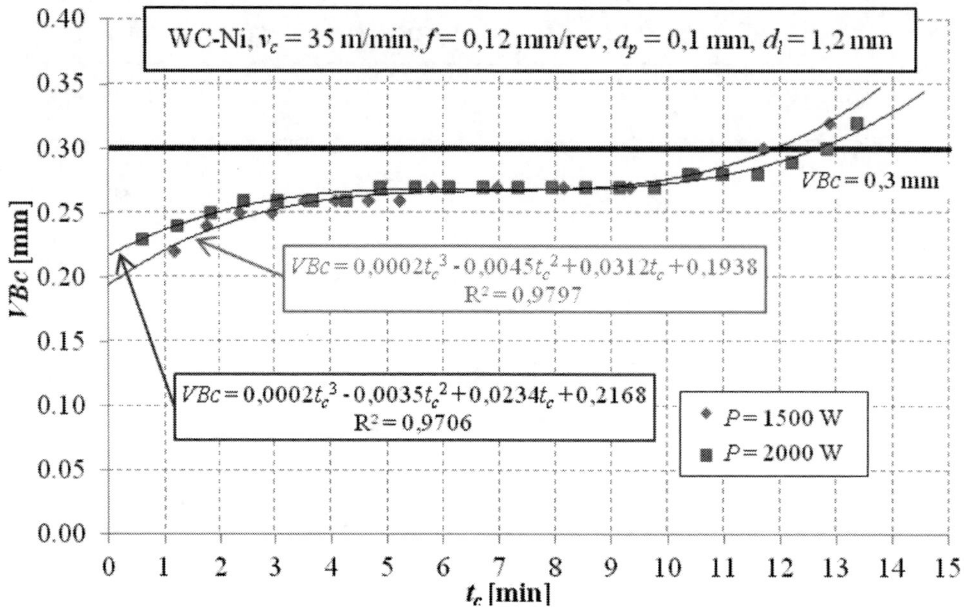

Fig. 6. Tool flank wear *VBc* as a function of cutting time t_c for a various laser power *P*.

The worst result was obtained with cutting speed v_c = 50 m/min. The best machining volumetric efficiency of removing a material was obtained in third test (Q = 0,611 cm^3/min) and increasing a cutting speed to v_c = 50 m/min allowed to obtain a volumetric efficiency of Q = 0,597 cm^3/min. However, this test was characterized by the most intensive tool wear. The influence of laser beam power P on machining efficiency Q has not been proven. For first and fourth test the value of machining efficiency was equaled to Q = 0,408 cm^3/min.

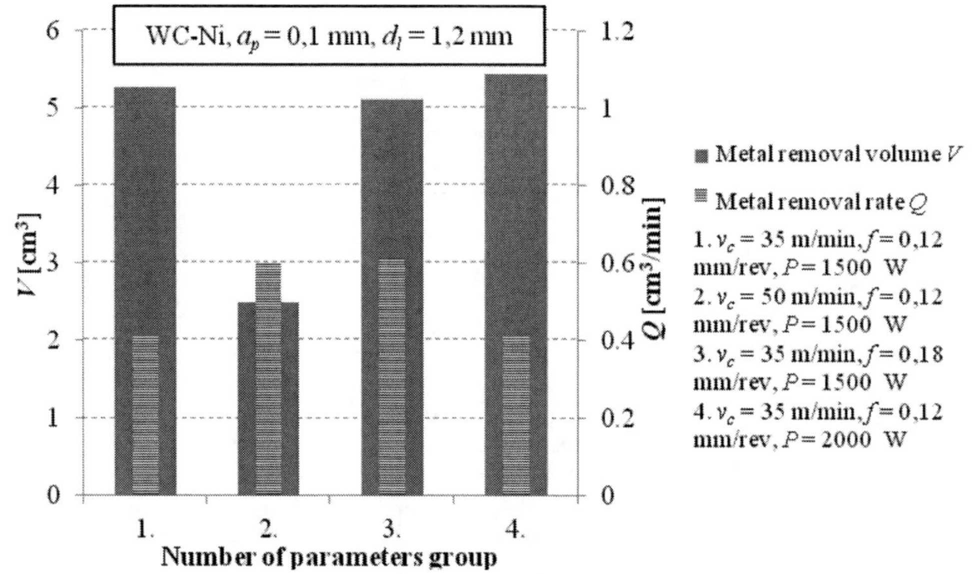

Fig. 7. Metal removal volume V and metal removal rate Q for a various machining and laser parameters.

4 Summary and conclusions

In this article the research of PCD tool durability during laser assisted turning was presented. The method of improving durability of tools when machining hard-to-cut laser-cladded cemented carbide was demonstrated and made by changing cutting parameters and laser beam power. Obtained results may be useful for selection of cutting parameters for machining cemented carbides in further research.

The results led to following conclusions:

- an increase of cutting speed causes an increase in PCD tool wear intensity,
- an initial wear period of edge with geometry as used in this research is very short,
- an increase in power of laser beam which assists machining allows to improve cutting edge durability,
- both feed rate increase to $f = 0{,}18$ mm/rev and cutting speed to $v_c = 50$ m/min caused a 50% enhance in machining volumetric efficiency Q in comparison with initial parameters $f = 0{,}12$ mm/rev, $v_c = 35$ m/min,
- an increase of machining volumetric efficiency Q of cemented carbides while laser assisted machining should take place by increasing feed rate f because it has minor influence on tool life,
- 50% increase of feed rate f led to 32% decrease of tool life,
- 43% increase of cutting speed led to 67% decrease of tool life,
- total volume of removed material V in set tool working time T does not depend on feed rate value,
- laser beam power P does not affect machining volumetric efficiency Q,
- obtained PCD tool durability proves that Technolase® 30S material is hard to cut.

Acknowledgments: This scientific research work is supported by National Centre for Research and Development (NCBiR) of Poland grant No. LIDER/005/141/L-5/NCBR/2014. Project: "Laser assisted machining of sintering carbide deposited by laser cladding"

References

1. S. Wojciechowski, P. Twardowski, M. Pelic, R. Maruda, S. Barrans, G. Krolczyk, Prec. Eng., **46**, 158-165 (2016)
2. S. Wojciechowski, R. Maruda, P. Nieslony, G. Krolczyk, Int. J. Mech. Sci., **119**, 360-369 (2016)
3. D. Przestacki, P. Szymanski, S. Wojciechowski, Composites Part A: Applied Sci. Manuf., **91**, 370-379 (2016)
4. D. Przestacki, Procedia CIRP, **14**, 229-233 (2014)
5. P. Kieruj, D. Przestacki, Adv. Manuf. Sci. Technol., **38**, 4, 61-70 (2014)
6. D. Przestacki, T. Chwalczuk, S. Wojciechowski, Int. J. Adv. Manuf. Technol., 1-12 (2017)
7. D. Bartkowski, A. Bartkowska, Int. J. Refract. Met. Hard. Mat., **64**, 20-26 (2017)
8. J. Amado, M. Tobar, A. Yáñez, V. Amigó, J. Candel, LiM, Phys. Procedia, **12**, 338-344 (2011)
9. S. Zhou, X. Dai, X. Zeng, App. Surf. Sci., **255**, 8494-8500 (2009)
10. I. Tabernero, A. Lamikiz, S. Martinez, E. Ukar, J. Figueras, Int. J. Manuf. Tech. Manag., **51**, 465-470 (2011)
11. P. Twardowski, Metrol. Meas. Sys., **18**, 1, 105-114 (2011)
12. www.technogenia.com
13. N. Coppini, A. Diniz, M. Bonandi, E. De Souza, E. Baptista, Procedia CIRP, **8**, 368-373 (2013)
14. M. Zhong, W. Liu, Y. Zhang, X. Zhu, Int. J. Refract. Met. Hard Mat., **24**, 453-460 (2006)
15. S. Song-hua, Y. Xiao-min, L. Yue-long, H. Yi-zhu, K. Shin, J. Iron Steel Res. Int., **13**, 3, 74-78 (2006)
16. D. Przestacki, M. Jankowiak, J. Phys. Conf. Series, **483**, 1 (2014)
17. C. Brecher, M. Emonts, C. Rosen, J. Hermani, LiM 2011, Phys. Procedia, **12**, 599-606 (2011)

DOI: 10.1051/matecconf/201712103012

Measuring unbalance-induced vibrations in rotating tools

Martin Kimmelmann[1,*], and *Thomas* Stehle[1]

[1]Institute for Machine Tools, University of Stuttgart, 70174 Stuttgart, Germany

Abstract. Unbalances in a tool cause vibrations of the spindle and the machine itself and lead to a waviness of the machined workpiece surface. This paper presents an experimental and analytical procedure for optically measuring the unbalance-induced displacements of the tool centre point (TCP). Therefore, a new method is introduced to determine the dynamic vibrations of a tool by comparing the geometrical profile of the tool with the dynamical profile at a high rotational speed. The necessary steps for measuring the signals and calculating the underlying dynamic vibrations of the tool are presented here. Afterwards, the unbalance-induced vibrations of a milling tool are shown as well as their influence on the eccentricity of the rotation axis. With this newly introduced method it is possible to directly link the displacements of the tool under rotation to the waviness of the workpiece surface and the dynamic stiffness of machine tools.

1 Introduction

Unbalance in rotating components generally leads to a deviation of the shaft axis from the axis of rotation [1]. Rotating tools such as drills and milling cutters produce centrifugal forces depending on rotational speed, which can cause the tools themselves or the machine to vibrate. In order to keep these vibration effects caused by unbalance, especially in the bearing planes and the tool centre point, within tolerable limits, the tools must often be balanced [2]. The vibrations separately excited by unbalance lead to dynamic displacements of the tool, which may show on the workpiece surface [3]. In particular in the machining of wood and derived wood products, the tool's condition of unbalance plays an important part owing to the generally high cutting speeds. When unbalance causes great eccentricities here, it may happen that only one cutting edge of the tool produces the workpiece surface determining the quality. Although the other cutting edges are involved in the cutting process as well, only the cutting edge protruding furthest, however, produces the workpiece surface in this so-called single-edged finishing [4]. Concerning high-quality and productive machining processes, such as precision planing, it is therefore necessary to equalise the centre point path radii of the cutting edges by the so-called jointing method at operating speed in order to attain the desired workpiece quality [5].

Unbalance in tools is limited according to two points of view: the bearing loads caused by unbalance and the guarantee of tool safety. In DIN 69888, based on a simplified model

*Corresponding author: martin.kimmelmann@ifw.uni-stuttgart.de

for motor spindles, the load on the front bearing point caused by unbalance is limited to 1% of the dynamic basic load rating C_{dyn} of the bearing [6]. The requirements for balancing are laid down in ISO 15641 and EN 847 regarding high circumferential speeds on tools, typical of wood machining processes [7, 8]. The influence of unbalance on the machining quality and the tool wear in wood machining is being analysed in the IGF project 18381 N „Establishing the necessary balance quality of tools in wood machining", funded by the Federal Ministry for Economic Affairs and Energy (BMWi). The goal here is to establish economically sensible, required balance qualities for wood machining.

2 Procedure for identifying vibrations in the TCP

Unbalances in tools are centrifugal forces due to the rotation of the tool. According to Equation (1), the unbalance U here causes the centrifugal forces F_U, which are quadratic to the angular velocity ω. In practice, unbalance is mostly given as the product of the unbalance mass u and the eccentricity of the centre of mass e as well [1]. The unit of unbalance is usually given in gmm here.

$$F_U = U \cdot \omega^2 = u \cdot e \cdot \omega^2 \tag{1}$$

Unbalances here work as preloading forces on the tool depending on rotational speed. Due to these forces, the tool axis is eccentrically displaced in the direction of the radial force. In this way, the tool axis moves around the axis of rotation with an eccentricity e. In order to compare different kinds of rotors with each other, the balance quality G was introduced as parameter in ISO 1940 [9] and describes the circumferential speed of the tool axis around the axis of rotation, cf. Equation (2).

$$G = e \cdot \omega \tag{2}$$

As an example, figure 1 illustrates a circle rotating centrically as well as one rotating eccentrically around the eccentricity e. In the stationary coordinate system, a constant distance signal was recorded in the case of the centrically rotating circle and a signal depending on the angle φ and the eccentricity e was recorded in the case of the eccentrically rotating circle. The centre point curve of the eccentrically rotating circle in the X- and Y-directions can be calculated from the difference between the distance signals, according to Equations (3) und (4):

$$x(\varphi) = e \cdot cos(\varphi) \tag{3}$$

$$y(\varphi) = e \cdot sin(\varphi) \tag{4}$$

Based on this principle, a method for establishing the centre point curve of eccentrically rotating tools is presented in the following.

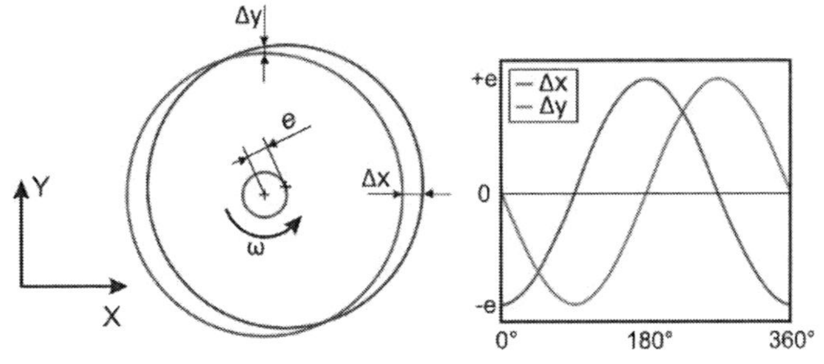

Fig. 1. Establishing the centre point curve of an eccentrically rotating circle

3 Method for measuring the eccentricity of the tool

A suitable method for examining how unbalances influence the eccentricity of the tool axis in machine fixtures is presented in the following, based on the principle described in Chapter 2. This method makes it possible to establish the eccentricity caused by unbalances of the tool by means of comparing two distance signals, one in the case of a slow rotation of the tool and one at operating speed. Figure 2 represents schematically a test tool rotating around the fixed coordinate system at the angular velocity ω. A sensor in the direction of the X-axis here measures the surface profile of the tool during rotation.

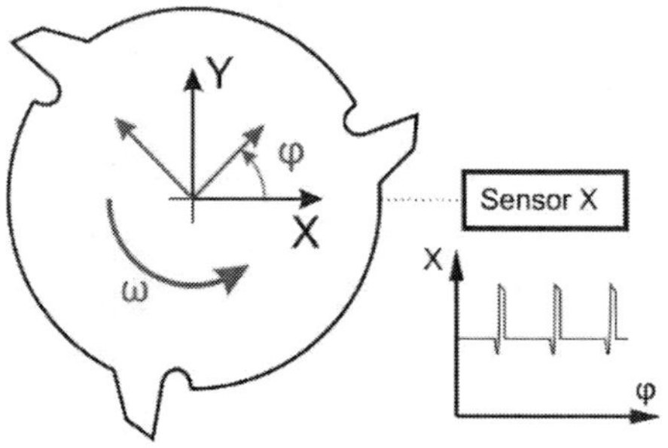

Fig. 2. Schematic test set-up for recording the surface profile of the tool

When the tool rotates slowly, it is hardly affected by centrifugal and unbalance forces. Thus, it can be assumed that the recorded surface profile in the area of the tool cutting edges is purely the tool geometry there. When the tool rotates at a higher rotational speed, unbalances lead to a displacement of the tool axis away from the axis of rotation, as described in Chapter 2. In order to establish the eccentricity of the tool caused in this way, a method for determining the tool eccentricity at operating conditions is described in the following, based on the comparison of the geometrical tool profile at $n \approx 0$ min^{-1} and the superimposed profile at operating speed.

Figure 3 represents schematically the procedure for the data processing in the form of a block diagram. These data are processed with the algorithm described in the following, starting from the discrete signals in the two recorded measuring signals at $n \approx 0$ min^{-1} and at operating speed. In order to process the signal proportions per tool revolution, striking topographies of the tool like teeth or chip spaces are used in the first step to clearly identify the angular position of the tool. The tooth flanks are identified in the algorithm for this, if a previously defined threshold is exceeded. Based on this, the signals measured can subsequently be saved in blocks, for example, per tool revolution. In the third step, the signals are synchronised by using striking profile areas in order to guarantee the phase balance of the tool profiles at slow and fast tool rotations. If the sampling frequency of the measuring device is not adjusted to the rotational speed in this method, the tool profile is scanned much more precisely at a low rotational speed than at the higher operating speed. In the fourth step, the signal at operating speed is interpolated to the same length as the signal at a low rotational speed in order to set off the signal proportions against each other. Then the signal blocks are put together again in the fifth step. The signals produced in this way can then be set off against each other so that a vibration signal caused by a tool unbalance can be calculated in the sixth step of the algorithm. The difference between the signals at operating speed and at a slow tool rotation is calculated for this. In the last step, the vibration signal is processed by filtering, and the frequency proportions in the signal are assessed with a fast Fourier transformation (FFT).

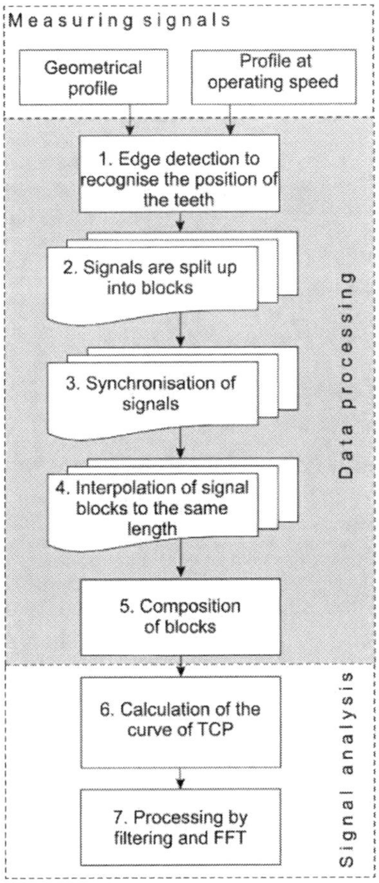

Fig. 3. Algorithm for signal processing in Matlab

4 Measuring the eccentricity of the test tool

The method described in Chapter 3 was tested with an inserted-tooth cutter by Leitz on a machining centre, type MAKA PE 170. A static unbalance of $U_{mess} = 355$ gmm was specifically generated in the test tool with a mass of $m = 5.9$ kg by means of inserting balancing screws and then measured with a balancing machine by Haimer Tool Dynamic.

Figure 4 shows the measuring set-up for recording the surface profile of the tool during rotation. The surface profile of the tool was measured both in the X-direction and the Y-direction of the machine axes by using two laser triangulation sensors, type Keyence LK-G32, with a sampling frequency of $f = 50$ kHz.

The profile recorded by the sensor was then converted into an analogue voltage signal and transferred to the signal analyser by Dactron, with which the information about the distance between tool and laser was saved in the ASCII format and afterwards evaluated with Matlab.

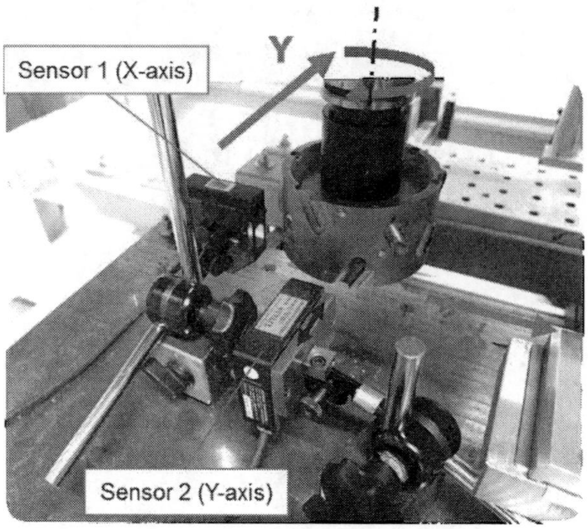

Fig. 4. Measuring set-up for establishing the profile of the test tool

The dynamic behaviour of the test tool was analysed with the method described in Chapter 3. The geometrical tool profile was recorded here for a rotational speed of $n = 100$ min^{-1}. Afterwards the rotational speed was increased from $n = 1{,}000$ min^{-1} up to $n = 5{,}000$ min^{-1} in steps of thousands in order to demonstrate the influence of varying unbalance forces. As an example, Figure 5 on the top compares the signal of the geometrical profile with that of the measured profile for one revolution of the tool. The great amplitudes of the signal here represent the chip space and subsequently the tooth. A qualitative comparison of the measured tool profiles already showed that tooth 3 has a greater centre point path radius due to the unbalance. Using the evaluation procedure described in Chapter 3, it was then possible to synchronise the signals, process them and calculate the difference between the profiles. Figure 5 at the bottom illustrates the vibration profile of the tool calculated in this way. A clear sinusoidal vibration, superimposed with high-frequency signal proportions, can be seen here.

The protruding peaks occurred periodically per tooth and could be attributed to synchronisation errors of the two signals due to the algorithm described and measuring errors owing to the measuring method based on triangulation. The sensors used in the measurement had a measuring range of ± 5 mm. As there are, however, greater differences in height at the tool owing to the cutting edges and the chip spaces, the measuring instrument got into an overload in the areas of the chip spaces. Upon the transition from overload back to the measuring range, there was an overshoot of the signal [10].

A Fourier transformation of the signals in the frequency range clearly showed a vibration frequency of $f = 83.3$ Hz, which corresponds to the rotational frequency of the tool at operating speed. The vibration amplitude for $n = 5{,}000$ min^{-1} here corresponded to an eccentricity of the tool of $e = 140$ µm. This corresponded to the displacement of the tool caused by the unbalance and could be seen as unbalance-related deviation of the shaft axis from the axis of rotation.

In figure 6, the curves of the TCP established in this way are shown in two directions for the signals recorded with the measuring set-up presented in Figure 5. The evaluated amplitudes of the first order showed that the deflection of the TCP in the X-direction of the machine was greater than in the Y-direction for all measurements. Hence, it could be assumed here that the TCP had an elliptical curve due to the anisotropy of the machine rigidity.

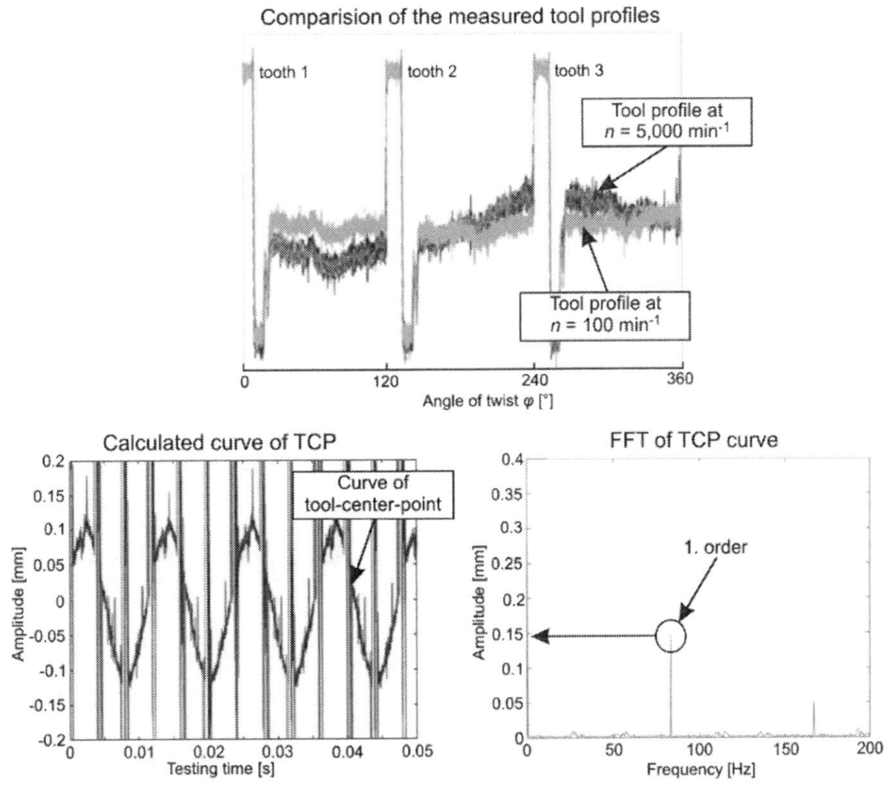

Fig. 5. Evaluation of the signals in sensor 2 at $n = 5{,}000$ min^{-1}

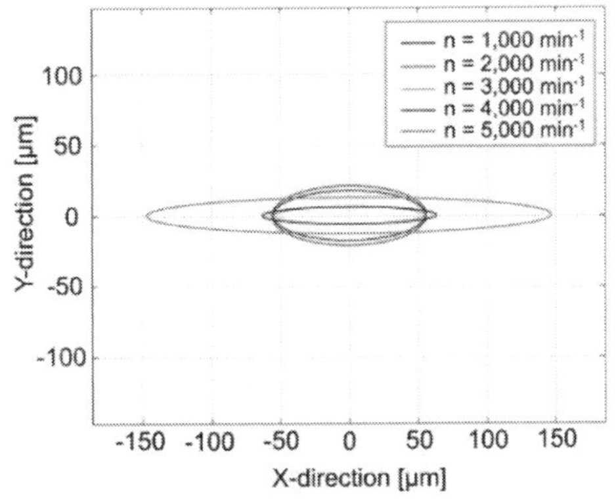

Fig. 6. Curves of the TCP at varying rotational speed

The eccentricity of the TCP could be measured with the method presented here. When comparing the curves of the TCP in Figure 6, it had to be taken into consideration that the unbalance forces increase quadratically with growing rotational speed. In this way, the measured eccentricity can be taken into account, for example, when calculating the waviness height or the path length of the tool's cutting edges to be expected. Figure 7 presents the flexibilities of the machine calculated from the ratio between the measured displacements of the TCP and the unbalance force of $U_{mess} = 355$ gmm measured on the balancing machine. A relevant influence of the rotational frequency on the machine behaviour could absolutely be detected here. Regarding the analysed rotational speeds, a minimum of the flexibilities in both directions of space could be found at a rotational speed of $n = 4{,}000$ min^{-1}. However, in view of arising unbalances it can absolutely be recommended to operate the tool at a rotational speed of $n = 4{,}000$ min^{-1} as the machine flexibility is expected to be lowest here. Since the tool also has a relatively great unbalance, the eccentricities are probably considerably lower in practice for a balanced tool as well.

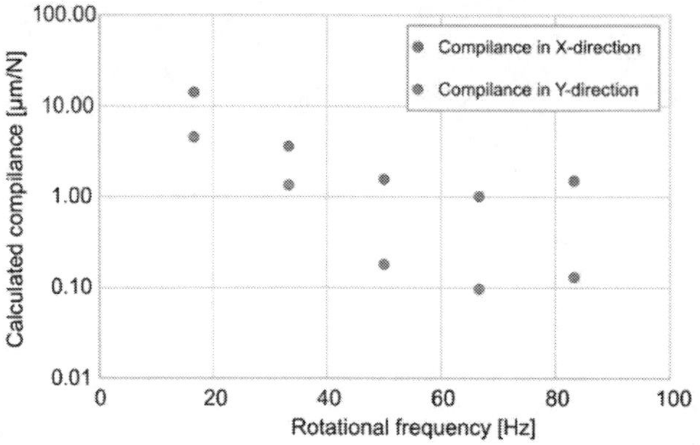

Fig. 7. Calculated flexibilities of the machine depending on rotational speed

5 Conclusions

In this paper, a method is presented for experimentally establishing the eccentricities of a tool caused by unbalances. It is based on comparing a measured tool profile at a low rotational speed and at the operating speed of the tool. It could be shown that unbalance-related displacements of the TCP can be reliably established with this method, which is applied to a 5-axis machining centre as an example. It was detected here that the tool showed a directionally dependent vibration behaviour, which could be attributed to different rigidities of the machine structure and had a direct influence on the eccentricities of the TCP. In addition, it could be shown that this method can be used to analyse the influence of the dynamic machine rigidity on the expected displacements of the tool. By comparing different designs of machines, it will be possible in this way to make recommendations on the permissible unbalance, based on the expected eccentricities of the tool.

References

1. H. Schneider: Auswuchttechnik, Springer-Vieweg, Berlin, Heidelberg, **8** (2013)
2. H. Schulz, T. Wuerz, CIRP Annals-Manuf. Techn.**47.1**, 321-324 (1998)
3. U. Heisel, A.Feinauer, Annals of CIRP **48.1**, 321-324 (1999)
4. U. Heisel, H.Krondorfer, Proceed.of the 13th Internat. Wood Machining Seminar, Vancouver, 253-264, (1997)
5. U. Heisel, J. Tröger; S.Martynenko, Symp. on Wood Machining, Properties of Wood and Wood Composites Related to Wood Machining, Vienna, Austria, 161-173, (2004)
6. DIN 69888, *Auswuchtanforderungen an rotierende Werkzeugsysteme*, (Beuth-Verlag, Berlin, 2008)
7. ISO 15641, *Milling cutters for high speed machining - Safety requirements*, (Beuth-Verlag, Berlin, 2001)
8. DIN EN 847-1, *Tools for woodworking - Safety requirements - Part 1: Milling tools, circular saw blades*, (Beuth-Verlag, Berlin, 2013)
9. ISO 1940-1, *Mechanical vibration - Balance quality requirements for rotors in a constant (rigid) state - Part 1: Specification and verification of balance tolerances*, (Beuth-Verlag, Berlin, 2004)
10. K. C., Fan, Computer integrated Manuf. Syst., **10.4**, 277-285 (1997)

DOI: 10.1051/matecconf/201712103013

Creating and evaluating replicas of surfaces machined by laser beam

Milena Kubišová[1,*], *Vladimír* Pata[1] , and *Libuše* Sýkorová[1]

[1]Tomas Bata University in Zlín, Faculty of Technology, Vavrečkova 275, 760 01 Zlín, Czech Republic

Abstract. In industrial practice, we often encounter a problem when it is necessary to measure and subsequently evaluate the surface quality of the part produced by the water jet, laser cutting. In that case, it is very often difficult to evaluate the final surface because of its weight or bulkiness. However, the literature sources recommend using of material known under the commercial name Dentacryl™. Although this material is still being used in technical practice, its utilisation for creating the replicas of the polypropylene surfaces which is produced by the laser beam is not recommended due to its limited viscosity and mainly due to its health risk for allergic people. Therefore, this article will describe a method of an application the two-component dental material Siloflex© for the creation of the replicas of polypropylene surfaces formed after CO_2 laser machining. Furthermore, the statistical approach for evaluation of the quality comparison of the original surfaces and produced replicas is necessary to surface quality of produced replicas.

1 Introduction

The main part of this article is the research of the preparation of replicas of surfaces and the determination of the methods impression and its evaluation. This special technology is very little explored from the surface quality control of products in which can brings new trends and methods of evaluation mostly in the section of the production of the moulds. For creating replicas of surfaces is historically the most commonly used the material known as Dentacryl™. In the nowadays modern concept of the quality, assessment is this material problematic inapplicable terms of both mechanical as well as security reasons.

For the production of replicas, we use the impression compounds, mainly used in dentistry. In this field, we can search not only the inspiration for the choice of material suitable for scanning surfaces but also knowledge in the preparation and the application itself.

Another independent issue consists of the discovery of suitable dental impression materials of the required properties (as discussed above) is followed to discover suitable mathematical and statistical methods to the assessment of the quality of the replicas.

* Corresponding author: mkubisova@ft.utb.cz

The proposed article will deal with a method of the production of the replicas using the dental materials SILOFLEX© including the technological process of the replicas. It will propose methods based on the statistical hypothesis testing to assess the degree of the agreement between the original surface and the replica. The scanning method of these two surfaces the original and the replica will be performed by using the contactless manner matrix CLI method [1].

2 Materials and Methods

2.1 Machining of Polypropylene Surfaces by CO2 laser

When creating the model surfaces on polypropylene (from now on referred to as PP) plates, the CO_2 laser ILS 3NM with the following characteristics was used: the maximum power of 100W, the maximum feed rate of 1524 mm.s^{-1} [8]. Next, the parameters of laser power and feed were changed - and Tab.1.

Table 1. Machining parameters.

f[mm.s^{-1}] / P[W]	**1.**	**2.**	**3.**	**4.**
1.	20/305	20/610	20/915	20/1220
2.	40/305	40/610	40/915	40/1220
3.	60/305	40/610	60/915	60/1220
4.	80/305	80/610	80/915	80/1220

2.2 Replica Creation

From all surfaces that were produced under cutting conditions (see Table 1), were formed the replicas using the dental material Siloflex©.

As a typical case used for the purpose of this article was choose the surface made at the cutting conditions P = 20 W and f = 1220 mm.s-1 (see Table1). The reason for choosing this sample was that the final surface showed maximum relief section and due to the biggest problems with separability of the replica from the sample. The last surface of the product and its replica (Fig. 2) is characterised by many surface defects, such as sinks, these defects complicate the replica production, scanning method, filtration type and the subsequent statistical evaluation [2].

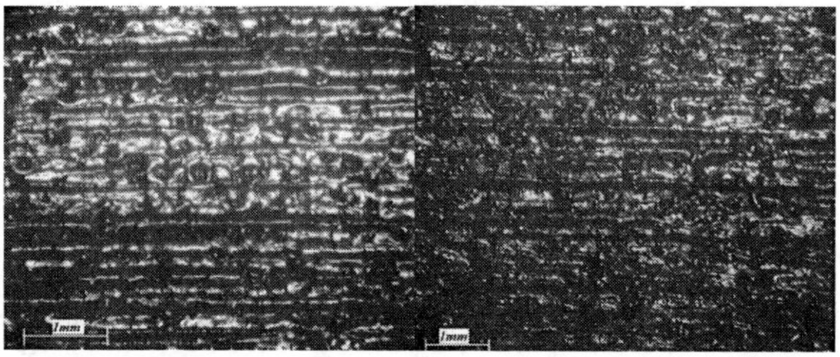

A. B.

Fig. 1. MS

DOI: 10.1051/matecconf/201712103013

2.3 Evaluation of the method

Firstly, original surfaces of dimensions 5 x 5 mm were scanned using 3D surface scanner Talysurf CLI 500 with a resolution of 0,25 μm in the X and Y axes. Sensitivity in Z axis was allowed to a maximum possible level 1 nm.

After scanning the individual values of X, Y, and Z parameters into a three-dimensional data matrix. The data was filtrated using a double Gaussian sturdy filter to ensure the correct calculation of amplitude parameters of surface quality of the originals and the replicas.

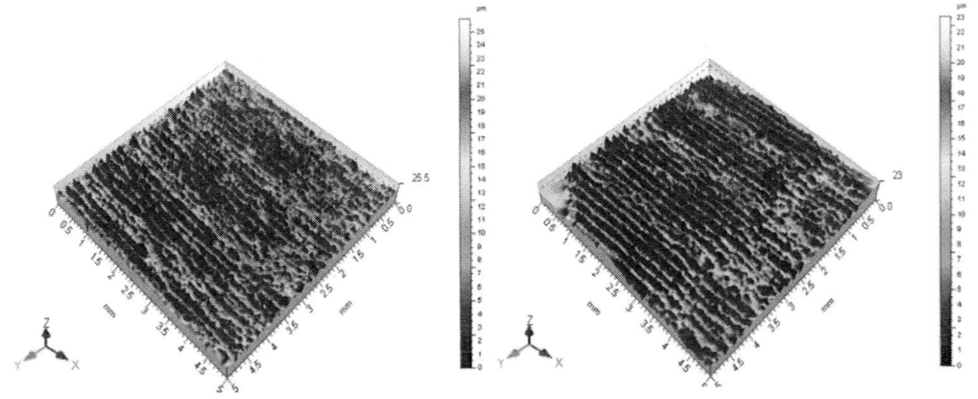

Fig. 2. 3D scanned: A) original B) replica under conditions: PPI 1000, P 20 % and f 80 %.

To specify whether the original and the replica are close, each to the other ISO 4287 was applied. We look at main parameters: Arithmetic mean deviation of the assessed profile Ra and a maximum height of profile Rz [3-4].

Table 2. Roughness parameters by ISO 4287 for original and replica

Original		Mean	Std. Dev.	Median	Min	Max
Rp	μm	4,35	0,64	4,40	2,70	6,09
Rv	μm	4,44	0,74	4,44	2,82	6,17
Rz	μm	8,79	1,29	8,88	6,10	12,10
Rt	μm	11,70	1,81	11,80	7,31	16,10
Ra	μm	1,87	0,29	1,87	1,21	2,53

Replica		Mean	Std. Dev.	Median	Min	Max
Rp	μm	3,93	0,792	3,91	2,56	6,44
Rv	μm	4,05	0,734	4,13	2,52	5,62
Rz	μm	7,97	1,450	7,960	5,12	11,4
Rt	μm	10,5	2,070	10,30	6,64	16,4
Ra	μm	1,67	0,269	1,73	1,01	2,25

After inventing methods, impression replicas were needed to prove that they can originate with a replica similar. For this demonstration was ingested to ISO 4287.

From the table 2, it is clear that the surface of the original and a replica are very similar. The difference between the original and the replica was calculated for parameters Ra 0, 2 μm and Rz 0, 82 μm. Evaluation of the degree of similarity based on cluster analysis [5-7].

2.4 Evaluation of the degree of similarity based on „ Cluster analysis."

In the case of cluster analysis was as the optimal method chosen Ward´s method, linking was the Euclidean distance.

It was found that the measured values disintegrate into two clusters, namely amplitude parameters of the originals and replicas see figure 3 and table 3.

 DOI: 10.1051/matecconf/201712103013

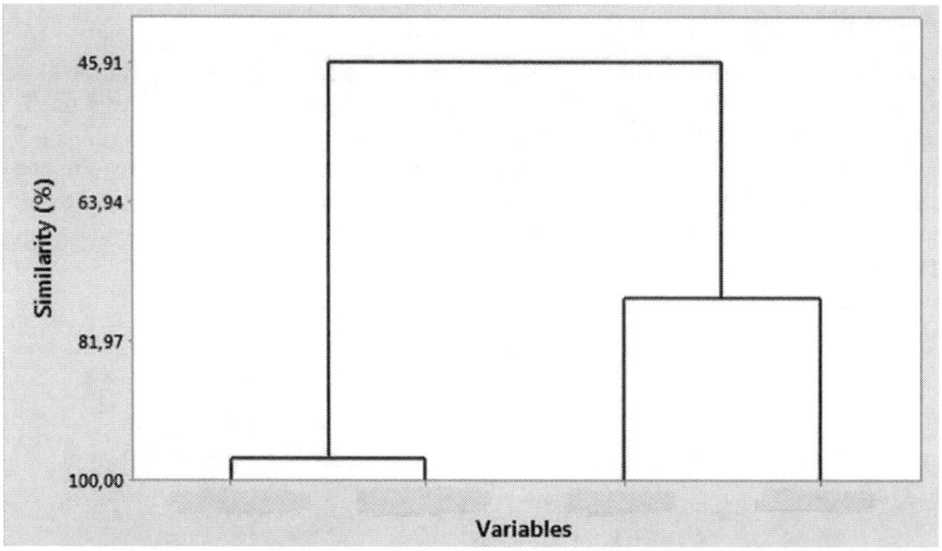

Fig. 3. Graph Cluster analysis

Table 3. Numerical results of Cluster analysis

Step	Number of clusters	Similarity level	Cluster
1	3	97,2496	Rz_Original Ra_Original
2	2	76,3666	Rz_Replica Ra_Replica
3	1	45,9051	Rz_Original Ra_Original and Rz_Replica Ra_Replica

To confirm the obtained results, Factor Analysis which has been optimised for the two main factors Varimax methodology were further applied. Biplot diagram (see Fig. 4), which again clearly demonstrates significant agreement from the above-displayed Cluster analysis, was selected as a typical demonstrates [2].

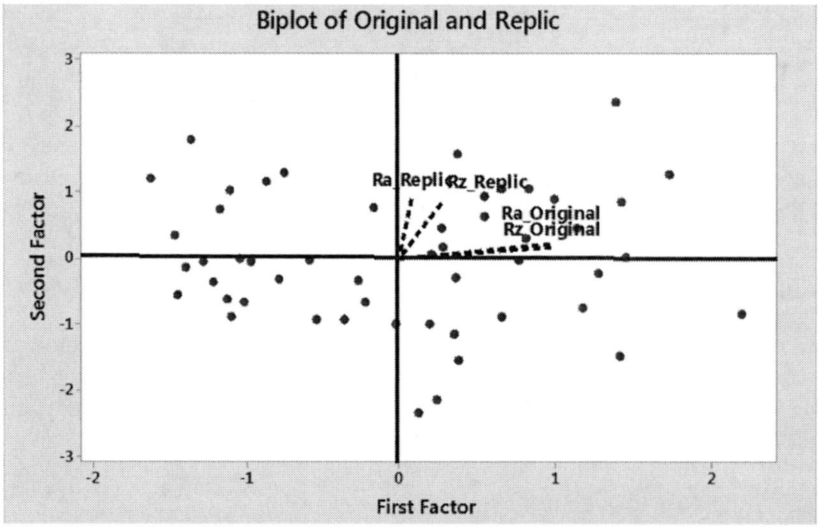

Fig. 4. Biplot diagram

Table 4. Unroasted Factor Loadings and Communalities

Variable	Factor 1	Factor 2	Factor 3	Factor 4	Communality
Rz_Original	0.879	-0.445	-0.055	0.162	1
Ra_Original	0.901	-0.397	-0.052	-0.166	1
Rz_Replica	0.703	0.498	0.508	0.005	1
Ra_Replica	0.591	0.675	-0.442	0.005	1
Variance	2.820	1.059	0.459	0.054	4
% Var	0.607	0.265	0.115	0.013	1

Table 5. Unroasted Factor Loadings and Communalities Varimax Rotation

Variable	**Factor 1**	**Factor 2**	**Factor 3**	**Factor 4**	**Communality**
Rz_Original	0.972	0.092	0.151	0.157	1
Ra_Original	0.959	0.128	0.184	-0.171	1
Rz_Replica	0.208	0.273	0.939	-0.003	1
Ra_Replica	0.122	0.959	0.256	-0.003	1
Variance	1.270	1.019	1.005	0.054	4
% Var	0.481	0.255	0.251	0.013	1

Table 6. Factor Score Coefficients

Variable	**Factor 1**	**Factor 2**	**Factor 3**	**Factor 4**
Rz_Original	0.561	-0.034	-0.105	3.015
Ra_Original	0.521	-0.046	-0.111	-3.087
Rz_ Replica	-0.172	-0.295	1.189	0.094
Ra_ Replica	-0.074	1.136	-0.313	0.097

3 Conclusion

That article describes the ability of a particular method of using dental silicone materials Siloflex© to produce replicas of removable difficult surfaces on which should be evaluated their quality.

For this purpose, is given by accurate procedure silicone mass in the manufacture of surface replicas and is applied the methodology of scanning the surface itself and the surface of the replicas using contactless scanner Taylor Hobson CLI 500.

This is followed by recommendations amplitude comparison of selected surface parameters using statistical methods multidimensional, namely Cluster analysis, Factor analysis using Varimax technique.

Finally, it can be stated that by comparing the amplitude parameters of similarities, their differences do not exceed 7%, for the purpose of laser machining surface material polypropylene can be considered the satisfactory result.

Acknowledgements: This work and the project is realised with the financial support of the internal grant of TBU in Zlín No. IGA/FT/2017/002 funded from the resources of specific university research.

References

1. B. Muralikrishnan, J. Raja, *Computational surface and roundness metrology* (Springer, New York, 2009)
2. M. Bilodeau, D. Brenner, *Theory of multivariate statistics* (Springer, New York, 1999)
3. Intelligent Laser System III-NM - *Operation Manual, Version 1.(*Laser Tools & Technics Corp., 2007)

4. R.D. Reiss, M. Thomas, *Statistical analysis of extreme values: with applications to insurance, finance, hydrology and other fields* (Birkhäuser Verlag, 2nd ed, Basel, 2001)
5. R. Stasiak-Betlejewsk, A. Czajkowska, MATEC Web of Conferences (2017)
6. V. Dițu, G. Oancea, R. Daicu, MATEC Web of Conferences (2017)
7. A.Mat Deris, A. Mohd Zain, R. Sallehuddin, S. Sharif, MATEC Web of Conferences (2016)
8. Yang, T. and Bellouard, Y., MATEC Web of Conferences (2015)

Generating of spatial formations composed of carbon atoms as fullerenes and nanotubes type by applying of pulsed electrical discharges to the formation of graphite films on metallic surfaces

Laurentiu Marin[1,*]

[1]Institutul National de Cercetare Dezvoltare pentru Chimie si Petrochimie ICECHIM Bucuresti.Bucuresti, Splaiul Independentei nr. 202, sector 6 Romania

Abstract. The present paper presents the results of experimental researchings concerning the technology for obtaining the graphite films on metallic surfaces by applying the method of electric pulsed discharges (DEI). It was established that the efficiency for the graphite film obtaining is according with the conecting mode for the workpiece subjected to treatment of electric pulsed discharges to the polarity of the electric pulse generator. Thin pellicles obtained after the electric pulsed discharges treatment, contain mostly carbon, but after of detailed and special analysis (termogravimetric analyses, electronic microscopy SEM and chemical analyses), it was found that different chemical forms of carbon other of graphite are generated. They are spatial formations made of carbon atoms the type of fullerenes and carbon nanotubes. This occurance explains the special functional properties of graphite films and that is to encapsulate small molecules such as N_2 in the case of fullerenes, or organic molecules in case of nanotubes with single wall.

1 Introduction

The obtaining of graphite film at micro and nano scale, on the surfaces of metallic pieces made of different metals or alloys, determines its diffusion in a superficial layer on the metallic surface of the piece, and is accompanied by the formation of carbides with high hardness and also spatial formations made of carbon atoms the type of fullerenes and carbon nanotubes. [1-5]. Graphite film formation also determine the reducing of surface roughness.

In accordance with the experimental results obtained by the authors [4], the application of graphite films on the surfaces of components that make up the kinematic assemblies leads to reducing the coefficient of friction of about three times.

Graphite films obtained using pulsed electrical discharge technology with graphite cathode, maintain their good resistance for long time at temperatures of 300-400 °C.

*Corresponding author: andree_marr@yahoo.co.uk

The procedure of surface treatment, which consists of depositing of graphite films by pulsed electrical discharges on the active surfaces of metal workpieces or subassemblies of some machine tools that are moving or there are in direct contact is a new and original method of surface processing which was not applied before. Specialized literature does not mention scientific interests or applications of the graphite film on metal workpiece. Most of technical and scientific communications which refer to surface treatments, have as subject these surface treatments which leading to improved the surface hardness or preventing the oxidation such as pack cementation, nitriding, carbonitriding,

This paper presents a part of the research activity of the doctoral thesis of the author Marin Laurentiu conducted under the coordination of mentors Prof Dr Hab Topala Pavel State University Alecu Russo Balti, Republic of Moldova and Prof Dr Hab Stoicev Petru Technical University of Moldova Chisinau

Applied researches and industrial experiments performed in the frame of the thesis have demonstrated that graphite film deposition is occur more efficient if the metal piece on that the deposition is done is connected to the positive pole – anode - and the graphite electrode who provide the graphite for deposit is connected to the negative pole – cathode - of the generator for electrical discharges in pulse. Under these circumstances the graphite films formed attains a thickness of about 7 microns [2, 5].

From above presented facts result that the submission of thin films of graphite on the surfaces of metallic pieces would solve a number of technical problems e.g. superficial wear resistance, reducing the coefficient of friction of the parts that make up the kinematic assemblies, avoidance of galling pieces.

2 Experimental research methodology

Experimental investigations have occurred under normal conditions of environment - 22 °C, 760 mmHg and 65-70% relative humidity. The working environment was the atmosphere. Electrical pulsed discharges have been applied such that the flow of plasma that carries vaporized graphite to provide so-called cold spots to avoid unintentional melting of the metal.

As a tool electrode graphite rods were used with cylindrical in shape with sectional area of 5-7 mm^2. The graphite rods used as tool electrode were the cathode. Target pieces on which the graphite were deposited were made of steel type OL45, OL 37, alloyed cast iron, bronze, brass. The target pieces have been connected to the positive pole of the pulse generator as anode.

In order to achieve graphite films by electrical pulsed discharges it was used an electric pulse generator [1]. It ensures the generation of electrical electrical pulsed discharges with the following parameters: the duration 10^{-6}-10^{-7} s which corresponds to the duration of „cold spots electrode", energy released in the interstitium W_S = 0 - 4,8J, the accumulated energy in the capacitor W_c = 0-12J, applied voltage to charge the capacitor U_c = 0 - 250V, capacity of capacitors in the range C = 100-600 μF with a step of 100 μF. The generator provides high voltage electrical impulses (U=12-24 kV), which enables ignition of electric discharges at values of the interstitium of S=0,05-2,5 mm. The frequency of electrical discharges was f = 0-50Hz.

The morphology of surfaces on that were applied graphite film by the method of electric discharges in impulse was studied by scanning electron microscopy SEM and the chemical composition by the method EDX.

3 The results of experimental research

3.1 Choosing the shape and the material of the electrode

During the process of applying graphite films by pulsed electrical discharge process, the shape of tool electrode is amended because graphite is consumed, which leads to the changing of operating parameters. In this way a set of operating parameters that ensure a predictable and controllable wear of electrode tool is required. In this purpose four values for capacitors 0,5J; 1J; 1,5J; 2J. have been set.

Tool electrodes with cylindrical, hemisphere and cone shape of the active area were used. This was done to identify the influence of active surface geometry of tool electrode on its depreciation. Figure 1 shows the dependence between the electrode tool wear of made of pyrolytic graphite, with active zone having a hemisfere shape and the accumulated on the capacitor. As can be seen in all cases the dependence electrode wear of is directly proportional to the accumulated energy from the capacitor.

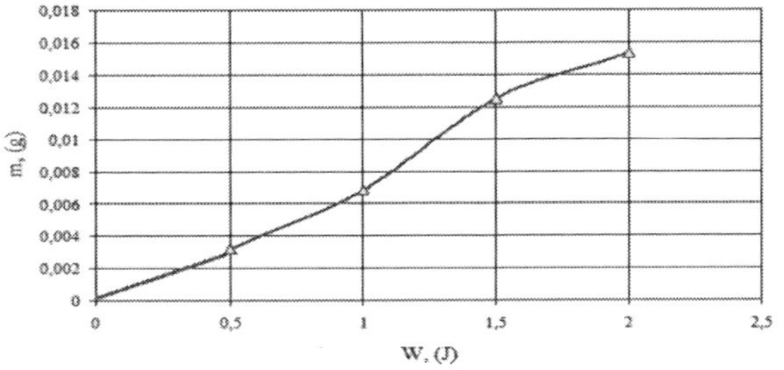

Fig. 1. The variation tool electrode wear depending on the accumulated energy in the capacitor. The active zone of electrode tool has the shape of a hemisphere

Figure 2 shows the dependence between the electrode tool made of pyrolytic graphite, with active zone having a sharp cone shape and the accumulated on the capacitor. As in the previous case the dependence electrode wear of is directly proportional to the accumulated energy from the capacitor. But when the shape of the active electrode tool is sharp, the slope of dependency graph wear of / energy is more pronounced which shows that the electrode consumption is faster through a sharp tip. This, leads to in default to the conclusion that the process for submitting graphite films over metallic surfaces it is more efficient when it use electrodes with the active area with sharp form.

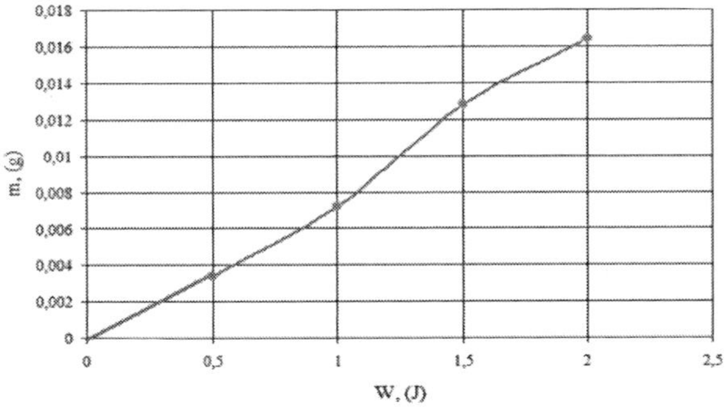

Fig. 2. The variation tool electrode wear of depending on the accumulated energy in the capacitor. The active zone of electrode tool has the shape sharp cone form

From the analysis of bibliographic sources [7, 14-15] and further experimental research it was found that removal of the vaporized material is more pronounced from the surface of the cathode compared to the similar phenomenon that would occur on the anode. Effectiveness of obtaining of graphite films when the tool electrode is connected to the anode pole is very low, for which reason, the electrode erosion phenomenon have not been studied in detail.

For the manufacturing of electrode a material with high resistance to erosion is required, because the electrode wear of involves changings of the interstitial space (space between the electrode tool and the metal surface on that deposition is occurs). Modification of interstitial space leads to the changing for operating parameters of the submission procedure which leads to failures of the processing technology.

According to the previously results obtained, [11-13] erosion stability of the materials is determined by thermosetting properties thereof and it is explained by the relationship (1).

$$\gamma = C\rho\lambda T_{nn}^2 \qquad (1)$$

C- thermal capacity;
ρ - density;
λ - thermal conductivity;
T_{nn}^2- melting temperature.

The material which the electrode tool it is made is pyrolytic graphite. Electrode shape is obtained after a splintering processing operations. The final form of the electrode tool is a cylinder with a diameter of 6 mm. If is necessary, the electrodes with smaller diameter up to 1 mm can be made. The shape of the active zone of the electrode tool is selected depending on the wanted processing parameters and is carried out following a splintering processing operations too.

When applying spots of graphite following pulsed electrical discharges on the metallic surface to be processed the necessary condition is that graphite spot to be applyed on a cold surface. The electrode shape that best satisfies this technological condition is cylindrical rod or disk.

Knowing the frequency of electrical impulses, the diameter of the interaction zone of electric discharge from the active surface area of the electrode tool the linear speed for submission can be determined with the mathematical relationship (2).

$$V = fd \qquad (2)$$

f– frequency of succession for working pulse;
d – diameter zone of interaction of plasma channel with the electrode surface.

The frequency of succession for working pulse is determined by pulse generator settings and d diameter zone of interaction of plasma channel with the electrode surface is calculated using the relationship (3):

$$d = \sqrt{\frac{4W}{QS}} \qquad (3)$$

or it is experimentally determined using the relationship (4).

$$W = \int_0^\tau i(t)\, U(t) dt \qquad (4)$$

In the relationship (3):

W - energy released in the interstitium;
Q - melting heat;
S – size of interstitium.

3.2 Scanning electron microscopy analysis (SEM)

The analysis of processed surface morphology under pulsed electrical discharges using pyrolytic graphite electrode tool it showed that physico-chemical changes were not exceed micrometer depth. Besides the main component of processed material - ie pyrolytic graphite the analyzes attest a significant amount of atomic carbon. This leads to the conclusion that on the processed surface is also formed binary compounds of carbon carbides which demonstrates the enhancement of the resistance for processed metallic surface to mechanical actions.

Fig.3. The morphology of graphite film deposited by pulsed electrical discharge on the surface of steel OL45

Analyzing the presented images that make up the Figure 3 it can be seen that the film of graphite contains a series of nanometric globular formations. This explains a number of interesting properties of graphite film which will be presented in Chapter thermo-gravimetric analysis.

Identification and method for obtaining of these spatial formations composed of carbon atoms by a procedure unspecified in scientific specialized literature is a novelty and represents the main scientific discovery of the PhD thesis sustained by the PhD student Marin Laurentiu.

3.3 Thermogravimetric analysis (TGA)

The presence of globular formations on the surface of graphite film identified by SEM imaging determined performing of additional analyzes.

A result of performing thermo-gravimetric analyzes of the graphite film deposited by pulsed electrical discharges process it has found a number of interesting addition of mass phenomena. [29].

They are reversible and occurred at the following values of the working temperature 222,99°C; 476,12°C and 614,73°C (fig. 5).

These mass additons occurred in conditions which thermo-gravimetric analysis were held in nitrogen environment N_2. This phenomenon is explained by the obtaining of spatial

DOI: 10.1051/matecconf/201712103014

formation of carbon fullerenes and carbon nanotubes [17-25, 27, 29] type who have the ability to encapsulate small molecules (N_2) (fig.6a).

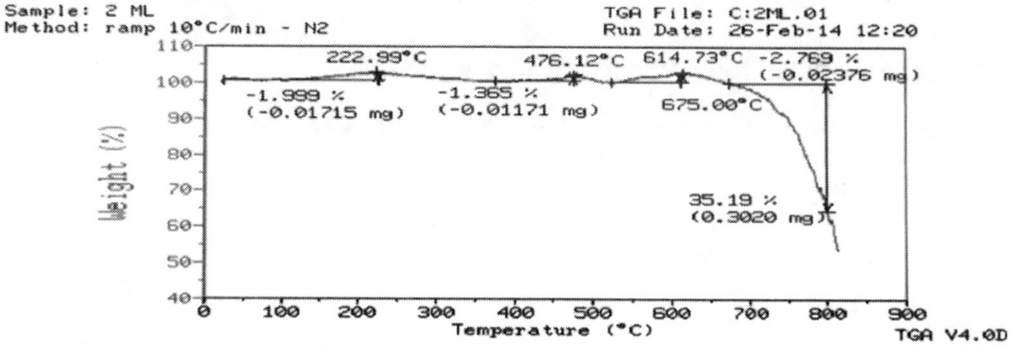

Fig.4. Thermogravimetric curve graphite of the film deposited by pulsed electrical discharges process

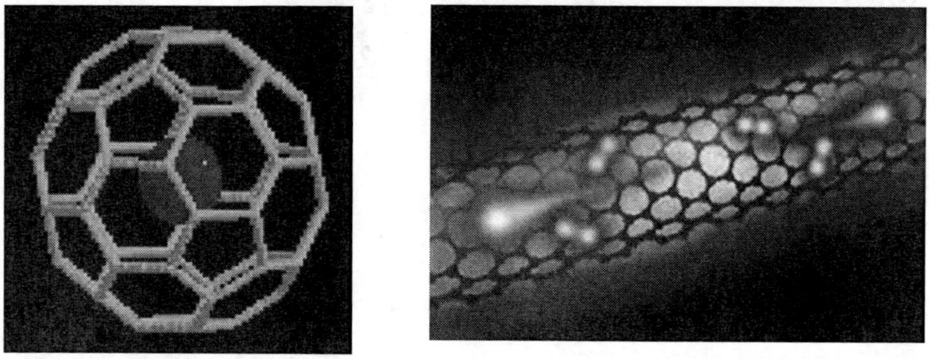

a) b)

Fig. 5. Graphic presentation of spatial structures of carbon: a)–fullerenes that encapsulates nitrogen molecules; b)– carbon nanotubes that encapsulate: OH^-, O–C and O–C=O

Following this analysis we determined that the film of graphite also contains a number of other compounds containing in their molecule atoms of oxygen OH^-, O–C and O–C=O. [30]. They are the result of oxidation processes that occur unavoidably during the graphite film submission considering that the process unfolds in normal atmosphere.

4 Conclusion

The submission of graphite films by electrical discharges in pulse in process is influenced by the geometry of the active part of the electrode tool.

Graphite films deposited by pulsed electrical discharges process have anticorrosive properties, following the submission of an inert airtight film on the surface of metal pieces, protecting them by aggressive actions of corrosive agents;

At the same time with graphite films deposited by pulsed electrical discharges process is also obtained a series of spatial formations composed of carbon atoms fullerenes and carbon nanotubes type which determine the mass addition phenomenon.

Identification and method for obtaining of these spatial formations composed of carbon atoms by a procedure unspecified in scientific specialized literature.

They represent the main scientific discovery of the PhD thesis sustained by the PhD student Marin Laurentiu.

References

1. V.P. Luneva, A.D. Verhoturov, A.V. Coziri, T.V. Glabet, V.N. Brui. *EPI*, **4**, 11-18 (2005)
2. F. H. Burumculov, P.P. Lezin, P.V. Senin, V.I. Ivanov. *The electric discharge tehnology aimed at retreading and hardening of the machining pieces and tools.* (USM "Ogareva", Saransk, 2003)
3. P. Topala. *The transfer of mass and diffusion processes on surface layers of engine parts during electrosparkle processing.* Saint-Petersburg, **2**, 234-242 (2007)
4. V. Beshliu. Technologies in machine building, **1**, 24-30 (2008)
5. A. Mihaliuc. EPI., **3**, 21-23 (2003)
6. P. Topala. Nonconventional technologies review, **1**, 129-132 (2007)
7. P. Topala, V. Beshliu. Bulletin of the Polytechnic Institute of Iasi, **54**, 105-111 (2008)
8. P. Topala, P. Stoicev, A. Epureanu, V. Beshliu. International Scientific and Technical conference Machinebulding and technosphere of the XXI century, Donetk, 262-266 (2006)
9. P. Topala, P. Stoicev. *Tehnologii de prelucrare a materialelor conductibile cu aplicarea descărcărilor electrice în impuls,* (Chişinău, TEHNICA – INFO, 2008)
10. P. Topala, V. Beshliu, S. Petru, O. Alexandr. International Journal of Modern Manufacturing Technologies, **2**, 97-102 (2013)
11. P. Topala, L. Marin, V. Beşliu. Advanced Manufacturing Technologies, 97-104 (2013)
12. Yu. V. Kurochkin, and Yu. N. Demin. Chemical and Petroleum Engineering, **37**, 404-408 (2001)
13. A. Mihaliuc. EPI, **3**, 21-23 (2003)
14. B. Ekmekci, O. Elkoca, A. Erden. ProQuet Science Journals, **36B**, 117-124 (2005)
15. V. Beshliu. Technologies in machine bulding, **1**, 75-82 (2008)
16. P. Topala, S. Mazuru, V. Beshliu, P. Cosovschii, A. Ojegov. *ModTech, Vadul lui Voda-Chisinau, Republic of Moldova,* 1093-1096 (2011)
17. G. Ruxanda, M. Stancu, S. Vizireanu, G. Dinescu, D. Ciuparu. Journal of Optoelectronics and Advanced Materials, **10**, 8, 2047-2051 (2008)
18. J.B. Howard, J.T. McKinnon, Y. Makarovsky, A.L. Lafleur, M.E. Johnson. Nature **352 (6331)**, 139–141 (1991)
19. J. Howard, A. Lafleur, Y. Makarovsky, S.Mitra, C. Pope, T. Yadav. Fullerenes synthesis in combustion. Carbon **30 (8)**, 1183 (1992)
20. H.W. Kroto, J.R. Heath, S.C. O'Brien, R.F. Curl, R.E. Smalley, Nature **318 (6042)**, 162–163 (1985)
21. Y. Zhao, Y.-H. Kim, A.C. Dillon, M.J. Heben, S.B. Zhang. Physical Review Letters **94 (15)**, 155504 (2005)
22. A. Hirsch, C. Bellavia-Lund. *Fullerenes and Related Structures (Topics in Current Chemistry).* (Berlin, Springer, 1993)
23. F.N. Diederich. Pure and Applied Chemistry, **69 (3)**, 395–400 (1997)
24. Y. Li. Chemical Physics Letters **335 (5–6)**, 524 (2001)
25. P. Topala, S. Mazuru, V. Besliu, P. Cosovschii, A. Ojegov. Proceedings of The 15th International Conference, Modern Technologies, Quality and Innovation. **2**, 1093-1096 (2011)
26. P. Topala, V. Beshliu, L. Marin, Advanced Materials Research, **1036**, 172-177 (2014)
27. P. Topala, V. Beshliu, A. Ojegov. *Сборник трудов XVIII международной научно-технической конференции, Донецк,* 240-245 (2011)
28. P. Topala; S. Mazuru; V. Besliu; P. Cosovschii; P. Stoicev. Proceedings of The 14th International Conference, Modern Technologies, Quality and Innovation, 635-638 (2010)
29. J. M. Moon, K. H. An, Y. H. Lee, Y. S. Park, D. J. Bae, G. S. Park. Journal of Physical Chemistry B, **105 (24)**, 5677-5681 (2001)
30. I. Tiginyanu, P. Topala, V. Ursaki. *Nanostructures and Thin Films for Multifunctional Applications,* (Springer, 2016)

Effect of fatigue testing on the properties of Glass-Epoxy composites using the acoustic tool

Younès Menail [1,*], *Abderrahim* El Mahi[2], *Bachir* Redjel[3], *Rachid* Berbaoui[2] and *Mustapha* Assarar[4]

[1]LR3MI, Université Badji Mokhtar, Sidi Ammar, BP 12, Annaba, Algérie
[2]LAUM, UMR CNRS 6613, Université du Maine, av. O. Messiaen, 72085 Le Mans, France
[3] Laboratoire de Génie civil, Université d'Annaba, Algérie
[4]Université de Reims Champagne Ardenne, LISM, EA 4695, IUT de Troyes, 9 rue de Québec, 10026 Troyes Cedex, France

Abstract. This study presents the experimental results of the influence of mechanical fatigue on composite material. The plates of Glass fiber with SR 1500 epoxy resin with SD 2505 composite were realized by vacuum molding. Experimental tests were carried out on a standard hydraulic machine INSTRON 8516. The machine is interfaced with a dedicated computer for controlling and data acquisition. The fatigue tests were performed using sinusoidal type of waveform at a displacement control with frequency of 10 Hz. The evolution of Young's modulus and strain based on fatigue gives us an idea about the resistance of the material. Degradation of mechanical properties was observed, and the results have showed that the Young's modulus of plates undergo only minor changes. In fact, the residual stiffness and residual strength decrease when the cycle number of fatigue increase (100 to 50000 cycles), indicating that the studied composites have experienced some forms of mechanical damage.The mechanical tests were backed by Acoustic Emission Monitoring (AEM) during the load cycle, in order to understand the nature of the failure process in the composites such as fiber breakage, matrix crazing, matrix debonding and delamination etc.

1 Introduction

The XXI century is a century of research, innovation and competition. This is the century of new materials. And polymer matrix composites are at the top of the list. Their characteristics, modes of varieties and their costs open all fields, including automotive, aviation, aerospace, shipbuilding, oil industry, civil engineering and many other fields.

The choice of fibers and resins depends on several parameters such as Workplace [1-2], cost, features required, technical moldings, life, etc., because these materials must meet specific conditions. Research in this area is vast and varied and researchers are expected to

* Corresponding author: menailyounes43@gmail.com

fully optimize the performance of these materials by simulations and practical research in situ.

These materials have invested among many other domains, petrochemical, whether on land or offshore. While the results of work in this area are not as numerous as in other areas, but the results show that the composite materials are required as key materials in the field of petrochemicals. They do not corrode and they have a very low coefficient of expansion compared to steel [3-4].

We want know if this material can be used in petrochemical, whether on earth and suffer significant temperature gradient or suffer substantial pressure for offshore use. [5] For this, we subject this material cyclic tensile fatigue then we submit to the effect of the wetland during well-defined periods.

To understand the phenomena experienced by this material, we used the acoustic tool to monitor different degradations that occur within these (matrix cracking, delamination, debonding and breakage of matrix and fibers) [6 - 10].

2 Experimental procedure

To carry out our work, we selected materials that have proven themselves in the market, and have made their molding with great care to have identical specimens, having the same characteristics.

2.1 Plates

To achieve our glass test plates E twill, we used a mixture of epoxy resin SR 1500 associated with 2505 SD amine hardener and respected percentages recommended by the supplier Sicomin Composites France, Table 1. After contact molding, by said vacuum bag for 6 hours under a vacuum of 0.3 bar technique, we obtained parallelepipedal plates. What we placed in an oven at 80 °C for 8 hours for polymerization. The specimens are cut thereafter the recommended dimensions (200x20x1) mm using a diamond disc mounted on a chainsaw. The weight content of fiber materials is determined by calculation and verified by loss on ignition.

Table 1. Material

Fibre	weaving	density [g/m2]	mass [%]	Resin	supplier	Dimensions [mm]	Layer	weight [g]
Glass E	twill	300	65	SR 1500 SD 2505	Sicomin France	200x20x1	4	7.2

2.2 Equipments

Tensile tests were conducted on a universal hydraulic machine INSTRON model 8516, equipped with a load cell of 100 kN, the following standard ASTMC365-57, represented in Figure 1. Steering and data acquisition are done by computer, to record the evolution of the stress in function of the deformation. The tests are performed at room temperature. The machine is driven at constant speed through. This speed was determined after a series of preliminary tests, which helped set it to 1 mm / min for all types of samples. The use of the same speed, regardless of the type of test, overcomes the viscoelastic effect of the resins, when comparing results from different tests.

Monitoring and acoustic data acquisition, are in parallel with the mechanical tests, thanks to another computer connected to the test with three sensors placed longitudinally. Software processes the data for acoustic monitoring.

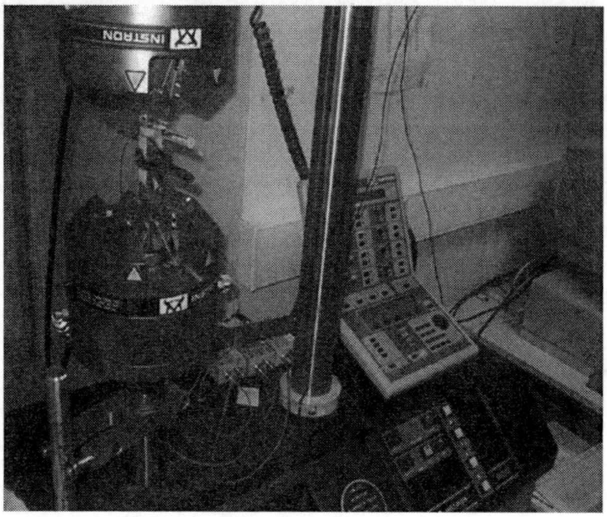

Fig. 1. Tensile machine and acoustic equipment

2.3 Environment

The various specimens were subjected to two different environments in this case, the ambient air and tap water Table 2. Series of test pieces subjected to the air ambient serves as reference. After fatigue, the specimens were immersed in the same time in different environments, at 100, 500 and 1000 h, before undergoing final tensile loading until failure. The various pieces are subjected to the ambient laboratory temperature, which varies depending on the season (15-25 ° C). After loading, they are placed in PET containers, undergoing both daylight and night light variations.

Table 2. Environment

Fibre	weaving	density [g/m2]	mass [%]	Resin	supplier	Dimensions [mm]	Layer	weight [g]
Glass E	twill	300	65	SR 1500 SD 2505	Sicomin France	200x20x1	4	7.2

2.4 Mechanical Testing

To determine the impact of tap water on the composite epoxy-glass material, we conducted two sets of mechanical tensile tests, static tests and fatigue tests.

Static tests allow us to determine the stress σ and the corresponding strain ε. These two parameters allow us to characterize the two materials.

2.4.1 Static tests

Three to five pieces of glass, as the case may undergo static tensile tests to determine both stress and strain parameters needed for testing fatigue.

2.4.2 Fatigue testing

From the results of static tests, we opted for the optimal values of displacement and amplitude fatigue, to achieve the fatigue testing.
The experimental protocol was done as follows figure. 2:

- Loading of the specimen at a constant tensile rate 1mm/mn under controlled 50% displacement at break static displacement.
- Fatigue with a sine wave form, at a frequency of 10Hz with an amplitude of 10% of the displacement at failure. In our case, we opted for a number of fatigue cycles 10000 cycles.
- Unloading the specimen after fatigue.
- Aging in tap water according to the case (and in the air for controls).
- And finally, a tensile strength with speed of movement of 1mm/mn.

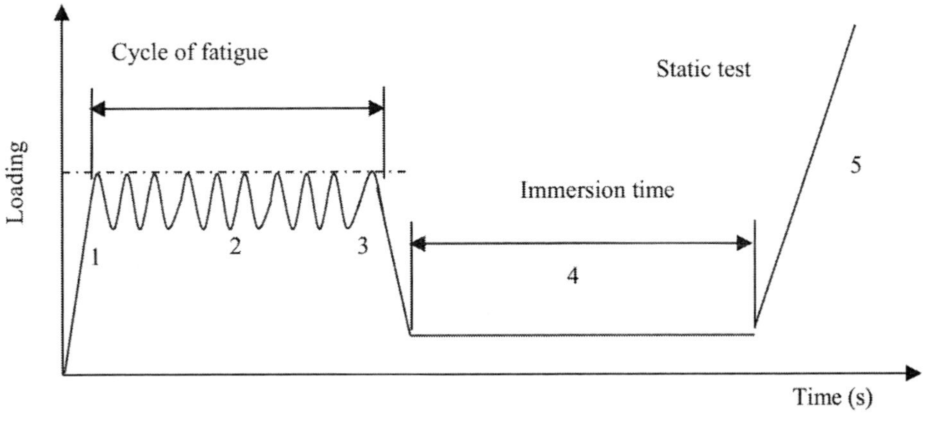

Fig.2. Experimental protocol

2.5 Immersion

Fatigues after specimens are immersed in tap water. A series of samples was left in the open air as a witness. The immersion times are of the order of 100, 500 and 1000 h. Thereafter the specimens are removed from the water according to a program to be broken in tension, as for the static test.

3 Results and discussion

The results obtained in our research work is divided into two, the preliminary results of static tensile tests, which serve as a basis and the results of stress tests, the subject of our work.

3.1 Static Tests

The curves of static tests for glass-epoxy material are manifested by a proportional increase between stress and strain. They indicate brittle-type behavior of the materials, which is manifested by a substantially linear variation of the stress in function of the deformation. The tensile force causes a progressive matrix cracking, the weakest of the composite element.

This cracking is followed by debonding and delamination, which further weaken the material. The effort is continuous, the material being weakened, the composite will sell quickly with a sudden drop in stress indicating fiber breakage and end of life of the material, Figure 3.

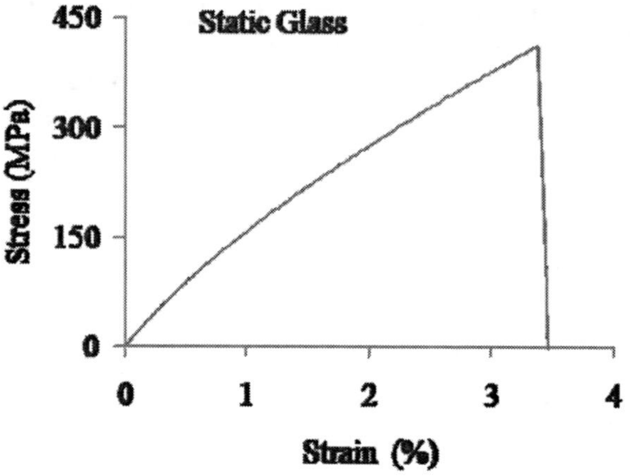

Fig.3. Stress-strain responses under tension loading GFRP

Three sensors connected to the plates allow us through specialized software to record and view live the effect of tensile stress on the materials. Consulted the literature does not accurately distinguish different ranges corresponding to different mechanisms of damage. De Groot and al. [11] Che and al [12], Kim and Lee [13] Karger-Kocsis and al. [14] Kotsikos and al. [15-16] and Benzeggagh and al. [17-18], all these authors and many others have classified the amplitudes into three or four groups, as appropriate, to involve the various failure modes of composite materials. They ranked the lowest to the highest, to associate the matrix cracking and delamination and debonding of those and finally fiber breakage. For we know that against degradation is gradually starting with the matrix cracking, followed by debonding, delamination at the fiber matrix interface, comes on and in parallel to complete delamination followed by fiber breakage. Each type of damage is characterized by its amplitude range. It should be noted that these overlap one another and on the other hand the various authors do not give the same amplitudes for the same damage [19-21].

We opted for the following average values:

- cracking of the matrix, amplitudes between 40 and 60 dB,
- debonding and delamination amplitudes between 60 and 70 dB,
- rupture of fibers amplitudes between 70 and 85 dB and more.
- the amplitude range between 0 and 40 dB is eliminated, since it corresponds to the noise generated by the machine, friction, background noise, etc..

These different amplitude ranges allow us to identify the different damage suffered composites during static and fatigue tests. The acoustic tool allows us to better visualize the degradation of materials and make a more accurate comparison.

The test results of the static tensile test show that the glass is broken after 230 seconds biasing (table 3. and figure.3.). During the first 50 seconds is that the matrix begins to crack under the effect of the traction, which represents less than 20% of the life of material. After this period, it is the different types of damage that occur until the final failure of the material. We can say that the damage is done gradually from the beginning to the end, starting with the cracking of the matrix followed by delamination and debonding, ending with fiber breakage.

Table 3. Physical and Mechanical properties of GFRP and KFRP

Fibre	Speed [m/mn]	Maximal stress [MPa]	Maximal strain [%]	Longitudinal module [GPa]	Transversal Module [GPa]
Glass	1	380	3.5	16	16

3.2 Testing of tensile fatigue

The tensile fatigue tests are made according to the protocol mentioned above, Table 4. Specimens undergo a load equal to 50% of the displacement at break static at a frequency of 10 Hz and an amplitude of 10% of the displacement at failure. We opted for a stress fatigue of the order of 10,000 cycles. Fatigue manifests itself as a matrix cracking, throughout the operation. Delamination and debonding are very low and the fiber breakage is very rare. Work Talreja [22] taken by Muc [23] explain the evolution of the damage in three stages: the collapse of the matrix, delamination and fiber breakage. We stopped fatigue testing to 10,000 cycles before the final break, immersing the samples in different environments.

Table 4. Experimental data

Fibre	Deformation [mm]	Rupture Force [N]	Rupture [mm]	Amplitude [mm]	Speed [m/mn]	Frequency [Hz]	Number of cycles	Immersion [h]
Glass	3.54	8000	1.76	0.35	1	10	10000	100, 500 1000

3.3 Effects of Aging

Fatigue after 10000 cycles, the specimens was immersed in tap water, as appropriate, for 100, 500 and 1000 hours, to undergo the same aging water. Whereupon, they undergo static tension until failure.

This choice aging times is based on a literature study to clearly identify the behavior of composite materials in wet conditions. Damage by cracking due to significant osmotic water uptake has been identified beyond 120 hours of aging. These results confirm the state of damage of the material. [24] At 20 ° C, the samples quickly reach equilibrium (35 days), 0.35% absorption and a model Frick can adequately describe the absorption [25] kinetics. At a temperature of 23 ° C, the interactions between moisture and fatigue damage can only be done locally, after capillary moisture in the crack tip [26]. Up to 50 ° C, the loss of properties due to plasticization of reversible phenomena of the matrix [27]. Different studies which we referred we have shown that osmotic origin of degradation of composite materials has been confirmed by preliminary tests in seawater and distilled water [28]. The attack of the resin in the laminate occurs between two folds and led to a decline of the most important mechanical properties [29].

At an exhibition in wet, it is known that the glass-epoxy composites can undergo three main types of damage:

- Reversible lamination process of the matrix resulting from water sorption in polyepoxide network,

- Macroscopic damage (matrix cracking, fiber debonding / matrix .....), either from the chemical degradation of the polyepoxide network or stress induced by differential swelling water sorption,
- Chemical attack of the reinforcement in the case of glass fibers, which results in a loss of tensile strength. This particular fiber degradation allows the creation of surface defects by heavily dependent on physico-chemical characteristics of the surrounding environment [30-33] ion exchange mechanism.

The results, given in Figure 4 are compared among themselves to highlight the effect of aging on epoxy-glass composite. For fatigue damage is brittle type for the series of samples tested. First, a quasi-linear variation of the stress in function of the deformation, then a sudden drop in the latter. This drop is due to fiber breakage, causing the total failure of the specimen.

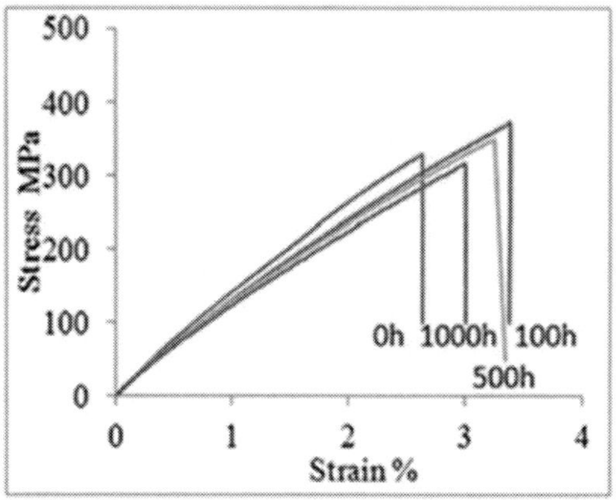

Fig 4. Evolution mechanical behaviour according to the immersion duration and fatigue test

A comparative study of the effect of fatigue and aging tap water allows us to make the following observation figure 4:

- For epoxy-glass test plates, the values of stress and strain are higher than that of the control samples remained in the open air.

In general, all the samples, regardless of their nature, experience a larger than those of the control specimens elongation. This is explained by capillary diffusion of moisture in the crack in the gaps left by fatigue by creating a differential swelling induced by water sorption. This gives the matrix lamination, giving it certain elasticity, it does not naturally.

The σ/ε ratio of stress to strain given by the tangent α, gives us an idea of material degradation due to fatigue and aging, Tables 5 and Figure 5 shows us that effect, indicated by a sharp drop of the material characteristics. We conclude that fatigue followed by aging alters the characteristics of the glass-epoxy specimens.

Table 5. Glass's stress and strain

Glass Tap water ($_{gtw}$) 10000 cycles		
σ [MPa]	ε [%]	σ/ε_{gtw}
σ_{gtw0}= 328,06	ε_{gtw0}= 2,64	124.26
σ_{gtw100}= 373,43	ε_{gtw100}= 3,38	110.48
σ_{gtw500}= 347,59	ε_{gtw500}= 3,24	107.28
$\sigma_{gtw1000}$= 315,58	$\varepsilon_{gtw1000}$=3,00	105.19

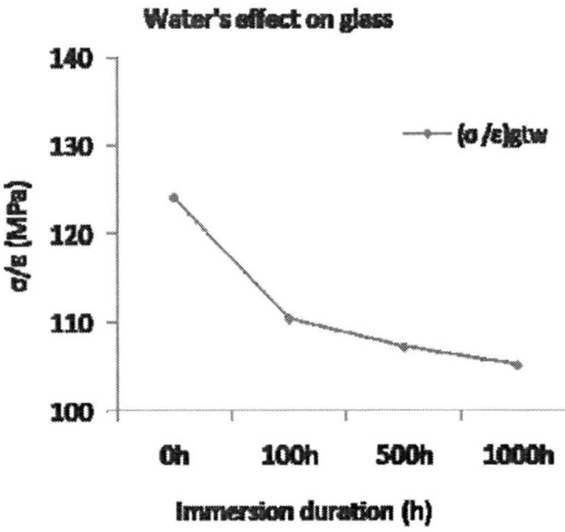

Fig. 5. Water's effect

The foregoing results show that the test pieces of glass-epoxy subjected to progressive aging after tensile fatigue undergo a gradual loss of resistance as a function of the immersion time.

For acoustic monitoring, we opted for a fatigue life of 10,000 cycles and immersion time of 100h, figure 6.

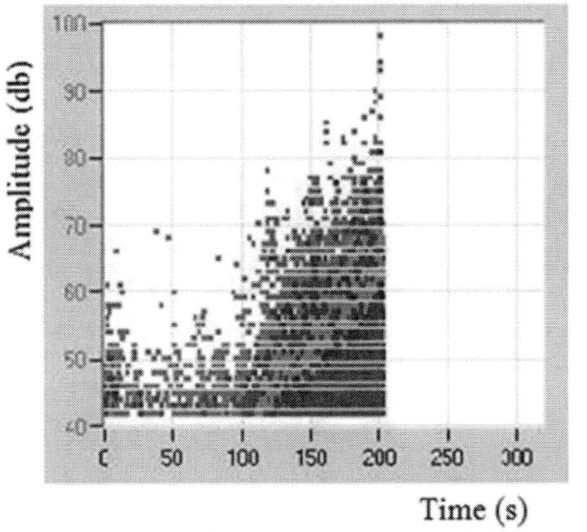

Fig. 6. Fatigue 10000 cycles and immersion 100 h

The glass plates exhibit a matrix cracking for 100 seconds, which correspond to 50% of the life of the material. We find that aging tap water gives the resin plates a greater elongation. The acoustic monitoring allows us to better visualize the different mechanical phenomena experienced by different materials. These results allow us to distinguish the various ranges of different damage. They also allow us to see that the same fatigue and even aging does not have the same impact on the epoxy resin, for glass. This may be due to the binding resin fiber which is not the same for materials.

4 Conclusion

This study aimed at investigating the influence of water ageing after fatigue on the mechanical proprieties of glass-fibre reinforced epoxy composite. The test specimens were subjected to fatigue for various numbers of 10000 cycles and aged in tap water for varying lengths of time (100, 500 and 1000 h) under acoustic control. Then the composite specimens were tested in tension in order to determine the evolutions of the strength and stiffness under local interactions of the humidity and fatigue.

From this investigating, the following conclusions can be drawn:

—The tensile mechanical proprieties are clearly affected by water ageing and fatigue, both on the shape of the stress-strain curves and on the properties' values.

— The specimens subjected to the tap water are more fragile than those which were preserved in the air.

— The residual stiffness and residual strength decrease when immersion time increases, indicating that the material has experienced some forms of mechanical damage and irreversible chemical degradation. The decrease in stiffness and strength are explained by the degradation of fiber / matrix interfacial by water absorption and the swelling of the matrix.

References

1. A.M. Visco, N. Campo, P. Cianciafara. Composites: Part A **42**, 123 (2011)
2. K. Imielinska, L. Guillaumat. Compos Sci Technol **64**, 2271–2278 (2004)
3. E. Poodts, G. Minak, A. Zucchelli. Composite Structures **97**, 222–230 (2013)
4. N. Godin, S. Huguet, R. Gaertner. Composite Structures **72**, 79–85 (2006)
5. P. Odru, B. Dewimille, F. Grosjean, and M. Huvey. Journées Scientifiques et Techniques AMAC (1996)
6. K. Ono. *Acoustic emission encyclopedia of acoustics* (New York: Wiley, 1997)
7. A.G. Beattie. J Acoustic Emiss **1/2**, 95–128 (1983)
8. A. Nielsen. The Danish Welding Institute Publication, **80**, 15 (1980)
9. A. Marec, J-H. Thomas, R. El Guerjouma. Mech Syst Signal Proc, **22**, 1441–1464 (2008)
10. H. Gu, S. Hongxia. Mater Des, **29**, 262–4 (2008)
11. P. J. De Groot, P. A. M. Wijnen, and R. B. F Janssen. Composites Science and Technology, **55**, 405-412 (1995)
12. O. Chen, P. Karandikar, N. Takeda, and T. Kishi. Nondestructive Testing and Evaluation, **8-9**, 869-878 (1992)
13. S-T. Kim, and Y. T. Lee. Materials Science and Engineering, A **234-236**, 322-326 (1997)
14. J. Karger-Kocsis, T. Harmia and T. Czigany. Composites Science and Technology, **54**, 287-298 (1995)
15. G. Kotsikos, J.T. Evans, A.G. Gibson, and J.M. Hale. Polym Compo, **20 (5)**, 689–696 (1999)
16. G. Kotsikos, J.T. Evans, A.G. Gibson, and J.M. Hale. Composites–part A **31**, 969–977 (2000)
17. M.-L. Benzeggagh, S. Barre, B. Echalier, and R. Jacquemet. AMAC Journées Nationales Composites, Paris, **8**, 703-714 (1992)
18. S. Barre, M.-L. Benzeggagh. Composites Science and Technology, **52**, 369-376 (1994)
19. S. Huguet, N. R. Godin Gaertner, L.Salmon and D.Villard. Compos Sci Technol, **62**, 1433–1444 (2002)
20. H. Nechad. *Thèse à l'Institut National des Sciences Appliquées de Lyon* (2004)
21. M. Bentahar. *Thèse de L'Institut National des Sciences Appliquées de Lyon* (2005)
22. R. Talreja. *Fatigue of composite materials: Analysis, Testing and Design, Siminar* (Technomic publishing Co Inc., Lancaster, Basel, Switzerland, 1990)
23. A. Muc. Computers and Stuctures, **76**, 211-218 (2000)
24. M.P. Foulc, A. Bergeret, P. Ienny, L. Ferry and A. Crespy. Matériaux (2002)
25. F. Dal Maso, B. Dewille , F. Grojean, M. Huvey, J. Martin et L. Montabord. *Annale des composites, AMAC, France,* 27-37 (1996)

26. S. Popineau, C. Rondeau-Mouro, C. Sulpice-Gaillet, and M. Shanahan. Polymer, **46**, 10733–10740 (2005)
27. A. Chateauminois, B. Chabert, J.P. Soulier, and L.Vincent. Annale des composites, AMAC, France, 823-832 (1990)
28. P. Castaing, and L. Lemoine. Polymer Composites, **16**, 5, 349-356 (1995)
29. Ph. Castaing, N. Tsouvalis, and L. Lemoine. IFREMER, Actes de colloques, **15**, 34 (1992)
30. E. Vautier, J.C. Abry, T. Bailliez, and Chateauminois. Journées Scientifiques et Techniques *AMAC*, France, 49-58 (1996)
31. A-HI. Mourad, AM. Beckry Mohamed, T. El-Maaddawy. Appl Compos Mater, **17**, 557–573 (2010)
32. A. Kootsooks, and AP. Mouritz. Compos Sci Technol, **64**, 1503–1511 (2004)
33. H. Gu. Mater Des, **30**, 1337–1340 (2009)

Behavior of some steels at vibrorolling

Gheorghe Nagîț[1], *Oana* Dodun[1,*], *Laurențiu* Slătineanu[1], and *Marius Ionuț* Rîpanu[1]

[1]Gheorghe Asachi Technical University of Iași, Department of Machine Manufacturing Technology, 700050 Iași, România

Abstract. The vibrorolling could be sometimes applied as a finishing method to diminish the height of the surface asperities. It is expected that the mechanical characteristics of the workpiece material affect in a certain extent the generation of the new profiles corresponding to the surface roughness and distinct values of the surface roughness parameters could be obtained. An experimental research was designed and achieved taking into consideration some vibrorolling process input factors and distinct steels as materials for the test samples. The experimental results were mathematically processed and empirical models were determined. These mathematical empirical models highlight the distinct behavior of the studied steels evaluated by means of the *Ra* surface roughness parameter.

1 Introduction

There are various reasons which need the finishing of metallic parts surfaces. Thus, in order to ensure a low value for the friction coefficient corresponding to surfaces which moves one other, usually a low surface roughness is necessary. In the case of interference fits, the surface roughness should be diminished up to certain values, but not too low, in order to not diminish the clamping force. Sometimes, the parts superficial layers must have certain specific properties, like high hardness, high crushing strength, corrosion resistance, certain internal stresses etc. In all such situations, adequate finishing methods should be applied in the final part of the technological route.

In order to obtain only low values of the surfaces asperities heights, abrasive finishing methods or finishing methods based on the use of metallic cutting tools could be applied. When the surface layer has to be characterized by an increased hardness compared with the rest of part, certain burnishing or rolling processes could be applied to metallic workpieces. With this aim in view, the surface layer is pressed by means of balls or rolls tools, so that the surface asperities are plastically deformed, and a low surface roughness could be thus obtained [1-5]. Additionally, the plastic deformation of the metallic crystals found in the workpiece surface layer could determine an increase of the hardness and, in this way, an increase of the mechanical strength corresponding to this layer. In such situations, the quality of the surface layer could be evaluated by means of the surface roughness parameters, of the surface layer hardness, of the stress state corresponding to the workpiece surface layer.

[*] Corresponding author: oanad@tcm.tuiasi.ro

The surface roughness could be defined as the assembly of surface asperities characterized by a ratio between width and height lower than 50 (some specialists take into consideration a ratio of 40). There are various parameters designed to be used when the problem of surface roughness evaluation is considered, but currently the arithmetic mean deviation of the profile is used.

The surface roughness parameter *Ra* could be determined as the arithmetical mean of the absolute values of the profile deviations from the so-called mean line of the roughness profile.

As above mentioned, the surface rolling supposes the plastic deformation of the surface asperities by means of balls or rolls, which rotates in contact with the surface layer and exert a certain pressure on this surface layer. In order to increase the effects of rolling, the specialists proposed processing methods in which an additional vibration motion is achieved by the deformation tool and this processing method was named *vibrorolling* or *vibroburnishing*. Even from its initial proposal, the vibrorolling was investigated by the researchers, in order to find combinations of process input factors values able to ensure convenient values for the process output parameters.

Thus, in the former Soviet Union, Odintsov [3] and Schneider [4] developed ample researches to investigate how various process input factors exert influence on process output parameters, such are the surface roughness or the surface layer hardness.

In the Polish university from Rzeszow, the professor M. Korzynski together his collaborators designed and achieved various equipment for vibroburnishing and investigated also aspects concerning the results of this processing method [5].

The objective of the researches presented in this paper is to highlight some differences between certain steels used in machine building, if workpieces made of these steels are affected by a vibrorolling process.

2 Phenomena specific to vibrorolling

In order to materialize a vibrorolling process, the processing scheme from Figure 1 could be applied. The case of vibrorolling by means of a universal lathe could be taken into consideration. A cylindrical workpiece clamped in a universal chuck and in a live center is trained in a rotation movement, specific usually to the turning process. On the lathe apron, a device able to generate the force necessary to the vibrorolling process and to ensure the additional vibratory motion to the vibrorolling tool is placed. Thus, a force *F* of known value is exerted on the workpiece surface. As a vibrorolling tool, a ball of known radius and made of a hard material is used. Together with the vibrorolling device, the lathe apron achieves a common longitudinal feed movement f_l.

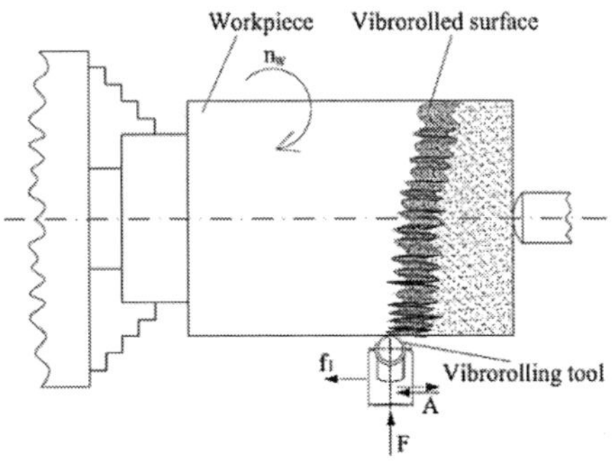

Fig. 1. Schematic representation of vibrorolling process applied to a cylindrical surface corresponding to a metallic workpiece

The vibrorolling device placed on the lathe apron must ensure also an additional vibratory movement of amplitude A and frequency f. If the force F is high enough, the vibrorolling tool will plastically deform the superficial layer of the workpiece, penetrating in the workpiece material on a depth a_p. If the vibratory motion does not exist, due to the rotation movement of the workpiece and the apron feed motion, the ball will follow a spiral trajectory on the workpiece surface. When the vibratory movement is also present, a more complex trajectory is achieved on the workpiece surface (Fig. 1) and sometimes, two consecutive routes followed by the ball on the workpiece surface could be partially superposed.

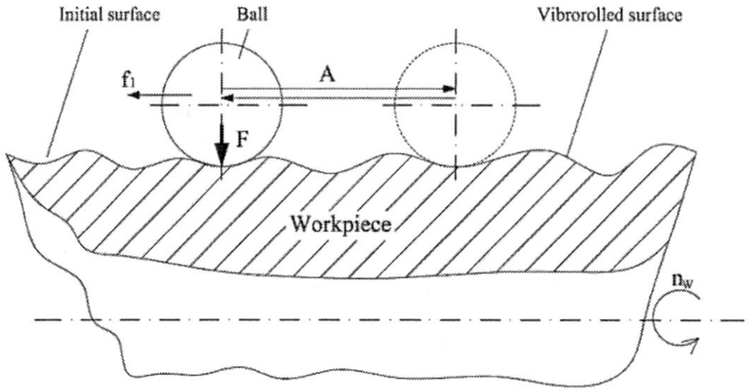

Fig. 2. Detail form the vibrorolling zone.

During the movements achieved on the workpiece surface and due to the force F, the surface asperities are affected by a plastic-elastic deformation process. As a consequence of the plastic deformation process, the height of the asperities diminishes (Fig. 2) and the surface roughness parameters could highlight this phenomenon. Due to the elastic component of the deformation process, after the ball passing on the workpiece surface, a certain recovery process develops.

The vibrorolling process will determine not only the decrease of the asperities height, but also a hardening of the workpiece surface layer and this result could be highlighted by measuring the evolution of the material hardness in the workpiece surface layer.

Within this paper, only the changes of the surface roughness as a consequence of the vibrorolling process application will be taken into consideration.

There are some main groups of factors able to affect the surface roughness parameters as a consequence of applying the vibrorolling process: 1) The physical and geometrical characteristics of the deformation ball (ball radius, ball surface roughness parameters, ball material hardness etc.); 2) The kinematic scheme of the vibrorolling process; in the case of vibrorolling cylindrical surfaces of metallic workpiece, the workpiece rotation speed and the apron feed speed could affect the values of the vibrorolled surface roughness parameters; 3) The physical and mechanical properties of the workpiece material (hardness, plasticity, tensile strength, relative elongation etc.). 4) The presence and the nature of media used in order to decrease the friction phenomena.

It is expected that as a consequence of the plastic deformation process by vibrorolling process applied to metallic workpieces surfaces, a diminishing of the values of the surface roughness parameters could be highlighted. Practically, under the action of the pressure exerted by the ball, the material corresponding to the asperities is affected by a process of re-distribution by the displacement of the material to the bases of the asperities.

3 Experimental research

In order to investigate the behavior of some steels as a consequence of applying the vibrorolling process, an experimental research was designed and materialized [6].

As scheme of the vibrorolling process that was illustrated in Figure 1 was used. A universal lathe was taken into consideration as equipment for training the workpiece in a rotation movement and for achieving the apron feed movement.

A device able to ensure a vibration movement with pre-established frequency and amplitude was placed and clamped on the lathe apron.

Table 1. Experimental conditions and results

Rotation speed n_{tp} of the test piece: n_{tp}=100 rev/min							
Longitudinal feed f_l =0.024 mm/rev							
Frequency f of tool oscillation motion: f_{min}=5.83 Hz; f_{max}= 15.66 Hz							
Diameter of ball used as a deformation tool: d_{min}=6.75 mm; d_{max}=15.85 mm							
Oscillations amplitude: A_{min}=0.6 mm; A_{max}=1.6 mm							
Force exerted by deformation element: F_{min}=100 N; F_{max}=600 N							
Exp. no.	**Process input factors**				***Ra*, µm, for distinct steels**		
	f, Hz	*R*, mm	*A*, mm	*F*, N	1C45	E 335	18 MnCr 4-4
1	5.83	6.75	0.6	100	0.07	0.05	0.04
2	5.83	6.75	0.6	600	0.02	0.03	0.05
3	5.83	6.75	1.6	100	0.65	0.61	0.55
4	5.83	6.75	1.6	600	0.25	0.30	0.40
5	5.83	15.85	0.6	100	0.03	0.02	0.02
6	5.83	15.85	0.6	600	0.01	0.01	0.02
7	5.83	15.85	1.6	100	0.24	0.23	0.20
8	5.83	15.85	1.6	600	0.10	0.14	0.20
9	15.66	6.75	0.6	100	0.04	0.03	0.02
10	15.66	6.75	0.6	600	0.02	0.02	0.03
11	15.66	6.75	1.6	100	0.42	0.50	0.46
12	15.66	6.75	1.6	600	0.11	0.16	0.18
13	15.66	15.85	0.6	100	0.02	0.01	0.01
14	15.66	15.85	0.6	600	0.01	0.01	0.01
15	15.66	15.85	1.6	100	0.12	0.09	0.08
16	15.66	15.85	1.6	600	0.06	0.07	0.10

The force necessary to the vibrorolling process was ensured by means of a spring, for which a previous calibration operation allowed establishing a correspondence between the applied force and the diminishing of the spring length.

A full factorial experiment with four independent variables at two levels was considered. One appreciated that within the experimental interval of independent variables variation, there will be a monotone variation of the response function (*Ra* surface roughness parameter) and, in such a situation, only two levels of the independent variables could be taken into consideration. On the base of some preliminary experiments, the values for the process input factors were established: these values, together with other initial experimental conditions, were mentioned in the first lines of table 1.

Sleeves shaped test pieces made of three distinct steels were prepared. The sleeves were positioned and clamped on a mandrel that at its turn was clamped in the universal chuck and in a live center. As steels for test pieces, there were used the steel 1C45, as a material widely used in the field of machine building and sometimes used as an etalon material when the problem of machinability evaluation is formulated, steel E335, as a material that does not need heat treatment, but able to ensure a certain high tensile strength, and the steel 18 MnCr 4-4, as a steel supposing the applying a thermo-chemical heat treatment (carburisation). The surfaces proposed to be vibrorolled were previously obtained by applying a finishing turning, in the same cutting conditions. Even in such a situation, due probably to their distinct machinability by finishing turning, a certain dispersion of values corresponding to the *Ra* surface roughness parameter was observed (*Ra*=1.27-1.56 μm). The surface roughness was measured before and after applying the vibrorolling process by means of a G4 type profilometer-profilograph (Russia).

The values of the input factors were included in the table 1. In the last three columns of table 1, the experimentally obtained values of the *Ra* surface roughness parameter were mentioned.

4 Processing and analyzing the experimental results

The experimental results were mathematically processed by means of a specialized software [7] based on the method of least squares. In this way, three power type empirical models were obtained. The power type empirical models were preferred since they offer a direct image concerning the influence exerted by each process input factor on the *Ra* surface roughness parameter. As indicator of adequacy of the power type mathematical empirical models to the obtained experimental results, the so-called Gauss criterion could be used.

The power type mathematical empirical models were the following:

$$Ra = 32.742 f^{-0.433} R^{-0.98} A^{2.138} F^{-0.530}, \tag{1}$$

in the case of steel 1C45 (the Gauss criterion having the value S_G=1.208144·10^{-3}),

$$Ra = 16.753 f^{-0.514} R^{-1.184} A^{2.411} F^{-0.293}, \tag{2}$$

in the case of steel E 335 (the Gauss criterion having the value S_G=2.484984·10^{-3}),

$$Ra = 4.754 f^{-0.655} R^{-1.090} A^{2.372} F^{-0.0282}, \tag{3}$$

in the case of steel 18 MnCr 4-4 (the Gauss criterion having the value S_G=4.080313·10^{-3}).

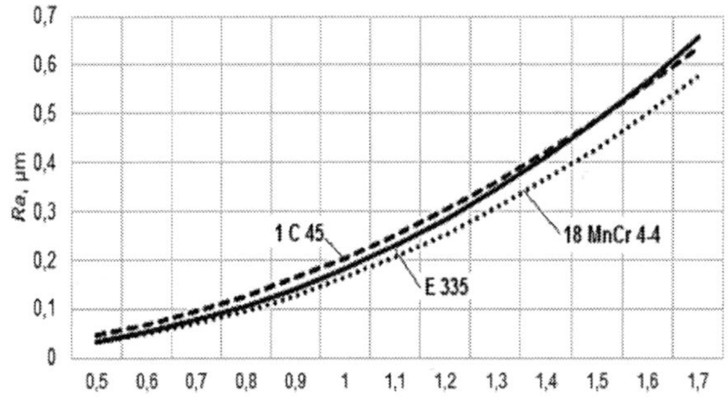

Fig. 3. Influence exerted by the amplitude *A* of the vibratory movement on the value of the *Ra* surface roughness parameter, in the case of applying vibrorolling process to test pieces made of three distinct steels (*f*=5.83 Hz, *R*=6.75 mm, *F*=100 N).

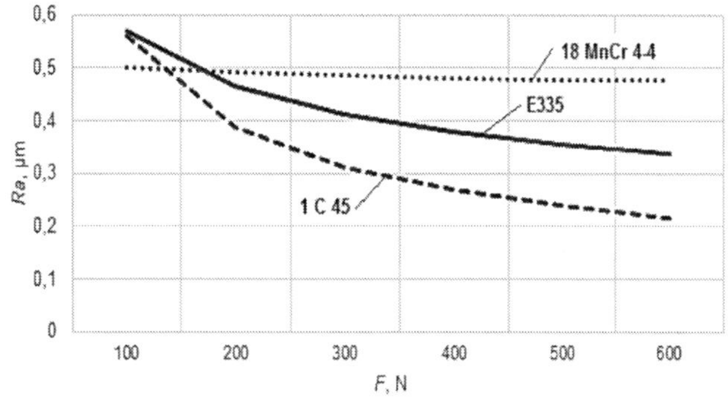

Fig. 4. Influence exerted by the force F on the value of the Ra surface roughness parameter, in the case of applying vibrorolling process to test pieces made of three distinct steels (f=5.83 Hz, R=6.75 mm, A=1,6 mm).

On the base of these power type mathematical empirical models, the graphical representations from Figures 4 and 5 were elaborated. The analysis of the empirical mathematical models and of the graphical representations allows statement of some remarks concerning the influence exerted by the process input factors on the Ra surface roughness parameter. Thus, one could notice that for all the three steels, the most important influence factor is the vibration amplitude A, whose exponent in all empirical models has the maximum absolute value. As a second influence factor, the ball radius R could be mentioned, since the exponent value has the second absolute value in all empirical mathematical models, compared with the absolute values corresponding to the other exponents. As expected, the increase of the frequency f, of the ball radius R and force F determines the decrease of the value of Ra surface roughness parameter. In the case of steel 18 MnCr 4-4, the influence exerted by the force F is not significant (Fig. 4), since the value of the exponent attached to F in the empirical model has a value very near to zero). The steel 18 MnCr 4-4 ensures the minimum value of the parameter Ra when the influence of the amplitude A is considered (Fig. 3) and a general higher value of the same parameter when the influence exerted by the force F is considered (Fig. 4); probably the presence of the alloying elements manganese and chrome generate such a behavior of this steel.

5 Conclusions

One of the processing methods applied in order to decrease the height of the surface asperities is the vibrorolling. Compared with the older burnishing process, the vibrorolling uses an additional vibratory movement of the burnishing tool. An analysis of the burnishing process input factors able to influence the height of the surface asperities showed that these factors could be classified in some groups, taking into consideration the geometrical and mechanical characteristics of burnishing tool, the values of the factors corresponding to vibrorolling kinematic scheme, the characteristics of the workpiece material, the presence and nature of eventual used processing media, the pressure exerted by the vibrorolling tool on the workpiece surface.

An experimental research aiming to highlight the influence exerted by some vibrorolling process input factors (frequency f and amplitude A of vibratory movement, ball radius R and force F) on the Ra surface roughness parameter was designed and materialized. By mathematical processing of the experimental results by means of a specialized software, three mathematical empirical models were determined.

As expected, one noticed that the increase of the process input factors f, R and F determine a decrease of the values corresponding to the Ra surface roughness parameter, while the increase of amplitude A generate an increase of the same parameter. In the future, there is the intention to extend the experimental research in the direction of a more detailed investigation of the microphenomena developed in the work zone corresponding to the vibrorolling process.

References

1. A.V. Ezhelev, I.N. Bobrowski, I.N., A.A. Lukyanov (in Russian). Fundamental research, **6** (2012)
2. V. G. Revenko, G. R. Alekseeva, N. F. Stepanov, N. L. Bogdashkina, and G. P. Chernova. Protection of Metals, **38,** 1 (2002)
3. V.G. Odintsov. *Finishing processing of details by diamond burnishing and vibroburnishing* (in Russian) (Mashinostroenie, 1981)
4. I.G. Schneider (in Russian). Stanki i instrument, **7**, 34-35 (1972)
5. M. Korzynski and J. Lubas. Bushing texturing by burnishing, in *Nonconventional finishing technologies*, (Polish Scientific Publishers PWN, Warsaw, 2013)
6. G. Nagîţ, *Theoretical and experimental contributions concerning vibroburnishing process. Doctoral thesis* (in Romanian) (Gheorghe Asachi Technical University, Iaşi, 1997)
7. G. Creţu, *Fundamentals of experimental research. Laboratory handbook* (Technical University of Iaşi, Romania, 1992)

Incremental deformation: A literature review

Daniel Nasulea[1,*], and *Gheorghe* Oancea[1]

[1]Transilvania University of Brasov, Department of Manufacturing Engineering, Mihai Viteazu No.5, Brasov, Romania

Abstract. Nowadays the customer requirements are in permanent changing and according with them the tendencies in the modern industry is to implement flexible manufacturing processes. In the last decades, metal forming gained attention of the researchers and considerable changes has occurred. Because for a small number of parts, the conventional metal forming processes are expensive and time-consuming in terms of designing and manufacturing preparation, the manufacturers and researchers became interested in flexible processes. One of the most investigated flexible processes in metal forming is incremental sheet forming (ISF). ISF is an advanced flexible manufacturing process which allows to manufacture complex 3D products without expensive dedicated tools. In most of the cases it is needed for an ISF process the following: a simple tool, a fixing device for sheet metal blank and a universal CNC machine. Using this process it can be manufactured axis-symmetric parts, usually using a CNC lathe but also complex asymmetrical parts using CNC milling machines, robots or dedicated equipment. This paper aim to present the current status of incremental sheet forming technologies in terms of process parameters and their influences, wall thickness distribution, springback effect, formability, surface quality and the current main research directions.

1 Introduction

In the last few decades, the competition between the product manufacturers becomes increasingly tighter. Because of the customers needs, the manufacturers and their supliers have to manufacture new products and to improve them as soon as possible. If in the 20th century the automotive manufacturers, for example, were launching a new car model in about 7...10 years, now, in the 21th century, they launch new models in about 4 years. That is the reason why the industy is now focused on the flexibles processes, the products are manufactured in small batches, targeting a minimum production cost.

Metal forming is one of the most important industry, 15-20% GDP comes from this industry [1], playing a significant role in manufacturing and development. The conventional processes from the metal forming industry like drawing, deep drawing, stamping, stretch forming, etc. allow to manufacture large and mass production batches, being very well developed with an high level of automatization and innovation, high productivity, but low flexibility. Due to the market competition, the products batches are smaller and smaller,

* Corresponding author: daniel.nasulea@yahoo.com

parts have more complex shapes and the customer requirements are to decrease the products price. In this case, the conventional sheet forming processes are not suitable because of the tool complexity and costs, and because of the high time from design to manufacturing (large die-preparaton time and costs).

In the last years, the researchs were focused on new processes and methods to meet the customer demands, innovative and flexible processes for small batches products with low costs and less time from design to manufacturing. One of the most used processes in metal forming is Incremental Sheet Forming (ISF). It is a relatively new process for sheet metal parts manufacturing, used for small batch production and also rapid prototyping. The conventional sheet metal forming processes depend on a dedicated tool (every part is manufactured with its own tool). The ISF is a flexible „die-less" process, using a universal, simple and low cost tool for more products. The process allows to reduce the time and costs for developing of new products, and it is suitable for small volume production and also for prototypes manufacturing.

In incremental forming, a simple shaped tool produces a local plastic deformation in the sheet, using layered manufacturing principles, converting the part geometry into many layers in forming direction, in each layer, the forming tool follows a two dimensional contour . This step is repeated layer by layer until the final shape of the part is obtained. All this process is numerically controlled by a machine, therefore, when a new shape of the part is required, only the CNC program has to be be changed and maybe also the forming tool, thus, the change are easy to do with small costs in comparison with conventional processes.

The predecessors of incremental sheet forming are the stretch forming and metal spinning process [2]. The ISF combines the advantages of these two predecessors with CNC technologies and has become one of the most flexible process in sheet metal forming industry.

2 Advantages and disadvantages of ISF

In comparison with conventional metal forming processes, ISF has some advantages. They are summarised bellow [1-4]:

- usual parts can be formed directly from CAD data with minimum dedicated tools;
- process does not require either positive or negative dies; it is a dieless process;
- changes in part design dimensions can be easily and quickly modified; the process has a high flexibility;
- making metal rapid prototypes is normally difficult, but easy with this process;
- small plastic zone and incremental nature of the process contribute to increase formability, making it easier to use low formability sheets;
- a conventional CNC milling machine or lathe can be used for this process;
- dimensions of the part is limited only by the working area of the used machine;
- necessary forces do not depends on the parts dimensions;
- process is quiet and relatively noise free.

The ISF process has a few disadvantages [1-4]:

- forming time is much longer than classical processes such as drawing;
- process is limited to small-size batch production;
- forming of right angles cannot be done in one step, requires a multi-step process;
- ISF process enables much higher strains.

3 Incremental Sheet Forming working principle

The tooling design for ISF process is simple and it supposes low costs. The main elements needed to implement an ISF process [5] are ilustrated in figure 1. There are five basic elements according to the type of ISF: a blank holder with or without a backing plate, the sheet metal (blank), a clamping plate and a forming tool. As shown in figure 1, the sheet metal blank is clamped on the backing plate or blank holder. The forming tool have three degrees of freedom (translations on x, y and z axis), and rotation around its own axis, following a path programmed by a CNC software. The tool moves around the outer contour from top to bottom, layer by layer decreasing with the predefined vertical increments (steps in z direction) in contact with the sheet metal, applying force and deforming the blank until the final desired shape was obtained. Therefore, using a simple forming tool and a CNC controlled path can get simple but also complex sheet metal parts. The process is characterized by deformations of a small surface in the contact area between the tool and the blank, and can achieve greater deformation than the conventional processes.

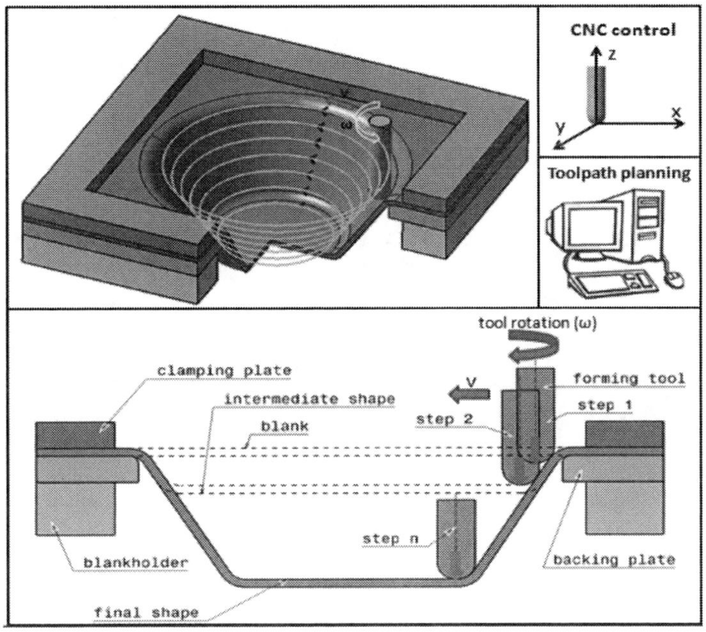

Fig. 1. Principles of ISF processes.

4 Application of ISF processes

The ISF processes are used in many domains of metal industry and a lot of situations [3, 5]:

- aerospace industry: instrument panel, body panel, passenger seat cover, and so on;
- automotive industry: door and hood panels, engine cover, and so on;
- electronic industry : cellular phones, hard disk drives, sensors;
- boat industry - ship hull plates;
- products with high level of customization: denture plate, ankle support, metal helmet, health care, and so on;
- low volume or unique products, prototypes or products after-sales;
- architectural - custom made formwork, panels;
- reflectors and casings for lighting;

- products of national security and defense.

The ISF have been implemented by industrial companies in the process of parts remanufacturing. An example is to produce hoods replacement parts for collection cars (Fig. 2) [6]. The hood was scaned with a 3D scanner, redesigned in CAD system and then remanufactured using the ISF technology. In this way a replacement hood is obtained. Another exemple was the production of some panels for a limited edition of well-known cars, taking some series production parts and reshape some new sharp edges or logo mark using ISF process (Fig. 3) [6].

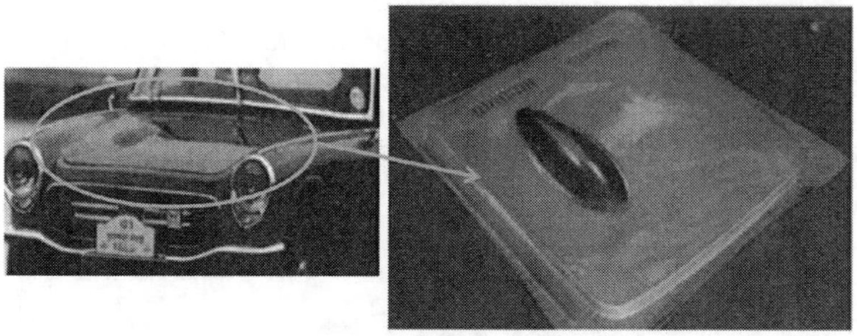

Fig. 2. Hood panel of collection car made with ISF process [6].

a) Series production car. b) Limited edition reshaped with ISF.

Fig. 3. Reshaping of automotive part [6].

5 Types of incremental forming processes

The ISF processes can be generaly classified taking into account various aspects. Based on the shape of the manufactured part, the ISF can be [4] :

- symmetric incremental forming (SIF) - for axis-symmetric manufactured parts;
- asymmetric incremental sheet forming (AISF) –for asymmetrical manufactured parts;

From the contact surface between the forming tool and the blank point of view, the ISF can be [4, 7]:

- single point incremental forming (SPIF) – one single contact point;
- incremental forming with counter tools – in this type of ISF, another tool offer support from the other side of the blank in the same area with the forming tool;
- two points incremental forming (TPIF): the deformation takes place in more then one point in the same time; it can be with partial die support or with full die support.

In the first two configurations above mentioned, only the tool is moving along the controlled path and the blank holder is stationary. In the third configuration not only the forming tool is moving, also the blank holder moves step by step along the tool axis direction, to ensure die support.

6 Equipments used for ISF

In many cases, the ISF process supposes the using of a CNC controlled machines. A minimum 3 axis CNC machine are needed to control the path of the forming tool along the predefined contours at different levels (Fig. 4). Milling machines are the most commonly used because they allow high speed, large working volumes and good stiffness [4]. For an axis symmetric part (SIF) a CNC lathe is suitable to perform the process in less time, but also a CNC milling machine can be used (Fig. 5) [8].

Fig. 4. ISF on CNC milling machine [4].

Fig. 5. ISF on CNC lathe [8].

More than that, some researcheres or companies also developed or porposed some machines, who was special builded for ISF process. In 1996, an industrial company developed the firs prototype machine and continued to develop it until 2002 when they started to provide specialized equipment for industry [6]. Another proposal was made by Allwood et al. for a single point incremental sheet forming [9].

Industrial robots are frequently used for ISF process. They have benefits like large working volume and much flexibility than a milling machine, but also some disadvantages, low stifness and low capacity to apply big forces [4, 10]. Using robots in incremental sheet forming was patented by Tuominen and he describes many configurations with one or two robots, for both SPIF and TPIF technologies [2, 11]. For ISF aro used also Stewart Platforms [4, 12], robotised cells [13] and hexapods [4].

7 Forming tools

The most usual tools in incremental forming are the solid tools. A wide variety of solid tools are designed and made by the users taking into account the used equipement. There are not yet standardized tools available on the market [4, 14]. The solid tools have usualy hemispherical shape and can be straight when it is used on CNC miling machine or for ISF machines (Fig. 6a) [15], and bended when it is used on a CNC lathe (Fig. 6b) [8]. A wide range of tool diameters are used, starting at small diameters of 6 mm and going up to large tool diameters of 100 mm for the manufacturing of large parts [14]. Tool diameter is an important factor, influencing on the formability, because of the contact area with the blank.

DOI: 10.1051/matecconf/201712103017

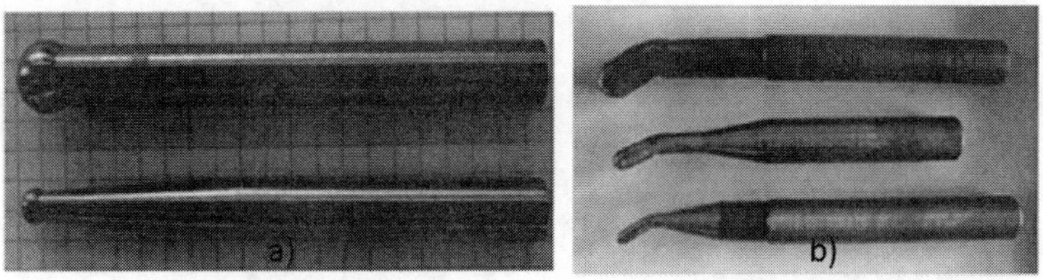

a) Straight tools [15]. b) Bended tools for lathe utility [8].

Fig. 6. Solid forming tools.

In the last years has been increased the interest of using the flexible tools for ISF process. Iseki [16] and Jurisevic et al. [17] studied the incremental forming using water jet, meen time Geiger and Vollertsen [18] used laser technology like forming tool. The laser technology can be also used to apply a local heating, to improve the material properties, for better formability [19]. Thus, new forming tools are used, water jet and laser forming, but they need future researchs.

8 Toolpath and tool parameter

In most of the cases the blank is fixed and the deformation is made by the forming tool that follows a programed path on the contour of the part. The path can be composed by a succession of planar curves that follow the part contour on each step level/increment, or a single spiral contour which makes a passing from one step to next in a smooth „spiral" way. To genarate the toolpath the CAD model of the part and a CAM software package are needed. Usually, any available CAM software package can be successfully used with same results. Obvious, the CAM softwares are used for cutting process, but can be also used for ISF if is chosed the right strategies. Thus, if the blank is deformed in multiple passes on each level, then one roughing strategy have to be used, because those commands are made to „mill" / deform all the material from inward to outward in multiple steps until the end of the part contour [4]. If the part shape allows to be deformed in a single step, then a finishing strategy can be used, because this is programed to make a single closed loop to the end part contour [4].

Basicaly, two types of toolpath are used: a succession of planar curves (part contour) on each level – level path (Fig.7a) [20], or a spiral toolpath (Fig.7b) [20];

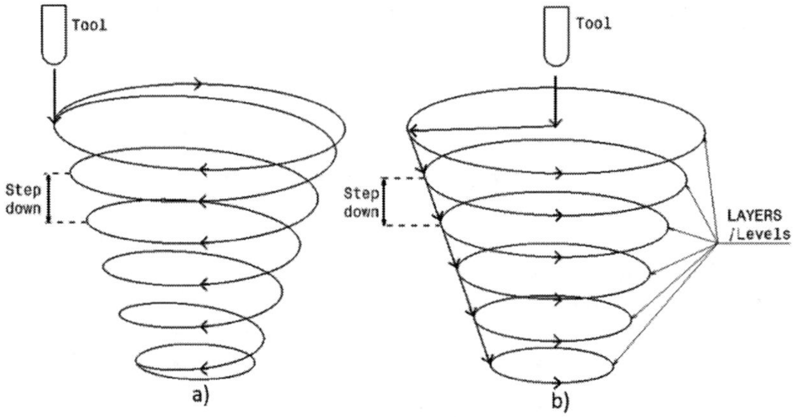

a) Spiral toolpath b) Succession of planar curves – level toolpath ;

Fig. 7. Tool toolpaths used in ISF processes [20].

Using the level toolpath, on the wall is vissible a trace where the tool passes from layer to layer, while the spiral toolpath is a continuous one where the pitch is equal with the step down and it moves with no traces on the part surface, being frequency utilised. For the level toolpath is needed 2.5 axis milling machine, while for the spiral path is needed minimum 3 axes milling machine. The final part obtained using an ISF process, strongly depends by the tool path generation. It is one of the most important stages which influence the quality, thickness distribution, roughness and deviation because of springback effect. Studies about tool path in ISF were made by Yamashita et al. [21] and Zhou [22]. The paremeters used for tool movement are the same like in the cutting processes, but the values are customized for ISF process. They are feed rates, spindle speed and step down. The feed rates [mm/min] have a higher value than in a cutting process [4]. Regarding the spindle speed, a ISF process can be done with or without spindle speed. To obtain a surface with high quality, the blank lubrification have to be used [4, 23]. The step down (the increment), in fact is the distance between layers in tool axis direction which is usually on z axis.

9 Materials

Incremental sheet forming is basically used for deforming product from metal sheets. The material is chosed according to the needed strength, weight and/or corrosion resistance. In the top of used materials in ISF are alluminum and its alloys. A a lot of researchs regarding diferent topics have been developed [24]. Because of their formability properties and light weight, the brass and copper are two materials frequently used for ISF process [24, 25]. Titanium alloys well-known for their corrosion resistance and high strength and magnesium alloys for their light weight are used in many industries, thus the parts can be also obtained using incremental forming processes [26, 27]. A lot of components from all the industries are made from steel sheets and its derivates, thus these materials are investigated if it is possible to be formed with ISF process, especially cold presing steel like deep drawing and extra deep drawind steel [24, 28] There are research to evaluate the possibility to incremental deform the polymer sheets, for a low volume with a low cost production [29].

10 ISF parts quality

10.1 Surface roughness

Surface quality or surface roughness is one of the important aspects in sheet forming. This is direct influenced by the forming strategy and process parameters like incremental step Δz, tool radius, draw angle, spendle speed [30]. To measure the surface roughness can be used many indicators Ra, Rz, etc., but the most useful indicator is Rz, mean peak to valley height [4].

In the literature, it was found that the surface roughness strongly depends on the incremental step size Δz, when it is increased the result is that surface roughness is also increased [30]. Figure 8 [4] ilustrates the various material ridge hight for various step depth. For part made from aluminium Al 3003, Hagan and Jeswiet defined a mathematical model between incremental depth and peak-to-valley roughness for a 45 degrees wall angle [31]. It was found that the roughness or the ridge height increases exponentially with increesing the incremental step size Δz. Another researcher, Bhattacharya et al. [32] discovered that the surface quality decreases with step depth increasing, but only till a certain angle. After this angle the surface quality is increased again.

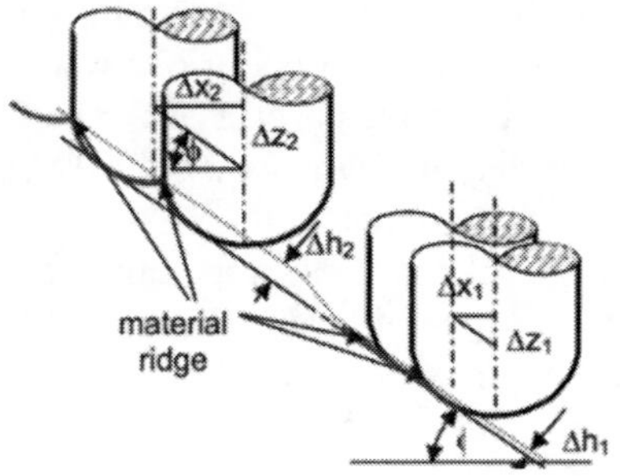

Fig. 8. Influention of step depth on ridge size (roughness) [8].

The tool diameter has also influence on surface roughness. Bhattacharya et al. [32] claim that the surface roughness is decreased when the tool diameter is increased. The tool material and the tool surface quality also influence the final manufactured surface quality. In this way, Oleksik et al. [33] reduced the tool roughness in order to improve the part surface finish, and there were obtained a comparable roughness of the final part with the initial roughness of the titanium alloy sheet.

Another parameter, wall angle or draw angle, influences the surface roughness. When the draw angle is increased, the stretching tension in the material is increased and the sheet metal is strongly stretched, helping to obtain a small value of surface roughness [32]. That meens the deeper products formed with ISF process have a smooth surface quality. It is important to note that the high wall angle can easily lead to failure [30].

Tool rotation or spindle speed has an important effect upon maximum profile height. It was observed that in case of tool with small rotation speed, the surface roughness is decreased. With decreasing, but not stopping the relative motion beetween blank and forming tool, a better surface finish can be obtained than using the high speed rotation [4]. Durante et al. [34] studied the influence of tool rotation on an incremental forming process and he found that the use of rotating tool gives a better surface roughness than a non rotation tool. The diference is about 10% in advantage of using a rotation tool [30].

10.2 Springback in ISF

A shortcoming for ISF processes is the dimensional accuracy of the parts because of the springback effect. Jeswiet et al. [4] appreciates that the most of the world ISF applications impose a dimensional accuracy of less than 1mm, even less ±0.5mm tolerance, while the ISF processes, as Bramley reports [23], the tolerance are ±1.5mm for symmetrical parts, and even more, ±2mm for the asymmetrical one. Springback effect is the main cause of the dimensional accuracy problems, being one of the most researched topics in ISF processes. It is the geometric change made by a part at the end of the forming process when the part has been released from the forces of the forming tool. In ISF the springback also appear nearby the deformed area, right after the tool just formed it. Therefore, springback affects the dimensional accuracy of the finished part. A lot of researchers investigated this topic and they found many solutions to improve the dimensional accuracy of ISF parts:

- it is indicated to use a backing plate (Fig.1) to support the sheet flange [4, 35];
- using a smaller tool diameter and incremental step depth, the springback can be reduced, as Ambrogio et al.[36] and Oleksik et al. [33] demonstrated;
- Duflou et al. [35] concluded that using a multi-pass toolpath strategy leeds to a better accuracy, minimising the sprigback effect. On the other hand Junchao et al. [37] find that increasing the number of stages the springback becomes higher using a single stage strategy, because of the springback accumulation during each forming stage;

- some researchers tryed to reduce the springback by correcting the toolpath of the forming tool to adjust the dimensional precision. It was done „in-process" measurements [38], or out of process was measured the discrepance from the CAD model and the manufactured part and then reload the process with corrected toolpath [4, 39]. Lu et al. [40] presented a new feature-based tool path special algorithm for a better geometric accuracy;
- using finite element methods (FEM) to define the springback deviation from the CAD model [39], before the forming process, and redefine the toolpath strategy the precision is improved.
- it was also proposed various models to overcome accuracy problems such as systems with flexible programmable back support structures [41], which are working with multiple and synchronised tools and high complexity.
- heating the sheet metal to improve the mechanical properties of the blank and the plasticity is also used. Duflou et al. [42, 43] uses laser assisted incremental forming with local heating for better formability and accuracy of the part obtained.

10.3 Wall thickness distribution

The deformation phenomenon in ISF process is achieved with stretching and redistribution of the material and the final part volume keeps constant with the initial sheet blank, same like shear spinning process. Because the sine law is used in spinning to predict the thickness distribution of the walls, it is taked into consideration also for incremental sheet forming. The sine law materialised that the deformation in this case is the projection of the deformed sheet metal onto the final part surface [14] – see figure 9 [4].

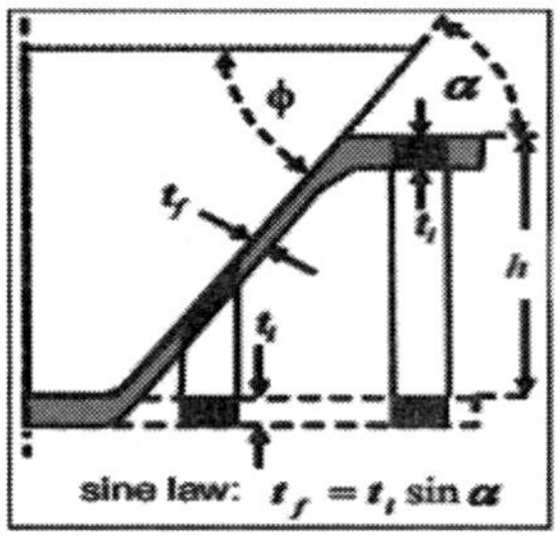

t_f – final part thickness;
t_i – initial sheet thickness;
$\alpha = \phi$ - draw angle;

Fig. 9. Equation for sine law [4].

In recent researchs, a single point incremental forming of a DC04 steel frustum cone on a CNC lathe machine, the sine low predicted the wall thickness in the deformed area with a precision less than ±0.1, without taking into consideration the surface near the fixing area who keeps his thickness [44]. Other researchers discovered that the thickness distribution is nonuniform in all formed area. Young and Jeswiet [4, 45] demonstrated this by measuring the thickness distribution on two wall with 30^0 and 70^0 inclination of a cone made by SPIF process, formed from an 1.21mm initial sheet thickness. The first portion near the flange is the most thinned portion of the wall and then there is section with a thickness of about 1mm. It was calculated the teoretical wall thickness predicted with sine law and compared with the real measurement. It was found that the sine law prediction is bigger than the real wall measured [4, 45]. Severe thinning appears at high draw angles when also the failure can occur, and the sine law can not be used to predict this unexpected effect. Bambach [46] developed a new model for the process kinematics of ISF to predict sheet thickness distribution. It was compared with experimental results and sine law prediction, and it was found that this new model make better thickness estimates than the sine law.

11 Formability

ISF processes are characterized by a small, localized plastic deformation, which is limited to the contact area between tool and blank. It can be said that the incremental forming processes are characterized by increased material formability [4]. It is well-known that the material behaviour and the formability can be described by the maximum value of the draw angle ϕmax [4, 47, 48]. When the ϕ angle increases, the walls of the part become increasingly thinner until the fracure occur. The ϕmax value is considered the maximum angle value up to wich the ISF parts can be processed in good conditions without fracture. Thus, the maximum draw angle ϕmax is a formability benckmark in ISF processes. Knowing the value of ϕmax for diferent material and sheet thickness, can be easy decided at the beginning, if the part can be made in one single pass without failure or shoud be taked other solution like multiple pass or local heating [4].

During the time was made experimental investigations to determine the maximum values ϕmax for diferent materials and sheet thickness and was developed a benchmark product that helps for an easy and sure way to find the ϕmax value. The benchmark product shape is a cone base diameter with standard dimensions [49]: diameter D0 = 72 mm, high H= 40mm, and the flange 10mm –figure 10 [49]. Keeping the base cone dimensions, the experiments must be repeated with various degrees of draw angle ϕ until the tearing occur. The safe angle value before tearing occurs is in fact the maximum draw angle ϕmax.

The formability in ISF is basicaly infuenced by four parameters: sheet thickness, size of the step down, Δz, spindle speed and tool forming diameter [4]. It was observed that with increasing the size of the step down Δz, the sheet formability decreases because of dificults deformation conditions [31, 50].

Talking about spindle speed, Micari [50] found that increasing the tool rotation (spindle speed) is influencing in a positive way the formability due to the local heating effect of the relative motion between the blank and the tool. Micari [50] also explains the tool diameter influence upon the formability -using a small diameter, the strain will be concentrated on a small surface under the forming tool, while using a big diameter, the strain tends to distribute on an extended area.

Thus, the conclusion regarding the tool size is that with decreasing the tool radius, the formability is increased [30, 50]. Hirt et al. [39] demonstrated in a study on DC04 material of three diferent sheet thickness, that the draw angle / formability are strongly influenced by the sheet thickness. Because the ISF principle is shear and distribution of the material, keeping the volume constant from blank to final part, with increasing the sheet thickness, thus the material volume increases, and can be obtained bigger deformation degrees, thus greater formability. Kim and Park [51] found similar conclusion about the sheet thickness influence upon the formability, using AA1050–O material.

Another way to describe the formability or the process limit is to consult the forming limit diagram (FLD). The FLD of conventional sheet metal forming processes are inapplicable for ISF [4]. Thus, some researchers developed a dedicated FLC for incremental forming processes. Filice et al. [47] developed a FLC for SPIF which has not only a negative slope but also a high strains achieved by comparison with conventional forming processes (Fig.11) [47]. Young and Jeswiet [4, 48] developed a composite FLD using more slope from diferent geometries parts.

For every geometry resulted an own FLD, witch was combined in one common FLD. Also here a high strain level was obtained. The FLD's developed helps for a better prediction of the formability or the strategy to be used.

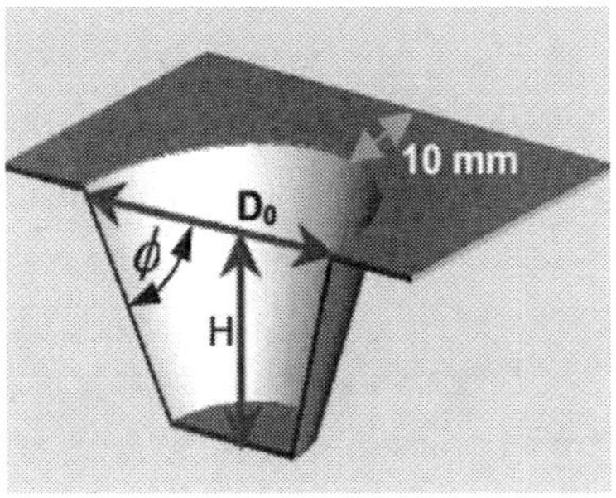

Fig. 10. Benchmark cone dimensions [49].

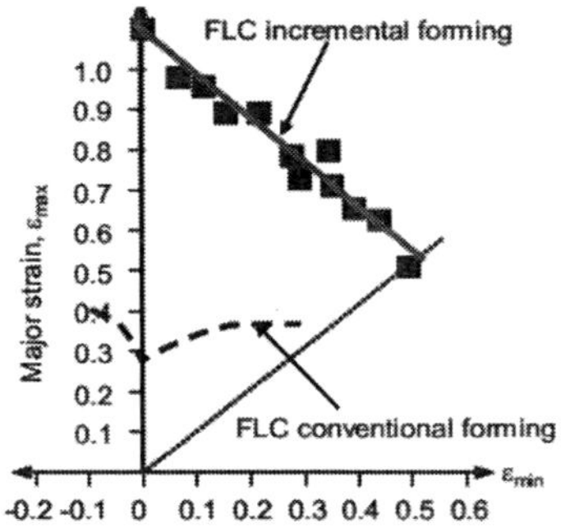

Fig. 11. FLC–ISF vs conventional processes [47].

12 Forces in ISF

Forming forces and strain resulted in ISF processes were studied because of their infuences and process limitations. Teoretical researchs was made by Blaga [52] in his PhD thesis based on the previous researchs made by many authours. It was succeed [52] to determine some teoretical mathematical models for strain and forming forces in ISF.

In ISF processes the tool applies force on the blank. As can be seen in figure 12, there are three components of the force [4, 53]: Fx or Fr (radial force) and Fy or Ft (tangential force) are in horizontal plane, and Fz or Fa (axial force) actions in tool axis direction. Both horizontal components Fx and Fy are repetitive on each level folowed by the tool from top to bottom of the part. It was concluded [54], that the most segnificant value of the force is for Fz, valid for both to SPIF and TPIF.

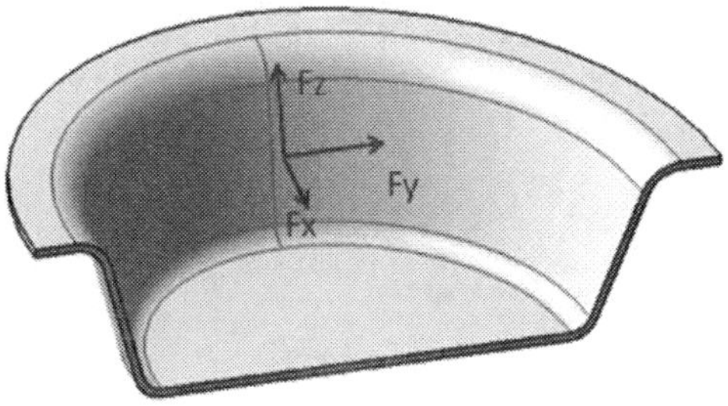

Fig. 12. Force components.

Duflou et al. [55] performed a series of experiments on a reference conical part manufactured by SPIF from five diferent materials. It is measured the forces for differents part configuration and it was concluded with the infuence of sheet thickness, draw angle, tool diameter and step depth Δz on total forces:

- when the sheet thickness is greater, then bigger forces are needed to form the parts;
- increasing the draw angle, the forming forces also increase;
- increasing in step depth Δz, the forces are slight increase, thus no a segnificant infuence. Therefore the incremental step can be increased to reduce the process time, but with high attention to fracture effect [4];
- increasing the tool diameter, the forming forces will have a high amplitude. Otherwise, the increase in tool diameter is used for a better surface finish of the final parts, but can by a limitation due to high forces and failure can occur [4].

Similar results were found by many authors [20], who research also the sheet orientation influence upon the vertical forming forces. To reduce the forming forces of SPIF process,

Duflou [42] used a laser beam for a local heating in the contact zone between tool and sheer. Thus, using the rights parameters for ISF processes, it can help to keep control in terms of forming forces.

13 Stress and strain

In ISF the tool follow a predefined toolpath and step by step it is stretching when passing over surface. Stress and strain occurs in ISF processes and they represent one of the most important criteria in failure prediction. Many reserchers describe one ISF model, in particular for SPIF, build upon membrane analysis [30, 56, 57], based on investigation of the small cantact area beetween tool and sheet. Jackson and Allwood [58] examinated the ISF mechanism using copper sheets. It was made experiments using both SPIF and TPIF to evaluate the strain distributions through the thickness of the sheets. The measured strain, valid for both configuration processes, shows that the material is streatched and shared in a horizontal plane (perpendicular to tool direction) and only sheared in tool axis direction [30]. The most significant strain component is sheared in the tool axis direction, but all strain components increase passing on every level untill the final part is finished. The wall thickness can not be precise predicted by the sine law in case of increasing stretching and shear perpendicular to the tool direction [30]. Usually in ISF, not only the sheet wich is touched the tool, but also the neighbourhood area to the contact zone is affected, and thus also this area will be analized. Commonly, to analyze the stress strain in ISF, a small element in the contact zone is chosen, as shown in figure 13 [59]. Here are presented the small element analised (in red) and the three stress components σr, σθ and σØ along their directions r, θ and Ø. Because the stress along θ direction causes the circumferential through-thickness shear, it is considered dominant shear efect, while the other two components can be ignored since they do not cause evident plastic deformations [60]. The conclusion regarding the stress evolution is that it increases as the forming tool moves along its own path going to finish, thus the stresses increase with respect to time [1].

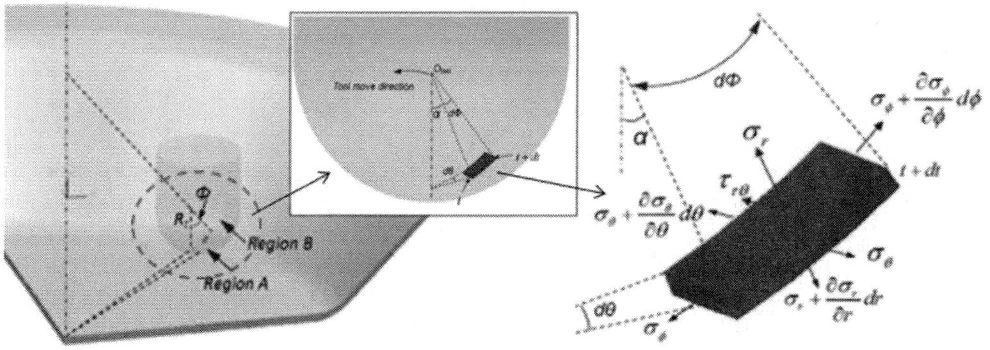

Fig. 13. Small element analized and stress components [59].

14 FEM simulation in ISF

Nowadays the finite element method (FEM) is one of the best tools used to improve the part specification from the virtual development phase. Because the traditionals experiments are usualy time consuming and expensive, to reduce the time and costs, the FEM simulation is also used for ISF processes. Usually are used dedicates FEM softwares like Abaqus, LS-Dyna and Ansys. The main topics studied using FEM simulation are forces analisys [60], springback amplitude and toolpath correction [61], deformation analisys [62], influence of the tool parameter and tool path on the deformation behaviour [30] and so on.

To implement a FEM simulation, must be folowed several steps [14]:

- design CAD model of the part and fixing devices;
- simulate and generate the toolpath in CAM software;
- design finite element assemble in FEM software;
- define materials and applying boundary conditions;
- running simulation and result processing.

Because in ISF, the tool is in continuous motion, the FEM simulation in such a model took a huge computation time. To improve this, explicit solver was used, but also time scaling and mass scaling of the model. Also the mesh parameters was analized [14] and was concluded that using a adaptive mesh or small element dimension mesh, better result will be obtained than using a big element dimension.

15 Future research directions

ISF processes have been made big steps forward because of its advantages. Worldwide, it has been researched basically topics as follow: forming tool, toolpath and tool parameters, materials, parts quality, formability, forces, stress/strain, and also FEM simulation tools. The ISF processes are not enogh stable and precisely to capture on a large scale the manufacturers' attention. Thus, future investigations are needed. In figure 14 are ilustrated future possible research topics.

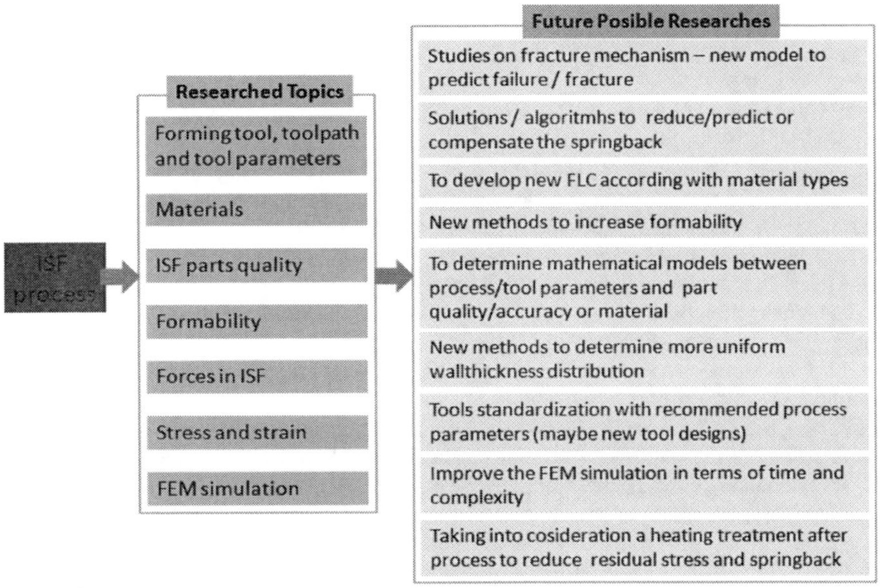

Fig. 14. Actual and future possible research in ISF domain.

16 Conclusion

Sheet incremental forming is a dieless forming technology which has been widely studied in the last years. Because it is cheap, easy to implement and it has low tooling cost and a high flexibility, the ISF processes can be used to manufacture unique parts, prototypes or test parts in small batch volumes. It can be also used for manufacturing complex shape parts from hard to form materials. The ISF is a time consuming process thus it is not suitable for high volume production. This paper presented the advantages and disadvantages of the process and the domains where it is applicable. It was made a short introduction in ISF processes, talking about the working principle, and the types of ISF. A wide range of equipment can be used to implement one ISF process: CNC lathe (only for axis-symmetric parts), CNC milling machine, industrial robots, Stewart platforms, hexapods, but also dedicates equipment created by various manufacturers. The main topics investigated by the researchers were discussed, and the results for each topic on ISF were presented: forming tool, tool parameters and toolpath, materials used to be deformed, ISF parts quality (surface roughness, wall thickness distribution, springback and accuracy), formability, stress and strain, and forces in ISF. In conclusion the Incremental Forming processes have been made big step forward in rapid manufacturing of sheet metal prototypes. Future researchs will be focused on: fracture mechanism, solution to reduce/predict or compensate springback, development of new FLC and methods to increase formability, mathemathical models between process/tool parameters and parts quality, wallthickness distribution, tools standardization, FEM simulation and maybe implementation of a heating treatment after process to reduce residual stress or springback.

References

1. Y. Kumar, S. Kumar, , Adv.in Mat. Forming and Joining 5th International and 26th All India Manufacturing Technology, Design and Research Conference, AIMTDR 2014, **29** (2014)
2. W.C. Emmens, G. Sebastiani, A.H. van den Boogaard, J. Mater. Process. Technol., **210** (8), 981 (2010)
3. M. Oraon, V. Sharma, AMAE Int. J. on Production and Industrial Engineering, **1**(1), 31 (2010)
4. J. Jeswiet, F., Micari, G. Hirt, A. Bramley, J. Duflou, J. Allwood, CIRP Annal Manuf. Technol. **54**(2), 88 (2005)
5. J. Jeswiet, M. Geiger, U. Engel, M. Kleiner, M. Schikorra, J. Duflou, R. Neugebauer, P. Bariani, S. Bruschi, CIRP J. Manuf. Sci. Technol., **1,** 2 (2008)
6. M. Amino, M. Mizoguchi, Y. Terauchi, T. Maki, 11th ICTP, **81,** 54 (2014)
7. L. Lihui, L. Kangning, G. Cai, X. Yang, C. Guo, G. Bu, Manufacturing Rev., **1**, 9 (2014)
8. I. Neagoe, A. C. Filip, M. A. Vasiloni, AMM., **371**, 168 (2013)
9. J. M. Allwood, N. E. Houghton, K. P. Jackson, AMR, **6-8**, 471 (2005)
10. H. Meier, O. Dewalt, J. Zhang, Steel Res. Int., **76**(2-3), 167 (2005)
11. T.Tuominen, *Method and apparatus for forming three-dimensional shapes in a sheet metal,* Patent EP 1 560 668 A1 (2004)
12. N. D. Joao Torrao, A. F. Jorge Ferreira, J. Ricardo Alves de Sousa, 11th Port. Conf. on Autom. Control, **321,** 292 (2014)
13. M. Callegari, D. Amodio, E. Ceretti, C. Giardini, *Industrial Robotics: Progrming, Simulation, and Application* (edited by Low Kin Huat, Intech Open Science, 493, 2006)
14. D. H. Nimbalkar, V. M. Nandedkar, IJERA, **3**(5), 39 (2013)
15. Le Van Sy, *Modeling of single point incremental forming process for metal and polymeric sheet,* PhD thesis, University Of Padova, Italy, 2009
16. H. Iseki, JSME INT J C-MECH SY, **4**, 486 (2001)
17. B. Jurisevic, K. C. Heiniger, K. Kuzman, IDDRG, 139 (2003)
18. M. Geiger, F. Vollertsen, Annals of CIRP, **42**(1), 301 (1993)
19. A. Mohammadi, H. Vanhove, A. Van Bael, J. Duflou, AIP Conference Proceedings, **1567**(1), 864 (2013)
20. Z. Liu, Y. Li, P.A. Meehan, Int. J. Precis. Eng. Manuf., **14**(11), 1891 (2013)
21. M. Yamashita, M. Gotoh, S-Y. Atsumi, J. Mater. Process. Technol., **199**(1-3), 163 (2008)
22. L. R. Zhou, Adv. Mater. Res., **239-242**, 940 (2011)
23. A. N. Bramley, *Incremental sheet forming process for small batch and prototype parts,* in Idee-Vision-Innovation, edited by F. Vollersten and M. Kleiner, Verlag Meisenbach, 95 (2001)
24. L. Pekka, V. Tapio, S. Mika, Physics Procedia, **78**, 312 (2015)

25. D. Fritzen, A. Daleffe, J. Castelan, L. Schaeffer, Key Eng. Mater, **554-557**, 1419 (2013)
26. A. Daleffe, L. Schaeffer, D. Fritzen, J. Castelan, Key Eng. Mater, **554-557**, 195 (2013)
27. R. Hino, K. Kawabata, F. Yoshida11th ICTP, **81**, 2330 (2014)
28. S. Kurra, S. P. Regalla, J. Mater. Res. Technol., **3**(2), 158 (2014)
29. P.A.F. Martins, L. Kwiatkowski, V. Franzen, A.E. Tekkaya, M. Kleiner, CIRP Ann. Manuf. Technol., **58**(1), 229 (2009)
30. S. Gatea, H. Ou, G. McCartney, Int. J. Adv. Manuf. Technol., **87**(1), 479 (2016)
31. E. Hagan, J. Jeswiet, J. Eng. Manuf., **218**(Part B), 1307 (2004)
32. A. Bhattacharya, K. Maneesh, N. Venkata Reddy, L. Cao, J. Manuf. Sci. Eng., **133**(6), 061020 (2011)
33. V. Oleksik, A. Pascu, C. Deac, R. Fleacă, O. Bologa, G. Racz, Int J Mater Form, **3**(S1), 935 (2010)
34. M. Durante, A. Formisano, A. Langella, F. M. Capece Minutolo, J. Mater. Process. Technol., **209**(9), 4621 (2009)
35. J. R. Duflou, B. Lauwers, J. Verbert, Y. Tunckol, H. De Baerdemaeker, 8th ESAFORM Conf. on Material Forming, 675 (2005)
36. G. Ambrogio, L. Filice, L. Fratini, F. Micari, Proc. of the 6th Conference ESAFORM., 175 (2003)
37. L. Junchao, P. Geng, J. Shen, Int. J.Adv. Manuf. Technol., **68**(9-12), 2637 (2013)
38. G. Ambrogio, L. De Napoli, L. Filice, M. Muzzupappa, Proc. of 7th ASME ICSDA, **3**, 339 (2004)
39. G. Hirt, J. Ames, M. Bambach, R. Kopp, Annals of CIRP, **53**, 203 (2004).
40. B. Lu, J. Chen, H. Ou, J. Cao, J. Mater. Process. Technol., **213**(7), 1221 (2013).
41. H. Meier, V. Smukala, O. Dewald, J. Zhang, Proceedings of the 12th International Conference on Sheet Metal, **344**, 599 (2007).
42. J. R. Duflou, B. Callebaut, J. Verbert, H. De Baerdemaeker, CIRP Ann. Manuf. Technol., **56**(1), 273 (2007).
43. J, R. Duflou, B. Callebaut, J. Verbert, H. De Baerdemaeker, Int. J. Mach. Tools Manuf., **48**(5), 543 (2008).
44. I.Neagoe, I. Neagoe, A. C. Filip, A. Manolescu, AMM., **657**, 142 (2014).
45. D.Young, J. Jeswiet, IMECHE part B, J. of Eng. Manuf., **218**(B11), 204 (2004).
46. M. Bambach, J. Mater. Process. Technol., **210**, 1562 (2010).
47. L. Filice, L. Fantini, F. Micari, Annals of the CIRP, **51**, 199 (2002).
48. D. Young, J. Jeswiet, IMECHE part B, J. of Eng. Manuf., **219**, part B, 1-6 (2005).
49. F. Micari, G. Ambrogio, *A common shape for conducting incremental forming tests*, 1st Incremental Forming Workshop, University of Saarbrucken, (2004).
50. F. Micari, *Single Point Incremental Forming: recent results,* Seminar on Incremental Forming, Cambridge University (2004).
51. Y. H. Kim, J. J. Pack, J. Mater. Process. Technol., **130-131**, 42 (2002).
52. A. Blaga, *Contributions to the incremental forming of thin metal sheets,* PhD thesis – abstract, "Lucian Blaga" University Of Sibiu, Romania (2011).
53. R. Aerens, P. Eyckens, A. Van Bael, J. R. Duflou, Int. J. Adv. Manuf. Technol., **46**, 968 (2009).
54. J. Jeswiet, J. Duflou, A. Szekeres, AMR., **6-8**, 449 (2005).
55. J. R. Duflou, A. Szekeres, P.Vanherck, AMR., **6-8**, 441 (2005).
56. M. Silva, M. Skjoedt, P. Martins, N. Bay, Int. J. Mach. Tools Manuf., **48**, 73 (2008).
57. P. A. F. Martins, N. Bay, M. Skjoedt, M. B. Silva, CIRP Ann. Manuf. Technol., **57**(1), 247 (2008).
58. K. Jackson, J. Allwood, J. Mater. Process. Technol., **209**(3), 1158 (2009).
59. B. Lu, Y. Fang, D.K. Xu, J. Chen, H. Ou, N.H. Moser, J. Cao, Int. J. Mach. Tools Manuf., **85**, 14 (2014).
60. F. M. C. Minutolo, M. Durante, A. Formisanom, A. Lagella, 2nd I*PROMS Virtual International Conference, 229 (2006).
61. J. Mo, F. Han, P. Gong, Int. J. Adv. Manuf. Technol., **64** (9), 1239 (2012).
62. H. Iseki, J. Mater., Process. Technol., **111**(1-3), 150 (2001).

DOI: 10.1051/matecconf/201712103018

Some aspects regarding the static behavior of basalt in the machine building industry

Ilie Octavian Popp[1,*]

[1]Lucian Blaga University of Sibiu, Department of Industrial Machines and Equipment, 550025 Victoriei 10, Sibiu, Romania

Abstract. Basalt is a material available in big quantities in nature. It has a low price and can be used in various domains. It can be relatively easily extracted from natural deposits and processed by means of melting. Melted and recrystallized basalt has demonstrated promising behavior with regards of its use in manufacturing, as reported in the literature. This work aims to present some aspects regarding the opportunity of using basalt in manufacturing the subassemblies of machine-tools. The research was made using grey cast iron and steel parts as reference for a comparative study. The researches presented in this work were focused on analyzing the behavior of basalt parts submitted to static stress. Testing methodology, specific measuring apparatus used, experimental test data records, data processing and resulting conclusions are presented in the paper.

1 Introduction

During its working regime, a machine-tool is subject to various loads which confer the predominant form of stress. The most important loading factors are:
- the weight of supporting and mobile assembly, of workpiece rigs acting as a static working load;
- the cutting forces, the running mobile assemblies, acting as a dynamic working load;
- frictional loss, the heat recessed while chipping cutting determine thermal deformation.

Structural elements of the machine-tool are subject to stresses in different ways due to all previously presented factors, consequently a lot of research work regarding the use of different materials for manufacturing machine-tools elements is reported in the literature [1-4]. There are quite few studies reporting basalt as building material for machine-tools elements. However, a literature survey [5-9] shows a tendency to use mineral casting instead of iron casting, due to their superior characteristics. A composition containing basalt, spodumenne and fly ash was found to have the highest flexural strength and lowest thermal expansion coefficient [1, 7].

The main purpose of the presented research was to demonstrate that when taking into consideration machine-tools beds and other structural parts, basalt can be accepted as building material. To achieve this goal, the stress behavior for basalt parts must be assessed. Consequently, one of the main contributions of this work is considered proposing basalt as

* Corresponding author: ilie.popp@ulbsibiu.ro

solution for structural parts in the machine building industry and to develop an experimental set of test for evaluating its static behavior.

Nowadays there is a technological limit of 800 mm as main dimension limit for basalt parts. Consequently, a complex machine structure from basalt (like housing, body, trunk, lug support, base) must be built using many basic parts (rungs (draw bar), plates) joined together.

2 Tests and simulations

2.1 General remarks

In order to assess the stability range of the components within the structures, static stiffness is often used as a characteristic input. For a given part, made from different materials, static stiffness represents the capacity to impede the external forces which tends to deform it. The static stiffness can be defined as the ratio between the force size which acts in assigned direction and the traveling size of respective structure [10].

It comes from stiffness definition the general approach mode used in test methods: taking measurement of displacement generated in closing point by a force well defined in size, direction and sense.

The analysis of the basalts behavior was performed on a representative structural element. Based upon structural machines analysis and technological limits of machining basalt parts, it was selected a natural embedded rectangle plate having dimensions: 250 x 250 x 40 mm. the shape and dimensions were considered by the author as very common for the structural elements encountered within machine building industry. The laboratory equipment used for unfolding the experimental tests includes: six channels Hottinger bridge, Hottinger force cell, signal convertor and Hewlett Packard graphic plotter (Fig. 1).

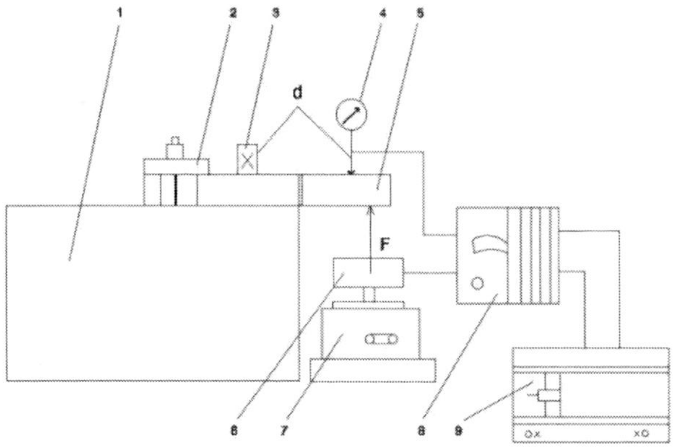

1. Support
2. Structure fasteners (bracket)
3. Magnetic support fixed to the structure
4. Inductive displacement transducer
5. Basalt structure
6. Power transducer (Hottinger)
7. Static loading device
8. 6-channel amplifier bridge (Hottinger)
9. Graphic x-y plotter (Hawlett-Packard)

Fig. 1. Sketch of the test stand

2.2 Experimental tests

The design on the test sample is presented in figure 2. The schematic representation from figure 2 depicts the loading forces positioned on the part, with their directions and application points, situated on the upper face of the part.

In order to load the test sample, two forces of 250 daN were applied consecutively as presented in figure 2. The application points were situated as follows: F1 eccentric at 40 mm from the edge of the part and F2 centric at 125 mm from the table's edge).

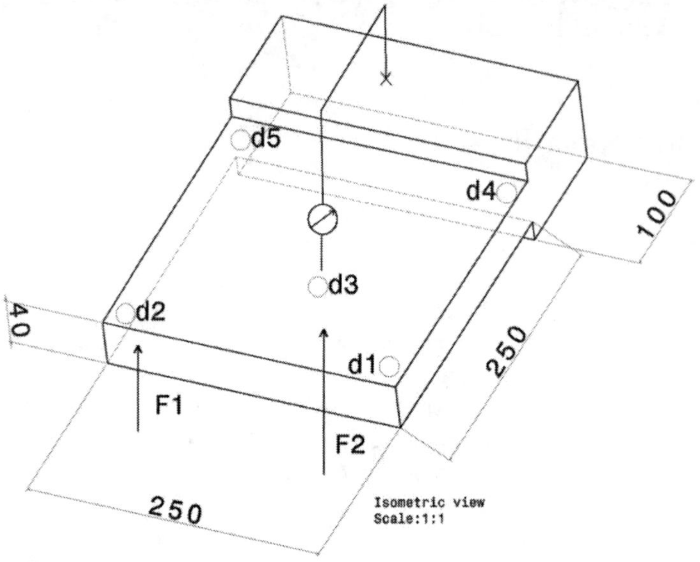

Fig. 2. Test sample (naturally recessed plate type element).

The deformations were measured in all the five point from figure 2, situated symmetrical on the upper face of the test sample. Specific methodology was used to determine the the indicators of directly static stiffness.

The synthetic results of performed tests are presented in table 1 which enclose the values of static elasticity of flexure recorded in all measuring points.

Table 1. Synthetic list of static flexure elasticity values measured

Measurement point	**Static elasticity (μm/daN)**	
	F1	**F2**
d1	0.1917	0.1215
d2	0.3186	0.1161
d3	0.1242	0.0675
d4	0.0216	0.0108
d5	0.0108	0.0108

2.3 FEM analysis

After running the experiments, the Finite Element Method (FEM) was used as theoretical mean for simulation, in order to perform a comparison between experimental and simulated results. The complete set of equations which generally describes the structure behavior has the following form [10, 11]:

$$[M]\{\ddot{u}\} + [C]\{u\} + [K]\{u\} = \{R\} \quad (1)$$

where:
[K] - stiffness matrix;
[M] - mass matrix; it contains the terms which defines the net weight of component parts or other implicated mass in dynamic behavior of structure;

[C] - damping coefficient matrix;
u} - branch point displacement vector;
{u} - branch point speed vector;
{ü} - branch point acceleration vector;
{R} - external loading vector.

Particularly on static analysis the speed and acceleration vector are nulled and the terms of loading vector are constant. Static behavior analysis is applied to evaluate the values of structure deformations and stresses. The analysis was by means MARC, a finite element analysis software package. The results obtained running this software include complete information about dimensional variation and about normal and tangential stress distribution.

The initial conditions were the following:

- sample plate dimensions 250 x 250 x 40 mm;
- material: basalt with elasticity modules E = 85230 MPa, contraction coefficient Λ = 0.33;
- load F = 2500 N applied in the two points indicated in figure 1.

The static elasticity modules value was experimentally measured using the electrical strain gauge method. Due to the geometrical symmetry and of the loading and supporting mode the analyzed structure was transformed in a finite element mesh containing hexahedral tridimensional solid elements resulting 1131 mesh point having 3393 degrees of freedom.

The results of FEM analysis for both tests are presented in figures 3 and 4, where in the left side of the figures are presented the displacements u_y [mm] and on the right side the stresses Φ_{11} [MPa].

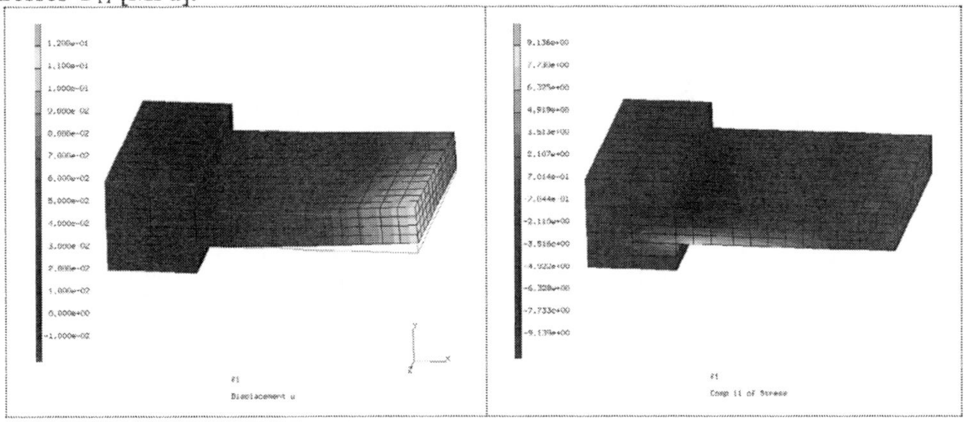

Fig. 3. Displacements u_y [mm] and o stresses Φ_{11} for F_1 [MPa].

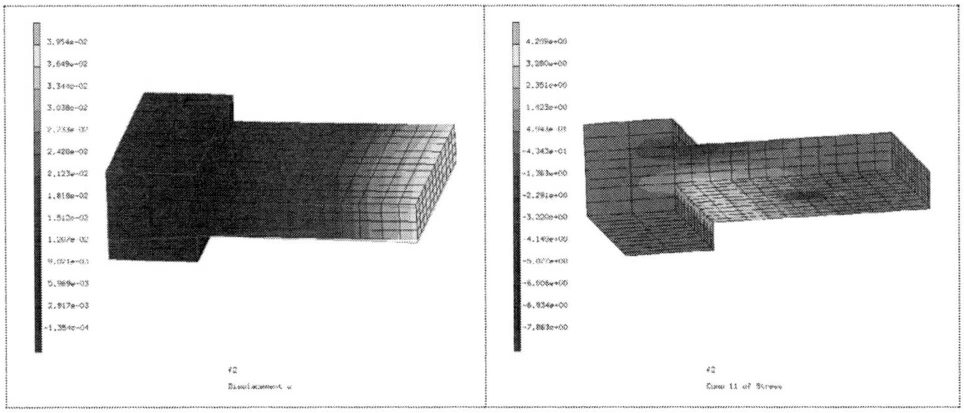

Fig. 4. Displacements u_y [mm] and o stresses Φ_{11} for F_2 [MPa].

The analytical solution for the main stress component on OX direction in F_2 loading case is:

$$\sigma_{11}=\frac{M}{W}=\frac{Fl}{\frac{bh^2}{6}}=\frac{125\times 2500}{250\times 40^2}6=4.687MPa \quad (2)$$

The maximum value of Φ_{11} determined by FEM is 4.209 MPa. The difference may be caused by the stress located phenomena in restraining zone where the geometrical structure is different from ideal structure. The displacement values are influenced by elastic modules E and structural geometry.

The analytical solution of d_3 point displacement equation is:

$$d_3=u_{y(3)}=\frac{1}{3}\frac{Fl^3}{EI_z}=\frac{1}{3}\frac{2500\times 125^3}{85230\times\frac{250\cdot 40^3}{12}}=0.014322mm \quad (3)$$

The FEM determined solution is u_y = 17,977 :m.

A comparison between measured displacements in each point, obtained by running the FEM analysis and by means of experiments is presented in table 2.

Table 1. FEM determined and measured displacements

Measuring point	Displacement u_y [:m]			
	FEM		experimental	
	F1	F2	F1	F2
d_1	50.780	31.143	47.925	30.375
d_2	85.516	31.143	79.651	29.025
d_3	31.142	17.977	31.051	16.875
d_4	6.95	2.925	5.413	2.715
d_5	3.015	2.925	2.712	2.715

3 Conclusion

Through the bibliographic study we have ascertained that the problem of the behavior of the basalt to the specific stress for its use in the manufacturing of the structural elements of the machine tools is not yet addressed.

This fact has led me to research on the use of basalt in the basic structure of machine tools, in the manufacture of structural parts as frames, pillars, housings, supports, etc.

In this paper is presented a part of the study of the static behavior of the basalt subjected to specific machine tool requirements, both experimentally on a basalt plate structure and an analytical model, for comparing the results.

The results of the experiments are quite similar with the results obtained by means of theoretical analysis and FEM simulations. The deformation amplitudes have shown maximum values in the near vicinity and direction of the lading forces application points. Static stiffness has a linear variation on restraint plate structures. When the sample part is subject to eccentric loading forces, the increase in the deformation values is significant (with a size factor of 2.5). If the sample is subject to centric loading forces, symmetrical deformation occurs.

Finally, comparing the test results with the one obtained for different materials, basalt demonstrated better behavior than steel or grey iron cast parts, which make it suitable for machine-building industry.

References

1. H.-C. Möhring, C. Brecher, E. Abele, J. Fleischer, F. Bleicher, *Materials in machine tool structures*, CIRP Annals - Manufacturing Technology, 64 (2), 725–748 (2015)
2. V. Ducatti, R. Lintz, J. Santos, Int. RILEM Conf. on the Use of Recycled Mat. in Building and Structures 925-934 (2004)
3. G.P. Zhang, Y.M. Huang, W.H. Shi, W.P. Fu, Int. J. of Machine Tools & Man., 699-706 (2003)
4. C. Gornic, *Bazaltul materie prima pentru piesele destinate masinilor unelte* (*Basalt raw material for machine tool parts)* (ICSITMU – Titan, Bucuresti, 1989)
5. N. Kepczak, W. Pawlowski, M&ME., **17**(4), 285–289 (2013)
6. C. Bruni, A. Forcellese, F. Gabrielli and M. Simoncini, J. of Mat. Proc. Tech., 493–499 (2007)
7. H. Haddad, M. Al Kobaisi, Composites **B**, 3061–3068 (2012)
8. F. Cortes, G. Castillo, Mat. and Design **28**, 1461–1466 (2006)
9. T. Erbe, J. Król, R. Theska, Annual Meeting of the American Soc. for Precision Eng. and Twelfth ICPE, **23** (2008)
10. C. Minciu, s.a. *Comportarea statica a masinilor unelte (Static behavior of machine tools),* (T.C.M.M. 11, Editura Tehnica, Bucuresti, 1995)
11. V. Nastasescu, *Metoda elementelor finite - Finite element method*, (Editura Academiei Tehnicii Militare, Bucuresti, 1995)

Systemic analysis of the caulking assembly process

Claudiu Rodean[1*], *Livia Dana* Beju[2], and *Paul Dan* Brindasu[2]

[1] Continental Automotive Systems SRL, 8 Salzburg Street, Sibiu, Romania
[2]Lucian Blaga University of Sibiu, Faculty of Engineering, 4 Emil Cioran Street, Sibiu, Romania

Abstract. The present paper highlights the importance of a caulking process which is nowadays less studied in comparison with the growing of its usage in the automotive industry. Due to the fact that the caulking operation is used in domains with high importance such as shock absorbers and brake systems there comes the demand of this paper to detail the parameters which characterize the process, viewed as input data and output data, and the requirements asked for the final product. The paper presents the actual measurement methods used for analysis the performance of the caulking assembly. All this parameters leads to an analysis algorithm of performance established for the caulking process which it is used later in the paper for an experimental research. The study is a basis from which it will be able to go to further researches in order to optimize the following processing.

1 Introduction

Nowadays technologies require fast and efficient processing which ensure high quality and low costs in addition with the process safety. Among the operations that ensure these needs it's included the caulking process. Caulking operation is an advanced process used to obtain permanent assemblies in various fields such as: automotive, aeronautics, medical or energy.

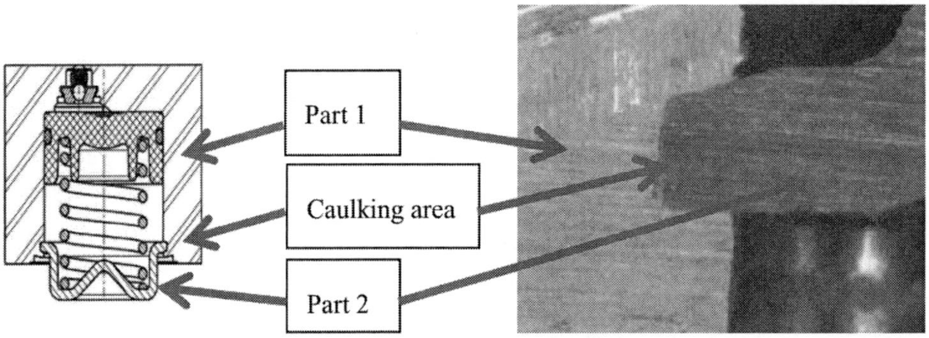

Fig. 1. Presentation of the caulking area.

* Corresponding author: claudiu_rod90@yahoo.com

Caulking is the deformation operation which consists of tightening press of a piece of material around another to avoid their relative movement (Fig 1). Part 2 is placed on Part 1 and a caulking tool strikes Part 1 in order to deform an amount of material in such a way, that the resulting assembly becomes non-removable.

Examples of domains where the caulking process is used are the brake systems (Fig. 2) and the shock absorbers (Fig. 3).

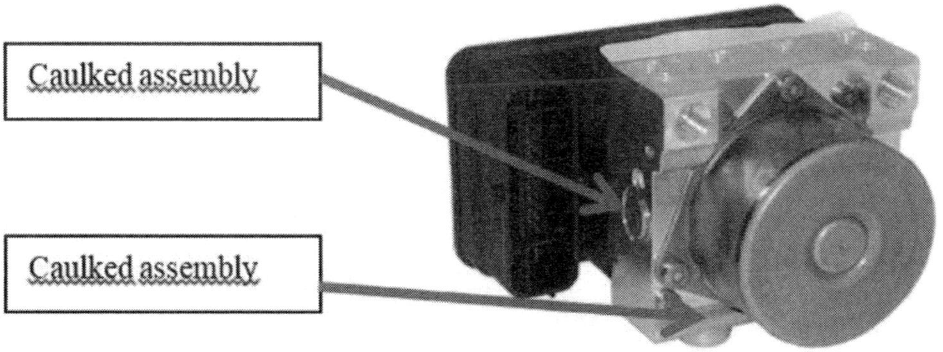

Fig. 2. Caulked assembly used at brake systems

Fig. 3. Caulked assembly used at shock absorbers systems

In the automotive industry there are certain products that require special plastic deformation. In this area is framed the caulking operation which is started to be used on a larger scale operation because of the advantages it presents:

- The operation can be performed on CNC equipment.
- No material loss occurs.
- Replaces more complex operations, which have a higher cost.
- Operation allows the obtaining of accurate geometrical parameters of the assembly.
- The simulation using finite element method allows the optimisation of the process.
- It is a low cost operation.

Although at present, the caulking process begins to be used more and more, the technical literature do not analyze this process very much.

Sin and Lee [1] evaluated qualitatively a caulking through numerical analysis and inspecting a plastically deformed shape. Following the caulking analysis, the structural responses were considered. In addition, three design variables related to the manufacturing process were defined, and the effects of design variables with respect to pull-out strength, caulking depth, and maximum stress were obtained. Finally, the optimum design maximizing the pull-out strength was suggested.

Park [2] studied an inner tie-rod (ITR) end-assembly module which is widely used in the steering systems of automobiles. The ITR module is composed of a ball-stud, ball-seat, and socket and manufactured by a caulking process. To ensure that the final ITR module is qualified for application, the caulking process of the ITR module was optimized to satisfy certain requirements, including the allowable displacement, pull-out force, tilting angle, and torque of the ball-stud. Accordingly, this paper investigates the deformation characteristics of both the caulking and pull-out processes using a finite element method analysis and clarifies the effect of the forming parameters on the allowable displacement, pull-out force, and tilting angle. The finite element method analysis results are also compared with experiments to validate the finite element method analysis.

Beck [3] study the heating tubes from the pressurized water reactor. The tightness is tested for the repaired tubes done using a roll-expanded plug.

Calcaletty [4] makes an analysis of the metrological leak detection method based on the application of mass spectrometry with tracer gas (generally helium).

Gorash [5] presents a numerical study involving the deformation of contact faces in the metal-to-metal seal in a typical pressure relief valve. The study method is the fluid pressure penetration technique.

An ideal caulking assembly must fulfil the following requirements:

- Precise geometrical parameters and high surface quality of the deformed volume of material.
- An adequate stress state along the entire caulking length.
- The assembling must have very good tightness.
- Assembly must withstand to fatigue.
- Low cost of the operation.

During a caulking assembly process, several problems may occur:

- The assembly does not acquire the imposed geometrical parameters.
- Sealing cracks appear along the caulking zone.

An assembly process can have several reasons which can cause these problems:

- The caulking force is out of range.
- The active area of the caulking tool is damaged.
- The device which holds the units is damaged.
- The press machine which drives the caulking tool can be damaged also etc.

The paper proposes to develop a method for the study of the performances of caulking assemblies by considering the entire caulking process as an Input-Output data system.

2 Caulking operation viewed as a system

Caulking process is a typical one which has input data such as geometrical and technological parameters and output data like geometrical parameters after caulking and values to be measured (Fig. 4).

Fig. 4. Caulking operation – systemic perspective

Due to the fact that there is a big list of requirements which it is asked by different customers, we intend to establish a list of technological demands which the caulking process must respect:
- Continuous deformation along the entire profile.
- Relative position between the parts and the caulking tool.
- Tool geometry.
- Caulking force.

In order to have a better understanding of the caulking process, a picture with all the geometrical parameters (before the caulking deformation process) is presented in figure 5.

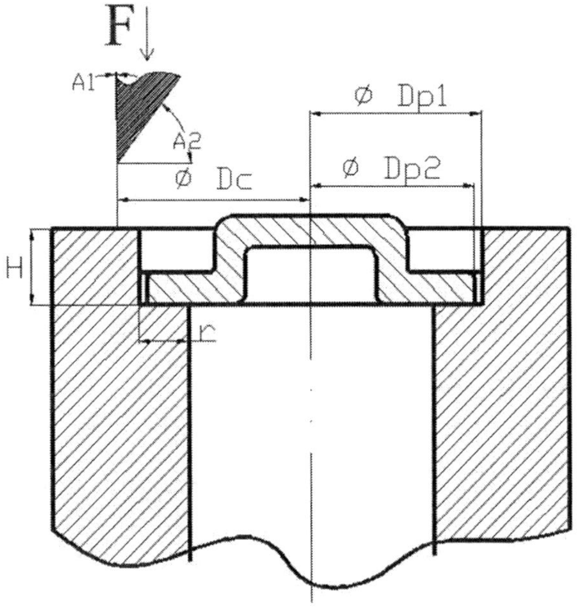

Fig. 5. Geometrical parameters of the caulking process (before deformation)

The geometrical parameters of the two parts, defined before the caulking process is done (the "Input data") are:
- H – the height of the material which is deformed during the caulking process.
- ØDp1 – the diameter of the part 1, which will be deformed in the caulking aria.
- ØDp2 – diameter of the part 2, which is going to be assembled in the caulking aria.
- ØDc – tool diameter.
- r– constructive parameter of part 1.
- A1 – clearance angle of the tool.
- A2 – rank angle of the tool.

All this parameters are independent one from each other and in this way they have a big influence on the caulking process.

Together with the geometric parameters, the caulking process demands also a list of technological parameters:
- V- The speed (velocity) of the caulking process.
- T - Tool pressing time.
- F - Caulking force.
- The accuracy of the devices: perpendicularity; linearity etc.

After the caulking process is performed, there are several output parameters which occurs.

The geometrical parameters can be seen in the following picture (Fig 6):

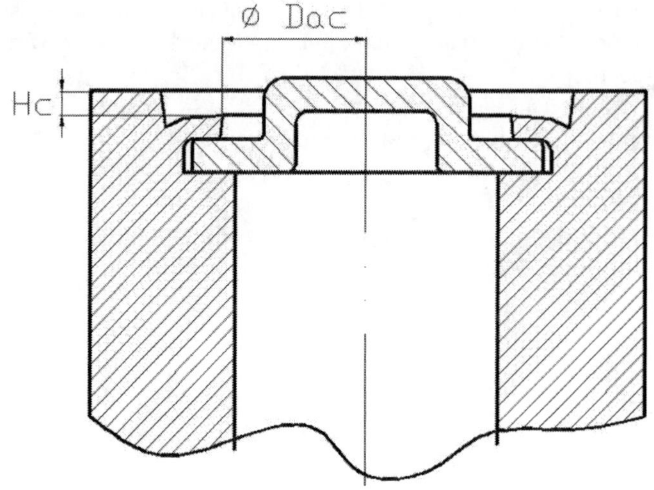

Fig. 6. Geometrical parameters after the caulking process

The output geometrical parameters are:
- Ø Dac – the actual diameter of the caulked material.
- Hc – the height of the caulked material which covers the second part.

Together with the output geometrical parameters, the caulking process includes also a list of output technological parameters:
- Sp_gas - the speed of the decreasing pressure inside the caulked assembly – gas conditions.
- Sp_liq - the speed of the decreasing pressure inside the caulked assembly – liquid conditions.
- Bp - the limit of the burst pressure.

Table 1. Caulking parameters

Input Parameters		**Output parameters**	
Name	**Measuring unit**	**Name**	**Measuring unit**
H	[mm]	Ø Dac	[mm]
ØDp1	[mm]	Hc	[mm]
ØDp2	[mm]	Sp_gas	Mbar*l / s
ØDc	[mm]	Sp_liq	Bar/s
r	[mm]	Bp	Bar
A1	[°]		
A2	[°]		
V	[m/s]		
T	[s]		
F	[KN]		

The definition of all the caulking parameters (Table 1.) allows us to perform a systematic analysis of the process.

3 The functional analysis of the caulked assembly

Starting from the requirements presented before, we aim to establish the assembly functions (Table 2.) and then to determine which ones are more important. This analysis allows us to design the research in order to optimize the caulked assembly.

Table 2. The functions of the caulked assembly

Requirement	Function
Precise geometrical parameters and high surface quality of the deformed volume of material	Precision
An adequate stress state along the entire caulking length	Resistance
Assembly must withstand to fatigue.	Reliability
The assembling must have very good tightness	Tightness
Low cost of the operation	Cost

The ranking of the functions is evaluated taking into consideration the importance of each one in the life-cycle of the final assembled product (Product lifecycle methodology).

The adopted method for this evaluation is "the triple cross method" (Fig 7).

	F2	F3	F4	F5	S	R		P [%]		
									F1	-precision
F1	1	1	1	2	5	0.125	F1	12.5	F2	-resistance
	F2	2	2	3	10	0.25	F2	25	F3	-reliability
		F3	2	3	10	0.25	F3	25	F4	-tightness
			F4	3	10	0.25	F4	25	F5	-cost
				F5	5	0.125	F5	12.5		
				Σ	40	1		100		
									Importance factor	

Fig. 7. Ranking functions; triple cross – method

Each function is compared with another one and receives some points in the following way:

- If a function is more important than another one it receives 3 points.
- If a function has the same importance with another one it receives 2 points.
- If a function is less important in comparison with another one it receives 1 point.

The ranking shows that the tightness, resistance and the reliability are the most significant functions for the caulked assembly.

4 The actual methods used for the evaluation of the caulked assemblies

As it was mentioned before, the sealing between the two parts which had suffered the caulking process is one of the most important requirements. Due to the fact that caulking process is very often used in brake systems from the automotive field, a simple inadvertence during the caulking process can lead to important damages to the functionality of the final product (Fig. 8).

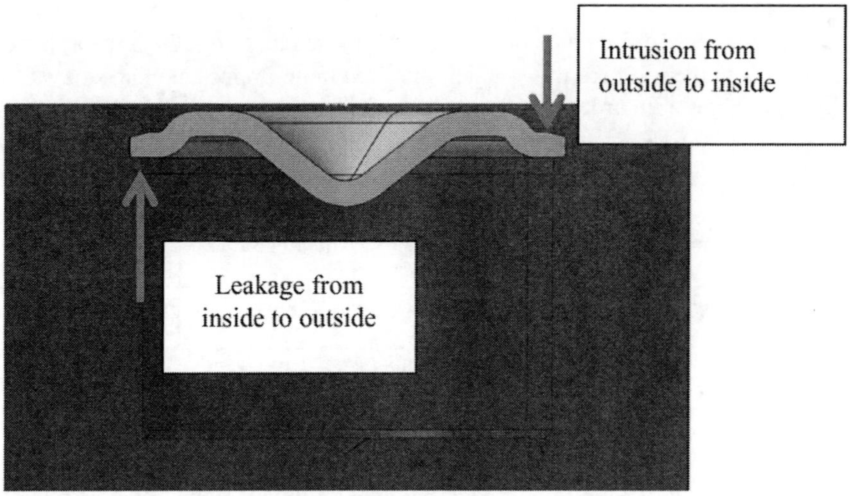

Fig. 8. Possible sealing problems at caulked assembly

There are two possible problems if the sealing was not performed in good conditions. A first one is the possible intrusion of gas or liquid from outside to inside the caulked assembly. The second one is a possible leakage of gas or liquid from inside to outside the caulking assembly.

In the caulked assembly the tightness is checked to assure that there is no leakage between the inside of the assembly and the outside.

Checking out the tightness of a product obtained after a caulking process implies the measurement of a fluid flow coming out of a defect or entering through it. Leak detection is a non-destructive test performed to verify the accordance of products and components with the prescribed specifications, by using techniques, which do not damage their characteristics or performances. The leak test is continuously implemented in a production process, to check the agreement of the goods to the tightness requirements.

Considering the functionality of the final product obtained after caulking process, the tightness requirements can be checked with liquids or gases. Some methods are presented next.

4.1 Method for testing the caulked assembly tightness using helium

The testing of a non-removable assembly is done using a special instrument which is based on the detection of helium.

The working principle of the helium leakage test device is as it follows. The caulking assembly is connected to the test device. Everything that exists inside the assembly is emptied using a vacuum pump.

Helium is flushed towards the caulked area to determine tightness. Helium can penetrate into very small holes and may reveal the quality of the caulked assembly (Fig. 9).

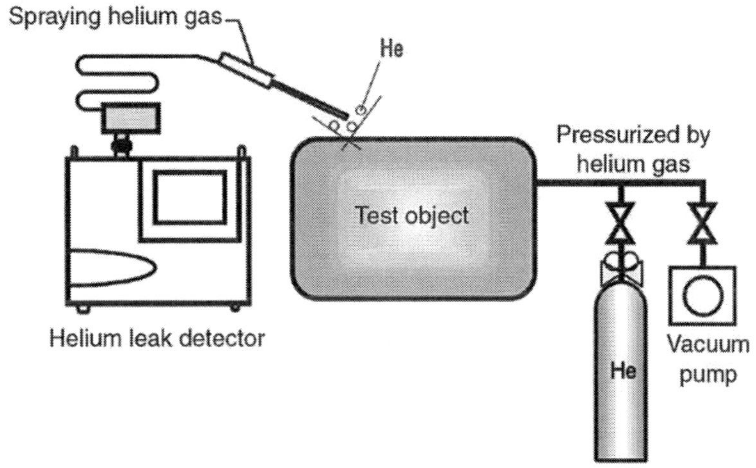

Fig 9. Helium leakage method

In normal conditions (proper conditions for the functionality of the final product) the left side of each curve must be in the right side of the pressure of 1bar.

Due to the fact that the caulking process was not performed in good conditions there was a leakage of the gas from inside the assembly to outside and in this case the functional curve (coloured with blue) was moved to the right side of the pressure of 1 bar. In these conditions the assembled product will not be able to perform its normal functionality.

The red curve from the picture shows how the functionality of the product must look when the caulking process is performed in good conditions.

The picture with two curves presents how important is to perform the caulking process in good conditions. Even if everything seems to be ok from the visual point of view, a simple pore in the caulking process can lead to clear damage of the final product.

In the following picture, examples of good and bad tightness can be seen, measured with the equipment presented before (Fig. 10).

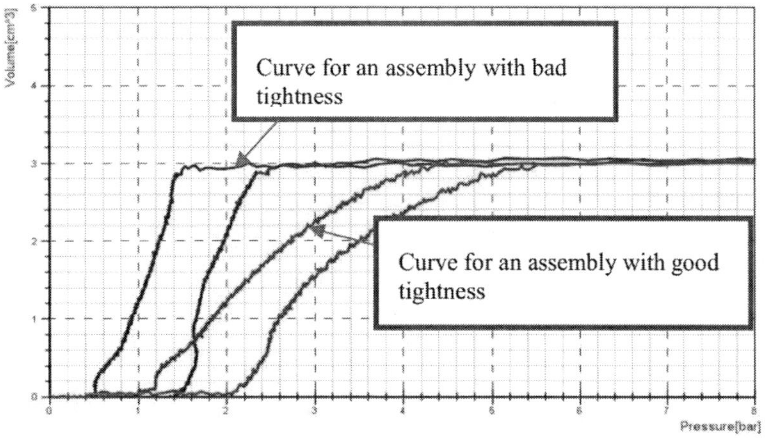

Fig. 10. Tightness evaluation using gas pressure

4.2 Method for testing the caulked assembly tightness using hydraulic pressure

Another method which can be used to check the tightness of the caulking process is the hydraulic tightness of the assembly.

This method consists of holding up a certain value of a pressure inside the caulking assembly and check if there is a pressure drop in a certain amount of time. This type of tightness test is used for assemblies which are supposed to run at low pressures.

In figure 11, the result of a hydraulic test (done after a caulking process) is presented. The parameters used for the test were a holding pressure of 22 bars and a holding time of 600 seconds. The graph shows that there was no pressure drop during the entire test.

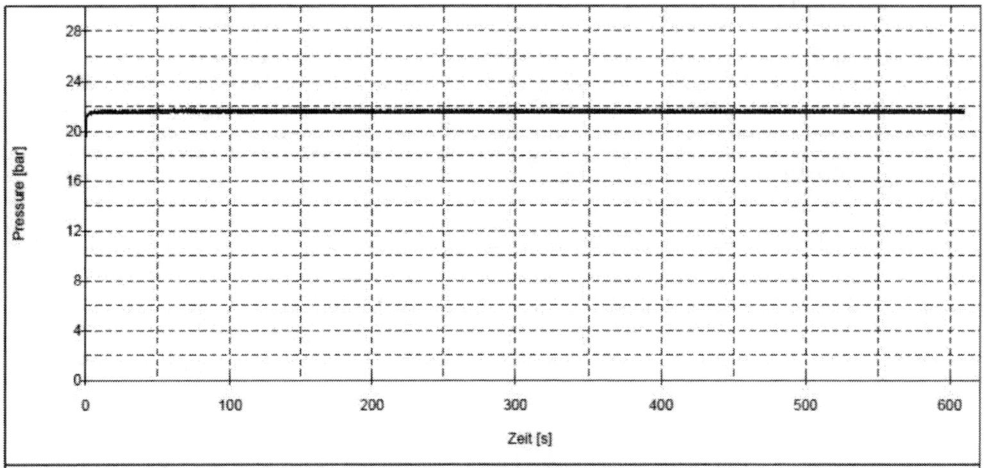

Preload_and_burst_test			
		Nominal	Measured
Maximal pressure	bar		22
Burst (0=No 1=Yes 2=Leak)			0

Fig. 11. Hydraulic tightness testing

4.3 Method for testing the burst pressure

There are cases when the caulked assembly must stand against big values of pressure. The best method to test if a caulked unit can meet the expectations of these kinds of big values is the burst pressure test.

Burst pressure tests are performed on dedicated test benches which are able to provide very big values of pressure inside the caulked assembly (test benches can go to values of 1500 bars – Fig 12).

 DOI: 10.1051/matecconf/201712103019

Fig. 12. Burst pressure test bench

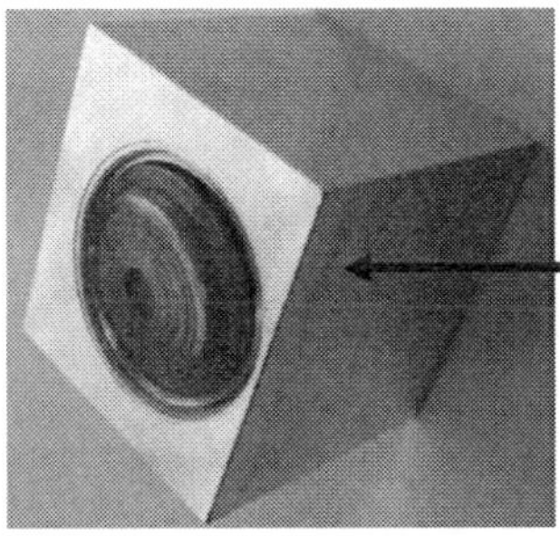

It is visible that the caulking process was not performed in specific conditions so the assembly failed at high values of pressure.

Fig. 13. Assembly failed at burst pressure test

A specific problem with the burst pressure value can be seen in the following picture (Fig 13): The assembly didn't resist to the testing pressure. In figure 14, the recording of a failed test is presented. After 20 seconds, the assembly failed.

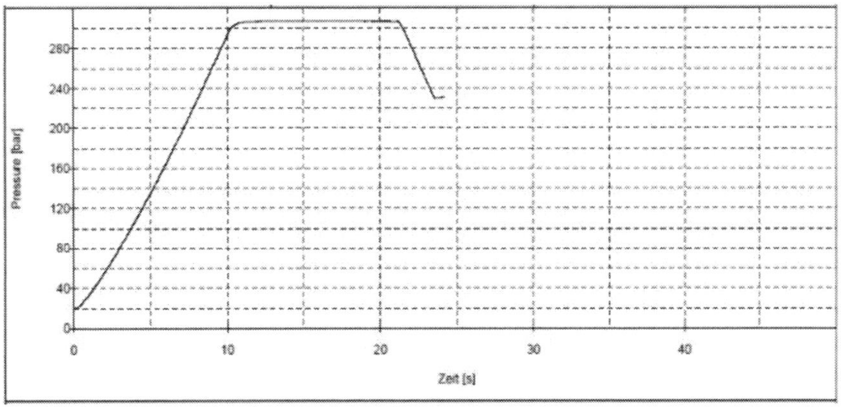

Fig. 14. Recording of a failed burst pressure test

4.4 Measuring the geometrical parameters of the caulked area

The amount of material which covers one of the parts and in this way creates the caulked assembly is also very important for the success of the process. This amount of material can be measured in many ways. The simplest one is the measurement of the height of the deformed material (Fig 15).

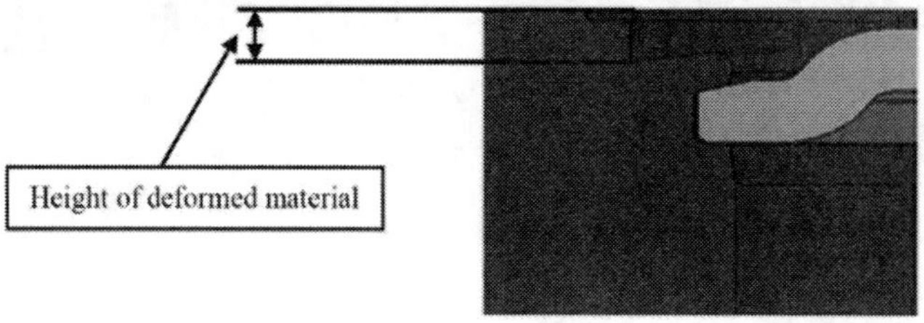

Fig. 15. Height of the deformed material at caulking process

It is highly important that this height of deformed material to be constant on all over the assembled area. A single zone where the height is smaller can lead directly to problems such as air leakage or low burst pressure value.

4.5 Comparative analysis of the testing methods for the caulked assembly

By analysing the testing methods presented before, it was possible to underline the input and the output data for each of them. The synthesis of this information is presented in table 3.

Table 3. Comparative analysis of the measurement methods used for the verification of the caulked assembly

Measurement methods	Input Parameters	Output Parameters
Measurement of caulked assembly tightness using helium	- ØDc; - H; - T; - F; - ØDp1; - ØDp2.	- Hc; - ØDac; - Sp gas.
Measurement of caulked assembly tightness using hydraulic pressure	- ØDp1; - ØDp2; - ØDc; - H; - T; - F	- Hc; - ØDac; - Sp liq.
Measurement of burst pressure limit	- ØDp1; - ØDp2; - ØDc; - H; - A1; - A2; - r; - T; - S; - F	- Hc; - ØDac; - Bp.

The presented methods analyse the tightness and the resistance of the caulked assemblies. None of the methods analyse the reliability of the caulked assembly. Therefore there is a need to develop a new test bench which will be able to test this function of the product.

5 Experimental researches

The analysis done before was applied to an experimental research. The aim of the research was to test the validity of the analysis made in this work and to develop a coherent methodology for testing the caulked assemblies. The input parameters are presented in table 4.

Table 4. Input parameter for the experimental research

Name	Values	Measuring unit
H	2.65	[mm]
ØDp1	25.63	[mm]
ØDp2	25.48	[mm]
ØDc	2.85	[mm]
r	0.82	[mm]
F	76	[KN]

The visual inspection reveals a good shape of the caulked assembly.

At the testing of the tightness using helium, the speed of the decreasing pressure inside the caulked assembly – gas conditions. (Sp gas) – result 7.8 * 10E-10 mbar*l/s. As a result, the assembly tightness function regarding gas pressure is assured.

At the testing of the tightness using– liquid (Sp liq) – the assembly was tested under a pressure of 20 bar for a time of 20 minutes – without any leakage drop. The assembly tightness function regarding liquid is assured.

The test for burst pressure (Bp limit test) – was done for the burst pressure limit and the value is ca. 363 bar. The assembly is assured from the burst limit point of view.

The experiment reveals the fact that for the geometrical parameters presented in the table all the requirements for the final caulked product are respected.

Experimental research revealed the algorithm necessary to fully evaluate a caulked assembly.

6 Methodology for testing the caulked assemblies

The functional analysis done before revealed that tightness, resistance and reliability are the most significant functions for the caulked assembly. Actual testing methods are focused on tightness and on resistance. The logical order of doing these tests is presented in figure 16. First the gas pressure test is done. If the assembly pass the test, then the liquid pressure test is done. If the assembly passes this test too, than the burst pressure test is done.

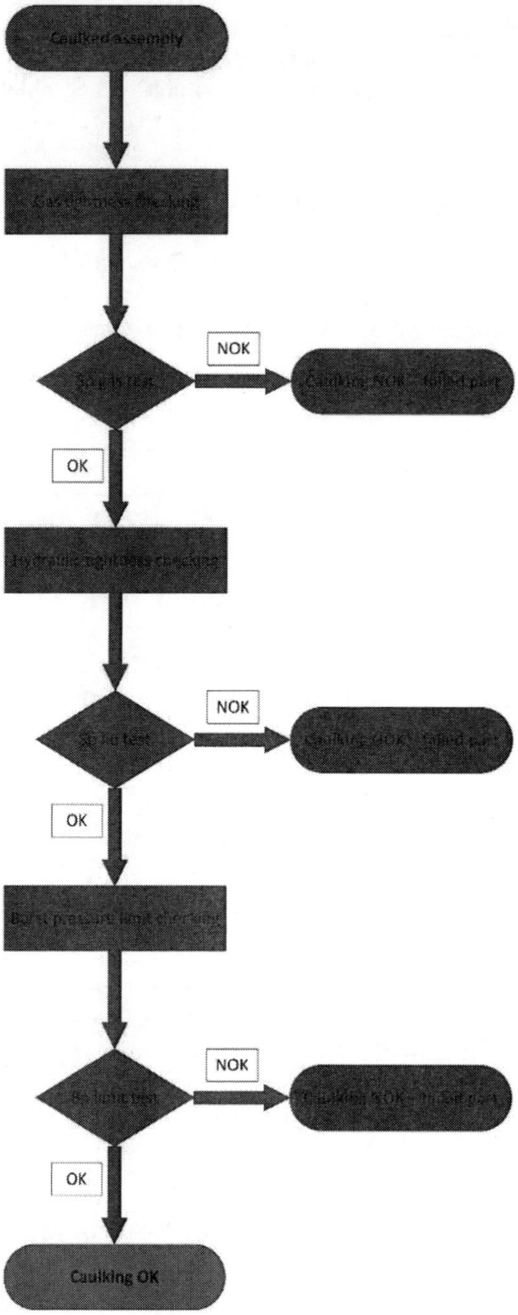

Fig 16. Algorithm of testing caulked assemblies

The study shows that the reliability function is not analysed. There is a need of developing a new stand for reliability testing.

7 Conclusion

The paper highlights the caulking process as a systemic process. The input data and the output data of the process were presented together with the methods of analysing the performance of the caulking assembly.

The actual knowledge and test equipment allowed us to perform an experimental research which reveals the validity of the measurements methods.

The performed analyse and the presented algorithm allows a complete and complex assessment of the performances regarding the caulking process.

In the future we propose to study the optimization of the input parameters in order to obtain precise assemblies for different materials and also for different dimensions of the parts.

This methodology can be used to achieve optimization of the process in order to process parts of various materials and also with different dimensions.

References

1. B.-S. Sin, K.-H. Lee, The Scientific World J., Article ID: 971679 (2014)
2. J. Park, J.Park, S. Choi, K. Na, Y. Kim, J. Mat. Processing Tech., **198**, 471, (2006)
3. J. Beck, R. Ziegler, N. Schonheit, Nuclear Eng. And Design, **263**, 179 (2013)
4. A. Celcatelli, M. Bergoglio, D. Mari, Vacuum, **81**, 1538 (2007)
5. Y. Gorash, W. Dempster, W. D. Nicholls, R. Hamilton, A. A. Anwar, J.of Loss Prevention in the Process Ind., **43**, 61 (2016)

Investigation of chips morphology after turning of materials applied in aerospace industry

Marek Rybicki[1,*], and *Piotr* Szablewski[2]

[1]Poznan University of Technology, Faculty of Mechanical Engineering and Management, ul. Piotrowo 3, 60-965 Poznan, Poland
[2]Pratt & Whitney Kalisz, ul. Elektryczna 4a, 62-800 Kalisz/ PWSZ Kalisz, Poland

Abstract. In the paper is presented analysis of chips morphology after turning of three groups of modern materials applied in aerospace industry: titanium alloys, stainless steels and heat resistant alloys. Various feed rates f and corner radiuses r_ε as well as recommended cutting speed $v_c$for given group of work material were applied in machining tests. Range of burrs on thinner edges of chips and serration on the chips surface occurrence were identified dependent on work material and uncut chip thickness (values of the f and r_ε). In the research height h and distance b between the burrs as well as distance a between serrated segments were measured. It was identified two kind of burrs connected with different physical phenomenon and specified h and b dimensions.

1 Introduction

During machining of materials in group M (stainless steels) and S (heat resistant alloys, titanium alloys) continuous chips are formed due to the materials properties. When cutting speed and federate is increased, chips types is changed to serrated or share localized type [1-4]. Connection of chips types with cutting parameters is attributed to changes of cutting temperature [1], what is confirmed by another chips types formation during hybrid, laser assisted or cryogenic machining of the materials [5-6].

The chip formation process can lead to fluctuation of cutting force, vibration and poor surface quality [2,7].

On edges of chips can be visible burrs of height and frequency depending on cutting parameters. Burrs forming at thinner edge of chips (on the non-free end)can affect machined surface roughness [4]. The burrs arise due to instability of chips formation when thickness of cut is smaller than minimal.

Authors [4] have revealed that at the critical cutting speed, both pitch $b$and height $h$of chip burrs reach the lowest value and serrated chips begin to appear above the speed. At the critical speed machined surface roughness had the smallest value.

*Correspondingauthor: marek.rybicki@put.poznan.pl

2 Range, conditions and technique of research

Machining was carried out on the CTX 310 ECOLINE turning machine made of DMG. During turning flood cooling/lubrication by emulsion: CIMTECH M21-02 made of Cimcool Industrial Products B. V. with concentration 10,45 %was used. Five work materials were machined:
- heat resistant alloy Waspaloy (EN: 2.4654; AISI: AMS 5706),
- heat resistant alloy Inconel 718 (EN: 2.4668; AISI: AMS 5662),
- heat resistant alloy Inconel X750 (EN: 2.4669; AISI: AMS 5667),
- stainless steel (EN: 1.4541; AISI: 321) and
- titanium alloy (Ti6Al-4 Grade5).

In turning tests were used Sandvik cutting inserts CNMG1204 with the SM chip breaker made of sintered carbide grade 1125. Applied holder DCLNL 2020K12 gave tool cutting edge angle $\kappa_r = 95°$.

Three corner radiuses r_ε with various range of feed rates, were applied (table 1). Recommended cutting speed v_c for given group of work material was applied in machining tests: v_c =35 m/min for heat resistant alloys, v_c= 150 m/min for stainless steel and v_c= 75 m/min for titanium alloy.

Table 1. Values of feed rates *f* applied for various corner radiuses r_ε

Corner radius r_ε [mm]	**0,4**	**0,8**	**1,2**
Feed rates *f* [mm/rev]	0,01	-	-
	0,02	0,02	-
	0,03	0,03	0,03
	0,05	0,05	0,05
	0,09	0,09	0,09
	-	0,17	0,17
	-	-	0,33

The chips were inspected on stereoscopic microscope SteREO Discovery V20 with AxioVision software made of Zeiss. Images of chips were made with 9 and 80 magnification. In the chips were measured following dimensions (fig. 1):
- burrs height on thinner edge of chip– *h*,
- distance between burrs on thinner edge of chip – *b*,
- distance between segments on rough surface of serrated chips – *a*.

It was assumed that burrs are large when its height h> 0,1mm (usually b> 0,2mm) and small when 0,02 mm $<h \leq$ 0,1mm (usually 0,05 mm$<b \leq$ 0,2mm).

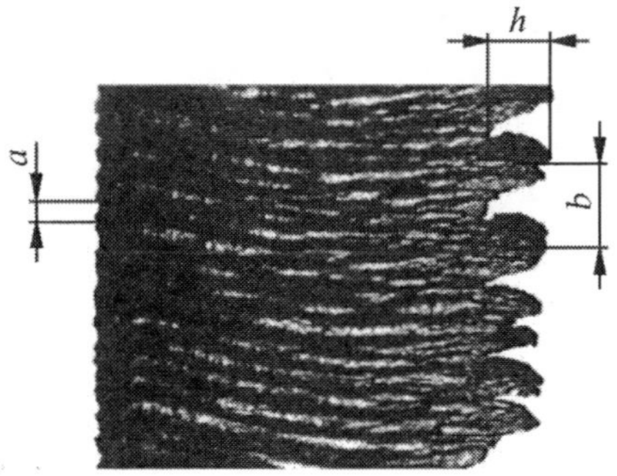

Fig. 1. Measured dimensions of chip: burrs height *h*, burrs distance *b* and segments distance *a*

3 Results and analysis of research

From the table 2 follows that large burrs on chips edges occurs for smaller federates f and higher corner radiuses r_ε (smaller thickness of cut) most often. Smaller burrs are usually connected with serration of chips surface. The serration is visible for stainless steel and titanium alloy only, what is connected with higher cutting speed v_c applied for the materials, the most probably [1-2].

Table 2. Occurrence of burrs on thinner edge of chip and serration of its surface for various work materials and machining conditions

Material	Chips properties	Large burrs[1]			Small burrs[2]			Surface serration		
	f \ r_ε	0,4	0,8	1,2	0,4	0,8	1,2	0,4	0,8	1,2
Waspaloy	0,01									
	0,02		X							
	0,03		X							
	0,05		X	X						
	0,09		X	X						
	0,17		X	X						
	0,33						X			
Inconel 718	0,01									
	0,02									
	0,03		X							
	0,05		X		X					
	0,09		X	X	X					
	0,17		X	X		X				
	0,33									
Inconel X750	0,01									
	0,02	X								
	0,03		X	X	X					
	0,05		X	X	X					
	0,09			X	X					
	0,17						X			
	0,33									
Stainless steel	0,01									
	0,02									
	0,03		X	X				X		
	0,05		X	X				X	X	
	0,09			X	X	X		X	X	X
	0,17					X	X		X	X
	0,33						X			X
Titanium alloy	0,01	X								
	0,02		X		X					
	0,03		X	X	X					
	0,05		X		X		X			
	0,09				X	X	X	X	X	X
	0,17					X	X		X	X
	0,33						X			X

[1] $h > 0{,}1$mm
[2] $0{,}02$ mm $< h \leq 0{,}1$mm

During machining of Inconel 718 with conditions: $r_\varepsilon = 0{,}8$ mm, $f = 0{,}17$ mm/rev was observed transient state. On chip formed in the conditions is visible all cases of chips morphology: irregular large and small burrs and beginning of the chip serration in form of surface waviness (fig. 2).

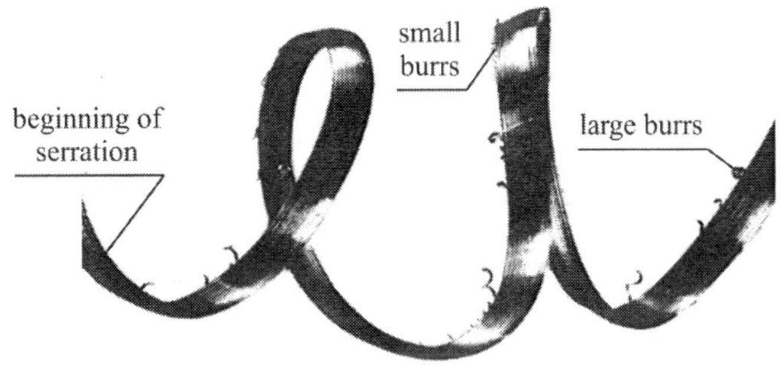

Fig. 2. Occurrence of large and small burrs as well beginning of segments formation on rougher surface of chip (Inconel 718, r_ε = 0,8 mm, f = 0,17 mm/rev)

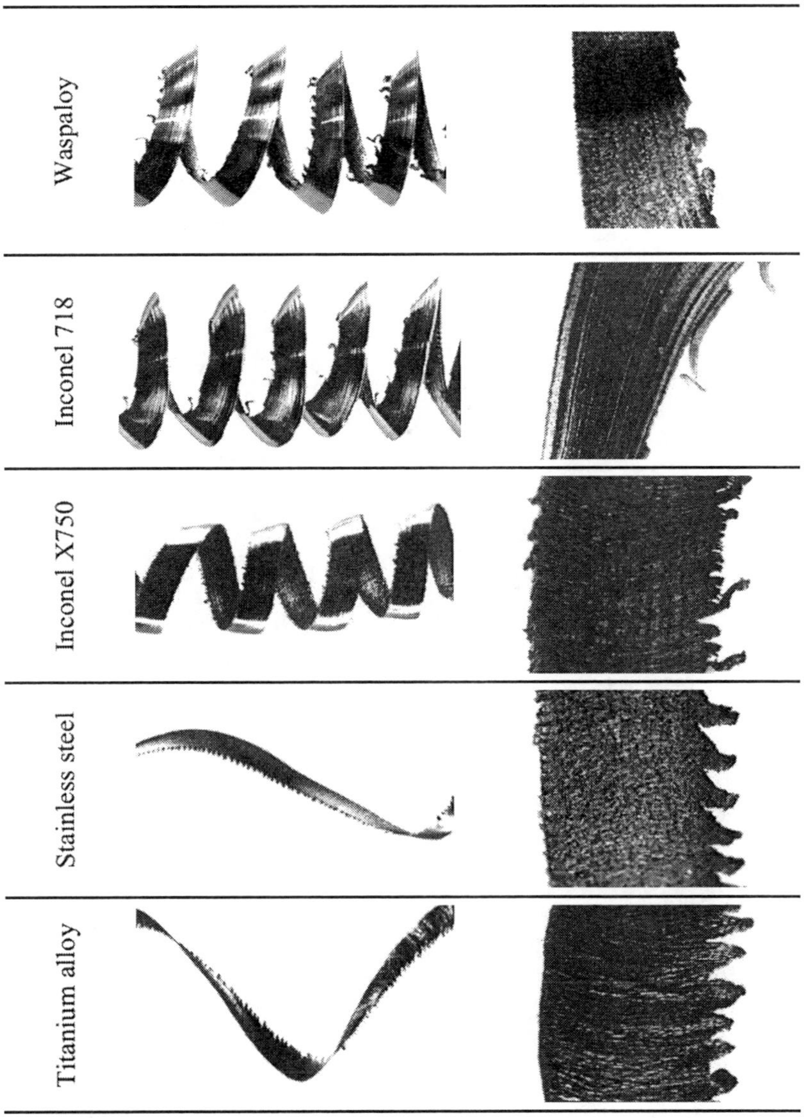

Fig. 3. Views of shape and rougher surface of chips after turning of various materials (r_ε= 0,8 mm, f = 0,03 mm/rev)

DOI: 10.1051/matecconf/201712103020

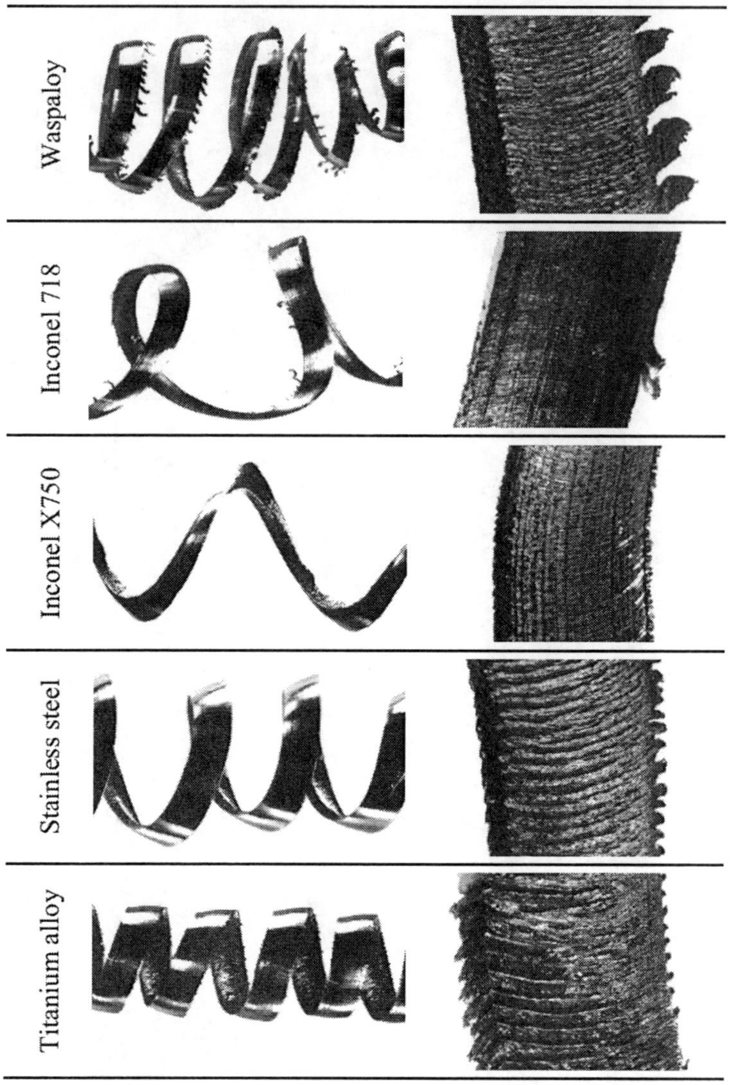

Fig. 4. Views of shape and rougher surface of chips after turning of various materials (r_ε = 0,8 mm, f = 0,17 mm/rev)

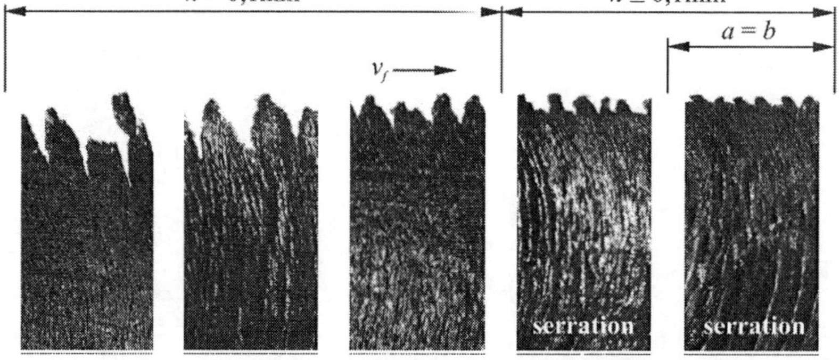

Fig. 5. Influence of federate on chips morphology during turning of titanium alloy (r_ε = 0,8 mm)

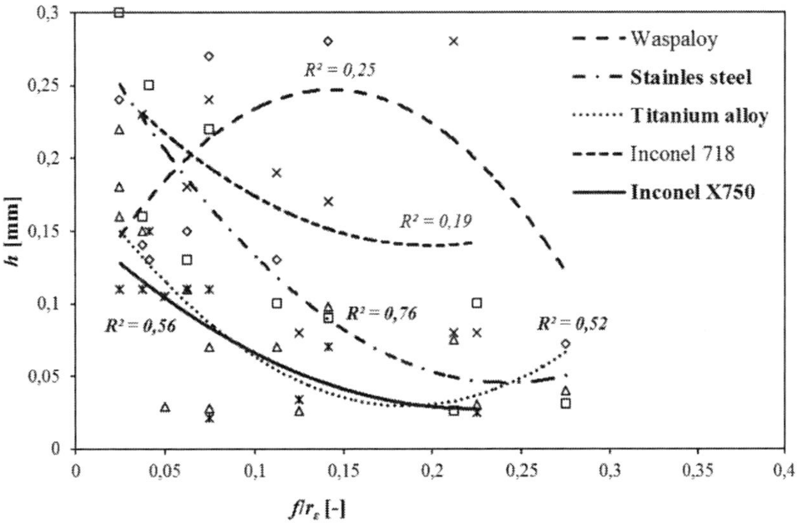

Fig. 6. Influence of ratio f/r_ε on chips burr height h after turning of various materials

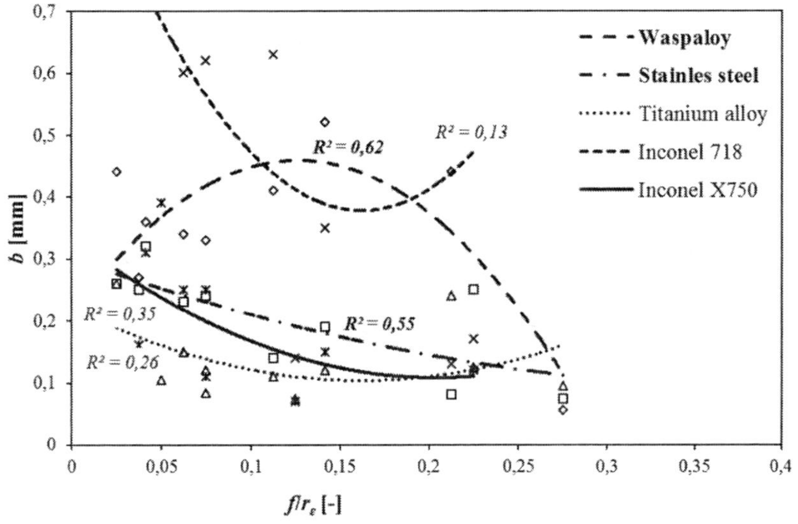

Fig. 7. Influence of ratio f/r_εon chips burr distance b after turning of various materials

At figures 3 and 4 is presented views of chips after machining with the most popular corner radius r_ε= 0,8 mm and two federate values: 0,03 mm/rev and 0,17 mm/rev respectively. From the figures follows that all the chips have tubular shape, except chips after turning of stainless steel and titanium alloy with small federate.

For smaller federate regular large burrs are visible for all of the materials and for higher federate large burrs is visible for the Waspaloy alloy only. Smaller burrs distance is equal to serration distance $a = b$ usually, what is especially visible after turning of stainless steel and titanium alloy (fig. 4,5).

At figures 6 and 7 is shown influence of the f/r_ε ratio on the chips burrs height h and distance b. The f/r_ε ratio is proportional to thickness of cut, affecting chip formation. Applied ratio allowed to relate the chips burrs dimensions to the specific combination of f and r_ε values easier, than in case of thickness of cut using. At the figures quadratic polynomial was used as a regression model, basis on research results [4] in which burrs height has minimum in function of cutting parameter v_c. From other research [2,3] follows, that federate has similar impact on chips burrs and serration, but lesser than v_c.

From the figure 6 follows that heat resistant alloys: Waspaloy and Inconel 718 are characterized by the highest burrs occurring independently of federate f and corner radius r_ε. The f/r_ε ratio has considerable impact on burrs height h for three of investigated materials: stainless steel, titanium alloy and Inconel X750. For the materials with increasing ofthe f/r_ε ratio both pitch b and height h of the burrs firstly decrease and then can slightly increase when the serration begin to appears. Below ratio f/r_εcorresponding to the smallest burrs height h the burrs arising is connected with smaller thickness of cut than minimal and under the ratio the burrs arising is connected with chips surface serration. After beginning of serration the continuous increase of the f/r_ε, causes that burrs connected with the serration becomes much more severe.

Comparing figures 6 and 7 is also visible that the highest burrs height h is connected with occurrence the highest burrs distance b.

3 Conclusions

The large burrs are usually arised for smaller federates f and higher corner radiusesr_ε, due to instability of chips formation in regions of the chips where thickness of cut is smaller than minimal. Whereas small burrs occureence are usually connected with the chips serration when the burrs distance is equal to serration distance $a = b$.

Waspaloy and Inconel 718 are characterized by the highest burrs, occurring independently of the f and r_ε values. For other investigated materials with increasing ofthe f/r_ε ratio (proportional to thickness of cut) both pitch b and height h of the burrs firstly decrease and then can slightly increase when the serration begin to appears. The serration is visibled for stainless steel and titanium alloy only, what is connected with higher cutting speed v_c applied for the materials than for heat resistant alloys, the most probably.Minimal values of the h and b dimensions is occured for ratio $f/r_\varepsilon = 0{,}15\text{-}0{,}25$ depending on work material. It corresponds to: $f = 0{,}06\text{-}0{,}1$ mm/rev for $r_\varepsilon = 0{,}4$ mm, $f = 0{,}12\text{-}0{,}2$ mm/rev for$r_\varepsilon = 0{,}8$ mm and $f = 0{,}18\text{-}0{,}3$ mm/rev for$r_\varepsilon = 1{,}2$ mm. It is optimal range of feedrate from the point of view of machined surface roughness the most probably, what need to be verified.

Acknowledgments: The paper is supported by The National Centre for Research and Development (Project No. PBS3/B5/36/2015).

References

1. C. Veiga, J.P. Davim, A.J.R. Loureiro, Rev. Adv. Mater. Sci. **34** (2013)
2. A. Daymi, M. Boujelbene, S. Ben Salem, B. Hadj Sassi, S. Torbaty, Archives of Computational Mat. Science and Surface Eng.,**1**, 2 (2009)
3. T. Mabrouki, L. Deshayes, J. -F. Rigal, K. Jurrens, R. Ivester, Proceedings: 7th CIRP Intl Workshop on Modeling of Machining Operations, (2004)
4. F. Ning, F.Wang, Z. Jia, J. Ma, Int. J. Machining and Machinability of Materials **15**, 3/4 (2014)
5. D. Przestacki, Procedia CIRP **14**, 229-233 (2014)
6. P.Szablewski, E. Weiss, T. Chwalczuk, Mechanik **89**, 8-9 (2016)
7. D. Przestacki, M. Jankowiak, Journal of Physics: Conference Series **483**, 012019, 1-7 (2014)

MATEC Web of Conferences **121**, 03021 (2017) DOI: 10.1051/matecconf/201712103021
MSE 2017

Effect of concentrated energy of laser beam on polymer material

Libuše Sykorová[1,*], *Vladimir* Pata[1], *Milena* Kubišová[1] , and *Jana* Knedlová[1]

[1] Tomas Bata University in Zlin, Faculty of Technology, Department of Production Engineering, Vavrečkova 275, 76001 Zlín, Czech Republic

Abstract. The article deals with the area of non-conventional technologies, specifically with the effect of concentrated radiant energy on the polymer material. There was studied the laser beam and its application to two different types of polymeric materials in details. PMMA, as a representative of the amorphous polymers and POM, which is a crystalline polymer, were used for experimental cutting as they frequently applied in industrial practice. The input technological parameters were changed during machining followed by evaluation of their interaction with the concentrated radiant energy of the laser beam. The results were subjected to statistical processing.

1 Introduction

The limiting factor of the level of the engineering production is the applied technology as it sets demands on energy consumption and thus determines the production energy intensity, raw material consumption and rate of their utilisation. Next, it also specifies practical, utility, and commercial characteristics of the products, thus it conditions the development of all industrial sectors.

The current production level requires the applied materials to meet the most demanding criteria in terms of a long durability, wear or economics. High demands on process ability of these materials are solved by the application of non-conventional technologies as the standard methods often do not meet requirements on speed and quality of the production. Therefore, the methods by which the better results in a shorter time are reached are used, especially laser machining [1].

This technology is highly accurate, fast, low-tech and ecological with high productivity and production efficiency. In this way, a laser is an optimal tool due to its properties. The laser beam can be applied to almost all kinds of material processing. Achievable dimensions of the structures depend on the type of laser, its parameters and material.

Laser cutting of material is affected by a variety of factors. The result of the optimisation process is to evaluate the impact of individual factors. It is necessary to emphasise the major effects, neglect less important once and generally to be able to set the parameters of the laser device so that the product capabilities are maximised [2].

* Corresponding author: sykorova@ft.utb.cz

2 Experiment

2.1 Preparation of specimen

From the obtained scientific knowledge it follows, that the character of surface machined by laser depends among others on the thermal conductivity of the polymer. Therefore, it was decided to test two different kinds of polymers, namely amorphous PMMA and crystalline POM. Commercial CO_2 laser Mercury L - 30 by firm LaserPro, USA was used for cutting of specimens. Changing power and feed rate of the laser system is possible. Ray of the laser could be focused on Mark diameter d = 185 μm. The laser is cutting with the software help of Corel Draw.

2.2 Machining of specimens on the CO2 Laser Mercury L - 30

The desired symbol was created in program Corel Draw, the width of the slot was 2 mm in all cases. Cutting parameters (output power and feed) were set, and these were changed gradually. Values of power and feed are presented as percent's from maximum power (P = 30 W) and maximum feed (fmax = 1066 mms^{-1}) in charts of parameters combination and graphs. These specimens were prepared from this sheet (proportions 130 x 61 mm, thickness 10 mm). The edges were milled and grinded.

Table 1. Adjusted parameters of machining

Percentage rate of feed f [%]	50	70	100					
Percentage rate of laser power P [%]	10	20	30	40	50	60	70	80

2.3 Dimension and profiles measuring of machined slots

Dependence of slot depth on the laser parameters – output power and feed - was measured. The dimension measuring of machined slots was realised on the optical microscope ZEISS 2772. Suitable microscope optics which ensured 117 x enlargements was set first of all. For the reason of statistical evaluation of measured values, the measurement of depth was realised in all slots five times. Slot depth (d in μm) in dependence on laser parameters combination is presented in following results tables. See Table 2, Table 3 and Table 4. Values of power and feed are shown as percent's from maximum power (P = 30 W) and maximum feed (fmax = 1066 mms^{-1}) in charts of parameters combination and graphs. Description 20/70 means 20% value from power 30W and 70% value from maximal feed 1066 mms^{-1}. These experimental results were evaluated and depicted in the graphs.

Table 2. Slot depth in μm for POM

P(%)	d(um)_f(50%)_POM	d(um)_f(70%)_POM	d(um)_f(100%)_POM
10	29	18	11
20	69	22	17
30	87	55	28
40	128	76	40
50	150	92	57
60	184	105	72
70	195	143	81
80	213	138	95

DOI: 10.1051/matecconf/201712103021

Table 3. Slot depth in μm for PMMA

P(%)	d(um)_f(50%)_PMMA	d(um)_f(70%)_PMMA	d(um)_f(100%)_PMMA
10	24	1	0,3
20	48	5	0,8
30	105	14	8,0
40	131	57	27,0
50	177	82	33,0
60	203	108	38,0
70	247	147	45,0
80	251	162	61,0

These experimental data were evaluated and depicted in the charts.

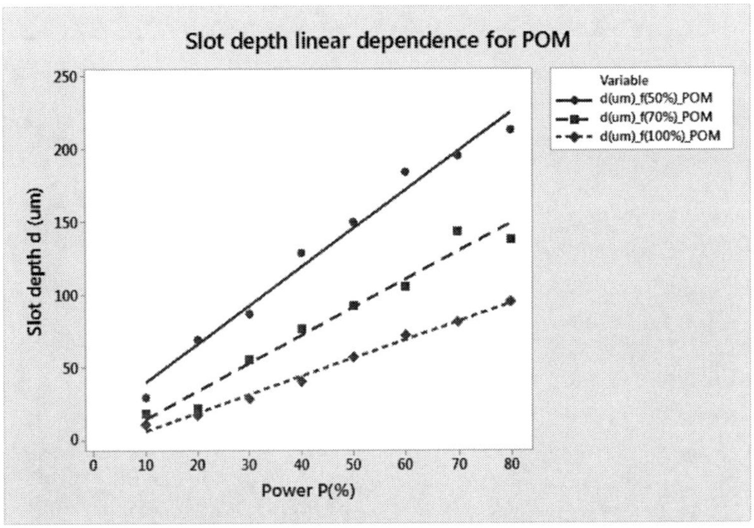

Fig. 1. Slot depth linear dependence for POM

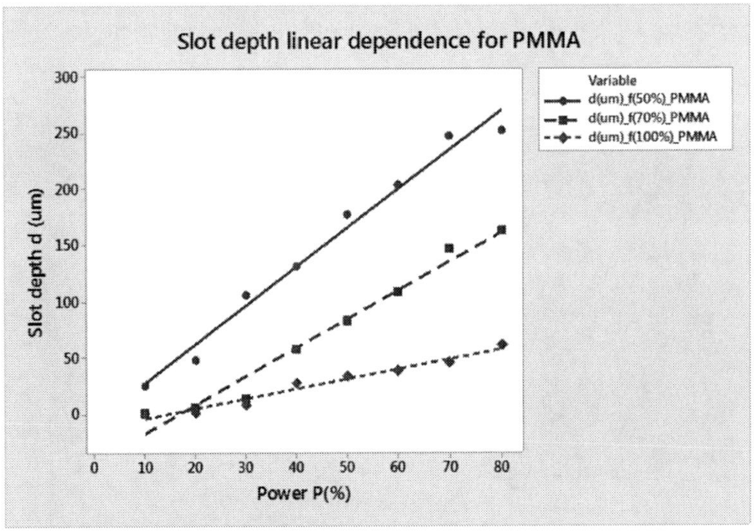

Fig. 2. Slot depth linear dependence for PMMA

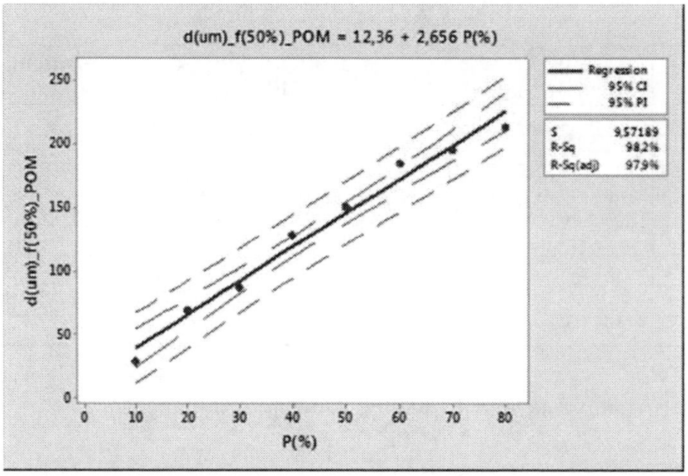

Fig. 3. Regression equation of slot depth dependencies on the laser parameters combination f (50%) for POM

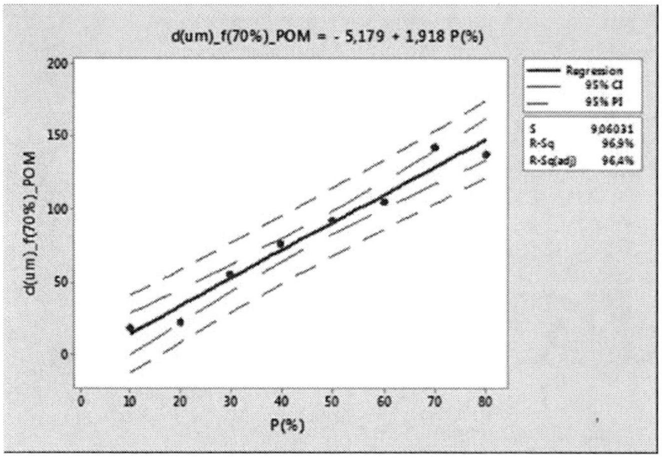

Fig. 4. Regression equation of slot depth dependencies on the laser parameters combination f (70 %) for POM

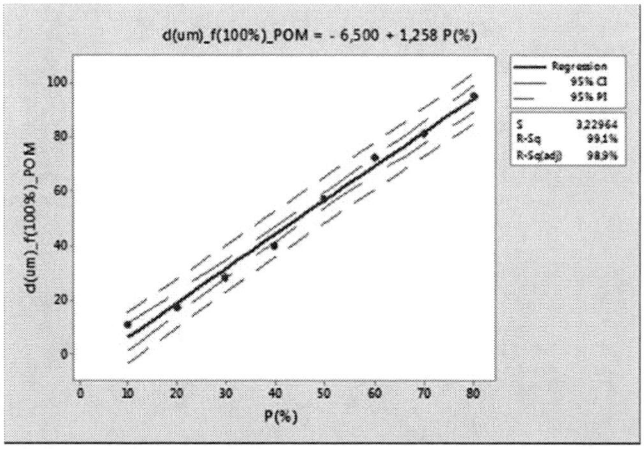

Fig. 5. Regression equation of slot depth dependencies on the laser parameters combination f (100 %) for POM

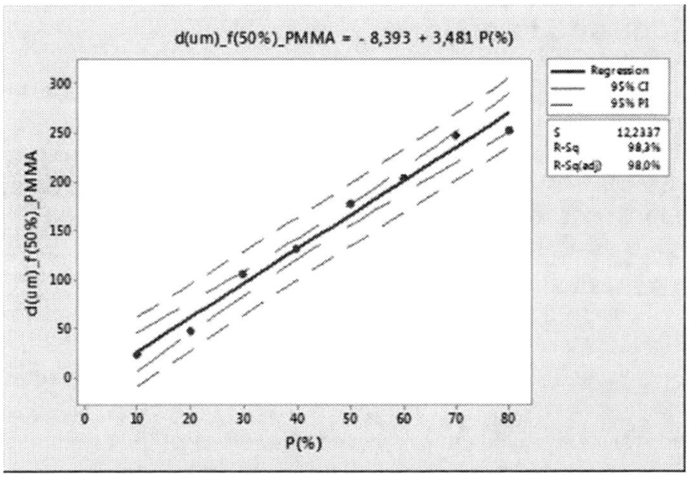

Fig. 6. Regression equation of slot depth dependencies on the laser parameters combination f (50 %) for PMMA

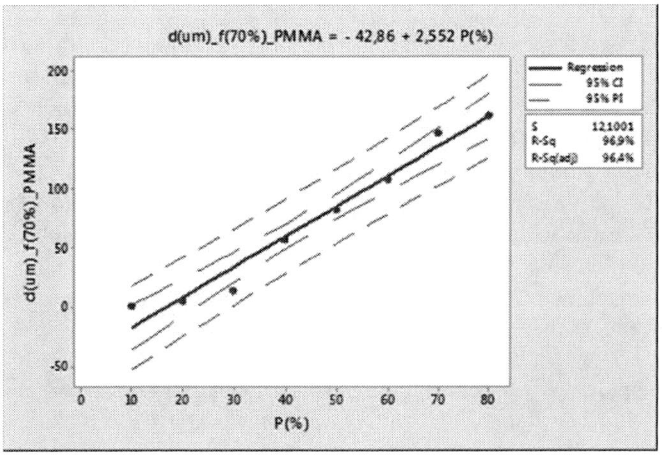

Fig. 7. Regression equation of slot depth dependencies on the laser parameters combination f (50 %) for PMMA

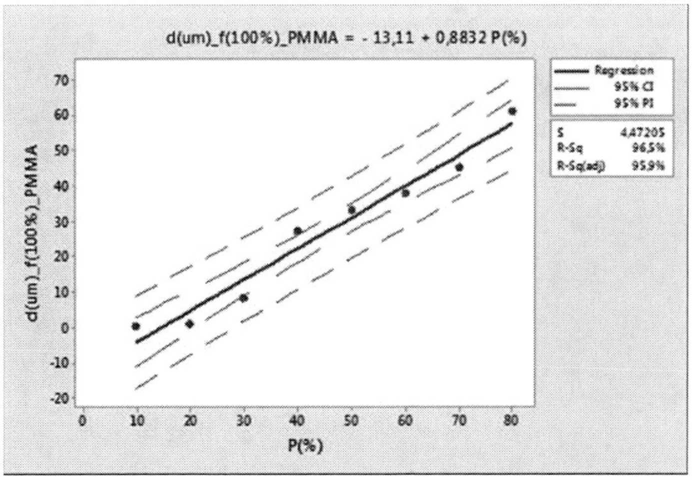

Fig. 8. Regression equation of slot depth dependencies on the laser parameters combination f (100 %) for PMMA

3 Conclusions and Intentions

All measured results for POM and PMMA were evaluated for the performance of P (10 - 80) % by the method of least squares using linear regression coefficient estimates.

For the relevant material and feed, there are listed below the corresponding regression equations, the standard deviations and determination coefficients including its adjusted form. (see Fig. 1 – 8)

As a result, it is evident that the slot depth grows linearly with the increasing values of power at given adjusted value of the feed. Biggest depth arises at low values of feed. This phenomenon was observed for both materials. Biggest slot depth was measured for PMMA material at the lowest speed of feed f = 50%, corresponding with f = 533 mms^{-1}. Compared to the PMMA, regarding the crystalline polymer POM, there was bigger interaction even at the maximal speed of feed f max = 1066 mms^{-1}. Based on the detailed examination, it can be said that there was not proved either the particular effect of the different configuration of the polymer structure (amorphous vs. crystalline) or effect of different thermal conductivity on the laser machinability.

The gradient of lines and correlation coefficients offer information about dependence character of slot depth on adjusted power and feed at visible presenting results. The value of gradient (k) is used for accurate control of stock removal (slot depth). The gradient of the line gives information about sloth depth at increasing of power about a definite number of percents. Correlation coefficient gives the relative quality of linear dependence [3].

Resulting structures can be very exact and with the high quality of surface in dependence on laser parameters and on the type of machining materials.

At the conclusion, it is possible to state that it is necessary to know output parameters combination of the particular laser system and properties of machined polymer materials for obtaining good results of micromachining by laser. It is possible to get the high accuracy of machined texture on these conditions. The result machining is different from the use of various kind of laser [4].

This work and the project is realised with the financial support of the internal grant of TBU in Zlín No. IGA/FT/2017/002 funded from the resources of specific university research.

References

1. R. Radovanovič, Strojírenská technology **9** (2004)
2. C. Felho, J. Kundrák, Characterization of Topography of Cut Surface Based on Theoretical Roughness, Key Eng. Mat., **581** (2014)
3. I. Hendl, *An overview of statistical methods for data processing* (Portal, Praha, 2008)
4. J. Jurko, Int. Conf. GC TECH Trencin (2002)

Effects of Core Softness and Bimodularity of Fibreglass Layers on Flexural Stiffness of Polymer Sandwich Structures

Oldřich Šuba[1,*], *Oldřich* Šuba jr.[1], and *Soňa* Rusnáková[1]

[1]Tomas Bata University in Zlín, Faculty of Technology, Department of Production Engineering, Vavrečkova 275, Zlín, 760 01, Czech Republic,

Abstract. This paper deals with the study of the flexural stiffness of the sandwich structures based on fibreglass and polymeric foams. The influence of geometrical and material parameters on the resulting effective flexural stiffness of the sandwich structure is being studied experimentally, analytically and by using FEM models. The effective modulus of elasticity of the sandwich-structured element is being studied and its theoretical and model dependencies on the flexibility of the foam core and bimodularity of the fibreglass layers are being investigated. The achieved results are compared with the experimentally observed values. This study shows that it is necessary to pay special attention to the issue of flexural stiffness of the walls when designing sandwich shell products in order to prevent possible failures in the practical applications of these types of structures.

1 Introduction

The production technology along with the description of the characteristics of the sandwich structures on polymer base achieve sustainable rise and are the subjects of various research. Technological process is relatively demanding but the resulting structure is characterised by the high flexural stiffness and load capacity at very low weight [1-5].

Particularity of the mechanical behaviour of the sandwich structures comes along with certain singularities and difficulties at the design compared to the standard structures. Regarding the standard – homogenous plastic or fibreglass structures, the ratio of stiffness to strength is considerably low; therefore in many cases the product design is done by limiting conditions in relation with the wall stiffness and not with its strength. In case of load where there is pressure membrane stress in the shell wall, the design is based on the limiting conditions of the loss of the stability of the structure shell which are related with the value of flexural stiffness of the wall. In such cases the sandwich alternatives are applied [6-8].

On the other hand, the sandwich component concept brings along also some problems related to its macrostructure [4, 6-7]. The result of the geometrical and material parameters of the sandwich wall is the layered and thus non-homogeneous element whose aspects of its

* Corresponding author: suba@ft.utb.cz

mechanical behaviour can be significantly different when compared to the conventional homogenous wall.

2 Flexural Stiffness of Sandwich Structures. Effective Modulus of Elasticity

For the pure flexural stiffness of the layered – composite cross section (excluded the effect of the core shear deformation), the modification of the technical theory of flexure can be expressed by

$$K_0 = E_{1+} J_R \tag{1}$$

where E_{1+} is the modulus of elasticity of the surface layer of the fibreglass layer tensile value, J_R represents quadratic moment of the reduced cross sectional area of the layered structure to the neutral axis - see Fig.1. So-called reduced cross sectional area of the non-homogeneous (composite) element is generally given by the individual layers width reduction to the selected referential (usually the biggest) value of the modulus of elasticity of the given structure. The Fig.1 calculates with the possibility of including the bimodularity of the compressive layers, i.e. Different values of modulus of elasticity in tension E_{1+} and compression E_1, when calculating the flexure stiffness.

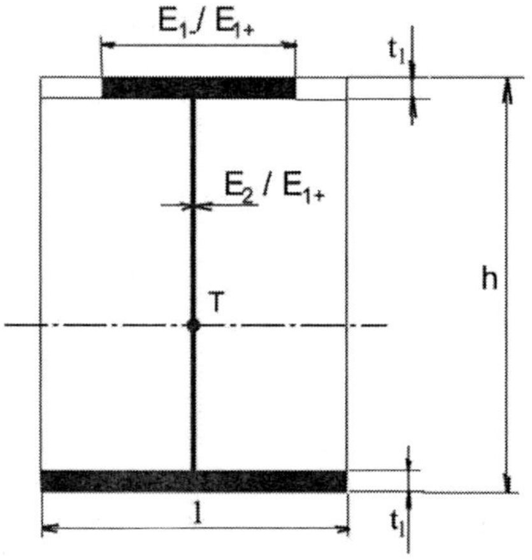

Fig. 1. Reduced cross-sectional area of the three-layered (sandwich) wall.

For the cases of practical evaluation of the experimental results, the concept of the effective modulus of elasticity of the sandwich structure was introduced. It is defined as the modulus of elasticity of theoretically homogenous (one-layered) wall thickness equal to the sandwich thickness *h*, having the same value of the flexure stiffness as the sandwich structure. Based on this condition, then according to (1)

$$E_{ef} = E_{1+} \frac{12 J_R}{b h^3} \tag{2}$$

The value of the „modulus of elasticity in flexure“ resulting from the standard bending test is apparently the experimentally given value of the effective modulus of elasticity of the sandwich wall.

3 Test: The effect of core shear compliance on the effective mechanical behaviour of sandwich elements

One of the aspects of the mechanical behaviour of the sandwich structure is the effect of the generally significant shear compliance, i.e. low values of shear modulus of elasticity of the lighter core on the resulting effective mechanical behaviour of the sandwich element. As it is shown schematically in Fig. 2., the overall deformation of the sample in the three-point bending test is done by not only the deformation due to the impact of the flexural – normal stress but also by the relatively high contribution of the shear deformation as the shear stress given by the cross forces in the cross-sectional areas of the sandwich element are, as known, transmitted only by the the layer of the light-weight core.

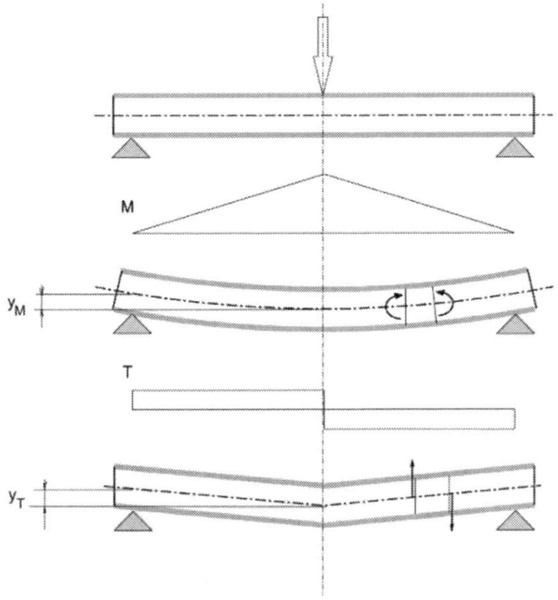

Fig. 2. Impact of core shear compliance on the overall deformation of the sandwich beam in the three-point bending test.

The equation (1), or (2) respectively, including only the deformation component by bending stress in the compressive surface layers gives, unlike to the homogenous cross sections, show always higher values of the stiffness when compared to the values resulting from the experimental three-point standard test. Analytical expression of the deformation, or stiffness respectively, according to the three-point test where the shear compliance of the core is included, is according to the Fig. 2 with application of the principle of energy given by

$$y_M + y_T = \frac{\partial A}{\partial F} \tag{3}$$

where work deformation A is given by the sum of the work deformation of the flexure moments and shear forces.

4 Experimental Testing of Sandwich Structures Stiffness

To verify the results of the numerical modelling, the sandwich structure with the outer coating of the fibreglass pre-impregnated fabric weighting 296 g/m^2 was used. As impregnating component of the prepreg the phenolic resin in amount of 42 weight percentage was applied. The core of the structures was made from polyethylene terephftalate (PET) foam with density of 100 kg/m^3, and the compression modulus of elasticity of 85 MPa.

The above-mentioned materials were used to produce the sandwich structures composed of one layer of prepreg from both, the upper and lower sides, and 20 mm thick PET core. It was produced by the method of the resin curing in the prepreg under the flexible foil in the curing own. Curing started by gradual heating for 30 mins to the temperature of 130°C, followed by dwelling at this temperature for 120 mins.

After removing from the mould, the sandwich panels were cut into testing samples for standard bending test. There were done both, three-point and four-point tests. Testing speed was set on 5 mm/min as recommended by the ASTM C393 standard.

5 Numeric Modelling of Sandwich Structure Bending tests

To study the dependency of the flexure stiffness of the sandwich wall on the stiffness of the weight-lighted core, the linear-elastic FEM models of the bending tests with the structures corresponding to the ones of the samples prepared for the bending tests were used.

The solution was done as linear-elastic analysis of the symmetrical half of the model for the three-point, or four-point bending test respectively. The FEM model for the three-point test is shown in Fig.3.

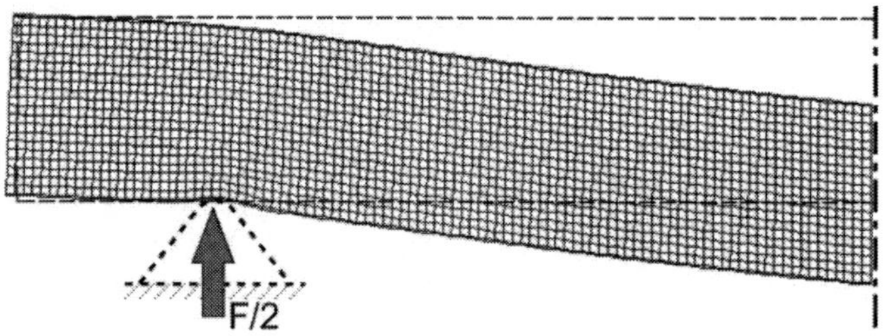

Fig. 3. Schema of the design model of the three-point bending test for the testing sample of the 20mm sandwich structure with the core module of elasticity of $E_2 = 85$ MPa and the supports distance of L = 150 mm. The character of the mesh deformation shows substantial proportion of the core shear compliance on the overall deformation of the sample.

6 The Resulting Stiffness of Sandwich Structure Determined by Three-point and Four-point tests

Comparison of the values obtained by model, analysis and experiment for the effective modulus of elasticity for the three-point test of the studied structure is shown in diagram in Fig. 4.

The effective modulus of elasticity values are taken for the coefficient of elasticity of covers in tension of 22000 MPa, which is determined by the manufacturer and it is, in fact, identical with the value obtained by the performed experiment.

It is obvious, that the dependencies E_{ef} determined analytically according to the equation (3) are located a bit under the values obtained by the FEM modelling.

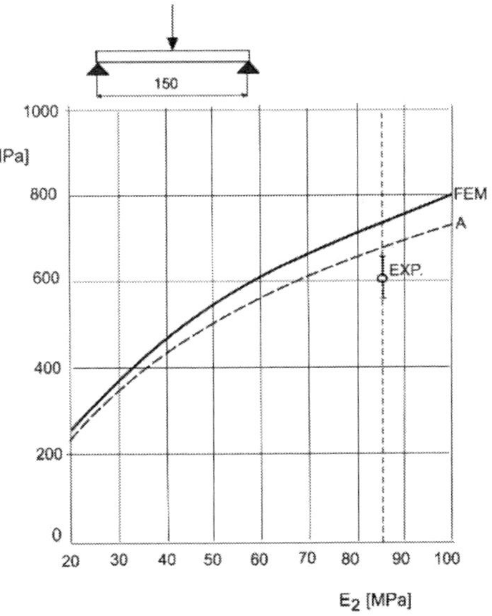

Fig. 4. Calculation dependencies for the effective modulus of elasticity during three-point testing on the foam core modulus of elasticity for the supports distance of L = 150 mm. Full curves – FEM, dashed curves – analytical calculation according to equation (3), EXP - range of experimentally obtained values E_{ef}.

The resulting differences between the experimentally obtained and numerical values of flexure stiffness of the studied structures can be attributed to the secondary aspects or imperfections respectively. Among those can be classified especially the problem of exact definition of the covers modulus of elasticity and their bimodularity, faults at determination of the shear modulus of core elasticity, as well as potential local indentation at the place of the supports. The impact of the core shear compliance is in the case of chosen 150 mm distance of the supports dominant. It will decline with the growing distance L between the supports and in case of the four-point test, it will be eliminated completely.

Comparison of the values obtained by model, analysis and experiment for the effective modulus of elasticity for the four-point bending test is shown in diagram in Fig.5.

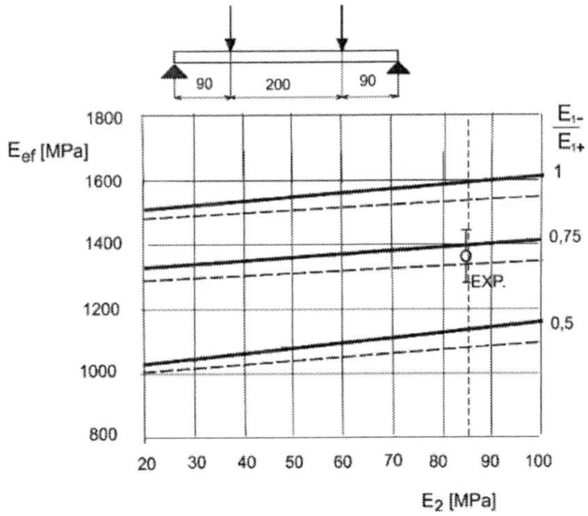

Fig. 5. Calculation dependencies of the effective modulus of elasticity during the four-point test on the modulus of elasticity of the foam core and bimodularity level of the compressive layers. Full curves – FEM, dashed curves – analytical calculation according to equation (3), EXP - range of experimentally obtained values E_{ef}.

The results show somewhat surprising values; there is decrease of the flexure stiffness of the structure in the dependence on the bimodularity level of the compressive layers. Next, the calculated and experimentally obtained values of E_{ef} are much higher in case of the four-point test when compared to the results of the three-point test. Regarding the studied structure, the pure flexure stiffness determined by the four-point test is approximately two times higher than the result of the three-point test with the same parameters. Experimental results point out certain bimodularity level of the studied samples.

7 Conclusion

The achieved results enable qualified evaluation of the effect of the core shear compliance, as well as of the bi-modularity level of the fibreglass compressive layer on the effective flexural stiffness of three-layered (sandwich) structures.

Based on the result of this study, it is necessary to pay attention to the design of sandwich products to prevent potential failure of these constructions in the practical applications.

Acknowledgement: This work and the project is realized with the financial support of the internal grant of TBU in Zlín No. IGA/FT/2017/002 funded from the resources of specific university research.

References

1. K. F. Karlsson, B. T. Åström. Applied Science and Manufacturing, **28**, Issue 2 (1997)
2. M. Mohamed, S. Anandan, Z. Huo, V. Birman, J. Volz, K. Chandrashekhara. Composite Structures, **123** (2015)
3. J. R. Vinson. *The Behavior of Sandwich Structures of Isotropic and Composite Materials* (1st ed. New York: CRC Press, 1999)
4. D. Zenkert. *The Handbook of Sandwich Construction*, (Nordic Industrial Fund, Worcestershire, EMAS Publishing, 1997)
5. A. Zinno, E. Fusco, A. Prota, G. Manfredi. Composite Structures, **92** (2010)
6. G. Belingardi, M. P. Cavatorta, R. Duella. Composite Structures (2003)
7. L. Liu, H. Wang, Z. Guan. Composite Structures (2015)
8. M. Giglio, A. Gilioli, A. Manes. *Computational Materials Science*, **56** (2012)
9. Y-M. Jen, CH.-W. KO, H.-B. Lin. International Journal of Fatigue, **31**, 3 (2009)
10. M. Garrido, J. R. Correia, T. Keller. Construction and Building Materials, **75** (2015)
11. B. Han, K.-K. Qin, B. Yu, Q.-CH. Zhang, CH.-Q. Chen, T. J. Lu. Composite Structures, **130** (2015)

MATEC Web of Conferences **121**, 03023 (2017) DOI: 10.1051/matecconf/201712103023
MSE 2017

Prediction of springback in V-die air bending process by using finite element method

Tomasz Trzepiecinski[1,*] , and *Hirpa G.* Lemu[2]

[1] Rzeszow University of Technology, Department of Materials Forming and Processing, al. Powst. Warszawy 12, 35-959 Rzeszów, Poland
[2]University of Stavanger, Department of Mechanical and Structural Engineering, N-4036 Stavanger, Norway

Abstract. Springback phenomenon affects the dimensional and geometrical accuracy of the bent parts. The prediction of springback is a key problem in sheet metal forming. The aim of this paper is the numerical analysis of the possibility to predict the springback of anisotropic steel sheets. The experiments are conducted on 40 x 100 mm steel sheets. The mechanical properties of the sheet metals have been determined through uniaxial tensile tests of samples cut along three directions with respect to the rolling direction. The numerical model of air V-bending is built in finite element method (FEM) based ABAQUS/Standard 2016.HF2 (Dassault Systemes Simulia Corp., USA) program. The FEM results were verified by experimental investigations. The simulation model has taken into consideration material anisotropy and strain hardening phenomenon. The results of FEM simulations confirmed the ability of numerical prediction of springback amount. It was also found that the directional microstructure of the sheet metal resulted from rolling process affects the elastic-plastic deformation of the sheets through the sample width.

1 Introduction

Springback is considered as the main problematic phenomenon that affects the quality of bent elements where the material undergoes geometrical changes. It occurs because of the elastic relief from the bending moment imparted to the sheet metal during forming [1]. Springback can be seen as an elastically-driven change of a shape of a deformed product which takes place during removal of the external loads that brought about the elastic deformation. Springback plays an important role in tooling and process designs in many industries, e.g., automotive industry, aerospace industry, and machine-building industry.

The amount of springback during unloading depends on three main groups of parameters [2-4]:

- *technological*: speed of deformation, lubrication, punch holding time, punch radius, repeat bending, and temperature,
- *geometrical*: bending radius, sheet thickness, sheet width,

* Corresponding author: tomtrz@prz.edu.pl

- *material*: material type, Young's modulus, yield stress, microstructure of the material, orientation of bending line according to the rolling direction of the sheet, ratio of yield stress to tensile strength.

In order to have better understanding of the springback phenomenon and the parameters influencing it, a number of research works have been previously reported. The effect of punch tip radius, sheet thickness, and sheet orientation on springback of CK67 steel sheets in V-bending process was reported by Bakhshi-Jooybari et al. [5]. They observed that springback decreases with an increase of sheet thickness. Furthermore, they concluded in their study that there exists no effect of bending angle according to rolling direction on the springback. Meinders et al. [6] investigated the effect of the geometry of punch and die on springback defect and indicated that the bulk material affects springback in bending operation. Blank holder force and effect of lubrication in springback analysis are studied by Chirita [7] and it was found that the greater the friction, the smaller becomes the springback. The effect of pad force and elevated temperature of sheet metal were studied by Duflou and Aerens [8]. It was experimentally demonstrated that local preheating of 25 mm thick steel plates result in reduction of bending force by 55%. Chalal et al. [9] observed that springback decreases almost linearly with an increase of sheet tension.

Among many parameters affecting springback, the Young's modulus is the main material property that influences the value of elastic springback of the material. In the analysis of sheet metal forming, it is common practice to assume that the elastic modulus remains constant. However, experimental investigations of Lems [10] revealed that elastic constants of a material may change during the plastic deformation. It was shown by Cleveland and Ghosh [11] and Yang et al. [12] that for steel materials the Young's modulus can decrease with plastic straining by up to 20% of its initial value. Additionally, Lems [10] and Morestin and Boivin [13] also indicated experimentally that the elastic modulus recovers to its initial value with time. According to Cleveland and Ghosh [11] and Wang et al. [14] the reduction of the Young's modulus can be seen as an extra nonlinear component of the total springback strain which is strongly dependent on the crystallographic texture evolution.

In many years, various experimental tests have been developed to study springback in sheet metal forming. The most popular and commonly used methods include: cylindrical bending, U-bending [15], V-bending [16], double-curvature forming [17], flanging [18] and other simple bending-reverse bending experimental methods [19].

Several analitycal methods are available for the prediction of springback [20]. The commonly used analytical models predict springback based on elastic unloading from a bending moment [2]. These models require the knowledge of the stress distribution within the part under bending. Other residual differential strain models only require the stress state on the outer and inner surfaces of the sheet.

The mechanical properties of anisotropic sheets depend on the chosen direction within the specimen. Accordingly, rolling process creates different features of the material microstructure and hence it results in planar anisotropy of sheet metals [21]. To account for these and other influencing factors and make nearly accurate numerical modelling of springback in sheet metal forming, finite element method (FEM) is commonly used. However, the achievable accuracy mostly requires the knowledge of sheet material model description. Thus, the selection of an appropriate material model is one of the crucial steps in preparing a FEM model for analysis of sheet bending [2].

Generally, material modelling can be divided into two parts: yield function and hardening law. Widely used yield function which can represent the behaviour of anisotropic steel sheets is the quadratic Hill'48 criterion [22]. To ensure an accurate springback prediction it is necessary to use hardening models which are able to describe the effects of strain hardening. Shu and Hung [23] reported that poor representation of the Bauschinger effect is one of the main reasons for poor springback prediction accuracy. Considering the contact behaviour, the Lagrange multiplier method and the penalty method are the main methods used to incorporate the contact conditions into a finite element formulation [24, 25]. The choice of an element type is also very important for accurate numerical prediction of springback using FE analysis. Depending on the geometry of the sheet metal specimen, different types of elements can be used to discretise the workpiece, i.e. 2D plane strain, shell, solid or solid-shell. To speed up the analysis time, shell elements are commonly used because they provide better accuracy of springback prediction, compared with 3D solid elements.

In this paper, the results of experimental and numerical investigations of springback of deep-drawing steel sheets commonly used in automotive industry are presented. Air bending experiments were carried out in a designed 90° V-die. The numerical model of air V-bending is built in FEM-based ABAQUS/Standard program and material anisotropy and strain hardening phenomenon have taken into consideration.

2 Material and method

The experiments are conducted on 40 mm x 100 mm DC04 steel sheets cut under the angle of 0° and 90° with respect to the rolling direction of the sheet. The mechanical properties of the sheet metals (Table 1) have been determined through uniaxial tensile tests of samples cut along three directions with respect to the rolling direction. The average values of mechanical parameters are determined using the formula:

$$X_{av} = \frac{X_0 + 2X_{45} + X_{90}}{4} \tag{1}$$

where: X – the mechanical parameter, the subscripts denote the orientation of the sample with respect to the rolling direction of sheet.

The anisotropy of plastic behaviour of sheet metals is characterized by the Lankford's coefficient *r*, which is determined by uniaxial tensile tests. Hollomon law describes the value of parameters of strain hardening phenomena as follows:

$$\sigma = C \cdot \varepsilon^n \tag{2}$$

where: σ – stress, ε – plastic strain, C – strain hardening coefficient, n – strain hardening exponent.

The values of the parameters *C* and *n* are determined from the logarythmic true stress – true strain plot by linear regression. The mean value of *n* exponent for the whole range of strain is usually assumed in the numerical simulations. Upon derivation of Eq. (2), the following expression for the strain hardening exponent is obtained:

$$n = \frac{d\sigma}{d\varepsilon}\frac{\varepsilon}{\sigma} \tag{3}$$

Based on the results of the uniaxial test, it was found that the value of the exponent *n* depends on micro-structure texture of the material and changes with effective strain value.

Air bending experiments were carried out in a designed semi closed 90° V-die (Fig. 1). The die assembly consists of a die with Rd = 10 mm rounded edge, and a punch with a nose radius of Rs = 10 mm. During the tests, three parameters were measured: bending force FB, punch bend depth under loading fl and punch bend depth under unloading ful (Fig. 2). The values of these parameters were registered by QuantumX Assistant V.1.1 program for processing the signals of both force and punch stroke transducers. The amount of springback was then evaluated as:

$$K = f_{UL} / f_1 \quad (4)$$

Table 1. Mechanical properties of the tested sheets.

Orientation	Yield stress σ_y [MPa]	Ultimate tensile strength σ_m [MPa]	Strain hardening coefficient C [MPa]	Strain hardening exponent n	Lankford's coefficient r
0°	182.1	322.5	549.3	0.214	1.751
45°	196	336.2	564.9	0.205	1.124
90°	190	320.9	555.2	0.209	1.846
Average value	191.02	328.9	558.57	0.208	1.461

where: ful – punch bend depth under unloading, fl – punch bend depth under loading.

Samples cut in the rolling (0°) and transverse (90°) to the rolling direction were used in V-bending test. The resultant of springback coefficient for each experiment is then obtained by averaging the values of springback for the three experiments.

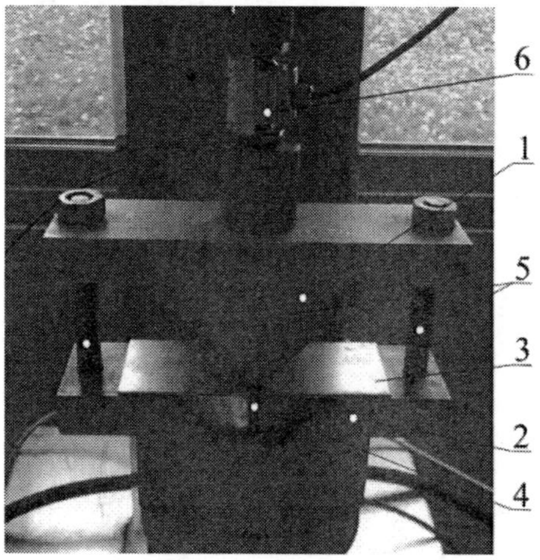

Fig. 1. The experimental setup: 1 – punch, 2 – die, 3 – sample, 4 – punch stroke controller, 5 – guide columns, 6 – force transducer.

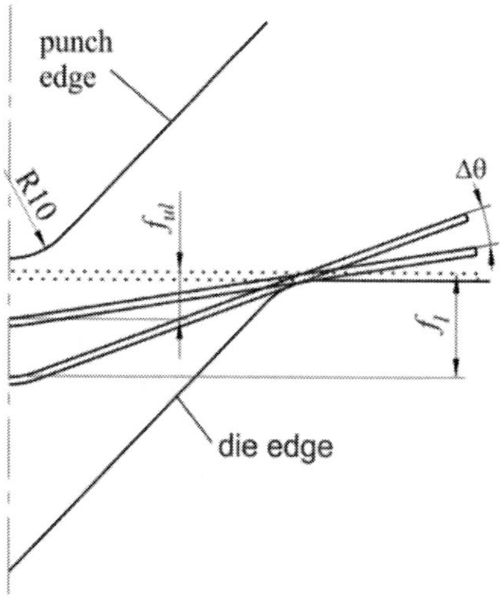

Fig. 2. The schematic of measurement of springback

3 Numerical model

The geometry of the blank and tools corresponds to the experimental set-up (Fig. 1). The analytical discrete rigid tools are meshed using 4-node 3-D bilinear rigid quadrilateral R3D4 elements. The blank was modeled with 4-node quadratic, doubly curved shell elements S8R [26] with five integration points through sheet thickness direction. Five integration points were employed because they are accurate to obtain satisfactory results [27]. The S8R shell type element is intentionally applied for the analysis of sheet metal forming processes [28], and this type of element accounts for the change of thickness in its output variables, unlike solid and plane strain elements. The blank model is composed of the different number of elements to analyze the mesh sensitivity. The tools consist of 9586 elements. The used boundary conditions allow moving the punch at Z-direction (Fig. 3). To determine an optimal mesh size the numerical analyses were carried out for four meshes with the number of elements: 84, 280, 1120 and 4400. Furthermore, the sensitivity analysis of the mesh size was done for four punch strokes f_g: 3, 6, 12 and 18 mm. After getting certain stroke, the punch is moved back and the springback was measured. In air V-bending process, there exists linear contact between the sheet and samples so the friction coefficient influences the springback amount in small scale. The friction coefficient between the sheet metal and tools was assumed to be 0.01 [29].

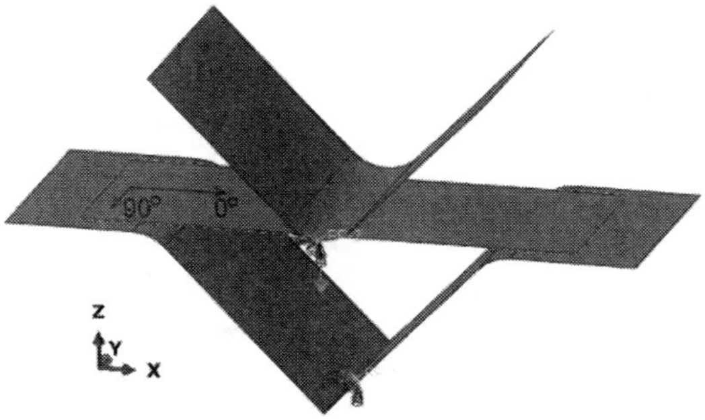

Fig. 3. Boundary conditions in V-bending model of the sample cut along the sheet rolling direction

The elastic behavior is specified in numerical simulations by the value of Young's modulus, E = 210 GPa, and of Poisson's ratio ν = 0.3. In the numerical model, the anisotropy of the material has been established using Hill (1948) [22] yield criterion (Eq. 5) with hardening behavior using Hollomon power-type law (Eq. 2).

$$\bar{\sigma}^2 = F(\sigma_{22}-\sigma_{33})^2 + G(\sigma_{33}-\sigma_{11})^2 + H(\sigma_{11}-\sigma_{22})^2 + 2L\sigma_{23}^2 + 2M\sigma_{31}^2 + 2N\sigma_{12}^2 \quad (5)$$

where, $\bar{\sigma}$ is the equivalent stress, and indices 1, 2, 3 represent the rolling, transverse and normal direction to the sheet surface. Constants F, G, H, L, M and N define anisotropy state of material and can be computed based on Lankford's coefficients [22].

The major advantage of the Hill (1948) function is that it gives an accurate description of yielding of steel sheets [30].

4 Results and discussion

Increasing the number of elements stabilizes the springback (Fig. 4) measured as the difference between punch bend depth under loading f_l and unloading f_{ul}:

$$K_S = f_l - f_{ul} \quad [\text{mm}] \quad (6)$$

The error of springback prediction K_s (Table 2) is determined using the formula:

$$E_S = 100 - \frac{100\left(f_l^{a+1} - f_{ul}^{a+1}\right)}{f_l^a - f_{ul}^a} \quad [\%] \quad (7)$$

where a is the level of the mesh density I, II, III and IV corresponded to the number of elements: 84, 280, 1120 and 4400, respectively.

Next, the absolute mean error E_s^{abs} is determined:

$$E_s^{abs} = \frac{1}{4}\sum_{i=1}^{i=4}\left|E_s^{(i)}\right| \ [\%] \qquad (8)$$

where i is the level of punch stroke i = 1, 2, 3, and 4 corresponded to punch stroke 3, 6, 12, and 18 mm, respectively.

The error E_s after increasing the number of elements from 1120 to 4400 is the smallest for all analysed mesh densities (Table 2).

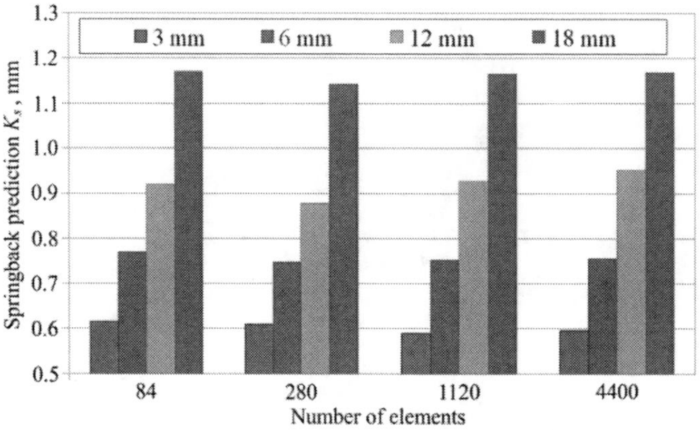

Fig. 4. Influence of the number of elements on amount of springback

Table 2. Error of springback prediction

Punch stroke, mm	The springback prediction error for the change in the number of elements E_s, %		
	84 → 280	280 → 1120	1120 → 4400
3	-0.81	-3.44	1.01
6	-2.98	0.52	0.41
12	-4.45	4.35	2.79
18	-2.3	1.93	0.33
Absolute mean error	2.64	2.56	1.14

The absolute mean error value E_s^{abs} for all analysed punch strokes f_g is equal to 2.64% (after increasing number of elements from 84 to 280), 2.56% (from 280 to 1120) and 1.14% (from 1120 to 4400). We also assume the accepted E_s^{abs} value of 1.5%. So, the number of elements 1120 is acceptable and then numerical models considered this number of elements.

Springback intensity is influenced by the orientation of the sample in terms of symmetric plane of the punch. It is conceivable that sample orientation due to rolling direction produces variations of the elastic-plastic properties of the sheet metal and residual stress after sample unloadning (Fig. 4). Although the distribution of the residual stress is similar for both analysed orientations, the difference in the value of maximum stress depends on punch depths and the difference is in the range of 3 - 5 MPa. The sheets produced in cold-rolling process exhibit directional elongation of grains in the rolling direction. According to Eq. (4), the high springback of the material denotes the lower values of springback coefficient. So, the samples bent parallel to the sheet rolling direction (0° in figure 3) exhibit higher value of springback. The variation of springback coefficient (K) value with the relative bending depth w/f_l (where w is bending length of a sample) is analyzed as depicted in figure. 5. As shown in the figure, the K value and w/f_l have almost linear dependence. The differences in K value between the results of both orientations of the sample are similar for all analyzed relative bending depths. However, the numerical models in all cases overpredict the springback coefficient value.

This is because, in the case of 0° rolling direction, during the draw bending process the grains are only subjected to tensile or bending stresses, but in the case of 90° rolling direction an important amount of deformation energy is used to change the orientation of grains [31]. The lower springback coefficient evaluated based on the punch depths under unloading and loading translates into higher value of springback angle $\Delta\theta$ (Fig. 2). It is understood that the springback angle is a function of yield strength to Young's modulus ($\sigma y/E$) [32], which is higher for transverse direction than rolling direction.

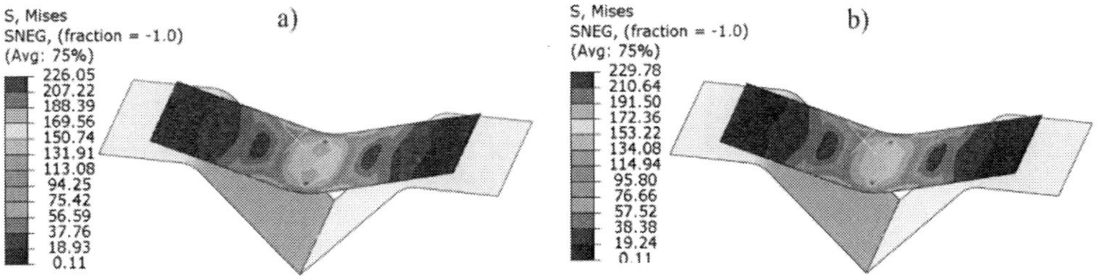

Fig. 4. The von Mises stress distribution after unloading samples oriented at 0° (a) and 90° (b), punch depth under loading 12 mm

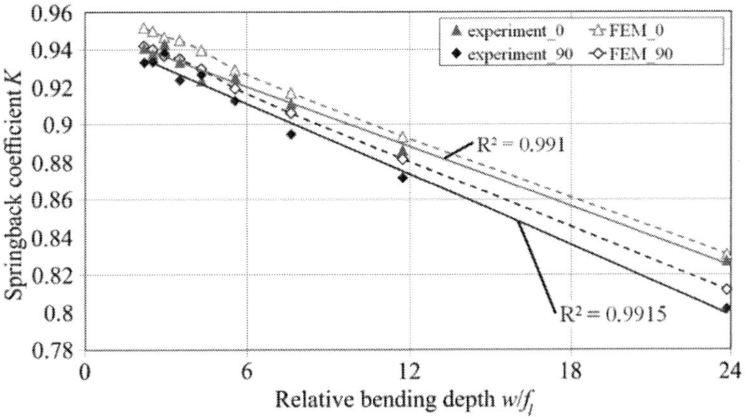

Fig. 5. Comparison of springback coefficient value determined experimentally and by FEM approach

4 Conclusions

The results of experimental and numerical investigations of springback of deep-drawing steel sheets commonly used in automotive industry are presented in this paper. The results indicated that orientation of the sample cut from anisotropic steel sheet has a great influence on springback. The test specimens were cut in the rolling direction (0°) and transverse direction (90°) of the sheet metal and the blank cut along the transverse direction exhibits greater springback than the blank cut in the rolling direction. These results were found both experimentally and by FE models. However, in all analyzed case, numerical models overpredict the springback coefficient value. A decreasing springback coefficient is observed when the relative bending depth (w/f_l) increases. The analysis of variation of springback coefficient value shows almost linear dependence between the spring coefficient (K) value and relative bending depth. In the case of high values of bending depth (f_l), there is an increasing plastic zone through the sheet thickness in the middle part of the sample that influences the low values of springback coefficient.

References

1. H.M. Huang, S.D. Liu, S. Jiang, J. Eng. Mater. Tech. **123** (2001)
2. I.A. Burchitz, Improvement of springback prediction in sheet metal forming (PrintPartners Ipskamp, Enschede, 2008)
3. I.A. Choudhury, V. Ghomi, Proc. Inst. Mech. Eng. B: J. Eng. Manu. **288** (2014)
4. H.K. Yi, D.W. Kim, C.J. Van Tyne, Y.H. Moon, Proc. Inst. Mech. Eng. C: J. Mech. Eng. Sci. **222** (2008)
5. M. Bakhshi-Jooybari, B. Rahmani B, V. Daeezadeh, A. Gorji, Mater. Des. **30** (2009)
6. T. Meinders, I.A. Burchitz, M.H.A. Bonte, R.A. Lingbeek, Int. J. Mach. Tool. Manu. **48** (2008)
7. B. Chirita, Arch. Civ. Mech. Eng. **3** (2003)
8. J.R. Duflou, R. Aerens, CIRP Ann: Manuf. Tech. **55** (2006)
9. H. Chalal, R. Sever-Gabriel, B. Tudor, Int. J. Mech. Sci. **59** (2012)
10. W. Lems, The change of Young's modulus after deformation at low temperature and its recovery (Delft University of Technology, Delft, 1963)
11. R.M. Cleveland, A.K. Ghosh, Int. J. Plast. **18** (2002)
12. M. Yang, Y. Akiyama, T. Sasaki, J. Mater. Process. Tech. **151** (2004)
13. F. Morestin, M. Boivin, Nuclear Eng. Des. **162** (1996)
14. J.F. Wang, R.H. Wagoner, W.D. Carden, D.K. Matlock, F. Barlat, I. J. Plast. **20** (2004)
15. D.K. Leu, Mater. Des. **94** (2016)
16. W. Xiong, W. Wang, M. Wan, X. Li, J. Manu. Process. **19** (2015)
17. P. Xue, T.X. Yu, E. Chu, J. Mater. Process. Tech. **89-90** (1999)
18. H. Livatyali, H.C. Wu, T. Altan, J. Mater. Process. Tech. **120** (2002)
19. J.T. Gau, G.L. Kinzel, Int. J. Mech. Sci. **43** (2001)
20. D.K. Leu, J. Mater. Process. Tech. **66** (1997)
21. F. Barlat, D. Banabic, O. Cazacu. Anisotropy in sheet metals. In: D.Y. Yang, S.I. Oh, H. Huh, Y.H. Kim (Eds.) Proc. 5th Int. Conf. And Workshop on Numerical Simulation of 3D Sheet Forming Processes NUMISHEET 2002, 515 – 524, (Jeju Island, Korea, 2002)
22. R. Hill, Proc. Royal Soc. Lond. **193** (1948)
23. J.S. Shu, C. Hung, Int. J. Mach. Tool Manu. **36** (1996)
24. B.D. Carleer, Finite element analysis of deep drawing (University of Twente, Enschede,1997)
25. G. Kloosterman, Contact methods in finite element simulations (University of Twente, Enschede, 2002)
26. F. Song, H. Yang, H. Li, M. Zhan, G. Li, Chinese J. Aeronaut. **26** (2013)
27. M. Hudovernik, D. Staupendahl, M. Gharbi, M. Hermes, A.E. Tekkaya, K. Kuzman, J.M. Slabe, Stroj. Vest. -J. Mech. Eng. **59** (2013)
28. H.G. Lemu, T. Trzepieciński, Stroj. Vest.- J. Mech. Eng. **59** (2013)
29. W. Frącz, F. Stachowicz, Manu. Ind. Eng. **2** (2008)
30. T. Trzepieciński, H.L. Gelgele, Int. J. Mater. Form. **4** (2011)
31. A. Albut, G. Brabie, Arch. Civ. Mech. Eng. **6** (2006)
32. R. Narayanasamy, P. Padmanabhan, J. Manuf. Eng. **3** (2008)

Peculiar behaviour of pure ENAW-Al99.5 sheet metal during uniaxial tensile test

Emil Yankov[1,*], and *Maria P.* Nikolova[1]

[1]University of Ruse, Dept. Material Science and Technology, Ruse, 8 Studentska Str., Bulgaria

Abstract. Due to the generalized use of aluminium (Al) and Al alloy a detailed investigation of the mechanical and technological behaviour of a wide range of sheet metals is particularly needed. Often during deformation the Al sheets fold or crack which makes their subsequent use harder. The aim of this study is to identify the causes of the changes taking place during deformation that leads to earlier breaks of the Al sheet material. Pure ENAW-Al99.5 in as-received condition is used for examination in order to determine its behaviour without interference of the alloying elements. After the uniaxial uniform tensile test significant differences in the mechanical strength of the samples cut out from the sheet material at three different directions - 0°, 45° and 90° angle to the rolling direction are observed. The hardness values measured at three different directions: alongside the long side of the samples, cross-section and top view are compared. The XRD and microstructure analysis confirm the grain and lattice reorientation after the tests compared with the initial structure of the sheet material. These results could be applied as a basis of comparison when examining other Al alloys used in aviation and mechanical engineering, chemical and agro-food industry, etc.

1 Introduction

The wide application of aluminium (Al) and its alloys in the modern mechanical engineering, car, aircraft, agro-food industry, etc., imposes a detail examination of the material behaviour during deformation and form-changing. For the purpose of improving its mechanical properties and plasticity Al is often alloyed with Mn and Mg and the materials thereof are used for higher levels of deep drawing. The behaviour of these alloys has been investigated by many authors [1-9] in order to determine the technological, mechanical, thermal, electro-chemical, etc. properties.

Over the years there have been signs that the pure Al is attracting less interest likely because of the low mechanical properties of the metal. In order to make a correct comparative analysis for the material behaviour, it could therefore be better the properties of the pure Al to be well defined. These properties afterward could be used as points of reference for future examination concerning the influence of the alloying elements during different types of testing.

* Corresponding author: eyankov@uni-ruse.bg

The uniaxial uniform tensile test of technically pure Al sheet material cut at three different directions - 0°, 45° and 90° in respect to the rolling direction is being subject to this investigation. The aim is to establish the mechanical properties in these three different directions and the reorientation of the crystallographic planes by means of XRD analysis. Additional evidence for the strengthening of the material is the change in the hardness values in three different directions of the tested samples.

2 Experimental Procedures

The material ENAW-Al99.5 (ENAW-1050A) was received in the form of sheets with the following dimensions: 1000×2000 mm and thickness of t_0 = 1.5 mm in annealed condition. Its chemical composition shown in Table 1 was measured by glow discharge optical emission analysis (GDOES), operated at 20 mA, 800 V and working pressure of 2.6 hPa.

Table 1. Chemical composition of ENAW-Al99.5 (ENAW-1050A).

Element	**Fe**	**Mn**	**Ti**	**Cu**	**Mg**	**Ti**	**Al**
Wt.%	0.00	0.021	0.011	0.079	0.0089	0.0042	Bal.

Table 2. Mechanical properties of ENAW-Al99.5 (ENAW-1050A) at 20°C.

Mechanical properties	R_m (MPa)	Elongation, A (%)			HV0.5
		A_5	On a basis of L_0 = 50 mm	On a basis of L_0 = 65 mm	
Value	74 ÷ 77.9	24 ÷ 50	18.9 ÷ 24.8	27.6 ÷ 37.8	24.7

The mechanical properties (Table 2) of the as-received material were examined on a universal testing machine Instron-3384, on the base of standard ISO 6892-1:2009. A Vickers Hardness tester 432 SVD by Wilson-Wilpert was used for the hardness measurements of the metal according to the standard ISO 6507-1.

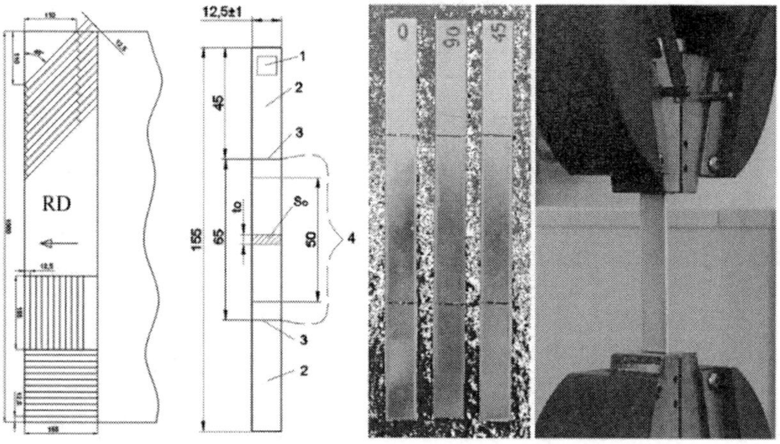

a) b) c) d)

Fig. 1. Sample preparation and testing: a) a scheme of cutting of the samples; b) shape and dimensions of the samples: 1) marking place; 2) gripping area of the clamps; 3) restraints; 4) tested area; t_0 – initial thickness, S_0 – initial cross-sectional area; c) overall appearance of the samples; d) a fixed sample in the clamps of the machine.

Samples were cut out from the sheet material at three different directions - 0°, 45° and 90° to the RD as shown in Fig. 1a. The specimens used for the tensile test were sawn on the base of standard ISO 6892-1:2009 (Fig. 1a, b, c) and cut from the sheet by a guillotine with digital laser positioning. The material was tested in as-received (called undeformed) condition at ambient temperature. The static uniaxial tensile test was performed on a universal testing machine Instron-3384 (Fig. 1d) on the base of the same standard, supplied with a software Bluehill-3. For each test a control measurement of the working length of testing, height and width was held and the values were set in the used test method via the software program. For the purpose of the testing method a constant strain rate of 5 mm/min until the load fell down to 100 N (or downturn of the maximum load below 90%) was used. The data were saved in an appropriate digital form for subsequent analysis.

The surface preparation for hardness measurements of all samples before and after the tensile test (TT) involved grinding with grit paper (from P400 to P2500) and mechanical polishing with non-abrasive paste. For the hardness measurements three mutually perpendicular planes from each specimens cut near the break down zone were prepared. A Vickers Hardness tester 432 SVD by Wilson-Wilpert was used for the hardness measurements of the metal according to the standard ISO 6507-1. In order to compare the hardness changes the researches were made on three different planes: 1) alongside the long side of the sample (plane I); 2) cross-section (plane II) and 3) top view (plane III) of each (undeformed and TT) sample with a load of 0.5 kg and dwell time of 10s. Ten sequential measurements by increments of 50 μm between the imprints were made for each plane.

The X-ray phase identification was performed with URD-6 diffractometer, applying Bragg-Brentano geometry and Cu-K_α radiation in the scanning range of 30-120° 2θ with a step size of 0.05° 2θ and counting time of 2.5 s/step. The Cu-K_α tube voltage and current were 30 kV and 20 mA, respectively. The lattice strain response was measured in the longitudinal and transverse directions of plane III of all samples. The qualitative phase analysis was held using Match!3 software. The texture analysis was performed by calculating the texture coefficient (TC) by the equation shown in [10].

3 Experimental Results

The representative indicator diagrams (Fig. 2a) are result of five samples testing from one the same angle of cutting sample (0˚, 45˚ and 90˚, respectively). The experimentally determined data indicate the load at maximal charge F_m for tensile extension at maximum charge ΔL_m, elongation at break ΔL_{br}, elongation at tensile strength at allongement A and tensile strength R_m. The diagram considered as representative is the one with the closest mechanical properties to the average results from each indicator diagram for a particular angle of cutting [11].

As shown in Fig. 2a, both curves of 0˚ and 45˚ angle of cutting samples practically coincide, except for the last stage of deformation where the degree of relative elongation differs by more than 2 mm. The mechanical strength of the 90˚ cut samples is lower than the others but the change in the trajectory is like that of the other two directions of cutting. These phenomena can be attributed to the internal trans-crystal displacement, favourable lattice reorientation, grain fragmentation and other crystal effects.

A quantitative comparison of the mechanical properties of the material according to the tensile strength - tensile extension diagram is not possible because the real changes in the transverse section width have not been taken into account. Therefore, the $F = f_1(\Delta L)$ diagram (Fig. 2a) is converted into true stress – strain diagram $\sigma = f_2(\varphi)$, Fig. 2b, that takes account of equatorial section changes at any time compared to the tensile stress applied. The diagram indicates similar trend for the stress changes of the 0˚ and 45˚ angle of cutting samples and their mechanical strength values are slightly higher than those of 90˚ cut samples. The true strain values do not exceed 41% which has been highlighted in previous examinations [3, 12]. The true stress of the material ranges from 110 up to 118 MPa.

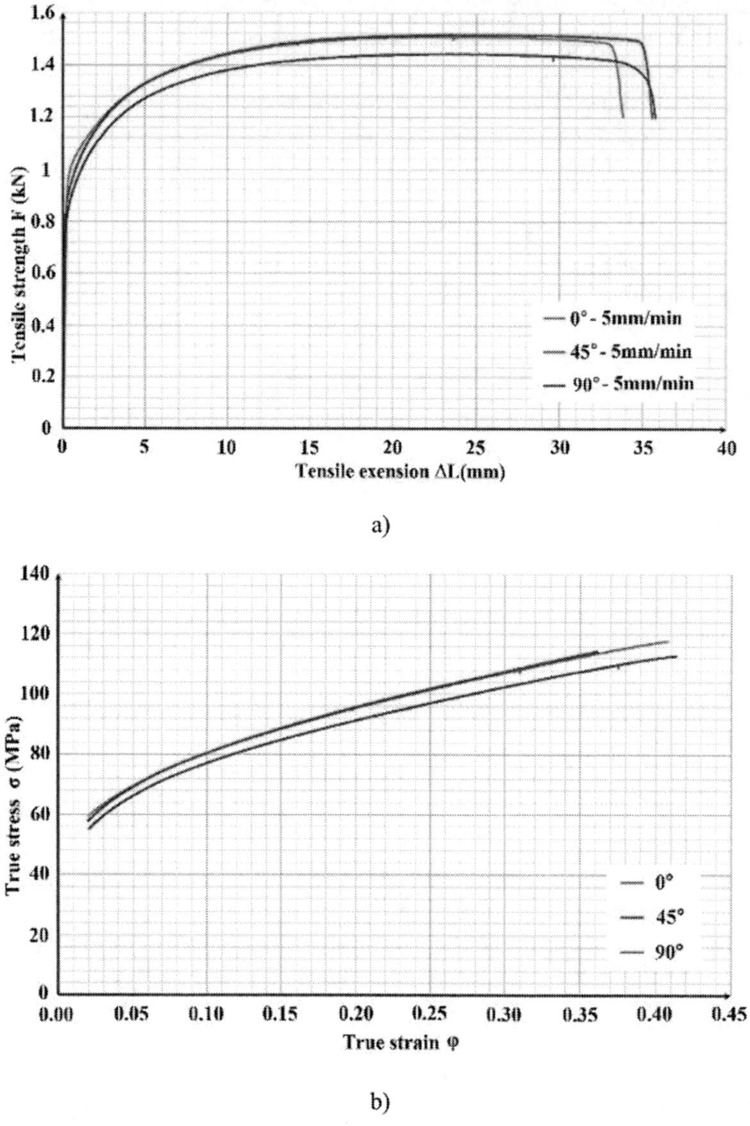

Fig. 2. Representative indicator diagrams of ENAW-Al99,5; t_0=1.5 mm: a) tensile strength - tensile extension, b) true stress – true strain.

To clarify the reasons for the differences in the mechanical strength, Vickers hardness measurements in orthogonal to each other planes and XRD measurements for analysis of the reorientation of the crystallographic planes could be used. The results of the average hardness values of planes I, II and III of the undeformed and TT samples cut at 0°, 45° and 90° angle to the RD are listed in Excel tables and graphically shown in Fig. 3. Ten measurements are averaged for each plane (I, II and III) of a tested specimen. As the number of the directions examined is six, the total number of the completed tables is equal to eighteen.

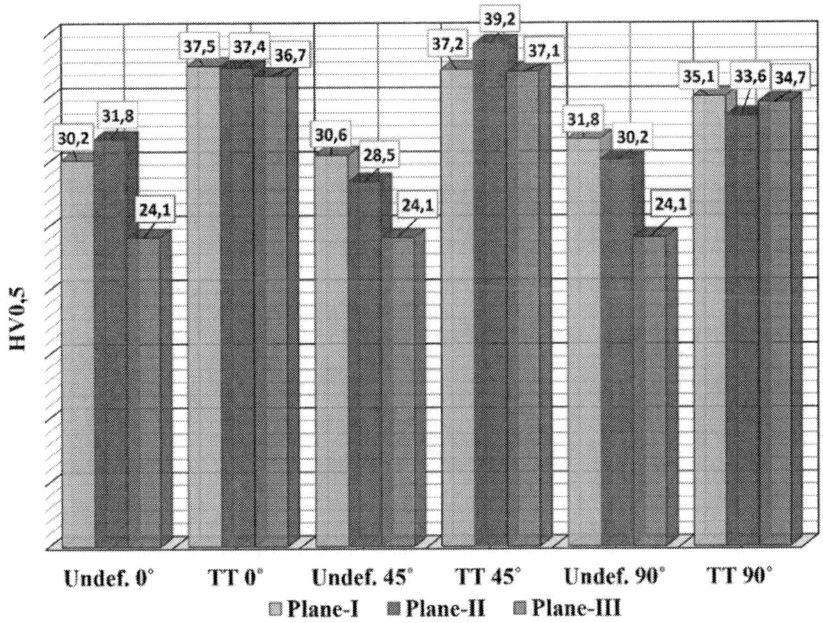

Fig. 3. Average hardness values before and after the tensile test of 0°, 45° and 90° cut samples measured at planes I, II and III.

In as-received condition of the 0° cut samples the lowest hardness values are measured in plane III and the highest – in plane II (Fig. 3). This difference is probably due to the initial pre-treatment – annealing of the undeformed metal that leads to surface softening while the sheet core remains not fully recrystallized. The observed phenomena will affect the trans-crystal movements during the uniaxial test. Conformation of this hypothesis would give a metallographic examination in planes I, II and III of all as-received and TT samples. The undeformed samples cut at 45° and 90° also show the lowest hardness value at plane III, while the highest values are measured at plane I. The reason for the higher hardness in plane I is the strengthening that runs perpendicular to the plane I within the 90° cut sample and the texture direction within the 45° cut sample that could be confirmed by XRD analysis.

After the tensile test the 90° cut sample demonstrates the lowest tensile strength values. This fact also affects the results of the hardness values after the test (Fig. 3) that are lower than the analogue values of the 0° and 45° cut samples. The highest are the hardness values measured at plane I of the TT 90° cut sample and the lowest are those in the perpendicular direction (plane II). This shows that the material has been subject to a hindered trans-crystal movements, but nevertheless, the hardness is increased by 10.5%, 11.26% and 43.6% in plane I, II and III, respectively (Fig. 4). The result gained shows that the sheet metal hasn't been fully recrystallized and the deformation takes place mainly on the surface.

For the TT 0° cut sample the highest hardness values are measured at plane I and the lowest – at plane III (Fig. 3). Calculating the hardness changes in respect to the values of the undeformed 0 ° cut planes (Fig. 4), the highest strengthening is observed in plane III (52.1%) and the lowest – in plane II (17.55%).

Despite the highest hardness values measured in plane I for the as-received 45° cut sample, after the tensile test the highest is hardness measured at plane II (Fig. 3). In contrast, the highest is the strengthening in plane III (53.62%) and the lowest – in plane I (21.83%), Fig. 4.

DOI: 10.1051/matecconf/201712103024

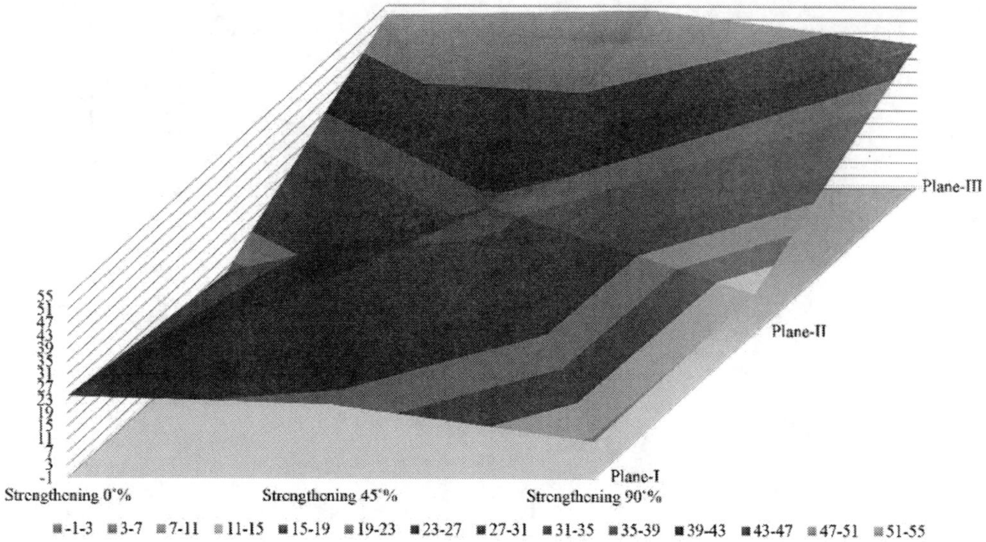

Fig. 4. Strengthening in percentage of the tensile tested samples (0°, 45° and 90°) applied to the undeformed samples for planes I, II and III.

The results thus obtained show that the highest is the strengthening in plane III for all three examined directions that is due to the softer sheet surface in as-received condition. For this reason, those planes of all samples are used for determining the changes in the crystallographic planes and texture of the material. The XRD analysis results obtained by a horizontal and vertical Bragg-Brentano scanning are summarized and shown in Fig. 5.

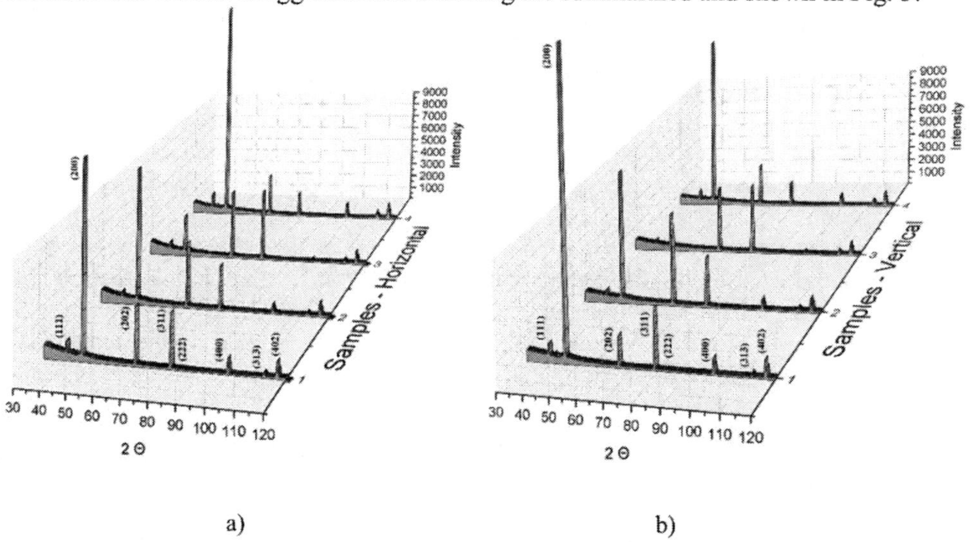

Fig. 5. Summarized diffraction patterns of ENAW-Al99.5 in initial state and after the tensile test of the samples in: a) horizontal direction and b) vertical direction of plane III. The numbered sample are as follows: 1- undeformed; 2 - 0° cut samples; 3 - 45° cut samples; 4 - 90° cut samples. The scanning geometry used corresponds to Bragg-Brentano (BB).

In as-received condition the highest is the intensity of the (200) plane both in horizontal and vertical directions, while the lowest is that of the (222) crystallographic plane (Fig. 5). According to the reference data the 100% reflection should be the (111) plane peak. This proves once more that there are residual traces of the initial deformation within the annealed metal. The inherited texture could explain the differences in the tensile strength values during the tensile test.

After the deformation of the 0° cut sample scanned in horizontal direction the highest is still the intensity of (200) plane, because the grains do not undergo reorientation but elongation in the tensile direction. Nevertheless, the (200) intensity after the deformation decreases with 31% with respect to the undeformed peak intensity (Fig. 6a), possibly caused by grain refinement, reorientation or other crystallographic reasons. Reductions in the intensities of planes (111), (311), (400), (313) and (402) with 35.33, 17.03, 36.6, 18.04, 12.01%, respectively are also observed for the account of an increase of (202) reflection intensity with 17.22%. In the vertically scanned 0° cut sample the (202) plane increases its intensity with 100.81% at the expense of the decrease of all the other planes intensities which is most relevant for (200) (53.17%) and (400) (32.27%) planes (Fig. 6b).

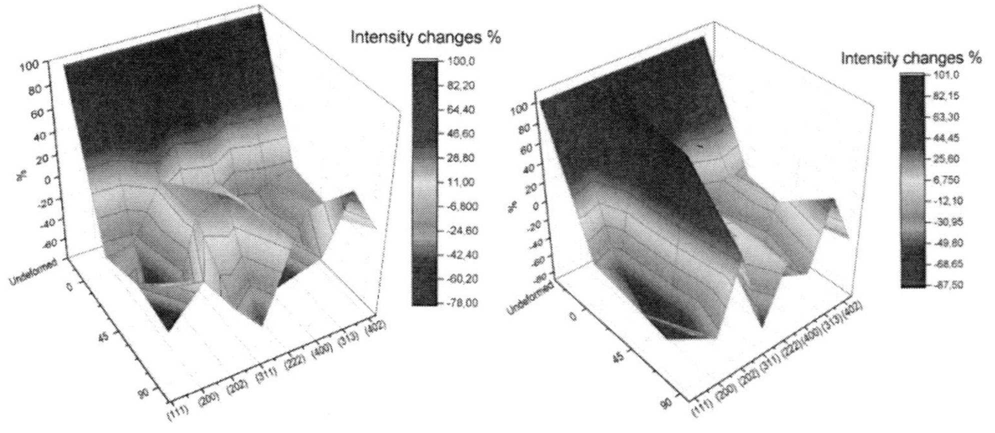

a) b)

Fig. 6. Intensity changes (in percentage) of TT 0°, 45° and 90° applied to the underformed sample measured at planes III in: a) horizontal and b) vertical direction.

For the 45° cut sample the highest are the intensities of (311) and (202) reflections that rise up to 30.32% and 19.75, respectively in horizontal direction and 11.79 and 90.18%, respectively, in vertical direction (Fig. 5). The rest of the crystallographic reflections decrease their intensities. Mainly affected are: (200) plane that reduces its intensity with 77.72% and 87.11% for the horizontal and vertical direction, respectively, and (400) plane with decreasing intensity of 76.09% and 77.79% for the horizontal and vertical direction, respectively (Fig. 6a, b). This significant difference in the reorientation of the crystals is due to the existence of favourable planes for trans-crystal movements applied to the direction of the tensile strength.

In the 90° cut sample the intensity of (200) plane in horizontal and vertical direction of scanning is higher than that of the TT 0° and 45° cut samples (Fig. 5). Nevertheless, the (200) plane intensity decreases with 11.72% and 38.27% in horizontal and vertical direction, respectively, with respect to the undeformed sample. The highest is the reduction of (311) reflection intensity – with 50.04% and 59.27%, in horizontal and vertical direction, respectively. This decrease is for the account of the increase of (313), (222), (111), (200) and (402) plane intensities with 42.41%, 18.26, 16.63, 11.72, 0.22%, respectively in horizontal direction (Fig. 6a) and (313) and (202) with 37.67 and 31.91%, respectively, in vertical direction (Fig. 6b). The strong reorientation of all planes in the horizontal direction is also caused by the direction of the applied tensile strength. Despite the unfavourable crystals' orientation, the (311) plane greatly reduces its intensity at the expense of (313) intensity increase. This reorientation is the only one possible, which accounts for the residual plasticity of the material cut at 90° angle to the rolling direction and tensile tested.

The results obtained from the XRD analysis of plane III scanned in horizontal and vertical direction indicate clearly the favourable crystallographic changes occurring during the tensile test. Those planes that increase their intensities have been less deformed and textured for the account of the highly deformed and textured ones. The latter have higher initial intensities that subsequently decrease.

In order to confirm the results the texture coefficients (TC) for all reflections are calculated compared to the standard integrated intensity and shown in Fig. 7. The undeformed material is strongly textured in (400) and (200) and less textured in (202) and (311) planes both in horizontal and vertical directions. After the tensile test the (400) and (200) planes of 0° and 90° cut samples remain the strongest textured ones, but they slightly decrease mainly as a result of the increase of the TCs of (202), (313) and (402) for the 0° cut samples and (202) and (313) for 90° cut samples scanned in vertical direction while the other coefficients of the planes slightly amend.

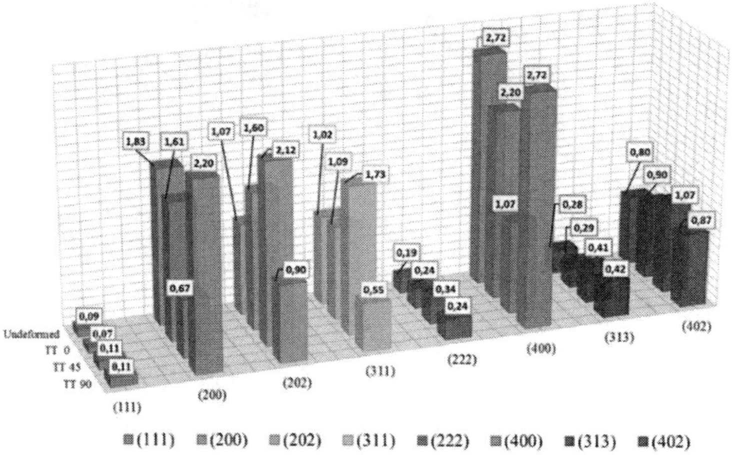

a)

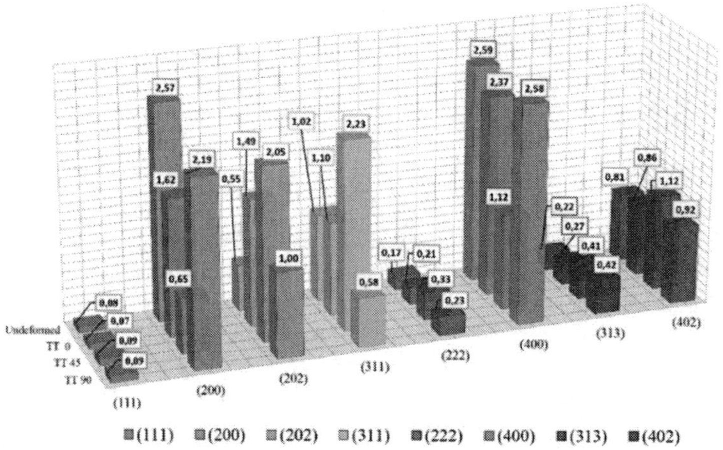

b)

Fig. 7. Calculated texture coefficients of the crystallographic planes of the undeformed and TT samples measured at planes III in: a) horizontal direction; b) vertical direction.

In horizontal and vertical direction of the TT 45° cut sample the strongly textured (400) and (200) planes greatly diminish their TCs for the account of changes of the TCs of the other planes. The strongest textured planes in horizontal direction are (202), (311), (402), (313) (222) and (111), while in vertical direction these are (311), (202), (402), (313) (222) and (111) planes.

The result of the calculated TCs reiterates the conclusion that the strongest changes occur in the crystallographic planes of the TT 45° cut sample. The tensile test curves of the 0° and 45° cut sample merely overlap despite the different nature of the crystallographic and texture changes. It therefore follows that the most favorable trans-crystal movements are in planes (202) and (311), because both planes strongly change their TCs applied to the basic values of the undeformed sample.

Metallographic observations in each plane (I, II and III) of each undeformed and tested specimen with a subsequent software analysis for determining the grain sizes, orientations, distribution, etc., are further needed for fully explaining the mechanisms of strengthening, trans-crystal movement, texturing that will disclose thoroughly the peculiar behavior of ENAW-Al99.5 sheet.

4 Conclusions

It has appeared from the various examinations that the annealed sheet material shows traces of residual texturing despite the heat treatment process that recrystallizes the surface material more than the sheet core. That's why the highest strengthening is observed in this softest surface zone – plane III for all three examined directions of cutting - 0°, 45° and 90° angle to the rolling direction. The highest positive reorientation in both horizontal and vertical direction of scanning is seen in (202) plane in the case of all tested samples, with the exception of the 90° cut sample scanned in horizontal direction where the TC decreases. The TC of the initially strongly textured (200) plane undergoes the strongest reduction after the tensile test of all samples scanned in horizontal and vertical direction, with the exception of the TT and horizontally scanned 90° cut sample where the TC slightly increases. The calculated TC clarifies the mechanisms of the trans-crystal movements that follow a decrease in the values of the textured planes for the account of slightly textured or non-textured planes with stocks of residual plasticity. The latter increases their TCs enabling greater deformation.

Additional microstructural analysis after the uniaxial tensile test in planes I, II and III of the undeformed and TT samples cut at 0°, 45° and 90° angle to the RD needs to be further fully exploited for clarification of the peculiar behaviour of the sheet material.

References

1. M. T. Di Giovanni, E. Cerri, D. Casari, M. Merlin, L. Arnberg, and G. Garagnani, Metall. Mater. Trans. A **47** (5), 2049 (2016)
2. A. Zykova, L. A. Kazantseva, I. A. Kurzina, V. Kh. Dammer, and A. V. Chumaevskii, Tsvetn. Met. (Moscow), **5**, 4 (2015)
3. T. A. Kalashnikova, S. Tarasov, A. V. Chumaevskii, F. Andrey, E. A. Kolubaev, V. E. Rubtsov, A. Eliseev, and V. A. Beloborodov, Advanced Materials with Hierarchical Structure for New Technologies and Reliable Structures, AIP Conf. Proc., 020079 (2016)
4. M. Akbarifar, M. Divandari, Conference: Iran International Aluminum Conference (IIAC2014), (2014)
5. X. Wu , N. Tao , Y. Hong , B. Xu , J. Lu , K. Lu, Acta Mater., **50**, 2075 (2002)
6. K. Dai, J. Villegas, Z. Stone, L. Shaw, Acta Mater., **52** (2), 5771 (2004)
7. R. Kapoor, J.B. Singh, J.K. Chakravartty, Mater. Sci. Eng. A, **496**, 308 (2008)

8. R.Z. Valiev, N.A. Enikeev, M.Yu. Murashkin, V.U. Kazykhanov, X. Sauvage, Scr. Mater., **63** (9), 949 (2010)
9. P. H. R. Pereira, Y. C. Wang, Yi Huang, T. G. Langdon, Mater. Sci. Eng. A, **685**, 367 (2017)
10. E. Yankov, M. Nikolova, MATEC web of Conferences, CoSME'16, **94**, (2017)
11. E. Yankov, *Sheet materials testing using biaxial hydraulic bulge test*, Dissertation, Ruse, 112 (2014)
12. S. Dumoulin, O. Engler, O. S. Hopperstad and O.-G. Lademo, Modelling and Simulation in Materials Science and Engineering **20** (5), 1-20

DOI: 10.1051/matecconf/201712104001

Augmented reality application for training in pipe defects ultrasonic investigation

Cătălin Gheorghe Amza[1], *Aurelian* Zapciu[2], *Diana* Popescu[2,*], and *Octav* Teodorescu[3]

[1]University POLITEHNICA of Bucharest, Technology and Welding Department, Splaiul Independentei 313, sector 6, Bucharest, Romania
[2] University POLITEHNICA of Bucharest, Department of Machine and Production Systems, Splaiul Independentei 313, sector 6, Bucharest, Romania
[3]Namicon SRL, Romania

Abstract. The paper presents the development process of an Augmented Reality (AR) application used for training operators in using ultrasonic equipment for non-destructive testing (NDT) of pipework. The application provides workers useful information regarding the process steps, the main components of ultrasonic equipment and the proper modality of placing, aligning and moving it on pipe and weld. Using tablet or mobile phone device, an operator can see on screen written details and images on standardized working method, thus offering assistance during the training process. Allowing 3D augmented visualization of ultrasonic equipment overlaid on the real-world environment consisting in pipes and welds, the AR application makes the NDT process easier to understand and learn, as the initial evaluation results showed.

1 Introduction

Although the first Augmented Reality (AR) application for training was developed in 1992 [1] for Boeing Company by Tom Caudell and David Mizell (who also coined the term), this field knew notable development only in the last ten years. This can be explained by the improvements brought to hardware equipment such as head mounted displays (HMD), hand held devices, stereoscopic glasses, mobile devices (tablets, phones), tracking systems, sensors and wearable technology. Moreover, the price decreased for all this equipment due to an increase in the number of producers and developers. This started AR technology consolidation process and made affordable the creation and use of AR applications for education and training in engineering [2-3], medicine [4-5], automotive and aerospace for training and maintenance [6-7], etc.

Literature [8] defines AR as a "human-machine interaction tool", which can supplement or overlay "a real-world environment with computer generated sensory inputs" [9]. There are several conditions to be fulfilled by an AR application: combines virtuality and reality, takes place in real time and objects are 3D represented in the virtual environment, which offers important advantages such as: enhanced perception of the real world, real-time

* Corresponding author: diana@mix.mmi.pub.ro

display of messages for users, integration with 3D CAD systems and interactive modalities to learn and experience the virtual worlds.

In this general context of interest and due to the usefulness of AR application in engineering training, the current paper focuses on presenting the development process of a new application for training operators in using ultrasonic equipment for non-destructive testing (NDT) of pipework.

A survey of the field showed that not many AR applications are available for ultrasonic NDT of pipeline. TWNKLS company specialized in AR created for Applus+ RTD a mobile application that allows operators to see augmented data on defects as coloured volumes superposed on the image of inspected pipe. 3D printed AR markers are placed as a ring around the pipe for generating images of dimensions, position and the nature of the defect. This application is dedicated mainly to maintenance work and not to training and it is based on 3D ultrasonic IWEX (Inverse Wave Field Extrapolation) technology [10]. Another application is reported in [11] and uses AR glasses for visualizing digital C ultrasonic scans overlaid on a tested object like they form "a unity with the structure". Thus, it simplifies testing results' evaluations, these being usually showed as 2D graphics which are harder to interpret, especially on curved surfaces such as pipes. Hybrid markers are considered by authors more appropriate for the application as they can provide both visual and sound characteristics. Another AR based NDT application (called HaroUT), developed using LabView and Microsoft Hololens is presented in. Ultrasonic delays are calculated by the software and the 2D and 3D plots generated are mapped on a virtual pipe.

The literature review pointed a niche in the field of training applications based on AR technology for ultrasonic inspection of pipework. Therefore, starting from the working flow for such type of non-destructive testing, a mobile application AR marker-based was developed. It allows operators to see, overlaid on a real pipe and weld, different displays of information regarding the operations to be performed before starting the inspection process. The correct placement and movements of ultrasonic probe (using an augmented 3D model) on the pipe and the identification of several types of defect by presenting scans images from a defects' database can be also learned using the AR application.

The developed application was initially tested by 5 researchers from University POLITEHNICA of Bucharest. Further tests will involve a larger group of participants for assessing the application usefulness, learning curve, interface friendliness based on a on a 5 points Likert scale questionnaire.

2 Materials and method

2.2 Ultrasonic non-destructive inspection

A typical ultrasonic testing system is composed of several functional units: transmitter/receiver, transducer and a display device. A transmitter/receiver is an electronic device that can produce high-voltage electrical impulses and turns them into high frequency mechanical vibration. Led by transmitter, the transducer generates high frequency ultrasonic energy. Ultrasonic energy is introduced and propagated through the material form of waves. When there is a discontinuity (such as a crack or air inclusion) in the sound waves, a part of the energy will be reflected back due to the defect. Reflected light signal is converted into an electrical signal by the transducer and displayed on a monitor.

By using different types of probes, ultrasonic NDT allows determining and positioning surface and internal defects of various sizes. The method involves the steps presented in Table 1.

DOI: 10.1051/matecconf/201712104001

Table 1. Ultrasonic NDT process steps

Step 1: Prepare the surface of the object to be explored: - Remove surface non-uniformities (notches, spatters etc.) and paint, dirt etc. The surface finish should be better that 6.3 μm - Apply couplings such as: oil without additive, glycerine or grease	Step 3: Establish the level of inspection Step 4: Defects' detection and localization: - Detect the existence of longitudinal defects - Localize discontinuities along material - Establish the depth at which the discontinuity is placed - Establish the existence of transversal discontinuities
Step 2: Preparatory operations for inspection - Choose the working frequency - Choose the angle of incidence for wedge probes - Establish the reference level - Establish the evaluation level - Establish the oscillograms characteristics to different type of defects	Step 5: Establish defects' shape, orientation and size Step 6: Evaluate the nature of the defect

Phased array is one of the ultrasonic testing methods characterized by the fact that computer-controlled excitation is applied to individual piezocomposite elements included in a probe. A steerable and focused ultrasonic beam is generated in phase by transducers which are pulsed independently. Table 2 presents different types of images created using phased array testing method, such as: B-scans lateral scanning, C-scans top scanning – based all on A-scans (amplitude vs. time). There are different ways to display information and the obtained images are dependent of the position of ultrasonic beam, the inspection direction and object sections that are inspected. These are used as 2D images augmented on the pipe during training, the trainee placing and moving the augmented probe on two directions perpendicular on the weld seam in order to examine the whole volume of welded joint.

Table 2. Types of phased array ultrasonic scans

B-scan is a 2D view of ultrasonic data presented as a cross sectional view of the part. 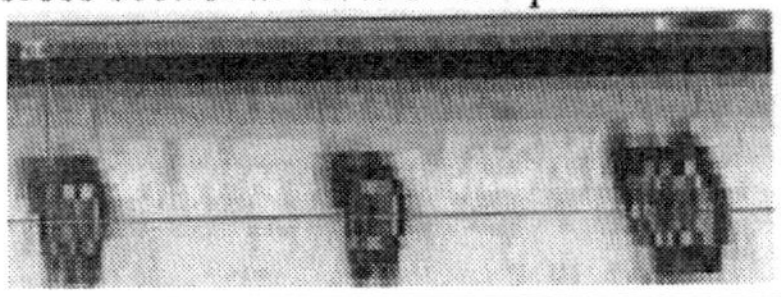	C-scan is a 2D view of ultrasonic data presented as top view or in-plane view of the tested part. 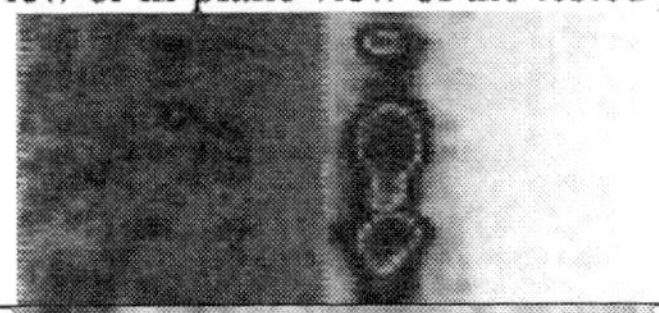
D-scan is a 2D view of ultrasonic data similar to B-scan and perpendicular to it. Both B-scan and D-scan are showing data at a predefined depth.	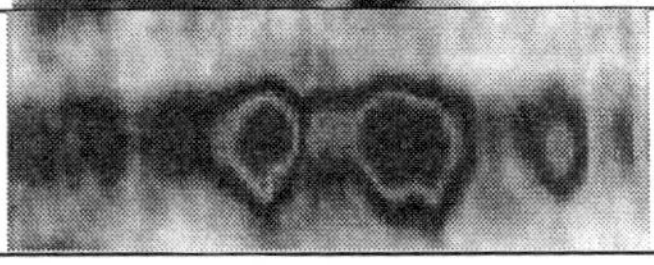

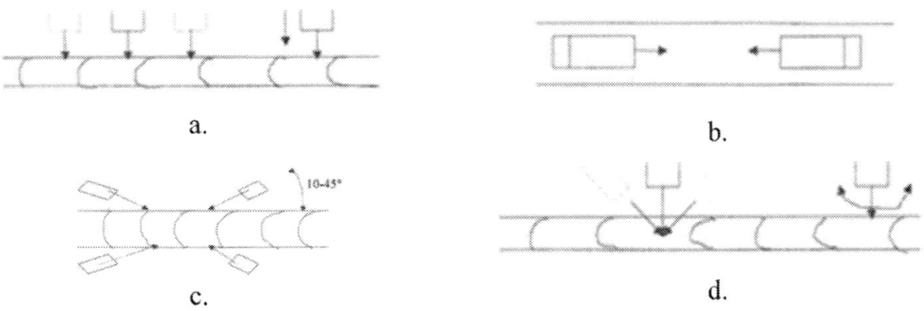

Fig.1. Examples of correct placement of the phased-array probe on the pipe for different indications

For longitudinal indications, the probe should be moved as in figure 1a, while for transverse indications, the scanning should be made as in figure 1b (for level ground at sound metal) and 1c (if the weld is ground level). For linear and volumetric indications, the probe should be rotated around axis as depicted in figure 1d.

During training using the developed application, if the probe is not moved by operators according to the information displayed on the screen, the 2D image will not be formed as augmented on the pipe.

Furthermore, during training, the correct succession of the ultrasonic NDT steps (table 2) is tested. An AR-marker is assigned to each stage and the operator should order them for obtaining the right sequence.

2.3 Development of the AR training application for ultrasonic NDT

The training application was built using Unity 5.3.4, a software development package which was preferred for its large database of scripting and 3D visual resources available open-source and for its extensive documentation.

The AR components of the application were built using Vuforia 6.2.10 SDK for Unity. The AR SDK utilizes the main video camera of a portable device (smartphone or tablet) and Unity prefabs to do image tracking, by identifying and tracking custom image markers. For this project, three image markers were created in graphic design software following the design criteria provided by Vuforia. The markers are 60 x 60 mm in size and have been verified to be suitable for use in image tracking applications using Vuforia's Target Manager developer tool. In order to better visualize the ultrasonic probe when using the AR application, an accurate 3D model was designed in CAD software, converted in a suitable format and then imported in Unity. Images of defect visualizations taken during real-world ultrasonic NDT testing are used to simulate the presence of defects in the application. This is achieved by meshing these images on the surfaces of a 3D model of a cylinder which is then made transparent using alpha compositing. The 2D visualizations of defects were provided by Namicon SRL and included in a database of defects which can be accessed by trainees.

For correct target detection, the application requires normal and stable lighting conditions. The application is used in an office/classroom environment that offers these conditions. However, in improper lighting conditions, the application can deploy the camera flash of the device through the setFlashTorchMode() function from the Vuforia API. Probe position tracking is enhanced by using the continuous autofocus mode (FOCUS_MODE_CONTINUOUS_AUTO). Moreover, the application uses embedded algorithms for extended tracking allowing improved tracking performances and sustained tracking even if the marker is partially out of frame. Vuforia SDK uses a predefined dimension for markers specified by the user. The dimension of the virtual model is set to be proportional to the marker dimension. This enables the autoscaling of the AR model when distance and viewing angle change. After the Unity project was fully developed, it was packaged using Android SDK 24.1 and installed on various Android devices for testing.

3 Results

The developed AR application is focused on acquiring knowledge and training skills:
- Knowledge on NDT process steps;
- Knowledge on applied ultrasonic NDT inspection steps by correctly placing and moving the probe relative to the inspected pipe;
- Knowledge on identifying different types of defect augmented on the pipe.

An end-user starting the AR application can observe on the first page the application logo and four buttons which give access to *Basics*, *AR Testing* and *About* sections, as well as an option to leave the application. *Basics* section presents the user with basic information about ultrasonic NDT, while *About* gives information regarding the authors and application development.

AR Testing section offers the user the possibility to watch a video tutorial showing how to use the application and its capabilities, as well as the possibility to start the main feature of the application – the AR ultrasonic NDT training process.

This feature can be accessed by touching the *Start AR Training* button in the *AR Testing* section. Touching this button switches the application into the training mode. Upon entering training mode, the main video camera of the device starts and the image tracking features are enabled. Training mode offers an interface which teaches the user the steps for performing ultrasonic NDT procedures.

The user can control the displayed information about NDT steps using three buttons (*Show next step*, *Show previous step*, *Reset process*). The type of NDT investigation to be pursued (*B-scan*, *C-scan*, etc.) can be selected from a dropdown menu. Another feature from the training mode is a simulation of the probe movements for each investigation type. This shows the user how the ultrasonic probe should be positioned and moved in order to correctly check for defects. This feature can be accessed by touching the *Show probe path* button. In order to assist with training, the application creates a virtual model of a pipe. The diameter of this virtual pipe can be modified using a slider. The training mode can be exit by touching the *Back* button.

The AR experience is achieved with the help of three embedded markers (fig.2):

1. *Namicon* marker which is used to determine the position of a pipe and creates a virtual equivalent of it;
2. *Scan* marker which creates a virtual defect;
3. *Probe* marker which is placed on the 3D printed probe in order to track its position.

To get the optimal training experience, *Namicon* and *Scan* markers should be placed on a physical pipe, but the training process can be carried out without a physical pipe, using the virtual pipe created by the application. The user handles the 3D printed probe while watching the screen of the device with the application in training mode. When the probe reaches the proximity of the Scan marker, the virtual defect becomes increasingly visible (thus simulating that the ultrasonic probe is detecting the defect).

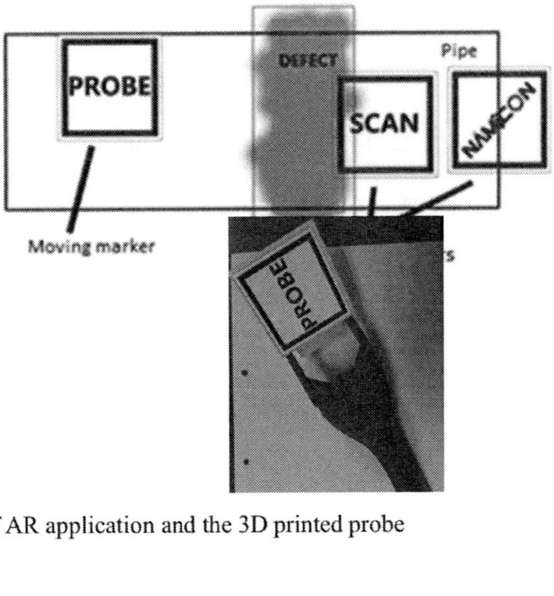

Fig.2. Schematic of AR application and the 3D printed probe

Figures 3-6 are captures from the application for illustrating some of the information presented above.

An initial test with a group of 5 researchers was performed and the participants rated 4,6 out of 5 for the application usefulness and 4,8 out of 5 for interface friendliness.

Several improvements were suggested and they will be considered in the next version of the application, together with feedback obtained from the industry via Namicon SRL.

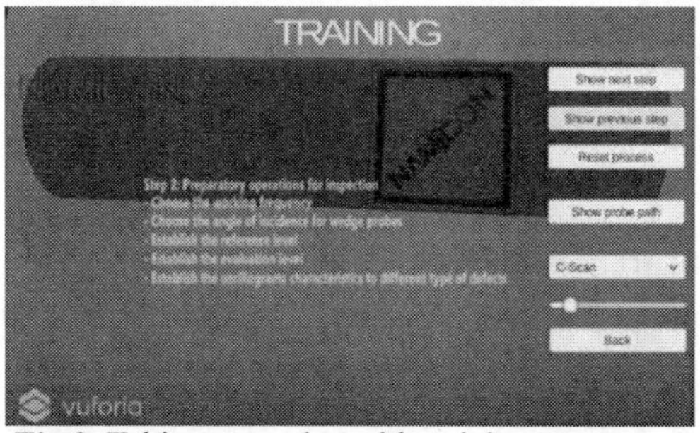

Fig.3. Tablet screenshot with training steps

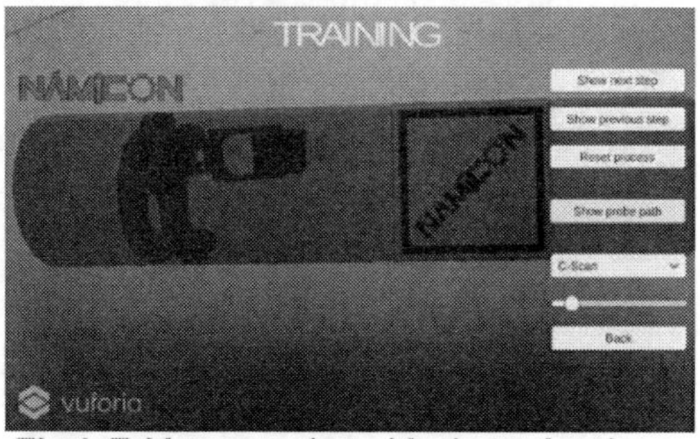

Fig.4. Tablet screenshot with pipe and probe moving path

Fig.5. Screenshot showing the probe coming closer to the defect

Fig.6. Screenshot of a clear image of a defect when the probe reaches the defect

4 Conclusions and further work

The AR application presented in this paper was developed as a necessity to improve the learning curve in NDT ultrasonic testing, in the same time decreasing the costs involved in the traditional training.

Vuforia SDK for Unity was used for developing the application. A 3D printed model was used as a cheap mock-up of the scanning equipment and probe.

Initial tests were performed by five researchers from University Politehnica of Bucharest. The application was considered useful. However, they suggested several improvements which will be included in the next version of the application. These improvements referred mainly to enhancing the displayed defects database, improve the display of the scan probe correct moving path and increase tracking capabilities.

Further work implies using the assessment data for improving the AR application. A larger group of participants, including also students, will be asked to test the application and give their feedback.

This work was supported by a grant of the Romanian National Authority for Scientific Research and Innovation, CNCS/CCCDI – UEFISCDI, project number PN-III-P2-2.1-BG-2016-0036, within PNCDI III.

References

1. T. P. Caudell, D. W. Mizell, Proc. IEEE Hawaii Intl.Conf. on Sys.Sciences, **992**, 659-669 (1992)
2. S. Webel, U. Bockholt, et al., BIO Web of Conferences, 00097, 1-4 (2011)
3. S. Iliano, V. Chimienti, G. Dini, Proc. of 4th CIRP Conf. on Assembly Technology and Systems, Ann Arbor (2012)
4. E. Zhu, et al., Peer J, **2**, e469 (2014)
5. A. Karthikeyan, P. Mani et al., Advances in computer science and information technology. Computer science and engineering, Springer, vol. **85**, 222–230 (2012)
6. G.S. Jo et al., Proc. of the Twenty-Sixth Annual Conference on Innovative Applications of Artificial Intelligence (2014)
7. S.J. Henderson, S.K. Feiner, IEEE Int. Symposium on Mixed and Augmented Reality (2011)
8. G. Dini, M. Dalle, The Fourth International Conference on Through-life Engineering Services, Procedia CIRP 38, 14 – 23 (2015)
9. G. Reinhart, C. Patron, CIRP Annals - Manufacturing Technology, 5-8 (2003)
10. K. Chougrani, N. Pörtzgen, 17th World Conference on Nondestructive Testing, Shanghai, China (2008)
11. J. Meyer, J. Rehbein, J. De freese, J. Holtmannspötter, 7 th International Symposium on NDT in Aerospace, We.2.A.5 (2015)

Design for Additive Manufacturing, to produce assembled products, by SLS

Nicolae Bâlc[1,*], and *Cristian* Vilău[1]

[1]Technical University of Cluj-Napoca, B-dul. Muncii, No. 103-105, Cluj-Napoca, Romania

Abstract. The paper presents a new methodology on how to design a product, which could be produced directly as an assembly, instead of manufacturing all individual parts, to be assembled after words, in a classical manner. A detailed case study is presented, on how the product's structure could be simplified, the number of the components being drastically reduced. The theoretical Design for Assembly (DFA) methodology was tested and validated, by manufacturing the physical products, using the Selective Laser Sintering (SLS) machine.

1 Introduction

The Additive Manufacturing (AM) technologies bring more alternatives both on design and manufacturing the products, which can be obtained directly assembled, without being necessary to produce separate parts and to assemble all of them later on, in a classical manner. It is not only that the assembly time and costs could be saved, but there are also other advantages, such as to simplify the structure of the product and reduce the number of the components, by eliminating the assembly elements.

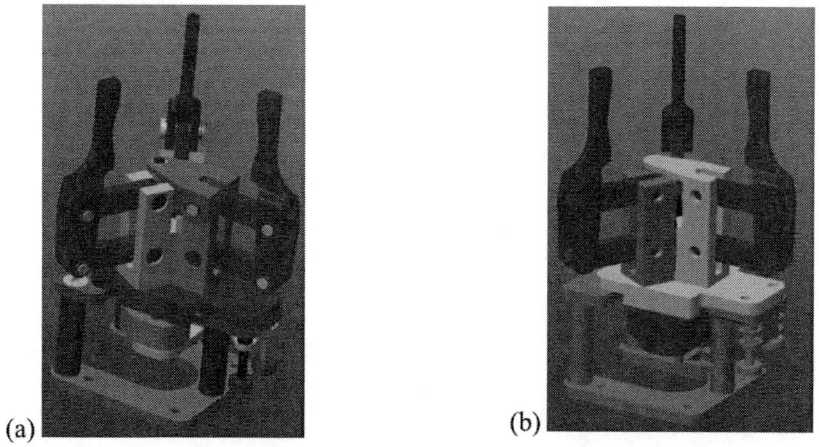

Fig. 1. Design for manufacturing assembled products, versus classical design.

* Corresponding author: nicolae.balc@tcm.utcluj.ro

Selective Laser Sintering (SLS) is a suitable technology to produce assembled products, as it uses powder materials, which could be eliminated easy at the end of the SLS process. The SLS process takes place in a solid faze, so the excess of powder could be removed during post-processing, if the assembly has been properly designed, with suitable gaps, in between the moving elements.

A special gripper, for a Kuka robot, was designed as part of the PhD thesis, in order to be flexible and able to handle both, rotational and prismatic parts [3]. This new gripper, illustrated in figure 1a, is composed by 61 elements, if the parts are to be made separately and being assembled at a later stage. If this gripper can be made by SLS, it's structure could became a lot simpler, and the new gripper could be designed for AM, the number of the components being decreased down to 19, as illustrated in figure 1b.

2 Design to manufacture assembled products

It is a concurrent engineering, as the authors paid attention in the same time, to the functionality of the gripper and to the new possibilities of the SLS manufacturing technology, in order to redesign the gripper, to be made by SLS, directly assembled. We managed to remove more than 2 thirds of the components, as the screws, nuts, pins, elastic rings and other assembly elements, are not necessary any more.

The functionality of the new design of the gripper should remain exactly the same, as the version of the gripper, designed in a classical manner. The base frame is going to be fixed onto the pneumatic linear motor. Besides of this frame, most of the other parts are mobile, each of them having a complex functionality purpose. The gripper is meant to be able to handle, both rotational and prismatic parts. That is why, many of its components need to adapt their position, from 2 to 3 jaws and back to 2.

Design for AM and virtual assembly require detailed analyses of the functionality of the new product [2]. For each component and/or assembly element which is a candidate for elimination, the designer needs to change the adjacent components of the product, in order to undertake the role of the eliminated parts.

The authors evaluated each component of the classical version of the gripper and the following three basic Design for Assembly (DFA) criteria were applied, in order to check if a part is a candidate for elimination [1]:

- Is this part moving (rotational or linear), with respects to the adjacent parts?
- Does this part have to be made from a different material, as compare to the adjacent parts?
- Does this part have to be separate, in order to be able to assemble or disassemble other parts?

According to the Boothroyd & Dewhurst methodology, if the answer is NO, to all the upper 3 questions, this part is a candidate for elimination. If the answer is YES to at least one of these 3 questions, then the analysed part has to remain separate.

The linear movement from the step motor is taken by the bracket, which is acting all 3 racks at once. Each of the racks is acting one of the 3 arms of the gripper. The rack is having a linear movement and it rotates the pinion (which has a gear sector on one side). An intermediary element with two cylindrical joints was designed, suitable for a maximum 65° rotating angle. The distance between the holes is depending of the maximum opening diameter, imposed to the gripper. The movement of the jaws is a parallel translation movement, obtained from the rotation of the pinion, around its primary cylindrical joint. The jaw is supported and guided by the intermediary element, connected onto the jaw by a cylindrical joint.

One of the 3 jaws has the supporting joint, fixed onto the central raceway of the gripper, which has a complex shape and belongs to the base plate. The other 2 jaws have intermediary elements fixed on special supports. Switching the gripper from 3 jaws gripping of the rotational parts, to 2 jaws gripping of the prismatic parts, is done manually, by linear transversal movement of the intermediary plate, which rotate the vertical supports, to align the 2 jaws.

3 Virtual assembly the new product, designed for AM

The "Creo Parametric" software was used to design the product to be made as an assembly, directly. Each of the 19 components of the new simplified structure of the gripper was designed in a separate part file. The first component called support is considered implicitly fixed, connected onto the pneumatic motor.

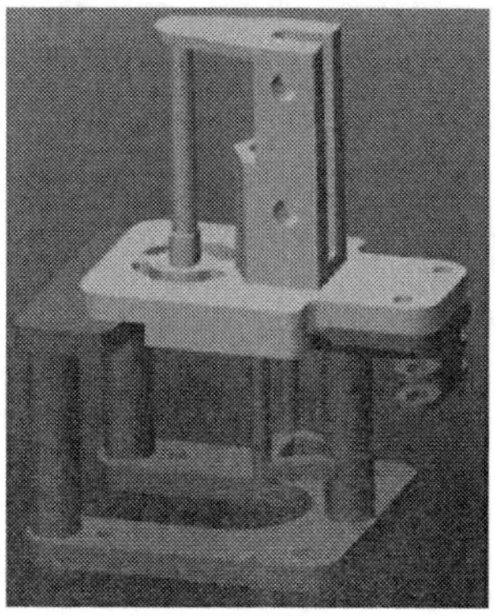

Fig. 2. DFA Considerations.

Fig. 3. Redesigned pinion.

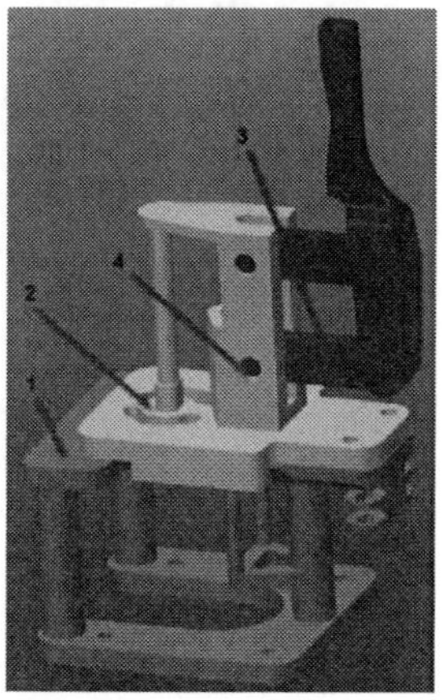

Fig. 4. The first jaw, assembled.

For the next component, which is the intermediary plate, the linear movement conditions were set up, to be acted manually. The length of the translation movement is correlated with the maximum allowed movement, from 3 jaws gripping to 2 jaws gripping.

The pinion illustrated in figure 3 comes next into the assembly and here the novelty and the difference appear, as compare to the classical procedure, when the parts were assembled one by one. Within the design for AM, the authors designed the pin, as being permanently joint (connected) to the pinion, so that no any separate bar is necessary anymore, for that cylindrical joint. Up to now, such an assembly could have been produced virtually, only. The new AM technologies open new doors and now such a virtual assembly could also be practically produced by SLS.

The research and practical experiments undertaken by the authors, allowed to find out the practical conditions to produce such an assembled product design. The gaps between the shaft and the surrounding support plate should be in accordance both, with the functional requirements and with the grain size and layer thickness of the SLS process. The gripper has been designed in two alternatives, with the space values of 0.1 mm and 0.2 mm, in between all its moving elements.

Figure 4 illustrates how the first jaw was virtually assembled, for the product to be made by SLS. The rack was assembled next and in a similar manner, all the new designed components were virtually assembled. The rack has a cylindrical guiding surface (within the cylindrical inner surface of the intermediate support) and could do a maximum 4 mm linear movement. For the gap between these elements, the 0.2 mm aperture was adopted, as well. This rack is also virtually assembled, as in a classical manner it would be impossible to be inserted. The rack (figure 5) has two continuous contact surfaces, the cylindrical area for self guiding (1) where it gets the movement onto, from the intermediary plate and the other cylindrical zone (2), acted by the linear motor.

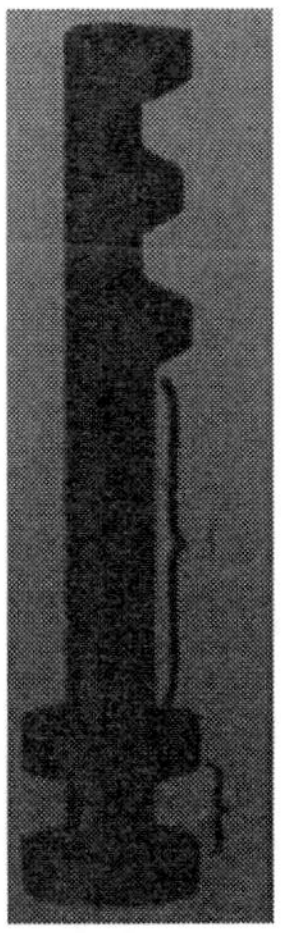

Fig. 5. The rack.

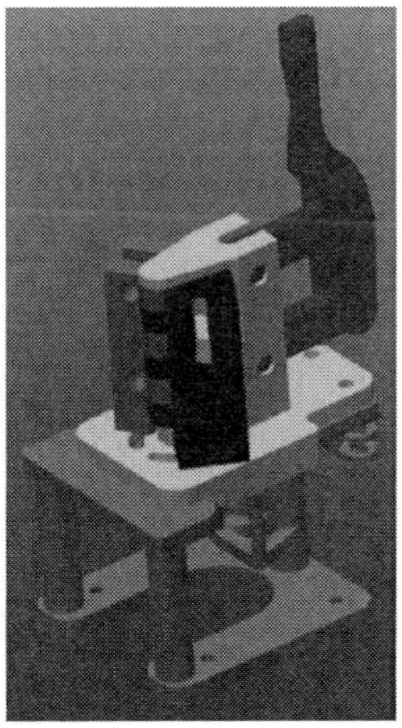

Fig 6. The joint supports.

Fig. 7. All 19 parts virtually assembled.

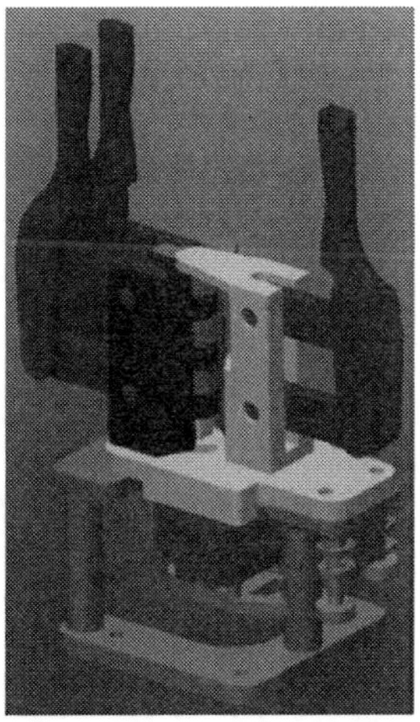

Fig. 8. Jaws for prismatic parts.

Two supports (green and blue - figure 6) were designed and assembled onto the same vertical shaft, with three intercalate zones, for cylindrical guiding. The rotation of these supports around their common shaft, is guided and limited by a cylindrical pin (which is joined onto the support), which is guided into a special slot, which was designed as a cavity into the intermediary plate. This slot has a special profile, representing the trajectory of the guiding element.

The intermediary elements of the other two jaws are added into assembly, so that the jaws could be assembled. Finally, the acting element and the two blocking bolts are assembled. Figure 7 illustrates all 19 components virtually assembled for cylindrical parts to be handled and figure 8 illustrates the gripper, when its jaws are aligned for prismatic parts to be handled.

4 Selective Laser Sintering of the assembled products

The practical case study was developed in order to test and validate the new DFA methodology developed by the authors. Two virtual assemblies designed with "Creo Parametric" software were saved as *.STL files and uploaded into the "Build set up" software of the SinterStation 2000 machine. Figure 9 illustrates the assemblies designed with 0.1 mm gap (left) and 0.2 mm gap (right). Both assemblies were further processed into the SLS machine's software (version 2.2.0.7). We also made experimental tests for higher gaps, such as 0.3 mm, but in such cases we obtained a product with bad functionality. That gripper was blocking, because of the too big gaps between the moving elements.

The powder material (Duraform PA 650) used for the experimental research, has 50 μm grain size. Both assemblies illustrated in fig. 9 were produced in the same time onto the Sinterstation 2000 SLS machine, available within the Department of Manufacturing Engineering, at Technical University of Cluj-Napoca.

The main technological parameters applied for the SLS process had the following values: laser power 4.5 W, layer thickness 0.1 mm, part bed temperature 170 °C and scanning speed 1.25 m/s.

The AM techniques and in particular SLS have many advantages, such as the fact that the cost of the product does not increase with the complexity of the shapes to be made, but still require strong experience, to produce an accurate and reliable product. For this case study, the authors calibrated the SLS machine, using a standard part. Also, a set of scaling factors was applied, in order to compensate the shrinkage of the parts made by SLS.

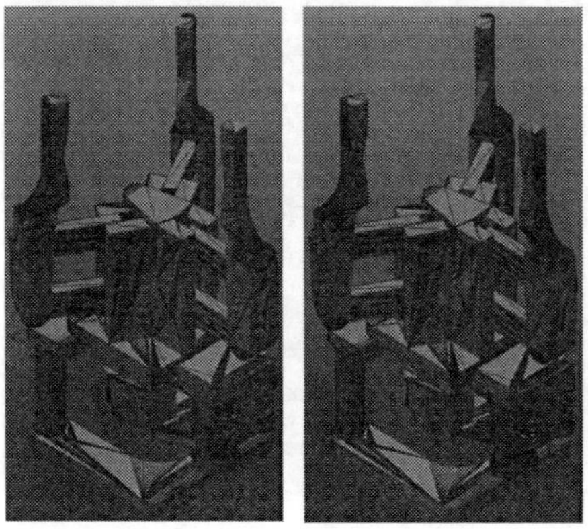

Fig. 9. Assemblies designed with 01. mm and 0.2 mm gaps.

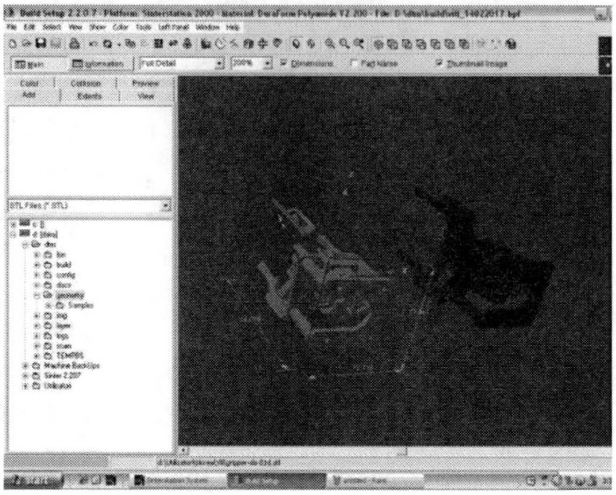

Fig. 10. Optimized orientation for the SLS manufacturing, for the assembled products.

It was very important the position in which these 2 assemblies have been orientated within the working space of the SLS machine. If the gripper would have been built into the horizontal position, there would have been some advantages (such as short manufacturing time and theoretically a correct geometrical shape), but the disadvantage would have been, too much heat incorporated within the product, leading to significant shrinkage (3D deformations).

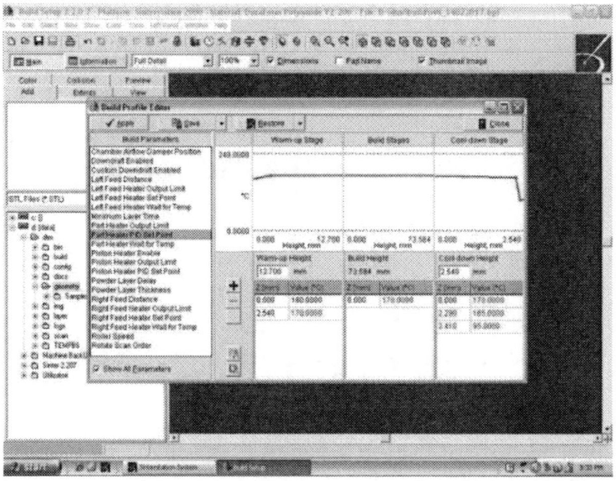

Fig. 11. Temperature cycles time, for SLS assemblies.

The authors found an optimized position of the gripper during its SLS manufacturing, at double angled position (15 ° rotation about the X axes and 15 ° rotation about the Y axes). In this way, we made the assembled grippers by adding layers with different shapes, leading to smaller heat incorporated within the product, smaller shrinkage, so a higher accuracy of the SLS products. In this case, the disadvantage was a higher manufacturing time, which is proportional with the high of the part, but this increase of time is not causing inconvenience for such a small volume production.

The building time was approx. 9 hours for the two grippers illustrated in figure 12, which were built together, in the same time, onto the SLS machine.

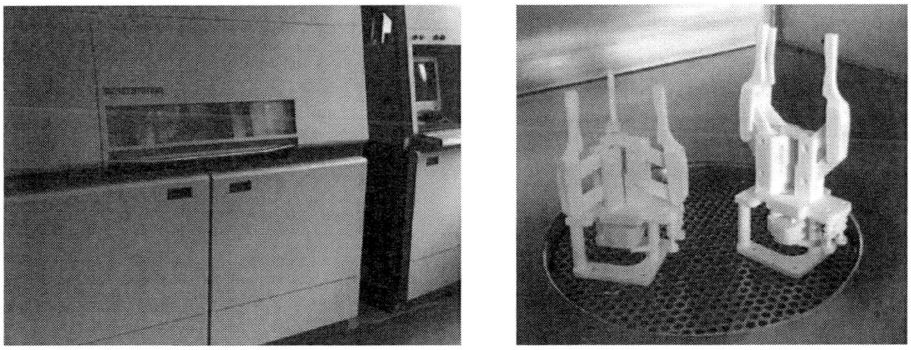

Fig. 12. Assembled products, made by SLS.

5 Conclusions

The new methodology on how to design a complex product, in order to be produced by SLS as an assembly, was tested and validated.

This practical case study shows both, the new possibilities and the advantages of the new design method. The number of components was decreased from 61 down to 19, as only the functional components have to be incorporated and all the auxiliary and assembly elements could be eliminated.

Static and dynamic analyses were undertaken for this version of the DFA gripper, to check what the maximum gripping force would be, as compare to a similar gripper, made out of laminated steel, manufactured part by part and assembled in a classical manner.

Due to the significantly lower mechanical properties of this plastic material (PA 650), the maximum weight of a part which could be safely handled, is 0.5 kg (at the maximum velocity of the Kuka robot).

The research was supported within the PCCA PECIFCO project (No. 115/2014), H2020 AMaTUC project (GA 691787) and the OpTi-DeP project (No. 101BG/2016).

References

1. G. Boothroyd, P. Dewhurst, W. Knight, *Product Design for Manufacture and Assembly*, Third Edition (CRC Press, USA, 2016)
2. S.B. Maidin, R.I. Campbell, E. Pei, Assembly Automation, **32** (3), 235-244 (2016)
3. Popan I.A., Balc N., Luca B., Popan A., Carean A, Applied Mechanics and Materials **808**, 339-344, (2015)
4. https://www.shapeways.com/tutorials/design_rules_for_3d_printing
5. https://www.3dsystems.com/resources/sls-plastics-design-guide

Automated system for data acquisition and monitoring

Sorin Borza[1,*], and *Iunia Cristina* Borza[2]

[1]University Lucian Blaga of Sibiu, Romania
[2]Microsoft Iberica, Paseo del Club Deportivo, 1Centro Empresarial La Finca – Edificio 1 28223 Pozuelo de Alarcón (Madrid), Spain

Abstract. The Environmental management has become, with the development of human society a very important issue. There have been multiple systems that automatically monitors the environment. In this paper we propose a system that integrates GIS software and data acquisition software. In addition the proposed system implements new AHP multicriteria method that can get an answer online on each pollutant influence on limited geographical area in which the monitors. Factors pollutants of limited geographical areas are taken automatically by specific sensors through acquisition board. Labview software, with virtual instrument created by transferring them into a database Access. Access database they are taken up by software Geomedia Professional and processed using multi-criteria method AHP, so that at any moment, their influence on the environment and classify these influences, can be plotted on the screen monitoring system. The system allows, the automatic collection of data, the memorization and the generation of GIS elements. The research presented in this paper were aimed at implementing multi-criteria methods in GIS software.

1 Introduction

With the development of human society has become obvious that It affects the environment and human activity is in turn influenced by to it. Passing through various stages of organization, quality protection environment and new integrated approach to environmental management become a vital necessity to ensure the prosperity of society and sustainable development.

Environmental management has become an essential part for any type of development regardless of scale that may occur environmental impacts. Environment monitoring system is an integrated system that continues to supervise the state of the environment and provides data on all of its structural components (air, water, soil, biocoenosis). The data acquired are processed by specific methods, and the information thus obtained is used for the final assessment of the environmental impact, for warning and alarm and control of the environment.

* Corresponding author: sorin.borza@ulbsibiu.ro

The main purpose of an environmental monitoring system regardless of scale covered or the number of components that includes is to provide an objective view as more realistic environment for the adoption of corrective measures control pollution and recovery [1].

Integrated monitoring aimed at achieving true-overall environmental quality of the stage at a time and historical trend two basic components - the biotic and abiotic environment - in their interconnection. From passive phase highlight the state of quality monitoring is currently an active intervention, control and evaluation of the effectiveness of regulatory measures, strategies profile and generation needs to promote clean technologies [1].

The integrated monitoring ensures database for both depth research (understanding, explanation and evaluation quantified the various ecosystems in terms of structure, processes and mechanisms operating within the hierarchy of ecological systems), and assessing the impact of actions undertaken to maintain the integrity of ecological systems [2].

Monitoring can be done for various purposes, including to establish "baselines, trends and cumulative effects" on the environment [3], to improve process modeling environment, to educate the public on environmental conditions, to support the decision-making environment for compliance with environmental regulations, to assess effects anthropogenic influences on the environment, or perform an inventory of natural resources [3]. Local level monitoring refers mainly to environmental observations near sources that supposedly can cause impacts. Apart from this restricted from another point of view the local concerns and national monitoring systems, relative to the overall system.

Environmental monitoring system works similarly to a double hierarchy system in business management and information system.

It is characterized by a functional interdependence between its five components, namely:

- supervision;
- evaluation of the actual state of the environment;
- prognostication state environmental;
- assessment of prognosis;
- adjust the quality of the environment.

Key indicators of the European Environment Agency (EEA) supports the monitoring and implementation of the priority objectives of EU environmental policy, is regularly updated [4]. They are bound by obligations of European Union resulting from the signing of agreements, treaties or international protocols targets assumed under various directives (eg Water Framework Directive, the Waste Framework Directive, the Habitats Directive, Birds Directive) or attempts of solving direct problems incidence in the quality of housing at community. Exist the 37 key indicators monitored by the European Environment Agency (Core Set indicators - CSI) [5].

In Automatic Environmental Monitoring System were used IEPE sensors and acquisition board, NI 9234 NI Mseries. The Drager Pac III apparatus has been used for the measurement of the monoxide carbon. For the sound measurements EXTECH 407780 was used.

The automatic taking over and the analysis of the data using the AHP method (Analytic Hierarchy Process) and of generating the element's classes necessary to elaborate the intelligent maps consists of hardware components: sensors of the acquisition board, apparatus used for measuring the polluting factors and also, of software elements, like: LabVIEW, Geomedia Professional, Access Database, presented in figure1.

DOI: 10.1051/matecconf/201712104003

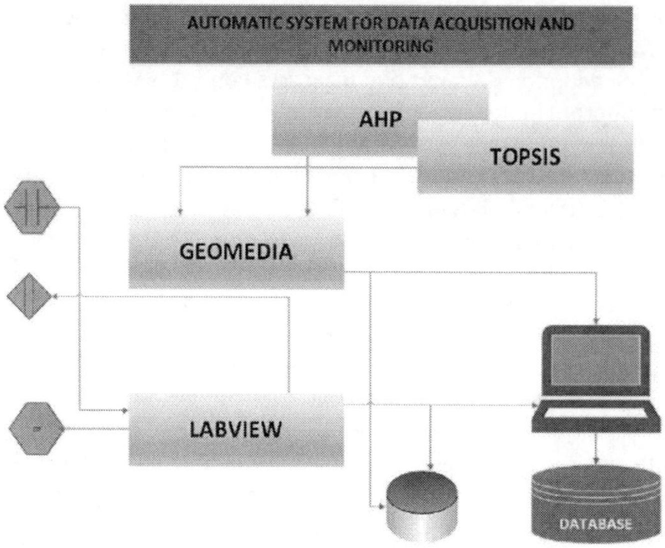

Fig. 1. The system for Acquisition and Monitoring Data

The MCDM methods are frequently used to solve real-world problems with multiple attributes. Recently, various methods have been developed to support decision making in various fields. Multi-criteria decision making (MCDM) is a well-known branch that deals with problems making decisions based on a number of criteria.

2 The Object Oriented Programming in LABVIEW

In object oriented programming with LabVIEW, the concept of "Dataflow" is extented [6]. In LVOOP, a class consists merely of a user defined data type together with methods that can be applied to values of that data type. Once could say that object oriented programming in Labview allows the developer to create object oriented wires.

As in other object oriented languages, a derived class may overload an abstract method of its base class. However, the override method must have exactly the same input and output parameters as the respective method of the parent class. The concept of dataflow requires objects to flow through their "class wires" from source to drain [7].

The fundamental differences between LVOOP and conventional object oriented languages prevent a straightforward *implementation* of design patterns that are based on the idea of objects as entities. However, many of those design patterns are useful for designing control systems. LabVIEW contains several built-in reference-based components, such as File I/O, TCP/IP Networking and ActiveX Automation. Using the LVOOP approach we can take advantage of the benefits of object-based design and development. Adding the class and object concepts from other object-oriented languages to LabVIEW, you can build your own components. By implementing classes, you can take advantage of the design and development practices used by developers who use standard OO languages. You will also be able to use standard OO analysis and design tools, such as Rational Rose and Microsoft Visual Modeler and apply principles described in literature about OO design and development.

The component-based approach to developing application augments the traditional data-flow programming used in LabVIEW applications. You will develop large applications by identifying and designing components that implement distinct parts of the functionality of the application [8].

3 The MCDM METHODS

The Multicriterial evaluation is used in cases where there are several alternatives, variations, locations or processes that have to be assessed by their total environmental load or quality. The common result of multicriterial evaluation methods is dimensionless number that indicates the degree of environmental load of alternatives that are valued. In addition to indicators that represent the environmental impact it is possible to include indicators that have economic, social, and technological character.

3.1 The AHP method

AHP (Analytical Hierarchy Process) is a structured technique for solving complex decision developed by Thomas L. Saaty [9]. This technique allows the decision maker to take the most appropriate decisions for him by failing necessarily the right decision.

This method is based on splitting the problem into simpler subproblems ranking / ordering that can be analyzed independently of each other. Once constructed hierarchy, decision maker evaluates its elements by comparing their successive pairs. These comparisons reflect the main advantage of the approach, namely that it necessarily requires the use of objective and clearly defined, allowing comparisons on the basis of purely subjective and ambiguous. All of these orderings are converted into numerical values, which will then allow the aggregation results obtained in such a way that the solution to the solution to the original problem.

An important component of the AHP method is a mathematical model by which priorities of elements are calculated (weighted), for elements that are on the same level hierarchical structure. AHP was successfully used in environmental impact assessment for determining of weights for impact categories [10]. AHP was used for verification of results gained by quantification of environmental aspects and impacts [11]. Summary of AHP method consists of converting subjective assessments to the relative importance of the criteria scores and weights. The AHP method was developed by Saaty [12] and is one of the most used methods MCDM.

In the method AHP each criterion is compared with the other subjective criteria considered by method. This results are compared in pairs, resulting are in scores and weights. For each criterion is taken into account Table 1:

Table 1. The Saaty table

Definition	Importance Intensity
Equally important. Compare criteria with itself	1
The Criterion is more important than the criterion they compare	3
The Criterion is strongly more important than the criterion they compare	5
The Criterion is very strongly more important than the criterion they compare	7
The Criterion is extremely more important than the criterion they compare	9
Intermediate values are allowed: 2, 4, 6, 8	

The intermediate values are allowed to shade difference between the two criteria. The total number of evaluations will be:

$$\frac{1}{2} \times n \times (n-1) \tag{1}$$

Using this method the following four steps:
Step 1. Defining decision criteria form a hierarchy of objectives. On the upper level is the objective of the problem, the decision criteria are intermediate levels and the lower level, alternatives / variants decision. A priority hierarchy is a system and organization of concepts in which each element, with the exception of the upper level, is subject one or more elements of the hierarchy. AHP hierarchy involves placing the higher purpose / primary objective, and then placing the decision criteria to intermediate levels, possibly broken down into sub-criteria, if necessary. The lowest are found alternatives decision to be ordered by preference decision maker / decision makers.
This is essentially an alternative multi-criteria decision modeling problem under deterministic. A graphical representation of this hierarchy is presented in figure 2.

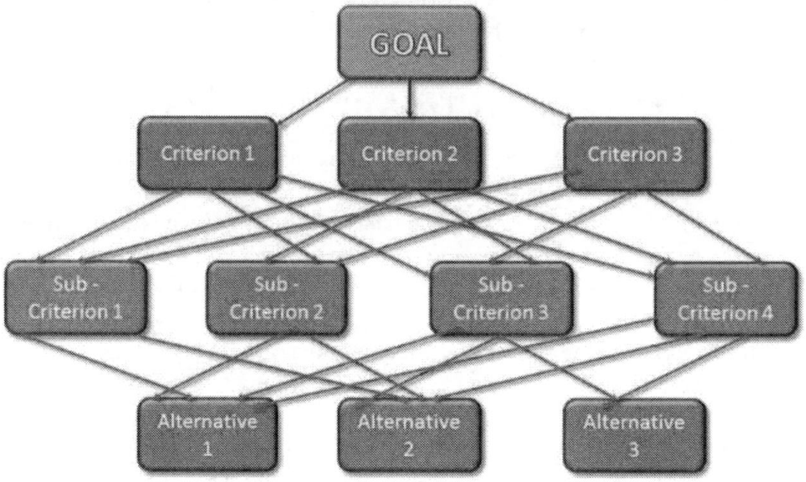

Fig. 2 General hierarchical model in AHP

Originally established hierarchy can be modified during the course of decision-making by identifying new nodes.
Step 2. Weight criteria, sub-criteria and alternatives according to their importance for higher level correspondent. To achieve this ordering using simple comparisons.
At this stage, the priorities established for the hierarchy nodes. The priorities are numerical values in the range [0,1] and represent the weights associated with nodes of said node relative to any group.
By definition, the share associated upper level / target is 1.00. For each level in the hierarchy relative weights sum will be 1.00.
The significance of these contributions is to show how to calculate them. A node with 0.20 weight will be twice as important than the 0.10 significance node. This way of quantifying the preferences that permits not only the ordering of preferences of a decision maker, but also quantify the intensity of this preference. Two concepts are closely related to the hierarchy of local priorities and global priorities. Local priorities are above and independent of the hierarchy that is node, while global priorities are obtained by weighting local priorities all levels weights directly above the level considered. Thus, the overall sum of the weights of the nodes-be is always equal to the overall share of the parent node. Although priorities goal, decision criteria and alternatives are closely interrelated, the AHP method is to take them as independent values [12].
These values of weights are then modified to obtain a decision matrix. These modifications are made by successive comparisons of the elements of the problem, two by two.
It compares each of the same hierarchical level elements which have the same parent node using table 1.

Step 3 Local priorities (weights) of criteria, sub-criteria and alternatives at same hierarchical structure level are calculated through appropriate mathematical model and afterwards they are synthesized in total priorities of alternatives.
Step 4 Implementation of the sensitivity analysis for final decisions.

$$A = \begin{bmatrix} 1 & a_{12} & \cdots & a_{1j} \\ a_{21} & 1 & \cdots & a_{2j} \\ \vdots & \vdots & \ddots & \vdots \\ a_{i1} & a_{i2} & \cdots & a_{ij} \end{bmatrix}$$

Fig. 3. Matrix A

The matrix *A,* figure 3, has special features (all of it's rows are proportional to the first row, and they are all positive and $aij = 1/aji$ is true) and because of that only one of it's eigenvalue differs from 0 and is equal to n.
If the matrix A contains inconsistent estimates (in practical examples almost always), weight vector w can be obtained by solving the equation $(A-\lambda max\ I)w=0$ with prerequisite that $\Sigma wi = 1$, where λmax is the largest eigenvalue in matrix A. Because of matrix A properties $\lambda max \geq n$, the difference $\lambda max - n$ is used in measuring timations consistency. With consistency index

$$CI = (\lambda max - n)/(n-1) \tag{2}$$

After consistency index, consistency can be calculated:

$$CR=CI/RI \tag{3}$$

In the method developed by Saaty [9], the calculation of the weights is based on a relatively complex mathematical apparatus, using matrix algebra tools. The results are values associated to eigenvector of maximum eigenvalue matrix. The calculations are quite complex, so it is necessary to use a dedicated program.

But in practice, we provide a simple method of calculation, which gives the same result with two decimal places:

- Calculate the geometric mean of each row of the matrix.
- It calculates the sum of the geometric mean.
- Normalized geometric mean.

4 Results and discussions

In the automatic system, measurements can be taken automatically by sensors connected to computer through data acquisition boards or measurements can be made directly with the apparatus dedicated to polluant factors. The data can be retrieved by means of the following IEPE sensors: SM 70 ozone fixed sensor, dust sensor g2yp1010 for dust particle, MQ-5 sensor for detect nitrogen oxide and NI MyRIO-1900 acquisition board.

This board allows connection to LABVIEW software environment and having a specific driver for connection. Environmental pollution data, figure 5, are captured by means of sensors and acquisition board placed in the Access database attached to the virtual machine. The MCDM AHP method, is implemented in Labview virtual machine and provides specific data for each measurement point.

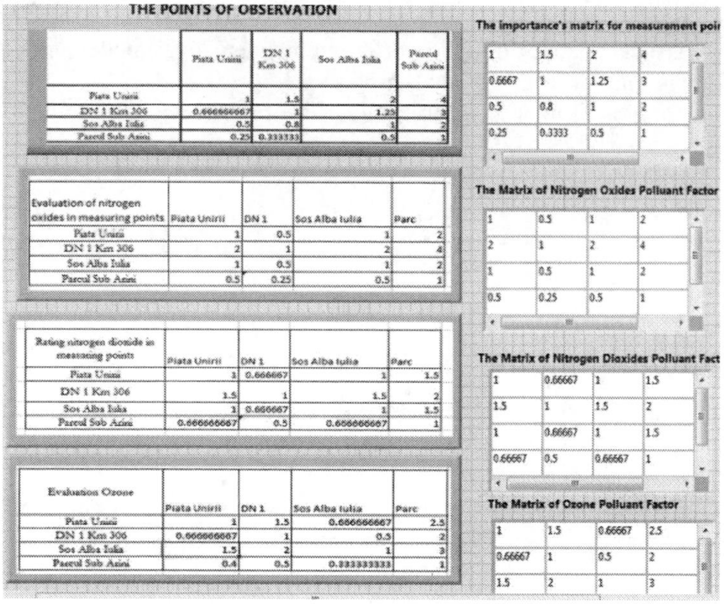

Fig. 5. The Input data for Virtual Instrument

The data is taken in Labview software, where was created a virtual instrument, whose panel is shown in Figure 6.

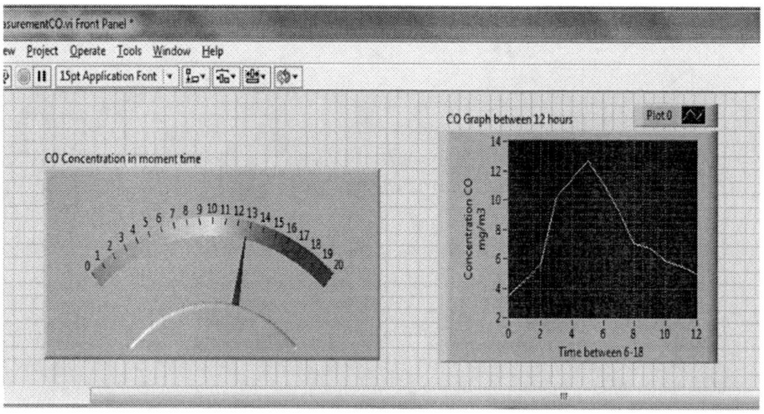

Fig. 6. The panel of virtual instrument that measures the concentration of carbon monoxide

Furthermore, the paper present the automatic acquisition, processing and analysis of environment data system, which is the analysis of the AHP multi-criteria method implemented within the system. The virtual instruments of our integrated system are based on objected-oriented programming. The input data is processed by using the LabVIEW functions. With the help of these functions the data is memorised in a database, in order to be further used in the realization of the objected classes for the generation of the intelligent maps of the Geomedia Professional Software.

The chemical factors which designed system can measure are: COx, PM10 dust and NOx.

The measurements were carried out in 4 different points from Sibiu: Union square, Alba-Iulia street, DN 1 306 km point and the Sub Arini park.

Virtual instrument design enables the implementation of multi-criteria analysis method works as follows:

- A subjective appreciation of the importance of each point stated above is made depending on the number of vehicles, number of people that are found in that concrete point at a certain time. The comparison matrix of pairs is realized, the weight of each observation point is calculated, the priority vector λ (lambda), CR and CI are being calculated. The diagram of virtual instrument are presented in figure 7.

- Depending on the measurements made for each of the polluting factors, the virtual apparatus will calculate the weight of each polluting factor from the observation points taken into account.

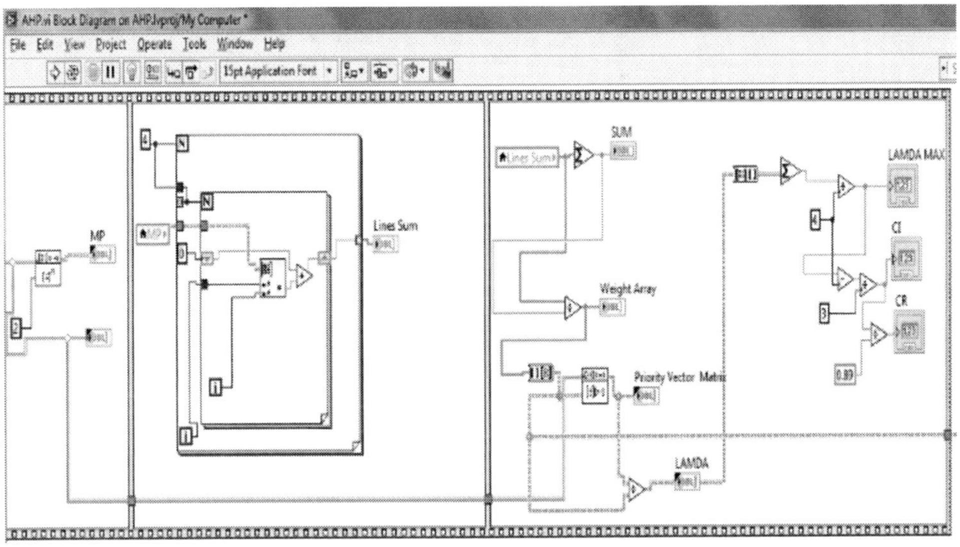

Fig 7. The diagram of VI for measurement points in AHP analysis

- Finally, the matrix of polluting factors is obtained. The matrix will be multiplied with the observation point's weights matrix. The virtual instrument allows the determination of the most and least polluted observation point, depending on the subjective appreciation made upon them. The final results are presented in the panel of virtual instrument created with Labview software, figure 8. Lambda and lambda max parameters to be calculated for obtaining CI and CR. Depending on their size choice was correct or not.

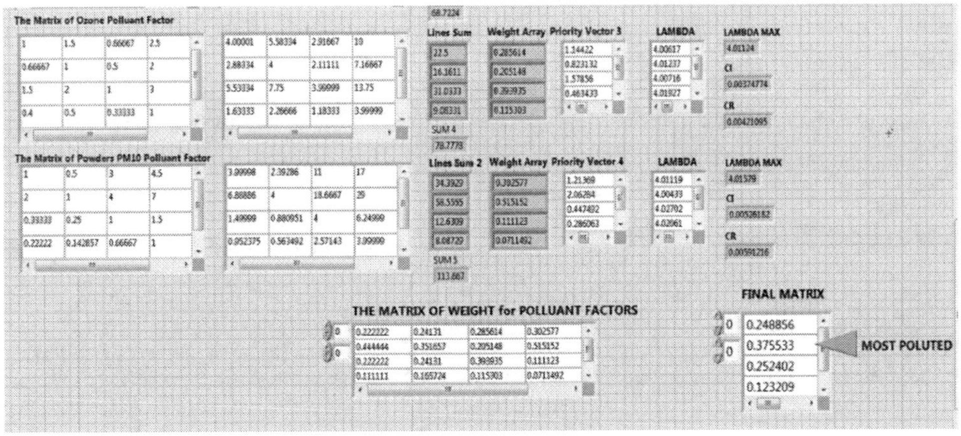

Fig. 8. The panel of virtual instrument with the matrix of weight for polluant factors and final matrix with results

The data is then retrieved from the Access database, in the software Geomedia Professional and intelligence map is generated by the polluted measuring point and the measurement point less polluted. For the examples presented, we can observe that the most polluted point is DN 1 at 306 km, point, shopping area, followed by Alba Iulia street, Unirii Square and, as expected, Sub Arini Park, figure 9. The data taken with Labview software are stored in the database corresponding to virtual instrument. This Access database is then used by GIS software to generate proper intelligence map. On this map appear different colored points where measurements were made. Calculations using AHP multicriteria method allow to present in the map the most polluted and least polluted point, depending on chemical factors measured.

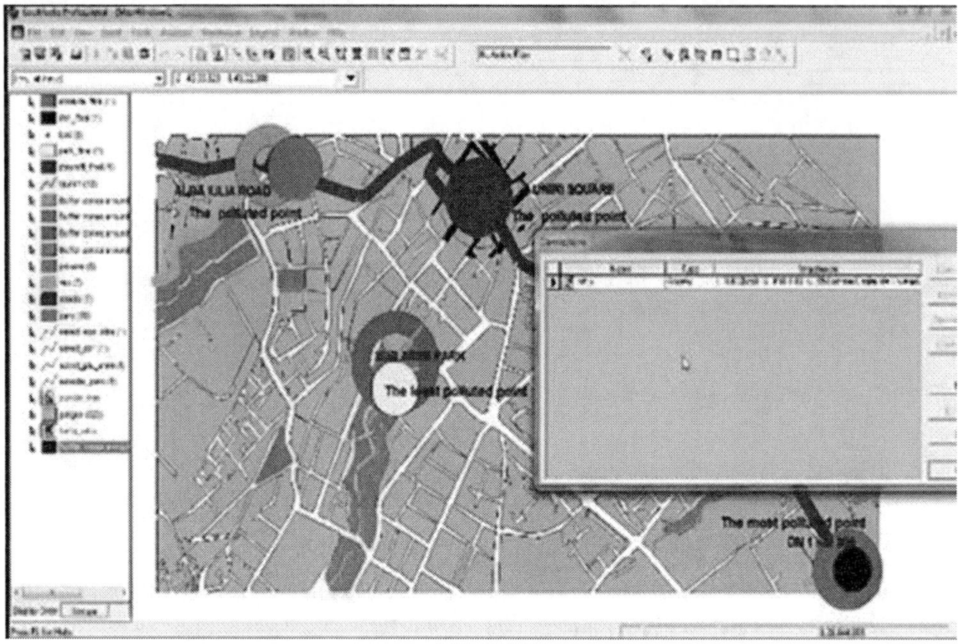

Fig. 9. The SIBIU map with the observation points and database connection

A very important thing to note is that the proposed automatic monitoring of environmental pollution factors system, allows obtaining environmental pollution situations in real time without human intervention. Data are obtained from the environment, is processed using virtual instrument, is stored in the common database with GIS software and then develops interactive map to view results.

In the future the system will be extended by using other MCDM analysis methods, like for instance the TOPSIS method. The system will be extended by implementing innovative methods based on genetic algorithms.

References

1. R. Mihaescu, *Monitoringul Integrat al Mediului, (*Cluj Napoca, 2014), http://enviro.ubbcluj.ro/studenti/cursuri%20suport/Carte_Monitoring_Radu_SITE.pdf
2. O., Ciolpan, *Monitoringul integrat al sistemelor ecologice*, (Ars Docendi, Bucureşti, 2005)
3. K. J. Mitchell, M. K., Johnson, M. Mather, Applied Cognitive Psychology Journal, **17**(1), 107-1119 (2003)
4. ***European Commission Communication from the Commission. Europe 2020. A strategy for smart, sustainable and inclusive growth. Brussels (2010).
5. C. Ioja, *Metode de Cercetare si Evaluare a Starii Mediului* (Ed. Etnologica, Bucuresti, 2013)
6. R. Bitter, T.Mohiuddin, M. Navrocki, *LabView Advanced Programming Techniques* (Boca Raton: CRC Press LLC, 2001)
7. D. Beck, H. Brand, *Proceedings of ICALEPCS07* (Knoxville, Tennessee, USA, 2014)

8. J. Jehander, *Graphical Object-Oriented Programming In LabVIEW – Tutorial - Developer Zone*, National Instruments, http://labview360.com/document/an/pdf/an143.pdf, (2005)
9. T. L. Saaty, Int. J. Services Sciences, **1**(1), 83-98 (2008)
10. B.G. Hermann, C. Kroeze, W. Jawjit, Journal of Cleaner Production **15** (18), 1787-1796 (2007)
11. Maliki, G. Owens, and D. Bruce, *IPCBEE 37* (IACSIT Press, Singapore, 2012)
12. T. L. Saaty, *The Analytic Hierarchy Process* (McGraw-Hill, New York, 1980)

Experimental modular stand used for studies of the High Tibial Osteotomy

Ileana Ioana Cofaru[1],*

[1]Lucian Blaga University of Sibiu, Computer Science and Electrical Engineering Department, 10, 550025, Emil Cioran Street, No. 4, Sibiu, Romania

Abstract. The optimization of the surgery strategies for the correction of the axial deviations of the human lower member presents a great importance both for the affected patients and for surgeons. This article seeks to develop a modular stand used mainly for experimental study of the High Tibial Osteotomy (HTO). In the article it is also shown the relation between the axial deviations of the human leg and their corrections using HTO. The modular devices were designed, executed and used in the experimental research.

1 Introduction

In this paper we will achieve the design and 3D modelling for an experimental stand used for the study of the HTO. High Tibial Osteotomy (HTO) is a surgery strategy which treats the axial deviations of the human lower member. A correct alignment of the bones of the human lower member is defined by collinearity of the line segments which unite the centre of the hip joint, the centre of the knee articulation and the centre of the ankle.

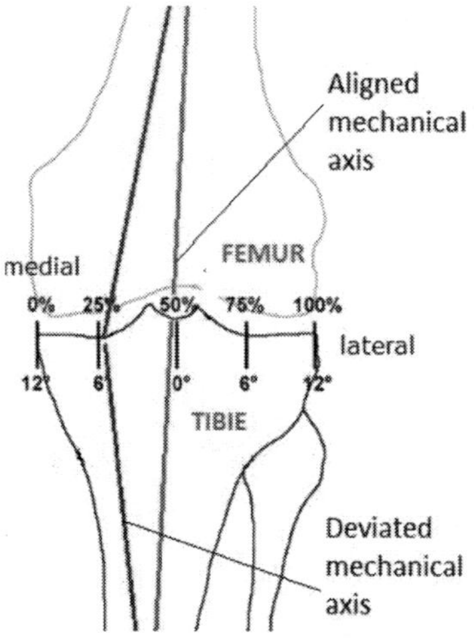

Fig. 1. Axial deviation of the mechanical axis in the knee articulation

* Corresponding author: cofaruioana@gmail.com

Any swerve from this axis becomes an axial deviation and it has important consequences on the health and the comfort of the patient affected by this deviations. In figure 1 it is shown both the knee articulation with the mechanical axis of the lower member both correctly aligned (green line) and one with the mechanical axis deviated (red line).

In the case where the mechanical axis is deviated we can observe that the mechanical axis of the tibia and of the femur aren't collinear, the axis of the leg appears broken and the loading does not take place in the centre of the knee articulations.

The main cause of this deviation is unicompartmental ghonartrosis, malady that unfortunately is more and more common today. The effect of this malady is the wear of the articular cartilage which leads to an incorrect loading of the knee and thus to the appearance of the deviations. From this moment there is a potentiality of the effects, meaning that as the deviation increases, the wear of the cartilage increases too and this usage involves a bigger deviation and so on.

One of the common procedures used in surgical practice for the correction of such axial deviations is the HIGH TIBIAL OSTEOTOMY (HTO). This procedure assume creation of the bone's wedge at the level of proximal human tibia in order to eliminate the axial deviations and to realign correctly the mechanical axis of the human lower member.

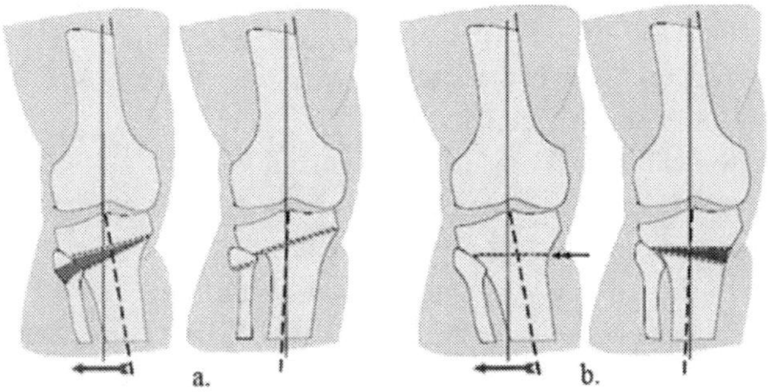

Fig. 2. High tibia osteotomy. a – with closing wedge, b – with opening wedge

In figure 2 it is shown a high tibial osteotomy and in it there are presented both the osteotomy with closing wedge and that with opening wedge [2,4] (see Figure 2a). It can be seen that in case of osteotomy with closing wedge two cuttings are performed, the resulting wedge is eliminated and then the cuttings are placed one over the other (see Figure 2a) achieving in this way the axial correction of the tibia. In case of osteotomy with opening wedge only one cutting is performed and the wedge is created through the angular opening of the two resulted surfaces. Regarding the area in which the cuttings are performed in Figure 2a is an osteotomy from the lateral (exterior) side and in Figure 2.b is one from the medial (interior) side. It has to be underlined that the approach from the medial side, both in the case of closing and in the case of opening, doesn't require the cutting of the fibula which represent an important advantage.

This surgical procedures is very interesting also from a biomechanical point of view studying several problems such as: the good cutting of the bone material, the precision in achieving the angle of the wedge [3], the state of the stress and strain resulting in the area where the bone is not cut. In this area the bone will be rotated resulting axial alignment.

This variety of problems open up the possibility to achieve ample biomechanical studies regarding HTO. These studies can be: analytical, numerical, and experimental.

In this article we will present the design and modelling of the modular experimental stand destined for the experimental study of the HTO.

2 Requirements, necessities and functions of the stand

The approach of the design of the experimental stand was achieved in a generalized manner. This was created modular in order to be used for the research of both HTO (closing wedge, opening wedge) and other orthopaedics surgeries.

For the design of the stand, specialty articles were studied in order to determine the current stage of research in this domain [5]. The other information came from analysing the necessities, from which we extracted the functions that the stand should accomplish. The constructive solutions were related to these functions.

The main objectives proposed for the development of the stand are:

- Establishing the needs and requirements that the stand should accomplish,
- The identification of the primary functions,
- The development of the implementation solutions,
- The 3D design of the stand

When it came to choosing the needs, a first step was the concretization of the current problems regarding the surgical strategies of osteotomy. Also, in order to establish the requirements, we took into account the opinions of surgeons, anaesthesiologists and last but not least of the patients.

By analysing all these aspects, the principal needs can be established. These are practically linked with the effects of specific surgeries. The stand must accomplish experimental studies regarding elements such as:

- the efficient planning before surgeries,
- enable the use of the knee functions as fast as possible,
- decreasing the stress states which appear in the tibia correction area,
- avoid further complications,
- equalise the distributions of the stress on the knee level,
- the bone fixation in correct position in order to achieve the correction of the mechanical axis,
- increase the safety of the surgery in order to prevent the developments of future infections.

Regarding requirements, at first we have to take into account the universality of the study, as it has to be applicable for every osteotomies type: opening, closing, medial, lateral, uniplanar or biplanar. Other important requirements are: achieving the fixation and orientation of the tibia and the possibility of closing and opening of the osteotomy wedge.

The constructive structure of the stand will be determined after the establishment and the analysis of the functions that the stand must carry out. These are:

- the orientations and fixations of the tibia both for closing and opening HTO,
- achieving the pushing in order to realize the closing of the osteotomy wedge,
- achieving the opening of the wedge for the opening osteotomy,
- the control of the contact surfaces position and of application of the forces in both situation from above;
- the orientation of the tibia-femur ensemble for experiments both for traction and compression stress,
- the possibility to adjust the position of the tibia and femur,
- achieving the controlled relative axial rotation between tibia and femur.

From the functions presented above, the last three are less important as far as osteotomies are concerned, but these were also took into account in order to confer a larger generalization of the stand. Also we considered for designing the stand the characteristics of the equipment and of the machines on which the stand will be assembled [6].

3 Design of the stand. Elaborations of the implementation solutions.

Starting from the functions shown in the last paragraph, we focused our attention on proposing the constructive solutions. Correspondence functions-solutions can be seen in the table 1.

Table 1. Functions and constructive solutions.

Functions	Constructive solutions
Tibia fixation	Spindle, Vice, Press, Clamp, Flange
Femur fixation	Bar, Grip Clamp
Push on tibia	Roll, Piercer
Angular positioning	„U" Frame
Segmentation of the tibia	Wedge, Clipper

The criteria that helped with the evaluations of the optimal solutions have been: the cost, the functionality, and safety in exploitation of the parts of device. In order to determinate the optimal solutions the share calculus method was used [1], resulting in this way a structure of the dispositive which uses as rotation elements "U" frames and for the orientation of the tibia the spindle. For pushing on the tibia in order to put together the bone segments (resulted after elimination of the wedge in the closing osteotomy) it was used a roll, and for creating an angular rotation of the bone segments after the elimination of the wedge in case of the closing osteotomy it was used a graduated wedge.

From the constructive point of view two big parts can be highlighted: an inferior ensemble responsible with the orientation and the fixation of the tibia and a superior ensemble responsible with the effective applications of the loadings.

The constructive solutions implemented will be shown in the following.

The inferior ensemble (see Figure 3) will accomplish the clamping and the positioning of the tibia. The tibia will be oriented and fixated in one shaft that permits a stabile fixation in three planes of the bone segments in a vertical plane related of the interior frame of the ensemble.

This ensemble consists form two shaped U frames, one placed in the interior of the other one. The exterior frame is fix and the interior frame is mobile and can rotate around the articulations materialized by a pin placed at the level of the frame's lateral sides.

This structure permits the tilt and lock of the interior frame at various degrees, until 120° in every way. The fixation of the assembly in a certain position is achieved through locking with two pins, which are inserted in the special holes situated on the lateral sides of the ensemble.

The shaft can be used with the tibia in the vertical position, horizontal position or tilted in any angle.

 DOI: 10.1051/matecconf/201712104004

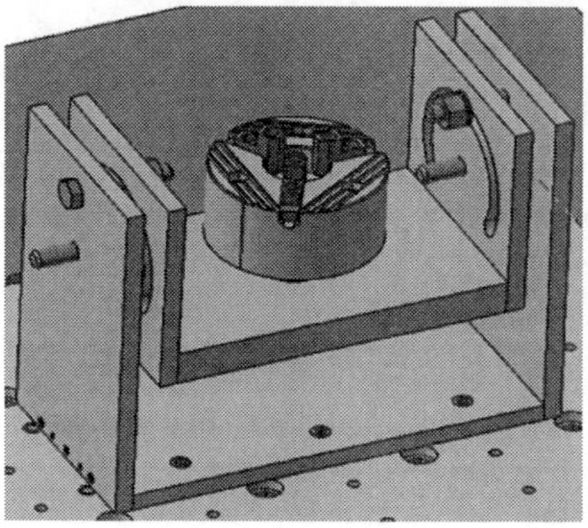

Fig. 3. High tibia osteotomy. a – with closing wedge, b – with opening wedge

The superior ensemble will differ depending of the experiments that are achieved. In principle two different modular devices will be made:

- for the study of the closing HTO,
- for the study of the opening HTO.

In figure 4 is shown an assembly view of the experimental stand used for the closing HTO. Apart from the inferior ensemble (described above) fixed on the table of the Instron universal tensile testing machine, we can also observe the superior ensemble fixed on the mobile traverse, situated in the superior area of the machine. The main functional part of the superior ensemble is the roll 1 which puts pressure on the superior side of the tibia in order to close the osteotomy wedge (see figure 4).

Roll 1 is articulated through a pin in fork 2, which is guided in the linked part 3, guidance that allows it to make anterior/posterior movements.

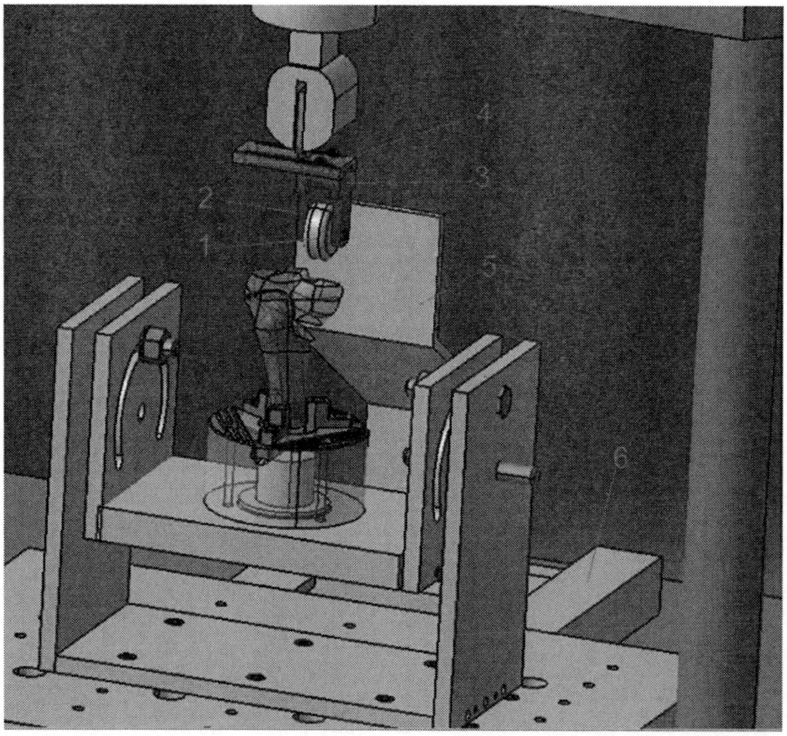

Fig. 4. Experimental stand used for studies of High Tibial Osteotomy with closing wedge

For achieving left-right movements, part 3 is guided in the guiding plate 4. On this plate there are a prismatic zone dimensioned in such a way that they can be assembled in the clamping devices of the machine. The two rectangular movements that the roll can accomplish are useful for determining both the point of forces application and medial-lateral position for achieving the loading after HTO. The fixing in the needed positions will be realised through screwed assembly. The tibia was clamped in the shaft in vertical positions. In order to assure a correct position of the tibia in the device, it was designed a little ensemble which had the role to improve the guidance of the tibia in the clamping moment. This is useful in order to orientate the cutting in the transversal plane and the pressure exerted by the roll to be distributed evenly on the action area. For this auxiliary ensemble, we have proposed a plate vertically orientated 5, which can approach or retreat from the U frame moving on the guiding plate 6. This ensemble will be mounted on the machine's table. The system is useful for the study of both the opening HTO and the closing HTO (see figures 4, 5).

The second experimental stand is destined for the studies of the opening HTO.

Analysing the assembly view (see figure 5), we can observe that the inferior ensemble used is the same as the inferior ensemble used in the previous stand. The difference is that the positioning of the ensemble isn't central with the symmetry centre situated in the axis machine but in the lateral part of the table's machine. Also the tibia that will be clamped in the shaft will not be fixed in the vertical position but close to the horizontal positions which allows the access of the graduated wedge 1 in the osteotomy area.

The superior ensemble is created from the same parts as is the closing HTO stand. The difference being that roll and fork were substituted in this case by the graduated wedge 1 guided by the linked part (see figure 5). The graduated wedge can execute movements both anterior-posterior and right-left.

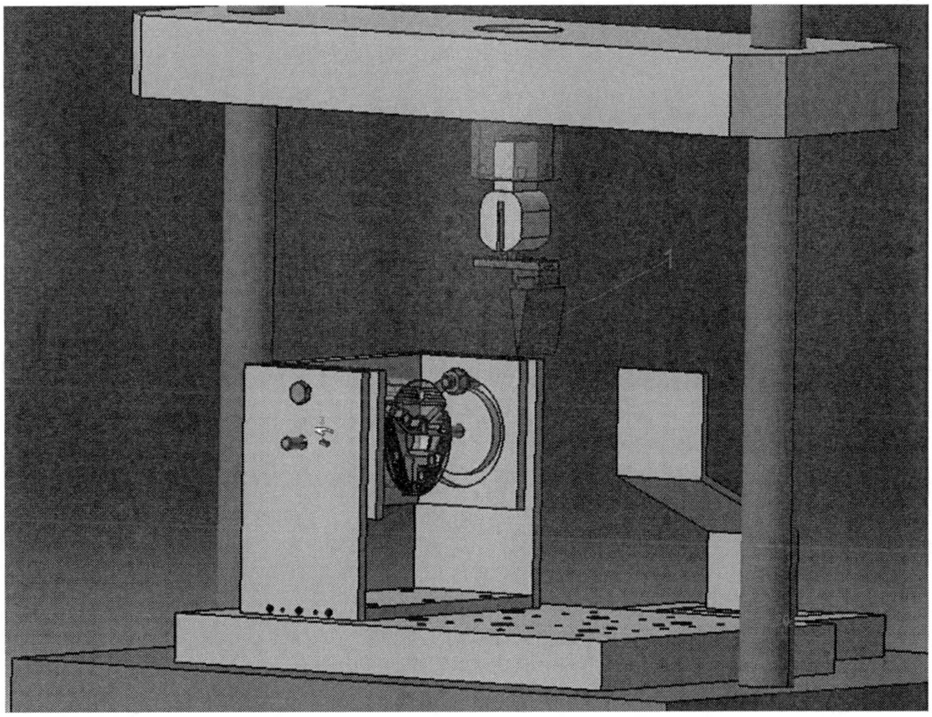

Fig. 5. Experimental stand used for studies of High Tibial Osteotomy with opening wedge

An important element in this case is the possibility and necessity of the graduated wedge to rotate around of the axis of the mobile traverse of the machine. The machine ensures this function. For correct positioning of the tibia we will use the same guiding systemas the one presented at the previous stand

From a functional point of view the tibia is oriented and fixated in the shaft and it rotates around its own axis until the posterior surface meets the guiding plates. The next step is the orientation of the HTO cutting in the vertical plane in order to approach this with graduated wedge.

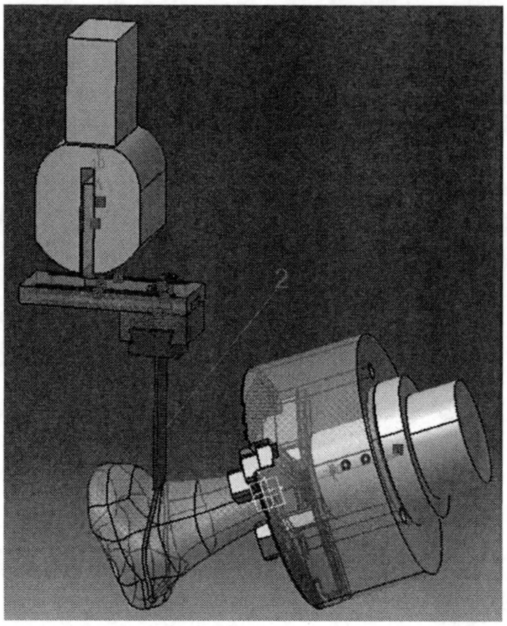

Fig. 6. The use of the searcher

For this purpose we will use the searcher 2 mounted in the superior ensemble instead of the graduated wedge (see figure 6). For moving the searcher in the HTO cutting, two translations on rectangular directions and one rotation in the horizontal plane are necessary. Also the interior frame together with the shaft and the tibia will perform a rotation until the searcher will be perfectly aligned in the HTO cutting (see figure 6).

This position will be conserved through the fixation screws form the U frames in the angular slots. After the orientation was achieved, the searcher 2 is replaced with the graduated wedge (see figure 5). The graduated wedge was designed and realized starting from the real tool that is used in the surgical techniques. This has engraved on the lateral surfaces small channels with the lengths between 4 and 18 mm. For achieving the required opening the graduated wedge is introduced in tibia until it touches the gradation corresponding to the needed angle.

Often in surgical practice it's made this approximation: 1 mm on the graduated wedge corresponds to a 1 degree angular correction.

For the both stands, the inferior ensemble and the superior ensemble are put in a cinematic chain through an extern machine. Concretely, for our experiments this machine will be an Instron universal tensile testing machine for traction, compression and buckling (Instron 5587, maximum force 300 kN) that can be found in our faculty.

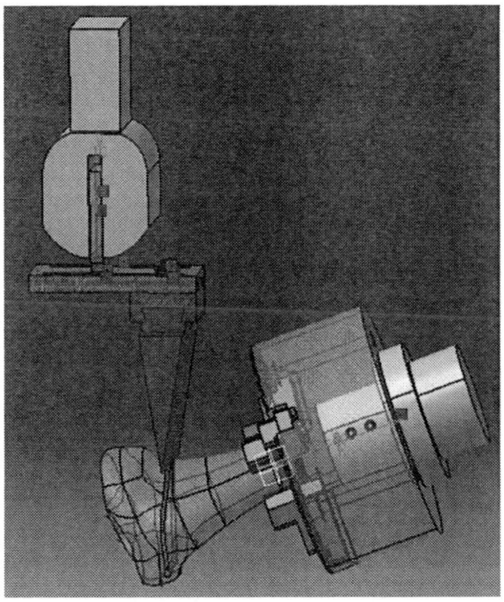

Fig. 7. The use of the graduated wedge

In figure 8, it is show the experimental stand destined for the study of the opening HTO, as this was developed and placed on the machine in our laboratory.

Fig. 8. Experimental stand used for studies of High Tibial Osteotomy with closing wedge

4 Conclusions and further work

In this paper, we have shown the designing of the devices used for the experimental study of the HTO and its variants. The designing was achieved starting from the necessity, requirements and functions of the stands. The modularization concept was used in order to cover a bigger area of research.

These stands designed and developed will be an experimental platform of research for the PhD Students or teachers implied in the research concerning the optimization of the surgical procedures, in this case from the orthopaedic and traumatology fields.

The experiments achieved with these stands are very useful for the improvements of the quality of this type of surgeries.

References

1. I.I. Cofaru, - *Researches regarding the biomechanics of the axial deviations of the human lower member and the development of the correspondent surgical devices*, PhD Thesis, Sibiu, (2013)
2. AA. Amis, Knee Surg Sports Traumatol Arthrosc., 21(**1**), (2013)
3. D. Pape, K. Dueck, M. Haag, O. Lorbach, R. Seil, H. Madry A., Knee Surg Sports Traumatology Arthrosc., 21(**1**), (2013)
4. I.I. Cofaru, Annals of the University Of Oradea, Romania, May, XXIII (2014)
5. I.I. Cofaru, P.D. Brindasu, N.F. Cofaru Applied Mechanics and Materials, 371 (2013)
6. M. Roman, R Fleacă, N. Cofaru, V. Oleksik, A. Pascu, C. Deac, I. Baier, The 11-th Congress of European Federation of National Association of Orthopaedics and Traumatology, Madrid, Spain, (2010)

Simulation of random events for adaptive control systems calibration

Mircea Viorel Drăgoi[1,*], and *Mircea Anton* Vasiloni[1]

[1]Transilvania University of Braşov, Manufacturing Engineering Department, 5 MihaiViteazul, Romania

Abstract. The paper deals with a mathematical model that simulates the random occurrence of events during cutting processes by milling. The evolution of certain parameters that typify the cutting processes depends on predictable and non-predictable variables. In this context, either the material hardness that varies in different sides of billet, or cutting depth, can act as non-predictable variables. In order to design a response in terms of cutting parameters to non-predictable variations of inputs, a simulation of such phenomena is very useful. A mathematical model that generates random events, both in terms of non-uniform frequency and intensity is here described. A virtual instrument built in LabVIEW generates (pseudo) random events based on a combination of random numbers, as the evolution of the simulated process to be much like a real one. Furthermore the user of virtual instrument can generate himself events at certain moments and of certain intensity. This can be a useful tool to study the algorithms of designing the response which should re-balance the process within adaptive control systems.

1 Introduction

Generally, the cutting processes are supposed to be deterministic ones, that is the input is predictable, and the output depends on input according to a certain function (e.g. the cutting torque is a function of feed-rate). In fact, not only the feed-rate influences the cutting torque, but other ones as well. Some of it, unfortunately, have a non-predictable evolution. Occurrence of random events, as variation of input parameters that are supposed to be constant, have a bad influence on cutting output, and they have to be balanced by acting on controllable inputs. A sample in these terms could be the variation of the hardness of the raw material, which may vary along the billet. Increasing the hardness causes an increasing cutting torque, which could be balanced by a decrease of feed-rate. Many other input may vary randomly, and their cumulated influence on the output should be attenuated.

Some steps before the experimental research is carried on, should be the theoretical one, and simulation. To design the correct reaction to a random event occurred during cutting process, a simulated evolution of input is needed to save resources.

In the scientific literature, some researches related to deterministic processes and random events are described. So, in [1] is presented a method so solve the industrial problems caused by random events in manufacturing systems. A study case has shown that

*Corresponding author: dragoi.m@unitbv.ro

random events may cause an increasing of production time of up to 20%. Applying the method to schedule production the authors cited propose, can lower the risk to increase production time to 0.5%. In [2] are described by means of simulation the possible effects of small probabilities (10^{-6}).

Random signals and their influence in opposition with deterministic signals in the electrical domain, in terms of collecting, transmitting and processing information are described in [3]. In [4] is stated that computer simulation is considered to be an experimental research that saves resources and can give even better output than the experimentation in the real system itself, if the computer model is appropriate. That encouraged the work to build a simulation model of the random events occurred during the cutting processes.

2 The concept

The random events are characterized by two main features: the intensity and the period of time they produce again. Both of them are non-predictable, and their combined evolution can describe the way a process supposed to be touched by random events progresses.

It is important to note that only when both the two mentioned parameters are irregular they can describe close to reality the random events that affect a process. Not any order of combinations of two random numbers can be considered a proper description of reality (a lifelike simulation). So, one can observe that several ways to randomize the events could be taken into account, as described in the next paragraphs. In this context, an event is that which causes a variation of the cutting torque measured during a cutting process.

To create models of events occurrence some virtual instruments (VIs) built in LabVIEW have been used. The engine of generating random events is the LabVIEW function *Random* that returns random values in the range of 0..1. Of course, these values are processed in an appropriate way to get the significance of intensity of event, or the period of time after it occurs. The type of processing can be selected by user of the VI, according to his needs and strategy of research.

The core of the algorithm to translate a random value returned by LabVIEW *Random* function in the interval [0..1] to a certain interval [min..Max], conserving the apportionment is formula (1).

$$Vx=R*(Max-min)+min \tag{1}$$

where, according to Fig. 1,
Vx – a value in interval [min..Max],
R – a random number in interval [0..1]
min, Max, the border limits of the interval to project [0..1] on.

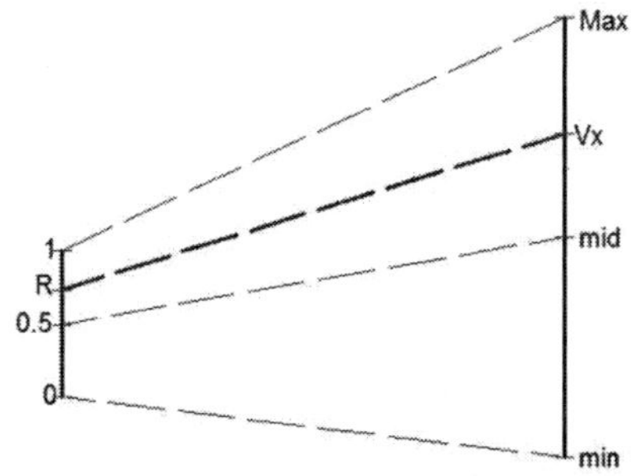

Fig. 1. Converting a random value into one placed in a certain interval, conserving the apportionment

Using it, any random value returned by *Random* function can be converted proportionally into a value in an interval of a given aperture and position, as shown in figure 1.

3 Some types of randomization

3.1 Completely randomized events

In this case, pairs of random numbers are repeatedly generated, and each pair is considered to be an event. Of course, such an evolution is not lifelike, because either the events succeed too fast (Fig. 2a), or, if events occur at random periods of time, their intensity is almost always the same and the difference of intensity too big (Fig. 2b). To bring simulation closer to reality, some "control" should be applied to the progress of events.

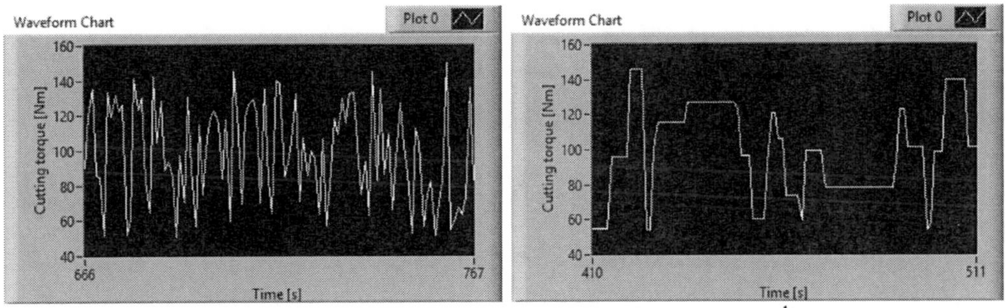

a b

Fig. 2. Completely randomized events

3.2 Selected randomized events

Because of in a regular repeatedly generation of random number pairs is not proper to model a real succession of events (variation of cutting torque), a selection of such pairs should be performed. To do that, some restrictions must operate. In this way, not any pair of random numbers (N_1, N_2) will be considered an event (as shown in Fig. 2a), but only those that meet a selection criterion. For instance, the criterion can be either N_1 or N_2 should be smaller than a threshold value. Say that N_2 controls the frequency of the events, that is the lower the threshold for N_2 is, the smaller the probability to occur the event (Fig. 3 a – threshold=0.3, b – threshold=0.1). The value of the threshold is an input of the VI, so the user can keep under control the events generation, without corruption of it's random feature.

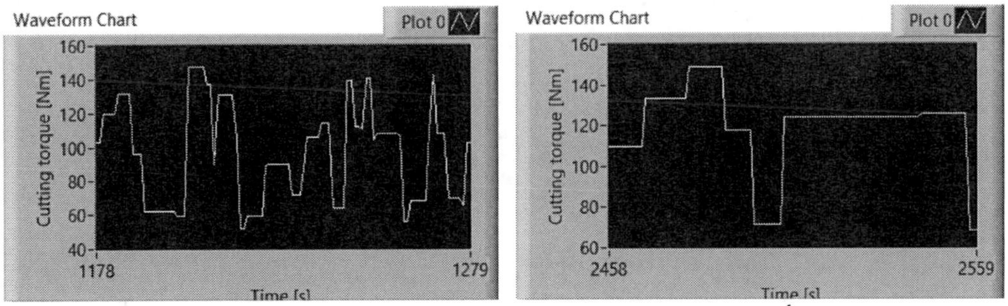

a b

Fig. 3. Selected randomized events

The evolution of the cutting torque shown in Fig 3 a, and b still does not fit to reality, because rarely the skipping of the cutting torque is so dramatic. In real conditions the variation of the torque is not so fast, that is the segments that represent cutting torque skipping in chart should be smaller. To bring simulation closer to reality, that is the skipping of the torque variation should have different values on one hand, and the bigger are the variations, the more rarely they occur, further processing of the randomly generated numbers must be performed. Now the randomly generated numbers do not represent the value of the cutting torque itself, but the variation of it, and it can be either positive or negative. The value of the variation is generated in an interval [-V .. V], and it is added to the previous value of the cutting torque. In Fig. 4 a, and b one can see simulations of cutting torque variations that are generated in these conditions. They display some consecutive increasing or decreasing which have different values, and the big skipping of cutting torque are rare, while the small ones occur oftener.

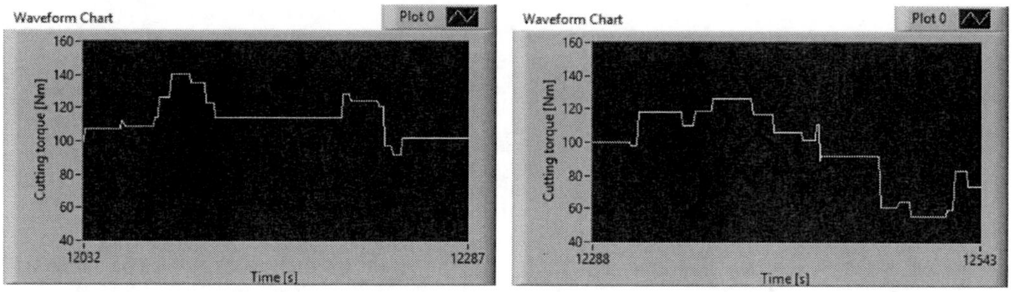

a b

Fig. 4. Selected randomized events. The big skipping are rare

3.3 Pseudo-randomized events

In practice there are situations when the measure monitored has a certain/expected way to evolve that might be affected by events (unexpected, random variations). To simulate such progress of a certain measure a more complex processing of the random numbers is needed. The events should be generated as deviations from the expected value of the cutting torque at a moment. That means a random number will be generated in a vicinity of the expected value. The vicinity may be smaller or larger, according to the user's needs. In fact, the random values are generated within a band around the theoretic values of the measure (in our discussion the cutting torque).

In Fig 5 a, and b, a random progress overlapped on a sine-shaped evolution is presented, for two widths of the band.

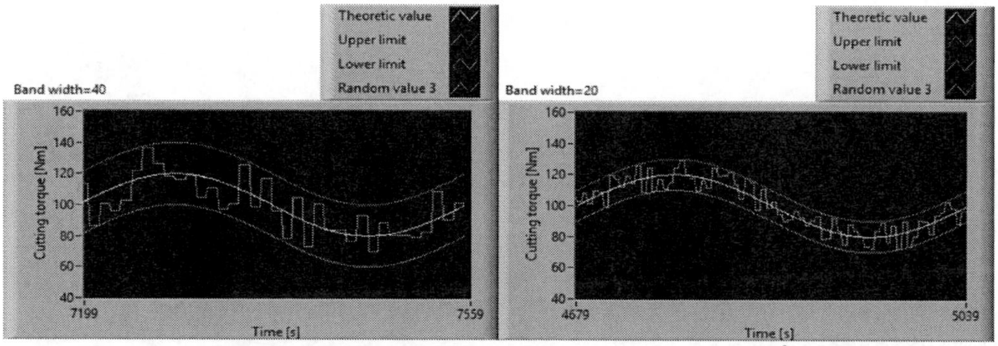

a b

Fig. 5. Pseudo-randomized events. Different band widths

If further refine the algorithm of generating pseudo-randomized events, new shapes can be obtained. When the process is supposed to be stable, and deviations from the expected values are very rare, but significant, insulated events can be simulated, as shown in Fig. 6 a, and b. Note, that in these cases, on the general sine-shaped evolution, a vibrating noise have been overlapped, as well.

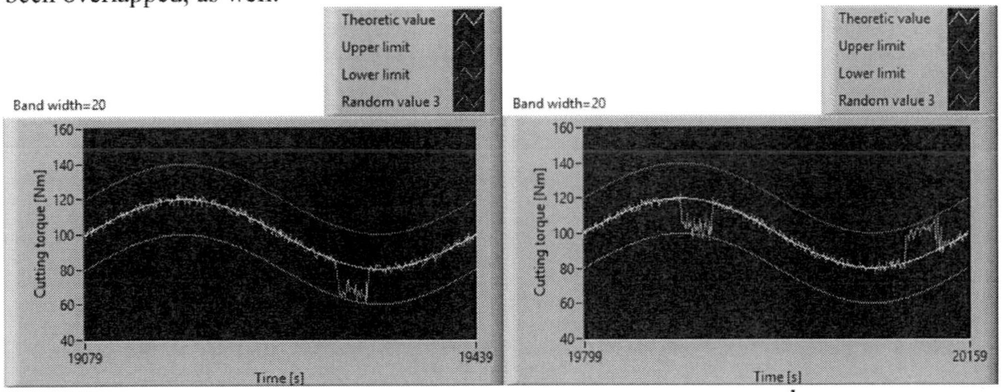

a b

Fig. 6. Pseudo-randomized events. Insulated events

3.3 Discussion on results of simulation

Combining in different ways two or more randomly generated numbers, a large variety of random events or sequence of random events can be generated. Many options are available for the user of this software tool, so he can describe in different ways the evolution of a process affected by random events. Several inputs, as the minimal and/or maximal time between two consecutive events, the minimal and/or maximal magnitude of event, the shape of predicted general evolution, and others, allow user to simulate events that occur at randomized periods of time, having random intensity, following a certain predicted shape, but altered by significant, insulated events, and many other. All of these keep the random occurrence of the events

4 Conclusions

The piece of software here described fits to the necessity to benefit from simulated evolution of processes that are subject of adaptive control. Calibrating an adaptive control systems is a difficult task, because a mathematical model has to be developed in order design correctly the reaction that has to be provided to an unbalanced process that was affected by an unpredictable event, in order to re-balance it, to bring it under control again. Using this simulator, the mathematical model that governs an adaptive control system can effectively and safety be tested and adjusted, without wasting material resources before putting it to work in real conditions,

We hereby acknowledge the structural funds project PRO-DD (POS-CCE, O.2.2.1., ID 123, SMIS 2637, ctr. No 11/2009) for providing the infrastructure used in this work.

References

1. G. Fleury, J. Y. Goujon, M. Gourgand, et al. J Intell Manuf **10**, 81 (1999)
2. S. Asmussen, *Proceedings of a Conference held at Hong Kong Baptist University, Hong Kong SAR, China,* 200, **1** (2000)

3. M. C. Jeruchim, P. Balaban, K. S. Shanmugan, Chapter *Simulation of Random Variables and Random Processes in Simulation of Communication Systems*, 189 *(2002)*
4. B. Wu, Chapter *Computer simulation in manufacturing systems analysis* in *Manufacturing Systems Design and Analysis*, 193 (1992)

Virtual model of an automated system for the storage of collected waste

George Enciu[1], *Adrian* Popescu[1], *Cicerone Laurentiu* Popa[1,*], and *Costel Emil* Cotet[1]

[1] University Politehnica of Bucharest, MSP Department, Spl. Independentei 313, Bucharest, Romania

Abstract. One of the problems identified in waste collection integrated systems is the storage space. The design process of an automated system for the storage of collected waste includes finding solutions for the optimal exploitation of the limited storage space, seen that the equipment for the loading, identification, transport and transfer of the waste covers most of the available space inside the integrated collection system. In the present paper a three-dimensional model of an automated storage system designed by the authors for a business partner is presented. The storage system can be used for the following types of waste: plastic and glass recipients, aluminium cans, paper, cardboard and WEEE (waste electrical and electronic equipment). Special attention has been given to the transfer subsystem, specific for the storage system, which should be able to transfer different types and shapes of waste. The described virtual model of the automated system for the storage of collected waste will be part of the virtual model of the entire integrated waste collection system as requested by the beneficiary.

1 Introduction

In the context of technological progress and the society evolution we must have a strict waste management so as to ensure the continuous development and protection of the planet's resources. For this, it is necessary that any company provide means for collecting waste, which should be as simple as possible, and with continuous accessibility. Such a means for the waste continuous collection represents the automated collection and storage system [1, 2-3].

In its current configuration, the waste collection system contains a RVM (Reverse Vending Machine) for plastic recipients, glass recipients and aluminium cans, different modules for collecting: plastic film, plastic bags, paper, cardboard, batteries, small and big electronics and appliances. These modules are not automated; two operators are receiving the collected waste. The operators are identifying and weighing the WEEEs and are transporting the WEEEs to the storage area. In this paper we intend to present a virtual model of an automated system for the storage of collected waste (plastic and glass recipients, aluminium cans, paper, cardboard and WEEEs) as a part of the virtual model of the entire integrated waste collection system.

[*] Corresponding author: laur.popa79@gmail.com

DOI: 10.1051/matecconf/201712104006

The automated storage system must permit the transportation and storage of all waste types (large home appliances, large blocks of compacted paper and cardboard and containers), with the possibility of quick access, without the intervention of human operators in the storage area.

The main advantages of using an automated waste storage system in the proposed application are the following
- reduced time for the storage and access to the stored waste;
- management of the stored waste;
- vertical storage availability which leads to reduced storage space.

The types of waste that will be stored via the automated storage system integrated into the application are divided into the following categories:
- Large home appliances: refrigerators, washing machines, TVs etc.
- Mid-size home appliances: microwave ovens, desktop computers, laptops etc.
- Containers that can contain small home appliances such as tablets, mouse, keyboards, phones etc. and pressed aluminium cans.
- Blocks of compacted paper and cardboard.

Examples of such systems that may be used in the software application for collecting and storing waste are presented in Figure 1.

Fig.1. Examples of automated storage systems

The designed storage system must be justified in terms of investment and operating costs by ensuring an adequate level of performance.

Therefore, the physical capacity specific to the storage system must be greater than the maximum volume of waste that is expected to be stored. The surface of the storage area designed will be of 9.8 m^2 and the height will be of 6 m.

The space afferent to the lane specific to the transport - transfer system which carries out the storage operation automatically, often occupies more space than the usable storage capacity.

Taking into account the performance that the storage system as part of the integrated system for the waste collection and disposal must meet and the types of parts that will be stored, the transport- transfer system specific to the storage system designed will need to meet the above demands.

Since using an automated storage and waste collection system has as main purposes to remove the human factor and a very good planning and use of storage space, respectively, we will use as storage strategy, at least in a first phase, the predetermined storing [4-5].

2 Equipment in the waste collection system

A possible variant of positioning the equipment into the waste automated collection system and the waste disposal station is shown in Figure 2.

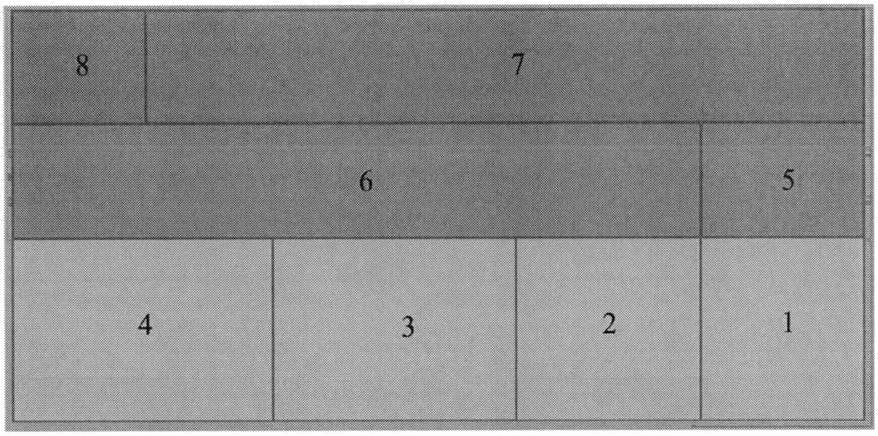

Fig. 2. Positioning the equipment in the waste collection and storage system

The correspondence between the positions shown on the Figure 2 sketch and the names of equipment used in the waste collection and storage system is presented in Table 1.

Table 1. Equipment in the waste collection system.

Position	Equipment name/zone
1	Input station for large home appliances wastes
2	Input station for small home appliances wastes
3	Input station for paper and cardboard wastes including the pressing machine for compacting and making the paper blocks
4	Input station for plastic recipients, glass recipients, cans, including the equipment for selecting and compacting the waste
5	Transport transfer system included in the automated storage system
6	Access lane for the transport transfer system
7	Waste storage proper area made up of a rack with 27 specific places for storage
8	Output station for stored wastes

Positions 1, 2, 3 and 4 correspond to entering into the system of the waste in correlation with the description in Table 1. Each station will be previously fuelled with a support pallet for positioning the parts that will be stored (large home appliances, compacted cardboard / paper blocks or containers with glass recipients, plastic recipients or aluminium cans).

The input station into the system for large home appliances wastes (Figure 3) is equipped with a conveyor allowing the support pallet positioning and blocking in a fixed position when the home appliances waste is fed to avoid theft. When the access for home appliances feeding is allowed, the access gate that separates the input station from the storage system is closed; it will open only after the waste identification and weighting and after the feeding access gate will close. For the input stations corresponding to positions 2,3, and 4 the feeding with small home appliances waste, cardboard / paper is performed through small access doors, outside the waste collection storage system and, respectively, glass recipients, plastic recipients and aluminium cans are received, identified and sorted using a RVM equipment.

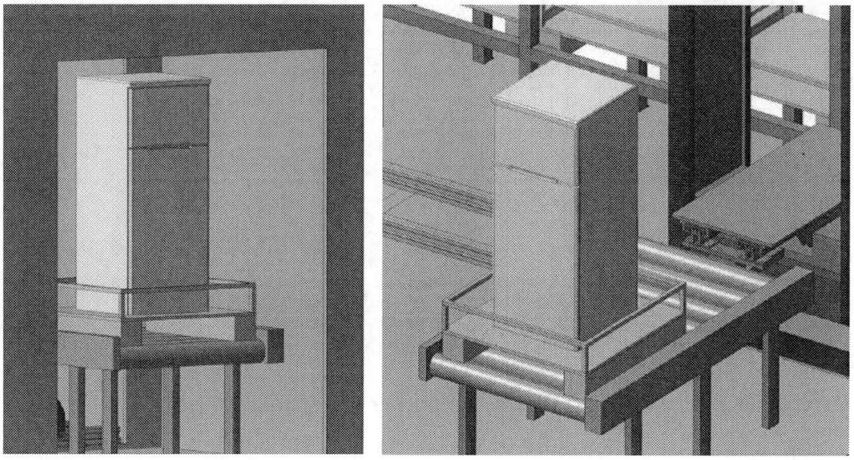

Fig. 3. Input station for large home appliances wastes

3 Wastes automated storage system

The waste storage system (Figure 4) comprises three main elements: storage structure, the transport-transfer system and the input-output stations [6-8].

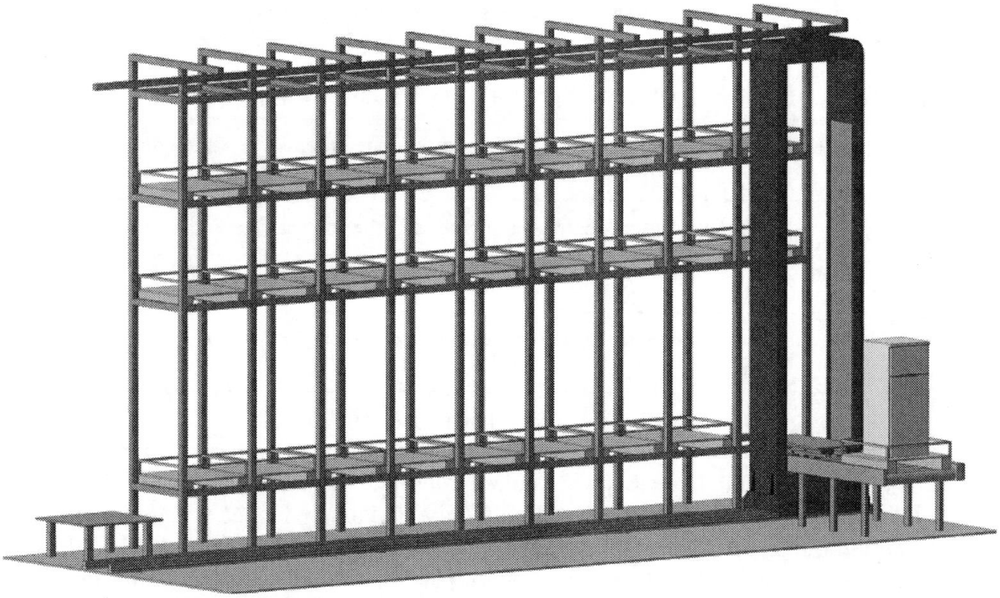

Fig. 4. Automated storage system

3.1 Storage structure

The storage structure consists of a shelving system used for the waste storage (Figure 5). The 27 specific storage places (3 levels with 9 places on each level) are of different sizes, so in base places there will be stored as appropriate, large electronic waste (refrigerators, washing machines, TVs etc.) and on the other places from levels 2 and 3 there will be stored blocks of paper or cardboard, containers with small electronic wastes, glass recipients, cans and plastic recipients. The storage places have a standard structure and to store different types of waste a support pallet is used. In the base places, there can be stored home appliances wastes with the maximum height of 2 m, and in the places afferent to levels 2 and 3, there can be deposited containers or paper/cardboard blocks with a maximum height of 1 m.

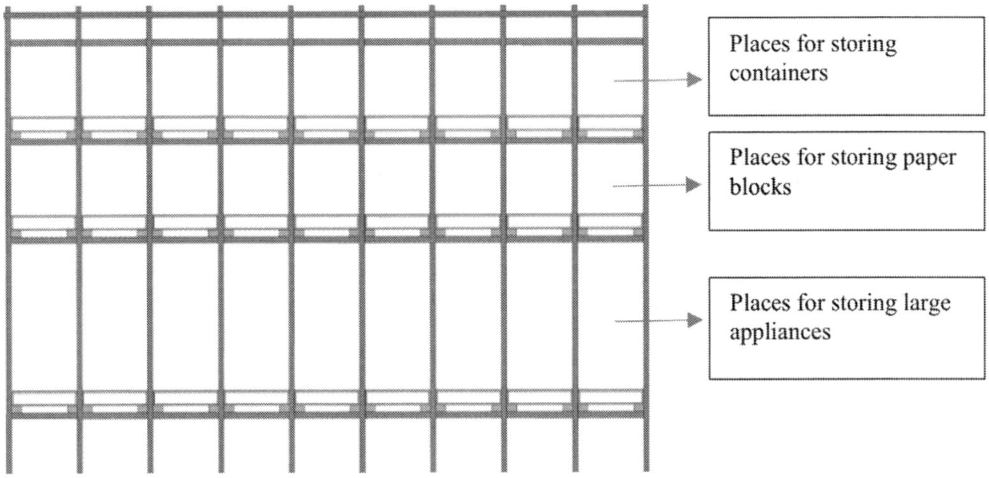

Fig. 5. Storage structure specific to the automated storage system

3.2 Transport-transfer system

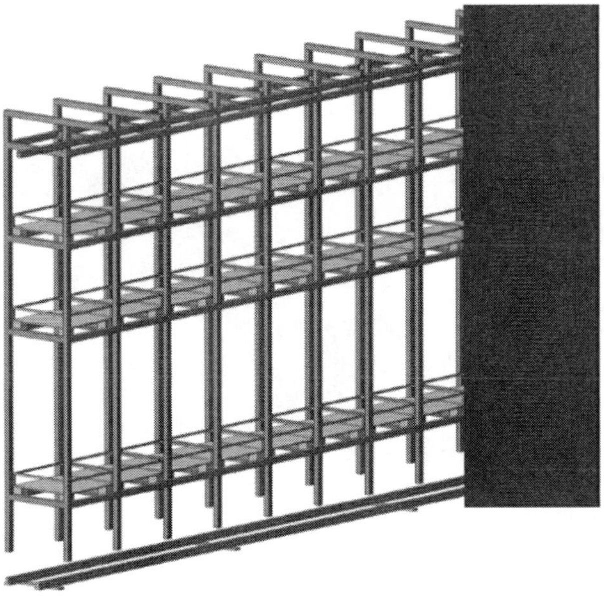

Fig. 6. The transport-transfer system

The transport-transfer system (Figure 6) does the wastes storage and taking over operations when they are evacuated. Picking up waste process automation performed by the transport-transfer system is available through a couple of sensors mounted on input posts which confirm the presence and the type of the waste that is going to be collected and stored. The transport-transfer system can move vertically and horizontally to position the transfer system in the input and output stations and in the places of storage. It is made up of a rigid beam that can be moved vertically to the transfer system. The whole ensemble moves on the ground on a rail system that crosses the storage lane and a beam stiffened by the storage structure does the guidance.

The transfer subsystem takes over the WEEEs, the containers with waste and the compacted blocks of paper/cardboard from the input stations and disposes them in the places specific to the storage structure for the operation of loading and, respectively, that of taking over the waste stored from the places of storage and positions them in the output station for the discharging operation (waste disposal).

The transfer system is made up of two telescopic arms that are driven by an electric motor.

Fig. 7. The transfer system specific to the storage system

A telescopic arm consists of 3 sections (Figure 8). Segment 1 is fixed to the structure supporting the transport -transfer subsystem, segment 2 can move in relation to segment 1 up to its halfway point and segment 3 is can move in relation to Segment 2 up to its halfway point.

The operation of the two mobile segments is done through the pinion-rack mechanisms and the movement guiding is done through the buckets that are found on all the three segments.

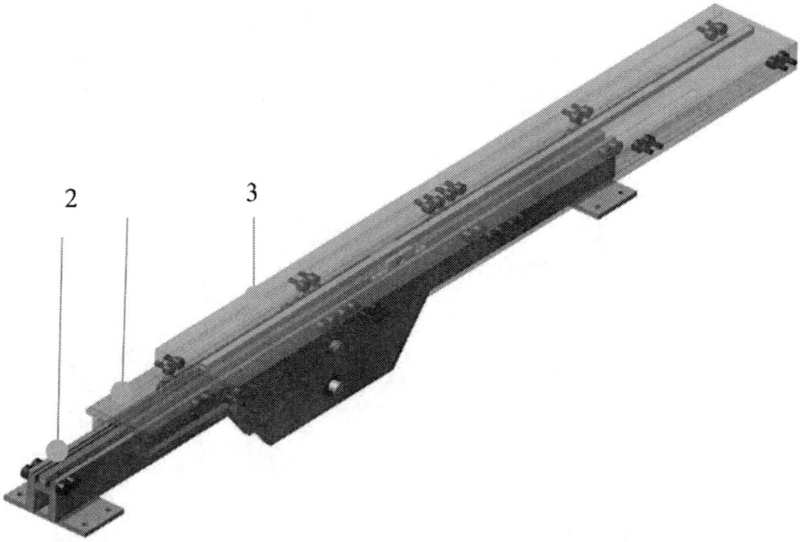

Fig. 8. The transfer system specific to the storage system

3.3 Input/Output stations

The input stations are specific to the 4 areas of feeding with waste (large home appliances waste - input station 1; small home appliances waste positioned in the container – input station 2; cardboard and paper compacted into blocks and tied - input station 3; glass recipients, plastic recipients and aluminium cans positioned in the container – input station 4). Through these stations the transport-transfer system takes over the wastes and stores them. The input stations are arranged laterally along the whole length of the access lane specific to the transport-transfer system.

The output or evacuation station from the waste collection and storage system is placed in the continuation of the storage structure on the opposite side to the entrance stations. Qualified personnel do waste disposal in a scheduled order during the discharge.

4 Conclusions

The proposed structure for the waste collection and storage system takes over and stores large home appliances waste, small home appliances waste, cardboard, paper, glass recipients, plastic recipients and aluminium cans. This was made possible by the use of a support pallet positioned in each input station before or together with the collection container.

The virtual model designed by the authors is meant to be a part of the virtual model of the entire integrated waste collection system. The storage system structure will be integrated (in correlation with the technical specifications of the other types of equipment) in the imposed usable area of 60 m^2 of the waste collection and storage system. The virtual model of the entire system will be the backbone of a digital twin, monitoring a modernized, completely automated waste collection system. This automated system, will considerably increase the waste collection level, will meet the requirements and the needs identified by the economic agent and will be ready to integrate the infrastructure of the smart city.

This work was supported by a grant of the Romanian National Authority for Scientific Research and Innovation, CNCS/CCCDI – UEFISCDI, project number PN-III-P2-2.1-BG-2016-0437, within PNCDI III.

References

1. B. Simon, M.B. Amor, R. Földényi, JCP **112**, 11 (2016)
2. I. Ervasti, R. Miranda, I. Kauranen WM **48**, 8 (2016)
3. F.C. Bergeron, RCR **115**, 13 (2016)
4. C.E. Cotet, C.L. Popa, G. Enciu, A Popescu, T Dobrescu, IJSMM **15**, 10 (2015)
5. C.A. Popescu, C.L. Popa, C.E. Cotet, G. Enciu, P DAAAM **27**, 7 (2016)
6. R. Manzini, M. Gamberi, A. Regattieri IJAMT **28**, 9 (2006)
7. C.A. Popescu, F.A. Nicolescu, G.C. Avram, A.M. Ivan, AMT **834**, 6 (2016)
8. G. Enciu, J P ICMaS **1**, 4 (2006)

MATEC Web of Conferences **121**, 04007 (2017) DOI: 10.1051/matecconf/201712104007
MSE 2017

Automated business process management – in times of digital transformation using machine learning or artificial intelligence

Daniel Paschek[1,*], *Caius Tudor* Luminosu[1], and *Anca* Draghici[1]

[1]Politehnica University Timisoara, Management Faculty, 300191, Timisoara, Romania

Abstract. The continuous optimization of business processes is still a challenge for companies. In times of digital transformation, faster changing internal and external framework conditions and new customer expectations for fastest delivery and best quality of goods and many more, companies should set up their internal process at the best way. But what to do if framework conditions changed unexpectedly? The purpose of the paper is to analyse how the digital transformation will impact the Business Process Management (BPM) while using methods like machine learning or artificial intelligence. Therefore, the core components will be explained, compared and set up in relation. To identify application areas interviews and analysis will be held up with digital companies. The finding of the paper will be recommendation for action in the field of BPM and process optimization through machine learning and artificial intelligence. The Approach of optimizing and management processes via machine learning and artificial intelligence will support companies to decide which tool will be the best for automated BPM.

1 Introduction

The ongoing industrial revolution from the implementation of mechanic production over the division of labour and assembly line work to the usage of electronic and automatic production were the first three named types of industrial revolution [1]. Actually we arrived at the 4.0 step of the industrial revolution – the so-called digitalization [2]. Behind the concepts Smart Factory, Internet of Things, intelligent manufacturing, Cloud Computing, Big Data, Machine to Machine communication and many more, new elements and trends affect the established economy [3]. Mc Kinsey advises therefore: "The traditional manufacturing business model is changing, and new models are emerging; incumbents must be quick to recognize and react to these new competitive challenges" [3].

Driven by the Internet, real and virtual worlds are growing ever further to form an internet of things and cloud computing to import [4]. Cloud computing is hosted on the internet and allows remote access to apps, services, and stored to manage processes, workflows or machines [5]. The internet of things is described by J. Brixel as: "The internet of things builds on this concept by using the cloud to store and automate processes in

* Corresponding author: paschi88@gmx.net

objects that are synced to the internet, like internet-enabled automobiles and remote home lighting and shade systems" [5]. In a nutshell, Industry 4.0 takes processes and the management of processes which are normally managed internally by people into the cloud where they can be adapted and analysed from anywhere [5].

In times of Big Data, constantly changing conditions in the supply chain and new customer requirements, process management and automation play a major role to reduce response time, to avoid back logs, dissatisfied clients and processing errors [6].

To enable a further step in process management and automation this research looks in detail at the possibilities of using machine learning or artificial intelligence for Business Process Management (BPM) in times of digital transformation. The opportunity of independent process management, via automated and learning computer programs promises in theory agility, flexibility and automation for a whole company as well as the process stake holders. The aim of this paper is to investigate the application these two mentioned developments in the field of Business Process Management. The outcome will show the effect as well as advantages and disadvantages of machine learning and artificial intelligence for process steering and optimization.

2 Scientific status of the research

2.1 Definition of Digital Transformation

The term digital transformation exists since the 1960s, with the wider availability of mainframe computers [7]. Over further development steps in digitalization like in 1980 the e-mail or 1990 the emergence of CRM tools and Business Intelligence (BI) to connected objects, 3D printing, the block chain technology and artificial intelligence (AI), the innovation in digital technologies and digital transformation changed the way of collaboration, execution at the level of the whole economy in a radical way [7].

In a nutshell, digital transformation means high-speed changes in the cooperation due to innovation and implementation of digital technologies and the integration into all aspects of human life and society at different dimensions [8].

2.2 Definition of Business Process Management

Business Process Management (BPM) is a management concept for controlling, adapting and optimizing business processes [9]. A concrete definition for BPM from the European Association of Business Process Management (EABPM) defined BPM as a systematic approach, to capture, shape, execute, document, measure, monitor and steering automatic and non-automatic processes to reach coordinated and sustainable company targets [10]. The aim of Business Process Management (BPM) is to improve the corporate performance by optimizing and managing business process of the company [11].

Business Process Management can be described as corporate business process optimization and management over single systems like ERP, CRM and SCM and the whole integrated network.

The assignment of a BPM-System is furthermore the coordination and execution of business processes step by step [11]. Through monitoring, identifying and evaluating of process issues or troubles, the BPM-System depicts where a process had trouble and enables the business to optimize their process issue [11].

 DOI: 10.1051/matecconf/201712104007

2.3. Definition of Artificial Intelligence

In the literature are two different approaches to define the artificial intelligence (AI). On one hand humans are drawn as comparison and on the other AI is characterized by the list of different application and work areas [12]. For example:

- "the theory and development of computer systems able to perform tasks normally requiring human intelligence…" [12];
- "The AI examines how to capture and understand the intelligent behaviour of computers, or how to solve problems by using computers that require interoperability "[12];
- "AI is a subdivision of computer science, which deals with the investigation of different problem areas like robotic, speech and flow text recognition as well as image and video processing [12].

In the scope of this paper AI can be defined as a section of informatics and applied computer science to pattern human proceedings of problem solving and transfer them to computers in order to invent efficient and new solutions as well as course of actions. Therefore, AI is a computer program running on any possible device(s) or data centre(s) with the skill to interact with its environment.

The definition of AI suffers from the fact that the terms "intelligence" and "intelligent human behaviour" are not well defined and understood yet [13]. In 1950 the British mathematician A.M. Turing (1912-1954) wrote the critical essay "Computing Machinery and Intelligence" on the subject of AI. He asks the question of whether a program can be intelligent and invented the so-called Turing test [13]. In this experiment, a test person communicates via a computer terminal with two non-visible partners, a program and a human being [13]. Finishing, the test person must decide which is which.

Two types of AI are distinguished in the literature like described in the following figure. It's important to understand the differentiation of AI for the following paper.

	Strong artificial intelligence	Weak artificial intelligence
Definition	• the form of artificial intelligence, which has the same intellectual abilities as human, or even surpasses him in it	• Weak AI is generally developed or used for specific application domains. • In a standard work on artificial intelligence, this is formulated as follows: "The assertion that machines could possibly act intelligently (called, weakness, act as if they are intelligent) is called the, weak AI 'hypothesis ..."
Capabilities and Domains	• Logical thinking • Making decisions in case of uncertainty • To plan • To learn • Communication in natural language • Use all these abilities to achieve a common goal	• Expert systems • Navigation systems • Voice recognition • Character recognition • Suggestions for corrections in searches

Fig. 1. Strong vs. Weak AI, own presentation based on [14]

Already today AI can be found in different types of weak AI in the human environment. Examples of this are object-oriented programming languages and graphical user interfaces, translation systems as well as "intelligent games" like chess can now be installed on almost all computers. Chess programs have recently defeated the best human players, like the IBM Deep Blue Super Computer [15]. Further samples of AI are voice recognitions and control systems like Cortana, Siri or Echo or intelligent bots and Agents who interact independently in phone conversations. The recent successes in AI are enabled by a very broad field of machine learning [15].

2.4. Definition of Machine Learning

Machine learning is a branch of AI and describes mathematical techniques that enable a system, that is, a machine, to independently generate knowledge from experience. [16] Machine Learning describes the development of algorithms based on empirical data as well as training data with the focus on the optimization of the results and the improvement of the predictions due to learning processes [16]. The term algorithm can be defined as detailed sequences of actions to solve a particular task [15]. The development of machine learning is due to the study of recognition patterns and computer learning theory [15]. The algorithmic implementation of machine learning takes place with monitored or unsupervised learning like shown in the following figure 2.

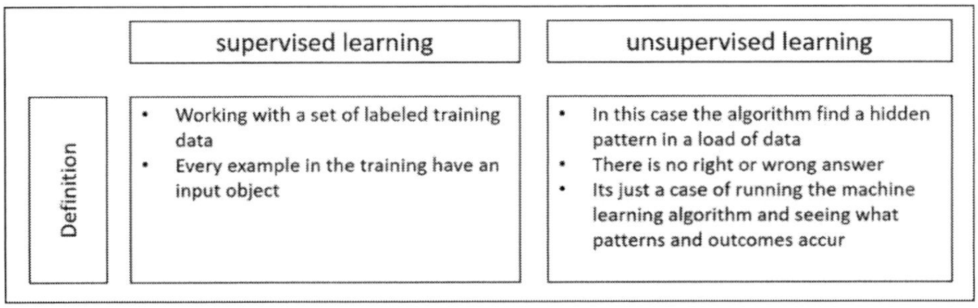

Fig. 2. Supervised and unsupervised Learning, own presentation based on [16]

In the past, the automatic processing of written and spoken language was attempted with the most comprehensive set of rules without any significant progress. [15] Nowadays, the methods of "deep learning" are used in areas of Natural Language Processing (NLP), above all for language recognition.

In part, the terms "deep learning" and artificial neural networks are used synonymously [15]. Deep learning works with artificial neural networks in order to achieve particularly efficient learning sequences like the language recognition “Siri” [15] The process eliminates the need for many neural network operations because the computer takes all intermediate steps. [15] The Scientist has only to present data to the neural network, such as images. How these are to be identified, the net finds completely by itself [15]. The neural networks are divided into several layers, like a filter that works from the coarse to the fine, thus increasing the likelihood of detection and output of a correct result [15]. Deep learning methods first learn low-level elements such as brightness values, then mid-level elements, and finally high-level elements like whole faces, [15]. These incremental enhancement is illustrated at the following figure 3.

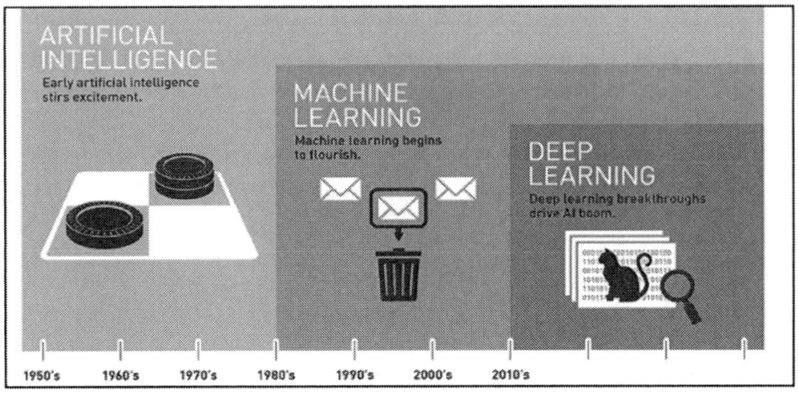

Fig. 3. Incremental enhancement, source: [17]

2.5. Evaluation of scientific status

The theory shows that methods of "deep learning" with neural networks are always meant when artificial intelligence is mentioned. The rapid increased progress of "deep learning" over the past years can mainly lead back to more powerful hardware for computing operations [15]. Furthermore the rising amounts of data which are important for the initial training of the neural Networks have positive effects [17]. After initial training, the "deep learning" of the procedure consists in always learning during the running application, which increases the recognition accuracy and results. From this can be deducted that next to the algorithm the big data plays a significant role for the deep learning and thus for the artificial intelligence.

3 The Research Methodology

For the proposed evaluation and analysis of using machine learning or AI a survey with interview characteristics and a shadowing with so-called digital companies was executed to identify the status quo and execution status at the economy to this topic. Based on the analysed facts and the AI approach a recommendation was announced to the survey participants and their companies for automated process optimization with machine learning or AI.

3.1 Detailed study description

For the analysis of the base data a field study and survey was carried out at 25 digital companies in Germany. A digital company is characterized by the preponderant use of information technology for their primary business activities within their value-added step. More than 75% of the survey questions were focused on the business process execution and the application of information technologies. Part of the survey was the possibility for an assessment of the daily work by shadowing the Process Management Team. Shadowing means to attend the participant at his daily business to have a look the work. For the study the anonymous survey method was used.

It was ensured that the subjects present their knowledge, experiences, expectations and there actually execution of process management. Target persons were mainly Chief Information Officers (CIO), Chief Technology Officers (CTO), Chief Execution Officer (CEO), Chief Digital Officer (CDO) and IT process managers. The base line is set up by 65 valid survey replies from a total of 75 requested ones. The survey embraced 30 closes and 5 open questions. The survey via interview proceedings took approximately 35 till 45 minutes per person.

Next to the survey a shadowing was requested at the selected companies. As observer were the process flows, the process management tasks, the processes itself as well as the necessary applications watched and analysed. The time duration of the shadowing differs via company between 4 and 8 hours.

Afterwards a comprehensive analyses and comparison of the collected data take place to identify the base line as well as deviations. Through the shadowing the analysis of the theoretical appreciation and the execution was possible.

3.3. Research and Interview results

In the following, the results of the survey are presented and the recommendations for using machine learning or AI are mentioned. 87% of the requested Managers of 25 companies allowed the survey and interview as well as the shadowing to analyse the base line of execute automated BPM.

The survey shows that every company and 100% of the respondents are affected by the digitalization. 59% of these are concerned very strong, 35% strong and only 6% stated low. These results indicate that the digital transformation and information technology plays a major role in the companies. This can be underpinned by the results which show in what state of digital transformation the survey participants are. Because approximately 38% are at the growth phase, 51% at the transition phase and only 11% at the beginning of the implementation of the digitalization.

Further relevant research results are listed in the following.

- 39% of the respondents worked in the IT branch, 19% at professional services,18% at the retail, 15 % at Logistics as well as 9 % at the bank sector
- The selected companies can classify in 52% family business and middle class business, 27% Start-up and Internet business and 21% enterprise business
- 51% of the respondent's rate data of the company, like process data as strategic assets and the foundation of the optimization. 39% assess data as feasible data and privacy problem and only 9% are sceptical and see data as cost drivers
- 100% of the attendees know and uses BPM in the company and 85% want more process optimizations and automated BPM applications
- only 12 % uses data warehouses, business intelligence systems or data lakes for data reporting and data management
- 75% of the respondents know the terms AI and machine learning but only 29% assess machine learning actually
- just 26% have already heart about deep learning and just 2% execute it

Very interesting are the survey results concerning the utilisation of machine learning now and in the future and AI like shown in the following figure 4.

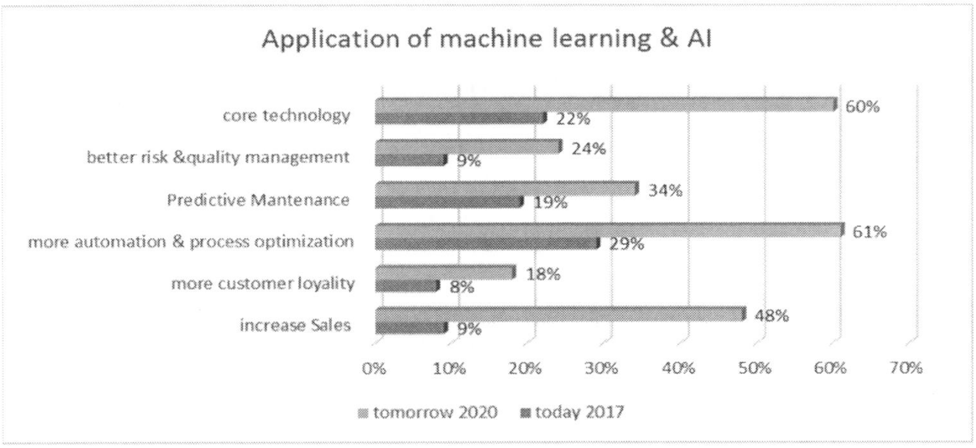

Fig. 4. Application of machine learning & AI, own presentation

It can be concluded that process optimization with 29% is the most important and assessed application for the participants. Other applications like machine learning as core technology for new business models or predictive maintenance are the second and third important areas of application. The outlook and estimation to the year 2020 shows the strengthened position of the desire to more process optimization and more automation. Also, the increased value of the application in core technologies shows the meaning of the continuing digitalisation. The biggest step in the field of machine learning can be conducted at the sales.

The findings of the shadowing are very different between the own assessment of the participants. As benchmark, the elevated data from the survey and interview will be set. As it can be seen in the following figure 5, most participants evaluate their use of BPM as up to date and their process automation and controlling as well set up.

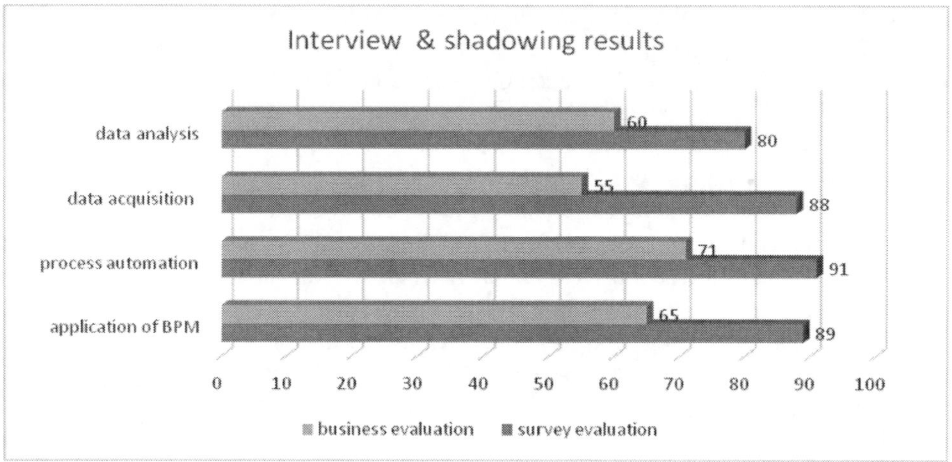

Fig. 5. Interview vs. shadowing results, own presentation

While the survey assessment generates 87% of all questions in figure five. The compared value while the shadowing is 15% lower by 62%. This shows the different exceptions and the big value for improvement. Particularly the process automation point shows a difference of 20% in theory and business. Such gaps can result in big problems in cases of missed BPM and missing continual improvement.

To avoid business faults, half automatic processes and the waste of resources, the base data regarding a special field like processes have to be set anew by the companies. In connection with detailed data analysis, deep learning and machine learning the optimization and automation opportunities can be pointed out.

3.3. Pre-Conclusion

For 80% of the participants deep learning, machine learning and AI are synonyms for the same concept. Like described in the theoretical part, machine learning is regarded as a subarea of artificial intelligence.

The results of the survey show the demand for machine learning and AI for the optimization and automation of the processes and the support of BPM. Furthermore, the participants evaluate the application of machine learning as very important in the future. Not only for process optimization, but rather as core competency for new business models and for an increase in sales support.

The results of the shadowing show two main findings:

1. the participants estimate their use of BPM and automated process management better than it is in reality
2. approximately 45% of the compiled data from the companies are not used or controlled effectively

This shows a big opportunity and space for improvements only in the evolution of process management. Big Data and data lakes are used in the business but the reporting and data analysis can be enhanced. Deep learning, machine learning and AI can be named as a recommendation to automate and optimize the process and they support the BPM through new algorithms and independent learning via constant data analysis.

4 Conclusion

Digitization is unstoppable and changes the market situation in many business sectors. Time-intensive manual workflows, based on e-mail, Excel or well and truly paper work reduce the productivity of companies. August Wilhelm Scheer, Managing Director of the Scheer Group stated already in 2015: "Everything that can be digitized will be digitized. Each process, product and business model becomes digital, so the revolution lies in the processes." [18] This quote underlines the opportunity for an optimized process management through digitalization and the execution of automated processes supported by AI. Therefore, targeted and consistent process automation is a critical success factor now and in the future as it can be seen by the evaluation of the survey results. It is becoming increasingly important for companies to use the power of their data and to make smart and profitbale decisions.

References

1. K. Schwab, *Die Vierte Industrielle Revolution*, (Publisher Pantheon Verlag, 2016)
2. Ö. Atiker, *In einem Jahr digital: Das Praxishandbuch für die digitale Transformation*, (Publisher Viley-VCH Verlag & Co.KGaA 2017)
3. http://www.mckinsey.com/business-functions/operations/our-insights/manufacturings-next-act Accesed 2015
4. http://www.hightech-strategie.de/de/Industrie-4-0-59.php Accesed 2017
5. https://www.salesforce.com/uk/blog/2016/09/what-is-industry-4-0.html Accesed 2016
6. http://ipt.ch/digitalisierung-mit-prozessautomatisierung-beginnen/ Accesed 2016
7. J.L. Leignel,T. Ungaro, and A. Staar, *Digital Transformation: Information System Governance*, (Publisher Wiley 2016)
8. D. Schallmo, A. Rusnjak, J. Anzengruber, and T. Werani, M. Jünger, *Digitale Transformation von Geschäftsmodellen: Grundalgen, Instrumente und Best Practices*, (Publisher Springer Gabler, 2016)
9. J.v. Brocke, and M. Rosemann, *Handbook on Business Process Management 2*, (Publisher Springer, 2015)
10. European Association of Business Process Management – EABPM, *Business Process Management*, Common Body of Knowledge, Gießen, (2009)
11. J.H. Schmelzer, W. Sesselmann, *Geschäftsprozessmanagement in der Praxis*.8th Edition, (Publisher Carl Hanser Verlag, 2013)
12. U. Lämmel, and J. Cleve, *Künstliche Intelligenz*, (Publisher Hanser, 2012)
13. R. Urwin, *Artificial Intelligence – The quest of the ultimate thinking machine*, (Publisher Arcturus Holdings Limited 2016)
14. http://www.informatik.unioldenburg.de/~iug08/ki/Grundlagen_Starke_KI_vs._Schwache_KI.html Accesed 2017
15. http://de.dice.com/nachrichten/aktueller-stand-kuenstliche-intelligenz/ Accesed 2016
16. J. Bell, *Machine Learning: Hands-On for Developers and Technical Professionals*, John Wiley & Sons Inc., (2014)
17. https://blogs.nvidia.com/blog/2016/07/29/whats-difference-artificial-intelligence-machine-learning-deep-learning-ai/ Accesed 2016
18. http://www.me-netzwerk.de/wp-content/uploads/2015/05/Industrie-4.0-mit-Scheer-BPaaS-_-ME-Forum-Berlin.pdf Accesed 2015

Methodology for computer-assisted optimization of waste flow

Cicerone Laurentiu Popa[1,*], *Costel Emil* Cotet[1], *Radu Constantin Parpala*[1], and *Tiberiu* Dobrescu[1]

[1]University Politehnica of Bucharest, MSP Department, Spl. Independentei 313, Bucharest, Romania

Abstract. The paper reports the development of a methodology based on computer simulations with the purpose to support decisions in designing the optimal architecture of different types of selective waste collection systems and recycling systems. The design of such systems is a complex task which involves both a very good knowledge of selective waste collection system equipment characteristics and of recycling processes, and the correct placing of the equipment along the flow so that to avoid underutilization of the structural elements and to avoid bottlenecks which generate low productivity or even blockages. The methodology is applied for three case studies in which different types of waste flow models are investigated: hybrid waste flows (windshields recycling), discrete waste flows (waste electric and electronic equipment collection) and continuous flows (industrial and automotive used oil collection and recycling). The architectures of these systems are optimized using the developed methodology in order to increase usage degree and productivity.

1 Introduction

In a very comprehensive view, the methodology proposed in this paper comprises optimization techniques for procurement, usage, handling, transformation and disposal of physical mobile entities like different types of waste in waste collection systems and recycling systems. The area of this methodology application is wide and implies concepts of the material flow theory [1]. An analysis of the literature in the field of MFT showed that the general focus is on presenting applications/case studies in different areas and, thus, obtaining specific conclusions, while the theoretical approaches and paradigms of MTF are considered only in few studies [2- 4].

The material flow theory (MFT) proposes a general scientific model without taking into account the nature of the described material flow. Even if some of the considerations made are covering the entire MFT area, the main focus of the paper does not concern the specific aspects of economic or social material flow, but only on the waste flow models area of applications. Therefore, in the present paper MFT will be used in a restricted sense, as a set of simulation and optimization rules for productivity duration and costs using various virtual waste flow models.

* Corresponding author: laur.popa79@gmail.com

2 Optimizing different types of waste flow models. A synoptic view

The proposed algorithm for optimizing recycling architectures performances should be adapted to the specific requirements of waste categories by specific types of selective waste collection systems and recycling systems. A waste processing architecture (WPA) can be defined as a set of structural elements interacting such that to contribute by specific processes to waste recycling. There are four main structural elements (work points, transport systems, transfer systems, buffers) and two auxiliary structural elements (mobile entities and human resources) in a waste processing architecture. Different types of waste flow models are considered: hybrid, discrete and continuous waste flow. For each one of these categories, establishing a WPA for a specific or a family of waste models involves designing and structuring the relationships between structural elements, processes, procedures and human resources. That is both defining these elements, placed and acting in a specific layout, as well as the material and information flow between them.

We agree to define here the role of the structural elements as follows. The work points are the structural elements covering all the technological stages necessary for the waste processing. Transport systems are moving the mobile entities between the long distance situated structural elements. The transfer systems are orienting and fixing the mobile entities, making as well the connection between near structural elements. Buffers are storing the mobile entities for a limited period of time. The mobile entities constitute the waste flow. The human resources are involved in two postures: operational (acting like a worker) and monitor (supervising the activity of the WPA).

The advances in three-dimensional computer-aided design (3D CAD software) made possible the creation and use of databases containing parameterized virtual models of all the structural elements that can be included in a WPA. These databases will be made available by producers (being already included in software applications) and their items can be put together in order to form a virtual model of a WPA. This virtual model should precisely reproduce the real system structural elements, in terms of overall dimensions and reciprocal positions and placement in the layout, being essential in optimizing the WPA. The modifications in the WPA needed after different simulation runs are easy to perform due to the parameterization of the virtual models, therefore decreasing the total optimization time.

The modelling and simulation algorithms for optimizing WPA using MFT are different according with two main categories as follows. A diffused WPA can be defined as systems with more than two work points connected by transport & transfer systems and using deposits at local or system level. A concentrate WPA can be defined as systems based on a single work point surrounded & assisted by transport, transfer & deposit facilities.

In a work point, one or several operations/activities can take place using processes and machines. In an attempt to decrease the overall production time, there are approaches which consider eliminating the transport & transfer times between work posts by combining in a single machine, two (or more) processes [1].

The three main processes categories are: subtractive – involving material removal, additive – involving material addition layer-by-layer and with material distribution. Additive and subtractive processes can be combined [1] within a single machine, completely computer-controlled, in order to benefit of the advantages of each technology, such as complex geometry, controlled geometry, functional graded material parts – for additive processes, and high precision and surface quality or very good mechanical properties – for subtractive processes.

From the point of view of the MFT, these hybrid processing systems, currently gaining more interest and importance, are considered as concentrated architectures, and can be modelled and optimized accordingly. As examples of hybrid processing can be mentioned:

solid recovered fuel systems [5, 6] or glass recycling systems [7]. The prosed algorithm to optimize WPA by increasing usage degree and productivity customizes general algorithms used in MFT. The main steps are starting from this point and are based on all the above information, as follows.

A preliminary processing architecture virtual modelling is done with the necessary parameters for the work points, buffers, transport and transfer systems. The trajectories of the mobile entities (parts and tools) are established. For this preliminary architecture, virtual prototype the material flow is simulating for a performance diagnosis report [8]. Using this report, the flow concentrators could be identified. There are two kinds of solutions for eliminating these concentrators by modifying the WPA model [9].

Functional remodelling potentiates the WPA performance by modifying the structural elements layout or functional parameters but is not introducing or eliminating structural elements.

Technological remodelling allows WPA changes also when structural elements should be added or removed. After functional or technological remodelling, another material flow simulation is performed for quantifying the increasing performance of the optimized WPA. The last stage is the profit validation based on the comparison of the investment in optimizing the WPA and the gain from the increased system performance.

3 Case studies

The methodology is applied for three case studies in which different types of waste flow models are investigated: waste electric and electronic equipment collection as an example for discrete waste flows, industrial and automotive used oil collection and recycling as an example for continuous flows, windshields recycling as an example for hybrid waste flows. First and second case studies are part of the same research project with the subject of selective waste collection. The third one refers to a research project in the field of waste processing.

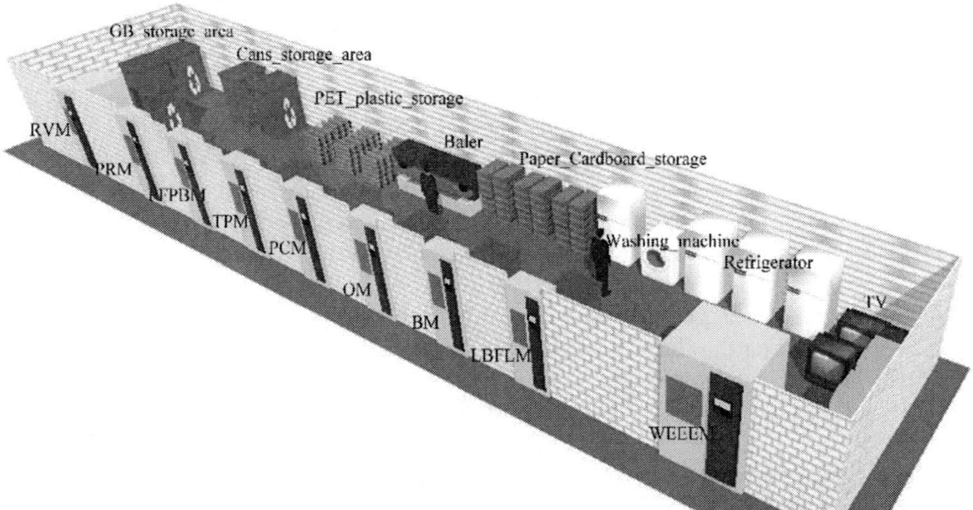

Fig. 1. The 3D model of the current selective waste collection system made in Witness Horizon

The three-dimensional virtual model of the selective waste collection system made in Horizon Witness is presented in figure 1, WEEEM (waste electric and electronic equipment module) being a part of it. The systems include also:

- RVM module (Reverse Vending Machine) for receiving, identifying and sorting the glass recipients, plastic recipients and aluminium cans;
- PRM (plastic recipients module) for plastic recipients over 5 litres;
- PFPBM - Plastic film and plastic bags module;
- TPM - Tetra Pack module;
- PCM - Paper and Cardboard module;
- OM - Oil module;
- BM - Battery module;
- LBFLM – Light bulb and fluorescent lamp module.

These modules are not automated. Two operators are receiving the collected waste. The operators are used for loading, weighing, voucher issuance and storage of WEEEs (including refrigerators, washing machines, TVs, phones, laptops, PC drives, printers, video consoles, microwave etc.).

The three-dimensional virtual model is used to perform a waste flow simulation in order to conduct a diagnosis. The waste flows from the system are optimized using the developed methodology to increase usage degree and productivity. For the WEEEM a technological remodelling was made to obtain an automated module. The module for collecting electrical equipment and electronic waste was divided in two separate modules:

- one for big electronics and home appliances. This will take over the refrigerators, washing machines, TV sets and after identification and weighing they will be transported by a conveyer to the transfer zone, from where they will be picked up by the AS/RS (Automated storage and retrieval system) and deposited in the storage area;
- one for small electronics and home appliances. This will take over the phones, laptops, PC drives, printers, video consoles, microwave and after identification and weighing they will be transported by a conveyor to the transfer zone, point from which they will be picked up by the AS/RS system and will be deposited in the storage area.

The data obtained after the optimization shows an increase in terms of usage degree and productivity

The oil module is included in the current waste collecting system, but is available only in a few locations where industrial and automotive used oil collection can be done. Now, the oil is collected in two different tanks (one for industrial an automotive oil and one for cooking oil). The operators are receiving the containers with oil from the customers and put it in tanks. After the preliminary simulation, the flow was have modified by including a RVM for oil in the system. Also, the oil recycling system was added in simulation as an example for continuous flow simulation.

The recycling process for cooking oil start with the identification of free fatty acids (FFA) percent, if FFA is larger than 2% then the FFA must be reduced by using an acid-catalysed esterification process. Using a trans esterification process based on sodium hydroxide and methanol reduces the oil viscosity. After trans esterification, the methanol is removed in a washing process. Once separated the biodiesel is purified. At the end the biodiesel is dried to remove all water remaining from the washing process [10]. The flow is shown in figure 2.

The results show possible usage increase degree of the system due to the fact of introducing the RVM machine. In this case the waste collection can be done 24 hours/7 days every week and is not limited to the presence of operators.

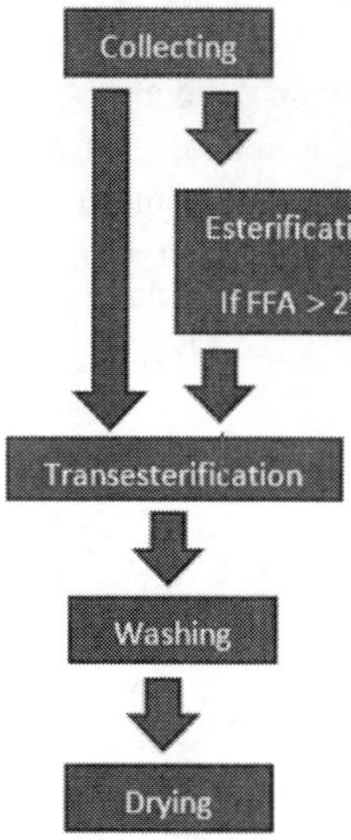

Fig. 2. Cooking oil recycling flow

The third case study refers to windshields recycling as an example for hybrid waste flows. The windshield glass material flow and the recycling installation virtual model made in Witness Horizon are shown in figure 3.

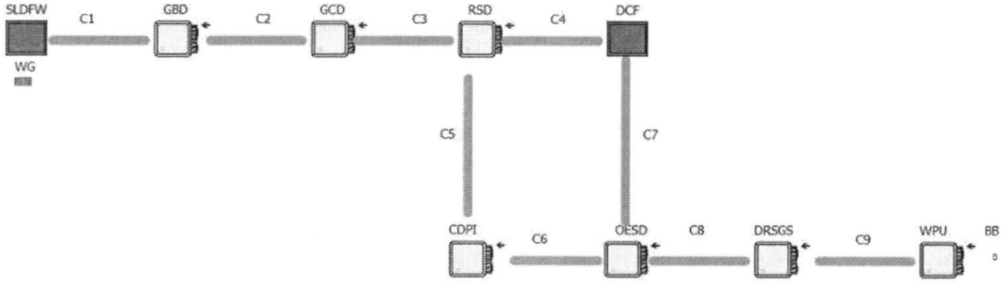

Fig. 3. The 2D virtual model of the windshield glass recycling installation made in Witness Horizon

WG (windshield glass) is introduced in the recycling system using the supply/loading device with glass waste (SLDFW). C1 conveyor will transport the WG to the glass-breaking device (GBD). From the Glass crusher device (GCD), the glass waste is transported to the rotating sieve device (RSD) using the C3 conveyer. Glass waste is transported to CDPI (Crushing device of pieces with impurities) by the C5 conveyer and PVB (laminated glass with polyvinyl butyral) waste is transported to DCF by the C4 conveyer. OESD (Optical-electrical sorting device) will sort the glass with impurities and using the C7 conveyer the waste will be transported to DCF (Catching device of fractions resulted by the recycling process). Clean cullets are sorted on DRSGS (device with rotating sieve for the glass sorting) depending on the desired granulation. On the Weighting and packaging unit (WPU) the granules are stored in big bags with 1000 kg capacity.

The preliminary simulation results show bottlenecks on GCD (22%) and RSD (17%). The functional remodelling was used to remove the bottlenecks and to increase the system productivity. The reports after system's optimization show that 3 big bags (BB) were obtained and at WPU during the packaging and another big bag are at 69% of capacity (approx. 690 kg). The quantity that the installation will process during one hour is 3.69 tonnes (an average of 26 tonnes per day – 7 hours).

4 Conclusions

Current needs and practice in industrial engineering determined us to adapt the general MFT and to develop specific algorithms for diagnosis and optimization for WPA.

As can be noticed, our approach accommodates the advances in digital fabrication and the development of new manufacturing processes/technologies to waste processing.

When designing the layout of WPA, virtual models of different structural elements can be used, while in simulating the material flow, data from processes simulations can constitute input parameters. In this context of using data from different types of applications for defining and optimizing WPA, data compatibility and data portability issues should not be neglected when choosing the software tools for supporting the optimal definition of a WPA.

The core of the proposed methodology assumes than in dealing with MFT applied to WPA, one should consider using a specific algorithm for optimizing the associated architecture by modelling and then simulating the material flow trajectory within the system. Such an algorithm, as detailed in the previous section, offers a structured modality to increase the productivity and profit in WPA in order to respond to different requirements.

Based on the results described above in this paper the authors established the main future research direction as improving performances analysis by developing new measuring tools and integrating the investment analysis in the material flow simulators.

This work was supported by a grant of the Romanian National Authority for Scientific Research and Innovation, CNCS/CCCDI – UEFISCDI, project number PN-III-P2-2.1-BG-2016-0437, within PNCDI III.

References

1. C. E. Cotet, D. Popescu, Enc. Inf. Sci. Tech., 3786 – 3794 (2014)
2. J. V. Steinaecker, G. Jürgens, Env. Inf. Sys. Ind. Pub. Adm., 356-366 (2001)
3. G. A. Swanson, Ent Inf. Sys., **2**, 21–31 (2008)
4. S. Xu, Inf. Sys. Fr., **10**, 601-609 (2008)
5. W. L Kepplinger, T. Tappeiner, Was. Man Re., **30**, 450-453 (2012)
6. K. E. Lorber, R. Sarc, A. Alexia, Was. Man Re., **30**, 370-383 (2012)
7. J. R Mueller, M. W. Boehm, C. Drummond, Was. Man., **32**, 1560–1565 (2012)
8. C. L. Popa, C. E. Cotet, V. Ionita, St. Gavrila, Proc. Eng., **100**, 334 – 339 (2015)
9. C. E. Coteţ, C. L. Popa, G. Enciu, A. Popescu, T. Dobrescu, Int. J.Simul. Model, **15**, 5-16 (2016)
10. N. H. Abdullah, S. H. Hasan, N. R. M. Yusoff, Int. J. Mat. Sci. Eng., **1**, 94-99 (2013)

Synergy of modeling processes in the area of soft and hard modeling

Robert Sika[1,*], and *Jakub* Hajkowski[1]

[1]Poznan University of Technology, Institute of Materials Technology, Piotrowo 3 street, Poland

Abstract. High complexity of production processes results in more frequent use of computer systems for their modeling and simulation. Process modeling helps to find optimal solution, verify some assumptions before implementation and eliminate errors. In practice, modeling of production processes concerns two areas: *hard modeling* (based on differential equations of mathematical physics) and *soft* (based on existing data). In the paper the possibility of synergistic connection of these two approaches was indicated: it means hard modeling support based on the tools used in soft modeling. It aims at significant reducing the time in order to obtain final results with the use of hard modeling. Some test were carried out in the Calibrate module of NovaFlow&Solid (NF&S) simulation system in the frame of thermal analysis (ATAS-cup). The authors tested output values forecasting in NF&S system (solidification time) on the basis of variable parameters of the thermal model (heat conduction, specific heat, density). Collected data was used as an input to prepare soft model with the use of MLP (Multi-Layer Perceptron) neural network regression model. The approach described above enable to reduce the time of production process modeling with use of hard modeling and should encourage production companies to use it.

1 Introduction

Starting from the 90s of the last century, a distinct trend can be observed in scope of use of dedicated methods and computer systems for modeling and simulation of production processes [1-6]. It is related to high technological complexity of these processes, as well as complexity of physicochemical phenomena accompanying them. Modeling of processes brings many benefits, such as selection of an optimal solution (the most beneficial solution by assumed criteria) [7], possibility of verifying assumptions before applying them in practice, as well as making it possible to eliminate errors (among others, by virtual application of Trial & Error method) [8]. Hereby, it needs to be stated that authors understand the model as a simplified object, which represents the real state. Behavior of a model or its properties are similar to a modeled phenomenon [9].

Considering the above mentioned statements, 2 types of models can be distinguished [10]: physical models and mathematical models. The first type exists in reality and is used

[*] Corresponding author: robert.sika@put.poznan.pl

mostly in universities or research institutes, can be a specially designed laboratory stand or a mockup in a decreased scale (e.g. electrical analogue used for physical modeling of stationary temperature field through the cross section of a pressure mold with filled cavity and cooling vents). The mathematical models, on the other hand, have a virtual shape and they are a modern alternative, which is very helpful in modeling phenomena and physical processes, often using dedicated computer systems. An activity consisting in use of mathematical models is known as the mathematical modeling (the results are presented using mathematical relationships) [11].

Subsequently, division of the mathematical modeling is related to a way of representing the nature of a process or phenomenon and it regards method of use of available data. The hard modeling is based on application of derivative/differential equations from the range of mathematical physics, known as early as in 19th century. The soft modeling, on the other hand, uses empirical cases gathered in data bases and operates on a basis of matching to the optimal variant, according to strictly defined criteria. Wider description of the problem, related not only to a model but also to coefficients of databases is presented in [12].

In this paper, the authors indicate possibilities of synergy of both groups of modeling (hard and soft), on example of aiding the hard modeling on the basis of tools applied in soft modeling. It was aimed at significant acceleration of obtaining end results from a hard model in some specific cases, not related to 3D modeling. Other cases of application of both soft and hard modeling in studies are presented in [13-14].

The tests conducted using dedicated computer systems (separately for hard and soft modeling) indicated high compatibility of prognoses. The above mentioned approach will help to reduce time of modeling of production processes with use of link between soft and hard modeling and it is an encouragement to apply such solutions in production companies and research institutes. In the further chapters of the paper, methodology of studies, their results and analysis are presented.

2 Thermal model – Calibrate

Initial simulation study was performed using the NovaFlow&Solid simulation system, in the Calibrate module [15]. This module is used to conduct one-dimensional simulation (1-D) of heat transfer for the basic shapes: sphere, cylinder, plate and thermal analysis cup (ATAS-cup). The last example was used for the studies – the ATAS-cup of dimensions of 60x60mm (geometric modulus = 1 cm), with a thermocouple located in the thermal center of the casting, which is at the height of 30 mm in the casting axis.

The Calibrate module allows obtaining the cooling curve and, what follows, determination of time of casting solidification and comparing it with a curve obtained from an experiment.

Then, in case of different results, it is necessary to change values of particular thermo-physical parameters of an alloy and/or mold. It will allow better matching of simulation results to the experiment.

In this paper, an assumption was made that the thermo-physical parameters of an alloy variate with temperature and are not varianted during the process. However, values of parameters of molding sand (green sand) were changed – they were assumed as mean, constant values (unchanged in function of temperature). It was a kind of simplification, which is often made for the molding sand, in case of testing of a simulation code.

3 Studies

3.1 Thermo-physical properties of the AC-42000 alloy

The authors proposed using the AlSi7Mg alloy (AC-42000), widely applied in the industry, for the simulation studies. The thermo-physical properties and chemical composition of this alloy are presented in Tables 1 and 2.

On the basis of the introduced chemical composition, internal thermodynamic solver – Thermo-Calc, built in the NovaFlow&Solid 4.7 system, determines characteristic temperatures (liquidus – T_{liq} and solidus – T_{sol}, beginning and end of solidification) [14]. For the AC-42000 material assumed in the studies and introduced concentrations of elements (see Tab. 2), the above mentioned temperatures had the following values: T_{liq}=613,67 [°C], T_{sol}=568,73 [°C].

Table 1. Thermo-physical properties for AlSi7Mg (AC-42000) alloy, dedicated to simulation studies as a function of temperature. Data from the Nova Flow&Solid database.

Temp. [°C]	**Thermal conductivity [W/mk]**	**Temp. [°C]**	**Specific heat [J/(kgK)]**	**Temp. [°C]**	**Density [kg/m³]**
25	151	100	880	20	2660
100	155	350	1050	100	2646
200	163	565	1200	200	2628
300	168	620	1100	300	2608
400	168	660	1100	400	2587
565	168			500	2566
620	100			550	2555
660	100			620	2374
				700	2354
				800	2330
Temp. [°C]	**C.L.E. [10⁻⁶/°C]***	**Temp. [°C]**	**Viscosity [m²/s]**	**Temp. [°C]**	**HTC [W/(m²K)]****
20	21,8	630	0,40	300	0,1
100	21,8	2000	0,48	565	1
200	23,7			590	7
300	25,1			600	8,5
400	26,5			620	10
500	28				
550	28,7				
700	34,2				
800	35,1				

*) C.L.E. – *Coefficient of Linear Expansion*
**) HTC – *Heat transfer coefficient*

Table 2. Chemical composition of AlSi7Mg (AC-42000) alloy.

Element	**Al**	**Si**	**Mg**	**Mn**	**Cu**	**Fe**	**Ni**	**Zn**	**Ti**	**Sn**
Contents [%]	91,68	7,00	0,30	0,30	0,10	0,30	0,05	0,15	0,10	0,02

3.2 Simulation studies - Calibrate (hard modeling)

In the simulation studies, different values of INput parameters were set and an OUTput parameter was obtained in form of time of alloy solidification in the cup. The parameters are presented below:

- IN_Par: Heat Conduction [W/mK], HC.
- IN_Par: Specific Heat [J/kgK], SH.
- IN_Par: Density [kg/m^3], D.
- OUT_Par: Solidification Time [s], ST.

Table 3 presents values of thermo-physical parameters (INput parameters) of the green sand and predicted value of solidification time for these parameters, obtained from the simulation in the Calibrate module (OUTput parameter). The input values for the particular records are assumed as constant mean values. Each variant of the simulation (Simulation number, Sim. no) contains different combination of values of parameters, which result out of physics of the modeled phenomenon.

Finally, 70 cases of simulation were performed, related to prognosis of the ST parameter for various settings of the HC, D and ST parameters, while results of 65 first courses were used for teaching of the soft model and 5 last cases were independently used to compare quality of prognosis. Time of realization of a single course of simulation, considering changing of settings of specific parameters, was approx. 120 seconds. Selected values of INput parameters (INput_par) and the OUTput parameter (OUTput_Par) obtained from the simulation are presented in Table 3.

Table 3. The values of input and output parameters used in the Calibrate simulation. Selected data records presented.

Sim. No.	INput parameters			OUTput parameter
	Heat Conduction [W/mK]	Specific Heat [J/kgK]	Density [kg/m^3]	Solidification Time [s]
1	0,83	558	1509	393
2	0,76	563	1539	411
3	0,91	588	1521	357
4	0,92	588	1529	355
5	0,96	562	1508	362
...	...	...	...	...
60	0,76	573	1546	407
61	0,82	575	1548	385
62	0,75	599	1540	401
63	0,97	568	1504	352
64	0,87	583	1502	374
65	0,98	587	1546	337

Fig. 1 presents a point relationship (XY scatter chart) of solidification time ST (determined using the software for hard modeling – NF&S, in the Calibrate module), depending on the INput variables. Additionally, each diagram contains the r-Pearson correlation coefficient, resulting out of the presented relationship. Only for the first case there is a strong relation between the parameters, indicating decrease of the solidification time ST with increase of the HC variable: $f = ST\downarrow(HC\uparrow)$. The other variables influence on the solidification time in a very unordered, non-linear manner. It is therefore difficult to indicate mutual interaction of the three mentioned parameters with the solidification time, outside the Calibrate module. The authors proposed further, expanded studies in direction of prediction of the OUTput variable ST, considering simultaneous share of all three parameters. The aim is to prepare a soft mathematical model in form of a neural network, which can be asked in a much shorter time than using the Calibrate module (instead 120 seconds it is less than 5 seconds). Moreover, the elaborated studies were used, by the way, to test quality of the hard model and its capability of generalizing.

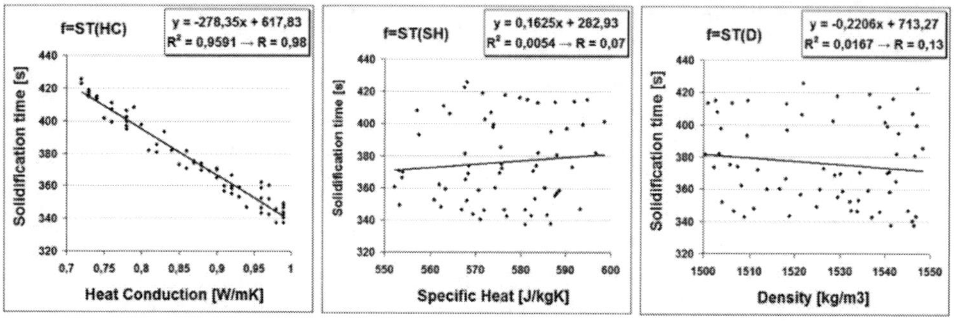

Fig. 1. Linear relation between the selected parameters of the Calibrate module. The Figure presents r-Pearson correlation coefficient (R)

The detailed results of studies regarding application of a neural network MLP-BFGS aimed at aiding of the hard modeling are presented in the next chapter.

3.3 The artificial neural network algorithm MLP-BFGS

To predict the OUTput variable ST using the soft modeling, the authors proposed use of a solution which is quite popular and tested in industrial practice – Multi Layer Perceptron (MLP) neural network [16]. The model was designed using the STATISTICA 10.0 network. The BFGS (MLP-BFGS, Broyden-Fletcher-Goldfarb-Shanno) algorithm was decided to be used, as it is an improved algorithm regarding the popular algorithm of Back Propagation (BP) due to consideration of variable set metric scale. [17]. The following assumptions regarding testing of the above mentioned algorithm were made:

- 80% are learning records, 10% are testing records and the remaining 10% are validating records.
- Learning of the network assumes use and testing of the following activation functions: linear, logistic, tanh and exponential.
- To compare values of prognosis of the neural network with results of the Calibrate simulation, 5 additional data records obtained as a result of Calibrate modeling, were used.

Finally, architecture of MLP neural network was used along with the activation functions mentioned above, while teaching of the network consisted in selecting weights in such a way, to make the network represent relations between the INput and the OUTput variables in the most correct way. The selected algorithm of the network adjusts weights of neurons in such a way, to ensure that for any values of the INput variables, the OUTput value of the neural network has as small difference from the OUTput value obtained from the Calibrate model as possible.

The BFGS teaching algorithm was used, with possibility of returning the 5 best suited models, considering used activation function and number of hidden layers. The applied error function, used for comparing network responses with real values, is the Sum of Square (SOS). With an assumed number of hidden layers (between 3 and 10) and a number of saved 5 of 20 MLP networks, the BFGS algorithm returned the networks containing between 3 and 9 hidden layers (see Tab. 2):

- Network model MLP-BFGS 3-9-1.
- Network model MLP-BFGS 3-8-1.
- Network model MLP-BFGS 3-9-1.
- Network model MLP-BFGS 3-5-1.
- Network model MLP-BFGS 3-3-1.

Table 2. Parameters of the best 5 MLP-BFGS models.

ID	MLP type	Prediction (learn)	Prediction (test)	Prediction (validation)	Algorithm	Error	Activation (hidden layer, hl)	Activation (output layer, ol)
1	MLP 3-9-1	0,9879	0,9996	0,9996	BFGS 9	SOS	Log	Linear
2	MLP 3-8-1	0,9896	0,9994	0,9987	BFGS 43	SOS	Log	Tanh
3	MLP 3-9-1	0,9886	0,9993	0,9992	BFGS 14	SOS	Log	Linear
4	MLP 3-5-1	0,9972	0,9999	0,9987	BFGS 238	SOS	Tanh	Tanh
5	MLP 3-3-1	0,9882	0,9996	0,9992	BFGS 15	SOS	Tanh	Linear

Table 3 presents summary of obtained results of prediction of the solidification time ST of the AC-42000 alloy for 5 cases of the best suited model, along with 5 additional validating records (not to be confused with 10% of validating records on the stage of the network learning), which are independent results of simulation in the Calibrate model. These results, as an empirical validation, were not taking part in teaching the network.

Predictions of the neural network were named as the MLP-BFGS_Model. Adjustment of a given network for the results obtained from the Calibrate model was estimated using the formula (1). The best matching is presented by the model no. 1 (MLP-BFGS_Model_No#1), for which 4 of 5 OUTput values are the closest to the real value. In this model, logistic function of activation was used for the hidden layer (hl activation) and linear function of activation was used for the output layer (ol activation). The worst matching was presented by the network no. 4 (MLP-BFGS_Model_No#4) with tangent function of activation for both hidden and output layer.

$$\forall\ MLP\_FIT_{ERROR}\ j : j \in \{1 \div 5\},\ \sum_{i=1}^{j_k} \frac{\left|W_{SMji} - W_{HMi}\right|}{\sigma_{W_{SMj}} + \sigma_{W_{HM}}} \qquad (1)$$

where:
MLP_FIT_{ERROR} – value of model matching error
j – model number (j=1,2...5)
i – validation record number (i=1,2...k)
k – number of all validating records
W_{SMi} – OUTput value estimated by the soft model (MLP-BFGS)
W_{HMi} – OUTput value estimated by the hard model (Calibrate)
σ_{WSM} – standard deviation for a set of OUTput values for a set of OUTput values estimated by a soft model for the validating record *i*
σ_{WHM} – standard deviation for a set of OUTput values estimated by a hard model for the validating record *i*

Table 3. Comparison of predictions of the OUTput parameters values ST obtained from the artificial neural network (5 best models) with predicted values in the Calibrate model.

Sim. no	INput parameters			OUTput parameter Solidification Time [s]					
	Heat Conduction [W/mK]	Specific Heat [J/kgK]	Density [kg/m³]	Calibrate Model	MLP-BFGS_Model				
					#1	#2	#3	#4	#5
66	0,94	569	1500	359	360	362	362	363	362
67	0,92	590	1543	352	352	354	351	353	351
68	0,95	563	1508	357	359	361	361	362	361
69	0,85	578	1504	381	381	381	381	380	381
70	0,83	572	1542	384	385	384	384	383	385
				MLP_FIT_{ERROR}	0,31	0,68	0,58	0,82	0,60

The network behavior was also examined with different number of neurons : (1-3), (1-5), (3-5), (3-8), but their accuracy was lower than 80%. It is also noteworthy, that with each new generation of a network with the same number of hidden layers, the network can return different models, with different activation functions. The authors noticed, that the most frequent case concerns generating of a logistic and a linear function. The authors propose future expansion of studies in scope of differentiation of values of model parameters (among other things – different number of hidden layers with different number of saved networks) and increase of records asking the network. It is therefore important to remember, that too high fitting of the INput data to the OUTput values may result in overfitting of the network.

Results of the conducted simulation indicate, that the MLP network with the BFGS algorithm is good in predicting of solidification times for a higher number of records (at least several dozens). As such, ready model of the network can be used to shorten time of calculations in the hard modeling simulation systems. In future, the authors plan to test other soft modeling methods, including regression trees, among others.

4 Summary

The paper presents study of possibility of predicting (in scope of HARD computer simulation) of output values on the basis of variable parameters of a heat model, taken from a database for casting a cup for the thermal analysis (Calibrate) and then comparing these predictions with results of a computer simulation. The initial simulation studies were conducted using the NovaFlow&Solid system in the Calibrate module and the STATISTICA system in the Neural Network module. The studies indicate large synergy of both areas.

The tests conducted using dedicated computer systems (separately for the hard and soft modeling) indicated high convergence of prognoses, illustrated in Tab. 3 and Tab. 4. The authors propose to expand future studies with differentiation of values of model parameters (among other things, different number of hidden layers with different number of saved networks), as well as increase of records asking the network. It needs to be considered, that too high fitting of INput data to the OUTput data will result in overfitting of the network.

The further studies concern problems of predicting the OUTput values on the basis of independent INput values using Regression Trees. In this project, this problem is merely suggested and authors propose to expand these studies in future.

References

1. M. Perzyk, J. Kozłowski, Arch. of Found. Eng., **16(4)**, 101, (2016)
2. J. Trojanowska, A. Karwasz, J. Machado, M.L.R. Varela, Effic. in Sustain. Supp. Chain, **2**, 119 (2017)
3. G. Vieira, L. Reis, M.L.R. Varela, J. Machado, J. Trojanowska, Rom. Rev. Prec. Mech. Opt. and Mechatr., **50**, 119 (2016)
4. A. Kujawińska, M. Rogalewicz, M. Piłacińska, A. Kochański, A. Hamrol, M. Diering, Metalurgija, **55 (4)**, 821 (2016)
5. P. Zawadzki, F. Górski, A. Hamrol, M. Kowalski, R. Paszkiewicz, Trans. of FAMENA, **35**, 69 (2011)
6. E. Dostatni, D. Grajewski, J. Diakun, R. Wichniarek P., Buń, F. Górski, A. Karwasz, Inter. Conf. Virt. and Augm. Real. in Educ., **75**, 348 (2015)
7. A. Kujawińska, M. Rogalewicz, M. Diering, Manag. and Prod. Eng. Rev., **7(2)**, 29 (2016)
8. P. Popielarski, Ignaszak Z., Drying and Energ. Technol., **63**, 181 (2016)
9. A.B. Rane, V.K. Sunnapwar, S.M. Khot, R. Vulavala, J. of Sim. and Proc. Model., **11(3/4)**, 305 (2016)
10. M.Perzyk, METRO, *Computer methods for analysis and control of production processes*, **course 10 (lec.1)**, 1 (2006)
11. I.H. Abdel-Halim Hassan, App. Mathem. Mod., **32(12)**, 2552 (2008)
12. Z. Ignaszak, J. Hajkowski, P. Popielarski, Def. and Diff. Forum, **336**, 135 (2013)
13. A. L. Pomerantsev, Chemom. and Intell. Lab., **79(1-2)**, 73 (2005)
14. J. Martinsson, F. Hagglund, J. E. Carlson, Ultrasonics, **48(5)**, 427 (2008)
15. J.O. Anderson, T. Helander, L. Hoglund, P.F. Shi, B. Sundman, Comp. Coup. of Ph. Diag. and Therm., **26**, 273 (2002)
16. A. Khotanzad, C. Chung, Neur. Comp. & App., **7(3)**, 249 (1998)
17. R.V. Abramov, J. of Comp. Phys., **228(1)**, 96 (2009)

Comparing calculated, experimental and finite element analysis punch forces for the same lenght punched contours but with different geometry

Stelian Uțuleanu[1, 2*], *Aurelian* Vlase[1], *Gheorghe* Sindilă[1], and *Nicolae* Căpățână[1]

[1]Politehnica University of Bucharest, IMST Faculty, Spl. Independentei, 313, Bucharest, Romania
[2]ICPE SA – Special Electric Machines, Splaiul Unirii 313, 030138 Bucharest, Romania

Abstract. When is calculated the punching force is taken into account only the length of contour and not its form. In this case, the pressing force for the contours with the same length, but with different geometries (circle, square, hexagon, triangle and shearing) may have different values. To demonstrate this hypothesis were made experimental tests which demonstrated this supposition. Also, this hypothesis was verified using the finite element analysis (FEA) in order to show the possible differences and to view the material behaviour during punching process. The calculated punching force, the experimental obtained punching force and the force determined using FEA were centralized into a table and were made graphics that show the variation of punch force according with the geometry of punched contour. The analysed materials are AISI 304, DC01, Al99.8 and Brass. The software used to make the finite element analysis is Deform 3D version 6.1 (service pack 2).

1 Introduction

Punching is a group of cold plastic processing methods based on wholly or partially material separation, after a closed or open contour [1-2].

It is used both for obtaining finished parts and semi-finished products for further processing operations and contains several methods like: cutting, punching, shearing, notching, trimming [2-3].

The required force F for shearing is the force used for total separation (punching, cutting, shearing) or for partial separation (notching) of a piece of waste. This force is determined using the relation [1-3]:

$$F = k \cdot l \cdot g \cdot \tau \text{ [N]} \qquad (1)$$

where:

* Corresponding author: stelian_u_86@yahoo.com

k=1,1... 1,3 is a coefficient that takes into account the anisotropy of mechanical properties of the material, the deviations from its thickness, the degree of wear of the cutting edges;
l [mm] – punched contour length;
g [mm] – material thickness;
τ [N/mm^2] - shear strength of the material.

After the fracture of material it is necessary to push it through the active plates. To calculate the force needed for this it is used the following relation [1, 3]:

$$F_{push}=k_{push}\cdot n\cdot F \text{ [N]} \tag{2}$$

where:
k_{push} is a coefficient that takes into account the type of material and its values are experimental determined ale cărui valori;
n is the number of pieces or waste that remain in the cutting plates from one punching to another. It is calculated with the next relation [1]:

$$n=\frac{h_{pl}}{g} \tag{3}$$

where:
h_{pl} is the thickness of active part of cutting plates. In our case is 3 mm.

The total punching force value is determined using the relation [1] :

$$F_{tot}=F+F_{push} \text{ [N]} \tag{4}$$

The centralized values of punch forces

Table 1. Centralised values of punch forces for different materials

		AISI 304	AISI 1008	Al99.8	Brass
k	min	1.10	1.10	1.10	1.10
	max	1.30	1.30	1.30	1.30
L		60.00	60.00	60.00	60.00
g		0.51	0.51	0.81	0.81
τ	min	490.00	310.00	90.00	200.00
	max	520.00	390.00	150.00	370.00
F	min	16,493.40	10,434.60	4,811.40	10,692.00
	max	20,685.60	15,514.20	9,477.00	23,376.60
k_{push}	min	0.010	0.010	0.009	0.006
	max	0.063	0.063	0.040	0.031
hpl		3.00	3.00	3.00	3.00
g		0.51	0.51	0.81	0.81
n	[hpl / n]	5.88	5.88	3.70	3.70
F_{push}	min	970.20	613.80	160.38	237.60
	max	7,665.84	5,749.38	1,404.00	2,683.98
Ftot	min	17,463.60	11,048.40	4,971.78	10,929.60
	max	28,351.44	21,263.58	10,881.00	26,060.58

The analysed contour length is 60 mm, with the gap of 0.07 mm between the cutting plates and the punch tools (Fig. 1 to Fig. 5).

 DOI: 10.1051/matecconf/201712104010

Fig. 1. Circle contour.

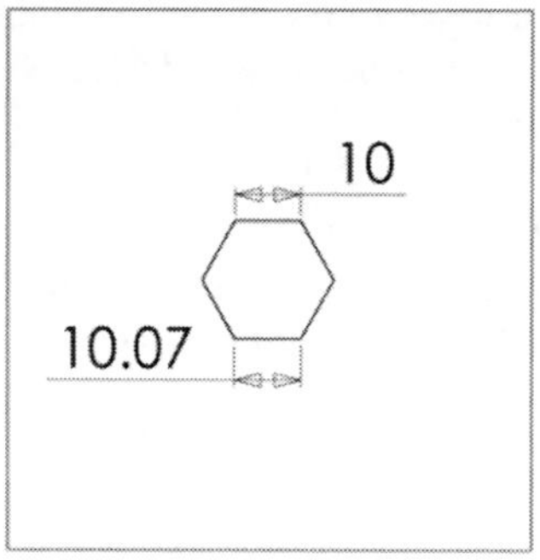

Fig. 2. Hexagon contour.

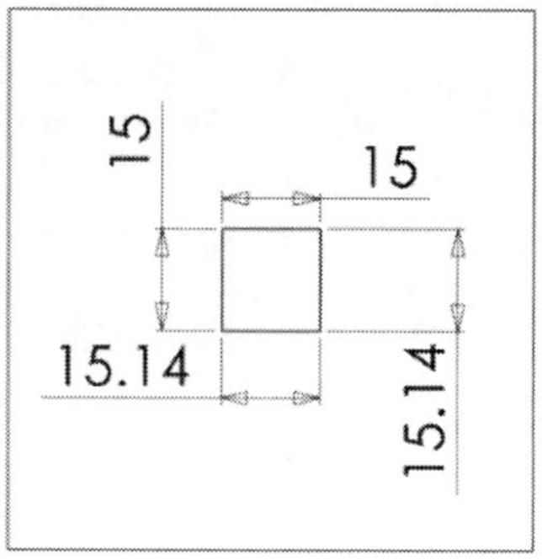

Fig. 3. Square contour.

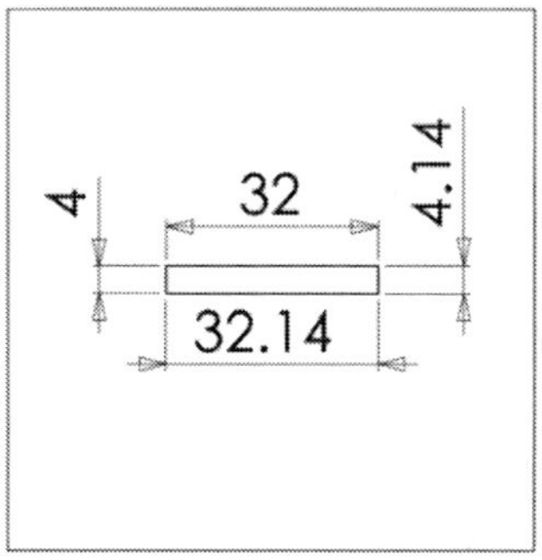

Fig. 4. Rectangular contour.

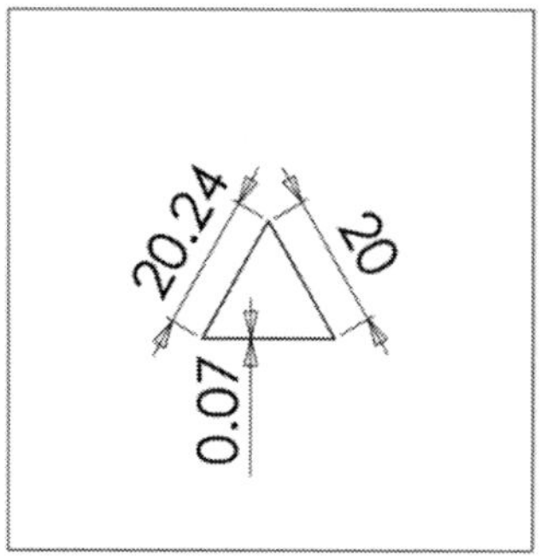

Fig. 5. Triangle contour.

2 Method

After were calculated the theoretical values of punching force they must be compared with the experimental and finite element analysis (FEA) results [4].

The software used for finite element analysis (FEA) was Deform 3D Service Pack 2. The number of mesh elements for workpiece was selected to the maximum of 100.000 elements, and the "Size ratio" as 3.

The materials used were imported from the library (AISI 304, AISI 1008, Al99.8 and Brass CDA 110) and the number of steps was set to 400 and step increment to 10.

The experimental values were obtained using a data acquisition system [5] that consists of mechanical press, force transducer, stroke transducer, one conditioner for each channel, a NI-USB-6008 data acquisition board, a server, the local area network (or Internet) and the computer from which is accessed the measuring stand.

The obtained results were centralised into a table and were made graphs to easily see the variations of punch force according with the geometry of punched contour.

In these graphs the stroke for experimental values was indexed with 9 mm to avoid the overlapping.

3 Results

The comparing results for ferrous materials are presented in the following graphs.

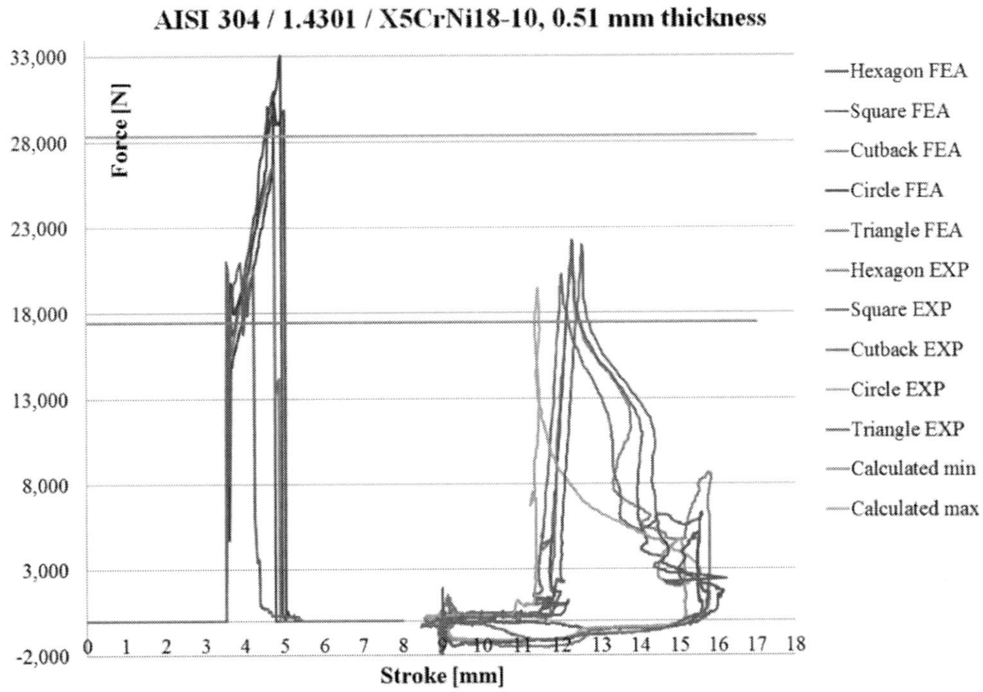

Fig. 6. Comparing results for AISI 304 material, 0.51 mm thickness.

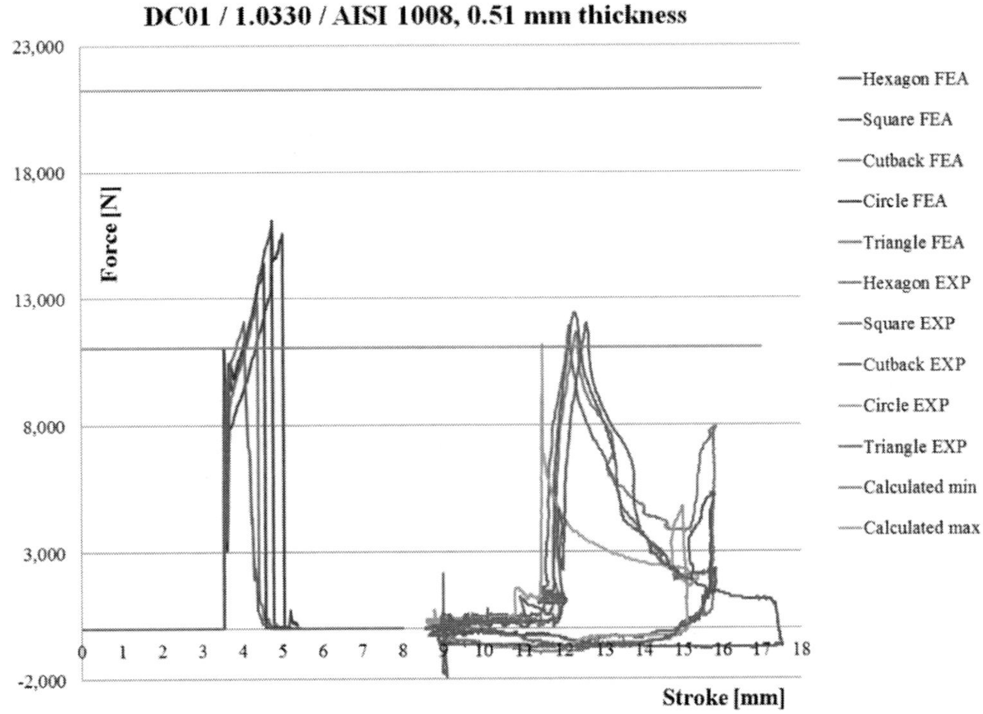

Fig. 7. Comparing results for AISI 1008 material, 0.51 mm thickness.

For non-ferrous materials the comparing results are the followings:

DOI: 10.1051/matecconf/201712104010

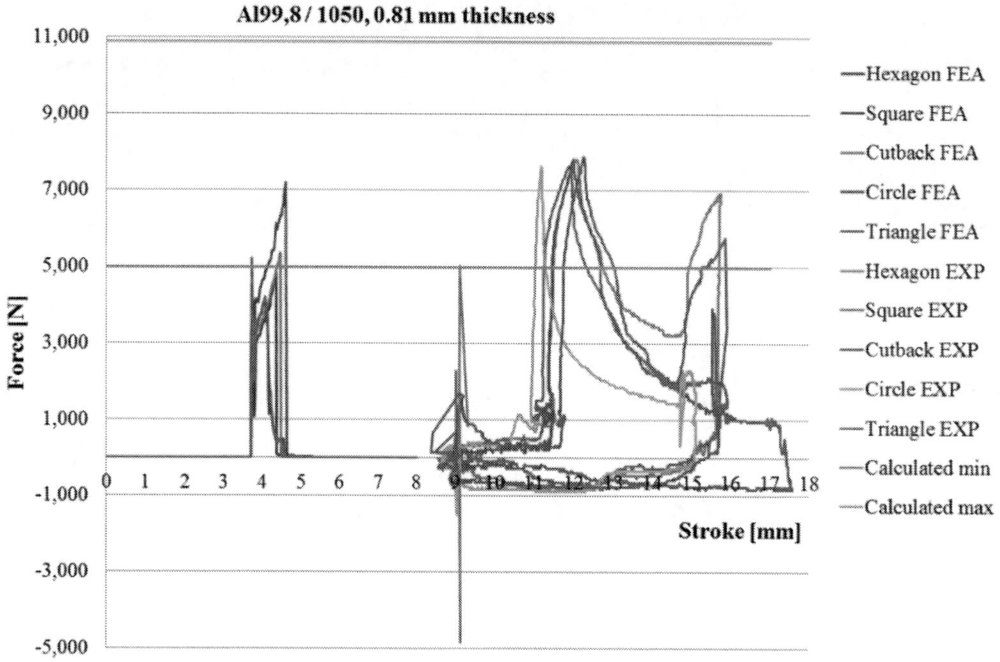

Fig. 8. Comparing results for Al99.8 material, 0.81 mm thickness.

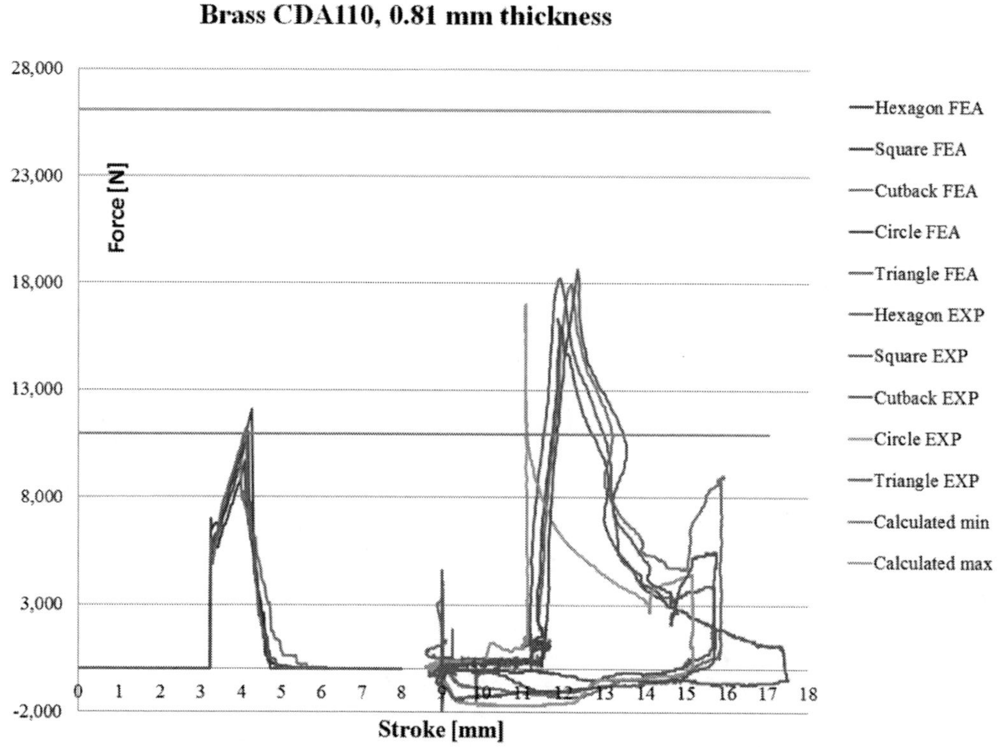

Fig. 9. Comparing results for Brass CDA 110 material, 0.81 mm thickness.

To easily observe the differences between the forces obtained for each type of contour geometry and materials was made a comparative table.

Table 2. Comparison of influence of punched contour geometry over the punch force.

		AISI 304	AISI 1008	Al99.8	Brass
Hexagon	F_{con} FEA	18,366.31	9,408.24	3,899.60	6,269.08
	F_{max} FEA	30,944.10	15,870.65	4,188.58	11,050.73
	F_{max} EXP	21,417.20	11,666.20	7,826.72	17,943.40
Square	F_{con} FEA	17,945.79	9,938.70	3,812.34	5,712.40
	F_{max} FEA	30,295.54	16,108.86	4,836.56	9,693.75
	F_{max} EXP	20,259.20	11,910.00	7,643.89	18,248.10
Shearing / cutback	F_{con} FEA	19,980.57	7,416.46	2,159.76	5,253.49
	F_{max} FEA	26,666.41	12,880.79	5,355.48	9,826.86
	F_{max} EXP	22,270.40	12,397.50	7,826.72	16,358.80
Circle	F_{con} FEA	17,917.21	11,029.96	3,951.62	6,999.85
	F_{max} FEA	33,089.05	14,386.40	7,185.62	12,095.67
	F_{max} EXP	19,406.00	11,178.60	7,643.89	17,029.20
Triangle	F_{con} FEA	21,030.08	6,854.26	5,209.33	6,002.73
	F_{max} FEA	21,030.08	12,089.80	5,209.33	11,261.86
	F_{max} EXP	21,965.70	12,031.80	7,704.83	18,674.70
Calculated force	F_{min} CALC	17,463.60	11,048.40	4,971.78	10,929.60
	F_{max} CALC	28,351.44	21,263.58	10,881.00	26,060.58

Where: "Fcon FEA" represent the force recorded when the top die hit the workpiece, "Fmax FEA" represent the maximum force recorded during FEA simulation, "F_{max} EXP" represent the maximum values recorded during experimental test, "F_{min} CALC" and "F_{max} CALC" represent the minimum and maximum calculated forces.

As can be seen in the graphs, there is differences between FEA and experimental results and the biggest difference is that the force is decreasing after the contact between the punch tool and material in the experimental case, while in FEA simulation the force is increasing.

But in both cases the geometry of punched contour has influence over the punch force.

References

1. Gh. Sindilă, *Cold plastic deformation manufacturing technologies,* (Publisher Bren, Bucharest, 2007)
2. S. Uţuleanu, A. Vlase, Gh. Sindilă, N. Căpăţână, ISCAME, **212**, Debrecen (2013)
3. C. Ciocârdia, Cold pressing technology, Ed. Did. and Ped., R.A., Bucharest (1991)
4. S. Uţuleanu, A. Vlase, Gh. Sindilă, N. Căpăţână, IManEE 2017, Iaşi (to be published)
5. S. Uţuleanu, A. Vlase, Gh. Sindilă, N. Căpăţână, ADEM, **172**, Drobeta-Turnu Severin, 2014

Improving the quality of the mechanical systems by specific methods in connection with a commercial entity

Claudiu Babis[1,*], *Andrei* Dimitrescu[1], *Gabriel* Iacobescu[1], *Gheorghe* Solomon[1], *and Dan* Nitoi[1]

[1]Politehnica University of Bucharest, Materials and Welding Technology Department, 313 Splaiul Independenței, 060042, Bucharest, Romania

Abstract. This paper aims is improving the quality of mechanical systems, using specific methods. In the first part of the paper, we will identify the causes of non-quality, using ISCHIKAWA diagram. Based on data collected from economic entity studied, using PARETO chart, we will identify the main action. In the second part of the paper, we will implement the specific quality improvement methods for mechanical systems, previously identified. In the last part of the paper, based on new data collected from ecnomic entity, it will make a statistical prediction quality for the next period.

1 Introduction

Product quality is the primary factor in the development of any organization, hence the importance of a quality management system. [1]

The concept of quality has evolved over the years, due to the industrial and information society development. In time, quality evolves from quality control, to quality assurance and then to total quality, taking into account both the social component (staff involvement, participation) and economic component (the organization's image, reducing quality costs).

The analysed company MGC INTERNATIONAL INVESTMENTS Ltd. has more then 1600 employees in several factories in Romania, with a production of 300 tons of finished product per week. The company produces and sales canned vegetables and fruit, canned meat and mixed.

The main objectives of the study are set out in the quality policy statement of the general manager: increasing the number of customers in the European market, reducing the number of complaints by 40% over the previous year and reducing production costs.

This study is significant because the analyzed company has the most important role in the labor market in the area.

* Corresponding author: claudiubbs@gmail.com

2 Sampling data

The Company has proposed implementing food safety management system in accordance with the requirements of SR EN ISO 22000/2005, which is described in the Handbook of Food Safety Management System (HFSMS).

Food Safety Management System (FSMS) described in this handbook sets out measures to demonstrate the application of the requirements of the aforementioned standard and the performance and capability of the organization in terms of providing consistent products with improved safety for consumers.

To operate legally and in good condition and meet the wishes of the beneficiaries in terms of quality of the finished product, proving the implementation of all laws and regulations in force, both in Romania and the European Union, the company has implemented the HACCP (Hazard Analysis and Critical Control Points) system, with the organizational procedures and implemented by this system by granting decisions by all team member.

They established checkpoints (CP) and critical control points (CCP) and were implemented proceedings and recording of data obtained at their level. This data is collected from the measuring and control instruments (AMC) and to the accompanying documents. All data is used to determine if it violates the law (temperature, microbiological parameters and physicochemical parameters). This data is archived electronically and on paper due to the possibility of establishing traceability of a finished product.

In developing the concepts of quality control, and quality assurance and subsequent total quality control were gradually developed a number of methods, tools and techniques that contribute to quality improvement. [2]

As no improvement action should not be employed without relying on data, this paper proposes a study of the relationship between cause and effect that they produce, and finding solutions to problems based on causes that led to negative effects on MGC INTERNATIONAL INVESTMENTS Ltd.

This study will be highlighted by cause and effect diagram (Ishikawa), shown in Figure 1 where the causes of non-complying products for human consumption correspond 6M: environment, labor, measurement, management, machinery, materials. [4]

The main advantage of Ishikawa diagram is highlights the 6 groups all causes of non-quality products. Another advantage is that for each group may consult various departments within the company. Main disadvantage is relatively long duration for proper preparation of this diagram.

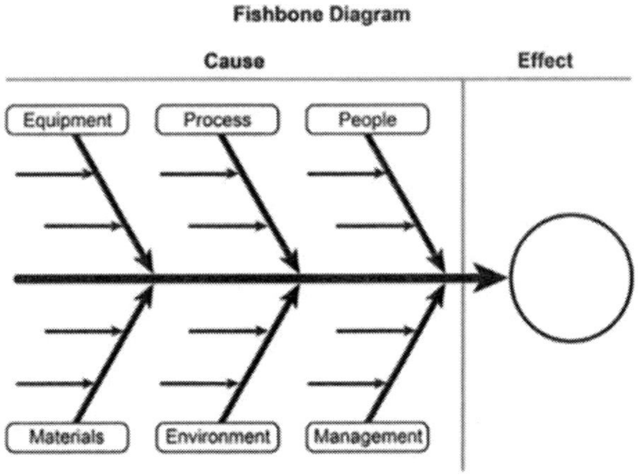

Fig. 1. Ishikawa Diagram

The analysis was conducted by quantifying the nonconformity products and analyzing the causes of non-compliant products for a period of 30 days.

Thus, resulting the following histograms for the non-compliant products in groups. It should be noted that these data were obtained from the department for products checking inside the factory prior to marketing.

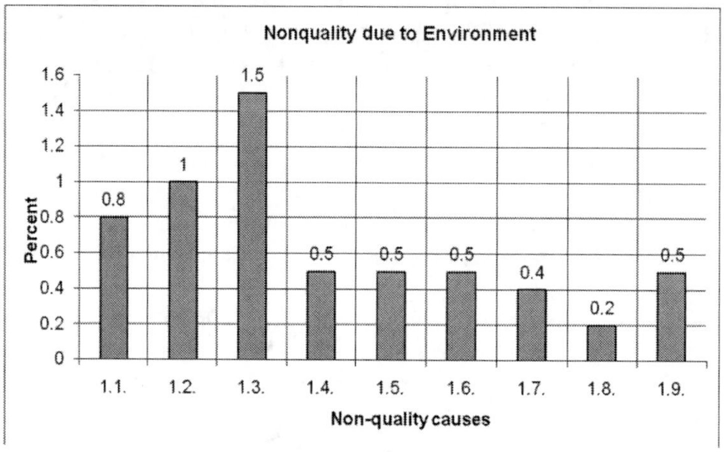

Fig. 2. Non-quality due to environment

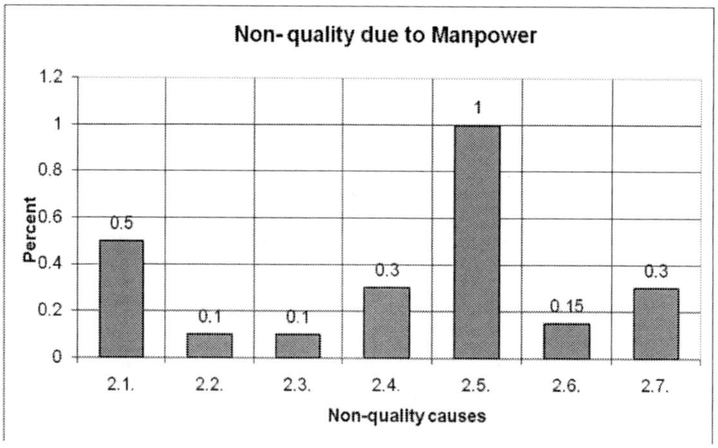

Fig. 3. Non-quality due to manpower

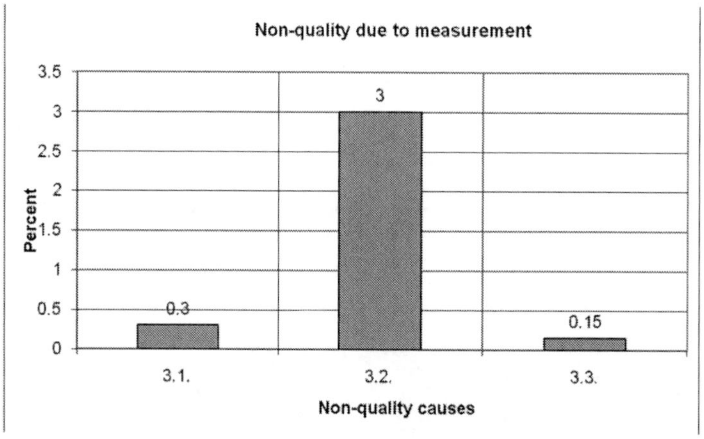

Fig. 4. Non-quality due to measurement

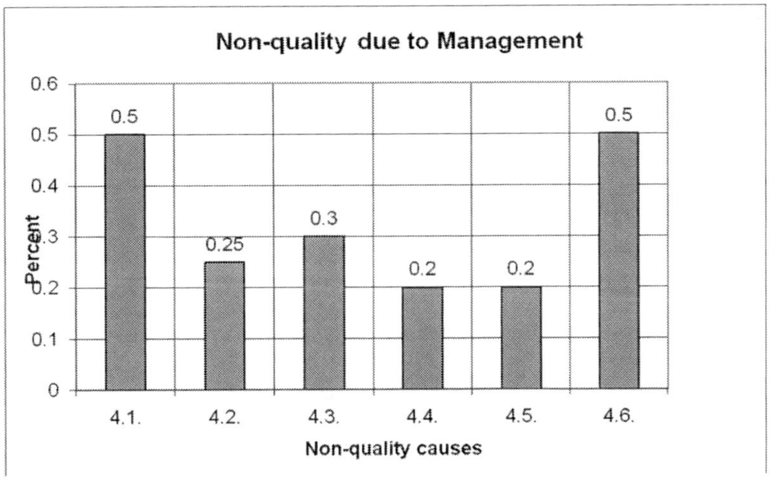

Fig. 5. Non-quality due to Management

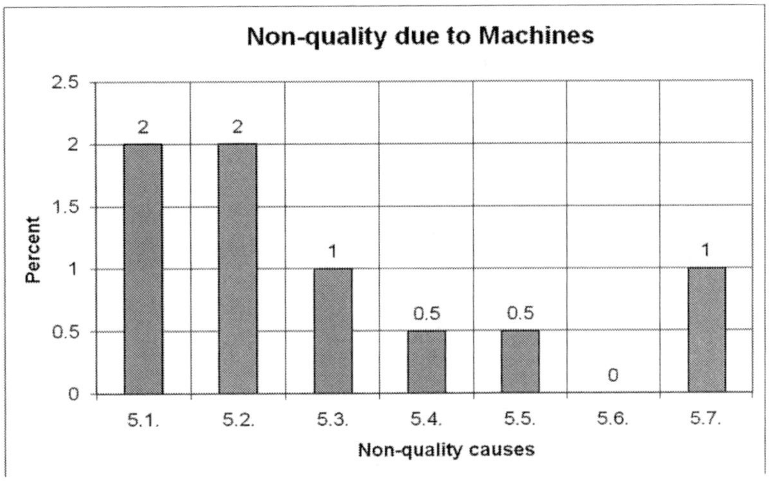

Fig. 6. Non-quality due to Machines

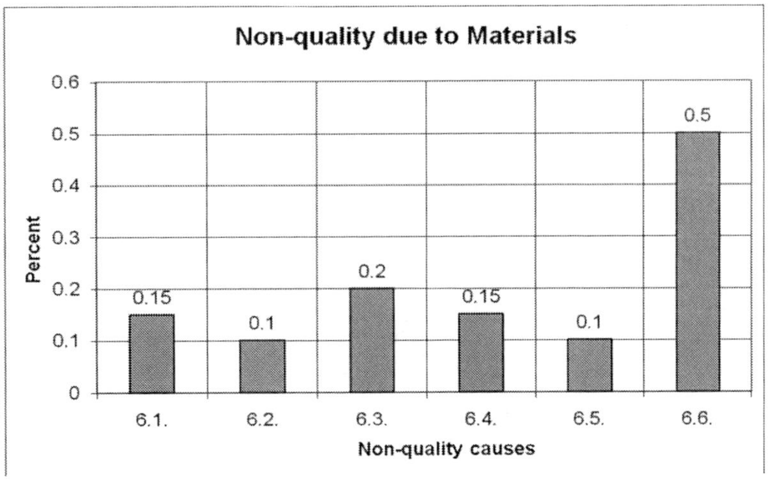

Fig. 7. Non-quality due to Materials

The table below highlights the percentages of influence of all causes of non-conformity of the finished product:

Table 1. Cases of non-conformity according to Ishikawa diagram

No.	Cause	Subcauses	Non-conformity [%]
1.	1. Environment	1.1. Lighting	0,8
2.		1.2.Ventilation	1
3.		1.3. Technical suspended routes	1,5
4.		1.4. Cleanliness	0,5
5.		1.5. Disinfection	0,5
6.		1.6. Equipment sanitation	0,5
7.		1.7. Staff hygiene	0,4
8.		1.8. Protection equipment	0,2
9.		1.9. Bacteria removal system	0,5
10.	2. Manpower	2.1. Good practices	0,5
11.		2.2. Staff training in the workplace	0,1
12.		2.3. Instructing	0,1
13.		2.4. Lack of financial motivation	0,3
14.		2.5. Training for equipment handling	1
15.		2.6. Working program	0,15
16.		2.7. Operators health check	0,3
17.	3.Measurement	3.1. Installations to measure the weight of raw materials, finished products	0,3
18.		3.2. Temperature measurement system	3
19.		3.3. Optimization of packaging - sensors	0,15
20.	4.Management	4.1. PO and PL procedures	0,5
21.		4.2. Employees information system - production stage	0,25
22.		4.3. Finished products transportation system	0,3
23.		4.4. Contracts with raw material suppliers	0,2
24.		4.5. Product sale system	0,2
25.		4.6. Marketing system	0,5
26.	5. Machines	5.1. Harvesters	2
27.		5.2. Transport	2
28.		5.3. Washing and sorting machine	1
29.		5.4. Processing	0,5
30.		5.5. Weighing and packaging machines	0,5
31.		5.6. Vacuum machines	0
32.		5.7. Maintenance	1
33.	6.Materials	6.1. Raw material suppliers database	0,15
34.		6.2. Providers Classification by quality / price	0,1
35.		6.3. Classification of suppliers based on timeliness of delivery	0,2
36.		6.4. Linking the two classifications	0,15
37.		6.5. Third party audits	0,1
38.		6.6. Archiving according raw material documents	0,5

3 Conclusions

Unable to act cumulatively to all causes of non-quality or because of organizational of financial reasons, it is necessary to decrease significantly the percentage of not complying products with minimum effort. Thus, actions have yield from all points of view.

To know in detail this phenomenon it must be applied another quality analysis which is called "Pareto Analysis".

Pareto analysis is a statistical technique for classifying tasks small in number but with significant effect. It is based on the Pareto principle (also known as the 80/20 rule), which states that 20% of resources generates 80% of total labor, or in terms of quality improvement, most of the problems (80%) have several key causes (20%). [3]

Pareto chart helps management interventions methodically through action plans built around the major causes of emerged problems, by being a strongly visually oriented, to make the decision. It is a tool for qualitative analysis because it focuses on process parameters in terms of frequency. Defined parameters can be events, errors, features, etc.

Of the 38 non-quality causes 20% representing 7.4 causes. The following eight causes were chosen as representative:

Table 2. Cases of non-conformity according to Pareto Analysis

No.	Cause	Subcause	Non-conformity [%]
1	1.Environment	1.2. Ventilation	1
2		1.3. Technical suspended routes	1,5
3	2.Manpower	2.5. Equipment handling training	1
4	3.Measurment	3.2. Temperature measurement system	3
8	5.Machines	5.1. Harvesters	2
9		5.2. Transport	2
10		5.3. Washing and sorting	1
11		5.4. Maintenance	1

After corrective actions listed above causes and establish new percentage of non-compliant products, where they exist we can start the depreciation of these corrective actions.

Using specific programs, we can predict when they will recoup their investments and especially what are the company's gross earnings. It should be noted that these calculations have real value only if it does not change the parameters, for example the average sales prices of products and raw materials or other taxes, contributions and excises.

For correct predictions, we suggest the use of higher mathematics regression curves with equal residuals and standard errors that tend to 1. These curves we aim to generate with the soft CurveExpert 1.4 after taking data from the company.

References

1. Bernard Froman, *Managementul Calitatii*, 77, (Ed. Tehnica, Bucuresti, 1998)
2. Henri Mitonneau, *O noua orientare in Managementul Calitatii*, 12, (Ed. Tehnica, Bucuresti, 1998)
3. Arthur L.J., *Rapid Evolutionary Development – Requirements, Prototiping & Software Creation*, (Ed. John Wiley&Sons, 2011)
4. Boca Dana, *Bazele managementului calităţii în afaceri*, (Ed. Risoprint, 2013)

Improvement of catheter quality inspection process

Mariusz Bożek[1], *Agnieszka* Kujawińska[2], *Michał* Rogalewicz[2,*], *Magdalena* Diering[2], *Paweł* Gościniak[1] and *Adam* Hamrol[2]

[1]Aesculap Chifa Ltd., Poland
[2]Poznan University of Technology, Chair of Management and Production Engineering, Poland

Abstract. Quality inspection is very often one of the most important stages of the production process, although it does not create any added value. Therefore, optimization of the related number of activities is of crucial importance. However, reduction should not be made arbitrarily, but preceded and documented by appropriate research. The article describes a study aimed at reducing the high cost of quality inspection as part of the manufacturing process of a diagnostic catheter at a medical company. The product is used for blood pressure monitoring and blood sampling by the Seldinger technique. A critical quality feature for the catheter is air/water tightness. Following a thorough analysis, some control points were eliminated, and others were improved. The resulting conclusion is that detection of defective components is the most beneficial for this specific production process if carried out during the 100 percent final quality inspection. The finding is based on the fact that the cost of producing the final device with a defective component is lower than a quality inspection run directly after each operation. The authors also managed to decrease the sample size for control charts used to supervise the adhesive connection strength.

1 Introduction

The key objective of a manufacturing process is the delivery of a product which meets customers' expectations. In practice, every manufacturing process is exposed to the effect of certain factors and obstacles which should be analysed and taken into consideration, but are difficult to be predicted at the stage of designing the process [1-4]. Such factors often cause non-compliance or defects in products, processes or equipment [5]. Undetected on an ongoing basis, non-compliant or defective semi-products are further processed or even reach the end user. In consequence, the manufacturer incurs additional costs related to repairs, claims, reduced efficiency or interruptions in the manufacturing process [6-7]. Therefore, proper assessment of the condition of the manufacturing process is important at the stage of designing the process.

[*] Corresponding author: michal.rogalewicz@put.poznan.pl

The condition of a manufacturing process is assessed based on the data obtained through quality inspection. Inspection results have to be compromised between accuracy and measurement time [8].

Quality inspection has many definitions. For the purpose of this paper, it is defined as the verification of the manufacturing process or product for compliance with the requirements of internal or external customers. Quality check is usually carried out through direct measurement or visual inspection, and provides data for interpretation of the condition of the manufacturing process or product. Therefore, the data should be a reliable source of information (utility of a quality control system can be assessed with the use of methods and procedures of the statistical analysis of measurement and control systems [9]). The feature being assessed may be expressed in any measurement scale (depending on the organization of the manufacturing process and the customer requirements). The data collected may be precise (the so-called hard data) or imprecise, incomplete and/or unreliable (the so-called soft data [10-11]). The inspection methods used should depend on the character of the monitored product feature or process parameter.

There are numerous quality inspection methods. Developed since the 1920s, they were initially constrained to the assessment of a finished product. Products which failed to meet certain requirements were discarded and returned to the manufacturing process or scrapped. Such an approach to quality control was criticised in the 1950s for generating high costs of repairs and scrapping to be incurred by the organization [12-15].

Subsequent decades brought the development of the process-based approach to manufacturing and the idea of "zero defects manufacturing". They both significantly changed the understanding of quality control. The underlying concept of the modern approach to quality control is that it should be carried out at selected (or even all) stages rather than at the end of the production process. The key objective of quality control is early detection of irregularities in the process and their causes, and application of corrective and preventive activities. Such an approach makes it possible to control the process quality and improves the opportunity to maintain the desired quality level [15].

The usual inspection method in repetitive manufacturing is the statistical inspection. Here, measurement of each produced unit is economically unjustified, and often technically impossible. An entire lot of products (the manufacturing process) is assessed on the basis of the results obtained from an analysis of a random sample.

Contrary to the statistical inspection, the 100 percent inspection method covers all the produced units. This method is time-consuming and applied only in unit or small series manufacturing. In series manufacturing, 100 percent quality inspection is implemented only in special cases, where the delivery of products incompliant with the requirements may have serious consequences [14, 16-17].

A vitally important aspect of quality inspection is planning. It should cover the definition of the objectives of quality control, time to carry out inspections, the inspected product features, methods of assessment, etc. First and foremost however, as a task which does not generate any added value, quality inspection should be justified in terms of the efficiency of the manufacturing process [18-19]. It should be noted here that even a perfectly planned and carried out quality inspection is not fully effective. Improper planning or poor organization of a quality inspection may generate waste of resources.

There are three types of quality inspection which are: incoming inspection, in-process inspection and final inspection [20]. This paper focuses on all of them – the paper presents the results of research related to the analysis of inspection operations in the process of manufacturing of the central venous catheter.

2 Problem statement

The research aims at reducing the costs of quality control in the process of manufacturing of the catheter used for central venous pressure monitoring and blood sampling by the Seldinger technique. The catheter comes in a kit with a stainless steel guide wire and a needle. The Seldinger technique of central venous pressure monitoring and blood sampling requires placement of the catheter in a blood vessel by means of the needle. A cannula is inserted into the vessel and the guide wire advanced through the cannula. Once the cannula is retracted, the catheter is inserted in the place where the guide wire was introduced. Next, the guide wire is pulled out and the catheter is secured by means of a wing [21]. The catheter consists of four components (Fig. 1): a capillary with a wing, a formed (sharpened) tip, a connector tube, a valve.

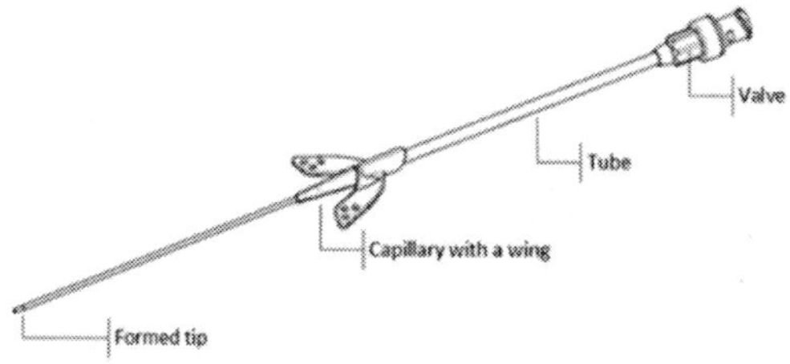

Fig. 1. The design of a diagnostic catheter.

The components of the catheter are connected in the process of bonding. The adhesive connections must be air/water tight and break proof. Considering the fact that leaks occurred in some of the products supplied to the customer in the past, the organization decided to implement additional inspections for air leaks in the manufacturing process. Moreover, following the introduction of certain alterations to the product design, an additional 100 percent acceptance inspection was implemented with respect to one of the components – the capillary with a wing.

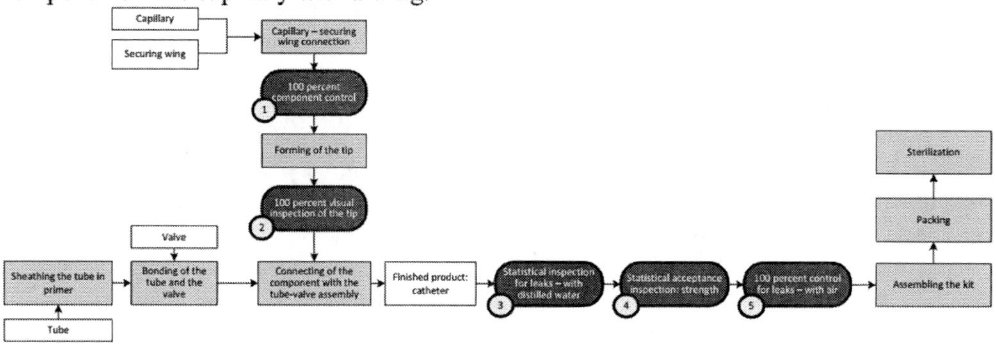

Fig. 2. Catheter manufacturing process map with control points.

In consequence, 5 inspections were implemented in the manufacturing process (Fig. 2):

- 100 percent component acceptance inspection – inspection of the diameter – non-destructive examination,
- 100 percent inspection of the edged tip – non-destructive examination,
- statistical inspection of adhesive connections for leaks, with the use of distilled water – destructive examination,
- statistical acceptance inspection of adhesive connections for strength – destructive examination,
- 100 percent inspection of the catheter for air leaks – non-destructive examination.

The implementation of additional inspections (Fig. 2 – control no. 1 and no. 3) made the manufacturing process more time-consuming and required the use of additional resources (material). After some time, the inspections were verified for validity.

3 Improvement of inspections

For several months, costs related to the inspections, as well as the efficiency and stability of the manufacturing process, were under analysis. The study was performed in five steps. The steps are described below, in an order corresponding to the numbering of inspections in Fig. 3. To improve inspection processes, DMAIC methodology was used [22].

3.1 Component acceptance inspection

The component acceptance inspection is a 100 percent inspection of the capillary bonded with the securing wing. The critical feature of the component is the diameter of the wing. A too large internal diameter of the wing opening is one of the reasons for leaks occurring in finished catheters. Twenty-two production batches of the component were examined to reveal that initially, the manufacturing process was characterised by a high percentage of incompliant products (in batch 2, more than 23% of the components were incompliant) (Fig. 3). However, starting from batch 3, the percentage of incompliant products diminished significantly. The mean percentage of incompliant products in the examined sample amounted to 1.09%.

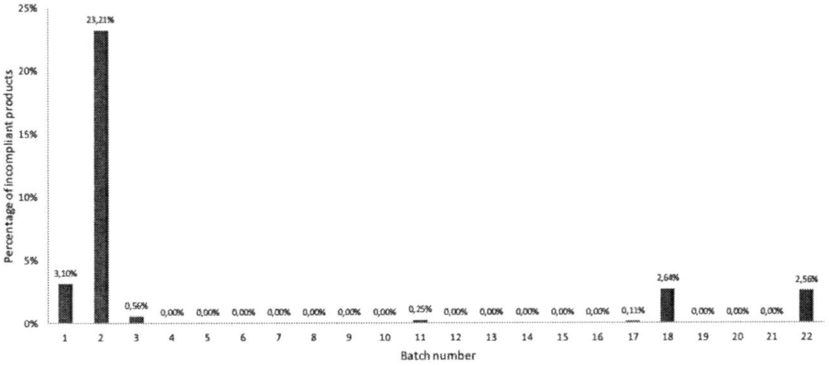

Fig. 3. Percentage of incompliant components (too large diameter) in analysed production batches (X axis – lot number, Y axis – percentage of incompliant components).

An analysis of the costs of inspection of the wing diameter in proportion to the total costs of manufacturing of the finished product revealed that the costs of inspection exceed the total costs of manufacturing of the catheter with a defective component by almost 25%.

The results of the analysis show that it would be reasonable to skip the first inspection, even more so considering that catheters with this type of defect will be excluded at the 100 percent final inspection for leaks.

3.2 Visual inspection of the catheter tip

Similarly to the inspection for leaks, this visual inspection is a 100 percent control. The geometry and manufacturing precision of the catheter tip are examined under the microscope. Five or six catheters are examined at a time for scratches, cavities, grooves, pits or jagged spots on the surface.

The inspection is carried out following the operation of forming the catheter tip, and is indispensable. Its cost amounts to 120% of the total costs of technological operations for the finished product.

3.3 Statistical inspection for leaks with distilled water

The 3rd type of inspection is the inspection for leaks with the use of distilled water. After contact with water, the catheter cannot be reused due to the risk of impurities in the water. Therefore, this type of destructive testing is carried out on a number of randomly selected catheters from a given lot.

The tested catheter is connected to the measuring device on one side. On the other side, a steel wire of an appropriate diameter is inserted and advanced to seal the measurement system. Next, distilled water is let into the system and pressure is monitored. Any drop of pressure indicates leaks in the catheter.

Considering that in the existing manufacturing process, the catheter was tested for leaks twice (inspection no. 3 and final inspection no. 5), the validity of the statistical inspection no. 3 was questioned (redundant destructive testing). Next, the effectiveness of the inspection for leaks with distilled water was verified against tests for leaks with air. In the first step, all catheters in the analysed lot were tested for leaks with air, and the leaking catheters were discarded. Next, the leaking catheters were tested again for leaks – with the use of distilled water. It turned out that in the process of testing with distilled water, only 35% of leaking catheters were detected. As a result, the researchers decided to skip the testing with distilled water due to its ineffectiveness.

3.4 Statistical acceptance inspection – evaluation of adhesive connections for strength

The 4th inspection in the manufacturing process is the statistical inspection of lots of finished products. Here, catheters are tested for strength of the adhesive connections with the use of a breaking force (destructive testing). The tube in the catheter is cut into two in order to examine the adhesive connection between the valve and the tube, and between the tube and the wing. The breaking force cannot be smaller than 18N. According to the inspection plan, 8 to 20 items are tested, depending on the size of the lot. For example, for a lot of 260 items, 13 randomly selected catheters are tested. A lot is considered compliant with the requirements if none of the catheters is defective.

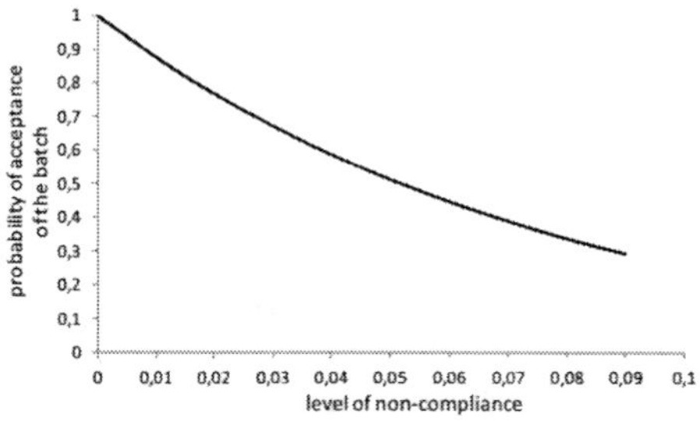

Fig. 4. The characteristic curve of the acceptance inspection plan for a sample of 13 items and Ac = 0.

It results from an analysis of the linear characteristic curve of the acceptance inspection plan for a sample of 13 items and the zero acceptance number (Ac = 0) (Fig. 4) that with the factual defectiveness of the process at 0.5%, the probability that a lot is considered compliant equals 94%. On the basis of the analysis results, the statistical acceptance inspection was considered inappropriate at this stage of the manufacturing process.

The downside of the discussed approach is that incompliance is detected only after the entire lot is produced. There is no possibility to discard the items where adhesive connections do not meet the strength requirements, because the testing is destructive. Therefore, a solution was proposed to monitor the strength of the adhesive connection during the manufacturing process, with the use of process control charts.

Mean value and range control charts were implemented for 13-item samples to monitor the valve-tube and wing-tube connections (Fig. 5). It results from an analysis of the graphs in Figure 6 and the performed calculations that the process is not stable, but shows a high capability in the short term. Further steps were taken to stabilize the process and optimise the size of the control sample.

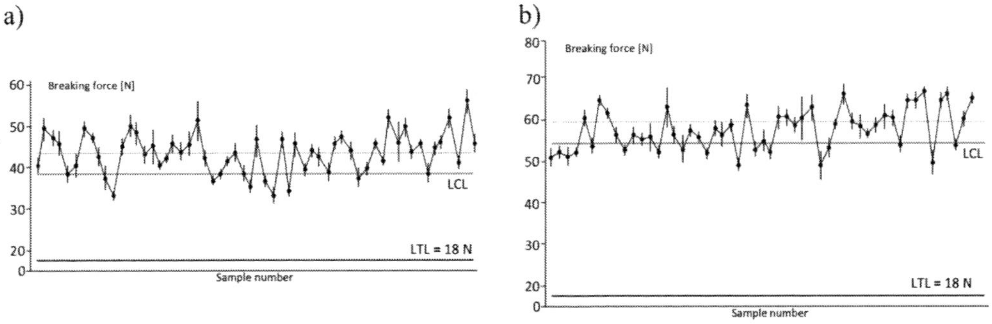

Fig. 5. Mean value chart for 13-item samples for the valve-tube (a) and wing-tube (b) connections.

The sample size was reduced from 13 to 3 items. Figure 6 presents a control chart for the valve-tube connection for a 3-item sample. In Table 1, calculations of probability are presented for the results of individual adhesive force measurements below the lower tolerance line (LTL). The LTL is the process safety indicator. The standard deviation was determined with the Hartley formula for a 3-item sample.

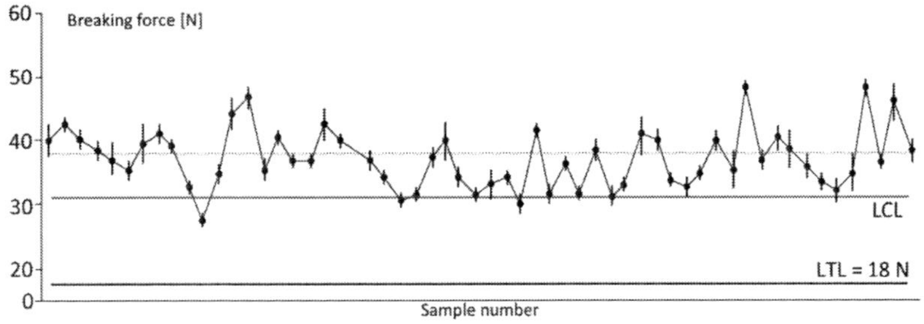

Fig. 6. Control chart for the valve-tube connection – 3-item sample.

Tab. 1. Probability of obtaining the adhesive force below the LTL for various values of the process mean deviation for the valve-tube and wing-tube connections (3-item samples).

Mean process deviation	Probability of obtaining the adhesive strength below the LTL	
	valve-tube	wing-tube
1	0.0006227	0.0000000
2	0.0129309	0.0000000
3	0.1096711	0.0000001

Owing to a high capability of both processes (1.41 for the valve-tube connection and 2.74 for the wing-tube connection), the chance of obtaining the adhesive force value below the LTL is very small, even at a significant deviation of the process mean. Further improvement of the process should aim at reducing its variability (stabilizing the process).

3.5 Inspection of adhesive connections for leaks – with compressed air

The fifth inspection is carried out after the capillary is connected to the valve-tube assembly. It is a 100 percent control, which means that each item in the lot is examined. For this purpose, compressed air at a pressure of 800 +5 mmHg is let into the catheter. Next, the compressed air inlet is closed and the pressure is monitored by means of a pressure calibrator. The control of catheters for leaks is indispensable. Owing to the use of compressed air, the testing is non-destructive.

4 Conclusions

The research aimed to analyse and – if justified and possible – reduce the costs of quality control in the process of manufacturing of the catheter for central venous pressure monitoring and blood sampling by the Seldinger technique. The research was conducted for about a year. The resultant savings are presented in Table 2.

Tab. 2. Quality control costs in the manufacturing process of the diagnostic catheter – before and after the implementation of changes.

Quality control cost items – manufacturing of the diagnostic catheter	Control value (percentage)	Savings*
A. 100 percent inspection – total cost	60.75%	reduced by 1.53%
Visual inspection of the tip	6.35%	no change
Inspection for leaks (with compressed air)	47.49%	no change
Component acceptance inspection	6.91%	reduced by 1.53%
B. Statistical destructive testing – total cost	39.25%	reduced by 31.98%
inspection for leaks (with distilled water)	4.26%	reduced by 4.26%
Inspection of adhesive connections for strength	7.10%	6.29%
Catheter (sample) manufacturing costs)	27.89%	21.44%
C. Total control costs	100.00%	33.51%
Total savings	**33.51%**	

* Savings after the implementation of changes – in percentage value, in proportion to the total control cost (C).

To sum up, the works covered the identification and research of control tasks. The inspections were analysed and actions were proposed for each control task (skipping the control task or implementation of improvement actions). Inspection no. 1 was proven to be economically unjustified and eliminated from the manufacturing process. Likewise, inspection no. 3 (tests for leaks with the use of distilled water) was found ineffective and unjustified, considering the fact that the next stage of the manufacturing process included a 100 percent control of air/water tightness with the use of air.

Therefore, in order to reduce the total control cost, skipping of inspection no. 3 was proposed. Inspection no. 4 (statistical acceptance inspection) was found to be inappropriate at that stage of the manufacturing process. Instead, as a result of the research, the use of mean value and range control charts was suggested. Moreover, the sample size was reduced from 13 to 3 items. As a result of the research, the total cost of quality control in the manufacturing process of the diagnostic catheter was reduced by more than 33%.

The results presented in the paper come from a scientific statutory research conducted at the Chair of Management and Production Engineering, Faculty of Mechanical Engineering and Management, Poznan University of Technology, Poland, supported by the Polish Ministry of Science and Higher Education from the financial means in 2017: 02/23/DSPB/7695.

References

1. A. Lewandowska, E. Więcek-Janka, 3rd Int. Conf. on Man. Sc. and Man. Inn. (MSMI 2016), AEBMR-Adv. in Econ. Bus. and Man. Res., Atlantis Press, **10**, 219 (2016)
2. J. Trojanowska, E. Pająk, 7th International Conference of DAAAM Baltic Industrial Engineering,Tallin, Estonia, **1**, 322 (2010)
3. J. Trojanowska, K. Żywicki, E. Pająk, Information Technologies in Environmental Engineering, Springer, 695 (2011)
4. K. Kalinowski, M. Zemczak, *Adv. in Intelligent Systems and Computing,* Springer, **368**, 475 (2015)
5. M. Jasiulewicz-Kaczmarek, *IFAC Conf. on Manufacturing Modelling, Management and Control (MIM)*, (IFAC Papers On Line, **49** (12), 674, 2016)
6. R. Dylewski, A. Jardzioch, I. Krebs, Man. and Prod. Eng. Rev., **7** (4), 97 (2016)
7. S. Kłos, Com. In Comp. and Inf. Sc., **639**, 42 (2016)
8. S.H. Mian, A. Al-Ahmari, Pi. Mech. Eng. B-J Eng., **231** (4), 565 (2017)
9. M. Diering, K. Dyczkowski, Proc. IEEE Int. Conf. on Industrial Engineering and Engineering Management, **228** (2016)
10. A. Stachowiak, K. Dyczkowski, Proc. IFSA World Cong. and NAFIPS, IEEE, 390 (2013)
11. E. Więcek-Janka, R. Mierzwiak, Grey Sys.: Th. and Appl., **5** (3), 302 (2015)
12. C.G. Drury, M.A. Sinclair, Hum Factors Ergon Manuf, **25** (4), 391 (1983)
13. A. Hamrol, D. Kowalik, A. Kujawińska,, Hum. factors ergon. manuf., **21** (2), 156 (2011)
14. M. Rogalewicz, R. Sika, Man. and Prod. Eng. Rev., **7** (4), 97 (2016)
15. A. Hamrol, Prod. Plan. Control, **11** (8), 797 (2000)
16. A. Kujawińska, K. Vogt, F. Wachowiak, *Technology Management for Sustainable Production and Logistics*, (EcoProduction, Springer, 193, 2015)
17. A. Kujawińska, A. Vogt, A. Hamrol, *Advances in Intelligent Systems and Computing*, HAAMAHA, **490**, 569 (2016)
18. B. Skołud, D. Krenczyk, *4th Int. Scientific-Technical Conf. MANUFACTURING*, in: Selected Conf. Proc. (Poznan University of Technology, December 2014, 35, 2016)
19. M. Suszyński, J. Żurek, S. Legutko, Tehnicki Vjesnik/Technical Gazette, **21** (6), 1229 (2014)
20. Y. Ngadiman, B. Hussin, A.T. Bon, N.A.A. Hamid, *ICMME 2016*, MATEC Web Conf., **95**:10007 (2017)
21. Z.C.J. Higgs, D.A.L. Macafee, B.D. Braithwaite, C.A. Maxwell-Armstrong, *The Seldinger technique: 50 years on*, Lancet, **366** (9494), 1407 (2005)
22. S. Mahato, A.R. Dixit, R. Agrawal, J. of Eng. Manufacture, **231** (4), 713 (2017)

Quality Management techniques embedded in Agile Project Development

Raluca Dovleac[1,*], and *Andreea* Ionică[2]

[1]Management Department, Doctoral School, University of Petrosani, Romania
[2]Management Department, University of Petrosani, Romania

Abstract. It used to be hard to make your dream come true and start a business but now, thanks to the dotcom boom and the rapid development of emerging technologies everybody has the chance to become the next Steve Jobs. Quality management tools and creativity techniques gave birth to the startup trend that transforms a brilliant idea into an innovative product. A common practice amongst startups is Agile development. In this paper, after analyzing the Agile approach and the tools used to ensure quality for every stage of a product's development life cycle and the results of using the Agile approach looking out for advantages and short comings as well, we proposed a PDCA inspired tool which embeds both quality and creativity techniques to help developers ensure quality for each stage of a product's life cycle as well as help them better organize their activities to reach the desired goals. We validated this approach by implementing it to the case of the Vitraly startup.

1 Introduction

The history of the word quality encapsulates plenty of definitions and aspects as well as case studies. As of lately, quality in all its forms has become a major factor taken into account by all companies and small businesses as well. In this sense, and in order to help companies integrate quality into their everyday business lives, standards, procedures, tools and techniques come to their aid to ensure a standardization and ease of implementation.

An interesting aspect emerges however from observing the correlations between the implementation and maintenance of quality throughout a product's lifecycle and the project management process of that product.

Another important aspect to factor in is the way that the agile approach and quality merge together to ensure quality products and/or services are delivered.

In our paper, we outlined the important aspects of quality, project management and agile development as well as proposed embedding new quality management and creativity tools into the agile development thus ensuring better results.

* *Corresponding author: raluca.dovleac@yahoo.com

2 Literature Review

From its birth in the late 13th century, when craftsmen began organizing into unions called guilds, quality evolved and developed around the needs of manufacturing companies and production plants initially, only to become part of the modern companies objectives later on.

Given the fact that there are multiple understanding of quality and quality systems [1], different strategies for quality tools emerged as well [2]. When it comes to classifying the quality tools available today, there are several approaches to the topic, a fundamental one covering "The Seven Basic Tools of Quality" also known as 7 QC Tools. The 7 QC Tools developed in Japan in the 1950's by Kaoru Ishikawa when the country was undergoing some major revolution in the field of quality and it was inspired by seven well known weapons of Benkei [3]. The tools are often referred to as The Seven Basic Tools of Quality because they can be implemented by any person with basic training in statistics and can be applied to solve complex quality related problems [4]. They include: The Pareto Chart, Histogram, Process Flow Diagram, Control Charts, Scatter Diagram, Check Sheets and Cause and Effect Diagram [5]. Carpinetti [6] argued that the most important techniques are: Quality Function Deployment (QFD), Failure Mode and Effect Analysis (FMEA), Six Sigma, 5S, Design of Experiments (DoE) and Statistical Process Control (SPC). After 1977, The Union of Japanese Scientists and Engineers (JUSE) appreciated the following tools and techniques to be the most essential for quality improvement: The Affinity Diagram, The Relations Diagram, The Matrix Diagram, Tree Diagram, Arrow Diagram, Decisions Diagram and The Factorial Analysis of Data [7].There are those who debate on classifying quality tools according to the type of inputs they require into: qualitative tools such as: Flow charts, Cause and effect diagrams, Multi- voting, Affinity diagram, Process action teams, Brainstorming, Election grids, Task lists etc. and quantitative tools such as Shewart cycle, Control charts, Scatter diagrams, Pareto charts, Sampling, Run charts, Histograms. Another perspective of classifying the quality tools focuses on the reactive, proactive approach, dividing the quality tools in: Reactive tools - Graphics, Histograms, Checklists, Pareto diagram, Brainstorming, Election Grids, Solutions matrix, and Proactive tools - Relations, Affinity, Tree, Matrix, Alternative, Arrow diagrams, Principal components analysis [8].

When it comes to the evolution of the Agile approach, it can be said that it underwent some significant changes and stages in its development. Studies such as those of Abrahamsson [9], Highsmith [10] and Cohen [11] focus on the aspect of Agile developing as a reactive response to the outdated traditional project management methods. Unfortunately, this approach considers Agile to have appeared only in the last couple of years but the agile ideas have been around for longer than that. In their paper, Larman and Basili [12] noted the usage of iterative and incremental approaches since the 1970's.

In his paper, Beck [13] suggested that the Waterfall model was first developed to help asses and define user needs by noting the user requirements. The Waterfall model tried to fix the problem of changing requirements by establishing well defined requirements and not allowing any changes to be made to it, but as practitioners observed, the requirements just couldn't be well defined in one swoop. [14] Incremental and iterative techniques evolved as a response to the rigidity of the user requirements, and tried to improve upon the Waterfall model by keeping the main aspects of it but introducing the notion of overlapping increments to help reduce the development time. [15] Now, the requirements are also noted before the development stage begins, but they are broken into chunks of standalone functionality known as increments. The development of each such increment can be overlapped with the others, creating a "multitasking" effect and saving time. And while the incremental development aimed to reduce the allocated development time, there are models that provide better handling capabilities for changing requirements and risk management. [16]

In the table below, we have outlined the main aspects and differences between the traditional approach towards project management and quality and the agile approach to these aspects.

Table 1. Key aspect of traditional project management approaches and agile development

Traditional project management tools	**Agile project management tools**
Rigid. Team members must adapt	Flexible. Can adapt to team members
Abstract	Visual and intuitive
Works best with large teams	Works best with small and medium teams
Reporting is an element of major importance	Constant feedback is important
Works best with teams that share a work Space	Can adapt to remote teams
Interest in the final result	Interest in each sprint release
Well established quarterly/semester/annual Meetings	Daily meetings
Schedules a series of events	Schedules releases and product versioning
Works with critical path calculations	Works with burn down charts
Formal management structure	Informal management structure
Dependencies between tasks can be dealt with on the go	Dependencies between tasks must be dealt with as soon as possible as they directly affect product releases
Quality within	
Traditional Quality Management	**Agile development**
Focused on delivering project objectives	Focused on constant improvement of delivered products
Adapts to established requirements	Adapts to customer changing demands
Comprehensive documentations is mandatory	Working software is more important than documentation
Progress is monitored through reports and periodical meetings	Progress is monitored through daily meetings and results
Defines working products criteria	Works with user stories
Reactive response to change	Proactive response to change
Sustainable development	Sustainable development
Complex solutions	Simplicity is essential
Considers that best decisions are made by professionals	Allows teams to self-organize and gives freedom to team members in choosing architecture, requirements and design
Gives power to team leader	Gives power to team members

3 Project lifecycles and agile systems

The lifecycle of a project is the period of time and the necessary steps to be made for a project to be completed. When talking about the lifecycle of a project, the following aspects need to be considered: the type of work that has to be completed, the expected deliverables, the people involved and criteria for the approval of each phase [17]. The lifecycle of a project management includes five phases: 1.Project Initiation: The stage in which the project and its scope is clearly defined, feasibility studies take place to decide whether or not the project will be launched. The purpose and requirements of the project will be noted in a project charter or a project initiation document (PID). 2. Project Planning: In this stage, the objectives of the project are outlined, goals are established, resources are allocated and estimates about deliverables are generated. Some of the documents of this stage include: Work Breakdown Structures (WBS), Gantt charts, Scope statements, Milestones and Risk management plans. 3. Project Execution: During this phase the deliverables are developed, resources are assigned, the project manager directs and manages the project execution, a tracking system is set up, tasks are completed and verified and the project plans are updated if necessary amongst other things. 4. Project Performance/Monitoring: This phase focuses on ensuring that the performance and progress is on track with the established goals, and in order to do so, project managers will use tools such as the key performance indicators (KPIs). It is recommended to pick 2 to 5 such KPIs to accurately determine the performance. Another important aspect to factor in at this stage is the budget and checking if the project is within the planned budget at all times. 5. Project Closure: This phase takes place when the deliverables have been made and all the tasks have been completed. During this stage, review meetings will take place to identify what went well and bad during the project development and to get feedback from the team members as well. [18].

When it comes to Agile, the life cycle also known as System Development Life Cycle (SDLC) is composed of six phases instead of five as was the case with traditional project management: Concept Phase, Iteration 0/Inception, Construction, Transition/Release, Production, and Retirement [13]. 1. Concept Phase: In this phase the business opportunity is defined, the viability of the project is determined and feasibilities studies are conducted. 2. Inception/Warm up: The phase in which the funding is obtained and general support is shown, the team is working with the stakeholders in order to determine the goal of the project, the team is built, the architecture for the system is modelled, and project estimates are established. 3. Construction phase: During this phase, the team is maintaining a close collaboration with stakeholders and the rest of the team, priorities for tasks are set and quality is ensured, at the end of every sprint there are deliverables available for testing and the deliverables are analyzed and designed. 4. Transition phase: The transition phase is also known as the "end game" because in this stage the solution is released for production, the final tests take place, the documentation is written, and the system is deployed. Other activities that can be included here are: reworks – if the product has any defects and training – for both users as well as team members. 5. Production phase: In this phase, all the focus is on keeping the system running and helping users adapt to it and be able to use it (can include help desk activities). During this phase a common practice is assigning operational personnel for running and monitoring the system. 6. Retirement: The retirement phase aims to remove a system release from the production process, and that happens mostly because of the new systems emerging and replacing the old ones, but other reasons can include: the system is no longer needed on the market, it has become obsolete or it is no longer supported [19].

4 Integrating the quality management and creativity tools in the agile system development life cycle (SDLC)

Based on all the research conducted, and the points detailed above, we proposed a new approach for the life cycle of agile project management inspired by the famous Plan Do Check Act (PDCA) method. Similar to it, our model has only four stages / phases: Initiation; Production; Delivery; Refactoring.

In the Initiation phase the following activities are conducted: creating the business model, value proposition, crowdfunding, market analysis, planning sprints and releases, finding the right team members, and a minimum of documentation.

The Production phase focuses on activities such as: development and task completion through reaching the established sprints; continuous testing, inspection & checkups with stakeholders and clients.

In the Delivery phase, the final version of the product is released to mass market and stakeholders; feedback is gathered and decisions regarding the future of the product are made. As it can be seen from fig. 1, this phase is important because it affects the future of the product. Once the phase is completed, there are three options available: closure, maintenance or refactoring. For the purpose of this paper, and because it is closer to reality, we have considered refactoring as the alternative that will be chosen.

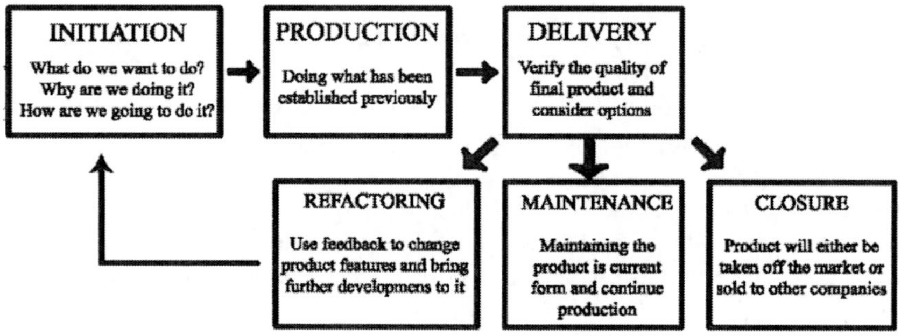

Fig. 1. Simplified System Development Life Cycle for Agile development projects

Optional Refactoring: The refactoring phase is perhaps the most particular one for the agile development because it links constant improvement with deliverables and it permits the team a flexibility in decision making. By going through the refactoring process, the team analyzes the feedback from the clients and decides what functionality to implement next taking into account their own ideas of course.

The main focus of this stage is deciding what to change about the current product and making sure that all the ideas are captured before making a decision. Once this stage is completed, the development process begins once more with the first stage of initiation.

This cycle can be repeated as many times as necessary in the agile development teams' environment.

Fig.2. Traditional quality management tools agile SDLC

Fig.3. Creativity tools adapted adapted for the for each stage of the agile SDLC

Given the fact that quality is a must for both traditional project management teams as well as agile ones, in fig. 2 we described what quality tools can be used, and how they can be adapted for each stage of the development process to ensure quality for each step of the process. When it comes to quality assurance in the agile environment, it can be said that even though the traditional quality management tools suffice, new tools have emerged that can be used for the same purpose. Creativity tools have been used lately as problem solution generators; hence they can complement the traditional quality management ones into finding the best option for quality assurance and control for each stage of any project especially in the case of agile development projects. [15] In fig. 3 we noted the main creativity tools which can be used for this purpose and depicted how each of them can be applied for the stages of the development process.

5 Case study and results

The proposed model has been implemented for the case of three startups: Vitraly, Check4Green [20] and Timetrak to observe and register the results, improvements or drawback if they occured. Next we will describe the implementation process for each stage of the system development life cycle as well as the results we achieved so far. Vitraly is a startup which brings the smart element to a regular mirror, turning it into a highly customizable mirror that acts as a "double agent". Besides displaying the world around it – like any other mirror would, it also displays widgets that the user wants such as: the clock, calendar, motivational quotes, weather and so on.

The team working on the product consisted of two engineers, two programmers and an economist. As was the case with the other two startups, for the project management of the Vitraly startup the free online project management platform called Freedcamp was used. For this, the team was created by inviting the members of the startup to join the Vitraly project group. In the initiation phase, the team held a brainstorming session combining this technique with the "Random Inputs" creativity techniques which requires the users to find as many correlations between unrelated items. Even though, a rough idea of a smart mirror was formulated already by the team during casual talks and interactions, a certain list of features and product capabilities wasn't clearly defined. During the brainstorming session, plenty of features for the smart mirror were identified such as for eg: it should display the temperature outside; it should display personalized messages that the user wants; it should allow the user to set alarms; it should turn off the widgets when the user is not around and display nothing, for power saving purposes, etc. All these features were noted down by the product owner and were later translated into user stories and tasks for each team member. During this stage, the team noted the lack of support for ideas, in the sense that the ideas captured during the brainstorming session couldn't be stored on the spot in the Freedcamp platform unless the team members inputted them in the discussion board afterwards which was time consuming and useless. After the ideas were selected and the team clearly defined the user requirements, these were noted on the platform by the product owner and later, tasks were derived from them by the scrum master and assigned to team members whilst defining the deadlines and allocated resources as well. Still in this stage, in order to better and easier define the business model, the team used the 40-20-10-5 method of problem solving which states that when dealing with a problem, a person must explain it in up to 40 words, then cut it down to 20 words, then 10 and lastly, to 5 words. These are the root cause of a problem and also the solution to that problem [21].

During the production phase, the product owner made sure that everything was going according to plan and with the help of the Kanban list view of the tasks that the Freedcamp platform offers, also supervised in real time the completion of the tasks and if these were in schedule with the established timeframes. Still during this stage, constant small releases were made to the public, more specifically, whenever a new widget was developed it was already deployed in order to gather the end user's opinions and adjust accordingly. Freedcamp didn't offer any kind of support for this activity, so once again, the changes in both customer demands and product features couldn't be noted only written as new tasks whilst marking the old tasks as completed or simply deleting them. Either way, a proper time estimate of how much time a task took if it was affected by the end user changes couldn't be made using this platform.

As soon as the product was finished it was launched on the market - encapsulating the delivery phase, through the help of a crowdfunding campaign in order to both get funding for further developing the product and producing more of it as well as to maintain a high level of feedback received from the customers who are more actively involved in the case of crowdfunding campaigns. Given the fact that people loved the idea but wanted improvements such as: making it available in smaller dimensions, more widgets and layouts that can be customized from the mobile phone, the team decided to step into the refactoring phase and decide what to deliver next though the help of the creativity methods noted in Figure3. Throughout the development lifecycle of this project as well as the other two: Check4Green and Timetrak, the usage of the free project management platform called Freedcamp proved to have some gaps in managing projects which are using the Agile approach. In order to fill these gaps and reach the established goals, the development team had to use a combination of quality management and creativity tools described above and noted in fig. 3.

5 Conclusions

Keeping in mind the principles of the Agile Manifesto, especially the one referring to less documentation and more valuable working code, we have designed a system development life cycle that is easier to understand, apply and monitor, inspired by the Plan-Do-Check-Act (PDCA) cycle. Given the fact that a product developed by an agile team does not necessary end after the delivery to the customer and that it can go through changes both during the production phase and the post-delivery one, the whole process can become cyclic a couple of times rather than having a well determined beginning and ending. The similarities between this and the PDCA cycle allowed us to define a system that depicts the lifecycle of an agile development environment more closely to reality and in accordance to the Agile Manifesto itself.

In order to use this model for a project, one must first define each stage of the development process and and know which stage they are currently facing then apply the following for each step. In the Initiation stage, the following quality tools can be applied: PERT chart; Benchmarking; Kano; Design of Experiments and Flow Charts. As for the creativity tools, we can have: Mindstorming; Brainstorming; Random Inputs and Inversion. In the Production stage, the quality tools used in the model are: Gantt Chart; QFD; Arrow Diagram; Eight disciplines and the creativity tools are: Kanban lists; Starbursting; Concept fan and Spaghetti Diagram. The Delivery phase encapsulates quality tools such as: Checksheets; Control Charts; FMEA and SMART matrix and the creativity tools as: Plus, minus, interesting; Brainstorming; Triz and 5 Why's. The last optional stage named Refactoring, quality tools such as: Decision Matrix; Histogram analysis; Idea Creation Tool; Success and Effect Diagram; Multivoting are used along with creativy tools such as: Six Thinking Hats; Future Wheels, Forced Associations and the 40-20-10-5 method.

Besides all this, keeping in mind the element of flexibility and creativity that the agile trend brings, we have incorporated creativity techniques for each stage of the development process so that it can be in line with the whole principle of this trend and adapt easily to any sort of agile development team. Throughout the whole production process and at the end of it, the promised quality will be monitored and delivered with the help of a combination of the traditional quality management tools integrated with the modern creativity tools used for problem solving purposed and quality assurance.

References

1. P.B.Crosby, *Quality is Still Free: Making Quality Certain in Uncertain Times* (McGraw-Hill,
2. 1996)
3. K. Liang, Mod.Appl.Scient, **4**, 9 (2010)
4. K. Ishikawa, Introduction to Quality Control (Springer Netherland, 1989)
5. A.Q. Press, *Seven Basic Quality Tools* (ASQ Quality Press, 2010)
6. R.H. Fouad, A. Mukattash, JJMIE, **4**, 6 (2010)
7. L. C.R. Carpinetti, T.Buosi, M.C. Gerólamo, Business Process Management Journal, **9**(4), 543-554 (2003)
8. A.C. Ionică, *Managementul calității* (Universitas, 2006)
9. R. Moțoiu, *Ghid pentru managementul calității și calitatea managementului anilor 2000* (Fiatest, 2010)
10. P.Abrahamsson, O. Solo, J.Ronkainen, J.Warsta, VTT technical Research Centre of Finland (2002)
11. J.Highsmith., *Agile Software Development Ecosystems*, (Addison-Wesley Longman Publishing Co., 2002).
12. D.Cohen, M.Lindvall,. P.Costa, Advances in Computers, 1-66 (2004)
13. C.Larman, V.R.Basili, IEEE Computer Society **36** (6), 47-56 (2003)
14. K..Beck, *Extreme Programming Explained: Embracing Change*, (Addison–Wesley, 1999)
15. K.Beck, *Embrace change with extreme programming*, 70–77, (IEEE Computer, 1999)
16. A.Cockburn, Cutter IT Journal, 6–12 (2002)
17. Cockburn A., Highsmith J., IEEE Computer, 120–122 (2001)
18. T.Kendrick, *Identifying and Managing Project Risk: Essential Tools for Failure-Proofing Your Project*, (Amacom, 2003)
19. R.K.Wysocki, *Effective Project Management* (Wiley, 2014)
20. M.Cohn, *Agile Estimating and Planning* (Prentice Hall PTR, 2005)
21. R.Dovleac, A.Ionica, M.Leba, Proceedings of Review of Management and Economic Engineering, 32-39 (2016)
22. A.Ionica, M.Leba, E.Edelhauser, Transformations in Business & Economics, **13**, 2B, 697-716 (2014)

Conceptual modelling of human resource evaluation process

Doina Olivia Negoiţă[1], *Anca Alexandra* Purcărea[1], *Andreea* Dumitrescu[1,*], and *Andronicus* Torp[1]

[1]Faculty of Entrepreneurship, Business Engineering and Management, Department of Management, University POLITEHNICA of Bucharest, 313 Independenţei Street, 060042, Romania

Abstract. Taking into account the highly diverse tasks which employees have to fulfil due to complex requirements of nowadays consumers, the human resource within an enterprise has become a strategic element for developing and exploiting products which meet the market expectations. Therefore, organizations encounter difficulties when approaching the human resource evaluation process. Hence, the aim of the current paper is to design a conceptual model of the aforementioned process, which allows the enterprises to develop a specific methodology. In order to design the conceptual model, Business Process Modelling instruments were employed - Adonis Community Edition Business Process Management Toolkit using the ADONIS BPMS Notation. The conceptual model was developed based on an in-depth secondary research regarding the human resource evaluation process. The proposed conceptual model represents a generic workflow (sequential and/ or simultaneously activities), which can be extended considering the enterprise's needs regarding their requirements when conducting a human resource evaluation process. Enterprises can benefit from using software instruments for business process modelling as they enable process analysis and evaluation (predefined / specific queries) and also model optimization (simulations).

1 Introduction

The evolution of the organizational structures, along with the increased complexity of roles and responsibilities undertaken by employees have generated the utter need for enterprises to develop an integrated system to exploit human resource talent [1]. Within this context, the concept of "human resource management" has emerged and continuously developed ever since, due to the strategic role of human capital in terms of professional and behavioural attributes [2-3]. The pressure to create and improve a system to focus on employees as a key resource for achieving organisational performance, has highlighted the importance of HR assessment.

The definition of "performance" takes different forms, depending on the subject it relates with, however, regarding human resource, the concept refers to the degree of accomplishment of a task in terms of "accuracy, completeness and efficiency" [4]. Hence, it is related to

* Corresponding author: dumidreea@gmail.com

specific performance standards and organization's expectations regarding employee's ability to achieve them.

2 Methodology

Based on the underlying concepts of human resource management, the authors have proposed a conceptual model of the aforementioned process with three alternative paths. The conceptual model is focused on the employees' performance evaluation sub-process, for which an in-depth analysis is provided. Therefore, the methodology of the current paper consists of workflow design using Business Process Modelling instruments - Adonis Community Edition Business Process Management Toolkit based on ADONIS BPMS Notation. The modelling software enables retrieving an integrated view of the human resource management, including: activities, risks, objectives, information and knowledge.

3 Conceptual model of human resource performance evaluation process

3.1 Integrated overview of human resource management model

The sequence flow contained by the human resource management model (Fig.1) [5] starts by performing the first activity *"Analyse the human resource requirements within the enterprise*". This task allows for continuing the process based on the current level of existing employees, which is assessed within the decisional node *"Existing employees (E) vs. Required employees (R)?"*. Hence, the enterprise should analyse their requirements in terms of human resource and follow the path which is according to their needs.

The first path corresponds to an equilibrium between the number of existing and required employees, which ends the process, respectively the End Event. When the number of current personnel exceeds the enterprise's needs in terms of human resource, then the second path is selected and the task "Dismiss employees" is executed. However, the enterprise may require more employees, which involves considering the third path, which comprises the following sub-processes: *"Recruit", "Select", "Guide", "Train" employees* and *"Evaluate employees"* [6-7].

The workflow of the latter sub-process is described in detail in section 3.2. Hence, based on the results emerging from human resource performance assessment, the model is divided into two paths, respectively the decisional node *"What are the evaluation results?"* is executed. Favourable results of the evaluation allow for keeping the employee within the enterprise, hence the opportunities for carrier development are thoroughly analysed in order to identify whether the employee is better suited for a higher job position, if available.

The decision of advancing the employee on the job hierarchy is retrieved from the decisional node *"Should the employee be promoted?"*, which retrieves the current status of vacancies and carrier development opportunities within the organisation. Therefore, either *"Promote employee"* or *"Maintain the employee on current position "*, the enterprise should provide training assistance in order to continuously improve the personnel's performance. The unfavourable results generated by the assessment sub-process mark the end of the model as the task *"Dismiss employee"* is fulfilled.

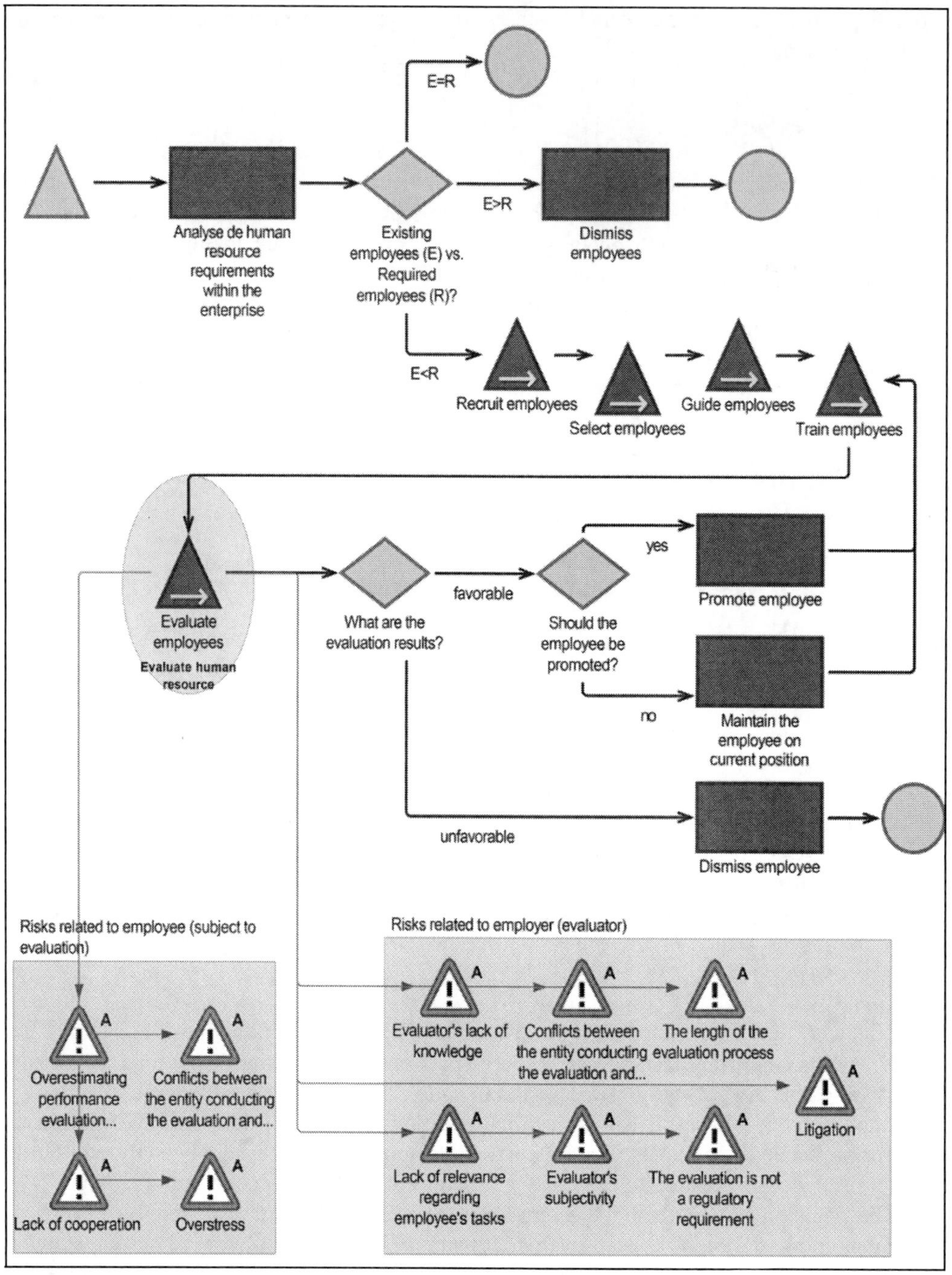

Fig. 1. Conceptual model of human resource management process.

When conducting the performance evaluation sub-process, both the employee and the employer are subject to different risks, which are depicted in Fig. 1 and described in Table 1. Understanding the risks related to the assessment activities has a crucial role, as it facilitates acknowledging the multiple factors which may influence the results of the human resource evaluation.

Table 1. Risks related to human resource performance evaluation [8-11].

No.	Risk	Description
R1.	Overestimating performance evaluation (employees' opportunism)	The employee has a tendency to change behaviour when the evaluation due date is approaching (the employee works longer hours; shows more interest in work results etc.).
R2.	Conflicts between the entity conducting the evaluation and the employee subject to evaluation	Employees consider the evaluation results as criticism, as opposed to an opportunity to identify and further improve current weaknesses. The evaluators focus on negative points. The employees are biased regarding the aim of the evaluation (they link the performance appraisal with a negative event as they consider the evaluation results are undervaluing the real job effort and accomplishments)
R3	Lack of cooperation	The employee is not motivated to contribute to the appraisal/ does not understand the importance (the added value) of the evaluation. Lack of communication between the assed and assessor / poor feedback (the employee does not understand the evaluation's results and/or the employee does not receive specific details regarding his current rating).
R4	Overstress	The employees become too preoccupied by the potential countermeasures which they would face if the evaluation results are unsatisfactory (dismissal, salary / incentives cut etc.)
R5	Evaluator's lack of knowledge	The evaluator does not apply the assessment methodology accordingly/ conducts the evaluation process poorly/ irrelevant results. The evaluator is not qualified for undertaking employees performance appraisals.
R6	Lack of relevance of evaluation criteria regarding employee's tasks	The evaluation methodology is not suited for particular job categories, which have specific assessment requirements (e.g. high risk jobs: air traffic controllers, mining, handling dangerous substances, services which involve exposure to dangerous behavior etc.)
R7	Evaluator's subjectivity	The system of appraisal (performance ratings) is too strict, hence the evaluators should provide input based on meetings and discussions. The evaluator favours an employee, while mistreating other employees (a biased review of employees performance due to lack of evaluator's independence).
R8	The length of the evaluation process	Process timeline is ineffective, as it is either too long (it becomes irrelevant as the employee becomes weary) or too quick (not enough time spent for establishing key and/ or emerging issues related to employees job performance)
R9	The evaluation is not a regulatory requirement	The performance appraisal is not performed on a regular basis, hence no continuity of the process.
R10	Litigation	The employee instigates lawsuit considering the performance appraisal is unfair/ based on discriminatory criteria.

Therefore, a dysfunction of HR assessment sub-process introduces flaws when designing employees' development programs. This may result in an inefficient training program due to lack of knowledge regarding the real features of the person's traits.

3.2 Conceptual model of human resource performance evaluation process

The performance evaluation sub-process starts due to trigger *"New job/ Periodical evaluation due time"*, which enables to *"Identify evaluation's objectives"* (Fig.2). The assessment objectives may be related to the labour system within the enterprise (people, physical environment, social connections, technology etc.), to new work methods/ techniques/ instruments etc.

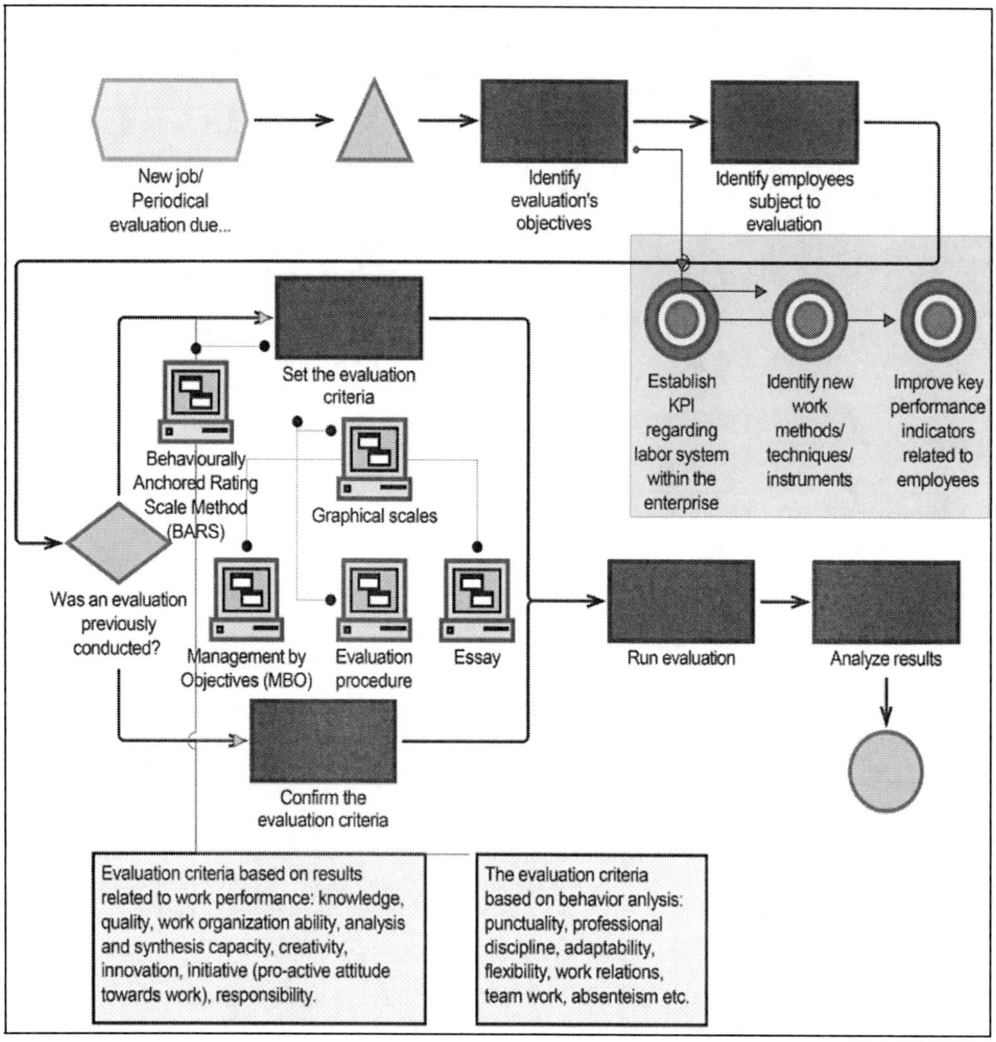

Fig. 2. Conceptual model of human resource performance evaluation process.

Based on the assessment objectives, a specific category of employees is selected in order to run the performance evaluation. Therefore, the entity conducting the process should choose and apply the proper methodology depending on the enterprises requirements. The evaluation criteria reveal different aspects regarding the human resource behaviour within the labour system, hence the choice regarding the methodology which should be used (Behaviourally Anchored Rating Scale Method–BARS, Graphical scales, Management by Objectives – MBO, essay, internal performance evaluation procedure etc.) is directly dependent on the type of information and knowledge required by the enterprise. The human resource evaluation methods have different applicability area, hence in Table 2 the authors have emphasized the limitations of the three main methodologies used for performance appraisal by performing a comparative analysis regarding the use of graphical scales, Behaviourally Anchored Rating Scale Method and Management by Objectives.

Table 2. Comparative analysis of human resource performance evaluation methodologies.

Criteria	Scales	BARS	MBO	Risk
Standardized questionnaire	x	x	-	Lack of precision (the questionnaire's content lack of flexibility)
Action-oriented	-	x	x	Time-consuming for developing plans for performance improvement
Based on critical incidents	-	x	-	Requires more effort in order to identify extreme (positive/ negative) behavioural cases
Evaluation criteria developed based on manager-employees interaction	-	-	x	Unrealistic targets for human resource performance (optimistic approach when defining evaluation objectives)
Based on real facts	x	x	x	Lack of knowledge regarding employee's potential behaviour in a specific scenario
Focused on providing rewards	-	x	x	The employees are focused more on evaluation results rather than work itself.

The risks related to human resource performance evaluation methodologies highlight the problems which enterprises may encounter when conducting this particular process, due to high diversity of factors which define the employees' work results - knowledge, analysis and synthesis capacity, creativity, innovation, initiative (pro-active attitude towards work), responsibility, punctuality, professional discipline, adaptability, flexibility etc.

4 Conclusions and future research directions

The enterprises' need for predicting the employees' performance has generated the high interest for scholars to develop HR assessment methodologies in order to prevent and/ or diminish the risks which may encounter both the employee and the employer when conducting performance appraisals. Therefore, the authors propose as future research direction the development of an innovative assessment method based on measuring the electromagnetic field of human being using *ElectroPhotonic Imaging device and software* [12].

The main benefits of this new method lie in the ability to run "on-the-job" measurements, hence conducting quantitative analysis, especially regarding highly stressful jobs. Therefore, this method will provide valuable input related to employees who require high stress resilience, an increased capacity to adapt to complex situations on a daily basis, a high level of emotional tension etc.

By examining the relationship between the results retrieved by *ElectroPhotonic Imaging device* and the employees' personality traits, the method allows for long-term performance enhancement and an overall organizational working system improvement.

References

1. E.A.M.E. Maycock, E. Allaputa, G.D., Waripanye, R. Geraghty, S., Chikafa, J. of Bus. and Man. **16,** 6 (2015), http://iosrjournals.org/iosr-jbm/papers/Vol17-issue6/Version-1/B017611421.pdf
2. J.S. Itika, *Fundamentals of human resource management* (University of Groningen, 2011), https://openaccess.leidenuniv.nl/bitstream/handle/1887/22381/ASC-075287668-3030-01.pdf?sequence=2
3. A.A. Purcărea, D.C. Şerban, O.I. Negoiţă , O. D. Negoiţă, Revue Européenne Du Droit Social, **20,** 3 (2013)
4. http://www.businessdictionary.com/definition/human-performance.html
5. A.A. Purcărea, *Human Resource Management -Training* (PUB, 2012)
6. R.L. Mathis, P.C. Nica, C. Rusu, *Managementul resurselor umane* (Editura Economica, 1997)
7. A.A. Purcărea, *Management modern* (Editura Niculescu, 1998)
8. http://www.simplehrguide.com/performance-management-risks.html
9. Delloite, *Risk committee performance evaluation* (Deliotte&Touche, 2014), https://www2.deloitte.com/content/dam/Deloitte/za/Documents/governance-risk-compliance/ZA_RiskCommitteePerformanceEvaluation_24032014.pdf
10. http://www.talentculture.com/conduct-a-performance-appraisal-with-minimal-risk/
11. http://legacy.juniata.edu/services/hresources/documents/WhitePaperPerformanceReviewEvaluations.pdf.
12. A. Torp, Z.I. Marosy, A.A. Purcărea, SEA **2,** 3 (5) (2014)

MATEC Web of Conferences **121**, 05005 (2017) DOI: 10.1051/matecconf/201712105005
MSE 2017

Experimental research regarding the study of surface quality of aluminum alloys processed through milling

Alina Bianca Pop[1], and *Mihail Aurel* Ţîţu[2,*]

[1]SC TechnoCAD SA, Vasile Alecsandri no. 72, code 430351, Baia Mare, România
[2]Lucian Blaga University of Sibiu, Victoriei Street no. 10, code 550024, Sibiu, România

Abstract. The aim of this paper is to present the necessary steps to conduct a design of experiment with application on the aluminum alloy surface quality obtained by end-milling process. The machined surface quality is influenced by the cutting process parameters, and especially by their variation. The research methods used in this paper are the Taguchi method - to determine the cutting parameters influence, on the surface quality, and the full factorial design of experiment to confirm the results obtained by Taguchi's method and to determine the percentage of the parameters influence involved in the process.

1 Introduction

The cutting process is the one of the most common processing techniques in the industry. There are a multitude of studies that are investigating it, both on the industrial side and also on the academic one.

In the end milling process on the workpiece a considerable number of factors are acting, like: the material microstructure, the parameters of the cutting tool geometry, the cutting regime or the tool wear etc. Due to this fact, to obtain a high surface quality, it is necessary to study the influence degree of the mentioned factors and also, to find the optimum methods to fulfill the intended purpose.

The experimental research involves the organization of a process or phenomenon under some well-established conditions, eliminating as much as it is possible the disturbance influences.

The overall objective is to make some measurements in order to determine the value or the intensity of physical measure of the considered phenomenon or process.

With this measurements can be deduced some connections between the factors and the laws that govern the phenomenon.

The research results, the cost and the research program duration, are depending on the methods choosing, on the investigation techniques, on the experimental stages organization and also on the results processing [1].

* Corresponding author: mihail.titu@ulbsibiu.ro

The purpose of this paper is to present the necessary steps of the design of experiments to study the influence exerted by the cutting process parameters on the surface quality machined by end milling of aluminum alloys, in this case the Al7136 alloy.

Similar research conducted on the design of experiments were carried out by [2] who determined the process parameters influence exerted on the impact forces, using the full factorial method.

Among the previous research are the investigation of the effect of the feed rate on the surface roughness and the reduction of processing time [3], and also the mathematical modeling of a regression equation of the surface roughness obtained by end milling [4].

2 Experimental Research Management

To conduct an experimental research, Montgomery recommends to follow the steps mentioned [5]:

- The recognition of the problem statement;
- The response variable selection;
- The choice of the factors (process parameters), levels, and ranges;
- The choice of the experimental design;
- The experiment performing;
- The statistical data analysis;
- Conclusions and recommendations.

2.1 The recognition of the problem statement

After the bibliography field completing, the research problem can be formulated as follows: „The study of the arithmetic mean deviation R_a of the surface profile, resulted by end milling process of the 7136 aluminum alloy, depending on the cutting process parameters."

2.2 The response variable selection

On this step, the response to analyze is established: the arithmetic mean deviation R_a of the surface profile.

2.3 The choice of the factors (process parameters) involved in the process

Considering the paper purpose, the factors to study are:

- Cutting speed (v [m/min]);
- Cutting depth (a_p [mm]);
- Feed per tooth (f_z [mm/tooth]).

Table 1. The established values of the cutting process parameters

Cutting regime	**Cutting speed [m/min]**
	495 530 570 610 660 710
	Cutting depth [mm]
	2 2.5 3 3.5 4 4.5 5 5.5 6
	Feed per tooth [mm/tooth]
	0.04 0.06 0.08 0.11 0.14

The choice of levels for each machining parameter was made in accordance with the SECO Tools manufacturer indications and also according to the technological possibilities of the tool, cutting inserts and CNC machine.

These factors are exerting their influence on the analyzed response, and their settled values to performing the experiments are shown in Table 1.

2.4 The choice of the experimental design

The experiment type to conduct the research is the full factorial design. With the established cutting parameters and their settled levels, a number of 5x6x5=150 experiments, are resulting.

After the experiments performing, the measurements will be collected.

The total number of R_a measurements results by the each replicated experiment. In order to conduct a data statistical analysis, each experiment requires a number of 3 to 7 replications [5].

In this research, the statistical analysis will be performed based on 7 measurements, made for each experiment. It follows a number of 150x7=1050 required measurements.

Based on the aspects mentioned above, first the Taguchi method shall be used to determine the influence levels of each parameter involved in the studied process (screening).

Starting with the process parameters values indicated in table 1 (settled based on the technological possibilities of the tool and of the cutting inserts), to conduct the research based on a complete factorial experiment, to use the Taguchi method it will be consider just 2 levels (the minimum and the maximum values) of each control factor, as it is shown in table 2.

Table 2. The control factors and their settled levels

Abbreviation	Parameter		Value 1 minimum	Value 2 maximum
A	1	Cutting speed	A1=495,00 [m/min]	A2=660,00 [m/min]
B	2	Cutting depth	B1=2 [mm]	B2=4 [mm]
C	3	Feed per tooth	C1=0,04 [mm/tooth]	C2=0,14 [mm/tooth]

The cutting depth values were chosen according to the limiting values of the CNC machine, whose technological possibilities are limited to a maximum rotational speed of 15.000 rev/min, with a safety rotational speed of 13.500 rev/min and the limiting safety on cutting depth of 4 mm.

Further, based on the results from Taguchi method, the full factorial design will be used in order to obtain the percentage influence of each studied parameter on the surface roughness.

2.5 The experiment performing

The effective experiments begun with the study of 7136 aluminum alloy behavior, according to the proposed objectives under the variation of the cutting regime parameters.

The experiment was performed using a standard set of tools for aluminums machining - 16 mm End milling cutter milling with 100% tool engagement - SECO R217.69-1616.0-09-2AN, holding two indexable cutting inserts XOEX090308FR-E05, H15.

The machine used for the milling tests is a HAAS VF2 CNC.

The optical microscope Micro-Vu VERTEX 310, was used to make an optical determination and evaluation of the finished surface.

To analyze the surface roughness for different machining conditions, it was used the portable surface roughness tester - Mitutoyo SURFTEST SJ-210, with a resolution of 0.002 μm.

DOI: 10.1051/matecconf/201712105005

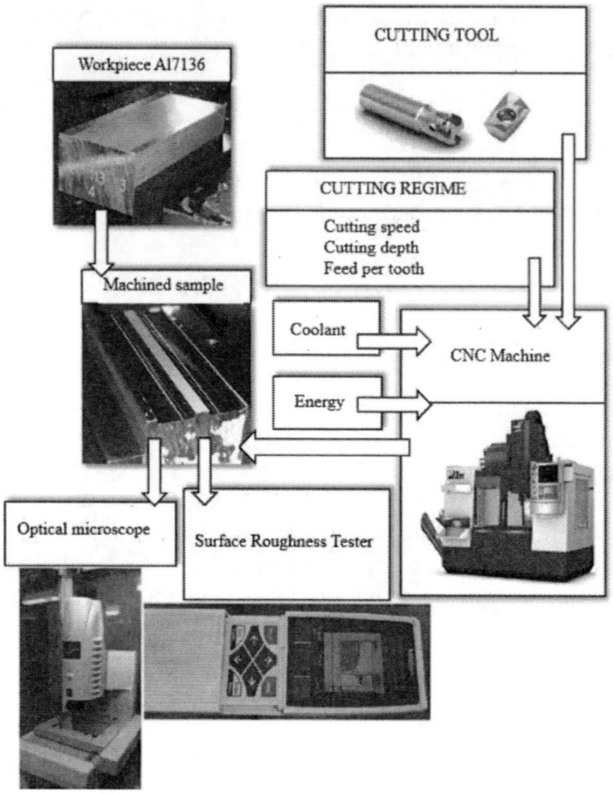

Fig. 1. Experimental stand [6]

With this experimental stand, can be fulfilled the main necessary requirements to achieve the sets objectives and also to obtain the experimental data on which can be analyzed the cutting parameters influence (cutting speed, cutting depth and feed per tooth) exerted on the machined surface roughness (fig. 1).

In the first phase, based on the Taguchi's method, data presented in table 2 will be used and a number of 8 experiments will be perform and for each experiment a set of 3 measurements will be associated, thus aiming to reduce the measurement error (Table 3).

Table 3. The values of 8 experimental setup with the measured roughness Ra values

No. exp.	Control factors			Surface roughness results Ra[μm]			
	v [m/min]	a_p [mm]	f_z [mm/tooth]	R_{a1}	R_{a2}	R_{a3}	R_{amed}
1	495	2	0,04	0,083	0,063	0,063	0,070
2	495	2	0,14	0,173	0,168	0,124	0,155
3	660	4	0,04	0,44	0,341	0,456	0,412
4	660	4	0,14	0,115	0,11	0,098	0,108
5	495	4	0,04	0,073	0,064	0,075	0,071
6	495	4	0,14	0,394	0,385	0,247	0,342
7	660	2	0,04	0,079	0,055	0,088	0,074
8	660	2	0,14	0,12	0,095	0,181	0,132

For a better understanding of the roughness resulted values it was used the arithmetic average method.

Given the pursued aim - to determining the factors levels to optimize the minimized quality criteria (as the roughness is lower it is better), after the Taguchi's method applying it was determined the configuration that consists of the most favorable factors levels to optimize the quality criteria: A2 B1 C1 A1B1 A1C2 B1C1.

Further, in order to determine the influence percentages of the cutting parameters involved in the process the full factorial experiment was used. In this situation, the input data consist of all cutting parameters values shown in Table 2.

A number of 6x5x5 = 150 experiments were carried out, which were 7 times replicated and a total of 150x7 = 1050 measurements was carried out.

Given these aspects, the percentages of the cutting process parameters influence exerted on the surface roughness were determined (Table 4).

The table 4 results shows that the cutting speed has the greatest influence on the surface quality obtained by end milling.

Table 4. The percentage influence of the process parameters exerted on the surface roughness [4]

Source	Freedom degrees	Square sum	Percentage contribution
Regression model	7	199,482	86,12%
Cutting speed A	1	194,322	83,89%
Cutting depth B	1	0,040	0,02%
Feed per tooth C	1	1,157	0,50%
A x B	1	3,093	1,34%
A x C	1	0,316	0,16%
B x C	1	0,366	0,14%
A x B x C	1	0,188	0,08%
Error	827	32,156	13,88%
Fitting error	143	18,735	8,09%
Pure error	684	13,421	5,79%
Total	834	231,638	100%

2.6 The statistical data analysis

The statistical data analysis has a great impact on the experimental research, because can certified that the studied process values are real and they are not affected by the system errors or measurement errors.

The statistical analysis of the experimental data consists of:
1) Verifying the aleatory data character;
2) Verifying the normality distribution of the experimental data;
3) Identifying the affected data by aberrant errors.

After performing the statistical data analysis, the experimental data modelling can be carried out in order to obtain those mathematical models which are best describing the studied process.

With the R_a measurements of the established experimental domain, a regression equation can be determined, which describe the surface roughness obtained by end-milling according to the involved process parameters values.

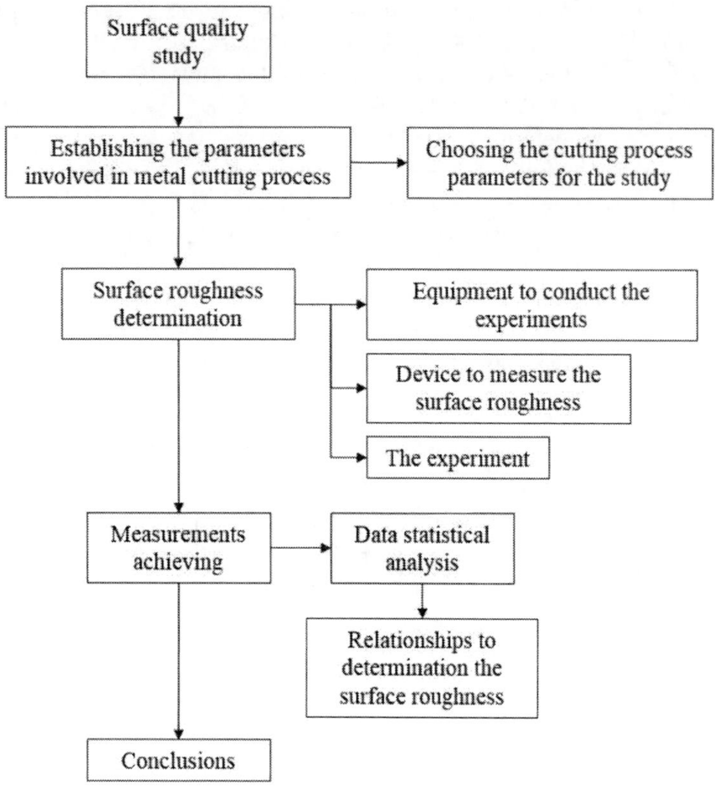

Fig. 2. Chart stages planned in the design of experiment to study the surface roughness.

In figure 2 is presented the planned stages of the design of experiment to study the surface roughness.

3 Conclusions and recommendations

To carry out the experimental research, it is necessary to follow a series of steps, broadly presented in scientific literature. The experimental research aim, is to improve the studied process with the preliminary obtained results.

The surface roughness is depending on certain process parameters, specifically on the cutting speed.

In the first phase three process parameters were determined: cutting speed, feed per tooth and cutting depth.

For the established experimental domain a number of 150 experiments it was necessary to carried out. In order to achieve the statistical data analysis of the experimental results obtained for each experiment, seven measurements were made; a number of 1050 measurements was resulted.

To determine the influence levels of cutting parameters was used the Taguchi's method. Then, to determine the influence values of the cutting parameters on the surface roughness was used the full factorial design.

So, using Taguchi's method for the process screening, the number of necessary experiments decreased from 150 to 8 experiments, on this way savings time and money. The statistical data analysis of the obtained results represent an important part of the design of experiment, after which the necessary steps to obtain a regression equation can be made.

References

1. V., Năsui, A., Coteţiu, R., Coteţiu, M., Lobonţiu, N., Ungureanu, *Bazele cercetării experimentale a actuatorilor electromecanici* (Editura Universităţii de Nord, Baia Mare, 2007)
2. N., Medan, M., Lobonţiu, M., Bănică, MATEC Web of Conferences **94**, (2017)
3. A.B. Bonţiu Pop, SCIENTIFIC BULLETIN, Serie C, Fascicle: Mechanics, Tribology, Machine Manufacturing Technology, XXVIII, 12-17 (2014)
4. A. B. Pop, M. A. Ţîţu, IOP Publishing, IOP Conf. Series: Materials Science and Engineering **161** (2016)
5. D., Montgomery, *Design and Analysis of Experiments* (Eighth Edition ed. Hoboken: John Wiley & Sons, Inc, 2013)
6. A. B., Bonţiu Pop, *Calitatea suprafeţelor la aşchierea aliajelor de aluminiu cu freze cilindro-frontale*, (Teza de doctorat, Baia Mare, 2015)

Energy recovery from Municipal Solid Waste in EU: proposals to assess the management performance under a circular economy perspective

Elena Cristina Rada[1,*], *Lucian-Ionel* Cioca[2], and *Gabriela* Ionescu[3]

[1]DICAM, University of Trento, Trento, Italy, & DIIM, University Lucian Blaga of Sibiu, Sibiu, Romania
[2]DIIM, Lucian Blaga University of Sibiu, Sibiu, Romania
[3]DPUE, Politehnica University of Bucharest, Bucharest, Romania

Abstract. In 2015 the European Commission issued a package of documents on Circular Economy concerning an integrated revision of legislative proposals on waste management. The aim was to stimulate a European transition towards a circular economy concept, which is expected to foster competitiveness, sustainable economic growth and new jobs generation. Three indicators are proposed in this paper to contribute to the assessment of the energy recovery management performance from MSW in a scenario of circular economy: a) referring to MSW directly used (RMSW) or indirectly used (SRF) as input of thermochemical plants, an indicator can be the percentage of waste having LHV > 13MJ/kg; b) referring to the MSW directly or indirectly used as input of thermochemical plants, the percentage of waste having ash recovered; c) referring to food waste, percentage of this stream sent to anaerobic digestion. The above indicators, proposed and discussed in this paper, have to be integrated with other ones in order to complete the quantification of the role of MSW management in term of energy recovery under a circular economy strategy. It is not the aim of the present paper to give a comprehensive solution to this complex issue.

1 Introduction

Currently, world cities generate about 1.3-1.9 billion tonnes of solid waste per year. This volume is expected to increase to 2.2 billion tonnes by 2025. Globally solid waste management costs will increase from today's annual 205.4 billion $ to about 375.5 billion $ in 2025 [1]. Worldwide, of the MSW collected, 19% is recycled, 11% is used in energy recovery and the rest ends up in landfills or dumps [2]. In this context, there is a stringent need for the development and implementation of sustainable sociotechnical systems. Since 1975, when the European Union (EU) first introduced its waste hierarchy, our society has been undergoing to a transition process, from only manufacturing and production activities to a systematic and environmental thinking [3-4]. In the last decades, the linear 'take-make-

* Corresponding author: elena.rada@unitn.it

use-dispose' economy approach has been used, relying on the production of large quantities of low-cost, easily accessible material and energy, a model that has reached-up to the end of the line, due to its many environmental challenges: the resource depletion, waste generation and emissions [5-6]. The sustainability role is to address economic, social and environmental issues of present and future generations, in the perspective of a balanced and systemic integration of intra and intergenerational economic, social, and environmental performance [7]. To overcome the environmental challenges of the conventional linear economy model and moving towards sustainability, the concept of Circular Economy (CE) has become one of the most modern approaches to address economic growth and environmental sustainability, aiming to the implementation of closed loop material flow in the whole economic system [8]. In line with the 3R of the environment, the CE relays on the circular material flow involving recovery, reuse, recycling and remanufacturing of products and the use of raw materials and energy through multiple phases [9,10].

In the light of the above mention, in 2015 the European Commission proposed a set of recommendations on the current waste legislations linked with CE. The official aim is "to stimulate Europe's transition towards a circular economy which will boost global competitiveness, foster sustainable economic growth and generate new jobs" [11]. The EU programme of action contains measures involving the whole cycle: "from production and consumption to waste management and the market for secondary raw materials" [11]. The common action plan proposed by the EU for CE, considers specific key targets for all the member states such as: to reach up to a recycling common target of 65% of municipal waste and 75% of packaging waste by 2030; to reduce landfill to maximum of 10% of municipal waste by 2030; to prohibit landfilling of separately collected waste through the promotion of economic tools; to have clear and simplified definitions, methods and standards for recycling rates all over EU; to have concrete forms that promote the close-loop of the re-use material option; to stimulate EU towards a sustainable market place by offering greener products and service programs that support recovery and recycling systems.

If we consider the CE a priority in the EU waste management, the above mentioned "package" underlines the importance of reuse and recycling before energy recovery. The target indeed is to close the loop of product lifecycles by an increase and an optimization of and re-use. It is important to point out that this target is seen also in terms of favorable balances for both the environment and the economy.

In this paper, the effects of material recovery maximization are discussed in terms of effects on the strategies of energy recovery.

2 Role of energy recovery from MSW in a frame of material recovery maximisation

At the moment the linear economy is being gradually replaced by the CE that aims towards a zero waste framework [12-16] The main factors that are underlying the bases of any sustainable waste management are the reliable and up-to-date information and their trends regarding the quantitative and qualitative characteristics of the MSW stream, considering the demographic changes, the socio-economic and technological maturity, environmental policies, decision-makers and the interests of the parties involved in the process [17-19]. The most advanced and sustainable, Integrated Municipal Solid Waste Management Systems must maintain an equilibrium between the separately collected (SC) waste fractions and the Residual Municipal Solid Waste (RMSW), both of them being suitable for material and energy recovery. The environmental citizenship and waste services play an important role in the RMSW treatment options [20-21]. Furthermore, in some regions the political differences and unstable legislative framework can recede, obstruct or cause modifications from the original waste management plan, leading to the community confusion, slowing the implementation of the industrial recovery processes and usage of valuable secondary products that can be put on the market, delaying the environmental and economic impact.

The different waste collection systems (through containers- common or separate by waste fraction, door-to-door or drop-off collection) affect not only the quantity of the MSW stream collected but also its mass distribution. In these context, studies have identified three individuals' indicators [22-30]:

- The organic waste (green and biodegradable household waste) source separated;
- Packaging waste source separated: paper and cardboard, plastics, glass, metals (e.g. aluminium, iron), wood;
- RMSW (the mixed waste not source separated).

Therefore, the amount of MSW collected becomes a key variable in the development of sustainable waste management plans. The ideal theoretical scenario is to achieve complete SC of waste, by fractions, even thou this case seems to be impossible, which means that the RMSW is composed from both recyclable and non-recyclable materials, with or without energy potential. According the European Commission report (2015) on ''Separate collection schemes in the 28 capitals of the EU'', the EU-28 average SC rate by capital reaches out to 18.5% [31]. The efficiency of MSW selective collection has successfully been registered in some EU regions, reaching out to 65.7 % in 2011 and heading to 80% [32,33].

It is obvious that the increase of SC rate represents the decrease of the RMSW stream. In this sense the RMSW treatment options should be developed considering the above mentioned main factors that underline a sustainable waste management plan. Therefore, two aspects can be pointed out and are described below.

- The first one relates to the decreasing percentage of the organic fraction in the RMSW. This aspect is due on the one hand to the increase of the use of packaging, not always accompanied by an increase of selective collection of this fraction; on the other hand, the activation of organic fraction SC decreases its content in the RMSW. In the most extreme cases the percentage of organic fraction in RMSW may fall below 10%. It may be pointed out that in Romania the MSW is mostly collected in an unselective manner, with no SC of the organic fraction and a poor SC of household Packaging Waste that covers a 23% from total population, 46% in urban and 4% in rural areas [34-35].
- The second one is related to a fraction emerged recently, the diapers. Their use has grown significantly in some regions, also among the oldest people, and the absence of SC for this fraction makes their presence even more relevant in RMSW, where their percentage may reach today, in the extreme cases, even 20% and more if SC of other materials is high.

Considering the RMSW characteristics (mainly heating value, chlorine and mercury content) the waste stream could be defined potentially as Refused Derived Fuel (RDF) or Solid Recovered Fuel (SRF) apart from the need of homogeneity. The role of these fuels in the development of waste management in the CE frame will be discussed in the next section. The waste stream characteristics (mainly volatile solids, non-volatile solids, and water content, chlorine and mercury content) will dictate the selection of the available treatment options (combustion, gasification, pyrolysis, integrated thermal plants, aerobic/anaerobic mechanical-biological treatments etc., which have as main target energy recovery).

Waste to Energy is the predominant mass burning technology. In 2013, 520 plants were operational in Europe, treating around 95 million tons of MSW and commercial waste per year [36]. Incineration is a well-known option [37-38]. In conventional waste combustors the RMSW Lower Heating Value (LHV), in areas with high SC rate, is around 13.5 MJ $kg^{-1}{}_{RMSW}$ with peaks of LHV expected during operation that can reach 20 MJ $kg^{-1}{}_{RMSW}$ [39]. The energetic potential of the RMSW depends also of the organic fraction SC, that has a typically 70% water content, making this stream unsuitable for any thermal treatment. The high SC efficiency of the non-combustible fractions, decrease the slag formation during RMSW incineration. Over the years the improvement of the incineration operation conditions led to the reduction of the slag generation up to 20% from the initial waste stream flow treated on one hand and the volatile solids content on the other. Slag today can be converted into products for the building sector even through "cold" processes of sorting and grinding. The resulting material is included into concrete products for simplified uses. That is surely interesting from the circular economy point of view.

Anaerobic Digestion uses microorganisms to break down biodegradable material in absence of oxygen into methane and carbon dioxide. The resulted co-product is a nutrient-rich digestate that can be used as fertilizer. The biogas can be used directly as fuel, in combined heat and power plants or treated in order to improve its quality and obtain biomethane. The correct SC of Organic Fraction of Municipal Solid Waste (OFMSW) stream plays an important role in the anaerobic process conditions and methane quantity and quality production. Wastes with higher lignocellulosic content present lower anaerobic biodegradability and extractable substances (soluble in water and alcohol) have a positive effect on methane production [40]. In Europe, 57% of the organic wastes treated is for compost production to be used in agriculture, and the remaining 43% is used for biogas production under anaerobic conditions [41]. One of the innovative waste pre-treatments that increases the anaerobic digesters performances is electro-oxidation. The adoption of this approach as pre-treatment of waste must take into account the different lasting of the treatment (shorter for MSW and its fractions) [39]. Anaerobic digestion of OFMSW is followed by post-composting, making the approach interesting from the CE point of view.

Composting, an aerobic degradation, the organic fraction of MSW using aerobic microorganisms in the presence of oxygen. The aerobic process produces CO2, H2O, nitrates, sulfates and heat in addition to a stable solid material commonly used as fertilizer for agriculture crops [42-43]. The adoption of direct composting (that is without a pre-stage of anaerobic digestion) does not change the amount of compost produced.

Considering the latest results in the optimization of the anaerobic process, the option of producing both biomethane and compost (two products) must be preferable from the CR point of view.

For aerobic mechanical biological treatments, two methods can be used as pre-treatment of the RMSW: bio-drying (usually single stream) and bio-stabilization (double stream).

Biostabilization is adopted as pre-treatment before landfilling thus it seem not coherent with the CE principles [44]. Bio-drying belongs to the biological mechanical treatment options and is aimed to the dewatering of MSW thanks to the biological exothermal reactions (thermal drying, on the contrary, needs an external heat source as methane, in order to increase the lower heating value of the treated material) [45,46]. Particular, bio-drying, by reducing the water content of the waste, produces an increase in calorific value of about 30–40% [47]. In practice, this process concentrates the initial energy content in a lower mass. A small energy loss (2%-3%) referred to the initial LHV characterizes this process. After a mechanical post-treatment the obtained material can used be re-used as fuel in thermal treatment plants. The possibility to extract recyclable materials and generate a product (SRF) makes bio-drying interesting from the CE point of view.

Gasification is the thermochemical conversion of the raw material (solid or liquid) in the presence of substoichiometric air/oxygen or water vapours, which converts volatile solids (VS) into a combustible gas (syngas). The syngas is mainly composed of carbon monoxide (CO) and hydrogen (H_2). In comparison with traditional waste incineration, gasification has some potential advantages: the syngas is suitable for different application (combustion in traditional burners, connected to a boiler and a steam turbine, gas reciprocating engines or gas turbines or its conversion in valuable products as chemicals and fuels due to CO and H_2 content) [48]. This change of approach can give environmental advantages [49].

To this concern, the integrated systems based on the concept of coupling directly a gasification chamber and a combustor for syngas give three advantages:

- The possibility to better regulate the combustion as performed by a combustible gas (and not by a solid, as in the conventional combustion)
- The possibility to enhance the efficiency of the electricity generation from steam turbine thanks to lower problems of corrosion
- The possibility to vitrify the non volatile solids opening to solutions alternative to a landfill (coherently with CE)

Pyrolysis is a endothermic process, consisting of thermal decomposition of the material, in a free oxygen or air atmosphere, in order to obtain non-condensable gas products (such as CO, CO_2, H_2, CH_4 and other hydrocarbons), liquid products (oils, tars) and carbonaceous residue within inorganic solid particles (char). Pyrolysis gas has potential applications in the energy industry such as: a) direct use by its combustion to produce heat and/or electrical (via combustion spark-ignition engine or compression ignition); b) directly or with different fuel (e.g. coal) to produce individual gases including H_2, CH_4 or other volatile liquid fuels or bio-synthetic. In some applications, the hot gases resulting from the pyrolysis process may be used in the process for preheating the reactor or inert gas [50-51]. Depending on its composition and additional treatments the pyrolysis oil can be used as unconventional fuel. Char is an attractive substitute for coal, due to increased energy potential [52]. Unfortunelly, essentially no commercial application on MSW or RSMW pyrolysis have been registred, due the limited operational conditions of the process, were a more homogenous input flow is preferable.

Considering the present analysis in the circular economy context, the aim of these options is to improve the environmental and energy balances compared to traditional approches and to obtain a valuable product as output. The possibility of having a product as output can be seen surely positive from the CE point of view.

3 The role of Solid Recovered Fuel among the options of energy recovery for landfill minimisation

The waste derived fuel can be classified as MSW by-product in case it meets certain classification standards, having more stable physical-chemical properties than the constituent raw material and markedly cheaper than primary fuels, becoming more attractive for the energy intensive industrial sectors such as cement industry, pulp and paper industry, thermal power plants [53].

In Europe, the mechanical/biological treatment (MBT) is an increasing option for the RDF/SRF production for industrial purposes, but also as a pre-treatment before landfilling or as a pre-treatment before combustion in dedicated plants. The interest for this kind of treatment has become higher and higher not only in the old European Union (EU) members, but also in the recent entries, in order to comply with landfilling of biodegradable waste minimization. As already underlined, this scenario is not steady, as the EU Member States must introduce source-separation strategies that can affect the composition of the residual MSW to be treated in MBT plants.

The RDF is a fuel product, which has more uniform particle size distribution and higher energetic potential than untreated MSW. The main sources of organic and inorganic chlorine content in waste fractions are from 0.1 wt.% in wood to >6 wt.% in non-packaging plastics (dry basis), polyvinylchloride (PVC) from packaging, electrical wire insulation etc. in plastics and chloride salts (mainly NaCl) in kitchen waste [54]. In the WtE plants chlorine causes high temperature corrosion and low conversion efficiency. Zinc presence in the MSW seems to be in the range 3000–4000 mg/kg_{DS} [55]. Zinc, lead and tin lower the melting temperature of deposits on the superheater tubes amplifying corrosion [56]. Trace and major elements affect the composition of ash [57].

There were only two RDF classes in Italy before 2010 (RDF and RDF_Q), depending mainly on the calorific value (>15 and 20 MJ/kg_{RDF} respectively), moisture (<25% and 18% respectively) and ash content (<20% and 15% respectively, referred to dry matter). Today there are 125 SRF classes, since 2010, based on limit values for three fuel properties:
a) the mean value for net calorific value (NCV; as received);
b) the mean value for chlorine content (dry basis);
c) the median and 80th percentile values for mercury content (as received) [58].

Therefore the RDF needs improvements and shifting to a high quality combustible fuel (SRF) that is covered by specific requirements in the legislations. In Italy, the classification of fuels from municipal solid waste (MSW) has changed in recent years with the introduction of 30 technical documents, mainly in the range of norms UNI CEN/TS 15357–15747, some of which have been operative since 2006. These documents set out all the characteristics, definitions, sampling methods, parameters of interest and analytical methods for SRF, while the previous term "Refuse Derived Fuel" (RDF) was not given by any legal definition and it was interpreted differently across countries [59].

Some new concepts have been introduced and other aspects are now more detailed; e.g. the quality management system concept for the production of SRF from the reception of waste(s) up to the delivery of SRF has now a specific norm, that requires data regarding source and origin of waste, information on waste production processes, data on the waste composition (main materials), appearance of waste, European Waste Code, chemical analysis when necessary ; moreover, the sampling method of SRF from production plants, from deliveries or from stock in order to obtain the laboratory samples refers to 3 technical norms instead of one; the metals characterization is now more detailed for each element being represented by 3 technical norms, that cover a large number of parameters: Al, Ca, Fe, K, Mg, Na, P, Si, Ti, As, Ba, Be, Cd, Co, Cr, Cu, Hg, Mo, Mn, Ni, Pb, Sb, Se, Tl, V, Zn and metallic aluminium; the precursors of acid gases are now analyzed in details; not only chlorine is taken into account but also sulphur (S), chlorine (Cl), fluorine, (F) and bromine (Br); a specific technical norm concerns the assessment of the biomass content in SRF. This is an important aspect as related to some incentives.

Summing up, SRF seems to be an interesting product generated from RMSW (that should have only non-recyclable materials) in the frame of a CE concept.

4 Preliminary proposals for assessing the energy recovery management performance

Referring to MSW directly used (RMSW) or indirectly used (SRF) as input of thermochemical plants, an indicator of energy recovery management performance can be the percentage of waste having LHV > 13MJ/kg. In a region the MSW stream sent to energy recovery by thermochemical plants could be differentiated in sub-streams depending on the final destination and the pre-treatment adopted (if any). In both cases it is important to verify if the threshold value proposed by the EU for addressing the energy strategies in the waste sector is taken into account or not.

Referring the MSW directly or indirectly used as input of thermochemical plants, another indicator can be the percentage of waste having ash recovered. Slag can be potentially used as a pozzolanic material in cement industry and fly-ash can be potentially vitrified and used as secondary product [60-61]. The indicator check if these opportunities are followed.

Referring food waste (from selective collection), a third indicator can be the percentage of this stream sent to anaerobic digestion. As discussed above, the option of anaerobic digestion gives two final products: biogas and compost. This should be preferable to direct composting.

5 Conclusions

In the described frame, the sector of MSW management seems to be ready to comply with the principles of CE. What seems still potentially critical is the economic sustainability. From the circular economy point of view, the evolution of the sector of SRF, if based on non-recyclable waste, can help to exploit industrial plants, like the cement factories ones, for integrating the ash of the input into a final product, without the need of a new complex thermochemical plant. Unfortunately, often SRF is not yet recognized as a product (a real fuel) that deserves to be bought. On the contrary who generates SRF could have to pay a fee for its exploitation (if a dedicate plant is not directly available). In spite of the present situation, the perspective of CE towards RMSW management seems to be clearly oriented to options characterised by a double output: energy generation and valorisation of the non-volatile solids.

This work has been partly supported by University Politehnica of Bucharest, through the "Excellence Research Grants" Program, UPB – GEX. Identifier: UPB–EXCELENȚĂ–2016, Research project title Asigurarea alimentării cu energie în regim continuu printr-un Sistem hibrid Integrat cu utilizarea Deșeurilor cu potențial energetic, surselor Eolione și Fotovoltaice/ Continuous power supply through an integrated hybrid system using wastes with high energetic content, photovoltaic and wind energy, Contract number: 90/26.09.2016 (acronym:SIDEF).

References

1. D. Hoornweg, P. Bhada-Tata. What a waste: a global review of solid waste management. (2012).
2. http://www.atlas.d-waste.com/ (2017)
3. M.E.B. Seiffert, C. Loch, J. Clean. Prod. **13**(12**)**, 1197–1202 (2005)
4. I.D. Williams, Waste Manage. **40,** 1-2 (2015)
5. S. Huysman, J. De Schaepmeester, K. Ragaert, J. Dewulf, S. De Meester, Resour. Conserv. Recy. **120,** 46-54 (2017)
6. D. Ness, Int. J. Sust. Dev. World Ecol. **15(4)**, 288-301 (2008).
7. M. Geissdoerfer, P. Savaget, N.M. Bocken, E.J. Hultink, J. Clean. Prod., **143,** 757-768 (2017)
8. Y. Geng, B. Doberstein, Int. J. Sust. Dev. World. **15(3),** 231-239 (2008)
9. Z. Yuan, J. Bi, Y. Moriguichi, J. Ind. Ecol. **10(1–2)**, 4-8 (2006)
10. I.S. Jawahir, R. Bradley. Procedia CIRP **40**, 103-108, (2016)
11. http://ec.europa.eu/environment/circular-economy/, (2016)
12. R. Patrascu, E. Minciuc, D. Tutica, M. Norisor, G. Ionescu, P. Stefani. Quality - Access to Success **15(138),** 84-88 (2014)
13. J.Y. Park, M.R. Chertow, J. Environ. Manage. **137**, 45-53, (2014)
14. P. Ghisellini, C. Cialani, S. Ulgiat, J. Clean. Prod. **114**, 11-32, (2016)
15. N.F. Cruz, P. Simões, R.C. Marques J. Clean. Prod. **37**, 8-18, (2012)
16. E.C. Rada, M. Ragazzi., L.I. Cioca, G. Ionescu, E. Ranieri, E. Trulli, Quality - Access to Success **18**, 151-156 (2017)
17. M. Ragazzi, V. Torretta, G. Ionescu, I.A. Istrate. WIT Transact. Ecol. Environ. **176**, 235-244 (2013)
18. V. Torretta, G. Ionescu (2016). Manag. Enviro. Quality: An Int. J. **27**(4), 419-426 (2016)
19. E.C. Rada, C. Bresciani, E. Girelli, M. Ragazzi, M. Schiavon, V. Torretta, Sustainability **8**(9**)**, 840 (2016)
20. A. Castagna, M. Casagranda, A. Zeni, E. Girelli, E.C. Rada, M. Ragazzi, T. Apstol, UPB Sci. Bull., serie D., 75(4), 253-264 (2013|)
21. Onyanta, Georg. Compass, **10**(12), 499-513 (2012)
22. L. Rigamonti, I. Sterpi, M. Grosso Ecol. Indic. **60**, 1-7 (2016)
23. E.C. Rada, M. Ragazzi, G. Ionescu, G. Merler, F. Moedinger, M. Raboni, V Torretta, Energ Procedia **50,** 1037-1044 (2014)
24. S. Ciuta, T. Apostol, V. Rusu, Sustainability **7**(1), 916-931 (2015)
25. F. Ferreira, C. Avelino, I. Bentes, C. Matos, C.A. Teixeira, Waste Manag. **59**, 3-13 (2017)

26. G. Ionescu, E.C. Rada, L.I. Cioca, Eviron. Eng. Manag. J. **14**(11), 2537-2544 (2015)
27. S. Bungau, C. Bungau C, D.M. Tit, J. Environ. Prot. Ecol., **16**(1), 56-62 (2015)
28. D.E. Popescu, C. Bungau, M. Prada, C. Domuta, S. Bungau, D.M. Tit, J. Environ. Prot. Ecol., **17(3)**, 1011-1020 (2016)
29. V. Torretta, E.C. Rada, M. Ragazzi, E. Trulli, I.A. Istrate, L.I. Cioca, Waste Manag., **45**(SI), 152-160 (2015)
30. S. Bungău, R. Suciu, A. Bumbu, G. Cioca, D.M. Tit, J. Environ. Prot. Ecol., **16(3)**, 980-987 (2015)
31. European Commission report, Assessment of separate collection schemes in the 28 capitals of the EU. Reference: 070201/ENV/2014/691401/SFRA/A2 (2015)
32. M. Ragazzi, E.C. Rada, WIT Transact. Ecol. Environ. **109**, 613-620 (2008)
33. R. Ciudin, C. Isarie, L.I. Cioca, V. Petrescu, V. Nederita, E. Ranieri, UPB Sci. Bull. serie D, **76**(3), 215-222 (2014)
34. V. Petrescu, R. Ciudin, C. Isarie, V. Nederita, World Transact. Eng. Technol, Educ. **8**(1), 107-110 (2010)
35. http://ec.europa.eu/environment/waste/framework/pdf/RO_factsheet_FINAL.pdf (2015)
36. https://waste-management-world.com/a/time-out-for-waste-to-energy-in-europe (2013)
37. U. Arena, F. Ardolino, F. Di Gregorio, Waste Manag. **41**, 60-74 (2015)
38. R.L. Smith, D. Sengupta, S. Takkellapati, C.C. Lee, Resour. Recov. Conserv. **104**, 311-316,(2015)
39. E.C. Rada, I.A. Istrate, M. Ragazzi, Environ. Technol. **30**(7), 651-661 (2009)
40. R. Campuzano, S. González-Martín Waste Manag. (to be published)
41. F. Micolucci, M. Gottardo, C. Cavinato, P. Pavan, D. Bolzonella, Waste Manag. **48**, 227-235 (2016)
42. G.C Fitzgerald, Waste to Energy Conversion Technology (2013)
43. M. Ragazzi, P. Tosi, E.C. Rada, V. Torretta, M. Schiavon, Waste Manag. **34**(11), 2400-2406 (2014)
44. S. Ciuta, S. Antognoni, E.C. Rada, M. Ragazzi, A. Badea, L.I. Cioca. Sustainability **8**(12), 1311 (2016)
45. E.C. Rada, A. Franzinelli, M. Ragazzi, V. Panaitescu, T. Apostol, Chemosphere **68**(9), 1669 -1674 (2007)
46. M. Ragazzi., E.C. Rada, D. Antolini. Waste Manag. **31**(9-10), 2085-2091 (2011)
47. F. Adani, D. Baido, E. Calcaterra, P. Genevini, Bioresour. Technol. **83**(3), 173-179 (2002)
48. U. Arena, Waste Manag. **32(4)**, 625-639 (2012)
49. C. Marculescu, Energ. Procedia **6**, 558-564 (2011)
50. O. Bičáková, P. Straka, Int. J. Hydrogen Energ. **37**(16), 11536-11578 (2012)
51. C. Marculescu, I. Minca, F. Bacalum, Rev Chim. **66**(11), 1837-1840 (2015)
52. A. Demirbas, J. Anal. Appl. Pyrol. **72**(1), 97-102 (2004)
53. E. Kakaras, P. Grammelis, M. Agraniotis, W. Derichs, H.P. Schiffer, J. Maier, T. Hilber, T. Glorius, U. Becker, Therm. Sci. J. **9**(2), 17-30 (2005)
54. W. Ma, G. Hoffmann, M. Schirmer, G. Chen, V.S. Rotter, J. Hazard. Mater. **178**(1-3), 489-498 (2010)
55. F. Jones, D. Bankiewicz, M. Hupa, Fuel **117(A)**, 763-775 (2014)
56. B. Gawlik, E. Sobiecka, S. Vaccaro, G. Ciceri, Energ. Policy **35**(12), 6293-6298 (2007)
57. M. Nasrullah, P. Vainikka, J. Hannula, M. Hurme, Fuel **145**, 1-11 (2015)
58. E.C. Rada, G. Andreotolla, Waste Manag. **32**(6), 1059-1060 (2012)
59. L. Lombardi, E. Carnevale, A. Corti. Waste Manag. **37**, 26-44 (2015)
60. K.L. Lin, D.F. Lin, Cement Concrete Comp. **28**(9), 817-823 (2006)
61. Y.J. Park, J. Heo, J. Hazard. Mater. **91(1)**, 83-93 (2002)

DOI: 10.1051/matecconf/201712105007

Demerit control chart as a decision support tool in quality control of ductile cast-iron casting process

Robert Sika[1,*], and *Michał* Rogalewicz[2]

[1]Poznan University of Technology, Institute of Materials Technology, Piotrowo 3 street, Poland
[2]Poznan University of Technology, Chair of Management and Production Engineering, Piotrowo 3 street, Poland

Abstract. In many industrial areas the product quality can be unequivocally assigned to classes such as: "good", "bad" or "to repair". In case of casting processes, the product is approved to sales considering customer's requirements. Except for common characteristics, such as structure, compactness and mechanical properties, physical state of the product is also important. This state is assessed by checking occurrence of specific kind of defects. They are often conditionally accepted by a customer if they do not have any influence on functionality of the product (e.g. negative adhesive and cohesive phenomena, fatigue strength, thermal shocks). Authors' experience shows that current registering of the most frequently occurring defects and comparing them to customers' requirements can be very useful and help a quality engineer to control the casting process. They suggest using the Demerit Control Chart (DCC), according to authors' own methodology, in aspect of information about the castings accepted conditionally by a customer (DCC-recognition). DCC-recognition can be used to assess this quality by monitoring the value of just one aggregated measure for all kinds of defects instead of using a single attribute control chart for each of them. The test version of this tool considering severity of defects proved to be useful in one of the European foundries.

1 Introduction

Foundry is one of the most complex manufacturing technologies, comprising of physicochemical processes, including the metallurgical ones. The aim is to obtain a final product (a casting) by shaping it and achieving functional features according to a specifically designed process, considering phenomena of solidification and cooling of a liquid alloy [1-3]. Starting from the 21st century, large organizational and technical changes can be observed in the industry and, what is following, in the foundry branch. The processing technologies, including the foundry, are more and more modern thanks to wider use of innovative industrial machines and improvement of methods of production, control and assessment of processes [4-9]. There is an ongoing trend of supporting this branch of industry

* Corresponding author: robert.sika@put.poznan.pl

by implementation of innovative projects and processes, such as 3D printing [10] and investing into more and more technologically advanced devices for the casting production [11]. Unfortunately, a still existing problem is high cost of manufactured castings and problems with ensuring their proper quality. In such a context, it is very important to appropriately plan processes of realization of production, organization of these processes and their effective supervision and control.

Manufacturing a final product in a form of a casting consists in realization of many processes, characterized by diversified levels of complexity and influence on the final quality [12]. The multi-staged process and necessity of ensuring quality on each its level, often performed in very changeable conditions, make obtaining a zero defect casting practically impossible. That is why, different from some products manufactured in other machine industry branches (e.g. by machining), it is difficult to unequivocally define and assess quality of a casting [13-14].

A casted product is therefore accepted for sales with consideration of criteria defined by its recipient. Besides basic characteristics, such as compactness of the structure and mechanical properties (often measured on separately cast samples or cast-on samples), state of the product prepared for sending is important. Eventually, the final state of the casting is evaluated for occurrence of specific types of defects (on a surface or inside the casting). What's more, a defect can be often accepted by the customer under certain conditions, if it does not affect usefulness of the product in a negative way (negative adhesive phenomena, fatigue strength, thermal shocks) [13].In foundries, quality control tools are rarely used or not used at all and all activities are taken only after the quality problems already occurred [15]. The authors know situations, when in case of occurrence of a so-called acceptable defect in a casting, quality controllers, due to lack of proper monitoring tools, eventually qualify products as discards (not acceptable for sales).

Considering the technological complexity of casting processes described above and related problems with quality assurance, the authors propose a modification of a tool for aiding quality control of casting processes in a form of a Demerit Control Chart (DCC) [16]. It is used to supervise a number of defects occurring in manufactured series of castings, with consideration of their criticality. This specific control chart was adapted by the authors for use in the foundry branch and it is based on monitoring an aggregated statistics determined for defects, which occurred in a sample of castings taken from each manufactured series. In the further part of the paper, a concept of the DCC will be presented, along with assumptions, methodology of its application in foundry processes and results of its implementation in one of European foundries.

2 Application of DCC in foundry conditions

2.1 Idea of Demerit Control Chart

The control charts are basic tools of statistical process control and they are used for supervision of a selected characteristics of a process/product through monitoring of changes of a statistic determined out of a sample, taken from this process in set time intervals. Values of a selected statistic (e.g. mean values) are plotted on a sheet of a chart and compared to control limits, representing variability resulting from purely random reasons. Exceeding of control limits or a specific pattern of points appearing in a chart shows that the process is affected by non-random conditions which cause its maladjustment [17].

The control charts are designed for features evaluated in a numerical way (variable charts) or in an alternative way (alternative charts, attributes charts). The first type of chart is usually composed out of two paths: location and dispersion – examples are x_{av}-R or x_{av}-s charts. The charts for attributes are plotted with a value of a statistic related to a number of defects or defective units. An advantage of variable charts is possibility of controlling the process and correcting it on-line. The alternative charts do not bring such a possibility. The process evaluation is made post factum – only after it is finished and an appropriately large sample is collected. Despite that these charts require a definitely larger sample than the variable charts and are less informative, they allow current evaluation if defectiveness level of a supervised process does not exceed an acceptable one (this level is determined on the basis of a specifically prepared pilotage trial) [18].

One of types of charts evaluated alternatively are charts related to defects: *c* – quantity of defects per sample of constant size and *u* – quantity of defects per unit. Their important assumption, which is sometimes a limitation, is treatment of each defect detected in a collected sample with equal importance. If, for a given process or phenomenon, it is advisable to consider certain defects as more important than the other ones, which is a case in foundry processes, it is worthy to use a Demerit Control Chart (DCC).

The DCC sheet is plotted with a statistic aggregating all the defects in a collected sample and taking their different importance into consideration. In literature [16, 19], it is recommended to apply 4 different classes and weights for the defects: class A – very serious defects - weight 100, class B – serious defects - weight 50, class C - moderately serious defects - weight 10, class D - not serious defects - weight 1.

Appropriate selection of these weights is very meaningful for the DCC. It is important for it not to be too sensitive to defects of low seriousness, with still being sensitive to the most relevant ones. Usually, values of weights are determined for a given problem. Assumptions and conditions regarding the use of DCC in casting processes are presented in the next subchapter.

2.2 Assumptions for use of DCC in casting processes

Cooperation with experienced quality controllers from European cast iron foundries allowed to conclude, that concept of the DCC can be successfully translated into industrial conditions in a foundry. However, it is important to modify approach for use of this chart, to take specific conditions of casting production into account. The most important assumptions are shortly described in the next paragraphs.

First of all, assessment of defects of castings is still performed by visual testing in most cases (VT method). It means that a casting is evaluated for a given defect without use of additional measurement tools (more accurate). Additionally, considering classification of criticality, a division into 4 classes suggested in the literature may be not justified in case of casted products. At most 3 cases can be distinguished here (C1, C2 and C3):

- C1 – a casting has a defect of such a size and/or dispersion that is making it impossible to sell it.
- C2 – a casting has a defect of a size and/or dispersion, which allows selling it under certain, strictly defined conditions (it is necessary to perform an additional study by another controller, preferably from outside the company) – however, such cases occur relatively rarely.
- C3 – a casting has a defect of a size and/or dispersion which usually allows selling it (does not cause any mechanical failures during standard operation of a product).

The second important issue of the proposed approach is classification of criticality of defects and assigning them with appropriate ranges. The authors decided not to make constant assumptions of certain types of defects being more or less critical. A given type of defect is assigned to a specific class of criticality each time, because depending on its size and/or dispersion it can take different weights. The most frequently (approx. 70-80% per each series) it is a C1 class. In other cases, the castings may be classified by a specific defect as C2 or C3.

The authors propose to name this approach the "DCC-recognition" (this notion will be used in the further part of the paper interchangeably with the DCC acronym), as large emphasis must be put on proper "recognition" of a class of a defect, due to criticality of its size and/or dispersion. As already mentioned, this "recognition" should be performed for each defect which occurred separately, without assigning constant weights for a specific type of defect.

The third important issue is reliable mapping of an analyzed process and definition of types of defects which may occur during this process. The authors propose that in scope of preparation of design of the DCC-recognition, selection of a specific type of casting for supervision (e.g. the most frequently produced) should be made in the first place and then a study should be made to answer what types of defects are the most frequently repeated for the casted series of this product. It is worth to select at most several defects, which are repeatable. A problem of mapping of the analyzed process is described in greater detail in the chapter 3.

It is noteworthy, that authors of this paper do not define precise nomenclature of defects assigned to cast iron castings. There is a considerable number of existing divisions, standards and atlases of defects, as well as branch standards. The methodology presented in the further part of the paper is suitable for practically any classification of defects, while this work shows its use in casting of cast iron.

The last important factor related to implementation of the DCC in foundry is selection of size of a pilotage trial. An advantage of this chart is a fact, that size of each subsequent sample collected from the process can be variable, what is important in case of different sizes of casted series. The authors propose, for the needs of the DCC-recognition, to collect and accurately evaluate randomly taken 10% of castings of each series. In practice, when quantity of a series is high, not all castings are investigated with 100% accuracy. It seems that 10% of castings investigated accurately will be a good representative of the whole casted series. A lot of defective castings are discovered after machining.

2.3 Methodology of design of DCC-recognition

Considering the guidelines presented in the previous chapter, methodology of implementation of the DCC into foundry processes on example of castings out of ductile cast iron is presented. This methodology is universal and can be adjusted for use in other processes of similar character, while its particular steps consider assumptions presented in the chapter 2.2. More details about particular steps of the methodology are presented in the further part of the paper.

- STEP 1: Mapping of analyzed process and defining defects, which may occur – on the basis of history of problems with a given process or with similar processes (for new processes), it should be determined which defects may occur in the process.
- STEP 2: Determination of classes of criticality of defects and assigning them with weights – due to specificity of the casting process, 3 classes were assumed: C1 (weight 10), C2 (weight 5), C3 (weight 1).
- STEP 3: Collecting a pilotage trial from a process as far as possible in the most beneficial conditions – the pilotage trial will be a „master pattern", that is referred to results of the current process control.
- STEP 4: Determination of an average number of Demerit per unit and standard deviation related to it:

  An average number of Demerit per unit is expressed by the formula (1):

$$\overline{u} = w_{c1}\overline{u}_{c1} + w_{c2}\overline{u}_{c2} + w_{c3}\overline{u}_{c3} \tag{1}$$

DOI: 10.1051/matecconf/201712105007

For example, for the C1 class the following formulas should be used (2):

$$u_{j_{c1}} = \frac{\sum_{i=1}^{n_j} x_{i_{c1}}}{n_j} \tag{2}$$

$$\overline{u}_{c1} = \frac{\sum_{j=1}^{k} u_{j_{c1}}}{k}$$

where:
u_j – number of defects per casting in sample j
j – number of a specimen in a pilotage trial (j=1,2,...,k)
k – number of specimens in the pilotage trial
w_c – weight of defect belonging to the c class (according to the previous assumptions for the needs of foundry, the following values were assumed: w_{c1}=10, w_{c2}=5, w_{c3}=1)
i – number of a casting in a sample (i=1,2,...,n_j)
xi – number of defects in a casting
n_j – quantity of j sample of castings
Standard deviation related to the average Demerit number per unit is calculated using the following formula (3):

$$\hat{\sigma}_u = \sqrt{\frac{(w_{c1})^2 \overline{u}_{c1} + (w_{c2})^2 \overline{u}_{c2} + (w_{c3})^2 \overline{u}_{c3}}{\overline{n}}} \tag{3}$$

- STEP 5: Determination of control limits in a DCC and plotting points from the pilotage trial. The control limits are determined using the following formulas (4):

$$CL = \overline{u} \qquad LCL = \overline{u} - 3\hat{\sigma}_u \qquad UCL = \overline{u} + 3\hat{\sigma}_u \tag{4}$$

- STEP 6: If points in a chart indicate existence of maladjustment while collecting the pilotage trial – identification and elimination of its cause and re-calculating the control limits.
- STEP 7: Supervision of a process, consisting in collecting a sample from each series of castings, determining the D number per unit and plotting this value in a chart, as well as observing potential symptoms in a chart. The control chart is plotted with a statistic using the following formula (5):

$$u_j = \frac{w_{c1}u_{c1} + w_{c2}u_{c2} + w_{c3}u_{c3}}{n_j} \tag{5}$$

3 Application of DCC in a process of ductile cast iron casting

The studies were realized in an European foundry manufacturing cast iron castings for the machine industry. The net weight of a single product (without the gating system and the risers) is 3,45 kg and it is manufactured on a modern automated box-less line with a vertical mold division. Daily, the company (in a 3-shift system) produces up to several dozen types of castings, with one series between several hundred to even 1000 pieces of castings (depending on the weight of casting). An example shown below refers to a specific melt and assortment produced in series. In scope of the pilotage trial, aggregated part of occurring of 3 the most commonly present defects for this assortment is presented. According to nomenclature by the Polish Standard PN-85 H-83105 they are the W101 (mechanical damages), W202 (blows) and W222 (sand holes) [20]. Table 1 presents data related to preparation of the pilotage trial and implementation of the DCC-recognition.

Table 1. Data related to pilotage trial preparation and DCC-recognition chart application.

Differentiating feature	Data according to pilotage trial	Data according to DCC-recognition
Period of testing:	January – June 2016	July – August 2016
Number of series:	20 pcs.	7 pcs.
Number of castings in series (min / max / average):	260 / 1045 / 688 pcs.	285 / 995 / 584 pcs.
Weight of the casting net/gross:	3,45 / 4,8 kg	3,45 / 4,8 kg
Number of defective castings (C1 / C2 / C3):	90 / 16* / 30**	28 / 5* / 9**

*) cases **must be** approved for sale with participation at least 2 controllers

) cases **may be approved for sale with participation at least 2 controllers (depends on the quality control regime)

For the needs of testing the presented methodology, a computer tool was designed for improving of casting defects evaluation in an electronic version of the DCC-recognition -see Figure 1.

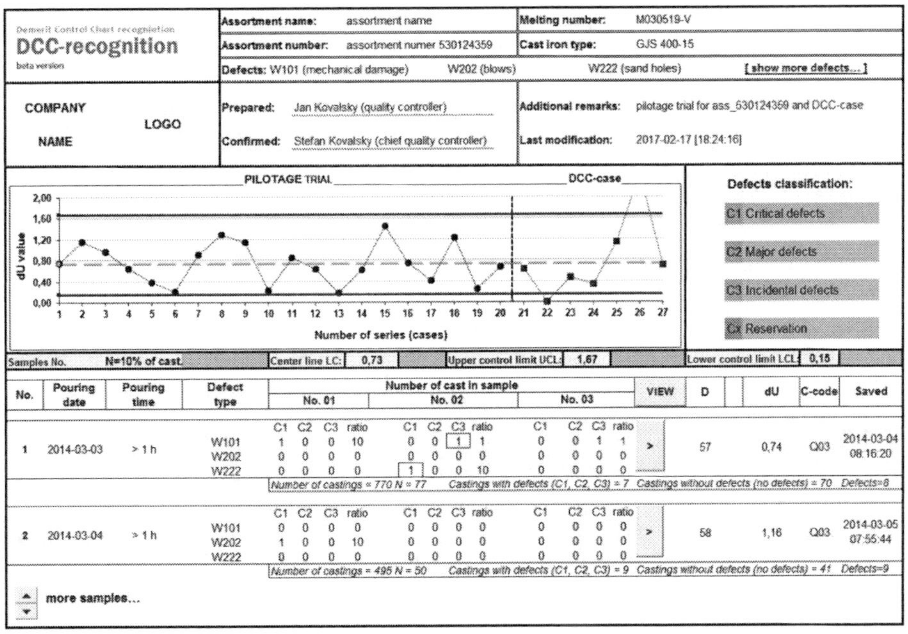

Fig. 1. DCC-recognition – an example of electronical control chart for assortment 530124359

The chart in its electronical shape allows displaying basic information about the currently casted series (assortment name, assortment number, melting number, cast type iron and information about defects) in a form readable for an employee. Moreover, it contains additional information, such as: a person responsible for the chart, data of modification, supplementary information. Central part of the chart is occupied by a diagram with a possibility of displaying last 30 values of samples. Below the diagram, there is a panel,

which allows recording cases for a randomly selected sample of each series (number of a casting in a sample, type of defect, class of defect, causes and preventive reactions) – see Fig. 2.
As it can be read from a sheet of the DCC (Fig. 1), first 20 samples were used to estimate basic characteristics of this chart (central line, upper and lower control limits). Formulas presented in chapter 2.3 were used. As presented, during the realization of the process for samples belonging to the pilotage trial, the process was stable – no maladjustment symptoms occurred. It means, that the control limits can be used for on-line supervision of the casting process. Another 7 measurements (21-27) are samples collected during realization of the process. For the series no. 6 (point 26), the chart indicated slight exceeding of the upper control limit. It is a valuable information for the quality controllers. It means that the number, type or criticality of occurring defects is higher than during collection of the pilotage trial and if no intervention is made in the process, the next series can present similar problems in production quality.

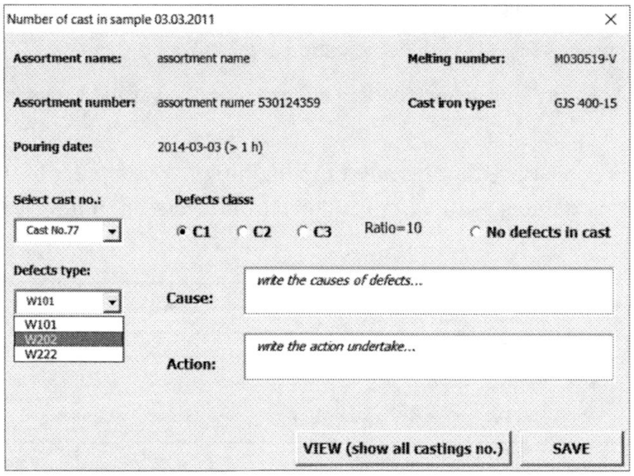

Fig. 2. Exemplary screen of the DCC-recognition – registering defect class for a sample.

4 Summary

The paper presents an example of implementation of a control chart for quality inspection and control of defects of casting (features evaluated alternatively) in the foundry industry. Although this type of chart does not allow current correction of the process, it is a valuable source of information for the quality assurance division, on problem of aggregated value of defects present for particular series of a specific assortment, with division into classes of criticality.

It turns out, that the proposed DCC-recognition allows, among other things, to systematize and order classes of defects occurrence by criticality. The authors think that the proposed methodology will be an encouragement for the foundries that unequivocally evaluate defects of castings and do not allow selling of products with acceptable defects. It is a matter which has been discussed for a long time between clients and suppliers (foundries). All conditions of selling of defective castings should be, obviously, settled with client beforehand (such as location of a defect, its size and dispersion).

A methodology was proposed and a tool based on it, in a form of an electronical chart (DCC-recognition), which is a new approach to monitoring of defects of casted products. Employees of the Quality Assurance division have a possibility of on-line supervision of aggregated criticality of occurrence of the most commonly observed defects in castings series, which makes it possible to take more accurate actions (differentiation between the class C1 – casting not sold – from the classes C2 and C3). As it turns out, average of 25% of cases of investigated castings can be directed for further selling.

References

1. 49th Cens. of Word Cast. Product., *Modest Grow. in Worldw. Cast. Mark.*, http://www.globalcastingmagazine.com/ (2016)
2. D.R. Cooper, World Found. Cong., **71**, World Foundry Congress, http://www.thewfo.com/contentfiles/pdfs/27.pdf (2014)
3. *Moulding – traditition and modernity* (in Polish), http://www.odlewnictwo.agh.edu.pl/ (2016)
4. A. Kujawinska, K. Vogt, F. Wachowiak, edited by: P. Golinska, A. Kawa, Technol. Manag. For Subst. Prod. and logist., Book Series: EcoProduction, 193 (2015)
5. M. Diering, K. Dyczkowski, A. Hamrol, 10th Int. Conf. on Soft Comp. Mod. in Ind. and Envir. Appl., Spring. Intern. Publ. Switz., **368**, 415 (2015)
6. M. Diering, K. Dyczkowski, IEEE Int. Conf. on Ind. Eng. and Eng. Manag., **2016**, 228 (2016)
7. G. Vieira, L. Reis, M.L.R. Varela, J. Machado, J. Trojanowska, Rom. Rev. Prec. Mech. Opt. and Mechatr., **50**, 119 (2016)
8. J. Trojanowska, K. Żywicki, E. Pająk, Inf. Technol. in Envir. Eng., **695** (2011)
9. P. Zawadzki, F. Górski, A. Hamrol, M. Kowalski, R. Paszkiewicz, Trans. of FAMENA, **35**, 69 (2011)
10. W. Kuczko, F. Górski, R. Wichniarek, P. Zawadzki, P. Buń, Adv. in Sc. and Technol. Res. J., **9**(26), 20 (2015)
11. R. Sika, Z. Ignaszak, Arch. of Found. Eng., **5**, 145 (2016)
12. I.G. Vidayev, N. Martyushev, A. Ivashutenko, A.M. Bogdan, Adv. Mat. Res., **880**, 141 (2014)
13. Z. Ignaszak, J. Ciesiółka, Weld. Over., *Conditions and prospects for non-destructive testing of castings before their exploitation – chapter II* (in Polish), **83(13)**, 36 (2011)
14. Z. Ignaszak, J. Hajkowski, P. Popielarski, Def. and Diff. Forum, **336**, 135 (2013)
15. M. Salem, S. Maik, J. Gottschling, D. Hartman, http://www.researchgate.net/ (2014)
16. L.H. Chen, J. Intell. Manuf., **16**(3), 349 (2005)
17. M. H. Omar, J. of Educ. Measur., **47**, 18, (2010)
18. T.R.Vijayaram, S. Sulaiman, A.M.S. Hamouda, M.H.M Ahmad, J. of Mat. Proc. Technol., **178**, 39 (2006)
19. M. L. Crossley, ASQ Quality Press, *The Desk Reference of Statistical Quality Methods. Demerit/Unit Control Chart*, 127 (2008)
20. PN-85 H-83105, *Castings. Classification and defects terminology* (in Polish),(1985)

MATEC Web of Conferences **121**, 05008 (2017) DOI: 10.1051/matecconf/201712105008
MSE 2017

Capability studies, helpful tools in process quality improvement

Carmen Simion[1,*]

[1]Lucian Blaga University of Sibiu, Department of Industrial Engineering and Management, 550025 Emil Cioran street 4, Sibiu, Romania

Abstract. Ability of processes to meet customer'quality requirements has become essential for providing competitive advantages such as cost savings, reducing the number of nonconfoming products or increasing customer satisfaction. This paper aims to conduct a capability study for a swaging process (that ensures an assembly dimension) of a key product characteristic (the outer diameter of a new part), related to the concept of capability and performance indices and how these metrics can be used and interpreted to become powerful tools for decision making. To achieve the goal of the paper, the following key aspects were analyzed: capability of the measurement system capability (gage R&R), production equipment (machine) capability and process capability/performance during the first serial production. The analysis was performed with Minitab® 17, the most commonly used software for quality improvement.

1 Process variation

As a matter of fact, in any process, regardless of how well designed or carefully maintained it is, a certain amount of inherent or natural variability will always exist [1]. This type of variation is the cumulative effect of many small causes, named common, chance or random causes. These types of causes produce a consistent and predictable type of variation over time. A process that is operating with only common causes of variation present is said to be a stable process or in a state of statistical control. Sometimes, dominant sources of variation may appear, that are not part of the process and they are named as special or assignable causes. These types of causes are not predictable and are sporadic in nature.

It is important to know what type of variation is present in a process, because the action one takes for the process improvement depends on the type of variation. If special causes of variation are present, they should be first identified and then eliminated; the next step is to reduce common cause of variation.

Competitive companies understand that satisfying the needs of customers and running a successful operation requires processes that are predictable and consistent. The variation which is present in a process (whether it is a business or a manufacturing process) is the main cause of the quality problems, so measuring variability and finding ways to reduce it, is essential for the success of any company.

* Corresponding author: carmen.simion@ulbsibiu.ro

In a manufacturing process, the observed process variation comes from two sources [2]: variation due to the differences between parts made by the process (part-to-part variation) called actual process variation and variation due to the measurement variation (measurement system or gage/gauge variation) i.e. imperfections in taking measurements.

There are numerous methods to control and reduce manufacturing process variation, e.g. statistical process control (SPC) tools, measurement system analysis (MSA) tools, six sigma methods, Taguchi´s design of experiments [3].

2 Capability studies

An important initial step in any process quality improvement is to quantify the capability of the product realization process (manufacturing process). Capability is defined [4] as the ability of the process to realize a characteristic that will fullfil the requirements for that characteristic. According to ISO standards, any organization shall measure the characteristics of the product to verify that product requirements have been met carried out at appropriate stages of the product realization process.

The capability study is an important concept to understand [5], which has gained wide acceptance around the world as a tool for process quality improvement. A process capability study is a set of calculations used to evaluate wheather a process (generally, a sistem) is statistical able to meet a set of specifications or requirements.

Automotive Industry Action Group [6] distinguishes between long-term and short-term capability studies; a short-term capability study (e.g., machine capability study) evaluates the parts manufactured in one continuous production run; a long-term capability study (e.g., manufacturing process capability study) evaluates the parts manufactured over a longer time which is representative of the variation encountered in series production.

Capability studies must be done not only to the machine on which the manufacturing process is running and to the manufacturing process, but also to the measurement system used in taking measurements of product characteristics.

Capability studies are performed for a new or changed production process (including assembly) in order to verify initially a preliminary capability and then over time to obtain additional inputs for controlling the process [7].

A capability study is summarized in indices; the purpose of these measures is to show the system's ability to meet its requirements, respectively to determine whether the system is capable of producing products which are within the customer specifications [8].

There are many capability indices that have been developed through time and they are divided into: capability indices of the machine, capability indices of the measurement system and capability/performance indices of the manufacturing process. The basic commonly used indices are presented in the following.

2.1 Machine capability study

Machine capability study is a short-term study and refers exclusively to the ability of a machine to produce parts. A machine capability study includes the following steps [9]:

- Manufacturing of a representative number of parts (minimum 50, if possible 100) in a continuous, uninterrupted production run;
- Measurement of the part characteristic(s);
- Calculation of capability indices.

Machine capability is summarized in Cm and Cmk indices: Cm represents the possible potential of the machine when no systematic factors exist; Cmk describes the practical state of the machine with reference to how centered mean value offsets are considering upper and lower specification limits. A machine is capable if Cmk $\geq$ 1.67.

2.2 Measurement system capability study

When one bases decisions on production data, the accuracy of that data is critical [2]. If the measurement system has too much variation, it will adversely influence the decisions. The measuring equipment (gage, gauge), the appraiser (evaluator, inspector) and the measuring procedure determine the variation due to the measurement system.

Information on how to conduct a measurement system analysis (measurement system capability study) are given in [2, 10]. An usual procedure is to do a repeatability and reproducibility (gage R&R) study. Repeatability is the variation in the measurements obtained by one appraiser while measuring the same item repeatedly; this is also called equipment variation. Reproducibility is the variation in the measurements obtained by several appraisers while measuring the same item repeatedly; this is called appraiser variation, too. Repeatability and reproducibility represent measurement system variability.

An ideal measurement system should not have any variation, but this is impossible; a common rule of thumb is that gage R&R variation should be less than 10 % of the smaller of either the process variation or tolerance (specification).

2.3 Manufacturing process capability study

The purpose of process capability study is to obtain a better understanding of process variation and to quantify how well the process can produce parts fulfilling the requirement.

Manufacturing process capability study can be done in short-term as well as in long-term by using control charts. In a short term study, the interest is in evaluating inherent process variation; many factors such as tools, operators, environment etc. do not change appreciably during study. Long term process capability study allows effects of many factors in the process such as operators, raw material, tools etc [11].

When one measures relatively less number of parts (about 50 parts are recommended) short-term capability is evaluated; for a long-term capability study, a minimum of 100-125 parts must be measured. A process capability study includes the following steps [9, 12]:

• Select and examined parts from series production in "rational" samples; for short-term study at least 10 subgroups should be evaluated; for long-term study at least 20-25 subgroups should be evaluated. The preferred subgroup size is n = 5 parts;
• Measure part characteristics and record the results along with production sequence;
• Statistical evaluation of the data: statistical distribution and stability;
• Calculate process capability and performance indices.

Most widely used indices are process capability indices Cp, Cpk and process performance indices Pp, Ppk [5]. The difference between these two indeces category is the way the process standard deviation is calculated. Process capability indices measures how capable the process is to satisfy specification limits by using the "within subgroup variation" which analyzes the shifts and drifts within subgroup. Higher Cp and Cpk value, more capable the process is. Many companies use Cpk $\geq$ 1.33 as benchmark value to determine whether the process is capable. Process performance indices Pp and Ppk consider "between (overall) subgroup variation" and they take into account the shifts and drifts between subgroups. Ppk $\geq$ 1.67 is a common value for most companies.

There are two critical assumptions to consider when performing process capability analyses. If they are not met, the resulting statistics may be highly unreliable [13]:

• The output data must follow a normal distribution. There are several analysis methods used to determine if the data are normally distributed or not: graphical methods as histogram and normal probability plot and analytical methods, goodness-of-fit tests.
• The process must be in statistical control (stable). Control charts are used to verify if a process is under statistical control.

 DOI: 10.1051/matecconf/201712105008

3 Case study

Before a serial production of a certain product type, the manufacturing process must first be omologated in order to better meet customer requirements and this calls for a preliminary capability study. The aim is to see that the manufacturing process is capable of production within specifications, which implies a certain minimum value for capability indices.

Preliminary capability study includes machine capability study, measurement system capability analysis and preliminary process capability study. After the process has been omologated (preliminary capability was demonstrated) production can be conducted. During production, the process will be regularly reviewed by using capability studies, to ensure an economic production level and of adequate quality.

To determine the general ability of the manufacturing process to meet customer specifications, capability indices of machine and gage and preliminary performance indices of process for all critical product characteristics should be calculated. Then from time to time, over the life of the product, process capability indices must be generated [14].

3.1 Company and analyzed product

In last years, Sibiu has attracted significant investment in automotive industry and some of the largest manufacturers of automotive parts in the world have opened production facilities in the area. One of these companies is CMP, expert in automotive anti-vibration systems.

CMP designs, develops and manufactures anti-vibration components, products which are part of the chassis systems for passengers and commercial vehicles. They can be categorised by functionality as products which are found in suspension systems (these are joints elements for front and rear axle mechanisms) and products which elastically support units in the structure of the vehicle, emphasising among them the engine mounts.

The case study presented in the paper analyzes the swaging process (a process that ensures an assembly dimension) for a new product "rubber bushing with mold bonded metal sleeves" (part number 14001). The critical quality characteristic of the product is the outer diameter, after the swaging process (Fig. 1).

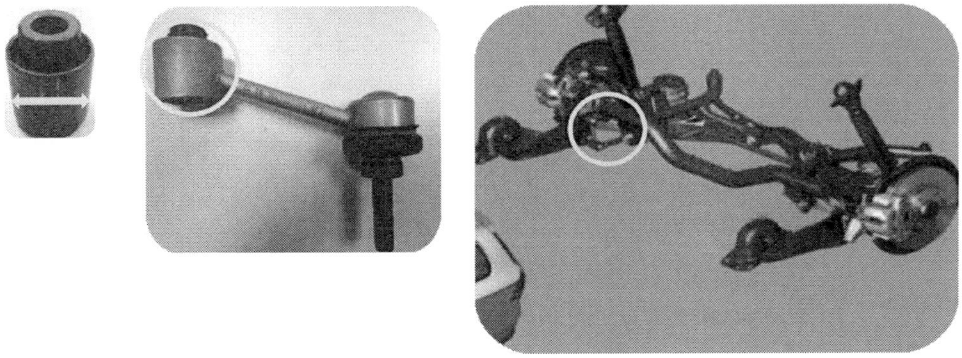

Fig. 1. Rubber bushing with mold bonded metal sleeves

The part is calibrated from initial diameter of Ø 27.3 +/- 0.1 to a final diameter of Ø 26 +0.45 / +0.25 on an automatic swaging machine with sectors, whose primary function is to reduce the outer diameter of the part with approximately 1 mm and also to compress the vulcanized rubber between the metal (inner and outer insert) sleeves of the bushing.

To achieve measurements, a digital caliper with a resolution of 0.01 mm was used (calibrated according to the internal calibration plan) which was initially tested for its capability, by using a repeatability and reproducibility study (gage R&R study).

Many Statistical Process Control software packages that calculate capability, produce control charts and perform measurement system analysis are available, one of the most common and simple being Minitab software. The analysis in this paper was performed with Minitab® 17, one of the most used software for quality improvement.

3.2 Results and remarks on capability studies

3.2.1 Measurement system capability

When quantifying process capability is important to assure that measurement system is capable. In this study, a R&R analysis was performed, respectively three inspectors measured three times with a digital caliper (resolution of 0.01 mm), the outer diameter of 10 "rubber bushing with mold bonded metal sleeves" (part number 14001), which had been selected to represent the full range of the manufacturing variation.

To assess measurement system capability (total gage R&R), the calculated specific metrics were calculated according to AIAG guidelines [2] and are presented in table 1.

- Variance component (VarComp) is used to assess the variation contributed by each source of measurement error relative to the total variation; in a good measurement system, the largest component of variation is part-to-part variation;
- %Contribution (of VarComp) is based on the variance component estimates; when %Contribution for part-to-part is high and for gage R&R as small as possible (less than 1%;), the measurement system is able to distinguish differences between parts;
- %Study Variation and/or %Tolerance; gage R&R may be calculated whether as a percent of study variation (%StudyVar) or as a percent of tolerance (%Tolerance) and must be less than 1% for a good measurement system. If the analysis is used for process improvement (reducing part-to-part variation), %StudyVar is a better estimate of measurement system capability and if it is used to evaluate parts relative to specifications, %Tolerance is a more appropriate metric;
- Number of distinct categories, tells how many separate groups of parts the measurement system is able to distinguish (it must be minimum 5).

In this study, according to AIAG guidelines [2], the measuring system contributes very little to the overall variation, as confirmed by gage R&R results from table 1, because:

Table 1. Gage R&R study -Xbar/R method

Source	VarComp	%Contribution (of VarComp)	Study Variation (6xStandDev)	%Study Variation (%StudyVar)
Total gage R&R	0.0000126	0.63	0.021264	7.96
Repeatability	0.0000126	0.63	0.021264	7.96
Reproducibility	0.0000000	0.00	0.0000000	0.00
Part-to-Part	0.0019691	99.37	0.266247	99.68
Total Variation	0.0019817	100.00	0.267095	100.00
Number of Distinct Categories (ndc)				17

- The variation due to the measuring system as a %Contribution is 0.63%, within acceptable range of less than 1%; therefore, 99.37 % of the total variation in the measurements is due to the differences between parts;
- The %StudyVar results (7.96) indicate that the measurement system accounts for less than 10% of the overall variation in this study. For this case study, 7.96% of the variation is estimated to be due to repeatability and 0% to reproducibility.
- The number of distinct categories is 17, which indicates the measurement system can distinguish between parts extremely well.

Based on AIAG guidelines [2] this measurement system is acceptable (adequate), meaning that the measurement system is capable to consistently record the same measurement of the outer diameter on this specific part.

3.2.2 Swaging machine capability

The study of the swaging machine capability is a preliminary analysis which refers exclusively to the ability of the machine to produce rubber bushing with mold bonded metal sleeves (part number 14001) within required tolerances.

Data of 50 rubber bushings were collected from a continuous, uninterrupted production run of 300 parts on the swaging machine; because it is a calibration with sectors, for each part were used two values, the minimum and maximum to determine an average value of the outer diameter. Design specifications for this characteristic are 25.25 and 25.45 mm.

The capability of the swaging machine has been performed in Minitab (Fig. 2).

Mean= 26,342
Standard deviation = 0,006468
Median = 26,34
Skewness = 0,2003
Kurtosis = - 0,7105

Cm=5.15
Cmk=4.74

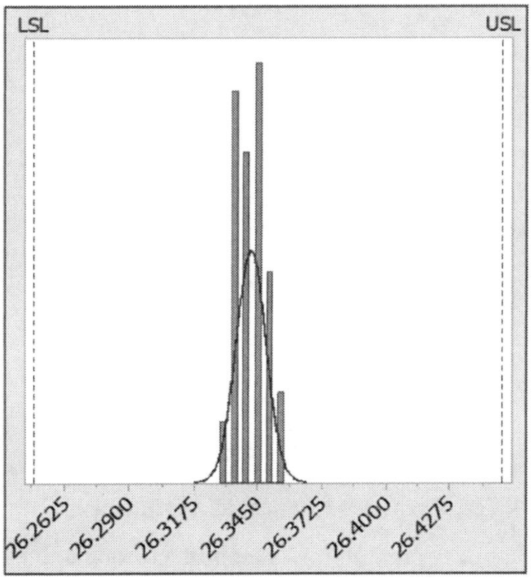

Fig. 2. Swaging machine capability

The capability of swaging machine is evaluated using the indices of capability Cm and Cmk, where Cm indicates variance and Cmk position. A normal requirement is that Cmk should be at least 1.67.

The capability of the swaging machine is very high (Cmk = 4.74); this means that it is a good machine with a small spread in relation to the tolerance with.

The slight difference between Cm (Cmk = 5.15) and Cmk indicate that the results are not perfectly centered but are slightly displaced to the lower specification limit.

High values of these indices are synonymous with good quality and high yield of the swaging machine under investigation.

DOI: 10.1051/matecconf/201712105008

3.2.3 Preliminary process capability

In the automotive industry, the Production Part Approval Process (PPAP) states "An acceptable level of preliminary process capability must be determined prior to submission for all characteristics designated by the customer or supplier as safety, key, critical, or significant, that can be evaluated using variables (measured) data."

Because the paper presents a process for a new product, a preliminary process capability study (short term study) was necessary to obtain early information on the performance of the swaging process relative to requirements. To determine the preliminary capability, 50 parts were randomly collected from the same parts produced for preparation of PPAP file.

Before estimating preliminary proces capability, two assumptions have been verified: the quality characteristic has a normal distribution and observations are random and independent of each other.

Figure 3 shows normal probability plot for the case study data, performed with Minitab: mean is 26.34, standard deviation is 0.006468, Ryan-Joiner test statistic value is 0.997 and P-value is greater 0.100 so more than the significance level (α=0.05) wich implies that the data is distributed normally.Thus, it was concluded that the data appears to be normal.

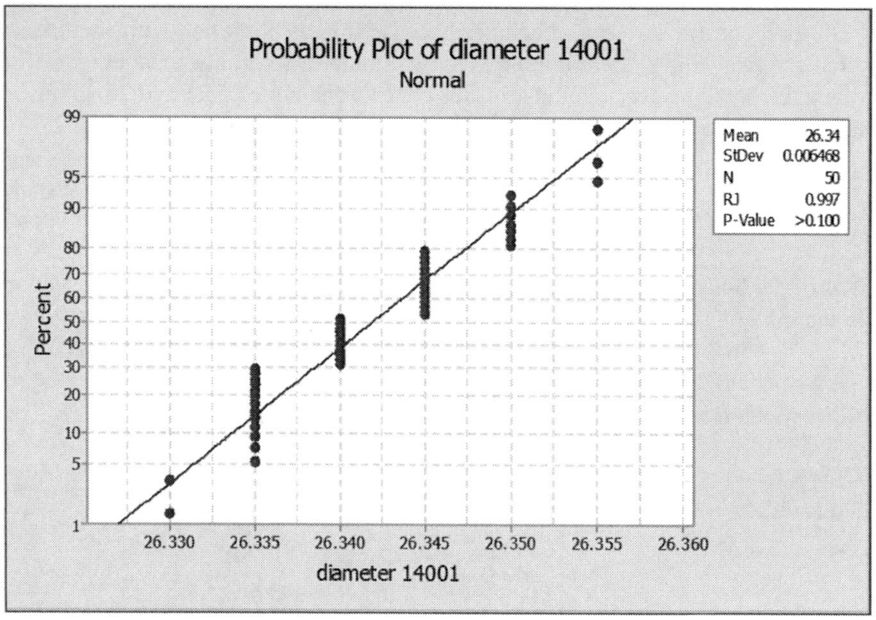

Fig. 3. Test for checking the normality assumption

Figure 4 presents the runs test, a nonparametric test for checking the assumption of randomness using Minitab software.

We see that the P-values for clustering, mixtures, trends and oscillation are greater than alpha value of 0.05 and the actual numbers of runs are close to the expected number of runs, so it was concluded that the data is independent and random.

After validating these critical assumptions, preliminary capability indices Pp and Ppk were calculated and were obtained the results shown in figure 5. It is noted that the swaging process has a preliminary capability that meets the requirements: Ppk = 4.11 > 1.67 and PPM = 0.

In addition, values of Pp (4.37) and Ppk (4.11) very close indicate that the process is well centered.

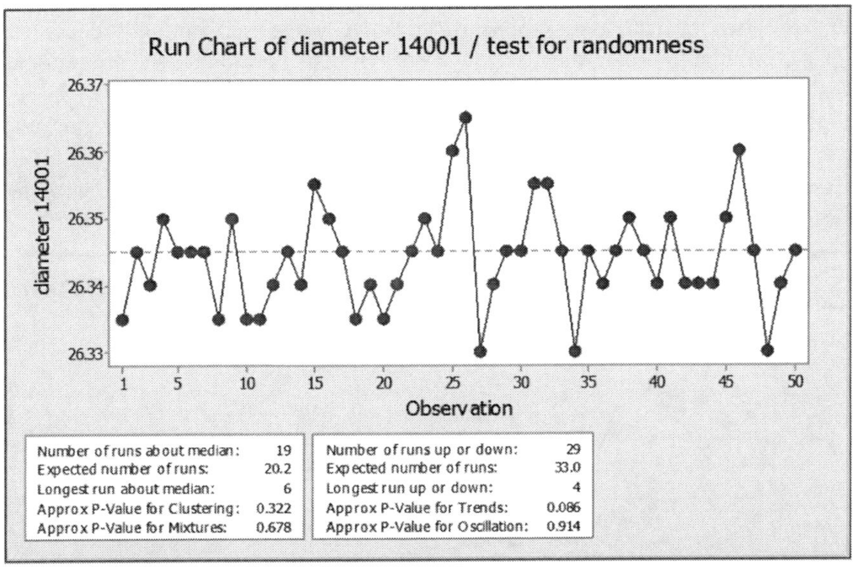

Fig. 4. Test for randomness

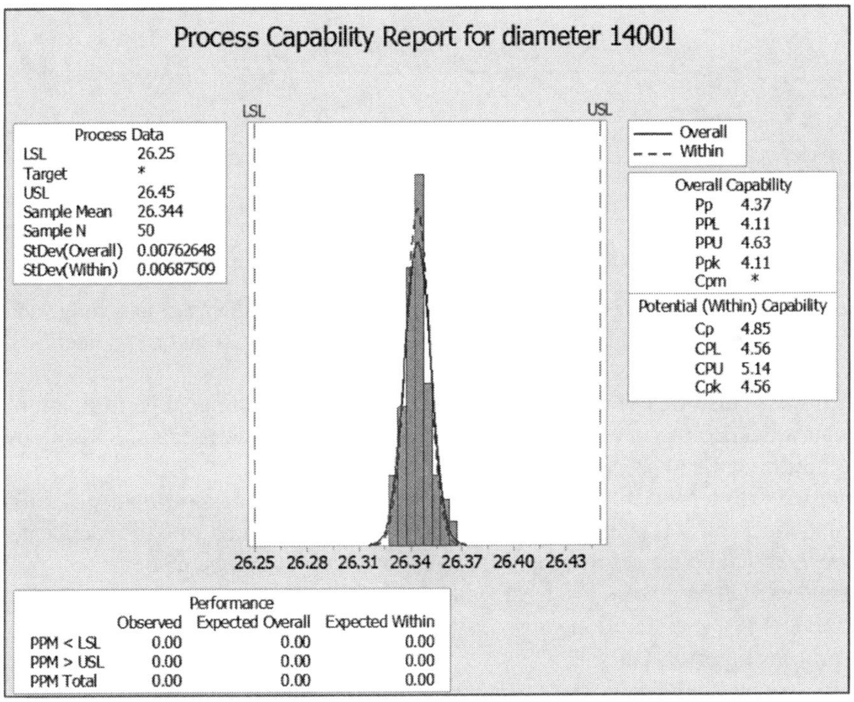

Fig. 5. Preliminary process capability

3.2.4 Process capability and performance

Capability and performance of the calibration process were analyzed during the first production serie for the rubber bushing with mold bonded metal sleeves (part number 14001), when there were vulcanized and then calibrated a number of 20 000 parts.

For this analysis, 24 subgroups have been collected each hour, during three different work shifts to include as many potential sources of variation; the subgroup size is 5 parts.

There are two critical assumptions to consider when performing process capability analyses [13]: the output data must follow a normal distribution and the process must be in statistical control (stable).

To ensure that the data is normal distributed, it was conducted again a normality test whose result was positive, the value of p being greater than 0.10.

To assess whether the data is statistically stable (in statistical control), with no special causes of variation present, a control chart by variables was completed (Fig. 6).

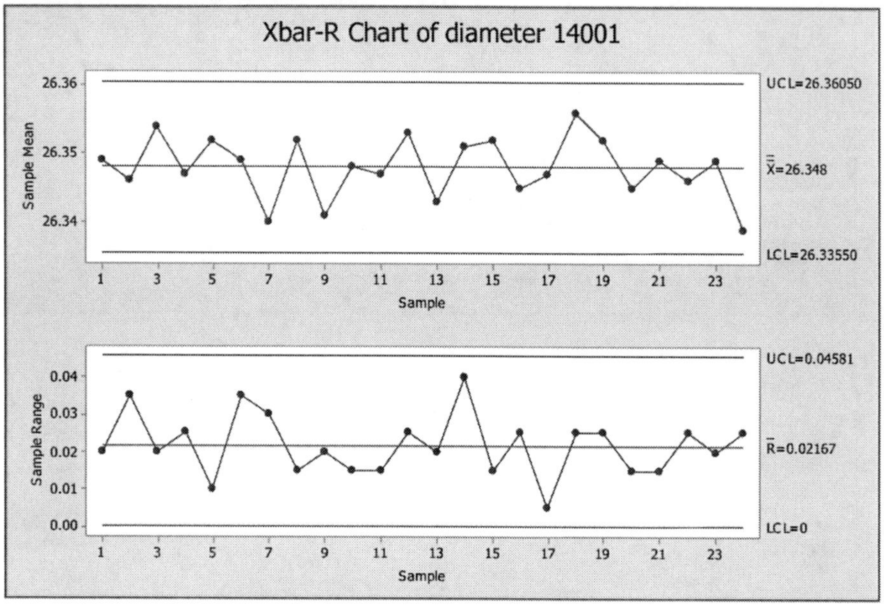

Fig. 6. Control chart by variables for case study data

From figure 6 it can be observed that all plotted subgroup mean and range values are within the control limits on both as XBar chart as well as R chart and no indication of trend, shift, run and clustering has been depicted.

So, the conclusion is that the swaging process is under statistical control and operating under the influence of only chance causes of variation, i.e. the process is stable over time both as adjustment and precision.

After validating the critical assumptions on the swaging process (normally distributed and under statistical control) overall capability analysis was performed, and the results are present in figure 7. Based on the results, the following observations can be made:

• The process is capable on short term, with capability indices: Cp=3.52/Cpk=3.45>1.33

• The process is capable on long term, with performance indices: Pp=3.50/Ppk=3.43>1.67, so the process is performant;

• The process meets inclusive the requirement for six sigma, Cpk≥2.00;

• Cp and Cpk values, respectively Pp and Ppk, very close indicate that the process is well centered;

• Cpk and Ppk values almost identical, show that process variation is due only to the random causes and no systematic variation occur, so the process is stable;

• No single index should be used to describe a process. For a better understanding of the process and to carry out correct decisions, it is strongly recomended that all for indices (Cp, Cpk and Pp, Ppk) be calculated and compared on the same data set;

• Parts per Million (PPM) is zero. This index indicates the ratio between the number of pieces exceeding the specifcation limits and a million produced units and his value represents the loss of quality due to special causes.

In conclusion, the results showed that we are dealing with a normaly distributed and stable process that comply with customer requirements both short-term and long-term.

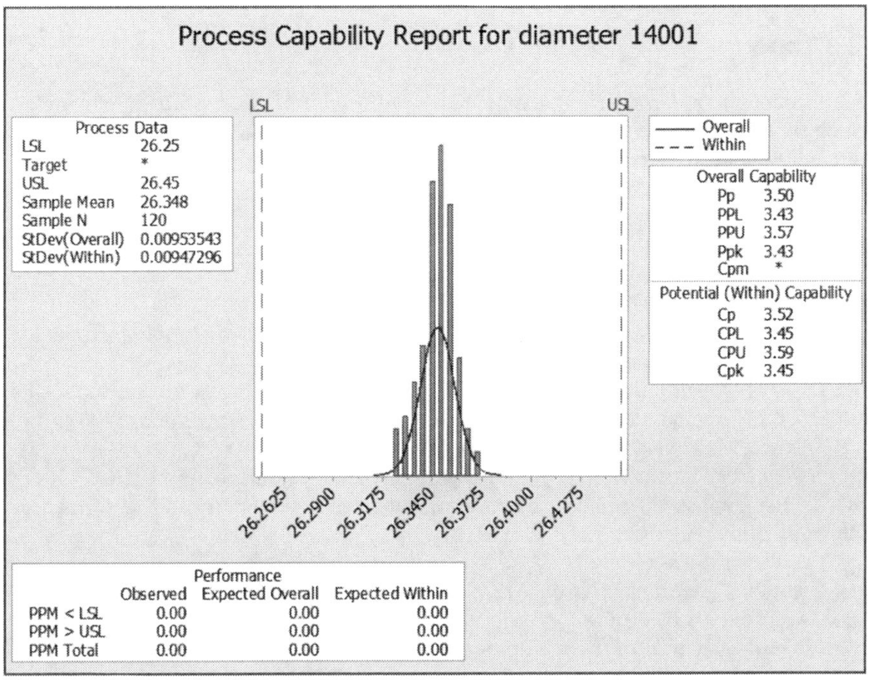

Fig. 7. Process capability and performance

If capability analysis is performed, Pp and Ppk indices will show approximately what happened in the past, but cannot be used to predict capability in the future. It will provide only a snapshot of the process at best. If, however, a system is stable, capability analysis through Cp and Cpk indices shows not only the ability of the system in the past, but also, if the system remains stable, predicts the future performance of the system.

4 Conclusions

The aim of this study was to highlight what powerful tools are capability studies for monitoring and ensuring the quality of products to satisfy the customer requirements, but also the importance of correct understanding, use and interpretation of capability indices.

This knowledge is necessary for controlling the process efficiently and effectively so that the products realized by the process fulfil the quality requirement.

No single index should be used to describe a process. For a better understanding of the process and to carry out correct decisions, it is strongly recomended that more indices be calculated and compared on the same data set.

Before quantifying the capability indices, validation of certain critical assumptions which allow the use of such metrics to have a meaningful interpretation, but that are frequently overlooked, was realized with the help of some quality statistical tools like normal probability plot, run chart and control charts by using the Minitab software.

The case study was conducted in a company activating in the automotive industry from Sibiu area, expert in automotive anti-vibration systems, to show the importance of capability studies in manufacturing process control. The automotive industry belongs to one of the largest industries and their manufacturing systems are placed before an increasingly fierce competition, within which the analysis of processes by the use of capability and performance indices came to play a key role since they assist to decide whether a manufacturing process is suitable to satisfy the customer quality requirements.

References

1. D. C. Montgomery, *Introduction to Statistical quality control, Sixth edition* (John Wiley & Sons, Inc., New York, 2009)
2. Automotive Industry Action Group, AIAG, *Measurement Systems Analysis, Reference Manual, Fourth edition* (Detroit-Michigan, USA, June 2010)
3. M. Mottonen, P. Belt, J. Harkonen, H. Haapasalo, P. Kess, The O. Ind. & Man. Eng. J. **1**, 29 (2008)
4. ISO 22514-1 Statistical methods in process management. Capability and performance - Part 1: General principles and concepts (2014)
5. https://elsmar.com/pdf_files/CPK.pdf, Accessed: 01.03.2017
6. Automotive Industry Action Group, AIAG, *Statistical Process Control, Reference manual, Second edition* (Detroit-Michigan, USA, 2005)
7. ISO/TS 16949 Quality management systems. Particular requirements for the application of ISO 9001:2000 for automotive production and relevant service part organizations (2009)
8. S. P. Wanare, M. V. Gudadhe, IJIET, **2** (2), 198 (2013)
9. https://wenku.baidu.com/view/12db0ffd04a1b0717fd5dd3e.html, Accessed: 01.03.2017
10. ISO 22514-7 Statistical methods in process management. Capability and performance - Part 7: Capability of measurement processes (2016)
11. http://www.world-class-quality.com/images/article/20080806154501_Initial%20process%20capability%20QW%20Feb07.pdf, Accessed: 05.03.2017
12. ISO 22514-4 Statistical methods in process management. Capability and performance - Part 4: Process capability estimates and performance measures (2016)
13. https://www.minitab.com/uploadedFiles/Content/News/Published_Articles/capability_analysis_1.pdf, Accessed: 05.03.2017
14. http://www.symphonytech.com/articles/processcapability.htm, Accessed: 05.03.2017

Considerations on oversized transportation in the UE(Community)

Mircea Bădescu[1], and *Carmen Maria* Purcar[1,*]

[1]Lucian Blaga University of Sibiu, Industrial Engineering and Management Department, 10 Victoriei Bdv, 550500 Sibiu, Romania

Abstract. The transportation problem is becoming increasingly complex at a national, European and global level. Achieving an efficient and safe transportation system with low environmental pollution, with a quick delivery is a requirement for any carrier in the 21st century. When talking about oversized transport, the challenge is even greater. Making a low loader go through several European countries has proved to be more complicated than we expected, due to the lack of uniformity of laws in this field. This paper seeks to find these different regularities characteristics from one European country to another and to propose developing procedures leading to the design and implementation of cheaper and faster oversized transport, with a lower environmental pollution and fewer human and material resources involved.

1 Introduction

Transportation is a technical and economic field in continuous development. As the production of goods is growing and desires of customers are becoming increasingly demanding, products must reach customers on tight deadlines, safely, with minimal costs, respecting the environmental conditions more restrictive.

The oversized transportation issue is more delicate. Special specifications are required for drivers, vehicles, infrastructure, special transportation permits, the maximum permitted dimensions, marking vehicles accompanying crews and more. Putting these specifications in full correlation, especially for the oversized transportation through many countries, and possibly multimodal regimen is a challenge that carriers and other economic factors, political and social must agree upon. To simplify the design and implementation of a low loader is a matter of creating an uniform communitarian law as well as creating European transport corridors, involving very diverse factors in decisions to implement such measures.

2 Current state

For a while there have been concerns in the European Comunity about the simplification of the process of oversized tranzit. The European Directive 96/53 establishes what are the

*Corresponding author: carmen.purcar@ulbsibiu.ro

maximum admitted dimensions of oversized cargo the maximum authorized weights, and the maximum authorized axle loads.

In 2006 the European Commission issued guidelines on good practices for oversized vehicle transit in Europe. These guidelines, jointly developed by European sector experts, member states and EU Commission, are aimed at facilitating the effective performance of oversized transport within the territory of the EU, the enhancement of operations relating to this kind of activities and the introduction of greater transparency in the area of OC transport. One of the main issues in the guidelines is the designation of transportation corridors for oversized cargo. These corridors should comprise economically important roads included in the E-way programme, other motorways, roads leading to industrial areas, ports, container terminals etc. Information on such corridors should be available for all parties dealing with oversized transportation: carriers, police agents, permit issuing authorities, bodies responsible for building and maintaining bridges, roads, tunnels etc. That way the oversized vehicle operations could be carried out on roads best suited for the purpose, belonging to the road corridor network where simplified application procedures are used [1].

Issues related to the development of transport corridors in the southern Baltic Sea for the economic development of the area were studied [1]. A network of information and possible routes of local transportation were created. In the study, oversized transport in the southern Baltic Sea was started to highlight economically developed areas, considering the main recipients of goods produced, along with prospects for medium and long-term development. There were taken into account all transport options adjoining their infrastructure and development opportunities for each. All have led to the inclusion of new corridors in the Berlin Moscow corridor. There have been analyzed oversized tranzit problems on smaller areas such as Lithuania [2] or Baltic Sea area. There have been created networks containing the possible and disponible routes from some areas using the GIS systems [3]. It has also been attempted to optimize the route for the oversized traffic taking into account the radii of curvature of the track intersections (4).

UAB Omega Technology (Norway) established in Klaipeda Science and Technology Park (KSTP), together with KSTP are creating the OTIN system (Oversize Transport Information Network). This system will help transportation companies get information about oversized cargo transportation and obtain necessary permits [5]. The oversized transportation strategy is developed in order to create a unified regional approach (South Baltic region) for oversized cargo transportation procedures and to identify four key aspects of development: transportation corridors, legal framework, economic impact, and improvements of infrastructure and technology. The system is intended to integrate with existing systems in Sweden and Germany thus significantly improving the accessibility of permits issuing services. In countries where information systems were not adapted to support web services, the extended version of the system will be used, allowing to simplify the communication between the authorities and carriers. Following the implementation of the project the development of the system both in functional and geographical context is planned.

The problem with transportation increasing in size and weight was combined with that of the multimodal transport [6]. The general trend is to increase the permissible parameters of means of transport, including weight, while cargo should reach destinations increasingly remote. However, the solution also aims for increasing the speed of transport, minimizing costs, enhancing traffic safety, means of transport and volume of cargo loaded. In the past few years in Europe many studies and scientific papers have been written and many opinions of experts on the impact of transport LHV (Longer and Heavier Vehicles) on the transport market have been expressed. The studies were coordinated by European and national authorities, research institutes, organizations and associations involved in transport. Of course the conclusions of these studies are not absolute, they are the following:

- compared with standard road trucks, LHV transportation reduce costs by 15%;
- LHV main transportion limitations come from the increase of fuel prices, several large salaries and maintenance costs;
- it is difficult to estimate the volume of freight transport over the LHV from rail operators on long distances at Community level;
- the highest costs related to the free movement of LHV are related to adapting infrastructure. These costs alone, will limit this transportion option for some countries.

Thus, the main transportation corridors to be used by LHV starts in Britain, through Benelux and Central Europe to Italy [6].

The oversized traffic can be regarded as a specific project logistics transport services, [7].

In some parts of Europe there have been examined more heavy tranzit routes and the optimal route was chosen according to certain restritive conditions through criteria analysis [8].

Risk evaluation and management are one of the key issues during the planning of oversized goods safe transportation and investments into transport infrastructure reconstruction. Usually it is international transportation and decision-making that requires through analysis of the problem both, on the national and international scale and only then the most rational decision (transportation route) can be made taking into consideration the effective risk lowering. The objective of the risk evaluation when investing into infrastructure reconstruction is to safeguard heavyweight or oversized goods transportation. When making the quantitative and economic assessment of the risk, the theory of chances and mathematic statistics are the most commonly used methods, because they are denoted to describe an event or process in case of ambiguous possibilities [8].

It can be concluded that there are serious concerns for optimizing Community oversized transport, with many factors involved. Computer networks and databases for easier access to information that facilitate the design and implementation of this type of transport were created as a consequence. There are proposals for the realization of European transport corridors oversized: the southern Baltic Sea, Berlin – Moscow, Britain to Central Europe and Italy.

3 Studies on implementing an oversized transportation system on the route Mokrice (Croatia) - Lulea (Sweden)

Documentation on oversized shipments in Europe was imposed, inter alia the need for such transport on the route Mokrice (Croatia) - Lulea (Sweden). From the oversized transportation organization, it has emerged that the design and implementation of a European heavy transportation system are not so easy because non-uniformity of rules and documents is required from country to country. These problems lead to high costs, prolonged transport, inefficient use of human resources, financial resources and means of transportation. We have analyzed the documents required for transit in countries like Romania, Hungary, Croatia, Austria, Slovenia, Sweden, Lithuania, Germany, Czech Republic, Poland or the directly documentation to the competent authorities of some of these countries.

Documentation in order to obtain special permits carriers has been put together and it can highlight a number of differences between ways of obtaining them, which constitute the major issues identified in carrying oversized transport in the EU:

- between the settlement terms of a special transport permits (between 3 days in Sweden and 15 days in Slovakia and Germany);

- between the number of crew accompanying, an economic performing operator who will make specialized accompaniment and the number of persons involved;
- between a transport permits format and language completing;
- a crew accompanying police from the road service area (Slovenia - without police);
- differences in costs for special transport permits
- differences in affordability in terms of the number of trucks used based on a permit:
  - one transport, one route, one assembly (Hungary, Romania);
  - any transport a limited period (3 months Austria);
  - for several ensembles in the same configuration defined by the number of axles of the tractor head and of the semi-trailer (Germany);
  - in Austria the authorization shall be issued for semitrailer (you can use any tractor);
- differences in terms of validity:
  - less than three months (Austria);
  - six month (Poland);
  - indicated by the carrier, within certain limits (Germany);
  - one transport, one route from one assembly;
- transit country is achieved only with crew accompanying certificate under the law of that country (in Sweden may be accompanying crew from another country, a car to maximum three trucks);
- documents accompanying vehicles;
- marking conditions of the motor vehicle;
- responsibility for assessing accessibility route (artwork, height, undercrossing elements of road infrastructure and utilities);
- the existence and use of appliances for height reduction;
- conditions eliminate any obstacles on trails (who requires authorization from whom);
- additional means of signalization road in case of accidental stationary roadway;
- the reasons for their refusals permits;
- conditionings trails:
  - avoid certain highways during holidays;
  - correlate with specific timetables for other terminals (multimodal transport);
- conditioning periods:
  - from Monday hour 9 am to Friday hour 3 pm;
  - between 10 pm and 6 am;
  - correlation with periods of traffic supported by other countries.
- coordination of transport made by:
  - carrier;
  - police;
  - traffic director (private Sweden).

Perhaps the solution of these drawbacks is the set up of a "Single office for obtaining special transportation permits".

In our case, several variants of routes were chosen, used depending on the circumstances, constraints, the beneficiary wishes, deadlines, gauge cargoes etc.

Also, the difficulty of such a project emerges from a comparison between "theory and practice" between the calculations preceding transportation project and collected on site or truck tracking system. Often there are a lot of contingencies situations along the transport operation.

DOI: 10.1051/matecconf/201712106001

4 Route description

Routes used were mostly known by drivers thanks to their experience. The first convoy was left to the most experienced drivers (transport recognition). To avoid problems, it was loaded with goods with insignificant exceeded gauge (the load was only 1 m longer than the loading, so overall 17.5 m).

Following this transport it was discovered that on the border between Hungary and Slovakia, exactly between the exit of the M15 (Hungary) and rise on the A2 (Slovakia) rehabilitation works were being performed of the road and its path was hence suited only for normal weight cars with minimum dimensions of 3,40 m width.

The first route was regarded as the optimal route because tolls for trucks were small, costs of transport permits and accompanying cars were not raised even if in Slovakia and the Czech Republic the time to reach them was relatively high. The restrictions of movement along this path were normal.

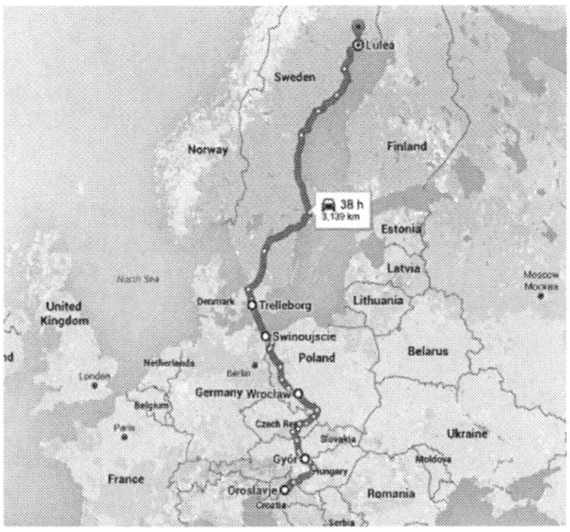

Fig. 1. Route 1 - Mokrice (Croatia) – Lulea (Sweden) via Gyor (Hungary) – Swinoujscie (Poland)

Following the findings of the first convoy yet another routewas adopted (Figure 2).

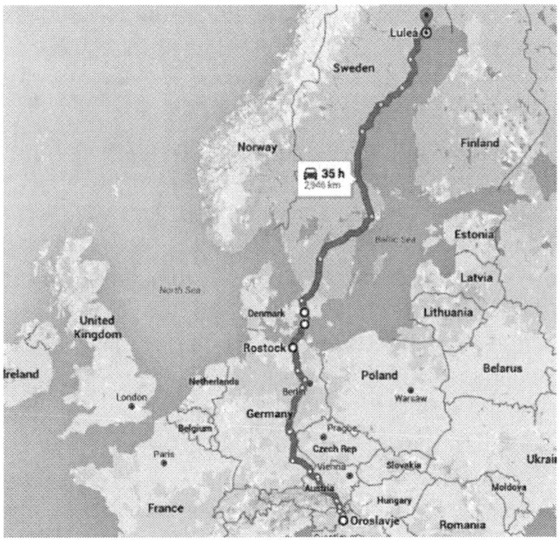

Fig. 2. Route 2 – Mokrice (Croatia) – Lulea (Sweden) via Austria-Rostock (Germany)

To browse this route the trucks has to pass through: Croatia, Slovenia, Austria and Germany reaching the port of Rostock in northern Germany, a place where they would board the ferry and continued their course to the port of Trelleborg in Sweden

From a financial standpoint this route was not the best due to high costs of obtaining special transportation permits and accompanying cars. Even if the distance is shorter than the first route (about 180 km), travel time was higher due to restrictions that impeded trucks to travel more than 600 km per night. Another factor that was considered a major impediment was represented by bureaucracy and the time for obtaining permits.

A particular problem was that the ferry reservation had to be made 3 days in advance. It was impossible to make a booking 3 days before because it meant that the place to be booked at the latest when the truck was loaded.

Along the way, many delays may have occured and this could have resulted in the loss of the reservation.

Fig. 3. Route 3 - Mokrice (Croatia) – Lulea (Sweden) via Austria - Czech Republic - Swinoujscie (Poland)

Also in Austria there is a succession of tunnels about to be closed in the opposite sense because their construction did not allow the passage of oversize trucks. Closing the tunnels was made after a specified timetable because the passage trucks overruns masses and dimensions is restricted during the day in Austria.

So, after the above was stated the need for a secondary route that were used when the seats were not found on ferryboat emerged (Figure 3). A third route would make a detour of Germany and of the tunnels in Austria. Transhipment ferryboat was made from Swinoujscie port (Poland) at the same port in Sweden, Trellerborg. Eventually it was found that this is the most effective route because of the existence of two daily ferry lines. Reservations must be made just a few hours before the arrival of the convoy to port. Also from a financial standpoint it was a mid-way solution with the best price - quality report in terms of transport permits, accompanying cars and ferries price, the difference in km being imperceptible.

Comparative presentation of the three routes is given in Table 1, showcasing the length, the costs and the limitations that apeared.

To choose between the three alternatives was considered appropriate performing a multi-criteria analysis [3], which takes into account the length of the route L, the duration D, the total cost C and the route constaints Con.

Table 1. Comparison between the three routes

Route no.	Transit countries	Length [km]/ Time [hour]	Cost permit transport	Cost accomp anying crew	Road tax	Limitation
1	via Gyor (Hungary) – Czech Republic - Swinoujscie (Poland)	3139/38	low	low	small	- work rehabilitation -road between Hungary and the Czech Republic – limitingwidth convoy to normal size
2	via Austria - Rostock (Germany)	2946/35	high	high	big	- Trucks advancing no more than 600 km per night - Ferry booking 3 days before - Transit tunnels in Austria
3	via Austria - Czech Republic - Swinoujscie (Poland)	2954/37	mediu	medium	medium	- ferry reservation a few hours before

The ranking criteria for carrier shows that the most important are D and Con while the last is L.

The calculation of the weights, acording to the multi-criteria analysis is presented in Table 2.

Table 2. The calculation of the weights

	L	D	C	Con	Σ points	Position	Weights
L	1/2	0	0	0	1/2	4	0,22
D	1	1/2	1	1/2	3	1,5	3
C	1	0	1/2	0	1,5	3	1,14
Con	1	1/2	1	1/2	3	1,5	3

The weight on the last column was calculated using the formula (empirical) Frisco:

$$\delta_i = \frac{p+\Delta p+m+0{,}5}{-\Delta p'+\frac{N_{crt}}{2}} \qquad (1)$$

where:
p is the sum of the points obtained on the line for the specific criteria;
Δp - the difference between the score of the considered element and theone from the last floor;
m - the number of outclassed criteria;
Δp' - the difference between the score of the considered element and theone from the top floor;
N_{crt} - number of criteria.

Next, each variant will be granted for each criterion. Notes are between 1:10 and a higher grade means abetter appreciation of the journey across the criterion. Note 10is the situation when option is perfect in terms of this criterion.

In Table 3 there are analyzed the three alternatives in terms of weights and grades of the four criteria opposite.

Table 3. The final score of the three alternatives

Criteria	Weight	Route 1		Route 2		Route 3	
		Note	Weight·Note	Note	Weight·Not e	Note	Weight·Not e
L	0,22	6	1,23	10	2,2	8	1,76
D	3	7	21	10	30	9	27
C	1,14	10	11,4	8	9,12	9	10,26
Con	3	7	21	5	15	10	30
			Sum=54,63		Sum=56,32		Sum=69,02

As one can see, the third route is the one who has the greatest score, so that in these circumstances it is reasonable to choose it.

So it can be said that the planning of the route to move combinations of vehicles has been complicated due to several factors (some unforeseen at the beginning):

- drivers experience;
- temporary restrictions of road works;
- restrictions on busy roads during certain times (highways in times of holidays);
- costs of transportation permits and accompanying cars;
- time (and bureaucracy) of obtaining transportation permits;
- advancement restrictions of trucks at night;
- time of booking to boarding the ferryboat (overcrowded that time of year);
- restrictions to the gauge cargo (not all vehicle assemblies were loaded equally);
- restrictions to the gauge during transit and tunnels in Austria.

From the above, it can be concluded that the establishment of an optimal route from the start, it is almost impossible. A route is suitable for a given load, for certain gauge restrictions, for certain costs, for a certain period of the week, and another is optimal within other constraints.

5 Conclusions

Following the study, there can be mentioned some conclusions:

1. From the legal perspective there are major differences between the component countries of the European Union regarding oversized transportation. These differences largely disappear if a "Single office for obtaining special transportation permits" would be created. But setting up such an office requires:

- Uniform legislation throughout the European Union;
- Reducing settlement terms of permit applications;
- The set up of accompanying crews certified in all European Community countries;
- Creating unified toll crossing permits;
- The use of combinations of vehicles of different configurations in the same Pass;
- Use of foreign languages in completing requests for authorization to transport;
- Establishing uniform standard application for transport permits;
- The presence of a representative of each country participating in such shipments.

2. In terms of routes, it is impossible to choose one optimal route. Therefore alternative routes were used for each type of cargo.

3. A car park properly equipped is a great facility in terms of transportation costs because various cargo transportation with minimal cost bring a great financial benefit to the carrier.

The information that the GPS system provides tracking of trucks is an option that carriers can not do without.
4. The GPS tracking of trucks, is an option that carriers can not do without.
5. Making a "Single office for obtaining special transportation permits" is a necessity that can ease the work and efficiency of transport carriers.

There is much talk of European corridors for carrying oversized transport, which is necessary and beneficial. We saw, however, that from the three routes used, each presented advantages and disadvantages, the most appropriate one variant or another, depending on certain temporary restrictions. There was performed a multi-criteria analysis, which takes into account the length of the route L, the duration D, the total cost C and the route constraints Con. Finally, the third route variant was found to be most suitable and could be a portion of a corridor linking southern to northern Europe.

The European Community's Large companies for oversized transport may suggest possible efficient routes of European corridors, alongside with companies that produce oversized products, their beneficiaries, the factors responsible for the environment, other economic and political factors.

References

1. A. Galor, W. Galor, B. Wisnicki, http://www.transportoversize.eu/en/articles/id/4030/ Accessed 20.02.2017
2. R. Palsaitis, A. Petraska, Transp. and Telec. J, **13**, 51 (2009)
3. Y. Li, T. Jerry, R. Danny et. Al. Mapping Oversized and Overweight Truck Routes with Procedure Based on Geographic Information Systems, **2291**, 8-16, (2012)
4. M. Lingkui, H. Zheenghua, H. Changquing, W. Zhang., T. Jia. Int. Jour. of Geo-Information, **4**, 2428-2445 (2015)
5. http://www.transportoversize.eu/en/articles/id/4026/ Accessed 15.11.2016
6. B. Wisnicki, W. Galor, http://www.transportoversize.eu/en/articles/id/4031/ Accessed 18.02.2017
7. I. Pisz, I. Lapunka. TranSopote Conference On Transport DevelopmentChallenges in the Twenty-First Century. Book Series: Springer Proceedings in Bussines and Economics, 139-160 (2016).
8. A Petraska, R. Palsaitis, N. Batarliene, et. al. Transport Means – Proceedings of the International Conference, 236-239 (2011)

Improving programming skills of Mechanical Engineering students by teaching in C# multi-objective optimizations methods

Adrian Florea[1,*], and *Ileana Ioana* Cofaru[1]

[1]'Lucian Blaga' University of Sibiu, Computer Science and Electrical Engineering Department, 550025, 4 Emil Cioran Street, Sibiu, Romania

Abstract. Designing an optimized suspension system that meet the main functions of comfort, safety and handling on poor quality roads is a goal for researchers. This paper represents a software development guide for designers of suspension systems with less programming skills that will enable them to implement their own optimization methods that improve traditional methods by using their domain knowledge.

1 Introduction

The teams dedicated for research, development and hardware-software deployment from industrial companies tend to become more heterogeneous, combining researchers with mechanical, electronic and informatics skills. Thus, possession of basic knowledge of programming by mechanical engineers is essential. Regarding the issue of suspension system optimization, in most approaches it is analysed just from mechanical viewpoint or mechanical accompanied by vision of transport and road vehicles engineers. In this case, the non IT specialists apply a series of software tools specific to computer science domain for simulation and optimization without regard or understand the particularity of optimization method, in whose kernel you could then intervene to customize them to bring a real scientific knowledge gain. In this paper we start the suspension optimization from the perspective of optimization methods, doing a research analysis from computers scientist viewpoint who wants to understand the engineering problem, to apply then a mathematical model and then learn to develop its own optimization tool, very parameterized, flexible and extensible.

Therefore, the methodology for modelling, simulation and optimization is tested on a mechanical problem, namely the passive suspension system of a Half-Car model (HCM) [1] with 4 Degrees of Freedom (4 DOF) because the suspension model design represents a challenging task for the vehicle designers due to multiple control parameters, complex and conflicting objectives, the system running under several existing constraints. In our previous work we analysed the quarter-car model (QCM) with two-degrees-of-freedom (2DOF) [2-3] in order to find optimal parameters of stability and to keep high standard of ride comfort under different exploiting conditions, minimizing the discomfort during

* Corresponding author: adrian.florea@ulbsibiu.ro

movement. Although the quarter car model is simple and widely used for dynamic performance analysis, it fails to capture the more realistic results of actual behaviour of the vehicle. Thus, in this work we analysed HCM because captures important characteristics of full car model. Although the passive suspension systems have some inherent constraints we consider them in our analysis due to their provided trade-off between the ride comfort level and vehicle stability at low cost. The passive systems consists in elastic (spring) and dissipative (damper) elements, their dynamic behaviour being described by the characteristics of the items referred. To have a good performance of handling and stability at different driving conditions (load, road profile, speed), the spring stiffness of suspension system should be very high, which determines the passenger to experience uncomfortable driving feeling. Conversely, if we expect that the suspension system to ensure a high level of ride quality protecting the vehicle and driver from harmful stresses, the spring stiffness should be decreased, which will determine the instability of vehicle when is drove.

This work can be seen also as a useful learning tool for students of Industrial (Mechanical) Engineering specializations that successfully combines in their training knowledge from different scientific fields such as: Computer Science (Artificial Intelligence, Programming Languages, and Distributed Computing), Mathematics (Differential Equations), Transportation Engineering (Vehicle Design), Mechanical Engineering (Vibrations) to solve a real problem, even critical in Romania, characterized by a developing infrastructure and poor quality roads.

The organization of the rest of this paper is as follows. In section 2 we review some related work solutions applied in suspensions optimization, whereas section 3 describes the dynamic suspension system in order to software implement. Section 4 presents the mathematical model, provides programming details on solving the system of differential equations by "*Runge-Kutta*" method, exhibits the application GUI and illustrates simulation results. Finally, section 5 suggests directions for future work and concludes the paper.

2 Related work

In [1] the authors investigate the vibration and ride features of a passive HCM model with 5 DOF vehicle suspension system excited by a sinusoidal surface. Their model parameters are kept fixed and made simulations in Simulink and Matlab for determining the amplitude of vibrations. The main differences between their work and this paper consist in the software vision of optimization methods that we performed (C# implementation of NSGA-II algorithm), the number of degrees of freedom and the random road profile that we tested.

In [2] the authors present the mathematical model of QCM suspension's system from OPEL cars and in [3] the authors apply multi-objective optimization techniques for designing the quarter car model on random and sinusoidal road profiles. Unlike these two previous works, in this paper we changed the model of suspension (analysing HCM) and also we have added two more degrees of freedom that expanded the differential equations system of order two.

Studying the vibrations induced by surface irregularities in road pavements in [4] the authors make some Matlab tests where they have shown that the road profile is a combination of a large number of longer and shorter holes of different amplitudes. Based on their equations we implemented the random road profiles in C#.

In [5] the authors present an optimization of a QCM 4-DOF vehicle's human with seat suspension system using weighted sum genetic algorithms (with fixed weights) to determine suspension parameters. Unlike their work we apply on HCM 4-DOF Pareto multi-objective optimization algorithms.

3 The Half-Car suspension system

The Half-Car model is a basic model used to simulate the vehicle's suspension system performance by investigating the dynamic response of cars when running on different surfaces. HCM includes an independent Front and Rear vertical suspension (see Figure 1). The HCM name is due to the fact that total vehicle mass is divided equally on two and might be seen as combination of two quarter car models. The spring has significant role in establishing the suspension function of the system, while the damper is used to dissipate kinetic energy of spring and supply a damping function. We consider that there is no effect of eccentricity of driver seat when located on the centre of gravity and the tires are always touching the ground.

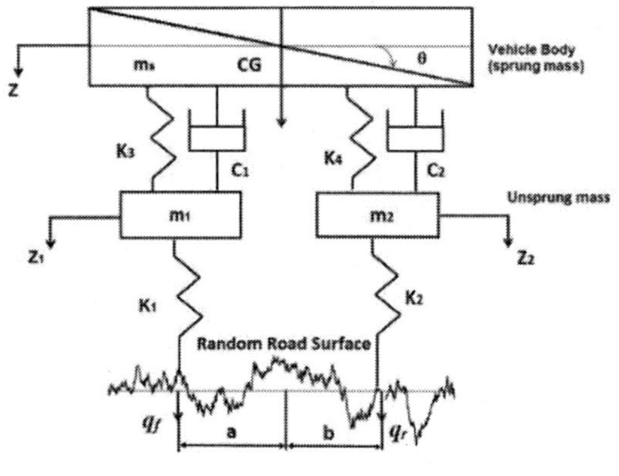

Fig. 1. The analytical model of half-car suspension system

We made the following notations:

- m_1, m_2, m_s: (in the following order) mass of the Front tire wheel, Rear tire wheel and the vehicle body (sprung mass).
- θ: pitching motion angle of car body.
- I_x: pitch moment of inertia.
- a: lateral distance between the body centroid and Front tire wheel.
- b: lateral distance between the body centroid and Rear tire wheel.
- c_1, c_2: Damping coefficient of the Front suspension and Rear suspension.
- k_1, k_2: Vertical spring stiffness of the Front tire, Rear tire.
- k_3, k_4: Vertical spring stiffness of the sprung body Front suspension and Rear suspension.
- Z_1, Z_2, Z_s: (in the following order) bounce motion of the Front tire wheel, Rear tire wheel and the vehicle body.

The four degree of freedom considered in this work are the three displacements: of car body, Front and Rear tire respectively, and the pitching angle of car body.

4 Modelling and software implementation of the suspension system

4.1 The mathematical model

In this section we described the mathematical model followed by C# software implementation of the half-car suspension model with four degrees-of-freedom (4-DOF) tested on a generic car. The purpose of modelling is to obtain a state space representation of the vehicle model. The mathematical model used for describing and studying of the suspension is provided by a system of four differential equations of order 2.

$$M \cdot \ddot{Z} + C \cdot \dot{Z} + K \cdot Z = F \cdot Q \tag{1}$$

In equation (1) M, C and K respectively, are mass matrix, damping matrix and stiffness matrix. The forces that are transmitted from the running unevenness surface are exhibited through Forces matrix (F).

Besides the aforementioned matrixes the system has as input the road excitation matrix Q, where $Q = (q_f \;\; q_{fr})^T$ (the height of potholes on front and rear wheel) and as outputs the generalized displacement vector Z, where $Z = (Z_1 \;\; Z_2 \;\; Z_s \;\; \theta)^T$.

$$M = \begin{pmatrix} m_1 & 0 & 0 & 0 \\ 0 & m_2 & 0 & 0 \\ 0 & 0 & m_s & 0 \\ 0 & 0 & 0 & I_x \end{pmatrix}; C = \begin{pmatrix} c_1 & 0 & -c_1 & -ac_1 \\ 0 & c_2 & -c_2 & bc_2 \\ -c_1 & -c_2 & c_1 + c_2 & ac_1 - bc_2 \\ -ac_1 & bc_2 & ac_1 - bc_2 & a^2c_1 + b^2c_2 \end{pmatrix};$$

$$K = \begin{pmatrix} k_1 + k_3 & 0 & -k_3 & -ak_3 \\ 0 & k_2 + k_4 & -k_4 & bk_4 \\ -k_3 & -k_4 & k_3 + k_4 & ak_3 - bk_4 \\ -ak_3 & bk_4 & ak_3 - bk_4 & a^2k_3 + b^2k_4 \end{pmatrix}; F = \begin{pmatrix} k_1 & 0 \\ 0 & k_2 \\ 0 & 0 \\ 0 & 0 \end{pmatrix} \tag{2}$$

The governing equations are:

$$\begin{cases} m_1 \cdot \ddot{Z}_1 + c_1 \cdot (\dot{Z}_1 - \dot{Z} - a \cdot \dot{\theta}) + k_1 \cdot Z_1 + k_3 \cdot (Z_1 - Z - a \cdot \theta) = k_1 \cdot q_f \\ m_2 \cdot \ddot{Z}_2 + c_2 \cdot (\dot{Z}_2 - \dot{Z} + b \cdot \dot{\theta}) + k_2 \cdot Z_2 + k_4 \cdot (Z_2 - Z + b \cdot \theta) = k_2 \cdot q_r \\ m_s \cdot \ddot{Z} + c_1 \cdot (\dot{Z} - \dot{Z}_1 + a \cdot \dot{\theta}) + c_2 \cdot (\dot{Z} - \dot{Z}_2 - b \cdot \dot{\theta}) + k_3 \cdot (Z - Z_1 + a \cdot \theta) + \\ \qquad + k_4 \cdot (Z - Z_2 - b \cdot \theta) = 0 \\ I_x \cdot \ddot{\theta} + c_1 \cdot (\dot{Z} - \dot{Z}_1 + a \cdot \dot{\theta}) \cdot a - c_2 \cdot (\dot{Z} - \dot{Z}_2 - b \cdot \dot{\theta}) \cdot b + + k_3 \cdot (Z - Z_1 + a \cdot \theta) \cdot a - \\ \qquad - k_4 \cdot (Z - Z_2 - b \cdot \theta) \cdot b = 0 \end{cases} \tag{3}$$

First two equations from system (2) describe the bounce motion of Front tire wheel and Rear tire wheel, respectively. The third equation illustrates the bounce motion of body vehicle while the last one shows the pitch motion of car. The system (2) will be transformed into 8 differential equations of first order by introducing new unknown functions. The next step aims solving the 8 kinematic equations of motion using C# DotNumerics.ODE [6] package (see section 4.2) by "*Runge-Kutta*" numerical time integration method.

4.2 The software design

The developed software application was implemented in Microsoft Visual Studio 2015 (C#), .NET Framework 4.5, using DotNumerics a numerical library for .NET [6, 7]. The class diagram of the project's solution is presented in Figure 2.

The class that contains the evaluation of the vehicle input variables and uses the "*Runge Kutta*" method [6] to solve the differential equations system in order to get the output variables of the system is the *SuspensionModelEvaluation* class. The principal method here is the *Evaluate* method, which receives as input parameters a *SuspensionDesignSolution* object and a *RandomRoadGenerator* object. First we briefly described these two classes and then we explain how *Evaluate* method works. Finally, it sets the displacement of the Front tire wheel, the displacement of Rear tire wheel and the body mass acceleration.

➢ *SuspensionDesignSolution* class contains the properties that characterize a solution:

- The design variables that we wanted to optimize (vertical stiffness of Front suspension (k_3), vertical stiffness of Rear suspension (k_4), vertical spring stiffness of the Front tire (k_1) and Rear tire (k_2), vertical damping coefficient of the Front suspension (c_1), vertical coefficient damping of Rear suspension (c_2)) as an array of type *DesignVariable*.
- The output variables that we want to optimize, such as *DisplacementSprungMassFrontTire* (Z_1), *DisplacementSprungMassRearTire* (Z_2), *bodyMassAcceleration* and their root mean square (RMS) version: *RmsDisplacementSprungMassFrontTire*, *RmsDisplacementSprungMassRearTire* and *RmsBodyAcceleration*.
- The *front number* (from all Pareto fronts) in which a solution is places, after evaluation of the results within NSGA-II algorithm.
- The *crowding distance* that is specific to the solution, also needed in NSGA-II.

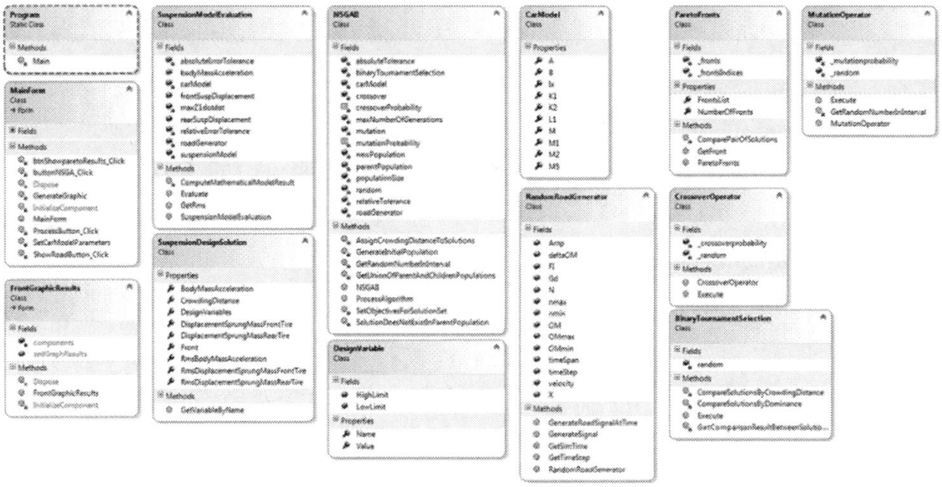

Fig. 2. The application classes

➢ *RandomRoadGenerator* is a class used for generating a random road signal on a specific length and with a certain velocity. The constructor gets a few parameters: the length of the road, the signal rate (step), the degree of roughness, the minimal and the maximal wavelength and the velocity.

- The method *GenerateSignal* calculates the road signal and returns an array of double values that represent the heights of the road on the specified length.
- Method *GenerateRoadSignalAtTime* is very important because it returns the road signal at a specific time. It is very useful when calculating the displacements and accelerations of vehicle model elements at a certain moment.
- Using *ZedGraph* functions package we realized 2D graphical representation of generated artificial road profiles. The *GenerateGraphic* method from principal *MainForm* class has the following parameters: a *ZedGraphControl* object, the title of the graphic, the X axis string, the Y axis string, the signal and the signal step.

➢ In the *Evaluate* method of the *SuspensionModelEvaluation* class we instantiate an *OdeExplicitRungeKutta45* object from DotNumerics library [6], establish the initial values of time (t0) and signal (y0), set the sampling interval of time array (t) and then with the method *Solve*(y0,t) we get as result a matrix that has the number of rows equal to the number of sampling values. In each row, there is an array of values that correspond to the outputs of the system that was solved with "*Runge Kutta*" method. The displacements and acceleration that we have interest in are saved in arrays and after iterating through all the samples, we calculate the RMS values of the outputs, which are saved in the *RmsDisplacementSprungMassFrontTire*, *RmsDisplacementSprungMassRearTire* and *RmsBodyAcceleration* properties of the solution of type *SuspensionDesignSolution*.

➢ The method *ComputeMathematicalModelResult* is used in the solving of differential equations system and it contains the real differential equations. It is sent as a parameter to OdeExplicitRungeKutta45 method.

4.2.1 Multi-objective optimization. Implementation of NSGA-II algorithm

Modified non-dominated sorting genetic algorithm (NSGA-II) [8] has been used for multi-objective optimization of a 4-DOF vehicle model which excited with random road profile. The conflicting objective functions considered were vertical acceleration of sprung mass, relative displacement between body mass and Front tire wheel and relative displacement between body mass and Rear tire wheel. NSGA-II is one of the most used algorithms in experiments due to its capacity of generating good solutions regardless of the problem [3]. Because the pseudo-code of NSGA-II algorithm is presented in [3, 8] here we are limiting to briefly describe how the main algorithm works:

- We have a parent population P_0 of chromosomes (design solutions) that is randomly initialized. The algorithm applies genetic operators (Binary tournament selection, crossover, and mutation operators) in order to create a child population Q_0 of size *N*. The old and new populations are combined together, then a sorting is done using non-dominance as criteria. Each solution is assigned a fitness equal to its *rank* (non-domination level where 1 is the best level). Thus, minimization of fitness is assumed. A combined population $R_t = P_t \cup Q_t$ is formed, where *t* is the generation number. The population R_t will be of size 2*N*. Then, the population R_t is sorted according to non-domination. The new parent population P_{t+1} is formed by adding solutions from the first fronts till the size exceeds *N*. The ranks and *crowding distance* (density of individuals surrounding a particular one) are used to guide the selection. An individual is better, if it has a smaller rank, or in case of equality, the one with the bigger crowding distance.
- In order to sort a population of size *N* according to the level of non-domination, each solution must be compared with every other solution in the population to find if it is dominated. This process is continued to find the members of the first non-dominated class for all population members. At this stage, all individuals in the first non-dominated front are found. In order to find the individuals in the next front, the solutions of the first front are temporarily discounted and the above procedure is repeated.
- First, for each solution we calculate two entities: (i) n_i, the number of solutions which dominate the *i* solution and (ii) S_i, a set of solutions which the *i* solution dominates. We identify all those points which have $n_i = 0$ and put them in a list. We call F_1 the current front.
- Then, for each solution in the current front we visit each member (*j*) in its set and reduce its n_j value by one. In doing so, if for any member the count becomes zero, we put it in a separate list *H*. When all members of the current front have been checked, we declare the members in the list F_1 as members of the first front. We then continue this process using the newly identified front *H* as our current front.
- To get an estimate of the density of solutions surrounding a particular point in the population we take the average distance of the two points on either side of this point along each of the objectives. This quantity distance serves as an estimate of the size of the largest cuboid enclosing the point without including any other point in the population (the crowding distance).

The implementation of the algorithm was done in *NSGAII* class. In the constructor we pass some parameters which are: the maximum number of generations of genetic algorithm, the population size, the road signal, the car parameters (as object of *CarModel* class) and the relative and absolute tolerance used in the evaluation of a solution. The most important method of NSGAII class is the *ProcessAlgorithm* method. It contains the main loop of the algorithm. The *GenerateInitialPopulation* method randomly initialized the first parent population. Then, each possible solution from the current parent population is evaluated to get the objective functions in method *SetObjectivesForSolutionSet*, which gets as input parameter the list of *SuspensionDesignSolution*. After this evaluation comes a *while* loop in which at each iteration from the current population is generated a new population, till the maximum number of iterations is reached. In this loop, the following actions are made:

- A new offspring population is generated from the parent population. They have the same size. Pairs of two parents are selected by the binary tournament selection mechanism (from *BinaryTournamentSelection* class, in which in the *Execute* method is done the random selection of two solutions from population, then these solutions are compared by their values of objective functions and crowding distance and the best solution is returned). Then, recombination with one crossover point is done in *Execute* method of *CrossoverOperator* class. After this, mutation is applied to both offspring resulted from crossover. In method *Execute* from *MutationOperator* class is done the mutation by randomly changing the value of one design variable.
- Offspring population is evaluated from the perspective of the 3 objective functions.
- A new population is generated by combining together the parent and children populations. This one has twice the size of the parent / children population.
- The new population is split in Pareto fronts. Fronts are obtained mainly by getting for each solution the number of solutions which dominate it and the set of solutions which the solution dominates. The implementation is made in the constructor of the *ParetoFronts* class.
- Each front is taken into consideration iteratively, from the first to the last one. Solutions from the fronts are copied in a new population one by one, front by front, till the new population has the size of the initial parent population. Each solution of the fronts that is added to the population is assigned a crowding distance in this moment.
- Best solutions from the resulting population (the ones from the first front) are saved in a list of solutions, which will be returned by the *Process* method.

4.3 The user guide and some simulation results

To run the optimization algorithm, a few steps are needed before to initialize the random road profile and the car model parameters:

- In the interface, the parameters for road profile can be selected and a graphic of the road is generated when pressing the "Show road" button.
- Then, the car model parameters must be set. The body mass, the Front wheel mass, the Rear wheel mass, the Front tire stiffness and the Rear tire stiffness can be edited from the interface. Others are hardcoded: pitch moment of inertia, distance between the body centroid and Front tire and distance between the body centroid and Rear tire.

It is possible to evaluate a single solution pushing the "Process this solution" button (see Figure 3) by editing also the specific design variables: vertical spring stiffness of the sprung body Front suspension and Rear suspension, damping coefficient of the Front suspension and Rear suspension.

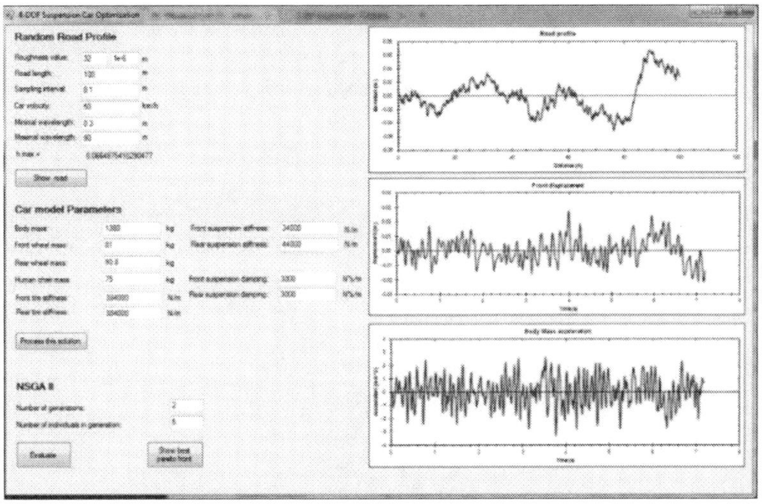

Fig. 3. Application's GUI: Random road profile parameters and simulation results

These values along with the car model parameters and random road signal are passed to the *Evaluate* method of *SuspensionModelEvaluation* class, and the graphics of the evolution of objective functions in time can be seen in the right part of the interface. Then, to apply NSGAII algorithm, the number of generations and the population size must be set and "*Evaluate*" button pressed. The results are written in a text file (see Figure 4). In this text file the best solutions from each generation are showed in this way: in the first row the values of the three objective functions and on the second row are written the values of the design variables that generated those results.

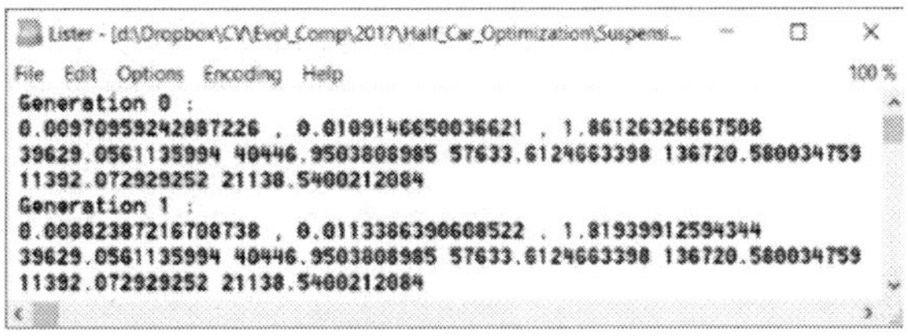

Fig. 4. NSGA-II simulation results

5 Conclusions and further work

This paper accomplished two-fold objectives: first, describe in details some C# functions useful for non-IT specialists to develop (custom) optimization methods of suspension systems and second, is concerned to improve 4-DOF HCM suspension's system by optimizing the stiffness and damping coefficients with the help of multi-objective methods (NSGA-II) in order to increase the ride comfort. Our preliminary results have shown that increasing generation number of genetic algorithm lead to obtaining better solutions.

References

1. M. S. Rahman, K. Muhammad, G. Kibria, Procedia Engineering **90,** 96-102 (2014)
2. L. Roman, A. Florea, I.I. Cofaru, Annals of the Oradea University Fascicle of Management and Technological Engineering **23** (3) 94-99 (2014)
3. A. Florea, I.I. Cofaru, L. Roman, N. Cofaru, Journal of Digital Information Management, **14**(6), 351-367 (2016)

4. M. Agostinacchio, D. Ciampa, S. Olita, *European Transport Research Review*, (Springer Berlin Heidelberg,2013)
5. S. Badran, A. Salah, Research Bulletin of Jordan, **2**, 42-51 (2012)
6. Santiago-Castillo Jose Antonio (2015) *DotNumerics*, http://www.dotnumerics.com/NumericalLibraries/DifferentialEquations/Default.aspx, (accessed 17 March 2017).
7. W.D. Passos, *Numerical Methods, Algorithms and Tools in C#* (CRC Press, 2009).
8. K. Deb, P. Amrit, A. Sameer, T. Meyariva, IEEE Transactions on Evolutionary Computation, **6**(2), 182 – 197 (2002)

DOI: 10.1051/matecconf/201712106003

Improving performance of roundabout intersections by optimizing traffic-flow speed

Marinela Inţă[1,*]

[1]Lucian Blaga University of Sibiu, Industrial Engineering and Management Department, 550025, Victoriei, 10, Sibiu, Romania

Abstract. Streamlining of urban traffic influence nowadays almost everything about our modern life. The main effect, which we all feel as the daily traffic participants, either as pedestrians or vehicles drivrers, is travel time to work and back. Roundabouts intersections have a number of advantages over intersections with traffic lights, they reduce the severity of accidents, traffic speeds to provide an "calmer" road environment. So, operating performance of roundabouts are complicated, influenced by the confluence operations. The paper aims at analyzing the current situation of traffic in two crowded roundabouts in Sibiu, comparing the results with the original data and proposing solutions to improve traffic conditions and especially traffic speed. The obtained results will be the basis for improving the overall capacity of roundabouts.

1 Introduction

1.1 Backgrounds

Transportation is an important factor in the development of society. Besides transport have a positive impact on economic development, they attract all the negative effects such as pollution of the environment, traffic congestion, accidents, etc.

Sibiu is one of the most prosperous cities in Romania and receives one of the highest rates of foreign investment. Thus, besides the traditional industrial city (engineering industry - Bilstein Compa, clothing industry - Mondostar, food industry - Scandia, after 1995 were created new companies in the field of automotive (Continental, Kuhnke relays, SNR bearings, Takata, Siemens).

With the development of industry, also population grew in Sibiu. As seen in the chart below, Sibiu's population has increased more than 11 times in a period of 163 years since they achieved first statistical data collected in censuses.

With this population growth, increased automatically and their need to travel. So, the number of registered vehicles has increased considerably in Sibiu, plus the ones in transit.

* Corresponding author: marinela.inta@ulbsibiu.ro

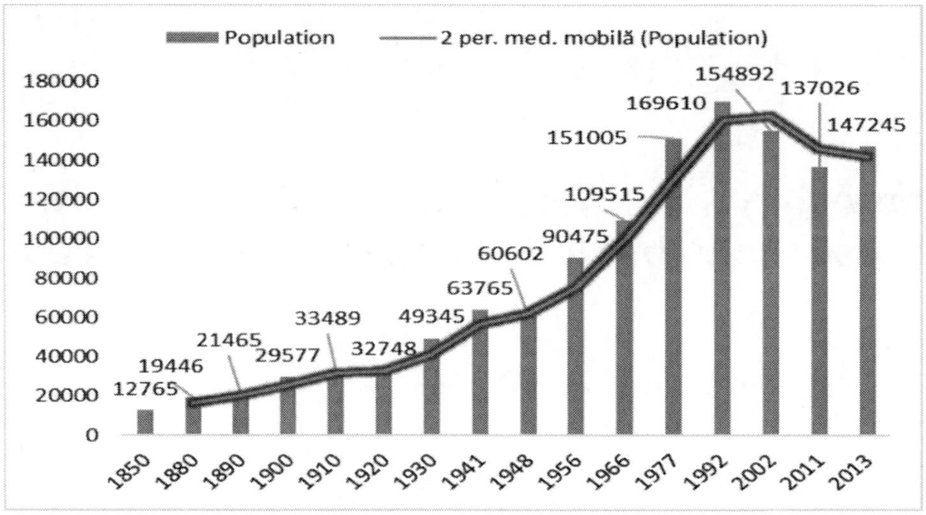

Fig. 1. Evolution of the population of Sibiu

If in 1985 the daily average annual vehicles circulating in Sibiu was 846, in 2010 their number doubled to 1752, and forecasts show that by 2025 the number of vehicles that will move could increase up to 3254, according to data provided by the National company of Motorways and National Roads in Romania [1].

1.2 Historic

The spread of roundabouts began when British engineers have redesigned circular intersection in mid 1960 and Frank Blackmore invented the mini roundabout to overcome capacity limitations of movement and protection measures [2].

Unlike circular intersections, roundabouts operate with rules to give priority to traffic already in the intersection and the traffic that leaves the intersection and the driver removes much of the confusion associated with circular intersections and waiting time for the drivers to intersections with traffic lights. Having roughly the same size as the intersections marked with the same capacity, roundabouts are significantly smaller in diameter than most circular intersections, separates entering and exiting traffic of the intersection with sites developed for pedestrians to encourage driving at lower speeds.

Sibiu is in full economic development, leading to the development of transport and automaticaly to the emergence of traffic congestion. This can be seen in the peak hours in the central areas and in areas of entry / exit in the city.

One of the main arteries of the city is the road of Alba Iulia that connects the central and western industrial area (exit to Sebes - Alba). This artery is made up of several types of intersections including roundabouts and large flow of vehicles creates problems in terms of fluency, resulting in increased congestion level.

2 Data collection

The biggest traffic congestion occurs on the road in the area formed by two roundabouts (Alba Iulia – Autogarii and respectively - Alba Iulia - Maramureșului - Malului). This is due to the following reasons:

- A zone of economic interest (linking with the Western industrial area, along artery are points of interest such as Banca Carpatica, store Lidl, DM, Kaufland, Peny market, gas stations, hotels, restaurants, etc;

 DOI: 10.1051/matecconf/201712106003

- Main access route to Sibiu International Airport;
- Positioned on both sides of the river Cibin.

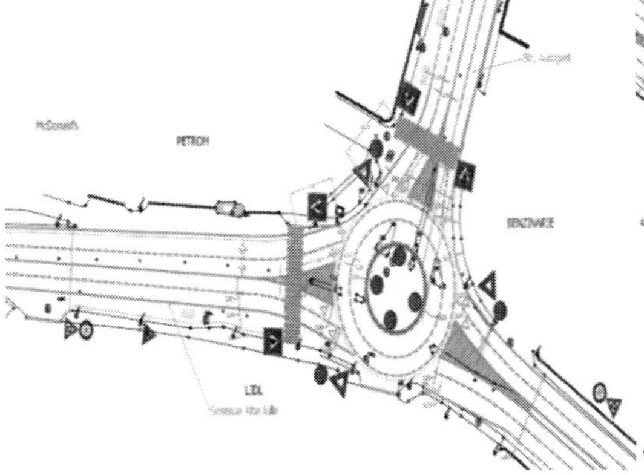

Fig. 2. Alba Iulia – Autogării

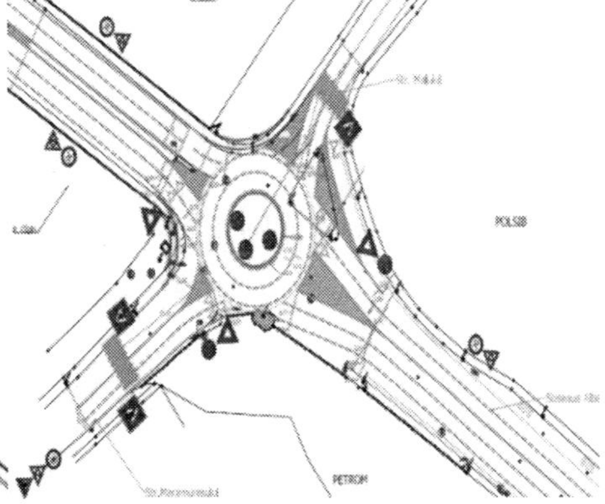

Fig. 3. Alba Iulia – Maramureşului – Malului

(source: http://www.sibiu.ro/index.php/primaria)

Table 1. Geometric elements of intersections.

	Roundabout Alba Iulia – Autogării	Roundabout Alba Iulia – Maramureşului – Malului
Inner radius	9,5 m	8 m
Outer radius	14,5 m	16 m
Number of annular strips	2	2
Width of the annular band	4-5 m	4 m
Entry lanes	2 lanes on each arm	2 lane on each arm
Entry lanes width	3,5-4 m	3-4 m

Although over time they have made improvements in terms of structure and type of intersection, and today can be seen queues and delays. Queues formed both morning and afternoon when most city residents are going or returning from work or school units, as well as on weekends when they go to supermarkets.

Traffic data were collected using traditional through manual spot counting of vehicles, then counting their record sheets per the category they belong.

These were collected through an observer team during a week to cover variations in traffic. The length of data collection was 15 minutes, three times per day for each junction. Under the traffic data, collection was followed:

- registration number of cars on each band;
- pedestrian traffic record;
- the length of the queue;
- the number of cyclists;
- frequency of the means of public transport.

It was verified that during data collection not appear major errors which, then to affect results because errors can lead to misinterpretations of traffic conditions. Before data collection there were studied the twoo roundabouts to determine the points of maximum visibility to allow observation of all intersection arteries.

Because traffic flows have a variable component, indentificating the vehicles in different categories was used for mixing standard vehicles through equalization coefficients. These coefficients represent the ratio of occupying road space by dynamic physical vehicle and vehicle occupancy standard - equal to unity - that is the car.

Thus, Figures 4 show the average standard vehicles volumule in the twoo roundabouts (Alba-Iulia- Autogării noted wih R1, Alba Iulia - Maramureşului – Malului noted wih R2) collected during one week.

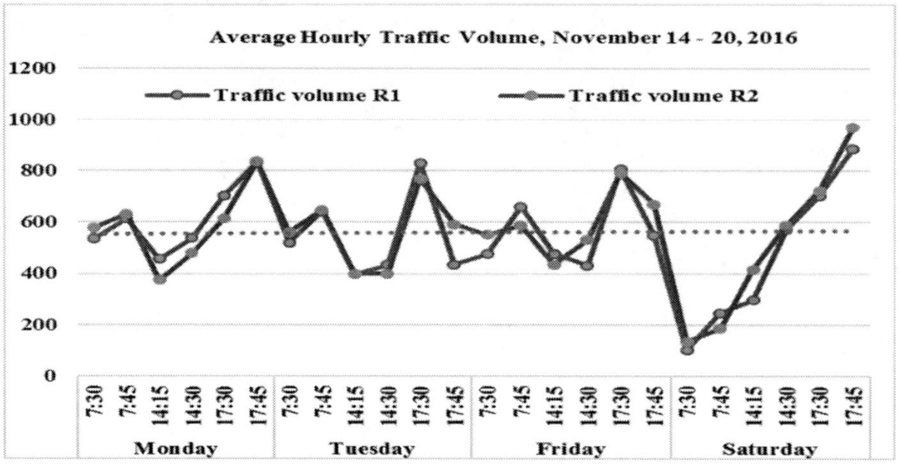

Fig. 4. Hourly and daily distribution of traffic volume data collected for two roundabouts

3 Traffic Data Analysis and levels of service (LOS)

After data collection, first were analyzed conflict points for the twoo intersection. Total conflict points of intersection being 33 and 35 respectively, divided into:
- divergence points - 9;
- convergence points - 7;
- crossing points - 17 and 19 respectively (7 points known crossing points vehicle - vehicle, 10 crossing points vehicles - pedestrian).

These data were then compiled Microsoft Excel data sheets calculation along with each roundabout, scheme, based on literature formulas were calculated and analyzed average speed on each arm, peak hour factor of each intersection.

3.1 Road traffic speed

Velocity is the main parameter of the road, because it characterizes purpose of travel and depends on the technical characteristics of vehicles, road network and capacity of vehicle drivers.

$$V = \frac{l}{t} \text{ [km/h]} \tag{1}$$

where:
l- length of road sector examined;
t – time in which the road was tranzited.
Average travel speed for every roundabout in each direction is shown in the figure below.
The difference between the speed between the three arms is caused by different diameters of the bands and confluence actions [3].

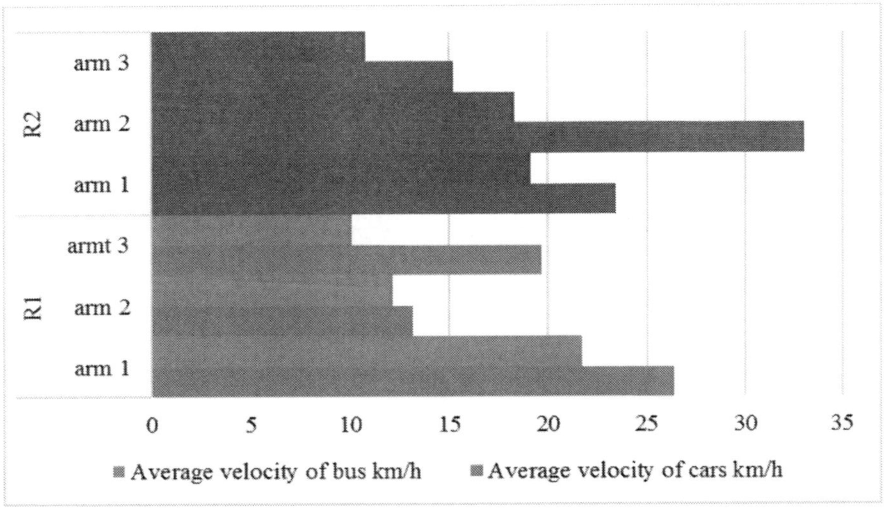

Fig. 5. Average velocity of vehicles (km/h).

These values indicate that vehicles confluence influences vehicle speed on band 1. Confluence vehicles have lower speeds because they usually decelerate before entering the roundabout.

3.2 Peak hour factor (PHF)

Peak hour factor in urban areas generally range between 0.80 and 0.98. Values lower than 0.80 means greater variability traffic flow, and values close to 0.98 represents small variations in traffic flows. Peak hour factor over 0.98 indicates a high volume of traffic.

Following application of the formula results for the first roundabout a peak hour factor of 1.23 and respectively 1.07 for the second sense.

To validate values obtained it were modeled and simulated with software Synchro Studio the twoo roundabouts junctions [4]. The sofware displays related reports of: intersection capacity utilization (ICU), level of service (LOS), delays of vehicles, total number of stops, travel time, HC, CO and NOx emissions and fuel consumption (Table 2 and Table 3).

Table 2. Str. Alba-Iulia & Str. Raului Performance by movement

Movement	EBL	EBT	EBR	WBL	WBT	WBR	NBL	NBT	NBR
Delay / Veh (s)	13.3	14.4	15	343.7	288.4	208.7	578.3	702	474.3
Stop/Veh	0.83	0.78	0.52	1.5	1.25	2.07	0.3	0.26	0
Avg Speed (kph)	22	21	21	5	4	14	2	2	2
Fuel Used (l)	0.3	5.1	1	2.1	7.9	2.2	2.9	4.2	0.7
HC Emissions (g)	0	15	2	0	22	1	5	0	0
CO Emissions (g)	11	356	57	43	435	67	103	67	11
NOx Emissions (g)	2	50	7	3	38	7	5	3	0

Following the completion of simulations of roundabouts using the program Synchro Studio, you may notice an ability to use the intersections of over 100% and a level of service H, therefore helps to create tails (congestion), low level of comfort and increased risk of accidents.

DOI: 10.1051/matecconf/201712106003

Table 3. Alba-Iulia & Str. Autogarii Performance by movement

Movement	**EB**	**WB**	**NB**	**SB**
Delay / Veh (s)	3.7	6.1	11.6	21.4
Stop/Veh	0.27	0.29	0.67	0.65
Avg Speed (kph)	22	27	8	14
Fuel Used (l)	3.2	2.8	0	3.7
HC Emissions (g)	9	12	0	15
CO Emissions (g)	198	279	1	315
NOx Emissions (g)	29	38	0	45

H level of service (LOS H) describes operations with control delays excessive, over 80 seconds to a vehicle. This level is considered unacceptable by most of the drivers, often due to over saturation of traffic that occurs due to excedes traffic capacity of intersections. These delays occur with a ratio V/Q great.

In other words, utilization capacity exceeded intersections and the traffic is not conducted in the normal range. Because the infrastructure does not allow major changes, only improvements that can be made are in the geometry of the intersection, which can have a positive influence on the traffic flow of the twoo roundabouts.

4 Methods of traffic decongestion

4. 1 Changes to the geometry of intersections

4.1.1 Roundabout Alba Iulia - Autogării

As a solution to increase traffic speed in the roundabout, they have made improvements to the intersection geometry, namely:
- arm Alba Iulia (the center) were added two lanes, one of which is exclusive left turn, and the other for, right turn having a length of 50 meters;
- arm Autogarii added an exclusive right turn band;
- arm Alba Iulia (to Turnisor) was added an exclusive lane for left turn with 20 meter.

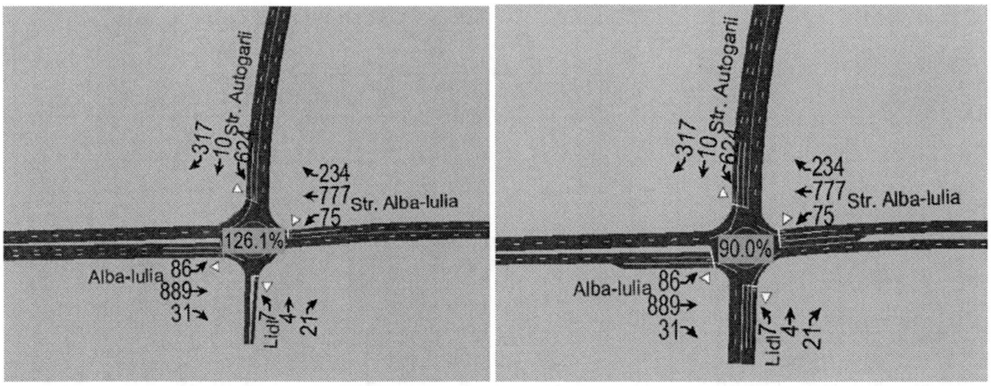

Fig. 6. Roundabout Alba Iulia – Autogării proposed solution

As seen intersection capacity utilization rate was reduced to 90% which means that traffic flows in normal parameters, and total delay time at the intersection was reduced by cca 40% reaching from 275. 6 to 164.1 seconds.

4.1.2 Roundabout Alba Iulia – Maramureșului - Malului

And our second roundabout noted that capacity utilization intersection exceeds 100%, which means that the movement is not conducted in the normal range.

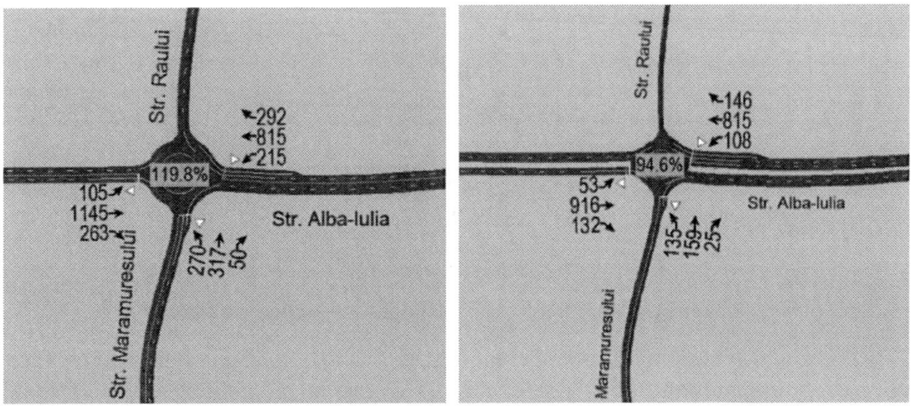

Fig. 7. Roundabout Alba Iulia – Maramureșului – Malului initial solution

And if this intersection were performed only changes the geometry and observed decrease in ICU parameter of the intersection to 94.6%.

4.2 Results

Following the implementation of the solutions presented in section 4 were obtained results.

- Roundabout Alba Iulia – Autogării

After implementing optimization, solution is seen that the average speed of traffic on the main roads between intervals are 20 to 30 km / h and 30-45 km / h.

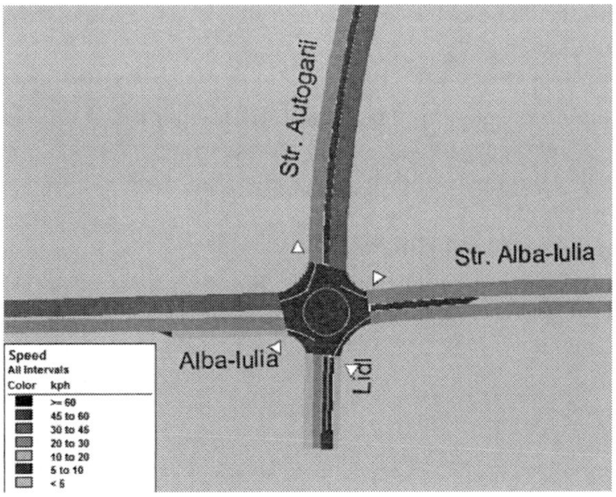

Fig. 8. Roundabout Alba Iulia – Autogării proposed solution

Speed being the smallest on the lanes from the Lidl store exit and the Autogarii is minor artery arm.

- Roundabout Alba Iulia – Maramureşului - Malului

If the initial situation average speed of traffic on each lane was low, especially on the arm of Maramures, after implementing the solution is observed that increased velocity being between 20 to 30 km / h.

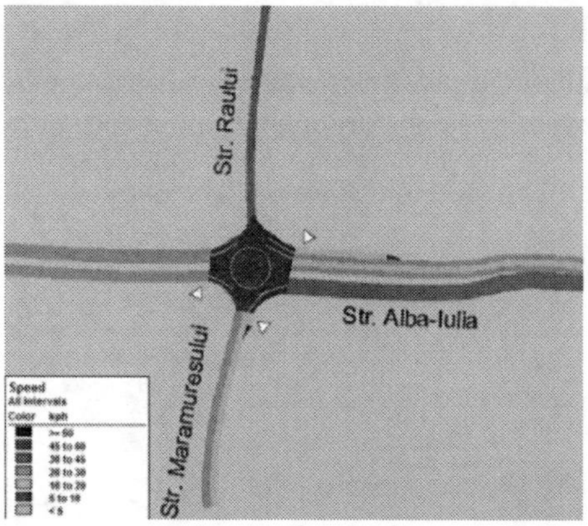

Fig. 9. Roundabout Alba Iulia – Maramureşului - Malului - the proposed situation

After geometric solutions implementations in the two intersections have reduced the number of stops and time of crossing. As disadvantage by implementing optimization, solution can not say that vehicle emissions have declined, which is higher due to increased traffic speed. hydrocarbons (HC) reached 18 g, the carbon monoxide (CO) being 1259 g and the oxide of nitrogen (NOx) is 88 g due to the continuous flow of traffic.

5 Conclusion

After analyzing the data collected from the two roundabouts intersections: Alba Iulia - Autogarii Road, Alba Iulia -Maramuresului – Raului road, resulted in large volumes of traffic moving toward the city center and industrial area with a low-speed, vehicles having frequent stops and starts, forming queues along the section.

It was concluded from the analysis and simulations performed Synchro that and pedestrian traffic around intersections influence in a considerable way, because in the area shopping centers are located.

Simtraffic software confirmed the results of calculations performed in the two junctions, which resulted in implementing simulation geometric changes. Following the simulation its implementation, improvement was observed in the range of peak hours' traffic, reducing queues and delays the vehicle stops. What has not been modified fuel efficiency, which has kept the value in both cases.

The results demonstrated that the proposed solution increases the average speed of traffic, reduce queues descongesting traffic, leading to a better development of it, but they could see a downside such as emissions made by vehicles, which increased.

References

1. www.recensamantromania.ro/wp-content/uploads/2013/07/sR_Tab_8.xls (accessed 20.10.2016)
2. Federal Highway Administration, *Manual of Uniform Traffic Control Devices for Street and Highways,* https://mutcd.fhwa.dot.gov/pdfs/2009/mutcd2009edition.pdf (2009)
3. R. Dowling, A. Kittelson. Transportation Research Board **447** (2013)
4. http://trafficware.infopop.cc/eve/forums/a/frm/f/9767036316, (accessed 04.01.2017)

MATEC Web of Conferences **121**, 06004 (2017) DOI: 10.1051/matecconf/201712106004
MSE 2017

GIS, integrated solution, for efficient administration of Hunedoara County

Eugenia Jornea[1], *Lucian* Lupu-Dima[1], and *Eduard* Edelhauser[1,*]

[1]University of Petroşani, 20 Universităţii Street, 20, România

Abstract. The local administration became in Romania a very important element for development. The requirements for the management activity in administration has changed as compared with what happened 10 years ago. The new issues raised by the complex needs and not always concurrent of the population, business agents and the trends in regional development are increasingly difficult to be managed with traditional tools. The adaptability and the scalability of Geographical Information Systems, represent an opportunity of the responsible from top management of the local administration that cannot be ignored. The intent of this study is to present a model to be followed for the use of GIS in management of the Hunedoara County Council.This model can change, the meaning of the idea of management at the local administration level by using the foresight as a support for decision making, through the access to the current and complete information and through the involvement of the GIS beneficiaries in providing updated data and in their transformation from simple beneficiaries of the management of the local administration in partners or even intervenients in the decisional process.

1 Motivation

Based on the statement made by Jack Dangermond of ESRI, which implies that: *"The application of GIS is limited only by the imagination of those who use it"*, we can let ourselves carried away by thoughts and ideas and to build almost any system.

Romanian territory is organized into territorial-administrative units: county, municipality, city, town, township and village. Bucharest the capital city of Romaniais organized by sectors. The county is lead under the laws of Romania, in a two-headed way, by the County Council and by the Government, through Prefecture. The County Council has the most important role and it also has the most appropriate tools for ensuring the proper administration of existing resources, the coordination of the various subordinate entities or of those with which it collaborates and also ofthe creation of the conditions of the materialization of the favourable opportunities for the county.

This paper work aims to analyse the possibilities of developing and to recommend a GIS solution for Hunedoara County, solution that can change the philosophy of the local and even of national administration. There are premises for such a system to simplify the

* Corresponding author: eduardedelhauser@upet.ro

decision making, to create the capacity of foresight and to generate resources for the system to be self-sustainable.

2 An evaluation of present

This evaluation aims to briefly present Hunedoara county and by that to suggest the complexity of a GIS for this county. A succinct look over the users of GIS in local administration or over GIS achievements in domains that can and should be used in the proposed systems proves the uniqueness of such GIS for Hunedoara county.

2.1 About Hunedoara, size, organization, institutions, resources

According to the public information displayed at National Institute of Statistics [3], Hunedoara county, situated in the central western part of Romania, represents 2.96% from Romanian territory, with a surface of 7063 square kilometres. In terms of administrative territory, the country includes 14 cities from which 7 municipalities and 55 communes with 457 villages.

In the monograph of Hunedoara [4] County we can find the information according with which the GDP of the county, represents 1.82% of the national GDP. One can also find in here interesting information regarding the structuring of economic agents by the number of employees.

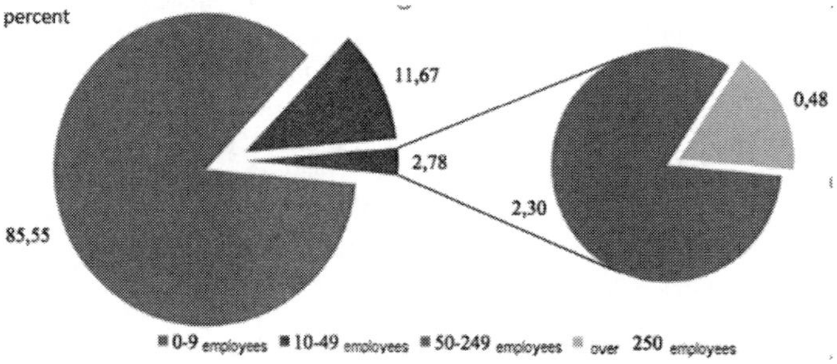

Fig. 1. The structure of active local units on size classes after the number of employees 2012 [4].

Note that Hunedoara is among the most important counties in terms of mineral resources exploitation, having three specialized companies: Energetic Complex Hunedoara, Min Vest Deva and National Society of Mine Closing Jiu Valley.

Analysing the public information offered by Hunedoara County Council [5], it is noticed that it has subordinated 16 institutions. The County Council is the majority shareholder of the two companies that provide alimentation with drinking water services, APA PROD S.A. Deva (with the administration of about 1000 km network) and ApaServ Jiu Valley S.A. Petrosani (with the administration of approximately 600 km network).

This brief presentation of the defining elements of the country has a goal the support of the affirmation that claims Hunedoara to be a county with great diversity and with special and complex needs.

2.2 Relevant examples of the use of GIS as presented in this paper

Following the analysis made by us we concluded that using GIS in administration is quite widespread in Romania nowadays. It can be stated, based on public information that Timisoara City Hall is the pioneer of public local administration, being the first institution of its kind, which implemented a GIS since 1998, the system becoming operational after five years and an expense of $ 1.2/inhabitant [6]. Oradea also started in 2003 the implementation of a GIS now according to public information the vectorization of the streets structure being resolved [7]. GIS is used in many localities such as: Bistrita, Satu-Mare, Iasi, Tecuci, Resita, Focsani, Cluj, Sibiu, Bucuresti. Utility companies have switched to using GIS in the current work, an example being Apa Nova Bucharest, which has become an example of best practice. Another example is Aquatim Timisoara that made an extra step, by extending the domain of activity of the company, achieving in 2013 from the ANCPI the right to perform specialized work in the field of cadastre, geodesy and cartography in Class S, authorization valid for 5 years.

Companies of the kind like those from: Braşov, Sibiu, Piteşti, Iaşi, Botoşani, Giurgiu, Alexandria can be mention as examples, with different GIS dimensions. In Hunedoara county one can find a GIS of green areas implemented in Petrosani and other 2 GIS of 2 societies of water and sewage.

According to ESRI Romania [8], in USA, over 80% from the administrative structures, from every level use GIS or to be more exact ESRI technology. The ESRI has created a model for the Local Administration as shown in figure 2.

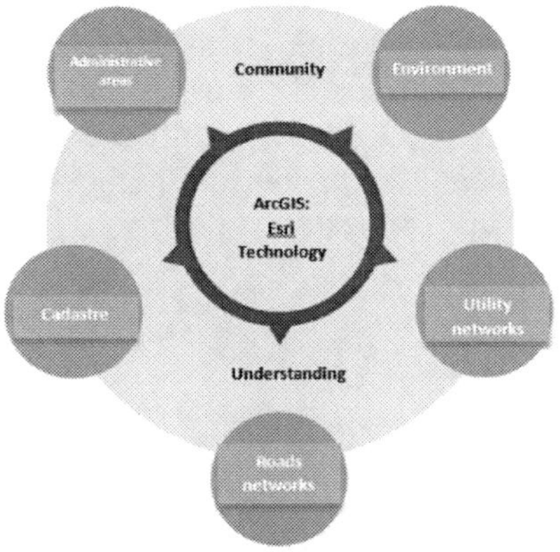

Fig. 2. Graphical model of GIS usage in Local Administration [8]

The model referred to, has the system focused on the economic development of the aria and on the citizen, as a source of information and as active participant in decision making. The system has a component for emergency situations, public safety, health and social services, administration of the terrains and cadastre, transport and utilities.

On the international level, there is enough evidence of the use of GIS as presented in this paper. A useful example for the component of the roads infrastructure can be found at Cepni and Aslan [9], tackled, by using GIS, aspects linked to the causality of accidents in traffic, approaching this topic of growing importance, in a complex mode that took into consideration aspects linked to technical component of vehicles, of road infrastructure yet without ignoring the psychological and socio economic component of the traffic participants. By using tools offered by GIS they modelled and designed risks maps, showing the distribution of accidents for 2015 (fig. 3). By using the analysis of collected information, the researchers show that when the routes, either that we talk about intersections or not, which represent high danger, are subjected to studies of geometric investigation, the number of fatalities is reduced. The use of such analysis system would contribute to the achievement of the National Strategy for Road Safety targets for the period 2015-2020, published by the Ministry of Environment [11].

Fig. 3. Accident at the road section. [8]

Directly related with the politics of local and regional development another relevant example is found in the thesis written by Sara Meerow and Joshua P. Newell [10] that shows how, by using GIS, solution for spatial planning for multifunctional green infrastructure. Using for the study Detroit city, it is proved that the planning of the green infrastructure moves from the inventory and keeping of record stage (situation found in green GIS from the most Romanian localities) to the one in which "green" is analysed, designed, made, maintained and developed by using the concept of Green Infrastructure Spatial Planning (GISP) model, a GIS-based multi-criteria approach that integrates six benefits: storm water management; social vulnerability; green space; air quality; urban heat island amelioration; and landscape connectivity. [10]

Fig. 4. Locations of major green infrastructure projects across the city of Detroit, [10]

There are many examples, but it is not the case to enumerate them all, because the goal is reached right now, the fact, that there are no limits in using GIS in this area of interest, being obvious.

3 GIS at Hunedoara County level

The difference between the classic of the attainment and of the implementation of a GIS and our preposition is that, that in this case the goal is the generation of the course of changing of the way a county is managed, involving all the actors that activate on the

Territory of Hunedoara County, either that they are localized in the county, or they are just in transit interact and this fact exceeds their choice and also offers to the decision making staff from the County Council the power of analysis and of processing of a GIS.

The efficiency of the administration becomes a result and ceases to be a purpose. A system, a mechanism, that makes this choice friendlier with themselves and with the environment of the county, that maximizes the strength and diminishes their weakness generated by inevitable divergent interests of the actors, previously mentioned, provides a natural increase of the efficiency of the managerial acts of the leaders of the county.

The fact that all those who have access to such system, have not only the status of beneficiary but also of information provider, system has a dynamic character, with permanent updated information and also competitive verified by users.

3.1 Proposition regarding the usage of GIS at Hunedoara County level

By using public information available now, some presented at the beginning of this paper, we have created by using ArcMap, a structure of layers to support, visually at this moment, this kind of project. We have chosen to use as layers, for the picture in figure 5, localities of the county, highlighting the localities where the population is over 2000 inhabitants suggestively, roads, railways, lakes, rivers, administrative territorial units. For the base layer "Corine Land Cover Europe 2006" was used.

Fig. 5. Hunedoara County (a part) with structure of layers.

In our opinion, such system is necessary to be implemented at the Hunedoara County Council, the argument being linked to what we showed at the beginning of this paper, regarding territorial organization of Romania, but also in direct connection with the structure of institutions subordinated to the Council and with unlimited life.

The specificity of such a project requires, in our view, the achievement of a correct staging of implementation. After getting over the first stage, that is already accomplished, the scope being "Hunedoara GIS", and that preceded the building of the Management System of Database, which is needed for the construction of the structure of involved actors. They will be structured after their role, contribution, commitment, benefits, constrains as follows:

- Project Leader - Hunedoara County Council, coordinator and system manager, the guarantor of the fulfilment of project objectives;
- Lead partners - city and town halls of all the localities of the county, state and private institutions, data providers and requirements from the design stage, throughout the operating of the system, direct beneficiaries of the system;
- Coercion partners - state entities, generating legal constraints and not only, who will be informed by the implementation of the system and requested to generate requirements, if the case;
- Volunteer partners - selected through public consultation, potential beneficiaries and generators of requirements in exhaustive completion.

This structure of partners will create a model of the project management that will be kept in the functioning time of the system, and then being treated as a project in continuous progress. Following the collecting of the requirements by GIS, a system of database management is designed, capable of satisfying all the requirements and at the same time of allowing the implementation of new requirements, in the context of continuous adding of information. It is anticipated that the acquirement of special data to be done in all eligible ways, taking into consideration as main channel, the automatic one, using remote sensing.

Regarding GIS structure, our opinion is that it will be constituted as a package of many GIS, in an arborescent style, those GIS coming to coalesce. This approach is unimaginable, as it is in fact the innovative element. For example, there will be GIS of localities, green, urban, of road signs, of cemeteries, which will be extended in plan from one locality to the other and they will have gaps of information, where the layer is not applicable. Although it is premature to appreciate that at this moment, we estimate that the number of layers will exceed 30.GIS of the institutions will cover the ascribed zone, as for example with APA PROD S.A. Deva and Apa Serv Valea Jiului S.A. Petroşani, which have identical domains of activity but different areas. Practically one can speak of one GIS with two subdivisions. The same way, the insertion of information and the requests from the operators of utilities, telecommunications, from the Road Company, Road Authority, Inspectorate for Emergency Situations, state or private institutions are approached.

Fig. 6. The visual model of the construction of layers as mini interconnected GIS (a Google Earth Pro capture was used).

Tools with which data are collected continue to be created and the data are uploaded into DBMS. Permanent verifications are conducted to identify potential errors and to permit the correction of possible discrepancies. The evaluation at this stage has stated a scope and some potential adjustments of DBMS, considering that in fact at this moment designing of DBMS is to be verified. Following the processing of data,outputs from the system are obtained such as maps and reports. All of these are built to meet the requirements of the partners from the project. At the same time, the analysis of the interdependencies from the requirements and of complementary factors from these, can conduct to results that exceed the expectations of the beneficiaries of the system.

3.2 Advantages of using GIS in the county

The implementation of GIS in the entire Hunedoara County, in the proposed form from the presented paper, would bring positive immediate results on many levels. At the level of the localities, local administration benefits of complete and permanently updated information, this way all that means the attribute of this administrative structure, from taxes to investment and development being tracked with ease. At the centralized level of the county the capacity of unitary administration of its resources is increased, by harmonious correlation of the development requests. At the level of direct interaction with the government, is made available for it, by means of prefecture, all the necessary elements. At the level of institutions for which the information of the kind speeds the reaction, such as Emergency Situations Inspectorate or those medical infrastructure, or those that protect the citizens. At the level of economic agents, which through rapid access at spatial information can design in the future, their investment plans, with a much smaller margin of error. At the level of population to which, such a tool, offers unexpected resources of interaction with entities of the county, but also means of information facile and current.

The most important advantage, that in fact is also the strong argument for the creation of such a system for Hunedoara county is the streamlining of the county administration, by means of all those shown previously, but also through the capacity of assimilation of all the applicable standards in the administration, management or related fields. GIS for Hunedoara county is a system that will be integrated with the informatic system of the County Council, the work interface with the management structure being ensured through a BI.

3.3 The possibility of extending and interconnecting the resulted system

Each of the benefits that the existence of GIS brings to Hunedoara county, can be detailed. Because from the technical point of view, the open character of such a system represents a central advantage, we choose to show it in detailed. A GIS of this kind, structured in the proposed form, offers a benefit we can call "*gold benefit*" since it can very easily interconnect with all kind of system, external to the county, either that it is about one of the neighbouring county or about a national one, in accordance with the INSPIRE [2] requirements. Actually, channels of informational connection adequate for almost any kind of situation are provided.

This idea started from the premise that at the county level all the needs will be covered. To support the open character of the system, we show some of the possibilities of extension and of interconnection:

- extension of the model to the level of the adjacent counties for the area of development from which Hunedoara is part, these managing to work unitary based on the same logic;
- interconnection with SNEP (*ghiseul.ro*), with the scope of the improvement of its performances;
- interconnection with the Unique Registry of Transparency of Interests, becoming a platform for the services of E-Consultancy, E-Debate or other actions of the Registry.

Each extension or interconnection will be constituted in exchange channels of information capable of feeding the system and to allow for it to serve better to the scope of the streamlining of the administration of Hunedoara county.

4 Conclusions

The idea of creating a GIS for a county, for Hunedoara county, is not a spectacular one and it can even be considered as a normality. The approach proposed in this paper is the innovative element and generator of change. Even if the idea is presented as a concept, it is feasible and sustainable.

This generate some elements of novelty:

- GIS as a complex structure of GISs;
- Principle of application of project management, in the implementation phase, based on the balancing of the roles of partners and on keeping of this principle during the functioning of the system;
- The use of the forecast as a support of the decision through the access to the current and to as complete as possible information.

Through this paper we have tried to make the first step, in the designing and implementing of what we believe to be the most important project of Hunedoara County. The suggested system exceeds the requirements of INSPIRE, because while the requirements of INSIRE are applied to the public institution, GIS for Hunedoara, brings together the public institutions, business, civil society and population and keeps intact the rigors of INSPIRE. The next step is the development of a complex study able to evaluate the existent and actual situation not only at the preparedness level of the territorial administrative units for the GIS implementation, but also in terms of their expectations from a GIS, namely the willingness to be an active part in the implementation process. This will represent, in fact, the premise from which a GIS implementation will start. The offered benefits, the lack of limits of development and expansion, are advantages that exponentially multiply. Premises for changing of the paradigm of management of Hunedoara County will be created.

The system can become an indispensable tool for each decision in the functioning of the Hunedoara County Council, allowing the propagation of these decisions and ensuring a better applicability for them and a control of their fulfilment.

The statement from the paper that such a system comes to be self-sustainable is substantiated by the way the information is to be available to the beneficiaries, under subscription regime or by using of the kind "pay for view". A GIS of such amplitude, after starting being used, comes to have a perpetual use, ensuring for itself the needed financial resources for the maintenance and for the further development, from any point of view.

References

1. INSSE - http://statistici.insse.ro/shop/?page=tempo3&lang=ro&ind=ADM101A (2016)
2. Parlamentul European, DIRECTIVA 2007/2/CE/14.03.2007, http://ogp.gov.ro/wp-content/uploads/2013/10/Directiva-INSPIRE.pdf
3. INSSEHd - http://www.hunedoara.insse.ro/main.php (2016)
4. BNR - MonografieJudetul Hunedoara, 26.06.2014, http://www.bnr.ro/DocumentInformation.aspx?idInfoClass=13567&idDocument=17760&directLink=1 (2016)
5. CJHd - http://www.cjhunedoara.ro (2016)
6. PrimăriaMunicipiuluiTimişoara – www.primariatm.ro (2016)
7. PrimăriaMunicipiului Oradea, www.oradea.ro (2016)
8. ESRI România – www.esri.ro (2016)
9. S. Cepni, M., O. Arslan,A International Scientific Conference On Applied Sciences – Antalya, Turkey (2016)
10. Sara Meerow, Joshua P. Newell, Landscape and Urban Planning, Volume 159, http://www.sciencedirect.com/science/article/pii/S0169204616302250 (2017)
11. MinisterulMediului - http://www.mmediu.ro/articol/strategia-nationala-pentru-siguranta-rutiera-pentru-perioada-2015-2020-si-a-programului-de-actiuni-prioritare-pe-perioada-2015-2020-de-implementare-a-strategiei-nationala-pentru-siguranta-rutiera-pentru-perioada-2015-2020/849, 2016)

 DOI: 10.1051/matecconf/201712106005

An empirical analysis of the relation between infrastructure and road accidents

Lucian Lobonţ [1,*], and *Lucian Ioan* Tarnu[1]

[1]Lucian Blaga University of Sibiu, Engineering Faculty, Industrial Engineering and Management Department, 550025, 4 Emil Cioran, Sibiu, România

Abstract. The concern for road safety in Europe is growing bigger. European Union by its means promote the aim of reducing the number of victims of traffic accidents by half in the period 2011-2020. In order to achieve this objective a lot of actions should be developed. One of our concern is the development of the infrastructure, mainly building motorways. Our research aims to study the relation between infrastructure – motorway versus highway and road accidents. Our findings show that is a great relationship between uses of motorways and reducing the number of accidents.

1 Introduction

The road transportation of freight and passengers counts for the most of the deaths by mean of transportation. Road traffic accidents are one of the leading causes of violent death in the European Union and at a global level. The actions promoted by the European Commission aims to reduce the number of victims of traffic accidents in the EU by half over the period 2011-2020.[1] From the many initiatives started to support the objective of reducing number of victims one of them is focusing on the safety of the infrastructure.

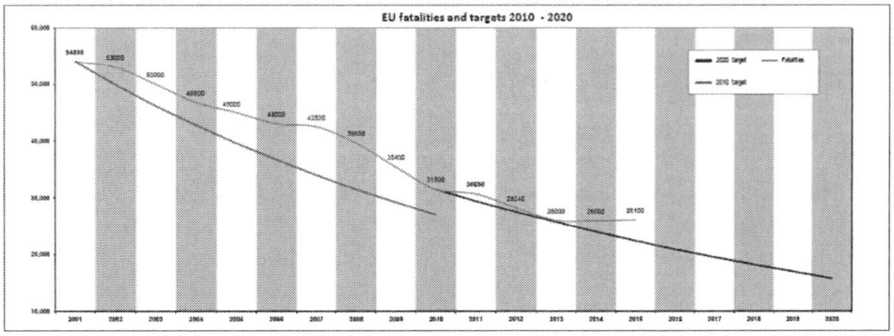

Fig. 1. Road safety evolution in EU – november 2016

* Corresponding author: lucian.lobont@ulbsibiu.ro

As is shown in the figure 1[2], even the fatalities are decreasing in an accentuated way the proposed target is still far from being achieved. Analyzing the situation of people killed in road accidents in Romania we can say that it is a positive evolution during the last years. (fig. 2) [2]. In reality when the numbers are compared with the other EU countries it can be seen that Romania is far behind other countries, for example second place after Bulgaria in 2015 with 95 killed/million inhabitants. The countries from the opposite part of the graph have only 30 killed/million inhabitants. (fig. 3) [2]

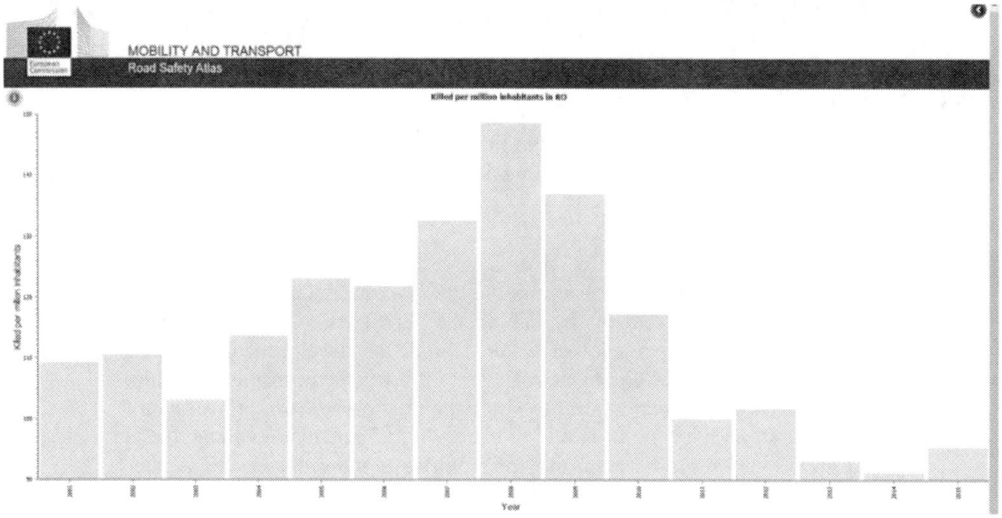

Fig. 2. Evolution of persons killed in road accidents in Romania

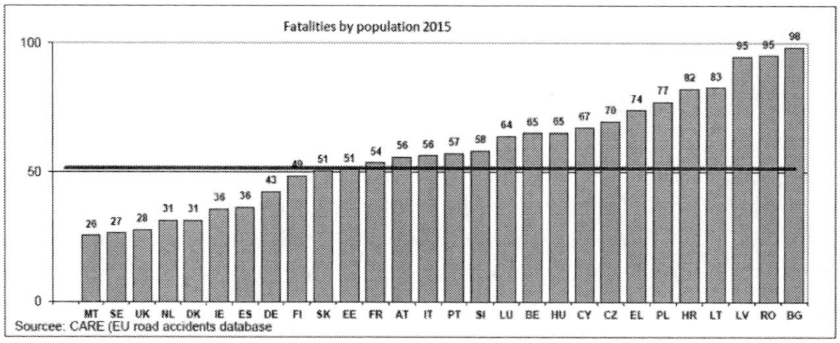

Source : CARE (EU road accidents database) or national publications

Fig. 3. Fatalities by population in EU countries

The previous researches regarding road traffic accidents is focused on different contributing factors that are connected with fatalities such as: uses of seatbelt [3], drinking, using of drugs [4,5], police inforcement [6]. Other researches [7] shows also that the risks of transportation depend on the characteristics of the road network such as road type, weather conditions, drivers' skills, and population concentration along the chosen routes.

Our research aims to study the relation between infrastructure – motorway versus highway and road accidents.

 DOI: 10.1051/matecconf/201712106005

2 Methodology

The analysis was conducted on a highly important road which connects the city of Sibiu in the center or Romania with the other cities situated on European corridor E81 and E68/DN1. European road E68/DN1 comprises the following: (Hungary) – Nădlac – Arad – Deva – Sebeş – Miercurea Sibiului – Sibiu – Braşov respectively E81 (Ucrain) – Halmeu – Livada – Satu Mare – Zalău – Cluj – Turda – Sebeş – Miercurea Sibiului – Sibiu – Piteşti – Bucureşti –Constanţa;

For our study we chose the roadways which connects the cities of Sibiu and village Cunţa at the territorial limit between Sibiu and Alba county. The road traffic between the two analyzed points can be done using a national road DN1 (highway) with 38,5 km length, represented with grey color in figure 4 or using the A1 motorway with 46,7 km length, represented with blue color in figure 4. According to Romanian legislation the speed limit is 100 km/h on highway, 130 km/h on motorway and 50 km/h in localities.

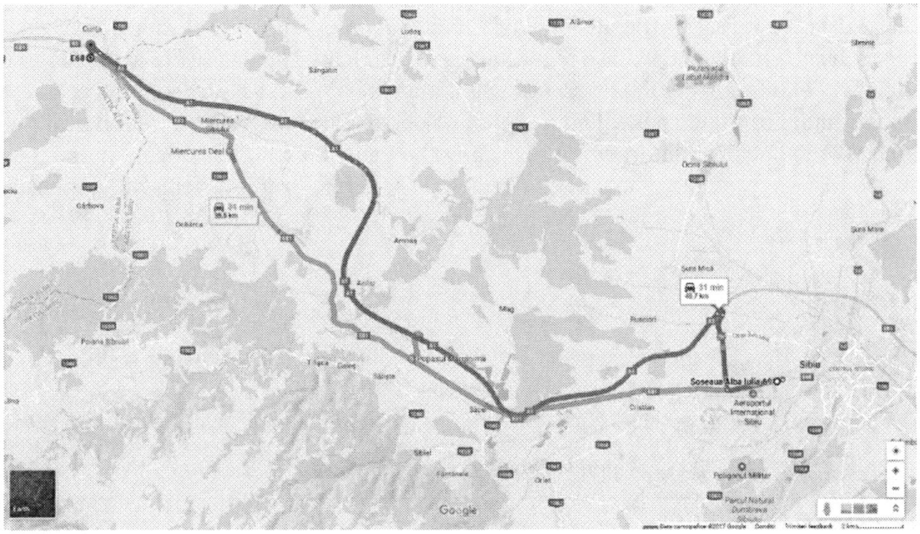

Fig. 4. Roadways chosen for the study. (image retrieved from Google Maps).

The analysis was conducted on two distinct periods:

a) 14.11.2014 – 07.09.2015 – within this time interval the motorway A1 was opened and the traffic was divided between motorway A1 and highway DN1. We estimate that most of the transit traffic was directed on A1, the highway DN1 being used for domestic traffic.

b) 07.09.2015 – 09.10.2016 – within this time interval, because of some malfunctions the motorway A1 has a closed section of 22 km (47,10 % from analyzed route). In this period all the transit traffic was diverted to the highway DN1. In the figure 5 is presented with dashed line the sector of the motorway which was closed.

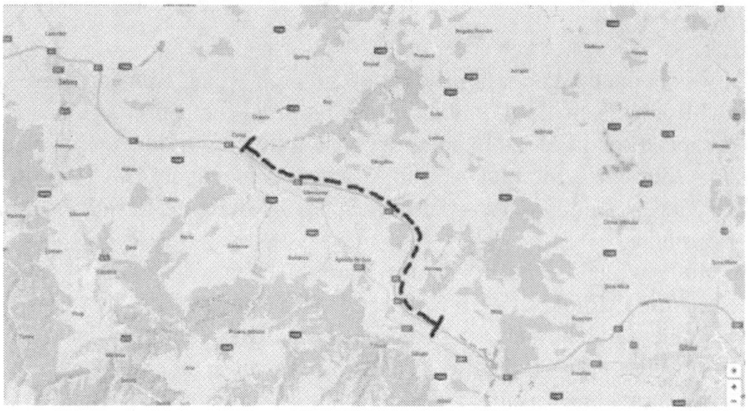

Fig. 5. Closed section of the motorway – dashed line (image retrieved from Google Maps).

The data regarding traffic accidents for the time intervals and selected roadways were exported from a national database regarding road accident. The data were analyzed in terms of quantity. The following data was extracted: number of total accidents, number of serious accidents, number of persons killed, number of persons heavily injured, number of persons injured. We extract also information regarding the main causes of accidents in the intervals analyzed and we presented the causes developing a Pareto graph which will emphasize the main causes.

3 Results

Retrieving the data for the two chosen intervals we obtained the following results: **Period 1: 14.11.2014 – 07.09.2015**
In the analyzed period, on DN1 Sibiu – Cunța, were 7 accidents as can be seen in Table 1.

Table 1. Number of accidents on DN1

	Total accidents (TA)	Serious accidents (SA)	Number of persons killed (PK)	Heavily injured (HI)	Injured (I)
Sibiu-Săliște	4	2	0	2	2
Săliște-Cunța	3	2	0	4	4
Total	**7**	**4**	**0**	**6**	**6**

The main accident causes (Table 2, Table 3):

Table 2. Main accident causes Sibiu - Săliște

Causes:	Total accidents (TA)	Serious accidents (SA)	Number of persons killed (PK)	Heavily injured (HI)	Injured (I)
Bicyclists	1	1	0	1	0
Careless lane change	1	1	0	1	0
Pedestrians on the road	1	0	0	0	1
Pedestrian crossing	1	0	0	0	1
Total	**4**	**2**	**0**	**2**	**2**

Table 3. Main accident causes Sălişte – Cunța

Causes:	Total accidents (TA)	Serious accidents (SA)	Number of persons killed (PK)	Heavily injured (HI)	Injured (I)
Inadequate speed at road conditions	3	2	0	4	4
Total	**3**	**2**	**0**	**4**	**4**

In the analyzed period, on A1 sector Sibiu – Cunța, were 14 accidents as can be seen in Table 4.

Table 4. Number of accidents on A1

	Total accidents (TA)	Serious accidents (SA)	Number of persons killed (PK)	Heavily injured (HI)	Injured (I)
A1 Sibiu – Deva (km 238-292)	14	8	1	10	11

The main accident causes on A1 - km 238-292 (Table 5)

Table 5. Main accident causes A1

Causes:	Total accidents (TA)	Number of persons killed (PK)	Heavily injured (HI)	Injured (I)
Distance between vehicle	4	1	5	4
Inadequate priority for other vehicles	3	0	1	4
Other causes that distract attention	2	0	1	1
Inadequate direction change	2	0	1	1
Inadequate speed at road conditions	1	0	1	0
Over speeding	1	0	1	0
Careless lane change	1	0	0	1
Total	**14**	**1**	**10**	**11**

Total number of accidents in the analyzed period on DN1 highway cumulated with A1 motorway is presented in Table 6.

Table 6. Total number of accidents (DN1+A1)

	Total accidents (TA)	Serious accidents (SA)	Number of persons killed (PK)	Heavily injured (HI)	Injured (I)
Total number accidents	21	12	1	16	17

The Pareto graph of the causes of accidents is presented in figure 6.

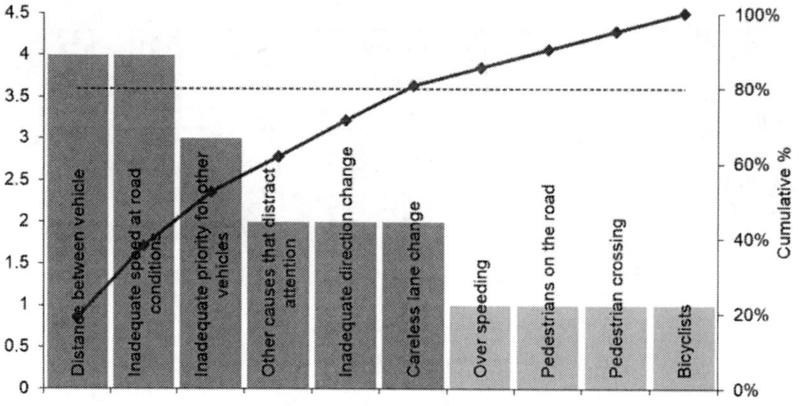

Fig. 6. Causes of accidents - period 1: 14.11.2014 – 07.09.2015

Period 2: 07.09.2015 – 09.10.2016

In the analyzed period, on DN1 sector Sibiu – Cunţa, were 44 accidents as can be seen in Table 7.

Table 7. Number of accidents on DN1

	Total accidents (TA)	Serious accidents (SA)	Number of persons killed (PK)	Heavily injured (HI)	Injured (I)
Sibiu-Sălişte	17	5	0	5	20
Sălişte-Cunţa	27	13	10	21	33
Total	**44**	**18**	**17**	**26**	**53**

Main causes of accidents are presented in Table 8 and Table 9.

Table 8. Main accident causes Sibiu – Sălişte

Causes	Total accidents (TA)	Serious accidents (SA)	Number of persons killed (PK)	Heavily injured (HI)	Injured (I)
Inadequate priority for other vehicles	4	1	0	1	6
Distance between vehicle	3	0	0	0	4
Careless lane change	2	0	0	0	3
Drivers of carts and animals	2	2	0	2	2
Inadequate direction change	2	0	0	0	2
Inadequate speed at road conditions	1	1	0	1	0
Pedestrian crossing	1	1	0	1	0
Other causes that distract attention	1	0	0	0	2
Drive without licence	1	0	0	0	1
Total	**17**	**5**	**0**	**5**	**20**

Table 9. Main accident causes Săliște - Cunța

Causes	Total accidents (TA)	Serious accidents (SA)	Number of persons killed (PK)	Heavily injured (HI)	Injured (I)
Inadequate speed at road conditions	19	9	8	18	23
Inadequate priority for other vehicles	2	1	0	1	6
Distance between vehicle	2	1	0	1	1
Other causes that distract attention	1	0	0	0	2
Sleeping driver	1	1	2	0	0
Inadequate overtaking	1	0	0	0	1
No priority for pedestrians	1	1	0	1	0
Total	**27**	**13**	**10**	**21**	**33**

The Pareto graph of the causes of accidents is presented in figure 7.

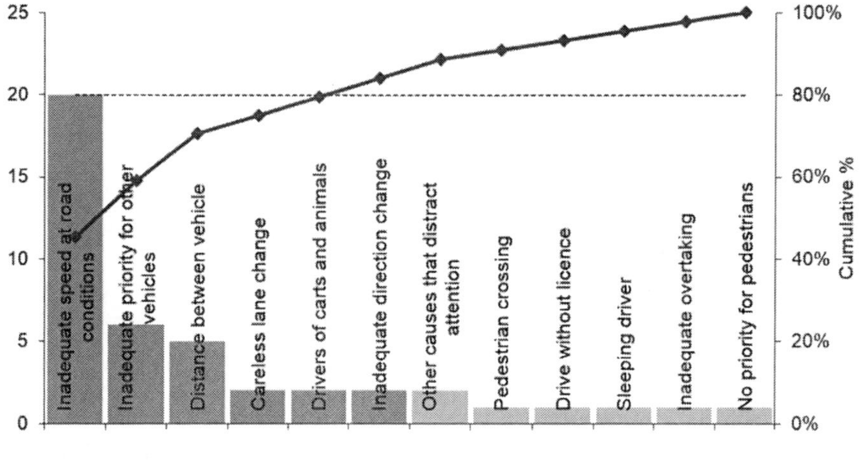

Fig. 7. Causes of accidents - period 2: 07.09.2015 – 09.10.2016

4 Discussion

The compared results from the two studied periods are presented in figure 8.

Even if there is a difference of 101 days between the studied periods, analyzing the results, substantial increases can be observed in all the areas, increases that cannot be attributed to the difference in the number of days.

Thus, in terms of total number of accidents the values are doubled. The number of serious accidents has grown from 12 to 18.

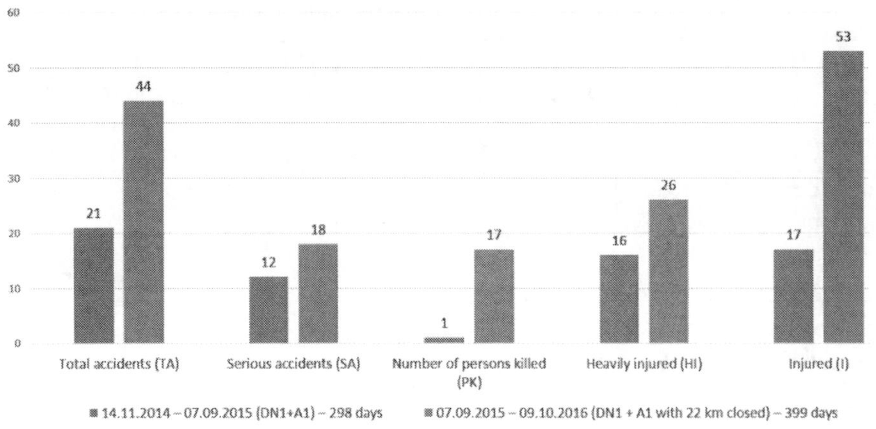

Fig. 8. Comparison of the accidents number in the two analysed periods

The consequences are visible through a terrifying increase in the number of deceased persons 17 persons within the 22 km of motorway were closed. There are also increases in the number of persons heavily injured and slightly injured.

Our findings, which show that the developing of the road infrastructure, especially building motorways, can help in a large decreasing of road accidents is supported also by other researchers such D. Albalate, and G. Bel, [8]

The main accident causes show different situations for the transit using national highway DN1. When the motorway was fully open there was a small number of interactions between cars and other traffic participants. After the motorway was closed the total number of interactions has grown resulting in the very big number of accidents. In order to sustain these and to emphasize the relationship between infrastructure and road accidents we can conclude analyzing the Pareto graphs that the inadequate speed, distance between vehicles and inadequate priority are the major causes of accidents but with the right infrastructure the fatal effects can be reduced.

5 Conclusions

While reports on road accidents recorded numerous causes for these accidents such as: inadequate speed at road conditions, inadequate priority for other vehicles, inadequate distance between vehicles, over speeding, pedestrians etc., a lot of these causes can be eliminated by developing the appropriate infrastructure, specifically building motorways. Consequently, the creation of appropriate conditions for making road transport by motorway construction impact not only by having a quick and easy transport of goods and people but also an increase in road safety and hence the saving of human lives.

It is important that the actions for reducing road accidents to be oriented not only to the obvious causes but to the fundamental ones.

References

1. *Towards a European road safety area: policy orientations on road safety 2011-2020,* (European Commission, Communication from the commission to the European Parliament, the council, the European economic and social committee and the committee of the regions, 2010)
2. European Commission, web site, retrieved from http://ec.europa.eu/transport/road_safety/specialist/statistics_en, (2017)
3. P.D. Loeb, Transp. Res. E Logist. Transp. Rev. **37**, 297–310 (2001)
4. D. Albalate, J. Pol. Anal. Manag., **27** (1), 20–39,(2008)

5. A. Delaney, K. Diamantopoulou, M. Cameron, *Strategic Principles of Drink-Driving Enforcement.* Report No. 249, (Monash University Accident Research Center, Melbourne, 2006)
6. E. O. Ryeng, Accid Anal Prev., **45**, 446-454 (2012)
7. V. Torretta, E. C. Rada, M. Schiavon, P. Viotti, Safety Sci. **92**, 1-9 (2017)
8. D. Albalate, G. Bel, SERIEs, **3**, 457 - 473 (2012)

AUTHOR INDEX

AUTHOR INDEX

AUTHOR INDEX

AUTHOR INDEX

EDP Sciences – Web of Conferences
17, Avenue du Hoggar
Parc d'Activité de Courtabouf
BP 112
F-91944 Les Ulis Cedex A
France

ISBN 978-1-5108-4705-7

CURRAN ASSOCIATES INC.
proceedings.com

8th International Conference on Manufacturing Science and Education (MSE 2017)

Trends in New Industrial Revolution

MATEC Web of Conferences Volume 121 (2017)

Sibiu, Romania
7 - 9 June 2017

Part 2 of 2

edp sciences

8th International Conference on Manufacturing Science and Education (MSE 2017)

Trends in New Industrial Revolution

MATEC Web of Conferences Volume 121 (2017)

Sibiu, Romania
7 - 9 June 2017

Part 2 of 2

Editors:

Ioan Bondrea
Carmen Simion
Marinela Inta

ISBN: 978-1-5108-4705-7

Printed from e-media with permission by:

Curran Associates, Inc.
57 Morehouse Lane
Red Hook, NY 12571

Some format issues inherent in the e-media version may also appear in this print version.

Printed by Curran Associates, Inc. (2017)

For additional information, please contact EDP Sciences – Web of Conferences at the address below.

EDP Sciences – Web of Conferences
17, Avenue du Hoggar
Parc d'Activité de Courtabœuf
BP 112
F-91944 Les Ulis Cedex A
France

Phone: +33 (0) 1 69 18 75 75
Fax: +33 (0) 1 69 28 84 91

contact-edps@webofconferences.org

Additional copies of this publication are available from:

Curran Associates, Inc.
57 Morehouse Lane
Red Hook, NY 12571 USA
Phone: 845-758-0400
Fax: 845-758-2633
Email: curran@proceedings.com
Web: www.proceedings.com

TABLE OF CONTENTS

PART 1

PART 2

Development of a drive management system by V2I communication

Alexandra Elisabeta Lörincz[1,*], *Marius Nicolae* Risteiu[1], and *Alin* Constandoiu[1]

[1] University of Petrosani, Departament of Computer and Electrical Engineering, Romania

Abstract. In this paper we propose a solution for drive management, by communicating between the vehicles involved in traffic and also with the road infrastructure, like traffic lights, road signs etc. All these elements make up a network in which information sharing is very quick and very safe. This system only warns the driver of dangers or some information that he missed on the road, like road signs. The system does not take control of the vehicle, but displays on an LCD display warnings, information on the operation of the car's engine (information received from the ECU) and information about the distance to the other cars around him.

1 Introduction

The continuous development in the automotive domain, lead to the implementation of new systems and technologies for the benefit of drivers but also for the passengers in vehicle. [1]

These systems are designed for specific reasons that are based on well-defined road traffic safety, easier management of vehicles by people and avoiding damage to property which greatly affects the economy because of accidents.

Avoiding these dangers led to development of certain systems center who warns the driver about a possible impact with another vehicle, pedestrian or cyclist but also warning the exit in cases unintentional vehicle off the road and an accident. These systems have been implemented even by some automobile manufacturers and the most advanced technologies include acquiring control of the vehicle in cases of extreme danger or when the driver announced this (setting automatic mode driving the car).[2]

To implement these systems on vehicles, manufacturers have developed specific communication networks between vehicles that allow transmission to other cars of basic information such as speed, direction of travel, exact location where there are vehicle messages to avoid traffic jams traffic, and weather information on a particular route.

This exchange of information made through various methods: radio communications, wireless communications, GSM communications or communications through satellites.[3]

Creating this system begins with the hardware, which includes radio module (RFM), module Wi-Fi (wireless), sensors etc. but the most important part is the software that applies a specific protocol that system. Protocol development allows us communication

* Corresponding author: lorinczalexandra@yahoo.com

DOI: 10.1051/matecconf/201712106006

between vehicles but also communication infrastructure that includes road traffic lights, road signs, etc.

All these elements make up a network in which information sharing is very quick and very safe. This system only warns the driver of dangers or some information that you do not know, he did not see or did not want to see them! (Road signs). The system not only takes control of the vehicle on an LCD display sends warnings, information on the operation of the car's engine (information received from the ECU) and information about the distance that has over other cars around him.

This system is created from devices placed on road signs and in-vehicle devices that creates a network and produce a useful exchange of information between them. [4]

1.1 Communication between vehicles and road infrastructure V2I

This communication between drivers and road infrastructure is beneficial in several ways:

- receiving information about road / highway which is the vehicle at the time;
- information on the maximum and minimum travel speed on certain portions of the road. This information is very useful for a leader traveling an unknown road or the possibility of another traffic law in some countries.
- information on dangerous roads with high accident
- heavy traffic in large cities who own many traffic lights and road signs
- informing the driver on the points of interest that are nearby.

Road infrastructure represents all roads usually open to public traffic, with road signs and decorations, that a person can move a vehicle from point A to point B. [5]

Romanian legislation requires specializing cataloging roads in three categories in the following Table1:

Table 1. The traffic signs

From a legal perspective			Traffic signs
The destination	Private utility roads		
	Public roads		
The movement	Open to public traffic roads		
	Close to public traffic roads		
Functional and administrative territorial	National interest roads	Highways	
		Express roads	
		National roads	
	County interest roads	County roads	
	Local interes roads	Village roads	
		Streets	

All these road signs are classified in eight specific categories:

- Road signs Warning: This category contains 59 different indicators;
- Road signs Priority: This category contains 6 road signs;
- Road signs Banning and restriction: This category contains 40 road signs;
- Road signs Ordering: This category contains 12 indicators;
- Road signs Guiding : This category contains 35 indicators;
- Road signs Information: This category contains 48 indicators;
- Road signs Tourist Information: This category contains 12 ;
- Road signs, additional panels: contains 23 indicators.

Each road sign includes one special device designed, to communicate with vehicles moving or stationary, information about its specification, Fig. 1.

All this communication is performed using a specific protocol created for this purpose.

Fig. 1. Comunication V2I [6]

1.2 V2V communication protocol

Department of Transport (DOT) of the United States compel car makers to include communications systems "vehicle-to-vehicle '(V2V) for reducing the number of road accidents.

V2V communication technology consists of motor vehicles between them by changing information useful in avoiding / preventing some accidents, but also early warning of the potential dangers of driver in traffic. This communication between vehicles is made using radio waves which I hope is opposed to cameras or certain sensors have a range of about 300 meters and are not blocked by some object in their area or another cars. Using this technology can prevent the driver from the dangers of traffic especially at intersections or when changing lane movement and travel.

V2V is also known as the VANETs (Vehicular Ad Hoc Networks). In 2001, it was mentioned in a publication as ad hoc networks can be formed by cars, which can help avoid blind spots, avoiding accidents etc. [7]

All this communication is performed using radio networks using special modules for sending and receiving data.

2 Proposed V2V and V2I communication protocol

This protocol is designed to warn the driver about road signs around it, the information you read in real time from the onboard computer of the vehicle on K-Line protocol and information on vehicles around it.

In Fig. 2 is present a block diagram of communication V2V and V2I.

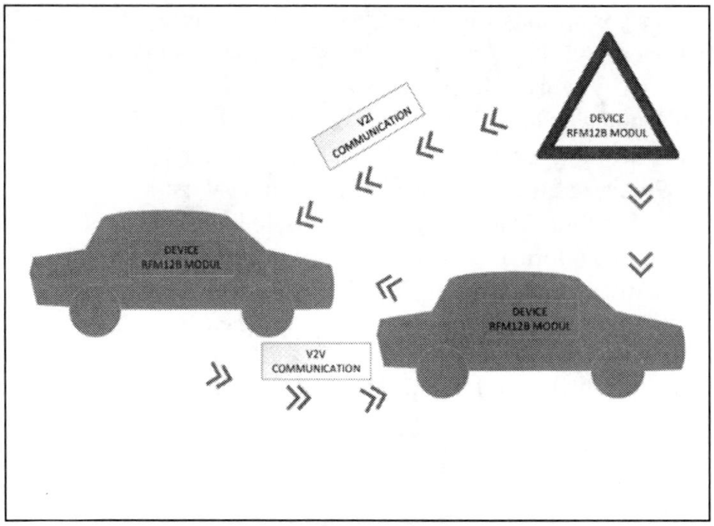

Fig. 2. Communication V2V and V2I.

2.1 Hardware System

The device of data transmission by moving or stationary vehicles, and road sign is placed on the sending data every second.

The module is use of a data transmission radio module (RFM12B) operating at a frequency of 915MHz, at a distance of 250m in the open air.

It creates STM32 with the microcotroller communication device, Fig.3.

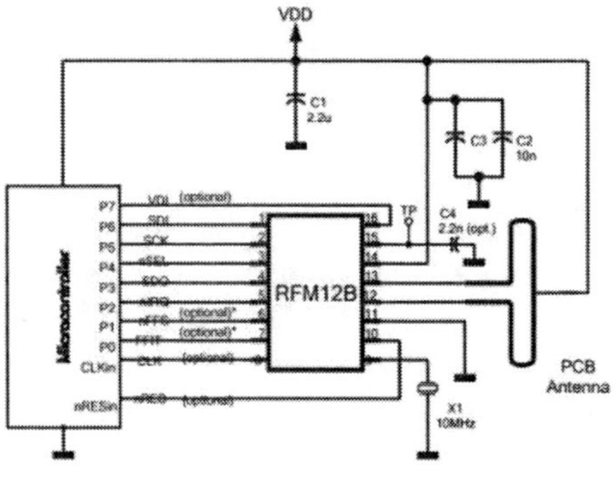

Fig. 3. Typical application with RFM12B.

Inside the vehicle is located this device RFM12B, set in slave mode and communicates with another vehicle (or more vehicles) which in turn holds the device (V2V communication) or communicate with road infrastructure (traffic signs, traffic lights etc.) holding devices RFM12B set in Master mode (V2I communication).

All the communication between the microcontroller and RFM12B module is done with SPI serial interface.

The Serial Peripheral Interface bus (SPI) is a synchronous serial communication interface specification used for short distance communication, primarily in embedded systems.

SPI devices communicate in full duplex mode using a master-slave architecture with a single master. The master device originates the frame for reading and writing. Multiple slave devices are supported through selection with individual slave select (SS) lines. [8]

2.2 Software System

Data sent from road signs to vehicles or directly between vehicles, forming a communication network where road signs are "Master", they only send data objects around them, and the vehicles are and "Master" and "Slave", they send priority their data and receive data from road signs or vehicles around them.

All this data is continuously exchange data and real-time data is received and transmitted between traffic participants.

The microcontroller communicates via SPI interface (Serial- Peripheral Interface) module RFM12B, which is a high speed serial interface that operates in full duplex mode.

The data stream transmitted automatically be filled with information specific vehicle or road sign.

This raw data consists of more bytes owning a priority and have a specific meaning if inficator received from a road or a vehicle.

In the next figure is presented the initialization function for RFM12B written in the microcontroller program (Fig.4).

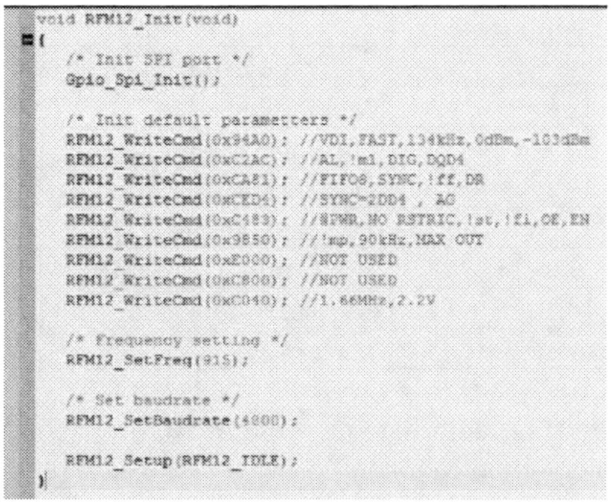

```
void RFM12_Init(void)
{
    /* Init SPI port */
    Gpio_Spi_Init();

    /* Init default paramentters */
    RFM12_WriteCmd(0x94A0); //VDI,FAST,134kHz,0dBm,-103dBm
    RFM12_WriteCmd(0xC2AC); //AL,!ml,DIG,DQD4
    RFM12_WriteCmd(0xCA81); //FIFO8,SYNC,!ff,DR
    RFM12_WriteCmd(0xCED4); //SYNC=2DD4 , AG
    RFM12_WriteCmd(0xC483); //@PWR,NO RSTRIC,!st,!fi,OE,EN
    RFM12_WriteCmd(0x9850); //!mp,90kHz,MAX OUT
    RFM12_WriteCmd(0xE000); //NOT USED
    RFM12_WriteCmd(0xC800); //NOT USED
    RFM12_WriteCmd(0xC040); //1.66MHz,2.2V

    /* Frequency setting */
    RFM12_SetFreq(915);

    /* Set baudrate */
    RFM12_SetBaudrate(4800);

    RFM12_Setup(RFM12_IDLE);
}
```

Fig. 4. Initializing communication with RFM12B module.

2.3 Message structure

The message structure is composed of header bytes, 1 byte for data length, 8 bytes of data, a checksum byte and 1 bytes stop.

Start Bytes to announce the receipt of an indicator, traffic lights or other vehicles of a new data message.

Stop Byte announces completion of the string received.

Bytes of data are divided into several categories.

- The first byte signify that comes directly from this message. It may be received from a road sign, a traffic light or in another vehicle. First Byte has the highest priority in continuing reading data from that string data.
- Second byet holds that a value according to the first byte. This value can range from 0 ~ 8 (warning signs, signs priority etc), where the first byte is the meaning of road signs.
- Third byte value has specific category of which the second byte. This category may vary from 0 ~ 59.
- Byte 4 has the same principles as the first byte of data. If there are two signs in the same place, they can communicate this to us. 4'lea byte can have the value of expressing an indicator, a traffic light or another vehicle.
- Byte 5 has the same features with byte No. 2.

- Byte 6 is strictly related to byte 5 and its value is specific category which includes byte 5.
- Byte 7 and 8 are used strictly for communication between vehicles, and mean values of speed or their location.

The checksum byte (CS) inserted at the end of the message block is defined as the simple 8 bit sum series of all the bytes in the message, excluding the checksum. [9]

The CS calculation is as follows:

If the message is:

<1><2><3>...<N>,<CS>

Where <i>(1<=i<=N) is the numeric value of the ith byte of message, then:

<CS> = <CS>N

Where <CS>i(i=2 to N) is defined as

<CS>i={<CS>i-1+<i>}Modulo 256

And <CS>1=<1>

Table 2. Message structure

Sequence	Description	Data
Byte 1	Message header 1	0xAA
Byte 2	Message header 2	0xAA
Byte 3	Message header 3	0x2D
Byte 4	Message header 4	0xD4
Byte 5	Message length	0x08
Bytes 6 to 13	Data (up to 7 bytes)	0xXX
Byte 14	Message stop 1	0xAA
Checksum Byte 15	Checksum	0xYY

To send a message using radio module RFM12B is need for some specific command this (Transmitter Register Write Command).

Table 3. Message block – Transmitter Register Write Command.

Bit	15	14	13	12	11	10	9	8	7	6	5	4	3	2	1	0
0xB8AA	1	0	1	1	1	0	0	0	t7	t6	t5	t4	t3	t2	t1	t0

With this command, the controller can write 8 bits (t7 to t0) to the transmitter data register. Bit 7 (el) must be set in Configuration Setting Command. [10]

An example of sending data from one road sign specified in the following figure, where SPI (Serial Peripheral Interface) has the following characteristics:

- MSB (**M**ost **S**ignificant **B**it) – First,
- 8 bit transfer,
- clock is low when inactive CPOL=0,
- data is valid on clock leading edge CPHA=0,
- enable line is active low.

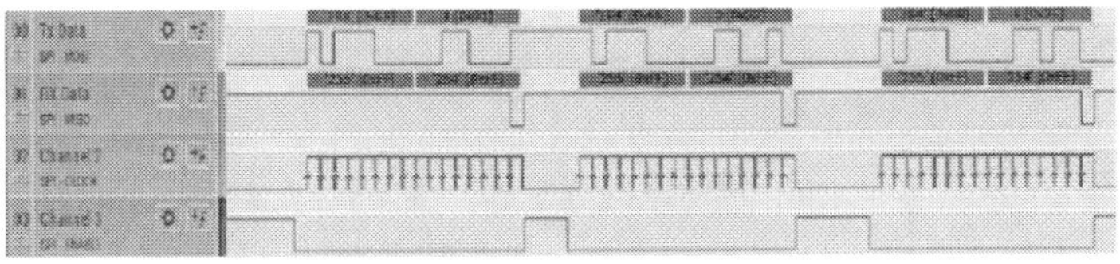

Fig. 5. The signals sent by the traffic signs.

The communication between the microcontroller and module RFM12B is performed using the SPI interface (**S**erial **P**eripheral **I**nterface).

In Fig. 5 are shown data signals:

- SPI MOSI – **M**aster **O**ut **S**lave **I**n (transmit data from Master device to Slave device)
- SPI MISO – **M**aster **I**n **S**lave **O**ut (receive data to Master device from Slave device)
- SPI CLOCK – Serial clock (from the Master device)
- SPI ENABLE – Slave select (active low, out of Master device)

First byte 0xB8 and 0x31, signify that the message is from a traffic sign, the second byte 0xB8 with 0x32 signify category traffic sign warning and the third byte 0xB8 with 0x36 signifies (the narrow road on the left).

The result of calculation described of the message is Checksum 0xA5.

After receiving this message by vehicle, the LCD will show the road sign specified.

3 Conclusions

This data communication helps alert the drivers from time with information about traffic, road signs, etc.

Composing hardware part is very simple, with a very low cost and easy to implement in any car.

V2V communication system proves to be the best system to achieve data transfer and information (operating at a frequency of 915MHz) compared with systems of vehicles carried by cellular networks, GPS etc. so is a high degree of flexibility for implementation. This communication is not restricted to different categories of roads or geographic areas (as restricting communication systems based on GSM or GPS).

Communication is done at a maximum distance of 250 meters between two cars in the open air.

In crowded cities are many factors that prevent communication to the maximum distance (buildings, trees, etc.), this can turn the disadvantage into an advantage, because speed is much smaller and the need to interpret information at a large distance is not necessary.

References

1. R. Dovleac, A. Ionica, M. Leba, Proceedings of Review of Management and Economic Engineering, 32-39 (2016)
2. https://www.nhtsa.gov/sites/nhtsa.dot.gov/files/readiness-of-v2v-technology-for-application-812014.pdf
3. www.termo.utcluj.ro/confstud07/lucrari/socea.doc
4. A. Ionica, M. Leba, Procedia Economics and Finances, 671-676, **23** (2015)
5. http://infrastructura-rutiera.blogspot.ro/
6. http://adevarul.ro/locale/targoviste/n-a-vazut-mai-mare-stop-judet-trei-tineri-ajuns-spital-1_50ad781c7c42d5a66395c02b/index.html

7. V. Oldenborgh, *System and Method for distributed data network having a dynamic topology of communicating a plurality of production nodes with a plurality of consumer nodes without intermediate node logically,* (2006)
8. http://digital.ni.com/public.nsf/allkb/862567530005F09C862566BE004E469D
9. *Road vehicles - Diagnostics systems-Part2. CARB requirements for interchange of digital information*, ISO 9141-2, First Edition (1994)
10. http://www.hoperf.com/upload/rf/RFM12B.pdf

Robotized semiautomatic motorcycle transmission development. Electronic and software design

Mihai Neghină[1], *Radu Emanuil* Petruse[2*], *Sebastian* Olteanu[3], *Ioan* Bondrea[2] , *Lucian* Lobonț[2] , and *Gabriel* Stanciu[1]

[1]Lucian Blaga University of Sibiu, Faculty of Engineering, Computer Science Department, Sibiu, Romania
[2]Lucian Blaga University of Sibiu, Faculty of Engineering, Industrial Engineering and Management Department, Sibiu, Romania
[3]Continental Automotive Systems, Sibiu, Romania

Abstract. In this paper, we propose an electrical design (implemented on a PCB board) and an accompanying software design for controlling the automatic gear change. The designs complement the mechanical solutions developed in Part 1. The paper also analyses the issues encountered during the intermediate steps of the development of the electronic module, which is expected to be small and adaptable enough to be installed on a motorcycle without changing its ergonomics. The control software runs on the Arduino Nano board and is built as a state machine with one idle state, five active states that cover different stages of the gear change and one error state for preventing malfunctions in case of an unexpected event. The sketch uses 5,760 bytes (or 18%) of program storage space and 706 bytes (or 34%) of the dynamic memory.

1 Introduction

This paper is based on a research project whose aim is to robotize a manual motorcycle gearbox. This paper is correlated with the paper "Robotized Semiautomatic Motorcycle Transmission Development. Mechanical Design and Integration". To control the mechanisms described in part one of this research, an electronical module must be developed. This module must be small and adaptable enough to be installed on a motorcycle without changing its ergonomics. Also, it must provide robustness in harsh electrical conditions of indoor and outdoor environment and it must assure a linear functionality no matter what loads are applied to it.

This module must be able to:

- Control two direct current motors capable of actuating the clutch and the gearbox lever.
- Separate the high current circuit, which is needed for the two DC motors, from the low current circuit needed for the microcontroller.
- To read feedbacks from the motorcycle's sensors and the robotized mechanism.
- Offer an intuitive human machine interface (HMI)
- Offer a high debugging capability

* Corresponding author: radu.petruse@ulbsibiu.ro

- Be adaptable to different mechanisms

The code compiled for the electronical module must be capable of offering a high reliability no matter what loads (e.g. user commands, false sensor readings etc.) are applied. It must run in a sequential manner and not enter in an infinite program loop or a dead-end (error). In order to show in which, state the program has reached or what problem has encountered, the module can communicate with the computer via USB.

2 Electrical development

2.1 Electrical schematics design

The schematic of the electronic module is made up of blocks, each one with its own purpose. Each block is arranged in a such way that the final populated printed circuit board is easy to debug and offers the best layout compatibility. The functionality of each block should not influence any other blocks, other than those connected to it. All the electronic components are chosen in order to be easy to mount and use. Next, each block will be explained as of its functionality and basics.

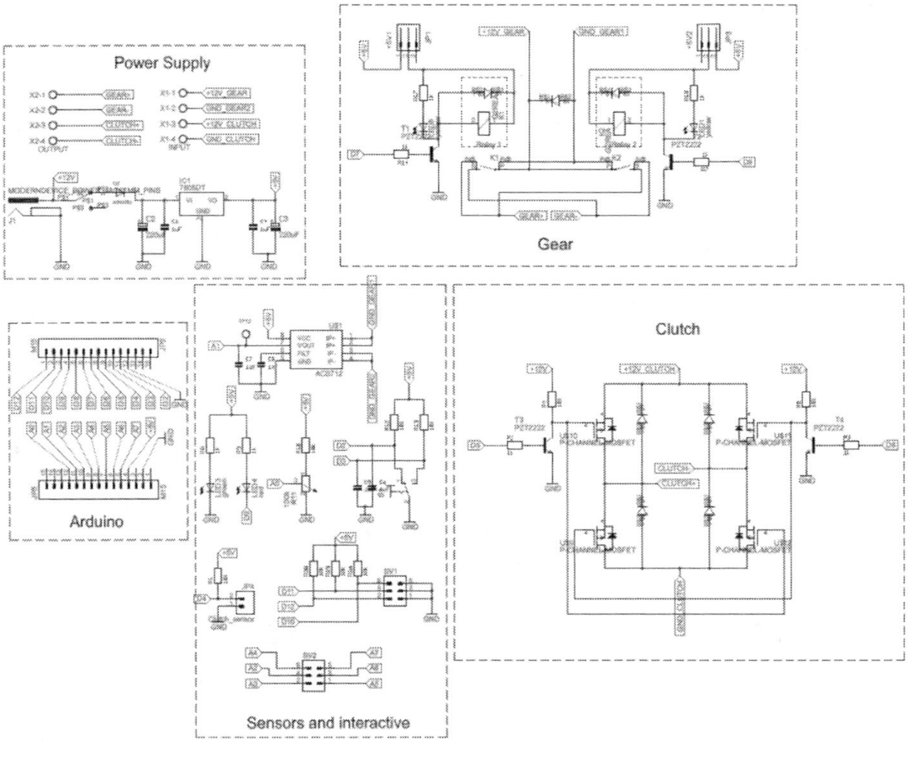

Fig. 1. Electronic schematic

2.1.1 Power supply

There are 3 main power supplies that come directly from the motorcycle battery:
+12V for supplying the logic circuitry from a barrel connector.
+12V_GEAR for supplying the gear shifter dc motor
+12V_CLUTCH for supplying the clutch actuator motor

All supplies are isolated from each other to remove the electric noise from the high current cables of the dc motors. The 3 ground connections are independent from each other.
The printed circuit board (PCB) terminal blocks (two 1x4 blocks) are used to connect cables for the 2 motors, which support a maximum current of 24 amperes. They are also used to grip the supply cables.

The barrel connector is chosen due to its easy assembly on the circuit board and because it can rotate around its interior positive pin.

A gliding switch is used to turn on/off the supply of the logic circuit. It can stand a maximum current of 2 amperes.

An LM7805 (DPAK capsule) voltage regulator is used to convert the ~12 volts from the battery to a constant 5 volts supply for the Arduino and the logic circuits. It offers a supply current of around 1 ampere without a heatsink mounted on top. It is accompanied with 2x220uF electrolytic capacitors and 2x1uF ceramic capacitors on the input and output to remove any unwanted voltage ripple and high frequency signals.

The two direct-current motors (for the gear shifter and the clutch) can reach 3000 rpm, at 12 volts, with a maximum power of 114 watts. Due to its high fly back voltage, to prevent it from suddenly interrupting the power supply, four fly back diodes are used to protect the H-bridge transistors and two fly back diodes are also mounted on the terminals of the two relays.

2.1.2 Arduino Nano

To control the integrated circuit is a ATmega328p microcontroller with a 32 Kbytes of Flash memory, 1 Kbytes EEPROM, 2 Kbytes SRAM that works with a maximum clock frequency of 20 MHz (16 MHz for Arduino Nano). There are two 8-bit Timers and one 16-bit Timer, six PWM channels, 8-channel 10-bit ADC, SPI, USART, I2C interfaces, Watchdog timer, On-chip analogue comparator, Interrupt and Wake-up in Pin Change. The uC can be supplied from 1.8 - 5.5 volts and can work in normal functions in a temperature between -40 -> +105 Celsius grades.

The Arduino's pin connection are set up as follows:

- D2 and D3 pins are connected to the rocker switch S1, which can offer external interrupts. Denounce capacitors of 1uF are used for better reading of signal.
- D4 is used to read the position of the clutch handle. Essentially, it is a Normal-Open micro switch.
- D5 and D6 pins are connected to the transistors that are controlling the clutch, via Timer1 PWM function.
- D7 and D8 pins are used to control the 2 relays. Special hardware control functions are not required in this case, so the program will be used for simple input/output states.
- D9 pin is used for red the LED light error. It can be controlled by PWM.
- D10 pin will be used for an eventual reading of the rotations per minute signal from the motorcycle engine.
- D11 and D12 pins will be used to read the 2 end stop micro switches for the dc motor mounted on the gear. There are pull-up resistors, so when an end stop is pressed, Arduino Nano reads a value of ~0 volts.
- A0 pin is used to read the value of the analogue voltage on the R11 potentiometer.
- A1 pin is used to read the electric current on the gear dc motor. It uses an ACS712 current sensor, which can read a maximum current of 5 amperes

2.1.3 Human-machine interaction

It is done via a rocker button and a potentiometer which calibrates the pulse width modulation for commanding the MOSFET transistors.
There are also 4 LEDs which offer visual information about the system's state:

- 2 orange to display the change in upper or lower gear
- 1 LED POWER green
- 1 red LED in case of error read by Arduino.

Each LED has a current limiting resistor of 1 kOhm. For the safety of the circuit and dc motors, an emergency STOP button is used on the positive supply of the dc motors. It is used in case of an unwanted operation of the motors. Two 4 amp fuses are mounted on the two positive cables of the two dc motors, for a current limitation safety. The 2 amps Schottky diode is used for the essential protection against reversed polarity connection.

2.1.4 Gear control driver

The gear dc motor is supplied via an H-bridge composed of 2 Double-pole Single-terminal Normal-open relays (Relpol RM84-2022-35-1005), which can support a maximum current of 8 amperes on the contacts, at a dc voltage level of 24 volts. The switching speed is around 7 milliseconds, fast enough to give a great speed for the motor command. They are selected due to their wide range of inductor voltage command (3.5 - 12.7 volts) and the possibility of making an H-bridge motor control using just 2 of them.

Due to the supply current of the relay inductor (83 mA) which Arduino's pins cannot provide, an NPN bipolar transistor (PZT2222 in SOT223 capsule) is used for each of the relay to drive their inductor directly from the 5-volt supply. Each transistor offers 600 mA of current.

Two LED indicators (LED1 and LED5) are used to indicate the supply of each relay. Two 1x3 pins (JP1 and JP3) are used to remove any supply voltage of the relay inductors, including the two LEDs.

2.1.5 Clutch control driver

Four P-channel MOSFETs (SUM110P04) are used to create an H-bridge to control the clutch dc motor. The gates of each transistor are connected to +12V signal directly from the dc barrel connector, as to stay open when no command from Arduino is received. The gate-source voltage (Vgs) always must be equal or higher than the drain-source voltage (Vds), so two NPN bipolar transistors (PZT2222) are used for switching from their +12V supply to ground when Arduino wants to close them.

One of the MOSFET transistors can withstand a maximum current of continuous -110 amperes on drain, at a voltage Vds of maximum -40 volts.
Two resistors of 10 kohm are used to limit the current through the two NPN bipolar transistors.

2.2 Layout design

2.2.1 Developing

The module consists of a 35 um thickness copper double-sided board which has all the routes and pads designed using Eagle Cadsoft (trial). The raw photoresistive pcb is developed using a local ultraviolet exposure station in order to obtain best quality of layout, especially for the

0.3 mm traces that are present on the board. Its populated components are mostly of Surface Mount Technology due to the need of pcb space and for better soldering.

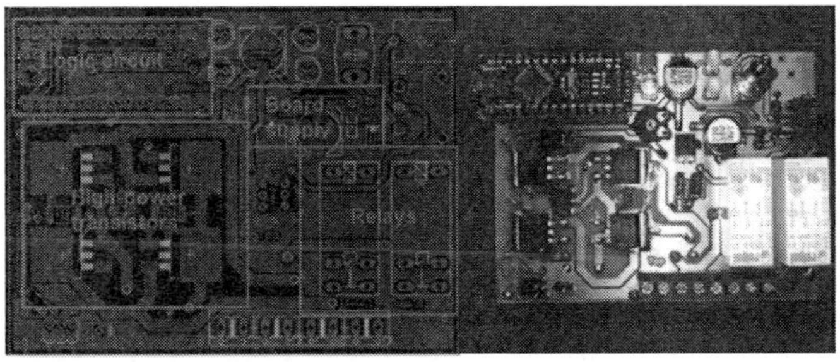

Fig. 2. Developed PCB

2.2.2 Electromagnetic compatibility

In order to limit mutual magnetic influence, traces are separated from each other with a minimum distance of 0.6 mm and the corners of each trace have a slight angle of 45 degrees in order to eliminate the possibility of radio emission. Also, the high-current traces are separated as much as possible from the sensitive low-power traces, and are partitioned in the low side of the board.

3 Control system software development

The code is written in C, using special function in Arduino IDE, specific for Arduino Nano development board. The control of the gear change mechanism is built as a state machine. Sketch uses 5,760 bytes (18%) of program storage space. Maximum is 30,720 bytes. Global variables use 706 bytes (34%) of dynamic memory, leaving 1,342 bytes for local variables. Maximum is 2,048 bytes.

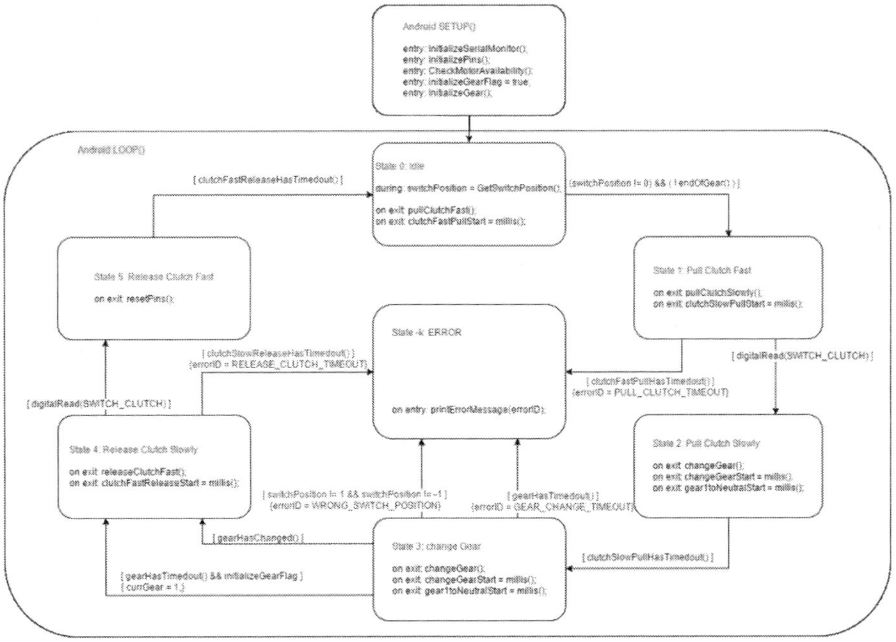

Fig. 3. Finite State diagram

Arduino Setup function

The setup function contains:

- The initialization of the serial monitor (for outputting messages)
- The initialization of the pins
- A check for the availability of the motor
- The setting of gear information. If initialization is requested, the application assumes worst case scenario (MAXGEAR) and attempts to go to neutral. If initialization is not requested, the application assumes that the gear is already in neutral.

Arduino Loop function

The state machine implemented in the Arduino Loop function contains the following states:

- State 0 (Idle) is the state of waiting for a command from the user. State 0 can be exited by a valid user command of changing the gear.
- State 1 (Pull clutch fast) acts on the clutch motor with maximum voltage. State 1 can be exited by reaching a certain position of the clutch (going to state 2) or by timer expiring (going to an error state).
- State 2 (Pull clutch slowly) acts on the clutch motor for a certain duration. State 2 is exited when the timer expires and goes to state 3.
- State 3 (Change gear) acts on the gear motor. State 3 is exited when the timer expires or when one of the end stops indicates that the gear has changed.
- State 4 (Release clutch slowly) acts on the clutch motor with moderate voltage. State 4 is exited when the timer expires (going to an error state) or by reaching a certain position of the clutch (going to state 5).
- State 5 (Release clutch fast) acts on the clutch motor with maximum voltage for a certain duration. State 4 is exited when the timer expires (going to state 0).
- Error states output an error message and prevent any further action.

State 0: Idle

State 0 (idle) is the state of waiting for a command from the user. While in State 0, the application checks the command switch pins, in order to determine whether there is a valid command. If no valid switch command is received, the application considers the user switch to be in the neutral position and remains in State 0.

In case a valid switch command is received and there is a possibility to change the gear in the direction requested by the user, then the cycle of changing the gear begins (the cycle goes through all positive states 1→2→3→4→5→0). Checking the possibility of changing the gear in the direction requested by the user prevents the cycle to begin in case that the motor already is in the maximum gear and the user requests a higher gear, or if the current gear is 0 (neutral) and the user requests a lower gear.

At the initialization phase, the application assumes that the current gear is in MAXGEAR (worst case scenario) and attempts to find the neutral. It does so by requesting lower gears automatically until gear 1 has been identified, and then requests once more a lower gear. During the initialization phase, the application runs several times through the cycle 0→1→2→3→4→5→0 without user intervention and any user command is ignored until the initialization phase has ended.

On exiting State 0, the application starts the timer for State 1, resets all pins and writes the pin for the fast phase of the clutch pull.

State 1: Pull clutch fast

The first phase of pulling the clutch acts on the clutch motor with maximum voltage. The application can exit State 1 in two cases:

- If the timer expires, in which case the application enters error state
- If a pin indicates that the phase has ended (application enters State 2)

On exiting State 1 towards State 2, the application starts the timer for State 2, resets all pins and writes the pin for the slow phase of the clutch pull.

State 2: Pull clutch slowly

The second phase of pulling the clutch acts on the clutch motor for a certain duration. The application exits State 2 when its timer expires. There is no error related to State 2 because there is currently no measurement to indicate whether the clutch has been fully pulled.

On exiting State 2, the application starts the timers for State 3 and resets all pins.

State 3: Change Gear

Depending on the positions of the gears relative to each other and the user command, the function sets the one or the other pin for controlling the gear motor. The implementation described in this paper is specific to the gears being positioned so that neutral is between the first and second gear. The initialization also takes into account this particularity of motorcycles.

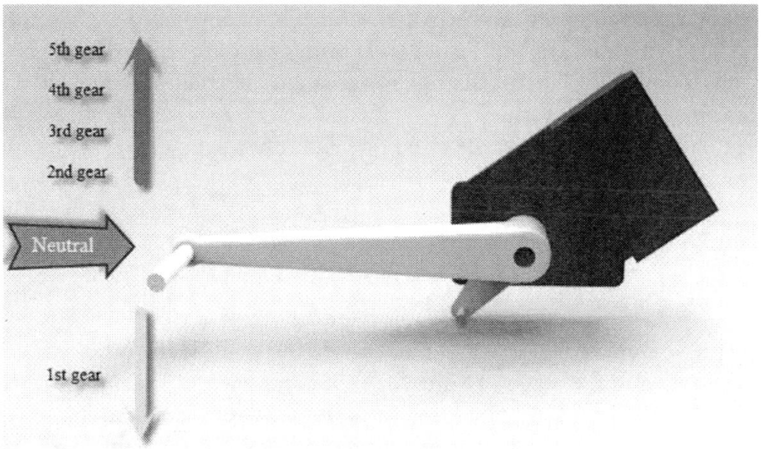

Fig. 4. Motorcycle gears

The application can exit State 3 in three cases:

- The gear has indeed changed (application enters State 4). The indication that the gear has changed relies on:
  - o either endstop pin is triggered (the endstop indications are not taken into consideration for the first few milliseconds to avoid unplausible triggerings), or
  - o If the user requested a change from gear 1 to neutral, 50ms have passed since entering State 3
- The timer expires in the initialization phase. The expiration of the timer is taken as an indication that the current gear is 1.
- The timer expires outside the initialization phase, in which case the application enters an error state

On exiting State 3 towards State 4, the application starts the timer for State 4, resets all pins and writes the pin for the slow phase of the clutch release.

State 4: Release clutch slowly

The first phase of releasing the clutch acts on the clutch motor with moderate voltage (moderate PWM duty cycle). The application can exit State 4 in two cases:

- If the timer expires, in which case the application enters error state
- If a pin indicates that the phase has ended (application enters State 5)

On exiting State 4 towards State 5, the application starts the timer for State 5, resets all pins and writes the pin for the slow phase of the clutch pull.

State 5: Release clutch fast
The second phase of releasing the clutch acts on the clutch motor for a certain duration. The application exits State 5 when its timer expires. There is no error related to State 5 because there is currently no measurement to indicate whether the clutch has been fully released.
On exiting State 5, the application resets all pins.
State: Error
When entering an Error state, the application outputs a message indicating the error and then moves to the ultimate Error State, where nothing happens (in order to not flood the serial monitor by sending one error message again and again).

4 Control software and electronic intermediate tests

During these tests, some issues with the PCB were discovered. The command for the 2 electrical operations was based on an Arduino development board which commands two SPDT-NO relays. These relays will form two H bridges for the DC motor which will change the gears and four MOSFET transistors for a future adaptation of an electric actuator dedicated for operating the clutch.
During the project development, some circuit design flaws were also encountered.

4.1 Layout issues:

4.1.1. PCB Version 1

After completing the electronic scheme, we reached the PCB development stage, drilling for pins and viases, then in the stage of soldering the components. The images bellow reflects the incomplete PCB in terms of populating the components on the board.
Due to unsuitable design of the PCB, the footprints of some components were wrongly made resulting in the difficulty of soldering their connection pins on the pads and viases of the PCB.

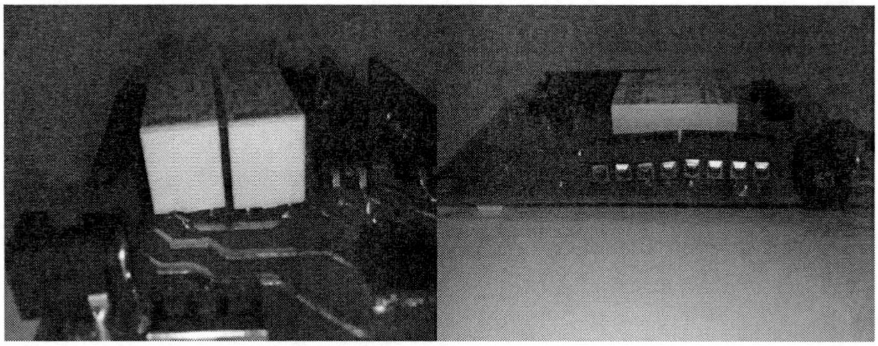

Fig. 5. Poorly welded relays and supply terminals for the board

One of the main reasons for the incorrect placement of the components is guiding current routes of the THT components on the top side of the board, therefore access to pads proved to be difficult. The THT components viases do not have connections between bottom and top side of the board.

Another problem of the circuit encountered during electrical tests is the poorly designed footprint of the voltage regulator.

DOI: 10.1051/matecconf/201712106007

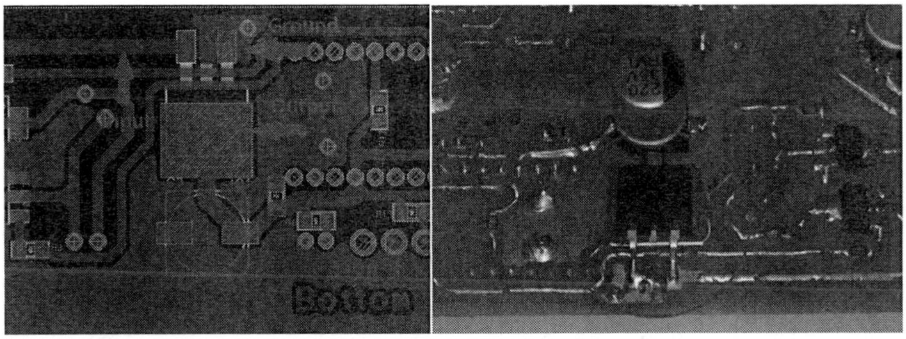

Fig. 6. Voltage regulator LM7805 - layout (footprint is incorrect)

4.1.2 PCB Version 2

P-channel MOSFET transistors were installed with the source to GND_CLUTCH and the drain to +12_CLUTCH, which led to an internal short-circuit on the internal diode.

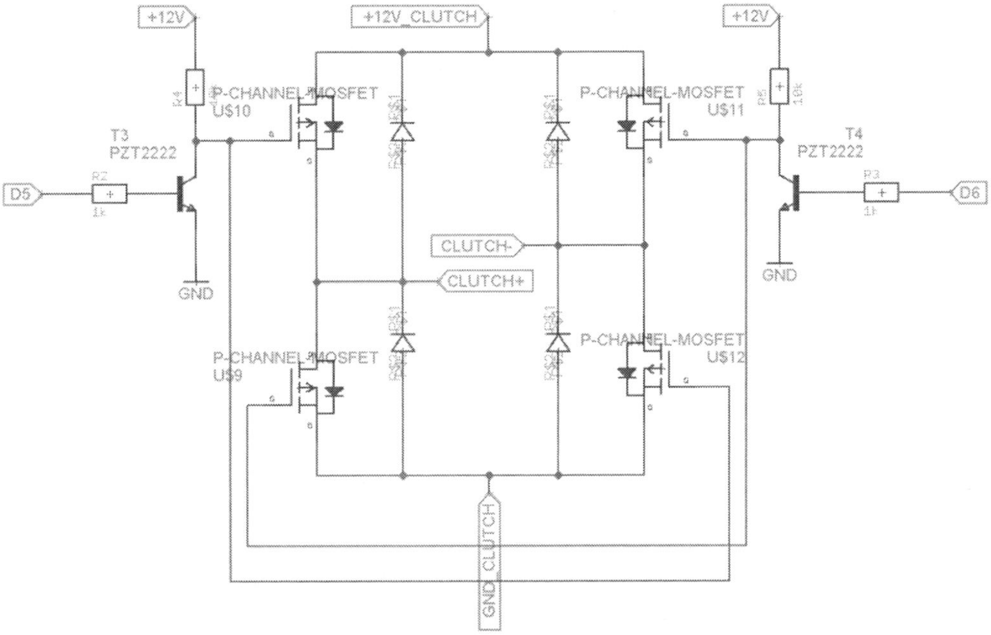

Fig. 7. Schematic with incorrect mounted transistors

Protection diodes D2, D3, D4, D5, D6, D7 and D8 did not have a corresponding footprint pin in the wiring diagram, so that every time you fed the relays and transistors, the short was made at the ground. The problem was later resolved by being mounted in the right direction.

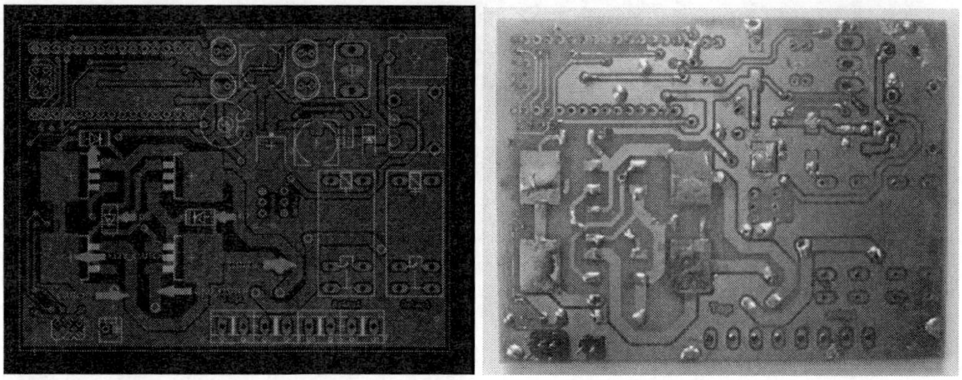

Fig. 8. Incorrect Layout Top

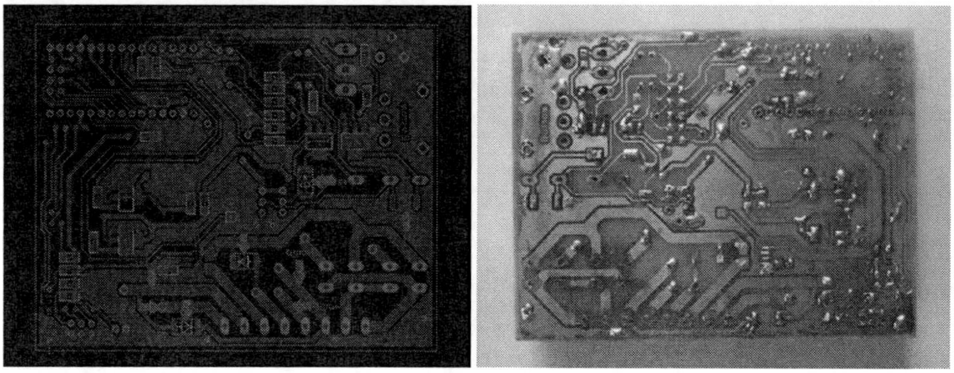

Fig. 9. Incorrect Layout Bottom

Due to not using correct fuse ratings (4a fuse, and we misleadingly used a 25a fuse) for the motors to work and removing power from the electronics, a short-circuit problem appeared during tests. Because the H-bridge for the clutch motor uses P channel MOSFET transistors, a voltage always must be existent on the gate of each transistor to keep it open. Once power was removed from the electronics, the voltage on the gate of the transistor fell to ~0.5, which made the transistor close its source-drain connection and short-circuit the whole 4 transistor H-bridge. This lead to a high current of almost 20 amperes passing through it, which heated up the transistors and the traces. One trace (GND_CLUTCH) heated in an extreme manner and it interrupted itself.

Fig. 10. Interrupted circuit due to short-circuit

5 Electrical functioning

5.1 PCB Version 1

From this point of view, we discovered that supplying for a short time the voltage regulator, voltage on the OUT pin fluctuated between 0.3 and 0.6 volts. Thus, we found that the footprint's pins from the program used to design the layout were not correct. Because of this reason, we discontinued to supply the rest of the circuit.

5.2 Version 2

The power supply was limited to a maximum current of 700 mA for safety. Reverse current diodes have allowed shorting +12V to ground due to the incorrect mounting, but without affecting their functionality, supporting up to 100 A.

After the correct install of the protective 1.5SMCJ120A diodes, we commanded the transistor's gate to rotate the motors, at which moment we observed their inappropriate behaviour due to incorrect mounting. The maximum current given went through the transistor's diodes and the voltage dropped from 12V to ~2V.

5.3 Version 3

Because the high electrical current once passed through the transistors, it heated them up at almost 100 degrees' Celsius temperature, but it didn't affect their functionality, as they are rated for a current of up to 100 amperes. They still performed well after resolving the trace interruption problem with solder.

6 Conclusions

The goal for the research is to eliminate the user's effort of changing the gears and make the gear-change as smooth, fast and autonomous as possible. In the current state in which the motorcycle is mounted on a stand, the project cannot be tested outdoors making this impossible to offer a clear idea of how the system should work in the purposed environment. But, despite all the encountered PCB layout problems, the module behaves as expected and can even be improved for better performance in the near future.

Acknowledgement: The research leading to these results has been supported by a research contract from Continental Automotive Systems Sibiu, Contract no. C-1028 form 17.10.2016.

References

1. M. Simon, *Programming Arduino: Getting Started With Sketches* (McGraw-Hill Education edition, 2011)
2. M. Banzi, M. Shiloh, *Getting Started with Arduino*, (Maker Media, Inc; 3 edition, 2014)
3. https://www.arduino.cc/en/Main/ArduinoBoardNano accessed on 17.01.2017

Individual risk evaluation and interventions for mitigation in the transportation of hazardous goods: a case study

Elena Cristina Rada[1,2,*], *Navarro* Ferronato[1], and *Vincenzo* Torretta[1]

[1]DiSTA, University of Insubria, Via G.B. Vico 46, I-21100, Varese, Italy
[2]DICAM, University of Trento, Via Mesiano 77, I-38123, Trento, Italy

Abstract. The transport of hazardous substances is an economic activity essential for goods' transference chain. However, the risk in transporting hazardous materials is related to the occur of accidents causing environmental damages and public health dangerous consequences. A quite recent Italian example is the Viareggio accident (2010), which involved a train with tank cars containing liquefied petroleum gas (LPG) which caused more than thirty deaths. This paper describes the safety state in the Varese district (an area of northern Italy with a very high population density and industrial activities), with the aim at comparing the current situation (considering the risks due to the transportation of hazardous materials on the main motorways and main national roads) with a potential scenario that introduces a few mitigating interventions, such as a partial conversion from road haulage to rail transport. This comparison can be accomplished by developing the existing intermodal platforms and implementing new ones in strategic areas.

1 Introduction

Transportation and management of dangerous goods is cause of risk for the population which live along the routes and around production areas [1-2]. The main guidelines for risk management in the chemical industry are provided by the European Union (EU) by the Seveso Directive [3], among others. However, there is not a unified presentations of industrial risks, while activities involving hazardous materials are not wide covered by existing directives and legislation. This means that, in EU, hazardous installations and activities are not automatically and always included in territorial risk assessments [3]. As a result, information transfer from decision makers to stakeholders, such as civil services, does not exist. This situation creates huge demands for territorial risk and vulnerability management tools, especially emergency preparedness planning [4]. Moreover, risk management instruments include a very critical item represented by risk communication and population involvement [5]. The main issues in hazardous goods transportation studies are data research and collection on traffic flows, running time, characteristics of neighborhood, the existent of rail bridges and tunnels along the way, among others [6-9]. Such items are required also by

* Corresponding author: elena.rada@unitn.it

customers which choose intermodal route mainly by lead-times [10]. Transport risk analysis presents more difficulties than risk analysis conducted, as instance, in a chemical industry since there are more uncertainties about potential accident dynamics and accident localization [11-12]. The main legislative references concerning the regulation of dangerous goods' transport are made by international agreements such as the ADR (European Agreement concerning the International Carriage of Dangerous Goods by Road), for road transport, and the RID (Rules concerning the Railway international transport of hazardous goods), for rail transport. Nevertheless, these agreements principally regulate container and loads features, classification and quantities of transportable goods and vehicles on board equipment. No indications are instead specified about the possible interactions among the vehicles and their loads, the infrastructures used and the surrounding environment. This approach, in the current set of rules, lack in measures aimed to manage and reduce transport risk such as rational restriction to the passage of transport of dangerous goods in areas characterized by high vulnerabilities for the population and the environment or the provision of indication about the most suitable modal split to adopt in a given territory.

The ADR is brief and simple. Its most important item is article 2, which states that, with the exception of certain extremely dangerous materials, hazardous materials may, in general, be transported internationally in wheeled vehicles, provided that two sets of conditions are met:

- Annex A regulates the merchandise involved, notably its packaging and labels;
- Annex B regulates the construction, equipment and use of vehicles for the transport of hazardous materials.

According to ADR and RID, nine are the main classes of dangerous goods: explosive substances and articles (class 1); gases (class 2); flammable liquids (class 3); flammable solids, self-reactive substances and solid desensitized explosives (class 4.1); substances liable to spontaneously combust (class 4.2); substances that emit flammable gases in contact with water (class 4.3); oxidizing substances (class 5.1); organic peroxides (class 5.2); toxic substances (class 6.1); infectious substances (class 6.2); radioactive material (class 7); corrosive substances (class 8); and miscellaneous dangerous substances and articles (class 9). Governance of hazardous materials transport requires complex decisions regarding regulation, routing, land use, and resources for emergency response [13-15]. Decision makers required to balance the needs of various actors, including public and private industries, and they should take a multi-hazard approach, considering the spatial dimensions of the problem and the vulnerability of their territory. Therefore, decision makers should have appropriate decision-support tools which facilitate the integration of data required [4]. In general terms, a decision support system (DSS) is a specific class of computerized information system that assists people in making decisions based on data gathered from a wide range of sources, and which could involve the stakeholders at all stages of the corresponding risk process [16]. DSS applications are not single information resources, such as databases, models or programs that graphically represent results and figures, since are a combination of integrated tools working together. The structure and design of a DSS can vary according to the skill and aptitude of the decision maker and the need of the decision-making process [17].

In this paper, the authors use a specific DSS to quantify risk due to the transport of hazardous substances through road, rail, pipeline and inland waterways [18]. The DSS allows the following evaluations [19]:

- population global risk (n. subject/year), considering the exposition to the potentially relevant incident;
- individual risk (yearly probability/person), with the hypothesis that the single person would be exposed during the entire duration of the dangerous event without protection.

This DSS allow comparing the risk distribution for dangerous substances conveyance with different methodologies. Such indications permitting the user to decide which course should be followed, minimizing the risk related to the transport of hazardous substances within the territory considered. The case-study refers to the Varese district, in the Lombardy region of northern Italy, which is an area densely populated and with a diffuse presence of industrial activities. The aim of the paper is the comparison of the current scenario (considering the risks produced by the transportation of hazardous materials on the main motorways and most important national roads) with a potential scenario obtained by introducing some mitigating interventions, such as a partial conversion from road haulage to rail transport, although road transport is the most frequently utilized means of transporting hazardous goods throughout Europe [20] particularly in Italy. As instance, in Lombardy, 87% of the total amount of transported hazardous goods is carried by road [21]. So, the development of intermodal capacities is of utmost importance in order to identify alternative solutions to road transportation [22]. For that purpose, the comparison provided by this study, accomplished by developing the existing intermodal platforms and implementing new ones in strategic areas, according to other similar experiences described in the technical literature [23] allow considering what transport methodology could reduce convey risk.

2 Materials and methods

Varese is an Italian province with about 880,000 inhabitants and a population density of 742 inh/km^2. The current economic state is mainly boosted by the numerous industries which are scattered within the whole area. Since Varese is one of the province closed to Milan and the big lakes of northern Italy, the transport facilities are developed and many time trafficked due to the many commercial activities and the tourism. Indeed, five important motorways cross through the province: A8 Milan-Varese, A9 Como-Chiasso, A8/A26 Gallarate-Gattico, A36 and A60, among other provincial and national ways (Figure 1), so that Varese is considered one important economic pole for the Lombardy region, where the province is inserted.

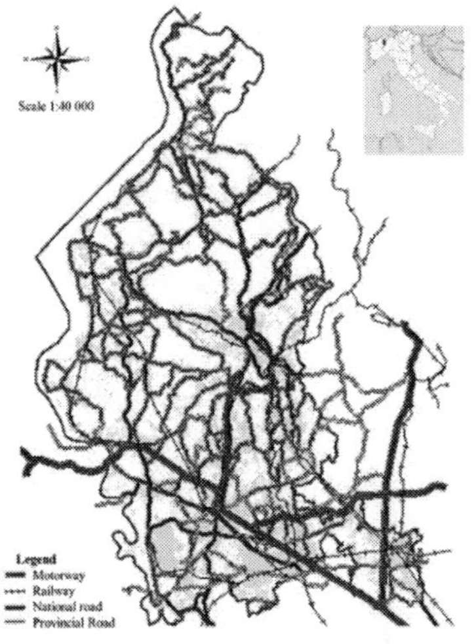

Fig. 1. Study area [24]

The DSS is fed by many types of information [19]:

- description of the transport stream (number of yearly loaded transport units per substance or category);
- description of the transport units;
- description of the transport routes;

- description of the number of accidents and level of traffic to determine accident frequencies;
- description of ignition sources;
- chemical characteristics and properties of transported substances;
- terrain classification of the surroundings of the transport route (urbanized, agricultural area, etc.);
- meteorological data;
- population living and working in the area surrounding the transport route (density of population);
- location of critical targets with high vulnerability (hospitals, schools, etc.).

All of these types of data feed the algorithms utilized for risk assessment [25].

The model and maps have been integrated into GIS software that can simulate the risk assessment for every road/rail/pipeline tract. When new data are generated, the system can be easily updated. The following are the main characteristics of the software:

- accessible standard language that is compatible with the most common operating systems;
- user interface for selection of roads, rail and pipeline streams;
- results can be geo-referenced into a technical regional map.

Users can select the area of interest and can move with a friendly pan operator. The software automatically shows the data associated with the selected tract, particularly the following:

- λ = incident frequency of the selected tract (number of accidents yearly per km);
- L = length of the tract (km);
- NT = hazardous traffic flows;
- A = impact area (km^2);
- DW = population in the environment associated with the selected tract.

Fifty-four substances (6 for each transport class) were considered, including hydrofluoric acid, ammonia, gasoline, benzene, liquefied oxygen, ethylene dioxide, methanol, sulfuric acid, and acetylene, along with their corresponding potential accident scenarios: flash fire, pool fire, UVCE, toxic release, BLEVE, and jet fire.

The first step corresponds to the identification of the transport stream with the characteristics of the hazardous flux. The most important parameters are λ (accident frequency), *L* (length of the tract) and *NT* (hazardous traffic daily flow). Data are visualized with GIS software. The user can select a road tract, and the software automatically identifies critical parameters. The quantified risk analysis is performed by considering different types of accident scenarios, taking into account that one substance can lead to more than one type of accident (for example, fire and toxic release). The evaluation of physical effects on human health is calculated using off-line software (TNO Effects) that can quantify the distance, in meters, of the threshold limits from the transport road line. The typical scenario (critical events) considered are the following [26, 27]:

- dispersion (release of substances that are toxic, flammable or dangerous to human health without fire);
- pool fire (flammable liquid);
- flash fire (vapor cloud fire without explosive effects);
- UVCE (unconfined vapor cloud explosion);
- BLEVE (boiling liquid expanding vapor explosion);
- jet fire (flammable pressurized gas).

A distinction is made between the following transport classes:

- flammable liquid transported in an atmospheric tank unit;
- flammable or toxic gas compressed or liquefied;
- compressed cryogenic flammable or toxic gas;
- compressed flammable gas.

The software was applied to all existing road and rail networks in the Varese district (using data on hazardous goods flows, specific incidence rates for every km of the networks, etc.).

3 Results and discussions

The present study facilitated the evaluation of risk and related potential consequences regarding the transportation of hazardous substances and goods in the territory of the Varese district. In particular, the study defined the distribution of individual average risk in the territory (yearly probability/person).

The most dangerous transportation infrastructures are:

- the Motorway A8 (Milan-lakes district), with an individual risk up to 3.713 E^{-4} prob. person/year;
- Como–Chiasso Motorway A9 (in the tract in Varese district) where the peak value of the yearly individual risk (Figure 2) on a 5 km way is 2.871 E-4;
- airport Milan Malpensa highway SS336 (1.268 E-4).

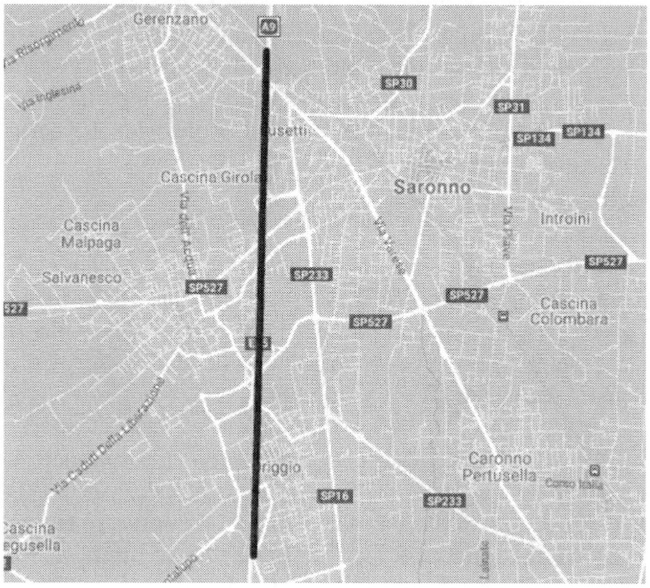

Fig. 2. Evidence of the 5 km of Motorway tract with the highest value of individual risk in the Como-Chiasso Motorway A9 (2.871 E-4)

The authors fixed the target of 1 E^{-4} as the individual risk maximum level. The mitigation intervention was the detour from road to rail of a part of the total amount of hazardous goods transported. Such adjustment requires considerable expansion of the existing intermodal pole in Busto and the realization of a new one. The combination of transport by road and by rail facilitates the following:

- transportation cost reduction (fewer personnel and lower energy and fuel consumption);
- the reduction of accident risk and environmental effects (in particular, air emissions and noise), increasing safety for people and protection for the environment.

The authors utilized the software in the current scenario, and we compared the results with those corresponding to different scenarios obtained by increasing the percentage of hazardous goods transported by rail (covering the same paths) instead of by roads, since transport risk decrease considerably [28]. In Figure 3 the results regarding the individual risk peak on the ordinate axis and the percentage detour on the abscissa axis, are reported.

The authors consider the point at which there is a significant change in the slope of the curve. After this point, we have a small reduction of risk level. The optimum value of the percentage of detour from road to rail is 40%. The increased capacity of the intermodal poles corresponds to this percentage.

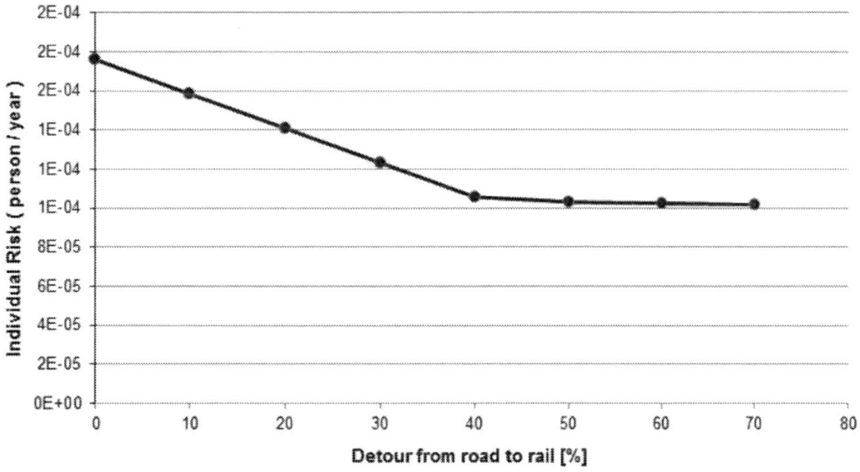

Fig. 3 Example of application of the risk curve to SS 336 (percentage vs. individual risk).

4 Conclusions

Decision-support systems have been found to be valuable tools for the management of dangerous goods and for crisis prevention, which increasingly rely on complex information technology architectures and powerful communication technologies. Such technologies are mature, but there are operational constraints related to the cost of the communication. Therefore, the challenge for the coming years is not to improve the available technology but to define a feasible and sustainable business model to reduce the operative costs of public controllers and decision makers and to stimulate competitiveness among private operators. Private sectors have put considerable efforts into improving safety and security of dangerous goods transportation, although improvements are still required. Public authorities need to improve control and monitoring activities, which requires large investments of public economic funds. Despite recent events, such as the Viareggio accident (June 29, 2009), rail transport remains the safest way to transport hazardous goods. Nevertheless, it is necessary to improve the safety level of rail networks and rail tanks. In general, rail transport makes use of proper housing and does not interact with other vectors (as road traffic does). Rail is a means of transport with higher safety levels and lower environmental impact than road transport, and it requires only 50% of the energy demands of road transport.

This study facilitated the evaluation of a new operative scenario corresponding to a significantly lower level of individual risk distribution in the territory (not higher than E-5). The main conclusions and recommendations of the study are as follows:

a) based on the current data, the optimal percentage of detour from road to rail transportation is 40%;
b) referring to (a), increasing the total intermodal capacity in the district territory from 1,000,000 to 3,000,000 tons year^{-1} of hazardous goods;
c) decrease of individual risk (always and everywhere lower than 10-5), with a maximum level of 9.365 E-5 evaluated in a tract of Motorway A8.

These figures demonstrate that railway can play a strategic role in the optimization of the hazardous goods transportation.

References

1. P. Cassini, J. Hazard. Mater. **61(1),** 133-138 (1998)
2. G.B. Băbuţ, R.I. Moraru, L.I. Cioca, Quality - Access to Success **16(147),** 85-91 (2015)
3. Directive 2012/18/EU of the European Parliament and of the Council of 4 July 2012 on the control of major-accident hazards involving dangerous substances, amending and subsequently repealing Council Directive 96/82/EC Text with EEA relevance.
4. C. Di Mauro, J.P. Nordvik, *Decision support system for crisis management planning* (EC, JRC, Ispra, Italy, 2010)
5. C. Di Mauro, S. Bouchon, V. Torretta, Chem Eng Transact, **26**, 297-302 (2012)
6. B. Fabiano, F. Curro, E. Palazzi, R. Pastorino, J. Hazard. Mater., **93(1)**, 1-15 (2001)
7. R.V. Gagliardi, V. Torretta, F. Conti, S. Stella, G.C. Bello, P. 3rd IASME/WSEAS Int. Conference, Crete, Greece, (2007)
8. V. Torretta, P. CBEE, Singapore (2009)
9. I.L. Balan, L.I. Cioca, V. Torretta, L. Talamona. Procedia Technol. **22**, 1028–1034 (2016)
10. M. Verma, V. Verter, Eur. J. Oper. Res. **202(3)**, 696-706 (2010)
11. S. Contini, P.M. Contini, V. Torretta, C.S. Cattaneo, M. Raboni, S. Copelli, Chem. Eng. Transact. **53**, 109-114 (2016)
12. S. Contini, S. Copelli, V. Torretta, M. Raboni, C. Sala Cattaneo, R. Chem. Eng. Transact. **33**, 487-492 (2013)
13. D.E. Popescu, C. Bungau, M. Prada, C. Domuta, S. Bungau, D.M. Tit, J. Environ. Prot. Ecol., **17(3)**, 1011-1020 (2016)
14. V. Torretta, E.C. Rada, M. Ragazzi, E. Trulli, I.A. Istrate, L.I. Cioca, Waste Manag., **45(SI)**, 152-160 (2015)
15. S. Bungău, R. Suciu, A. Bumbu, G. Cioca, D.M. Tit, J. Environ. Prot. Ecol., **16(3)**, 980-987 (2015)
16. A.V. Gheorghe, J. Birchmeier, D. Vamanu, I. Papazoglou, W. Kröger. Reliability Eng. Sys. Saf. **88(3),** 247-272 (2005)
17. V. Torretta, E.C. Rada, M. Schiavon, P. Viotti, Safe Sci. **92**, 1-9 (2017)
18. V. Torretta, M. Raboni, S. Copelli, A.G. Capodaglio, Environ. Eng. Manag. J. **13(10)**, 2643-2652 (2014)
19. V. Torretta, M. Raboni, S. Copelli, G. Urbini, Environ. Eng. Manag. J. **12(10)**, 2031-2039 (2013)
20. D. Ryland, J. Transport. Plan. Technol. **9(2),** 102-134 (1999)
21. ISTAT - *Conto nazionale delle infrastrutture e dei trasporti* (Rome, Italy, 2008)
22. *European transport policy for 2010: time to decide, White Paper*, Bruxelles, Belgium (2007)
23. Y.M. Bontekoning, C. Macharis, J.J. Trip, Transport. Res. Part A: Policy Pr. **38(1),** 1-34 (2004)
24. http://cartografia.provincia.va.it/maps/mapindex.html (2017)
25. P. Leonelli, S. Bonvicini. G. Spadoni, J. Hazard. Matater, **71(3)**, 423-437 (1998)
26. TNO-*Guideline for quantitative risk assessment 'Purple Book', CPR18E, Part 2.*, (The Hague, The Netherlands, 1997)
27. TNO Safety Software - *Effects version 7.5 - Modelling the effects of accidental release of hazardous materials*. (The Hague, The Netherlands, 2000)
28. M. Verma, Transport. Res. part C: Emerging Technol. **19(5)**, 790-802 (2011)

Soot and smoke emissions numerical evaluation for a direct injection (DI) diesel engine

Bogdan Radu[1], *Alexandru* Racovitza[1,*], and *Radu* Chiriac[1]

[1]University POLITEHNICA Bucharest Dept. of Thermodynamics, Engines, Thermal and Refrigerant Systems, Romania

Abstract. The reduction of Diesel internal combustion engines emissions is one of the major concerns of the engines manufacturers. Despite the fact that the efficiency of the gas post-treatment systems has been significantly improved, decreasing the smoke and the soot from the cylinder inside remains a main research goal. This work is proposing a theoretical study on these pollutants formation for different kinds of direct injection methods. By dividing the in-cylinder injection the heat release characteristic could be modified, leading to different temperature and pressure levels. Using exhaust gas recirculation (EGR) the reduction of the gas temperatures might also be decreased, limiting NO_x formation. To evaluate the level of the cylinder gas emissions formation a two-step procedure could be followed. First, by using a numerical calculation system the heat release characteristic can be highlighted concerning a Diesel engine with stratified injection; then, using an experimental relationship applying a large data base, the amount of the gas emissions can be subsequently provided. The authors propose some combinations between injection characteristics and EGR used fractions which could generate successfully results speaking in terms of NO_x, soot and smoke formation.

1 Introduction

The greatest concern referring to the use of commercial Diesel engines is to identify reasonable and applicable solutions in order to fit the limitations regarding the exhaust emissions. Therefore, the control of NOx, smoke and particulate matter concentrations is the basic target among the general activity of developing Diesel engines in terms of increasing their performance and efficiency.

Simulation and experimental research have proposed different techniques in order to cover the processes development inside the cylinder, from which air-fuel mixture formation, combustion and heat release play crucial role in the control of the motor fluid thermal status and consequently of the NOx genesis and exhaust. Thus, solutions like split

*Corresponding author: alexandru_racovitza@yahoo.com

injection and injection timing adjustment [1], premixed combustion [2], injector and nozzle geometric modifications [3], water or further more combinations of hydro-oxygenated fuels injection have been studied [4-7]. Separately or together with these, the Exhaust Gas Recirculation (EGR) method allows as well the reduction of the heat peak temperature reached inside the cylinder [8-9].

The more drastic limitations stipulated by the recent Euro 6 stages standards impose continuously reducing of Diesel engines emissions. Moreover, commercial vehicles with direct injection engines are limited at 0.005 g/km for Euro 5 and Euro 6. Starting with Euro 5b a particulate number standard (PN) has been established in 2011 for diesel engines, valued at $6x10^{11}$ per km and it was assigned also for gasoline engines in 2014 with Euro 6 [10-11].

The emissions formation has been subjected for simulation development studies more than three decades, since soft packages occurred as valuable theoretical research tools. Computational Fluid Dynamics (CFD) is also used for numerical analysis using data structures to better describe and solve problems linked to the engine fluid dynamics [12-13]. The obtained results presented in this paper are based on the AMESIM [14] numerical code, used in order to establish the necessary link between model result and the predictive experimental results of the engine.

2 Modeling the engine processes

In order to obtain the engine operating parameters as a result of diesel fuel injection variables when using a direct injection diesel engine, a specialized program AMESIM has been used [14]. This allows the creation of a diesel engine model with a multiple stage split injection (see Figure.1).

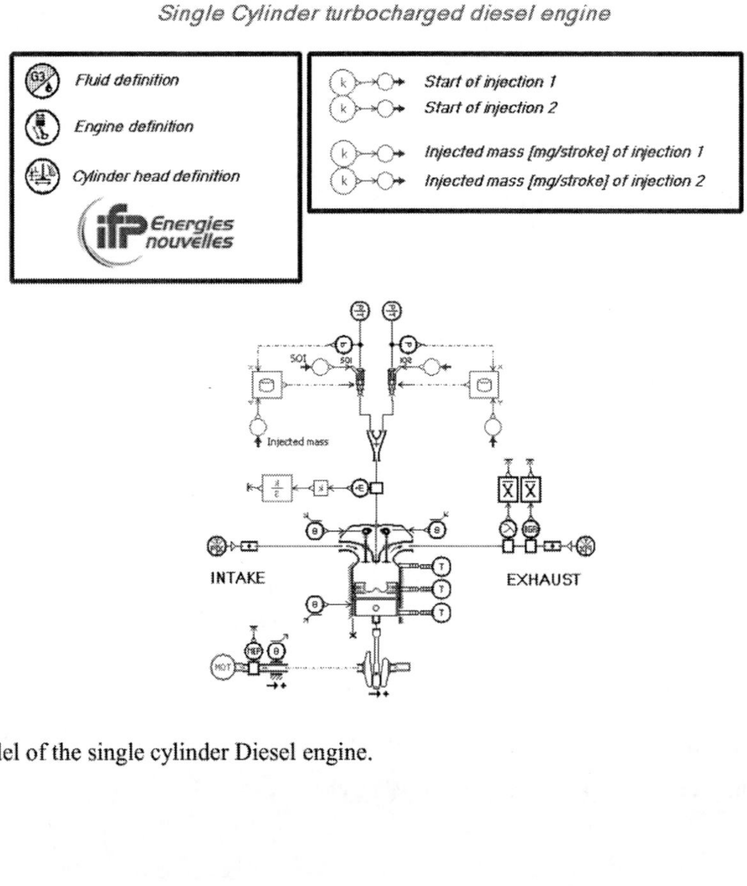

Fig. 1. Model of the single cylinder Diesel engine.

In order to simplify the engine model only one of the engine cylinders has been schematized. Regarding the gas exchange both inlet and exhaust are isobar processes. The structure elements defining the gas exchange systems are consisting parts of the IFP Engine Library [14] and allow the options for the engine operating parameters.

The main component of the system is the Diesel engine cylinder (see Fig.2) and contains 9 ports for the input and the output data transfer [15].

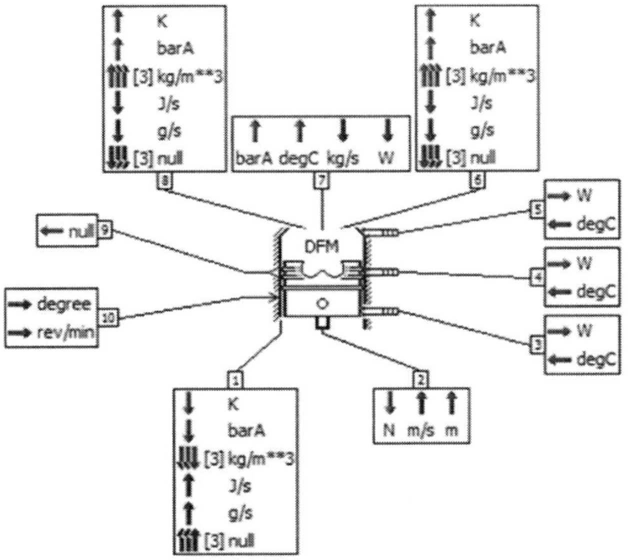

Fig. 2. The Diesel engine cylinder model

The combustion phase is using a model based on a previous work [16]. This considers a two stage scenario for the fuel burning process: the combustion of the preformed mixtures and the diffusive combustion. A first fuel injected mass covers in time the autoignition delay and the rest of the injected fuel is presumed to ignite immediately. For the autoignition delay time period the following Arrhenius type formula could be use:

$$\tau_i = (1 + 10 \times X_{\mathrm{BGR}}) A_t \times p^{-n} \times e^{\left(\frac{T_a}{T}\right)} \tag{1}$$

in which the only difference comparing to the classic form is the insert of X_{BGR} that defines the molar concentration of the burned gas before combustion and marks the effect of the exhaust recirculated gas accumulated together with the burned gas until the start of injection. Each unit of fuel injected mass should have its own description in terms of heat release.

The scenario of the preformed air-fuel mixtures formation could be expressed by five phenomena [15]:

1. The fuel is injected and vaporizes.
2. A mixing zone is formed, where the air is motioned into the fuel spray.
3. The preformed mixtures continue to form until the end of the autoignition period.
4. After this period the mixtures ignite, leading to the flame propagation through the reactive environment.
5. The combustion is reaching the cylinder volume.

In the case of split injection this scenario is repeating after each injected unit. Referring to those units injected after the combustion start the in-cylinder parameters are strongly affected, causing a significant reduction of the autoignition delay. Thus, for these following injection sequences the preformed mixtures should have a less important influence.

This above presented model [16] has been implemented in AMESIM.

3 The modeling results

In order to evaluate the engine performance a commercial Renault Energy dCi Diesel engine has been chosen as basic model. Its features are: 30 kW/cylinder rating power at 3500 rpm, 85 mm bore and 102 mm stroke. Its turbocharging pressure is 0.18 MPa, compression ratio 16 and for the chosen operating regime the relative air-fuel ratio is 1.42. At the beginning of the study the injection was considered an undivided process.

The optimized injection advance for to this regime is 18 deg CA (Crankshaft Angle degrees), corresponding to a power of 30.3 kW. The pressure diagram and the temperature on cycle variation are highlighted in Figure 3. The peak firing pressure is 15.8 MPa and the maximum temperature is 2450 K.

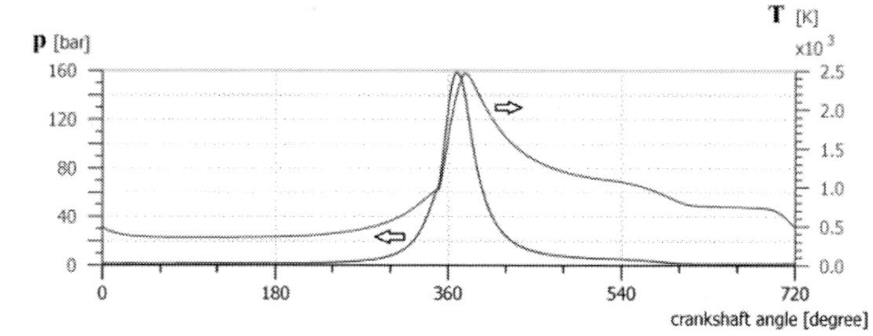

Fig. 3. Basic cycle pressure and temperature variations

The optimal situation has been chosen corresponding to the maximal values of the efficiency or of the rated power, as seen in Figure 4. This shows the fact that the 1% increase of the power is related to an increase of the cycle peak firing pressure from 13 MPa to 15.8 MPa, therefore the nitrogen oxides will significantly increase.

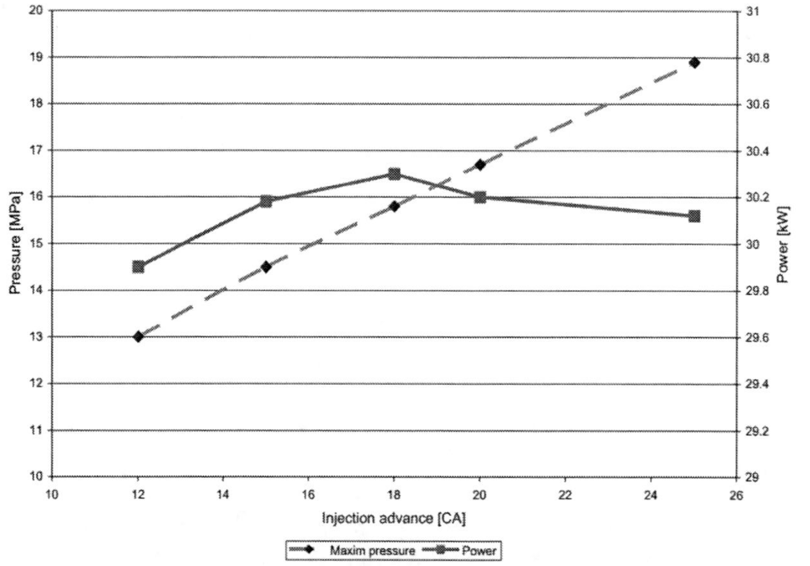

Fig. 4. Identifying the optimal injection advance value

From the point of view of the combustion development, for this situation characterized by 120 MPa fuel injection pressure the start of injection is at 342 deg and the end of the autoignition delay is at 348.8 deg. It can be also added that 5 % of the fuel mass burning corresponds to 352.7 deg, its 50% fraction corresponds to 365.4 deg and 90% fraction corresponds to 377.6 deg. The end of combustion is marked at 397 deg.

4 NO_x and soot emissions calculation

The theoretical calculation of nitrogen oxides and soot formation is a far complicated issue, due not only to their genesis mechanism, largely known, but mostly because to the fact that these reactions occur in very special zones of the combustion chamber. Because of that these calculations can not be performed using phenomenological models which consider only the global conditions existing inside the combustion chamber.

There are two empirical formulas [16] based on a large number of experiments in order to evaluate the two type emissions as a function of the cylinder conditions and the pressure diagram characteristics.

$$\text{Soot(filter smoke number)} = K_1 \times ZV^{e1} \times \alpha_1^{e2} \times \left(\frac{dp}{d\emptyset}\right)_{max}^{e3} \times \alpha_2^{e4} \times O_2^{e5} \times \left(\frac{m_{Luft}}{m_{gesamt}}\right)^{e6} \quad (2)$$

$$\text{NOx(ppm)} = K_2 \times ZV^{f1} \times \alpha_1^{f2} \times \left(\frac{dp}{d\emptyset}\right)_{max}^{f3} \times \alpha_2^{f4} \times O_2^{f5} \times \left(\frac{m_{Luft}}{m_{gesamt}}\right)^{f6} \quad (3)$$

In these expressions, the symbols have the following meaning:
- K_1, K_2 are the proportionality coefficients, each for each emission type;
- ZV is the autoignition delay;
- α_1 is the angle formed with the horizontal axis by the line between the combustion starting point and the highest point of the heat release characteristic;
- α_2 is the angle formed with the horizontal axis by the line between the combustion ending point and the highest point of the heat release characteristic (see Figure 5);
- $(dp/d\varphi)max$ is the maximum rate of the pressure increase, peaked from the pressure diagram;
- O_2 is the oxygen concentration from the exhaust gas;
- $(m_{air}/m_{fresh\ charge})$ is the ratio between the air quantity and the entire fresh charge;

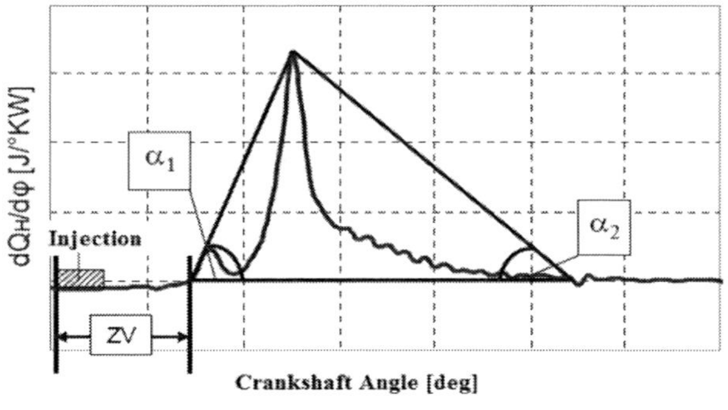

Fig. 5. The evaluation of α_1 and α_2 angles

Each of the above terms have an exponent (symbolized from a to f) and an index (from

1 to 6) numerically established to offer a good correlation with the experimental data. These 7 used terms [16] are listed in Table 1:

Table 1 Equation's 2 and 3 constant values

	K	1	2	3	4	5	6
Soot	1.74	-0.93	1.93	-0.06	-1.86	-1.37	-0.25
NO_x	177.3	0.19	1.67	0.86	-1.92	0.45	2.98

As a remark of the different sign referring to the terms 5 and 6, those depending on the cylinder air-fuel ratio, it means that decreasing one of them the other one increases.

5 Emissions evaluation at split injection

The evaluation for the emissions regarding the engine basic operating regime starts with the evaluation of the pressure diagram and the evaluation of the heat release characteristic (see Fig. 6).

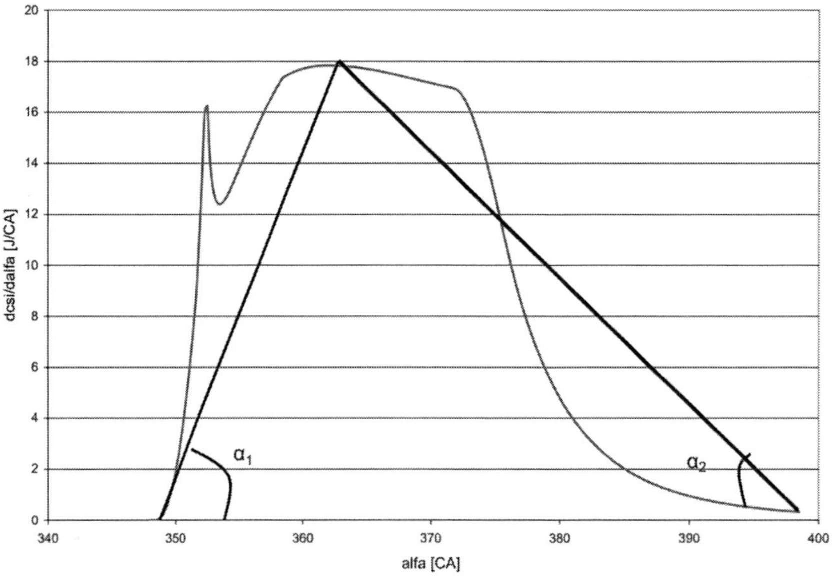

Fig. 6. Rate of heat release [J/CA]

The maximum rate of the pressure increase is 8.48 bar/deg CA, the autoignition delay is about 7 deg CA and the injection phase period is 48 deg CA. The terms used in the emissions calculation formula are (listed in Table 2):

Table 2 Equation's 2 and 3 values for basic regime

ZV	α_1	$(dp/d\varphi)_{max}$	α_2	O_2	$m_{air}/m_{fresh\ charge}$	NO_x	Soot
0.32	53.44	8.48	41.51	6.09	1	1218	0.779

The reduction of the analyzed emissions is a quite delicate problem because the nitrogen oxides emission could not be succeeded without the increase of the soot emission. Analyzing the basic undivided injection cycle it clearly appears that the main problem is that of the NOx. Therefore, some injection configurations could be tested so that the peak firing pressure could eventually decrease with less penalties on the engine efficiency.

For a second proposed engine regime, the injected fuel is divided into a 20% injection pilot corresponding to an advance of 8 deg CA and the rest of the injected fuel with an advance of 2 deg CA (see Fig. 7).

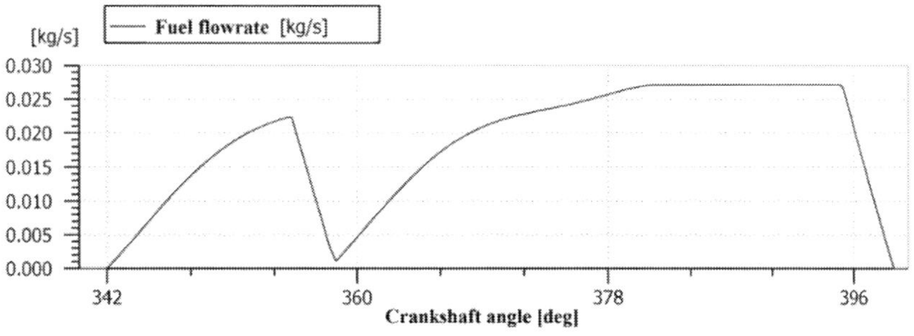

Fig. 7. The injection characteristic for regime no.2

The main injection phase starts approximately when the pilot injection ends and the ending point for the entire injection period is at about 400 deg. The engine performances applying this injection case are low, the rated power per cylinder decreasing from 30.3 kW down to 26.6 kW. The peak firing pressure reaches only 102 bar and the combustion process is expanding until 410 deg, as shown in Fig. 8.

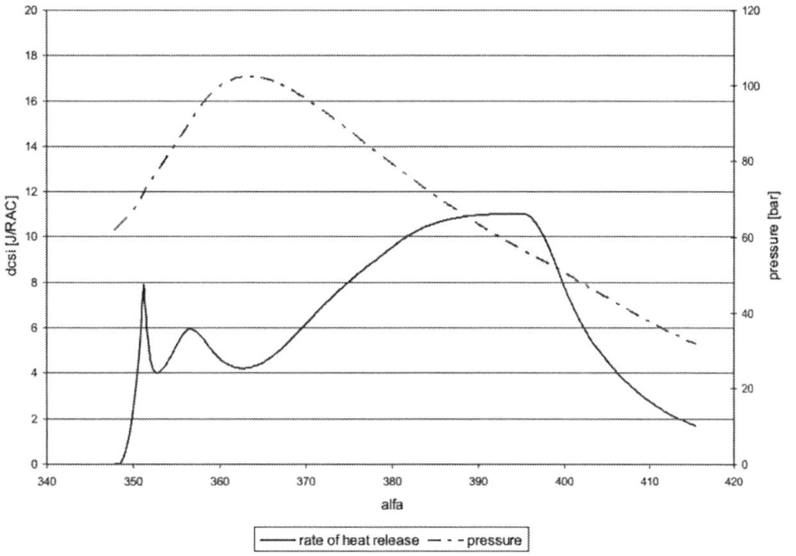

Fig. 8. Cycle characteristics for regime no. 2

In these conditions, the calculated emissions significantly drop, as listed in Table 3:

Table 3 Equation's 2 and 3 values for regime no.2

ZV	α_1	$(dp/d\varphi)_{max}$	α_2	O_2	$m_{air}/m_{fresh\ charge}$	NO_x	Soot
0.32	13.53	5.08	27.65	6.09	1	172	0.120

Maintaining the ratio between the pilot and the main injection fractions (20% for the pilot amount and 80% for the rest), in case of testing a third regime (Regime no.3) the

DOI: 10.1051/matecconf/201712106009

advance has been augmented in order to get higher the engine efficiency. Fig. 9 shows the new adopted injection characteristic.

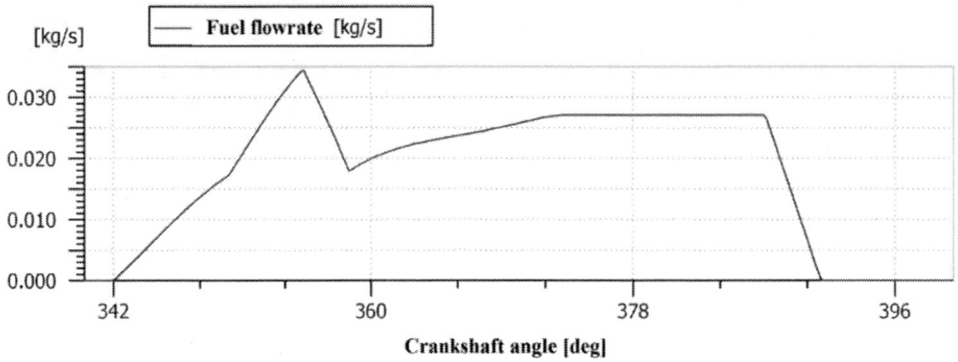

Fig. 9. The injection characteristic for regime no.2

The pilot injection has an advance of 18 deg CA and the main injection phase occurs with an advance of 10 deg CA. The obtained power in this situation is 27.8 kW, marking a drop of 6% comparing to the first regime. In this particular case the peak firing pressure is 117 bar. As for Regime no.2, Fig. 10 shows both variations for the cylinder pressure and for the temperature when analyzing Regime no.3 and Table 4 presents the values for the calculation terms regarding the emissions.

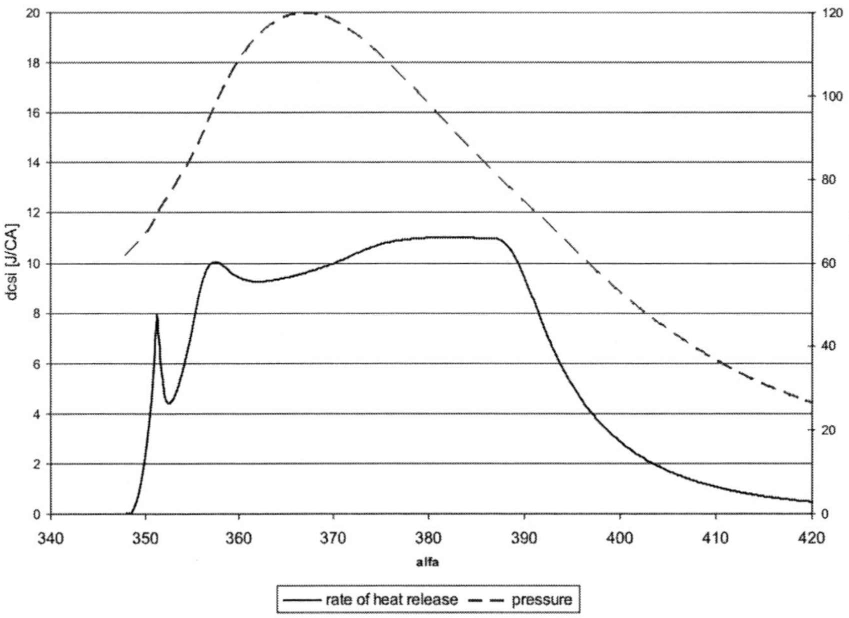

Fig. 10. Cycle characteristics for regime no. 3

Table 4 Equation's 2 and 3 values for regime no.3

ZV	α_1	$(dp/d\varphi)_{max}$	α_2	O_2	$m_{air}/m_{fresh\ charge}$	NO_x	Soot
0.32	18.33	5.08	26.95	6.09	1	301	0.228

6 Conclusions

Establishing the mechanism of the emissions formation following Diesel engine in-cylinder processes is very complex according to the chemical reactions and their kinetics. Therefore, this calculation is highly complicate and relatively inaccurate.

This paper combined two methodologies: first, the calculation of an engine cycle using the AMESIM soft procedures and secondary, the evaluation of the engine emissions using an empiric expression in order to track the most common values found in the experimental research data basis.

The proposed method can define a heat release characteristics, which can combine a good yield with relatively low pollutant emissions. These data, coupled with models of gas after-treatment systems, provides solutions for framing the analyzed engine in pollution norms.

This work could lead to an acceptable approximation of soot and NO_x formation trends when using different operating parameters of the fueling injection process. It opens also the path of research for the autoignition behavior when using different kinds of fuels, such as M, E fuels categories or Diesel-Biodiesel blends.

References

1. G.S. Hebbar, Int. Jour. of Mech. Eng. and Rob. Res., **3(4),** 471 (2014)
2. N.Raeie, S.Emani, O.K. Sadaghiyani, Elsevier Prop.Pow. Res.,**3(2)**, 59 (2014)
3. M.Fiebig, A.Wiartalla, B.Holderbaum, S.Kiesow, J. of Occup. Med. and Tox., **9(6)** (2014)
4. S.Prassad, J.Gonsalvis, V.S.Vijay, Elsevier En. and Power, **5(1A),** 28 (2015)
5. R.Chiriac, A.Racovitza, P.Podevin, G.Descombes, Elsevier Int. J. of H Energy, **40,** 15856 (2015)
6. A.Atmanli, Elsevier Fuel, **176**, 209 (2016)
7. J.H.Zhou, C.S.Cheng, C.W.Leung, Elsevier Int. J. of H Energy, **39**, 4611 (2014)
8. J.Hussein, K.Palaniradja, N.Alagumurth, R.Manimaran, Alexandria Eng. J., **51** (2012)
9. D. De Serio, A. De Oliveira and J. R. Sodré J of Phys, **745** (2016)
10. ***Emission Standards» European Union» Cars and Light Trucks. DieselNet. January (2015)
11. ***Regulation (EC) No 715/2007. 2007-06-20, 5, Retrieved 2015-10-29
12. J.O'Connor, M.Musculus, SAE Paper 2013-01-0917 (2013)
13. S.Emani, S.Jafarmadar, Elsevier Prop. and Power Research, **2(2),** 162 (2013)
14. ***AMESIM IFP Engine Library Rev.14
15. C. Barba, C. Burkhardt, K. Boulouchos, M. Bargente, SAE Paper, 2000-01-2933 (2000)
16. Wenzel, S.,P., *Modellierung der Russ und NO_x Emissionen des Dieselmotors*, (Dissertation zur Erlangung des akademischen Grades Doktoringenieur, Magdeburg Universitat, 2006)

Solutions to improve the decision-making process for dynamic allocation of resources in military pre-conflict environments

Ioan-Danut Balos[1,*], *Catalin* Cioaca[2], and *Mircea* Boscoianu[1]

[1] Transilvania University of Brasov, Department of EIM, 5 M. Viteazul, Brasov, Romania
[2] "Henri Coanda" Air Force Academy, FSM Department, 160 M. Viteazul, Brasov, Romania

Abstract. The complexity of the international security environment compels us to analyse the dynamics of the civil-military decision making process regarding the optimal allocation and efficient use of resource units (UoR). The main purpose of this paper is to improve the civil-military decision-making process by analysing from a dynamic perspective the influence of uncertainty on UoR allocation, thus providing a fast intuitive image of resource ratio. To achieve this goal, risk and uncertainty are defined in terms of knowledge (total or partial) of possible future events or potential outcomes of the action/ decision. The decision alternatives capitalize the results obtained by the optional (the real options analysis in conditions of uncertainty) and flexibility (the prediction of transition moment according to the life cycle of resources). Moreover, these findings could have a positive effect across the full range of economic sectors.

1 Introduction

From a strategic point of view the complexity of the international security environment compels us to analyze the dynamics of the civil-military decision making regarding the optimal allocation and efficient use of resource units (UoR). The statistical analysis of conflict shows that a non-conflict state can be considered a special case of the conflict state in the same way as peace is a preliminary stage that accumulates future confrontations sources [1]. Currently, according to Armed Conflict Database, there are 9 ongoing global conflicts of high intensity, 19 of medium intensity and 13 low-intensity conflicts.

For this scientific approach, risk and uncertainty are defined in terms of total or partial knowledge of possible future events or potential outcomes of the action/ decision [2]. Regardless of the activity and the levels of applicability, decisions are characterized by the fact that they require time and effort, cause concern and anxiety and are a means of exerting influence.

The classical decision making process is based on the backward approach because it is trying to solve a problem with a chosen alternative before understanding what one might hope to gain by solving that problem [3]. The real options analysis (ROA) provides a

* Corresponding author: danut.balos@gmail.com

theoretical justification of requests for additional UoR in a way that differs from the classical model (based on UoR flows).

Military decisions that seek to gain small and visible benefits with potentially negative and invisible side effects, lead to the disruption of a system that has survived just by using unpredictable shocks for continuous regeneration.

Disruption in the military system can take the form of actions with unintended consequences, repeated with a more sustained intensity in order to correct the side effects [4], which leads to disregarding the available resources or those that can be effectively mobilized in a timely manner.

The complex behavior of systems is a combination of "explosive" (the system contains only positive feedback, such as the emergence of a new weapons system) and "inert" (generated by the preponderance of negative feedback sites, such as defense spending) [5].

Each military system should answer the following questions: How are investments projected in order to have the necessary resources units available? How many units of resources are needed to take advantage of the first move (the initiative)? What are the solutions to minimize losses of units per resources? What are the upper and lower limits to investing in producing new units of resources?

The allocation of the resources process is used for reaching military strategic objectives through UoR distribution from the resource nexus, but the issue is generally presented in a quasi-static manner. Resource units which are the subject of military decision-making process usually include: weapons systems, categories/ types of forces, installation and network elements and defense research and development.

The main purpose of this paper is to improve the military decision making process by analyzing in a dynamic vision the influence of uncertainty on the UoR allocation. The model provides an intuitive image of the timing and intensity of investment transition, using binomial portfolio of real options.

2 The cost-performance analysis in defence procurement

The cost – performance analysis is specific to certain fields such as the airspace industry and defence industry, and is based on the hypothesis that the essential benefit are from the cost benefit analysis is given by performance in a wider sense.

The allocation of resources in order to reduce the risk of being surprised involves the development of risk assessment instruments whose results contribute to the identification of risk scenarios, the links between initial events and consequences, the probability of occurrence and mitigation measures [6].

Decision makers often, in the field of military operations, underestimate the risk of occurrence of initial events that have major impact and are difficult to predict [7].

At the strategic level, the objectives of the military action are influenced by political, military, economic and social risks. The most widely used risk assessment methods are focused on supporting the total military operation plan [8]. However, these techniques have a major limitation: the assumptions used by analysts can be highly subjective.

Risk management in defence procurement processes must be integrated with management requirements and configuration [9]. Thus, there have been identified four categories of risks (technical, financial, time and programmatic), having as main sources: changes in mission, technical inconsistencies, imposed budgets and tight deadlines [10]. Based on the results of the risk analysis, the level of confidence is established by estimated costs of the system associated with the lifecycle and the updated costs during the development of the acquisition program (Fig. 1).

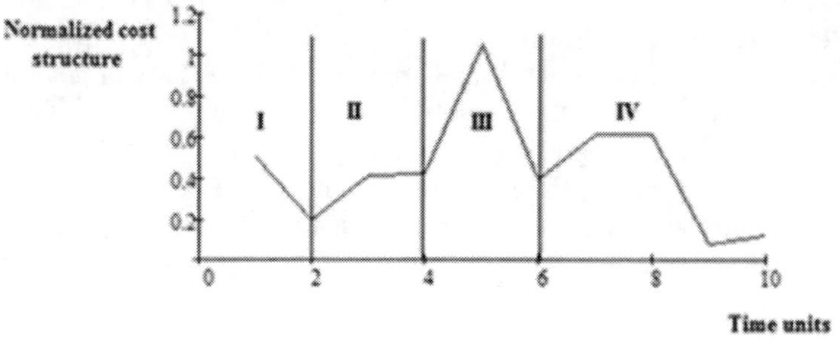

Fig. 1. The cost structure on lifecycle phases: I – studying and selecting concept; II – technological development; III – production and installation; IV – operating and support

The cost of the entire life cycle is structured as follows: I and II represent research and development costs; II and III correspond to acquisition costs/procurement expenses, IV represents operating costs and modernization. In terms of design, engineering, construction, testing, implementation and maintenance the acquisition of military equipment for endowment structures of forces requires a critical evaluation of the needs and capabilities of different available alternative systems and equipment from the opponent and associated costs. Modernization aims is to improve the operating performance, to cancel/ reduce the obsolescence, or to respond to new demands (by default it contains revitalizing and extending the resource).

The distribution of costs based on the lifecycle stages of military products differs from other products in that the duration and intensity of the consumption during certain stages are volatiles; also there are financial constraints as well as certain products characteristics, which all result in altering the altitude to investment on long term. Additionally, the various dynamics of evolution affects the strategic decision making process.

Cost assessments represent the input for the cost-performance analysis, which is developed in order to: identify the objectives of the mission and performance requirements; to support the performance allocation necessary for the functions of the system to work; to ensure selection of requirements regarding system design and to provide criteria for the selection of alternatives; to provide analytical confirmation that the project meets the end user requirements; to support system and process verification; to supports decisions on changes and upgrades throughout the lifecycle.

The real options approach is designed primarily to assist the management and investment selection in order to maximize the value of the organization [11-12]. Real options are completed and integrated solutions (using financial theory, economic analysis, management, decision theory, statistics and econometric modelling) aimed at improving decision making for military investment projects in the pre-crisis phase [13].

3 Decision model development

The need to optimize management strategies and to provide some flexibility to decision makers about the changes in the external environment of the organization should be based on the ingredient of optionality in order to quickly adapt to the actions (movements) of the competitors/ opponents.

Military decision makers constantly adapt and revise their strategies depending on the conditions of the operational environment. Also, it should not be overlooked the fact that the allocation of UoR is partially irreversible; some of them can easily be fully recovered, but others may only be partially recovered or even entirely lost. The regeneration resources process (from manufacturing to modernization) becomes a difficult process to assess in a conflictual environment.

For the development of a decision-making model, in which the use of UoR must become efficient, there the following assumptions are considered:

Hypothesis 1: two options are considered: the offensive (A), corresponding to the procurement cycle, which allows for the development of the capabilities needed for the achievement of the benefits from future opportunities; and delaying the decision of the offensive action (M), corresponding to the modernization cycle that allows for keeping the option open.

The option of going the offensive action encourages UoR allocation. In the case of irreversible allocations, there is an opportunity cost that represents the value of the postponing option. A and M have opposite effects, but they both contribute to increasing the level of uncertainty.

Hypothesis 2: for a dynamic options management, the envelope borders of the efficient use of a UoR nexus from weapons category are defined as follows: the lower limit corresponds to the NATO target of 2% of GDP for national defense; and the upper limit corresponds to Global Fire Power Index (GFPI). The upper limit also incorporates the technological progress. The study is designed for the complete lifecycle of a product.

Hypothesis 3: the increase in uncertainty is inversely proportional to the escalating stress and is a feature of the pre-conflict status. Currently, to manage resources, the military decision-making system is based on the concept of "military equipment static valuation", based on the achievement of some cumulative criteria necessary to the military equipment procurement programs that are subject to valuation, capitalization and depreciation [14].

The importance of strategic flexibility is given by the growing differences between solution and the classical model (e.g. the flexibility of the postponement decision when there is uncertainty regarding the future environmental conditions). If it would sacrifice option M (although UoR is usually irreversible), then the opportunity to make a specific decision "A" when new information is available would be lost. This mechanism has a significant impact on UoR timing allocation.

Hypothesis 4: the combination of effects of options A and M depend not only on their ratio (W) within the binomial portfolio P (A, M, W), but also on the interactions between A and M on various levels of uncertainty.

Ideally, "W" should be 0 or 1, but due to the inherent transition process, take values lying within the interval (0, 1), as follows: the variation W_A and W_M follows a logistic curve, while their sum any given time is 1.

The concept of "portfolio" first appears in the context of these transition processes and is characterized by frictions and delays. The accumulative of UoR for offensive is not encouraged because the cost of delay option (C_M) increases simultaneously with the uncertainty.

The main interest is to identify the critical threshold for when UoR request appear [15]. The mechanism through which the uncertainties delay process of resource allocation disregards the fact that this allocation process could provide unpredicted possibilities for growth.

If the value of option A evolves simultaneously with uncertainty, the increase in uncertainty diminishes the critical threshold of UoR requests, thus inducing another type of relation between uncertainty and allocation of UoR.

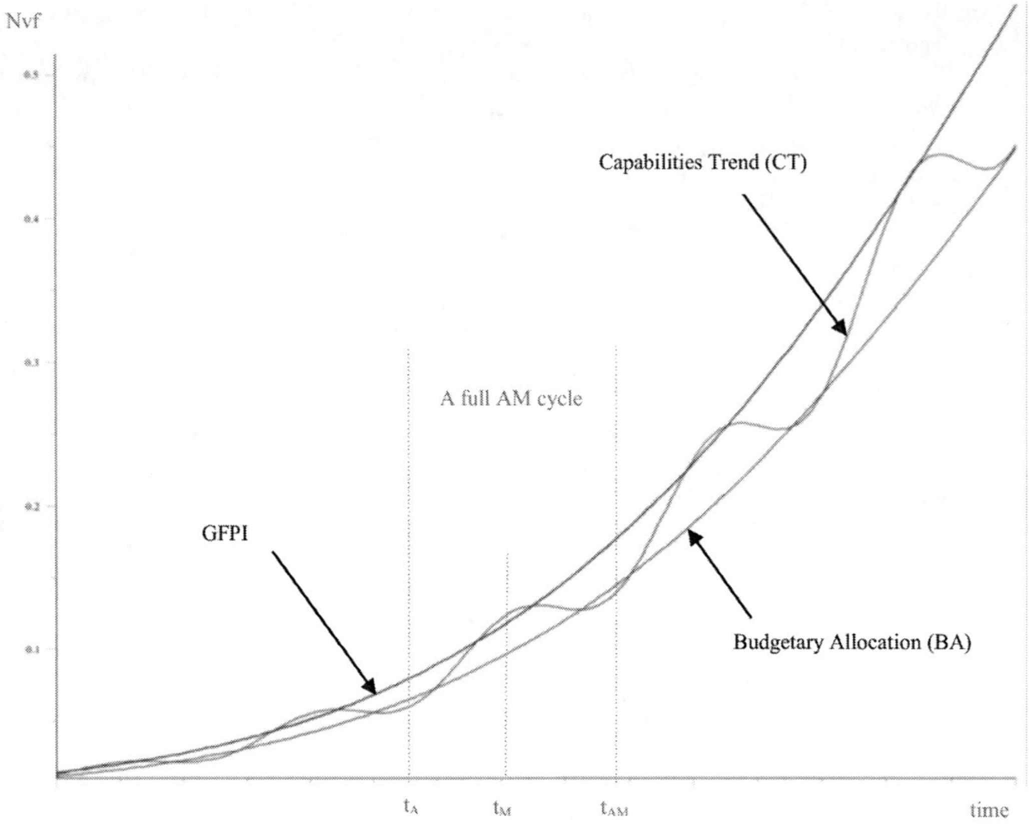

Fig. 2. The dynamics of the investments and military capabilities

In order to simplify the understanding of the current approach, the evolution of limitations can be approximated by means of a logistic function (or normal distribution function or a crisis evolution function). For the analyzed part, the functions corresponding to GFPI and budget allocations (BA) are defined in equation (1) and (2), and the evolution of military capability (f_{CT}) in equation (3).

$$f_{GFPI}(t) = \frac{1}{2\pi\sigma_1} e^{-\frac{(t-m_1)^2}{(\sigma_1)^2}} \tag{1}$$

$$f_{BA}(t) = \frac{1}{2\pi\sigma_2} e^{-\frac{(t-m_2)^2}{(\sigma_2)^2}} \tag{2}$$

$$f_{CT}(t) = Re\left(e^{-\frac{2\pi t j}{t_{AM}-t_A}}\right) f_{dl}(t) + f_{md}(t) \tag{3}$$

In a graphical (Fig. 2) and intuitive representation of the normalized value of the analyzed functions (Nvf), the probable values of expectations (m_1, m_2) and the standard deviation (σ_1^2, σ_2^2) which quantify the dispersion of a specific data values sets, can be viewed as strategic objectives that incorporate a high degree of uncertainty. A completely ideal cycle AM, is characterized by duration (t_{AM}-t_A), initial moment (t_A) and endpoint (t_{AM}). The ratio expectation-variance determines the frequency of the decisions oscillations and thus, the duration of the AM cycle. The evolution of the military capabilities under this circumstances may be expressed by using the equation (3) which shows the dependence of these on the upper limitations (expected GFPI) and lower BA through the $f_{dl}(t)$ - the function of variance limitations (the width of the decision tunnel) and $f_{md}(t)$ - the mean function of the limitations (trend achieved).

The immediate strategic choice between A and M has strong implications. A previous decision may position favorably the obtaining of an advantage and/ or of future resource savings.

4 Conclusion

As part of the planning processes, the volume, the structure and the way the resources are allocated for defence in order to fulfil the fundamental of defence and national objectives security are established.

The decision to launch the offensive action must be made either quickly in order to make use of the gained advantage, or later on when the environment becomes less uncertain.

Within the planning, programming, budgeting and evaluation system, the risks that may arise (respecting deadlines, procurement contracts, shortage of capabilities, necessary financial resources in the short / medium and long term, exterior condition) are taken into account and a reserve buffer-type reserve is provided in order to solve unexpected situations. Errors in this regard could result in failure to ensure the critical mass of defence resources for in a wider sense or unnecessary waste that could have been used for economic development. In this respect, it is imperative to create dynamic balance equilibrium able to fluidly support the necessary investments by means of multi-annual budgets.

The defence resources concept, as approached in a wider sense, including the correlation between budget allocation and capability requirements, as well as the exploitation of the results yielded by the engineering system within the phases of the procurement process management, represent important contributions to the dynamic planning of defences resource in volatile and turbulent environments.

The issue of the resources allocation was analysed based on the principles of complex engineering systems, allocation strategies design by integrating the concept of Dynamic Capabilities for Defence and correlating it with the option thinking concept expressed through the real options analysis paradigm.

Based on the results, the strategic decision-maker can better understand the spectrum of possible decisions and the mode of action in case of major shocks or turbulence. The method can be adapted at tactical level or over shorter periods of time, but the focus on the preparation and the preplanning sequences, should be maintained.

Future research directions include how operational and decision-making interferences comparatively evolve in multiple scenarios incorporating additional categories of resources.

References

1. H. Grotius, *On the Law of War and Peace*, (SPH, Bucharest, 1968)
2. T. B. Folta, J. P. O'Brien, Strat. Mgmt. J., **25**, 121-138, (2004)
3. R. Keeney, *Value-Focused Thinking: The Foundation for Decision Quality*. (Available on: http://www.spe.org/ogf/print/subscribers/2012/08/08_12_OGF_10_Culture.pdf, 2012)
4. N. Taleb, *Antifragile: Things That Gain from Disorder* (Random House, New York, 2012)
5. B. Arthur, *Complexity Economics: A Different Framework for Economic Thought*, Complexity Economics, (Oxford Univ. Press, 2013)
6. R. Moraru, G. Babut, L.I. Cioca, Archives Of Mining Sciences **55 (4)**, 873-878, (2010)

7. C. Cioaca, C.G. Constantinescu, M. Boşcoianu, R. Lile, Env. Eng. Manag. J., **14 (6)**, 1399-1408 (2015)
8. A. C. Karmperis, A. Sotirchos, I. Tatsiopoulos, K. Aravossis, J. of Comp. Model, **4 (1)**, 67-81 (2014)
9. T. J. Andersen, J. Manage. Stud., **41(8)**, 1271–1299 (2004)
10. N. Iancu, *Contributii la dezvoltarea ingineriei sistemelor de armament terestru*, (PhD Thesis, Military Technical Academy, Bucuresti, 2006)
11. F. Moreno-Hines, *Real Options Applications for Public Sector Investment and Acquisitions*, MITRE Technology Program (Available at: http://www.mitre.org/news/events/ tech07/1.html, 2007)
12. J. Mun, *Real Options Analysis: tools and techniques for valuing strategic investments and decisions*, (John Wiley and Sons, New Jersey, 2002)
13. C. Cioacă, M. Boscoianu, App. Mech. Mat., **436**, 32-39 (2013)
14. *** Department of Defense (DoD), Financial Management Department of Defense Office of Inspector General September 30, 2005 Report on Development of the DoD Baseline for Military Equipment (D-2005-114), Office of Inspector General, Arlington (2005)
15. H. Smit, L. Trigeorgis, R&D Manag., **36 (4)**, 403–419 (2006)

MATEC Web of Conferences **121**, 07002 (2017) DOI: 10.1051/matecconf/201712107002
MSE 2017

Study on heredity value in communication skills, for improving individual performance in the workplace

Mihaela Laura Bratu[1,*], and *Dan* Miricescu[1]

[1]Lucian Blaga University of Sibiu, Emil Cioran Street, No. 4, 550025 Sibiu, Romania

Abstract. Effective communication is the wish of every intelligent organization. In today's society, modern communication means saved time, reduced costs, multiplied resources. Communication can be improved early in the development. Heredity of communication can be an excuse, a brake but also a mean of focus personal resources early in child development. To establish the link between communication and heredity was conducted a study at a nursery school in Sibiu. It have been questioned a number of 98 adults and 45 children, using questionnaires for the evaluation of social and communication skills, attention to detail and distributive, and tolerance to change, carried out by researchers from United Kingdom. Adults who have significant problems of communication, have children with communication problems. The results slightly above average obtained by some adults do not always correlate with slightly different results of the children. Another conclusion is that communication ability has a social and cultural coordinated. The correlations have been established between the responses of children and adults, but cannot identify precisely, slightly above average for the cases, which is the incidence of occurrence of communication problems.

1 Introduction

In Romania, the so called MTO approach (Man, Technology, Organization) has been one of the main theoretical frameworks for the enhancement of systematic approaches to integrate human, organizational and technological factors in the context of industrial activities [1]. Risk management is carried out by people, acting individually or in groups. The human factor introduces an additional degree of complexity in the risk assessment. Emotional Intelligence provides the means by which attitudes can be managed [2]. The MTO approach brings a lot of advantages from economic and time point of view, but also creates a lack of comunication in tehnological process. So, the human factor can be in the same time very productive, but also „less intelligent" in the emotional way.

Human resource is present within any organization, so occupational health and safety (OHS) is becoming a pillar in sustainable development of the enterprise. Integrating OHS in corporate social responsibility (CSR) leads to an interesting approach that shapes and solves

* Corresponding author: ybratulaura@gmail.com

a number of current concerns. The integrated approach of the two concepts creates a framework for sustainable work [3]. Organizations, besides the strategic objectives that they have, have also social responsibility for the human resource, to enhance individual compentence, to develop personals skills.

Effective communication is one of the desiderata of any performance, successful organization. Communication quality means success, represents performance, speed of implementation of tasks, represents the optimal use of time and financial, human, material resources [4].

In the labour field are coming individuals with different personality traits and with different communication skills. As a result of the practice, in can be put the following questions: when is the optimal time to develop communication skills, when it can start the personal development of the individual. As earlier is starting, the resistance to change is small and the costs are lower. Advanced societies invest precisely for this reason first of all in the education of children. All it doesn't learn in school period, it will be recovered with costs much higher by employers.

This research has aimed the identification of adults and children with problems of communication and referencing data in order to identify hereditary factor of communicative proficiency.

The present study focuses around four areas that are associated with effective communication: social skills, communication skills, attention to details/distributive attention, tolerance to change [5].

Social skills makes interacting with others easier to. The lack of basic skills such as a smile, greeting, eye contact with the other side, can lead to the inability of the manager to establish a bridge of communication with employees. In the case of complex skills needed to manage conflict, employees can be put in a situation very stressful, because of a lack of social skills. They will be forced to work on the development of social skills consciously.

Communication skills increase processing speed, response time, speed of reaction, leading to understanding and execution tasks as they are envisioned by the superiors. The lack of these skills results in inability to convey accurate information and clear, to relate on horizontal and veritical, to participate in the working meetings to devise improved products etc.

Attention represents a mechanism of orientation and focus of a consciousness on a task, a question, a problem. In its superior forms, attention can be learn, improved, arranged, through the will and thought. There are people who see the smallest details and can track multiple tasks at the same time, but there are employees who are unable to perceive the task as a whole but only certain stages. The manager's role is to know the possibilities of each individual employee and to adapt the communication to the level of attention of the interlocutor.

Changing tolerance means accepting change, adapting to new situations, understanding and changing the behavior according to new parameters. This feature of one's personality is beneficial for employees who work in dynamic environments, constantly changing.

There are employees able to achieve tasks, performance at the required level, but have not developed the capacity for communication, which is a barrier, a barrier in professional development. Often, the performance of the employees are above average, giving evidence of his prodigious intelligence, that becomes an obstacle to communication forms. This type of specific behaviors employees manifest Asperger disorder, specific to gifted persons.

The Autism and Developmental Disabilities Monitoring Network (ADDM) and the Centers for Disease Control and Prevention (CDC) report that in 2002 approximately 10% of all 8-year-olds in the United States had an Autism Spectrum Disorder (ASD). If these findings are correct, by next year 10% of all 18-year-olds in the United States will have an

Autism Spectrum Disorder [6]. So, in 2017, 10% of the new hire young people can manifest communication problems.

So, the following questions arise:

1.Employee communication problems have a hereditary component;

2.To what extent can we intervene in childhood to improve the ability of individuals to communicate.

Research conducted at the level of the adults.

The purpose of the questionnaire was to investigate to what extent children who have high scores on applied questionnaires were coming from families with normal adults from the point of view of intelligence, which exhibit autistic-like symptoms.

The questions of the questionnaire covers five areas: social skills, communication skills, imagination, attention to detail, attention to distributive and tolerance change.

The questionnaire includes 50 statements, with four variants of answer: totally agree, partially agree, strongly disagree, totally disagree. Depending on the response, the statement shall be graded with 0 or 1 point, 1 designating the specific behaviors of Asperger syndrome, manifested totally or partial. Each statement describes a typical autistic behavior (asperger's).

People who replied to the questionnaire have been the female gender in 65%, 66% of these having ages between 30 and 40 years.

As a result of questionnaire, 16% of adults fall into normal spectrum, indicated by questionnaire. However, it cannot be asserted, for reasons such as psychological distress, culture, habits, that the remaining persons have a problem, but only manifests communication difficulties. 3% of the population has a high score of 37. Men have higher average scores than women. Immediately above the average of 16, lies 47% of women and 76 percent of men. However, there are no men with higher score then 24.

It is considered that with increasing age, people have a greater chance of improving communication, thanks to self-knowledge and social adaptation effort.

Dimension social skills that includes 14 items refer to social interaction, social rules, adapted to the needs of others. Items describe specific behaviors of Asperger syndrome, social expenses, incapable of following social rules, etc.

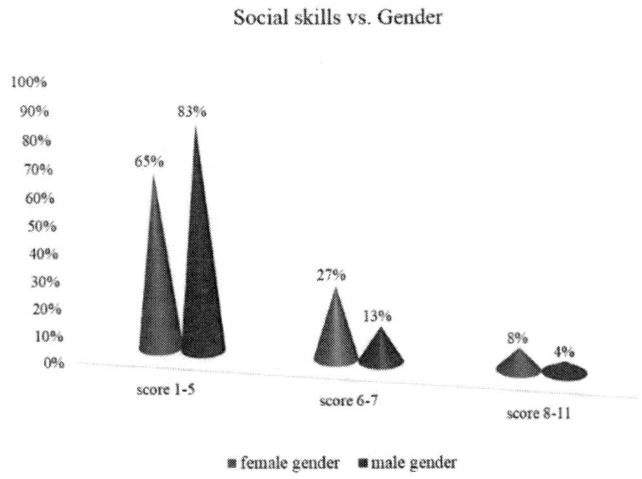

Fig. 1. Social skills vs. Gender

Social skills deficits are at 35% of women and only 17% of men.

Communication skills refers to the understanding of game words, verbal stereotypes, nonverbal communication, the interpretation of verbal and nonverbal messages. Within this questionnaire graded deficiencies of this dimension.

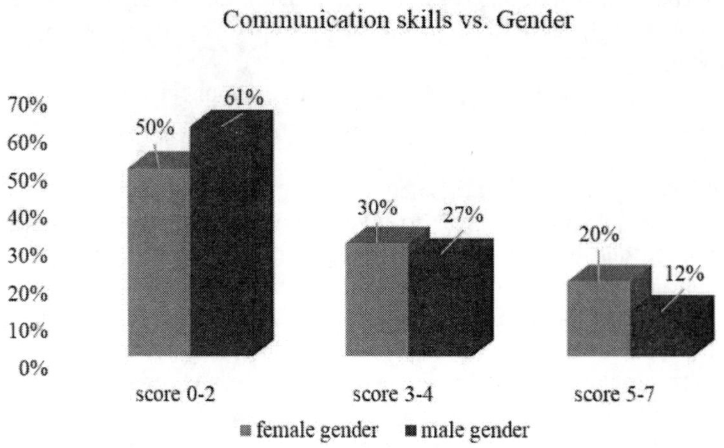

Fig. 2. Communication skills

Analyzing the scores from questionnaires show that women have more communication problems than men, surprising conclusion can be said. Might as a possible cause is the item to which they related, comparing with others female, so is appearing inhibition and low self esteem.

People who are diagnosed with Asperger's Disorder have trouble building mental images, mental visualization and imaging operation constructs.

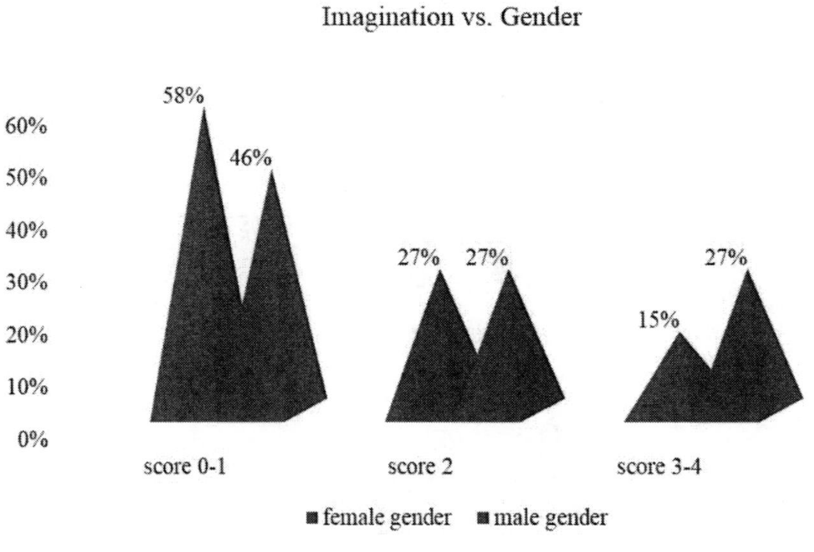

Fig. 3. Imagination vs. Gender

Men have difficulty in operating the imaging constructs against women, by twice the percentage (27% vs 15%).

Some details are captured by people with Asperger obsessively, while others are not noticed at all.

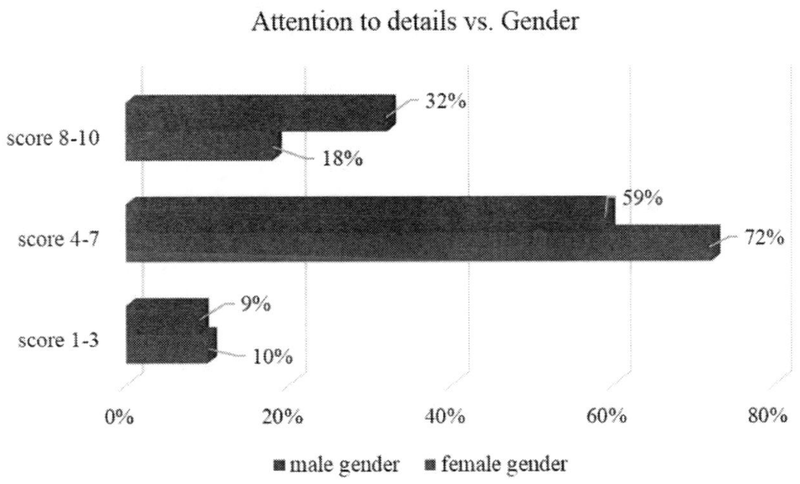

Fig. 4. Attention to details vs. Gender

Men generally given less weight than women to details, scores over 8 specific behavior is an indication of asperger.

People with Asperger become sad or anxious when removed from familiar places or when something changes in their daily routine. People female exhibit lower tolerance to change - 71 % compared to men - 52 %.

Research conducted at the level of children

Subjects were 45 children aged between 3 and 7 years at a kindergarten in Sibiu. Communication and social skills were tested with Scala A.S.A.S., part of the volume" Asperger's Syndrome: A Guide for Parents and Professionals" by Tony Attwood. Scala as authors and A.J. Attwood M.S. Garnett.

The questionnaire was filled first by parents and then for children with high score, was completed by the teacher also.

Each question or statement has a scale from 1-6, with 1 being the normal level expected of a child 's age at this level.

Following the questionnaire, scores over 22 are those of concern in terms of research. In this percentage falls 6% of girls and 18 % of boys. In what regards communication skills, a minimum score achieved on this dimension was 6 and the maximum 26. 7% of children had scores extreme, these being girls and boys in equal percentages.

2 Discussion and conclusions

11% of the questioned population has obtained high scores on the questionnaire AQ, which means that they are facing communication and adaptation difficulties. Comparing with the average obtained in university in the UK, it appears that the population surveyed, the average value of 16 passes around 24. A possible explanation, in my opinion, could be social and cultural values promoted by the history. In Romania, we have " Ewe " as a masterpiece and Romanian shepherd's image as a symbol of freedom, that people have so much confidence in them, they are afraid to have different views. Because of that the scores could be higher. So, communication is a social and cultural component. In Romanian space, communication is conducted harder, there are many barriers, many inhibitions, fears, lack of exercise in expressing their ideas.

Correlations can be made between data, resulting that adults with communication problems have a better chance to have children who are experiencing the same difficulties. The parents are deeper problems, the more likely their children to inherit.

At the organizational level it is important to objectively identify individual opportunities and develop employees to increase performance. Improving employment inclusion is likely to reduce mental health problems for adults with Asperger syndrome and reduce the burden on those who play a broader supporting role [7].

Like anyone, those with Asperger's can have particular strengths and qualities and it is important to look at some of the attributes which someone with Asperger's brings to the workplace. These can be as follows:

- excellent memory;
- precision and attention to detail;
- mathematical and technical skills;
- a preference for following instructions and abiding by rules;
- sticking to structured programmes;
- direct, open and honest;
- reproduction, imitation and mimicry;
- high level of vocabulary;
- dealing with projects which have a clear beginning, middle and end [8].

References

1. L.I. Cioca, R.I. Moraru, G.B. Babut, Proceedings of the 15th International Conference The Knowledge-Based Organization, Nicolae Bălcescu Land Forces Academy, Sibiu, Romania, 26-28 November, **2**, 43-48 (2009)
2. L.I. Cioca, L. Ivascu, 3rd International Conference on Social Sciences and Society (ICSSS), 27-28 December, **32**, 74-79 (2013)
3. L.I. Cioca, L. Ivascu, S. Rus, Proceedings of the 4th Review of Management and Economic Engineering International Management Conference, The Management between Profit and Social Responsibility, 18th – 20th September, Technical University of Cluj-Napoca, Todesco Publishing House, Romania, 2-7 (2014)
4. R. Moraru, G. Babut, L.I. Cioca, Archives of Mining sciences, 55 (4), 873-878 (2010)
5. T. Attwood, Asperger's Syndrome: A Guide For Parents And Professionals (Jessica Kingsley Publishers, London 1998)
6. S. Lesco, Employment and Asperger Syndrom, (Available at: www.autism.com/advocacy_lesco)
7. J. Richards, Personnel Review, **41** (5), 630-646 (2012)
8. M. Matusiak, Is It possible my Colleague Has Asperger's Syndrome?, (Available at: https://livingautism.com/possible-colleague-aspergers-syndrome)

Occupational emerging risks affecting international virtual project team results

Iulia Dumitraşcu-Băldău[1,*], and *Dănuţ Dumitru* Dumitraşcu[2]

[1] Lucian Blaga University of Sibiu, Department of Management, Calea Dumbravii 17, Sibiu, Romania
[2] Lucian Blaga University of Sibiu, Department of Industrial Engineering and Management, Emil Cioran 4, Sibiu, Romania

Abstract. The expansion of internet access, high-speed connection services, collaborative work platforms and tools, allowed employees to interact virtually offering companies the possibility to develop projects around the world, reducing operational costs and gain competitive advantage. Realizing the advantages and disadvantages of developing a project team in an international virtual work environment, requires adopting specific strategies to construct an effective team and ensure the project success. One of the most important disadvantages that we identified is that the new work environment brings new risks for both team members and managers. So, it becomes mandatory to identify and analyze the occupational emerging risks and their impact on the productivity of virtual team members, in order to prevent them efficiently and to ensure the safety and health of employees in a virtual working environment. This paper aims to highlight the necessity for project managers and organizations, to include in their specific project strategies, an efficient occupational risks management in the virtual workplace, to obtain a continuously improved virtual working environment, so to achieve a high performance from virtual employees.

1 Introduction

If until the advent of the Internet, outsourced collocated project teams, were communicating by phone, fax, or through numerous face-to-face extremely expensive meetings, subsequently the Internet has enabled a substantial reduction of these costs and even their elimination, transforming these dispersed teams into virtual project teams, allowing organizations to convert technology into a competitive advantage. "Organizations around the world are increasingly involving in projects developed across borders, building virtual teams, given the many advantages they offer" [1]. "Many companies will be less concerned with their internal capabilities and more focused on the value they can draw from external networks" [2]. Companies can use virtual project teams also to align environmental protection agreements, because these teams can gain the title of "green job" taking into account that one of their advantages is to reduce the carbon footprint [3], what emerges

* Corresponding author: iulia.dumitrascu@ulbsibiu.ro

indirectly from the definition of the virtual team: "a virtual team is one that does much of its work across distances facilitated by technology, as opposed to doing much of its work face-to-face contexts"[4]. 2020 EU Strategy [5] proposes three priorities, among which the first two, are in perfect accordance with the definition of the virtual teams. European Commission defines green jobs as "covering all jobs that depend on the environment or are created, substituted or redefined (in terms of skills sets, work methods, profiles greened, etc.) in the transition process towards a greener economy" [6]. European Agency for Safety and Health at Work (EU-OSHA) considers that "green jobs can be understood as contributing, in some way, to the preservation or restoration of the environment" [7]. These 'green-jobs' definitions, can be identified to some extent with the work in virtual environment or, with international virtual project teams.

2 Defining the problem

The emergence of virtual work environment, following the development of information and communication technology, has also led to the emergence of new occupational health and safety risks-OSH. "The new and emerging OSH, may have different origins, in particular: new technologies, new production processes, new forms of work and emerging forms of work" [8]. "ICT-related changes in the world of work, more so than the technology itself, brings about not only great opportunities but also a number of health and safety risks"[9]. Considering the importance that EU-OSHA is giving to "green-jobs" having dedicated a special section within emerging risks on the website [7], the present information regarding the virtual working environment and the risks involved, is very scarce in the literature and is not taken into account in the studies undertaken so far by EU-OSHA. Instead, we identified that there is an interest in studying them, in the future [10].

3 Hypothesis

We believe that identifying these risks can assist project managers in establishing effective strategies for preventing them, so the results of international virtual project teams, will not be affected. The European Parliament noted that globalization has negative effects on safety and fairness at work [11]. The best way to deal with these risks is the prevention, and the virtual team members' preparation in this respect is fundamental, given that they work remotely. Awareness of these risks and their severity by employees through specific trainings, setting rules and making available the necessary tools to work in the virtual environment, are some of the responsibilities of the manager and the organization that is creating such project teams. "Glover predicts big changes in the way that companies are organised, with less focus on developing internal capabilities among their own workforces and more on bringing together capabilities from a number of different sources, such as consultants and freelancers, in a virtual workplace" [2]. "It is likely to be hard to assess OSH risks consistently and regularly for mobile or remote workers. They may suffer from a lack of OSH monitoring and a poorer access to OSH services. A virtual workforce could find it difficult to know that OSH information and services relevant to them exists and it could also be very difficult for OSH regulators to influence such a workforce" [12].

4 Emerging risks affecting international virtual project team results

EU-OSHA defined emerging occupational risks as "any new and growing occupational risk" [13]. "Currently, the attention focuses on lesser-known risks, which are interpreted as emerging, although some of these were present before. They acquire their emerging nature due to one of the following conditions: the risks are better measured, quantified and evaluated or the society is sensitizing to the nature and potential consequences of those risks" [14]. "In 2002, EU- OSHA has set up a "Risk Observatory". There were identified and considered four different types of risks, to provide a more complete view of the work: a) physical risks; b) organizational, social and human risks; c) biological risks; d) chemical risks, without excluding multifactorial risks that may be identified in several of the categories mentioned"[15].

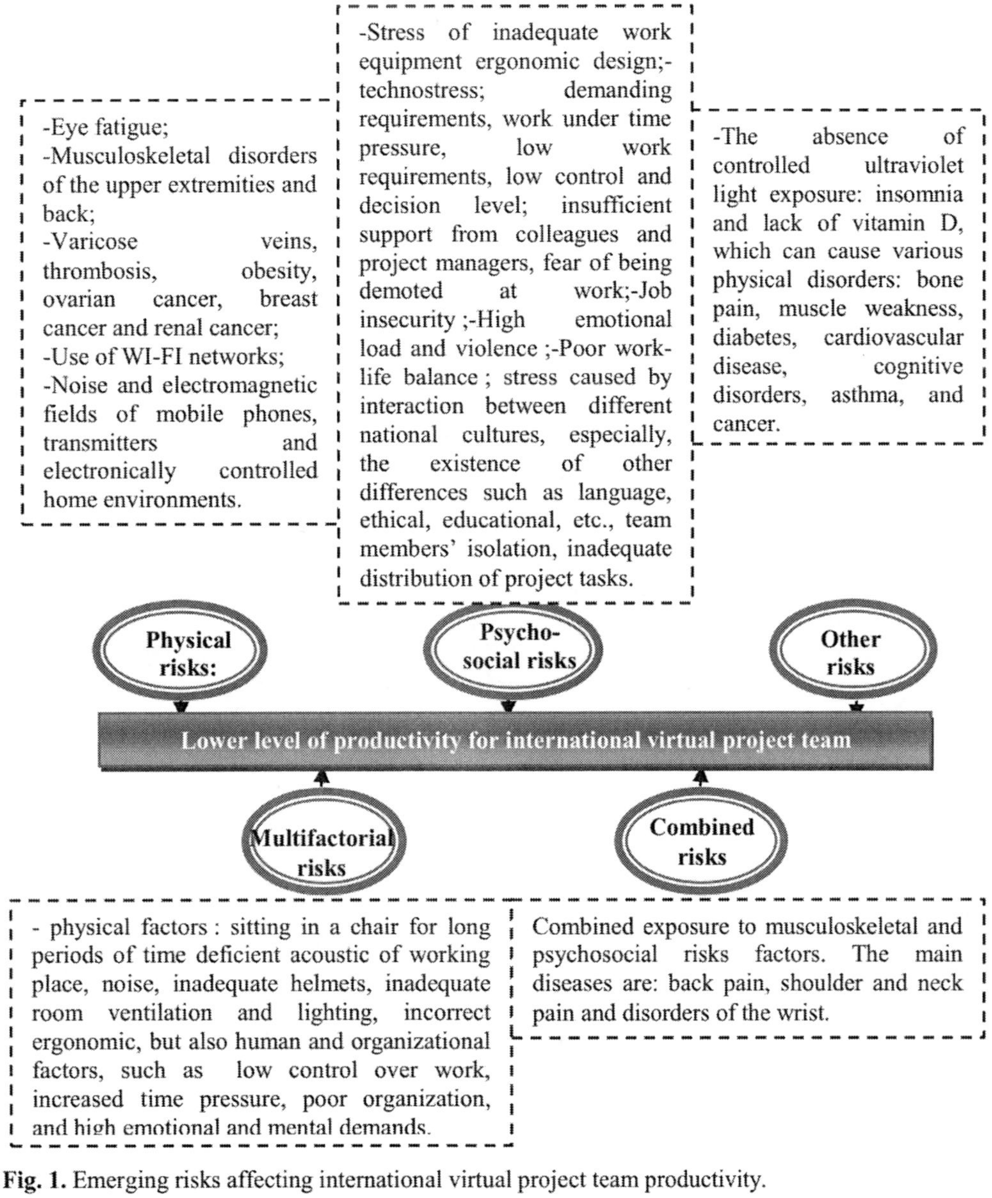

Fig. 1. Emerging risks affecting international virtual project team productivity.

We identified a number of risks, which can be assimilated to virtual work environment, establishing a link between office work risks, removing the ones related to the physical presence of the other team members, and adding the risks generated by the lack of their presence, the risks related to the work from home, those specific to teleoperator's work, or those related to visual display unit users. In addition to these risks in the literature, we propose specific risks for the international working environment. We've created Figure 1, in order to offer a schematic vision of these risks.

Given that virtual project team members communicate via technology-driven tools, they spend most of their time in front of a digital screen, which causes eye fatigue or accommodative asthenopia. The "lack of physical activity at work, involve both, sitting, or standing for long periods of time"[15]. "An increase in the use of computers and in productivity demands leads to more time spent in a fixed body posture"[16]. The widespread use of non-ergonomic mobile devices such as smart phones, tablets, laptops, can also lead to worsening musculoskeletal disorders. "Global mobile data traffic grew 63 percent in 2016. Global mobile devices and connections in 2016 grew to 8.0 billion, up from 7.6 billion in 2015. Of all IP traffic (fixed and mobile) in 2021, 50% will be Wi-Fi, 30% will be wired, and 20% will be mobile" [17]. Studies also suggest the need for a better understanding of the physical risks regarding noise and electromagnetic fields of mobile phones, transmitters and electronically controlled home environments [18]. Psychosocial risks identified in the literature [15], including those identified by EU- OSH [16] are mentioned in Figure 4.1. Addressing psychosocial risks in terms of poor mental health, presents additional challenges due to the complexity in measuring and assessing causation of mental disorders [19]. There are different theoretical models of psychosocial risks and their effects, and the best known is the model of R.A. Karasek [20], demand-control. This model identifies two psychosocial risks: high demands at work and poor control activity at work. This control activity relates primarily to the possibility of choosing tasks, how to do the tasks to achieve the desired results, working time and breaks. "This is particularly important to meet the demands of the workplace involving a high volume, fast pace and deadlines" [19]. "Common mental disorders such as depression and anxiety are the results of most epidemiological evidence available regarding decreased control over the activity at work" [20]. In terms of job security within international projects, it has a degree quite low, because contracts are signed for a limited period of time, during the project. "Workers employed based on insecure and flexible contracts with unpredictable hours and workload, are more prone to occupational diseases" [21]. Whittaker thinks it is possible that we will see a continuation of recent trends towards increased self-employment and use of non-standard employee contracts, including temporary working [2]. Also, into psychosocial risks category may be included the "technostress", which can occur when the employees experience regarding IT failures is not able to produce improvements or to face pressures produced by performance monitoring through technology [23]. This type of stress is specific to virtual international project teams, given the need of continuous improvement and adapting skills to the needs of the project, the technology used and also to the sheer volume of information. "Since ICT has the ability to allow a 24/7 economy, this can lead to stress during the work time, and experts have expressed their concern indicating that the widespread use of hand devices, such as smart phones, tablets, could lead to an imbalance between private and professional life, working hours in excess and insufficient time to relax" [24]. "It becomes increasingly difficult to achieve a good work-life balance due, in part, to work pressure but also driven by the 'fear of missing out' syndrome" [25]."The use of computers and smart phones with internet which provides easy access to all kinds of information but also the expectation from colleagues, supervisor and clients that one is always available, always reads e-mails and can always be contacted. Technology has a great impact on workers' task characteristics and hence on their health and well-being" [16]. As a result of excessive working time, we can add the risk that those employees may recourse to the use of drugs to enhance performance in work. In international projects, the high level of diversity, can cause communication difficulties, distrust, conflict, team members isolation, inadequate distribution of tasks and their lack of synchronization with time project, all leading to mental disorders such as depression and anxiety. Regarding combined risks, "it has been shown that exposure to musculoskeletal risks combined with psychosocial ones, affect workers at a greater extent than when taken separately" [26]. Concerning, multifactorial risks, the main problems observed are: musculoskeletal disorders, varicose veins, nose and throat diseases, voice disorders, fatigue, and stress" [27]. Also, if in the literature we find the exposure to ultraviolet radiation as a physical risk, we can mention that doctors consider the absence of controlled ultraviolet light exposure, as being an important risk, due to the disturbance of the circadian rhythm, which may lead to insomnia and also the lack of vitamin D, which can cause various physical disorders. This is caused by long hours in an office with limited or no access to sunlight.

5 Risk Assessment for virtual teams in international projects, using OiRA tool

"If risks are not assessed or properly dealt with, a suitable risk management process cannot be started and appropriate preventive measures are unlikely to be found or put in place"[28]. "What's more, case studies show that good OSH management in a business is linked to improved performance and profitability. Not only does poor safety and health cost companies money, but good OSH pays dividends. Businesses with higher safety and health standards are more successful and sustainable" [29]. "Studies estimate that for every euro invested in OSH, there is a return of 2.2 Euros and that the cost–benefit ratio of improving safety and health is favourable. The economic advantages of good OSH for businesses large and small are significant. To give just a few examples, good workplace safety and health: Improves workers' productivity, Cuts down on absenteeism, Reduces compensation payments, Meets the requirements of public and private sector contractors" [29].

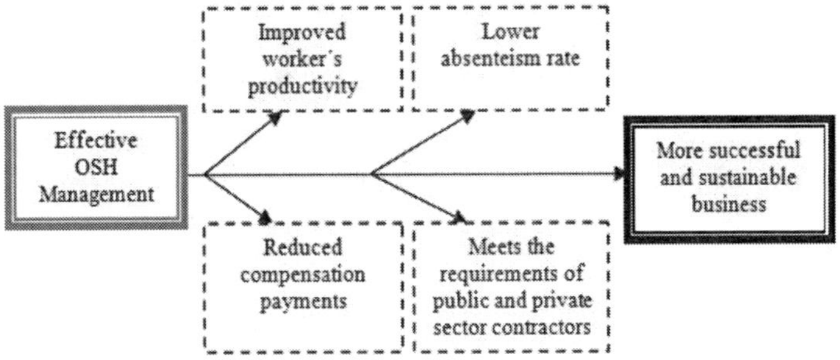

Fig. 2. Effective OSH Management Results.

We used OiRA instrument - Online interactive Risk Assessment, to assess risks in a virtual environment in international projects. OiRA is a web platform that enables the creation of sector risk assessment tools in different languages in a standardised way, developed and maintained by EU-OSHA and it is based on the Dutch risk assessment instrument RI&E [30]. There is no specific tool in Romanian or specific to Romania. As an activity sector or workplace, ICT or virtual work environment is not found in the default list. For which we used the generic tool in English. This risk assessment tool is mainly intended for micro (less than 10 workers) and small (less than 50 workers) enterprises/organisations, which corresponds to the virtual project team ideal size [31].

OiRA generic tool includes 11 categories of risks, of which we excluded 3 of them: dangerous substances; machine safety; personal protective equipment and health and safety signs, considering they are not appropriate for the virtual work environment. Following the risk assessment, we created Table 5.2, where we included all the identified risks due to the expertise on international virtual working environment. The priority of the risk is calculated by the system, according the answers given regarding the probability of the risk occurring, how often are the workers exposed to that risk and the severity of the damage.

Table 5.2. OSH risk assessment for virtual teams in international projects, using OiRA tool.

Nº	Category	Risks
1	Help with in-house emergencies	7 Present Risks: *7 Low priority risks:* • It´s possible that it´s not clear for some workers what to do in case of emergencies. • No special training has been provided for in-house emergency officers with regard to special risks. • Emergency equipment and resources laid down in the emergency planning is not or might not be sufficient. • An in-house emergency plan doesn´t exist. • There is no written document on the duties of in-house emergency officers. • In-house emergency officers did not receive adequate training. • There are no or not enough in-house emergency officers.
2	Organization workplaces	13 Present Risks: *5 High priority risks:* • No area for relaxation (canteen or other catering area) separated from the workplace is provided. • There are places where work is conducted and smoking is allowed or where tobacco smoke enters working areas. • Some equipment in the enterprise might be connected to extension cords and/or might not have a residual current operated circuit-breaker being tested at least once a quarter. • The enterprise's electrical installation is not inspected regularly. Inspection intervals of 5 years are sometimes not respected. • Work on the electrical systems including first connection of machinery and equipment is not always performed by a competent person. *2 Medium priority risks:*• Toilets, urinals, washbasins and changing rooms are not cleaned daily. • It´s not well or not at all indicated where fire prohibitions apply. *6 Low priority risks:*• Entrance doors are not periodically maintained and inspected (annually). • Not all fire extinguishing facilities are inspected regularly.• The required fire extinguishing facilities are not always present, recognisable and accessible in all areas.• There are no maps indicating where a person is and which escape routes can be used.• Not all emergency exits are clearly indicated and accessible.• Emergency exits might not be adequate.
3	Physical stress and visual display unit work	15 Present Risks: *12 High priority risks:* • The office is lacking windows through which daylight comes in. • There are problems with regulating properly the temperature in the office. • Storage space is not sufficient. • There are VDU work places where the minimum area provided per VDU is smaller than 6 m2. • No measures have been taken to protect the eyesight and vision of VDU workers. • Not everyone knows how to adjust his or her workstation to the correct height. • Lighting and VDU equipment are not well matched. • There is reflection nuisance from VDUs. • In the case of computer work sessions lasting at least two hours, the workstation (desk, chair, etc.) is not adapted to the worker's height. • After 2 hours of computer work, tasks can´t usually be interrupted by breaks or by performing other non-VDU-related work. • Normally workers can not alternate between working in a standing, sitting or walking position. • Working in an unfavourable posture is not always prevented. *1 Medium priority risk:* • Directly incident sunlight cannot be excluded at the work station. *2 Low priority risks:*• Printers and copiers that are used a lot are positioned close to workers. • Pushing and pulling with heavy loads is not prevented.
4	Undesirable behaviours and work pressure	9 Present Risks: *5 High priority risks:*• Working overtime is regularly occurring. • Resources needed for work are not always available in time and/or do not work well. • Tasks are not evenly distributed in all functions. • With regard to functions in your enterprise there is an imbalance concerning work rate, work supply and time pressure. • There is no special training for employees with contact with customers/patients/ visitors on how to deal with undesirable behaviour. *4 Low priority risks:*• There is no formal complaints procedure for employees who want to report undesirable behaviour. • There is no confidential adviser employees can appeal to in case of problems. • No attention is paid on prevention of undesirable behaviour. • There are no arrangements concerning undesirable behaviour between employees.
5	Working	2 Present Risks: 2 High priority risks:• The same rules apply to all employees, no differences are made for pregnant women and employees who gave birth. •

	hours and rest periods	Agreements concerning on-call services are vague and/or don't follow clear rules.
6	Environmental factors	4 Present Risks: *3 High priority risks:* • The amount of (day) light in the workplaces is not always adequate and/or incoming sunlight sometimes cannot be shaded. • There is inadequate ventilation in the space. • The temperature at the workplace might be suboptimal; workers are dissatisfied with the temperature. *1 Low priority risk:* • There are workplaces where draughts occur.
7	Occupational health care and organization of work	22 Present Risks: *17 High priority risks :* • It's possible that some employees who perform work on the premises of external parties have not received sound instruction on health and safety. • Employees might not always be properly protected against risks that may arise on site. • Employers (e.g. temporary employment agencies) of hired workers do not always have a copy of the risk assessment. • In assigning tasks, attention is not always paid to vulnerable groups. • There are no arrangements with a registered occupational health officer concerning the periodic occupational health examination. • There are no clear arrangements about what to do if someone breaks the health and safety rules. • It is not always ensured that everyone performs his or her work properly and safely. • Employees do not always strictly follow the enterprise's safety instructions. • Employee are not regularly informed and trained on the (safe) performance of their work. • Workers don't have access to possible working conditions consultation, not even if required by them. • The prevention officer might not all the health and safety risks in the enterprise and /or might lack knowledge on how to limit them. • Prevention tasks concerning working conditions have not been assigned to employee(s) known as prevention officers. • Topics like working conditions and absenteeism and not regularly included in work discussions. • Managers have not always good skills and might sometimes not have sufficient scope to monitor employees effectively. • In general not much attention is paid to a good mix of preparatory, executive and organisational tasks within the different jobs. • Occupational diseases are not systematically reported by the occupational health officer. • There are no clear arrangements concerning monitoring of absence through sickness, neither with the safety, health and welfare service nor with a registered occupational health officer. *5 Low priority risks :* • There is no clear procedure in place for reporting accidents with serious consequences immediately to the respective authority. • Accidents are not recorded and the causes are not investigated any further. • A reintegration plan is not drawn up for workers who are sick for more than 8 weeks. • Problem analysis and Action Plan are normally not drawn up for workers who are ill for more than 6 weeks. There are no clear arrangements concerning the person monitoring the sick employee and how this should be done.
8	Additional risks	4 Present Risks: 4 High priority risks : • Job insecurity • Tehnostress • Interaction between different national cultures. • Lack of direct sun exposure.

6 Conclusions and future lines of research

Following the bibliographic study and using the OiRA risk assessment instrument, we believe that the main risks that may affect, in a significant way, the international project team members performance in the virtual environment, are the psychosocial risks, and stress being one of the most complex phenomena needed to be studied and analyzed in accordance with the characteristics of virtual work environment and international project.

While recognizing the role of individual dispositions and the overall life circumstances, it has been demonstrated that stress caused by work-related factors can significantly affect the employees' proper functioning both within work and outside. Its symptoms include emotional problems (irritability, becoming solitary, feeling exhausted), cognitive (difficulty concentrating and making decisions, negative thinking) and behavioural (negligence, abuse of alcohol, drugs, medication) [32]. Great Britain's Health and Safety Executive, estimated that in 2015/2016, the total number of lost working days due to work-related stress, anxiety and depression was 11.7 million days, an average of 23.9 days lost in each case [33]. Absenteeism and poor performance attract significant economic implications. The total cost of psychosocial risks and stress related to work activity, for business and society as a whole- including health care, disability, early retirement, low productivity, high staff turnover and other direct and indirect expenses- is estimated at several billions [32].

Given the implications of the emerging occupational risks' effects, it becomes clear that is needed more information and tools adapted to virtual work environment to assess and achieve an effective risk management. The main contributions of this paper are represented in the following figures: Figure 1. Emerging risks affecting international virtual project team productivity; Figure 2. Effective OSH Management Results; Table 5.2. OSH risk assessment for virtual teams in international projects, using OiRA tool. Also, this paper may represent a starting point in conducting an empirical research on emerging risks in virtual international project teams, in order to validate the identified risks, and based on that, it can be developed a specific tool for the virtual work environment in international projects, using OiRA generator tool. Also, it can be used a multi-criteria analysis used in the integration process of the organizations [34], to the virtual environment, in order to take efficient risk management decisions.

References

1. K. Fisher, M. Fisher, *Manager's guide to Virtual Teams*, 1st edition (McGraw-Hill Education, Madison, Wisconsin, 2011).
2. Institute for Public Policy Research, *Technology, globalization and the future of work in Europe*, (2017).
3. Dumitraşcu, D.D. Dumitraşcu, Revista de Management şi Inginerie Economică, 14 (4), 691-701(2015).
4. P.E. Brewer, *International Virtual Teams: Engineering Global Success (*IEEE PCS Professional Engineering Communication Series, Wiley-IEEE Press, 2015).
5. European Commission, *Communication from the Commission Europe 2020- A strategy for smart, sustainable and inclusive growth* (Available at: http://ec.europa.eu/eu2020/pdf/COMPLET%20EN%20BARROSO%20%20%20007%20-%20Europe%202020%20-%20EN%20version.pdf, 2010)
6. European Commission, Commission staff working document – Exploiting the employment potential of green growth, (Available at: https://www.google.ro/url?sa=t&rct=j&q=&esrc=s&source=web&cd=1&cad=rja&uact=8&ved=0ahUKEwiRuIaQ2oPUAhXpFZoKHXNnAp8QFggkMAA&url=http%3A%2F%2Fwww.cedefop.europa.eu%2Ffiles%2FSWD_green-growth_EN.pdf&usg=AFQjCNGde1gxl-fshwTknZRSoyxPdqLamQ, 2012)
7. *** https://osha.europa.eu/en/emerging-risks/green-jobs, Accesed: martie 2017
8. *** https://oshwiki.eu/wiki/OSH_in_general, Accesed: martie 2017
9. C. Degryse, *Digitalisation of the economy and its impact on labour markets* (Working Paper. ETUI aisbl, Brussels, 2016).
10. N. Stacey, S. Bradbrook, J. Reynolds, and H. Williams, *Review of drivers and trends of change in information and communication technologies and work location*, (Available at: https://oshwiki.eu/images/f/ff/Summary_report_Key_trends_and_drivers_of_change.pdf, 2016).
11. European Commission, EU Employment and Social Situation: Quarterly Review (European Commission, Brussels, 2013).
12. Ministry of Social Affairs and Health Finland, *Global health security agenda pilot assessment of the United Kingdom*, (Available at: https://www.gov.uk/government/uploads/system/uploads/attachment_data/file/456984/IndependentReport_GHS_acc.pdf, 2015).
13. EU-OSHA, Topic Centre Risk Observatory, *European Risk Observatory Report*, Luxembourg: Office for Official Publications of the European Communities, (2007).

14. L. I. Cioca, R. I. Moraru, *Managementul Riscurilor profesionale psihosociale*, (Editura Universitatii "Lucian Blaga" din Sibiu, 2010).
15. D. Reinert, E. Flapsoler, A. Hauke, E. Brun, Safety Science Monitor, **11** (3), 1 – 17, (2007).
16. *** https://oshwiki.eu/wiki/Monitoring_new_and_emerging_risks
17. *** Cisco Visual Networking Index: Global Mobile Data Traffic Forecast Update, 2016–2021 White Paper, (Available at: http://www.cisco.com/c/en/us/solutions/collateral/service-provider/visual-networking-index-vni/mobile-white-paper-c11-520862.html, 2017).
18. European Agency for Safety and Health at Work – EU-OSHA, Scoping study for a foresight on new and emerging occupational safety and health (OSH) risks and challenges, Luxembourg: Publications Office of the European Union, (2014).
19. Kyaw-Myint, Su Mon, L. Strazdins, M. Clements, P. Butterworth, L. Gallagher, Safety Science **93**, 143 – 151, 2017.
20. R. *A.*, Jr. Karasek, Administrative Science Quarterly, **24 (2)**, 285-308 (1979),
21. N. Clumeck, C. Kempenaers, I. Godin, M. Dramaix, M. Kornitzer, P. Linkowski, F. Kittel,. Community Health **63 (4)**, 286–292, (2009).
22. ILO 'XIX World Congress on Safety and Health at Work - Global trends and challenges' (ILO, Geneva), (2011).
23. Dewe and Kompier, Foresight Mental Capital and Wellbeing Project. Wellbeing and work: Future challenges, The Government Office for Science, London, (2008).
24. European Agency for Safety and Health at Work, *EU-OSHA Multi-Annual Strategic Programme (MSP) 2014-2020*, EU-OSHA, Bilbao, (2013).
25. N. Stacey, S. Bradbrook, J. Reynolds, and H. Williams, *Review of drivers and trends of change in information and communication technologies and work location*, (Available at: https://oshwiki.eu/images/f/ff/Summary_report_Key_trends_and_drivers_of_change.pdf, 2016).
26. R.H. Westgaard, **31 (6)**, 569-580, (2000).
27. J. Cramer, R.P. Ellegast, T. von der Heyden,M. Liedtke, W.Pfeiffer, R. Stamm, *Arbeitsumgebung und Ergonomie*. (Call Report 4. Hrsg: Verwaltungs-Berufsgenossenschaft, Hamburg, 2001).
28. *** https://client.oiraproject.eu/nl/generic/generic-tool-in-english/start
29. D. Bräunig, T. Kohstall, *Calculating the international return on prevention for companies Final report (*Available at: *http://publikationen.dguv.de/dguv/pdf/10002/23_05_report_2013-en--web-doppelseite.pdf, 2013).*
30. *** https://oiraproject.eu/en/what-oira
31. K. Ferrazzi, *Getting Virtual Teams Right*, (Available at: https://hbr.org/2014/12/getting-virtual-teams-right, 2014.)
32. European Agency for Safety and Health at Work, *Second European Survey of Enterprises on New and Emerging Risks (ESENER-2) Overview Report: Managing Safety and Health at Work, European Risk Observatory*, Luxembourg: Publications Office of the European Union, (2016).
33. HSE, *Work related stress, anxiety and depression statistics in Great Britain, (*Available at: *http://www.hse.gov.uk/statistics/causdis/stress/stress.pdf,* 2016).
34. M. Cioca, L.I. Cioca, L. Duţă, Studies in Informatics and Control, **20 (2)**, 129-134, (2011).

A new approach in Supply Chain Processes – Smart Logistics

Olga Bucovetchi[1,*], *Andreia Elena* Simioana[1] , and *Radu D.* Stanciu[1]

[1]University POLITEHNICA of Bucharest, 313 Spl. Independentei, 060042 Bucharest, Romania

Abstract. Globalization is progressing steadily and cannot be stopped. Through existing technologies, this process will be accelerated further. Entire professions and industries, as well as government and politics, must constantly adapt to this process. Many things can be created quickly, which, in turn, creates more vulnerability. These changes are supported by different technologies, which also change rapidly, not to mention the perception of time influencing people's awareness. Things are faster and sometimes more fragile. In this context, supply chain is perceived as a mean of expressing globalization. The present paper emphasizes the switch of paradigm from classic logistics to smart logistics. The authors analysed several prestigious reports (released by World Economic Forum, Oxford Martin Commission for Future Generations, IEEE Computer Society, etc.) to highlight the global trends in supply chain processes. The main objective of the research is to shape the near future of logistics to prepare the production and transportation industry for change.

1 Introduction

Globalization means transforming the economy, culture, innovation and trade in new global synthesis, a new debate about how our civilization could develop into something better. Globalization refers to the close collaboration between nations in a free trade world, sharing global concerns such as climate, energy, economy, technology, security and democracy. Globalization will support several new advanced systems. Industries around the world are working together to find better ways of doing business. The main objective of the research is to shape the near future of logistics to prepare the production and transportation industry for change.

The paper presents, in great, a literature research towards mega trends revealed by prestigious worldwide entities. This study highlights mostly theoretical things about the future of globalization regarding Supply Chain Management.

The literature research shall represent the starting point for a deeper analysis in order to help practitioners to reshape their businesses, to embed these trends within the organizational behaviour to be more globally competitive.

* Corresponding author: olga.bucovetchi@upb.ro

2 Research methodology

After comprehensive trend analyses, a wide-ranging literature review was conducted and summarized [1-4]. Existing and new concepts, techniques to meet current challenges were investigated, analysed and positioned. Essentials have been identified and priorities were set as well and findings have been summarized and analysed.

It is important to note several aspects of risk concepts generated by these reports. First, updates, maintenance, and having a consistent implementation are difficult to carry out. The relationships between risks are also difficult to recognize and maintain, especially when they are difficult or impossible to understand. Right now, the global current risks maps are the most comprehensive and concrete, and have the most sustainable overviews, but these also examine risk at a global level, and not necessarily at smaller levels, where the risk(s) may have more significant impacts. Over the years, risk has shifted and some dependencies have now been documented and are visible. Some of these risks have been quantified and documentation is available to help mitigate these risks as they occur. There is a communal consensus that global risk assessment is an important instrument to promote risk sensitivity and awareness. However, the responsibility as to who is responsible for identifying, dealing with, or communicating risk factors is often unclear and inconsistent, and the consequences are not fully outlined. It is important to note that there is a lot of know-how and new technologies that are required to master risks as they occur. The dynamics and collaboration of risk management are hard to visualize and the different ideas of quality affinity are a challenge. Many resources are required for risk mitigation and can only, unfortunately, be provided by just a few. The related yet different assessments of countries, regions, or cultures are not documented. Because implementation is not always consistent, new risks can be created. It is also always important to note that people are still unpredictable and it is difficult to have risk that is manageable.

3 Main findings

The following trends provided by research institutes and globally focused organizations are currently leading the way. These have shaped the development of new transport management systems to gain a higher market share in logistics area.

The risk landscape developed and published by the World Economic Forum (WEF) was used as a source for the research because it includes over 10 years of documented global risks and are regularly updated by an accepted body. This risk landscape is very comprehensive, as the perception of risk over time changes and differences are perceived and implemented

Many companies compete through storage and transport services provided to meet a growing number of online orders, which is today the main driver of logistics. As a result, received more orders and a greater diversity ordered naturally requires a larger deposit. Fortunately, some carriers have a plan to improve warehouse utilization. The easiest way to make a deposit effectively is through a proper inventory management, but it just means more efficient stored products, and products not delivered.

To honour the high volume of orders and to have satisfied customers, retailers must embrace new technological trends and to understand the current state of logistics in electronic commerce to prepare for the future needs of commerce. [5] They must invest in devices that eases the work and soon in robotics, Amazon changing landscape in the supply chain and delivery robotics.

Transport industry is going through some remarkable changes now; these changes are leading up to a certain long-term impact that will change goods' movements in the world.

Transportation management systems (TMS) will be centralized in Cloud. TMS solutions must work with all aspects of transport, from inventory management to status updates of carriers in transit. For any TMS, one of the main tasks is to gather real-time information from all parties involved in the transport cycle. Given this, visibility becomes one of the most crucial factors in increasing the efficiency of TMS solutions, thus information gathering becomes even more important.

Transportation Management is headed by Internet of Things (IoT), which supports carriers by offering ways to retrieve information, self-service and automatic updating through portals. Thus, companies can access information from any location and take advantage of the latest information from all partners in the supply chain network. This TMS solutions help generate higher output in terms of efficiency and reduction in operational costs.

Logistics flows, payments and transactions will need to satisfy requirements related to the enormous trade economy based on innovations. Many of these chains will be automated and will work online to receive orders from one end of the world, to honour others at the other end and produce elsewhere. These transactions will be further accelerated by the visibility offered by internet, wireless connection in real time and a new digital global financial standard.

Within the book "New Mega Trends" [6], Frost & Sullivan expert, Sarwant Singh claims that "SMART is the new green". Furthermore, he predicts "the failure of green technology to show a manageable and predictable return on investment that will lead to a shift towards intelligent products, services and even cities that 'behave' green by adapting and changing to the environment around them. The megatrend of the last decade — Green products — will be replaced in this decade by Smart products and services". A smart product is characterized by an intelligent sensing technology that is increasingly being integrated with Internet technologies, thereby allowing the product to react to and communicate with the changing environment around it. This leads to optimal operations and improvement in terms of efficiency.

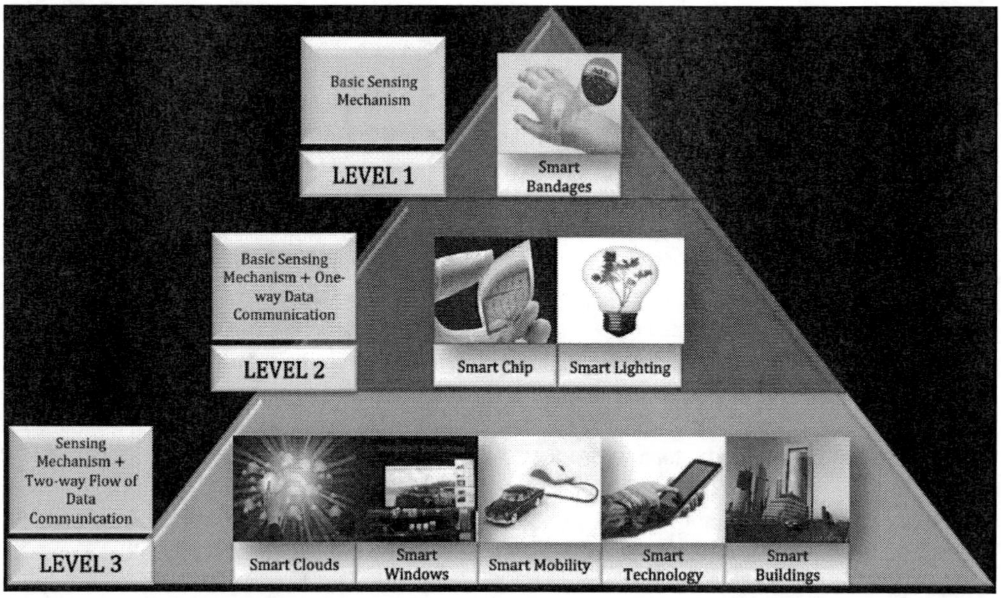

Fig. 1. SMART products, [7]

As one may observe from figure 1, there is a large number of SMART products, which can be structured in 3 levels, depending on their connectivity, from basic sensing mechanism to two-way flow of data communication. If analysing the first two levels, one may highlight the time gap between the sensor signing the problem or the misbehaviour of an apparatus and the moment of parameters' recovery, while on the thirds level of technology, one can speak about autonomous decision-making process – technology recovering by itself.

Currently logistics is controlled by simple transport processes; modern logistics will be composed of data, this concept will become essential for success, and the Internet started to become mountaintop informational data, which with the help of IT systems are analysed for insights and better transport operations. Every device that is connected to Internet via Wi-Fi or a wired connection generates more information for enterprises and carriers that are used in real-time by them. Meanwhile, technology is increasing to meet demand and current pressures on the industry

Virtual reality will be used in coming years in storage services; if nowadays scanning is done using mobile packages, in the coming years will be done with the help of glasses which visually scan all packages. On average, a person can hold approximately seven bits of information in short-term memory, but with these glasses and the data on parcels will immediately scanned and sent to a database. As a consequence, the working time of human workforce will be reduced, the inventory will be done automatically, this providing instantly updated information about the free space in storage.

Recently, global attention was directed toward unmanned aerial vehicles, (UAV), which are already used for military purposes in Pakistan (US Army) for operating aid in the development of Africa and delivery packages in Germany (DHL Packet).

Besides military purposes, the world's leading online retailer Amazon and the largest logistics company in the world, DHL have already started using these devices for delivering parcels.

Using drones in logistics represents a double-edged sword even if they have the advantage of a reduced delivery-time: on the one hand because of the cost of delivering - very high, and on the other hand in terms of the number of devices that would be needed to cover the need of delivery. Moreover, it's a matter of concern when taking into consideration the air safety. Airspace would very much clutter, and the smallest error on these devices could have dramatic effects and can even shoot down a plane full of passengers.

To avoid such incidents, large companies increasingly invest in testing several types of unmanned aerial vehicles, as those organizations are confident that these devices will be the future of efficient logistics

4 Conclusions

Supply chain industry leaders have access to more information today than ever in history. Thanks to data collection programs, software and applications for data entry, it is an enormous amount of data that has the potential to provide a revolutionary perspective on how to improve the efficiency of the business model. As a result, business leaders can reap a significant return on investment through detailed analysis of these data. Unfortunately, Big Data is not understood in its true potential, how the supply chain and help you would need profited as soon as possible.

Big Data offers insight to realize reduced costs in the supply chain and a more strategic approach to transport management. Analysis of data collected provides a chance to see how the fare and the route can be optimized and consolidated to improve efficiency and product flow.

World Economic Forum states within the Global Risk Report 2017 [1] that: "technology, in particular, may be where we can turn for innovative solutions to today's risks and challenges. It may also produce additional risks which must be assessed and factored in to thinking at business, government, and international levels. New and sometimes revolutionary structures such as Artificial Intelligence systems represent such opportunities for streamlining our everyday lives, but also bring with them many potential dangers. These include mismanagement, design vulnerabilities, accidents, unforeseen occurrences and malicious use that post risks to security and safety of individuals."

Safety and sustainability is a priority for carriers [8]. A safety problem could lead to negative reactions with rapid financial failures. Therefore, automated systems can help increase safety by eliminating the possibility of human error from the equation. In other words, transport will be organized to prevent accidents at work by using automated systems. There was a time when logistics was composed of simple transport processes. But modern logistics is data-driven, and this concept will become essential for success

References

1. *** http://www3.weforum.org/docs/GRR17_Report_web.pdf ,Accessed in March 2017
2. *** http://www.engr.washington.edu/news/trend, Accessed in March 2017
3. *** http://www.un.org/ga/search/view_doc.asp?symbol=A/RES/70/1&Lang=E, Accessed in March 2017
4. ***http://www.oxfordmartin.ox.ac.uk/downloads/commission/Oxford_Martin_Now_for_the_Long_Term.pdf , Accessed in March 2017
5. D. Badea, and S.M. Petrişor, Proceedings of the 16th International Conference The Knowledge Based Organisation, Sibiu, 479-484 (2010)
6. S. Singh, New Mega Trends: Implications for our Future Lives, (Palgrave Macmillan, 2012)
7. *** http://www.investinbsr.com/ipaforum/wp-content/uploads/Iain-Jawad-IPA-Forum-2014-Presentation.pdf , Accessed in March 2017
8. L. Ivascu, M. Tămăşilă, I. Tăucean, L.I. Cioca, M. Izvercian, Proceedings of the 5th Annual International Conference on Cognitive-Social, and Behavioural Sciences, 259-267 (2017)

Assessment procedure for the soft skills requested by Industry 4.0

Gabriela Beatrice Cotet[1,*], *Beatrice Adriana* Balgiu[1], and *Violeta – Carmen* Zaleschi (Negrea)[2]

[1]University POLITEHNICA of Bucharest, Training for Teaching Career and Social Sciences Department, 060042, Romania
[2]University POLITEHNICA of Bucharest, Training 060042, Romania

Abstract. In the context of an increasingly more performant global and informational environment, as the one found in the virtual organization, the process of selecting and evaluating the human resources holds a particular important role. The premise behind this article is that of the need to adapt the human resource performance to the requests of the industry 4.0 being supported by the organizational culture. Following the analysis of the state of the art literature we concluded that there is no clear procedure for the assessment of the constellation of skills and personal qualities – soft skills requested by industry 4.0. These capabilities should complete professional technical hard-skills; the procedure should provide a map for the constellation of capabilities necessary to adapt and perform, in specific industry 4.0 activities. Current psychological evaluation systems consider only some of such requested capabilities not always the most relevant. To solve this problem, we tried to identify and evaluate what we considered as the core of the complex various skills required. Our research was centred on the use of a psychological instrument for evaluating transversal capabilities. The capabilities map needed for the evaluation and selection of the human resource fit to work in the industry 4.0 environment was designed after the application of this complex system of evaluation on successive series of students from the University POLITEHNICA of Bucharest.

1 Introduction

The industrial revolution it was represented along the time by challenges offered by steam power in the nineteenth century, electricity in the early twentieth century, and automation in the 70s. All these waves of technological advancement reduced overall employment, however, today, the new jobs emerge and the demand for new skills grows. In our days, it's talk more and more about how fast technology advances and how these changes transform our life at both socially and professionally.

* Corresponding author: beatricenica@gmail.com

Cyber-physical systems and the Internet of Things, big data and cloud computing, robotics, artificial-intelligence based systems and additive manufacturing are the integrated components of fourth industrial revolution.

A new challenge that concern employment is on the horizon, the fourth wave of technological progress: industrial growth of new digital technologies, the rise of new digital industrial technologies that are collectively known as Industry 4.0.

The 4th wave is an industrial revolution or we can talk about a global revolution?

2 The fourth industrial revolution as Industry 4.0

Industry 4.0 represents a significant transformation of the entire industrial production through the unification of digital technologies and the Internet with conventional industry. Klaus Schwab briefly explains the paradigm of the fourth industrial revolution as evolution since the discovery of steam power for cars, through the discovery of electricity for mass production followed by digitization for industrial automation and finally Industry 4.0 the fourth revolution eliminating the limits [1].

The Fourth Industrial Revolution intends to leverage differences between the physical, digital, and biological sphere. Therefore, it aims to integrate: Internet of Things – IoT, Cyber-Physical Systems – CPS and, Artificial-Intelligence IA, cloud computing, robotics based systems and additive manufacturing (figure 1). Fourth Industrial Revolution develops in robotics, 3D printing, artificial intelligence, nanotechnology and biotechnology. Such advancement will radically remodel the labor market worldwide generating a paradigm change in defining the skill background needed to address these challenges [2].

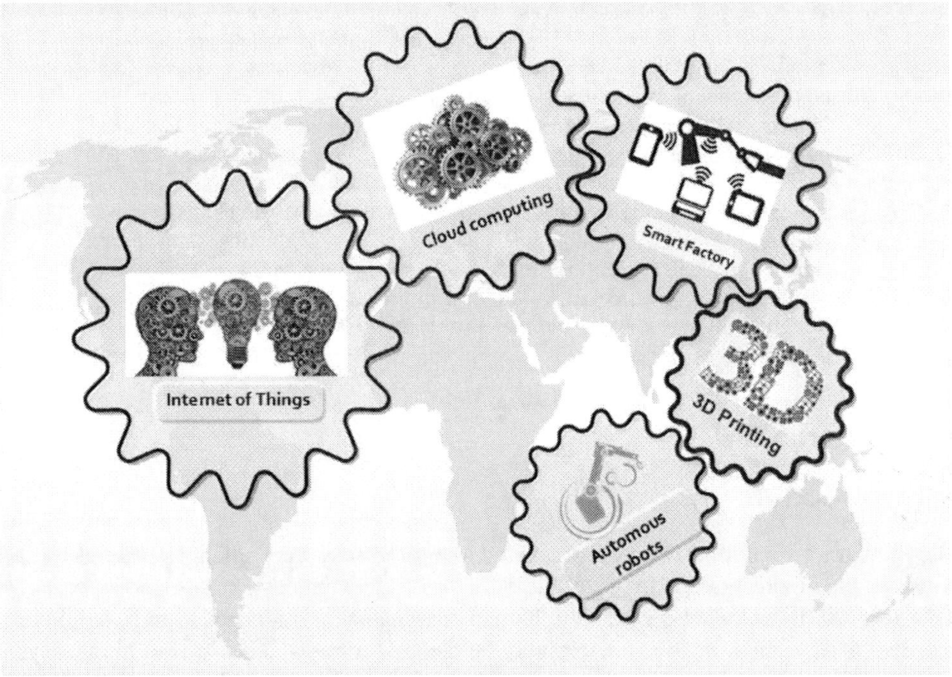

Fig. 1. Industry 4.0: a vision of the future industrial production

Schwab requires for responsible leaders and citizens to "together shape a future that works for all by putting people first, empowering them and constantly reminding ourselves that all of these new technologies are first and foremost tools made by people for people." [1]

The questions that arise are: How fast the technology advances and transforms our lives? How will this next wave of industrial evolution play out? How will job profiles involve? And what types of skills will be demanded?

The answers to these questions are critical to education, industry leaders and public politics as they have to seek to take full advantage of the opportunities arising from Industry 40 by ensuring that an appropriately skilled workforce, there are.

If we want to adopt Industry 4.0, the manufacturers will be able to increase their competitiveness, which will enable them to expand their industrial workforce at the same time that productivity increases. These capabilities identified as soft-skills can be developed during their academic studies and favourite underlie the formation of a complex series of competencies, including: teamwork and networking capacity; ability to communicate; personal effectiveness; self-development; creative and innovative thinking; leadership skills.

3 The constellation of skills and personal qualities – soft skills

As "soft skills" are presented not precisely defined aptitudes, but rather a cluster of personality traits which have a synergistic effect, contributing decisively to the personal and professional effectiveness. As a "cluster of skills and personal capabilities" soft skills describe the attitude of each of us, our compatibility with others and how we manage social interactions mostly in professional environement. This cluster should become "mandatory" to select human resources from the perspective of Industry 4.0. Unlike "hard skills", which describe a specific set of technical skills, "soft skills" are recognized as transversal competences, being found at the junction between the professional and social skills. It is often said that hard skills help you to succeed at an interview, but you need soft skills to keep the job and reach to professional fulfilment [3].

Interest in high performance, good interrelation, and high professional self-development potential should shape the future employee profile, which will face the new challenges coming from Industry 4.0.

Whereas one of the means by which we can "control" these parameters is psychological evaluation of candidates as part of the selection process, we consider that we have to define and build a standard profile to match training needs, identification and development of soft skills. Identification and evaluation the cluster of psychosocial skills are required still by formative university years, in order to answer the question: are prepared students, future employees, to address the new challenges? The rapid evolution of technologies requires a cluster of psychosocial skills - soft skills - to act holistically to resist "technological wave."

Five years from now, over one-third of skills (35%) that are considered important in today's workforce will have changed [4].

They will be brought together advanced robotics, autonomous transport techniques, artificial intelligence and human-machine learning, advanced materials, biotechnology and genomics. Current demands coming from leading global employers are redefining skills and workforce strategy for the future.

In this paper we propose the new concept of "constellation of skills and personal qualities", referring to the soft skills and capabilities demanded by Industry 4.0. To define this new concept we start from the archetypal Jungian constellation. In our proposed concept around of three capabilities-stars, revolve a series of psychosocial traits: empathy, inter and intrapersonal skills, effective communication, adaptability, self-power, etc. In these constellations, creativity is the core of the top three skills demanded.

With the avalanche of new products, new technologies and new ways of working, employees should become more creative in order to benefit from these changes. Emotional intelligence, which doesn't feature in the top today, is rated in our model as the second on the top skills needed.

The last of the three top skills is proactive thinking applied to new technologies. It assesses how human resources connect with the technological revolutionary model in order to implement change rapidly and sustain it in the global economy.

4 A map of the future - Case study

Are we prepared to face this Industry 4.0 reality? Is the young generation ready to cope with these challenges not only economic, industrial, global but especially social?

We must be prepared. Change won't wait for us: business leaders, educators and governments all need to be proactive in up-skilling and retraining people so everyone can benefit from the Fourth Industrial Revolution. The methodology described here, centred on a map of skills and personal qualities necessary to assess the specificity required by industrial environment 4.0. could be a useful tool in this reconversion process.

4.1 Methodology

This methodology is centred on the achievement of a map for the constellation of skills and personal qualities necessary to assess the specificity required by industrial environment 4.0. First we defined the conceptual backbone of this map. Then, we tested it by applying an instrument of psychological evaluation ABCD-M. This psychological assessment test has been applied to third-year students from three subjects: Robotics, Machine-tools and Logistics. The total number of students was 90 (Machine-tools 38, Logistics 19 and Robotics 33).

4.2 A "Pencil" for designing a map for the constellation of soft skills

The questionnaire that was used for the students' assessment was an ABCD-M complex personality test based on a psycho-lexical approach system compatible with Big Five.

ABCD-M personality questionnaire (the Romanian version of Big Five personality test) was created and benchmarked on Romanian population by M. Minulescu (2008). The survey is made of five supra-factors, each of them including other five facets: Extraversion (with the facets Activism, Optimism, Humour, Interpersonal skill, Personal self-assertion); Maturity (with the facets Respect, Adaptability, Friendship, Inhibition force, Force of the self); Agreeableness (Altruism, Romanticism, Affective warmth, Empathy, Honesty); Conscientiousness (Will, Spirit of perfection, Sensibility, Planning, Auto-discipline); Self-actualisation (Thoroughness, Tolerance, Refining, Independence, Creativity).

The total number of items is 151 each of them appreciated on a scale with 5 answering variants, from totally true to totally untrue. The consistency coefficient of the scales and the subscales varies between a minimum of .68 and a maximum of .76. We have chosen the respective questionnaire since it proposes a broad perspective on investigating personality [5-6].

To established a nuanced structure for our map of constellation of skills and personal qualities that better reveals soft skills, considering the three capabilities-stars defined above we had chosen: Interpersonal skills (E4), Personal affirmation (E5), Respect (M1), Power of ego (M5) Empathy (A4), Perseverance (C1), Spirit of perfection (C2), Self-discipline (C5), Refined (AC1), Refining (Ac 3), Independence (AC4) and Creativity (AC5).

One could notify that from the three capabilities stars only creativity could be identified as it is in this list. The other two are resulting from the other skills based on several considerations (e.g. between self-esteem as "the power of ego" and performance, there is a strong connection, Empathy, Interpersonal skills and Personal affirmation imply situational awareness and the ability to choose the best answer that determine the best outcome for all stakeholders, Refined – implies an open attitude towards its completeness of a sophisticated person etc.)

4.3 Results analysis

The interpretation of results is a diagnosis illustrating the differences between the three specializations. This interpretation followed next protocol:

- Diagnosis of the five dimensions
- Diagnosis of the characteristics that define the constellation of skills and personal qualities-soft skills

In terms of low values and high values, the values obtained are situated closer the high values, especially for students Machine-tools specialization.

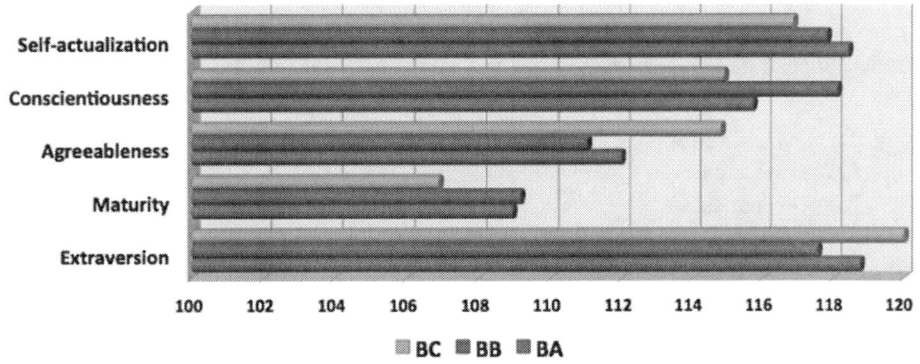

Fig. 2. Comparative values for all the five dimensions of ABCD-M

From the perspective of psychological dimensions considered relevant for describing soft skills (Conscientiousness and self-actualization), you may notice the following (figure 2 and 3):

- For dimension Conscientiousness - higher average values were recorded by students from robotics specialization, followed by machine-tools and manufacturing systems and logistics specialization.
- For dimension Self-actualization – we have on the first place the students from machine-tools and manufacturing systems, followed by robotics specialization and logistics specialization

Analysis of the five dimensions requires developing strategies to educational level (personal development, role playing, coaching, etc.) allowing the training and development of skills profiles as close specificity required by the new challenges.

A more nuanced differentiation arising of the processing and interpretation of data on the five dimensions subscales.

We considered relevant to describe features that are subordinated constellation of skills and personal qualities – soft skills, these subscales: interpersonal skills (E4), asserting personal (E5), respect (M1), strength of self (M5), empathy (A4), will (C1), spirit of perfection (C2), self-discipline (C5), Intellectual curiosity (AC1), refining (Ac 3), independence (AC4) and creativity (AC5).

Interpretation was accomplished in relation to the low values and high values. The first step was to create the schedule for each specialization. The reason for this assessment was given by the desire to identify critical values that would require establishing an intervention strategy.

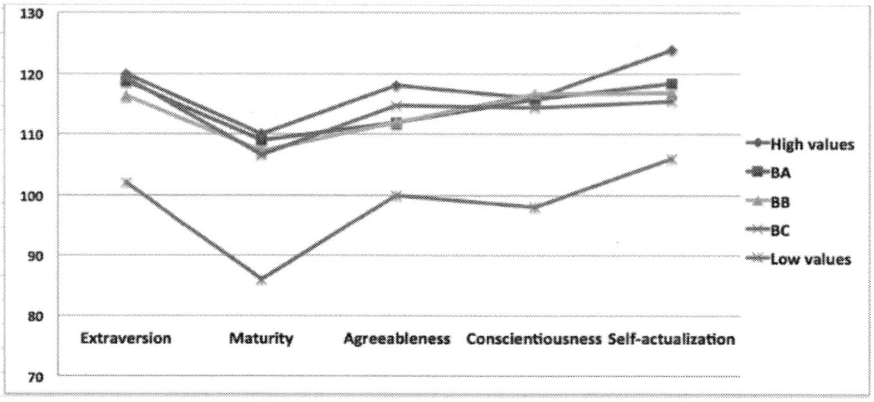

Fig. 3. Comparative average values reported at low and high values for all the five dimensions of ABCD-M

Where BA = Machine-tools ,
BB = Robotics,
BC = Logistics.

Average values for Machine-tools undergraduates (less refinement) are situated closer to the high values of the graph. Low value for Refinement indicates a closed attitude toward self-accomplishment, a low intellectual personal cultivation (figure 4).

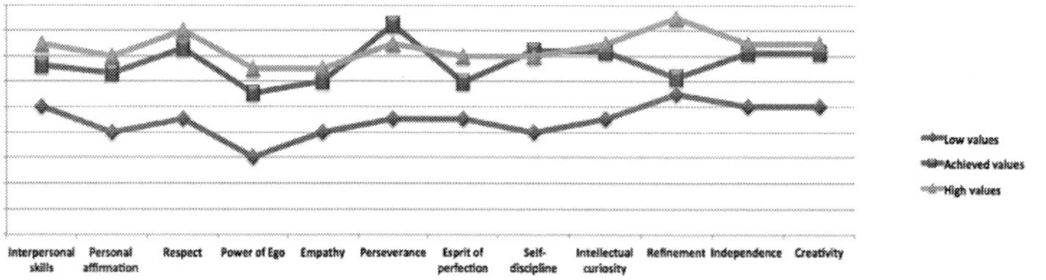

Fig. 4. The values for secondary scales of ABCD-M for Machine-tools' students

For Robotics undergraduates, average values are closer to high values, except for Empathy, Sense of perfection and Refinement (Figure 5).

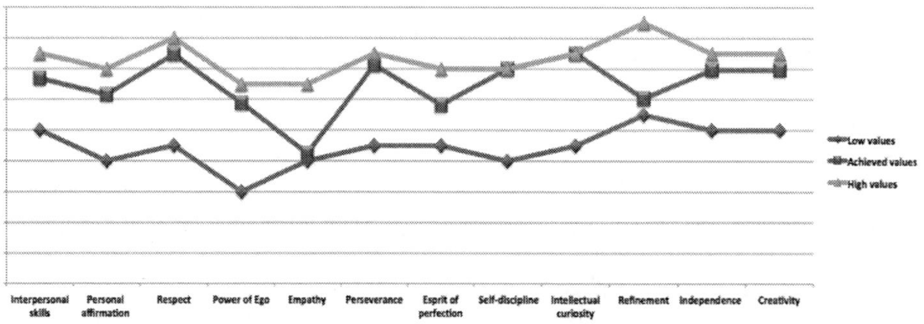

Fig. 5. The values for secondary scales of ABCD-M for Robotics' students

The low values represent:

- Low Empathy - people less intuitive, less concerned with others, selfish, devoid of understanding for others, cold and disinterested
- Low Esprit of perfection - superficial people, a lower accuracy in action, personal development, in pursuit of their goals.
- Low Refinement - closed attitude towards their accomplishment and low intellectual personal cultivation;

For Logistics undergraduates we have average values that are closer to the high values, except for Empathy and refinement. A description of personal characteristics to low levels is made above (Figure 6).

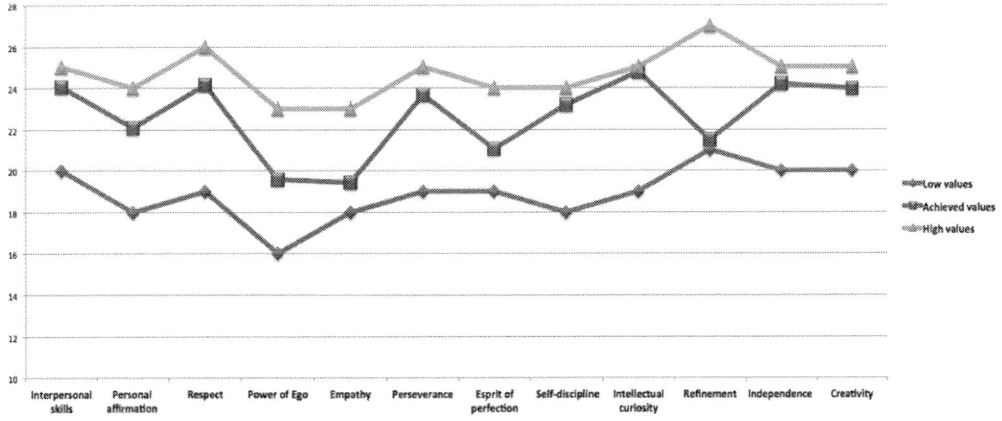

Fig.6. The values for secondary scales of ABCD-M for Logistics' students

The graph shown below captures average values obtained for subscales revealed differentiated specializations (Figure 7 and Figure 8).

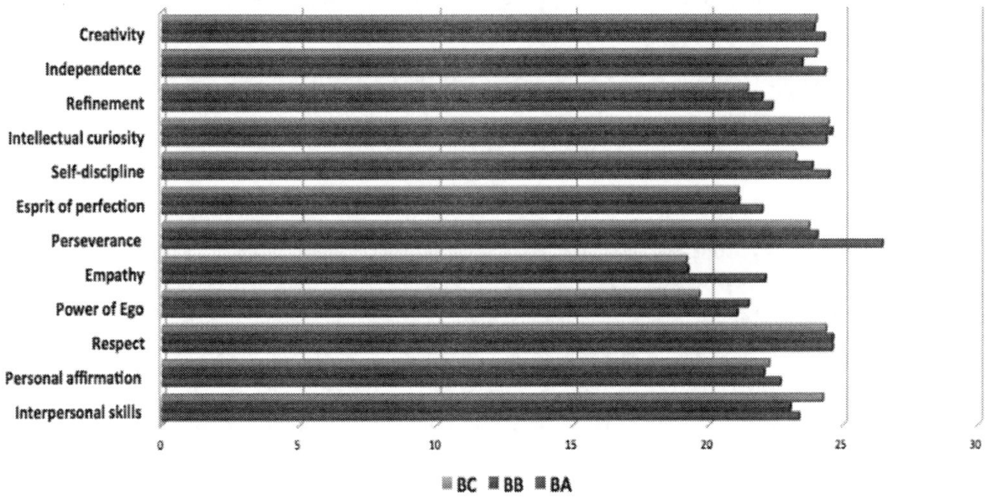

Fig. 7. Comparative values of ABCD-M secondary scales for different specializations

The results obtained at ABCD-M questionnaire shows the average values closer towards high values of interpreted grid. The final analysis of the results for all five dimensions correlated with the relevant subscales imposes rethinking development programs, personal training for future challenges.

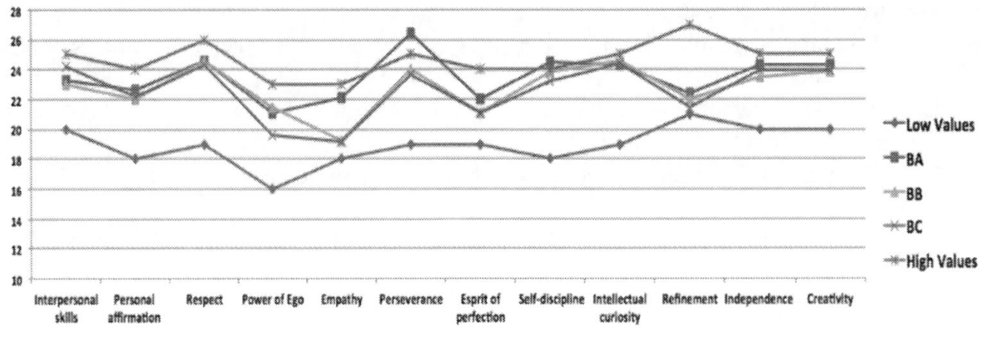

Fig. 8. Comparative average values reported at low and high values

5 Conclusion

Current psychological evaluation systems consider only some of the requested capabilities and not always the most relevant for smart manufacturing. To solve this problem, we tried to identify and evaluate what we considered as the core of the complex various skills required.

Our research was centred on the use of a psychological instrument for the assessment of these capabilities.

The constellation of skills and personal qualities map needed for the evaluation and selection of the human resource fit to work in the industry 4.0 environment was designed after the application of this complex system of evaluation on successive series of students from the University POLITEHNICA of Bucharest.

This work has been funded by University POLITEHNICA of Bucharest, through the "Excellence Research Grants" Program, UPB – GEX. Identifier: UPB–EXCELENŢĂ–2016 Research project title: Diagnosis of psychological skills associated with transversal skills required by multinational industrial companies - DiAPaSS, Contract number: 369/ Nr 60/26.09.2016

References

1. K. Schwab, *The fourth industrial revolution*, (World Economic Forum, 2016)
2. ***http://www.weforum.org/agenda/2016/01/how-can-we-embrace-the-opportunities-of-the-fourth-industrial-revolution
3. ***https://www.bcg.com/expertise/industries/engineered-products-infrastructure/man-machine-industry-4.0.aspx,
4. ***https://www.weforum.org/reports/the-future-of-jobs
5. M. Minulescu, *ABCD-M, Manual tehnic şi interpretativ*, (Editura Sinapsis, Cluj-Napoca, 2008).
6. W.K.R. Hofstee, B. De Raad, & L.R. Goldberg, Journal of Personality & Social Psychology, **63 (1)**, 146-163. (1992)

Integrated planning: a sustainable urban mobility concept for the Jiu Valley

Mihai-Ionut Danciu[1,*], and *Sabina* Irimie[2]

[1] Politehnica University, 300223, Traian Lalescu, 2A, Timisoara, Romania
[2] Petrosani University, 332006, Universitatii, 20, Petrosani, Romania

Abstract. Former industrial areas are characterized by transport infrastructures that exceed the national average in terms of development, spatial coverage and initial investments. Today, when production is replaced by services, retail and consumption, these networks are seen as amenities that have the ability to address the main local issues: the vulnerable economic flows, the lack of accessibility in poor or remote areas, urban shrinkage and the lack of coordination in spatial planning. This study reviews the current status of sustainable urban mobility plans in each of the six towns that form the Jiu Valley and correlates them with the local development strategies and plans. The main positive and negative aspects are highlighted to create an objective overview regarding the vision for mobility in Jiu Valley, as a whole. In the final part, we propose an integrated planning methodology for the entire urban area, envisioning a sustainable productive urban area, formed by accessible communities with strong identities, expressing a circular economic metabolism integrated in the regional development strategy.

1 Introduction

According to the latest legislative changes in Romania [1-2], planning is regarded as the main activity that can determine the economic and territorial development in all levels, from the state down to the local government. There are two types of planning documents: strategic plan and spatial development plans. The former are the creation and responsibility of various types of practitioners, from urbanists to economists and local administration experts. The latter is created by urbanists and architects that coordinate a multidisciplinary team of experts. Hierarchically speaking, the most important are the strategic plans, followed by the spatial development strategies. The latter must be drafted according to the former. Also, in order to receive financial European aids, every project must be mentioned in the local development strategic Plan.

In the Jiu Valley, there are six local administrative territorial units (ATU): Petrila, Petrosani, Aninoasa, Vulcan, Lupeni and Uricani. Each of them approved a local development strategy that corresponds to the 2007-2013 European financial exercise, but not all of them established a strategy for the 2014-2020 interval. Also, only some of the six cities are planning to renew their General Urban Plan (GUP), even though some of the current

* Corresponding author: mihaiidanciu@yahoo.com

GUPs are approved around 2000, when mining was still the main driving force for economic activity in the area. The rules for the application of Law no. 350/2001 on regional planning and development and updating of planning documents [3] state that every new GUP must be based on substantiation studies of several types, including the Sustainable Urban Mobility Plan (SUMP) and the Study regarding the Evolution of Economic Activities (SEEA). In theory, they must be the documents that give the main directions regarding the development of economy and mobility in every ATU. Practically, as we will see in this study, only one town (Petrosani) acknowledges the necessity of these studies and their essential role in their sustainable development. This is also the one municipality that needs to become the centre of the new mobility-oriented paradigm. Drawing a line as a conclusion in the failure of strategic local planning, this study proposes an objective overview concept that engages the most important and feasible assets of the Valley: the transport infrastructure and the concentration of human activities along the main transport network.

2 Methodology

The research uses qualitative methods to evaluate the strategic planning of each local administration in the Jiu Valley and finds gaps between planning and the actual needs of the local socio-economic context. Then, it uses the information received after a regular request was placed, in January 2017, at each mayor's office, enquiring for the status of each of the relevant documents in strategic local planning, in order to present their capacity to respond to the current spatial development challenges. Following the guidelines of OECD [4], and the demonstration regarding the sustainability of rail transport in both local and regional contexts, the study comes in the final part and proposes an alternative that checks the main sustainability goals for the Jiu Valley.

3 Mobility and Urban Development Strategies

3.1 Strategic documents

The XXIst century is dominated by mobility [5], a concept involving people, goods, natural elements, ideas and technologies. This is one of the emerging sign of the fourth industrial revolution, one that creates the so-called Industry 4.0, a contemporary trend that brings together automation and data exchange to create new manufacturing technologies. Considering these, urban development becomes a matter of negotiation between citizens, local administration and stakeholders and the regulation of the processes that occur inside the urban area and in relation to the surrounding environment. The city becomes an organism in itself, governed by a metabolism that can either develop its structure (organic development) or let it decrease (smart shrinking or destruction).

The main purpose of strategies in the context of the European Union is sustainability, through the meanings of protection, strengthening and further development of the cities [6]. European cities are acknowledged as possessing unique cultural and architectural qualities, a multiculturality that supports social inclusion of the vulnerable groups and great possibilities for economic development. Cities are the centre of knowledge and the sources for innovation and growth.

In this context, the role of the Sustainable Urban Mobility Plan (SUMP) is to consider each coherent urban core as the centre of a functional urban area and to introduce coherence in cooperation between sectors, policy areas, levels of administration, citizens and other stakeholders. This strategic plan is not just a concept regarding the development of the mobility in the Functional Urban Area, but it is also a section through all the strategic documents that have an impact in the urban area, from the national to the local level, from the socio-economic point of view to the territorial (or physical) field. This is the reason why we consider this as being the most important document that sustain the intentions of local authorities. In addition to this, we consider the Study regarding the Evolution of Economic Activities (SEEA), a founding study for the General Urban Plans that have to be revised, by 2018, in every city. It is the most important economic document that establishes the main drivers for local sustainable economic development, based on the historic background, the current level of development and the economic trends that can impact the local context. Taken together, SUMP and SEEA establish the main directions for urban development and should be the main documents consulted when local authorities decide to renew the urban plans and the local development strategies.

3.2 Contemporary urban mobility trends

In the field of mobility, the most efficient transit systems are the light rail for inner city areas and the commuter train in suburbia [4]. These two have to be combined with small scale means, as pedestrian and bicycle corridors, in order to be the most efficient (related to costs, time and impact on the environment) in complex urban systems. Railroad was introduced in central and eastern part of Europe in the second half of the XIXth century, mainly because of the Austrian politics regarding the extraction and processing of resources from the main dependent areas of the empire. Railroads were built on the perimeter of settlements, where the stations were also located, usually in relation to a main thoroughfare or a street connection to the centre of the settlement. Also, the new industrial developments occurred in this part of the settlements, usually behind the station and in direct relation to the rail lines.

Today, new trends in mobility recognize the situation presented above as being favourable for a new type of development, based on density and mixed use functions. According to the European Commission [6], the future of European mobility resides in high speed lines and new urban cores (NUC) developed around the main passenger stations. The former traditional railway station becomes the new city centre, creating a link between the traditional city centre and the former industrial or backstreet area, now in a complex state of redevelopment. These new areas are a mixture between new buildings (mainly offices and small scale production) and restored historic facilities.

Also, a new concept emerged in Western Europe: the tram-train system (TT). It uses the advantages of both tramway (short distances and high accessibility) and commuter trains (relation between urban settlements and the hinterland and higher speeds) to reconnect functional urban cores to the surrounding territory. By using the same gauge and electricity supply system, the tram-train system starts to be implemented even in the Eastern part of Europe, as we can see in the projects of Timisoara and Szeged [6].

Overlapping the two concepts (tram-train and new urban cores around the main rail stations), we can easily extract the third most important trend influencing our subject: Transit Oriented Development (TOD). As it is named, it is a type of development focused around transit, bringing new residential buildings, business hubs and a valuable public leisure space close to the transit stations. It resides upon the basic characteristic of urban areas: density. The higher the density of urban functions, the more effective is the use of space and public spending, and in the same time developing sustainable in economic and social terms.

In conclusion for this chapter, we can underline that urban development is strongly related to mobility, the former usual railway stations becoming the centre of the urban regeneration, one based on efficiency in the use of public funds.

4 Current strategic planning in the Jiu Valley

In the context of a former mining region, as the Jiu Valley will be after 2018, the most important role in local development resides upon the leadership of local administrations [7]. To test the transparency of local administrations, in January 25th we placed a request at each local administration, asking for the main local development documents. We asked to receive the information via e-mail, or via mail, based on the Law 544/2004 regarding the free access to public information. The results are presented in table 1, below.

Table 1. Result of the basic request for local development strategy (LDS) 2014-2020, SUMP and SEEA, from each local authority in the Jiu Valley

No. of request	Town	Answer medium	Days after the request was placed	Answer for SUMP	Answer for SEEA	Answer for LDS
10980 / 2017	Vulcan	e-mail	27	invited to study documents, but not specified their nature		
2186 / 2017	Petrila	e-mail	1	There is no SUMP	There is no SEEA but offered some information	There is an LDS, posted on the website
875 / 2017	Petrosani	conventional mail	`26	SUMP in approval phase	No answer for SEEA	(there is an LDS, posted on the website)
365 / 2017	Uricani	-	-	-	-	-
870 / 2017	Lupeni	e-mail	22	There is no SUMP	There is no SEEA but offered some information	Received LDS via email
264 / 2017	Aninoasa	e-mail	1	There is no SUMP	There is no SEEA but offered some information	There is no LDS yet

There are two local administrations that proved to be very responsive, Petrila and Aninoasa. Uricani City Hall did not respond to our request. As we expected, even if the legal term to renew the General Urban Plans in Romania is January 1st, 2018, three out of the six local administrations did not even begin the process. Vulcan did not specify what is the current state of the documentation, Uricani did not answer and the Petrosani local administration seems to be the most preoccupied with this purpose, as they are in the final stage of approval.

5 A vision for Transit Oriented Development in the Jiu Valley

It is clear that local authorities in the Jiu Valley do not regard strategic management as a necessary instrument for the sustainable development of the Jiu Valley. Even if some local authorities made some steps in the elaboration of LDS, only one of them, Petrosani, decided for an integrated vision for the entire urban area. This municipality collaborated closely to us to create the vision of an open city, one that is now left behind any intention for implementation. In this situation, benefitting from this experience, we propose an integrated system that respects all of the three conditions for a development based on sustainable mobility in the Valley.

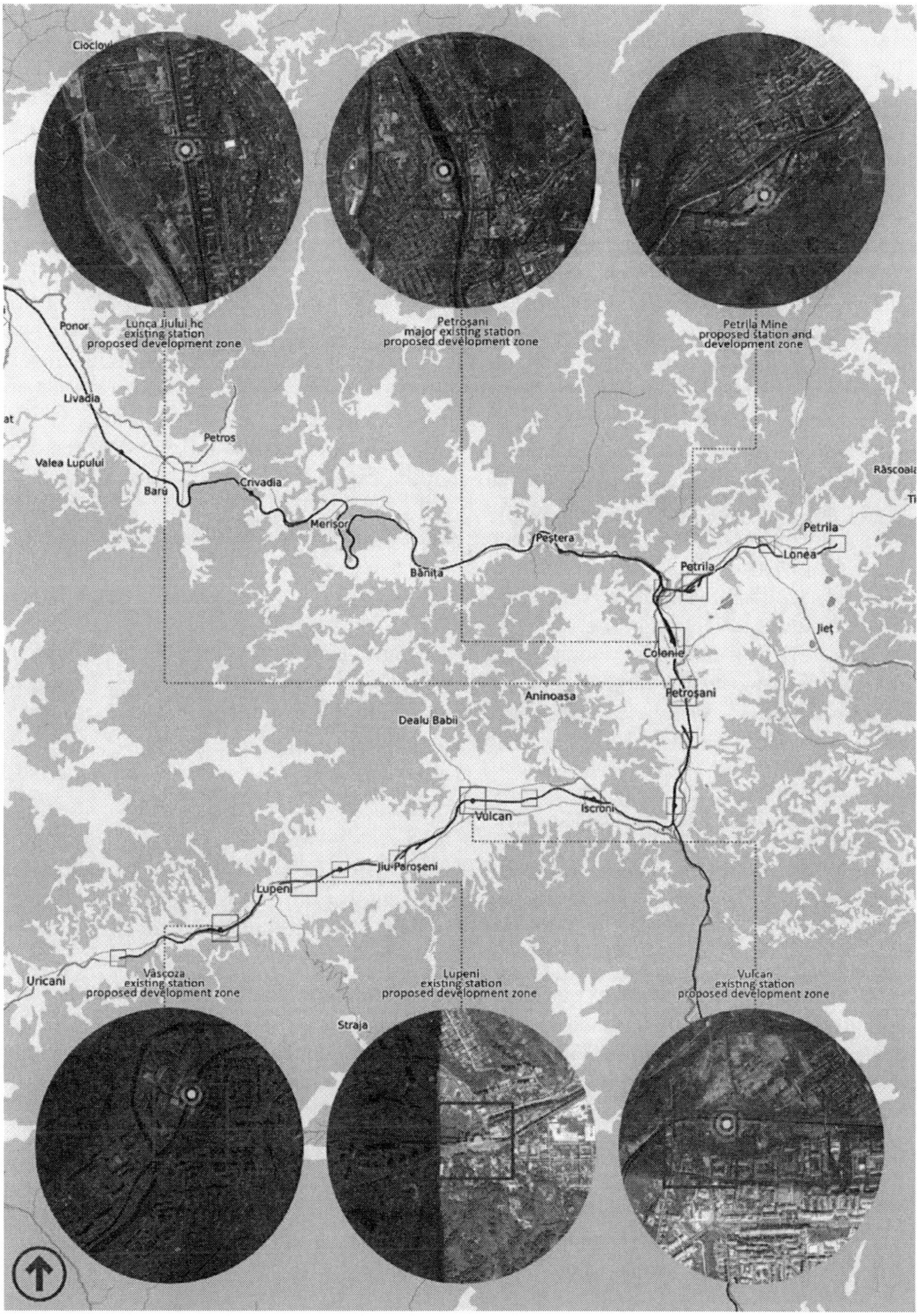

Fig. 1. Main development areas, related to the existent and proposed tram-train stations in the Jiu Valley. Two of the major ones are in Lupeni and two in Petrosani.

The development concept that we propose consists of creating an integrated administrative unit along the Jiu Valley, developed around the new and old city centres that emerge in the regeneration of the urban areas around the railway station and the creation of new ones. These main clusters will concentrate the efforts for economic development, being sustained by a main transit system that uses the existing, underused railroad infrastructure. The whole Valley becomes a linear city, limited by an outstanding natural system, traditional rural settlements and modern ski resorts. This is a vision that can streamline the public investment in revitalization of the former mining region and can drive it to survive and adapt to the 4th Industrial Revolution. In this scenario, we see the cities aligned along the unique railroad, taking passengers from Uricani to Petrila in maximum 30 minutes. This needs an administrative body born out of the local administration, in partnership with the railroad company and high capacity local investors.

Petrosani is the focal point, and the most delicate situation. Here, the main station is already the main transportation hub of the Jiu Valley. It benefits from the proximity of the single most important public space in the Valley, Victory square, and of large demolished areas behind it. The main station, itself, has the qualities to be declared a historic monument.

The main objectives for this station are to:
- protect the potential heritage building, restore and revitalize it;
- develop a new business cluster behind the train station, in what is today one of the most degraded neighbourhoods in Petrosani and concentrate here at least a quarter of the city's office spaces and small scale manufactures;
- bring the main bus station close to the train station;
- provide better facilities for passengers, pedestrians and bicyclists;
- restore the urban piazza behind the train station, between the station building and the new office building.

Each of the other train stations need to become a local development hub, layering functions and events, delivering a greater density in the new urban cores. These hubs become new productivity cores, based on the proximity of the transit system and the existence of refurbished former industrial and administrative buildings.

6 Conclusion and discussion

In order to implement such a vision, there is the need to put together all the six local administrations and create a unique administrative unit that can manage the entire Jiu Valley. The urban development must be concentrated along the main railroad, making it efficient enough.

The greatest investments must come in the reshaping of the degraded urban contexts around the main train stations and engage local stakeholders in a creativity cluster that can drive the interests of the entire Valley, with a strategic leadership and a vision that stretches over at least 30 years from now on.

References

1. RP – Romanian Parliament, *Law 350/2001 regarding urban planning and the territorial development* and its further amendments (2016)
2. RP – Romanian Parliament, *Law 50/1991 regarding the approval of constructions* and its further amendments (2016)
3. RG - Romanian Government, Emergency Ordinance 100 for the amendment and the completition of Law 350/2001 *regarding urban planning and the territorial development and Law 50/1991 regarding the approval of constructions* (2016)
4. OECD, *OECD Regional Outlook, 2014 – Regions and Cities: Where Policies and People Meet.* (2014) http://dx.doi.org/10.1787/9789264201415-en, retrieved 15.04.2017

5. A. D. Little, *Future of urban mobility* (2011), http://www.adlittle.com/downloads/tx_adlreports/ADL_Future_of_urban_mobility.pdf, retrieved 21.04.2017
6. European Comission, *High Speed Europe* (2010), http://ec.europa.eu/transport/sites/transport/files/themes/infrastructure/studies/doc/2010_high_speed_rail_en.pdf, retrieved 11.04.2017
7. S., Irimie, M.I., Danciu, Proceedings of the 10th International Management Conference, "Challenges of Modern Management", 2016, Bucharest, Romania, http://conferinta.management.ase.ro/archives/2016/PDF/4_5.pdf, retrieved 05.04.2017

Actual state of knowledge in the field of Supply Chain Management

Oana Dumitrascu[1,*], and *Constantin Manuel* Hila[1]

[1]University of "Lucian Blaga", Faculty of Engineering, 550025, Sibiu, Emil Cioran street, No. 4, Romania

Abstract. This paper aims to analyse the actual state of research in the field of supply chain management and to identify eventual gaps of knowledge and potential research directions in the field concerned. Concepts such as supply chain, supply chain management, procurement management, sales management, outsourcing, quality standards, key performance indicators and data visualisation are analysed and defined. Defining conceptual elements of the approached field, analysing different national and international studies, converge to identify eventual differences concerning authors' opinion, problems, research gaps and potential research directions. It has been found that the specialty literature analyses in general subjects such as supply chain management, supplier performance, suppliers' selection and evaluation, suppliers' development, performance management and do not take into consideration also the supply chain management performance, so that various gaps concerning the specialty literature in the field of supply chain management performance evaluation have been identified. Causes, effects and hypothetical solutions of the identified problems have been highlighted.

1 Introduction

Due to a previous work, the state of art in the field of supply chain management is presented through a bibliographic study that is conducted from various secondary sources. Through the benchmarking method best practices in this field are identified. Based on this research, specific terms, concepts, definitions related to this field, typology and orientation of existing specialised studies in the scientific literature were analysed. The main goal of this paper was to analyse the actual state of the scientific knowledge in the field of supply chain management and to ascertain the eventual gaps of knowledge in this area, identifying potential unexplored research directions.

In the first part of research, the epistemological approach and conceptual elements were presented. Concepts such as supply chain, supply chain management, procurement management, sales management, outsourcing, performance management and data visualisation were analysed and defined.

The concept of supply chain management is seen as a philosophy with a purpose to integrate the independent activities, material and human resources from the point of origin

* Corresponding author: oana.dumitrascu@ulbsibiu.ro

to final destination. Thus, the Supply Chain Management (SCM) can be defined as an integrated, strategic coordination of the organisation's functions, processes and transactions within the supply chain, with an overall goal of improving long-term performance and gain a competitive edge on the market, [1].

Procurement by definition represents the process of obtaining goods and services from preparation and processing of a requisition through to receipt and approval of the invoice for payment. Purchasing is responsible for acquiring all the materials needed by an organisation. Purchasing is the function responsible for issuing purchase orders and initiating the flow of materials, [2].

Sales management is presented as a complex unitary process, attached to which is an extended structure of specific activities that are aimed at the problems of leadership – coordination, forecast – planning, scheduling, organizing, contracting – selling products, training, monitoring and checking the status – realisation of the activity, analysing and evaluating results, [3].

The phenomenon of outsourcing is increasingly encountered today. By outsourcing we understand the transfer of employees, know-how, assets or the transfer of various services from one company to another in order to streamline the process and thus reduce costs, placing also focus on companies' specialisation within their maximum competence field.

It is the authors' opinion that Supply Chain Management leads to operational effectiveness and therefore is critical that organisations manage it properly. Without a proper management of strategic and operational performance of the supply chain, organisations will not be able to achieve their strategic objectives.

Data visualisation lays at the intersection of the study of human cognition and perception, advances in technology, graphical interfaces, extensive implementation of standards for internet applications, and the ongoing increase of interest and experience in analytics and data discovery, [4]. In our point of view, data visualisation plays a very important role when it comes to evaluate supply chain management performance. This can be evaluated by means of specific key performance indicators visualised in dashboards or scorecards, that improves supervising and conducting the entire supply chain management process and increases its performance.

When managed in a proper manner, supply chain performance ensures the strategic objectives derived from the organisations' strategy. This involves measuring, communicating, benchmarking performance, setting performance targets, managing issues and risks, leading improvements and changes, and fostering stakeholder.

In the bibliographical study various sources were used, such as [5-13].

In the epistemological approach and research state of scientific knowledge, specific terminology and conceptual elements are treated. The supply chain management concept is analysed, highlighting the characteristics of the procurement and sales management.

Concerning the concept of supply chain management and its performance evaluation a lack of knowledge in this field can be identified. The specialty literature includes various studies on supply chain management, supplier development, supplier performance evaluation, performance management, but does not link these research directions to the supply chain management performance evaluation, so that a certain knowledge gap in the national and international literature is identified. Concerning the authors' opinion, it is highly recommended that the specialty literature in this field should be developed by linking these fields together.

Many studies aim to analyse the supply chain management. It is shown how supply chain management has become a very important tool set concerning the solution that are offered for supply chain planning, execution, coordination, and also collaboration. [14]

Another research direction is the one focusing on supplier development. Proch, Worthmann and Schluchtermann (2016) point out how to coordinate supplier development in supply chains. The [16] analyses Toyota's supplier development practices. Dalvi and Kant (2015) analyses the crucial benefits of supplier development, such as effective supply chain management, efficient communication, strategic benefits, delivery performance, competitive advantages, cost reduction and supplier's performance. [15-17]

Another research direction focuses on performance evaluation. Some studies focus on supplier performance evaluation and selection. Schmitz and Plattz (2004) indicate the importance of performance evaluation of suppliers. [18]

Some studies analyse also the supply chain management performance. In [6] shows that a good implementation of supply chain management can lead to positive outcomes such as on-time deliveries, product quality, reduce wastage and inventory and low operating costs. [6]

Taking into consideration the dynamic field of supply chain management and analysing various international bibliographical sources, it can be affirmed that presently there is a vast number of theoretical models and principles companies can follow and implement. Current studies focus mainly on the process of logistics in general, taken as a whole, or on the supply management in particular. Usually, performance aspects, their evaluation process, neither their methods of implementation in different environments and organisational cultures are not taking in consideration. In the authors' point of view, analysing diverse national bibliographical sources, a lack of specialty literature can be noticed, in the field of performance evaluation of supply chain management, but also in the field of supply chain management, supplier selection or supplier development. A few studies focus also on the evaluation of performance within the supply chain management system.

The second part of the report presents the bibliographic research from various nationally and internationally studies. Diverse bibliographical references have been consulted: [14, 19, 5, 15, 16, 17, 20, 21].

Both in the international literature and in the national various studies, scientific articles and books deal on the one hand presenting the goals and advantages of supply chain management and on the other hand analysing the aspects of supplier development and the influence factors on the supplier selection.

The third part of the research consists in identifying certain gaps, problems regarding the current state of scientific knowledge and analysis of causes, effects, hypothetical solutions of problems and defining of new objectives for future research.

2 Analysis of causes, effects of problems in the field of supply chain management specialty literature

Both in the international literature and in the national various studies, scientific articles and books deal on the one hand presenting the concept of supply chain management and on the other hand on presenting supplier performance, suppliers' evaluation, selection and performance management. Current studies focus mainly on the process of logistics in general, taken as a whole, or on the supply management in particular and do not take performance aspects, their methods of implementation in different environments and organisational cultures in consideration. In our opinion an important gap concerning the specialty literature in the field of supply chain management performance can be observed, so that it is recommended to develop different studies, including key performance indicators, dashboards and scorecards in the field of supply chain management. Also, certain studies in the field of supplier development including the purchasing research should be developed.

It has been found that the specialty literature treats in general subjects such as supplier performance, suppliers' selection and evaluation, suppliers' development, performance management or supply chain management in general and do not focus also on the evaluation of performance within the supply chain management system.

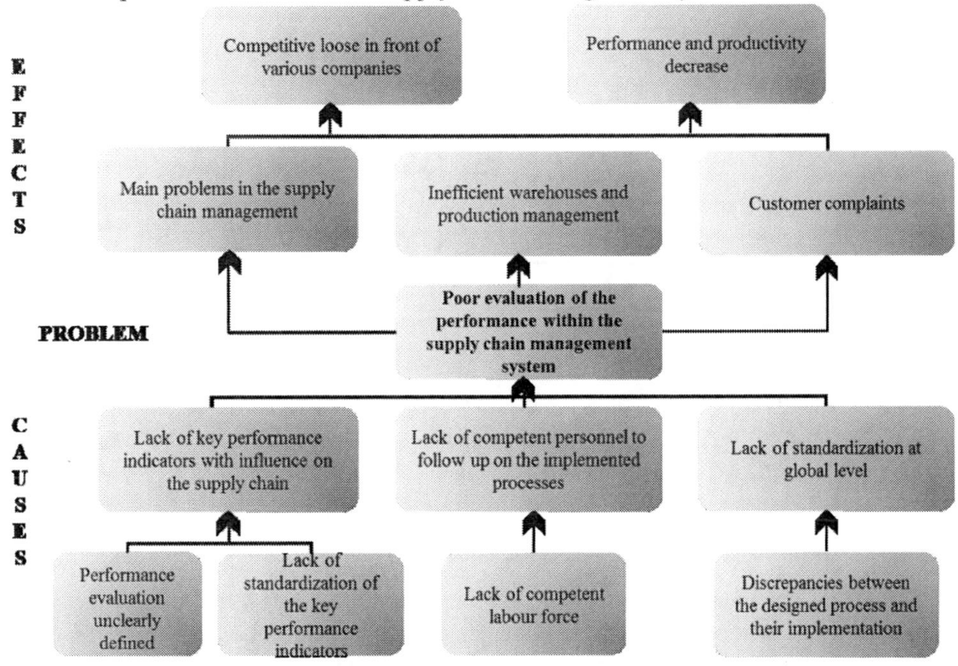

Fig. 1. Problem tree for identifying the causes and effects of the main identified problem

The presented problem tree (Figure 1) presents several causes that are leading to a poor evaluation of the supply chain management system, aspect which lead to the incapacity of performance measurement and thus to decreased productivity and efficiency.

In the authors' point of view starting from the issues identified in the problem tree, we can distinguish the main problem, the poor evaluation of the performance within the supply chain management system.

Certain issues were identified such as lack of performance indicators in the field of supply chain, caused by an unclearly defined performance evaluation and a lack of standardization of KPI's, lack of competent personnel to follow up the implemented process, due to the lack of competent labour force and due to lack of a global level standardisation, due to certain discrepancies between the designed process and their implementation.

A poor evaluation of the supply chain management performance can cause main problems in the supply chain management, can conduct to customer complaints and inefficient warehouses and production management that can lead to performance and productivity decrease and competitive loose in front of other companies.

3 Analysis of hypothetical solutions of problems regarding the supply chain management specialty literature

After analysing the problem and relationship between means and objectives, objectives are deducted and formulated with the role of finding potential solutions.

 DOI: 10.1051/matecconf/201712107007

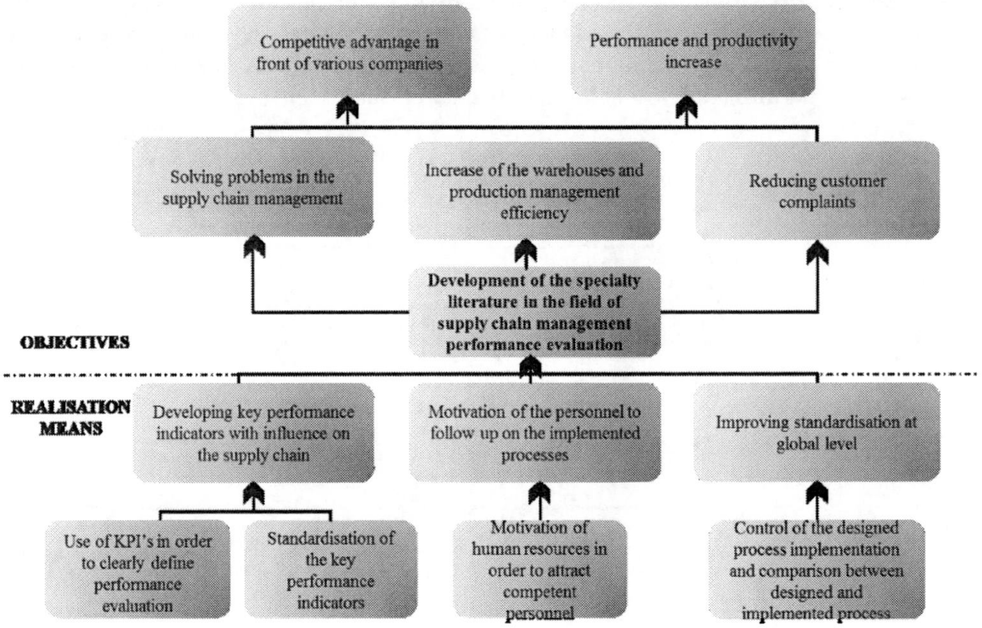

Fig. 2. Development of the relationship between objectives and realisation means

In our point of view, the main objective regarding the development of the specialty literature in the field of supply chain management performance evaluation can be achieved by developing key performance indicators with influence on the supply chain. The use of KPI's in order to define performance evaluation but also the standardisation of the KPI's could lead to this. The development of the specialty literature in this field could be reached as well by an increased motivation of the personnel in order to follow up on the implemented processes and by an improved standardisation at a global level, due to a control of the designed process implementation and comparison between designed and implemented process. These aspects will lead to an increase of the warehouse and production management efficiency, will reduce customer complaints and solving problems in the supply chain, that will increase the performance and productivity and bring a competitive advantage in front of various companies.

4 Conclusions

The work developed facilitated the analysis of the current state of research and of the conceptual elements specific of the field concerned by consulting numerous bibliographic sources, studies and researches, which led to the identification of problems, causes, effects and hypothetical solutions. Specific problems and potential solutions have been identified. This work forms the basis of future research, on the basis of the objectives and hypotheses set.

As specific objectives of a future research that converge to the achievement of the overall objective we can distinguish:

- Determination of the main problems in the supply chain management;
- Determination of the influence of supply chain management on quality, production, sales and procurement;
- Demonstrating specific correlations between supply chain management, performance management, key performance indicators, quality standards, lean planning, six sigma, just in time, Kanban and data visualisation;

- Comparison of the supply chain management systems from various companies;
- Analysing the existing performance evaluation system based on key performance indicators in the studied companies;
- Improvement of the specialty literature in the interest domains, aspect that will fill the observed gaps by determining the link between the interest domains which lay within the supply chain management system.

The general objective of the future research is to bring new contributions regarding the performance evaluation of the supply chain management system, with a positive effect on improving the standardisation of the processes. This will contribute to the performance improvement of specific processes by increasing the capacity of performance evaluation, but also the adaptability of the process as to allow an easier implementation according to the needs and specifications.

References

1. J. Mentzer, S. Min, Z. Zacharia, Journal of Retailing, **76 (4)**, 549-568 (2000)
2. R. Monczka, et al., *Purchasing and Supply Chain Management*, 34-36, (South-Western Publisher, Mason, 2010)
3. S. Crişan, *Managementul desfacerii*, (Sibiu, 2013)
4. D. Stodder, *Data visualization and discovery for better business decisions*, The Data Warehousing Institute best practices report, (2013)
5. A. Agus, JGSM, **5 (1)**, 134-145 (2011)
6. A. Agus, JOEBM, **3 (11)**, 1046 – 1053 (2015)
7. J. M. Aitken, *Supply Chain Integration within the Context of a Supplier Association* (Ph.D. thesis, Cranfield University, UK, 1998)
8. P. Baily, et al., *Procurement principles and management*, 10th edition, (Pearson Education Limited, Harlow, 2008)
9. M. Christopher, *Logistics & supply chain management*, (Pearson Education Limited, Harlow, 2016)
10. S. Few, *Data visualization for human perception*, 2nd edition, (The Interaction Design Foundation, 9, 2014)
11. L. Johnson, F. Fearon, *Purchasing and Supply Management: with 50 supply chain cases*, 13th ed., (2006)
12. V. Lupşe, O. Cosma, Revista Informatica Economică **2 (38)**, 120 – 123 (available at: http://revistaie.ase.ro, 2006)
13. T. Uta, I.D. Ion, *Analiza pieţei de furnizare şi selecţia furnizorilor*, Al treilea simpozion ECR Academic Partnership România, (available at: http://www.ecr-uvt.ro., 2009)
14. S. Bansal, *Technology Scorecards: Aligning IT investments with Business Performance*, 31-33, (John Wiley & Sons, Inc., Hoboken. NJ, USA, 2009)
15. M. Proch, K. Worthmann, J. Schluchtermann, European Journal of Operational Research, **256** (2), 412–429, (2016)
16. P. Marksberry, Benchmarking: An International Journal, **19 (2)**, 277-298, (2012)
17. M. Dalvi, R. Kant, Asia Pacific Journal of Marketing and Logistics, **27 (4)**, 653-654 (available at: http://www. emeraldinsight.com/doi/abs/10.1108/APJML-11-2014-0161, 2015)
18. J. Schmitz, K. Platts, International Journal of Production Economics, **89 (2)**, 231-243, (available at: https://www.researchgate.net/publication/4917445_Supplier_logistics_performance_measurement_indications_from_a_study_in_the_automotive_industry_International_Journal_of_Production_Economics_892_231-243, 2004)
19. C. Martin, *Logistics & Supply Chain Management*, Fourth Edition, (Pearson Publisher, Harlow, 2011)
20. A. Sanchez, M. Perez, International Journal of Operations & Production Management, **25** (7), 681-700, (2005)
21. Gallego, L., *Review of existing methods, models and tool for supplier evaluation*, (Linkopings Universitet, Institute of Technology, 2011)

The connection between absorptive capacity and open innovation in managerial perspective

Cristina Feniser[1,*], *Florin* Lungu[1], and *Javier* Bilbao[2]

[1]Department of Management and Economic Engineering Department, Faculty of Machine Building, Technical University of Cluj-Napoca, Romania
[2]Applied Mathematics Department, Engineering School, University of the Basque Country, UPV/EHU, Spain

Abstract: The growing number of research dedicated to open innovation has determined a growth of company's interest towards external sources of knowledge and of the interactions with these resources in order to obtain an added value. Absorptive capacity, together with the outside-in dimension of open innovation represent interconnected concepts. The present study examines their connection both in theory and in their representation in SMEs managers' perception. The study that forms the base of the research is a survey conducted on companies' managers who worked on open innovation projects. Although few studies dealing with the subject existed up until now, the vision of the managers was chosen to be examined due to the impact of a managerial decision over a company's ability to learn from an exterior environment. In the future, open innovation is thought to be integrated in the management of innovation, this being the reason we view it as of utmost importance for the deepening of the managerial perspective regarding its influences and component elements.

1 Open Innovation and absorptive capacity: literature overview

Open Innovation and the capacity of absorption are both very well-known notions in contemporary literature dedicated to the management of innovation, but the connection between them have been systematically treated very few times until now. The growing number of research dedicated to open innovation has determined a growth of company's interest towards external sources of knowledge and of the interactions with these resources in order to obtain an added value. Chesbrough (2003) describes open innovation as "*a model that assumes that companies can and should use external ideas as well as internal ideas, and internal and external path towards the market, as companies are turning to technological advancement*" [1]. The growing importance of external technological sources is due to the shortening of life cycle of technology, the development of emerging technologies, of cost-sharing practices and of associated risks with science based technology. Cooperation between firms regarding learning and innovation is found more often lately, mostly because of fast technology-based changes in the business world. The

* Corresponding author: cristina.feniser@mis.utcluj.ro

quick rise of technological knowledge along with growing costs of R&D lead to an improbability that any companies could keep all their production capabilities and knowledge in-house. Increasing specialization also facilitates situations in which companies occupy small positions in the space of knowledge.

In small and medium enterprises innovation becomes a business strategy in the moment when the product is a generic one, when the difference between the competition is only in the price. Such a strategy is necessary now more than ever due to technology complexity and of the shortening life span of products. Thus, in order to differentiate and to create value in a market that puts pressure on the existing products, companies choose innovation as a strategy.

Lately, the concept of capacity of absorption is more and more utilized in explaining a wide range of phenomena, from technology transfer in between states to the efficiency of international alliances. At an organizational level the term is used to examine the process of innovation and the result of accumulating knowledge within the organization to deliver a sustainable and competitive advantage. The capacity of absorption appeared as a concept in the 80s when different studies underlined the base role of gaining and applying recent knowledge in order to obtain competitiveness. The term was used by Kedia and Bhagat (1998) in the setting of technology transfer in between countries and they related it with the companies' receptivity to the evolution of technology [2]. Cohen and Leginthal are the first authors to have given a concise definition and have constructed a more general theoretical framework for the application of the capacity of absorption in the business domain. The definition they gave is "*the company's ability to recognize the value of a new external information, its assimilation and application in commercial purposes*" [3]. (Cohen and Levinthal, 1990). This definition includes some very essential aspects. The first is the multidimensional aspect of the notion, including three fundamental capacities related to the newly discovered knowledge, namely: recognizing the value, assimilation and application. The second aspect refers to the connection between the capacity of absorption and the previous knowledge which implies certain abilities and a common language, something that shows the cumulative character of the capacity of absorption in that its present development will allow for a more efficient accumulation in the future.

Zahra and George (2002) expanded the notion of capacity of absorption through distinguishing between potential and realized capacity of absorption. The first one requires acquiring and assimilating knowledge transfers and the second one helps its application in the development and refinement of routines that help transformation and exploitation [4].

The capacity of absorption is a functional concept and it implies different capabilities depending on the company and its domain of activity. But the main elements that make the presence of the capacity of absorption possible are:

- *Basic knowledge:* which offers the capacity of identification, communication, understanding and evaluation of new knowledge and information;
- *The organization and internal processes* which motivates, coordinates and organizes people and resources, and distributes and integrates knowledge;
- *The connections and relational mechanism with the external environment* are those that create communication channels and collaborating relationships;
- *Strategies* which offers direction, coherence and lead to the acquisition and application of knowledge.

Literature dedicated to this theme highlights the increasing role of knowledge and learning in companies and it recognizes the changes that take place in generating, distributing and application patterns of knowledge in the industrial domain. After analyzing this literature Lane and Lubatkin (1998) revises the definition of the capacity of absorption and defines it as: a company's ability to utilize external knowledge using three processes:

- *Explorational learning:* recognizing and understanding the knowledge of the outside environment that has potential to add value to the company.
- *Transformational learning:* the assimilation of valuable knowledge,
- *Exploitational learning:* using the knowledge assimilated in creating new knowledge and commercial results [5].

2 Methods

This study is based on a survey research focusing on the idea of individual perception of managers on open innovation in small and medium enterprises and was obtained by applying the research to the managers of 118 companies that were in the Top Companies in the County of Alba, who in a previous research have declared that they have implemented open innovation programs in the companies' activities they lead. Due to the nature of the problem of a questionnaire, the questionnaire was conceived for auto-application, in order to remove the effect of influence by the presence of an operator and to diminish the effect of interaction needed with one. Another reason why we applied this type of research was the fact that auto-administration implies time saving and less resources. The data was collected through e-mail and telephone.

3 Results

After analyzing the obtained responses, the most important reasons why managers choose to get involved in open innovation projects were identified, which are: *collection of new development ideas, a better understanding of the clients' needs and expectations and access to technologies and new technical knowledge used in industry*.

The advantages perceived by managers due to open innovation projects are:

- Finding new and innovating ideas;
- The development of new and innovating products and services;
- Reducing costs and risks sharing for R&D;
- Entering new markets;
- Increasing reputation;
- Shortening the length of time to enter the market.

Based on the elements that compose the capacity of absorption identified in revised literature, we have identified five determinative factors in SMEs' activity:

- The managers' and employees' experience, both in the activity that the company carries such as open innovation projects. Most research that studied the capacity of absorption have highlighted the importance of human, educational and experience resource.
- The technology used: considered as another indicator of the capacity of absorption due to the equipment used, its level of performance is related to the achievement of complex products;
- Organizational capabilities: they are key elements in competition of SMEs;
- Activities of learning and innovation: research/development activities are a common indicator of the capacity of absorption;
- The connections with entities from the external environment.

The managers were asked which mechanisms they search for opportunities to improve technological capabilities. The mains sources of innovation come from the supplier, 75% of the respondents use this mechanism, followed by clients, for 68% of the cases included in the analysis and the internet has a frequency of 62% in the total obtained responses. The connections with researching organizations have a significant proportion of 29% in the economical responses. The least used sources of innovation by managers are consultants (6%) and industrial or trade associations (14%).

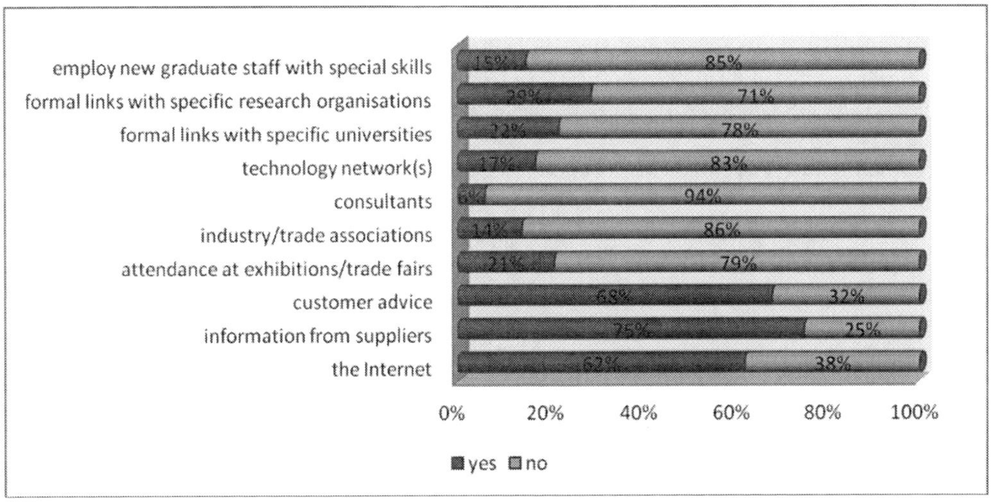

Fig. 1. Mechanisms used to identify technological and non-technological opportunities

Figure 2 illustrates which contributors are the main one for SMEs which are included in the analysis. Just as the variable analyzed previously suggests, referring to the mechanisms of identifying opportunities, the managers indicate as their main partners in innovation projects their suppliers, their clients and universities or other centers of research. This result is due to the smaller costs that this type of partnership implies, a significant aspect in the case of SMEs with a smaller budget for this type of projects.

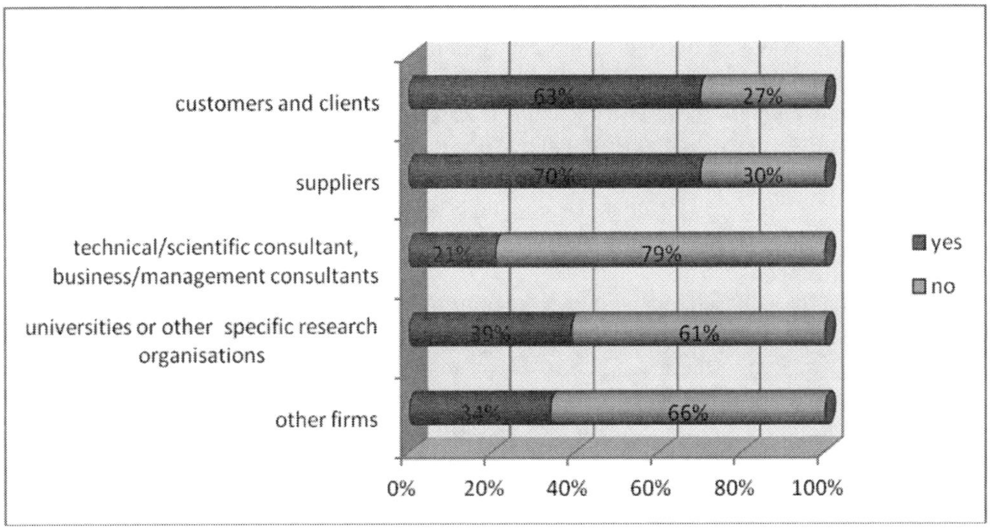

Fig. 2. Does the firm cooperate with other organizations?

4 Conclusions

Absorptive capacity and innovation are interconnected, in a relationship of interdependence and not as a substitute relationship. Absorptive capacity is of utmost importance in the analysis of the success some firm have had regarding others, namely relating to value creation and attraction of exterior sources, like adopting new technologies partnerships with others for innovation. An open innovation project doesn't end once a new product has been launched, this is just the beginning. The companies involved acquire new abilities and, thus, become more competent and more profitable. These types of initiatives always generate new opportunities, and these are finally translated in the improvement of the capacity of absorption. On the other hand, the market is very dynamic and a new product can be taken by competition in a relatively short time and new strategies always have to be taken into account.

The competition of SMEs doesn't solely depend on the employers' skills, but also on the relationship capital that the company disposes of because this strengthens the negotiation position and makes the company vigilant towards other companies of the same domain.

Through a research conducted on the companies that were the Top Companies in the County of Alba in 2010 according to the Chamber of Commerce, it shows that innovation on products and services is a strategical priority after productiveness and profit. One of the research's conclusions was that for small and medium enterprises the implementation of this type of activity greatly depends on the owner, his knowledge and abilities, on his experience and on his value system. Among the advantages identified by managers in open innovation projects are: the development of new and innovating products and services; reducing costs and risk sharing for R&D; access to new markets; improving the company's image.

We chose to deepen the managerial perspective on open innovation and the capacity of absorption of companies in order to understand how the perceptions and decisions of a manager can improve the capacity of a firm for learning from exterior sources. In the future, open innovation is thought to be integrated in the management of innovation, this being the reason we view it as of utmost importance for the deepening of the managerial perspective regarding its influences and component elements.

The rise of economic value, besides aiding new technologies development through innovation, also means an efficient use and broadcast. The capacity of absorbing and utilizing exterior knowledge is of utmost importance for SMEs. The fact that exterior knowledge from others exists is insufficient. The company needs to be able to absorb and implement new technologies. The capacity of absorption is an important dimension for the capacity of innovation which is responsible for the identification, evaluation, acquiring and integration of external knowledge - and it has a relevance only in a general strategy and capacity for innovation. The changes that take place once globalization happens, which require a specialized knowledge based on economy increase the importance of this dimension of innovation.

The research is based on a survey regarding the topic of individual perception of managers on open innovation in small and medium enterprises and was obtained by applying the research to the managers of 118 companies that were in the Top Companies in Alba County, who in a previous research have declared that they have implemented open innovation programs in the companies' activities they lead. Due to the nature of the problem of a questionnaire, the survey was conceived for auto-application, in order to remove the effect of influence by the presence of an operator and to diminish the effect interaction needed with one. Another reason why we applied this type of research was the fact that auto-administration implies time saving and less resources. The data was collected through e-mail and telephone.

References

1. H. Chesbrough, *Open innovation; The new imperative for creating and profiting from technology* (Harvard Business School Press, Boston, 2003)
2. B. L. Kedia, R.S. Bhagat, Acad. Manage. Rev.,**13**, 559–571 (1988)

3. W.M. Cohen, D.A. Levinthal, Admin. Sci. Q., *Absorptive Capacity: a new perspective on learning and innovation,* **35**, 128–152 (1990)
4. S. Zahra, G. George, Acad. Manage. Rev, **27(2)**, 185–203 (2002)
5. P. J. Lane, M. Lubatkin, Strat. Manage. J., **19(5),** 461–477 (1998)

Characteristics of the Romanian CIF Imports during the Period 01.01.2016 - 31.10.2016

Wiegand Helmut Fleischer[1,*]

[1]Lucian Blaga University of Sibiu, Faculty of Engineering, Department for Industrial Engineering and Management, 550024 Sibiu, Romania

Abstract: The aim of this paper is to analyze the Romanian CIF imports during the period 01.01.2016 - 31.10.2016. My research goal is to identify the most important section of CIF imports during the analyzed period and the main structural modifications in evolution of the Romanian imports. The expected results are to know the most important import section of the Combined Nomenclature, the structural modification in evolution of these industries and the most important partner countries.

1 Introduction

For a better understanding of the Romanian economy, it is important to analyze the Romanian commercial relations. The characteristics of the Romanian CIF imports will give us important results regarding the main industries and the most important partner countries.

Under the Incoterms 2010 standard published by the International Chamber of Commerce, CIF means "the seller pays for the carriage of the goods up to the named port of destination. Risk transfers to buyer when the goods have been loaded on board the ship in the country of Export." [13].

The CIF price (i.e. cost, insurance and freight price) is "the price of a good delivered at the frontier of the importing country, including any insurance and freight charges incurred to that point, or the price of a service delivered to a resident, before the payment of any import duties or other taxes on imports or trade and transport margins within the country" [14].

2 Sections

In the period 01.01. - 31.10.2016 CIF imports amounted to 55671.7 million euro and increased with 6.4% as against the period 01.01. - 31.10.2015 [10].

In the structure of exports, six sections of the Combined Nomenclature hold 71.8% of total imports, as follows (table 1).

The main structural modifications came up in the evolution of imports by CN sections in the period 01.01.-31.10.2016 as against the period 01.01. - 31.10.2015, consists of:
- XVII – increasing with 17.4 percentage points;

* Corresponding author: wiegand.fleischer@ulbsibiu.ro

- VI – increasing with 7.1 percentage points;
- VII – increasing with 6.8 percentage points;
- XVI – increasing with 6.0 percentage points;
- V – decreasing with 13.6 percentage points;
- XV – decreasing with 1.0 percentage points [10].

Table 1. CIF Imports during the period 01.01. - 31.10.2016 [10]

Section of the Combined Nomenclature (CN)	Value - million euro -	Structure in % as against total imports	In % as against 01.01. - 31.10.2015
XVI Machinery and mechanical appliances; electrical equipment; sound and image recorders and reproducers	15558.2	27.9	106.0
XV Base metals and articles of base metals	5682.5	10.2	99.0
XVII Vehicles and associated transport equipment	5680.1	10.2	117.4
VI Chemicals products	5647.1	10.1	107.1
VII Plastics, rubber and articles thereof	4162.0	7.5	106.8
V Mineral products	3273.5	5.9	86.4

Imports of section XVI holding 27.9% of total imports in the period 01.01. - 31.10.2016, holding the first place and registered an increase of 6.0% as against the same period in 2015.

In this section, the chapters are:

- Chapter "Machinery and mechanical appliances; electrical equipment; sound and image recorders and reproducers" holds 8627.7 million euro imports CIF, representing 55.5% in total section, 15.5% in total imports CIF and registered an increase of 10.0% in comparison with the period 01.01. – 31.10.2015;
- Chapter "Boilers, turbines engines, mechanical apparatus and devices, parts thereof" holds 6930.5 million euro imports CIF, representing 44.5% in total section, 12.4% of total exports FOB in the period 01.01. – 31.10.2016 and registered an increase of 1.4% in comparison with the period 01.01. – 31.10.2015 [10].

The second place is held by the imports of section XV with an amount 5682.5 million euro and representing 10.2% in total imports CIF and registered a decrease of 1.0% in comparison with the period 01.01 - 31.10.2015.

In this section, the chapter are:

- Chapter "Products of pig-iron, iron and steel" holds 1704.1 million euro of the imports CIF and registered an increase of 1.7% in comparison with the period 01.01. – 31.10.2015;
- Chapter "Pig-iron, iron and steel" holds 1699.4 million euro imports CIF and registered an important decrease of 8.7% in comparison with the period 01.01. – 31.10.2015;
- Chapter "Aluminium and articles thereof" holds 705.2 million euro imports CIF and registered a decrease of 2.4% in comparison with the period 01.01. – 31.10.2015;
- Chapter "Miscellaneous articles of base metal" holds 559.4 million euro imports CIF and registered an increase of 10.8% in comparison with the period 01.01. – 31.10.2015;
- Chapter "Copper and articles thereof" holds only 530.3 million euro imports CIF, but registered also a decrease of 0.4 % in comparison with the period 01.01. – 31.10.2015;
- Chapter "Tools and implements, cutlery of base metal" holds 265.6 million euro imports CIF and registered an increase of 13.5% in comparison with the period 01.01. – 31.10.2015;
- Chapter "Zinc ads articles thereof" holds 64.2 million euro imports CIF and registered an important increase of 4.4% in comparison with the period 01.01. – 31.10.2015;
- Chapter "Other base metals; cermets" holds 60.3 million euro imports CIF and registered an important increase of 11.9% in comparison with the period 01.01. – 31.10.2015;
- Chapter "Nickel and articles thereof" holds 37.9 million euro imports CIF and registered an important increase of 0.7% in comparison with the period 01.01. – 31.10.2015;

- Chapter "Lead and articles thereof" holds 36.3 million euro imports CIF and registered an important increase of 59.0% in comparison with the period 01.01. – 31.10.2015;
- Chapter "Tin and articles thereof" holds 19.8 million euro imports CIF and registered an important increase of 60.3% in comparison with the period 01.01. – 31.10.2015 [10].

Imports of section XVII hold the third place as weight in total imports CIF with 5680.1 million euro, registering an increase of 17.4% as against the period 01.01. - 31.10.2015.

In this section, the chapter are:
- Chapter "Vehicles, tractors and other ground vehicles" holds 5346.8 million euro of the imports CIF and represent 94.1% of total section and 9.6% of total imports CIF in the period 01.01. – 31.10.2016;
- Chapter "Railway or tramway locomotives and rolling-stock" holds 131.5 million euro imports CIF and registered an important increase of 135.7% in comparison with the period 01.01. – 31.10.2015;
- Chapter "Aircrafts, spacecrafts" holds 178.2 million euro imports CIF and registered an important increase of 69.8% in comparison with the period 01.01. – 31.10.2015;
- Chapter "Ships, boats and float structures" holds only 23.7 million euro imports CIF, but registered also an important increase of 104.1 % in comparison with the period 01.01. – 31.10.2015 [10].

Imports of section VI, hold the fourth place as weight in total imports CIF with 5647.1 million euro, registering an increase with 7.1% as against the period 01.01. - 31.10.2015.

In this section, the chapter are:
- Chapter "Pharmaceutical products" holds 2280.9 million euro of the imports CIF and registered an increase of 8.9% in comparison with the period 01.01.-31.10.2015;
- Chapter "Miscellaneous chemical products" holds 811.7 million euro imports CIF and registered an increase of 11.5% in comparison with the period 01.01. – 31.10.2015;
- Chapter "Essentials oils and resinoids; perfumery, cosmetic or toilet preparations" holds 503.4 million euro imports CIF and registered an increase of 15.4% in comparison with the period 01.01. – 31.10.2015;
- Chapter "Organic chemicals" holds 479.7 million euro imports CIF and registered an important increase of 18.6% in comparison with the period 01.01. – 31.10.2015;
- Chapter "Fertilizers" holds 422.2 million euro imports CIF and registered an increase of 21.7% in comparison with the period 01.01. – 31.10.2015;
- Chapter "Tanning or dyeing extracts; tannins and their derivates; dyes, pigments and other dyestuff; paints and varnishes; putty and mastics; inks" holds only 367.6 million euro exports CIF, but registered also an increase of 5.5 % in comparison with the period 01.01. – 31.10.2015;
- Chapter "Soap; washing and lubricating preparations, dental wax, modeling pastes" holds 350.0 million euro imports CIF and registered an increase of 11.6% in comparison with the period 01.01. – 31.10.2015;
- Chapter "Inorganic chemicals" holds 222.4 million euro imports CIF and registered an increase of 7.6% in comparison with the period 01.01. – 31.10.2015;
- Chapter "Albuminoidal substances; modified starch; glues; enzymes" holds 121.4 million euro imports CIF and registered an increase of 1.4% in comparison with the period 01.01. – 31.10.2015;
- Chapter "Explosives; pyrotechnic products; matches; pyrophoric alloys; certain combustible preparation" holds 56.5 million euro imports CIF and registered an increase of 6.4% in comparison with the period 01.01. – 31.10.2015;
- Chapter "Special woven fabrics; tufted textile fabrics; lace, tapestries; trimmings, embroidery" holds 29.7 million euro imports CIF and registered a decrease of 7.7% in comparison with the period 01.01. – 31.10.2015 [10].

Imports of section VII, hold the fifth place as weight in total imports CIF with 4162.0 million euro and registered an increase with 6.8% as against the period 01.01. - 31.10.2015.

In this section, the chapter are:

- Chapter "Plastics and articles thereof" holds 3124.9 million euro of the imports CIF and registered an important increase of 7.3% in comparison with the period 01.01.-31.10.2015;
- Chapter "Rubber and articles thereof" holds 1037.1 million euro imports CIF and registered an increase of 5.1% in comparison with the period 01.01. – 31.10.2015 [10].

Imports of section V Mineral products – hold the sixth place as weight in total imports CIF with 3273.5 million euro, registering a decrease of 13.6% as against the period 01.01. – 31.10.2015.

In this section, the chapter are:

- Chapter "Mineral fuels and oils; bituminous substances; mineral waxes" holds 2994.9 million euro of the imports CIF and registered a decrease of 14.6% in comparison with the period 01.01.-31.10.2015;
- Chapter "Ores, slag and ash" holds 169.1 million euro imports CIF and registered a decrease of 4.7% in comparison with the period 01.01. – 31.10.2015;
- Chapter "Salt; sulphur; stone; plaster; lime and cement" holds 109.5 million euro imports CIF and registered an increase of 2.7% in comparison with the period 01.01. – 31.10.2015 [10].

Table 2. Structure of CIF imports by main groups of goods per month (value in million euro) [1-10]

Period	Section of the Combined Nomenclature					
	XVI Machinery and mechanical appliances; electrical equipment; sound and image recorders and reproducers	**XV Base metals and articles of base metals**	**XVII Vehicles and associated transport equipment**	**VI Chemical products**	**VII Plastic, rubber and articles thereof**	**V Mineral products**
01.2016	1303.9	450.0	450.0	488.0	334.4	225.1
02.2016	1487.0	536.6	536.6	626.9	421.4	250.0
03.2016	1593.4	620.2	620.2	649.9	449.0	320.2
04.2016	1515.8	570.9	570.9	559.2	429.2	292.5
05.2016	1583.5	566.9	566.9	545.7	415.3	368.2
06.2016	1601.2	587.0	587.0	526.8	418.8	341.3
07.2016	1488.1	579.6	579.6	554.1	414.1	326.4
08.2016	1482.5	512.0	512.0	570.5	400.8	447.9
09.2016	1719.4	660.7	660.7	557.2	441.3	324.6
10.2016	1783.4	596.3	596.3	568.0	437.9	377.3
$\sum$	15558.2	5680.1	5680.1	5647.1	4162.0	3273.5

In the period 01.01.-31.10.2016, as against the same period in 2015, imports from the other 27 European Union (EU28) countries increased with 6.4%, registering a weight of 77.3% in total imports of Romania [10].

The European Union countries are the most important import partners of Romania and is very import to focus on this commercial relations but also on the Romanian production in order to reduce the imports and to improve the trade balance of Romania.

Partner countries holding the first 10 places in total amount of imports in the period 01.01. - 31.10.2016 (representing 68.4% of total imports = o. t. i.) were the following:

1. Germany (20.7% o. t. i.),
2. Italy (10.3% o. t. i.),

3. Hungary (7.3% o. t. i.),
4. France (5.6% o. t. i.),
5. Poland (5.1% o. t. i.),
6. China (5.1% o. t. i.),
7. Netherlands (4.1% o. t. i.),
8. Turkey (3.8% o. t. i.),
9. Austria (3.6% o. t. i.),
10. Russian Federation (2.8% o. t. i.) [10].

In the period 01.01. – 31.10.2016, CIF imports amounted to 55671.7 million euro, with 6.4% more compared with the same period in 2015.

3 Conclusions

The most important import section of the Combined Nomenclature in the period 01.01.- 31.10.2016 of the Romanian CIF imports are:

- XVI - 27.9% o. t. i.;
- XV - 10.2% o. t. i.;
- XVII - 10.2% o. t. i.;
- VI - 10.1% o. t. i.;
- VII - 7.5% o. t. i.;
- V - 5.9% o. t. i. [10].

By implementing renewable energy projects, the cost of energy imports can be substantially reduced and this would have a beneficial effect on the trade balance of Romania [12].

Also, by implementing the most advanced technologies, we can save energy and increase productivity. For this reason, it is very important to import advanced technologies from countries with best performance in the technologic sector and not import outdated technologies.

We recommend the reviving of the agriculture and animal husbandry in Romania, so that the domestic production should meet, to a great extent, the need of the Romanian market and the surplus should be exported under favorable conditions [12].

In order to stabilize the trade balance of Romania, we recommend the decrease in imports as a result of the quantitative and qualitative improvement of the domestic production and a more efficient awareness of the consumers in Romania regarding the importance of supporting the Romanian industry and the Romanian entrepreneurs by purchasing Romanian products [11].

Romania`s economy has the potential to grow from year to year and to produce most of the products needed by its own population. To realize this important objectives, it is impetuous necessary to improve the efficiency of the production activity and the quality of the Romanian products and services. Thus we have the possibility to reduce the imports, to reach a balanced and even positive trade balance.

References

1. Institutul National de Statistica, *Buletin Statistic de Comert Internațional*, **1/2016**, 54-61 (2016)
2. Institutul National de Statistica, *Buletin Statistic de Comert International*, **2/2016**, 54-61 (2016)
3. Institutul National de Statistica, *Buletin Statistic de Comert Internațional*, **3/2016**, 54-61 (2016)
4. Institutul National de Statistica, *Buletin Statistic de Comert Internațional*, **4/2016**, 54-61 (2016)
5. Institutul National de Statistica, *Buletin Statistic de Comert International*, **5/2016**, 54-61 (2016)
6. Institutul National de Statistica, *Buletin Statistic de Comert International*, **6/2016**, 54-61 (2016)
7. Institutul National de Statistica, *Buletin Statistic de Comert International*, **7/2016**, 54-61 (2016)

8. Institutul National de Statistica, *Buletin Statistic de Comert Internațional*, **8/2016**, 54-61 (2016)
9. Institutul National de Statistica, *Buletin Statistic de Comert International*, **9/2016**, 54-61 (2016)
10. Institutul National de Statistica, *Buletin Statistic de Comert International*, **10/2016**, 54-61 (2016)
11. W.H. Fleischer, Procedia Economics and Finance, **16**, 256 (2014)
12. W.H. Fleischer, *Relatiile economice romano-germane si Uniunea Europeana – Intre controverse si realizari*, 283-284 (Editura Alma Mater, Sibiu, 2011)
13. ***, http://en.wikipedia.org/wiki/Incoterms (2017)
14. ***, https://stats.oecd.org/glossary/detail.asp?ID=332 (2017)

Characteristics of the Romanian FOB Exports during the Period 01.01.2016 - 31.10.2016

Wiegand Helmut Fleischer[1,*]

[1]"Lucian Blaga" University, Faculty of Engineering, Department for Industrial Engineering and Management, 550024 Sibiu, Romania

Abstract: The aim of this paper is to analyze the Romanian FOB exports during the period 01.01.2016 - 31.10.2016. My research goal is to identify the most important section of FOB exports during the analyzed period and the main structural modifications in evolution of the Romanian exports. The expected results are to know the most important export section of the Combined Nomenclature, the structural modification in evolution of these industries and the most important partner countries.

1 Introduction

For a better understanding of the Romanian economy, it is important to analyze the Romanian commercial relations. The characteristics of the Romanian FOB exports will give us important results regarding the main industries and the most important partner countries.

Under the Incoterms 2010 standard published by the International Chamber of Commerce, FOB means "that the seller pays for delivery of goods to the vessel including loading. The seller must also arrange for export clearance. The buyer pays cost of marine freight transportation, insurance, unloading and transportation cost from the arrival port to destination. The buyer arranges for the vessel, and the shipper must load the goods onto the named vessel at the named port of shipment according to the dates stipulated in the contract of sale as informed by the buyer. Risk passes from the seller to the buyer when the goods are loaded aboard the vessel" [13].

The FOB price (free on board price) of exports and imports of goods is "the market value of the goods at the point of uniform valuation, (the customs frontier of the economy from which they are exported). It is equal to the CIF price less the costs of transportation and insurance charges, between the customs frontier of the exporting (importing) country and that of the importing (exporting) country" [14].

2 Characteristics

FOB exports during the period 01.01. - 31.10.2016, amounted to 47767.3 million euro and increased with 4.1% as against the period 01.01. - 31.10.2015. [10]

* Corresponding author: wiegand.fleischer@ulbsibiu.ro

In the structure of exports, six sections of the Combined Nomenclature hold 71.6% of total exports, as follows:

Table 1: FOB Exports during the period 01.01. - 31.10.2016 [10]

Section of the Combined Nomenclature (CN)	Value - million euro -	Structure in % as against total exports	In % as against 01.01.-31.10.2015
XVI Machinery and mechanical appliances; electrical equipment; sound and image recorders and reproducers	14107.6	29.5	111.3
XVII Vehicles and associated transport equipment	8455.7	17.7	110.2
XV Base metals and articles of base metals	3729.6	7.8	93.6
XI Textiles and textiles articles	3438.2	7.2	103.2
II Vegetable products	2783.0	5.8	109.9
V Mineral products	1678.8	3.5	75.0

The main structural modifications came up in the evolution of exports by CN sections in the period 01.01. - 31.10.2016 as against the period 01.01. - 31.10.2015, consists of:
a) increasing weight for sections: XVI - with 11.3 percentage points, XVII – with 10.2 percentage points and II - with 9.9 percentage points;
b) decreasing weight for sections: V – with 25.0 percentage points, and XV - with 6.4 percentage points [10].

Exports of section XVI are in an amount of 14107.6 million euro, representing 29.5% of total exports in the period 01.01. - 31.10.2016, hold the first place and registered an increase of 11.3% in comparison with the period 01.01 - 31.10.2015.

In this section, the chapters are:
- Chapter "Machinery and mechanical appliances; electrical equipment; sound and image recorders and reproducers" holds 8800.1 million euro exports FOB, representing 62.4% in total section, 18.4% in total exports FOB and registered an increase of 10.0% in comparison with the period 01.01. – 31.10.2015;
- Chapter "Boilers, turbines engines, mechanical apparatus and devices, parts thereof" holds 5307.5 million euro exports FOB, representing 37.6% in total section and 11.1% of total exports FOB in the period 01.01. – 31.10.2016 [10].

The second place is held by the exports of section XVII with an amount 8455.7 with 17.7% in total exports of total exports in the period 01.01. - 31.10.2016 and registered an increase of 10.2% in comparison with the period 01.01 - 31.10.2015.

In this section, the chapter are:
- Chapter "Vehicles, tractors and other ground vehicles" holds 7376.1 million euro of the exports FOB and represent 87.2% of total section and 11.1% of total exports FOB in the period 01.01. – 31.10.2016;
- Chapter "Ships, boats and float structures" holds 760.2 million euro exports FOB, but registered also an increase of 11.4 % in comparison with the period 01.01. – 31.10.2015;
- Chapter "Railway or tramway locomotives and rolling-stock" holds 134.5 million euro exports FOB and registered an increase of 0.2% in comparison with the period 01.01. – 31.10.2015;
- Chapter "Aircrafts, spacecrafts" holds 184.8 million euro exports FOB and registered an important increase of 11.9% in comparison with the period 01.01. – 31.10.2015 [10].

Exports of section XV hold the third place as weight in total exports FOB with 3729.6 million euro, registering a decrease of 6.4% as against the period 01.01. - 31.10.2015.

In this section, the chapter are:
- Chapter "Products of pig-iron, iron and steel" holds 1348.2 million euro of the exports FOB and registered a decrease of 1.8% in comparison with the period 01.01. – 31.10.2015;

- Chapter "Pig-iron, iron and steel" holds 1034.8 million euro exports FOB and registered an important decrease of 2.4% in comparison with the period 01.01. – 31.10.2015;
- Chapter "Aluminium and articles thereof" holds 769.1 million euro exports FOB and registered a decrease of 8.0% in comparison with the period 01.01. – 31.10.2015;
- Chapter "Miscellaneous articles of base metal" holds 294.6 million euro exports FOB and registered an increase of 1.0% in comparison with the period 01.01. – 31.10.2015;
- Chapter "Copper and articles thereof" holds only 165.1 million euro exports FOB, but registered also a decrease of 6.9 % in comparison with the period 01.01. – 31.10.2015;
- Chapter "Tools and implements, cutlery of base metal" holds 70.8 million euro exports FOB and registered a decrease of 2.3% in comparison with the period 01.01. – 31.10.2015;
- Chapter "Lead and articles thereof" holds 28.4 million euro exports FOB and registered an important increase of 13.8% in comparison with the period 01.01. – 31.10.2015;
- Chapter "Other base metals; cermets" holds 11.7 million euro exports FOB and registered a decrease of 4.5% in comparison with the period 01.01. – 31.10.2015;
- Chapter "Zinc and articles thereof" holds 3.3 million euro exports FOB and registered an important decrease of 26.1% in comparison with the period 01.01. – 31.10.2015;
- Chapter "Tin and articles thereof" holds 1.9 million euro exports FOB and registered an important increase of 17.2% in comparison with the period 01.01. – 31.10.2015;
- Chapter "Nickel and articles thereof" holds 1.5 million euro exports FOB and registered an important decrease of 43.2% in comparison with the period 01.01. – 31.10.2015 [10].

Exports of section XI, hold the fourth place as weight in total exports FOB with 3438.2 million euro, registering an increase with 3.2% as against the period 01.01. - 31.10.2015.

In this section, the chapter are:

- Chapter "Not knitted or crocheted clothing and accessories" holds 1752.3 million euro of the exports FOB and registered an increase of 1.1% in comparison with the period 01.01.- 31.10.2015;
- Chapter "Knitted or crocheted clothing and accessories" holds 598.4 million euro exports FOB and registered an increase of 1.1% in comparison with the period 01.01. – 31.10.2015;
- Chapter "Man-made staple fibres" holds 188.4 million euro exports FOB and registered an increase of 3.8% in comparison with the period 01.01. – 31.10.2015;
- Chapter "Wool, fine or coarse animal hair, horsehair yarn and woven fabric" holds 127.7 million euro exports FOB and registered an increase of 8.7% in comparison with the period 01.01. – 31.10.2015;
- Chapter "Man-made filaments" holds 114.0 million euro exports FOB and registered an increase of 10.6% in comparison with the period 01.01. – 31.10.2015;
- Chapter "Impregnates, coated, covered or laminated textile fabrics; technical articles of textiles" holds 74.4 million euro exports FOB and registered an important increase of 0.5% in comparison with the period 01.01. – 31.10.2015;
- Chapter "Cotton" holds only 56.2 million euro exports FOB, but registered also an important increase of 33.9 % in comparison with the period 01.01. – 31.10.2015;
- Chapter "Knitted or crocheted fabrics" holds 20.3 million euro exports FOB and registered a decrease of 9.6% in comparison with the period 01.01. – 31.10.2015;
- Chapter "Wadding, felt, special yarns: twine, cords, ropes, articles thereof" holds 46.1 million euro exports FOB and registered an increase of 10.7% in comparison with the period 01.01. – 31.10.2015;
- Chapter "Other ready-made textile articles" holds 325.1 million euro exports FOB and registered an increase of 8.2% in comparison with the period 01.01. – 31.10.2015;
- Chapter "Special woven fabrics; tufted textile fabrics; lace, tapestries; trimmings, embroidery" holds 56.3 million euro exports FOB and registered an important increase of 34.6% in comparison with the period 01.01. – 31.10.2015;

- Chapter "Silk" holds 65.0 million euro exports FOB and registered an important increase of 6.2% in comparison with the period 01.01. – 31.10.2015;
- Chapter "Carpets and other textile floor coverings" holds 12.7 million euro exports FOB and registered an increase of 1.9% in comparison with the period 01.01. – 31.10.2015;
- Chapter "Other vegetable textile fibres; paper yarn and woven fabric of paper yarn" holds 1.3 million euro exports FOB and registered an important increase of 37.4% in comparison with the period 01.01. – 31.10.2015 [10].

Export of section II Vegetable products - hold the fifth place as weight in total exports FOB with 2783.0 million euro, registering an increase of 9.9% as against the period 01.01. - 31.10.2015.

In this section, the chapter are:
- Chapter "Cereals" holds 1671.4 million euro of the exports FOB and registered an increase of 1.7% in comparison with the period 01.01.-31.10.2015;
- Chapter "Seeds and fruits; industrial and medicinal plants; straw and fodder" holds 946.5 million euro exports FOB and registered an important increase of 33.5% in comparison with the period 01.01. – 31.10.2015;
- Chapter "Edible vegetables, roots and tubers" holds 75.7 million euro exports FOB and registered a decrease of 3.2% in comparison with the period 01.01. – 31.10.2015;
- Chapter "Edible fruits" holds 40.6 million euro exports FOB and registered an important decrease of 37.1% in comparison with the period 01.01. – 31.10.2015;
- Chapter "Products of the milling industry; malt; starch" holds 22.5 million euro exports FOB and registered an important increase of 31.5% in comparison with the period 01.01. – 31.10.2015;
- Chapter "Coffee, tea, mate and spices" holds 20.5 million euro exports FOB and registered an important increase of 44.2% in comparison with the period 01.01. – 31.10.2015;
- Chapter "Live plants, flower products" holds only 2.5 million euro exports FOB, but registered also an important decrease of 23.8 % in comparison with the period 01.01. – 31.10.2015;
- Chapter "Vegetable plaiting materials; vegetable products not elsewhere specified" holds 2.4 million euro exports FOB and registered an important increase of 89.2% in comparison with the period 01.01. – 31.10.2015;
- Chapter "Lac, gums, resins and other vegetable saps and extracts" holds 1.0 million euro exports FOB and registered a decrease of 29.1% in comparison with the period 01.01. – 31.10.2015 [10].

Table 2: Structure of FOB exports by main groups of goods per month (value in million euro) [1-10]

Period	Section of the Combined Nomenclature					
	XVI Machinery and mechanical appliances; electrical equipment; sound and image recorders and reproducers	**XVII Vehicles and associated transport equipment**	**XV Base metals and articles of base metals**	**XI Textiles and textiles articles**	**II Vegetable products**	**V Mineral product s**
01.2016	1240.0	690.8	338.0	320.1	193.3	164.2
02.2016	1400.6	962.2	363.7	355.9	217.0	164.6
03.2016	1521.3	834.2	394.1	338.9	183.8	162.9
04.2016	1395.1	879.4	373.0	362.9	138.2	170.3
05.2016	1383.9	906.2	378.4	330.5	141.3	188.1
06.2016	1462.9	942.7	396.1	364.6	228.6	175.3

07.2016	1301.8	762.1	366.2	397.1	466.0	171.1
08.2016	1309.5	692.9	330.3	286.2	415.8	169.8
09.2016	1551.1	909.0	401.2	360.4	456.8	107.2
10.2016	1506.0	904.7	395.6	333.5	342.1	205.1
∑	14107.6	8455.7	3729.6	3438.2	2783.0	1678.8

Exports of section V Mineral products – hold the sixth place as weight in total exports FOB with 1678.8 million euro, registering an important decrease of 25% as against the period 01.01. – 31.10.2015.

In this section, the chapter are:

- Chapter "Mineral fuels and oils; bituminous substances; mineral waxes" holds 1614.2 million euro of the exports FOB and registered an important decrease of 24.8% in comparison with the period 01.01.-31.10.2015;
- Chapter "Salt; sulphur; stone; plaster; lime and cement" holds 33.8 million euro exports FOB and registered an important decrease of 23.5% in comparison with the period 01.01. – 31.10.2015;
- Chapter "Ores, slag and ash" holds 30.8 million euro exports FOB and registered an important decrease of 34.0% in comparison with the period 01.01. – 31.10.2015 [10].

In the period 01.01-31.10.2016, in comparison with the period 01.01. - 31.10.2015, exports to the other 27 European Union (EU28) countries increased with 4.1%, registering a weight of 71.6% in total exports [10].

Partner countries holding the first 10 places in the total amount of exports in the period 01.01. -31.10.2016 (representing 64.7% of total exports = o. t. e.) were the following:

1. Germany (21.6% o. t. e.),
2. Italy (11.7% o. t. e.),
3. France (7.1% o. t. e.),
4. Hungary (5.2% o. t. e.),
5. United Kingdom (4.4% o. t. e.),
6. Bulgaria (3.1% o. t. e.),
7. Turkey (3.1% o. t. e.),
8. Spain (3.0% o. t. e.),
9. Poland (2.9% o. t. e.),
10. Cehia (2.6% o. t. e.). [10].

In the period 01.01. - 31.10.2016, FOB exports amounted to 47767.3 million euro, with 4.1% more compared with the period 01.01. - 31.10.2015.

3 Conclusions

The most important export section of the Combined Nomenclature in the period 01.01.2016 - 31.10.2016 of the Romanian FOB exports are:

- XVI - 29.5% of total exports;
- XVII - 17.7% of total exports;
- XV - 7.8% of total exports;
- XI - 7.2% of total exports;
- II - 5.8% of total exports;
- V - 3.5% of total exports [10].

In order to stabilize the trade balance of Romania, we recommend the increase in exports through better use of the competitive advantages and the focus on the best performing industries of our country [11].

In order to obtain an improvement of the Romanian FOB exports, we recommend following measures:

1. Improvement of the Romanian infrastructure. Ensuring an infrastructure similar to those of the countries in western Europe (particularly the Bucharest - Nadlac highway) would result in an increase in foreign investments in general, because the poor infrastructure of Romania represents one of the most important impediments to foreign direct investments and creates logistical problems that should not exist if we take into account Romania's favorable geopolitical position;
2. Encouraging the economy through concrete measures to support the investments. Thus, we recommend the preservation of the flat rate tax at the current level or even its decrease by 10% according to the Bulgarian model.
3. Furthermore, for the investments that involve large capital consumption and provide a higher number of jobs, additional tax incentives should be offered;
4. The adoption of the European Union legislation in Romania is not sufficient; it must be accompanied by detailed application rules designed so that their implementation, their accomplishment should be easy;
5. Ensuring a legislative stability and predictability in Romania, in the sense that regardless of government, there should exist a medium-term strategy related to taxes and the absolutely necessary changes in this area should take effect from 1 January the following years;
6. Focusing on the industries where we would have competitive advantages and which can cope with the European and global competition. The analysis of Romania's exports in the recent years has shown that the production of auto parts and the car production (especially the Dacia brand, part of the Renault group, or Ford) are sectors that can make a significant contribution to balancing the Romanian trade balance [12].

Romania`s economy has the potential to grow from year to year and to increase its exports. To realize this important objectives, it is impetuous necessary to improve the efficiency of the production activity and the quality of the Romanian products and services. Thus we have the possibility to reach a balanced and even positive trade balance.

References

1. Institutul National de Statistica, *Buletin Statistic de Comert International*, **1/2016**, 46-53 (2016)
2. Institutul National de Statistica, *Buletin Statistic de Comert International*, **2/2016**, 46-53 (2016)
3. Institutul National de Statistica, *Buletin Statistic de Comert International*, **3/2016**, 46-53 (2016)
4. Institutul National de Statistica, *Buletin Statistic de Comert International*, **4/2016**, 46-53 (2016)
5. Institutul National de Statistica, *Buletin Statistic de Comert International*, **5/2016**, 46-53 (2016)
6. Institutul National de Statistica, *Buletin Statistic de Comert International*, **6/2016**, 46-53 (2016)
7. Institutul National de Statistica, *Buletin Statistic de Comert International*, **7/2016**, 46-53 (2016)
8. Institutul National de Statistica, *Buletin Statistic de Comert International*, **8/2016**, 46-53 (2016)
9. Institutul National de Statistica, *Buletin Statistic de Comert International*, **9/2016**, 46-53 (2016)
10. Institutul Național de Statistica, *Buletin Statistic de Comert International*, **10/2016**, 46-53 (2016)
11. W.H. Fleischer, Procedia Economics and Finance, **16**, 256 (2014)
12. W.H. Fleischer, *Relatiile economice romano-germane si Uniunea Europeana – Intre controverse si realizari*, 283-284, (Editura Alma Mater, Sibiu, 2011)
13. *** http://en.wikipedia.org/wiki/Incoterms, (2017)
14. *** https://stats.oecd.org/glossary/detail.asp?ID=1009, (2017)

MATEC Web of Conferences **121**, 07011 (2017) DOI: 10.1051/matecconf/201712107011
MSE 2017

Education, premise for a workforce reconversion trough a Professional Reinvention

Eugenia Jornea[1,*], *Lucian* Lupu-Dima [1], and *Eduard* Edelhauser [1]

[1]University of Petroşani, Universităţii Street, No. 20, Romania

Abstract The diversity of resources, made in the 1980-1990 years from Hunedoara County the most industrialized county of Romania. Hunedoara is today one of the most affected by unemployment, poverty and crime. Today Hunedoara is a county that decreased with 140,000 inhabitants in the last 20 years and 75,000 jobs lost, all of these bearing the seal of deindustrialization. An accurate professional framework, could allow the attracting of investors. A first step should be a study of the age and socio professional structure of the population of Hunedoara County, could permit in the next years on focusing to the education of undergraduate and graduate level workforce in the area.

1 Introduction

Professionals play a key role in meeting health, education, agricultural and water goals: engineers, doctors, nurses, teachers – the whole range of professionals who will be vital to any approach to development. So higher education is not just for people to invest in their own human capital and become part of the elite, but for development to happen, there's going to have to be a lot of professionals trained.[8] Earnings differences are substantial across EU-28 member states even after controlling for price level differences and earnings dispersion. In particular, looking at the 2011 data, inequality in monthly wages is substantially higher in Luxembourg, Netherland or Germany (2800-4000) than in Romania or Bulgaria (330-340). Inequality in income and wages has been increasing in almost all highly industrialized countries over the last decades. This path has been particularly reinforced in the recent crisis, although the impact has not been the same across Europe. If new technologies increase the productivity of highly skilled workers more than for low skilled laborers, wages of highly skilled workers will increase faster.

While technology can boost inequality, better education of laborers can work in the opposite direction.[4] On a local level, Hunedoara county shows a very difficult situation, for example people of Jiu Valley investigated in a study made by a local researcher, reflect a continuous worsening of the quality of life until now of their way of life from material point of view, from 3.7% of those dissatisfied before 1990 to 42% between 2000 -2009. [9].

* Corresponding author: eduardedelhauser@upet.ro

2 Analysis of the age structure of the population of Hunedoara County, as well as its professional social profile

The role of education and training plays a major role in the formation of human capital in supporting economic competitiveness, employment levels of the workforce and not least in personal and professional development. Hunedoara County's population is steadily declining for more than 20 years. Official data presented by the Department of Statistics (Table 1) show that over the past 20 years, Hunedoara county population decreased by 13%, from 544,036 in 1995 to 469,853 in 2016.

Table 1 Population, by age, 1 January (1995-2016), [11]

Years	Total	By age groups		
		0-14 years	15-59 years	60 years and over
1995	544036	111244	349625	83167
2000	526797	96190	339661	90946
2001	523662	92945	338698	92019
2014	477675	61935	310262	105478
2015	473635	60344	305715	107576
2016	469853	59281	301025	109547

Table 2 The average number of employees, on activities of national economy (2010-2014), [12]

Years	2010	2011	2012	2013	2014
Total County	108851	106870	108047	106384	106877
Agriculture, forestry and fishing	2275	2236	2356	2302	2908
Industry	42967	41812	42755	42256	41485
Constructions	7491	8195	9775	8410	9392
Wholesale and Retail	19914	20524	19740	19559	19162
Transport and storage	4970	4949	4798	4117	4129
Hotels and restaurants	2222	2145	2071	2535	3596
Information and communications	768	777	730	771	627
Financial and insurance activities	1438	1301	1270	1182	1139
Real estate transactions	321	271	297	344	267
Public Administration	4444	3891	3726	3828	3974
Education	7626	7186	6783	7112	7028
Health and social assistance	8280	7861	7563	7452	7354
Entertainment activities, cultural and recreation	563	626	707	735	907
Other activities of national economy	769	715	630	707	693

Hunedoara adult population decreases, from 349,625 people in 1995 between 15 and 59 years, to 301,025 people in 2016. The number of children was halved, and the persons over 60 increased by 30%. According to statistics, in 1995, lived in Hunedoara County 111,244 children aged between 0 and 14 years, and in 2016 their number was halved reaching 59,281. In other words, on average, each year, Hunedoara County had 2,500 children less. Worse is that, while the number of children and young people decreases, the increase is only for people over 60 years. In statistics, the number of seniors has increased from 83,167 in 1995 to 109,547 in 2016 (an increase of nearly 30%).

On the other hand, Hunedoara County employees involved in the activities of the national economy from 2010 to 2014 varies depending on the fields. According to INS statistics (Table 2) we can observe the decrease of employees from 2010 until 2014 in the field of industry with 1,482, in the field of transport with 841, in the field of information and communications with 141, in the field of financial activities with 299, in the field of public administration with 470, in the field of education with 598, in the field of health and social care with 926, while an increase was recorded only in the areas such as agriculture with 633, construction with 1,901, hotels and restaurants with 1,374 and entertainment, cultural and recreational with 344.

Table 3 The total number of unemployed registered in 2015 and 2016, [2]

Years	Total No. unemployed	of which women	Total No. unemployed paid	Total No. unpaid unemployed	Unemployment rate (%)	Male unemployment rate (%)	Female unemployment rate (%)
2016	10789	5411	4170	6619	5,79	5,42	6,21
2015	11131	5535	3348	7738	5,88	5,42	6,43

Before 1989 Hunedoara was considered one of the most industrialized counties. Now, the county is trying to escape from the label of "the jobless county".

In table 3 we can see that in 2016 the unemployment rate was 5.79%, one of the largest in Romania, yet it is gratifying that there is a decrease in the unemployment rate compared to 2015 by 0.09%. According to employment workforce agency statistics, Hunedoara has over 10,789 unemployed. The restructuring of the mining sector increase the unemployment since the area is one mono industrial, and all related activities are related with mining.

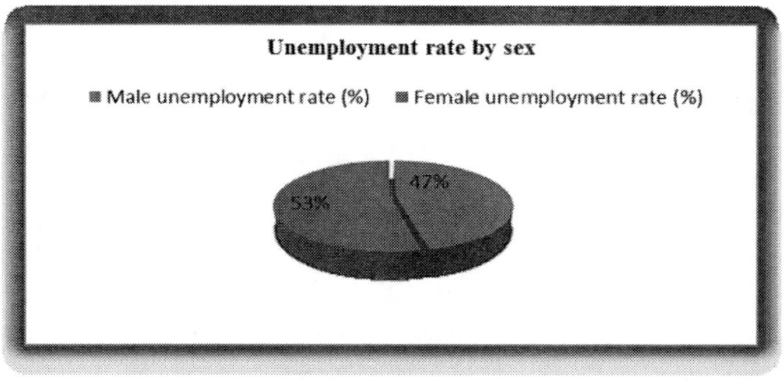

Fig. 1. Unemployment rate by sex 2016 in Hunedoara County

In the figure 1, which refers to unemployment by gender at 31 December 2016, we can see that the female unemployment rate exceeds with 6% the male unemployment in Hunedoara County.

In figure 2 we have presented the total number of registered unemployed in 2015 and 2016 years, and we can see a decrease from year to year and we can hope that this decrease will occur in the future also, due to increase of labor supply and professional qualification people from Hunedoara County.

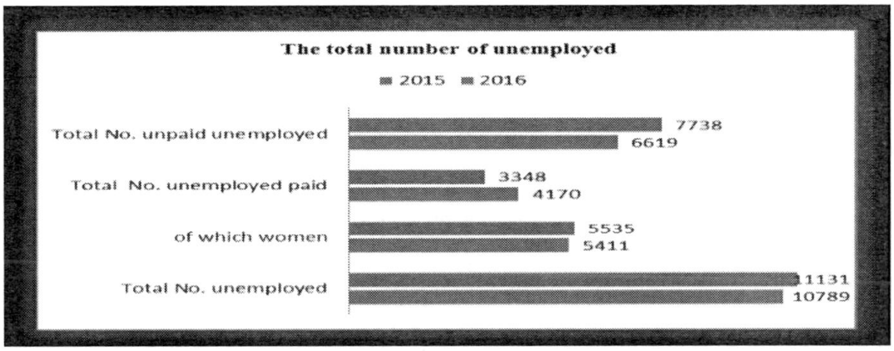

Fig. 2. The total number of unemployed registered in 2015 and 2016 in Hunedoara County

A possible solution of improving the situation in Hunedoara County, could be to restructure the economy by replacing addiction to mining with the creation of a variety of industries and business qualifications.

3 Statistical analysis of the training programs for the workforce

Among the most important factors that contribute to a sustainable development of a society, there are economic factors and social factors. By economic, social and political factors, in Hunedoara County the mining sector was held on a distinct position on national level. The financial crisis of the recent years has had an impact on the job market. Many jobs that once brought a stable income disappeared little by little from the market. Employers have changed their requirements and potential employees have to adjust on the fly, so unemployed and some employees had to refocus on the job market. [7]

The main objectives training programs for the workforce are:

- Obtaining a professional qualification
- Adaptation of the employee to the new job or workplace
- Updating knowledge
- Retraining determined by socio-economic restructuring
- Acquisition of advanced knowledge, modern methods and processes needed to achieve professional activities. [10]

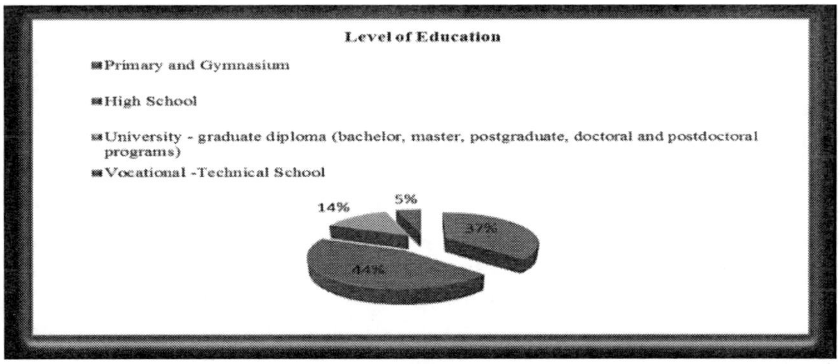

Fig. 3. Classifying the population according to the last level of education of educational institutions graduated in Hunedoara County in 2014 [13]

In terms of training of the workforce, the classification of the active population according to the educational level of the last educational institution graduated in 2014 in Hunedoara County (Figure 3) indicate a high percentage of people who have graduated high school (44%) followed by those with primary and secondary schools (37%), university education (14%) and technical and vocational school (5%). Values are relatively similar to those at the national level.

The high level of population that graduate high school compared to primary, secondary and university education level, raises the question of the need for post-secondary training programs to prepare high school graduates for the labor market. To break the deadlock some people had the courage to begin to make a change and have started business in the wood processing, representing every year a quarter of the county budget, while others are preparing to become specialists in accessing European funds, the so called EU programs.

The management of the National Mine Closures Company (SNÎMVJ) organized training programs for miners. There were over 50 people involved in training programs in the field of expert in accessing European structural and cohesion funds and were 3 classes, the 3 branches mining. Classes were held at Petrila and Uricani Paroşeni. Also miners that are dismissed can follow training programs to learn English or German or to qualify in the profession - craft mechanic.

According to statistics dated 22/02/2017 provided by the Local Agency for Employment Hunedoara, we can observe in figure 4 that most vacancies are for unskilled workers at assembly, fitting parts with 92 positions, after which the top closest followers vendors with 85 positions, truck driver with 33 positions, just as many places available for textile articles assembler, 18 jobs are granted to unskilled worker in the textile industry, 15 jobs for merchandisers, 12 jobs for trade workers and only 4 jobs for security agents. [15]

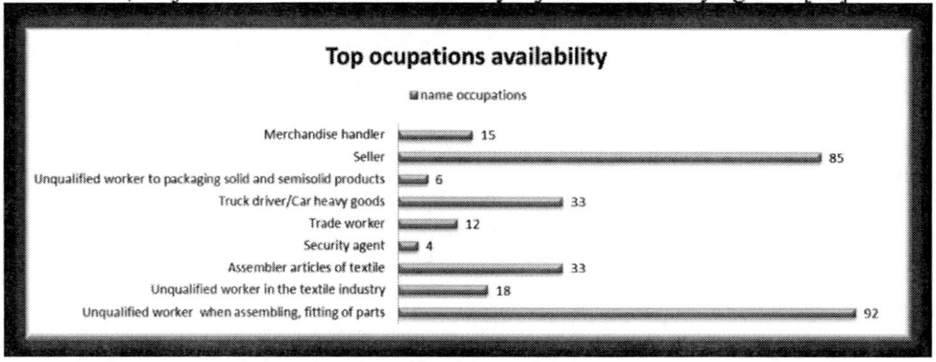

Fig. 4. Top occupations availability in Hunedoara County in 2017

Retraining or reconversion is similar to a reinvention of professional skills. The training programs could be a solution when your last job was eliminated. Of course to recover the situation in Hunedoara County, represent a major problem, besides humane resource development it is needed a strategy for development of the infrastructure of the area as a tourist attraction and diversification of economic activities.

Under the economic context of Romania and especially of the difficult situation the mining areas within Hunedoara and Gorj counties are faced with, the University of Petroşani represent a powerful center for improvement, continuous training and reconversion of the labor force in these counties through an European Founded Project (POSDRU/5.1/S/82/59756) during 2010-2013 years, when we have focused on digital competencies professional training that shall be absolutely necessary in terms of the information society that EU wants to implement. [6] The Project's objective was the increasing the employment rate of the labor force in mono-industrial areas in mining from the Jiu Valley and Oltenia, by attracting young unemployed and long term unemployed, on the labor market, by developing modern and innovative methods for electronic e-learning type of training. [17].

4 Trends in new occupations and professions that might be required in the labor market of Hunedoara County

Hunedoara County is identified as a single industries or mono industrial county. Highly dependent population of the mining industry, this territory was identified as a fragile area characterized by industrial decline. Performing an analysis of all aspects of the current situation and economic transformations of recent years in Hunedoara County, we can deduced that this county still has chances of recovery and becoming a county with a performance and diversified economy, competitive regionally and why not nationwide. [7]

Hunedoara County must take a series of actions for economic diversification, and support the training in the fields of production systems capable of engaging a large percentage of the workforce. A general objective for the deadlock, as the restructuring of the economy is replacing its dependence on mining as a result of the creation of a variety of industries and business qualifications. In this regard, metallurgy and textiles are already directions viable, also automotive and tourism and IT sector shows high growth potential, while being regional areas of interest. Another means of achieving the overall objective is to encourage private sector development, including actions such as creating a favorable business environment by focusing on SMEs. According to the Post-privatized Foundation Report on SME sector in Romania in 2013, specified that the share of SMEs of all registered companies is 99.7% and the number of employees of SMEs in Romania is 65.9 % of total employment in enterprises, we can say that the future of graduate job may be largely an SME. [3]

Hunedoara County loses chance for investments because it is not prepared infrastructural. Business support infrastructure is unevenly distributed in the county and its absence is highlighted especially in areas with reconversion problems. It is necessary to focus attention on the creation and investments in business incubators and the establishment of business parks. At the moment, there are registered in Hunedoara County 3 Industrial Parks - Hunedoara Industrial Park, Business Incubator Simeria and Călan Industrial Park (including business incubator) offering new opportunities to attract and retain investors in the county.

Another objective of the development strategy of Hunedoara County would support the development of tourism as a new sector of economic activity. Hunedoara County is one of the regions in Romania with most tourist spots, but unfortunately this area is not exploited. There is a lack of an unified management of protected areas that allow the practice of tourism in these areas, lack management knowledge, law and ignorance of foreign languages by those involved in networks of rural tourism so alpine areas are used to a minimum potential, lack of information and tourist maps, lack of access roads to tourist attractions, lacking specialized human resources in the field, lack of development organizations tourism. Developing a well thought out and implemented management would solve these problems, but of course with raising capital.

To develop a management, a very important current and future has the IT field. The European Commission launched a coalition for skills and jobs in the digital sector for career development and improving daily life.

The stated purpose is to reduce the existing digital skills shortages at all levels, from highly specialized IT to those who need all citizens to live, work and participate in the digital economy and society. The lack of digital skills are an obstacle to innovation and growth. In Europe, in all sectors remain vacant jobs because of the number of people who cannot use a computer. European Commission stated that 2020 to 750,000 jobs in IT will not be occupied because of this. A situation which is recorded in the context of an unemployment rate of nearly 20 percent among young Europeans aged between 15 and 24 years. More generally, almost 45 percent of Europeans have only basic digital skills. The goal is that by 2020 national coalitions exist in all Member States because until the same year, 2020, one million young people are unemployed to undertake training to fill the vacancies in the digital sector.

It would also be funded retraining the workforce with practical measures to support SMEs face particular challenges in terms of attracting and retaining talent in the digital domain. The plan is a modernization of education, to give all students and teachers the opportunity to use digital tools and materials in teaching and learning. [16]

5 Conclusion

Hunedoara County is now branded as a single industries county and having an industrial decline in recent years, so could be called a fragile area. In this paper based on economic, social and educational information, we can conclude that the recovery of Hunedoara County could be possible by developing and implementing development strategies, which should be on medium and long term.

The transition to a sustained economy must support, on the one hand, on retraining the workforce, and on the other hand, on attracting new investment. Therefore, it is necessary that county and local authorities work together to develop a joint plan of economic reorientation. Hunedoara County loses chance for investments because it is not prepared infrastructural. Business support infrastructure is unevenly distributed in the county and its absence is highlighted especially in areas with reconversion problems. It is necessary to draw attention to investment and on training the population Hunedoara on various fields in accordance with market demand jobs. T

he strengths of Hunedoara county, which would help change the economic course would draw attention to the development of tourism as a new sector of economic activity, focusing investments on business incubators and business parks, encouraging private sector-development of SMEs, and on course to pay particular attention to IT, which is one current and future opportunity.

References

1. *** http://www.ajofmhd.ro
2. *** http://www.anofm.ro/statistica
3. ***http://webcache.googleusercontent.com/search?q=cache:ruaY9dLak0wJ:mcr.doingbusiness.ro/uploads/51c2bb33b342cSINTEZA_Raport%2520IMM%2520-%25202013.docx+&cd=1&hl=ro&ct=clnk&gl=ro
4. C. Dreger, E. López-Bazo, *Wage and Income Inequality in the European Union*, IP/A/EMPL/2013, (Available at: http://www.europarl.europa.eu/studies, 2015)
5. E. Edelhauser E., A. Ionica, M. Leba, Transformations in Business & Economics, **13 (32B)**, 742-759, (2014)
6. E. Edelhauser, L. Lupu-Dima, A. Ionică, M. Leba, Annals of the University of Petrosani, Economics, **XII (3)**, 57-66, (Universitas Publishing House, Available at: http://www.upet.ro/annals/economics/2012.php, 2012)
7. E. Edelhauser, *Sisteme informatice în managementul resurselor umane*, (Universitas Publishing House, 2011).
8. G. Kruss, S. McGrath, I. Petersen, M. Gastrow, International Journal of Educational Development, **43**, 22-31,(Available at: https://www.elsevier.com/atlas/story/people/higher-education-is-key-to-economic-development , 2015)
9. R. Pleşa, Annals of the University of Petroşani, *Economics*, **XII (3)**, 187-196, (Available at: http://www.upet.ro/annals/economics/pdf/2012/part3/Plesa.pdf, 2012).
10. *** http://www.gov.ro
11. *** http://www.hunedoara.insse.ro/main.php?lang=fr&pageid=409
12. *** http://www.hunedoara.insse.ro/main.php?lang=fr&pageid=479

13. *** http://statistici.insse.ro/shop/
14. *** http://date.edu.ro
15. *** http://www.card-profesional.ro
16. *** http://www.romania-actualitati.ro/coalitie_pentru_competente_si_locuri_de_munca_in_sectorul_digital-97031
17. *** http://www.upet.ro/proiecte/

Evaluation of the organizational and technical level of a company transitioning into the sixth innovation mode

Valery Maslennikov[1], and *Karina* Chernitsova[1,*]

[1]Department of Theory of Management and Business Technologies, Plekhanov Russian University of Economics in Moscow, Moscow, Russia

Abstract. Relevance – during the transition to the sixth technological order, the importance of quantitative evaluating company's organizational and technical level rises, along with prospects for its improvement through innovation. The goal – examining the existing approaches to evaluating the organizational and technological level of companies in the high-tech business sector, and propose possible approaches to creating a dynamic real-time assessment of the named level. Methods – a comparative analysis was carried out of the approaches employed in the quantitative evaluation of a company's organizational and technical level in terms of their compliance with the transition to the sixth technological order. Results – the study showed that evaluating the technical level of high-tech companies is possible based on information about the influence of factors that can be grouped into three categories: technical, economic and organizational factors. Conclusions: It was established that the developed method allows identifying a company's technical level, assessing a company's competitiveness and formulating a strategy for production development.

1 Introduction

Assessing the impact of innovation on the technological level is viewed from different perspectives. The focus is placed on the differences of innovation in large, medium and small enterprises [1 - 2]. On an equal basis, universities are considered as a source of innovation for technology development [3 - 6].

That said, it has been proven [7] that the patents and technology resulting from research activities do not fully reflect the needs of enterprises in technological development. A similar position is held by other researchers in innovation [8-16]. This is due to the fact that scientists' ideas about innovations are implemented in their understanding of research results, declare Dilek Karahoca and Ayça Kurnaz. In this light, it has also been proven that an intuitive perspective of the researchers with respect to their results always supports the quality of the results of innovation, and therefore, it is difficult to apply the results on the dissemination of innovation in industrial enterprises [17].

*Corresponding author: kalinina309@yandex.ru

A company's organizational and technical level and staff training are profoundly interrelated processes. Therefore, the concept of structural training is needed for the organization of a continuous process of the reorganizing the existing technological level of production structures, while the availability of educational trajectories allows to transform challenges and shortcomings into opportunities [18].

The technological gap ultimately induces an increase in the inequality of individual industries occurs in many countries. An article by Fernando Perera-Tallo presents a model of growth in which technological changes lead to an increase in the share of enterprises income [19]. Cristiano Antonelli [20] determines the technological correspondence between the elasticity of production and costs, which is required to evaluate the technological level. Authors focusing on the wavelet transform [21] offer their own approaches to the macro-economic methods of quantifying changes in the economy in terms of technological changes.

A review of the literature suggests that there are not enough publications to assess the technological level of enterprises, industries, regions and countries during the transition to the 6th technological mode, while poorly conceptualized and methodologically weak statistical base is provided.

In this paper, we investigate approaches to assessing the technological level and innovative activity of individual European countries; and it is shown that the current methodology does not fully fit for the microeconomic level, which is made up of industrial companies under the 6th technological mode.

Purpose of the study was to formulate the main approaches to the evaluation of the technological level of industrial companies for assessing the changes related to the implementation of innovative development strategies in the transition to the 6th technological mode.

The novelty of the work is to develop a methodology for assessing the state of the technological level of an enterprise in relation to the influence of the organization of management, taking into account the deterioration of equipment.

The study uses the available statistical data from the European Commission and the authors' results of studies conducted in high-tech sectors of the Russian Federation.

2 Relevance of the research.

The current level of the Russian economy is characterized by a decrease in production and technological capabilities. Over the last 8-10 years, there has been a four-fold reduction in production investment, resulting in the depreciation of fixed assets closer to 50%; at the same time, the renewal coefficient was 3.4%, thus predetermining Russia's technological backwardness. The average age of equipment exceeded 20-30 years, i.e. twice as much as in the developed countries, which was one of the reasons for the rapid degradation of the country's technological structure.

The technological gap is also associated with a low-cost expenditure, implemented by Russian companies on research and development: on average, 65% of them spend more in this area of less than 1% of their turnover. In this case, the structure of expenditure on technological innovation is dominated by market research and industrial design (50%).

At the same time, the transition to the 6th technological system requires launching sectors of higher technological redistribution (mechanical engineering, chemical and petrochemical complexes). Today, Russia's industrial policy strategy should be based on investment and technological development, which will see the emergence of new industry leaders at its core. These industries must ensure the progressive and rapid development of the national economy in the medium and long term.

Research methods: A comparative analysis of approaches to the quantitative evaluation of the organizational and technical level of companies in terms of their compliance with the transition to the sixth technological order.

Main part. Economic science still has not yet developed a solid understanding of the process of modernization based on the evaluation of the technological level, which is seen primarily as investments in R & D and human capital.

To determine the technical and economic level of production in Russia and abroad, a number of techniques are used that allow to evaluate development of the country by implementing private and general indicators on technique indexes and indicators of competitiveness development – Growth Competitiveness Index (GCI) provided by the World Economic Forum in the Global Competitiveness Reports [22], as well as indices and indicators of the international rankings of the innovative activity of countries on the European Innovation Scoreboard (EIS), provided by the EU [23] (Table 1).

Table 1. Comparison of the ratings of the innovative and technological levels of countries occupying the first 10 places in the technological and innovation level rankings

Country	**Availability of latest technologies**		**European Innovation Scoreboard**		**Discrepancies in normalized estimates**	**Place**
	Value	**Normalized estimate**	**Value**	**Normalized estimate**		
1	**2**	**3**	**4**	**5**	**6**	**7**
Switzerland	5,8	1	0,79139	1	0	1
Sweden	5,4	0,931034483	0,70398	0,889548769	0,041485714	3
Denmark	5,3	0,913793103	0,70027	0,884860815	0,028932289	2
Finland	5,5	0,948275862	0,6494	0,820581508	0,127694354	5
Germany	5,5	0,948275862	0,63155	0,798026258	0,150249604	8
Netherlands	5,5	0,948275862	0,6315	0,797963078	0,150312784	9
Ireland	5,1	0,879310345	0,60854	0,768950833	0,110359511	4
Belgium	5,2	0,896551724	0,60211	0,760825889	0,135725836	6
United Kingdom	5,4	0,931034483	0,60177	0,760396265	0,170638218	10
Luxembourg	5,2	0,896551724	0,59824	0,755935759	0,140615966	7

In 2016, Switzerland occupied the first five places in the ranking (0,79139), Sweden (0,70398), Denmark (0,70027), Finland (0,6494) and Germany (0,63155).

A comparison of the estimates of the innovation and technological levels of the countries occupying the first 10 places in the technological and innovation level rankings shows that the direct connection between innovative activity and technological level is characteristic only for the leading countries that rank first in the discrepancy between the normalized estimates.

For other countries, such a relationship is not detected, indicating a multidirectional change dynamic at the micro-level.

The following indicators are used as change indicators on the macro-economic level:

- the share of innovation-active enterprises in the total number of organizations,
- the share of innovative products in the total output,
- the share of expenditure on research and development in relation to the volume of product output (research intensity).

At the same time, the 2016 European Innovation Scoreboard does not provide the assessments of the levels of technological innovation for the development of innovative strategies at the enterprise level of individual industries that are united by a common technology platform.

The technological level of industry enterprises depends on their innovation strategy, and is largely determined by the investment firms in their own research and development as well as their effectiveness. This connection determines the technological modernization of the material and technical bases of industries.

The study showed that the technological levels of high-tech enterprises should be considered as the ability of a technical subsystem to work in a fault-free operation with a given range of parameters under the influence of disturbances on the quality of raw materials, etc.

On changing the processes being influenced by the innovation, a multivariate study is usually conducted, while selection of the best process is carried out at a minimum economic cost of the planned production output volume.

Based on the above, an innovation level determining a change in the technological processes is formed in the following composition:

1) organizational and technical level of production;
2) operations management level.

For any production unit, the composite index for the organizational and technical level of production is comparable, regardless of objects or the type and quality of the products, the applied technology, processes and their organization.

The composite index shows how much of the actual economic cost makes up the cost of manufacturing the same product in the same volume with the full use of the basic elements of production, i.e. without internal production loss of material and labor resources.

The reciprocal value indicates how much the actual economic cost of producing a specific product exceeds the necessary costs in the absence of losses, while the difference between the denominator and the numerator makes up the internal production reserves for their reduction.

The operations management level is a relative characteristic that allows evaluating how efficiently and fully the basic elements of production are used under the planned product output program and under the terms and conditions of existing and new technological levels.

In this case, the composite index of the efficiency of use of the main production process of the organizational and technical level of production and operational management for production under changing technological processes shall be determined (I_{OT}):

$$I_{OT} = \frac{C_{T_{new}}}{C_{T_{actual}} + \Delta C_{PL} + \Delta C_{OM}} \qquad (1)$$

$C_{T_{new}}$ and $C_{T_{actual}}$ – respectively, the specific economic costs of production without taking into account losses associated with the organization of the production process, i.e., when fully utilizing the equipment and working hours, RUR.,

ΔC_{PL} – additional costs associated with an increase in the technological level of production, RUR.;

ΔC_{OM} – additional costs associated with the operational management of the production of a new technological level, RUR..

In the company "Mechanical Engineering Technology" the production cost with full use of equipment and working time ($C_{T_{new}}$) amounted to 150 million rubles in 2014.

As a result of increasing the technological level of production, the production cost with full use of equipment and working time ($C_{T_{actual}}$) amounted to 200 million rubles in 2016. This required additional costs associated with raising the technological level of production (ΔC_{PL}) in the amount 3 million rubles. Also, additional expenses related to the production operational management of a new technological level (ΔC_{OM}) 0.5 million rubles were realized.

As a result, the organizational and technical level of production and operational management was 0.737 in 2014-2016, which means lower production costs due to increased efficiency in the use of equipment, technology and operational management.

Evaluating the organizational and technical level is necessary for supplementing the monitoring of the physical deterioration of the equipment, as the new level of technology leads to the rapid development of resource capacity.

The selected equipment deterioration criteria are linearly independent and enable to comprehensively assess the development of equipment resources of innovation-active enterprises with respect to changes in the technological level.

The indicator weights are determined expertly based on the importance of a particular type of equipment for the entire production complex of the enterprise. Use of the following scale for rating the depreciation of equipment is proposed (Table 2).

The physical deterioration of the equipment as a whole is estimated based on the deterioration of a *key element* with a physical deterioration that is assigned the highest weight, for example, equal to 0.5. Knowing the other less significant elements of the equipment with physical deteriorations that affect the costs to a lesser extent, the weight may be calculated, for example, equal to 0.3. The least expensive and relatively easy-to-replace elements of equipment may have a physical deterioration weight in the overall assessment of the physical deterioration, of, for example, equal to 0.1.

Therefore, the integrated resource development index for the equipment when assessing an enterprise's technological level is calculated by the formula:

$$R = 0{,}5\,K_1 + 0{,}3\,K_2 + 0{,}1\,K_3, \quad (2)$$

where Ki — development index of the i-th equipment resource, selected for the rating; the integral index values range from 0 to 1.

An estimation example of the equipment life-cycle integrated indicator for the key elements wear: it is required to determine the physical wear coefficient for the mechanical press. The annual depreciation rate for A = 7.7%. The chronological age is 12 years.

Table 2. Scale for assessing the physical deterioration of equipment

Depreciation rate	Equipment condition	Description of equipment condition
0-5	New	New, newly installed equipment still to be put into operation, equipment in excellent condition
5-17	Very good	Used equipment that has been completely repaired or restored, in very good condition
17-33	Good	Used equipment that has been completely repaired or restored, in good condition
33-50	Normal	Used equipment that is requiring some kind of repair or replacement of individual small parts, such as bearings, brasses, etc.
50-67	Relatively usable	Used equipment that is in a condition suitable for further work but is requiring significant repair or replacement of major parts such as the engine and other critical parts
67-83	Poor	Used equipment that is requiring major repairs, such as replacing the working bodies of basic units
83-95	Unusable	Used equipment that is unsuitable for further use
>95	Broken	Equipment in respect of which there is no reasonable prospect of resale, except for the cost of basic materials that can be extracted from it

The equipment key elements were replaced in the seventh year of operation and accounted for 15% of the press components. Less significant elements of equipment (25% of parts and assemblies) on the 9th year of operation have been replaced with new ones.

Least expensive and relatively easily replaced equipment elements (60% of parts and assemblies) were replaced by the 12th year of the press operation.

Calculation:

1. Determine the press normative service life as a value opposite to the depreciation rate: 100% / 7.7 = 13 years. Since the chronological age of the press is 12 years and no major repairs have been made, it can still remain in operation.
2. 15% of details and units have a chronological age: 12-7 = 5 years and referred by experts to a good condition (10%); K1 = 0.15 x (1-0.1) = 0.135.
3. 25% of details and units have a chronological age: 12 - 9 = 3 years and referred by experts to a satisfactory condition (35%): K2 = 0.25 x (1-0.35) = 0.1625
4. 60% (100% -15% -25%) of parts and assemblies have a chronological age: 12 years and are classified by experts as conditionally suitable (60%): K3 = 0.60 x (1-0.6) = 0.24.
5. The resource development integral index of the mechanical press will be: R = 0.5 x 5 x K1 + 0.3 x 3 x K2 + 0.1 x 12 x K3 = 2.5 x 0.135 + 0.9 x 0.1625 + 7.2 x 0.24 = 0. 3375 + 0.1463 + 1.728 = 2.218 years.
6. The physical wear coefficient of the press will be: 1 - 2.2 / 13 x 100% = 83%.

Thus, the mechanical press will be able to work without a noticeable loss of quality for 2.2 years of service, taking into account the resource consumption, provided the operation conditions are met by the recommended manufacturer, and the repair and maintenance work will be performed on time and qualitatively.

3 Conclusion

The advantage of the provisions put forward is the validity and reliability of the private and general indicators that are used to assess the organizational and technical level of production and their strict adherence to the essence of the more efficient use of production resources as a result of changes in the technological level of production.

In this system, there is a relationship between the current and non-recurring costs of producing goods. The system can be used not only as a means of analyzing, planning and controlling the organizational and technical level of production, but also as a means of research and involvement in the use of the existing reserves in production. The increasing of the production technological level is proposed to be evaluated taking into account the three groups of factors: technical, economic and organizational

1. Technical factors include:

- Development and application of modern production processes
- Application of advanced methods of processing parts
- The use of machines and equipment with modern means of active and objective quality control
- Application of an automated production management system
- Application of the computer-aided design (CAD) system for technological equipment
- Control automation over the network schedules implementation for the design and production of technological equipment

2. Economic factors are a phased advance financing of the work for “technological” production preparation, the creation of a stimulating fund for the new technology development (Start-up into production), etc.
3. Organizational factors - development and deepening of production specialization; certification of the technological process quality and manufacturing of technological equipment, improvement of the auxiliary production organization; improvement of the relationship between auxiliary and main production; expansion of intra-factory, intra-industry cooperation, etc.

References

1. G. Ahuja, C. Morris Lampert, Strategic Management Journal, **22 (6-7)**, 521–543, (2001).
2. C. Corradini, P. Demirel, G. Battisti, Small Bus. Econ., **47**, 163–177, (2016).
3. M. Guerzoni, T.T. Aldridge, D.B. Audretsch, S. Desai, Research Policy, **43 (10)**, 1697–1706, (2014).
4. Lissoni, F., Montobbio, F., Cahiers du GREThA, **2012-24**. (Available at: http://cahiersdugretha.u-bordeaux4.fr/2012/2012-24.pdf, 2012)
5. A.M. Petruzelli, Technovation, **31 (7)**, 309–319, (2011).
6. V.V. Maslennikov, L.G. Skamaj, N.V. Kozlov, Setevoe vzaimodejstvie vuza, hozyajstvennyh obshchestv i vysokotekhnologichnyh predpriyatij. - Vestnik Universiteta (Gosudarstvennyj universitet upravleniya). № 26. p. 191-199, (2011).
7. Srecko Nesic, Ulrich Krupp, Wilhelm Michels. An Investigation on Configuration status of Science and Technology Resource in the Coal Industry of Heilongjiang Province Original Research Article Energy Procedia, Volume 5, Pages 2167-2171, (2011).
8. L. Fleming, O. Sorenson, Strategic Management Journal, **25**, 909–928, (2004),.
9. W.B. Arthur, Research Policy, 36 (2), 274–287, (2007).
10. S. Breschi, F. Lissoni, F. Malerba, Research Policy, **32 (1)**, 69–87, (2003).
11. M. Coffano, G. Tarasconi, Center for Research on Innovation, Organization and Strategy, CRIOS Working Paper n.1, (2014).
12. P. Criscuolo, B. Verspagen, Research Policy, **37 (10)**, 1892–1908, (2008).
13. A. Datta, L.M. Jessup, Technovation, **33 (10-11)**, 355–367, (2013).
14. M. Du Plessis, B. Van Looy, X. Song, T. Magerman, Data production methods for harmonized patent statistics: Patentee sector allocation, (Eurostat Working Paper-Annual Report, available at: http://ec.europa.eu/eurostat/documents/3888793/5835901/KS-AV-06-001-EN.PDF , 2010).
15. F. Malerba, M. Mancusi, F. Montobbio, Review of World Economics, **149 (4)**, 697 – 722, (Available at: http://dx.doi.org/10.1007/s10290-013-0167-0, 2013),
16. J.M. Wooldridge. *Econometric Analysis of Cross Section and Panel Data* (second edition), (MIT Press, 2010).
17. Dilek Karahoca, Ayça Kurnaz, Procedia - Social and Behavioral Sciences, **(131)**, 211-217, (2014).
18. Antonio Andreoni, Structural Change and Economic Dynamics, **29**, 58-74 (2014).
19. F. Perera-Tallo, Journal of Macroeconomics, **52**, 23-38, (2017).
20. C. Antonelli, Structural Change and Economic Dynamics, **38**, 15-24, (2016).
21. F. de Melo, V.V. Maslennikov, E.V. Popova, T.L. Bezrukova, I.V. Kyksova, Asian Social Science, **11 (20)**, 66-73, (2015).
22. Global Competitiveness Report (Available at: http://reports.weforum.org/global-competitiveness-report-2015-2016/competitiveness-rankings/, Accesed martie 2017).
23. European Innovation Scoreboard (Available at: http://ec.europa.eu/growth/industry/innovation/facts-figures/scoreboards_en, Accesed martie 2017).

Barriers in organizing the Romanian firms in responsibility centers

Gina-Maria Moraru[1,*]

[1]Lucian Blaga University of Sibiu, Engineering Faculty, Industrial Engineering and Management Department, 4 Emil Cioran St, 550025 Sibiu, Romania

Abstract. In the current turbulent business environment, the big and medium enterprises must organize their activity in order to face new challenges. One of the organizational forms that can stimulate the performances of the firms is the one in responsibility centers. This paper addresses the issue of the responsibility centers starting from the possibility of using benchmarking before and after the implementation of this organizational form. After highlighting the advantages of the responsibility centers, we presented a study made in Romanian big and medium organizations, regarding the barriers in the design and implementation of this organizational form. The main causes that determine Romanian managers to avoid the organization in responsibility center were analyzed through several techniques specifically used in quality management. After this analyze, we proposed several measures for overcoming these barriers.

1 Introduction

One of the modern ways of organizing large or medium-sized companies is the management center, called responsibility center by some authors. In [1], the two designations are used interchangeably. In essence, the company's budget is divided into sub-budgets corresponding to each division that consists of a control center, characterized by the following features:
- Setting up a management center is based on a field of activity, on a particular activity or process within the company;
- One or several specific missions for each center, substantiated by clear-cut quantitative and qualitative targets;
- The existence of a stock of available resources, necessary to reach these objectives;
- A center manager;
- All areas of activity within the company are covered by the management centers, but with no overlapping activities;
- The ability of the center to adjust to the changes in the business environment (called "self-stabilization" in [1], [2]);
- The ability of the center to develop its own stable sub-structures (called "self-organization" in [1], [2]).

*Corresponding author: gina.moraru@ulbsibiu.ro

2 Advantages and disadvantages of the responsibility centers

Among the managerial advantages of this type of organization, we mention the technical role of the budgets, which confers the organization around management centers the capability of establishing the foresight management of the company, and the behavioral role of the budgets, allowing the empowerment and motivation of the human resource in the long term [1].

To identify the disadvantages of this form of organization, we must first correlate the functioning of this structure, the type of organizational culture that it fosters [3] and the way in which it stimulates the use of production factors. At first glance, the functioning seems to be ideal, as any process irregularities can be detected and corrected at an early stage. We are informed of the exact way in which the tangible resources of the company are employed [4], by whom and to what effect. This "mechanism" of the management centers seems to work flawlessly, with clock-like precision. The organizational culture stimulated within the company is based on order, discipline, respect, waste avoidance and, above all, efficiency. Essentially, quantitative results are particularly important in this form of organization, but qualitative ones are usually left in the background. The freedom of the individual is subject to be limited by figures that mark actual achievements, and this leads to the risk of suppressing change initiatives and creativity acts.

By analyzing in detail a company organized around management centers (Fig. 1), we can notice that a clear-cut delineation of these centers and of their targets raises barriers to the possible creative interaction between groups and individuals belonging to different centers. Creativity is a new factor of production mainly used by research and development teams (Fig. 1). In this type of organization, production as well as the personnel and marketing functions have many duties related to material and financial purposes but very few creative duties. The creative results of the research and development departments usually cross the other departments, without gaining further innovation.

Another potential danger caused by this form of organization is the exacerbation of the importance of purely financial values. For instance, by tracking the financial profitability of the company, the management may adopt measures which are harmful to the natural environment, but which improve the operation of the "mechanism" of management centers, and increase their economic indicators and momentary customer satisfaction.

However, we believe that the primary disadvantage of this form of organization is the danger of the excessive branching of the company's activities, especially those generating profit. Similar design, production, accounting, etc. departments, carrying out almost identical activities, yet working at full capacity only for short periods of time, may unnecessarily appear within different management centers. Their operation in parallel may prove to be – in the long run – unprofitable or insufficiently profitable for the organization, even if it is profitable for each individual center.

I deliberately set aside for the final section of this brief analysis a few of the advantages of the organization around management centers, advantages that are directly or indirectly related to the possibility of using regular benchmarking (Fig. 1) in order to improve the results of the organization and all changes in its system (circular dotted arrow in Fig. 1). For a start, each center provides, through its self-regulation mechanism, effective and permanent management. From the perspective of management planning, the organization around management centers supports the efficient implementation and running of the operational plans within its links; thus, a context that is favorable to the performance management is created. In order for budgetary procedures to connect the strategy to short-term actions, benchmarking – especially internal – is an extremely effective and useful solution available to center managers and to the senior management of the company, facilitating the strategic management of the organization. Once the three levels of control are correlated (management, performance and strategic), the optimal corrections needed in the system are easier to find.

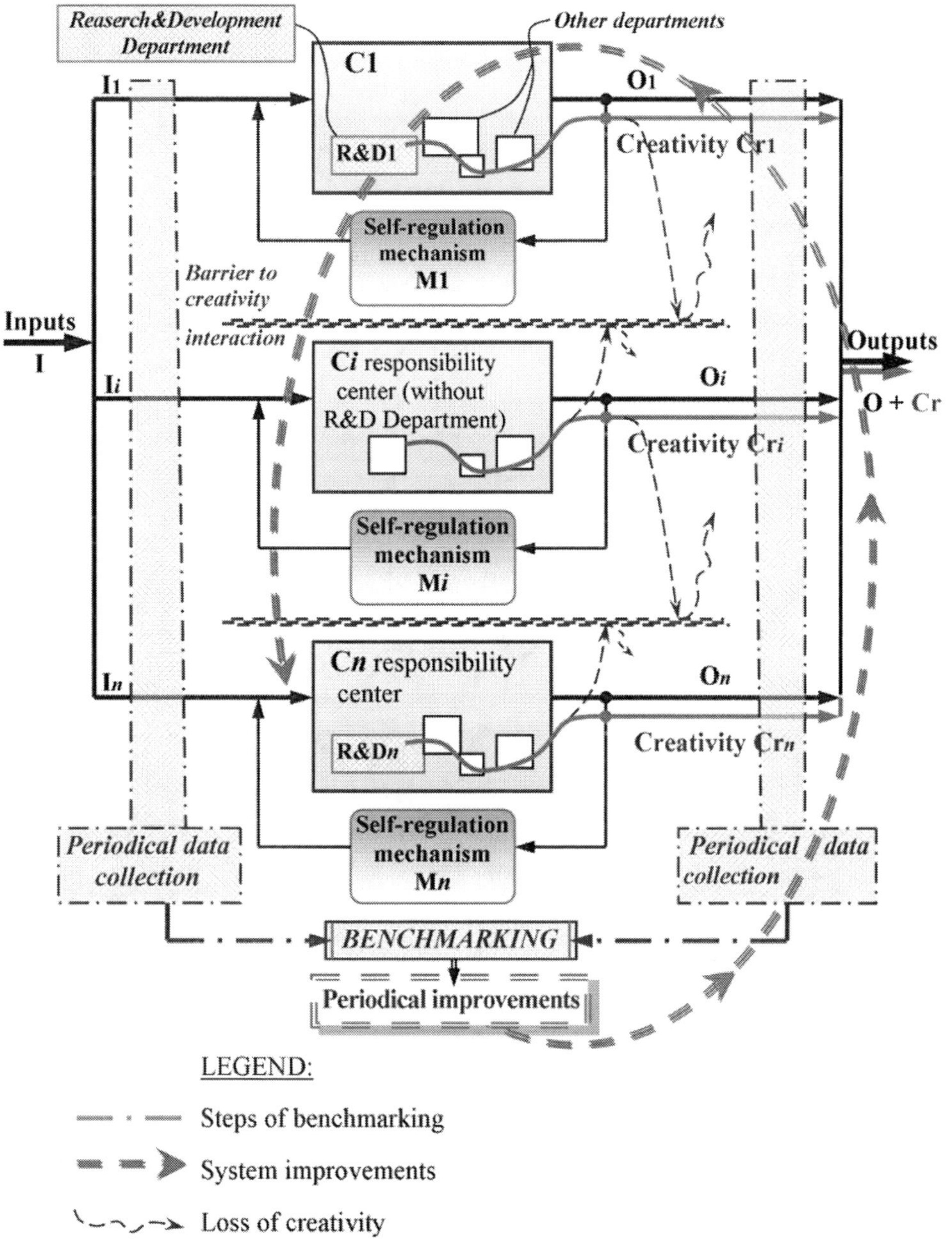

Fig. 1. Benchmarking in a company that is organized in responsibility centers

Another principle necessary for the proper functioning of management centers needs to be observed: correctly motivating the employees, within the framework of a partially decentralized human resource management. Costache Rusu and Monica Voicu point out: "The methods of motivation within the management center vary and depend largely on the competence and creativity of the one who devises them" [2]. This reality holds one of the most significant strengths of the organization around management centers, which can also constitute one of its most important weaknesses: getting the most out of individual creativity by every manager, *with* or *without* the interaction of these individual creativities at group level [5]. The system is successful in the long run only if all managers creatively work together, destroying the barriers between the centers (Fig. 1).

3 Study in Romanian companies

In order to discover the causes who determine the Romanian managers to reject the firm reorganization in responsibility centers, we firstly made a focus-group with 6 members of the staff from large and medium companies/enterprises and 3 academics, specialists in finance-accounting and/or management. During debate, we selected the most frequently mentioned causes of the problem. These 25 causes are represented in the Ishikawa diagram in Figure 2. They were grouped by the 9 participants to the focus-group in 6 main categories: materials, methods, people, environment, equipments and money. The participants assessed the severity of these causes with 1 point for a low severity, 3 points for a medium severity and 9 points for a high severity (see Table 1).

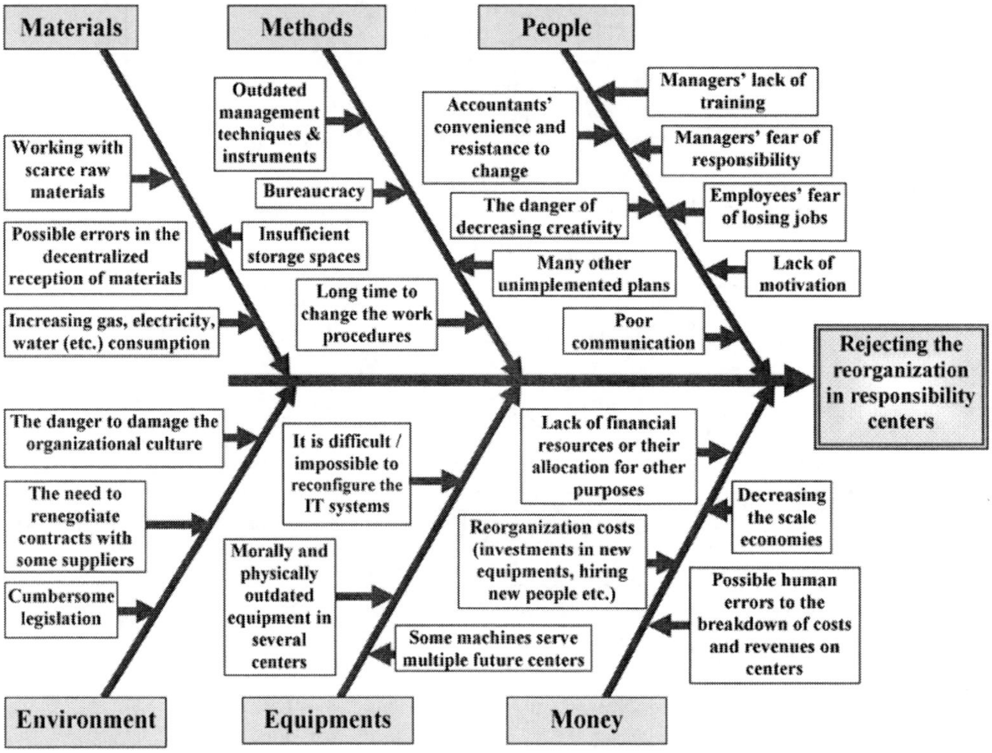

Fig. 2. Causes of rejecting the firm reorganization in responsibility centers – the Ishikawa diagram

Secondly, we gave questionnaires to 33 managers and specialists in various large and medium enterprises in Romanian industry. We asked respondents to mark any of the 25 issues in the questionnaire, if they consider them potential causes of the problem. We presented in table 1 the scores accorded by the respondents to the analyzed causes of the problem, in descending order. Using these results, we built the Pareto chart of the main barriers that prevent the organization of Romanian firms in responsibility centers (Fig. 3).
Thirdly, we proposed the following measures for overcoming the first 6 barriers: achieving a sustainable financial plan for at least 3 years, clarifying to the employees their importance in all changes made in the company and the management intention for keeping them in the firm, making periodical training for managers, finding ways to respect deadlines and

developing of collaborative programs between centers and departments for enhancing the creativity in the firm.

Table 1. Scores accorded by the respondents to the causes of problem.

Code	Cause	Frequency	Severity	Score	Percentage	Cumulative percentage
Money 1	Reorganization costs (new equipments, hiring people etc.)	32	9	288	19.20%	19.20%
People 1	Employees' fear of losing jobs	29	9	261	17.40%	36.60%
People 2	Managers' lack of training	22	9	198	13.20%	49.80%
Money 2	Lack of financial resources or their allocation for other purposes	19	9	171	11.40%	61.20%
Methods 1	Many other unimplemented plans	18	9	162	10.80%	72.00%
People 3	The danger of decreasing creativity	12	9	108	7.20%	79.20%
Methods 2	Outdated management techniques & instruments	18	3	54	3.60%	82.80%
People 4	Accountants' convenience and resistance to change	14	3	42	2.80%	85.60%
People 5	Lack of motivation	12	3	36	2.40%	88.00%
Equipments 1	Morally and physically outdated equipment in several centers	3	9	27	1.80%	89.80%
Environment1	Cumbersome legislation	21	1	21	1.40%	91.20%
People 6	Managers' fear of responsibility	2	9	18	1.20%	92.40%
Methods 3	Bureaucracy	17	1	17	1.13%	93.53%
Equipments 2	It is difficult or impossible to reconfigure the IT systems	5	3	15	1.00%	94.53%
Materials 1	Insufficient storage spaces	4	3	12	0.80%	95.33%
Methods 4	It is difficult to change the work procedures	12	1	12	0.80%	96.13%
People 7	Poor communication	11	1	11	0.73%	96.87%
Environment2	The need to renegotiate contracts with some suppliers	3	3	9	0.60%	97.47%
Money 3	Decreasing the scale economies	3	3	9	0.60%	98.07%
Materials 2	Possible errors in the decentralized reception of materials	2	3	6	0.40%	98.47%
Materials 3	Working with scarce raw materials	2	3	6	0.40%	98.87%
Equipments 3	Some machines serve multiple future centers	6	1	6	0.40%	99.27%
Money 4	Human errors to the breakdown of costs and revenues on centers	2	3	6	0.40%	99.67%
Environment3	The danger to damage the organizational culture	1	3	3	0.20%	99.87%
Materials 4	Increasing gas, electricity, water (etc.) consumption	2	1	2	0.13%	100.00%

DOI: 10.1051/matecconf/201712107013

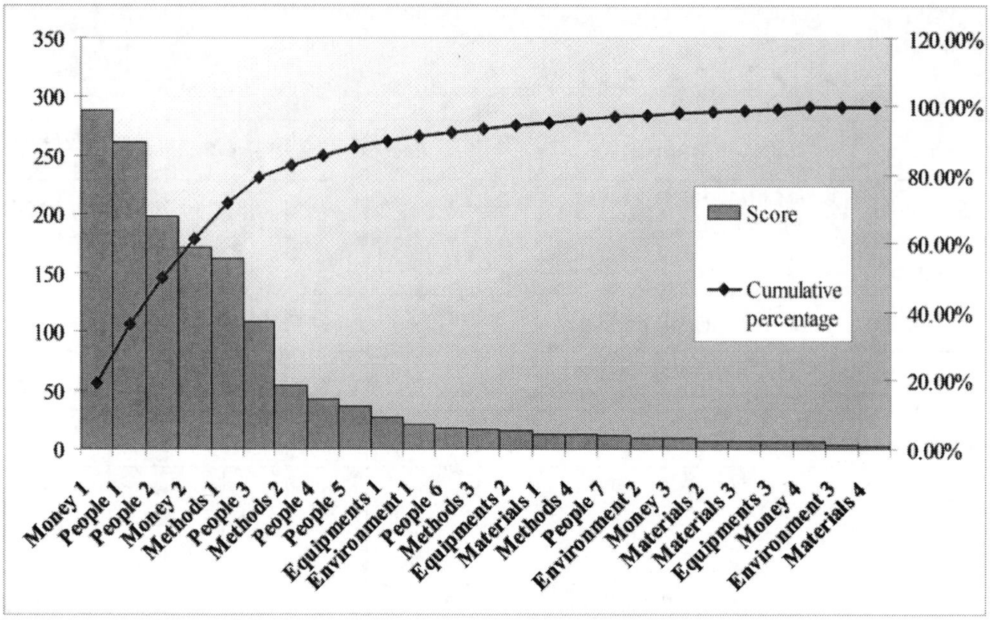

Fig. 3. Barriers in organizing Romanian firms in responsibility centers – the Pareto chart

4 Conclusions

One of the organizational forms that can stimulate the performances of the big and medium enterprises is the one in responsibility centers. The most difficult to overcome barriers for reorganizing the Romanian firms in responsibility centers are: the costs of this process, the employees' fear of losing their jobs, the managers' lack of training, the lack of financial resources or their allocation for other purposes, the existence of other unimplemented plans and the danger of decreasing creativity in the firm.

In order to reorganize the big and medium companies in responsibility centers, managers have to:

- Achieve a sustainable financial plan for at least 3 years;
- Clarify to the employees their importance in all organizational changes and the management intention for keeping them in the firm;
- Make periodical training for managers, in accounting and other economic fields;
- Find ways to respect the deadlines for all ongoing programs;
- Implement collaborative programs between centers and departments for enhancing the creativity in the firm.

The future research regarding the possibilities of reorganizing the big and medium firms in responsibility centers will focus on several factors that customize this process from one case to another [4]:

- The size and complexity of the company and the specificity of the processes, according to its industrial field;
- The components of the strategy and their correlation with the interests of owners;
- The financial situation of the organization in correlation with the technical endowment (of the firm and of each department) and with the intellectual capital [6];
- The organizational culture, its elements and availability for changes;
- The complexity and characteristics of human resources, including their professional aspirations.

This work was supported by the grant LBUS-IRG-2016-02, co-financed by the Lucian Blaga University of Sibiu, Romania.

References

1. O. Nicolescu, I. Plumb et al. (coordinators), F. Badea (author of §1.3), *Abordari moderne în managementul si economia organizatie* **2**, §1.3, 39-57 (Ed. Economica, Bucharest, 2003)
2. C. Rusu, M. Voicu, *Managementul pe baza centrelor de responsabilitate* (Ed. Economica, 2001)
3. H. Bouquin, *Comtabilite de gestion* (Economica, Paris, 2004)
4. C. Obreja, Annals of "Stefan cel Mare" Univ. Suceava. Fasc. Faculty of Ec. & Public Adm. **8**, 162-169 (2008)
5. G.M. Moraru, *Creativity Management and Benchmarking Promoting Performance and Competitiveness in Modern Organizations*, PhD Thesis, "Lucian Blaga" University of Sibiu (2011, unpublished)
6. L.M. Gogan, A. Draghici, Proceedings of The 6th Int. Conf. MSE 2013, Sibiu, 461- 464 (2013)

DOI: 10.1051/matecconf/201712107014

An integrative approach of the marketing research and benchmarking

Gina-Maria Moraru[1,*]

[1]Lucian Blaga University of Sibiu, Engineering Faculty, Industrial Engineering and Management Department, 4 Emil Cioran St, 550025 Sibiu, Romania

Abstract. The accuracy of the manager's actions in a firm depends, among other things, on the accuracy of his/her information about all processes. At this issue, developing a marketing research is essential, because it provides information that represents the current situation in organization and on the market. Although specialists devote the marketing research exclusively to the organizational marketing function, practice has shown that it can be used in any other function of the company: production, finance, human resources, research and development. Firstly, the paper presents the opportunities to use the marketing research as a management tool in various stages of creative thinking. Secondly, based on a study made from secondary sources of economic literature, the paper draws a parallel between marketing research and benchmarking. Finally, the paper shows that creative benchmarking closes the management - marketing - creativity circle for the benefit of the organization and community.

1 Introduction

The accuracy of the manager's actions in a firm depends, among other things, on the accuracy of his/her information about all processes. At this issue, developing a marketing research is essential, because it provides information that represents the current situation in organization and on the market. Therefore, we consider that marketing research is an efficient tool in the study of the external environment of the company, but also in the study of its internal environment. There are many organizations that have implemented an internal marketing plan, in order to educate, stimulate and lead the workforce to higher levels of performance [1]. Implementing the internal customer concept stimulates the employees to encourage the customer to buy again the products or services of the company [2]. Internal customer satisfaction helps the organization to achieve sustainable competitive advantage over other firms [3].

Although specialists devote the marketing research exclusively to the organizational marketing function, practice has shown that it can be used in any other function of the company: production, finance, human resources, research and development [4].

*Corresponding author: gina.moraru@ulbsibiu.ro

2 Marketing research and creative thinking

Many companies in Romania and around the world comply step by step with procedures of different types of marketing research when they:
- Are looking for new ideas for products among stakeholders (research and development).
- Are studying employees' behavior, perception or attitudes towards certain measures of change (human resources).
- Examine possibilities offered by the labor market, to employ staff (human resources).
- Are following the reorganization of production based on modern criteria (the organizational production function).
- Seek to detect all unprofitable activities which should be outsourced (finance).

We will demonstrate how marketing research can support the development of the newest resource of the company: the creativity.

Tony Proctor presents several of the theories of creative thinking [5]. We chose two of the most representative theories, which can use various types of marketing research as management instruments. We sketched in Figure 1 several white ellipses representing the Donald G. Marquis theory of creativity (1969) and four rectangles with Arabic numbers representing the theory of Graham Wallas (1926).

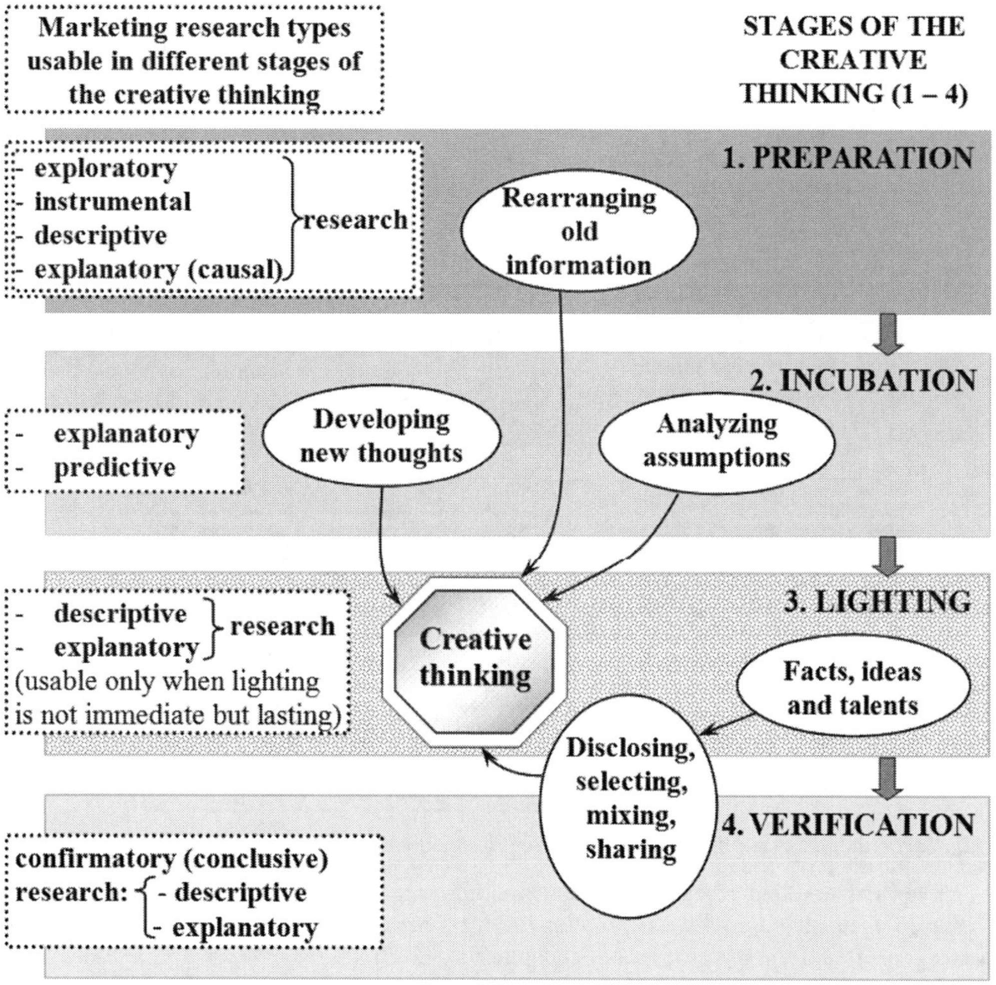

Fig. 1. Marketing research as a management tool in stimulating the creative thinking [4]

We suggest in the left side of Figure 1 the possibility of using different types of marketing research in different stages of the creative thinking. We refer here to the human creative thinking, to the expert system (computer) creative thinking, or to the thinking of a group of specialists aided by a computer system, too.

3 A parallel between marketing research and benchmarking

According to Philip Kotler, the marketing research is structured into four stages: defining the problem and research objectives, developing the research plan, implementing the plan (with phases of collection, processing and analysis of data/information), and the stage of interpretation and reporting of the results [6]. Placing these stages alongside those of benchmarking, we notice that marketing research shows some similarities with benchmarking (Fig. 2).

Benchmarking is the method/process that determines who is the best in a specified field, who set the standards and what these standards are. Many managers have detected benchmarking "as the main tool delivering improvements through various management systems" [7].

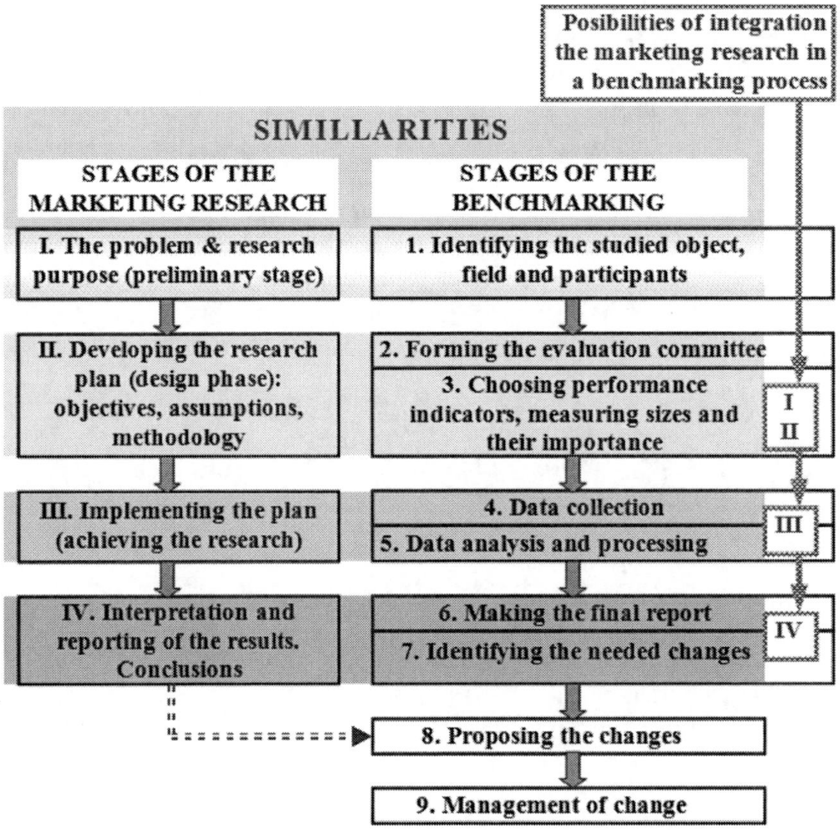

Fig. 2. Possibilities to use the marketing research in a modern benchmarking process

Even if marketing research and benchmarking are different processes, the similarities between their stages allow assimilation of marketing research - as extremely valuable management tool - in the process of benchmarking (Fig. 2).

The right side of Figure 2 has suggested a possible assimilation by Roman numerals indicating the place that you can handle benchmarking stages to ongoing research.

The fact that there is an external and an internal benchmarking, but there also is an internal and an external marketing too, make it even easier the adoption of marketing research as an effective tool in benchmarking processes.

Besides effectiveness, marketing research is recognized as an instrument capable of operating in extremely complex environment to measure various organizational phenomena. There are four essential characteristics of data and values measured through a correctly carried out marketing research, characteristics that gives those data accuracy [8]. These characteristics offer many advantages to the use of marketing research in benchmarking. Thus, a benchmarking that use marketing research as a management tool, compared to a usual benchmarking process, will benefit a larger proportion from:

- The accuracy of the data/sizes with which the process is operated. Random errors of benchmarking decrease or are even canceled.
- The validity of the measurements made. In our opinion, the degree to which the measurement procedure does not generate a systematic or a random error increases.
- The generalization of the results of measurements of the entire population statistical samples. It greatly reduces the cost of benchmarking process if, for instance, not all employees will be surveyed a section, but only a representative sample.
- The sensitivity of the discovered input-output relationships. With the help of marketing research, the benchmarking reveals with a greater precision the factors that actually participate in determining the cause-effect reactions in the organization.

After studying the specialty literature ([9-13]), we proposed a summary of the most important advantage of the creative benchmarking (Fig. 3). Using the marketing research as an instrument, the creative benchmarking has become today a strategic method used more and more often at the national or global level in economy.

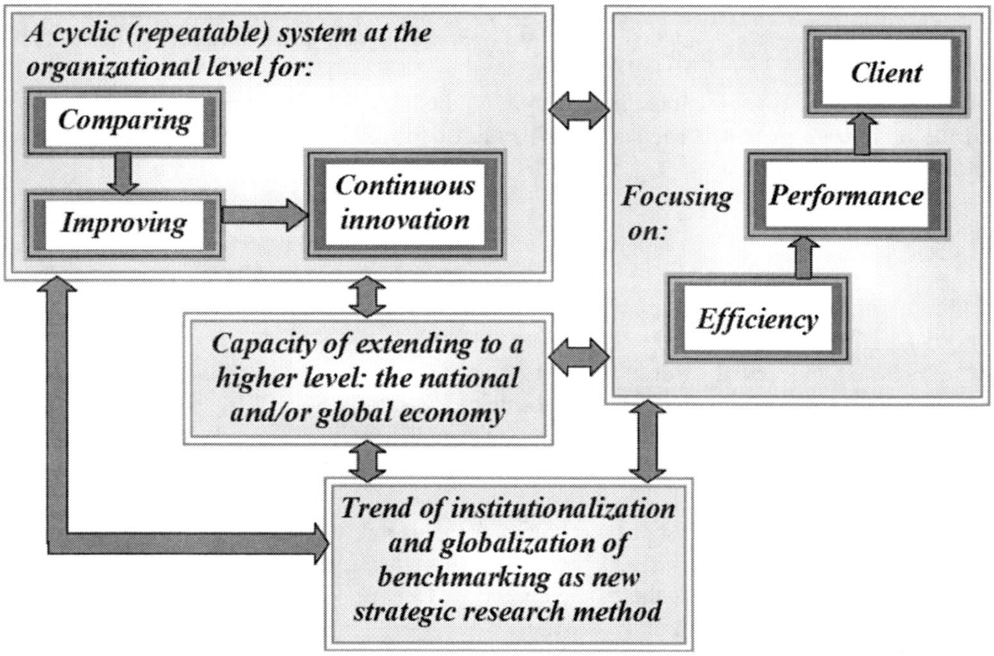

Fig. 3. A summary of the most important advantages of the modern creative benchmarking

A modern creative benchmarking has four essential characteristics: it is selective, has clear objectives, is focused on client and based on the organizational human values [4].

4 Case-study

One of the results of a larger research made in the Romanian academic environment on creativity management and modern benchmarking shows that 31.8% of the academics and 40% of the PhD students in management accept the need for a perfect balance between three of the factors that contribute to the value creation in the company: the creativity management, the marketing concept and the use of the modern management methodology, including the modern benchmarking [4]. These percents are the majority of respondents' opinions.

The study was made on 44 academics and 50 PhD students in management and was based on questionnaires.

Some of the respondents suggested to an open question that an integrative approach of these three elements in company will help the manager to lead more efficiently the teamwork – sustaining other studies presented in the managerial literature [14] – and will sustain the organizational effort for achieving new performances.

22.7% of the academics and 20% of the PhD students accorded more importance to the creativity management than to the other two analyzed factors, but they accepted anyway the necessity of integrating the three mentioned factors. 18.2% of the academics and 20% of the PhD students accorded more importance to the marketing concept, while the same percentages of respondents accorded more importance to the use of the modern management methodology, including the modern benchmarking. 9.1% of the academics accorded more importance to an integrative approach of two from the three above mentioned factors [4].

The respondents' answers at many open questions have suggested the idea that all participants to a creative benchmarking process will find new solutions to improve their activities and processes. However, many respondents noticed many similarities between the marketing research and modern benchmarking, concerning their stages, deployment mode and impact on the organizations. In this context, the increase of efficiency and performance of the organizations that participate to a modern benchmarking process will have positive effect in their community.

5 Conclusions

This paper recommended the use of the marketing research as a management tool in various stages of creative thinking, in order to base the new ideas developed in an organization on real, accurate and actual information. The paper suggested some possibilities to use the marketing research in a modern benchmarking process, too.

The similarities between marketing research and benchmarking allow their integrated use in organization, in order to discover all important and actual information that can lead the firm to performance.

Moreover, the entire society has to win after the use of the new type of benchmarking at a national or global level (Fig. 3). Focusing on efficiency, performance and client, the creative benchmarking closes the management - marketing - creativity circle for the benefit of the organization and community.

This work was supported by the grant LBUS-IRG-2016-02, co-financed by the Lucian Blaga University of Sibiu, Romania.

References

1. S. Mishra, Int. J. Bus. & Mgmt. **5**(1), 185-193 (2010)
2. G. Herrington, W. Lomax, Kingston Bus. School, Occasional Paper Series **34** (1999)
3. K.P.M. Byju, Global J. Mgmt. & Bus. Studies, **3**(5), 519-524 (Research India Pub., 2013)
4. G.M. Moraru, *PhD Thesis: Creativity Management and Benchmarking Promoting Performance and Competitiveness in Modern Organizations*, Lucian Blaga Univiversity of Sibiu (2011, unpublished)
5. T. Proctor, *Elemente de creativitate managerial* (Teora, Bucureşti, 2000)
6. Ph. Kotler, G. Armstrong, *Principiile marketingului* (Teora, 2000)
7. J. Moriarty, C. Smallman, Benchmarking Int. J. **16**(4), 484-503 (Emerald, 2009)
8. I. Catoiu, *Cercetari de marketing,* 171-178 (Uranus, 2002)
9. J. Alexis, *Metoda Taguchi in practica industriala. Planuri de experiente* (Tehnica, 1999)
10. R.L. Ackoff, Systemic Practice & Action Res. J. **6**(6), 581 (1993)
11. O. Nicolescu, *Management comparat* (Ed. Ec., 1997)
12. O. Nicolescu, I. Verboncu, *Management* (Ed. Ec., 1996)
13. M.S. Attiany, J. Bus. Studies Quarterly, **5**(4), 41-51 (2014)
14. D. Miricescu, Rev. Manag. & Economic Eng. **14**(1), 65-77 (2015)

Information support in the assessment of female students as potential entrepreneurs

Paulina Siemieniak[1], and *Jakub* Pawlak[1,*]

[1]Poznań University of Technology, Faculty of Engineering Management, ul. Strzelecka 11, 60-965 Poznań, Poland

Abstract. On world's average, every third business is run by a woman. Typically, these businesses are small and limit their growth, they rarely employ staff, have shorter surviva[l rates, lower level of capitalization, and are financed with foreign capital to a lesser extent in comparison with businesses run by men. In addition, women are motivated to start their own business due to the phenomenon of a "glass ceiling", problems finding a job after returning from parental leave, and the need to balance work and parental duties. Running their own business opens up many opportunities for women. It seems, therefore, appropriate to examine the factors that will help the next generation of women realize their entrepreneurial potential. The article presents the opinions of management students on the usefulness of different measures in supporting the initiation and conduct of business in the SME sector. The results were compared with assessments made by female entrepreneurs. This helped identify areas in which the expectations of students differed significantly from the reality. The authors of this article believe that such a discrepancy between the ideas and the realities of running your own business may be the cause of the lower survival rate of businesses run by women and their limited participation in the overall number of entrepreneurs. Based on these results, the authors propose directions for the education of potential female entrepreneurs.

1 Introduction

The experience of modern market economy shows that small and medium-sized enterprises play a significant role in the development of both regions and countries. Carrying on one's own business is a factor in improving the situation on the labour market, for example through reduction of unemployment. According to the literature on the subject, entrepreneurial determinants include both the personal resources represented by the entrepreneur as well as environmental factors, which tend to be specific to a certain geographic area. As regards entrepreneurs themselves, they include psychological traits, knowledge, skills, and experience. In the case of women's entrepreneurship, particularly significant is the impact of their socio-cultural environment, for example the level of social approval for self-employed women. Entrepreneurial determinants in the economic environment, on the other hand, are those factors which, among other things, fulfil an

* Corresponding author: jakub.pawlak@put.poznan.pl

informational function, such as the local government, funding agencies, research institutions, business incubators or consulting companies.

The situation of women in the labour market is special because of the need to reconcile professional activity with the exercise of their familial role. One of the forms of employment characterized by higher flexibility is setting up and running one's own business. Although the share of self-employed women in Poland is relatively large compared to the rest of Europe, it still does not match that of self-employed men. Many studies point to the existence of numerous restrictions on women's entrepreneurship and the coexistence of its relatively high potential. Developing and utilizing existing capabilities is not only a solution to unemployment, but also enables self-fulfilment of individuals and economic development of the region [1].

Understanding the specific nature of women's entrepreneurship seems to be particularly important due to the fact that the number of women starting their own business has continued to grow in recent decades. Therefore, this trend can be expected not to reverse. However, because female-owned companies are generally smaller and encounter more trouble surviving on the market, the increase in their number may result in the increasingly common problems faced by SMEs [2]. Therefore, this article focuses on one of the factors that support women's entrepreneurship, which is gathering information on starting and running one's own business.

2 The essence of the concept of 'entrepreneurship'

Many definitions of entrepreneurship found in the literature on the subject characterize it as a broad, multi-dimensional, interdisciplinary phenomenon.

Entrepreneurship as a psychological category means those psychological characteristics of an individual which allow them effective implementation of their intentions. They include initiative and vigour, resourcefulness, ingenuity and a high need for achievement. In the psychological sense, entrepreneurship is the result of innate characteristics, such as temperament or intelligence, and characteristics acquired in ontogenetic development, such as character, motivation, and attitude [3]. Entrepreneurship as a feature (or a set of psychological traits) facilitates the taking of initiative in the field of entrepreneurial activities (starting an economic activity), however, it is not a sufficient factor to originate them [4]. This is where the concept of the so-called "latent entrepreneurial potential" comes to the aid. It is created in an individual by five fundamental dimensions, which are the need for achievement, a sense of internal control, creativity, leadership skills, and the ability to resort to one's intuition [5].

Personality traits that allow an individual to implement their intentions are referred to as personal entrepreneurship. Entrepreneurship is often seen as a core competence, which should be owned by every working individual (including workers). However, even from the perspective of competence systems it is very difficult to list all the properties which a person considered to be entrepreneurial should have [6].

Entrepreneurship in the economic sense is based on starting an entrepreneurial activity and taking the associated risks in order to achieve certain, mainly material, benefits. According to one of the first definitions by J. B. Say, an entrepreneur is someone who "transfers economic resources from the lower area to the area of higher productivity and higher yield" [7]. Entrepreneurship is, therefore, the basis for the creation of a company and its development, and is inseparably linked with the desire to make profit. Thus, in this dimension entrepreneurship is a way of action focused on development, successful management of a company, innovation, and expanding to new markets.

Although entrepreneurship as a feature facilitates the taking of initiative in terms of starting and developing a business, it is not a sufficient condition for such behaviour to occur. We must emphasize the difference between an individual's entrepreneurial potential, which is a set of characteristics, and the intention – the desire to become an entrepreneur [5]. Possessing the psychological predispositions to run a successful business does not have to mean that an individual has the motivation to undertake such activities.

3 Knowledge as one of the determinants of women's entrepreneurship

In addition to conditions that apply to all individuals, one can assume the existence of a certain group of specific determinants of women's entrepreneurship. They include knowledge, self-confidence, partnership in sharing household duties, and resolving the issue of institutional child care [2].

Knowledge and education are an important factor influencing entrepreneurship [8]. Research conducted at the University of Szczecin among Polish students and students living in selected countries of Central and Eastern Europe shows that discouragement arises mainly due to the lack of information on the formal side of starting and running one's own business. Students of economic universities as well as residents of large cities (37%) are more convinced about their knowledge (42.9%) than those studying at other universities (18.5%) and those living in smaller towns (20.8%) [9]. Among entrepreneurs, only a small share participate in courses or training that would enable them to broaden their knowledge and acquire new skills. The vast majority of them sees no need to do so [10].

A source of knowledge, besides formal education, is also the experience in running a business. On the basis of the results of Zięba's research one can draw the conclusion that although women have a higher level of education, they lack knowledge and experience in business. As a result, companies run by women have a lower survival rate [11].

The example of parents who are business owners, or at least help in running a family business, allows to gain experience and has a positive impact on the attitude of young people towards creating their own business. One could even talk about the existence of a relationship between setting up their own business and the mother's or father's entrepreneurial activity. About half of all entrepreneurial attitudes originate in the home [5]. According to the results of research by students of the Poznan University of Technology, experience in running a business is acquired more often by young men than women [12]. Men are, furthermore, more confident about their own business (47%) than their female counterparts (23.3%). Interestingly, male students list personal qualities as the most important factor in running a business (47.1%), while female students consider it to be experience (43.3%) [12]. It can, therefore, be assumed that the reason for the lower participation of women in entrepreneurship is the lack of faith in their own abilities.

Independent of gender, barriers in running one's own company cited by business owners can be classified into one of the three groups: economical, educational, and social [13]. Barriers to the development of small and medium-sized enterprises vary, depending on the stage of life of the company. Some difficulties will appear at the time of its launch, and other in the maturity phase [14]. The biggest limitations in the start-up phase are:
- Psychological barriers in deciding to start a business
- Insufficient experience and knowledge
- Lack of capital
- Bureaucratic problems with registration

Deficits in the field of information are of particular importance in the development phase of the company. The barriers experienced by SME owners in the growth phase of their business include [14]:
- Lack of knowledge about the market and the ability to navigate it
- Lack of technological and economic knowledge and of sales skills

- Lack of access to material resources (premises, machines)

At the stage of maturity, entrepreneurs may be restricted by [14]:

- Lack of financial resources for investments
- Changes in the market
- Low profitability
- Inadequate skills
- Limited access to technical infrastructure

Entrepreneurs, regardless of gender, are limited by lack of knowledge, mainly in the field of accounting, bookkeeping, labour law, economics, and civil law [13]. One can, therefore, speak about the existence of educational barriers in running one's own business. This limitation can be overcome through training and a developed advisory system. Limited access to training is one of the elements which hinder the economic activity of enterprising individuals. Entrepreneurs need information that would enable them to overcome the administrative and legal barriers, and in particular information to enable them to deal with problems such as [14] changes in tax legislation, levies imposed by the state (taxes, fixed costs, social and health insurance), access to information and business consultancy, and training and further education. Information support is also helpful in overcoming market barriers (economic barriers such as difficulties in obtaining a loan, growing competition, and a shortage of demand) [13-14].

4 Sources of information in the opinion of female students as potential entrepreneurs on the basis of own research results

The information that entrepreneurs can find on the Internet when using portals for small and medium-sized enterprises relates to financing, technology, laws, principles applicable on the EU market, and consulting bodies providing support for SMEs [16]. Some sites also make it possible to establish cooperation, find an investor, sell real property, etc. An important role in this context is played by portals for businesswomen, stressing the importance of social support (e.g. in the form of mentoring) and transferring contacts between the users of the real reality.

A particular source of knowledge are online discussion boards that allow interactive communication of those involved in setting up or running one's own business [16]. In Poland there are organizations for entrepreneurial women of different ages, with different family situations, education, and income. Each of these institutions has a website containing the information needed to conduct business and to allow the exchange of information. Women can learn from them about training, counselling, empowerment of women in the socio-economic life, returning to work after childbirth, promoting entrepreneurship among mothers of up to three years, and the possibility of using modern technology in conducting their own project [17].

Contemporary Polish economy is evolving towards a knowledge-based economy. Traditional sectors are replaced by the service sector and new technologies, which requires constant upgrading of professional qualifications. In the case of female entrepreneurs with small children, this may be hindered due to lack of time or problems with providing care [2]. Women starting and running a company need knowledge and skills that can be acquired at different stages of their lives, and which relate to many different areas.

At the earliest stage, when the most common entrepreneurial attitudes are usually formed, competencies are acquired through school education and concern two areas: running a business and the industry sector of interest. Initiating and developing an economic activity will depend on the ability to utilize legal advice and various types of training.

In order to determine the importance of information support in promoting women's entrepreneurship, 25 factors were distinguished that led to the creation of two categories:
1) Sources of information as elements of support
2) Areas of consultancy and training in promoting entrepreneurship

Next, a study involving 384 women from the Polish province of Wielkopolska was conducted [1]. Quota sampling of respondents was used, and on the basis of the results obtained in 2013 and 2014 the usefulness of different types of assistance was determined. The results were compared with an assessment of the same factors made by 174 students of two economic majors at the Poznan University of Technology: a) Engineering Management and b) Management. To enhance the representativeness of the results, two seminar groups were randomly selected from each year and tested. The choice of the study group was based on the belief that these female students are a group of potential entrepreneurs because of their knowledge and skills that prepare them to run their own businesses [1].
Below are the factors in information support of entrepreneurship (Table 1):

Table 1. Factors in information support of women's entrepreneurship

Information source	**Area of information**
Competitions promoting entrepreneurship	Legal advice on EU support
Social campaigns promoting entrepreneurship	Legal advice in the field of infrastructure investments
Billboards	Legal advice on state aid
Media (radio, TV)	Legal advice on the knowledge and development of new technologies
Specialized websites	Legal advice on subsidizing jobs
PARP website	Legal advice on tax law
MRR website	Industry-specific training in setting up and running a business
Social networks	Industry-specific training in export strategy
Books	Training courses, for example in implementing new technologies, handling equipment or new trends
Trade press	Training in social skills and business skills
Seminars and conferences on entrepreneurship	Knowledge and skills in the field of running one's own business acquired during school education
Family, friends	Knowledge and skills in the field of their own industry sector acquired during school education
Participation in trade fairs and/or exhibitions	----------------------

Source: Own elaboration.

Both female entrepreneurs and potential female entrepreneurs (students) responded by checking on a 5-point scale, where 1 meant completely useless support and 5 very useful support.

We investigated the motivation of students to start their own business. Almost half (48.3%) declared the desire to start their own business, 14.9% had no such intention, while 36.8% did not know at the time.

The results obtained allowed identification of the factors of greatest importance in the promotion of women's entrepreneurship. At the same time, they enabled a comparison of the students' ideas about running a business against the realities known to business owners.

Industry-specific portals and specialized websites were ranked top by female entrepreneurs in terms of importance in supporting entrepreneurship. In addition, they considered friends and family to be a useful source of knowledge. In the opinion of the female students, the most valuable sources of information are portals and specialized websites. This was followed by media, press and industry-specific books (cf. Table2).

Table 2. Usefulness of sources of information on entrepreneurship. Comparison of ratings by female entrepreneurs and by potential entrepreneurs

Information sources	**Potential entrepreneurs**	**Female entrepreneurs**
	Arithmetic average:	**Arithmetic average:**
Competitions promoting entrepreneurship	3.74	3.33
Social campaigns promoting entrepreneurship	3.82	3.50
Billboards	3.61	3.35
Media (radio, TV)	4.05	3.79
Specialized websites	4.38	4.14
PARP website	3.89	3.79
MRR website	3.89	3.65
Social networks	3.99	3.64
Books	3.66	3.80
Trade press	3.99	4.07
Seminars and conferences on entrepreneurship	3.89	3.79
Family, friends	3.82	4.08
Participation in trade fairs and/or exhibitions	3.80	3.89

Source: Own research.

The students evaluated the suitability of different areas of support higher than female business owners. The entrepreneurs valued most the industry-specific knowledge and skills gained during their school education. The students highly appreciated the role of industry-specific training in establishing and running a company (cf. Table3).

Table 3. Usefulness of areas of information about entrepreneurship. Comparison of ratings by female entrepreneurs and by potential entrepreneurs.

Areas of information	**Potential entrepreneurs**	**Female entrepreneurs**
	Arithmetic average:	**Arithmetic average:**
Legal advice on EU support	4.51	3.79
Legal advice in the field of infrastructure investments	4.26	3.42
Legal advice on state aid	4.06	3.56
Legal advice on the knowledge and development of new technologies	4.34	3.56
Legal advice on subsidizing jobs	4.28	3.79
Legal advice on tax law	4.44	3.90
Industry-specific training in setting up and running a business	4.63	3.91
Industry-specific training in export strategy	4.09	3.44
Training courses, for example in implementing new technologies, handling equipment or new trends	4.38	3.84
Training in social skills and business skills	4.31	3.81
Knowledge and skills in the field of running one's own business acquired during school education	4.17	3.93
Knowledge and skills in the field of their own industry sector acquired during school education	4.22	4.05

Source: Own research.

It is worth noting that the generation of potential entrepreneurs attributed particular importance to the media (Internet, television, radio) as sources offering the most useful information on entrepreneurship (cf. Table 2). Meanwhile, business owners, in addition to knowledge gained from the Internet (industry-specific portals and specialized websites), appreciated their family and friends as a source of knowledge. In addition, female students valued knowledge about starting and running a business, while business owners – knowledge of their own industry sector. Thus, the older and more experienced generation of businesswomen is more focused on professional skills, while students realize the importance of knowledge in the field of starting and running a business.

5 Summary and conclusions

The results of this study reveal the existence of gaps that occur between the assessment of support by students and by female entrepreneurs. The opinions expressed by business owners lead to the conclusion that the functioning forms of support are not as useful as theory or the common sense of potential entrepreneurs would suggest. This discrepancy between expectations and the actual state may result in disappointment and abandonment of entrepreneurship upon confrontation with reality.

Based on the results obtained, one can infer that there is support business owners are not familiar with but which is becoming increasingly popular among the younger generation of students of economics. Lack of knowledge or subjectively understood lack of need may be the reason for not using the help offered by institutions providing information support.

References

1. P. Siemienak, *Metody wspierania przedsiębiorczości kobiet*, Praca doktorska. Politechnika Poznańska (2014)
2. P. Siemieniak, T. Łuczka, *Przedsiębiorczość kobiet. Wybrane aspekty ekonomiczne i psycho-kulturowe,* Poznań, Wydawnictwo Politechniki Poznańskiej (2016)
3. J.F. Terelak, *Psychologia menedżera*, Warszawa, Difin (1999)
4. H. Bieniok, *Kształtowanie postaw przedsiębiorczych młodzieży w rodzinie, szkole i uczelni*, in: *Kształtowanie postaw przedsiębiorczych a edukacja ekonomiczna*, (Ed: P. Wachowiak, M. Dąbrowski, B. Majewski, Fundacja Promocji i Akredytacji Kierunków Ekonomicznych, Warszawa, 2007)
5. R. Athayde, Entrepreneurship. Theory and Practice, Vol. *33*, Iss. *2*, March, pp. 481-500 (2009)
6. A. Marszałek, E-mentor, nr **3** **(40)**, (2011) *http://www.e-mentor.edu.pl/ar-tykul/index/numer/40/id/841* (as assessed on 10.10 2012)
7. P. F. Drucker, *Innowacja i przedsiębiorczość. Praktyka i zasady*, (Warszawa, Państwowe Wydawnictwo Ekonomiczne, p. 30, 1992)
8. T. Łuczka, M. Rembiasz, Horyzonty Wychowania, *15* **(34)**, 27-47. DOI: 10.17399/HW.2016.153402 (2016)
9. T. Bernat, J. Korpysa, M. Kunasz, *Przedsiębiorczość studentów studiów dziennych–raport z badań*, Szczecin, Uniwersytet Szczeciński, Katedra Mikroekonomii (2008)
10. J. Wasilczuk, N. Daszkiewicz, *Polscy właściciele/menedżerowie wobec jednolitego rynku europejskiego*, in: Małe i średnie przedsiębiorstwa w obliczu internacjonalizacji i integracji gospodarek europejskich. Przykłady Włoch, Francji, Polski i Czech, ed: P. Dominiak, J. Wasilczuk, N. Daszkiewicz, Gdańsk, Scientific Publishing Group (2005)
11. K. Zięba, *Płeć właściciela/menedżera jako czynnik kształtujący rozwój małej firmy – aspekty teoretyczne i wyniki badań empirycznych*, in: Strategie zarządzania mikro i małymi przedsiębiorstwami, ed: A. Bielawska, A. Szopa, Zeszyty Naukowe Uniwersytetu Szczecińskiego, **585**, Ekonomiczne Problemy Usług, **50** (2010)
12. P. Siemieniak, Zeszyty Naukowe Uniwersytetu Szczecińskiego, **588**, Ekonomiczne Problemy Usług, **51**, ed: A. Bielawska, T. Łuczka (2010)
13. E. Rollnik-Sadowska, *Przedsiębiorczość kobiet w Polsce*, Warszawa, Difin (2010)
14. A. Kamińska, *Regionalne determinanty rozwoju małych i średnich przedsiębiorstw*, Wydawnictwo, Warszawa, Difin (2011)
15. T. Łuczka, *Makro- i mikroekonomiczne determinanty struktury kapitału w małych i średnich przedsiębiorstwach*, Poznań, Wydawnictwo Politechniki Poznańskiej (2013)
16. M. Rembiasz, P. Siemienak, J. Pawlak, *Internet as a tool to promote women's entrepreneurship. Selected aspects, Research Reports* (to be published) (2016)
17. M. Rembiasz, Zeszyty Naukowe Politechniki Poznańskiej. Organizacja i Zarządzanie **68**, 167-179 (2016)

MATEC Web of Conferences **121**, 08001 (2017) DOI: 10.1051/matecconf/201712108001
MSE 2017

Linear pneumatic actuator

Mihai Avram[1,*], *Constantin* Nițu[1], *Constantin* Bucşan[1], and *Bogdan* Grămescu[1]

[1]Politehnica University of Bucharest, Dpt. of Mechatronics and PrecisionMechanics, Bucharest, Romania

Abstract. The paper presents a linear pneumatic actuator with short working stroke. It consists of a pneumatic motor (a simple stroke cylinder or a membrane chamber), two 2/2 pneumatic distributors "all or nothing" electrically commanded for controlling the intake/outtake flow to/from the active chamber of the motor, a position transducer and a microcontroller. There is also presented the theoretical analysis (mathematical modelling and numerical simulation) accomplished.

1 Introduction

The actual stage and the future development of pneumatics make it a high technology domain. The successes guaranteed by the results of some fundamental and especially applicative "fascinating" researches [1], in the direction of diminishing the negative effects produced by the physical properties of the working fluid: low viscosity and high compressibility. These are the main difficulties in building high performance pneumatic systems.

The informatisation of the pneumatics systems is an important qualitative step ahead for the domain [2]. The concept of pneutronics was developed by the synergetic joining of three domains: pneumatics-electronics-informatics. So, pneutronic actuating systems were born.

The configuration of such a system may vary from simple actuating circuits to complex structures, controlled with programmable automatons or computers. The basic elements of such a system are the proportional pneumatic servo-systems [3 - 5]. In the structure of the system are also included sensors and transducers, electronic circuits for signal processing, A/D and D/A converters, controllers or microprocessors.

Unfortunately these systems are very expensive and they are rarely used. This is the reason why sustained efforts are made in order to reduce the costs [6, 7].

In this paper a pneumatic linear actuator is proposed. It is in fact a pneutronic positioning system that does not use proportional devices.

2 The structure of the intelligent pneumatic actuator

Figure 1 shows the original principle scheme of the actuator to be developed. It consists of the following elements:

*Corresponding author: mavram02@yahoo.com

BPA – the pneumatic supply block;
MPL – the short stroke pneumatic motor;
BEC – the electronic command block;
Tr – the position transducer.

The pneumatic supply block *BPA* delivers into the active chamber of the motor a pressure proportional with an electric command signal: $P_c = f(x_c)$.
For the short stroke pneumatic motor one can choose between special short stroke motor, a chamber with a membrane or a corrugated tube.

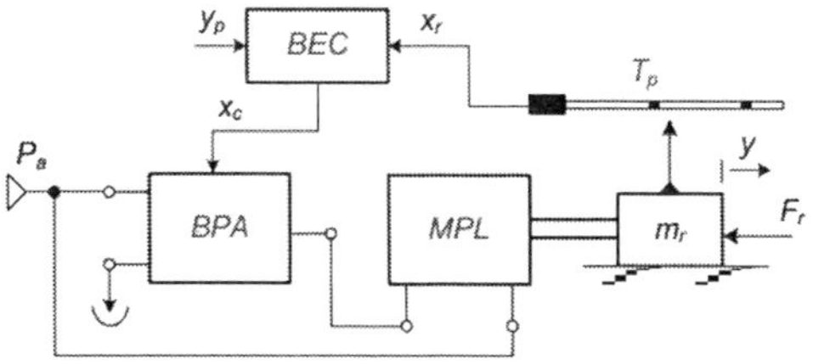

Fig. 1. The original principle scheme of the actuator.

The functional scheme of the actuator is shown in figure 2.

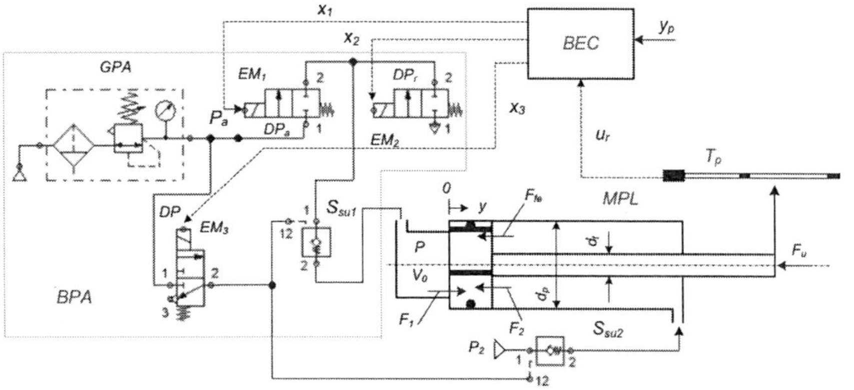

Fig. 2. The functional scheme of the actuator.

The pneumatic supply block consists of the following devices:

- the 2/2 pneumatic control valves DP_a and DP_e, with preferential position, electrically controlled;
- two one-way valves, relievable, S_{su1} and S_{su2}, pneumatically controlled;
- the 3/3 pneumatic control valve DC, with preferential position, electrically controlled;
- the air supply group.

The control valves DP_a and DP_e connect the active chamber of the rod less pneumatic linear motor MPL to the pneumatic supply or to the atmosphere; the chamber has the volume V.

The rod chamber of the motor is permanently connected to the pressure supply. When the control valve DP_a is commanded the pressure P increases; when the pressure force F_1 becomes greater than the force $F_2 + F_{fe}$ (the pressure force and the elastic friction force respectively), the mobile assembly of the motorstartsmoving. The position transducer Tp gives information on the position of the mobile assembly to the electronic command block BEC; at this level, the actual position is in real time compared with the programmed position; according to the result of this comparison, digital command signals are or are not generated for the control valves DP_a and DP_e.

The logic of these signals is the following:

- if $y < y_p - \varepsilon$ then $x_3 = 1, x_1 = 1$ and $x_2 = 0$;
- if $y_p - \varepsilon \le y \le y_p + \varepsilon$ then $x_3 = 0$, $x_1 = 0$ and $x_2 = 0$;
- if $y > y_p + \varepsilon$ then $x_3 = 1$, $x_1 = 0$ and $x_2 = 1$,

where: *"1"* corresponds to the case when the electromagnet of the control valve (DP, DP_a or DP_e) is commanded, and *"0"*when the electromagnetic not commanded.

3 The theoretical analysis of the intelligent pneumatic actuator

The static and dynamic performances of a system can be determined theoretically or experimentally. In this stage, without a physical model of the actuator, only a theoretical analysis can be performed.

Taking into consideration the functional scheme and imposing the maximum stroke, the useful force and the supply pressure, the devices composing the system can be pre-dimensioned and then they can be chosen from the producers catalogues.

The next step is the elaboration of the mathematical model and then the simulation of the system functioning. So, the first information on the dynamic behavior of the system is obtained.

Mathematical modeling and numerical simulation are laborious; require knowledge from many domains besides pneumatics (numerical analysis, automated systems theory, computer programming, informatics, optimizing methods etc.).

A pneumatic system is characterized by a large number of variables and interactions that are difficult to quantify, so a rational and systematic approach strategy for modeling and simulation is required.

The mathematical model of a pneumatic system is highly non-linear [8], [9], [10], and the numerical simulation is used to determine the capability of the system to perform a given task, in terms of dynamic behavior.

The equations of the mathematical model are the following:

a. The mobile assembly moving equation is given by:

$$m_r \cdot \frac{d^2y}{dt^2} = F_1 - F_2 - F_{fe} - F_0 - F_u \tag{1}$$

where:

$$F_1 = P \cdot S_1$$

$$F_2 = P_2 \cdot S_2$$

$$F_0 = P_0 \cdot S_t$$

$$F_{fe} = \mu \cdot \pi \cdot d_p \cdot b_p \cdot (P_2 - P) = k_1 \cdot (P_2 - P)$$

$$S_1 = \frac{\pi}{4} \cdot d_p^2$$

$$S_2 = \frac{\pi}{4} \cdot \left(d_p^2 - d_t^2\right)$$

$$S_t = \frac{\pi}{4} \cdot d_t^2$$

In these conditions, equation (1) can be written:

$$\frac{dv}{dt} = \frac{1}{m_r} \cdot \{(S_1 + k_1) \cdot P - [(S_2 + k_1) \cdot P_2 + S_t \cdot P_0 \cdot + F_u]\} \qquad (2)$$

where:

$$\frac{\mathrm{dy}}{\mathrm{dt}} = \mathrm{v} \qquad (3)$$

b. The pressure P differential equation is given by:

$$\frac{\mathrm{dP}}{\mathrm{dt}} = \frac{\chi \cdot \mathrm{R} \cdot \mathrm{T_a}}{\mathrm{V_0} + \mathrm{S_1} \cdot \mathrm{y}} \cdot \dot{\mathrm{m}} - \frac{\chi \cdot \mathrm{P} \cdot \mathrm{S_1}}{\mathrm{V_0} + \mathrm{S_1} \cdot \mathrm{y}} \cdot \mathrm{v} \qquad (4)$$

where:

$$\dot{\mathrm{m}} = \frac{\mathrm{K} \cdot \mathrm{P_a} \cdot \mathrm{S_n}}{\sqrt{\mathrm{T_a}}} \cdot \mathrm{N}\left(\frac{\mathrm{P}}{\mathrm{P_a}}\right) \qquad (5)$$

$$\mathrm{N}\left(\frac{\mathrm{P}}{\mathrm{P_a}}\right) = \begin{cases} 1 & \text{if } 0 \le \mathrm{P}/\mathrm{P_a} \le 0{,}528 \\ \mathrm{a} \cdot \left[\left(\frac{\mathrm{P}}{\mathrm{P_a}}\right)^{\frac{2}{\chi}} - \left(\frac{\mathrm{P}}{\mathrm{P_a}}\right)^{\frac{(\chi+1)}{\chi}}\right]^{\frac{1}{2}} & \text{if } \frac{\mathrm{P}}{\mathrm{P_a}} > 0{,}528 \end{cases} \qquad (6)$$

where the coefficients are given by [1]:

$$\chi = 1{,}4\ [-]$$

$$\mathrm{R} = 287.04 \left[\frac{\mathrm{m}^2}{\mathrm{s}^2 \cdot \mathrm{K}}\right]$$

$$\mathrm{K} = 0.04042 \left[\frac{\sqrt{\mathrm{K}} \cdot \mathrm{s}}{\mathrm{m}}\right]$$

$$\mathrm{a} = 2.6143\ [-]$$

When functioning, the following cases can be found:

1. If $R = (S_1 + k_1) \cdot P - [(S_2 + k_1) \cdot P_2 + S_t \cdot P_0 + F_u] < 0$ than the displacement$y = 0$ and a filling process of a fix volume V_0 takes place, which produces the increase of the pressure in this volume to the value:

$$P_{lim} = \frac{(S_2 + k_1) \cdot P_2 + S_t \cdot P_0 + F_u}{S_1 + k_1} \qquad (7)$$

In these conditions the speed $v = 0$, and the pressure differential equation (4) becomes:

$$\frac{dP}{dt} = \frac{\chi \cdot R \cdot T_a}{V_0} \cdot \dot{m} \qquad (8)$$

The initial pressure in the volume $V_0$is the atmospheric pressure.

1.1. First, the flowing regime through the nominal area$S_n$of the control valve$DP_a$is a sonic one, according to the pressure ratio value: $P \le 0.528 \cdot P_a$. In this case, the mass flow rate is given by:

$$\dot{m} = \frac{K \cdot P_a \cdot S_n}{\sqrt{T_a}},$$

and the pressure differential equation becomes:

$$\frac{dP}{dt} = \frac{\chi \cdot R \cdot T_a}{V_0} \cdot \frac{K \cdot P_a \cdot S_n}{\sqrt{T_a}} = c_1 \cdot P_a \quad (9)$$

where:

$$c_1 = \frac{\chi \cdot R \cdot \sqrt{T_a} \cdot K \cdot S_n}{V_0} [s^{-1}]. \quad (10)$$

Integrating equation (9) the pressure is obtained:

$$P = c_1 \cdot P_a \cdot t + c_2$$

The integration constant$c_2$is found from the condition that at$t = 0$the pressure in the volumeV_0 is $P = P_0$and its value is$c_2 = P_0$.
So, the pressure is given by:

$$P = c_1 \cdot P_a \cdot t + P_0 \quad (11)$$

The pressure in the volume V_0will reach the critical value $P_{crt} = 0.528 \cdot P_a$(corresponding to the end of the sonic flowing regime) at the time:

$$t_{crt} = \frac{0.528 \cdot P_a - P_0}{c_1 \cdot P_a} [s] \quad (12)$$

1.2. From this point the flowing regime becomes subsonic. The equation (8) becomes:

$$\frac{dP}{dt} = \frac{\chi \cdot R \cdot T_a}{V_0} \cdot \frac{K \cdot P_a \cdot S_n}{\sqrt{T_a}} \cdot a \cdot \left[\left(\frac{P}{P_a}\right)^{2/\chi} - \left(\frac{P}{P_a}\right)^{(\chi+1)/\chi}\right]^{1/2}$$

This equation can be written as:

$$\frac{dr}{dt} = c_3 \cdot \left[r^{2/\chi} - r^{(\chi+1)/\chi}\right]^{1/2} \quad (13)$$

where:

$$c_3 = \frac{\chi \cdot R \cdot \sqrt{T_a} \cdot K \cdot S_n}{V_0} \cdot a = c_1 \cdot a \quad (14)$$

$$r = \frac{P}{P_a}$$

Equation (12) can be written:

$$\frac{dr}{r^{1/\chi} \cdot \sqrt{1 - r^{(\chi-1)/\chi}}} = c_3 \cdot dt$$

Using the substitution $1 - r^{\frac{(\chi-1)}{\chi}} = z^2$ the equation can be integrated and the solution is given by:

$$z = -\frac{\chi-1}{2\cdot\chi}\cdot c_3\cdot t + c_4,$$

from which the value of the pressure is obtained:

$$P = \left[1-\left(c_4-\frac{\chi-1}{2\cdot\chi}\cdot c_3\cdot t\right)^2\right]^{\frac{\chi}{\chi-1}}\cdot P_a \quad (15)$$

The constant$c_4$is obtained from the condition that at $t = t_{crt}$ the pressure in the volume V_0 is $P = P_{crt}$ and its value is given by:

$$c_4 = \sqrt{1-0.528^{\frac{\chi-1}{\chi}}} + \frac{\chi-1}{2\cdot\chi}\cdot a\cdot\left(0.528-\frac{P_0}{P_a}\right) \quad (16)$$

The pressure P has the valueP_{lim}at $t = t_{lim}$. Imposing the condition for these values to check the equation (15), the equation of $t_{lim}$is obtained:

$$t_{lim} = \frac{2\cdot\chi}{(\chi-1)\cdot c_3}\cdot\left[c_4-\sqrt{1-\left(\frac{P_{lim}}{P_a}\right)^{\frac{\chi-1}{\chi}}}\right][s] \quad (17)$$

2. If $R > 0$ then the mobile assembly moves and a filling process of a variable volume V takes place. In this case the dynamic behavior of the system is described by the mathematical model consisting of equations (2), (3), (4), (5) and (6).

The numerical simulation of the system functioning was performed using Matlab Simulink.

3. The following constructive and functional parameters values where imposed:

$d_p = 40\ [mm]$, $d_t = 28\ [mm]$, $b_p = 4\ [mm]$, $l_0 = 20\ [mm]$, $\mu = 0{,}3[-]$, $F_u = 100\ [N]$

$d_n = 3\ m[m]$, $P_a = P_2 = 4\ [bar]$, $T_a = 293{,}15\ [K]$.

The programmed position value was imposed$y_p = 200\ mm$.

Figure 3 shows the simulation model. Running the model, the following characteristics were obtained:

- $P = P(t)$ – figures 4a and 4b;
- $y = y(t)$ – figures 4c and 4d;
- $\dot{y} = \dot{y}(t)$ – figures 4e and 4f.

The effective start moment is$t = t_{lim} = 0.0102$ s, as shown in the diagrams from figures 4.a, 4.c and 4.e. In this moment the pressure in the chamber becomes $P = P_{lim} = 3.5\ bar$.

When the distance to the programmed position reaches a pre-settled value, the command of the control valve DP_a is cancelled, the air supply of the volume V stops, and the mobile assembly continue to move due to the inertia.

The speed of the mobile assembly decreases to the value$v = v_f = 0.1 m/s$. When the condition$y \geq y_p - \varepsilon$is fulfilled, the movement stops, due to a braking system (not discussed until now).

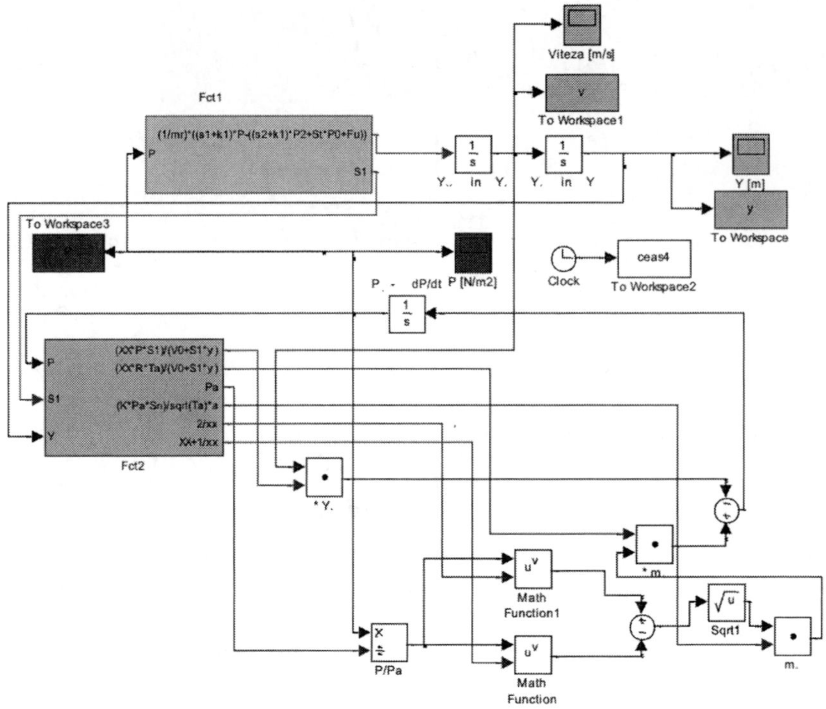

Fig. 3. The simulation model.

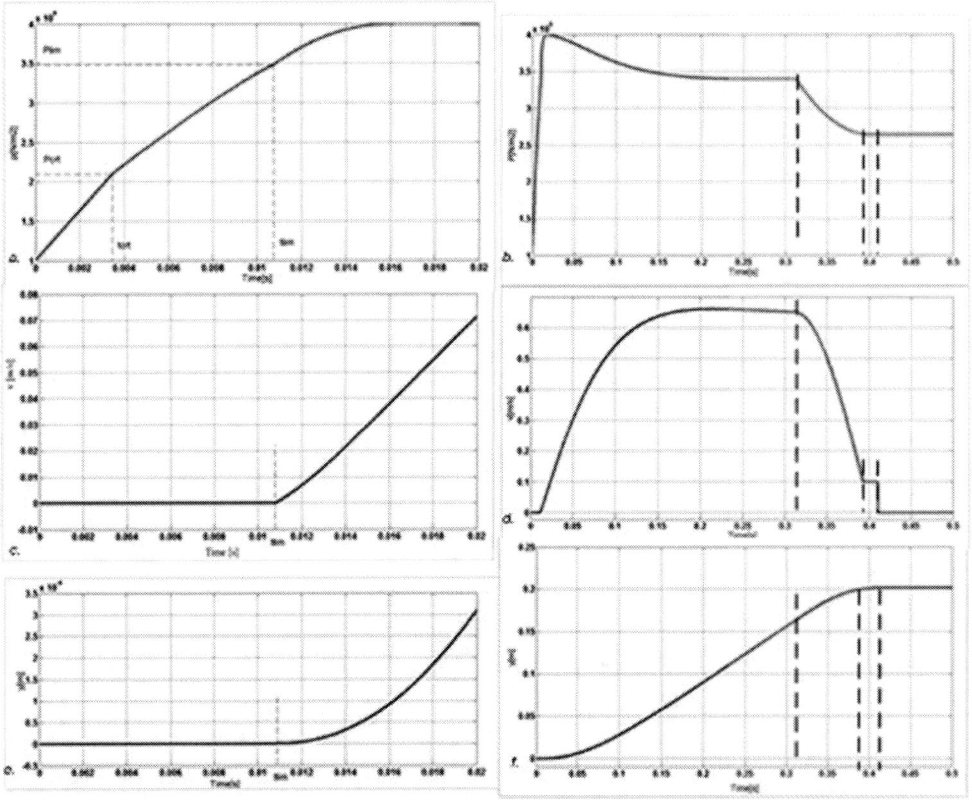

Fig. 4. The characteristics obtained by running the model.

4 Conclusions

The proposed pneumatic linear actuator allows the positioning of a pay load in any point of the working stroke with an imposed accuracy. The structure of the actuator is original, very simple, using common devices, most of them being standardized. Using two 2/2 control valves, "all or nothing" electrically controlled, an accurate control of the pressure in the working active volume can be obtained and the desired position can be reached.

The theoretical analysis of the system, performed on the basis of the developed mathematical model, confirmed the anticipated behavior of the actuator, according to the results obtained by integrating the model using the graphical programming environment Matlab Simulink.

The future approach will be the construction of the experimental model of the actuator. The command system will use a microcontroller and the working programs will integrate different variants of command and control algorithms. The results will be presented in a future paper.

References

1. G. Vogel, E. Mühlberger, *L'univers fascinant de la pneumatique* (Würzburg Vogel, 2001)
2. M. Avram, C. Bucsan, *Sisteme de actionare pneumatic inteligente* (Ed. Politehnica PRESS, Bucuresti, 2014)
3. M. Avram, D. Duminica, C. Udrea, V. Gheorghe, *Hidronică şi pneutronică - Aplicaţii* (Ed. Universitară, Bucureşti, 2008)
4. M. Avram, *Acţionări hidraulice şi pneumatice – Echipamente şi sisteme clasice şi mecatronice*, (Ed.Universitară, Bucureşti, 2005)
5. G. Belforte, A.M. Bertetto, L. Mazza, *Pneumatico – curso completo* (Ed. Technichenuove, Milano, 1998)
6. K. Ahn, S. Yokota, Mechatronics, **15**(6), 683 (2005)
7. R. B. Van Varseveld, G.M. Bone, IEEE/ASME Transactions of Mechatronics **2**(3), 195 (1997)
8. C. Valdiero, C.S. Ritter, C.F. Rios, M. Rafikov, Proceedings of the 9th Brazilian Conference on Dynamics Control and their Applications, *Serra Negra, Brasil,* 198 (2010)
9. Z. Varga, P. Keski-Honkola, **25,** 01096 (available at http://www.epj-conferences.org or http://dx.doi.org/10.1051/epjconf/20122501096, 2012)
10. T. Kimura, S.Hara, T. Fujita, T. Kagawa, Control Engineering Practice **5**(10), 1385 (1997)

MATEC Web of Conferences **121**, 08002 (2017) DOI: 10.1051/matecconf/201712108002
MSE 2017

Specialised sensorial block

Mihai Avram[1,*], *Constantin* Bucsan[1], and *Tudor Cătălin* Apostolescu[2]

[1]Politehnica University of Bucharest, Department of Mechatronics and Precision Mechanics, SplaiulIndependentei nr. 313, Bucharest, Romania
[2]Titu Maiorescu University of Bucharest, Faculty of Informatics, Department of Informatics, CaleaVacaresti nr. 187, Bucharest, Romania

Abstract. In driving systems using working fluids, for monitoring the fluid's parameters such as pressure, flow and temperature appropriate transducers are present in different areas of the system. Such an approach has certain disadvantages with regard to the installation and powering of the transducers as well as the transmission of the signals supplied by the transducers to the system's electronics command block. Both hydraulic and pneumatic driving systems are considered. The paper proposes a sensor block which integrates the above mentioned sensors and eliminates the previously outlined disadvantages. The proposed block can be designed for an easy installation in a driving system containing a pneumatic linear motor; basically for a standardised motor one of the caps is replaced with the proposed block. An example is also shown for the integration of the proposed block in the structure of a pneutronic system.

1 Introduction

Production systems in the future will be based on mechatronic assemblies with autonomous functioning; these are also known as intelligent components. Data processing is increasingly handled at a local level with a higher number of functions being directly integrated into modules. These communicate in a network and can organise and self-configure for receiving commands from the superior control level. These aspects outline a new stage of industrial development, called Industry 4.0, with components defined by 4 „i" properties:

- intuitive operation;
- intelligent: due to functional integration, such as a programmable logical controller with communication abilities;
- internet: local network communication and Internet connection abilities, by using WLAN or industrial Ethernet;
- integrators: modules that log on to the main computer, communicate available capabilities and so their actions can be planned in the production process.

Preconfigured controllers and adaptable interfaces minimise installation times for the system's installation, configuration and integration. Intelligent components have a "plug and produce" capability. This principle is similar to a USB computer interface where

* Corresponding author: mavram02@yahoo.com

connected devices can log on and communicate directly with a computer. In future enterprises, individual components, production subsystems and modules will be able to similarly log in to the production system [1].

In Industry 4.0, smart factories represent a combination of the following components [2, 9]:

- Internet of Services, or Cloud Computing, that allows for the usage of applications, data bases, records, storage and computing power for all those connected to a network;
- Cyber-physical systems, for digital and physical connectivity;
- Internet of Things (IoT) is the most important and active of these components, where connected devices use the Internet without any human interference.

As with all aspects concerning production, hydraulic and pneumatic technologies will also benefit from the Industry 4.0 standards. A study of the Rexroth Bosch Group shows that most controllers for hydraulic equipment are pre-programmed with the necessary software for passing from "path control" to "force control", in order to achieve synchronization on the production line. This gives hydraulic equipment a decentralised intelligence which allows for autonomous operation and adjustment to the changing process parameters [3].

As for pneumatic equipment, FESTO has been simulating process and factory automation for some years now and offers a wide variety of products for control technologies and software by using smart components to simplify and shorten the configuration and operation of engineering processes [1].

As for the development of equipment and hydraulic and pneumatic driving systems a strong trend toward computerisation can be noticed in order to achieve real time system and equipment control using programmable controllers. This obviously involves that the system contain sensor elements for the physical parameters needed in the application. Also, a low energy consumption is required and the possibility to intervene on the functioning system in order to reduce idle times caused by stops.

2 Analysis of the options for integration of sensors in the structure of pneumatic and hydraulic equipment

Below we have 3 scenarios which led the authors to the idea of developing intelligent sensor modules that integrate more transducers.

Scenario 1 refers to pneumatic/hydraulic systems where different measurements need to be made, such as: pressure, pressure difference, flow and temperature. Transducers need to be placed at several points in the system. A more efficient option would be grouping these transducers into a single module, universally and efficiently developed so that it involves a single power source and a single communication protocol with or without wire connection.

Figure 1 shows a hydraulic positioning system from the Automation and Robotics Lab in the Department of Mechatronics and Precision Mechanics at the Politehnica University of Bucharest [4]. The load positioning is done with micrometric accuracy with a specifically designed software. Real time information is processed from a series of pressure, flow and position sensors which allow for interference compensation. The system is driven by a PC and interfaced with Field Point input/output modules (figure 2). The command program is developed using the graphic programming language Lab VIEW.

This system uses different transducers located along the circuit that are produced by different manufacturers pressure transducer T_P, flow transducer T_q, position transducer $T_{poz,1}$.

The main challenge consists in coupling these transducers to the hydraulic system and the driving system. The existence of a block module that includes all these sensors with a single power supply and communication interface could result in a simpler system design process and an optimal control for the transducers block.

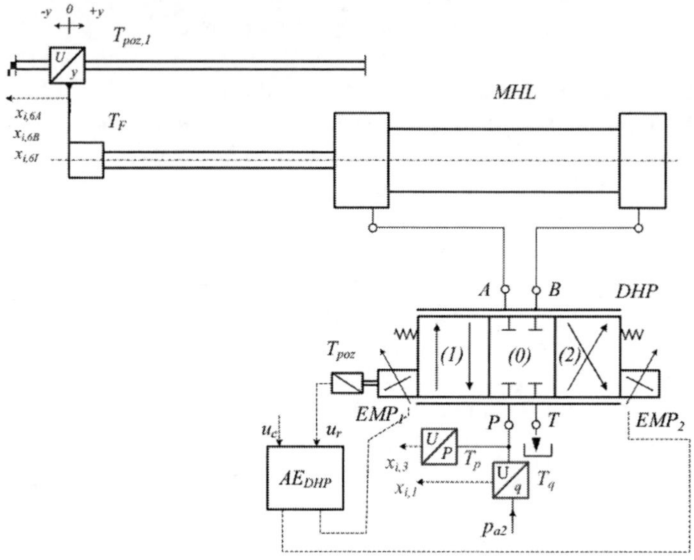

Fig. 1. Hydraulic positioning system.

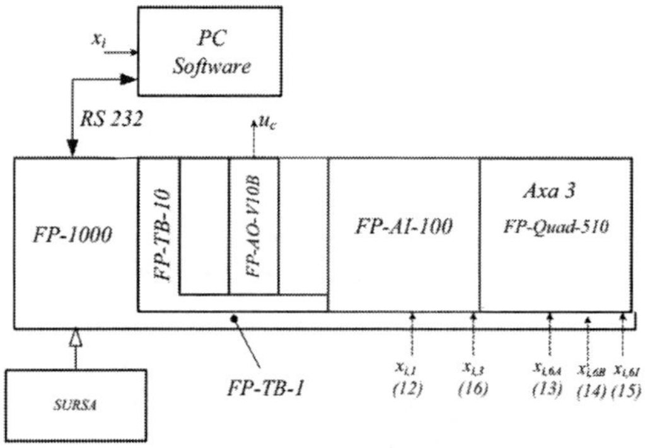

Fig. 2. PC Interfaced with Field Point input/output modules.

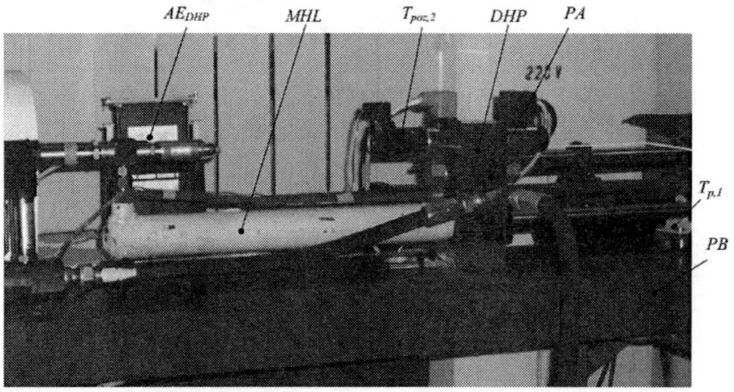

Fig. 3. Experimental model developed by the authors.

Scenario 2 involves proportional equipment. A reoccurring problem for this equipment is represented by hysteresis. The global hysteresis can be significantly reduced by automatic adjustment for the position of the mobile fixture [5]. An inductive position reaction is generally used. A practical example for a proportional distributor could be the one produced by Rexroth (figure 4), this being equipment*4 WREE 6 EA08* [3].

Fig. 4. Rexroth proportional distributor

Similar situations can be encountered for other manufacturers in the area of pneumatic systems such as FESTO and SMC.

Scenario 3 follows the construction of hydraulic motors used in positioning systems. These have position transducers placed on the outside of the motor (resistive, temposonic), or integrated (magneto-strictive). These solutions are available from several manufacturers but the costs are significant and also, they require specialised command modules. Using a different command system than the one imposed by the manufacturers can prove difficult and costly. An example for a hydraulic motor produced by Rexroth [3] is shown in figure 5,with the magneto-strictive transducer 8integrated in the motor's construction and the distribution equipment 7placed on the motor's stator.

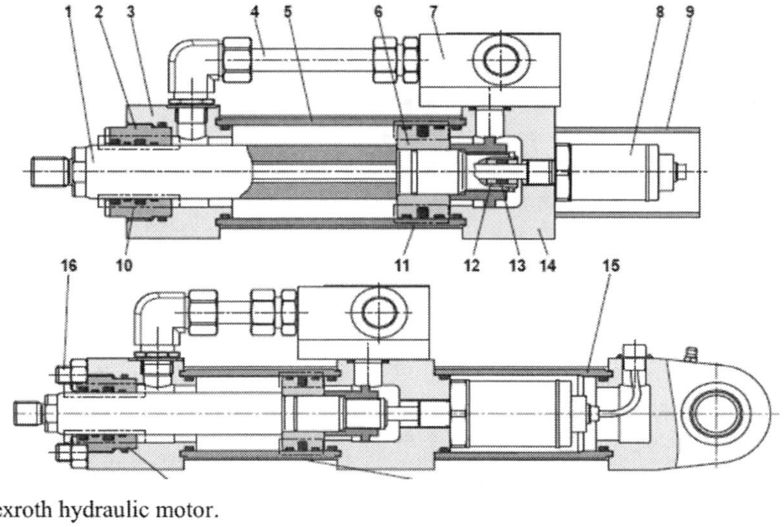

Fig. 5. Rexroth hydraulic motor.

If one wishes to use the motor with a different controller than the one indicated by the manufacturer problems arise regarding the interface and communication protocol. This is another reason for which an intelligent module that can adapt to any motor with a simple, clear and universal interface could prove to be a very efficient solution.

3 Theproposed solution

The authors have developed the structure for an intelligent sensor module, with integration capabilities for most pneumatic equipment and systems with the imposed performance criteria.

Figure 6 shows the basic schematics for the proposed sensor module.

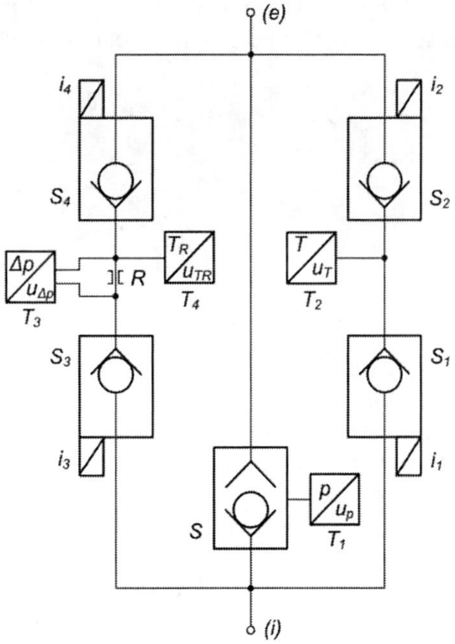

Fig. 6. Schematics for the proposed sensor module.

The sensor block consists of the controlled valves S_1, S_2, S_3 andS_4, that according to the command currents i_1, i_2, $i_3$andi_4, select the work fluid's direction of flow so that its main parameters can be measured: pressure p, temperature T and mass flow $\dot{m}$.

Flow$\dot{m}$is calculated based on values obtained by direct measurement of the pressure difference Δpon the calibrated section$R$and of the fluid's temperature T.

The formula based on which these calculations are done is [6]:

$$\dot{m} = \frac{K \cdot P_a \cdot S_n}{\sqrt{T_a}} \cdot N\left(\frac{P}{P_a}\right) \tag{1}$$

where:

$$P = \begin{cases} P_i \; for\ the\ flow\ direction\ (i) \to (e) \\ P_e \; for\ the\ flow\ direction\ (e) \to (i) \end{cases}$$

$$N\left(\frac{P-\Delta p}{P}\right) = \begin{cases} 1 & if\ 0 \le \frac{P-\Delta p}{P} \le 0{,}528 \\ a \cdot \left[\left(\frac{P}{P_a}\right)^{\frac{2}{\chi}} - \left(\frac{P}{P_a}\right)^{\frac{(\chi+1)}{\chi}}\right]^{\frac{1}{2}} & if\ \frac{P-\Delta p}{P} > 0{,}528 \end{cases} \tag{2}$$

The following relevant coefficients are considered:

$$\chi = 1{,}4\ [-]$$

$$R = 287{,}04 \left[\frac{m^2}{s^2 \cdot K}\right]$$

$$K = 0{,}04042 \left[\frac{\sqrt{K} \cdot s}{m}\right]$$

$$a = 2{,}6143\ [-]$$

So for the flow direction from *(i)* to *(e)*, transducer T_1measures the pressure p_i. If valve S_1 is commanded, the work fluid flows through and transducer T_2 measures the temperature T_i. If valve S_3 is commanded, the work fluid flows through the resistance R and transducer T_3 measures the pressure drop Δp on this resistance, and transducer T_4 measures the fluid's temperature in the resistance's area T_R. Considering these values the flow q for the work fluid can be calculated.

Table 1 shows a synthesis of the system's functioning.

Table 1. Synthesis of system's functioning.

Flow direction	i_1	i_2	i_3	i_4	p	T	$q=f(\Delta p,T_R)$
$(i) \rightarrow (e)$	1	0	0	0	p_i	T_i	-
	0	0	1	0	p_i	-	q_i
$(e) \rightarrow (i)$	0	1	0	0	p_e	T_e	-
	0	0	0	1	p_e	-	q_e

The construction for the controlled valve is shown in figure 7a.

The valve is composed of the body 1, the cover 2, where the valve's seat is machined, the valve 3, solidary with the rigid centre 4 of the cut-out metallic membrane 5, the cylindrical helical spring 6, that presses the valve on the seat, and the electromagnet 7.

The cut-out metallic membrane is shown in figure 7b.

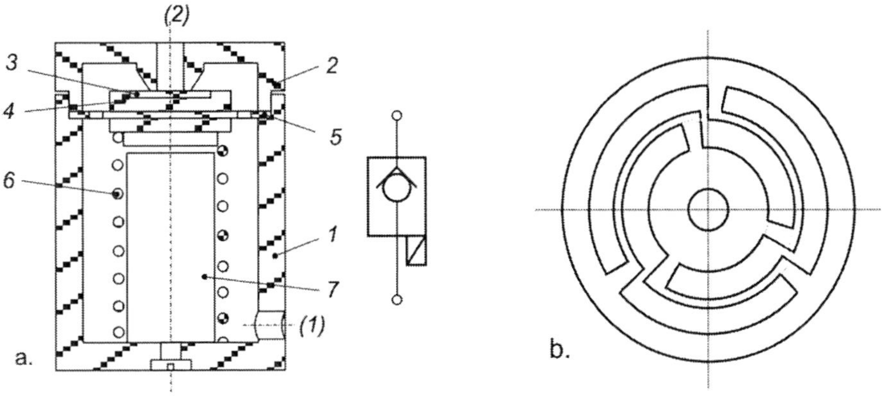

Fig. 7. a. Construction of the controlled valve; b. Cut-out metallic membrane.

If the work fluid enters through hole *(1)*, its pressure acts as the spring in the direction in which the valve presses on the seat, so the controlled valve is closed.

If a command current is applied on the electromagnet this pulls the rigid centre of the membrane overcoming the force in the spring and so the controlled valve is opened allowing the fluid to exit through hole *(2)*.

If the work fluid enters through hole *(2)*, its pressure creates a force that overcomes the force in the spring, the rigid centre of the membrane with its valve will move away from the chair and the fluid will flow through the controlled valve and will exit through hole *(1)*.

Based on the schematics in figure 6 the authors designed the experimental 3D model. The model should be easily integrated in the construction of a linear pneumatic motor where it would replace one of its covers.

4 An example of integrating the sensor block in a pneutronic system

Figure 8 shows a pneutronic system containing the following equipment:

- *MPL* – Linear pneumatic motor with magnetic couplings and integrated position transducer with one of the lids replaced by the proposed sensor block;
- *DPP*–proportional pneumatic distributor;
- *BEC* – electronic command block.

The system allows for multiple functions, such as:

- positioning a load with specific precision at any point of the work run [5];
- data acquisition for functional parameters (pressures, temperatures, mass flows) and their comparison with data obtained from numeric simulation; this ensures verification for the developed mathematical model;
- adjustment of load speed at the desired value and keeping that speed constant even with the resistant force varying over time.

The proposed system has a simple structure, is a multifunctional system and the command electronics is implemented in the system's construction.

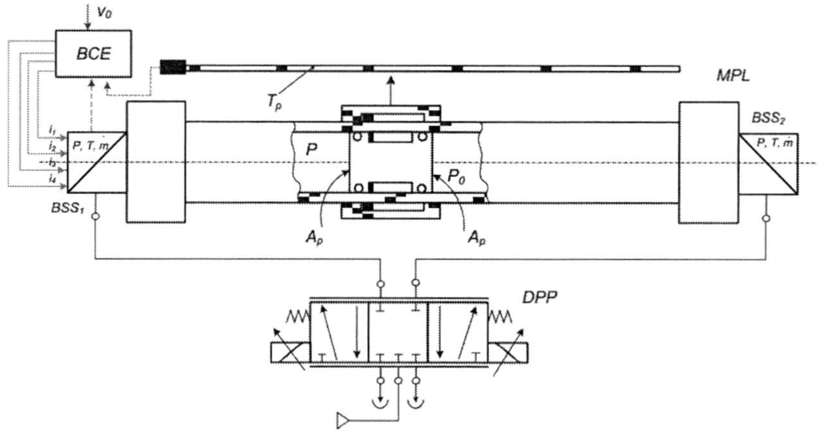

Fig. 8. Pneutronic system.

An example for an algorithm developed for the last function is shown next – load speed adjustment at the desired value and keeping constant that speed even if the resistant force varies over time.

This can be done by modifying in real time the control value i for the proportional pneumatic distributor *DPP* so that the desired speed can be maintained.

In order to determine the necessary correction$\Delta i$the starting point is the mathematical model for the linear pneumatic motor *MPL*. This model consists of the following equations:

- the motion equation of the mobile subassembly:

$$M_r \cdot \frac{dv}{dt} = (P - P_0) \cdot A_p - F(t) \tag{3}$$

where: M_r – reduced mass of the mobile subassembly (piston, mass, load),
$F(t)$ – resisting force.

- the differential equation of the pressure in the motor's active chamber:

$$\frac{dP}{dt} = \frac{\chi \cdot R \cdot T_a}{V_0 + A_p \cdot x} \cdot \dot{m} - \frac{\chi \cdot P \cdot A_p}{V_0 + A_p \cdot x} \cdot v \tag{4}$$

Considering a constant speed movement: $v = v_0 = ct.$ şi $\frac{dv}{dt} = 0$the motion equation becomes:

$$P - P_0 = \frac{F(t)}{A_p}.$$

By deriving:

$$\frac{dP}{dt} = \frac{1}{A_p} \cdot \frac{dF(t)}{dt} \tag{5}$$

By replacing relation (5) in relation (4):

$$\dot{m} = \frac{V_0 + A_p \cdot x}{\chi \cdot R \cdot T_a \cdot A_p} \cdot \frac{dF(t)}{dt} + \frac{P \cdot A_p}{R \cdot T_a} \cdot v \tag{6}$$

Following measurements values are determined at any time $P = P_i$, $T_a = T_i$and position x. Using relation (6) the necessary flow for obtaining the desired speed can be determined $v = v_0$. At the same time by measuring the real mass flow $\dot{m}_r$for powering the active chamber, corrections for the control values of the pneumatic proportional distributor can be established:

$$\Delta i = \frac{|\dot{m} - \dot{m}_r|}{k_{DPP}} \tag{7}$$

Where$k_{DPP}$is the proportionality factor for the proportional pneumatic directional valve.

5 Conclusions

Placing the necessary transducers for monitoring the work fluid parameters (pressure, flow and temperature) for an integrated sensor block in the construction of the execution element eliminates the disadvantages associated with transducer installation, powering and signal transmission from the transducers to the electronic command block for the respective system. The authors have developed a multifunctional driving system containing the proposed sensor module. This system has a particularly simple structure with the command electronic implemented in its build and the command aspects ensured by a controller based electronic system.

References

1. ***www.festo.com, Accesed 18.05.2017
2. ***http://www.hydraulicspneumatics.com/other-technologies/iot-intends-revolutionize-hydraulics-and-pneumatics, Accesed 8.02.2017
3. ***www.boshrexroth.com, Accesed 1.03.2017
4. M. Avram, C.Bucsan, *Sisteme de actionare pneumatice inteligente*, (EdituraPolitehnica PRESS, Bucuresti, 2014)
5. M. Avram, D. Duminica, C. Udrea, V. Gheorghe, *Hidronică şi pneutronică - Aplicatii*, (EdituraUniversitară, Bucuresti, 2008)
6. M. Avram, *Actionări hidraulice si pneumatice - Echipamente si sisteme clasice si mecatronice*, (EdituraUniversitară, Bucuresti, 2005)
7. M. Avram, C. Bucsan, V. Constantin, Buletinul Institutului Politehnic din Iaşi, **LVI(**LX), 1 (2010)
8. S. Hesse, *Blue Digest on Automation, Sensors in Production Engineering, Festo AG & Co,* (Available at: http://www.festo-didactic.com/int-en/seminar-search/?fbid=aW50LmVuLjU1Ny4xNy4yOA&search=download, 2001)
9. G. Belforte, G. Eula, Journal of Control Engineering and Applied Informatics, **14**(4), 70 (2012)

Considerations regarding the use of technological equipment for indexed and continuous multi-axes machining

Radu-Eugen Breaz[1,*], and *Melania* Tera[1]

[1]Lucian Blaga University of Sibiu, Department of Industrial Machines and Equipment, 550025 10 Victoriei, Sibiu, Romania

Abstract.CNC machine-tools are becoming more and more complex, with regards of kinematic capabilities, driving systems, HMI interfaces and controllers. Consequently, the prices of these machines are also increasing, being directly linked with the machine characteristics. Multi-axes kinematic is nowadays seen as the possibility of controlling the motion of the machine elements not only on translational axes, but also on rotational ones. There are mainly two types of CNC milling machines with multi-axes kinematic: indexed 3+1 and/or 3+2 axes and continuous 4 and/or 5 axes machines. This work will present some aspects regarding indexed multi-axis and continuous multi-axes machining.

1 Introduction

The structure and kinematic of CNC machine-tools greatly influences the machining performances. The most important parameter, by this point of view is the overall number of controlled axis of the machine. While 3-axis machine-tools are the most encountered equipment, nowadays 4- and 5-axis CNC machine-tools are gaining ground in front of them. However, there are still a lot of aspects to be taken into consideration to balance the performances and costs.

While 3-axis CNC machine-tools machining capabilities are quite enough for most of the parts within machine-building industry, sculptured surfaces usually require machining performances which can be achieved only by means of 4- and 5-axis CNC machine-tools. However, higher performances require higher cost, related to both equipment and workforce, consequently a comprehensive analysis is needed before taking the decision of using either 3-axis or 5-axis CNC machine-tools.

The work presented in [1] presented a method of performing 5-axis machining on a 3-axis machine-tools, by using a rotary-tilting indexing table. The approach presented in [2] proposed a method of using a 3-axis CNC machine-tool, which has also the capability to perform to supplementary indexed rotations (seen as ½ axis), for machining sculptured surfaces. Other researchers have targeted the identification of the deviations of 5-axis CNC machine-tools with tilting-rotary table [3]. To reduce the costs related with buying a new 5-

*Corresponding author: radu.breaz@ulbsibiu.ro

axis CNC machine-tool, some authors proposed methods of equipping 3-axis CNC machine-tools with additional axis [4].

Another problem related with the use of CNC machine-tools is the collision avoidance. While for 3-axis CNC machine-tools the collisions may occur only between tool and workpiece (and also the fixing device), for 5-axes CNC machine-tools the situation is somehow more complex.

On multi-axis machining, the collision may occur between tool and workpiece but also between tool, workpiece and moving machine elements (slides and or tilting-rotary tables). The more complex the geometry of the part is, the more collision occurrence possibilities are expected, so various methods for avoiding them are presented in the literature [5].

In [6] a method of building a 3D geometric and kinematic model of a 5-axis CNC machine-tool was presented. The use of such a model allow the user to identify and eliminate the collisions during the simulation phase. Consequently, the above-mentioned model must be used in conjunction with a CAM software package, which can simulate the machining process.

2 The geometric and kinematic model of the machine

The part presented in figure 1 has five pockets on each face which have to be machined. On a 3-axis CNC milling machine the part can be machined by re-positioning each face, meaning that the part must be un-fixed and fixed again before and after each operation. Also, the technological origin (which is set in the program under G54 address) must be changed for each operation, a fact which usually involves complex and accurate adjustments. Consequently, five setups will be required.

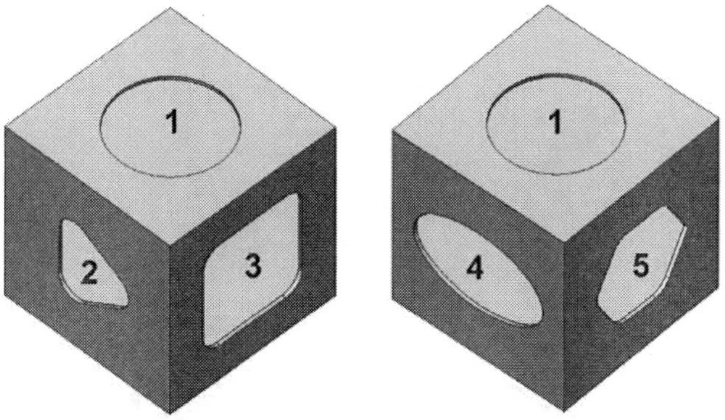

Fig. 1. Part for 3+2 indexed machining.

On the other hand, on a 5-axis CNC milling machine, the part can be machined using a single setup, by means of a 3+2 axis machining process. It means that only translational movements (on X, Y and Z axes) will be used during the machining process, while the rotational movements will be used only outside the machining process, just for repositioning the part (indexation), without the need of un-fixing it from the machine table. To illustrate the process of 3+2 machining, the DMF 260-11 DMG Mori-Seiki 5 axis CNC machine-tool was taken into consideration (fig. 2). The machine can rotate the machining head around Y axis (on B axis), while the part can be rotated around Z axis, using a rotary table (on C axis). Both indexed (3+1 and 3+2) axis and continuous 5 axis machining operations can be performed on this machine.

Fig. 2. Workspace of DMF 260-11 DMG Mori-Seiki 5-axis machine-tool.

To machine the part, a CAM software package was used. In a first approach, during the process of generating the NC code, no model of the machine was used. The second approach took intro a consideration a 3D geometric and kinematic model of the machine, presented in figure 3. This model was built by the authors of this paper. A 3D CAD program was used for building the geometric model, while the tools provided by the CAM software package were used to define the kinematic of the machine.

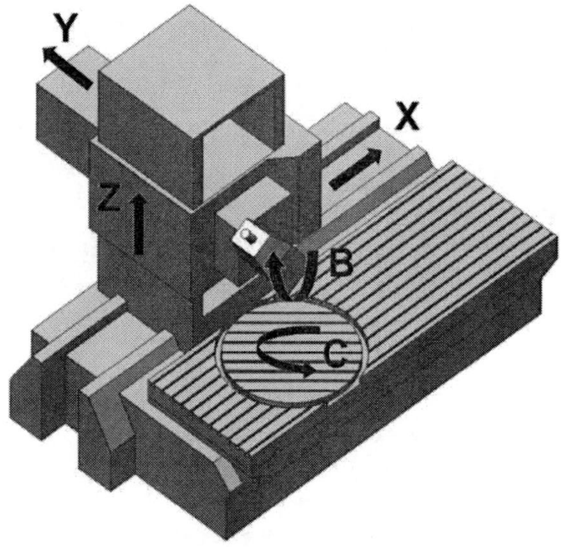

Fig. 3. 3D geometric and kinematic model of DMF 260-11 DMG Mori-Seiki 5-axis machine-tool.

In order to machine the part, five different operations were needed. Aside the first operation, all the other four operations required indexation movements on axes B and C. A scheme of each operation and the values of the B and C angles are presented in table 1. The fixing device (vise) was also taken into consideration for simulating the machining process on the DMF 260-11 DMG Mori-Seiki 5-axis machine-tool.

Table 1. 3+2 indexed machining of the considered part.

Operation	Indexation	Scheme
Millingofface1	B=0° C=0°	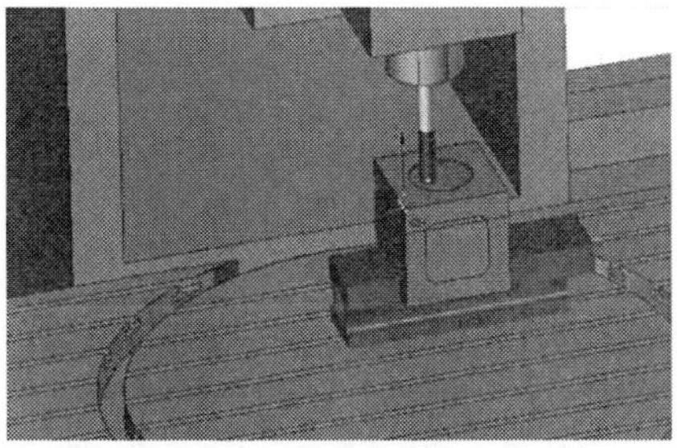
Millingofface2	B=-90° C=0°	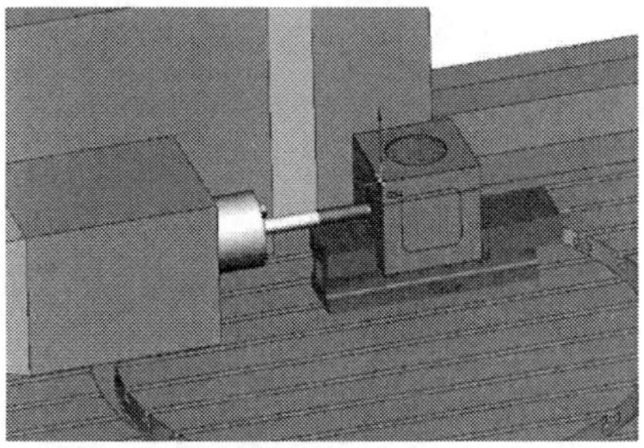
Millingofface3	B=-90° C=90°	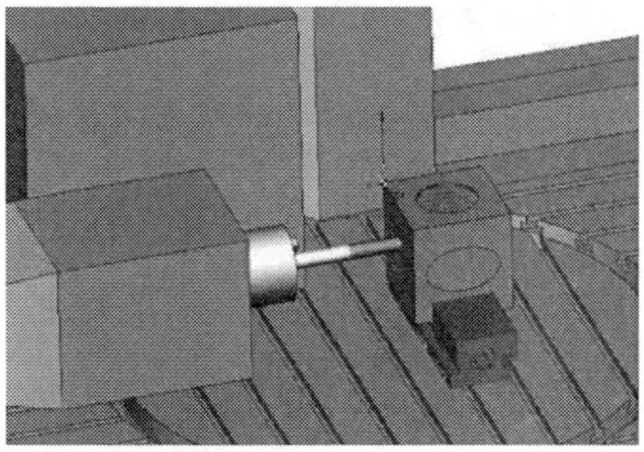
Millingofface4	B=-90° C=180°	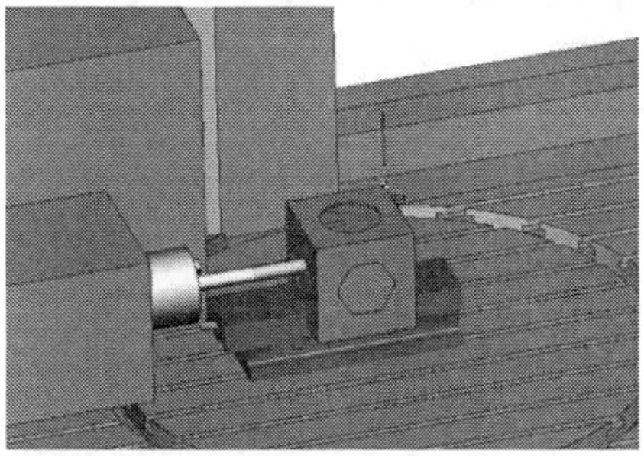
Millingofface5	B=-90° C=270°	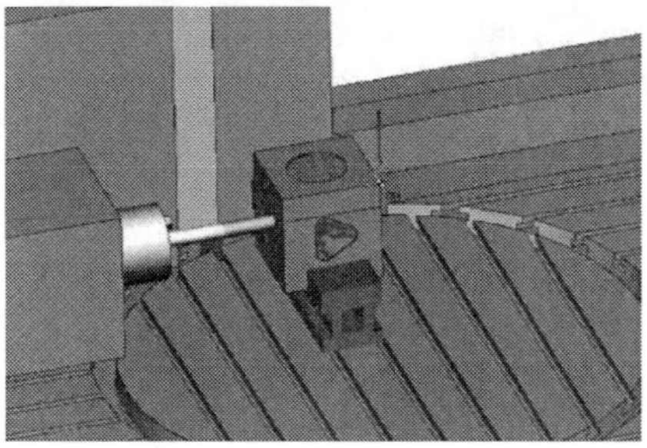

Figure 4 shows how the use of the 3D geometric and kinematic model can identify a collision between a moving machine element (working head rotating on axis B) and the fixing system (vise), while the simulation performed without the use of the model is not able to identify it, because the movements of the machine elements are not simulated.

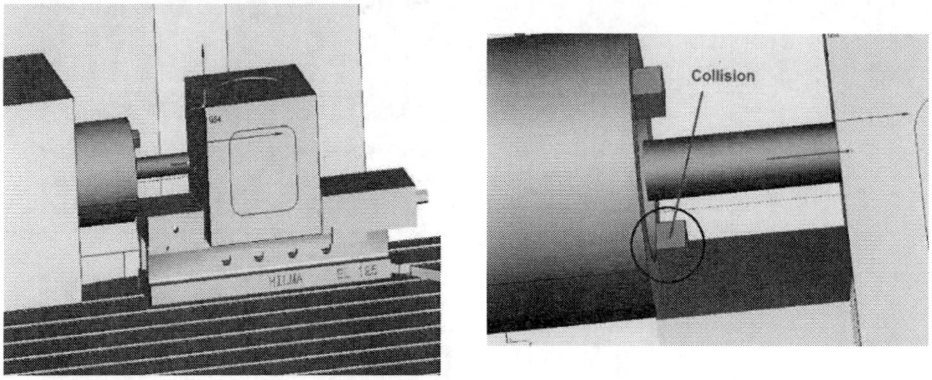

Fig. 4. Collision between a moving machine element and the vise.

3 Comparison

Table 2 presents a comparison between the use of 3-axis CNC machine-tools, indexed (3+2)-axis CNC machine-tools and continuous 5-axis CNC machine-tools. Paper [7] presents a decision-making tool which assists the user in the process of choosing between 3-axis and 5-axis CNC machine-tools.

Table 2. Comparison

Features	Multi-fixture 3-axisCNC machine-tools	Indexed (3+2)-axisCNC machine-tools	Continuous5-axisCNC machine-tools
Fixingthepartonthemachine	-	+	+
Setting-uptheoriginofthepart	-	+	+
Complexityoftheparts	-	+	++
Machiningaccuracy	-	+	+
Costsofthemachine	+	-	--
Costsoftheworkforce	+	-	--
Costsoflogistics	+	-	--

It is here noticeable that no solutions have only advantages or only drawbacks. However, if machine-building industry is targeted, one can say that indexed (3+2)-axis CNC machine-tools could represent the best compromise. Sculptured parts, which require 5-axis continuous milling are not usually encountered in this industry. Moreover, development and acquisition costs related to technological equipment may be dramatically reduced by simply equipping 3-axis CNC machine-tools with rotary-tilting tables and transforming them into indexed (3+2)- axis CNC machine-tools.

This solution, while allowing the machine-tool to perform rotational movements, is much cheaper, because it does not imply costly upgrades to the CNC controller, which are needed for 5-axis continuous machining.

4 Conclusion

The work presented in this paper has demonstrated how multi-axis CNC machining can benefit from using 3D geometric and kinematic models during the simulation process. Also, some considerations regarding the advantages and drawbacks of using CNC machine-tools with different kinematic capabilities were presented. Finally, it was stated that a kinematic upgrade of a 3-axis CNC machine-tool in order to transform it in a (3+2)-axis CNC machine-tool could be a better solution than using a 5-axis CNC machine-tool for indexed machining.

References

1. S.H. Suh, J.J. Lee, J. Manuf. Sci. Eng **120**(1), 120 (1998)
2. Z.C. Chen, Z. Dong, G.W. Vickers, Comput. Ind., **50**, 319 (2003)
3. M. Tsutsumi, A. Saito, Int. J. Mach. Tool. Manu., **44** (12–13), 1333 (2004)
4. S.H. Suh, J.J. Lee, S.-K. Kim, Int. J. Adv. Manuf. Technol., **14** (12), 865 (1998)
5. S. Ding, M.A. Mannan, A.N. Poo, Comput Aided Des., **36**, 1281 (2004)
6. A. L. Chicea, R. E. Breaz, O. Bologa, Applied Mechanics and Materials, **809-810**, 1004 (2015)
7. O. Bologa, R.-E. Breaz, S.-G. Racz, M. Crenganiş, Procedia Comput. Sci., **91**, 184 (2016)

Colour based sorting station with Matlab simulation

Victor Constantin[1,*], *Dana* Rizescu[1], and *Ciprian* Rizescu[1]

[1]DMMP, Politehnica University of Bucharest, Romania

Abstract. The paper presents the design process and manufacturing elements of a colour-based sorting station. The system is comprised of a gravitational storage, which also contains the colour sensor. Parts are extracted using a linear pneumatic motor and are fed onto an electrically driven conveyor belt. Extraction of the parts is done at 4 points, using two pneumatic motors and a geared DC motor, while the 4th position is at the end of the belt. The mechanical parts of the system are manufactured using 3D printer technology, allowing for easy modification and adaption to the geometry of different parts. The paper shows all of the stages needed to design, optimize, test and implement the proposed solution. System optimization was performed using a graphical Matlab interface which also allows for sorting algorithm optimization.

1 Introduction

Automatic sorting is one of the most common tasks needed in automated systems [1]. The advantages of such a system are high working speed, high precision, low costs as well as the possibility of sorting through a vast number of objects. Current solutions to this problem are based on several principles, mostly adapted according to what the sorting is done: colour, volume, weight, material type – based on this the measured parameter can be optical, geometric, and magnetic [2].

Such systems can be based on electric, pneumatic or hydraulic actuators according to the particularities of the objects being sorted: weight, volume, rigidity, risk of explosion, working in clean environments, etc. For example, electric systems are preferred for angular displacements while fluid-based actuators are widely used for linear movement [3, 4]. There is a growing need for training in the use of such systems, as almost all sorting is now done automatically [5].

The paper shows the methods employed by the authors for the design and of a colour-based sorting station.

The basic schematic of the system is shown in figure 1. It is comprised of a gravitational storage for cylindrical parts, which also contains the colour sensor, HDJD-S822-QR999, alongside a supplemental LED light source.

The parts are taken out of the gravitational storage one at a time by means of a linear pneumatic motor and pushed onto a constant speed conveyor belt, actuated by a servo DC

* Corresponding author: victor.f.constantin@gmail.com

motor. The sorting station has 4 pre-set sorting positions where the parts are stopped depending on their colour, which has been previously determined in the gravitational storage. Removal of the parts from the conveyor is done by means of two pneumatic motors, a servo dc motor as well as the conveyor itself (end of the conveyor). This was done to exemplify different means of performing this operation.

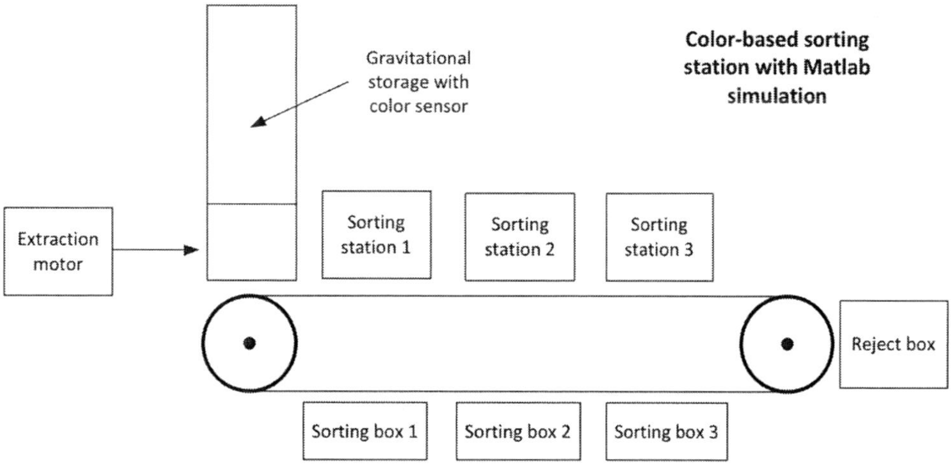

Fig. 1. Basic schematic of the system.

Figure 2 shows a logic diagram of the algorithm used for the sorting machine.

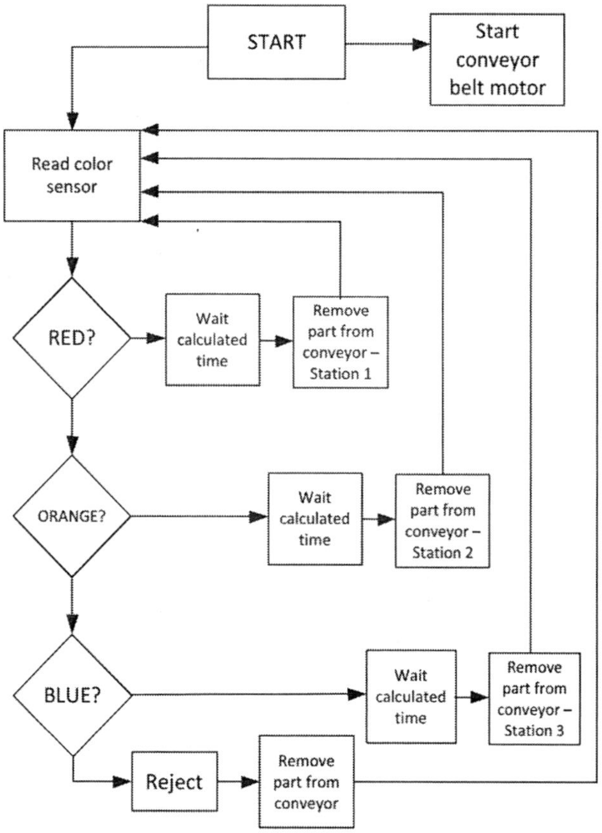

Fig. 2. Sorting station algorithm.

DOI: 10.1051/matecconf/201712108004

2 Design of the sorting station

Design of the components of the system was done using Solidworks 3D software, used to obtain STL models of the parts needed to 3D print each component. Using this method allows for easy modifications to be made if the sorting pieces are changed.

The gravitational storage (figure 3a- STL model and 3b- printed model) is comprised of three parts. The first one is cylindrical and is used for actual storage of the parts. Pieces 2 and 3 are attached to the storage and are necessary for the colour sensor, including isolating it from exterior light sources. Colour of the parts represents of the early versions of the system.

Design of the gravitational storage allows for easy modification to be made to the system if for example, the shape of the sorted parts is changed.

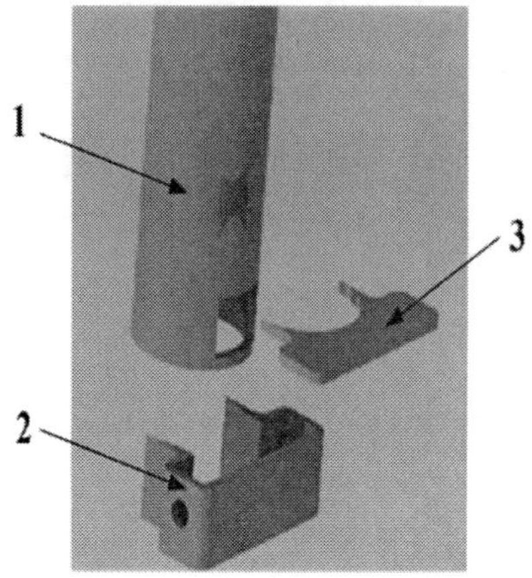

Fig. 3.a. 3D model of the parts.

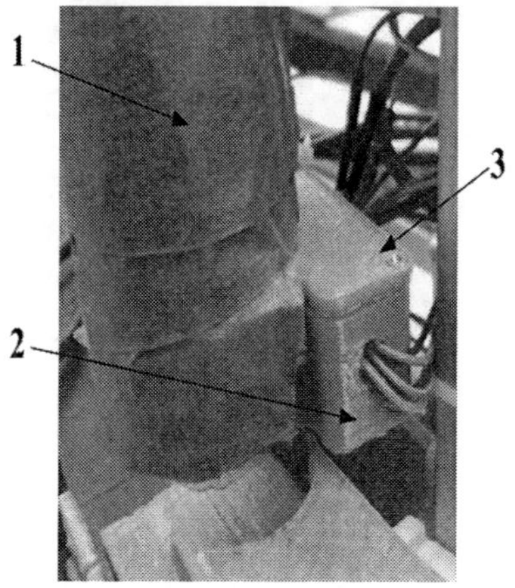

Fig. 3.b. Printed model.

The pieces are extracted from the gravitational storage using part 4 mounted on the rod of motor 6 and then are pushed along part 5. Figures 4a and 4b show the 3D model and the printed version of the parts.

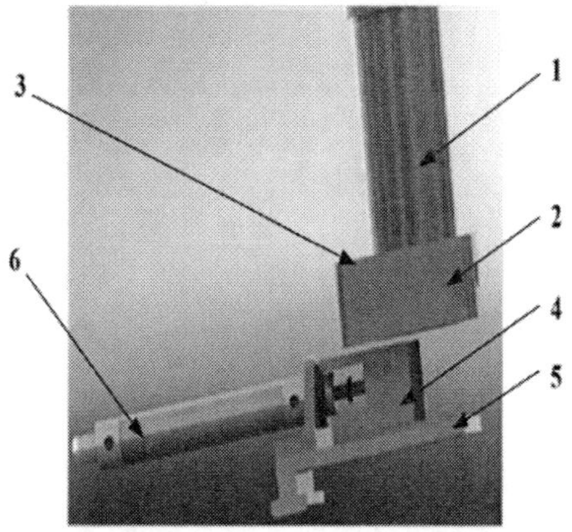

Fig. 4.a. 3D Model of the extractor.

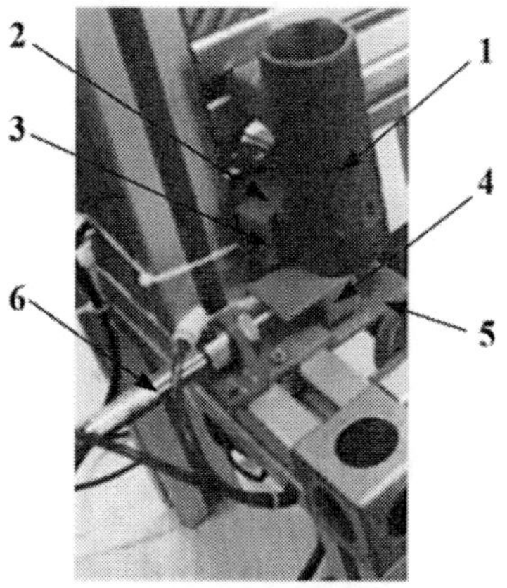

Fig. 4.b. Assembled extractor.

Conveyor 9 is connected to servo dc motor 7 by coupling 8. The system is also equipped with a tensioner mechanism 10. Parts are removed from the conveyor using linear motors 12, equipped with cams 11 or by means of a electric motor 13. The electronic system used to control the system is housed in box 14.

The 3D model of the system is presented in figure 5a and the physical model in figure 5b.

Fig. 5.a. 3D rendered model of the system.

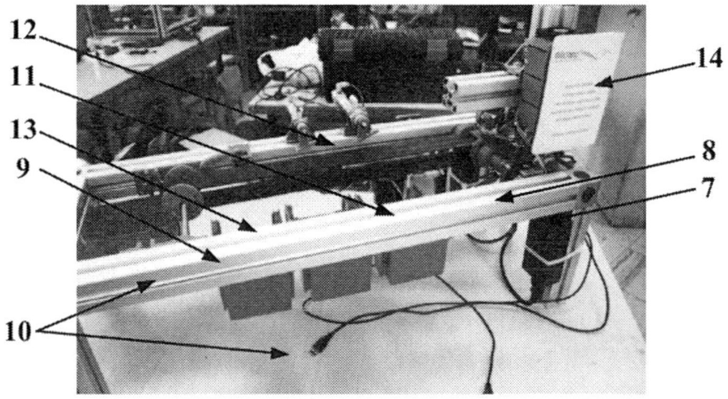

Fig. 5.b. Assembled model.

3 The control system

The pneumatic control system of the sorting station is presented in figure 6. It is comprised of a single acting cylinder A, needed to remove the parts from the gravitational storage, as well as single acting cylinders B and C (12 in figure 5), needed to remove the parts from the conveyor belt. All pneumatic motors are controlled by 3/2 pneumatic way valves, electrically actuated.

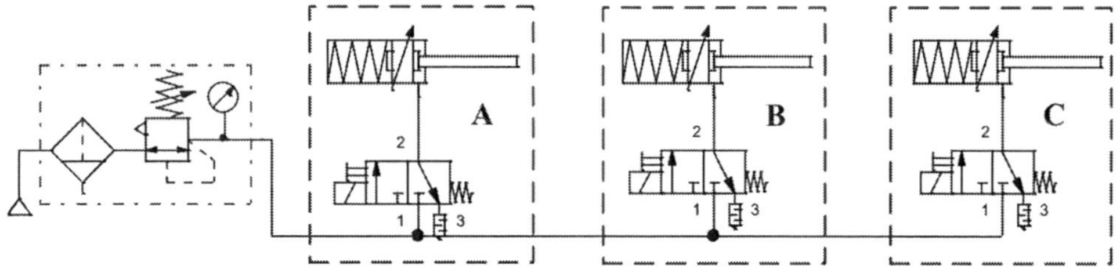

Fig. 6. Pneumatic control schematic.

The electronic control system is structured around two Atmel microcontrollers, one responsible for control of the servo dc motor needed for the conveyor and one used to read and send control signals to all other parts of the system. A separate microcontroller is used for the conveyor motor to ensure constant speed and also be able to implement a accelaration/deceleration interrupt-driven algorithm. This is neded since removal of the parts from the conveyor is done in a time-sensitive loop, rather than implementing a proximity sensor algorithm. This was done to ensure low cost and low maintenance of the system. The second microcontroller is used to read analog voltage values from the colour sensor – between 0 and 5V for red, green and blue respectively. The power level of the control system is interfaced using transistor relays. General schematic for the electronic circuit is presented in figure 7.

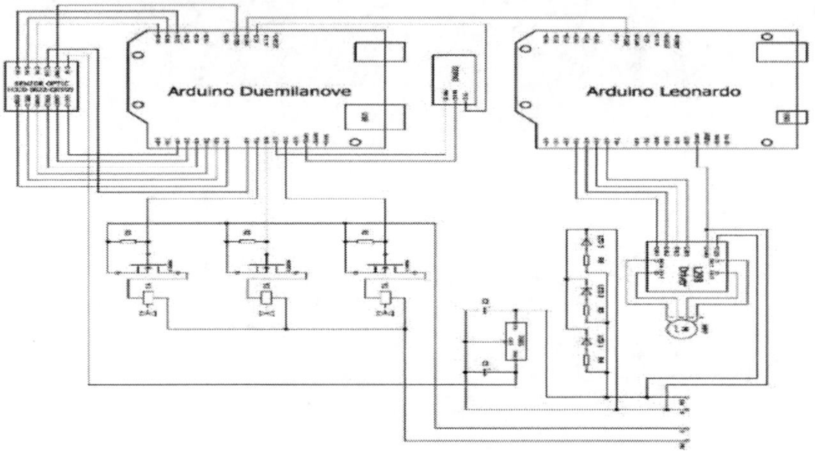

Fig. 7. Electronic circuit.

4 Simulation

In order to study the working principle of the sorting station, the authors have developed a graphic interface using Matlab programming environment (figure 8).

The interface recreates the system and allows for different conveyor speeds to be selected, as well as the distance between the sorting positions. In order to start the simulation the user must first select the overall length of the station, by using the „Maximum Length L[mm]" pop-up selector. The speed of the system is also selected by using the „Transport belt speed v [mm/s]" selector. Simulation is started by using the appropriate button and in the new window a random generated part will appear in the storage position. The part is transported using the conveyor to the appropriate position, according to its colour. The application returns travel time to sorting position as well as the speed of the conveyor.

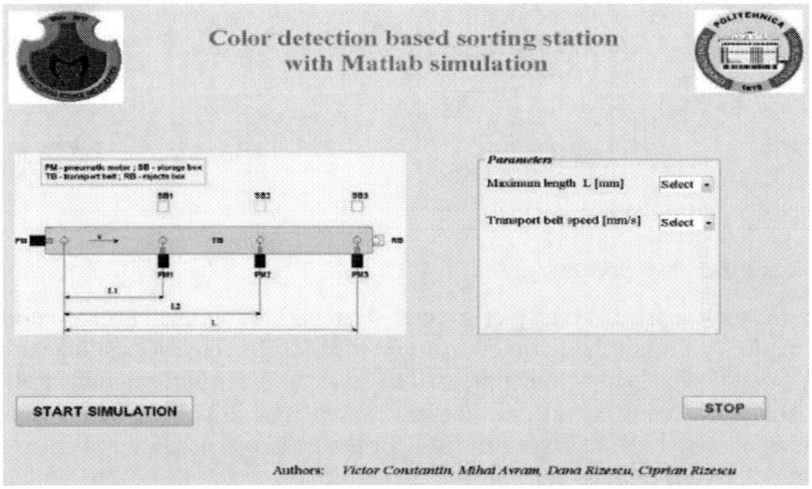

Fig. 8. Matlab interface.

Figures 9a through 9f show the result for different colours. As can be seen in figure 9a depicting the sorting of a red part, time elapsed (1.73s) from the storage PM to the sorting position PM1 is shown in red under the appropriate station. Figures 9c and 9b show sorting of a black part, which is not removed from the conveyor by one of the motors but is stored in a box at the end. Figures 9e and 9f show the sorting process and time for a blue and orange part.

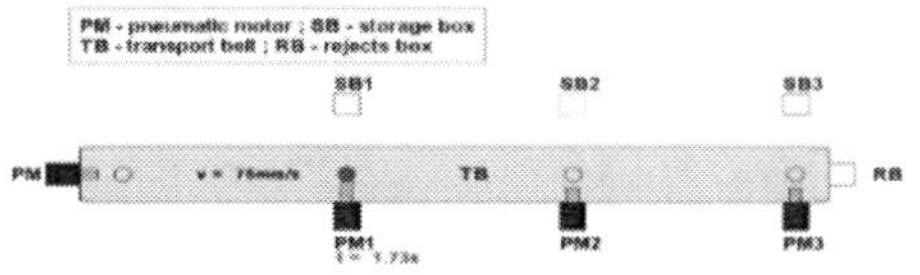

Fig. 9a. Red part at sorting station.

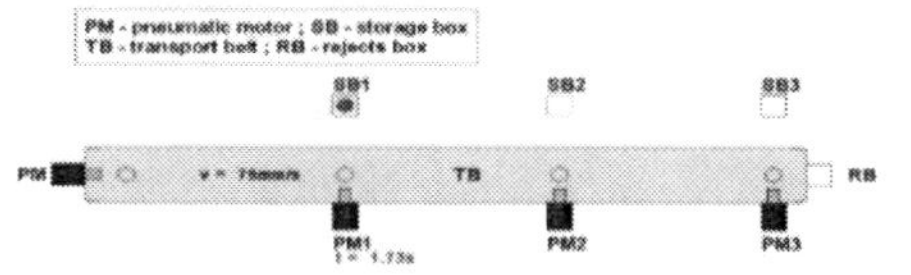

Fig. 9b. Red part in sorting box.

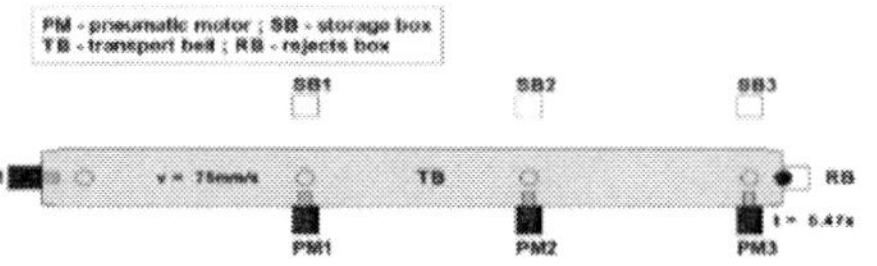

Fig. 9c. Black rejected part at end of conveyor.

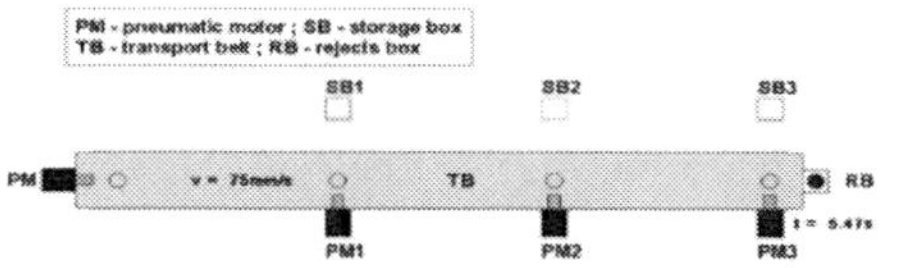

Fig. 9d. Black rejected part in sorting box.

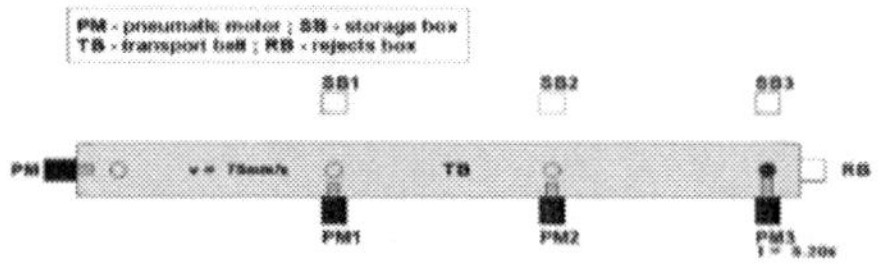

Fig. 9e. Blue part at sorting station.

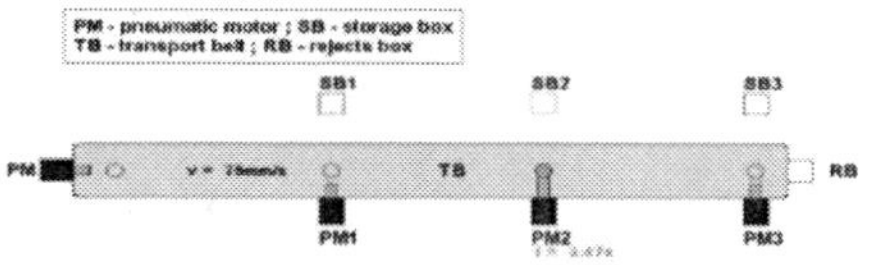

Fig. 9f. Orange part at sorting station.

5 Conclusions

The proposed system has been designed, simulated and tested by the authors. During these stages different modifications were made to the initial design to ensure that it will function within desired parameters but also to ensure that the system could be easily replicated or adapted if needed. Simulations done in Matlab programming environment have confirmed results obtained by testing the installation.

Acknowledgement: This work has been partially supported by University Politehnica of Bucharest, through the "Excellence Research Grants" Program, UPB – GEX. Identifier: UPB–EXCELENȚĂ–2016 "Modul universal de testare și control pentru acționări fluidice și electrice", 39/26.09.2016 (acronym: MUTCAFE).

References

1. Gh. Prisăcaru, M. Bercea, B. Gramescu, V. Ciupe – *Mecatronica aplicata* (Ed. Academiei Oamenilor de Stiinta din Romania, București, 2011)
2. M.R. Satpute, S.M. Jagdalem, International Conference on Inventive Computation Technologies, ICICT 2016; Coimbatore; India, 1, **1** (2016)
3. S.V. Krichel, O Sawodny, Proceedings of 2011 International Conference on Fluid Power and Mechatronics, FPM, Beijing, China (2011)
4. A.A. Ata, A. Rafeek, H. Yusof, Journal of Intelligent and Robotic Systems: Theory and Applications, **43(**1), 99 (2005)
5. A.L. Bencsik, M. Lendvay, Proceedings of IEEE 9th IEEE International Symposium on Applied Computational Intelligence and Informatics SACI, Timisoara, Romania, 213 (2014)

DOI: 10.1051/matecconf/201712108005

Assistive humanoid robot MARKO: development of the neck mechanism

Marko Penčić[1,*], *Maja* Čavić[1], *Srđan* Savić[1], *Milan* Rackov[1], *Branislav* Borovac[1], and *Zhenli* Lu[2]

[1]Faculty of Technical Sciences, University of Novi Sad, TrgDositejaObradovića 6, 21000 Novi Sad, Serbia
[2]School of Electrical Engineering and Automation, Changshu Institute of Technology, Hushan Road 99, 215500 Changshu, People's Republic of China

Abstract. The paper presents the development of neck mechanism for humanoid robots. The research was conducted within the project which is developing a humanoid robot Marko that represents assistive apparatus in the physical therapy for children with cerebral palsy.There are two basic ways for the neck realization of the robots. The first is based on low backlash mechanisms that have high stiffness and the second one based on the viscoelastic elements having variable flexibility. We suggest low backlash differential gear mechanism that requires small actuators. Based on the kinematic-dynamic requirements a dynamic model of the robots upper body is formed. Dynamic simulation for several positions of the robot was performed and the driving torques of neck mechanism are determined.Realized neck has 2 DOFs and enables movements in the direction of flexion-extension 100°, rotation ±90° and the combination of these two movements. It consists of a differential mechanism with three spiral bevel gears of which the two are driving and are identical, and the third one which is driven gear to which the robot head is attached. Power transmission and motion from the actuators to the input links of the differential mechanism is realized with two parallel placed gear mechanisms that are identical.Neck mechanism has high carrying capacity and reliability, high efficiency, low backlash that provide high positioning accuracy and repeatability of movements, compact design and small mass and dimensions.

1 Introduction

Children with non-progressive neuromuscular disorders, such as cerebral palsy [1], can have a significant deficit of motor functions and skills, body posture, as well as intellectual and emotional disorders [2]. The primary problem is a defect in the initiation and execution of the movement [3], considering that the motor activities represent the basis of human functioning.

The paper presents the development of neck mechanism for humanoid robots. The research was conducted within the project, which develops humanoid robot Marko that represents assistive apparatus in physical therapy for children with cerebral palsy – Fugure 1a. The

*Corresponding author: mpencic@uns.ac.rs

conventional medical treatment involves exercises that are monotonous and exhausting for children, wherefore children lose interest for the work. Since the success of therapy depends directly on the time that child spent exercising, it is necessary to find a way to further motivate the child.The aim of the project is to offer the children a partner – robot, with which it will be more interesting for them to exercise and who will be able to demonstrate an exercise that the child cannot perform. It is expected are that the effects of the therapy will be increased through additional motivation and encouragement to the patient to exercise more. It should be noted that the robot is sitting the entire time on the mobile platform in the form of a horse and that the legs of the robot are intended solely for demonstration of movements during treatment, not for walking.

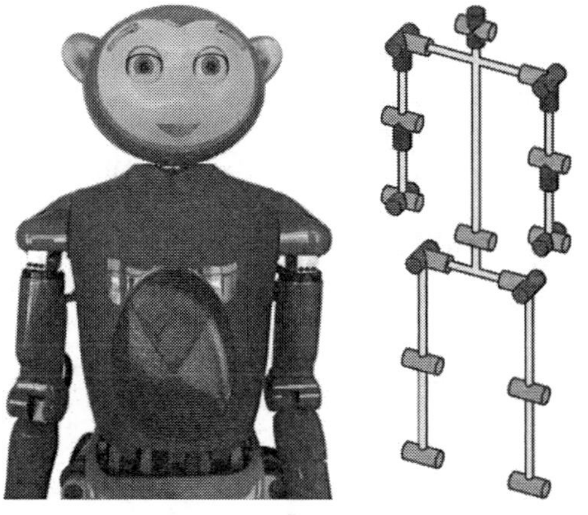

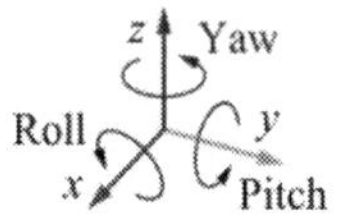

Joint	DOF
Eyebrows	3
Eyeballs	4
Eyelids	4
Neck	2
Arms	14
Hands*	10
Waist	1
Legs	8
Total	46

(a) (b)

Fig. 1. Photography of assistive humanoid robot Marko (a) and its kinematic structure (b).

Figure 2b presents kinematic structure of the robot, which has a head with 11 DOFs, neck with 2 DOFs, two arms with 14 DOFs, two hands which should have 10 DOFs [4], waist with 1 DOF [5] and two legs with 8 DOFs [6], which makes the total 46 DOFs. In addition to verbal communication, Marko is able to intuitively realize nonverbal communication by moving the characteristic elements of the face – eyebrows, eyes, eyelids, and with the use of LEDs – ears and mouth are changing colors depending on the situational context of the robot.Therefore 3 DOFs eyebrows are realized, which is2 DOFs for the roll axis and 1 DOF for the yaw axis – lifting both eyebrows, 4 DOFs eyeballs, which is 2 DOFs for the yaw axis and 2 DOFs for the pitch axis, and 4 DOFs eyelids for the pitch axis –opening/closing the upper and lower eyelids.

1.1 Human neck and its mobility

Neck is multi-joint structure that connects the head to the chest. It is a part of the spine and consists of irregular bones that are placed one on top of the other and called vertebrae. The movements of the cervical spine are enabled with intervertebral joints and intervertebral discs that are placed between two adjacent vertebrae. Due to the elastic structure of disc, it is possible to perform locomotor activities of the neck and better amortization. In neck movements a large number of muscles and ligaments are involved. Neck provides stability and head posture, enableshead movements, transmits the load and absorbs shocks. Therefore, it has to be strong, stable and sufficiently movable. The movements between individual vertebrae are of small range, however, the mobility of the neck in whole is large, because it represents the sum of numerous small movements that are performed in 7 joints.

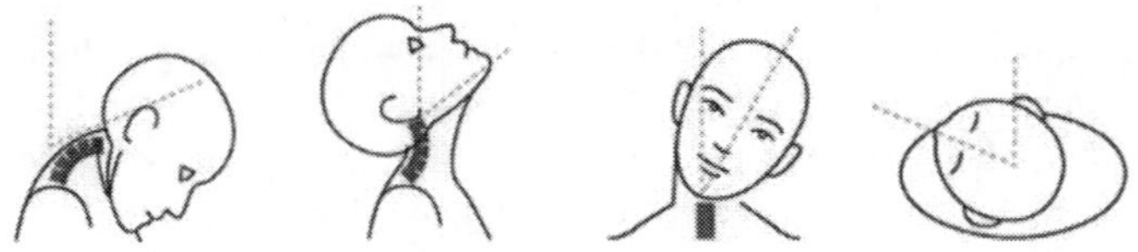

(a) (b) (c) (d)

Fig. 2. The human neck movements: (a) bending forward – flexion, (b) bending backward – extension, (c) left/right side bending – lateral flexion and (d) left/right torsion – rotation.

In the cervical part of spinal column, there are movements flexion – bending forward (Fig. 2a), extension – bending backward (Fig. 2b), lateral flexion – side bending (Fig. 2c) and torsion – rotation (Fig. 2d). The mobility of the cervical spinefor the movement of flexion-extension is approximately 100°. Head and neck movements are inseparable, so the overall mobility of the head and neck in the direction of flexion-extension is 130÷150°. Total lateral flexion movement of the neck is approximately ±30°, and if the movement of the head is added, then the total lateral bending is approximately ±50°. Rotation movements are approximately ±80° [7-9].

2 State of the art

There exist two basic groups of robots that can move the head independently from the trunk by activating the neck. The first group of robots has a neck with 2, 3 or 4 joints where each joint has 1 DOF – rigid structures, while the second group has a neck with 3-7 joints – spherical, spherical and rotational or universal and rotational, which is 9-19 DOFs – elastic structures.

Neck with 2 DOFs – movements about the yaw and pitch axes, have ASIMO [10], HUBO [11], HRP-4 [12], EveR1 [13], NAO [14], AILA [15], TORO [16], Justin [17], Pepper [18] etc.Neck with3 DOFs – movements about the yaw, roll and pitch axes, haveWABIAN-2 [19], Albert HUBO [20], HRP-4C [21], Affetto [22], iCub [23], Probo [24], EveR-2 [25], Geminoid F [26], Actroid-F [27] etc.Neck with4 DOFs – yaw, roll, upper and lower pitch movements, haveWE-4RII [28], ARMAR-III [29], ROMAN [30], KOBIAN [31], SAYA [32], BARTHOC [33], Romeo [34], Flutist Robot WF-4 [35], Twente Humanoid Head [36] etc.Neck structures with 2, 3 or 4 DOFs usually consist of rigid and low backlash mechanisms that are interconnected – harmonic drive, cable-driven mechanisms, spindle drive, low backlash gears, etc. In addition to high reliability and carrying capacity, the advantage of these mechanisms is low backlash that provide high positioning accuracy and repeatability of movements.

Biologically inspired neck structures with 3-7 cervical vertebrae and large number of DOFs – movements about the yaw, pitch and roll axes, have Kenta [37], Kenji [38], Kotaro [39], Kojiro [40], Kenzoh [41], Kenshiro [42] and Kengoro [43].Each vertebra has points for attaching tendons which enables independent motion of each joint. Between each two vertebraethere isviscoelastic element – disc of silicone rubber and tension springs – ligaments. The height of each rubber disc is slightly greater than the distance between adjacent vertebrae and wherefore discs generate the pressure on the joints and ligaments tighten them.

During neck bending, the elastic elements generate force that opposes gravity and thereby help actuators.

The advantage of elastic elements is in achieving a significant force at the time of releasing the accumulated energy. In addition, flexible neck allows that the head movements are more natural and human-like.

3 Neck mechanism

Since the robot head has the shape of an ellipsoid, the basic requirement is an adequate mobility of the head, respectively the neck about the pitch and yaw axes. In order to minimize driving torques it is necessary to decrease mass of the segments as much as possible. In addition the simple and reliable mechanisms are requiredthat have a high carrying capacity, high efficiency, low vibrations, low noise and low backlash that provide high positioning accuracy that enables high accuracy and repeatability of movements which is essential for motion control [44].

3.1 Forces and torques

The robot head is approximated by a sphere of mass 3.5 kg whose center of gravity is on distanceof 0.1 m above the neck joint. The neck movements should be as natural as possible and therefore it is adopted that their duration does not exceed 1 s, which as a consequence have a certain dynamic effects. For that reason, based on the formed 3D model of the robot upper body a dynamic simulation was performed, and the driving torques of the neck mechanism are determined.

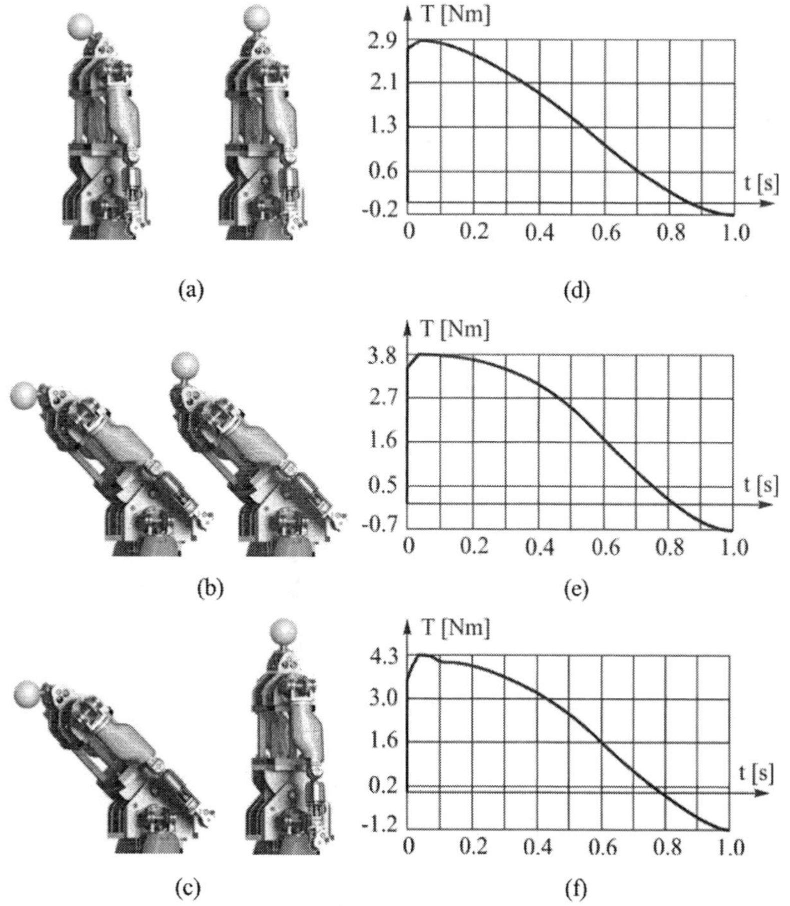

Fig. 3. Dynamic simulation of neck: (a) flexion of neck –50° to 0°, (b) flexion-extension of the neck –50° to +50° where the trunk is fixed in position of –45°, (c) simultaneous movements of neck and trunk respectively flexion of the neck –50° to 0° and flexion of the trunk –45° do 0°, and (d)-(e) graphics of torque changes for the appropriate range and movements of the neck mechanism.

Three cases were examined depending on the type of movements and range of motion. The first is the neck flexion for range of motion from –50° to 0° (Fig. 3a), the second is flexion-extension of the neck for range of motion from–50° to +50°where the trunk is fixed in position of – 45° (Fig. 3b), while the third case involves the simultaneous flexion of neck and trunk in the range of–50° do 0°for neck and–45° do 0°for trunk(Fig. 3c).Profile of velocity is such that maximum values are reached in the middle of the movement – cubic polynomial trajectory. Based on the dynamic analysis the maximum torque value for dimensioning of the neck mechanism is 4.3 Nm.

3.2 Mechanical design

Realized neck has 2 DOFs and enables movements in the direction of flexion-extension 100°, rotation ±90°, and the combination of these two movements. It is based on the differential mechanism with three bevel gears of which the first two are driving and are identical, and the third one which is the driven gear. If driving gears have the same circumferential velocity and the same direction of rotation, head moves about the pitch axis – flexion and extension of the neck – Figure 4b, and for the opposite direction of rotation, head moves about the yaw axis – rotation of the neck – Figure 4c. If the circumferential velocities of driving gears are different a combination of these two movements is obtained – Figure 4d.

(a) (b) (c) (d)

Fig. 4. The neck movements: (a) upright position, (b) flexion 50°, (c) left rotation 90° and (d) flexion 50° + left rotation 90°.

Figure 5 shows the structure of the neck mechanism. It consists of three subassemblies– two driving and identical, and one driven to which the robot head is attached. The driving subassembly consists of two support plates within which are placed the actuator and the elements for the power transmission and motion – bearings, shafts, helical gears and driving spiral bevel gear.

The actuator consists of a brushless DC motor, planetary gearhead with 3 stages and a magnetic encoder. Driving helical gear is directly attached on the gearhead shaft. Driven helical gear and driving bevel gear are placed on the same shaft – driving shaft. Driven subassembly has a housingwhich is used for mounting the two other subassemblies. Driven shaft, bearings and output links of the differential mechanism– driven bevel gear to which is attached the head support, are placed within the housing. Spiral bevel gears provide low backlash because several teeth are always in mesh. Low backlash at gearing with helical gears is achieved by precise machining of support plates and gears preloading – center distance is with negative tolerance.

Figure 6 shows a photograph of neck mechanism and in Table 1 its configuration is given. The neck width is 164 mm, length 84 mm, a height of 92 mm, while the mass is 2 kg.

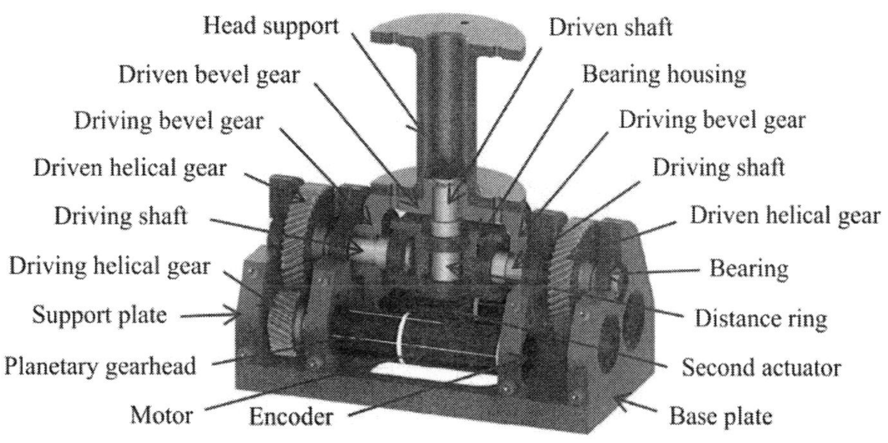

Fig. 5. Neck mechanismwith details – cross section.

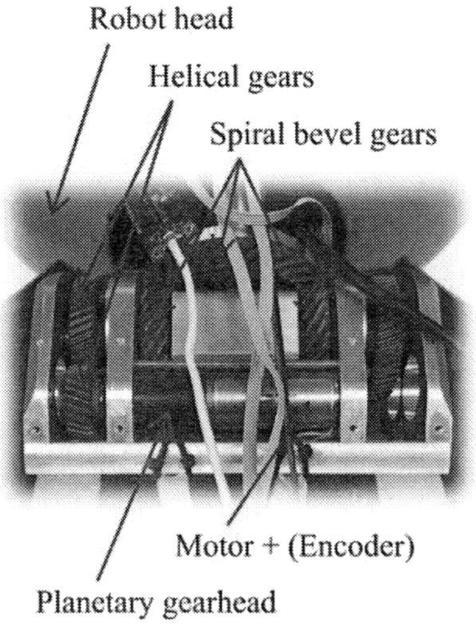

Fig. 6. Neck mechanism of the robot Marko.

Table 1. Neck mechanism configuration.

Flexion-xtension and Rotation – neck joint		
Maxon motor EC-max 22 (2 pcs.)	Voltage [V]	24
	Power [W]	25
	Torque [mNm]	22.7
	Speed [rpm]	10400
Planetary gearhead GP 22-HP (2 pcs.)	Reduction	157
	Torque [Nm]	3
	Efficiency [%]	59
Encoder MR-ML (2 pcs.)	Channels	3
	Counts per turn	512
Gearing with helical gears (2 pcs.)	Reduction	1.84
	Module [mm]	1
	Helix angle [°]	20
Gearing with spiral bevel gears (2 driving &1 driven)	Reduction	1.625
	Module [mm]	1.5
	Spiral angle [°]	44

4 Conclusion

This paper presents the development of the neck mechanism for humanoid robots. Since the robot head has the shape of an ellipsoid, the basic requirement is to enable the mobility of the neck about the pitch and yaw axes. Based on the kinematic-dynamic requirements a dynamic model of the robots upper body is formed.

Dynamic simulation for several position of the robot is performed and the driving torques of neck mechanism are determined. Realized neck has 2 DOFs and enables movements in the direction of flexion-extension 100°, rotation ±90° and the combination of these two movements. It consists of a differential mechanism with three spiral bevel gears. First two are driving and identical, while the last one is driven gear to which the robot head is attached.

When driving gears have the same circumferential velocity and the same direction of rotation, head moves about the pitch axis – flexion and extension of the neck, and for the opposite direction head moves about the yaw axis – rotation of the neck. If the circumferential velocities of driving gears are different then a combination of these two movements is obtained. Power transmission and motion from the actuators to the input links of the differential mechanism is realized with two parallel placed gear mechanisms that are identical and consist of helical gears. Neck mechanism has high carrying capacity and reliability, high efficiency, compact design and low backlash that provides high positioning accuracy that enables high accuracy and repeatability of movements, which is essential for motion control.

This work was funded by the Ministry of Education and Science of the Republic of Serbia under the contract III 44008 and by the Provincial Secretariat for Science and Technological Development under the contract 114–451–2116/2011.

References

1. C. Bayon, R. Raya, S. Lerma-Lara, O. Ramirez, I. Serrano, E. Rocon, Transl. Biomed. **7**(1), 1 (2016)
2. F. Miller, *Physical Therapy of Cerebral Palsy*, (Springer, New York, 2007)
3. I. Novak, S. McIntyre, C. Morgan, L. Campbell, L. Dark, N. Morton, E. Stumbles, S.A. Wilson, S. Goldsmith, Dev. Med. Child. Neurol. **55**(1), 1 (2013)
4. S. Savić, M. Raković, M. Penčić, M. Nikolić, S. Dudić, B. Borovac, Proc. of the Int. Conf. on Electrical, Electronic and Computing Engineering – IcETRAN, Zlatibor, SR, (ROI1.4.) 1 (2016)
5. M. Penčić, M. Čavić, M. Rackov, B. Borovac, Proc. of theInter. Conf. on Accomp. in Mechanical and Industrial Engineering – DEMI, Banja Luka, BA, (to be published, 2017)
6. M. Penčić, S. Savić, M. Čavić, B. Borovac, Z. Lu, Proc. of theInt. Sci. and Expert Conf. – TEAM, Trnava, SK, 23 (2016)
7. M. Mijač, M. Blagotić, Lj. Djordjević, G. Teofilovski-Parapid, *Humans Anatomy – Osteology* (in Serbian), (Savremenaadministracija, Belgrade, 1998)
8. D. Nikolić, *Kinesiology* (in Serbian), (College of Health Studies, Ćuprija, 2007)
9. P. Opavsky, *Introduction to Sports Biomechanics* (in Serbian), (Author's edition, Belgrade, 1998)
10. Y. Sakagami, R. Watanabe, C. Aoyama, S. Matsunaga, N. Higaki, K. Fujimura, Proc. of the IEEE/RSJ Int. Conf. on Intelligent Robots and Systems, Lausanne, CH, ,**3** *2478* (2002)
11. I.-W. Park, J.-Y Kim, J. Lee, J.-H. Oh, Adv. Robot. **21**(11) 1305 (2007)
12. K. Kaneko, F. Kanehiro, M. Morisawa, K. Akachi, G. Miyamori, A. Hayashi, N. Kanehira, Proc. of the IEEE/RSJ Int. Conf. on Intelligent Robots and Systems, San Francisco, CA, USA, 4400 (2011)
13. H.S. Ahn, D.-W. Lee, D. Choi, D.-Y. Lee, H.-G. Lee, M.-H. Baeg, Inter. J. Human. Robot. **10**(2), 1350017 (2013)
14. D. Gouaillier, V. Hugel, P. Blazevic, C. Kilner, J. Monceaux, P. Lafourcade, B. Marnier, J. Serre, B. Maisonnier, Proc. of the IEEE-ICRA Int. Conf. on Robotics and Automation, Kobe, JP, 769 (2009)
15. J. Lemburg, J. de GeaFernández, M. Eich, D. Mronga, P. Kampmann, A. Vogt, A. Aggarwal, Y. Shi, F. Kirchner, Proc. of the IEEE-ICRA Int. Conf. on Robotics and Automation, Shanghai, CN, 5147 (2011)
16. J. Englsberger, A. Werner, C. Ott, B. Henze, M.A. Roa, G. Garofalo, R. Burger, A. Beyer, O. Eiberger, K. Schmid, A. Albu-Schäffer, Proc. of the IEEE-RAS Int. Conf. on Humanoid Robots, Madrid, ES, 916 (2014)
17. C. Ott, O. Eiberger, W. Friedl, B. Bauml, U. Hillenbrand, C. Borst, A. Albu-Schaffer, B. Brunner, H. Hirschmuller, S. Kielhofer, R. Konietschke, M. Suppa, T. Wimbock, F. Zacharias, G. Hirzinger, Proc. of the IEEE-RAS Int. Conf. on Humanoid Robots, Osaka, JP, 276 (2006)
18. J. Lafaye, D. Gouaillier, P. B. Wieber, *Proc. of the IEEE-RAS Int. Conf. on Humanoid Robots, Madrid, ES, 336* (2014)
19. Y. Ogura, H. Aikawa, K. Shimomura, H. Kondo, A. Morishima, H.-O. Lim, A. Takanishi, Proc. of the IEEE-ICRA Int. Conf. on Robotics and Automation, Orlando, FL, USA, 76 (2006)
20. I.-W. Park, J.-Y Kim, B.-K. Cho, J.-H. Oh, Rob. Auton. Syst. **56**(1), 95 (2008)
21. S. Kajita, K. Kaneko, F. Kaneiro, K. Harada, M. Morisawa, S. Nakaoka, K. Miura, K. Fujiwara, E.S. Neo, I. Hara, K. Yokoi, H. Hirukawa, *Cybernetic Human HRP-4C:* In book: C. Pradalier, R. Siegwart, and G. Hirzinger (eds.) Robotics Research: 14th ISRR. STAR, **70**, 301 (2011)
22. H. Ishihara, M. Asada, Adv. Robot. **29**(18), 1151 (2015)
23. N.G. Tsagarakis, G. Metta, G. Sandini, D. Vernon, R. Beira, F. Becchi, L. Righetti, J. Santos-Victor, A.J. Ijspeert, M.C. Carrozza, D.G. Caldwell, Adv. Robot. **21**(10), 1151 (2007)
24. K. Goris, J. Saldien, B. Vanderborght, D. Lefeber, Int. J. Hum. Robot. **8**(3), 481 (2011)
25. H.S. Ahn, D.-W. Lee, D. Choi, D.Y Lee, M.H. Hur, H.Lee, W.H. Shon, Proc. of theAnnual Conf. of the IEEE-IECON Industrial Electronics Society, Melbourne, AU, 104 (2011)

26. A. Becker-Asano, H. Ishiguro, Proc. of the IEEE-WACI Workshop on Affective Computational Intelligence, Paris, FR, 1 (2011)
27. M. Yoshikawa, Y. Matsumoto, M. Sumitani, H. Ishiguro, Proc. of the IEEE-ROBIO Int. Conf. on Robotics and Biomimetics, Phuket, TH, 2378 (2011)
28. K. Itoh, H. Miwa, M. Zecca, H. Takanobu, S. Roccella, M. C. Carrozza, P. Dario, A. Takanishi, In book: T. Zielinska, C. Zielinski (eds.) ROMANSY 16: Robot Design, Dynamics and Control. CISM, **487,** 255 (2006)
29. T. Asfour, K. Regenstein, P. Azad, J. Schroder, A. Bierbaum, N. Vahrenkamp, R. Dillmann, Proc. of theIEEE-RAS Int. Conf. on Humanoid Robots, Genova, IT, 169 (2006)
30. N. Schmitz, J. Hirth, K. Berns, *Proc. of theIEEE-RAS Int. Conf. on Hum. Robots, Daejeon, KR, 579* (2008)
31. M. Zecca, Y. Mizoguchi, K. Endo, F. Iida, Y. Kawabata, N. Endo, K. Itoh, A. Takanishi, Proc. of theIEEE-ROMAN Int. Symp. on Robot and Human Interactive Communication, Toyama, JP, 381 (2009)
32. T. Hashimoto, H. Kobayashi, N. Kato, Proc. of theIEEE-FUZZ Int. Conf. On Fuzzy Systems, Taipei, TW, 766 (2011)
33. M. Hackel, S. Schwope, J. Fritsch, B. Wrede, G. Sagerer, Proc. of theIEEE/RSJ Int. Conf. on Intelligent Robots and Systems, Edmonton, CA, 2443 (2005)
34. N. Pateromichelakis, A. Mazel, M. A. Hache, T. Koumpogiannis, R. Gelin, B. Maisonnier, A. Berthoz, Proc. of theIEEE/RSJ Int. Conf. on Intelligent Robots and Systems, Chicago, IL, USA, 1374 (2014)
35. K. Chida, I. Okuma, S. Isoda, Y. Saisu, K. Wakamatsu, K. Nishikawa, J. Solis, H. Takanobu, A. Takanishi, IEEE-ICRA Int. Conf. on Robotics and Automation, New Orleans, LA, USA, **1,** 152 (2004)
36. R. Reilink, L. C. Visser, D.M. Brouwer, R. Carloni, S. Stramigioli, ISR **4**(2), 107 (2011)
37. I. Mizuuchi, R. Tajima, T. Yoshikai, D. Sato, K. Nagashima, M. Inaba, Y. Kuniyoshi, H. Inoue, Proc. of the IEEE/RSJ Int. Conf. on Intelligent Robots and Systems, Lausanne, CH, **3,** 2527 (2002)
38. I. Mizuuchi, T. Yoshikai, Y. Nakanishi, M. Inaba, Proc. of theIEEE/RSJ Int. Conf. on Int. Robots and Systems, Edmonton, CA, 692 (2005)
39. I. Mizuuchi, T. Yoshikai, Y. Sodeyama, Y. Nakanishi, A. Miyadera, T. Yamamoto, T. Niemela, M. Hayashi, J. Urata, Y. Namiki, T. Nishino, M. Inaba, Proc. of the IEEE-ICRA Int. Conf. on Robotics and Automation, Orlando, FL, USA, 82 (2006)
40. I. Mizuuchi, Y. Nakanishi, Y. Sodeyama, Y. Namiki, T. Nishino, N. Muramatsu, J. Urata, K. Hongo, T. Yoshikai, M. Inaba, Proc. of theIEEE-RAS Int. Conf. on Humanoid Robots, Pittsburgh, PA, USA, 294 (2007)
41. T. Izawa, M. Osada, N. Ito, S. Ohta, J. Urata, M. Inaba, Proc. of theIEEE-ROBIO Int. Conf. on Robotics and Biomimetics, Phuket, TH, 2384 (2011)
42. Y. Asano, T. Kozuki, H. Mizoguchi, Y. Motegi, M. Osada, T. Shirai, J. Urata, K. Okada, M. Inaba, Proc. of the IEEE-RAS Int. Conf. on Humanoid Robots, Osaka, JP, 811 (2012)
43. M. Kawamura, S. Ookubo, Y. Asano, T. Kozuki, K. Okada, M. Inaba, Proc. of theIEEE-RAS Int. Conf. on Humanoid Robots, Cancun, MX, 814 (2016)
44. M. Penčić, M. Čavić, B. Borovac, Proc. Of the Int. Conf. on Power Transmission – BAPT, Ohrid, MK, 115 (2016)

Using brain-computer-interface for robot arm control

Sebastian-Daniel Rosca[1], and *Monica* Leba[1,*]

[1]University of Petroşani, Department of Computer and Electrical Engineering, 332006, Petroşani, Romania

Abstract. In this paper, we analyze the principles of brain computer interface that convert the electrical activity of the human brain into commands for a computer of any embedded device. For this reason, we present the existing devices and applications from the area of brain-computer interfaces with advantages and disadvantages. Further, we propose a solution for brain control of a robotic arm. We develop the model and simulate the entire system functioning, both the robotic arm control and the brain signals processing. The final purpose of our research is to achieve a brain-computer system that controls a robotic arm that can replace a human arm.

1 Introduction

Over the time, people have fantasized about the ability to communicate and interact with the machine by simply thinking or creating devices that read thoughts. These ideas have captured people's imagination in the form of ancient myths and science fiction stories.

In recent years, control devices „by the power of the mind" is a very controversial topic, but highly researched, both in the last generation gadget, such as smart phones, laptops and tablets and even televisions intelligent, but also in medicine, to be used by people with disabilities for which these technologies could be the only way of communication with the external environment. Today, brain-machine interfacing (BMI) development [brain-computer interfacing or (BCI)] has a rapid growth of research because it provides a unique way of communication between a human and a machine (or device) without any neuro-muscular intervention [1].

Brain-computer interfaces are often used to help paralyzed people or with motor functions diminished by connecting to a robotic arm; it's important that these people should not be fixed in one place, as it happens when they are connected to the interface brain-computer through a cable, but they can move around the room - thus defending the interest for a wireless connection. The basic idea of a BCI system is to measure electrical activity, magnetic or any other physical manifestations of brain activity and convert them into commands for a computer or any other device (wheelchair, neuroprosthesis). A BMI is a system based on digital signal processing tools and machine learning how to identify and predict the user's cognitive state from corresponding brain signal [1].

*Corresponding author: monicaleba@upet.ro

DOI: 10.1051/matecconf/201712108006

The brain signals are recorded either invasively or noninvasively [1]. Noninvasive means are widely used by most BMI/BCI researchers for their simplicity in user interface, although invasive means provides better performance in terms of accuracy and precision. The most widely used noninvasive recording technique is electroencephalography (EEG - where the signals are recorded by electrodes placed on the scalp) because it is inexpensive, portable, easily available and has high temporal resolution [1].

2 Applications

Neurospeller application ensure a way for communication for people that suffer from severe communication disabilities gained from neurological diseases and offers a new mode for human interaction to improve their life [2]. BCI applications for communication purposes are based on a virtual keyboard [3] that is shown on the display that offers the user possibility to select a letter from the alphabet by means of a Brain-Computer Interface (BCI) [4].

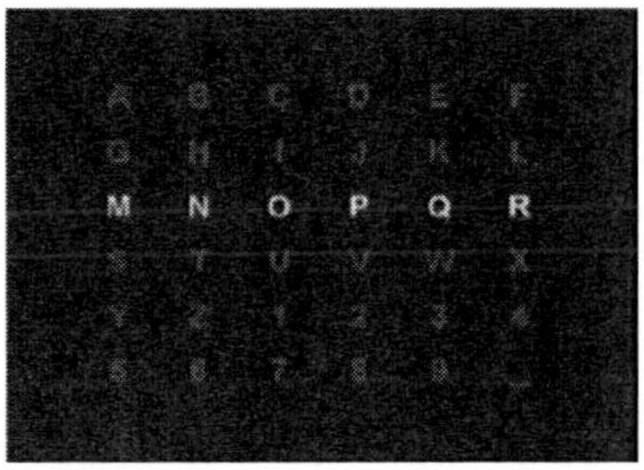

Fig. 1. User display in P300 Speller [3].

Wheelchair Control with Mental Tasks could be represent a real help for person that suffer from hemiplegia because they don't have a full control of their body and they cannot use a joystick of an electric wheelchair [5]. This technology offers for these person a chance to obtain independence and don't have any more to request help from a companion to push their wheelchair [6].

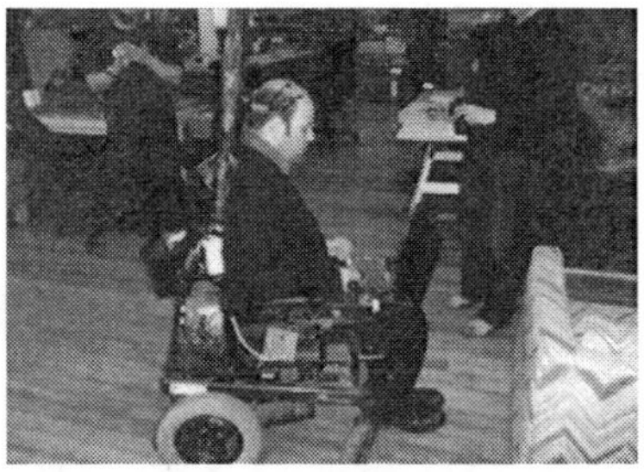

Fig. 2. BCI - Brain-Controlled-Wheelchair [5].

Fig. 3. BCI-based navigation in a complex VE [9].

Brain-controlled video gaming technology can be used both for medical and non-medical purposes [7]. As possible purposes can be: to allow the persons with severe mobility deficiency (e.g. spinal cord injuries) that cannot use a keyboard, a mouse, or a joystick to play video-games; to recover the lost functions result from an disease or a trauma at the level of brain (e.g. stroke or neurodegenerative diseases); for relaxation and entertainment to bring experience to a new level [8].

Brain-controlled Robotic Prosthetic Limboffers a solution for people who have suffered limb amputations resulted in the development of Brain-Computer Interface for controlling robotic limb prosthesis through EEG signals that offer a better solution than the existing ones (e.g. control of robotic human prosthesis by electromyographic sensors placed on the muscles of healthy limb to replicate the movements) and can be helpful for completely paralyzed peoples [10].

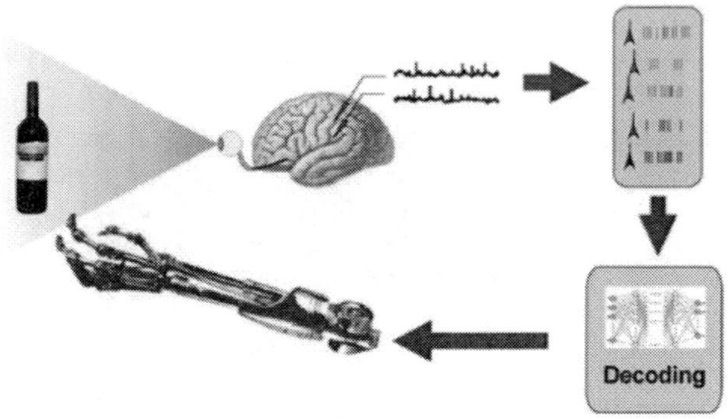

Fig. 4. Diagram of a neural prosthetic system [11].

3 Software

BioSig is an open source software library developed by Graz University of Technology for biomedical signal processing. BioSig provides open source software tools for many different areas applications: neuroinformatics, brain-computer interfaces, neurophysiology, psychology, cardiovascular systems and sleep research [12].

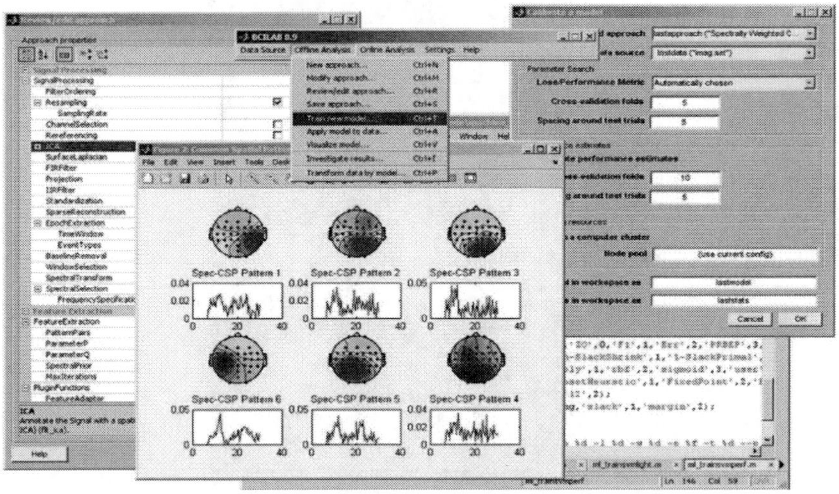

Fig. 5. BCILAB Sample GUI.

OpenViBE is developed by INRIA and is a software platform used to designing, testing and using brain-computer interfaces. OpenViBE is a free open source software for real-time neurosciences (real-time processing of brain signals) because it can be used to acquire, filter, process, classify and visualize brain signals in real time. Some of OpenViBE application fields are: medical (assistance to disabled people, real-time biofeedback, neurofeedback, real-time diagnosis), multimedia (virtual reality, video games), robotics and all other application fields related to brain-computer interfaces and real-time neurosciences.

BCILAB is a MATLAB toolbox and EEGLAB plug-in for the design, prototyping, testing, experimentation with, and evaluation of Brain-Computer Interfaces (BCIs) and other systems in the same computational framework. The toolbox has been developed by C. Kothe at the Swartz Center, inspired by the preceding PhyPA BCI toolbox created by C. Kothe and T. Zander at the Chair for Human-Machine Systems, Berlin Institute of Technology [13].

4 Brain controlled robotic arm

The system developed in this paper, which is still in research and needs improvements, aims to replace lost functions due to amputations for the human arm. The system includes a robot arm with five fingers and BCI headset with five channels.

In accordance with the real model of a human arm that provides seven degrees of freedom, the proposed robot arm provides 7 degrees of freedom for arm and 16 degrees of freedom for hand [14].

The robot allows seven types of movement along the x, y, and z axes, as follows:3 degrees of freedom for the shoulder joint (movement right-left, up-down and forward-back); 2 degrees of freedom, 1 for the elbow joint (movement forward-back/flexion-extension) and 1 for forearm (pronation/supination); 2 degrees of freedom for the wrist (rotation between the shoulder–elbow and elbow–wrist parts).

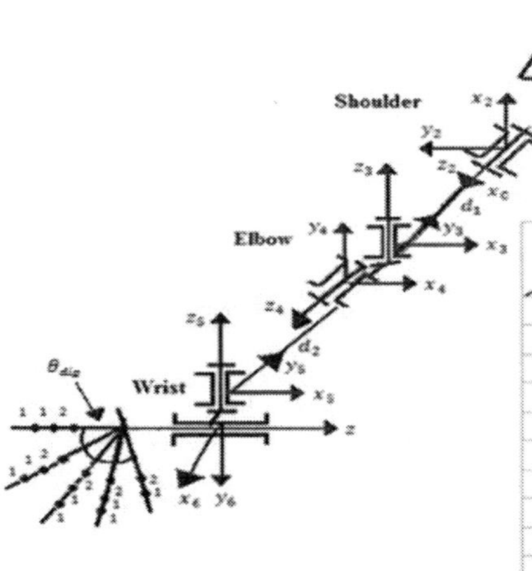

Coord. / i	a_i	α_i	d_i	θ_i
1.	0	$-\pi/2$	0	θ_1
2.	0	$-\pi/2$	0	$-\pi/2+\theta_2$
3.	0	$-\pi/2$	d_1	$-\pi/2+\theta_3$
4.	0	$-\pi/2$	0	θ_4
5.	0	$-\pi/2$	d_2	θ_5
6.	0	$-\pi/2$	0	$-\pi/2+\theta_6$
7.	d_3	0	0	$-\pi/2+\theta_7$
8.	0	$-\pi/2$	0	$-\pi/2+\theta_8 \pm \theta_{dig}$
9.	d_4	0	0	$-\pi/2+\theta_9 \pm \theta_{dig}$
10.	d_5	0	0	$\theta_{10} \pm \theta_{dig}$
11.	d_6	0	0	$\theta_{11} \pm \theta_{dig}$

Fig. 6. Robot arm outline.

We apply the Denavit-Hartenberg formalism to determine the direct kinematic model of the robot arm in order to model and simulate the movement in MatLab-Simulink.

Using the parameters identified in the Denavit-Hartenberg table from figure 6, we determine the movement matrix for each finger, based on the formula:

 DOI: 10.1051/matecconf/201712108006

$$T_{i-1,i} = \begin{bmatrix} cos\theta_i & -sin\theta_i cos\alpha_i & sin\theta_i sin\alpha_i & a_i cos\theta_i \\ sin\theta_i & cos\theta_i cos\alpha_i & -cos\theta_i sin\alpha_i & a_i sin\theta_i \\ 0 & sin\alpha_i & cos\alpha_i & d_i \\ 0 & 0 & 0 & 1 \end{bmatrix} \quad (1)$$

The mathematical model for the robot arm was next simulated in MatLab-Simulink. In figure 7 there is presented the shoulder joint formed up by three revolute simple joints linked by bodies of 0 lengths. We adopted this solution in order to point out the structure of the shoulder joint and to be able to accurately control the three revolution movements for this element. All the other joints are similar to this.

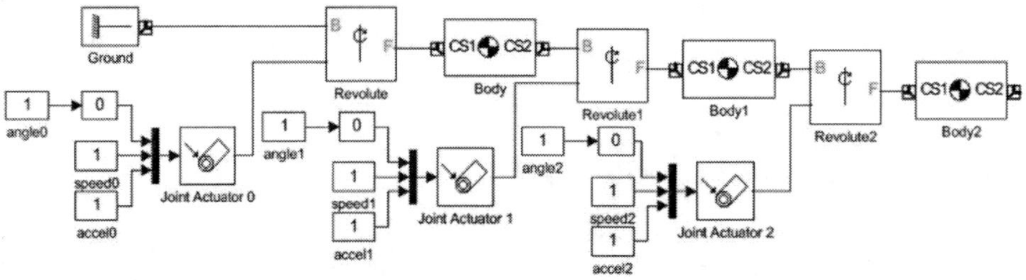

Fig. 7. MatLab-Simulink model of the shoulder part of the robotic arm.

The robotic hand is equipped with five high-end servomotors for advanced robotic applications that can operate the 16 joints of the fingers, either individual or in the same time for the gripping operations.

Entire parts of the robotic arm were projected in computer aided design software and were printed from Polylactic acid plastic (PLA) using a 3D printer. The BCI headset that was used in our research is Emotiv Insight.

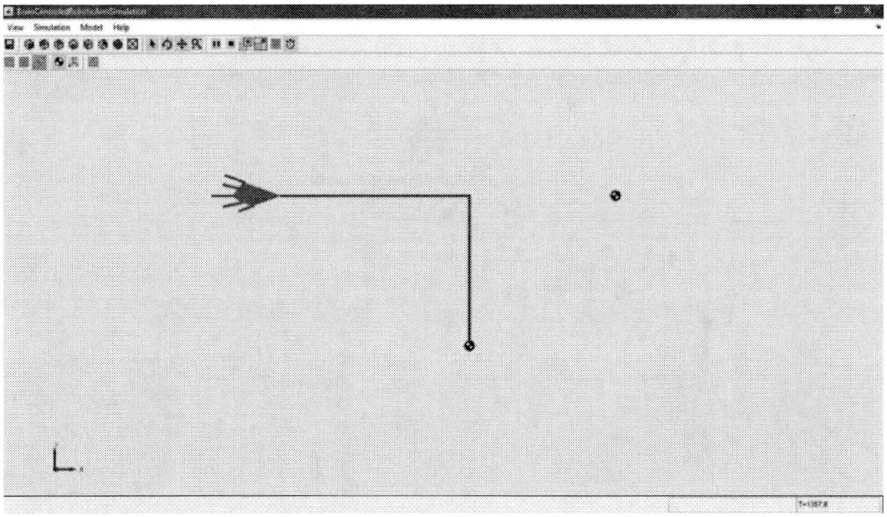

Fig. 8. Simulation of robotic arm (90°).

The Emotiv Xavier Control Panel interface that we use for control the human robotic arm through mental commands is a revolutionary interface for human-computer interaction. The interface allows 13 types of movement of a virtual object. After that movements are training, the Emotiv Insight headset create a pattern based on EEG principle and signals are convert into a command that can be understood by a computer.

We proposed to use 4 mental commands for control the human robotic arm in four complex movements. The mental commands chosen are the follows:

- lift mental command (fig. 9) for up the shoulder along the z_1 axis (0°-90°) and flexion the ellbow along y_3 axis (0°-90°) (fig.8,9)
- drop mental command for down the shoulder joint along the $z_1$axis (90°-0°) and extension the ellbow along y_3 axis (90°-0°)
- rotate right mental command for close the human robotic hand
- rotate left mental command for open the human robotic hand

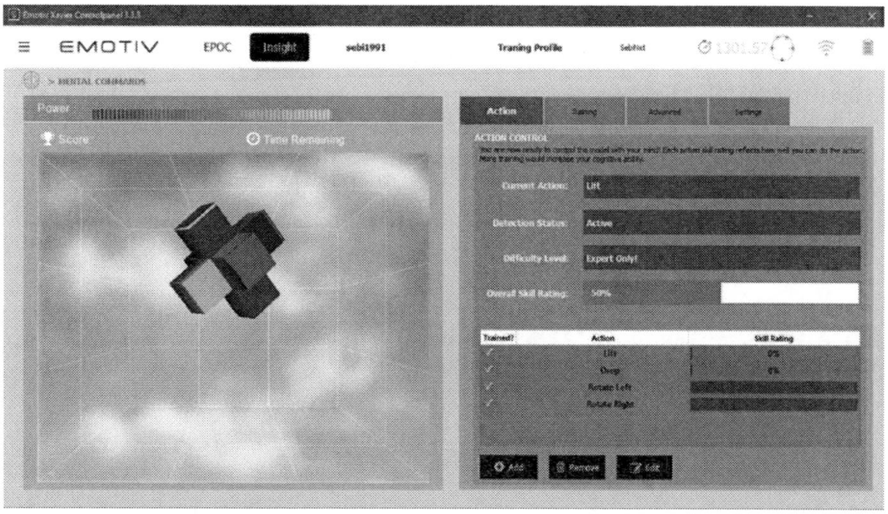

Fig. 9. Interface Emotiv Xavier Control Panel for lift mental command.

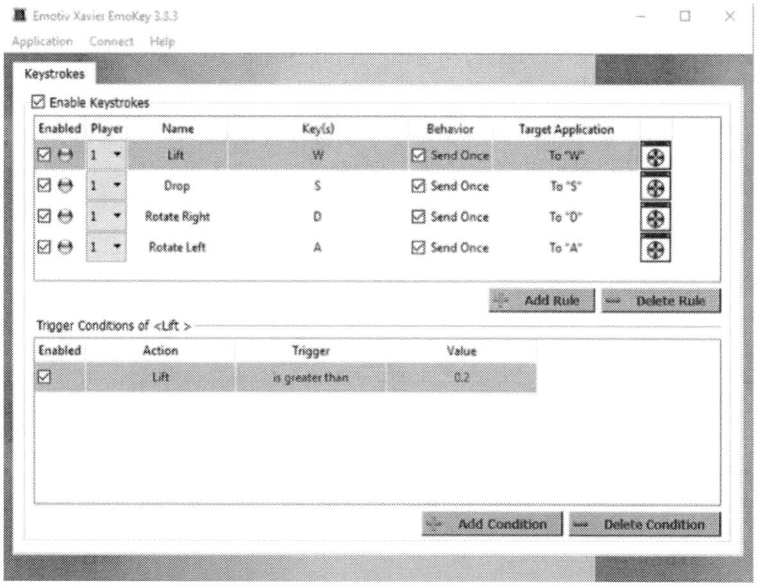

Fig. 10. Emotiv XavierEmoKey user interface.

In figure 8 is presented the simulation results for a lift command.

In figure 9 is presented the corresponding brain computer interface. As can be seen the mental command for arm lift resulted from the trainning, based on EEG and focus signals.

Emotiv XavierEmoKey user interface (fig.10) converts the results obtain after the measurements of mental command in predefined key sequences based on logical rules – Mapping EmoKey defined by the user. We assigned a set of rules for convert the mental commands in keystroke event.

5 Conclusions

Brain-computer interface is a revolutionary technology that opens a path from a new era of human computer interaction both by the fact that it addresses a large number of target groups but also that it can be a real help for making daily activities much easier.

In this research we propose a robot arm that has all the joints' movements specific to the human arm. We developed the Denavit-Hartenberg model for this robot arm and simulated it in MatLab-Simulink in order to validate its use.

Then, there was developed a simple brain control interface based on a commercial BCI headset in order to prove the possibility to brain control such a robot arm using 4 mental commands, 2 of them for the shoulder movement and 2 of them for the hand.

We propose as further research to implement within the brain-controlled robotic arm a series of pressure sensors for each finger to control the objects grip force. Also, we propose to develop a method to real-time adapt the brain-controlled robotic arm movements with the goal of improving response time and performing natural movements.

References

1. S. Bhattacharyya, S. Shimoda, M. Hayashibe, IEEE Transactions on Systems, Man and Cybernetics: Systems, **46**(7), 957 (2016).
2. L. F. Nicolas-Alonso, J. Gomez-Gil, Sensors 12(2), 1249 (2012), (Available at: http://doi.org/10.3390/s120201211)
3. M. Thulasidas, C. Guan, J. Wu, IEEE Transactions on Neural Systems and Rehabilitation Engineering, **14**(1), 24 (2006)
4. R. Fazel-Rezai, B. Z. Allison, C. Guger, E. W. Sellers, S. C. Kleih, A. Kübler, Frontiers in Neuroengineering, **5**(14), (Available at: http://doi.org/10.3389/fneng.2012.00014, 2012)
5. ***http://www.instructables.com/id/Brain-Controlled-Wheelchair/, Accesed 8.03.2017
6. T. Kaufmann, A. Herweg, A. Kübler, Journal of Neuro Engineering and Rehabilitation, **11**(7), 15 (Available at:http://doi.org/10.1186/1743-0003-11-7, 2014)
7. D. Coyle, J.Principe, F. Lotte, A. Nijholt, IEEE Transactions on Computational Intelligence and AI in games, **5**(2), 77 (2013)
8. C. G. Lim, T. S. Lee, C. Guan, D. S. S. Fung, Y. Zhao, et al, PloS one, **7**(10), e46692., (Available at: doi:10.1371/journal.pone, 2012)
9. F. Lotte, Proceedings of the 6th International Conference on Foundations of Digital Games. ACM Jun 2011, Bordeaux, France, 325, (2011).
10. C. Guger, W. Harkam, C. Hertnaes, G. Pfurtscheller, Proc. AAATE 5th european conference for the advancement of assistive technology, 3, (1999)
11. C. Pedreira, J. Martinez, R. Q. Quiroga, (Available at: https://www2.le.ac.uk/departments/engineering/research/bioengineering/neuroengineering-lab/Publications/PID924857.pdf, 2009)
12. *** http://biosig.sourceforge.net/, Accesed 1.05.2017
13. ***https://sccn.ucsd.edu/wiki/BCILAB, Accesed 11.02.2017
14. A. Ionica, M. Leba, Procedia Economics and Finances **23,** 671 (2015)

Method for analysing the deformation and stress of vertebra assembly

Alina Rodica Spanu[1,*] and *Daniel* Besnea[1]

[1]University Politehnica of Bucharest, Mechatronic and Precision Engineering Department, 313 Splaiul Independentei, Sector 6, Bucharest, Romania

Abstract. The paper presents a method for testing the deformation and stress values of the vertebra assembly when variable forces are applied on the body surface. The working points have very well-known positions and they could be changed, so the compression force, torsional or bending couples could act. The system is actuated by an electrical stepper motor controlled with Arduino. The force and displacement sensors were used.

1 Introduction

The lumbar spine is one of the main sections of human body due to its important role of providing the capability of maintaining the vertical position during walking, working and some other activities. Generally speaking, it is a part of our body with static and dynamic influence as well as the anatomic role. A per cent of 60 – 85% of population is afflicted with spine pains and this becomes a phenomenon even for young people, 40 -45 years old for instance [1].

The main problems that could appear causing some troubles and pains are degeneration process due to aging, some loads exceeding the normal values during working and leisure time or accidental situation causing injuries.

Regarding the forces acting on spine during movement or stationary periods, we have to mention at first the gravity, which is proportional with body mass. Meantime, this mass value influences the inertia force developed during accidental fallings amplified by the acceleration. Supplementary forces could appear when we move some objects by humping them, so the load becomes mechanical couple depending on the distance between the spine axis and the object. Ergonomic aspects, such as chair position in front of the table as well as chair shape, are brought about some distress too especially when we are sitting a long period without pause for standing or walking. At last we should mention the sedentary life at home watching TV set or doing something else.

There are many studies that analyse the influence of all these factors on the spine deformation, so the conclusion is the lumbar section is one of the most affected. The lumbar spine is the section of spinal column between the thorax and the sacrum. It consists of five vertebrae named L1 to L5 with their posterior elements and articular facet joints, of intervertebral discs, ligaments and the surrounding muscles.

* Corresponding author: spanu_alina@yahoo.com

The lumbar vertebrae are quasi cylindrical with a lateral width of 40-50 mm and sagittal depth of 30-35 mm. The height of a lumbar vertebra is about 25-30 mm. The lumbar vertebrae are thicker anteriorly than posteriorly. The intervertebral discs separate the adjacent vertebrae. They are quasi cylindrical with a lateral width of 40-45 mm and sagittal depth of 35-40 mm. The height of the lumbar disc is about 10 mm. The cartilaginous endplates separate the nucleus and annulus from the vertebral bodies and they cover almost the entire surface of the adjacent vertebral bony endplates. The plates have a mean thickness of 0.6 mm [1].

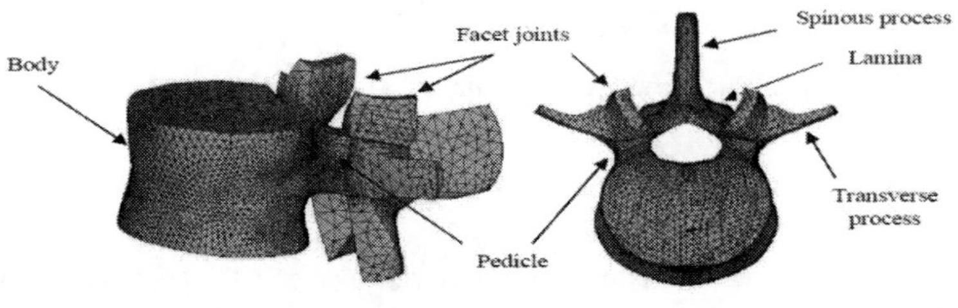

Fig. 1. Anatomy of the vertebra of the lumbar spine [2].

As we may infer from figure 1, there are the vertebra body, spinous process, transverse process, lamina and pedicle and facet joints. The layers of cortical and cancellous bone of the body are integrated in the same type of material only for computing the deformations.

The particularity of each person should be taking into account by various types of apparatus gathering biomechanical data by image processing and finding the vertebra dimensions using computation methods finally.

The study [2] presents Finite Element Analysis (FEA) of the three – dimensional model of L2 vertebrae, made by using SolidWorks environment that has been done with ANSYS 11.0 considering hyperextension loading. The material has linear and isotropic properties and tetrahedral solid element was used for the entire model. The applied force in axial direction was 460 N for a person of 80 Kg and the level of stress was analysed.

The paper [3] presents the experimental study made on the vertebra under compression with variable force. There are two steel plates, up and below the vertebra and variable pneumatic force is acting by varying the supply air pressure. The force and occurring displacement are simultaneously monitored.

The goal of this paper is to analyse the deformations and stress level for the assembly of two vertebrae considering the values for mechanical couple by applying variable forces over the cylindrical body surface of the vertebra. We suppose there are the real working conditions during lumbar spine deformation, if the factors presented above are acting. We may analyse the working point positions on the body surface as the result of person movements. An experimental setup was done for pointing out the force variation and the positions of its applying points while the deformation was registered and analysed.

2 Three dimensional model of the vertebra

In order to analyse the stress and deformation whether the forces and couples are acting, it is necessary to make the 3D model of the vertebrae by using Catia V5R17 environment using the Generative Shape Environment due to the very complex spatial shape.

We aim to provide a realistic model as possible as we may asses it and the result is presented in figure 2.

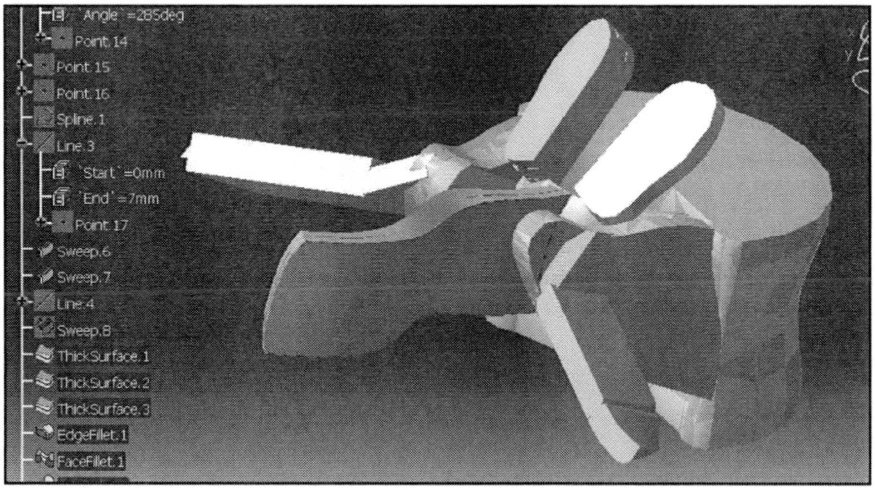

Fig. 2. The three dimensional model of the vertebra.

The next step was the rapid prototyping stage applied to the vertebra model by using the 3D printer – MakerBot Replicator. The main advantage of using the new technology of Additive Manufacturing is that it allows the user to manufacture the product directly from the 3D model in a very short period. Moreover, the spatial shapes could be very complex such them we aim to use for our experimental setup. The entire process is a cheaper one and the accuracy of the vertebra dimensions is good enough for later measuring.

After we have designed the model, it is transferred to the sectional processor and divided into a series of computed triangles. In figure 3 it is presented the manufactured model of the vertebra.

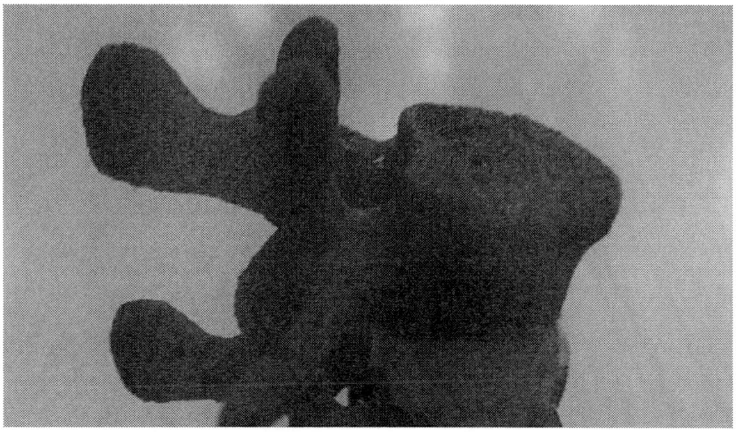

Fig. 3. The manufactured model of the vertebra.

In order to study the tensions and internal deformations during the forces are acting on the vertebra, we have design the profile of spatial cam – follower mechanism. The rotational cam is actuated by an electrical steper motor controlled with Arduino and the translational follower is developing the compression force acting on the vertebra.

The movement law is given by the equations below comprising the phase of starting the process, meaning the time t is in the range of t_1 0 ÷ 1.2s and it is followed by the phase of ending the process, when the time is in the range of t_2 1.2 ÷2.4s. Taking into account the way the maximum acceleration value influences the internal forces of this mechanism, we have imposed this value $y''(t) = 0.8$ m/s. The minimum value of acceleration has the main role in establishing the contact surface of spatial cam and follower, so that it may be computed with the equation:

$$y''_{min} = -(t_1 / t_2) \cdot y''_{max} \quad (1)$$

The function of movement for the starting phase of spatial cam mechanism is given by the equation below [5]:

$$y = 1 / (t_1 \cdot (3 - t_1 - t_2)) \cdot (3 \cdot x^2 - x^3 / t_1) \quad (2)$$

Meantime, we have computed the function for the last phase when the follower has to be stopped, as we may infer from the equation below:

$$y = 1 - (3 \cdot t_2 \cdot (1 - x)^2 - (1 - x)^3) / (t_2^2 \cdot (3 - t_1 - t_2)) \quad (3)$$

The third dimension for describing the trajectory of the contact point between the cam and the follower is computed knowing the variation of angle Φ, whose value is starting from zero degrees and its maximum value is 24^0 in order to achieve the maximum and minimum value for ordinate y. It is expressing the level difference between the horizontal plane and the working plane.

Consequently, we have presented in figure 4 the variation of movement law y(x) and in figure 5 the variation of cam profile depending on x, y and Φ.

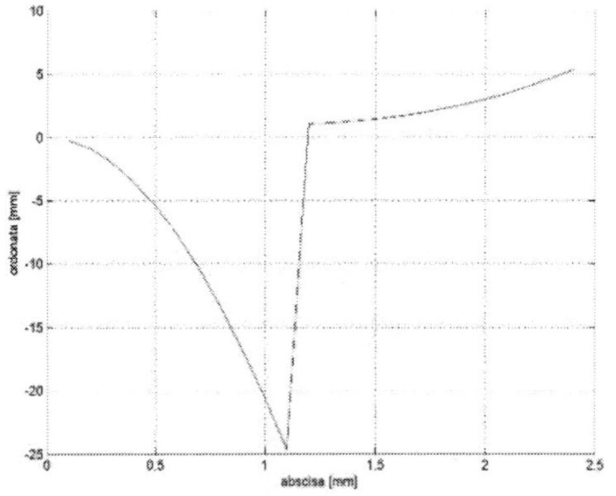

Fig. 4. The variation y as x function describing the movement law.

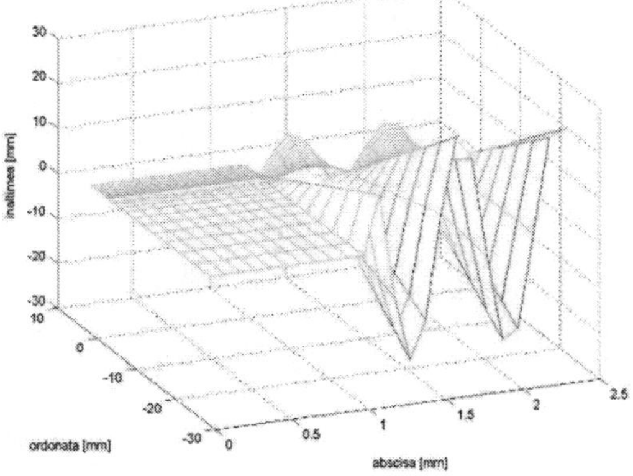

Fig. 5. The variation of cam profile.

3 Experimental set-ups

Our goal is to analyse the stress and deformation values of the vertebra assembly 1 by measuring them, so that we have designed the system with three dimensional model presented in figure 6. It could provide a variable force applied in a well-known point position by using the electrical stepper motor 2 placed on a guiding subassembly 3 having an imposed accuracy.

The cam 4, actuated by the motor, provides the force acting on the vertebra body surface. Its working point position could be changed by motor displacement or by changing the vertebrae position, because of the particularity of placing it on the table 5.

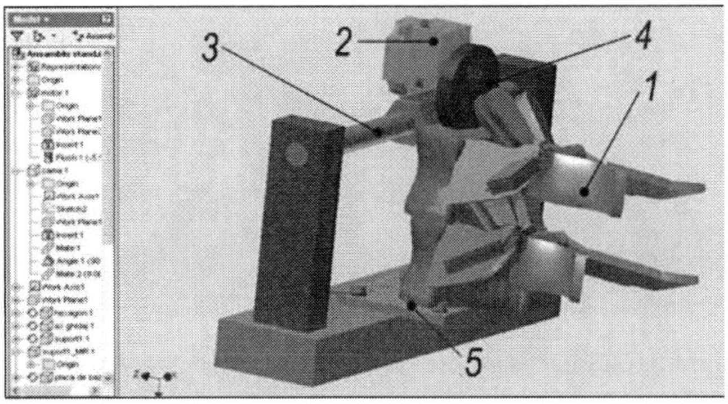

Fig. 6. The experimental setup of the system with variable force.

The main advantages of this solution are the variable and controlled force applied on the surface and the possibility of changing the working point position.

The most important force is the compressive normal one, but it is acting together with sagittal and lateral shear forces, so that the flexion/extension movement appears. Meantime, the bending and torsional mechanical couple values are generated. Consequently, the deformations of each element of the assembly appear and there are amplified by the disc with elastic properties between the two vertebrae.

The figure 7 presents the experimental set-up that we have made in order to measure the acting forces and the corresponding displacement of vertebra body. The system is actuated by the electrical stepper motor 1, which could be moved along the X axis by the guiding system 2.

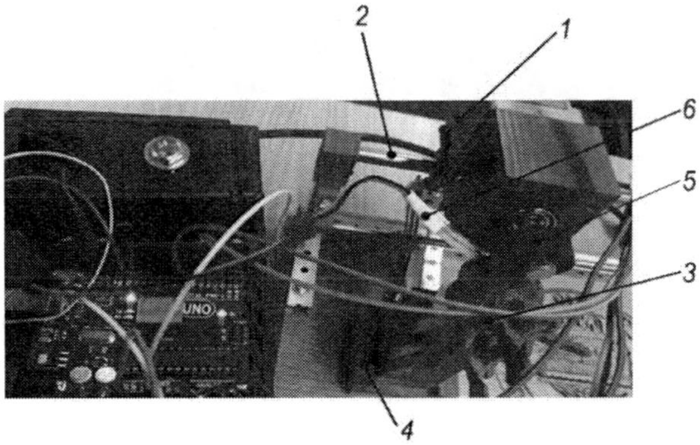

Fig. 7. The experimental set-up made for measuring the vertebra deformations.

The vertebra assembly 3 is put on the part 4, so it could be rotated around Z axis. On the motor shaft is the cam 5 whose computed profile generates the variable forces applied on the points with very well-known coordinates. There are force and displacement sensors 6 for measuring the vertebra deformation, if flexion/extension movements are acting. We have used the FSR 400 force sensor, a very accurate one, with 4 mm its active diameter whose resistance is depending on how much pressure is being applied. We have used a 3 axis accelerometer/3 axis gyro with I2C interface accesses for detecting the vertebra assembly deformation.

Fig. 8. The internal tension values according to Catia V5R17 Analysis.

In figure 8 we have shown the computed values of tensions applying the finite element analysis. In figure 9 we have presented the vertebra deformations when the compression force is acting.

Fig. 9. The internal translational displacement of the vertebra.

3 Conclusions

The paper presents a method for deformation and stress measurement that could be applied for vertebra assembly. The main advantages of the proposed method are variable forces and mechanical couples provided by using a cam with computed shape actuated by controlled electrical stepper motor.

The three–dimensional model was designed according to the imposed demands and it was manufactured by FDM technology. We may determine the maximum values for deformation as force function.

The future work will be focused on improving the experimental set-up solution considering compressing and torsional couple acting in the same time for a vertebra that has a material with properties ressembling with the human bone tissue.

References

1. M. Kurutz, L. Oroszvary, *Finite Element Modeling and Simulation of Healthy and Degenerated Human Lumbar Spine, Finite Element Analysis - From Biomedical Applications to Industrial Developments*, (ED D. Moratal, 2012).
2. A. Zulkifli, A. K. Ariffin, M. M. Rahman, IJAME **3**, 256 (2011)
3. K. D. Bouzakis, S. Mitsi, N. Michailidis, I. Mirisidis, G. Mesomeris, G. Maliaris, A. Korlos, JMNJ, **4**(2), 152 (2004)
4. J. Noailly, H. J. Wilke, J. A. Planell, D. Lacroix, J.B., **40**(11), 2414 (2007).
5. I. Simionescu, V. Moise, A. Spânu, *Mecanisme,* (I.P.B., 1991)

Adaptive robotic end-effector with embedded 3D-printed sensing circuits

Aurelian Zapciu[1,*], *George* Constantin[1], and *Diana* Popescu[1]

[1]University POLITEHNICA of Bucharest, Department of Machine and Production Systems, 313 Splaiul Independentei, sector 6, Bucharest, Romania

Abstract. The paper presents the development and testing of an adaptive robotic end-effector used for manipulation of sensitive objects such as fruits and vegetables. The end-effector uses Fin-Ray-structured 3D-printed fingers with embedded conductive 3D-printed sensing circuits, which give the end-effector capacitive touch sensing and bend sensing capabilities. The conductive 3D-printed circuit is connected to a control circuit consisting of a low-current DC power source and a microcontroller. As the end-effector finger is subjected to various forces and other external stimuli, changes in the electric signals that run through the conductive circuit of the end-effector finger are detected by the microcontroller. The electric signal is processed in order to provide real-time information about contact detection, finger position or gripping force. This information was used for process monitoring purposes and as feedback for the end-effector actuator.

1 Introduction

Research in the field of additive manufacturing (AM) focused recently on improving materials and processes. Thus, processes, such as PolyJet, allowing the build of realistic multi-color, multi-materials parts [1] were developed for use in many fields, especially for visualization, marketing or training purposes, but improvements were as well brought to the existing AM processes in terms of accuracy, resolution, reproducibility, adaptability to new materials, printed part quality and mechanical properties [2]. Currently 3D-printing can produce flexible metamaterial parts [3], parts with controlled microstructures and properties [4] for smart materials applications, as well as functional parts and assemblies from materials with electric current conductivity that can be used in a large range of applications and industries [5].

Robotics is one field which benefits from these advances. The possibility to 3D print intelligent end-effectors that can adapt to objects' shapes and dimensions, similar to a human hand, opens new perspectives in manipulating, for instance, sensitive or irregular objects like fruits or vegetables, as the universal gripper based on jamming granular material [6] does. In this context, this study is focused on developing and testing an end-effector with three 3D-printed fingers based on Fin-Ray structures [7-8]. It embeds 3D-printed conductive sensing circuits, which provide the end-effector capacitive touch sensing and bend sensing capabilities.

* Corresponding author: aurelianzapciu@yahoo.com

DOI: 10.1051/matecconf/201712108008

Adaptive or soft grippers are designed and optimized for different tasks as presented in [9], their ability to adapt to vary surfaces and sizes offering the versatility required in diverse applications ranging from grabbing biological samples [10] to thumb rehabilitation [11]. Although all based on Fin-Ray effect, the differences in design (dimensions, number of fingers, number and type of joints, etc.) determine different grip strength which is an important factor to consider as it is highly related to the end-effector functionality and application constraints. Moreover, there are different actuation solutions and modalities to monitor the fingers position and the contact between object and gripper, our solution being based on establishing a control circuit between the conductive 3D-printed circuit in each finger, the power source and a microcontroller. The changes in electric signals when the finger is subjected to a load are detected in real time by the microcontroller and used as feedback for the actuator.

The use of Fused Deposition Modeling (FDM) process for manufacturing end-effector fingers with embedded conductive sensing Circuits already proved feasibility [12], this research focusing on developing and testing a whole gripper.

2 Method and materials

The multiple approaches to gripper adaptability presented previously, while functionally sound, experience difficulties when it comes to accessing data from the environment and providing feedback. To address this, third-party vision or sensor systems are used driving up system complexity and costs.

The approach taken by the authors uses novel conductive polymers to create electrically conductive circuits which are integrated in end-effector parts using multi-material 3D printing. Previous work [12] focused on identifying methods and physics which could be applied in the development of such an end-effector, while the current paper considers integrating multiple of these principles in a single prototype.

First, the experimental adaptive end-effector (fig. 1a) under development integrates capacitive touch sensors, which are used to sense the contact between the fingers and the target object. This effect is achieved using a plate made out of conductive polymer, which forms the electrode of a capacitor, fabricated on top of an insulating plate. When a target object enters this electrode's electrostatic field, it changes its capacitance. The second capability is sensing the bending of the end-effector finger under load. As the material bends, the outer radius of a conductive plate becomes longer or shorter than its nominal length under no load. The conductive polymer is subjected to elongation or compression

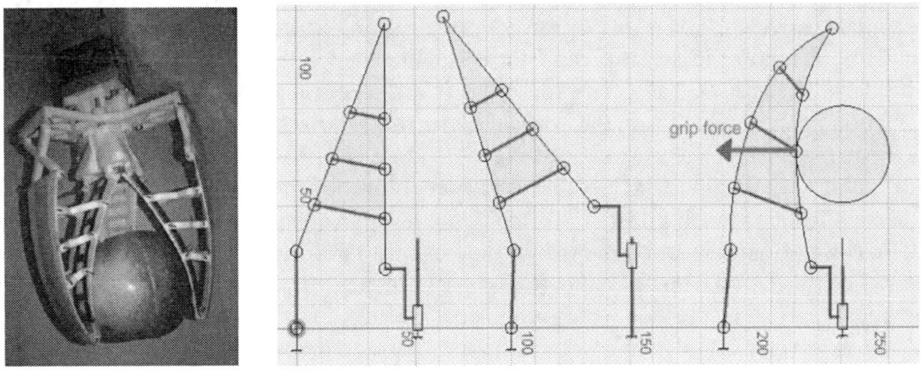

a b

Fig. 1. Developed adaptive end-effector: *a* – end-effector during testing; *b* – end-effector geometry and forces acting on a finger.

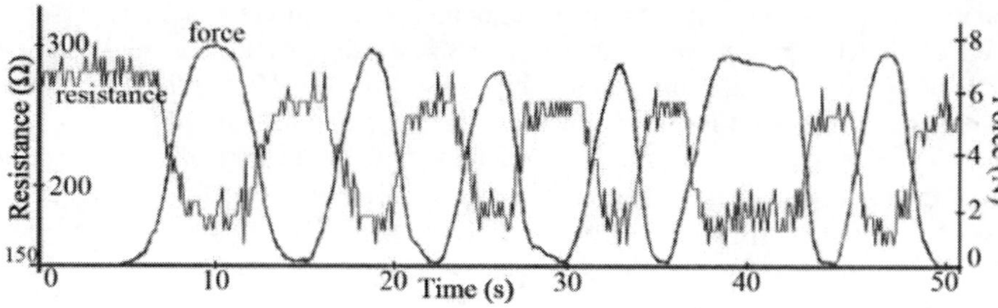

Fig. 2. Resistance of a 3D printed and applied force depending on time.

strain and microscopic cracks form in the structure, lowering or raising the resistivity of the conductive circuit. This effect is visible in fig. 2 showing resistance, which decreases when the gripping force of an end-effector finger increases.

2.1 Materials

Throughout the experiments, four different materials were used for manufacturing the gripper parts and the parts for the experimental stand. Unfilled ABS polymer filament was used to create the structure of the end-effector fingers and form the insulating plates of the integrated capacitive and bending sensors. For the conductive parts of the end-effector fingers, using an ABS-based material allows for a strong bond between part layers and prevents delamination during bending. Another important aspect of multi-material 3D printing is thermal expansion coefficient compatibility, as too big of a difference between the thermal expansion coefficients of different materials can create internal mechanical strain which can destroy the part during the cooling process. The chosen material was 3D Prima's Conductive ABS. The dual-material parts were fabricated using a QidiTech Avatar IV 3D printer. For the other components of the testing device, which have higher geometrical complexity and require higher dimensional accuracy, ABS P400 and ABS P400 Support were used. These components were fabricated using a Mojo 3D printer (Stratasys).

2.2 Determining a suitable build orientation

Considering how loads are applied in gripping applications and the geometry of the end-effector finger (fig. 1b) it was concluded that the ideal part orientation during fabrication is horizontal. However, given that electrical properties also vary with layer orientation, and for this application the sensing capabilities of the final product are just as important as mechanical strength, it was determined that it was necessary to test the second best part orientation as well, which is in the vertical plane, with horizontal layers covering the length of the part.

In order to determine which part orientation allows for the optimal electrical properties of this application, a series of four 80×15×4 mm test parts (fig. 3a) were fabricated and then tested. The test parts were orientated so they can be fabricated without the use of support material, with 0.8 mm thick conductive material plates orientated in the horizontal or in vertical plane. The conductive plates are insulated using a 0.4 mm thick unfilled ABS layer. For the parts fabricated in the vertical plane, the conductive layers were printed using parallel filament strands along the length of the part (fig. 3b) or in a zigzag pattern (fig. 3c). For the parts fabricated in the horizontal plane, the conductive layers were printed using an orthogonal crosshatch pattern with parallel (fig. 3d) and 45° (fig. 3e) road orientations.

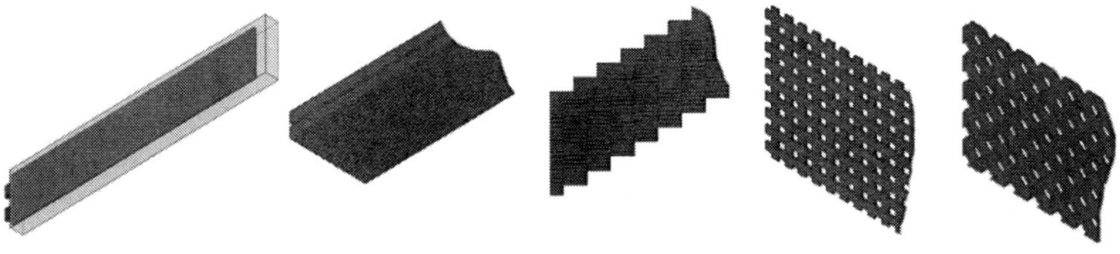

a b c d e

Fig. 3. Patterns used to test build orientation: *a* – test parts; *b* – parallel filament; *c* – zigzag pattern; *d* – crosshatch parallel; *e* – crosshatch 45° orientation.

Test parts were connected to an Arduino UNO microcontroller. The resistance of the conductive circuit is determined using the controller's integrated voltmeter and a simple voltage divider circuit where the measured resistance is wired in series with a known-value resistor. The known resistor is sized to match the resistance of the conductive circuit under no load, which was measured manually. The capacitive sensing capabilities are tested using a timing circuit that uses the controller's integrated timer and a 5 MΩ resistance. Capacitance variance is tracked using arbitrary units, as determining the absolute value of the capacitance is not needed in this touch-sensitive application.

In order to test the integrated bending sensor, the parts were bent on a testing rig (fig. 4). Resistance was measured for each 5 mm increments in displacement in the median plane between 0 and 20 mm. 5 sets of measurements were taken and the values were averaged.

The results of this preliminary test are shown in fig. 5. The test samples that were fabricated with the conductive layers in the vertical plane (square and star markers) performed poorly. The resistance of the conductive circuit does not vary sufficiently to provide a reliable determination of finger bending. The test samples that were fabricated in the horizontal plane both provided satisfactory results. The conductive circuit printed with an angular orthogonal pattern (up-pointing triangle) is preffered over the one that is parallel with the length of the part (down-pointing triangle) as the experiment show a bigger drop in resistance.

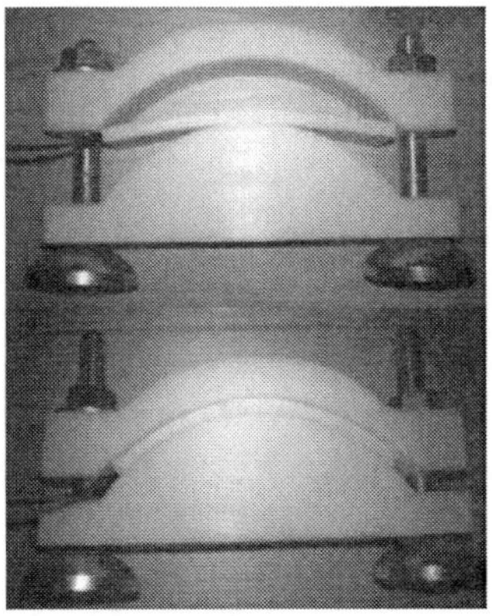

Fig. 4. Experimental stand for bending tests of finger plates.

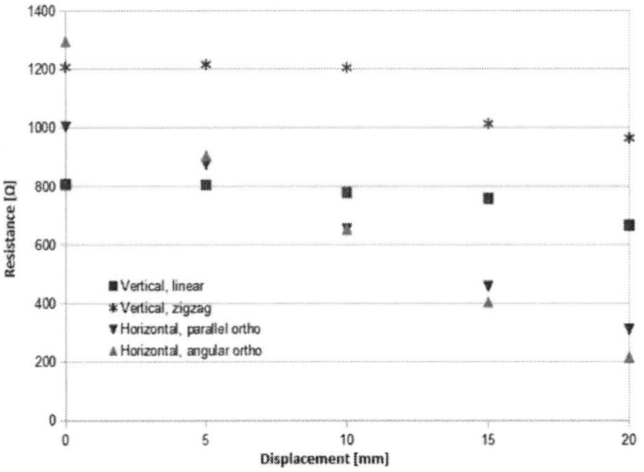

Fig. 5. Results of the bending test (vertical linear, fig.3b; vertical zigzag, fig.3c; horizontal parallel orthogonal; fig.3d, horizontal angular orthogonal, fig.3e).

2.3 Adaptive gripper prototype design

A prototype for an end-effector was designed using Autodesk 360 Fusion (fig. 6). The prototype uses three Fin Ray-type fingers, each composed of two sides with integrated conductive circuits. In total, six different conductive circuit patterns were designed and 3D printed. As determined from previous testing, the end-effector finger components were printed in a horizontal orientation, using orthogonal cross-hatch at a 45°angle. Each configuration was printed in a single process, using a dual-extruder QidiTech Avatar IV 3D printer.

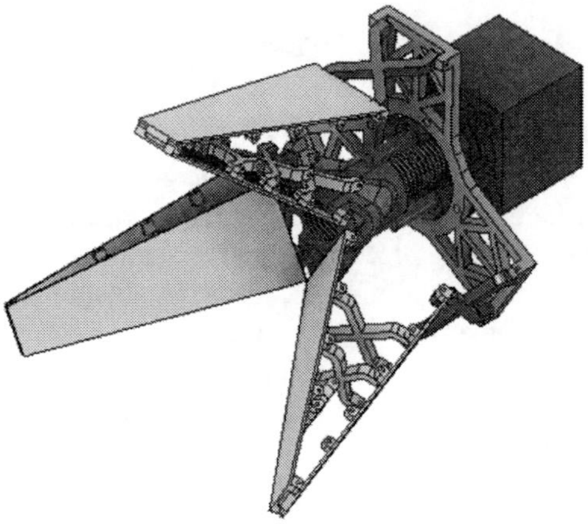

Fig. 6. Virtual prototype of the adaptive gripper.

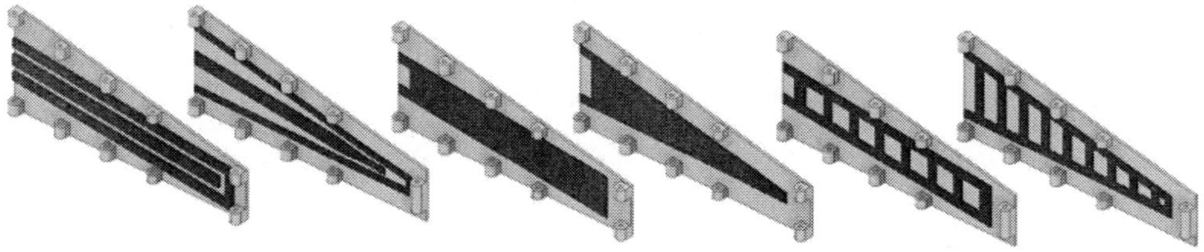

a b c d e f

Fig. 7. Various conductive plate designs 3D printed: *a* – v1, parallel stripes; *b* – v2, inclined stripes; *c* – v3, rectangular area; *d* – v4, triangular area; *e* – v5, rectangular ladder; *f* – v6, trapezoidal ladder.

2.4 Experimental setup

The testing setup is shown in fig. 8. The three Fin-Ray end-effector fingers were connected using steel wire to the main body which has an internal M16 thread. A 3D printed screw fixed to the shaft of the stepper motor drives the main body. The main body slides along a system of three steel round guides.

The end-effector prototype has a maximum opening of 110 mm at the tip of the fingers and 30 mm at the base. It can grasp firm round objects up to 80 mm in diameter and soft objects up to 100 mm in diameter. A stepper motor is connected to an Allegro A4988 stepper motor driver and powered by a 12V DC power supply. The entire system is controlled by an Arduino UNO controller connected to a laptop computer. The controller is powered through the USB port of the laptop computer. Data retrieved by the controller is sent via serial protocol to the laptop computer, plotted in real-time on the screen and saved in CSV format for further analysis. For this particular prototype, a stepper motor drive was preferred due to the simplicity of controlling it using the Arduino UNO controller, but all the results presented in this paper can be replicated with pneumatic or other types of actuators as well, as the sensing system does not depend on the driving method.

In order to determine which conductive circuit design performs better and to verify the feasibility of applying this concept in a real-world setting, a 12-hour endurance test was carried. The gripper is closed to grasp a tennis ball. The grasp is held for 2 seconds, after which the gripper is opened and the tennis ball is released. After 1 second, the test cycle is repeated. In total, one cycle lasts 8 seconds. After 12 hours and more than 5,000 test cycles, the orientations of the three end-effector fingers are reversed and another 12-hour test is carried. In total, each conductive circuit design was tested for 5,000 cycles while positioned on the inside part of the gripper finger (on the face that comes in contact with the target object) and 5,000 cycles while positioned on the outside part.

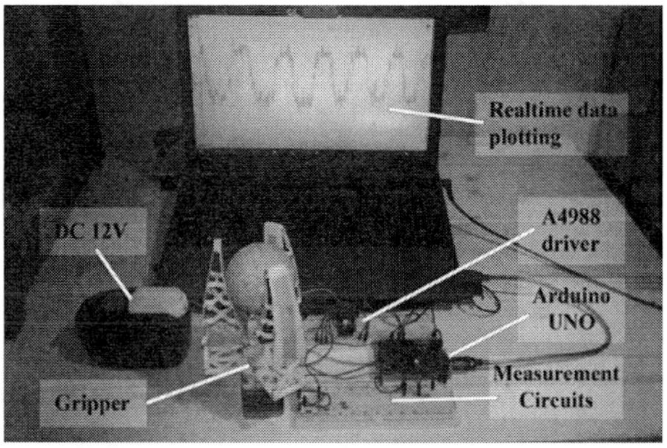

Fig. 8. Experimental setup for testing the adaptive end-effector prototype.

3 Results

Four out of the 6 conductive circuit designs - v1, v2, v5, v6 -showed good repeatability during the 24-hour continuous testing (more than 10,000 work cycles). Figure 9 shows the changes in the resistance of these 4 bend sensor designs during a work cycle. For better visualization of the performance, the graphs show the measured resistance increase or decrease in relative percentages, 0% being the resistance of the unloaded conductive circuit.

Results show that placing the finger face which contains the embedded bend sensor towards the target object produces an increase of resistance of up to 100% the initial value, allowing for a proportional determination of the amount the finger bends. When the bend sensor is positioned on the face that is away from the target object, there is a drop in resistance which is nominally less significant and the results are harder to use.

Following the endurance tests, an automated grasping procedure was setup using a feedback loop from the two embedded sensing circuits. Signal input from the capacitive touch sensor is used as a trigger to start closing the gripper and signal input from the bend sensor is used to stop the closing process when a threshold is met. The design variant chosen for this test was v1, which has shown the most reliable performance during the endurance test. The finger is mounted with the bend sensor on the inner face, as this positioning showed the biggest resistance difference. To further increase the reliability of the bend sensor readings, a signal smoothing algorithm was used (moving average). For this test, an apple attached to a string was used. The apple is lowered towards the opened gripper. When the apple touches the surface of the fingers, the change in capacitance is sensed and the microcontroller commands the stepper motor to close the gripper. As the gripper closes, the fingers start bending around the apple. When signal from the bend sensor reaches the preset threshold, the stepper is stopped and the grasp is held.

DOI: 10.1051/matecconf/201712108008

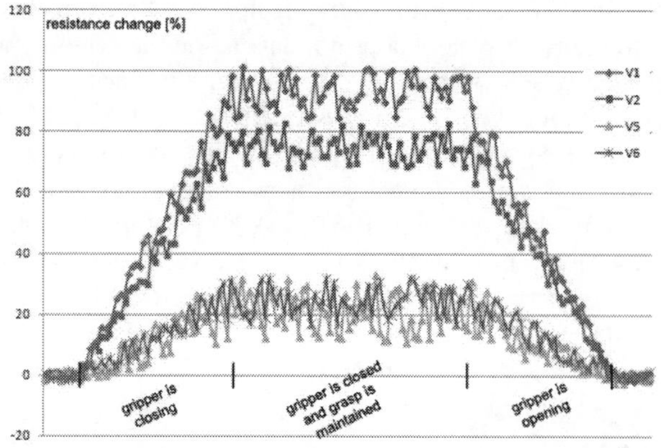

a

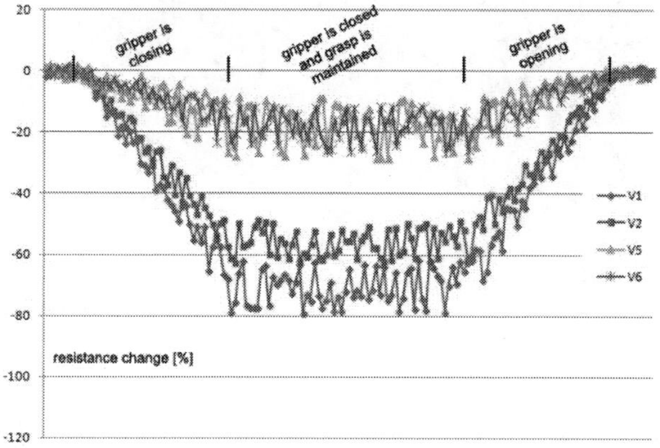

b

Fig. 9. Relative conductive circuit resistance during work cycle: *a* – bend sensor embedded in on the finger face coming in contact with the target object; *b* – bend sensor embedded on the opposite face.

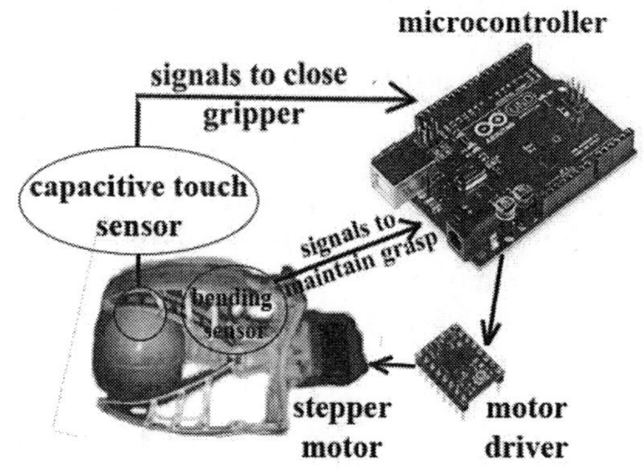

a

b

Fig. 10. Automated grasping test: *a* – working principle; *b* – automated grasping an apple.

4 Conclusions

The current paper was focused on developing and testing an adaptive end-effector with three fingers with Fin-ray geometry (Festo) manufactured using new conductive polymers printed inside the fingers' plates – multi-material 3D Printing.

The development process consisted in designing and analyzing different types of conductive circuit patterns, designing each finger and then the whole end-effector, as well as manufacturing these parts from different materials using filament deposition based 3D-printing process. For the conductive plates, four types of deposition patterns were tested and comparatively analyzed in order to determine the optimal build orientation, i.e. the one ensuring the best electrical properties for the application.

Thus, two experimental stands were designed, manufactured and used. They were needed for: (1) finding suitable build patterns and orientation for 3D printing end-effector fingers with embedded bend sensing capabilities, (2) determining which conductive circuit design performs better and (3) checking the gripper prototype's performance in a 12-hour endurance test (successive closing and releasing fingers for grasping a tennis ball, 8 seconds/cycle).

The results of the bending test showed that the conductive circuit printed horizontally in an orthogonal pattern rotated by 45° ensures the best bend sensing capabilities, being preffered to the vertical builds.

The results of the endurance test showed good repeatability for 4 of the 6 tested conductive circuit designs. The design that uses a line pattern (v1) had the most stable signal throughout the 24 hours of continous testing. This sensor design was then succesfully used in an automated grasping procedure that uses input signals from the embeded capacitive touch sensor and bending sensor.

References

1. A.T. Gaynor, N.A. Meisel, C.B. Williams, J.K. Guest, J. Manuf. Sci. Eng., **136,** 061015 (2014)
2. J. Scott et al., *Additive Manufacturing: Status and Opportunities*, (Science and Technology Policy Institute 2012)
3. A. Ion et al., Proceedings of UIST '16, *Tokyo, Japan* (available at: http://alexandraion.com/wp-content/uploads/2016UIST-Metamaterial-Mechanisms-authors-copy.pdf, 2016*)*
4. S. Tammas-Williams, I. Todd, SMM-11342 (available at: http://dx.doi.org/10.1016/j.scriptamat.2016.10.030, 2016)
5. T.M. Llewellyn-Jones et al., Smart Mater. Struct., ***25****(02LT01)*, (available at: http://iopscience.iop.org/article/10.1088/0964-1726/25/2/02LT01, 2016)
6. E. Brown et al., PNAS, **107** (44), 18809 (2010)
7. P.G. Madden, G.V. Lauder, Silas Alben, J. R. Soc. Interface, **4** (13), 243 (2007)
8. ***https://www.festo.com/net/en-gb_gb/SupportPortal/press.aspx?DocumentId=367236, Accessed 20.9.2016
9. W. Crooks, G. Vukasin, M. O'Sullivan, W. Messner, and C. Rogers, Front. Robot. AI, **3(**70) (available at: doi: 10.3389/frobt.2016.00070, 2016)
10. K.C. Galloway, K.P. Becker, B. Phillips, J. Kirby, S. Licht, D. Tchernov, et al., Soft Robot, **3**(1), 23 (2016)
11. P. Maeder-York, T. Clites, E. Boggs, R. Neff, P. Polygerinos, D. Holland, et al., J. Med Dev, **8**(2), 020933 (2014)
12. A. Zapciu, G. Constantin, Manufacturing Systems, **11**(4), 201 (2016)

MATEC Web of Conferences **121**, 09001 (2017) DOI: 10.1051/matecconf/201712109001
MSE 2017

Energy management strategies in the oil industry for chemical and thermal environment pollution reduction

Lazăr Avram[1], *Tudora* Cristescu[1], and *Monica Emanuela* Stoica[1,*]

[1]Petroleum-Gas University of Ploiesti, Drilling, Extraction and Transport of Hydrocarbons Department, 39 Bucuresti Blvd., Ploiesti, Romania

Abstract. This paper depicts certain thermodynamic processes possibly involved in the oil industry specific activities operation and/or proper equipment maintenance.
On this basis we present a series of courses of action in performing processes like oil storage and transportation, aiming the thermal energy consumption reduction.

1 Introduction

Chemical pollution reduction, respectively reducing of pollutant emissions, is achieved by reducing fuel consumption. Applying energy management can attain this. For the thermal pollution reduction, energy management is considering the recovery of secondary energy resources, rehabilitation of buildings and equipment.

The ultimate goal of energy management is to maximize energy efficiency, and involves the systematic application of techniques and procedures developed and refined over time. Improving energy efficiency is a strategic objective of the national energy policy, due to the major contribution that makes to the achievement of security of supply, sustainability and competitiveness, to save primary energy resources and reduce emissions of greenhouse gases [1].

An important part of material resources resides in the energy resources. They became notorious after the oil crises that occurred in industrialized countries. Thus, concepts were developed, like renewable energy, alternative energy, energy management and energy efficiency.

Directions implementing energy management strategy address the following issues:

- Technical;
- Organizational;
- No cost, low cost, high cost;
- Psychological component;
- Flowchart.

An activity in a given contour in an organized manner, aims to achieve maximum benefit. Improving energy efficiency is a requirement resulting from this general requirement. Schematically, these issues are as follows:

* Corresponding author: monicastoica20022002@yahoo.com

Given contour $\Rightarrow$ an activity in an organized manner $\Rightarrow$ Obtaining maximum benefit $\Rightarrow$ profitable activity $\Rightarrow$ Increasing energy efficiency.

Energy costs (energy bills) are part of the total costs involved in the smooth conduct of the work performed inside the contour under analysis. Lowering the energy costs contributes to slide of total costs and thus to develop the profits obtained.

Energy expenses $\Downarrow$ $\Rightarrow$ **Total expenses** $\Downarrow$ $\Rightarrow$ **Benefit** $\Uparrow$

Unprofitable activities require minimizing costs. Reduction of expenses involves lowering energy costs.

The concept of energy efficiency has a specific character and content only if it is linked to:

- A well-defined outline;
- An activity that is carried out in an organized manner therein.

The concept of energy efficiency refers to the energy performance involving diminishing the size of energy bills or specific energy expenditure [2].

Increasing energy efficiency in an industrial application requires determination, consistency and professionalism techniques and energy management procedures.

Oversight and improving management activities is done by going through the four stages of the cycle Plan-Do-Check-Act (P = plan; D = do, C = check, A = act) called "wheel of Deming", shown in Figure 1.

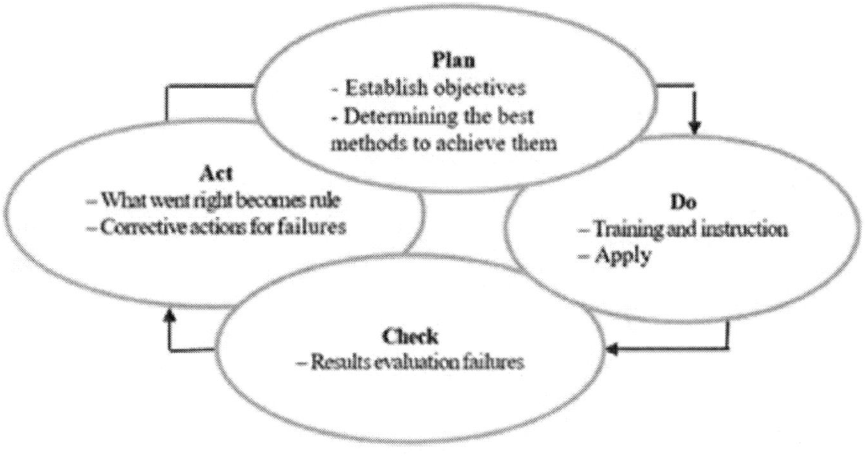

Fig.1. Cycle Plan-Do-Check-Act

Management techniques and procedures aim at:

- identification;
- allocation;
- harnessing maximum efficiency of material, human and financial resources within an organization.

Energy management procedures involve [3 - 5]:

- in-depth knowledge of specific activity;
- monitoring each of the flows of energy that have entered into and came out of the contour and establishing the links between them.

The activities carried out within the considered contour relate to:

- analysis of energy efficiency in an area;
- staff reaction to the size of energy bills;

- the operation and effectiveness of the tracking and transmission of information systems on energy consumption within the boundaries given.

Analysis of energy efficiency in a given perimeter begins with an accurate depiction of the qualitative and quantitative matters of energy supply to the activities in the same area:

- determining the nature of energy carriers entering the contour of the balance sheet;
- determining the order of magnitude of consumption for each category of energy carrier;
- determining the payment method for each of them.

The first aspect of the analysis is the size of the energy bill and how it is constituted. Staff reaction to the size of energy bills is another aspect that relates to the attitude in relation to the energy bill; it can fall under one of the following situations:

- energy bills are paid on time without any analysis or internal controls;
- monthly energy bills are compared to readings (records) monthly metering devices mounted at the entrance to the contour of the balance sheet;
- readings (records) are reported monthly to the volume of activity, calculating specific global consumption of energy;
- there is a data acquisition system (not necessarily authomathic), conducting at least weekly monitoring of energy consumption of the major internal and consumers reporting them to their share of the volume of work;
- an automated / computerized monitoring system is implemented and operates continuously evaluating the efficiency of energy utilization, and possibly other material resources usage.

The attitude of the executive management and other personnel regarding the energy usage efficiency is reflected by [3-4]:

- awareness of the importance of the problem;
- the quality and effectiveness of the monitoring system;
- how to capitalize the results obtained;
- the expected reaction from each of the levels of authority to the size and evolution of energy costs.

The third issue is the functioning and effectiveness of information on energy consumption tracking and transmission within the boundaries given.

The analysis includes design, material base and the importance given to system within the organization. The following matters are followed:

- the reading method and frequency of the measuring instruments, especially those that are the basis of measurement, even more those which are the bases for billing;
- the reading transmission method (using templates, through electrical signals or a computer network, etc.);
- information processing module (model algorithm, sizes calculated);
- frequency of writing this report (daily, weekly or monthly) and addressee;
- the impact of reporting and how decisions are made on energy efficiency.

Energy bill reduction requires the commitment of all staff and an effective system for monitoring, processing and capitalizing on data [3-5].

Using secondary energy sources, reusable and renewable recyclable energy sources (RRES), Table 1, together with improving energy efficiency (EEI) can contribute to diminishing energy consumption, ie to reduce greenhouse gas emissions and, consequently, prevent dangerous climate change [6].

Secondary energy resources, Table 2, in a process or aggregate represents all forms of energy used in the process or technological unit. They result from a primary technological process, as by-products thereof, falling within the category of energy loss in the process or unit respectively. Part of the power of secondary energy resources sunk, dissipates in the environment. The causes of secondary energy resources are imperfections of the energy processes occurring in the contour considered. When the energy content of secondary energy resources is high, the outline of the processes that occur has a low energy yield [7].

Reusable energy resource is part of the secondary energy resource that can be recovered in terms of economic efficiency. This rate depends on the stage of development of this technology and the costs, at that time, of the energy and materials used [8 - 9].

Table 1. The main categories of renewable energy sources

Renewable energy resources category	Renewable energy resources examples
Wind	Wind with constant speed 6-10 m / s
Hydroenergy	Rivers
Sun	Solar thermal -electric
Biomass	Biodegradable fraction of products, waste and residues from agriculture (including plant and animal substances), forestry and wood processing, as well as the biodegradable fraction of industrial and municipal waste.
Geothermal	Hot springs, geysers
Photovoltaic	Solar radiation

Table 2. The main categories of secondary energy resources and form of usable energy [6]

Category of secondary energy resources	Usable energy form	Secondary energy resources examples
Thermal	Sensible and/or latent heat	- Waste gas resulting from pirotechnological processes from the industries: metallurgical, chemical, construction materials, burning of industrial and urban waste; - Waste hot technology (slag, coke); - Used steam; - Moist air discharged from industrial plants and drying plants
Combustible	Chemical energy	- Waste gas resulting from chemical processes, furnaces, coke ovens, converters, refining, coal uplifting; - Leaching of pulp and paper; - Waste wood; - Agricultural waste
Overpressure	Potential energy (overpressure)	- Furnace gas; - Gases resulting from combustion plants under pressure; - Solutions or fluids of technological aggregates pressure as steam, compressed air

2 Case Study

2.1 Determination of pollutant emissions

We considered a firm that produces thermal energy. The company holds four thermal plants that run on natural gas. The main emissions of pollutants discharged from chimneys of combustion installations are SO_2, NO_x, particulates and CO_2. The determination is based on measurements made with specialised equipment. In our company case periodic analysis is performed regarding the discharged flue gases in the atmosphere.

The amount of pollutants discharged into the atmosphere is determined by relations of the form:

$$E = B*Q_i*\varepsilon \tag{1}$$

where: E is the quantity of pollutants discharged into the atmosphere over a period of time, [kg]; B is the quantity of fuel consumed during that period [kg]; Q_i - The lower calorific value of the fuel [kJ / kg]; ε - Emission factor [kg / kJ].

Taking into account the lower calorific value of the fuel gas used relation (1) becomes:

$$E = B*35815*\varepsilon \tag{2}$$

The emission factor is the amount of pollutants discharged into the atmosphere per unit of heat introduced with the fuel in the combustion plant. When using several kinds of fuel, the amount of pollutant is determined by summing the amounts calculated for each of them. Emission factors can be determined by calculus or estimate using the specific references [2].

In the case under analysis, the burned gases contain no sulfur or combinations thereof, SO_2 emissions are zero. For pollutant NO_x emission factor, in the case of combustion units with a thermal power of 50 ... 100 MW_t [2] and as fuel natural gas is ε_{NOx}= 130g / GJ (value corresponding to a boiler load of 100 %).

Note that in the case studied, the power of the boiler hot water falls under the mentioned levels.

For pollutant CO2 emission factors are those adopted at present in countries of the European Economic Community; for natural gas is recommended ε_{CO2} = 50 000 g / GJ.

For hot water boilers within thermal power plants belonging to the company analyzed the emissions calculated for an hour are shown in Table 3.

Table 3. Boiler emissions during one hour

Emissions	**Unit**	**BAR 1740**	**BAR 1930**	**HR2 1100**	**HR2 1400**	**BAR 1930**
NOx	kg	0.20	0.24	0.47	0.38	0.43
CO_2	kg	78.08	93.71	179.83	145.77	166.58
Total	kg	78.28	93.95	180.29	146.15	167.01

In Table 4 the corresponding emission values for two consecutive years for each boiler in the analyzed company are mentioned.

In year 2, compared to year 1, there was a decrease in the quantity of emissions; the main reason resides in the diminishing of fuel consumption because of low number of users and applying specific measures on fuel economy.

Table 4 Thermal plants emissions

Emis.	Unit	CT2		CT3		CT4		CT5	
		Year 1	Year 2	Year 1	Year 2	Year 1	Year 2	Year 1	Year 2
0	1	2	3	4	5	6	7	8	9
NOx	kg	2490.12	1912.59	973.75	742.56	957.55	404.74	1016.66	1075.84
CO_2	kg	957739	735613	374520	285600	368289	155669	391023	413786
Total	kg	960229	737526	375494	286343	369247	156074	392040	414862

Table 4 (continuation)

Emissions	Unit	Total	
		Year 1	Year 2
10	11	12	13
NOx	kg	5432	4161
CO_2	kg	2089348	1600311
Total	kg	2094780	1604472

2.2 The effect of a habitat thermal insulation on heat dissipation

Assessment of energy efficient energy performance of buildings and installations is made taking into account climatic parameters [10].

The study proposes estimate of dissipated heat flow, the flow of fuel gas or assessing the amount of CO2 emitted and the price of fuel used for heating.

The survey is conducted for a flat wall size 4x25 m, built of concrete, insulated on the outside with rigid polystyrene boards and mineral wool inside.

Two cases were considered: non-insulated and insulated walls and two seasons - spring and winter.

It is noted that by isolating a wall, fuel consumption is reduced by approximately 50% and heating costs will correspondingly decrease. Also, the amount of carbon dioxide emitted into the atmosphere is decreased in proportion to the amount of fuel consumption, from previously very high levels in the case of non-insulated wall, [11-12].

Table 5. Estimated heat flux and flow of fuel gas [12]

Wall	Season	k	$\dot{Q}$	$\dot{V}_N$	$\dot{V}_N$	CO_2 emis.	Fuel cost
		W/(m^2 K)	W	m_N^3/s	m_N^3 /month	m_N^3 /month	RON/month
Non insulated	winter	0,5474	273,7	$7{,}64\cdot10^{-6}$	20	39,26	14,47
	spring	0,5474	54,74	$1{,}5\cdot10^{-6}$	4	7,85	2,894
Insulated	winter	0,2532	126,62	$3{,}5375\cdot10^{-6}$	9	17,67	6,5115
	spring	0,2532	25,32	$7{,}0738\cdot10^{-7}$	2	3,93	1,447

3 Conclusions

In order to increase energy efficiency the following are taken into consideration:

1. The integrated analysis of Contour Energy.

2. It is necessary to coherently and pragmatically monitor the material and energy flows in the production activities, transport and heat usage.
3. It is recommended to correlate energy costs with the activity carried on in the contour profile analyzed in order to increase competitiveness.

The synthesis of these data characteristic to be known, discussed and compared with similar data and discussed with the personnel involved directly or indirectly in order to establish appropriate ways of improving management activity not only in energy efficiency field but also in other areas.

References

1. *** *Law 121/2014 regarding energy efficiency* (Romania, 2014)
2. *** PE 902/86, *Normative regarding the making and analysis of energy balance sheets* (ICEMENERG, Bucureşti, 1995)
3. A. Badea, A., *Bazele termoenergeticii* (Ed. Universitatea Politehnica din Bucureşti, Bucuresti, 2003)
4. C. Răducanu, *Bilanţuri termoenergetice* (Ed. Universitatea Politehnica din Bucureşti, Bucuresti, 2004)
5. *** *Monitorul Oficial al Romaniei*, 628 (2008)
6. *** apps.upm.ro/cip/docm/m9.pdf, accessed 12.03.2017
7. V. Athanasovici, *Resurse energetice secundare – Resurse energetice refolosibile*, (Editura Tehnică, Bucureşti, 1989)
8. T. Cristescu, *Termotehnica*, (Ed. Petroleum - Gas University of Ploiesti, 2009)
9. M. Marinescu, A. Chisacof, P. Răducanu, A. Motorga, *Bazele termodinamicii tehnice* (Editura Politehnica Press, Bucureşti, 2009)
10. H. Petran, *Parametrii climatici necesari determinării performanţei energetice a clădirilor noi şi existente, dimensionării instalaţiilor de climatizare a clădirilor şi dimensionării higrotermice a elementelor de anvelopă a clădirilor* (Bucuresti, 2011)
11. S. Suditu, M. Pătărlăgeanu, T. Cristescu, Bul. UPG, **LVII** (2005)
12. T. Cristescu, M. Pătărlăgeanu, S. Suditu, S. Proceedings of the Thermotechnics Conference **2**, 74 (2007).

DOI: 10.1051/matecconf/201712109002

Research on the internal pressure behavior of metal gas distribution pipelines with different types of tubing defects

Stefan Mihai Filip[1,*], and *Eugen* Avrigean[2]

[1]Delgaz Grid S.A., Investment Management, Project Team Sibiu, Romania
[2]Lucian Blaga University of Sibiu, Faculty of Engineering, Department of Industrial Machinery and Equipment, Sibiu, Romania

Abstract. The paper aims to approach an important subject related to natural gas distribution networks which, depending on the expansion of the localities, are composed of intercommunicating pipes, pressure reducing stations and branch connections fittings. The urban networks are the most complex ones and the rural areas networks are the simplest. However, irrespective of their installation, they must meet the safety operating requirements as much as possible. According to standards, all these components must be tight and pressure resistant. In this regard, we intend to approach a very important issue related to the behavior of the tubular steel material showing corrosion and/or material defects, and to the internal stress caused by the gas pressure on the walls of the tubing material.

1 Introduction

The distribution pipelines are usually installed underground at a depth at least equal to the local freezing depth. If we consider classifying the territory of Romania according to the freezing depth standard (STAS 6057-77), in the municipality of Sibiu where we focus the current research, the freezing depth is 900-1000 mm.

The distribution system also includes above-ground pipelines installed in accordance to ANRE approved projects.

The underground distribution pipelines are made of steel or high-density polyethylene, while the above-ground ones are made of steel only.

The distribution pipeline routes are rectilinear as much as possible, along streets or thoroughfares. It is not accepted to install them under or along the tram lines or railroads, or under channels communicating with the neighboring buildings, without special preparation. It is forbidden to pass the distribution pipelines under buildings or their annexes.

The technical regulations for the design, manufacture and operation of gas supply systems (Official Gazette. Part I, no. 255 bis of 2009, as subsequently amended and supplemented), establish the minimum distances between the distribution networks and the different obstacles, buildings or on-route installations.

* Corresponding author: stefan.filip@hotmail.com

DOI: 10.1051/matecconf/201712109002

2 Analyses and Classification of Defects

The gas distribution piping and its equipment can have a number of defects and imperfections that influence operating safety. The residual mechanical strength and the safe operation of the gas distribution pipelines are essentially determined by the presence of defects. These defects (imperfections with significant negative influences on the correct operation and the bearing capacity of a pipeline) and imperfections (abnormal configuration, size, microstructure, etc, present in the wall of a pipeline that do not inadmissibly affect the bearing capacity of the pipeline) can be classified as:
a. geometric imperfections and defects – deviations in size and shape that significantly change the cross-sectional configuration of the pipe: *local gouges and deformations* - dents or signs of interaction with external forces.
The local deformations (Fig.1.) are less severe defects or imperfections, since they do not modify the pipe wall thickness, producing only local variations in the curvature of its cross section. The local deformations influence the gas flow in the pipes and may cause major difficulties in carrying out cleaning operations and the internal inspection of the pipes, by blocking the advancement of the intelligent control installations.
The gouges consisting of the local plastic deformation of the pipe and the removal of material by cutting effect are high severity defects or imperfections. When material is removed, the upper layer of the gouge is hammer hardened, so that the presence of these imperfections or defects can result in triggering the processes of cracking and brittle fracturing.

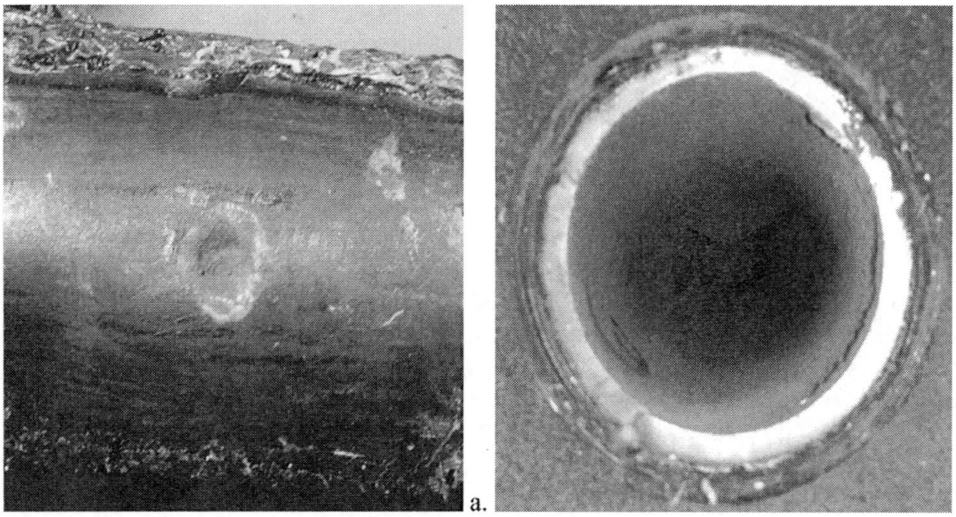
a. b.

Fig. 1. Local deformation defects detected on pipes

b. metal loss is the imperfection or the defect consisting in the thinning of the pipe wall due to metal loss in the presence or absence of a corrosion process (fig. 2).
c. fissures and cracks (Fig. 3) are the most harmful defects that cause powerful effects of mechanical stresses concentration and noticeably decreases the bearing capacity of the pipelines. These can be generated during the operation of the pipelines, due to the phenomena of fatigue, stress corrosion, hydrogen embrittlement, etc., or they may be present in the pipe walls since its start-up (cracks and similar defects caused by pipe lamination or cracks in the welded joints).

Fig. 2. Metal loss defect due to corrosion

In case of the metal loss type defects subjected to the action of external physical factors (the interaction of the pipe wall with different objects or the slide of the land in the immediate vicinity of the pipeline route) the problems that arise are not clarified and there is no consistent and complete unanimously accepted methodology for assessing the severity of these types of defects.

This category of defects is not classified in certain technical requirements for estimating the residual resistance of the pipelines, and the current norms do not contain a methodology for estimating the severity of these types of defects. [4]

Fig. 3. Cracking defect caused by landslide

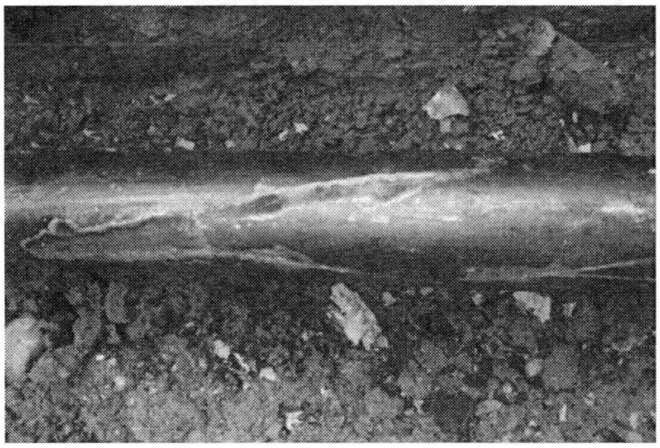

Fig. 4. Metal loss defect detected on a pipeline wall

In order to assess residual mechanical resistance and the failure probability in operating the pipelines having metal loss defects, it is required to take the following steps:

- classifying the defects, determining the main causes for their appearance on the pipe walls, and establishing the means for detecting them;
- determining the characterization of these defects in terms of size, configuration and changes they cause on the mechanical characteristics of the tubular material;
- establishing the measures to avoid the causes leading to the occurrence of these defects;
- determining the means of correcting the metal loss type defects. [2]

In addition to the metal part defects, the pipelines may also have insulation defects:

- due to incorrect design of the insulating layer and/or failure to follow the applying technology on the insulation ramps or on-site, which happens frequently with the pipes insulated with bitumen and mounted in corrosive soils;
- occurring in pipe sections where the insulation has been pierced during handling in the trench or laying on unprepared surfaces;
- caused by improper weld insulation, according to technical regulations;
- due to foreign bodies in the insulating material;
- occurring in pipelines coming into contact with drains or other pipes, which are not applied a reinforced insulation or are not laid in protective tubes;
- arising from insulation damage caused by interventions to other utilities, when the gas pipeline is covered without restoring the insulating layer;
- caused by probing with tools that damage the protective layer and even the tubular material, like drill bits of the asphalt drilling machines to detect gas leaks;
- due to the improper application of the insulation on the branch tees;
- occurring in the inside, where the pipeline is buried at a greater depth, reaching the maximum camber; where the water is no longer be sucked in the gas flow;
- due to the pores in the tubing material that the insulation cannot cover. [1]

3 Researches on the Internal Pressure Behavior

We considered important to know the behavior to the internal pressure of the pipes that have material and corrosion defects, as shown in figures 5. b) and c). For both types of defects we analyzed the same wall thickness.

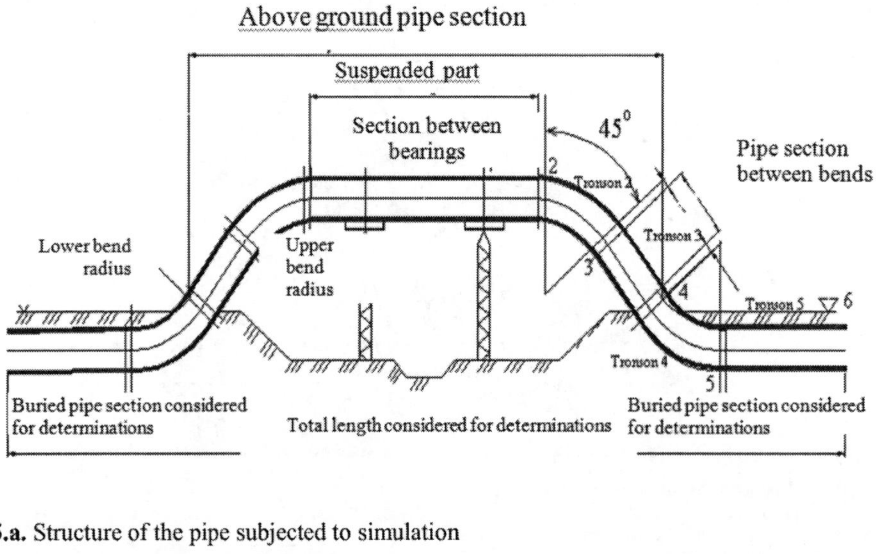

Fig. 5.a. Structure of the pipe subjected to simulation

The determinations were made by the finite element method on a steel pipe, made of type L235GA material, according to STAS 10208, material that is also called 1.0458 or ST 37.0 according to the former standard DIN EN 12007-3, which crosses over a river and has a structure as shown in Figure 5.a.

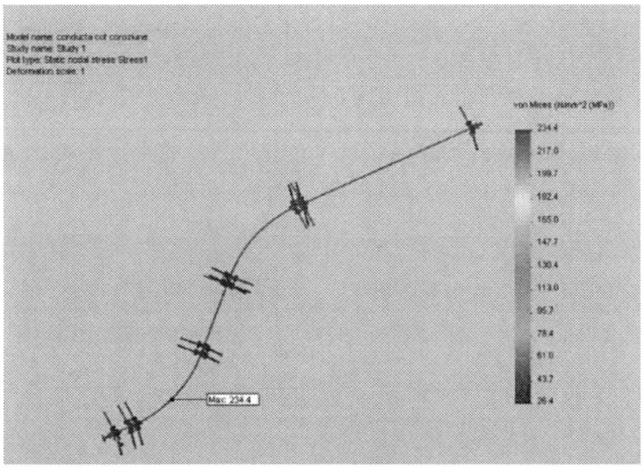

Fig. 5.b. Simulation of a corrosion defect

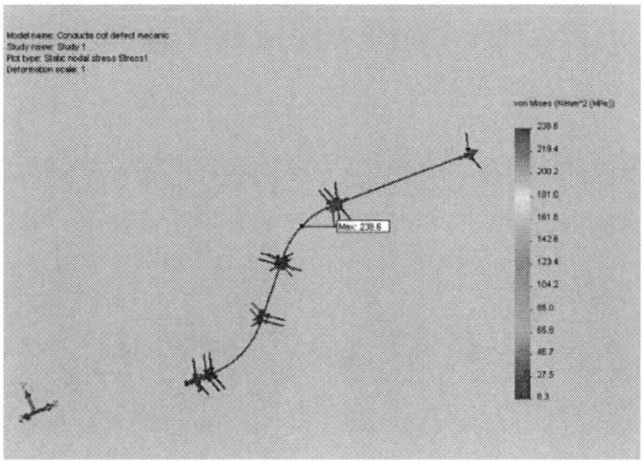

Fig. 5.c. Simulation of a material defect

There have been 6 determinations performed for the values of the wall thickness and the inner pressure shown in Table 1.

The design for the assembly for which the calculus was conducted is shown in figure 5.a), where the positions of the welded joints are marked.

We considered useful that the determinations should be simulated on a pipe section that is part of an assembly of a river overpass, on the grounds that all the defects can be simulated on this type of network; which has led to the simulation of the two types of defects: one corrosion defect found in section where the pipe exits from the ground (considered the section most exposed to corrosion) and one material defect, simulated on the upper bend of the overpass. [5]

After modeling the pipe section subjected to determinations in Catia (a software described in Chapter 4), we simulated the two defects on the section. We did a simulation of the defects for the wall thicknesses shown in Table 1, for every size we applied an increasing pressure up to the fracture of the pipeline in the defect area, and we also showed the pressure for each defect size in Table 1.

The geometry of the corrosion and material defects is shown in Figures 6 and 7.

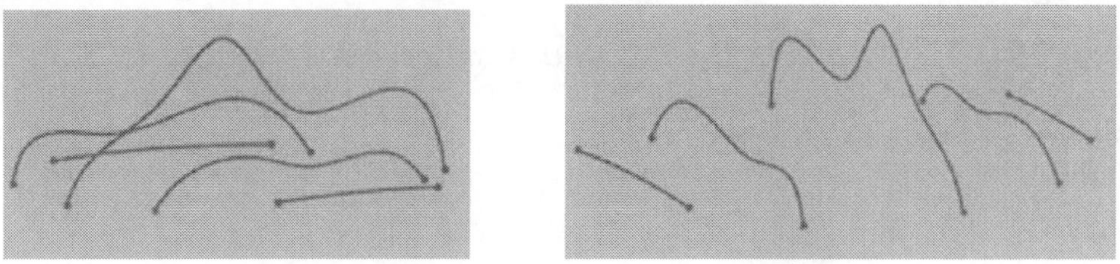

a. b.

Fig. 6.a. Geometry of the corrosion defect - Curbs defining the defect – side view – a; longitudinal view – b.

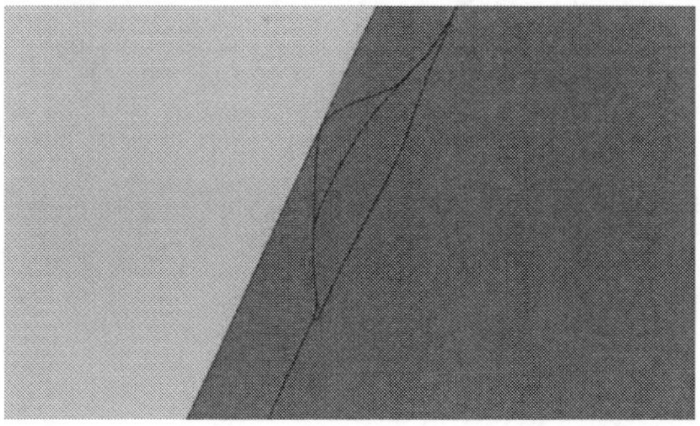

Fig. 6.b. Geometry of the corrosion defect – cross section of the defect in the pipe wall

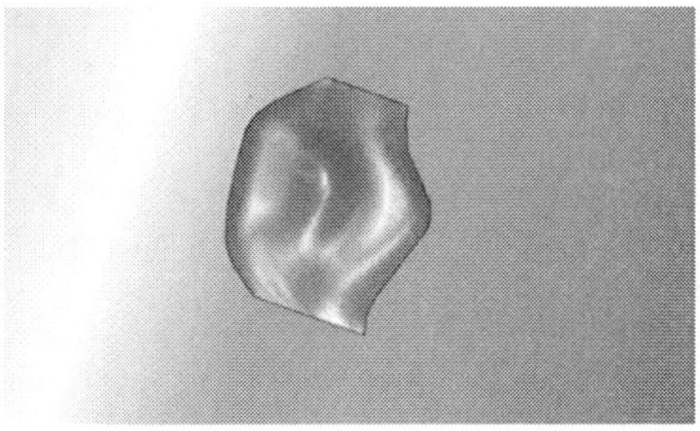

Fig. 6.c. Geometry of the corrosion defect – front view of the defect

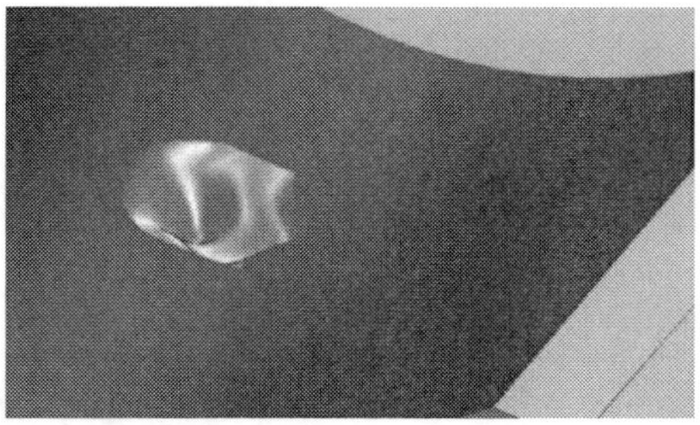

Fig. 6.d. Geometry of the corrosion defect – solid view of the defect

Fig. 7. Image of the material defect

Table 1. Yield pressure of the tubing

Wall thickness		Yield Pressure [MPa]	
[mm]	%	Material defect	Corrosion
4,5	100	181	181
2,25	50	88,9	104
1,125	25	57	103
0,45	10	54,4	90
0,225	5	52	86,5
0,1125	2,5	33,1	84,8

By determining the geometry of the corrosion defect, we tried to simulate the size of a real defect as much as possible (Figure 3), only on the area affected by the pipeline fracture, and not on the entire corroded area. This defect was located on the lower vertical bend which is buried in the ground.

The material defect simulates the damage caused on a pipeline by the tooth of an excavator bucket, not only as a deformation action but as a geometric shape of the blunt body that produces the defect. Assuming that the greatest influence of the operating pressure of the pipeline is in the area of the direction changes, in our case the vertical bends of the overpass, the simulation of the material defect was conducted in this area. [3]

The calculated values of the pressure at which the pipe has yielded in the defect area depending on its depth are shown in Table 1 and the variation graph in Figure 8. [6]

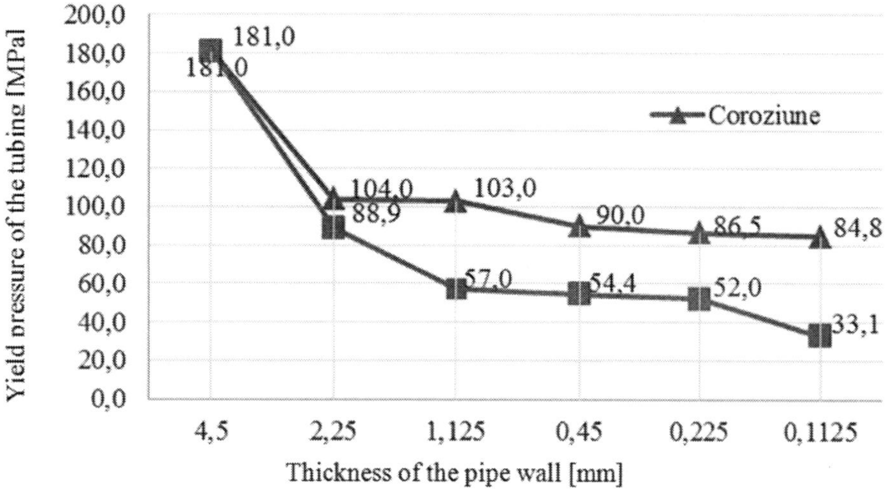

Fig. 8. The piercing pressure according to wall thickness

4 Conclusions and Recommendations

Following the theoretical and experimental analyses, we have concluded the following:

- comparing the behavior of the two types of defects under the influence of the internal pressure of the pipeline, we have deduced that, contrary to the initial considerations, in the case of the corrosion defect, the yield pressure of the tubing is higher, under the same wall thickness conditions. We attribute this to the fact that it is possible that the influence should be due to the defects being simulated on the outer side of the pipe and not on its inner side, which generates a new hypothesis for future scientific research, related to the simulation and the determination of the maximum tubing yield pressures when there are certain inner material tubing defects;

• the maximum yield pressure corresponds to the lack of material and corrosion defects;

• in case there are material or corrosion defects, the effect is felt when the wall thickness has been reduced by half;

• the minimum yield pressure corresponds to the material defects because they have a larger expansion than the corrosion defects and a geometry that determines the concentration of stresses. [7]

• taking into account these conclusions and determinations, we might suggest as effective measures to help and lead to a better and safer operation, an efficient corrosion protection, both from a passive and an active point of view. Also, another very important aspect in this regard is ensuring an optimal handling of the pipes both during transport and installation.

References

1. C. Calin, S. Filip, *Corrosion protection – less defects, more operational safety* (3R International, 2008)
2. Ghe. Zecheru, Ghe. Drăghici, A. Dumitrescu, and P. Iukhymets, *Design of Composite Material Reinforcing Sleeves Used to Repair Transmission Pipelines* (University of Ploieşti Publishing House, Ploiesti, 2014)
3. I. Bondrea, E. Avrigean, *Proiectarea constructiva si tehnologica asistata de calculator* (Lucian Blaga University Publishing House, Sibiu, 2002)
4. I. Tudor, *Materiale speciale si coroziune* (University of Ploiesti Publishing House, Romania, Ploiesti, 1992)
5. M. Frăţilă, Acta Universitatis Cibiniensis, XVIII (1996)
6. Ş. Filip, Report no. 2, *Cercetări privind evaluarea analitică şi numerică a rezistenţei mecanice a conductelor care prezintă pe tubulatură defecte de tip lipsă de material* (University Petroleum-Gas, Ploieşti, 2009)
7. Ş. Filip, *Contribuţii privind evaluarea stării de degradare a reţelelor urbane de distribuţie a gazelor naturale* (University Petroleum-Gas, Ploieşti, 2016)

Considerations on sand control in natural gas wells

Ion Foidaş[1,*], and *Dan-Paul* Ştefănescu[2]
[1]Lucian Blaga University of Sibiu, Faculty of Engineering, 4 E. Cioran st., 550025 Sibiu, Romania
[2]Lucian Blaga University of Sibiu, Faculty of Engineering, 4 E. Cioran st., 550025 Sibiu, Romania

Abstract. Mechanism of sand production in gas wells is complex and influenced by every operation performed into the well, starting from the opening of the production interval by drilling, and continuing with completion and putting into production as well as with the exploitation regime. Sand production along with fluids from gas reservoirs creates a series of potentially dangerous and costly problems. For selecting the most appropriate methods of sand-control all the data and information related to properties of reservoir rock, the history of sand production, the potential well flow rates as well as HSE and costs need to be evaluated. The conclusion of the authors is that the best results in sand control is achieved when the methods are applied before the production of sand may become an issue.

1 Introduction

One of the problems that natural gas producers are facing during wells operation is the appearance of sand from the production unit into the formation fluids. This happens when the well produces from a weakly consolidated formation, where detachment of solid particles (called sand) from the rocks is possible, those particles being driven by layer fluids and getting into the boreholes and even at surface where they can lead to erosion of the production equipment.

The sand that entered into the well can deposit on the bottom and, in time, it can cover the perforations, forming sand plugs in the tubing and in the annulus between the tubing and production casing, as well. Accumulation of sand into the well decreases and even stops the influx of formation fluid. In this case, the production should be stopped and the wellbore cleaned, not always a successfully operation, but incurring always significant costs.

On the other hand, after sand floods from formation into the well, inside there will be disorders, forming caverns behind production casing, which could damage the casing and lead to the collapse of the impermeable rock above the production layer. Consequently, water layers can be opened that deteriorate the productive formation and as a result, production difficulties may occur.

The technologies used to stop problems caused by sand carrying over from the productive formation are diverse and continually enhanced. All these technologies except one (production restricting) imply consolidation of the formation from the area adjacent to the productive interval, to prevent sand movement because of the extracted fluids or of layer - well differential pressure.

* Corresponding author: ion.foidas@ulbsibiu.ro

1.1 Sand control by restricting gas flow rate

Restricting the gas flow rate is the simplest method that can be applied to prevent the occurrence of sand floods into the wells. In some cases, it can be a successful alternative, but in most cases, it is not a sustainable or economical method.

Thus, the natural gas wells that produce from poorly consolidated formations cannot be operated at nominal rates but at much lower rates not to produce sand floods from the formation. For each case, the maximum allowable flow rate a well produces can be determined, without having problems due to formation sand (Figure 1).

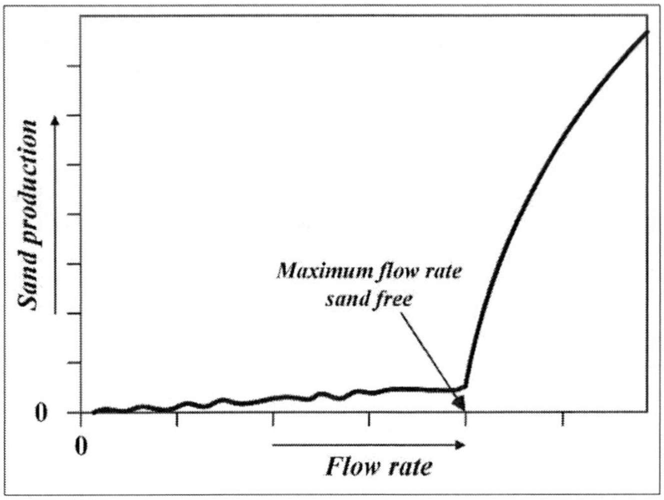

Fig. 1. Influence of the well flow rate on the sand production from a non-consolidated formation

As an indication, one can consider a maximum acceptable amount of sand produced by a well to be 8 mg/ m^3 for the wells producing gas with flow rate higher than 2,000 m^3/ day and 16 mg/ m^3 in wells with flow rate lower than 2,000 m^3/ day.

Maximum flow rate allowable for a well is determined using an equation based on the condition that the actual maximum gas velocity in the formation-well influx area should not reach the value where the rock grains (sand) are driven into the wellbore:

$$Q_m = 2\pi r_s h v_m \quad (1)$$

where Q_m- the maximum flow rate under the conditions of the well-layer interface to which rock grains begin to enter into the wellbore;
- r_s - the well radius;
- h - thickness of the production layer;
- v_m - maximum gas filtration velocity in the influx formation-well area at which the granules begin to penetrate the rock into the well bore.

Because gas flow rate is usually expressed under standard conditions, applying the state gas equation, relation (1) becomes:

$$Q_{m_{st}} = 2\pi r_s h v_m \frac{T_{st}}{T_z}\frac{p_{dp}}{p_{st}}\frac{1}{Z} \quad (2)$$

where $Q_{m_{st}}$ - is the maximum gas flow rate in standard conditions;
- p_{dp}- perforations dynamic pressure;
- Z- gas compressibility factor;
- T_z- reservoir temperature.

Equation (2) can be written as:

$$Q_{m_{st}} = \frac{W p_{ad}}{Z} \tag{3}$$

where $W = 2\pi r_s h v_m \frac{T_{st}}{T_z}$ - velocity factor, which contains the maximum filtration velocity;

- $p_{ad} = \frac{p_{dp}}{p_{st}}$ - dynamic pressure at the perforations, dimensionless expressed relative to standard pressure.

From equation (3) one can calculate the velocity factor W:

$$W = \frac{Q_{m_{st}}}{p_{ad}} Z \tag{4}$$

The velocity factor must be maintained constant during the production of the well, not to cause sand inflow into the well.

According to the equation (3) if the velocity factor *W* has a constant value throughout the entire life of the well, the admissible flow rate decreases proportionally to the dynamic bottom pressure.

In practice, for determining the velocity factor, a special equipment can be used which allows the real time monitoring of solid particles from the well, operating based on the principle of sensing the noise produced by the collision of sand to the inner wall of a pipe which conveys the produced gas. Thus, the solid particles in the gas stream which hit the inner wall of the pipe generate ultrasonic signals that circulate through the pipe wall and which can be sensed by an acoustic sensor, from where they will be picked up and processed to determine the exact amount of sand coming from the well.

The sensor's location is preferred to be at a bend where fluid flow changes its direction of movement so that the noise from the collision of sand with the inner wall to be maximum.

The data collected by the sensor are input in a dedicated software which filters them and processes them, providing information about the amount of sand coming from the well according to production data and well actual conditions.

Introducing the gas volume factor expression (b_g) in equation (2) it becomes:

$$Q_{m_{st}} b_g = 2\pi r_s h v_m \tag{5}$$

For wells producing from weak consolidated formations $Q_{m_{st}} b_g$ should be kept constant during exploitation in order to prevent the appearance of sand into the well [1].

For safe well exploitation, the maximum allowable flow rate is considered excessive and therefore the well will be operated with a lower flow rate, 0.9 usually from the last flow rate at which the well did not carry off rock particles.

In most cases, however, restricting the flow of production is not sufficient to stop the sand movement; the degree of formation consolidation, type and amount of the cementing materials and the amount of water that is produced by the well are also significant factors in sand carrying off. These factors may cause sand floods into the well, although production flow rate has been severely restricted. As a conclusion, we can say that even if the restriction of flow rate has as effect, in a first stage, preventing the appearance of sand into the well, this success may be short-lived, depending on how other factors act on formation sand during well exploitation.

1.2 Sand flood control by mechanical control methods

The mechanical sand control methods consist of wells downhole completion/recompletion with mechanical devices capable to block the influx of sand into well. These devices are slotted liners, wire-wrapped screens, prepacked filters, sintered filters, expandable sand screen etc.

The devices most commonly used are the wire-wrapped screens, made of a piece of tubing or pipe, of various lengths, over which spacers are welded, on which a special wire with up to 2.5 mm diameter is wound. The tubing is provided with circular holes of 8 - 10 mm or with slits with relatively large widths for fluids access (Figure 2).

The wire is made of stainless steel with a trapezoidal or triangular cross-section, and is wrapped with the narrower side inwardly positioned in order to ensure the filter's self-cleaning action [2]. The wire is welded at every connection point with the spacers in order to maintain an equal distance between coils/ gauges.

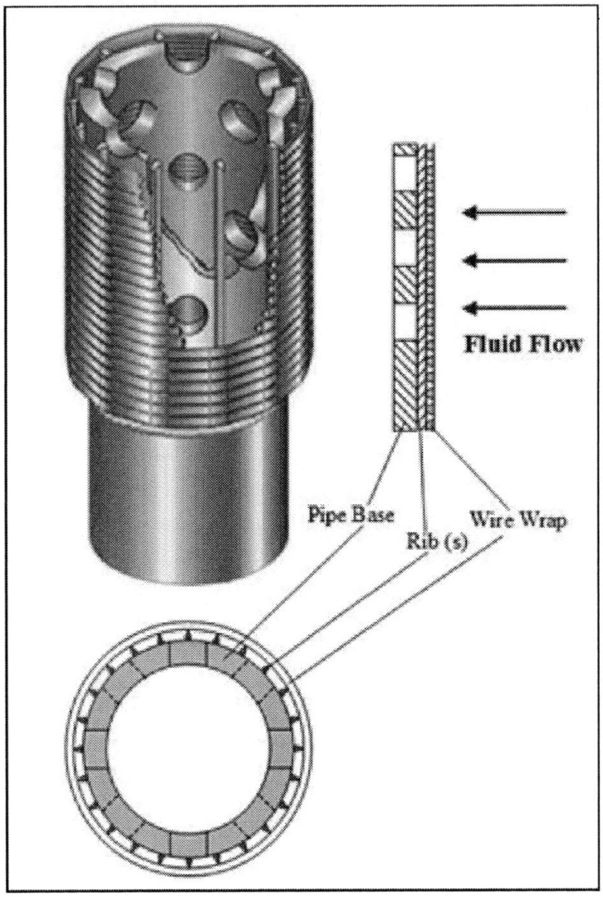

Fig. 2. Wire-Wrapped Screen.

The classical method of controlling sand floods consists of gravel packing the metal filter, which is fixed against perforation by using a packer.

The purpose of the wrapping is to provide an extended artificial filtering area in order to stop efficiently the sand carried - over by the fluid flow but also to create a flow resistance as small as possible (Figure 3). To achieve these two main purposes, a gravel for packing with larger dimensions should be used but also being capable to block the smallest particles from the formation.

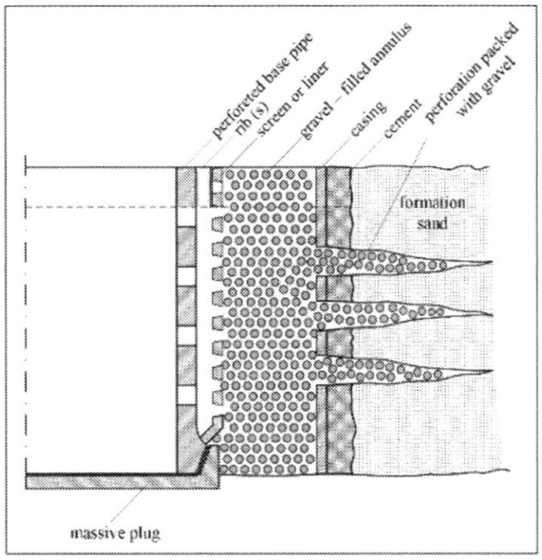

Fig. 3. Detail of internal gravel pack.

The filter wire is designed to retain gravel sand packing particles, which is why the distance between screen gauges must be less than the diameter of the smallest granules of gravel sand packing. Therefore the filters selection is done based on particle's size of the gravel sand packing, which is also chosen depending on sand layer's characteristics.

Sand layers characteristics are obtained after carrying out the particle size analysis of a sand sample derived from the productive formation, following these steps:

- weigh just a sample of sand and sieve, consecutively, through a set of sieves that have progressively smaller openings;
- determine the weight of sand passed through each sieve and is expressed as a percentage of the total sample weight;
- in a coordinate system plot the points, having as ordinate, the cumulative percentage of the amount of sand remained on each sieve and/or the percentage of the amount of sand passed through a sieve of a certain size (percentage), characterized by sieve's mesh diameter, indicated on the x-axis to a logarithmic scale.

For example, below, are presented the results of particle size analysis of a sand layer sample derived from a well in Glavăneşti structure (Bacău County) [3].

Data on sieve's mesh size and data on sand quantity, which has passed through each sieve, are shown in Table 1, and their graphical representation is shown in Figure 4.

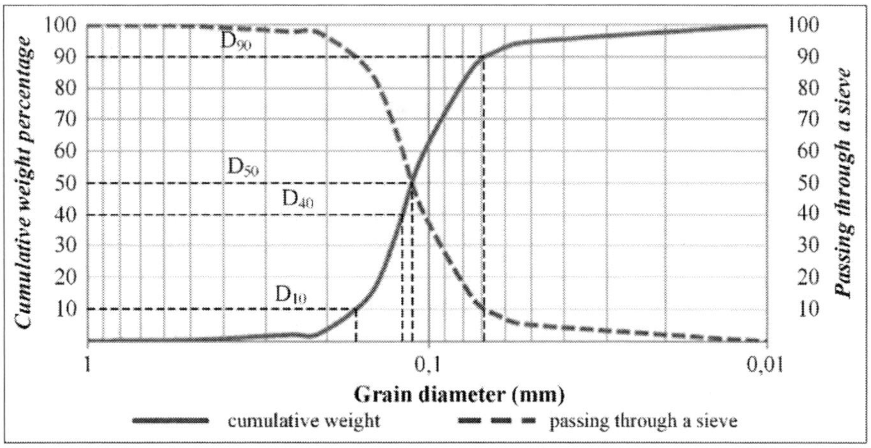

Fig. 4. The grading curve of sand from Glăvăneşti structure.

Table 1. Particle size analysis results

Size sieve	Weight on sieve	Cumulative weight	Cumulative weight	Passing through a sieve
mm	grams	grams	%	%
1	0	0.00	0.00	100.00
0.85	0.01	0.01	0.10	99.90
0.5	0.02	0.03	0.30	99.70
0.42	0.04	0.07	0.70	99.30
0.3	0.09	0.16	1.60	98.40
0.25	0.05	0.21	2.10	97.90
0.21	0.04	0.25	2.50	97.50
0.15	1.2	1.45	14.50	85.50
0.125	2	3.45	34.50	65.50
0.106	2.33	5.78	57.80	42.20
0.075	2.84	8.62	86.20	13.80
0.063	0.56	9.18	91.80	8.20
0.053	0.27	9.45	94.50	5.50
0.01	0.55	10.00	100.00	0.00
TOTAL	10.00			

From the grading curve is shown that the sample's larger granules have a diameter below 1.0 mm because through the sieve with mesh of 1 mm, 100% of the sample passes, and the smallest granules have a diameter of about 0.01 mm.

By the grading curve traced like this, the sand layer can be characterized. First of all, four points are located on the grading curve to give the following characteristic sizes (Fig. 4):

- D_{10} = maximum size of the largest 10% of the sample;
- D_{40} = maximum size of the largest 40% of the sample;
- D_{50} = maximum size of the largest 50% of the sample;
- D_{90} = maximum size of the largest 90% of the sample.

From these characteristic sizes, the following grading characteristics are defined:

- ✓ Effective size, D_{90}
- ✓ Median diameter, D_{50}
- ✓ Uniformity coefficient, $Uc = D_{40} / D_{90}$

Depending on the uniformity coefficient, the formation sand may be classified as shown in table 2.

Table 2. Classification of formation sand [4]

$U_c < 3$	well sorted, highly uniform sand
$3 < U_c < 5$	uniform sand
$U_c > 10$	moderate/poorly sorted sand
$5 < U_c < 10$	poorly sorted highly non-uniform sand

The gravel pack sand may be selected once the formation sand has been characterized. There are a number of criteria used [4]:

a. Saucier's Method

D_{50} (grain) = 4 to 8 times D_{50} (sand)

b. Schwartz Uniformity Method

D_{10} (grain) = 6 x D_{10} (sand) for $U_c < 5$ uniform sand

D_{40} (grain) = 6 x D_{40} (sand) for $U_c > 5$ poorly sorted highly non-uniform sand

D_{70} (grain) = 6 x D_{70} (sand) for $U_c > 10$ moderately uniform sand

Usually Saucier method is used to determine the variation limits of sand sizes used to gravel packing, modified as follows: amplify by 5, namely by 6 the size of 50% of the formation sand, afterwards chose the gravel sand with the median grain size ranging between the two limits, from the commercial offer.

Gravel pack sand is supplied in different sizes, the most used are: 40 - 60, 20 - 40, 12 -20; the first number represents the mesh, though which all grains fall downwards in the sieve, where all particles are gathered.

Table 3 shows the main properties of these gravel sand types:

Table 3. Properties of gravel pack sands [4]

Sand size	Range gravel size	Median gravel diameter	Average permeability	Screen opening size	Weave
Mesh	mm	mm	D	mm	
40 - 60	0.420 – 0.250	0.338	45	0.200	Coarse
20 - 40	0.838– 0.420	0.640	121	0.300	Medium
10 - 20	1.980 – 0.838	1.220	500	0.600	Fine

In stand-alone (without gravel packing) screen applications, the screen acts as porous barrier to the formation sand. For effective sand control, the screen opening is sized relative to the formation sand.

In this case the rule of the thumb can be applied, which means to size the screen opening to control the D_{10} - tenth largest accumulative percentile of the formation particles [5].

2 Case studies - results further to gravel pack operations on Hurezani-Piscu Stejari field

Hurezani - Piscu Stejari gas field is located in the Getic Depression in the contact area with the Moesian Platform, along the pericarpathian fault.

Based on production tests, hydrocarbon was found in the following formations: Pontian, Meotian and Sarmatian. The production units on this gas field group a number of layers of marl deposits, sand and sandy limestone, poorly consolidated, that give the fault the following physical properties [3]:

- permeability = 0.65 mD;
- porosity = 20%;
- fracturing gradient = 0.17 bar/m;
- median sand size D_{50} = 0.09 mm.

The wells drilled on the field have to be exploited with restricted flow rates (3.5 mm or 4 mm choke) not to drag formation sand.

A series of consolidation operations have been performed to increase production performances of this field that resulted in increasing monthly production from the entire field, as may be seen in figure 5.

Gravel packing has been performed at production units at an average depth of 1,800 m, using filters of 2 3/8" x 0.008" gauge assembled in sections of 28 m.

For the gravel pack a gravel sand of 40-60 mesh was used in quantities between 550 - 1,600 kg/well [3]. After performing the gravel pack operation the wells were able to produce through larger chokes (5 mm and 6 mm) without having sand production.

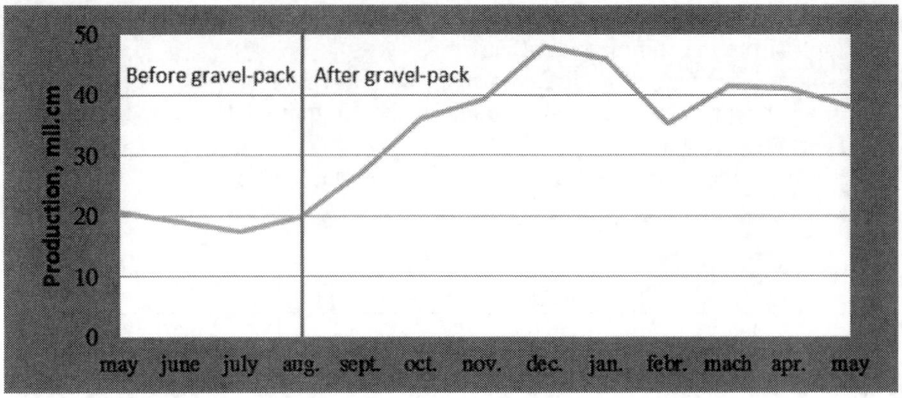

Fig. 5. Impact of gravel-pack operations on monthly production

3 Conclusions

Natural gas wells that produce from poorly consolidated formations cannot produce at maximum flow rates but at lower flow rates, not to cause sand floods in the layer. To improve the performance of such wells it is mandatory to use technologies that impede formation sand from entering in the well.

One of the most efficient sand control method is gravel packing. This method can be successfully applied by accurately sizing the gravel sand, sizing the filter according to the formation sand size and by using a procedure able to rigorously comply with the requirements of a gravel packing operation.

Acknowledgment: The authors would like to thank Romgaz for facilitating the access to technical information and documentaries on the subject matter of this article.

References

1. P. Svoronos, I. Oltean, *Proiectarea exploatării zacamintelor de gaze* (Editura Tehnica, Bucuresti, 1979)
2. N. Macovei, *Deschiderea stratelor productive* (Editura Universității Petrol-Gaze, Ploiești, 2008)
3. **** Romgaz Documentation*
4. *** *Schlumberger Sand Control Engineering Manual, Volume **1*** (Schlumberger, 1992)
5. *** *Baker Oil Tools Sand Control Screen Application Guide* (Baker Oil Tools, 2008).

MATEC Web of Conferences **121**, 09004 (2017) DOI: 10.1051/matecconf/201712109004
MSE 2017

Studies regarding the use of a neuro-fuzzy mathematical model in order to determine the technological parameters of the polyethylene pipes butt welding process

Alina Gligor[1,*], *Marcel-Letitiu* Balan[2], *Letitia-Corina* Balan[1], and *Horatiu-Daniel* Crisan[3]

[1] Lucian Blaga University of Sibiu, Faculty of Engineering, 4 E. Cioran St, 55025, Sibiu, Romania
[2] E-on Gaz, 83 Stadionului st., 551105, Medias, Romania
[3] Romgaz, 4, P-ta C.I. Motas, 551130, Medias, Romania

Abstract. The paper analyzes the possibility to use a neuro-fuzzy type mathematical model, with the final goal of establishing the welding parameters for new types and dimensions of pipes and fittings. Anticipating the developing dynamic of polyethylene-made elements, especially pipes and fittings, starting from the current situation when already a wide range of pipes and fittings with different wall thicknesses and nominal working pressures is produced and commercialized, and taking into account also new development, it was considered necessary to find out the welding parameters for any new pipe type and dimension. The usage of existing welding equipment for new pipe dimensions is impossible without a preliminary set of welding parameters: pressure, temperature, time. Based on experimentally validated data for discreet values of the characteristic welding parameters, there was generated, using mathematical laws and functions, a new model that can estimate the necessary values of the welding parameters for any value within their variation range. As a result, the mathematical model created using neuro-fuzzy techniques allows the obtaining of the correct value for certain parameters (e.g. required welding pressure) for any values of the input variables pipe diameter and pipe thickness.

1 General aspects

The paper analyzes the possibility to use a neuro-fuzzy type mathematical model, i.e. a logical system that is an extension of the multi-criteria evaluation logic, with the final goal of determining the welding parameters for new types and dimensions of pipes and fittings.

Anticipating the developing dynamic of the polyethylene materials, especially pipes and fittings, starting from the state of the art when already a wide range of pipes and fittings with different dimensions, wall thicknesses, nominal working pressure are produced and commercialized and looking towards new ones (intermediate diameters with specific wall thicknesses), it was considered necessary to find out the welding parameters for any new pipe type and dimension.

Using the existing welding equipment for new pipe dimensions becomes impossible without having a preliminary setup of the welding parameters: pressure, temperature, time.

* Corresponding author: alina.gligor@ulbsibiu.ro

Based on some experimentally validated data for some discreet values of the characteristic parameters of the welding process, there was elaborated, using mathematical laws and functions that are implemented in the software system, a new model which can estimate the necessary values of the welding parameters for any value from their variation range, more precisely the values for the equipment setup, starting from the known and existing values.

The result is that that the mathematical model realized using neuro-fuzzy techniques gives the possibility to obtain the correct value for some parameters (e.g. welding pressure required) for any values of the input variables: pipe diameter and pipe thickness.

2 Construction steps of a preliminary fuzzy model

As was previously mentioned, the authors attempted to analyze the possibility of using fuzzy logic methods for the elaboration of a mathematical model regarding the butt welding process of the polyethylene pipes.

The advantages of using this method are:

- the fuzzy logic allows to obtain precise conclusions, based on poor, unprecise entry data (information);
- in the elaboration of the multicriterial fuzzy model, especially for the fuzzy type rule set, the experience of field professionals can be beneficial; this experience can successfully replace the use of highly complex mathematical algorithms;
- the fuzzy model has a high flexibility, resulting in its changing and refining being made with ease;
- both the gathering of information necessary for the application of the multicriterial decision fuzzy model and its usage can be done in a very short time, it can be described even as a real-time decision;
- nowadays there exists a large range of software programs that make the deployment of models based on fuzzy logic possible, very easily, without the need of advanced programming knowledge.

The elaboration of a fuzzy model requires going through the stages that will be presented below.

3 Application of the Fuzzy Logic Methods

The next step in approaching the problem is the application of the fuzzy logic methods for the data related to the existing input [1]. It should be taken into account that the structure of the mathematical model governing this problem is not known and one of the goals of this research is in fact to determine this model.

The fuzzy approach is complicated by the fact that it is difficult to establish the form of the affiliation functions for the proposed fuzzy model.

Most industrial processes have a multidimensional character. The mathematical modeling of these processes is very difficult, and in many cases impossible due to the absence of necessary information about the involved processes and about the interconnections between the components. Often, the information about processes are just examples of numerical data and linguistic forms of rules.

One of the methods for eliminating those difficulties is the use of both the theory of the fuzzy sets and of the artificial neuronal networks [2]. By applying the neuronal networks methods for the weak points of the fuzzy control and at the same time, the facilities of the fuzzy sets for the problems of the neuronal networks, a synthesis of both methods has been achieved. The method of the neuro-fuzzy intelligent shaping is a combination of the best performances of fuzzy logic and of the artificial neuronal networks.

The basic idea of this approach is to use the techniques for the fuzzy model to learn, based on initial data input/output, in such a way as to calculate optimal affiliation functions, expressing the best interdependence of the input and output data.

The most important features of the theory of the fuzzy sets and artificial neuronal networks, for the purpose of their use in models of industrial processes, are shown in table 1.

Table 1. Characteristics of the fuzzy sets theory and of artificial neuronal networks for use in industrial process models.

CHARACTERISTICS	NEURO	FUZZY
Advantages	- learning possibility based on numerical data files and generalization of the gained knowledge	
Disadvantages	- convergence of the learning process	- absence of the functioning adaptation of components

In a typical neuro-fuzzy structure following parts can be clearly distinguished:

- The fuzzy component, that consists in converting the value of a variable or quantitative values in a proper representation of the fuzzy sets;
- The artificial neuronal network component;
- The unfuzzyfication component that, using the method of the center of gravity or other methods, converts the outputs that the neuronal network expressed through appropriate fuzzy sets, into a simple numerical form.

3.1 The construction of the mathematical model for the process

The method studies the construction of a mathematical model for the studied process, based on the experimental data. The multivariable input character of the process and the fact that the material is exposed to some transformations allow to determine some analytical relations that characterize the process [3].

The goal is to build, based on experimentally validated data, for certain discrete values of the process parameters and using advanced modeling techniques, a model for estimating the required values of the parameters for any value within their variation range.

3.1.1. Steps in the model construction

A. In a first stage an attempt was made to construct a fuzzy model, but the complexity of the requirement to achieve this by conventional methods made it practically impossible. Therefore, it was suggested to create a model based on the neuro-adaptive fuzzy methods.

Thus, as input parameters for the process there have been chosen:

- the diameter;
- the nominal pressure;
- the pipe thickness.

And as an output variable there has been chosen the welding pressure.

From the set of experimental data there was separated a data set for the drive shaft and a data set for verification.

An initial fuzzy model has been built that reflects the system's behavior only to a limited extent. Then, based on the drive shaft and verification data, there was created a refined fuzzy model using artificial neuronal networks and an autolearning process, until the modeling error reached acceptable values. From this stage, it can be said that fuzzy model reflects the real consequences of the process, being also possible to work on it, using characteristic fuzzy methods.

In order to carry out the activities mentioned above, a specific interface and subprograms was created in the software package Matlab using the Fuzzy Logic Toolbox modules and Neural Network Toolbox software.

Starting from the collected input/output data set, two sets were generated, a first set used as the drive shaft and the second set used as verification data. The two sets of data are shown in tables 2 and 3.

Table 2. Experimental training data set

Nominal pipe diameter [*mm*]	Nominal fluid pressure [*bar*]	Pipe wall thickness [*mm*]	Welding pressure [*bar*]
75.0000	2.5000	2.0000	2.8000
75.0000	4.0000	2.9000	3.9000
75.0000	10.0000	6.9000	8.9000
90.0000	3.2000	2.8000	4.6000
90.0000	6.0000	5.1000	8.2000
90.0000	16.0000	12.5000	18.3000
110.0000	2.5000	2.7000	5.5000
110.0000	4.0000	4.3000	8.6000
110.0000	10.0000	10.0000	18.9000
125.0000	3.2000	3.9000	8.9000

Table 3. Verification data set

Nominal pipe diameter [*mm*]	Nominal fluid pressure [*bar*]	Pipe wall thickness [*mm*]	Welding pressure [*bar*]
75.0000	3.2000	2.4000	3.3000
75.0000	6.0000	4.3000	5.7000
75.0000	16.0000	10.3000	12.6000
90.0000	2.5000	2.2000	3.6000
90.0000	4.0000	3.5000	5.7000
90.0000	10.0000	8.1000	12.5000
110.0000	3.2000	3.5000	7.0000
110.0000	6.0000	6.3000	12.3000
110.0000	16.0000	15.1000	27.0000
125.0000	2.5000	3.1000	7.1000

Figure 1 shows the graphical user interface used for determining the mathematical model with the graphical representation of the data on the hard drive. Entering the input and output data into the model's structure is done from two data files (with an extension of "."), which will be then loaded from the computer memory.

It is important to assign the readings into dimensional series or into series of parameters in such a way that the modeling remains valid and the mistakes of interpretation or confusion due to process variables are eliminated. The attributes remain unchanged, regardless of whether the data is for the drive shaft data set or for the verification data set.

To plot the drive shaft data, there was used the graphical symbol "o", while for the representation of the verification data there has been used the symbol "+". Figure 2 shows both data sets.

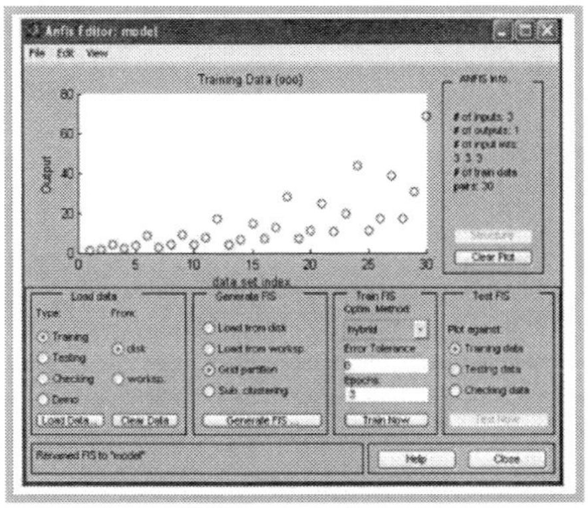

Fig. 1. Graphical interface used for the construction of the neuro-fuzzy model and graphical representation of the data on the hard drive

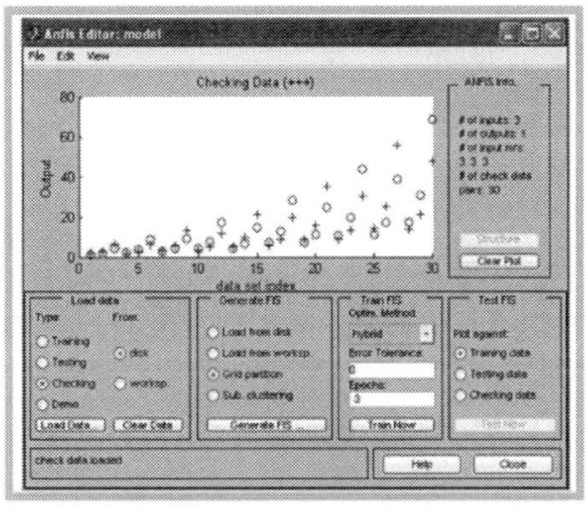

Fig. 2. Graphical representation of the data for the drive shaft and of the verification data.

B. The next step in the construction of the model was to determine a preliminary fuzzy interference structure, based on which to begin the neuro-fuzzy process.

As indicated above, there was chosen a fuzzy model with three input variables (diameter, nominal pressure, thickness) and an output variable (welding pressure).

The preliminary fuzzy structure is shown in figure 3.

In this preliminary phase, three variation ranges for each variable have been set and the triangular alternative was selected for the affiliation functions.

The distribution of the variation ranges for the diameter input variable and the form of the affiliation functions is shown in figure 4.

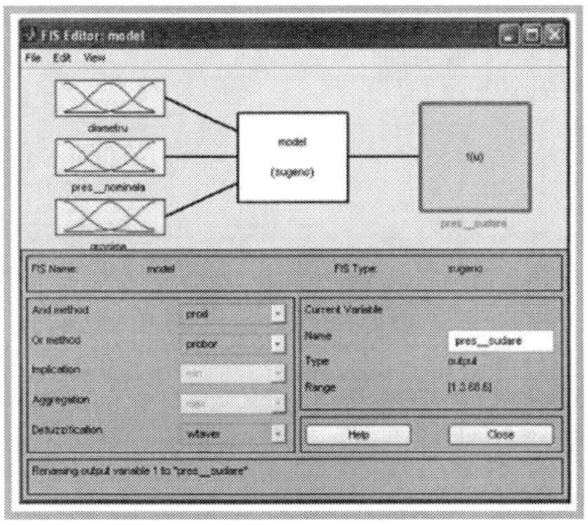

Fig. 3. Preliminary structure of the fuzzy logic system

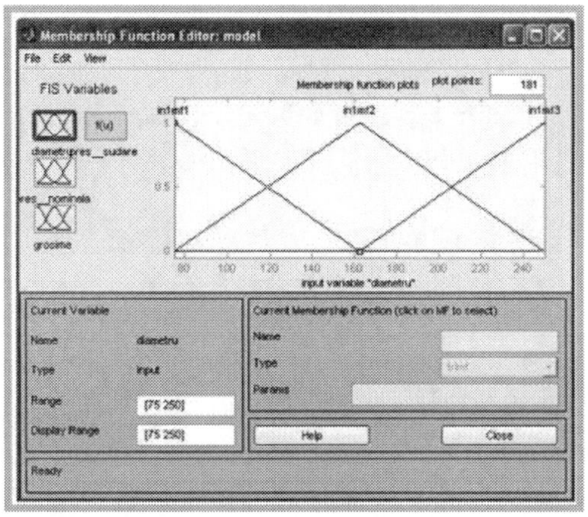

Fig. 4. The function of belonging to the diameter input variable

C. A further stage in the generating the model consisted in unfolding the training process of the neuro-fuzzy system until achieving an acceptable error. The stages of the training process bear the name of ages. For the initial phase of the training process there has been chosen a number of 80 ages. Figure 5 shows the evolution of the training error, which achieves an acceptable minimum after 55 ages of training. After exceeding this value it should be noted that the training error begins to increase, which led to selecting the value of 55 as an optimal number of ages (stages of training). The value of the error after completing the 55 ages is 0.016554, a value that in this case has been considered as acceptable.

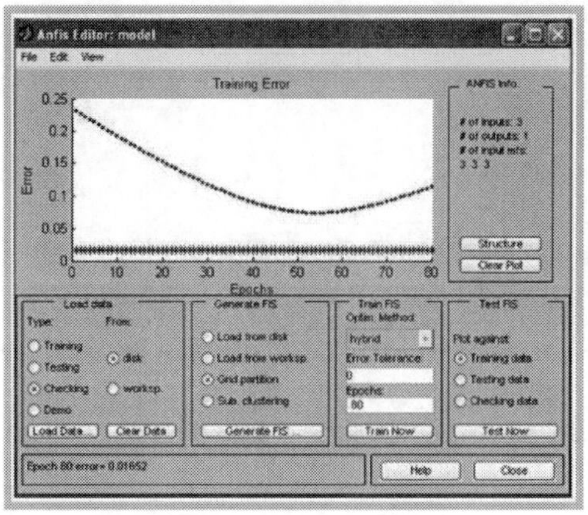

Fig. 5. Evolution of the training error

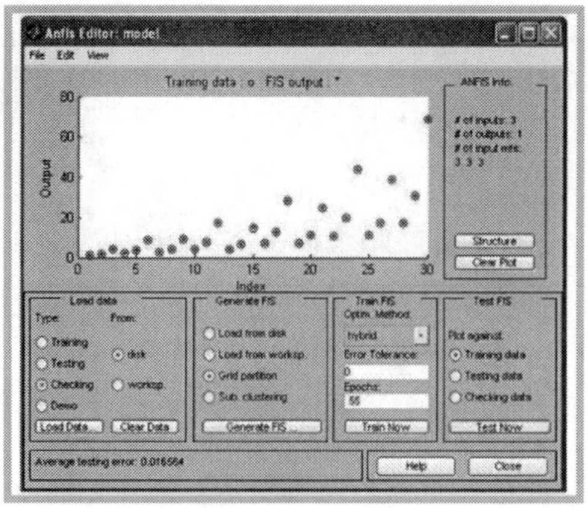

Fig. 6. Predicted data for the model "*" compared to the training data "o"

As training error has been defined the difference between the output value from the sets of training variables, and the output value generated as a result of the fuzzy inference.
D. The next step in the model's construction focused on the quality assessment, through the so-called checking operation. The checking required to compare the values of the experimental output with those obtained by using the neuro-fuzzy model for the same values of the input data.

Figure 6 shows the overlapping of two sets of values, (for displaying the experimental data there was used the symbol "o", while for displaying the data obtained by running the mathematical neuro-fuzzy model there has been used in the symbol "*").

The tree structure of the neuro-fuzzy system results as shown in figure 7.

A preliminary overview at this stage revealed the existence of 3 variables, 5 levels of the neuronal networks organization, 74 neurons and 27 fuzzy rules.

Figure 8 presents the neuro-fuzzy structure, which at this stage is used to calculate the output for any intermediate values of the variables, in accordance with its intended purpose.

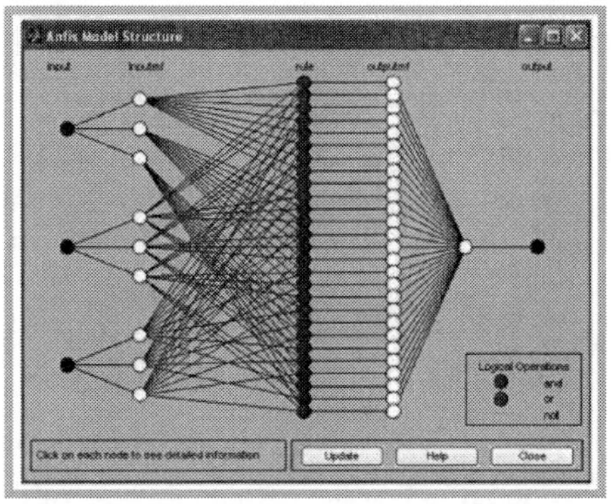

Fig. 7. Tree structure of the fuzzy model

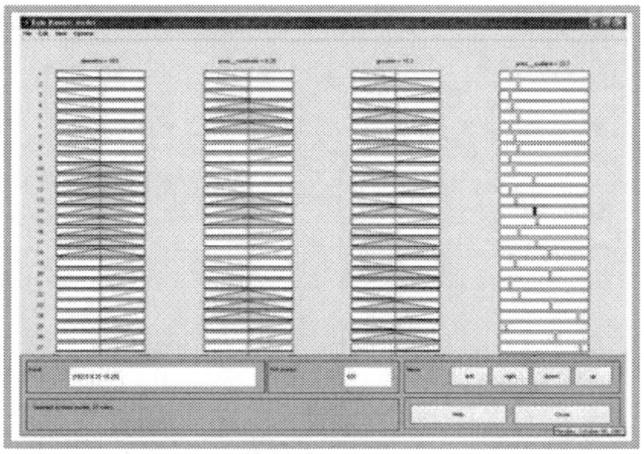

Fig. 8. The neuro-fuzzy model

The main problem which occurs here refers to the share of each input variable in the construction of the mathematical model [4]. A preliminary estimation has revealed that the influence of the variable "nominal pressure" is reduced in comparison to that of the other two, "diameter" and "thickness". However, due to the advanced possibilities of the neuro-fuzzy method used, the variable in question has been taken into account when setting up the model.

However, based on running the mathematical model, it can be noticed that the value for the variable output is only very slightly influenced by large variations of the input value "nominal pressure". For example, for a variation of the nominal pressure from 3.23 bar to 15.6 bar, the welding pressure varies from 22.4 bar to 29.6 bar.

3.2 Refining the model

As a result of the findings described in the previous paragraph, in the model's refining the nominal pressure has been eliminated as an input variable. The running of the neuro-fuzzy system required in this case a number of 35 ages, less than the value of 55 for the previous model with three input variables.

The evolution of the values of the training error (the difference between the output value from the training data set and that obtained by running the neuro-fuzzy model) and of the verification error (the difference between the output value from the verification data set and that obtained by running the neuro-fuzzy model), for the structure of the new model are shown in figure 9.

The structure of the employed fuzzy system was customized by the association of linguistic variables for each input variable. Thus, to the diameter input variable there were associated the variables SMALL, MEDIUM, LARGE, and for the thickness input variable there were associated the linguistic variables LOW, MEDIUM, HIGH.

Figure 10 presents the rules used by the fuzzy system. It should be specified again that generating the rules was carried out in automated mode through the neuro-fuzzy process, but the adaptation and their refining for the studied welding process were made by the authors.

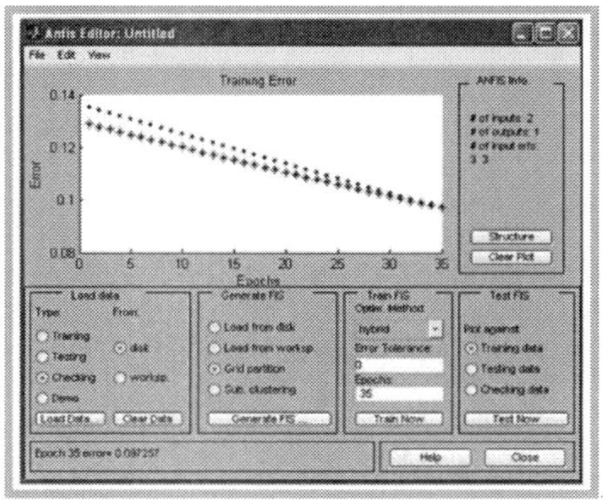

Fig. 9. Evolution of the training and verification errors

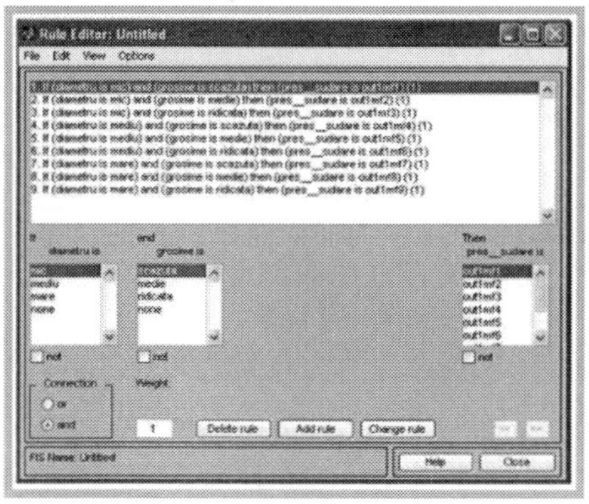

Fig. 10. The fuzzy system rules

Figure 11 presents the adjustment surface of the fuzzy system, which expresses in a graphical form the dependence between the input variables (diameter and thickness) and the output variables (welding pressure).

The neuro-fuzzy model in its final form is shown in figure 12. The model interface allows both the interactive graphical section of the values of the input variables and the accurate input of their values. In both cases the output value is calculated and displayed in real time.

4 Conclusions

The generation of the mathematical model for the analyzed welding process has included following stages:

- generation of a preliminary fuzzy model, which combined the user experience with the mathematical system made available by the fuzzy logic; the model's purpose has been to provide some introductory statements on the subsequent modeling strategies;

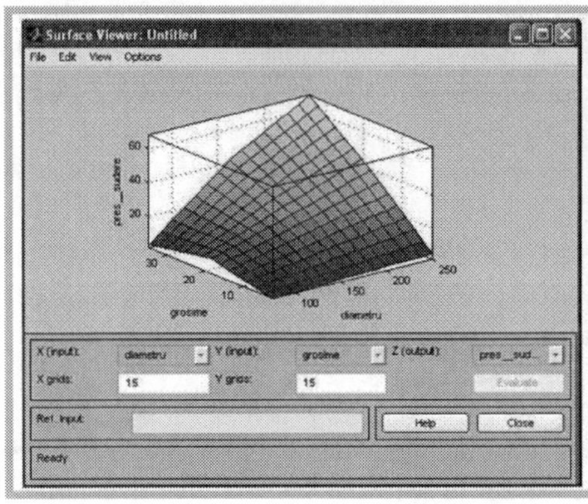

Fig. 11. The fuzzy system adjustment surface

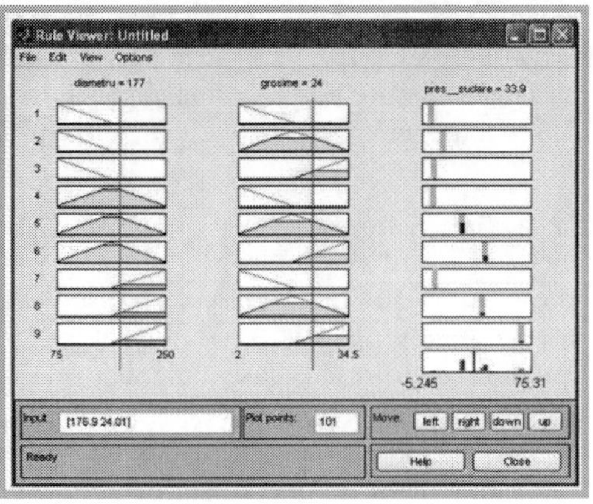

Fig. 12. The neuro-fuzzy model after refining

- creation of a mathematical model whose techniques are based on the neuro-fuzzy concepts, including as an input variable the nominal pressure, although preliminary indications suggested that the size of the output (welding pressure) is very slightly influenced by this; running the developed mathematical model has confirmed this hypothesis;
- refining the previously obtained mathematical model, by eliminating the input variable pressure and adjusting the fuzzy rules and the variation ranges of the quantities, in light of the experience and researches carried out by the authors.

It can be asserted that the mathematical modeling carried out gives the possibility to obtain the correct value of the pressure required for welding, for any pairs of input variables: pipe diameter – pipe wall thickness. Also, the proposed model can easily take into account other input variables for the process.

References

1. D. Conner, EDN **31** (1993)
2. M.H. Hassoun, *Fundamentals of Artificial Neural Networks* (MIT Press, Cambridge, MA., 1995)
3. M. Jamshidi, N. Vadaiee, T.J. Ross, *Fuzzy Logic and Control: Software and Hardware Applications* (Prentice-Hall, Englewood Cliffs, 1993)
4. J.F. Agassant, P. Avenas, J-Ph. Sergent, P. Carrean, *Polymer Proccessing Principales and Modeling* (Hanser Publishers, Munchen, 1991)

Computer-aided study of the gas flow within a natural gas storage and the associated collection network

Alina Gligor[1,*], *Cristian* Deac[1], and *Marius* Bibu[1]

[1]Lucian Blaga University of Sibiu, Faculty of Engineering, 4 E. Cioran st., 550025 Sibiu, Romania

Abstract. The current paper presents a computer-aided analysis of the way in which a natural gas storage created in a depleted gas deposit and the associated gas collection network are operated, both during the extraction phase and during the injection phase. In a first step, an analysis is carried out on a two-dimensional, nonstationary model for the gas storage, based on the deposit's geometry and the rock properties, but also on the gas motions. Also, a dynamic simulation of the injection and extraction processes is unfolded, allowing the study and optimisation of elements such as well placement and gas flows through the wells with the goal of maximising the storable gas quantities. A second model is then generated for the gas collection network associated with the natural gas storage, using the simulation software Simone. Its main benefit is allowing the simulation and modelling of nonstationary flows through the network. The two models are closely interconnected, in that for the extraction stage, the output from the first model is used as input for modelling the collection network, while for the injection phase, the results from modelling the collection network are used as entry data for the gas storage model.

1 Introduction

Natural gas storages are facilities built with the purpose of holding, during the warm season, the largest possible amount of gases which can then be used for heating or other purposes during the cold season [1]. The processes of injecting gas into a storage space and then extracting gas from that storage are complex and involve the consumption of relatively large amounts of energy.

Gas storages realised in depleted gas deposits are complex structures that can occupy wide areas and are operated by means of several gas wells. An efficient management of these storages involves the constant monitoring of the processes taking place and an advanced knowledge of the gas pillow's spatial distribution in the storage space.

While a review of the relevant speciality literature has revealed several studies that concern the modelling and optimisation of processes unfolding in gas deposits or reservoirs [1- 4], most of these refer to either general problems or to the specific problem of storing carbon dioxide in depleted gas deposits [3 - 4]. Furthermore, it should be mentioned that no specific studies have targeted the reservoir in the area described by this paper; also, processes unfolding in the gas collecting network are usually treated separately.

* Corresponding author: alina.gligor@ulbsibiu.ro

 DOI: 10.1051/matecconf/201712109005

The current paper describes the manner of linking two numerical models that allow the modelling of an injection process: on the one hand a numerical nonstationary two-dimensional model for the simulation of storage-extraction processes in time and space that allows the obtaining of a dynamic spatial image of the gas distribution in the storage and on the other hand a model for the gas collecting network, created with the help of the professional simulation software Simone allowing to model the transit of a non-stationary gas flow through the collection network.

2 The model of the gas collecting network

The first step in realising the combined model was to generate the model of the gas collecting network leading to the gas deposit using the numerical simulator Simone, developed by the SIMONE Research Group s.r.o. from the Czech Republic; this software uses as inputs the network configuration as well as the storage process dynamics.

In order to allow the validation of its results, the authors have chosen to allow the model's input data to reflect the reality on the ground, the pipe dimensions, for example, being based on GIS data from an actual gas storage located in Romania. Thus, the studied collection network includes on the one hand adduction pipes that link the gas wells to the collection points (groups) and on the other hand adduction pipes that link the collecting pipes to the gas delivery/reception points on the National (Gas) Transportation System (NTS). Specifically, the analysed network consists of a compressor station, 5 well groups, 8 collecting pipes, 19 adduction pipes and 19 wells.

Figure 1 presents the flow diagram of the studied collection network for the gas deposit.

Fig. 1. The diagram of collection pipes in the studied network

In order for the model to be realistic, the working scenarios defined in the program needed to contain some constraints due to the unsteady gas flow in the national transportation system. An important constraint that arose was the limited transport capacity of the collecting network. The flow rates injected/extracted into/from the gas storage must not exceed the transport capacity of the collecting network.

The compressor station is designed to raise the pressure of the gas taken from NTS pipes to a level that allows it to be introduced into the gas storage. At the points where the gas leaves the collection system and enters the well, regulators allow the defining of injection parameters. These parameters depend on the well's configuration (diameter, length) and on the depleted deposit's properties (permeability, porosity, deposit pressure around the well).

Based on the created gas collecting network model, the authors created in a next step dynamic simulation scenarios for the gas injection. Here, the parameters such as gas flows and pressures injected daily into the storage were used as boundary conditions for the storage model; the model allowed to calculate for each considered day the spatial distribution of the injected gas volumes, the total amount of gases injected up to that day as well as the pressure field in the storage.

The total duration of the considered gas injection stage was of 132 days, but due to the large amount of data needed to be fed into the scenario and due to the calculation possibilities offered by the available software and hardware, this time span was divided on scenarios with a duration of 10 days each (except for the last scenario that lasts only 2 days), the next scenario starting from the parameters resulting at the end of the simulation of the previous scenario. The full injection cycle resulted from the totality of the 14 scenarios for that process.

3 The Numerical Model of the Gas Storage

The numerical model of the gas storage can be used to assess the gas amounts that should be injected and to accept these amounts into the storage or, to the contrary, to reject the amounts that would surpass the holding possibilities of the storage.

For the construction of this model too, the authors have started from the real geometry of the considered gas storage. Since it can be considered that the upper and lower limits of the storage, respectively, have a relatively flat profile and that the thickness is largely constant, a two-dimensional model has been chosen, reflecting only the extent on a horizontal plane.

As stated, the numeric model was designed to reflect the storage's real geometry. Therefore, the model's integration field had to be identical with the shape of the storage. This has been achieved by defining a so-called informatic field map with the purpose of allowing the running of calculations for the case of irregularly shaped integration fields.

The integration process is done at the same time with reading the field map. With the help of the information that is stored here, calculations can be carried out according to the type of points: some of the points require integration calculations, others require the definition of boundary limit conditions or limit conditions on wells, respectively etc.The modelling of processes occurring in the studied gas storage uses flow equations for a flat storage, defined in Cartesian coordinates [2, 3]:

$$\frac{\partial}{\partial x}\left(\frac{kh}{\mu z}\frac{\partial^2 p^2}{\partial x}\right)+\frac{\partial}{\partial y}\left(\frac{kh}{\mu z}\frac{\partial p^2}{\partial y}\right)+M_s hRT=2mh\frac{\partial}{\partial \tau}\left(\frac{p}{z}\right) \tag{1}$$

where: x, y, z - spatial coordinates; p – pressure; k – permeability; h - height of the storage; μ - dynamic viscosity of the gas; Ms - mass flow of the source; R - gas constant; T - temperature; m - porosity;

For the analysed storage in the depleted gas deposit, the chosen method was a Du Fort-Frankel integration with two temporary steps, because it offers unconditional stability and a high processing speed. The employed interactive scheme is the one indicated in equation 2:

$$u_{ij}^{n+1}=\frac{2\alpha_x}{1+2\alpha_x+2\alpha_y}(u_{i+1,j}^{n}+u_{i-1,j}^{n})+\frac{2\alpha_y}{1+2\alpha_x+2\alpha_y}(u_{i,j+1}^{n}+u_{i,j-1}^{n})+ \\ +\frac{1-2\alpha_x-2\alpha_y}{1+2\alpha_x+2\alpha_y}u_{ij}^{n-1}+2BM_s \tag{2}$$

The integration process requires the defining of initial conditions and of limit conditions. These allow in turn to define various scenarios applied to the gas storage model [5]. Therefore, the difference between the gas injection stage simulated on the model and the simulated extraction stage is made by the specific limit conditions.

The process of injecting gas into the deposit starts from an initial gas pressure (and amount) in the storage, called a "gas pillow" (Fig. 2). It can be assumed that the pressure is initially distributed evenly throughout the storage.

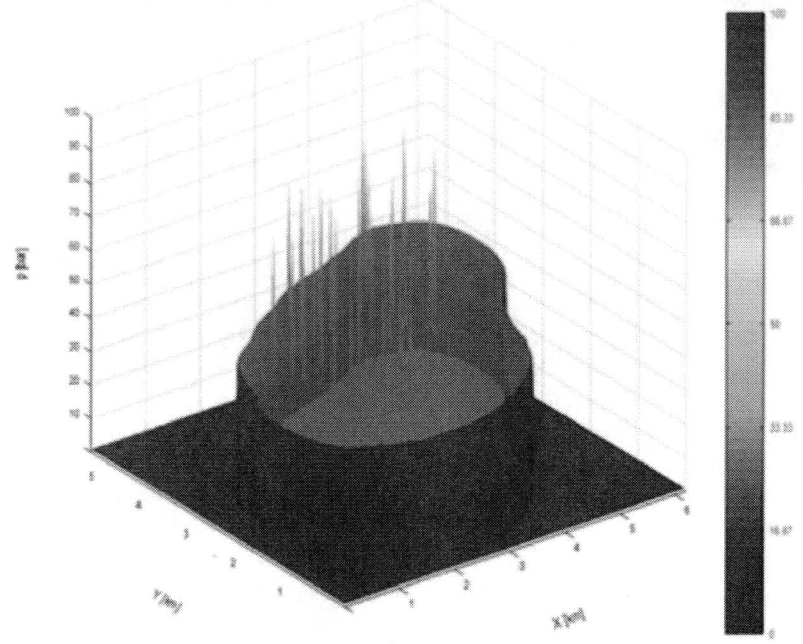

Fig. 2. Gas pillow and storage wells

The initial conditions are determined at the beginning of the integration process. Since the flow equation is defined based on pressure, the initial conditions for this model can be written as:

$$p_{ij}^{0} = p_{i,min} \begin{cases} i = 1,2 \dots nx \\ j = 1,2 \dots ny \end{cases} \tag{3}$$

Limit conditions are used to define the manner in which the integration process unfolds. In this case, the authors have employed, two types of limit conditions: boundary limit conditions and limit conditions on wells.

Boundary limit conditions are determined in the boundary points of the integration field (points that limit the area of the considered gas storage). Since for gas storages, the boundaries have to be considered impermeable, the speed of the gas in the boundary points is zero. Considering a flat (two-dimensional) storage area, there are only two sets of boundary conditions: one for the gas speed on the direction of the Ox axis and the other for the gas speed on the direction of the Oy axis (Equations 4 and 5).[3]

$$v_x = -\frac{k}{\mu}\frac{\partial p}{\partial x} \cong -\frac{k}{\mu}\frac{\left(P_{i,j}^{n+1}\right)_f - P_{i+1,j}^{n+1}}{\Delta x} = 0 \tag{4}$$

$$v_y = -\frac{k}{\mu}\frac{\partial p}{\partial y} \cong -\frac{k}{\mu}\frac{\left(P_{i,j}^{n+1}\right)_f - P_{i+1,j}^{n+1}}{\Delta y} = 0 \tag{5}$$

where:
x, y - coordinates of the points within the storage area;
p – pressure;
k – permeability;
μ - dynamic viscosity.

Using the relations (4) and (5), the boundary conditions have been rewritten as follows:
$(P_{i,j}^{n+1})_f = P_{i\pm1,j}^{n+1}$ on the Ox direction
$(P_{i,j}^{n+1})_f = P_{i,j\pm1}^{n+1}$ on the Oy direction
where the index f refers to boundary, i.e. $\left(P_{i,j}^{n+1}\right)_f$ represents the pressure of the point with the coordinates i,j located on the boundary of the domain.

For the injection stage, the limit conditions in the gas wells describe the injected flow moving through the well. The flow is defined by the well dimensions, as well as by the pressure and permeability of the storage-containing layer. The flow then allows the generation, within the model, of the injection and extraction pressures that provides the actual transportation of that flow through the gas well.

In order to inject a mass flow of gas ṁ into a well located at the coordinates (i,j), the existence of a pressure $\left(P_{i,j}^{n+1}\right)$ in that well. This value can be determined from the limit conditions, based on the law of mass balance. This leads to the expression of $\left(P_{i,j}^{n+1}\right)$ given in equation 6:

$$P_{i,j}^{n+1} = \frac{\dfrac{m}{\frac{k}{\mu}h} + \left(P_{i+1,j}^{n}\rho_{x_1} + P_{i-1,j}^{n}\rho_{x_2}\right)dy + \left(P_{i,j+1}^{n}\rho_{y_1} + P_{i,j-1}^{n}\rho_{y_2}\right)}{\left(\rho_{x_1} + \rho_{x_2}\right)dy + (\rho_{y_1} + \rho_{y_2})dx} \tag{6}$$

For the extraction stage, the limit condition would not change, but the flow ṁ extracted from the well would need to be introduced with the minus sign.

In the case of a gas storage in the settlement stage, that is, gases are neither not injected nor extracted, the pressure in the storage tends to homogenize itself.

The limit condition for this case is then given by the sum of the average pressures from nearby points:

$$P_{ij}^{n+1} = \frac{\left(p_{i+1,j}^{n} + P_{i-1,j}^{n} + P_{i,j+1}^{n} + P_{i,j-1}^{n}\right)}{4} \tag{7}$$

4 Results of the simulation for the stage of gas injecting stage

4.1 Results for the simulation on the gas collecting network model

After running the simulations with the model defined as presented above, there have been obtained several results that, after comparison with the data from the real gas storage, allowed the validation of the conceived model and confirmed its accuracy. In the following there are presented the most relevant ones results. mainly as graphs in order to be more suggestive.

Figure 3 presents the variation of flows through the surface system, after 20 days of injection.

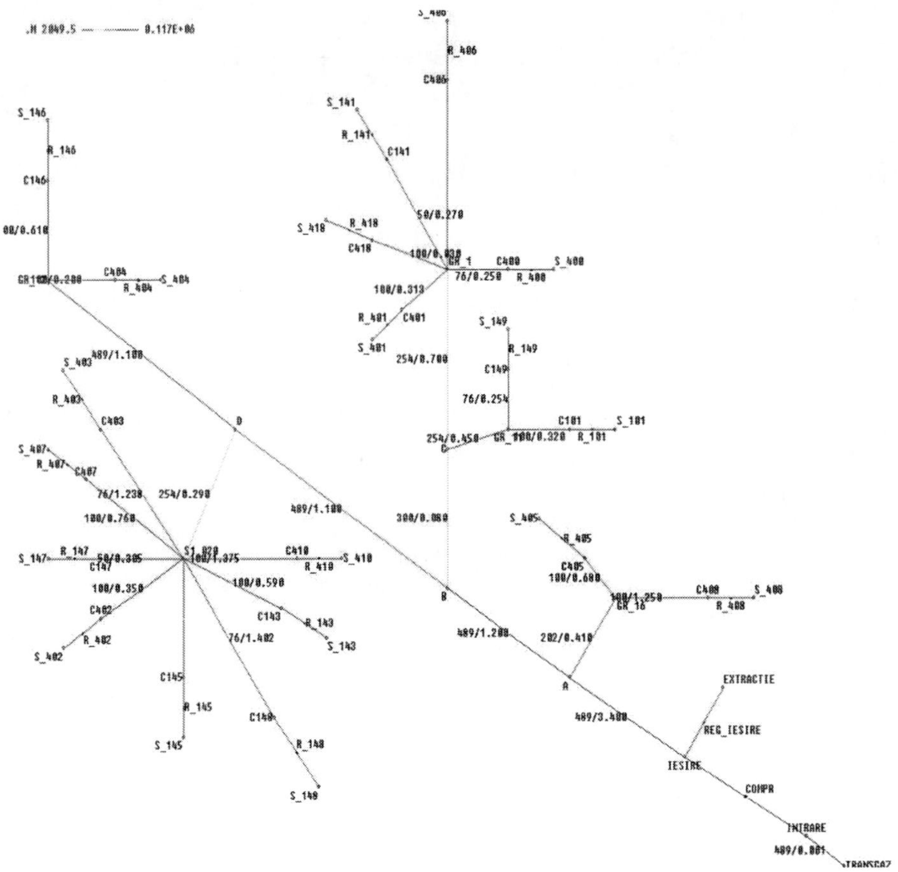

Fig. 3. Variation of the gas flows in the collecting network after 20 days

In order to synthetise the results of all simulation scenarios considered, there were created graphs of the main parameters for the whole injection period. Thus, figure 4 presents the variation of the gas compression pressure. It can be seen that in this case, the compression pressure is approximately constant.

Figure 5 presents the variation of the total gas flow injected into the gas storage. Beneath the daily variations, there can also be noticed a descending trend - as the injection period increases, the gas amount entering the storage is smaller, because the pressure in the deposit increases.

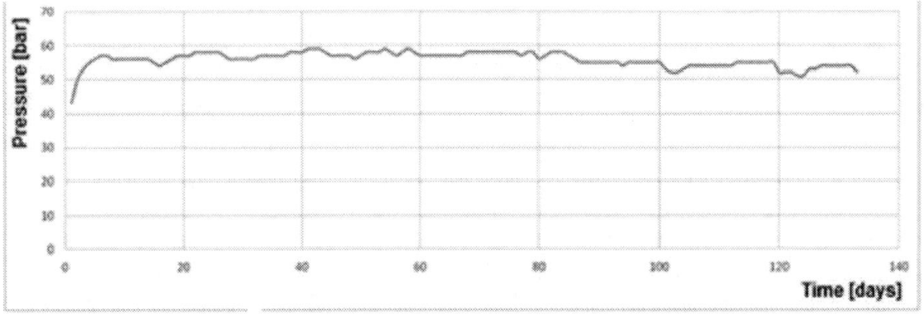

Fig. 4. Variation of the gas compression pressure during the injection stage

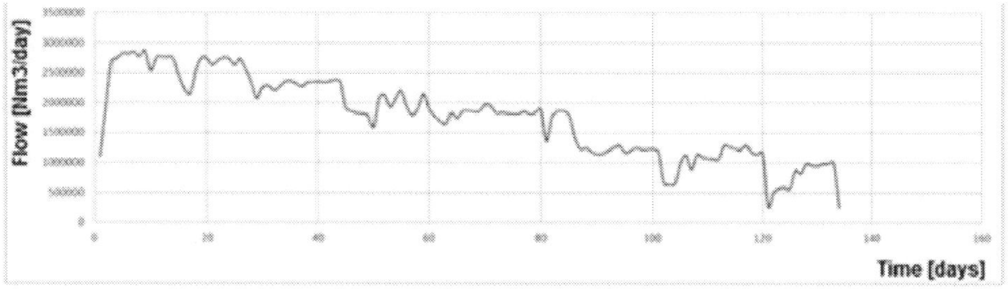

Fig. 5. Variation of the injected gas flow

The total flow injected into the deposit is divided on each well. Figure 6 presents the variation of flows on 5 of the existing wells.

Even if the general descending trend is similar to the one in figure 5, when analysing each well in part, it can be noticed that the injected flow differs very much from day to day. This needs to be correlated with the pressure variation in the wells, a variation exemplified in figure 7 for other 5 wells.

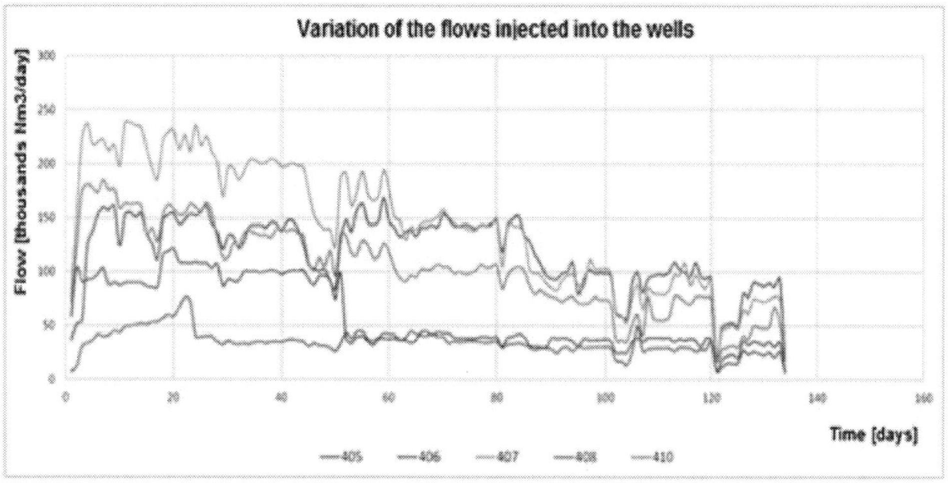

Fig. 6. Variation of the flow injected into the wells

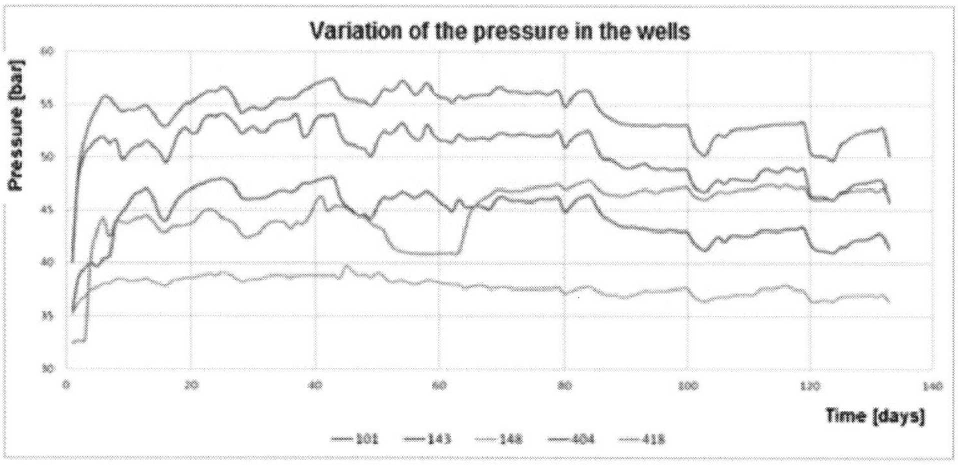

Fig. 7. Variation of the pressure in the wells

DOI: 10.1051/matecconf/201712109005

4.2 Results for the simulation on the gas storage model

Using the storage model for each day of injection, the flows and pressures obtained from the simulation of the transport network of the storage were imposed as limit conditions for the wells on which the injection is done. Using these conditions, a complex simulation was realised for the entire duration of the injection process.

Figure 8 presents the average daily pressures and the cumulative values for the entire period, for all 19 wells considered. The unevenness is due to the variation of the injected daily flows. This graph also presents the average pressure in the deposit (yellow curve) and the total amount of injected gases (blue curve). The slope of this blue curve depends on the daily flows that are injected, so if in the last part of the injection process (after 90 days) the slope decreases, it is because the daily flows that are injected are also smaller.

Figure 9 presents the spatial distribution of the pressure at the end of the injection stage. The peaks on the surface correspond to the pressures in the wells. However, it can be also noticed that at the end of the considered injection period, due to the current location of the injection wells, there remains a large part of the original deposit volume which is not filled by the stored gases (green area). This can be corrected by placing new wells in the areas that show a deficit in stored gas.

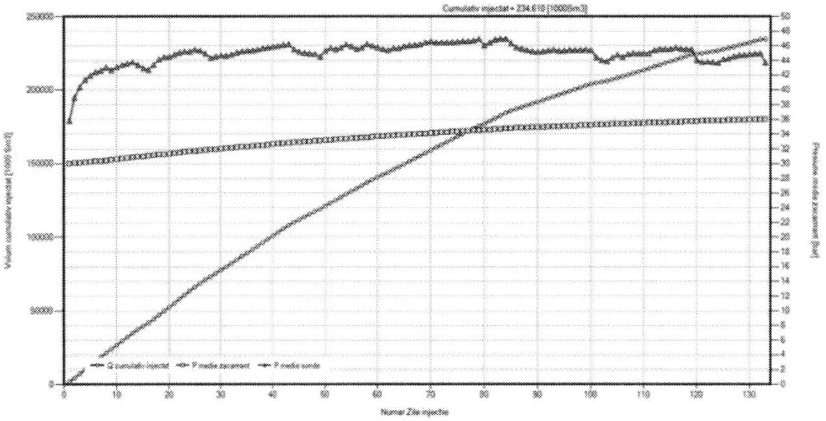

Fig. 8. Average and total values for the injection process

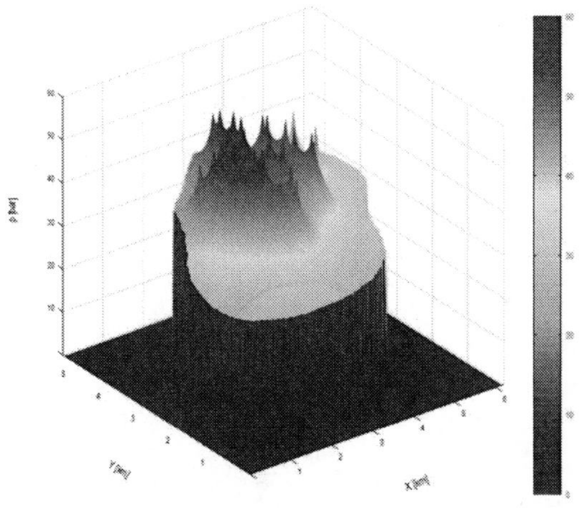

Fig. 9. Pressure values throughout the storage area at the end of the injection cycle

After injection there follows a settlement stage during which pressures homogenize themselves. In Fig. 10 the pressures in the storage are represented at the end of the settlement stage of 14 days. One can notice the homogenization of pressures in the field of injection wells, but that here too, the storage is not filled evenly with gases even if during the settlement stage a part of the gases has migrated towards the boundary of the storage.

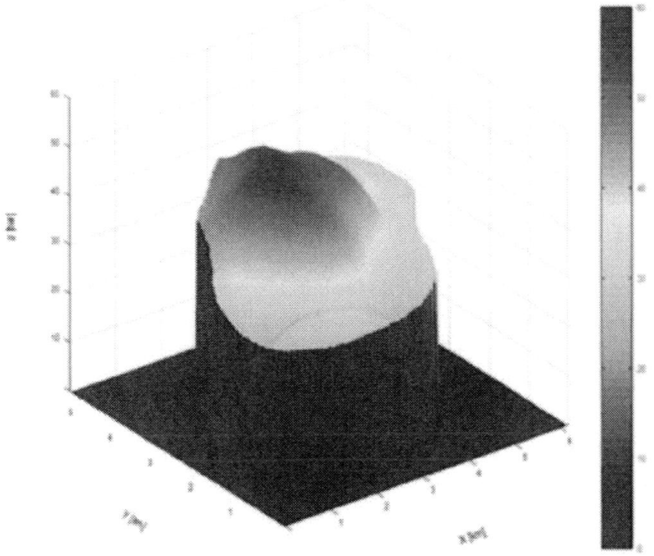

Fig. 10. Pressure values throughout the storage area at the end of the settlement stage after injection

5 Conclusions

As has been presented in this paper, the developed numerical simulation allowed to define the settings required for the daily usage of operating elements such as controllers, valves, compression station etc. that facilitate the transportation of the gas through the gas collection network from the compression station to the well for the injection phase and reverse for the extraction phase under optimal pressure conditions.

By using a dynamic model of the collecting network for the field/ deposit, it was possible to simulate full injection / extraction cycles in real conditions.

The model of both the collection network and the gas storage is especially useful for monitoring their behaviour, which is carried out by comparing the acquired SCADA information with the expected values. If larger differences between individual data points are detected, or if data evolution trends emerge that differ from the projected ones, corrections can be determined following quick simulation processes.

The developed model can simulate also the period between two cycles during which gases are neither injected nor extracted (settlement period).

For the gas storage, the developed model determines the values of parameters such as pressure, gas saturations, gas volumes for any given point of the storage and for any moment during the analysed period. For example, the pressure matrix can be determined precisely and can be represented as a 3D graph.

The numerical model described here also facilitates operations such as the placement of new wells (extraction points) in order to achieve a more homogeneous distribution of the pressures within the storage.

References

1. M. Albulescu, S. Neacsu, C. Eparu, S. Bambo, S. Suditu, Termotehnica, **XV**, S1 (2011)
2. D. He, A. Jia, G. Ji, Y. Wei, H. Tang, Petroleum Exploration and Development, **40,** 1 (2013)
3. I. Crețu, *Modelarea numerica a miscarii fluidelor in medii poroase* (Technical Publishing House, Bucharest, 1980)
4. C. Eparu, R. Radulescu, D. Stoica, Annals of the Ovidius University, Chemistry series, **24**, 2 (2013)
5. C. Eparu, S. Suditu, A. Prundurel, Annals of the Ovidius University, Chemistry series, **24**, 2 (2013)
6. S. Neacsu, C. Eparu, Journal of Eastern Europe Research in Business and Economics, **2013** (2013)
7. S. Neacsu, *Comprimarea si lichefierea gazelor* (Romconvert Publishing House, Ploiesti, 2002)
8. S. Neacsu, C. Eparu, Proceedings of the 21st International Business Information Management Association (IBIMA) Conference, *Vienna, Austria* (2013)
9. S. Neacsu et al., Environmental Engineering and Management Journal, **9**, 10 (2010)
10. S. Neacsu, S. Suditu, C. Popescu, Proceedings of the 1st International Conference on Power and Energy Systems, *Chania, Greece* (2013)
11. S. Neacsu, S. Suditu, D. Stoica, Annals of the Ovidius University, Chemistry Series, **24**, 2 (2013)
12. M. Nikolaou, Computers and Chemical Engineering **51** (2013).
13. H. Zangeneh, S. Jamshidi, M. Soltanieh, International Journal of Greenhouse Gas Control, **17** (2013)
14. *** Simone 5.7, http://www.simone.eu/simone-company-about.asp, accessed 21.02.2017

Solutions to operate transmission and distribution gas networks

Sorin Neacsu[1], and *Cristian* Eparu [2,*]

[1]UPG Ploiesti, DETH Department, 39 Bucharest Blvd, Ploiesti, România,
[2]NetGas R&D, 28 Bucharest Blvd, Ploiesti, România

Abstract. In order to respect the actual and future regulations, besides SCADA, there is a need for further modern operating solutions for the transmission and distribution gas network. The paper presents the newest operating principles and modern software solutions that represent a considerable help to operate the transmission and distribution gas networks.

1 Introduction. AdmoduNet simulator, real time operating solution for gas transport and distribution systems

The modernization of natural gas transport and distribution systems has necessitated the introduction of SCADA systems. These are offering at each moment a snapshot of the situation of the transport system. The information is discontinuous because of the large area of the transport and distribution network and the big number of pipes. The number of points where data are acquired are limited due to costs.

The transport and distribution process is a continuous process and in order to get a precise image of the parameters at any moment in time it is necessary that, besides SCADA systems, a dynamic simulator calibrated for the respective network to be in place [1]. This simulator can complete the picture of parameters in the entire transport or distribution network based on the real data received. In this way, the parameters that are to be measured, like line pack for pipes or the entire network [2], spatial and time continuous distribution of transport parameters (pressure, flow rate, temperature) and also gas quality (composition, calorific value) can be determined [3].

Having a nonstationary simulator calibrated on a transport or distribution network allows making simulations for preparing the next day plan of transport (base plan). The simulator can also be used to predict the flow in the system for bigger time intervals: days, weeks, months and years. Another important function is the possibility of coming up, very quickly, with a solution in case of accidents that can deviate from the normal functioning of the system (base plan).

NetGas R&D has developed the AdmoduNet simulator for gas transport and distribution and oil and gas gathering systems. The simulator contains calculation engines for monophase (gas, liquid) and multiphase systems.

The simulator concentrates all calculation effort on servers and has a multiuser type web interface so that it can be used by multiple departments in a company, even from mobile devices.

* Corresponding author: sneacsu.ph@gmail.com

 DOI: 10.1051/matecconf/201712109006

AdmoduNet has performant proxy models for all regulation points in the network (compressors, control valves, slug catchers) that allow simulating any desired command.

2 Necessary data for a base plan

The transport program is dynamic due to the gas capacities provided by the gas market. In order to define these quantities, AdmoduNet has a business module through which the equilibration of demand and offer from the market is possible. It should be mentioned here that the connections between the gas suppliers and the customers form an entire different transport network called commercial network. The business module makes the daily equilibration of this commercial network (fig. 1), converts commercial agreements into point-to-point Demand/Supply transactional relationship configurations and enables synoptic Business Rules visualization, converts Business Rules and current supply/demand constraints into operational conditions for process simulation

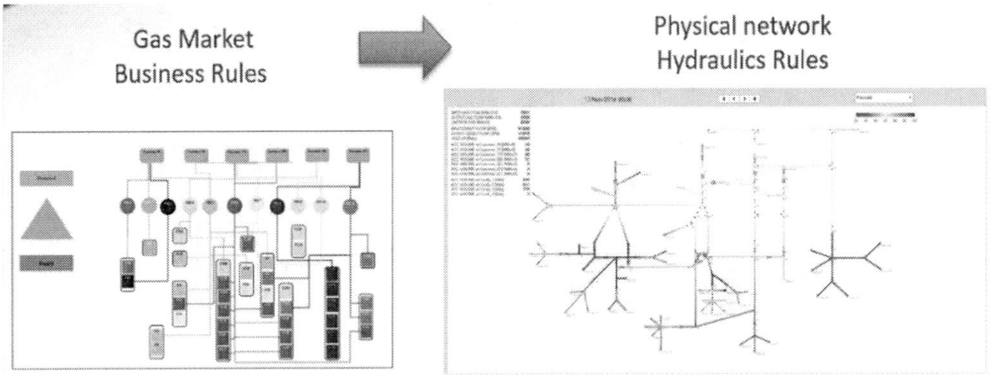

Fig. 1. Differences between commercial and physical network

The consumption volumes used for network entry and exit points are global values for 24 hours that came from the gas market. For these values to be properly used in dynamic simulations, there are modeled with hourly profiles defined for each consumption point from historical data [4] (fig. 2).

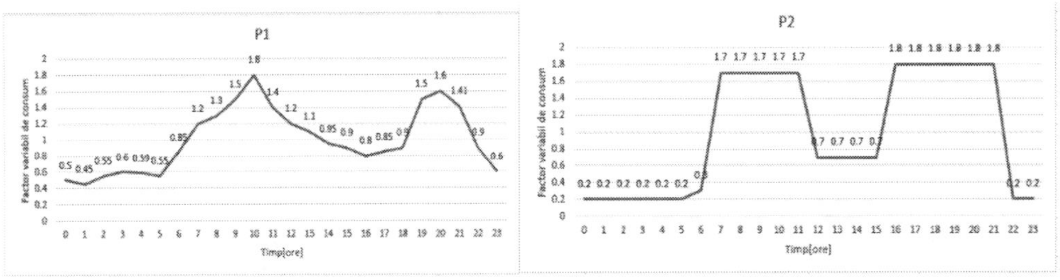

Fig. 2. Hourly consumption profiles for a locality and an industrial consumer

A performant profiling module that has the ability to define a consumption profile based on temperature, day of the week and other major influences, is used in order to determine the small clients' (non-monitored) consumptions (fig. 3). This module is used for modelling the consumption behavior of an area in order to get reliable prognosis for the gas volumes needed.

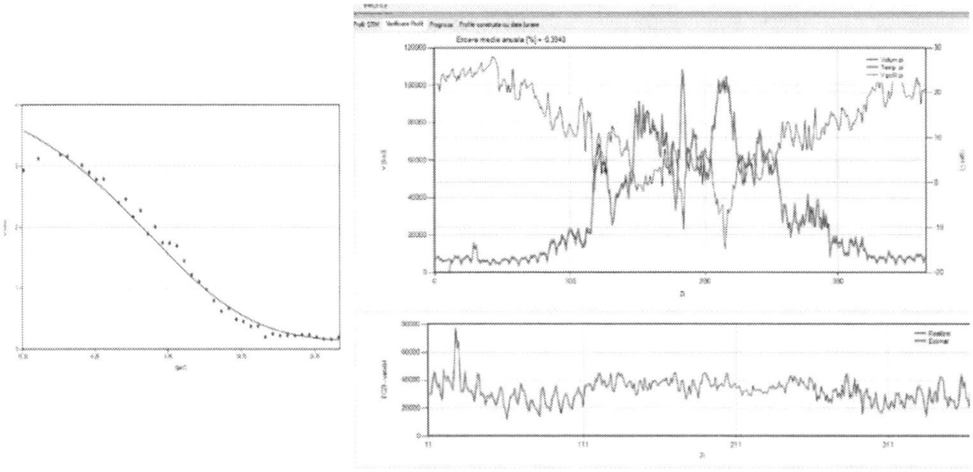

Fig. 3. Profile use and gas volume nomination

3 Dynamic simulations with AdmoduNet

Once the transport system load is defined, by establishing the gas values at the entry and exit points, necessary simulation can be made starting from an initial state of the transport system. Figure 4 presents a graphic representation of the results of a simulation.

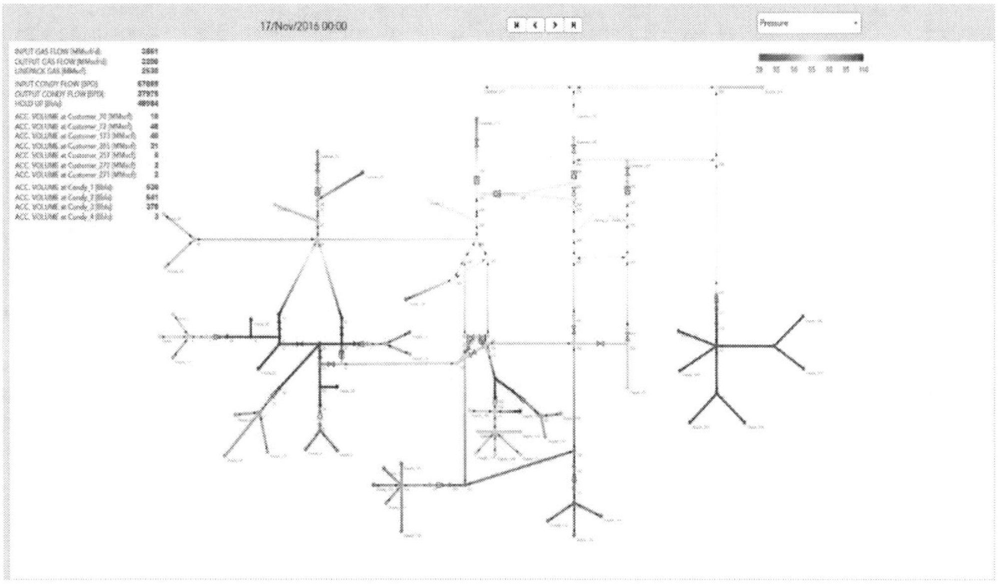

Fig 4. View of pressure variation in the network

The parameter chosen to be presented in this case is pressure. The coloring of the pipes is based on a color code. In the left upper part of the picture global values for the network are presented for a specific moment in time. We can see the evolution in time for any defined parameter by clicking the middle buttons.

The software allows different types of results visualization directly from the main tab.

In figure 5 one can see the possibility of graphically viewing different parameters and gas composition variations in space (every node of the network) and time.

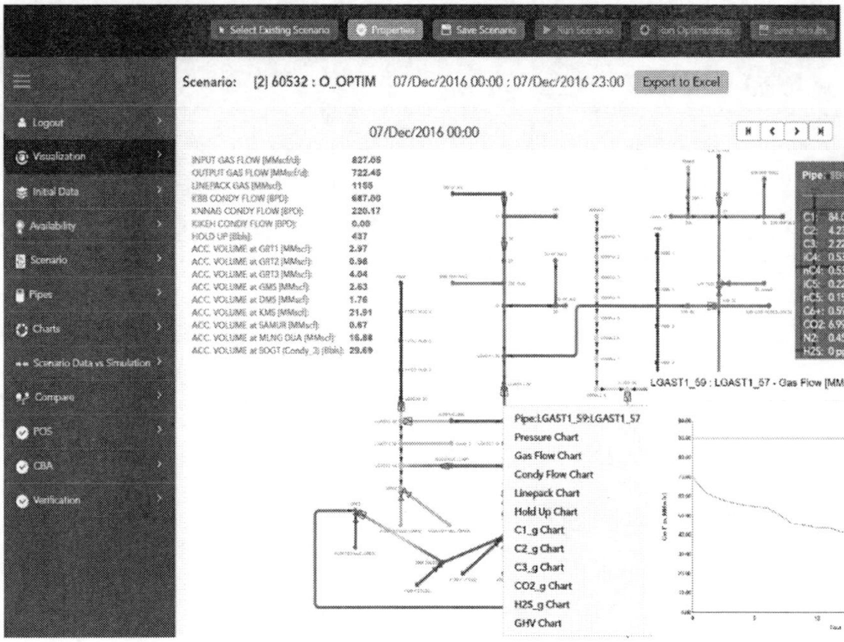

Fig 5. Results visualization

Figure 6 presents different possibilities of graphic representation for gas volume delivered to the client, variation of pressure on a path that connects an entry point to an exit point and priority of supply, where we can see from which supply point came the gas delivered to a customer. This last calculation is the base for cost business analysis.

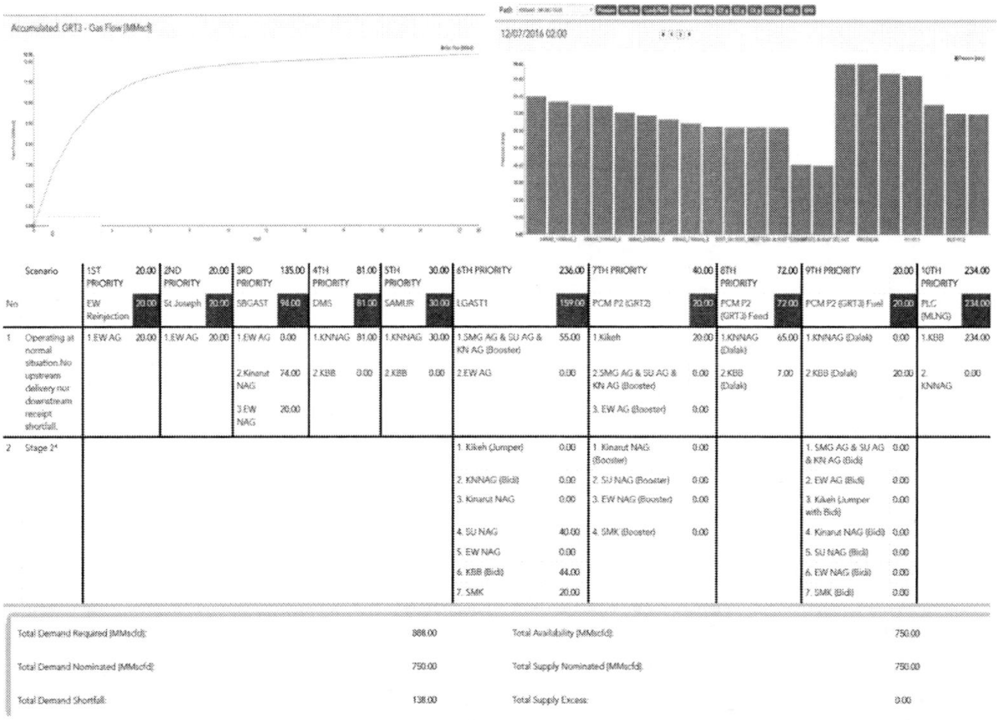

Fig. 6. Results analysis and visualization

4 Simulation types available with AdmoduNet

Forward simulation starts from gas volumes at supply points and determine the quantity and quality of gas that will be delivered to the customers, taking into consideration the restriction and parameters for control elements [5].

Reverse simulation provides the best operating scenario starting from the gas delivery target on volumes and quality and determines the platforms and network set-points of control elements and operational conditions. Because multiple solutions can exist, this type of simulation is used together with an optimization module that can be configured based on the clients' needs.

Auto-simulation (Monitoring) is used in the dispatch center by the operators; it is a very important predictive application that will give the operators the necessary comfort to make decisions in ~~a~~ real time, solving thus future problems that can appear in the system. The monitoring application automatically simulates the transport process every hour for the remaining time ahead in the monitor time horizon, compares the results simulated with the initial base plan and also identifies the deviation from plan in the remaining time ahead.

In figure 7 one can see a predictive exception based on monitoring which shows what will happen in the future as an exception to your plan. What is shown here is a line which is the plan line/base, a line that was created the day before and then, every hour, it shows you what will happen in the next hours, more precisely what may deviate from that plan. It is a predictive exception surveillance, different from any other surveillance known so far, that is actually showing a deviation only after the detection of a problem.

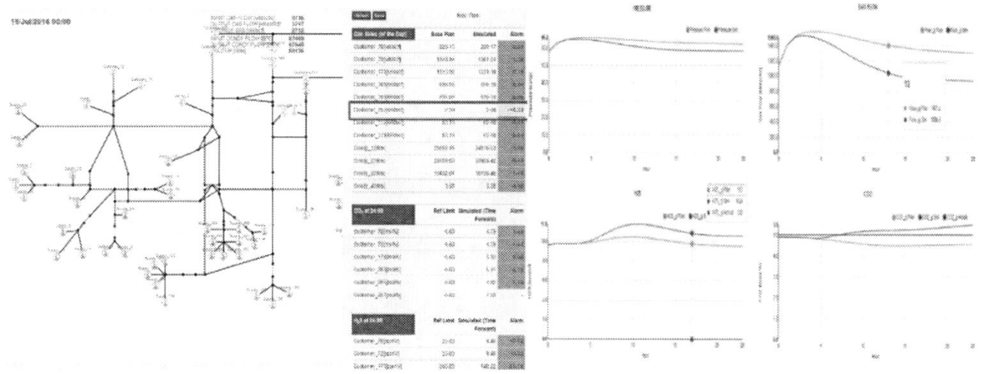

Fig. 7. Monitoring dashboard and graphs

By fulfilling the objective of simulation (supplied and delivered quantities, gas quality, pressure regime in the network) we obtain the operation mode, represented by the totality of commands needed to be executed and the moment in time when these commands must be executed for all the control elements in the network (control valves, compressors).

In the case of gas distribution system, where only large consumers (B4-B6) are monitored, NetGas R&D prepared a similar solution with AdmoduNet that can be used to make the equilibration of the distribution network every 24 hours, based on data from a limited number of control points in the network.

5 Conclusions

Transport and distribution systems are spread on large areas and have many pipelines. SCADA systems give information from a limited number of points in the network. In order to determine the flow parameters in the entire network, a dynamic simulator has to be used, a simulator that continuously determines the parameters values in the whole network starting from the monitoring data.

From the dynamic simulation, a very important measure for the equilibration process and its variation is obtained.

The use of a performant dynamic simulator allows the operation process for the next time interval to be defined.

In case of various incidents, by using the optimization module, low-priced operational solutions can be easily and promptly found.

Certification of correct solution results of AdmoduNet simulator presented in this paper is represented by Petronas using the gas and condensate gathering system from offshore platforms.

References

1. S. Neacşu, C. Eparu, *Proceedings of the 21st International Business Information Management Association (IBIMA) Conference*, (Vienna, 2013)
2. C. Eparu, R. Rădulescu, D. Stoica, Ovidius University Annals of Chemistry, **2**, 83 (2013)
3. S. Neacsu, S. Suditu, C. Popescu, *EDU '10* (Iwate Prefectural University, Japan, 2010)
4. C. Eparu, S. Neacşu, D. Stoica, *Proceedings of the 21st International Business Information Management Association (IBIMA) Conference* (Viena, 2013)
5. A.J. Osiadacz, *Simulation and Analysis of Gas Networks* (Gulf Publishing, 1987)

Intergranular degradation of stainless steel tubular components used for kerosene transportation. Metallographic analysis

Ion Palamarciuc [1,2], *Dan Gelu* Galusca [1], *Monica* Ratoi[3], *Brian* Mellor[3], *Cătălin Gabriel* Dumitraş[4], and *Maricel* Agop[5,*]

[1] Gheorghe Asachi Technical University of Iasi, Faculty of Materials Science and Engineering, Bd. D. Mangeron no. 41, 700050, Iasi, Romania
[2] Alexandru Ioan Cuza University of Iasi, Faculty of Physics, Bd. Carol I, 700506, Iasi, Romania
[3] University of Southampton, Faculty of Engineering and Environment, United Kingdom
[4] Gheorghe Asachi Technical University of Iasi, Faculty of Machine Manufacturing and Industrial Management, Bd. D. Mangeron no. 67, 700050, Iasi, Romania
[5] Gheorghe Asachi Technical University of Iasi, Faculty of Machine Manufacturing and Industrial Management, Department of Physics, Bd. D. Mangeron no. 67, 700050, Iasi, Romania

Abstract. The chemical interactions between the kerosene compounds and the stainless steel structure of the tubular components, through which kerosene is transported, develop a degradation phenomenon, causing intergranular contamination from the surface to the inner part (process based also on mass transport, i.e. diffusion at low temperatures). This is the most dangerous effect of chemical contamination, affecting the active section of the tubular components, which can reach a critical section, leading to catastrophic consequences in the operation of a jet engine. Metallographic analysis of the tubular components of such a jet engine, manufactured from stainless steel 12H18N10T, through which kerosene has been transported, come to confirm the intergranular contamination.

1 Introduction

Fuel system components of turbo engines, as well as lubricating oil systems are subjected to repeated thermal shocks, which hampers their performance and operating time.

Pressure increasing in the fuel system (65 .... 90 bar) leads to temperature increase, which favors improvement of combustion efficiency. Also, the temperature of the flue gas increases, temperature restricted by conditions for the safe operation of gas turbines. Cooled gas temperature should not exceed 1198 °C at turbine entry and 730 °C at the turbine output, not to damage the blades. Also, the oil temperature must not exceed 150 °C [1-2]. Maintaining the temperature of lubricating oil and flue gas temperature below the maximum permitted levels is achieved using heat exchangers that use kerosene as coolant [3].

* Corresponding author: m.agop@yahoo.com

In conditions of high temperature and pressure, kerosene becomes a supercritical fluid - is neither gas nor liquid - but has properties of both aggregation states [4]. This kerosene property ensures improved performance of engines, but also produces kerosene oxidation, thermal decomposition (pyrolysis) and the deposits formation [5], processes strongly influenced by temperature, pressure, fuel composition and surface support [6]. The formation of deposits on the fuel system components is inevitable and can have catastrophic consequences [7]. Therefore, the study of the mechanisms that lead to the "generation" and evolution of deposition is very important.

In this paper, such a mechanism, specifically, intergranular contamination is confirmed by metallographic analysis of the tubular components of such a jet engine, manufactured from stainless steel 12H18N10T, through which kerosene has been transported. The originality of this research consists in the fact that the samples were collected from a jet engine and the pipes have been used in real conditions, while in literature we can find information based on laboratory operating conditions.

2 Experimental procedures and results

2.1 Samples

The analyzed pipes have been collected from the fuel system of a jet engine. They were used for transporting TS-1 jet fuel at high pressure (maximum 64 bar) and different temperatures (minimum -60 °C and maximum 150 °C). The pipes are made from 12H18N10T austenitic stainless steel with the external diameter of 16 mm. On the inner surface of the pipes we could observe carbon depositions and we prepared two types of samples: as received and ultrasonically cleaned in distilled water in order to detach the deposited layer for further analysis.

2.2 Metallographic analysis

The microstructure of the samples has been analyzed in cross section and longitudinal section by optical microscopy after metallographic preparation and etching in 10 % oxalic acid. In cross section we could observe minor surface defects (pits), as seen in Figure 1.

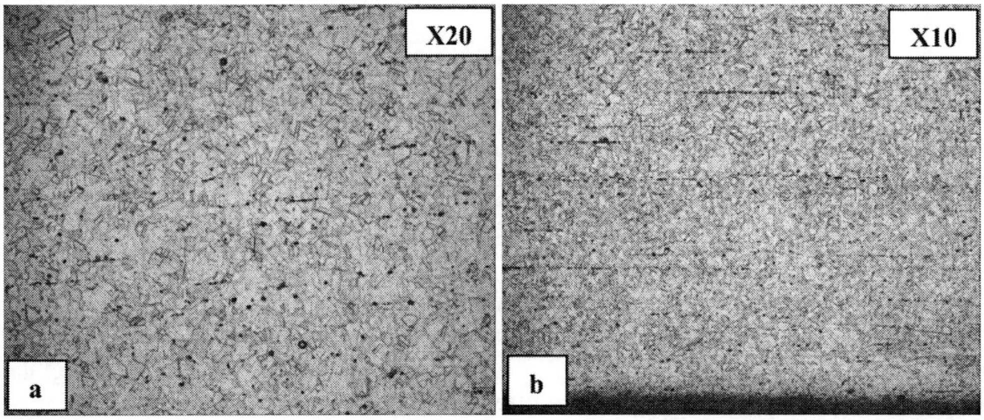

Fig. 1. Microstructure of the steel in cross section: (a) in the middle of the pipe wall, (b) at the external wall

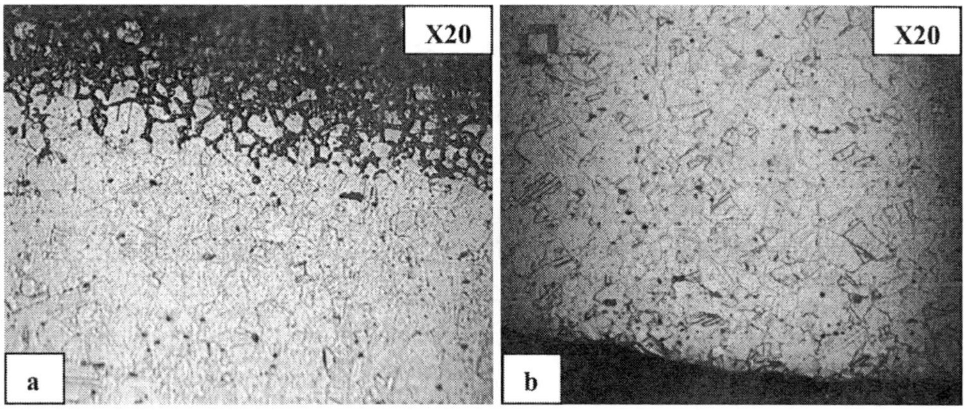

Fig. 2. Microstructure of the steel in longitudinal section: (a) at the internal wall, (b) at the external wall

However, only in longitudinal section, on the internal wall we could observe the intergranular corrosion attack (Figure 2a).

2.3 3D and 2D surface characterization

The cylindrical geometry of the samples, the non-uniform and dark color of the depositions did not allow us to perform any exact quantitative measurements.

This required the use of focus-variation microscopy techniques, so we characterized the surfaces with an Alicona Infinite Focus G4 microscope and used its IF-Measure Suite 5.1 software for image processing. Figure 3 illustrates the 3D image of the carbon deposition on the steel surface and the defects emerging to the surface.

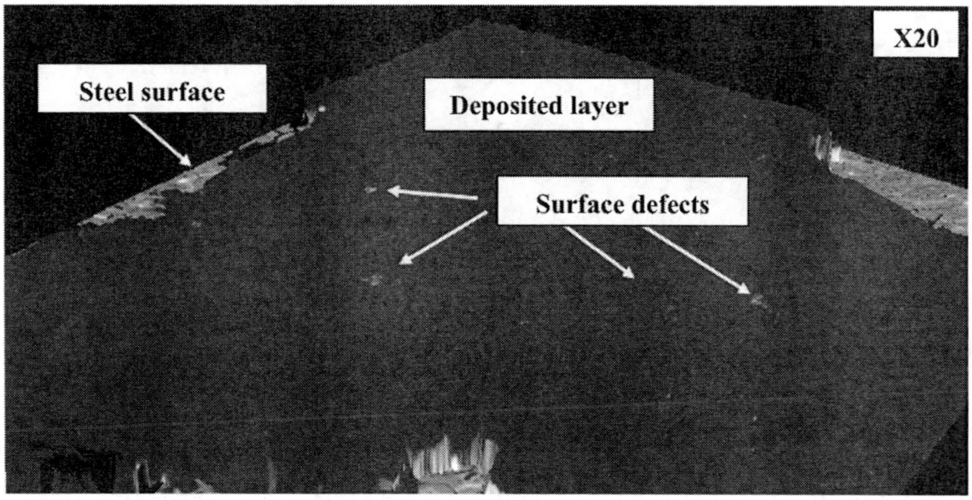

Fig. 3. 3D image of the carbon deposition on the surface of the steel

At higher magnification we detected surface defects such as corrosion pits. After extracting the surface profile along a linear defect (that is parallel to the direction of the fluid flow) we could measure the depth of the defect. It can be seen (Figure 4) that the depth of the defect in the analyzed area is about 3 µm.

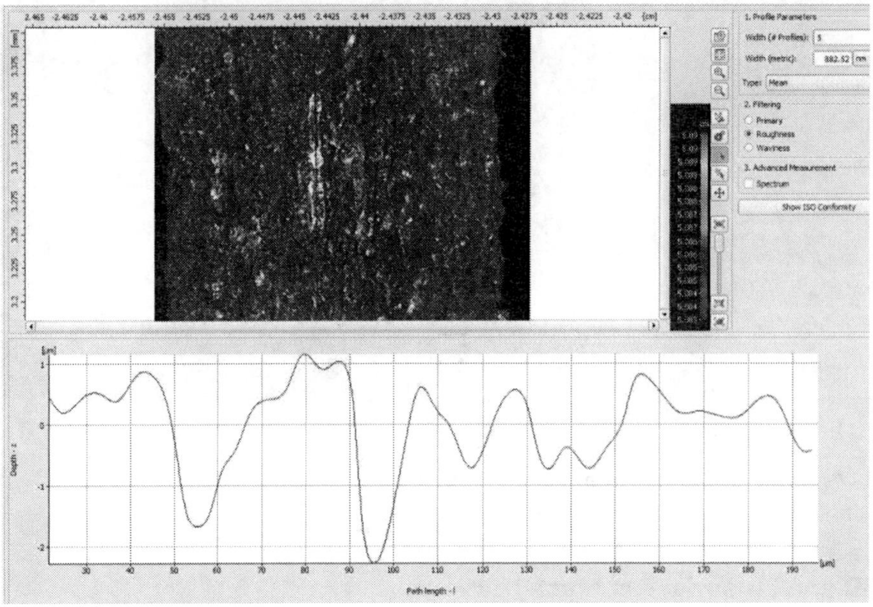

Fig. 4. Surface profile along the defect

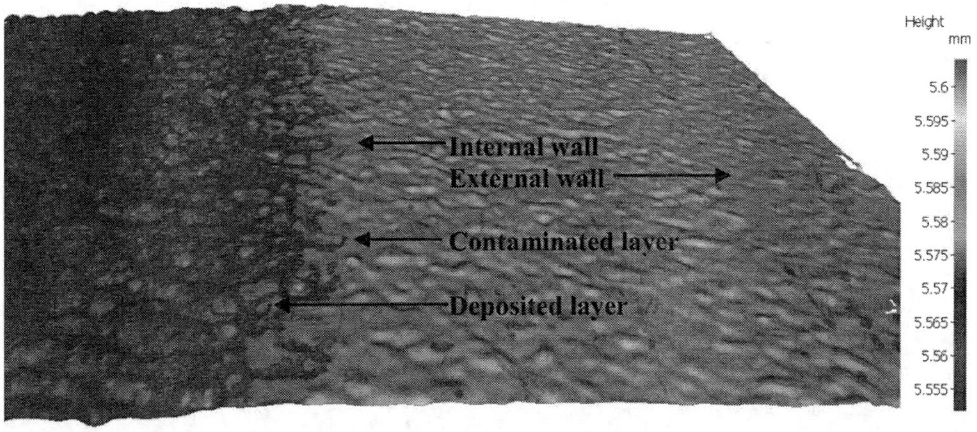

Fig. 5. Surface topography of a longitudinal cross section of a pipe

From Figure 5 it can be seen that two types of interfaces were formed: an interface between the steel and the deposition and another interface between the deposition and the transported fluid. The first one represents a contaminated layer (containing corrosion products), anchored in the microstructure of the steel and the second represents a deposited layer.

The interaction between kerosene components and steel is causing changes in the chemical composition also [8, 9] and it is directly correlated with the effects on the mechanical and physical properties. For example, penetration of sulfur into the steel surface, through the grain boundaries is a process that increases the dislocations density (the result of steel embrittlement) and is amplified by the chemical interaction leading to brittle fracture.

3 Conclusions

Our metallographic analysis of the tubular components of such a jet engine, manufactured from stainless steel 12H18N10T, through which kerosene has been transported, come to confirm the intergranular contamination.

This is the most dangerous effect of chemical contamination, affecting the active section of the tubular components, which can reach a critical section, leading to catastrophic consequences in the operation of a jet engine. We note that a similar effect occurs in the controlled release of drugs by diffusion complex at low temperatures (critical, subcritical and supercritical) [10-13] and plasma discharge [14-16].

This work was supported by the strategic grant POSDRU/159/1.5/S/ 133652, co-financed by the European Social Fund within the Sectorial Operational Program Human Resources Development 2007 – 2013.

References

1. A. Ersin, A. T. Shih, S. M. Jones, M. S. Reveley, J. T. Luxhøj, J. K. Evans, J. Risk Res. **18**, 428 (2015)
2. *** *Operating and maintenance manual of the jet engine AI-25* (Moscow, 1980)
3. H. C Mongia, Proceedings of the Turbine Technical Conference and Exposition (2013)
4. I. Gainar, *Jet engines* (Ars Docendi, Bucuresti, 2000)
5. S. P. Heneghan, S. Zabarnick, Fuel **73**, 35 (1994)
6. E. Grant Jones, L.M. Balster, W. J. Balster, Energy Fuels **12**, 990 (1998)
7. T. Edwards, W. Harrison, S. Zabarnick, M. DeWitt, C. Bentz, Proceedings of the 40th AIAA/ASME/SAE/ASEE Joint Propulsion Conference and Exhibit, 3886 (2004)
8. I. Palamarciuc, *Cercetări privind efectul contaminării cu produse petroliere asupra caracteristicilor fizico-chimice ale oțelurilor utilizate în construcția unor componente tubulare*, (Universitatea Tehnica "Gh. Asachi", Iași, 2017).
9. I. Palamarciuc, D. G. Galusca, J. Manhart, N. Okolie, Proceedings of Metal 2015 Conference. Ostrava, Czech (2015)
10. V. A. Paun, M. Popa, J. Desbrieres, C. A. Peptu, S. V. Dragan, G. Zegan, G. Cioca, Mat. Plast. **53**, 590 (2016)
11. V. A. Paun, L. Ochiuz, M. Hortolomei, A. Creteanu, I. Stoleriu, C. M. Ghiciuc, G. T. Serban, G. Zegan, G. Cioca, Mat. Plast. **53**, 699 (2016)
12. M. Stanciu, I. S. Zaharie, L. G. Beraz, G. Cioca, Acta Endocrinol. **12**, 485 (2016)
13. E. S. Băcăiță, C. Bejinariu, B. Zoltan, C. Peptu, G. Andrei, M. Popa, D. Magop, M. Agop, J. Appl. Math., 653720 (2012)
14. S. Gurlui, M. Agop, M. Strat, G. Strat, S. Băcăiță, A. Cerepaniuc, Phys. of Plasmas **13**, 063503 (2006)
15. M. Agop, D. Alexandroaiei, A. Cerepaniuc, S. Băcăiță, Chaos Solitons Fractals **30**, 470 (2006)
16. S. Gurlui, M. Agop, M. Strat, G. Strat, S. Băcăiță, Jpn. J. Appl. Phys. **44**, 3253 (2005)

Brownfield management opportunities to reduce the back pressure effects on the gas wells

Dan-Paul Stefanescu[1,*], and *Diana-Andreea* Lupu[2]

[1]Lucian Blaga University of Sibiu, Faculty of Engineering, 4 E.Cioran, Sibiu, Romania
[1]Exploration-Production Division, SNGN Romgaz SA, P-ta I.C.Motas 4, Medias, Romania
[2]Stimulation-Rehabilitation Production Department, E&P Division, SNGN Romgaz SA, P-ta I.C.Motas 4, Medias, Romania

Abstract. Gas mature fields are associated with challenges to optimize the hydrocarbon flow from reservoir to the sales point in a cost effective manner due to declining well productivity. Laslau Mare field is a mature gas field in Transylvanian basin (Mures County) developed in 1970s and is producing ~99% methane with low water-gas ratio. As any brown field, the state of depleted reservoir will generate several constraints for gas flow from formation to surface facilities and further to delivery point. During the exploitation has been observed that the operation conditions are facing with unstable pressure in the system due to low demand. Therefore, the back pressure effect will affect the wells in terms of inability to unload the bottomhole accumulated liquids and the reservoir will suffer a higher pressure drawdown. The best fit-for-purpose solution to overcome the above challenges is represented by installation of group compressor. Laslau Mare field has 3 group compressors running and shows positive results, especially when external pressure fluctuates continuously. This paper explain the challenges seen in 2016 in Laslau Mare field with back pressure effects and how the compression overcome them, and also other solutions that should be considered to optimize the well production.

1 Introduction

Gas fields from Transylvanian Basin have an advanced state of depletion considering the reservoir conditions after more than 50 years of exploitation and the intensive approach in terms of production. During exploitation, it has been found that is necessary to operate the fields in a rational and efficient manner in order to achieve attractive recovery factors. Rehabilitation concept has been developed which means a new approach of production management for field development or redevelopment plan.

Laslau Mare field is one of the field which is in a rehabilitation process subjected to many optimization projects executed in order to increase the recovery factor and to reduce the production decline.

Gas fields operation was faced throughout 2016 with many challenges, field operation conditions changed in terms of dynamic surface parameters represented by field pressure and

* Corresponding author: dan.stefanescu@romgaz.ro

the national grid pressure at the sales point from Production System to National Transportation System.

Dynamic surface parameter variations plays an essential role on the wells performance and behaviour and influences the flow condition from porous-permeable formation to surface facilities.

Demand on natural gas market being at this moment an uncontrollable factor at national level has been dropped significantly during warm period of 2016. Uncertainty conditions on natural gas market demand have a secondary effect in field operation and obviously achieved production doesn't match with forecasted production.

Extending recovery factor by implementing rehabilitation projects involves those technical and economical solutions that can improve the productive performance of the reservoir and to face all challenges caused by external factor.

These technical and economic solution must target both improving productive performance of the reservoir and gas production infrastructure.

2 Problem definition. Field challenges

Laslau Mare field management faced during 2016 with high external pressure causing high impact in production and reservoir backpressure causing liquid loading.

The reason of higher field pressure is mainly due to higher sales pressure and/or Filitelnic/Tigmandru compression stations don't work at full capacity. Filitelnic and Tigmandru compression station are located at the interference point of production-transportation system and are used to increase discharged pressure into network system.

It it has been observed during operation period there are different malfunctions in the system which could generate local pressure drops and is even more accentuated by reservoir pressure depletion and pipeline pressure drops. The flow gas through pipelines with different lengths and diameters is introducing a new pressure drop in the system which will led to increased production losses.

Jones 'Gas inflow performance relationship (IPR) equation characterized the pressure losses and it is very useful for gas reservoir exploitation design:

$$p_r^2 - p_{wfs}^2 = a \cdot Q_g^2 + b \cdot Q \tag{1}$$

$$AOFP = \frac{b \pm \sqrt{b^2 + 4a(p_r^2)}}{2a} \tag{2}$$

$$a = \frac{3.16 \cdot 10^{-12} \beta \gamma_g T Z}{h_p^2 r_w} \tag{2.1}$$

$$b = \frac{1.424 \cdot 10^3 \mu T Z \left[\ln\left(0.472 \frac{r_e}{r_w}\right)\right] + S}{kh} \tag{2.2}$$

where:
AOFP = Absolute Open Flow Potential
Q = gas flow rate [mcf/d]
a = turbulence term
b = Darcy flow term
p_r = reservoir pressure [psia]
p_{wfs} = sandface flowing bottomhole pressure [psia]
β = turbulence coefficient [ft^{-1}]

γg = gas specific gravity [dimesionless]
Z= supercompressibility [dimensionless]
μ = viscosity [cP]
re = drainage radius [ft]
rw = wellbore radius [ft]
S = skin factor [dimesionless]
hp = perforated interval [ft]
kh = effective permeability of net vertical formation thickness [mD*ft]

In this case, under permanent pressure drawdown conditions linked with sales contract agreement it is necessary an additional effort for gas compression.

To effectively plan and design for gas well liquid loading problems, it is essential to be able to predict accurately when a particular well might begin to experience excessive liquid loading. Turner and Coleman developed correlations to predict the critical velocity assuming the droplet model (Turner). In practice, the critical velocity is generally defined as the minimum gas velocity in the production tubing required to move droplets upward. Lowering the surface pressure by compression also increases the actual gas velocity. Turner adjusted the theoretical equation for velocity upward by 20% and the equation is:

$$V_t = \frac{1.59\sigma^{\frac{1}{4}} \cdot (\rho - \rho_g)^{1/4}}{(\rho_g)^{1/2}} \tag{3}$$

where σ - surface tension [dynes/cm], ρ - density [lbm/ft^3]

Typical values of surface tension is 60 dynes/cm for water and 20 dynes/cm for condensate.

Although critical velocity is the controlling factor, one usually thinks of gas wells in terms of production in multiples of scf/d rather than velocity in the wellbore. Critical velocity equation is easily converted into a more useful form by computing a critical well flowrate:

$$Q_g = \frac{3.067 \cdot P \cdot V_t \cdot A}{(T+460) \cdot Z} \left[\frac{MMscf}{d}\right] \tag{4}$$

Where:
A -Tubing cross-sectional area, ft^2
T,P-surface temperature ^{0}F, surface pressure, psia

For lower surface tubing pressures, Coleman has developed similar relationships for the minimum critical flow rate for both gas and water. In essence, the formulas from Coleman are identical to Turner but without the adjustment to fit Turner's data.

Wells flowing at just below the critical velocity before compression can be increased to rates beyond the critical velocity with compression, thereby alleviating liquid-loading problems. Because the critical velocity is directly proportional to surface pressure, compression installations can be sized on the critical velocity calculation. For example, the compression station can be sized on a surface pressure that maintains the tubing production velocity at some percentage above the minimum critical velocity. With estimates of the decline of the reservoir, the installation size can be optimized to produce the most gas over the longest period.

System nodal analysis tools are ideally suited to analyse the effects of reducing the surface tubing pressure with compression because allows for the simultaneous comparison of the effects of various surface pressures on the well productivity. System nodal analysis uses single or multiple flow correlations and theoretical models developed for various components of reservoir, well completion and surface equipment systems to calculate the pressure loss associated with each component in the system. This information is then used to evaluate well performance under a wide conditions, which will lead to optimum single well completion and production practices, and to make the analysis of the effects of liquid loading on the gas wells. In the same time, is also used to clarify the detrimental effects of excessive surface production tubing pressure. Increased surface pressure adds back pressure on the reservoir at the sand face. The added back pressure reduces the gas production and lowers the gas velocity in the tubing, which also reduces the efficiency with which the liquids are transported to the surface.

In general, lowering the surface pressure of a well by using compression will result in more production. This is true for flowing wells and with nearly all lifted wells. Depending on the individual well, a well production can be increased over a range from only few percent to several times the current production of the well. For flowing gas wells, lowering the surface pressure can significantly increase the production, prolong the life of the well and increase reserves.

Lowering the surface pressure of a gas well has two effects both of which are beneficial for liquid loading.

- The bottomhole flowing pressure decreases, increasing production rate and gas velocity throughout the wellbore
- The required critical rate to remove liquids decreases because of the reduced pressure.

3 Production optimization. Group compressor performance

Brownfield field management involves a field development plan in order to explore options and targets the optimal development for individual wells. This process is performed by specialist teams who analyse, evaluate, manage and integrate multiple disciplines, subsurface and topsides complexities, uncertainties and opportunities into a single development plan for field, maximizing the overall field recovery and asset value.

Production optimization is the main objective and challenge for specialist teams and can include besides artificial lift, the subject of this paper, many other methods such as stimulation through re-perforation, wash acidizing or matrix acidizing, and fracturing operation, outlined by a surveillance program of data gathering to minimize the high level of the uncertainty of field parameters.

Starting from the main objective of rehabilitation project to increase the recovery factor, a preliminary study was performed for compression implementation in the field. After risk management was evaluated, the project team executed a pilot project of using compression in Laslau Mare field. Based on this project, in 2014 one screw compressor was installed in one of the five groups.

The incremental production has been observed during compression time and a feasibility study was conducted to evaluate the technical and economic aspects for the field by installing screw compressor in the groups. Since the feasibility study confirmed the benefits of this project, four screw compressor were installed in groups.

Any project which is carried out in a rehabilitation phase, has the main objective to increase the production, reduce the production decline and add value by running at optimal parameters. After installation, expected incremental has not been achieved due to operational field condition and other technical issues regarding the running time of compressor.

The real incremental production has been observed when the operational field condition changed in terms of field pressure. Due to lower demand for gas during warm season in Romania, the national grid pressure increased above expected or forecasted parameters, thereby an external pressure was recorded in the field. This external pressure it was also affected by turning off Filitelnic and Tigmandru Compressor Stations which are located at interference of production-transport.

Considering this uncontrollable and unpredicted situation on the market, screw-compressor installation in the field has been shown to be a successful application.

It is observed that every 1 bar higher of field pressure, field gas production decreases around 20 kscm/d but this assumption it was no more available in 2016 when operation field condition changed. Mainly, the cause of higher field pressure is the back pressure given by national grid transportation system or because the compression station is not working at full capacity. Many calculation using system nodal analysis were made to highlight the production losses due to high field pressure. Unlike the previous case, in 2016 the following situation corresponded to production losses caused by field back pressure effects:

Table 1. Production losses/bar at high field pressure

Production losses (kscm/bar)	Field pressure interval (bar)
28	7-9
28	9-12
33	12-13
38	13-15

To overcome the above potential losses caused by field back pressure effect, the running time of field compressor was essential. Three compressors installed in the groups shown the effective incremental in current conditions (fig.1):

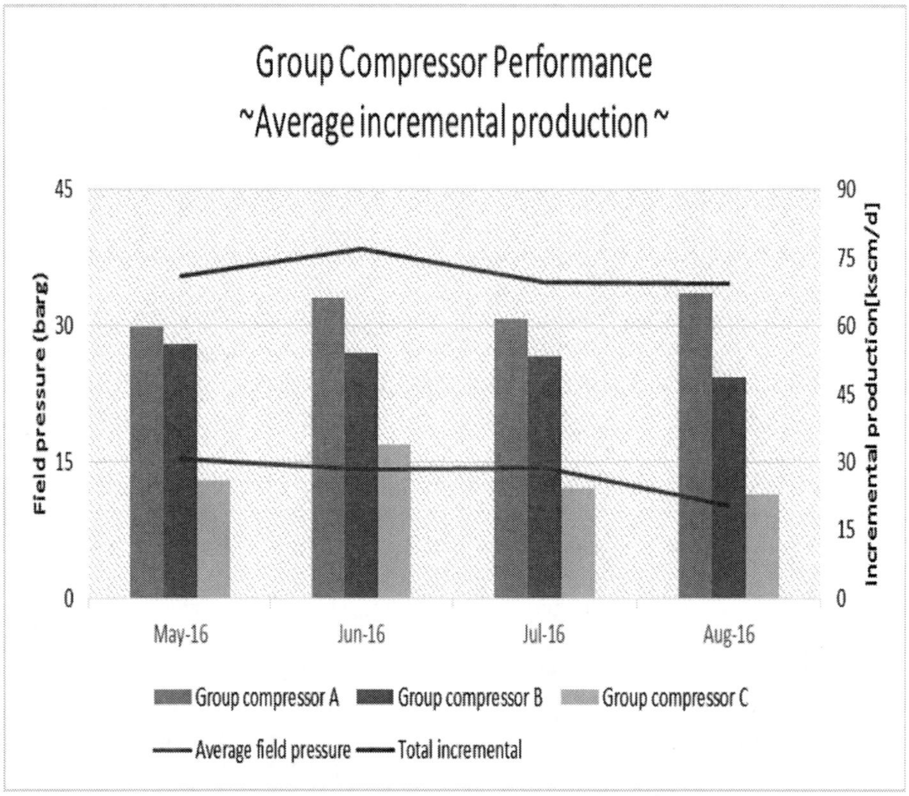

Fig. 1. Incremental production from group compressor during warm season in 2016

The wells connected to compressors were tested previously whether are reactive to pressure changing or not.

Through nodal analysis system the well performance was simulated by using multiphase simulator PIPESIM (Schlumberger Software), in other words to get the expected incremental gas production by sensitizing the wellhead pressure values.

To calculate the incremental production for each well, the baseline was defined considering the same field operating conditions. Group compressor A had average baseline production 25.6 kscm/d with 5 wells tied in, group compressor B has 2 wells connected to it with an average baseline production of 30 kscm/d and group compressor C had one well connected with an average baseline production of 26.75 kscm/d.

Definitely, selected wells were high reactive to pressure changes and their potential production under compression it was considered when candidate selection was done. Also, another important parameter is the maximum capacity of group compressor which is 60 kscm/d and limits the number of well that can be connected to it.

Running parameters for group compressors presented in table below were monitored in daily basis to maintain the optimum condition for compression and prevent any failure of equipment.

Table 2. Running parameters for group compressors

Group Compressor	**Suction pressure (Barg)**	**Discharge pressure (Barg)**	**Total gas compressed (kscm/d)**
A	2.2	18.2	60.7
B	2.3	14.4	60.0
C	3.1	16.2	55.5

Figure 2 presents the well performance during warm season from each group compressor.

As we can observe in Figure 2 the well performance is directly depended on the wellhead pressure, the gas production increases when the wellhead pressure is decreases and conversely. Therefore, reducing wellhead pressure will give insignificant impact of gas production. The wellhead pressure is lower as the result of decreasing group pressure, so the gas production increases subsequently. This condition surely extends its production life cycle and improve the final gas recovery factor.

Compression at a group or field level would help to dewater multiple wells with one piece of equipment. Compression assists the well to dewater by increasing the drawdown from the reservoir. The increased drawdown translates into higher flow rates and increased gas velocities. If the drawdown is great enough the well could be able to unload itself without the assistance of afore mentioned technologies.

Compression proved to be a good option for Laslau Mare because it is not as dependent on fluid chemistry as soaping is and operates over a larger range than velocity strings do.

Group compression is advantageous in that it will assist more than one well at a time. Having multiple inlets also means that maintenance or intervention on a well will not cut off the gas supplied to the compressor, so it will be able to function continuously.

Also, from an operations standpoint no physical changes will occur at the wellhead, so operators will be able to go about their normal work without noticing any difference.

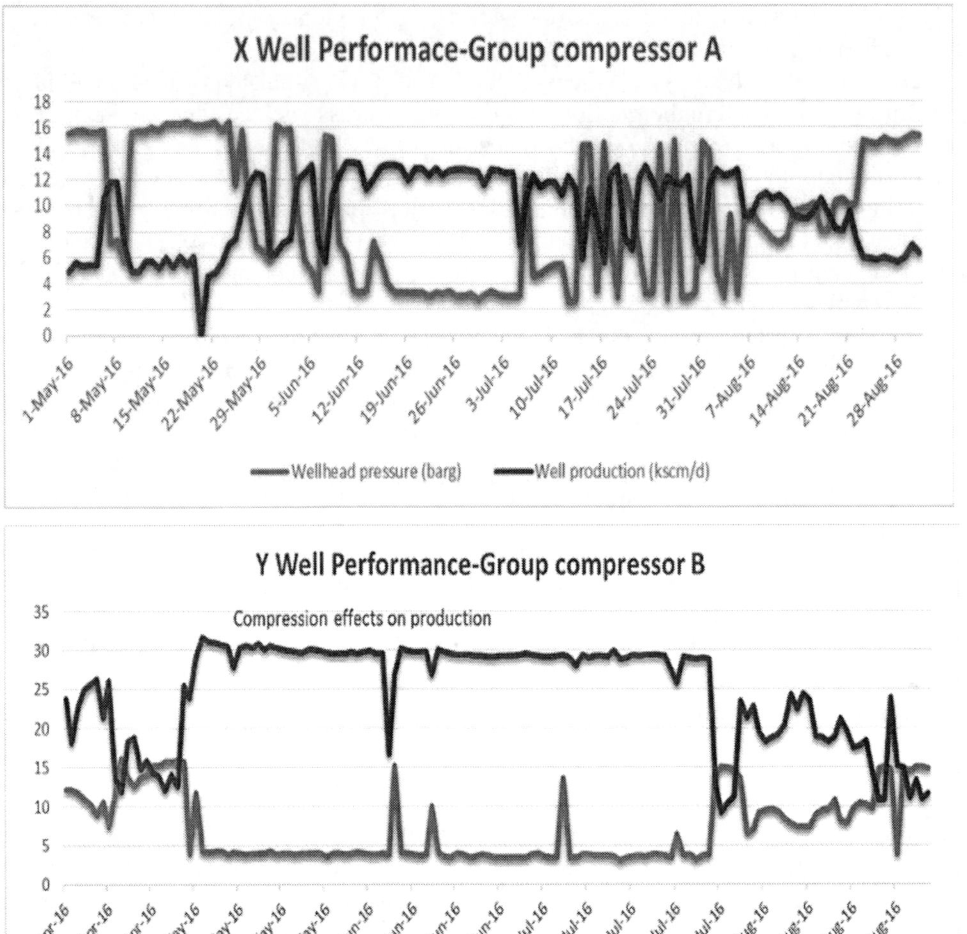

Fig. 2. Example of well performance under group compressor during warm season in 2016

Fig. 3. Screw medium compressor installed in the group

4 Conclusions

Laslau Mare field, similar to others mature gas field from Transylvanian Basin, being a depleted reservoir is directly depended on external pressure from National Transport System and is having liquid loading problems. Group compressor application in the field has proved to be extremely beneficial especially when gas demanded decreases correlated with field back pressure effects.

Group compressor application can be considered a mitigation measure for field back pressure effects on the gas wells, for liquid loading problems created by external pressure and definitely to reduce the drawdown pressures.

One advantage of group compressor besides the ones mentioned, is that can compress multiple wells in the same group as long as the volume doesn't exceed 60 kscm/(d·unit). It doesn't require changes at wellhead and the operation is uninterrupted as the unit will be standby in the group at any time. Nevertheless, disadvantages to the technology are that it requires a high capital expense. Also the equipment is rather large and will need a sizeable footprint of proper foundation. The compressor will require maintenance and consume either fuel gas during operation.

The maintenance program is very important to make sure is running in optimum conditions and the flat time is reduced.

Having the group compressor installed in a mature gas field which is running at maximum capacity, undesirable effects on wells performance created by external conditions are mitigated, production incremental is a certainty benefit, the life cycle of the field is extended, production decline is decreasing and definitely the recovery factor is increasing..

References

1. D.P Stefanescu, *Introduction to the mature natural gas reservoir rehabilitation-Theory and Case studies,* **375-377** (Editura Universitatii Lucian Blaga, Sibiu, 2012)
2. S.B Coleman, H.B Clay., D.G McCurdy, H.L. III Norris, Journal of Petroleum Technology **43**, 329 (1991)
3. J.Lea, H.V Nickens, M.Wells, *Gas Well Deliquification* (Gulf Professional Publishing, 2003)
4. *** *Matrix Engineering Manual-Well Performance,* **119** (Schlumberger, 1998)

Underbalance well completion - a modern approach for mature gas fields

Argentina Tătaru[1,*], and *Nicolae Bogdan* Simescu[2]

[1] Lucian Blaga University of Sibiu, Sibiu, Romania
[2] SNGN ROMGAZ SA, Mediaş, Romania

Abstract: The exploitation of natural gas fields from Transylvanian Basin started a century ago. The majority of these fields were discovered and developed in the last century, from 1950 to 1970. So, these reservoirs have over 50 years of production historical. These are mature fields with a very low reservoir pressure. Now, for some of these reservoirs the pressure is 10 – 20 % of initial values. The biggest challenge for a production company is to make completions and recompletions in depleted reservoirs wells. At the beginning was not a problem to do a workover in these wells because the completion fluids were lost in the reservoir and the pressure helped the wells to clean up. Now, because the reservoir pressure is very low, it takes time for well to be cleaned. From time to time, some net pays have to be bypassed because the fluid can be lost in the reservoir. So the best method to do recompletion works in these wells is to work underbalance or even with the well under pressure. This paper presents some of the technologies used by Romgaz to accomplish this goal.

1 Introduction

Romgaz operates mainly in the natural gas mature fields located mostly in the Transylvanian Basin. There are reservoirs which are producing from over one century. In present, some of these reservoirs have dropped the pressure to 10 – 20 % of initial values.

These reservoirs are producing using natural gas elastic drive mechanism. Reservoir history matching shows us that once these fields are put in production, the pressure is dropping irreversibly. This phenomenon can be described by the so-called natural decline pressure curve.

In these depleted fields, the overbalanced workovers at higher pressures can lead to undesirable phenomena such as completion fluid invasion near the wellbore zone. In turn, this can lead to a drop in the well productivity and can produce even well blocking.

Conventionally gas wells are drilled, cased and cemented using fluids with high density, $\rho > 1$ kg/m^3, which because of their weight generate a hydrostatical pressure higher than the pressure of porous-permeable media. This phenomenon is called overbalance.

Working overbalance is a safety measure in well completion. This prevents technical accidents, like eruptions. But this safety measures must take in consideration the actual reservoir conditions not the initial ones.

* Corresponding author: argentina.tataru@romgaz.ro

Over time there were studied and tested the possibilities of avoiding and minimizing as much as possible the formation damage phenomenon due to reservoir contamination with various types of fluids in balance or underbalance conditions.

In the green fields the invasion phenomenon is not too accentuated and normally the wells could clean up from the undesirable fluid contamination in short time.

In mature fields things look much different. Besides reservoir contamination are present partial and total loses in the reservoir which can lead to formation damage. So to avoid these situations, the state – of – the art technology offers solutions for every type of workover. If there are needed completion fluids these could contain bridging elements which can avoid or reduced reservoir losses. Perforating operations are mostly made in balance or underbalance conditions, even with empty well, for avoiding salt water or completion fluids to flood the perforations. Tubing handling can be done with well under pressure, so the perforated area is not possible to be damaged by various fluids types. Partly reservoir abandonment can be achieved by introducing expandable plugs trough the tubing with well under pressure.

Taking this into account, the company' specialists have to use the last technologies applied to mature fields. So in many cases, the operations during workover are made at balance conditions taking in consideration the actual reservoir pressure if it is necessary to kill the well. In this case to minimize productive layer contamination most often is used filtered reservoir salt water or completion fluid.

2 Underbalance well completion alternative solution for traditional well completion

2.1 Snubbing unit

Science people invented equipment for tubing handling under pressure, in order to eliminate any productive layer contamination with different type of fluids. This equipment is called "snubbing unit" which can perform tubing handling without killing the well.

The equipment is provided with a blowout preventer system which seals the tubing and a hydraulic system which permits lifting and lowering the tubing string in well (see fig. 1 and 2).

Unfortunately this unit cannot perform the operation in wells with any type of wellhead. It can be used only on wells with API Christmas Tree (oil type). The workover under pressure is not possible for non API Christmas Tree (gas type) because the tubing is screwed in tubing hanger so you the wellhead must be lifted and rotated if the tubing must be lifted up.

Fig. 1. Snubbing unit working in the field

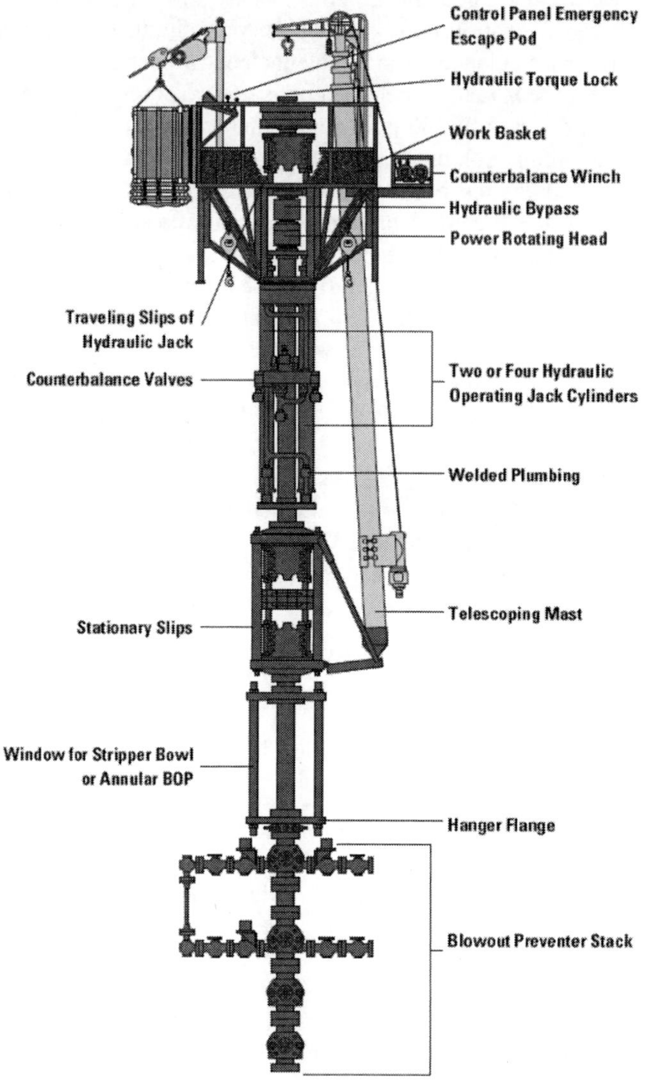

Fig. 2. Snubbing unit scheme

The snubbing unit can perform the operation of repositioning the tubing in well (pull out from the perforated zone or run in the perforated zone), with the well still under pressure, by setting up a plug in tubing. After the tubing is out from the perforations zone it is possible to perform cased hole logging operations like PLT, thermometry or throughout tubing perforation jobs.

The snubbing unit can perform also the following operations: tubing pulling out of the hole, underbalance deep penetration perforation or casing logs as MIT or RBT.

After the desired operations are performed the snubbing unit is used to reinsert the tubing back into the well with the well under pressure and if is needed can be used to introduce tubing strings together with a packer which can be set up.

Romgaz performed all these operations with the snubbing unit in different situations.

So, there were performed the following operations like tubing repositioning like tubing run in hole for a better liquid unloading and tubing pulling out from the bottom of perforations zone to an upper zone for a better communication between top perforations and tubing.

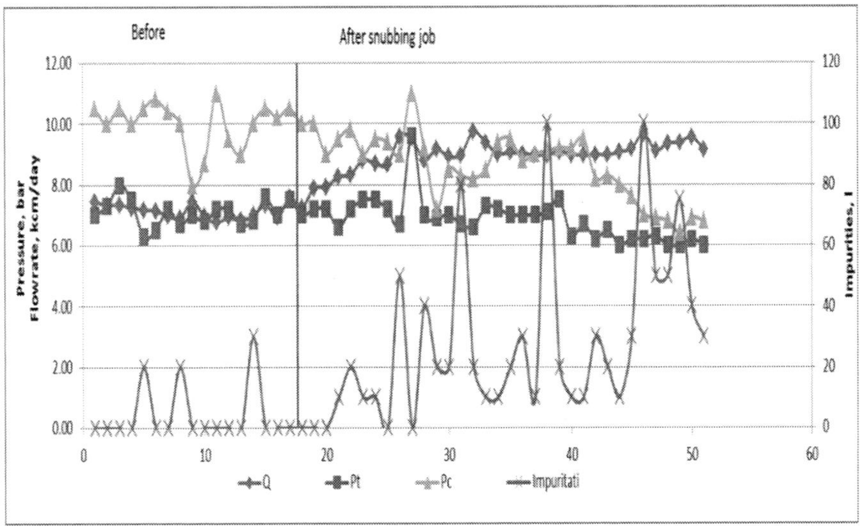

Fig. 3. Production behavior of a well before and after the snubbing job

Chart 3 presents the production behaviour of a well which had the tubing set up with 50 m above the top of perforations and a perforated interval of 100 m. After analysing we decided to introduce the tubing in the middle of the perforated area for a better unloading of the liquids accumulated in the wellbore. The presence of these liquids are lowering the well productivity. After the tubing was set in the perforations zone there were a gain 3 kcm/day and the well was lifting the reservoir water by itself.

Chart 4 presents another example of production evaluation for well were have been performed several operations. First the tubing string was pulled over the perforations and the well was reperforated underbalanced with 2" guns. After this job, we observed that the flowrate has been doubled from 2 ksm/day to 4 kcm/day. Because we consider that this job was not a real success, we decided to pull out of the hole the entire tubing string and we have reperforated this well with 4" deep penetration casing guns. After the repeforation job, we obtain an 11 ksm/day flowrate.

The main advantage for all these performed operations was the possibility of handling the tubing under pressure so the introduction of fluids, in order to kill the well, was no longer necessary.

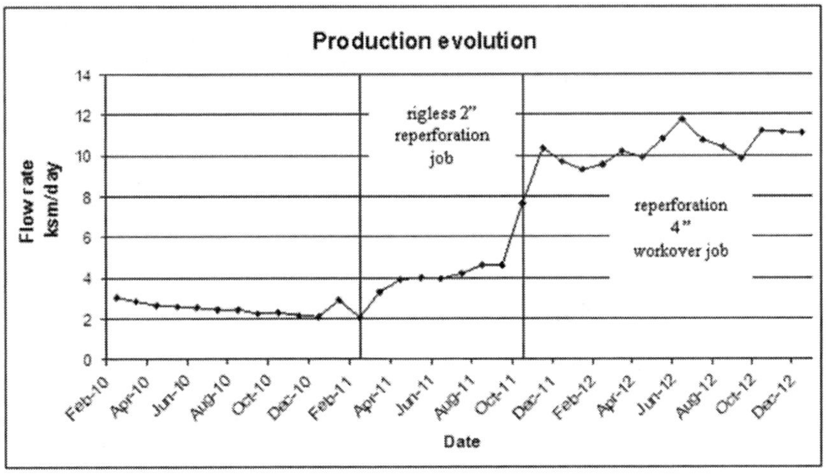

Fig. 4. The production behavior for a reperforated well using the snubbing unit

DOI: 10.1051/matecconf/201712109009

2.2 Underbalance perforation

If in the past all the perforation operations were performed with the well filled up with fluid, now sealing systems are developed allowing performing underbalance or even with empty well perforation jobs.

There are several advantages for underbalance perforation such as:

- Avoiding the phenomenon of fluid invasion near wellbore zone;
- Less time spent for well testing or well set up into production;
- Less time = lower operational / job costs.

To perform the underbalance perforating operations are taking safety measures such as:

- If it is planned to do a perforation job with deep penetration guns like 3 3/8" or 4" guns it is necessary to set up a 7" safety valve upper the tubing-hanger;
- Above this valve it will be set up the pressure retriever string, wireline blow out preventer and / or other safety devices;
- If it is planned to do a trough tubing perforation job then on top of the Christmas tree it will be set up the pressure retriever string, wireline blow out preventer and / or other safety devices.

Using these safety measures there could be performed perforating or re-perforating operations with the well under pressure. In recent years there are performing more often these types of operations due to mature reservoir conditions.

2.3 Water shut off

Since the investigations regarding the actual fluid saturation in the open hole for most of the wells in mature fields are not updated and so accuracy is not very high, there can be situations after the perforation job to open accidentally a depleted or water flooded layer. Such situations can happen and they can be solved by performing different investigations like thermometry or production logs to see the water influx.

After these logging operation were performed there were took measures for closing (isolating) the opened aquifers and putt the well back in production.

Fig. 5. Rigless underbalance perforation job

There are many solutions to solve this type of issues but doing this job with the well under pressure is a challenge. It depends from where the water is coming. If it is coming from the lower zones, one good option is to set up a non-retrievable trough tubing inflatable plug with the mention that it is very difficult to do a reentry for the lower part of the well. For the upper zones a good option is to set up a packer or a casing patch to isolate the water influx.

As follow we present one case where a well was perforated underbalanced 70 m split in 4 intervals. After these layers were opened there were recorded a gas influx and large quantities of water. It was performed a thermometry which shows that there is a big water influx from the lower part. (see results in picture 6)

This well had an opened interval between 1090 and 1230 m and the reservoir pressure was 16 bar. So we decided to do an underbalance water shut off job using a trough tubing inflatable plug to avoid other water loses in the productive layers. The whole job finished in 5 days (setting up the plug and putting 2 times a quantity of cement above the plug).

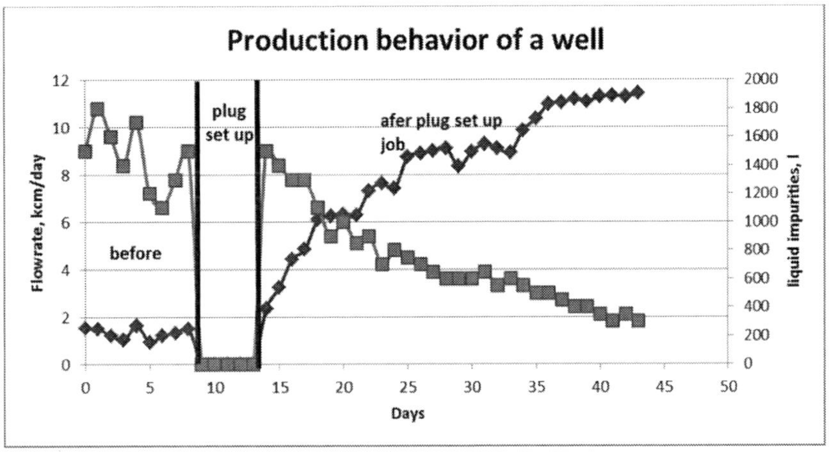

Fig. 6. Production behavior of a well were a water shut off was performed

These operations were successful and currently the well is producing with a flow of 11 kmc/day and 300 liters salt water/day. The results can be seen in chart 6 were presented the values recorded before and after the water shut off job.

3 Conclusions

In Transylvanian Basin are several major mature reservoirs. In these fields is better to apply the underbalance completion technology instead the traditional completion technology to avoid formation damage and to improve flow conditions as much as possible.

Especially in mature fields it is preferable to perform the underbalance completion because compared to a traditional workover it takes less times for the well to clean up.

Working underbalance in this type of reservoirs is mandatory because in most cases killing fluids are hard to recover and it can take months for the flow rate to reach the same value as it was before the workover involving high costs operations like nitrogen kick off and coil tubing jobs.

Learning from our experience performing these types of operations in mature fields we suggest to run a saturation log before a re-perforation job or adding new perforations in the wells where this operation are possible.

As we have seen, performing snubbing operations is not a major risk for well abandonment. In most of the cases, these types of operations help us to maintain or increase the production.

In the future, we will continue to perform underbalance jobs in the rehabilitation process.

The big advantage is that we can optimize the production rate, minimizing the formation damage and maximizing the recovery reserves. Using this technology the wellbore is isolated from undesired flow of water which can be easily controlled.

References

1. F. Şuţoiu, A. Tătaru, and B. Simescu, AGH Drilling Oil and Gas **33**, 2 (2013)
2. I. Falk, B. Simescu, SPE Conference, 14 October, Medias, Romania (2014)
3. D. P. Ştefănescu, B. Simescu, AGH Drilling Oil and Gas **36**, 2 (2016)
4. *** https://oilandgastechnologies.files.wordpress.com, accessed 14.02.2017

MATEC Web of Conferences **121**, 09010 (2017) DOI: 10.1051/matecconf/201712109010
MSE 2017

Hazardous gas areas on high-pressure gas pipelines in Poland

Łukasz Zabrzeski[1,*], *Łukasz* Drążkiewicz[2], *Krystian* Liszka[1], *Mariusz* Łaciak[1], and *Piotr* Paszylk[1]

[1]AGH University of Science and Technology, Faculty of Drilling, Oil and Gas, Department of Natural Gas Engineering. Al. Mickiewicza 30, 30-059 Cracow, Poland,
[2]The Main School of Fire Service, Faculty of Fire Safety Engineering, Słowackiego 52/54 St, 01-629 Warszawa, Poland

Abstract. Natural gas transmission is strongly connected with fire and explosion safety. Generally, explosion hazard occurs, when at least 5% of natural gas is present in the atmospheric air. As high-pressure gas transmission infrastructure includes many potential sources of gas releases, both operational and accidental, many different cases should be considered. To properly face the potential hazard of explosive atmosphere formation, the hazardous zone should be estimated. There are some guidelines, describing the process of hazardous zones calculating. This paper compares calculations of such hazardous based on Polish and European standards which are also valid in Poland. Calculations are focused on releases that may occur on the safety block and relief valve systems.

1 Introduction

The natural gas transport with high pressure pipelines is a key part of the gas system. This energy carrier is a mixture of all light hydrocarbons having lower density than air and which are flammable. This causes that the mixture of natural gas and air may be explosive. This happens when the molar participation of blue fuel in air totals to about 5 to 15%. The origin of explosions, their course and possible consequences and extent can be found, e.g. in [1] and [2].

The transmission network consists of a number of elements which are crucial for its proper operation. The explosion hazard will be present in places where these elements are localized, starting from pipelines to gas compression stations, but with different probability and range. This hazard can be counteracted with prevention measures and one of them is distinguishing the explosion hazard zone and estimation of the explosion hazard [3, 4]. Depending on the classification, they distinguish zones in which the explosive atmosphere may be formed to a different degree and with different probability. The problem lies in the fact that the zones can be established individually, in view of the localization and the steps specified in respective guidelines. They slightly differ in content, but always give the user some freedom as to the size and classification of zones.

* Corresponding author: lukasz.zabrzeski@agh.edu.pl

2 Principles of determining the zones

The explosion hazard is defined as potential occurrence of explosive atmosphere generated when flammable substances mix with the air, and also outbursts which may result in injuries, death or material damage [5]. As already mentioned, the explosive mixture is formed when about 5 to 15% of natural gas is present in the air. These are Lower Explosive Limit (LWL) and Upper Explosive Limit (UEL).

There are a few papers containing guidelines specifying how to design explosive hazard zones. Among the most important ones are Technical Standard of the Chamber of the Natural Gas Industry ST-IGG-0401 of 2015 [6], British standard of 2013 issued by the Institution of Gas Engineers & Managers IGEM/SR/25 Edition 2 [7] and standard PN-EN 60079-10-1 of 2016 [8], based on standard IEC 60079-10-1:2015 RLV [9]. All they recommend the explosive zones to be designed individually for a specific case. They offer guidelines to follow, and which should not be ignored. There are also ready formulae, on the basis of which the shape and range of explosive zones can be obtained. Importantly, in each case the zones should be classified as zone 0, zone 1 and zone 2. Basing on standard [8] and IGG standard [6], the zone can be identified depending on the probability of occurrence of explosive atmosphere or degree of emission and efficiency of ventilation. In the British standard, the zones depend on the probability of occurrence of explosive atmosphere, giving the ventilation system efficiency classes from poor to more than adequate. In both cases in zone 0 explosions take place continuously over a long time, or with high frequency. In zone 1 the probability of occurrence of explosive atmosphere is associated with regular operation of the installation. In zone 2 probability of occurrence of explosive atmosphere during regular operation of the installation is low; such effects are rare, if at all, and short. Among the degrees of emission, we can distinguish continuous, primary and secondary degrees, in the order of decreasing frequency of occurrence and duration of emission of flammable substances. Ventilation is defined on the basis of its availability and reliability. It should have appropriate quality especially when the second degree emissions tend to pass to higher groups.

Unlike its other counterparts, standard presented in [8] gives recommendations on how to calculate characteristic emission parameters, without giving ready formulae for the range of the zones. The only direct guideline is a plot on the basis of which the distance from the emission source can be safely determined in a graphic form, if the characteristic of the emission, i.e. density of medium, yield and lower explosive limit are known.

3 Character of source of emission

When determining explosive zones attention should be paid to the type of medium, conditions in which it mixes with the air, what is the pressure of emission and character of its source. Standards [6] and [7] mainly concentrate on natural gas, while standard [8] also provides guidelines for substances of higher density. The character of the source is defined by its surface, shape and spatial localization. In Poland the gas sector relies generally on IGG standard [6]. The character of emission and surface of source of emission are the main parameters here. It was assumed that gas propagates in natural-turbulent or jet mode. Unlike [7] and [8] there is no division to the vent pipe terminations. The direction of the release pipes is extremely important because of the high pressure involved. The zone extent will be larger towards the propagating emission than in the case of a vertical pipe. As to the rooms, the British standard [7] and standard [8] also define the effect on the quality of ventilation of neighbouring objects and also wind speed for natural ventilation (British standard [7] also includes a map of Great Britain with wind speed isolines).

4 Sealing elements

An important assumption should be made that in the lack of possibility to measure the surface of the opening (especially when assuming unexpected emissions resulting from leaky sealing elements), it equals to 0.25 mm^2 (with exception of [8], where it depends on type of sealed parts – fixed or moving). The division is also made for various pressures. Here the focus is on high pressure intervals. A spherical shape of the zone is assumed. The probability of occurrence of explosive atmosphere is very low therefore we have to do with zone 2.

In standard IGEM/SR/25 each type of sealing elements has its own guidelines determining possible shapes of zones, with a sphere as the basic one. For comparison's sake, the radius of a sphere defining the zone for sealing elements at gas pressure of 5 MPa equalled to 1.6 m in IGG standard [6]; according to IGEM [7] the radius totalled to 1.0 m (assuming undisturbed natural ventilation).

5 Release pipes

For the sake of showing the differences in the obtained results, emphasis was put on the emission from release pipes in high pressure pipelines, depending on the applied standards. The calculations were based on a British standard [7], IGG standard [6] and standard [8]. The release pipe (100 mm of diameter) is a source of emission; its orifice is placed 3 meters above the terrain surface. Natural gas is discharged from this pipe to the atmosphere at a pressure of 5 MPa. Natural gas has a molar mass of 16.5 kg/kmol and temperature 5°C. All the guidelines suggest a different shape for the zone: British standard [6] – 2 cylinders of different size, and a sphere; IGG [6] – a sphere and a cone; standard [8] – the shape should be defined by the person who defines the explosion hazard.

Standard [8] provides shapes, depending on the gas pressure and certainty of direction of propagation, as in figure 1: for low and high pressure in the lack of certainty as to the emission direction – sphere; for high pressure at known emission direction – cylinder or a truncated cone decreasing the radius towards the source of emission. The calculated dimensions for given emission parameters are given in tables 1-3, and are schematically presented in Figure 3. The shape of the zone calculated after [8] has not been accounted for in the figure because authors of the standard did not provide respective formulae, so it can be equal to the calculated one or smaller, depending on the conditions. Thus defined hazardous distance can be treated as maximum for high pressure natural gas emissions from release pipes (jet flux). This value is smaller than the obtained radius of a sphere obtained with the presently applied IGG standard [6].

The range of the hazardous zone in the form of a sphere was calculated on the basis of [6] and equation (1), where p_r – working pressure of gas in the source of emission [MPa], F – cross section of orifice, being a source of gas emission [mm^2]. The dependence of the radius of the sphere on the gas pressure and diameter of the release pipe is additionally presented in Figure 2. The hazardous zone at the base of the cone [6] was calculated with formula (2), where d is a diameter of the release orifice [mm].

The size of the zone in [7] was calculated on the basis of tables (in [7]), dedicated to the respective parameters of emission source. The hazardous zone in [8] was based on a plot of the range vs. character of source of emission, which for non-choked release can be determined with equations (3) and (4), where W_g is mass release rate of gas [kg/s], ρ_g is gas density [kg/m^3], k is the safety attributed to LFL (typically between 0.5 and 1.0), LFL is lower flammable limit (for natural gas approximately 5%), C_d is discharge coefficient (dimensionless) which is characteristic of the release openings and which accounts for the effects of turbulence and viscosity (for rounded orifices 0.95 to 0.99), S is the cross section of the opening hole), through which the fluid is released [m^2], p is pressure inside pipe [Pa],

M is molar mass of gas or vapour [kg/kmol], Z is compressibility factor (dimensionless), R is universal gas constant (8314.3 $J \cdot kmol^{-1} \cdot K^{-1}$), T is absolute temperature of the gas [K], γ is polytropic index of adiabatic expansion or ratio of specific heats and p_a is atmospheric pressure (101 324.7 Pa). Values of coefficients k and C_d were assumed as equal to 1. Coefficient Z was calculated analogously as in [10], making use of Peng-Robinson equation and Newton-Raphson method, whereas adiabatic exponent γ with the method as in [11].

$$Z_R = 0{,}33 \cdot F^{0{,}5} \cdot (p_r + 0{,}1)^{0{,}5} \quad (1)$$

$$Z_S = 0{,}175 \cdot d \quad (2)$$

$$Release\ characteristic = \frac{W_g}{\rho_g \cdot k \cdot LFL} \quad (3)$$

$$W_g = C_d \cdot S \cdot p \cdot \sqrt{\frac{M}{Z \cdot R \cdot T} \frac{2 \cdot \gamma}{\gamma - 1} \left[1 - \left(\frac{p_a}{p}\right)^{\frac{1-\gamma}{\gamma}}\right] \left(\frac{p_a}{p}\right)^{\frac{1}{\gamma}}} \quad (4)$$

Table 1. Extent of hazardous range, after [6]

Standard ST-IGG-0401:2015			
Parameter	**Symbol**	**Value**	**Unit**
Radius of sphere - zone 2	Z_r	66,05	m
Radius of base of cone - zone 2	Z_s	17,50	m
Radius of sphere – zone 1	Z	1,00	m

Table 2. Extent of hazardous range, after [7]

Standard IGEM/SR/25 Edition 2			
Parameter	**Symbol**	**Value**	**Unit**
Height of zone 2 below orifice, being result of lower dispersion	X_L	0,20	m
Height of zone above release pipe	X_h	120,00	m
Radius of zone 2 from release orifice, being result of lower dispersion	X_k	12,00	m
Radius of zone 1 from release orifice	X_s	1,00	m
Radius of zone 2 from release orifice	X_r	30,00	m

Table 3. Maximum extent of hazardous zone, after [8]

Standard PN-EN 60079-10-1			
Parameter	**Symbol**	**Value**	**Unit**
Range of zone (height of cylinder or radius of sphere)	*r*	45,61	m

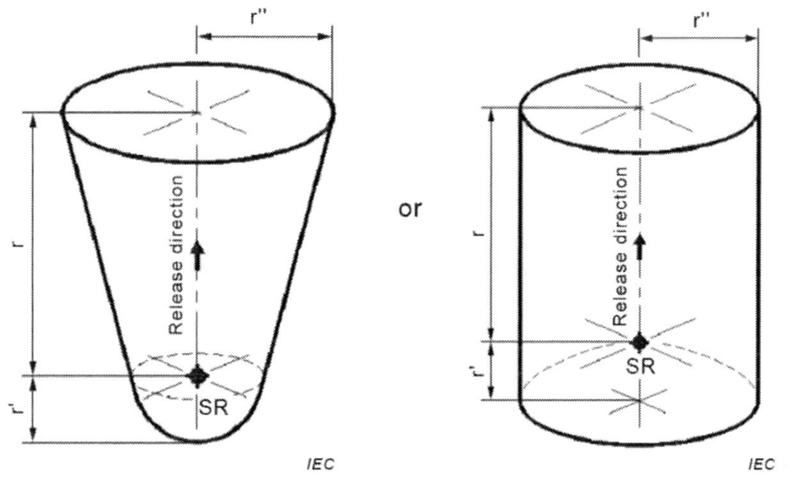

Fig. 1. Proposed hazardous zones shapes in case of high pressure of gas, where *r* is main extent of the hazardous area to be defined considering the estimated hazardous distance, *r'* and *r''* are secondary extents of hazardous area to be defined considering release behaviour, *SR* is release source. [8]

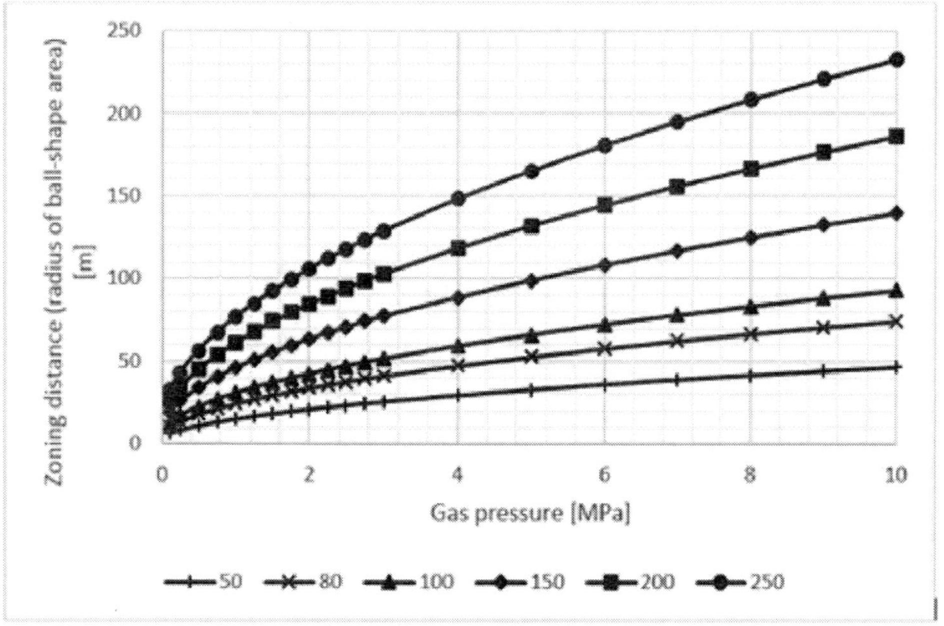

Fig. 2. Extent of hazardous zone 2 – radius of sphere, for various diameters of release pipes [mm], depending on gas pressure, after [6]

DOI: 10.1051/matecconf/201712109010

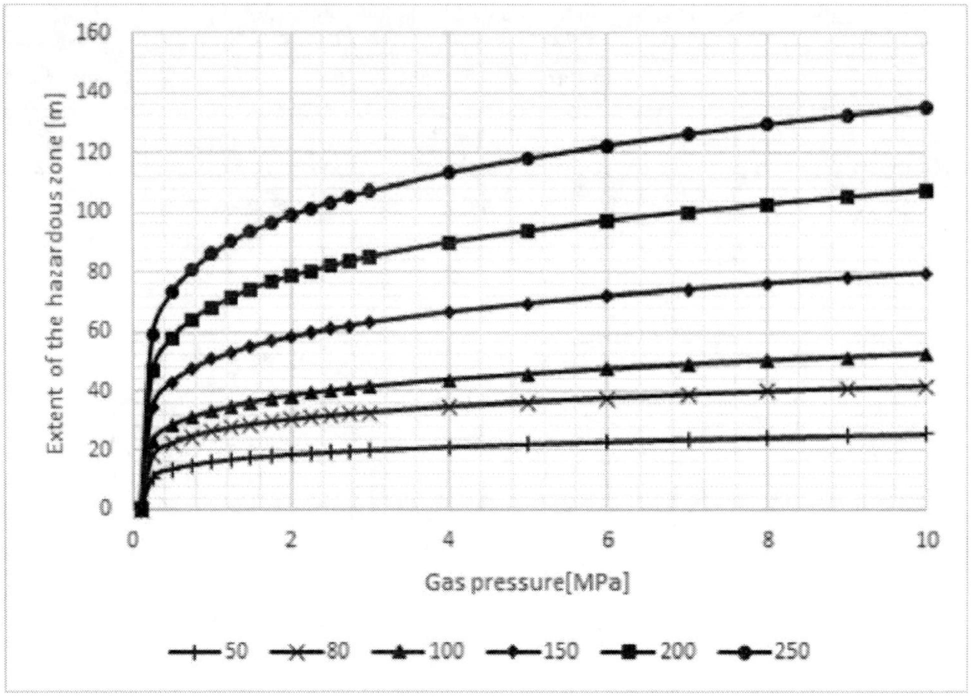

Fig. 3. Extent of hazardous zone or area where flammable gas may occur for different nominal pipe diameters, depending on gas pressure in pipe, after [8]

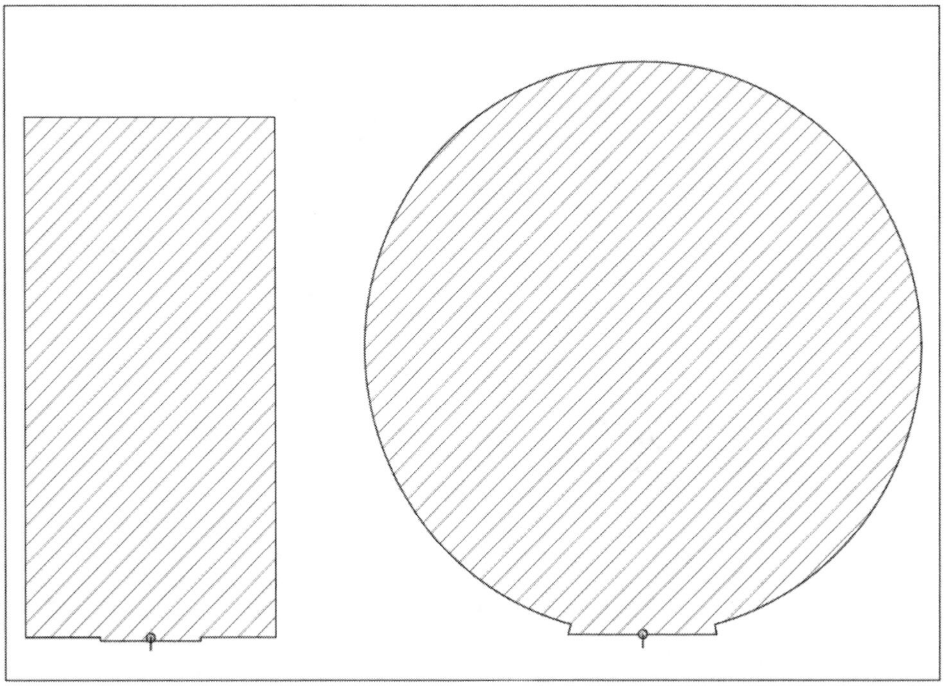

Fig. 4. Hazardous zones for analysed example of release pipe after standard IGEM [7] (left) and IGG [6] (right)

6 Discussion and conclusions

The analysis of the results of calculations performed for the assumed case in line with respective guidelines reveals that each approach brings about different results. Values calculated after IGG standard [6] are definitely highest. Lower results were obtained for standard [8], which became applicable in Poland in 2016, whereas the lowest for the British standard [7]. It should be emphasized that standards are not obligatory according to the Polish law, though the evaluation of the gas explosive hazard is obligatory. Hence, ready-made guidelines are provided for the hazard zones designers. They are part of the assessment or own methodics, which should be approved by the respective authorities.

The hazardous zones can be evaluated in a number of ways and they can significantly differ in their shape and extent, as seen in figures 2-4. Attention should be paid to the difference in the dependence between the extent and the pressure. The influence of the diameter of the release pipes on the zone area in IGG standard [6] and standard [8] is similar, the increase of the extent of the zone in [8] for pressures lower than ca. 1 MPa is considerably higher than in [6]. For higher pressures the situation is reverse – the increase of hazardous zones in [8] is small (8% for DN100 and pressures 4 MPa and 6 MPa), whereas in [6] it is maintained on a relatively high level (increase of 22% for DN100 and pressures 4 MPa and 6 MPa).

Hazardous zones in natural gas installations should be determined individually in each case. Importantly, none of the factors which may affect the shape of the zones should be ignored. The analysis of the calculations reveals that the difference in calculations for horizontal extents may significantly differ (in the analysed example – over twice; according to IGG the horizontal extent totals to 66 m, according to IGEM it is 30 m). The differences result from assumptions taken in line with the respective guidelines. Unfortunately there are no documented research works, which would prove the congruence of any of them with reality. It is worth observing that the extents calculated according to the Polish guidelines are very high, and the atmospheres are not likely to form on such large areas. Therefore it is crucial to find such shapes of hazardous zones as to maintain appropriate safety level at real probability of hazardous zones formation, and the zones will not be redimensioned.

This work has been prepared within the statutory research of Department of Natural Gas Engineering, Drilling, Oil and Gas Faculty, number: 11.11.190.555

References

1. D. A. Crowl, *Understanding Explosions* (Wiley, 2003)
2. W. E. Baker, P. A. Cox, J. J. Kulesz, R. A. Strehlow, P. S. Westine, *Explosion Hazards and Evaluation* (Elsevier, 1983)
3. M. Łaciak, *Bezpieczeństwo eksploatacji urządzeń, instalacji i sieci gazowych* (Tarbonus, Kraków-Tarnobrzeg, 2013)
4. T. Sawicki, Bezpieczeństwo Pracy: nauka i praktyka, **11**, 1 (2005)
5. M. Woliński, G. Ogrodnik, J. Tomczuk, *Ocena zagrożenia wybuchem*. (Szkoła Główna Służby Pożarniczej, Warszawa 2007)
6. *** ST-IGG-0401:2015 *Sieci Gazowe Strefy Zagrożenia Wybuchem Ocena i Wyznaczanie*. (Izba Gospodarcza Gazownictwa, Warszawa 2015)
7. *** IGEM/SR/25 Edition 2 with Amendments August 2013 Communication 1748. *Hazardous area classification of Natural Gas installations*. (The Institution of Gas Engineers and Managers, 2010)
8. *** PN-EN 60079-10-1 *Atmosfery wybuchowe Część 10-1: Klasyfikacja przestrzeni Gazowe atmosfery wybuchowe*. (Polski Komitet Normalizacyjny, Warszawa 2016)
9. *** IEC 60079-10-1:2015 RLV *Explosive atmospheres - Part 10-1: Classification of areas - Explosive gas atmospheres* (International Electrotechnical Commission, 2015)

10. T. Blacharski, P. Janusz, M. Kaliski, Ł. Zabrzeski, AGH Drilling, Oil, Gas **33** 2 (2016)
11. Ł. Zabrzeski, P. Janusz, K. Liszka, M. Łaciak, A. Szurlej, AGH Drilling, Oil, Gas (to be published)

Research experiences on the reuse of industrial waste for concrete production

Alessandro Abbà[1,*], *Sabrina* Sorlini[2] , and *Maria Cristina* Collivignarelli[1]

[1]University of Pavia, Department of Civil and Architectural Engineering, via Ferrata 1, 27100 Pavia, Italy
[2]University of Brescia, Department of Civil, Environmental, Architectural Engineering and Mathematics, via Branze 43, 25123 Brescia, Italy

Abstract. The aim of this study was to assess the feasibility of concrete production using different kinds of industrial wastes as "recycled aggregate". The wastes studied in this work were: fly ashes and slags from Electric Arc Furnace (EAF) steel plant; foundry sands produced from foundry dies; slags from lead processing; Waelz slags; solid residues from municipal solid waste incineration (MSWI) plant (with mass-burning kiln and fluidized bed reactor); sludge from industrial wastewater treatment plants. Good compressive strength (similar to natural concrete) was achieved after 28 days of curing by concrete mixtures obtained with the partial replacement (from 7% to 40% by weight) of natural aggregates with slags from lead processing, foundry sands, Waelz slags and bottom ashes from MSW incineration. The worst mechanical and leaching behaviours were shown by concrete samples containing EAF fly ashes and sludge from industrial wastewater treatment. For the residues with the best performance, concrete products (kerbs and flat tiles) were casted. Their mechanical and leaching characterization has shown that the reuse of these residues for concrete product is feasible.

1 Introduction

Concrete is a composite material that consists of different constituents such as binding materials, water, aggregates and admixtures. Among these ingredients, aggregate plays a very crucial role in concrete as it represents about 60–75% of total concrete volume. The extensive increase of industrialization and urbanization processes progressively increased the exploitation of natural resources for the production of concrete in the construction industry. Concrete has a very important role in the economy development of a country due to its large volume utilization. In the world, it uses approximately 20 billion tons of raw materials (coarse aggregate) each year [1, 2].

The use of low impact, renewable and recyclable building materials and the reusability of industrial by-products for construction purposes should be prioritised in the design, construction and management to achieve environmentally sustainable infrastructures [3]. Moreover, according to European [4] and Italian [5] legislation, a growing interest on the

* Corresponding author: abba78@unipv.it

possibility of using industrial wastes in concrete was observed, since this allows to: (i) save natural resources as conventional aggregates; (ii) reduce energy consumption due to mining activities; (iii) preserve environment; (iv) avoid the increase of landfill volume.

Many wastes produced by industrial activities can offer interesting characteristics for their potential reuse as construction materials. About this, the residues from metallurgical industry, municipal solid waste incineration (MSWI) and sludge from wastewater treatment plants (WWTPs) could be interesting for reuse in different fields also due to the quantity involved.

As concerns the metallurgical industry, according to the Euroslag Organization, 45.3 million tons of ferrous metallurgy slag was produced in 2010 [6]. The steel slag amount (derived from steel production) is equal to 29.6 and 4.1 million tons in Europe and in Italy, respectively [7]. Moreover, the recycling of Electric Arc Furnace (EAF) dusts produced in the European Union could lead to the generation of nearly 800,000 tons per year of Waelz slag [8]. Furthermore, in Europe, around 18 million tons of foundry sands is left over every year and in many cases big landfills do not have enough capacity to deposit those large amounts of surplus foundry sand [9].

About the municipal solid waste incineration (MSWI) sector, the solid residues produced in Europe can be estimated in 18 and 4 million tons for bottom ash and fly ash, respectively [10].

Finally, as regards the sewage sludge from wastewater treatment plants (WWTPs), the production in Europe will rise to just under 13.0 million tons in 2020 [11]. Actually, especially in Italy, the main routes for sludge disposal and recycling is the agricultural reuse, which could involve critical issues related to heavy metal accumulation and odour emissions [12].

The aim of this work is to evaluate the potential reuse of different industrial wastes as "recycled aggregate" in construction materials and to assess the mechanical and environmental characteristics of concrete containing these residues.

2 Material and methods

The wastes studied as aggregate replacement for concrete production are:
- fly ashes from electric arc furnace (EAF) steel plant (Steel FA);
- slags from electric arc furnace (EAF) steel plant (Steel SLAG);
- foundry sands produced from foundry dies (Foundry SAND);
- slags from lead processing (Lead SLAG);
- Waelz slags (Waelz SLAG);
- bottom ashes from MSWI plants with a mass-burning kiln (BA(MSWI-A)) and a fluidized bed reactor (BA(MSWI-B));
- exhausted sands from MSWI fluidized bed reactor (ES(MSWI-B));
- sludge from an industrial WWTP treating wastewater from galvanic industry (WWTP SLUDGE).

Some kinds of solid residues were used after pre-treatments mainly based on crushing and sieving operation in order to obtain a granular distribution suitable for concrete production; as concerns the solid residues from MSWI, iron removal and washing treatments were also applied with the aim to reduce the content of pollutants. The specific processes applied for waste pre-treatments are reported in Table 1.

Moreover, concrete mixtures were casted with the partial replacement of natural aggregate with different amounts of solid residues (see Table 2). For each residue, a reference mixture was casted only with natural aggregates, in order to compare the performance.

Table 1. Residues analyzed in the work and pre-treatments applied.

Residue	**Pre-treatment**	**Residue name**	**Granular distribution [mm]**
Fly ashes from EAF steel slag	-	Steel FA	0.05 - 0.125
Slags from EAF steel plant	Crushing and sieving	Steel SLAG	0 - 31.5
	Crushing, sieving and ageing (for 3-4 months)	Aged Steel SLAG	0 - 31.5
Foundry sands	-	Foundry SAND	0.125 - 0.5
Slags from lead processing	-	Lead SLAG	0.5 - 31.5
Waelz slags	-	Waelz SLAG	-
	Ageing (for 6 months)	Aged Waelz SLAG	-
Bottom ashes from MSWI mass-burning kiln	Sieving and iron removal	BA(MSWI-A)	0 - 10
			10 - 20
			20 - 30
	Sieving, iron removal and washing	WBA(MSWI-A)	0.063 - 2
			2 - 10
			2 - 20
Bottom ashes from MSWI fluidized bed reactor	Sieving and iron removal	BA(MSWI-B)	0 - 10
Exhausted sands from MSWI mass-burning kiln	-	ES(MSWI-B)	0 - 2
Sludge from industrial WWTPs	Drying and crushing	WWTP SLUDGE	0 - 1

Table 2. Concrete mixtures mix design

Type of residue	**Amount [kg m^{-3}]**			**Water/ cement ratio**
	Residue	**Natural aggregate**	**Cement (type)**	
Steel FA	200	1800	270 (32.5R)	0.65
Aged Steel SLAG	2580	-	380 (32.5R)	0.43
Foundry SAND	200	1800	270 (32.5R)	0.65
Lead SLAG	200	1800	220 - 300 (32.5R - 42.5R)	0.65
Waelz SLAG*	200 - 600	1330 - 1730	250 - 400 (32.5R - 42.5R)	0.48 - 0.54
Aged Waelz SLAG	200 - 400	1530 - 1730	300 (32.5R)	0.54
BA(MSWI-A)	400	1270	300 (32.5R)	0.4
WBA(MSWI-A)	200 - 1700	0 - 1500	300 - 400 (32.5R - 42.5R)	0.4
BA(MSWI-B)	200 - 400	1100 - 1300	320 (32.5R - 42.5R)	0.75
ES(MSWI-B)	100 - 200	1300 - 1400	320 (32.5R - 42.5R)	0.75
WWTP SLUDGE	200	1800	270 (32.5R)	0.65

* superplasticizer addition (3 kg m^{-3}).

As concerns the tests carried out, initially the residues were characterized in order to define their chemical and physical properties according to technical norms for the use of aggregate in concrete production [13-15].

Moreover, the mechanical characterization of concrete mixtures was provided by analysing the compressive [16] and tensile strength [17].

The environmental compatibility was assessed by means of leaching tests, according to UNI EN 12457-2 for granular material [18] and UNI 10802 for concrete mixtures [19].

3 Results and discussion

3.1 Characterization of residues

The chemical composition of the residues analysed is reported in Table 3. Significant concentrations of Fe, Ca, Si, Al and Mg were detected in the steel and Waelz slag. As concerns the potential environmental impact, the steel slag showed high concentrations of Cr_{tot}, V and Ba; in the foundry sand and lead slag, Cu and Pb were very relevant. Regarding the solid residues of MSWI, the content of metals depends both on the chemical composition of the MSW fed to the incinerator and on their volatility. The washing process was able to reduce the pollutant concentration, with the exception of Fe, Mn, Ba and As. The exhausted sand from fluidized bed reactor showed higher content of Zn, Cr_{tot} and Ni with respect to the bottom ash, probably due to the several recirculation of the sand inside the burning chamber during waste combustion, that increased the concentration of pollutants contained in the sand itself.

Table 3. Chemical composition of residues [7, 20-23]. n.d.: not determined.

Parameter		**Steel FA**	**Steel SLAG**	**Aged Steel SLAG**	**Foundry SAND**	**Lead SLAG**	**Waelz SLAG**	**Aged Waelz SLAG**	**WWTP SLUDGE**
		0.05-0.125	**0-31.5**	**0-31.5**	**0.125-0.5**	**0.5-31.5**	**-**	**-**	**0-1**
Ba	[mg/kg]	n.d.	1,526	1,322	-	-	n.d.	n.d.	n.d.
Pb	[mg/kg]	96.8	53.3	47.2	1,860	60,100	12,000	13,000	428
Cu	[mg/kg]	653	365	248	1,314	3,015	6,000	5,600	5,867
Cd	[mg/kg]	43.6	<1	<1	54.2	381	<1	<1	3.8
Cr_{TOT}	[mg/kg]	n.d.	4,393	6,346	n.d.	n.d.	6,950	3,940	n.d.
CrVI	[mg/kg]	< 1	n.d.	n.d.	<1	<1	<1	<1	<1
As	[mg/kg]	2.5	<1	<1	n.d.	535	59.3	49.9	1.6
Hg	[mg/kg]	<1	<1	<1	n.d.	<1	<1	<1	<1
V	[mg/kg]	n.d.	461	487	n.d.	n.d.	n.d.	n.d.	n.d.
Se	[mg/kg]	n.d.	n.d.	n.d.	n.d.	n.d.	11.2	11.0	n.d.
Ni	[mg/kg]	n.d.	n.d.	n.d.	n.d.	n.d.	1,200	500	n.d.
Zn	[mg/kg]	n.d.	255	215	n.d.	n.d.	57,500	63,000	n.d.
Fe	[mg/kg]	n.d.	277,000	276,000	n.d.	n.d.	436,000	424,000	n.d.
Ca	[mg/kg]	n.d.	132,039	154,235	n.d.	n.d.	113,500	123,500	n.d.
Si	[mg/kg]	n.d.	1,568	3,253	n.d.	n.d.	36,700	32,000	n.d.
Al	[mg/kg]	n.d.	36,475	33,133	n.d.	n.d.	19,400	11,300	n.d.
Mg	[mg/kg]	n.d.	42,671	38,239	n.d.	n.d.	20,500	19,200	n.d.
Na	[mg/kg]	n.d.	1,841	1,593	n.d.	n.d.	7,800	2,500	n.d.
K	[mg/kg]	n.d.	447	312	n.d.	n.d.	1,000	400	n.d.
Parameter		**BA (MSWI-A)**			**WBA (MSWI-A)**			**BA (MSWI-B)**	**ES (MSWI-B)**
		0-10	**10-20**	**20-30**	**0.063-2**	**2-10**	**2-20**	**0-20**	**0-2**
Ba	[mg/kg]	420	93	123	552	149	45	n.d.	n.d.
Pb	[mg/kg]	683	145	93	94	46	29	589	96
Cu	[mg/kg]	850	153	114	112	79	42	1,700	1,500
Cd	[mg/kg]	3	0.5	0.4	0.6	0.4	0.3	<0.3	2.9
Cr_{TOT}	[mg/kg]	169	150	80	70	24	4.4	39	415
CrVI	[mg/kg]	n.d.	n.d.	n.d.	n.d.	n.d.	n.d.	<0.3	<0.3
As	[mg/kg]	3.4	1.4	1.3	10	<0.3	<0.3	<0.3	<0.3
Hg	[mg/kg]	<0.3	<0.3	<0.3	<0.3	<0.3	<0.3	n.d.	n.d.
V	[mg/kg]	16	2,3	2,2	9.2	3	2.9	n.d.	n.d.
Se	[mg/kg]	<0.3	<0.3	<0.3	<0.3	<0.3	<0.3	n.d.	n.d.
Ni	[mg/kg]	35	34	23	32	47	32	70	11
Zn	[mg/kg]	1,173	936	433	145	96	84	185	1,661
Fe	[mg/kg]	10,400	2,500	4,200	10,450	n.d.	n.d.	n.d.	n.d.
Al	[mg/kg]	18,400	5,600	7,460	10,420	n.d.	n.d.	n.d.	n.d.

FA: fly ash; BA: bottom ash; WBA: washed bottom ash; ES: exhausted sand.

According to technical norms [13-15], the aged Steel SLAG and the WBA(MSWI-A) could be used as aggregate for concrete production. The ageing treatment (for 3-4 months) allowed to reduce the volumetric expansion of fresh Steel SLAG and to avoid cracking phenomena in concrete mixtures [7]. Moreover, the washing process applied to bottom ash from MSWI mass-burning kiln was able to reduce the contents of acid-soluble sulphates (that could involve concrete expansion) and water-soluble chlorides [22].

The different particle density (i.e. for steel SLAG much higher than natural aggregates) and water absorption (1-2.5% for aged Steel SLAG and 1.6% for WBA(MSWI-A)) should be carefully considered for concrete mix design.

3.2 Characterization of concrete mixtures

Regarding the concrete mixtures casted with the residues from metallurgical industry, the results of compressive strengths after 28 days of curing are summarized in Figure 1.

It can be noted that the partial replacement of natural aggregate with lead SLAG, also with a low cement dosage (270 kg/m^3 of CM32.5R or 220 kg/m3 of CEM42.5R), allowed to obtain compressive strengths close to 20 MPa. Moreover, the use of aged steel SLAG as complete replacement of natural aggregates led to good mechanical performance [7].

As concerns the Waelz SLAG, the ageing treatment seemed not to be effective in terms of compressive strength enhancement; furthermore, the increase of residue dosage (from 200 to 600 kg/m^3) allowed to obtain compressive strengths higher than 35 MPa, with the use of CEM 42.5R (300 kg/m^3) [21].

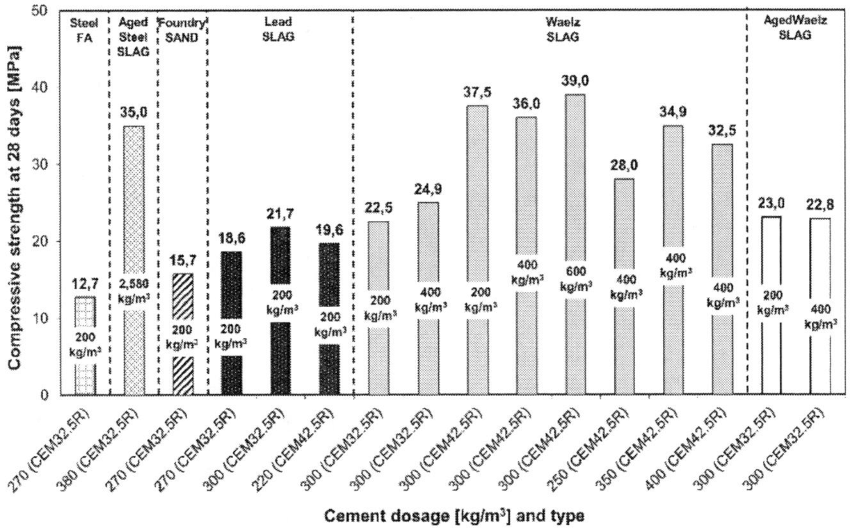

Fig. 1. Compressive strength of concrete mixtures casted with residues from metallurgical industry.

As concerns the residues derived from mass-burning kiln (MSWI-A), the effect of washing was relevant (Figure 2). Using dosages of 400 and 600 kg/m^3, the compressive strength values were higher than 15 MPa. Moreover, the use of the CEM42.5R, instead of the CEM32.5R, allowed to obtain an increase of concrete strength and could reduce the concrete swelling [22]. The dosage increase of residues from fluidized bed reactor (MSWI-B) led to an enhancement of compressive strength. The results obtained suggested that, although the use of these residues could negatively affect the concrete performance, an increased percentage of bottom ash and exhausted sand could contribute to improve the grading curve towards the achievement of lower porosity [24].

DOI: 10.1051/matecconf/201712110001

Finally, the mixtures casted with sludge as a partial replacement of natural aggregate (200 kg/m^3 only) showed a low compressive strength.

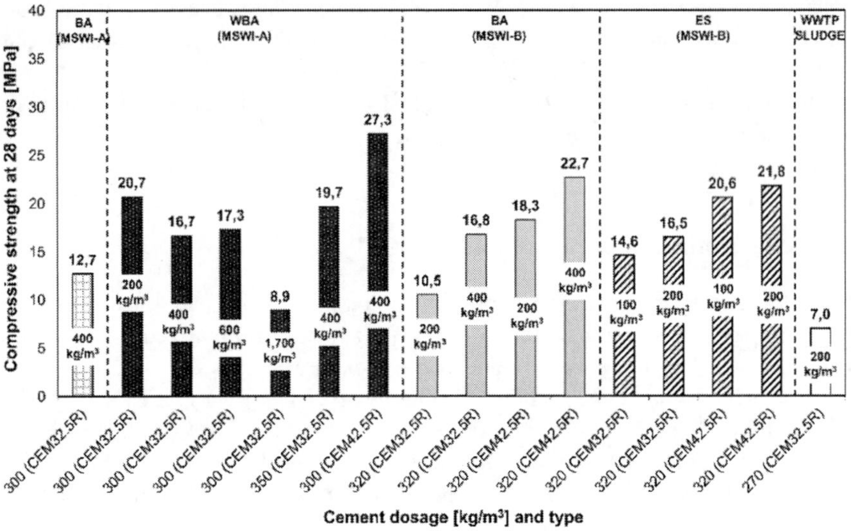

Fig. 2. Compressive strength of concrete mixtures casted with residues from MSWI and sludge.

The results of concrete mixtures mechanical characterization showed that the use of aged steel SLAG and Waelz SLAG do not affect the tensile strength. Furthermore, the dosage of bottom ash from fluidized bed reactor had a negative effect on tensile strength; this phenomena could be significantly reduced with the use of a better quality cement (CEM42.5R instead CEM32.5R).

3.3 Release of pollutants from residues and concrete mixtures

The results of the leaching test (according to UNI EN 12457-2) carried out on the steel SLAG and solid residues from MSWI are shown in Table 4.

According to the Ministerial Decree 2006/186 [25], the unwashed BA from mass-burning kiln (MSWI-A) and the residues from fluidized bed reactor (MSWI-B), especially the exhausted SAND, were not suitable for recovery in simplified procedures [23]. This behaviour could be explained, as said above, as a consequence of several recirculation of the sand in the burning kiln during the combustion.

As concerns the BA from fluidized bed reactor, a slight exceeding of total chromium normative limits could be reduced by means of a washing or ageing treatment. In fact, the washing treatment allowed to obtain residues that could be reused according to the simplified procedure. The concentrations of chromium and copper in the leachate obtained from the exhausted sand were higher than both the concentrations measured for the unwashed bottom ashes and the limit values for waste recovery in simplified procedures.

The results of leaching test for the steel SLAG (Table 4) were good: the only parameter that exceeds the limit value was total chromium [21]. As expected, as well as the MSWI residues, the release of pollutants increased in the finest granular fraction due to the higher specific surface.

The environmental compatibility of concrete mixtures, evaluated by means of leaching test for monolithic materials, has shown that the use of 400 kg/m^3 of washed bottom ash for mass-burning kiln and bottom ash from fluidized bed reactor allowed to obtain release of pollutants similar to the reference concrete mixtures (see Table 5). The release increasing in the crushed concrete samples was probably due both to the increase of the material specific surface and to the cement that significant content of chromium, barium and sulphate [22].

The leaching tests performed on concrete casted with the use of lead SLAG and Waelz SLAG have not shown any problems in terms of pollutants release [7, 21].

Table 4. Results of leaching test for solid residues (the values highlighted are higher than limits) – UNI EN 12457-2. All results are expressed in mg/L with the exception of pH. n.d.: not determined.

	Steel SLAG		Aged Steel SLAG		BA(MSWI-A)			WBA(MSWI-A)			BA (MSWI-B)	ES (MSWI-B)	Limit value (M.D. 186/06)
	0-4	16-31.5	0-4	16-31.5	0-10	10-20	20-30	0.063-2	2-10	2-20	0-10	0-2	
NO_3^-	<0.001	<0.001	<0.001	<0.001	n.d.	n.d.	n.d.	0.17	0.43	0.43	n.d.	n.d.	50
F^-	0.6	0.1	1.3	0.2	0.15	0.05	0.50	0.73	0.33	0.47	0.6	0.1	1.5
SO_4^{2-}	4.3	1.7	4.5	1.1	175	102	99.4	31.17	29.2	32.9	36.5	28.8	250
Cl^-	<0.001	<0.001	<0.001	<0.001	**239**	60.5	79.4	58.6	34.1	46.8	13.4	28.7	100
Ba	0.7	0.3	0.2	0.1	0.051	0.070	0.028	0.135	0.060	0.095	0.022	0.116	1
Cu	<0.001	<0.001	<0.001	<0.001	**0.195**	**0.103**	**0.111**	0.031	0.008	0.004	0.005	**0.245**	0.05
Zn	<0.001	<0.001	<0.001	<0.001	<0.001	<0.001	0.011	0.007	0.002	0.001	0.019	0.031	3
Co	<0.001	<0.001	<0.001	<0.001	n.d.	n.d.	n.d.	0.001	0.001	<0.001	n.d.	n.d.	0.25
Ni	<0.001	<0.001	<0.001	<0.001	<0.001	0.006	<0.001	<0.001	0.001	0.001	0.006	0.002	0.01
V	0.17	0.09	0.13	0.072	n.d.	n.d.	n.d.	0.030	0.017	0.008	n.d.	n.d.	0.25
As	<0.001	<0.001	<0.001	<0.001	0.008	<0.001	<0.001	<0.001	<0.001	<0.001	<0.001	**0.50**	0.01
Cr_{tot}	0.012	0.015	**0.073**	0.008	0.027	0.015	0.022	0.010	0.002	0.002	**0.082**	**0.45**	0.05
Pb	<0.001	<0.001	<0.001	<0.001	0.003	0.016	<0.001	<0.005	<0.005	<0.005	0.003	<0.001	0.05
Mo	n.d.	n.d.	n.d.	n.d.	0.085	0.039	0.025	0.021	0.020	0.009	<0.001	<0.001	-
COD	9	12	<5	<5	n.d.	n.d.	n.d.	20.7	10	12	n.d.	n.d.	30
pH	10.5	10.1	10.5	9.7	n.d.	n.d.	n.d.	11.1	9.8	10.3	n.d.	n.d.	5.5-12

Table 5. Results of leaching test for concrete mixtures. All results are expressed in mg/kg (CN^-, As, Be, Se, Co and Hg are lower than instrumental detection limit for all extractions).

	Reference mixture		WBA(MSWI-A) 400 kg/m³		Reference mixture		BA(MSWI-B) 400 kg/m³		ES(MSWI-B) 200 kg/m³	
	Monolit.	Granular	Monolit.	Granular	Monolit.	Granular	Monolit.	Granular	Monolit.	Granular
F^-	0.50	8	0.75	7	3.38	0.8	3.53	2.5	3.04	0.6
Cl^-	64.9	89	25.7	49	28.1	70	453.8	80	50.3	100
NO_3^-	2.38	3	1.19	3	53.3	140	14.50	170	3.91	100
SO_4^{2-}	21.8	32	28.4	25	114.6	1,750	223.7	2,160	198.9	1,310
Ba	0.11	1.14	0.38	5.84	0.20	6.05	0.20	5.15	0.13	5.9
Cr_{tot}	0.04	0.17	<0.04	0.015	0.06	0.54	0.16	0.47	0.12	0.46
Ni	<0.04	0.015	<0.05	0.01	0.01	<0.01	0.03	<0.01	0.54	<0.01
Pb	<0.16	0.245	0.18	0.375	0.24	<0.01	0.23	<0.01	0.21	<0.01
Cu	<0.07	0.01	<0.02	0.015	0.07	0.11	0.08	0.16	0.05	0.14
Zn	0.19	0.085	0.51	0.12	0.12	0.38	0.42	0.86	0.26	0.48
Mo	0.02	0.02	0.04	0.04	-	-	-	-	-	-
COD	<37.5	45	<50	60	85.4	300	79.4	800	77.8	400

Monolit: test carried out on monolithic samples ; Granular : test carried out on granular samples

4 Conclusions

The reuse of solid residues derived from metallurgical industry and from MSW incineration as "recycled aggregates" in construction sector could be an interesting alternative to landfill disposal.

An experimental work was carried out on concrete samples casted with different residues in partial replacement of natural aggregates. Mechanical tests were performed to determine the compressive and the tensile strength; leaching tests were also carried out both on granular residues and concrete mixtures.

Good mechanical characteristics are shown for concrete mixtures produced with lead slag, aged steel slag, Waelz slag, washed bottom ash from mass-burning kiln, bottom ash and exhausted sand from fluidized bed reactor; on the other hand, concrete samples produced with steel fly ash, foundry sand and WWTP sludge showed worse mechanical characteristics.

The environmental compatibility of steel slag and washed bottom ash was acceptable. Moreover, the leaching test performed on concrete mixtures casted with washed bottom ash from mass-burning kiln, bottom ash and exhausted sand from fluidized bed reactor showed a release of pollutants similar than reference concrete.

The authors wish to thank "PBR s.r.l." (Maclodio, Brescia), "MISE s.r.l." (Tromello, Pavia), "Bosetti s.r.l." (Roè Volciano, Brescia) and "Megazzini s.r.l." (Bressana Bottarone, Pavia) companies and the Laboratory of Department of Civil and Architectural Engineering (University of Pavia) for their collaboration during the experimentation.

References

1. M. Behera, S.K. Bhattacharyya, A.K. Minocha, R. Deoliya, S. Maiti, Constr. Build. Mater. **68**, 16 (2014)
2. M. Calkins, *Materials for Sustainable Sites: A Complete Guide to the Evaluation, Selection, and Use of Sustainable Construction Materials* (Wiley, New Jersey, 2009)
3. K. Lawania, P. Sarker, W. Biswas, Mater. **8**, 17 (2015)
4. Directive 2008/98/EC of the European Parliament and of the Council of 19 November 2008 on waste and repealing. Official Journal of European Union (2008).
5. Ministerial Decree 14 January 2008. Italian law on the Official Gazette 29 (2008)
6. Euroslag, *Position Paper on the Status of Ferrous Slag* (2012)
7. L. Rondi, G. Bregoli, S. Sorlini, L. Cominoli, C. Collivignarelli, G. Plizzari, Compos. Part B-Eng. **90**, 8 (2016)
8. D. Mombelli, C. Mapelli, S. Barella, A. Gruttadauria, U. Di Landro, Process. Saf. Environ. **94**, 12 (2015)
9. *** http://life-foundrysand.com/, accessed 8.03.2017
10. Eurostat, http://appsso.eurostat.ec.europa.eu/nui/show.do?dataset=env_wasmun&lang=en, accessed 8.03.2017
11. Milieu, WRC, RPA, *Environmental, economic and social impacts of the use of sewage sludge on land Final Report Part I: Overview Report* (2010)
12. M.C. Collivignarelli, A. Abbà, S. Padovani, M. Frascarolo, D. Sciunnach, M. Turconi, M. Orlando, Environ. Eng. Manage. J. **14**, 10 (2014)
13. *** UNI EN 12620 (2008)
14. *** UNI 8520-1 (2015)
15. *** UNI 8520-2 (2016)
16. *** UNI EN 12390-3 (2009)
17. *** UNI EN 12390-6 (2010)
18. *** UNI EN 12457-2 (2004)
19. *** UNI 10802 (2013)
20. S. Sorlini, C. Collivignarelli, *Fifth International Conference on the Environmental and Technical Implications of Construction with Alternative Materials, Wascon Conference 2003* (San Sebastian, 2003)
21. S. Sorlini, C. Collivignarelli, G. Plizzari, M. Delle Foglie, Proceedings of the International RILEM Conference on the use of recycled materials in buildings and structures, Barcelona, Spain (2004)
22. S. Sorlini, A. Abbà, C. Collivignarelli, Waste Manage. **31**, 9 (2011)
23. M.C. Collivignarelli, A. Abbà, S. Sorlini, M. Bruggi, J. Mater. Cycles Waste Manage. (to be published)
24. A. Abbà, M.C. Collivignarelli, S. Sorlini, M. Bruggi, Compos. Part B-Eng. **58**, 8 (2014)
25. Ministerial Decree 5 April 2006 n. 186, Italian law on the Official Gazette 115 (2006)

Analysis of the road traffic-induced pollution in two areas of Sibiu, Romania

Cristian Deac[1,*], and *Marius* Bibu[1]

[1]Lucian Blaga University of Sibiu, 4 E. Cioran St., 550025, Sibiu, Romania

Abstract. The current paper details a computer-aided research of the emission levels in two areas of the city of Sibiu, Romania. The research started from the numbers of motor vehicles passing through key segments of those two areas, located at entry points into the city of Sibiu. Using the software package Synchro Studio 7 and SimTraffic 7, the targeted crossroad areas were modelled and key indicators for the road traffic efficiency and for the emission levels were determined. These emission levels were then compared to real measured pollution values, which allowed both to assess the accuracy of the modelling and the influence of other pollution sources on the emission levels in those areas.

1 Introduction

Over the last few decades, the world population – and especially the urban population – has increased rapidly. This has resulted on the one hand in significantly more road traffic in the cities, with all the associated problems for the people's movement, but on the other hand also in higher levels of pollution, both in terms of noise and in terms of noxious gaseous emissions.[1]

In many cases, the problems created by pollution are enhanced by the fact that due to budgetary constraints, but often also due to practical reasons, the road infrastructure of cities is unable to adapt quickly enough to the increased demographic pressure. This problem becomes even more visible in larger towns/cities, especially when these have a certain local or regional importance.

The location analysed in this paper, the city of Sibiu, is located in central Romania and has a stable population of around 150000 inhabitants. It must be added however, that the studied city is an important industrial and commercial centre and also has a rich culture and history, so beneath the actual inhabitants there is also a significant temporary population, consisting of tourists, businessmen, workers etc., living, working or at least transiting Sibiu. All this contributes to a significant pressure on the road structure of the city, which, despite sustained efforts over the years, continues to lead to difficulties in driving through the city, but also leads to an increased pollution.

In the current paper, the authors analyse the traffic conditions and the associated noxious emissions in two key areas of Sibiu. Furthermore, using specialised software instruments, the authors offer possible solutions for an optimisation in the sense of decreasing the road traffic levels in these areas and also of decreasing the pollution levels.

* Corresponding author: cristian.deac@ulbsibiu.ro

2 Study areas, research means and method

For the current paper, the authors have focused on two areas in Sibiu: on the one hand the area around the Unirii square in the centre of Sibiu, which represents an important crossroads area, where the north-south road traffic intersects with the east-west road traffic through the city (fig. 1) and the area around another important crossroads, this time along the Lunga street, an important artery in the northeastern part of Sibiu, which collects the traffic between the Terezian neighbourhood of Sibiu and the city centre but also the traffic between Sibiu and the other larger town in the area, Medias (fig. 2).

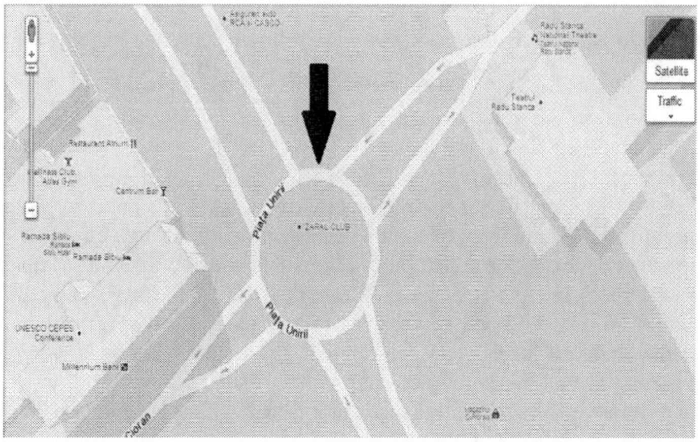

Fig. 1. The Unirii square and crossroads in central Sibiu

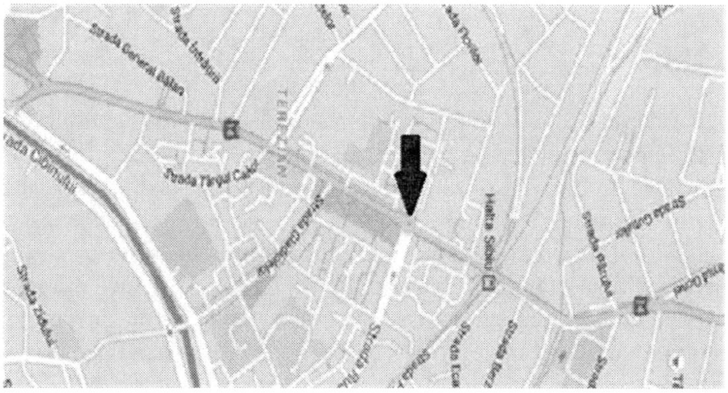

Fig. 2. The Lunga street in northeastern Sibiu, with the analysed crossroads highlighted

The first area is characterised by an intense road and pedestrian traffic for following main reasons:

- The area is of high economical and administrative interest (in or near to it there are important institutions such as banks, hotels, shopping centres, the local Police headquarters, hospitals etc., but also residential areas)
- It is crossed or bordered by several main roads that connect the various neighbourhoods in the city but also funnel the traffic towards the exits from the city and towards the surrounding cities.

The second area is located in the Terezian neighbourhood of Sibiu. It starts relatively close to the city centre, near the Cibin market and ends at the exit from Sibiu towards Medias and towards the national road DN 14. It too is bordered by residential areas, but contains also the headquarters of several important institutions or companies.

In both areas, the current road geometry seems to be well organised, but during peak hours their traffic capacity is visibly surpassed and there is also a significant pollution mainly stemming from exhaust gases.

The analysis described in this paper intends to determine the precise traffic and pollution conditions and to find optimal solutions for reducing or possibly eliminating the mentioned problems.

The main activities that were carried out for this study included:

- analysis of the two areas;
- collection of data regarding the traffic conditions;
- modelling of the crossroads area in a specialised software and simulation of the traffic conditions;
- using the simulation software to calculate the average noxious emissions in the studied areas, based on the previously gathered traffic data;
- determining optimisation solutions for the traffic;
- analysis of the impact of the proposed solutions on the pollution in the two areas.

For collecting the data on traffic conditions, following methods can be used [2]:

- manual collection using a team of human observers
- analysis of video recordings
- collection using specialised equipment (sensors, GPS etc.)

It is important to select appropriate moments and frequencies for the collection of the traffic data. Also, these data have to be accompanied by information on factors such as the meteorological conditions, on the traffic intensity and the traffic's complexity. [2, 3]

For the current study, the data on the number of vehicles passing through the targeted areas was collected by means of video recordings, on several days, using time periods of 15 minutes each. After that, the recordings were reviewed and the vehicles were counted manually [4].

Beneath the actual car counting on each individual lane, several other elements were recorded, in order to facilitate the subsequent computer-aided modelling of the traffic in the crossroads area [4]:

- the road geometry and road equipment present (traffic lights, traffic signs etc.);
- the traffic light cycles and the corresponding durations of each colour;
- the pedestrian traffic;
- the length of waiting queues on the various directions;
- the number and frequency of the public transportation means etc.

The number of vehicles passing through the considered areas had to be "normalised" using the provisions of the Romanian standard SR 7348-2001 [5] in order to account for the different types of vehicles (small cars, motorcycles, trucks, buses etc.) noticed there, resulting in a number of so-called equivalent vehicles, calculated according to formula (1).

$$N_{equiv} = N_1c_1 + N_2c_2 + \ldots + N_ic_i \qquad (1)$$

where:

N_{equiv} – number of equivalent vehicles for the considered traffic flow and time period;

$N_1, N_2, \ldots, N_i$ – number of vehicles of a certain type in the flow;

$c_1, c_2, \ldots, c_i$ – correction coefficients corresponding to each type of vehicle.

The counting revealed only relatively few motorcycles and trucks (because Sibiu has a well-organised ring road that is used by most of the heavy traffic without entering the city itself), so these vehicle categories have been neglected in the actual vehicle count.

The traffic data were then input into the specialised software package Synchro Studio 7. For this, the authors first modelled the geometry of the two areas and then the traffic-related data were added. The main element determined using this traffic simulation was the so-called crossroads usage capacity. This was then used as base for comparisons with the proposed optimisation solutions.

Also, the software package SimTraffic 7 was used to simulate traffic in the considered crossroads based on the data collected previously and to determine the main traffic and pollution statistics.

The optimisation of the traffic in the analysed crossroads was based on slight modifications to the crossroads' geometry, but also on modifications to other elements.

The noxious gaseous that were targeted for this analysis were hydrocarbons (HC), carbon monoxide (CO) and nitrogen oxides (NOx). The presence of hydrocarbons and carbon monoxide in vehicle exhaust is due to incomplete combustion processes, whereas NOx results when atomic oxygen formed through the thermal dissociation of molecular oxygen reacts with nitrogen molecules in the air [6].

3 Results and discussion

The data obtained after counting the cars passing through the Unirii crossroads during the morning hours is presented in table 1, while the data obtained after counting the cars passing through the Lunga-Rusciorului crossroads is presented in table 2.

Table 1. Number of vehicles determined for the Unirii crossroads, on each entryway (road)

Road / Time range	**1 Emil Cioran str.**			**2 Unirii square (descending)**			**3 Corneliu Coposu Blvd.**			**4 Piața Unirii (ascending)**		
Direction of movement	←	↑	→	←	↑	→	←	↑	→	←	↑	→
7:30 - 7:45	64	45	71	88	278	35	174	90	135	37	283	155
7:45 - 8:00	78	67	95	125	486	44	263	124	183	75	386	242
8:00 - 8:15	41	39	65	55	249	28	139	81	74	58	229	129
8:15 - 8:30	21	29	27	41	182	21	113	63	85	31	163	86
Number of buses (#/h)	-	-	-	13	12	-	12	-	13	-	12	12
Number of equivalent cars (7:30-8:30)	204	174	258	355	1237	128	731	358	523	201	1103	654

Table 2. Number of equivalent vehicles determined for the Lunga-Rusciorului crossroads, on each entryway (road), during the morning hours

1. Lunga str. (outwards)				2. Rusciorului str. (towards Lidl)				3. Lunga str. (towards Cibin market square)			4. Lunga str. 2 (towards Cibin market square			5. Rusciorului str. (towards city centre)		
↖	↰	↑	↗	↖	↰	↑	↗	↖	↑	↗	↖	↑	↗	↖	↑	↗
18	2	117	49	55	2	28	103	7	101	13	1	3	2	24	19	12

The number of equivalent vehicles indicated in the last row of tables 1 and 2 was determined using equation 1 for each particular flow on a street (on each branch of a crossroads) and for each direction.

The simulation of the traffic in the Unirii crossroads and in the Lunga-Rusciorului crossroads, respectively, in the initial state, carried out using the software Synchro Studio 7 lead to the calculation of the crossroads usage capacity as visualised in figures 3 and 4.

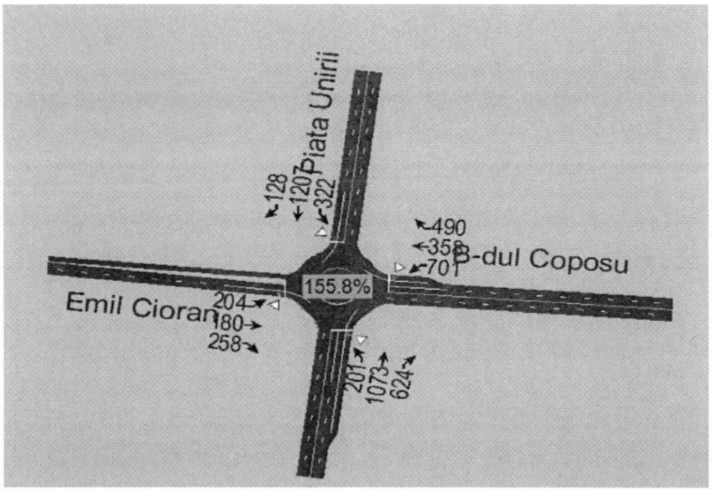

Fig. 3. Calculation of the crossroads usage capacity for the initial state of the Unirii crossroads using Synchro Studio 7

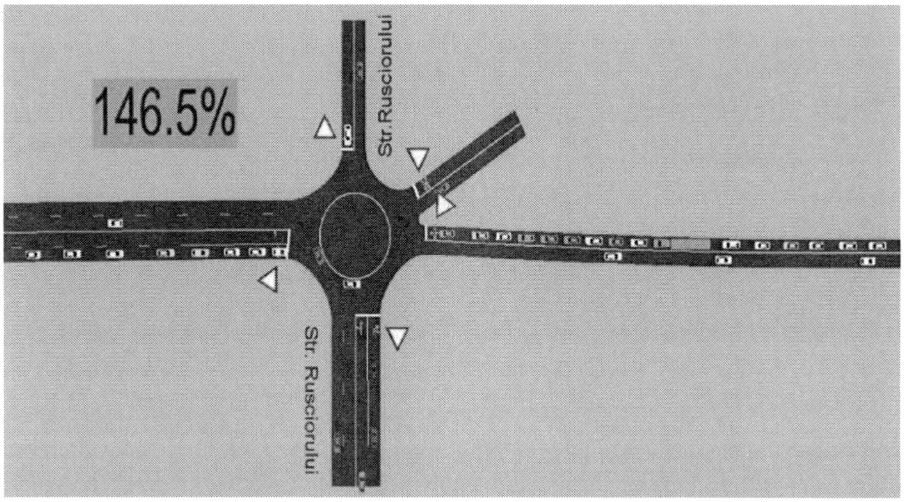

Fig. 4. Calculation of the crossroads usage capacity for the initial state of the Lunga-Rusciorului crossroads using Synchro Studio 7

As can be seen in figures 3 and 4, the crossroads usage capacity exceeds in both cases 100% by far, indicating that traffic is not unfolding under normal conditions.

Figure 5 presents the pollution statistics for the Unirii crossroads in its initial state, as calculated by the SimTraffic 7 software, while figure 6 indicates both the main road traffic and the pollution statistics for the Lunga-Rusciorului crossroads area in its initial state, also as calculated by the SimTraffic 7 software.

SimTraffic Performance Report
Baseline

2: Emil Cioran & Piata Unirii Performance by movement

Movement	EBL	EBT	EBR	WBL	WBT	WBR	NBL	NBT	NBR	SBL	SBT	SBR
HC Emissions (g)	0	0	0	0	0	0	0	0	0	0	0	0
CO Emissions (g)	34	36	38	141	66	80	43	153	119	57	206	27
NOx Emissions (g)	2	2	2	3	1	2	1	3	2	1	3	0

2: Emil Cioran & Piata Unirii Performance by movement

Movement	All
HC Emissions (g)	2
CO Emissions (g)	1044
NOx Emissions (g)	24

Total Network Performance

HC Emissions (g)	19
CO Emissions (g)	1946
NOx Emissions (g)	99

Fig. 5. Pollution statistics for the Unirii crossroads in the initial state, as determined with the SimTraffic 7 software.

3: Str. Lunga & Str. Rusciorului Performance by movement

Movement	SWL2	SWL	All
Total Delay (hr)	0.0	0.0	34.6
Delay / Veh (s)	7.6	7.0	323.9
Avg Speed (kph)	15	16	5
Fuel Used (l)	0.1	0.2	34.0
Fuel Eff. (kpl)	6.5	6.8	1.5
HC Emissions (g)	0	0	24
CO Emissions (g)	5	13	887
NOx Emissions (g)	1	2	55

Total Network Performance

Total Delay (hr)	35.0
Delay / Veh (s)	327.2
Avg Speed (kph)	9
Fuel Used (l)	44.5
Fuel Eff. (kpl)	2.8
HC Emissions (g)	50
CO Emissions (g)	2115
NOx Emissions (g)	158

Fig. 6. Road traffic and pollution statistics for the Lunga-Rusciorului crossroads in the initial state, as determined with the SimTraffic 7 software.

An analysis of the data in figures 5 and 6 shows that the three targeted pollutants have similar concentrations in the two cases, owing to vehicle numbers in roughly the same order of magnitude. The HC and NOx emissions are relatively low, but CO emissions are at an alarming level, which along with the general discomfort created by traffic congestion conditions call for serious and immediate corrective actions. Inhaling carbon monoxide at high concentrations can be harmful and even fatal, because CO prevents the transportation of oxygen through the body [6].

In order to optimise the road traffic in the area of the Unirii square, the authors suggested adding a traffic lane on a length of 50 m for the Unirii square (descending branch), allowing only right turns and one supplementary lane each for the Unirii square (ascending branch) and for the Coposu boulevard, at whose ends only left turns are allowed.

The optimisation possibilities of the Lunga-Rusciorului crossroads are limited by the impracticality of modifying the main problem element, namely the railroad overpass that is located just next to the crossroads, towards the exit from the city. Therefore, some alternative possibilities have been conceived:

- one-way traffic on the Rusciorului street, on two lanes, from the city centre towards the Lidl supermarket;
- adding another traffic lane on the Lunga street, for the outward direction (this can be done at the expense of a small portion of a park located in the area);
- deviation of a part of the vehicle flow entering the city and moving towards the city centre, to the nearby Cerbului street.

After implementing these changes in the model created in Synchro Studio 7, the crossroads usage capacity changed for the Unirii square as shown in figure 7, while the new crossroads usage capacity for the Lunga-Rusciorului crossroads is shown in figure 8.

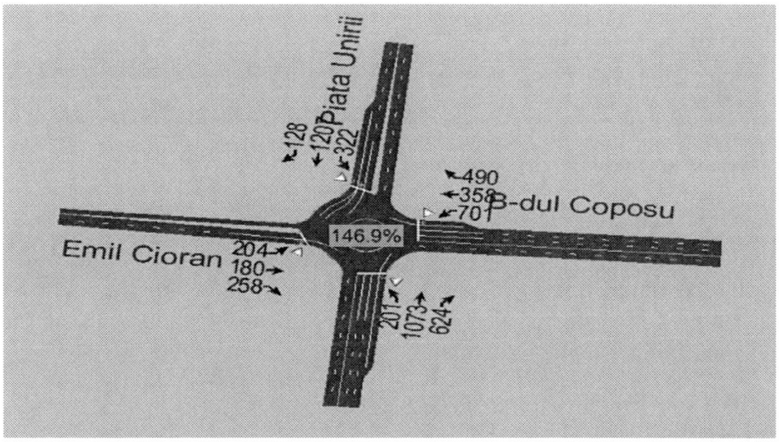

Fig. 7. Calculation of the crossroads usage capacity for the proposed changes in the Unirii crossroads

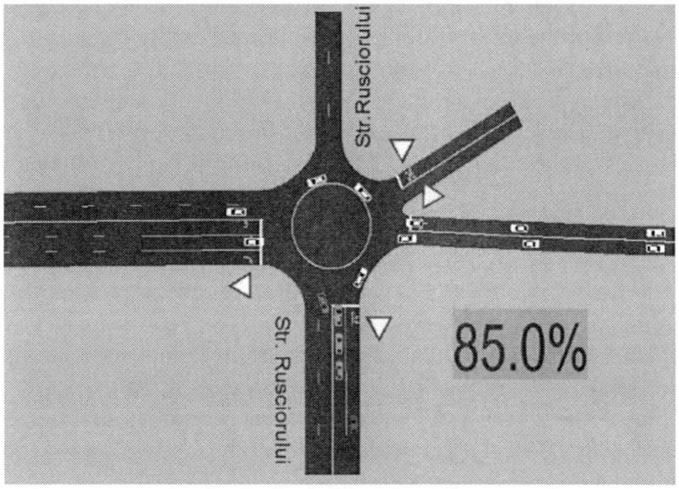

Fig. 8. New crossroads usage capacity for the Lunga-Rusciorului crossroads.

It can be seen that for the Unirii square, the proposed changes lead only to a moderate but still beneficial decrease of the crossroads usage capacity, whereas for the Lunga-Rusciorului crossroads, the decrease is significant.

However, as can be seen in figure 9, the improved vehicle flow in the Unirii crossroads is not also beneficial for the emissions, which even increase somewhat compared to the initial state. Therefore, in this case a compromise needs to be found between a higher traffic speed and higher pollution levels.

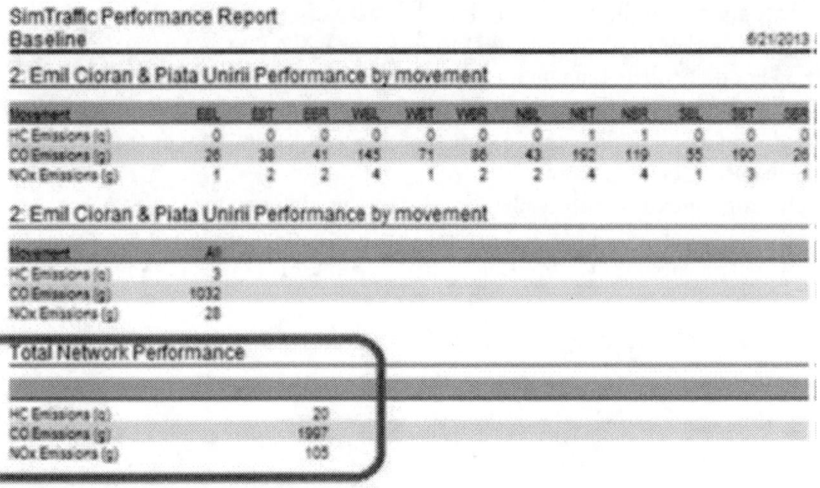

SimTraffic Performance Report
Baseline 6/21/2013

2: Emil Cioran & Piata Unirii Performance by movement

Movement	EBL	EBT	EBR	WBL	WBT	WBR	NBL	NBT	NBR	SBL	SBT	SBR
HC Emissions (g)	0	0	0	0	0	0	0	1	1	0	0	0
CO Emissions (g)	26	38	41	145	71	86	43	192	119	55	190	26
NOx Emissions (g)	1	2	2	4	1	2	2	4	4	1	3	1

2: Emil Cioran & Piata Unirii Performance by movement

Movement	All
HC Emissions (g)	3
CO Emissions (g)	1032
NOx Emissions (g)	28

Total Network Performance

HC Emissions (g)	20
CO Emissions (g)	1997
NOx Emissions (g)	105

Fig. 9. Road traffic and pollution statistics for the Unirii crossroads in the optimised state, as determined with the SimTraffic software.

Owing to the much improved crossroads usage capacity in the Lunga-Rusciorului crossroads, pollution levels there also decreased significantly when applying the suggested improvements in the simulation software.

4 Conclusions

The current paper presented a study on the traffic and pollution conditions in two important crossroads areas from the city of Sibiu, Romania, one in the city centre and the other one towards the exit from the city towards the northeast, but both located on very busy roads.

In both cases, the proposed optimisation solutions succeeded in reducing the crossroads usage capacity, in the case of the Lunga-Rusciorului crossroads the reduction being impressive. However, the simulations carried out on solutions that increased the traffic flow have yielded mixed results in terms of reducing the road traffic-related pollution.

Therefore, any solutions targeting the improvement of traffic conditions in the city of Sibiu either need to take into account more factors and/or bolder changes even to the infrastructure (which however would lead to dramatically increased investment costs for these improvements) that would allow a simultaneous reduction of traffic problems and pollution problems, or they need to take into account compromise solutions with only modest gains – or even losses – in terms of road speed and/or pollution levels.

In future, the authors intend to extend the number of areas analysed in this manner and to work together with the local authorities and with other specialists in order to find solutions for better environmental conditions along Sibiu's main roads.

References

1. M. Wiering, J. van Veenen, J. Vreeken, and A. Koopman, *Intelligent traffic light control* (Institute of Information and Computing Sciences of the Utrecht University, 2004)
2. J. Barceló, M. Delgado, G. Funes, D. García, J. Perarnau, and A. Torday, SIAM News, **40**, 9 (2007)
3. D. Popovici, V. Boboc, I. Galusca, *Sisteme de transport şi trafic urban* (Editura Societăţii Academice „Teiu Botez", 2004)
4. M. Inta, A. Muntean, Revista Stiinta si Inginerie, 2015, http://stiintasiinginerie.ro/category/revista/2015/volumul-28/ , Accessed 21.02.2017

DOI: 10.1051/matecconf/201712110002

5. *** *SR 7348-2001 Road works. Equivalation of vehicles for determining the traffic capacity* (ASRO, 2001)
6. T.J. Wallington, E.W. Kaiser, J.T. Farrell, Chem. Soc. Rev. **35** (2006)
7. Scottish Environment Protection Agency, http://apps.sepa.org.uk/spripa/Pages/SubstanceInformation.aspx?pid=4, Accessed 17.03.2017

Is agro-tourism eco-friendly?

Ramona Giurea[1,*] , *Marco* Ragazzi[2], *Moise Ioan* Achim[3], and *Mihai-Victor* Zerbes[4]

[1] Department Engineering and Management, Lucian Blaga University of Sibiu, 550024, Sibiu, Romania & Department of Civil Environmental and Mechanical Engineering, University of Trento, 38123, Trento, Italy
[2] Department of Civil Environmental and Mechanical Engineering, University of Trento, 38123, Trento, Italy
[3] Department of Exact Sciences and Engineering, "1 Decembrie 1918" University of Alba Iulia, 510009, Alba Iulia, Romania
[4] Department Engineering and Management, Lucian Blaga University of Sibiu, 550024, Sibiu, Romania

Abstract. The paper will highlight the impact of agro-tourism on environment in Romania, focusing on the Sibiu County. Thanks to the results obtained from this analysis, research and information collection, the paper will give some answers regarding the question: are agro-tourisms environmentally friendly? The authors identify potential causes of environmental pollution from agro-tourism and some resources that could be economically used. The negative and positive aspects on the increase of agro-tourism activities in the proposed region will be also discussed taking into account there impact on the environment. Based on the obtained results, the authors will propose a method for the management of resources in order to arrive to an eco-friendly system for the proposed case-study. Future developments and applications on other EU case-study will be also discussed in this paper.

1 European framework on environment

The fundamental of our health, our economy and our well-being is the environmental quality. Nevertheless, it faces several serious challenges as climate change, unsustainable consumption and production, as well as various forms of pollution [1-3]. A part of this pollution is also connected with the agro-tourism production and consumption: wastewater, solid waste, energy consumption, etc. [4,5]

The concern for environmental protection appeared on the European Agenda in the early 1970s [6]. Environmental policy of the European Union (EU) was created by the Treaty of the European Community with the aims to ensure sustainable environmental protection measures. The Treaty of Maastricht (1993) made the environment an official EU policy area [7]. Over the past decades, the EU has put in place a broad range of environmental legislation through Environment Action Programme (EAP) that sets out priority objectives as: the protection of nature, stronger ecological resilience,

* Corresponding author: ramona.giurea@ulbsibiu.ro

sustainable, resource-efficient and low-carbon growth; and the fight against environment-related threats to health [8].

In Romania, after the communist period (1989), began a long process of social, political and economic transformation, whose essence lies in the transition from a planned economy to a free market. So by Government Decision the Ministry of the Environment was established – a state controlled authority, which organizes the institutional framework, develop, guide and improves environmental activities on a national scale [9]. The main areas of expertise of Ministry of Environment are: air quality and environmental noise, management of protected areas, environmental infrastructure, waste management, sustainable development, climatic changes, water management, forestry sector management, etc [10].

The European Union has some of the world's highest environmental standards, developed over decades. Environment policy helps the EU economy become more environmentally friendly, protects Europe's natural resources, and safeguards the health and wellbeing of people living in the EU. The EU environment policy is based on Articles 11 and 191-193 of the Treaty on the Functioning of the European Union [11].

Apart from demands required at European and national level, there are a range of international standards adopted by European Union and implicitly of Romania. The principal standard in Romania is Environment ISO 14001 and in Europe EU Eco-Management and Audit Scheme (EMAS) [12-13].

Tourism plays a major role in the EU economy. According to the European Commission, it is the third largest socio-economic activity in the EU [14]. In the scientific literature we can find the term agro-tourism defined as a subset of rural tourism [15], having the roots on the use of the resources from the rural regions [16-17]. However, rural tourism is an intricate and extended phenomenon and the effect is directly proportional with each individual territory and the relationships between tourism products and local resources [18]. Agro-tourism includes various tourist activities, such as accommodation (bed and breakfast, rural lodgings, farm campsite); catering (evening meals); leisure activities (pedagogical farms, sports, horse-riding, farm visits) [19].

The existences of an economic interest and social benefits of agro-tourism is increasing [20-21], specific literature has considerable gaps in which concern environmental consequences [22-23], those gaps being a considerable opportunities to improve the specific processes of the field. Starting in the context of environmental actions, based on analysis, mainly case studies, we can observe that the agro-tourism can produce both positive and negative effects on the environment; for example, some positive effects on environment are: water and energy resources, landscape, biodiversity, reduced use of fertilizers and pesticides in productive processes and improvement of the quality of foods [24].

2 Proposed methodology

The methodology used in this article by the authors is inspired from the DMAIC model used by organizations in the six-sigma projects [25]. The overall methodology of the study (Fig. 1, Fig. 2) is detailed explained in an article [26], concerning "*The modelling of the improving environmental aspects process and of the associated impacts in industrial organizations*" which illustrates the key steps carried out in the model building.

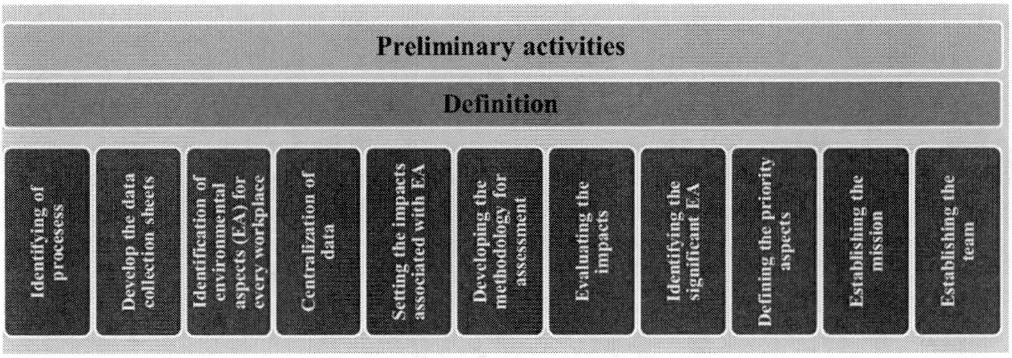

Fig. 1. Phase of the propose DMAIC model – Preliminary activities

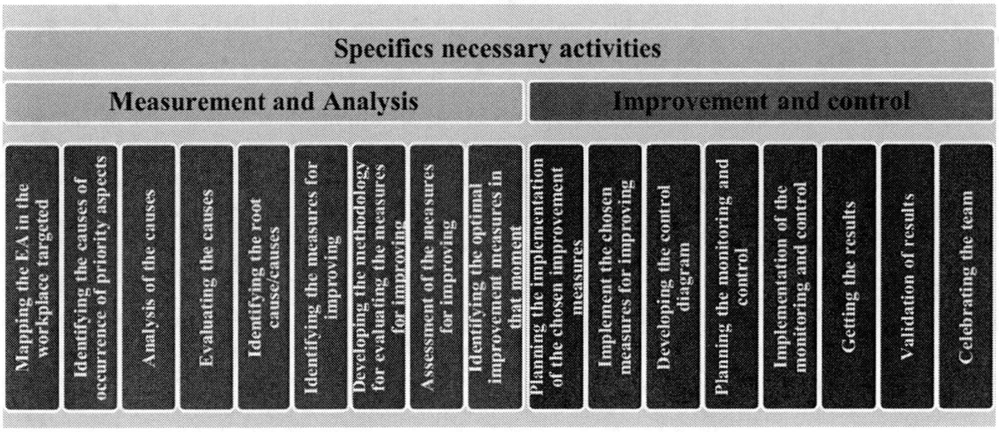

Fig. 2. Phases of the propose DMAIC model - Specifics necessary activities

For implementing the proposed model, the authors took into account the main activities in an agro-touristic structures, summarized in Table 1.

Table 1. Main processes in an agro-touristic structure

Process		Included activities		Input	Output	Code
Name	Code	Name	Code			
Transport	T	Gas supply	gs	Gas	Fuelled automobile	TGF
		Carrying out documents	cd	Documents	Documents completed	TDD
		Transport	tr	Gas	Transportation	TGT
		Maintenance	me	Documents, Inspect. equipment	Automobile maintained	TIA
Check in/out	C	Carrying out documents	cd	Personal documents. Specific formulation	Check in/out	CPC
		Delivery keys	dk	Keys	Check in/out	CKC
Accommodation	A	Takeover room	tr	Check in	Check out	ACC
Meal	M	Meal preparation	mp	Equipment and products	Food prepared	MEF
		Serving food	sf	Meal prepared	Meal served	MPS
		Cleaning table	dm	Disposal meals	Release tables	MDR
		Cleaning dishes	cm	Dirty dishes	Clean dishes	MDC
Leisure activities	L	Hike	hk	Transport	Visit	LTV
		Equipment	ha	Equipment	Household cared	LEH
Hygiene/ cleaning	H	Common space	cs	Equipment and cleaning solutions	Common space cleaned	HEC
		Kitchen and bathrooms	kb	Equipment and cleaning solutions	Kitchen and bathrooms cleaned	HEK

Process Name	Process Code	Included activities Name	Included activities Code	Input	Output	Code
		Guest rooms	**gr**	Equipment and cleaning solutions	Guest rooms cleaned	HEG
		Household	**hh**	Equipment and cleaning solutions	Household cleaned	HEH
Acquisition/ purchases	P	Analysis and drafting documents	**ad**	Documents	Documents completed	PDD
		Transport	**tt**	Gas	Transportation	PGD
		Reception	**re**	Documents	Documents completed, Product reception	PDP
		Storage	**st**	Documents	Storage products	PDS
Maintenance work	W	Maintenance equipment	**me**	Documents, Inspection equipment	Equipment maintained	MDE
		Automobile	**at**	Documents, Inspection equipment	Automobile maintained	MDA
		Equipment and installations household	**ei**	Documents, Inspection equipment	Equipment and installations maintained	MDI
Agriculture/ farming	F	Mechanized works	**mw**	Specific equipment	Tillage	FST
		Manual/handiwork works	**mh**	Specific equipment	Tillage	FSH
Zootechnics	Z	Caring	**ca**	Specific equipment	Groomed animals	ZSG
		Feeding	**fe**	Specific equipment	Fed up animals	ZSF
		Slaughtering	**sl**	Specific equipment	Meat, fur	ZSM
Pomiculture	P	Other manual works	**om**	Specific equipment	Fruit growing	PSF
		Spraying	**sp**	Specific equipment	Splash trees	PSS
		Pruning work	**pw**	Specific equipment	Fruit growing	PSP

To achieve the objective of this research, the authors, firstly, focused on identifying the main environmental aspects and impacts associated with them. In this way, the data about the identification and analysis of the aspects are highlighted in Table 2.

Table 2. Identification of environmental aspects and their impacts assessment

Process	Included activities	Input/ Output	Environment aspect	OC *	EI	Impact assessment					Impact classification	
						L	F	N	C	TS	I	S
T	**gs**	TGF	Exhaust emissions	N	AP	5	3	1	3	45	I	-
	cd	TDD	Waste paper	N	SP	3	5	1	1	15	I	-
	tr	TGT	Exhaust emissions	N	AP	3	3	1	3	27	I	-
	me	TIA	Waste paper	N	SP	3	1	3	3	27	I	-
C	**cd**	CPC	Waste paper	N	AP, SP	3	5	1	3	45	I	-
	dk	CKC	Waste paper	N	AP, SP	3	5	1	3	45	I	-
A	**tr**	ACC	Waste paper, organic paper	N	AP, SP	1	5	3	3	45	I	-
M	**mp**	MEF	Organic waste	N	SP, WP	3	5	1	5	75	-	S
	sf	MPS	Organic waste	N	SP, WP	3	3	3	1	27	I	-
	dm	MDR	Organic waste	N	SP, WP	3	3	3	1	27	I	-
	cm	MDC	Organic, chemical waste	N	AP, SP, WP	3	3	1	3	27	I	-
L	**hk**	LTV	Exhaust emissions	N	AP	5	3	3	1	45	I	-
	ha	LEH	Organic, paper waste	N	SP, WP	5	3	1	1	15	I	-
H	**cs**	HEC	Chemical waste	N	AP, SP, W	1	3	5	3	45	I	-
	kb	HEK	Chemical waste	N	AP, SP, WP	1	3	5	3	45	I	-

Process	Included activities	Input/ Output	Environment aspect	OC *	EI	Impact assessment					Impact classification	
						L	F	N	C	TS	I	S
	gr	HEG	Chemical waste	N	AP, SP, WP	1	3	5	3	45	I	-
	hh	HEH	Organic, chemical waste	N	AP, SP, WP	5	3	3	1	45	I	-
P	ad	PDD	Paper waste	N	SP, WP	3	3	1	1	9	I	-
	tt	PGD	Exhaust emissions	N	AP, SP, WP	3	1	3	3	27	I	-
	re	PDP	Paper waste	N	SP, WP	3	3	1	1	9	I	-
	st	PDS	Organic, Paper waste	N	SP, WP	3	3	1	3	27	I	-
W	me	MDE	Paper waste	N	SP, WP	3	1	1		3	I	-
	at	MDA	Paper waste,	N	SP, WP	3	1	3		9	I	-
	ei	MDI	Paper waste, electronics and electrical waste	N	AP, SP, WP	5	1	3		15	I	-
F	mw	FST	Organic waste	N	SP, WP	3	3	5	3	135	-	S
	mh	FSH	Organic waste	N	SP, WP	3	3	5	3	135	-	S
Z	ca	ZSG	Organic waste	N	SP, WP	3	3	1	5	45	I	-
	fe	ZSF	Organic waste	N	SP, WP	3	3	1	5	45	I	-
	sl	ZSM	Organic waste	N	SP, WP	3	3	1	5	45	I	-
P	om	PSF	Organic waste	N	SP, WP	3	3	1	5	45	I	-
	sp	PSS	Organic waste	N	SP, WP	3	3	1	5	45	I	-
	pw	PSP	Organic waste	N	SP, WP	3	5	1	5	75	-	S

Legend: WP-water pollution; AP-air pollution; SP-sol pollution; OC-Operating condition; EI-Environment impact; L-Existing of the legal requirements and other environmental requirements regulating environmental impact; F-Frequency of occurring environmental impact; N-The nature of the natural resource/the pollutant /the waste to which it relates those impacts; C-Data on the amount of natural resource/pollutant /waste implied by those impacts; I-Insignificant; S-Significant

The values obtained in Table 2 were analysed using Pareto diagram as follows:

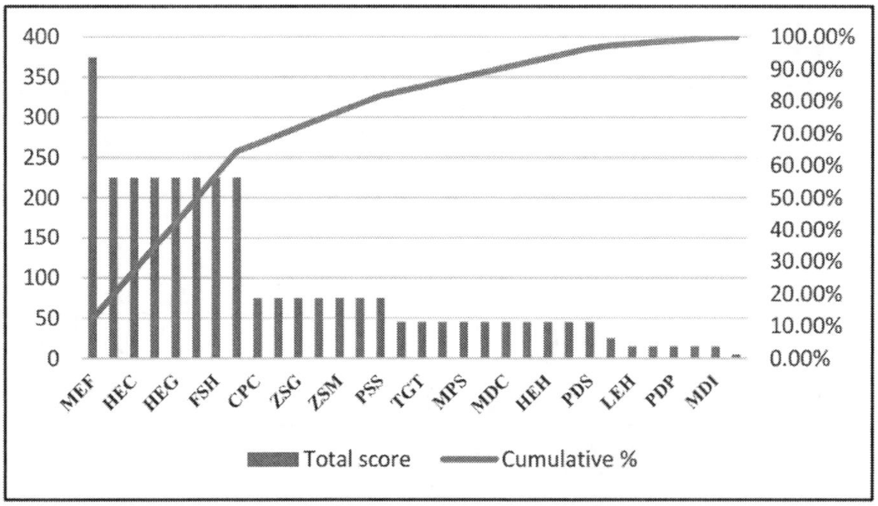

Fig. 4. Overview on analysed values by the instrumentality of Pareto diagram

As can be seen in this chart, 25% of cases generate about 70% of the problem. To illustrate how these cases are approached further, below we present the analysis of the first three causes. Therefore, were identified alternative amelioration presented in Table 3.

Table 3. Identification of alternatives for the top three causes

Cause	Alternative 1	Alternative 2	Alternative 3
MFE Meal preparation Equipment and products Food prepared	Outsourcing services through catering	Removal of the meal by providing the kitchen. The tourist will prepare his own meal	Increasing the consumption of their own products by prohibiting the consumption of products from outside the agro-touristic structure
ACC Takeover room Check in Check out	Creating a special zone for selective collection of waste in agro-touristic structures	Creating a special zone, serving the meal in agro-touristic structures	Optimizing water consumption and electricity by implementing sensors for light and water
HEC Equipment and cleaning solutions Common space cleaned	Acquisition of eco cleaning products	Some alert rules for dirty clothes through posters with environmental protection	Outsourcing services through specialized firms. Traditional cleaning

The alternatives identified will be evaluated using matrix for selecting the alternatives. To apply this tool is necessary to define a number of criteria as below.

The evaluation criteria previously developed alternatives are: a. total cost; b. impact on the problem; c. relation The cost / benefits; d. resistance / impact to change; e. time of implementation; f. the uncertainty about the effectiveness.

To evaluate alternatives to improve relative to these criteria the authors used as an instrument of quality matrix (Table 3) for selecting the alternatives where used the following notations: 3 - very favourable impact; 2 - medium favourable impact; 1 - weak favourable impact.

Table 4. The matrix for selecting the alternatives

Cause	Alternative for improving	Criteria selection						Total
		a	b	c	d	e	f	
MFE Meal preparation Equipment and products Food prepared	Outsourcing services through catering	2	3	2	2	3	1	14
	Removal of the meal by providing the kitchen The tourist will prepare his own meal	3	1	2	2	2	3	13
	Increasing the consumption of their own products by prohibiting the consumption of products from outside the agro-touristic structure	**2**	**3**	**2**	**3**	**3**	**2**	**15**
ACC Takeover room Check in Check out	Creating a special zone for selective collection of waste in agro-touristic structures	2	3	2	2	3	1	14
	Controlling serving the meal in the room by creating a special zone	2	2	3	2	2	1	12
	Optimizing water consumption and electricity by implementing sensors for light and water	**1**	**3**	**3**	**3**	**2**	**1**	**13**
HEC Equipment and cleaning solutions Common space cleaned	Acquisition of eco cleaning products	1	3	2	3	2	1	12
	Some alert rules for dirty clothes through posters with environmental protection	**3**	**2**	**3**	**3**	**2**	**2**	**15**
	Outsourcing services through specialized firms	2	3	2	3	2	1	13

After identifying suitable options, using matrix for selecting the alternatives, a plan of measures will be developed to improve compliance with the requirements of Table 5.

Table 5. Plan of measures to improve compliance

No.	Cause	Corrective action	Resources needed	Responsible	Term
1.	MFE Meal preparation Equipment and products Food prepared	Increasing the consumption of their own products by prohibiting the consumption of products from outside the agro-touristic structure	M: preparing from the own products U: The chef	The owner	**Daily**
2.	ACC Takeover room Check in Check out	Optimizing water consumption and electricity by implementing sensors for light and water	M: sensors for light and water U: the electrician, the plumber	The owner	**2 month**
3.	HEC Equipment and cleaning solutions Common space cleaned	Some alert rules for dirty clothes through posters with environmental protection	M: posters with environmental protection U: the staff	The owner	**1 day**

Legend: M – material resources, U – humane resource

With the control elements we have to ensure that corrective actions planned are used and maintained as required in Table 6, highlighting those items graphed control.

Table 6. Corrective actions planned

Variable	Increasing the consumption of their own products by prohibiting the consumption of products from outside the agro-touristic structure	Optimizing water consumption and electricity by implementing sensors for light and water	Some alert rules for dirty clothes through posters with environmental protection
How to measure	Tracking stock	Tracking consumption	Visual
Were	Storage area	At the electric and water counter	Rooms
Reference	Sheet Storage	The bill	The amount of washing
Who measures	The chef	The owner	The staff
Who decides	The chef	The owner	The owner
What makes	Remake stock	Checking the sensors	Checking the posters
Who checks	The owner	The owner	The staff
Frequency	Dailey	Monthly	Dailey
Were records	Register	Register	Register

3 Conclusions

In order to comply with the international environment requirements, the authors recommend that SMEs in the agro-tourism sector also, to assume the techniques, tools and modern methods of control, monitoring and improvement of environmental aspects. In this context the authors presented in the present paper a conceptual model that is meant to facilitate the purpose defined in the paper title.

Currently, there are numerous studies, research and methodologies that consecrate optimally chosen can contribute to achieving specific environmental objectives. For example, to establish the strategy to improve the waste can use other tools and methodologies from other fields. In this paper the authors attempted to clarify the relationship between the agro-tourism activities and environment. The extent to which succeeded this, it is up to the potential users of the model [27].

References

1. *** http://eurlex.europa.eu/summary/chapter/environment.html?root_default= SUM_1_CODED= 20&locale=en, accessed 6.03.2017
2. M. Schiavon, M. Redivo, G. Antonacci, E.C. Rada, M. Ragazzi, D. Zardi, L. Giovannini, Athmos. Environ. **120** (2015)
3. E.C. Rada, WIT Transact. Ecol. Environ. **191** (2014)
4. E.C. Rada, C. Zatelli, P. Mattolin, WIT Transact. Ecol. Environ. **180** (2014)
5. R.D. Ionescu, M. Ragazzi, L. Battisti, E.C. Rada, G. Ionescu, WIT Transact. Ecol. Environ. **176** (2013)
6. European Union, https://europa.eu/european-union/about-eu/history/1970-1979_en, accessed 6.03.2017
7. *** http://eur-lex.europa.eu/legal-content/EN/TXT/?uri=uriserv%3Axy0026, accessed 04.03.2017
8. European Commission, http://ec.europa.eu/environment/action-programme/, accessed 04.03.2017
9. *** http://legislatie.just.ro/Public/DetaliiDocumentAfis/1725, accessed 04.03.2017
10. *** http://www.gppbest.eu/?page_id=46&lang=ro, accessed 04.03.2017
11. European Union, *Consolidated Versions of the Treaty on European Union and the Treaty on the Functioning of the European Union, Charter of Fundamental Rights of the European Union* (Publications Office of the European Union, p. 132, 2010)
12. ***, https://www.iso.org/iso-14001-environmental-management.html, accessed 04.03.2017
13. European Commission, http://ec.europa.eu/environment/emas/index_en.htm, accessed 04.03.2017
14. European Parliament, M. Juul, *EPRSTourism and European Union – Recent trends and policy developments* (European Parliament, 2015)
15. P. Sharon, C. Hunter, K. Blackstock, Tourism Manag. **31,** 6 (2008)
16. M. Cawley, D.A. Gillmor, Annals of Tourism Res. **35,** 2 (2008)
17. D. Hall, M. Mitchell, L. Roberts, *Tourism and the countryside: Dynamic relationships,* In: Hall D, Mitchell M, Roberts L, editors. *New Directions in Rural Tourism* (Aldershot, 2003)
18. I. Frochot, Tourism Manag. **26,** 3 (2005)
19. European Parliament, http://www.europarl.europa.eu/thinktank/en/document.html?reference=EPRS_BRI(2016)58 1978 , accessed 04.03.2017
20. N. G. McGehee, K. Kyungmi, G.R. Jennings, Tourism Manag., **28,** 1 (2007)
21. C. Tew, C. Barbieri, Tourism Manag. **33,** 1 (2012)
22. N. Polovitz Nickerson, R. J. Black, S. F. McCool, J. Travel. Res. **40,** 1 (2001)
23. G. Veeck, D. Che, A. Veeck, The Professional Geograph. **58,** 3 (2006)

24. L. Mastronardi, V. Giaccio, A. Giannelli, A. Scardera, https://agriregionieuropa.univpm.it/ it/content/article/31/40/ agriturismo-e-sostenibilita-ambientale-primi-risultati-di-unanalisi-aziendale, accessed 6.03.2017
25. C.V. Kifor, C. Oprean, *Ingineria Calității. Imbunatațirea 6 Sigma* (Editura Universitatii "Lucian Blaga" din Sibiu, 2006)
26. M.V. Zerbes, L.I. Cioca, "Quality – Access to Success", **1,** S1 (2017)
27. S. Bungau, R. Suciu, A. Bumbu, G. Cioca, D.M. Tit, J. Environ. Protect. Ecol. **16,** 3 (2015)

DOI: 10.1051/matecconf/201712110004

3D modeling of olive tree and simulating the harvesting forces

Dan Ovidiu Glăvan[1,*], *Theoharis* Babanatsas[1], and *Roxana Mihaela* Babanatis Merce[1]

[1]University "Aurel Vlaicu" of Arad, Faculty of Engineering, 310130, Arad, Romania

Abstract. The paper presents the results of the study regarding the influence of shaking forces on olive tree harvesting systems. Shaking forces can be released through several methods. Important is the end result, namely the shaking force and the cadence of shaking speed. Mechanical and automatic harvesting methods collect more olives than traditional methods but may damage the olive trees. In order to prevent this damage, we need to calculate the necessary shaking force. An original research method is proposed to simulate shaking forces using a 3D olive tree model with Autodesk Inventor software. In the experiments, we use different shaking forces and various shaking speeds. We also use different diameters of the olive tree trunk. We analyze the results from this experiment to determine the optimal shaking force for harvesting olives without damaging the olive tree.

1 Theoretical considerations

Harvesting through manpower requires employing a large number of people for a limited time period, which could lead to social problems in addition to high labor costs [1, 4]. Due to the economic and social importance of olive trees, it is necessary to use an olive-harvesting robot or an olive-harvesting automatic system, thus eliminating a big number of the human operators, who are employed in large numbers for a limited time, and increasing the speed and the volume of work [2, 3, 5, 11]. In order to use an automatic system for harvesting, we must calculate the necessary force of vibration to shake the olive tree [4, 7, 12].

For this we can use a 3D model of an olive tree based on real dimensions. We use the average diameter dimensions, which is 30 cm at a height of 120 cm in the position of the shaking dispositive and the average density of olive tree wood in raw condition is ρw - 1.09 g/cm3 [5, 8, 10].

2 Experimental conditions

Before creating the 3D model, we were traveling in Greece in an olive orchard in region of Halkidiki that was designed to involve planting trees.

* Corresponding author: glavan@fortuna.com.ro

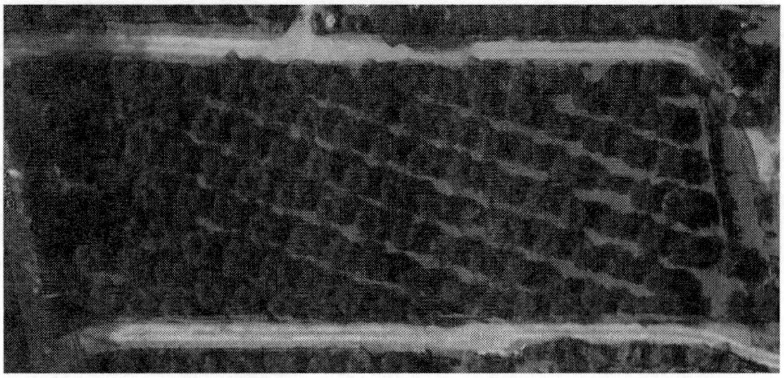

Fig. 1. Olive orchard in region of Halkidiki

Afterwards we took measurements of more than 110 olive trees (height, thickness, perimeter and number of branches). The perimeter was taken at 120 cm from sol where is the position of the shaking system and of the vibration forces. Once we measured more than 110 olive trees we determined the average sizes table 1.

Table 1. Olive tree measurements.

Olive tree	Perimeter of olive tree (cm) at high of 120 cm from sol	Height of olive tree (cm)
1	62	353
2	59	322
3	67	370
4	65	341
5	63	335
6	68	340
...	...	...
112	59	294
113	58	289
114	64	296
Average dimensions	62	335
Minimum dimensions	53	258

We used this data to create a 3D model. We used one 3D model: figure 2, with two sizes, one with the smallest size we found, and one based on dimensional average.

Fig. 2. 3D olive models rendering in Autodesk Inventor.

DOI: 10.1051/matecconf/201712110004

Afterwards, we simulated different vibration forces in Autodesk Inventor analysis study, figure 3. The vibration forces were at a high of 120 cm in the olive tree, in positions of the shaking system [6, 9].

Fig. 3. Autodesk Inventor analysis.

3 Results and discussions

In order to determine the necessary vibration forces we can use certain software such as Inventor or Solidworks, where we can create the 3D model and simulate the forces. In this case, we used Autodesk Inventor with different vibration forces and statically force, table 2.

All results were satisfactory on highest forces without breaking the olive tree.

Table 2. Autodesk Inventor vibration results on olive tree wood.

Parameter	Minimum value	Maximum value
Volume	176715000 mm^3	
Mass	134.303 kg	
Von Mises Stress	0.000176446 MPa	0.776895 MPa
1st Principal Stress	-0.0196544 MPa	0.716821 MPa
3rd Principal Stress	-0.648668 MPa	0.0204877 MPa
Displacement	0 mm	0.593403 mm
Safety Factor	15 ul	15 ul
Stress XX	-0.132117 MPa	0.134893 MPa
Stress XY	-0.0538508 MPa	0.233049 MPa
Stress XZ	-0.102971 MPa	0.0967356 MPa
Stress YY	-0.622679 MPa	0.681943 MPa
Stress YZ	-0.0697603 MPa	0.249304 MPa
Stress ZZ	-0.296653 MPa	0.272747 MPa
X Displacement	-0.000801034 mm	0.314307 mm
Y Displacement	-0.0455565 mm	0.0453073 mm
Z Displacement	-0.00139634 mm	0.501298 mm
Equivalent Strain	0.0000000128804 ul	0.0000634389 ul
1st Principal Strain	-0.0000697494 ul	0.00000219895 ul

3rd Principal Strain	-0.000014206 ul	0.0000145032 ul
Strain XX	-0.000014206 ul	0.0000145032 ul
Strain XY	-0.00000579099 ul	0.0000250616 ul
Strain XZ	-0.0000110733 ul	0.0000104027 ul
Strain YY	-0.0000669544 ul	0.0000733271 ul
Strain YZ	-0.00000750186 ul	0.0000268095 ul
Strain ZZ	-0.0000318973 ul	0.0000293271 ul

The harvest forces taken into consideration in the present paper were validated by the virtual model simulated in Autodesk Inventor, results being part of a previous paper already presented in another Conference.

The researches will continue by creating a real model of harvesting device with adjustable forces that will confirm or not our presumptions and aloud us to establish closer values for forces closer with the real situation.

References

1. Debojit, Biswas Jain Hitesh, K. Arora Manoj, R. Balasubramanian, International Journal of Earth Sciences and Engineering, **4,** 6 (2011)
2. M.Z. El Attar, M.N. El Awady, M. Rashwan, M.A.I. Genaidy, Proceedings of the 10th MSAE Conference (2004)
3. D. Mnerie, D. Tucu, A. Golimba, G. Anghel, Proceedings of the 36th International Symposium "Actual tasks on agricultural engineering", Opatija, Croatia, 617, **36** (2008)
4. C. Peri, *Olive handling, storage and transportation* (Wiley-Blackwell, 2014)
5. C. Peri, *The extra virgin Olive Oil Handbook* (Wiley-Blackwell, 2014)
6. D. Tucu, Proceedings of the 39th International Symposium "Actual tasks on agricultural engineering", Opatija, Croatia, 171, **39** (2011)
7. D. Tucu, Proceedings of the 40th International Symposium "Actual tasks on agricultural engineering", Opatija, Croatia, 549, **40** (2012)
8. D. Tucu, Proceedings of the 40th International Symposium "Actual tasks on agricultural engineering", Opatija, Croatia, 405, **42** (2014)
9. D. Tucu, A. Filipovici, Proceedings of the 40th International Symposium "Actual tasks on agricultural engineering", Opatija, Croatia, 397, **42** (2014)
10. D. Tucu, A. Golimba, D. Mnerie, Proceedings of the 40th International Symposium "Actual tasks on agricultural engineering", Opatija, Croatia, 447, **38** (2010)
11. D. Tucu, A. Golimba, T. Slavici, Proceedings of the 40th International Symposium "Actual tasks on agricultural engineering", Opatija, Croatia, 455, **38** (2010)
12. K. Wang, J. Yang, G. Shi, Q. Wang, Proceedings of the Fourth International Conference on Natural Computation, ICNC '08, Jinan, China (2008)

DOI: 10.1051/matecconf/201712110005

Reducing the environmental impact of a gas operated cogeneration installation

Sabin Ioan Irimie[1], *Sorin Mihai* Radu[1], *Dan Codruț* Petrilean[1,*], and *Andrei* Andras[1]

[1]Departament of Mechanical Engineering, Industrial Engineering and Transportation, 332006, University of Petrosani, Romania

Abstract. The paper deals with a case study for carried out in order to justify the installation of a gas cogeneration system fitted with an internal combustion engine, system used to supply electricity and heat to the users of an energy user which exceeds 1000 toe/year. 2 scenarios were therefore compared based on the energetic analysis:

-the actual scenario in which energy is supplied from the National Energetic System.

- the scenario in which a combined heat and power plant is used.

The amount of energy saved yearly, the specific fuel consumption and the environmental impact were determined by the comparative study. The diagrams representing the variation of the performance indicators according to the operation period were also created. The usefulness of the paper consists in the creation of the yearly optimum installation operation time chart.

1 Introduction

Cogeneration defines those processes in which the useful electric energy (or mechanical energy) and the thermal energy are simultaneously produced through fuel burning. Cogeneration systems are efficient alternatives to central heating power plants and traditional central heating plants. Cogeneration is a completely developed technology which has been widely spread and introduced in the industrial sector. For several years now, it has been widely used by the services sector due to its undeniable advantages.

Cogeneration is a potential source used for the efficient generation of electricity as long as they are allowed by the thermal industrial consumption and centralised heating. Nevertheless, cogeneration is the only way of combined production of electricity and heat using fossil fuels considering a general efficiency of over 80%.

Compared to the separate heat production (in central heating plants, CHP) and electricity (in central heating power plants, CHPP), the fuel economy resulted from the combined production of heat and electricity may reach 32-34% (Figure 1).

The fuel economy may be observed in a reduced invoice received by final beneficiaries of heat and electricity. Considering as well the case in which modern cogeneration

* Corresponding author: dcpetrilean@yahoo.com

competes with the new and advanced electricity generation technologies (such as, steam-gas combined installations, with efficiencies comprised between 55-58%), the realised economy is comprised between 15-18%. It is estimated that at the present fuel price the proportion of fuel expenses represent 50-80% of the cost of supplied energy, being therefore the most important cost in the proportion of electricity and central heating.

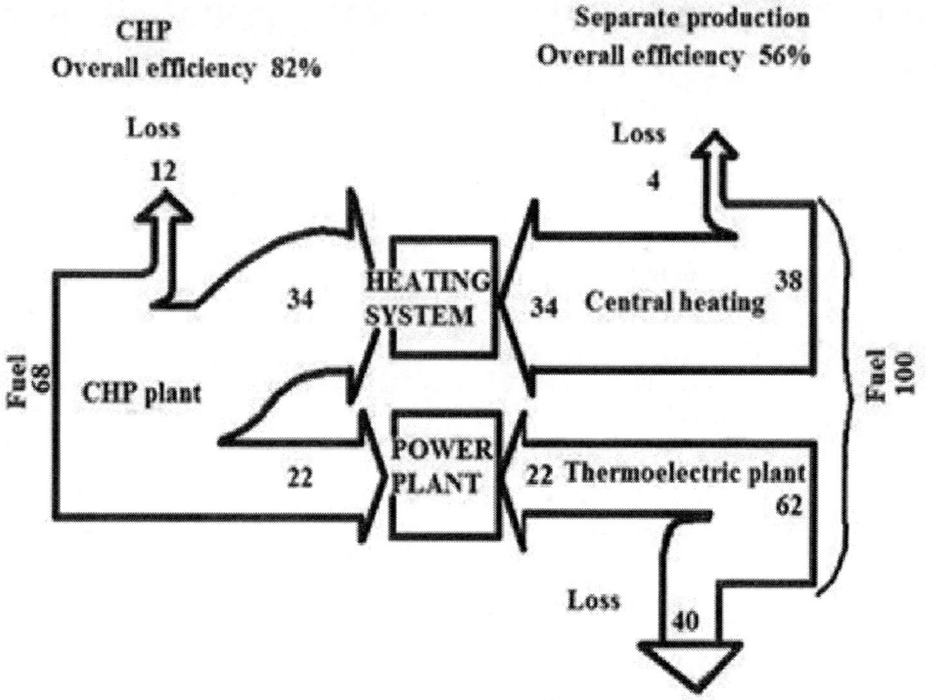

Fig. 1. Comparison between energetic and fuel fluxes considering separate electricity and heat production and respectively cogeneration

The emission of greenhouse gases into the atmosphere represents an important cause of global warming and climate change (especially CO2), the energetic sector being therefore the main contributor (roughly around 80%). The polluting emissions per useful energetic unit may be significantly reduced through cogeneration, corresponding therefore to the economy of unused fuel, compared to the separate production of heat and electricity.

The ecologic, social as well as economic advantages of cogeneration are materialised through: flexibility in using fuels, contributing therefore to the stability of the price of heat, the creation of jobs for the local labour market, avoiding transport costs for electricity as cogeneration plants fit perfectly within the modern concept of electricity production and distribution, increasing therefore the safety in supplying electricity locally and reducing the risk of breakdowns which might be caused by a major breakdown of the national electricity supply system. [1].

The international importance of cogeneration has fluctuated during the past years: it has appeared as a practical technology during the 1980s developing rapidly (58% of the electricity generated in the USA by different industries was produced by cogeneration plants) following a decreasing trend around the 1950s (15% of the total generated potential) continuing its course reaching 5% in 1974. The increase of fuel prices, environmental pollution problems, and the focus of energy policies on efficiency determined an increasing trend both on a national as well as on an international scale.

The speciality literature found in the field follows the fluctuations of the technologic trend as nowadays there are papers which deal with theoretical, methodological as well as instrumental aspects regarding its exploitation [2 - 5] as well as papers with an increased applicative character [6 - 8].

The purpose of the present paper consists in creating a basis, focusing on a case study, for to the necessary steps required in order to justify the use of a gas cogeneration installation with a combustion engine destined for the supply with electricity and heat to the users of a certain company. The usefulness of the paper is justified by future financing plans which focus upon the increase of energetic efficiency and the reduction of the environmental impact through cogeneration. The method used consists of comparing the quantity and quality performance parameters for the 2 scenarios, namely the classical thermal energy capitalisation scenario and the cogeneration one. A decision concerning the installation of a cogeneration unit with a gas operated combustion motor may be taken based on the comparison and on the energy usage chronogram. Nevertheless, he should be skilled enough to make the difference between "what is happening?" and what "should happen?" in the work environment [9]. An energy usage chronogram shall be created as in [10].

2 Case study. Presentation

The analysed system makes reference to a company which is supplied with natural gas in order to ensure its demand for technological steam and hot water used for heating and consumption. In order to render the energetic source more efficient, to reduce the expenses as well as the environmental impact, the acquisition and installation of a cogeneration unit with a combustion engine is therefore proposed. (Figure 2).

Arguments for choosing the type of installation: The main component of the installation is a diesel engine which operates an electricity generator. The electric power of the unit is comprised between 25 and 5000 kW, the fuel used being gasoil. 70% of the heat from the burnt exhaust gases (at 500°C) can be recovered by cooling them down to 200°C. The heat of the cooling water and that of the lubrication oil (at approximately 100°C) may be fully recovered. The heat delivered to the consumers consists in hot water and steam.

The main indicators characteristic to the energetic balance of a combustion engine system are presented in Table 1:

Table 1. The characteristic indicators of the energetic balance of a combustion engine system [11]

	Otto	**Diesel**
Fuel consumption	100%	100%
Electricity production (α_e)	35%	40%
Heat production (α_t)	50%	43%
where: burnt gases	20%	21%
cooling water	30%	22%
Total efficiency	85%	83%
Central heating index -y-(cogeneration), [J_e/J_t], [kW_e/kW_t] [kWh/Gcal]	0.6...0.93 700... 1082	

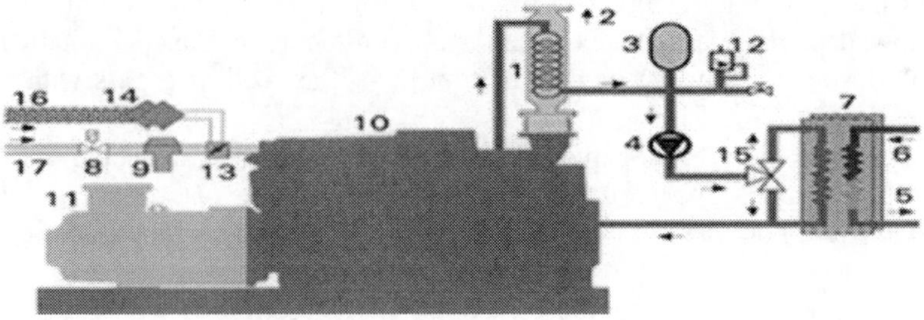

Fig. 2. Basic diagram of the cogeneration installation using an electricity generator group with a combustion engine, 1 – water – burnt gases heat exchanger; 2 – exhaust pipe; 3 – expansion tank; 4 – pump; 5 – hot water outlet; 6 – cold water inlet; 7 – water – water heat exchanger; 8 – closing valve on the gas fuel circuit; 9 – pressure regulator; 10 – internal combustion motor; 11 – electricity generator; 12 – safety valve; 13 – carburettor; 14 – air filter; 15 – thermostat; 16 – combustion air inlet; 17 – gas inlet.

The main economic indicators of different types of installations with cogeneration (internal combustion engine – CE, gas turbine – GT, steam turbine- ST) are presented in Table 2:

Table 2. The main economic indicators of different types of cogeneration installations

	CE	GT	ST
Specific investment, [EURO/kW$_e$]	1500...400	1200...530	1500...1000
Specific maintenance cost, [EURO/MWh$_e$]	20...5	7...4	2.3...1.5
Investment recovery period, [years]	2...3	2...3	3.5...4
Life span, [years]	10	15	30

The main principles considered while conceiving the solutions: dimensioning and operating the cogeneration is determined by the thermal criteria, in order to be cost efficient (with energetic economies sufficient enough to compensate the investment effort), they need to be used as more as possible considering as well higher loads (an operation span of over 4000 h/year may be considered as one of the basic rules which need to be respected).

The energetic performances of the gas operated electricity cogeneration group are presented in Table 3:

Table 3. Energetic performances of the gas operated electricity cogeneration group

Measure	Value
Gas thermal power	1446 kW
Gas flow	137 m^3_N/h
Mechanical power at the shaft of the motor	620 kW
Electric power of the electricity generator	600 kW$_e$
The available thermal power of the steam generator	348 kW$_t$
The available thermal power for heating and hot water	320 kW$_t$
Electric efficiency	41.5 %
Thermal efficiency	46.2 %
Cogeneration efficiency	87.7 %
Cogeneration index	0.898 kW$_e$/kW$_t$

The obtained results are brought forward following the realisation of an hourly energetic balance:

- Used energy	1446 kWh – 100 %
- Useful energy	1268 kWh – 87.69 %
- Losses in the electricity generator	20 kWh – 1.383%
- Losses with the exhausted burnt gases	94 kWh – 6.501 %
- Losses with the reduced potential water	41 kWh – 2.835 %
- Radiation losses (engine, generator)	23 kWh – 1.591 %
- Total lost energy	178 kWh – 12.31 %

Calculation data

A. ACTUAL SITUATION

- Natural gas consumption for the thermal generators (steam and hot water) $C_{gas} = 174\ m^3_N = 1839\ kWh_t$;
- Thermic steam flow 930 kWh_t;
- Thermic hot water flow 730 kWh_t;
- Total thermic flow of the thermal generator 1660 kWh_t.

B. THE SCENARIO CONSIDERING THE USE OF A COGENERATION ELECTRICITY GENERATOR GROUP OF 600 kW_e

- Natural gas consumption for the electricity generator 137 $m^3_N = 1446\ kWh_t$;
- Electricity 600 kWh_e
- Thermic steam flow 348 kWh_t;
- Thermic hot water flow 320 kWh_t;
- Total thermic flow electricity generator group 668 kWh_t;

Additional calculation data

- The yearly operation period may be estimated within the interval: $\tau_{an} = 4000 \div 8760$ hours/year
- The price for one kWh of electric energy 0.304 lei/kWh [12]
- The price for one kWh thermal energy with gas 0.128 lei/kWh
- The price of one m^3 of gas 1.35 lei/ m^3
- The energetic equivalent of one m^3 of gas 10.57 kWh
- Tonne of oil equivalent 1 t.e.p. = 11.627907 MWh
- Average value of the factor of CO_2 emissions for gas fuels in Romania $f_{CO2} = 0.535$ kg/kWh
- Investment cost 2 700 000 lei
- Maintenance cost 30 lei/hour.

3 Results

Based on the herein mentioned results, calculations for different operation periods during a year and for different loading degrees were carried out, obtaining the following performance parameters: the yearly economic actual energy, the yearly effective economy considering the cogeneration scenario, the investment recovery period and the reduction of the environmental impact.

In order to give an example, the operation period was set within the following range 4000-8660 hours / year, while the required load by the users considered 3 situations: 100% (668 kWh), 80% (534.4) kWh and 50% (334 kWh).

DOI: 10.1051/matecconf/201712110005

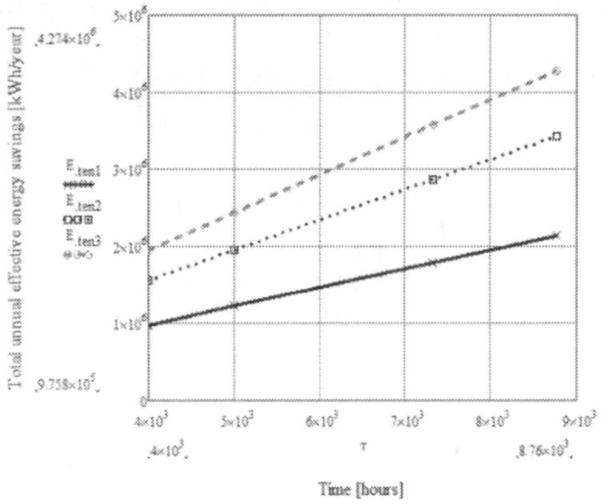

Fig. 3. The scenario in which the yearly actual economy depends on the yearly operation period, corresponding to the loads: 100%, (668 kWh), corresponding to the loads 100% (668 kWh), 80% (534.4) kWh and 50% (334 kWh)

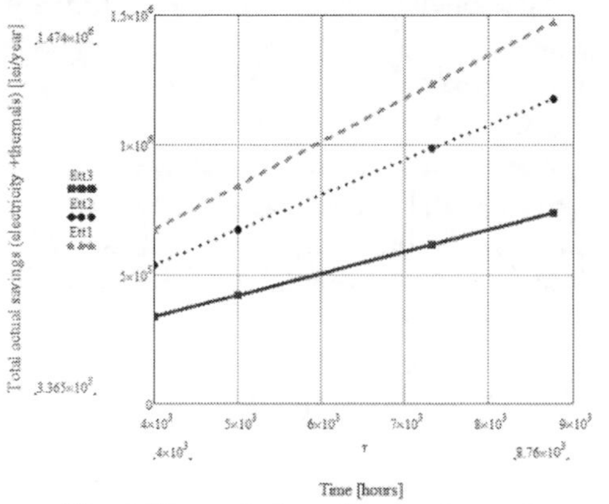

Fig. 4. The variation of the total actual electric and thermal energy economies depending on the yearly operation period

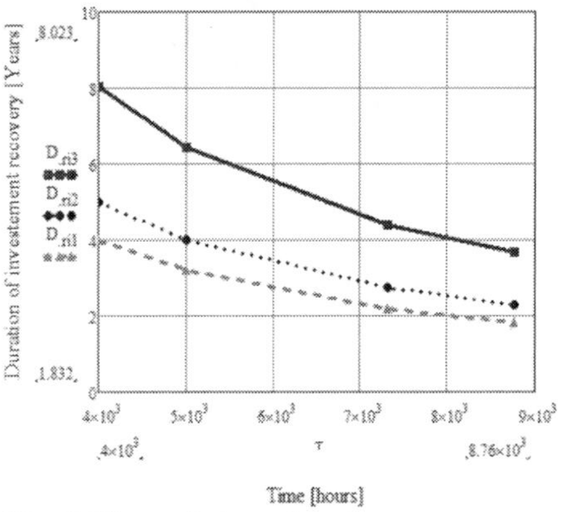

Fig. 5. The variation of the investment recovery period depending on the yearly operation period corresponding to the loads 100% (668 kWh), 80% (534.4) kWh and 50% (334 kWh)

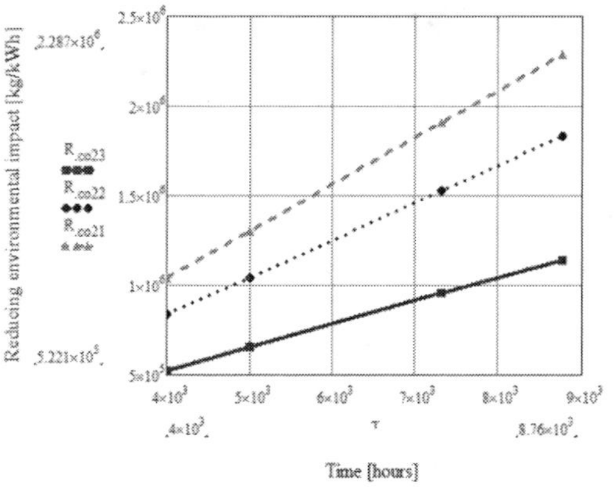

Fig. 6. Environmental impact reduction depending on the yearly operation period corresponding to the loads 100% (668 kWh), 80% (534.4) kWh and 50% (334 kWh)

The previously presented diagrams highlight the energetic, economic and environmental performances depending on the yearly operation period and the loading demand of the final beneficiaries, namely the users.

Figures 3-6, highlight the fact that the optimum performances are obtained for larger operation periods. Considering Figures 3, 4 respectively 6, the values obtained for an 80% load may be increased considering the increase in efficiency for the said load.

Considering Figure 5, a correction may be introduced according to which, towards the maximum limit of the operation period, the investment recovery might become more favourable due to the increase of the loading efficiency, namely that of 80%.

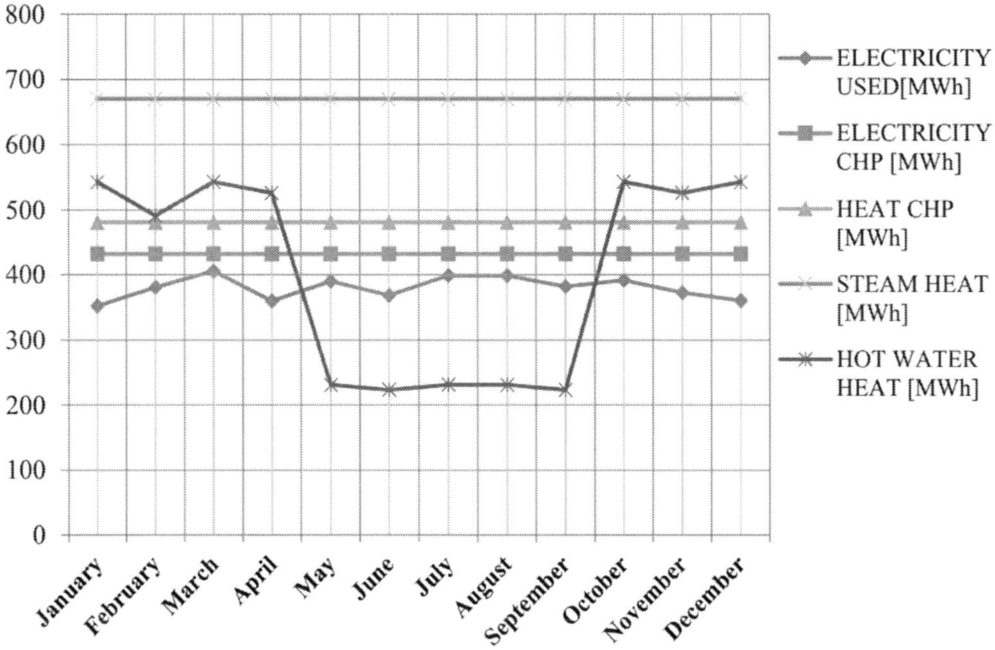

Fig. 7. Energy usage chronogram

4 Conclusions

The energy usage chronogram for the case dealt within Figure 7 highlights the following:
- the available electricity at the cogeneration installation fully covers the demand of the company;
- during the cold season, the thermal energy available at the cogeneration installation fully covers the hot water demand, the steam demand being covered by the steam generator existent within the company;
- during the warm season there is a thermal energy surplus which can be capitalised through the water supplying the steam generators or through the use of a recovery steam generator.

The solutions mentioned capitalise the cogeneration advantages determined by the quasi-continuous yearly operation.

The solution used to render the cogeneration group more efficient is proposed considering the energy usage chronogram.

References

1. A. Leca, C. Cremenescu, *Serviciul public de încălzire centralizată. Propuneri de eficientizare*, (AGIR Publishing, 2008)
2. V. Athanasovici, *Tratat de inginerie termică-Alimentări cu căldură - Cogenerare* (AGIR Publishing, 2010)
3. B. Kolanoswski, *Small-scale Cogeneration handbook* (Lulu Press, Inc, 2013)
4. J. Orlando, *Cogeneration Design Guide* (American Society of Heating, Refrigerating and Air-Conditioning Engineers, Inc., 1996)
5. T. Lipper, H. Setetter, P. Krzyslak, Proceedings of the Xth Int. Conf. "Steam and Gas Turbines for Power and Cogeneration Plants", Karlovy Vary, Czech Republic, 178, **1** (1994)
6. *** *Studiu de fezabilitate pentru centrala termică de cogenerare pe bază de biomasă lemnoasă* (National Research Institute for the Development of Electric Engineering ICPE-CA, 2010)

7. *** http://www.sistemecogenerare.ro/continut/tehnologii-cogenerare, Accessed 14.03.2017
8. *** *Cogenerarea de mica si medie capacitate OPET* (Organization for the Promotion of Energy Technologies, 2002).
9. R. I. Moraru, G. B., Babut, L. I. Cioca, Environmental Engineering and Management J., **13**, 6 (2014).
10. D.C. Petrilean, S.I. Irimie, J. of Environmental Protection and Ecology, **16**, 3 (2015).
11. D. C. Petrilean, S. I. Irimie, J. of Environmental Protection and Ecology, **17**, 4 (2016).
12. V. Pasculescu, S. M. Radu, D. Pasculescu, T. Niculescu, Proceedings of the 13th SGEM GeoConference on Science and Technologies in Geology, Exploration and Mining, Albena, Bulgaria, 417, **1** (2013).

Environmental assessment of the Sibiu County, Romania: proposal for sewage sludge and OFMSW management

Gloria Trombin[1], *Marco* Ragazzi[1], *Claudiu* Isarie[2], *Rodica* Ciudin[2], and *Vincenzo* Torretta[3,*]

[1]Department of Civil and Environmental Engineering, University of Trento, Trento, Italy
[2] Department of Industrial Engineering and Management, Lucian Blaga University of Sibiu, Sibiu, Romania
[3]Department of Theoretical and Applied Sciences, Insubria University of Varese, Varese, Italy

Abstract. The aim of the work is to assess current solid waste management (SWM) and wastewater treatment (WWT) in the county of Sibiu, Romania. Sibiu is a region where industrialization and tourism are spreading constantly, while pollution monitoring and the introduction of circular economy (CE) principles are still lacking. The environmental issues are mostly due to the absence of reliable sanitation systems, in particular for what concerns sewage sludge treatment. Moreover, Organic fraction from selective collection of municipal solid waste is still missing, since it is disposed of directly to landfill, increasing environmental pollution and contributing considerably to global warming. As a result, SWM and WWT should be improved at the same time, as sludge requires specific handlings in order to be safety disposed. The objective of the study is to investigate the introduction of affordable technological facilities able to treat sludge and the organic fraction of municipal solid waste, recovering, at least, the energy required for the working phase of the treatment plants. Anaerobic digestion resulted viable, but the organic fraction full stream collection seems to be the only way to have a significant energy recovery. From the financial point of view, the feasibility of this co-treatment option could count on specific opportunities that Romania has in the frame of the European Union context.

1 Introduction

The declared aim of the EU Sustainable Development Strategy is to work towards improvements of the life quality, both for the present and the future [1]. Despite the progress in the recent years, it is a fact that Romania's economy still relies on intensive consumption of resources. The society and the administrations have yet to develop a shared vision, and so the natural capital faces the risk of damages that may become irreversible [2].

The relentless deterioration of natural resources, which affect many emerging countries due to economic improvements boosted by national policies [3], and the need of their conservation requires a good waste and wastewater management, an increasing recovery of wastes and their reduction [4,5,6]. In this sense, according to European regulations, the waste

* Corresponding author: vincenzo.torretta@uninsubria.it

hierarchy of prevention, re-use, recycling, recovery for other purposes such as energy and disposal should be applied [7].

Wastewater and solid waste are to be considered as new sources of energy and materials that can be used in a circular economy perspective, also for developing and rural regions [6,8, 9]. European directives transposed into Romanian legislation led to a new approach to the waste problem, focusing on the need of natural resources protection, costs management reduction and effective solutions to reduce pollution [10-12]. The general framework waste regime in Romania stated on the *Legea nr. 211/2011* on waste regime [13].

For the creation, organization and operation in the common interest of sanitation services of a national waste management system, Romania formed administrative-territorial units called Asociatia de Dezvoltare Intercomunitaria (ADI) [14], and developed waste management plans at national, regional and county level.

In the case study the focus is on municipal solid waste (MSW), which is generated by: households, commercial activities and other sources whose activities are similar to those of households and companies. The current management of urban waste generates negative impact on the environment both in the presence of household waste and industrial waste, since most of the MSW is disposed of in landfills. Landfills are a source of air pollution (emissions of methane), surface water pollution and degradation of soil quality through the percolation of hazardous liquids that cause low soil fertility, also at great distances [15-18].

The main problems identified in the environmental assessment of Sibiu consist of:

- waste management, with focus on Organic Fraction of MSW (OFMSW);
- sludge management.

Nowadays in the city of Sibiu the OFMSW is not recovered in any way, and it is instead disposed of in landfills, management which should be avoid, as stated in the Council Directive 1999/31/EC. However some proposals where put forward [19].

In addition to that, as reported in the next paragraphs, the sludge from the wastewater treatment station are just stored in the area adjacent the WWTP and not recovered.

For all these reasons, it is necessary to find another way to manage the OFMSW of the city of Sibiu and the WWTP sludge. Thus, the aim of this work is to offer practical (and replicable) proposals to improve urban waste management, focusing in particular on its organic fraction, integrating it with the appropriate sewage sludge management, in order to reduce the environmental impacts and to recover resources and energy. This integrated approach is considered an opportunity both as national strategy and at local level [20,21].

2 Materials and methods

Sibiu is located in the southern part of Transylvania, and extends for an area of 5,432.5 km^2 (2.3% of the country), with a population of about 425,000 inhabitants, 66% of which are urban citizens and the others live in rural areas. The capital of the county, Sibiu, has about 148,000 inhabitants. The status quo of the economy of the Sibiu County is analysed in the *Annual report of the State of the Environment*, which is published every year [22]. At the end of 2012, the county could count 10,031 active companies.

The main activities in this area are trade, manufacture, scientific profession, agriculture and livestock. Sibiu has also a significant agricultural potential thanks to the total of about 117,000 hectares of arable land, and about 108,000 hectares of pasture land [7]. The forest department in Sibiu also plays an important role, with an area of about 112,000 ha of land. The GDP/per capita is 6,837 €, higher than the national one (5,454 €). The unemployment rate is low and around 5.4% with an average monthly salary of just over 300 €. There is an environmental monitoring network divided by different areas: air quality, drinking water, wastewater and waste, which are included in the annual report [2, 22].

In 2014, the County had the following functional facilities for MSW management:

- 3 transfer and sorting stations for household waste
- 1 sorting station for household waste
- 1 transfer and compost station
- 1 station sorting and composting of household waste
- 1 composting station

Table 1 illustrates the situation of household waste collected separated or mixed. The waste was landfilled in the Sibiu landfill. Selective collection (SC) is one of the key steps of a modern waste management, in order to transform what is recyclable into useful products. In the Sibiu County, SC was extended with the implementation of five waste management projects financed by PHARE CES 2003-2005 [23]. Following the closure of the non-compliant landfills in Sibiu, an extensive awareness campaign was conducted in order to connect the sanitation services operator to the households, especially in the rural areas. For biodegradable waste, awareness activities in schools were also organized, distributed individual compost units, in order to decrease the amount of waste sent to the landfills.

As a matter of a fact, in 2016, in the area of the city of Sibiu, there are no bins for SC. The whole waste is collected together mixed in just one type of bin. There are two companies in charge of collecting waste. The citizens can choose to make use of the service of either company for their waste. Most of the projects about SC started in 2014 lasted for only few months, because of the arising of technical and political problems. Nowadays, since 2016, SC is in a law impasse after the awarding of the contract to a firm. Another company challenged the result of the competitive procedure and now it is currently awaiting trial.

Table 1. Household waste (collected separately and mixed) (2008-2012).

	Year				
	2008	**2009**	**2010**	**2011**	**2012**
Household waste [ton]	90,242	128,271	100,636	85,421	87,715
MSW Collected separately	1,795	1,715	2,940	4,481	3,901
MSW Recycled	1,463	1,715	3,275	2,869	2,972
Biodegradable MSW	65,012	46,661	49,468	38,253	40,146
Level of connection with sanitation services [%]					
Sibiu County	61.82	79.88	77.01	78.24	71.18
Urban area	77.83	78.42	77.91	84.00	73.05
Rural area	29.01	82.77	75.18	67.35	67.47

The wastewater treatment station has a capacity of 225,000 Equivalent Inhabitants (EI), but nowadays it serves 180,000 EI. It is a conventional activated sludge treatment plant, with processing of sludge with thickening and anaerobic digestion. Currently the sludge from the wastewater treatment plant in Sibiu is disposed of on the left bank of the Cibin river. Here the natural drying of the sludge takes place. The total area of the deposit is approximately 2 hectares, divided in 1.35 ha of the deposit itself and 0.65 ha of access and perimeter roads [24]. The daily quantity of sludge collected by the station is shown in the Table 2.

Table 2. Flow and moisture of the sludge [19]

Parameter/Stage		**Value**
Daily sludge flow	[m^3/day]	180
Hourly sludge flow	[m^3/hour]	8.7
Final polymerization step	[H_2O content %]	75.0

The sludge is left on the area near the station for an unstated time and presently there is not a plan to recover them. In 2016, the managers of the WWTP claim that this is not an imminent problem. The main reason is the lack of capital that they would need in order to launch a new project. In Table 3 the characteristics of the digester are reported.

Table 3. Characteristics of the digester.

Parameter		**Value**
B_r: Maximum organic loading rate	kgvvs $m^{-3}day^{-1}$	4
V_f: Available digester volume	m^3	900
T_f: Required hydraulic retention time	day	40
Q_{ks}: Daily amount of digester sludge	m^3/day	15

The proposal of this study is to recover the OFMSW and solve the problem of the sludge management using the digester as a co-digestion unit to treat these two kinds of waste, as suggested by other studies [4, 25-28].

Potentially, this would allow producing the required biogas for the thermal drier in which the waste would be modified in order to make the resulting material appealing for a co-combustion in cement kilns. A flowchart of the process is shown in Figure 1 [29]. Two different operative scenarios have been selected: case A and B:

- **Case A:** the starting idea is to collect the OFMSW from the city centre of Sibiu, starting from the markets of vegetable and fruit, and then from the restaurants and bars. The choice of these places where to begin the collection is based on their convenience. In fact the markets' waste is mainly OFMSW, and the two markets produce big quantities of it. The two fruit and vegetable markets are Cibin and Rahova. They are in the area around the city center. Other than the markets, the OFMSW, can be easily collected from the pubs, bars and restaurant.
- **Case B:** it is taken into account the possibility to improve SC of the OFMSW from civil residences. The waste production per capita is of 0.83 kg inh^{-1} d^{-1}, where the weight percentage of the organic fraction is equal to 70% and a moisture of 70%. From the calculation, it results a total quantity of OFMSW equal to 86 t/d.

Presently, in the historic center of Sibiu 93 pubs/bars/restaurants of medium small dimension, and 2 big ones are located. Each of the smaller services is supplied with 2 bins of a 120 l volume in which the OFMSW is gathered. All of these bins are then collected weekly by the waste company in charge.

The same happens for the two big restaurants, which have five bins, because of their intense OFMSW production. The compactness of the waste in these bins can be estimated as the one in the markets' bins, and so of around 300 - 350 kg/m^3.

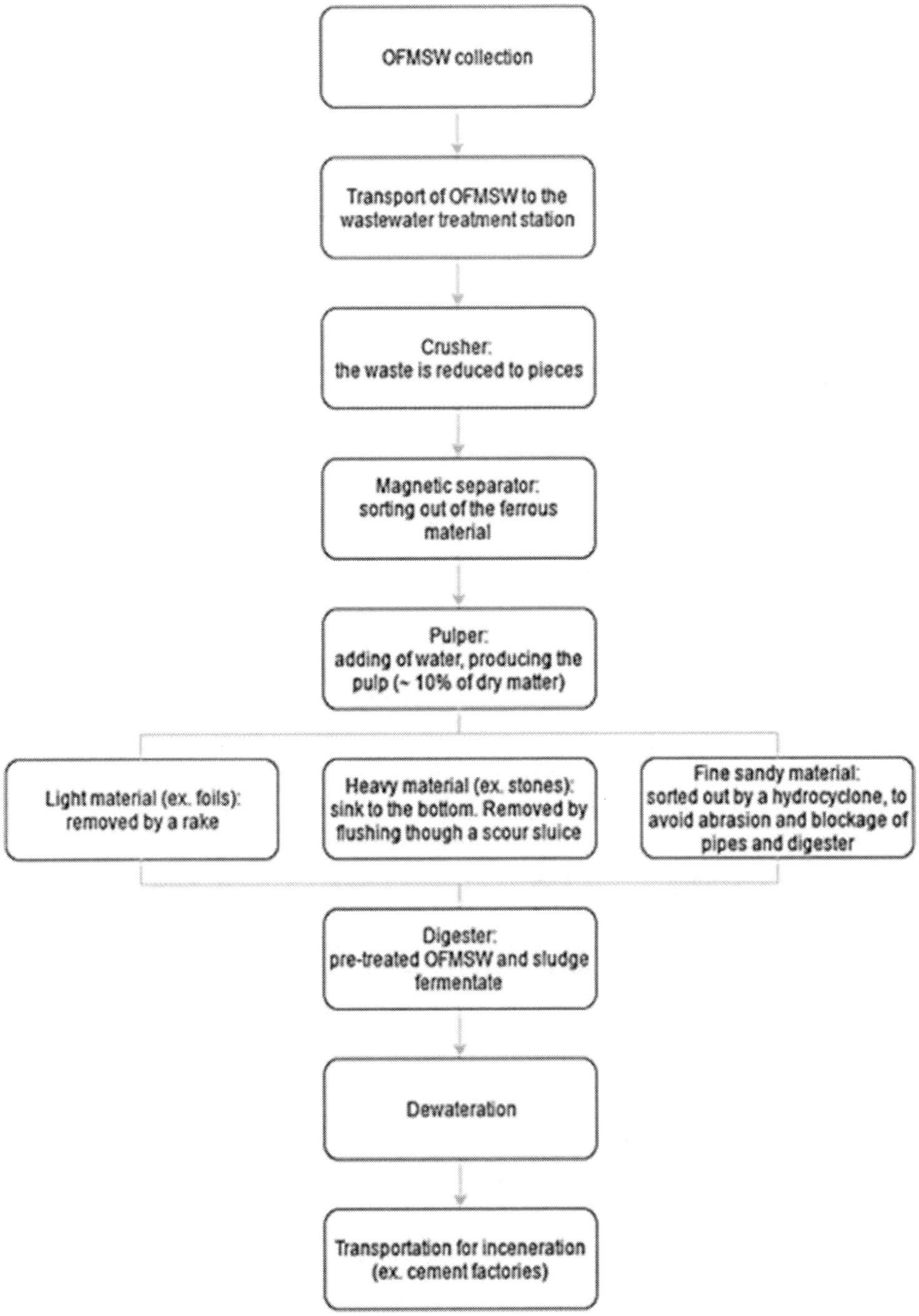

Fig. 1. Flowchart of the co-digestion and drying of the waste.

The total quantity of OFMSW that could be collected from the historical center of Sibiu is about 17 ton/month, considering the lowest compactness, and circa 20 ton/month, considering the highest. This estimation is done calculating only the 60% of the bin as full, doing an average throughout the year. Adding the fraction from the markets and the one from the pubs/bars/restaurants the total quantity of OFMSW per month is between 40 and 53 ton/month. The data referring to the Case A are reported in Table 4.

Table 4. Data related to the two markets, bars and restaurants.

Source	Nr of bins	Capacity m^3	Volume m^3/month	Density kg/m^3		Weight ton/month	
Cibin Market	-	-	1.1	300	350	20	
Rahova Market	-	-	1.1	300	350	20	
TOTAL FROM MARKETS						33	
Bar – Restaurants (B-R)							
B-R of big dimesion	5	0.12	2.1	300	350	0.43	0.50
B-R of small dimesion	2	0.12	0.96	300	350	0.17	0.20
TOTAL FROM B-R						40	53

3 Results

The current situation in Sibiu shows no critical issues in the regards of air quality and it does not prevent an additional combustion of biogas. The current output from the wastewater treatment plant does not present any issue, since there is a good pollutants removal. Of great importance is the analysis of the nitrogen pollutants in the wastewater output from the station, because the addition of a thermal drying increases the load of nitrogen to the system because of the condensation of moist air as an output of the dryer. From the drinking water side, there are no critical environmental issues; in fact the study shows a total separation of the sector from the purification of waste water due to the type of sources used. As a conclusion, there are no critical environmental issues that would stop the construction of the new system.

Moving to the results connected to the proposal, Table 5 reports the amount of OFMSW in the two scenarios, to be treated in co-digestion with sludge, and its characteristics.

Table 5. OFMSW in the two scenarios.

		Case A	Case B
Sludge:			
EI served by the WWTS	EI	180,000	180,000
Sludge production	grs/inhab d	100	100
Sludge density	g/l	1,000	1,000
Sludge moisture		97%	97%
Daily sludge input	t/d	180	180
OFMSW:			
Daily urban waste production	kg inh^{-1} d^{-1}		0.83
OFMSW percentage in the UW			70%
Residents in Sibiu	inh		147,245
Daily OFMSW production	t_{OFMSW}/d	1.77	87.32

		Case A	Case B
OFMSW moisture		70%	70%
Total quantities in input:			
VS in input	t_{VS}/d	4.28	28.67
NVS in input	t_{NVS}/d	1.65	2.93
TS in input	t_{TS}/d	5.93	31.59
Water in input	t_{H2O}/d	175.84	235.72
Biogas conversion:		50% of VS	50% of VS
Total quantities in output:			
VS in output	t_{VS}/d	2.14	14.33
NVS in output	t_{NVS}/d	1.65	2.93
TS in output	t_{TS}/d	3.79	17.26
Water in output	t_{H2O}/d	175.84	235.72
Moisture in output		98%	93%

The following tables show the comparison between the current dehydration and the target for the sludge, if thermally dried, the additional production of biogas by co-digestion and the energy balances, for the additional stream of OFMSW.

Table 6. Comparison between the current dehydration and the target level.

		Case A	Case B
Mechanical dehydratation :TS in output	-	25%	25%
Mechanical dehydratation :Substance quantity in output	t/d	15.15	67.94
Dryer: Target Moisture of the output material		10%	10%
Dryer: Water in input	t/d	11.36	50.95
Dryer: Water in output	t/d	0.42	1.89
Dryer: Water to be evaporated	t/d	10.94	49.06
	kg/h	455.99	2,044.37

Table 7. Addional production of biogas by co-digestion.

		Case A	Case B
Biogas yield	Nm^3/t_{OFMSW}	150	150
Daily biogas quantity	Nm^3_{biogas}/d	265	12,832
CH_4 quantity in biogas	-	60%	60%
CH_4 quantity	Nm^3_{CH4}/t_{OFMSW}	90	90
Daily CH_4 quantity	Nm^3_{CH4}/d	159	7,699

Table 8. Dryer energy demand [30] and the energy production in the co-generator engine (CHP).

		Case A	Case B
Dryer energy demand	kWh/kg_{H2O}	kWh/d	
Electric energy demand	0.03 – 0.5	328-1,642	1,472-7,360
Thermal energy demand	0.8-0.9	8,755-9,849	39,252-44,158
Energy production in the CHP	kWh/Nm^3_{biogas}	kWh/d	
Electric energy production	1.8-2	477-530	23,098-25,665
Thermal energy production	2-3	530-795	25,665-38,497

The demand of thermal energy is greater than its production in co-generation. On the other hand, the production of electric energy covers the demand. For the lack of thermal energy, it is necessary to do a partial co-generation using only a fraction of the biogas generated from the co-digestion (supposing that the role of the present sludge stream is neutral). The remaining part of the biogas is thus used to generate thermal energy using a heater. An alternative could be the drying of only a percentage of sludge. Typically, the final solid content in the sector is 90%, but, through a partial bypass of the drier, it is possible to modulate the final solid content down to a target level suitable for balancing the thermal energy demand. This must be done taking into account the final destination of the output. Co-digestion of OFMSW and sludge could suffer from a bad quality of sludge in terms of heavy metal content, if a biological destination would be planned. The alternative could be a combustion in industrial burners as the ones of cement factories. In this case the problem of heavy metal presence in the sludge could be less stringent, but the case of partial drying should be deeply analysed as the lower heating value of the dried output could be no longer compatible with an industrial furnace.

4 Conclusions

The improvement of the waste management systems and sewage sludge can be obtained through an assessment of the current environmental loads and the possibility that the environment has to support other treatment plants. A decisive step in the implementation of sustainability principles is the application of Circular Economy. In this frame, the current situation assessment on the waste and sewage sludge from WWTP management in a developed area in Romania has been described and the main environmental pressures were analyzed in order to find out any criticality.

This was followed by the setting of a proposal to adjust the current management system of waste, with a focus on organic waste and sewage sludge. The feasibility of interventions is affected both by the environmental and financial feasibility. After the environmental assessment no critical issues were found. Regarding the financial aspect of the project, it is linked to the availability of funding opportunities that in Romania is much easier than in other countries of the European Community (e.g.: Structural Funds 2014-2020, European Regional Development Fund and Operational Program for Big Infrastructure).

Some specific issues related to the composition of sludge should be deeply analysed taking into account the final destination of the co-digested output.

The authors acknowledge the co-tutors dr. Elena Cristina Rada and prof. Lucian Ionel Cioca, and the Erasmus+ Programe for the support during the development of the research and during the stage in Romania.

References

1. *** http://ec.europa.eu/environment/eussd/, accessed 04.03.2017
2. *** *Raport anual privind starea mediului în judetul Sibiu* (Ministerul Mediului, Apelor si Padurilor, 2014)
3. M. Ragazzi, R. Catellani, E.C. Rada, V. Torretta, X. Salazar-Valenzuela, Sustainability, **6**, 1 (2014)
4. E.C. Rada, M. Ragazzi, S. Villotti, V. Torretta, V. Waste Manag. **34**, 5 (2014)
5. E.C. Rada, M. Ragazzi, G. Ionescu, G. Merler, F. Moedinger, M. Raboni, V. Torretta, Energ. Procedia, **50** (2014)
6. G. Ionescu, E.C. Rada, L.I. Cioca. Environ. Eng. Manag. J. **14,** 11 (2015)
7. *** Directive 2008/98/EC of the European Parliament and of the Council of 19 November 2008 on waste and repealing certain Directives (European Council, 2008)
8. V. Torretta, F. Conti, M. Leonardi, G. Sustainability, **4** (2012)
9. R.D. Ionescu, M. Ragazzi, L. Battisti, E.C. Rada, G. Ionescu, WIT Transact. Ecol. Environ. **176** (2013)
10. S. Di Leo, M. Salvia, Renew Sustain. Energ., Rev. **68** (2017)
11. A.N. Iuga, IOP Conf. Ser. Mate. Sci. Eng., **161,** 1 (2016)
12. C. Iacoboaea, O. Luca, F. Petrescu, Theor. Empirical Res. Urban Manag., **8,** 4 (2013)
13. *** http://ec.europa.eu/environment/waste/framework/, accessed 04.03.2017
14. *** http://www.adiecosibiu.ro/despre-noi.html, accessed 04.03.2017
15. V. Torretta, N. Ferronato, I.A. Katsoyiannis, A.K. Tolkou, M. Airoldi, Sustainability, **9** 1 (2017).
16. *** Council Directive 1999/31/EC of 26 April 1999 on the landfill of waste (European Council, 1999)
17. E.C. Rada, I.A. Istrate, V. Panaitescu, M. Ragazzi, T.M. Cirlioru, T. Apostol, Environ. Eng. Manag J. **9**, 4 (2010)
18. L.I. Cioca, L. Ivascu, E.C. Rada, V. Torretta, G. Ionescu G. Sustainability. **7** (2015)
19. R. Ciudin, C. Isarie, L.I Cioca, V. Petrescu, V. Nederita, E. Ranieri, E., UPB Sci. Bull., Series D, **76,** 3 (2014)
20. G. Silvestre, A. Bonmati, B. Fernandez, Waste Manag, **43** (2015)
21. B. Morero, R. Vicentin, E.A. Campanella, Waste Manag., **61** (2017)
22. *** *Raport anual privind starea mediului în judetul Sibiu* (Ministerul Mediului, Apelor si Padurilor, 2013).
23. *** http://www.adrcentru.ro/Detaliu.aspx?t=PFPhareImpleID=32, accessed 04.03.2017
24. E. Trulli, V. Torretta. Environ. Technol., **36,** 12 (2015)
25. G. Andreottola, M. Ragazzi, P. Foladori, R. Villa, M. Langone, E.C. Rada, E.C., UPB Sci. Bull. Series C, **74,** 1 (2012)
26. P. Arnò, S. Fiore, V. Verda, Energy, **122** (2017)
27. V. Cabbai, N. De Bortoli, D. Goi, Waste Manag., **49** (2016)
28. G. Silvestre, A. Bonmatí, B. Fernández, Waste Manag. **43** (2015)
29. M. Krupp, J. Schubert, R. Widmann, Waste Manag. **25,** 4 (2005)
30. *** http://www.huber.de/, accessed 04.03.2017

Illumination of a production facility for communication and electric power cables

Cristian Tsakiris[1], *Cristin* Bigan[1,*], and *Valentin* Panduru[1]

[1]Ecological University of Bucharest, Faculty of Environment and Management Engineering, Romania

Abstract. Sight represents 90% of our sensory everyday life. In specific work conditions, like long distance communication and electric power cable production, sight is by far more important. Work personnel must distinguish between many colours cables in verifying the cable delivered finally after the machine assembling. Errors are very costly since any non-conformity results in an entire production stock rejection. The luminosity in the production facility is estimated by similitude of different room scales and the necessary of artificial illumination. The results will be computed per the level of illumination indicated for production. A scaled-up estimation will be for the production facility, and is corrected with respect to the production equipment situated between the center of the room, occupied by the directing staff, and the facility edges. Equipment and machines are placed in a visible way from the central area occupied by offices. Results are analysed to improve facility design in terms of rectangularity ratio, colour of the walls, ceiling and floor and light sources emplacement.

1 Introduction

Sight serves human perceiving the environment reality in proportion of about 90 % of all completed sensorial communication way. Information received by sight is enhanced in specific work conditions, where reactions to visual appearance direct to important decisions, like in driving.

Working with machines represents a peril of being injured by the machine assembles movement. In a phone cable production process, checking cable parameters on the working posts requires good sight abilities, besides good skills 1. Possible hazard management represents a minimum requirement for maintaining the operation of the production 2.

The right light characteristics that must be in the work area can be satisfied by adequate source lights, well positioned. For normal subjects working, visual perception, contrast sensibility and speed reaction are subject to the absence of shadows, of brilliance or of inadequate light intensity.

Illumination is expressed quantitively in lumen, which represents the light intensity expressed in candles propagated in a solid angle of 1 radian. It is of interest the light flux arrived on a target surface, after possible absorption, reflection or refraction. The light flux is measured in lux, that is 1 lumen reported to 1 square meter. Indirect illumination avoids brilliance of surfaces that makes sight discomfort 3.

* Corresponding author: cbigan@yahoo.com

Illumination sources used in the present are halogen bulbs, fluorescent tubes and LED (Light Emitting Diode) tubes, most recently. The halogen bulb replaces the former incandescent bulb, so would primarily serve in household destination; however, it will be considered in the electric sources comparison for the phone cable industrial hall production. The fluorescent tube has an oscillating light source linked to the alternative electric current frequency; this light oscillation is too high to be perceived by the human eye sensor, but induces eye fatigue for repeatedly unconditioned reflex adaptation to this oscillation; fluorescent tubes contain mercury, which is essential for a good light emission. The LED tube is formed of a semiconductor that have its electrons ready to jump on higher energetical levels when excited, emitting light at the same time, in the visible spectrum; the emission is not oscillatory and can be controlled to be on a constant wave length, i.e. constant chromatic 4.

1.1 General trends in illuminating facilities

Illumination quality grows in its importance together with work intensity, sophistication of machine technology and computerized stock evidence. At the same time, the energy consumption is desired to be lowered, even if the source illumination cost would increase, because of electric energy becoming more and more expensive.

The ABB and Philips companies arrived recently at the conclusions that LED bulbs are best for proper illuminations and energy savings, as well. Compared with their old system (not specified, perhaps a not far generation of fluorescent tubes), there is a 12 % reduce in power consumption 5. The illumination level is maintained at a 500-lux level, by combining artificial light with day light. Movement sensors switch off the light gradually, when no more detecting movement after a while. In this adaptive lightening way, a 50 % energy reduce was obtained.

Diehl Metall Schmiedetechnik, company that produces metal-plastic components, arrived to use LED illumination system from the manufacturer Philips, as well 6.

ETAP company from United Kingdom promotes its production of LED illumination tubes for large spaces and high ceilings that can be for production or storage facilities, or for hypermarket products on shelves good illumination display 7.

Studies of optimal illumination for an existing hall without windows were made in the Transylvanian University from Brasov, Romania. A statistical regression model was used for different source illumination intensity. Again, the conclusion was that if replacing the existing fluorescent tubes with LED tubes, the electric energy consume would reduce to half 8.

In this paper case, the trend of LED tubes usage should be investigated if it may be a good choice for the phone cable hall production facility ICME ECAB S.A. situated near Bucharest, Romania.

1.2 Phone cable production

Phone cables are made of 2 very thin wires, of 0.5 mm diameter, wrapped in alternative position in a protective sheath. In the case of phone apparatus connectors to cable there are 2, 4, 6 or 8 wires to be fixed there 9.

Every wire has a specific colour that encodes its functionality. Assembling is most often made by alternatively looking carefully in the manual operations guide. Workers must be able to distinguish relatively small type characters on paper, see univocally the wire colours, operate correctly their torsion or verify it after being done by machine and finally fix everything.

The phone cable lots are rolled on tambours of enormous size, of about 2-3 km of cable. Tambours are of different colours depending on the type of phone cable produced. When the beneficiary receives the goods, selects randomly one tambour category and tests with their apparatus a cable portion randomly, as well. If the test does not fully pass the beneficiary specified requirements in their preliminary offer request, all the same kind tambours are rejected and penalties may apply to the cable producer.

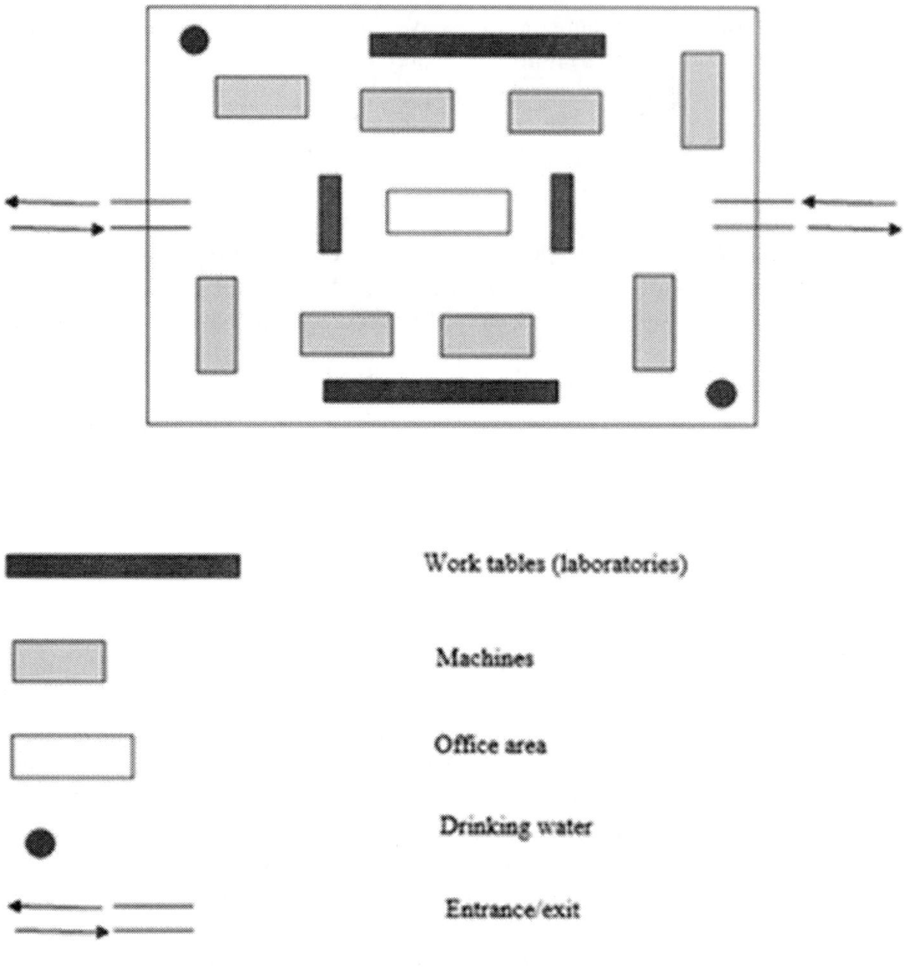

Fig. 1. Phone cable hall production

In these conditions, light required for production purposes must be of a good quality, having the recommended intensity and chromatic characteristics. Comparing the described source lights, oscillatory intensity of fluorescence tubes, even if not detectable by the human sensors may generate fatigue during work and in long practice professional medical problem. LED light sources are still expensive, but have guaranteed constant intensity and wave length of the light. Halogen ECO light sources have been produced as an alternative to classic electric bulbs with incandescence, but probably remain to be used competitively for small interiors. Compact fluorescent tubes may have the light oscillations attenuated, but they are not reported to give a better qualitative light than the LED source.

1.3 Cable hall production facility

The hall has 150 m of length L and 100 m of width W. It is equipped with all necessary machines to produce desired phone cables, which are packed in large roles (Fig. *1*). The principal machines are many extruders for different cables. Cables are twisted in roping machines, in two steps, arriving to 100 elementary cables. Tambours containing up to 2 km of rolled rope are pushed manually on tracks towards the testing tables, after being put on by cranes. Tables are equipped with testing apparatus of cable fulfilling required parameters; they are serving for studying internal work manuals, as well.

Production is continuously supervised from the hall center, where there is office facility for technical and management production supervision. From there, white collars have a comprehensive view to the blue collars work and machine functioning, and to any personnel access and exit. The office area has 30 meters' length, l and 10 meters' width, w.

2 Illumination method and projection

Illumination is made sectorial, according to local activities. Not all the machines may be required for use at the same time. Production may work continuously, day and night. To minimize the necessary artificial light requirements, the hall facility was provided with glass transparent walls from ground up to the transversal beam under the ceiling. The ceiling is partially of concrete, partial of glass, to let maximum day light to enter. Artificial light is not provided to adjust automatically to the required light level, however is commanded on the base of decades of experience of working in that facility.

There are analysed one by one the illuminating devices available for providing the necessary light all over the hall production. The calculus steps start in estimating the necessary luminosity for good work in the facility. One of the work nature is of machine working, supervising and checking or repairing, necessity of know-how handling, but as well fixing small landmarks in cable manufacturing machines. For this kind of operations, the necessary source light intensity is between 300 and 750 lx. Other operation is of making precise measurements for testing the fabricated cable. For laboratory kind operations, a higher source light intensity between 500 and 1000 lx is required, stipulated by on rule normative. Since both types of operation occur equally frequently, a mean light intensity i_p of 625 lx would be set to fulfil the requirements. This illumination level is not affecting the office area, which is isolated by non-transparent roof and by glass walls from the rest of the hall; there, the required level is between 220 and 370 lx for working both on paper and on LCD screen, luminosity assured by separate desk and low ceiling illumination; considering increased visual tasks during being in service, an i_o of 320-lx light intensity will be selected for a majority human subjects eye comfort 10.

The illumination surface S will be divided between production area, S_p and office area, S_o:

$$S_o = l \cdot w = 30 \cdot 10 = 300\, m^2 \tag{1}$$

$$S = L \cdot W = 150 \cdot 100 = 15.000\, m^2 \tag{2}$$

$$S_p = S - S_o = 15.000 - 300 = 14.700\, m^2 \tag{3}$$

The total light Q is the sum of the light for the production area Q_p and the light for the office area Q_o:

$$Q_p = i_p \cdot S_p = 625\, lx \cdot 14.700\, m^2 = 9.187.500\, lm \tag{4}$$

$$Q_o = i_o \cdot S_o = 320\ lx \cdot 300\ m^2 = 96.000\ lm \quad (5)$$

$$Q = Q_p + Q_o = 9.283.500\ lm \quad (6)$$

The rectangularity ratio does not influence the light requirement since its source propagates almost vertically by spreading enough the cross area light flow to cover entirely the floor. In this way, shades practically are not present, light arriving on all machine surfaces that are placed equally distanced between office area and hall edges 11.

As it is in present reality, the hall production walls are entirely of transparent glass. This may be replaced with another glass type, which has special properties of entirely permitting day light to enter and letting little to pass from the hall outside, especially during dark light occurrence. Besides light preserve, nobody from outside can look in the hall facility. Glass walls are doing economy tasks for electric energy consuming, for lightening reasons and to filter at the same time in the very sunny day lights, avoiding greenhouse effect especially during hot days.

The floor and the ceiling would be preferred in light colours that are most reflective in terms of light. The ceiling would be painted of an anti-static white. The floor would be of massive granite resistant and vibration absorbent polished surface covered by an eye comfortable light brown colour 12.

Light sources are mounted on the ceiling since the light intensity projection refers to the light source position to be on the top position, up oriented to reflect the light down without shadows. There are four types of light sources considered for comparing: LED, fluorescent tube, compact fluorescent tube, and ECO halogen light source. All sources will be considered to maximum possible power to reduce their number to the minimum. The case of light sources mixing is excluded first to avoid eye shock caused of abruptly adapt necessity to light variations, and of technological sources mounting and maintenance 13. A different light source may be considered for the office area, if benefit found. The total light necessity computed in relation (6) will be considered, rounded to an upper number as multiple of 25 to have integer result on hall length L and width W. Any light source is supposed to be replaced at about two of third times the reported functioning time to avoid production dysfunctionalities because of lower light performances, and to avoid short circuits due to longer use associated with fire hazards 14. The consumed electric energy is considered at $c_{elen} = 0{,}40\ lei/kWh$, TVA included.

2.1 LED tube

LED source light is available to a maximum power of $w_{LED} = 12\ W$ for each source, maximum producing 1070 lumen. The number of LED source light required is:

$$n_{LED} = \frac{Q}{1070} \cong 8700 \quad (7)$$

One LED is reported to cost $c_{LED} = 49{,}30\ lei$ and to last about $t_{LED} = 50.000\ h$ 15. The mounting and maintenance cost for each LED is estimated at $m_{LED} = 5\ lei$. The hourly cost of illumination P_{LED} would be:

$$P_{LED} = n_{LED} \cdot \left[\frac{c_{LED} + m_{LED}}{t_{LED} \cdot \frac{2}{3}} + \left(\frac{w_{LED}}{1000}\right) \cdot c_{elen} \right] = 8700 \cdot \left[\frac{49{,}3 + 5}{5 \cdot 10^4 \cdot \frac{2}{3}} + \left(\frac{12}{1000}\right) \cdot 0{,}4 \right] = 55{,}94\ lei/h \quad (8)$$

The surface covered by each LED is:

$$S_{LED} = \frac{S}{n_{LED}} = \frac{15000}{8700} = 1{,}72\ m^2 \quad (9)$$

2.2 Fluorescent tube

Fluorescent tube source light is available to a maximum power of $w_{flt} = 36\ W$ for each source, maximum producing 3350 lumens. The number of sources light required is:

$$n_{flt} = \frac{Q}{3350} \cong 2800 \quad (10)$$

The unit cost is $c_{flt} = 19{,}04\ lei$ and to last about $t_{flt} = 30.000\ h$ 16. The mounting and maintenance cost for each piece is estimated at $m_{flt} = 10\ lei$, since the manipulation is more difficult and there is a peril of mercury contamination in case of breaking by mistaken. The hourly cost of illumination P_{flt} would be:

$$P_{flt} = n_{flt} \cdot \left[\frac{c_{flt} + m_{flt}}{t_{flt} \cdot \frac{2}{3}} + \left(\frac{w_{flt}}{1000}\right) \cdot c_{elen} \right] = 2800 \cdot \left[\frac{19{,}04 + 10}{3 \cdot 10^4 \cdot \frac{2}{3}} + \left(\frac{36}{1000}\right) \cdot 0{,}4 \right] = 44{,}39\ lei/h \quad (11)$$

The surface covered by each fluorescent tube is:

$$S_{flt} = \frac{S}{n_{flt}} = \frac{15000}{2800} = 5{,}36\ m^2 \quad (12)$$

2.3 Compact fluorescent tube

The maximum power is of $w_{cft} = 25\ W$ for each source, maximum producing 1520 lumens. The number of sources light required is:

$$n_{cft} = \frac{Q}{1520} \cong 6125 \quad (13)$$

The unit cost is $c_{cft} = 44{,}18\ lei$ and to last about $t_{cft} = 10.000\ h$. The mounting and maintenance cost for each piece is estimated at $m_{cft} = 5\ lei$. The hourly cost of illumination P_{cft} would be:

$$P_{cft} = n_{cft} \cdot \left[\frac{c_{cft} + m_{cft}}{t_{cft} \cdot \frac{2}{3}} + \left(\frac{w_{cft}}{1000} \right) \cdot c_{elen} \right] = 6125 \cdot \left[\frac{44{,}18 + 5}{10^4 \cdot \frac{2}{3}} + \left(\frac{25}{1000} \right) \cdot 0{,}4 \right] = 106{,}44\ lei/h \quad (14)$$

The surface covered by each fluorescent tube is:

$$S_{flt} = \frac{S}{n_{flt}} = \frac{15000}{6125} = 2{,}45\ m^2 \quad (15)$$

2.4 Eco halogen bulb

The maximum power is of $w_{ech} = 105\ W$ for each source, maximum producing 1900 lumens. The number of sources light required is:

$$n_{ech} = \frac{Q}{1900} \cong 4900 \quad (16)$$

The unit cost is $c_{ech} = 5{,}95\ lei$ and to last about $t_{ech} = 2.000\ h$. The mounting and maintenance cost for each piece is estimated at $m_{ech} = 2\ lei$, since it is a very simple work procedure. The hourly cost of illumination P_{ech} would be:

$$P_{ech} = n_{ech} \cdot \left[\frac{c_{ech} + m_{ech}}{t_{ech} \cdot \frac{2}{3}} + \left(\frac{w_{ech}}{1000} \right) \cdot c_{elen} \right] = 4900 \cdot \left[\frac{5{,}95 + 2}{2000 \cdot \frac{2}{3}} + \left(\frac{105}{1000} \right) \cdot 0{,}4 \right] = 235{,}02\ lei/h \quad (17)$$

The surface covered by each fluorescent tube is:

$$S_{flt} = \frac{S}{n_{flt}} = \frac{15000}{4900} = 3{,}06\ m^2 \quad (18)$$

3 Results and discussion

The estimation hourly cost for lightening the industrial production hall helps to easily eliminate the halogen ECO bulbs and the compact fluorescence tubes that have sensibly much greater exploit costs than LED tubes or common fluorescent tubes. However, it is to consider the eye fatigue from oscillating intensity light emission for the fluorescent tubes and the mercury intoxication peril in case of mishandling them correctly. The LED tube exploit price is not far from the fluorescence tube and would be preferred even if it is of about 25 % more expensive. The LED tube illumination has the advantage of a higher emplacement density having the illumination task of deserving only 1.72 square meters of ceiling; this is a value that would avoid object placed on the ground and human body shade and that can be focalised on the designed surface without interference between two next situated LED tubes. At the same time, the illumination from LED tubes is most recommended for desk office tasks, so they can be used in all the hall facility.

It is noticeable that the cost advantage of using LED tubes instead of fluorescent tubes is not so evident, as presumed. The obtained result is dependent on right data to be used for. The final decision would be influenced by technological and economic reasons that would give in each case one or another result between the two nearly priced electric sources, LED tubes and fluorescent tubes.

4 Conclusions

A phone cable manufacturing facility was studied regarding the illumination options in the required parameters of technology specificity. The hall production facility was designed in terms of active areas emplacement inspired from information and experience of profiled engineers who had worked in such a facility for their entire active life. Four current used light sources, LED tubes, fluorescent tubes, compact fluorescent tubes and ECO halogen bulbs were compared in terms of hourly cost if assuring the all calculated necessary light for this specific production activity. The analysis proved useful in technically eliminating the non-economically efficient source lights, despite of their lower acquisition cost. It is foreseen a deepening of the study by obtaining additional information directly from the plant's leadership.

References

1. V. Lefter, A. Manolescu, A. Deaconu, *Ergonomie* (Editura Economica, 2010)
2. R.I. Moraru, G.B. Băbuţ, L.I. Cioca, Environmental Engineering and Management Journal **13**, 6 (2014)
3. A. Draghici, *Ergonomie* (Editura Regie, 2009)
4. *** https://www.magazin-leduri.ro/beneficiile-iluminatului-led-vs-iluminat-fluorescent/, accessed 12.03.2017
5. ABB, www.abb.com/lowvoltage, accessed 12.03.2017
6. Diehl Metall Scmiedetechnik, http://www.diehl.com/en/diehl-group.html, accessed 12.03.2017
7. ETAP NV, www.etaplighting.com, accessed 12.03.2017
8. C. Chis, Student Scientific Research Yearbook of the Transylvania University Brasov **65-68**, (2015)
9. *** http://cabluri.net/telefonice/, accessed 12.03.2017
10. *** http://www.misiuneacasa.ro/uploads/files/normative4.pdf, accessed 12.03.2017
11. *** http://www.rabalux.com/ro_RO/calculator-al-nivelului-de-iluminare, accessed 12.03.2017
12. A. Muresan, http://www.tex.tuiasi.ro/biblioteca/carti/CURSURI/Prof.%20Dr.%20Ing.%20Augustin %20Muresan/Curs%2014.%20Evaluarea%20culorii.pdf
13. *** http://ebooks.unibuc.ro/StiinteADM/enache/24.htm, accessed 12.03.2017
14. L.I. Cioca, R.I. Moraru, Archives of Mining Sciences **57**, 1 (2012)
15. *** https://www.magazin-leduri.ro/eficacitatea-si-durata-de-viata-a-becurilor-led/, accessed 12.03.2017
16. *** http://lightsmart.ro/blog/tuburi-led-versus-tuburi-fluorescente-consideratiuni-generale/, accessed 12.03.2017

Contributions to risk assessment in the departments of welding mechanical engineering companies

Gheorghe Amza [1,*], *Dan Florin* Nitoi[1], *Valentin Dan* Petrescu[2], *Zoia* Apostolescu[1], and *Oana Roxana* Chivu[1]

[1] Polytechnica University of Bucharest, Splaiul Independentei, no. 313, sector 6, Bucharest, Romania
[2] Lucian Blaga Universiy of Sibiu, 10 B-dul Victoriei, Sibiu, Romania.

Abstract. The risk assessment professional should cover every activity and every workstation in an enterprise engineering, considering each component of the production system (the system working), each duty, work equipment and working environment. This represents an extremely complicated and complex problem because of production system that is a powerful polluting technology, especially of the atmosphere and soil. The formation of gas welding process is the result of electrodes burning, fluxes, and development of bath fused metal the welded seam. In welding processes, human operators are exposed to smoke and toxic gases, arising from the welding process, which can be dangerous to health. Many acute intoxication caused by to excessive exposure or short exposure to severe smoke and gases resulting from the welding process were studied the course of time. The paper focuses on smoke inhalation and calculate the score of danger each chemical agent score volatility of process. This conducts to collective protection so ultimately results in a score of inhalation S_{inh} = 1000, or resulting risk moderate, requiring a set of protective and preventive measures that are proposed in this regard.

1 Introduction

Occupational risk assessment must cover each activity and each shop of a machine building company, taking into account each component of the production system (work system), every workload, work equipment and work environment. This represents an extremely complicated and complex problem because the production system of such an enterprise is particularly complex and complicated. Involvement of workers from each job and, where appropriate, of their representatives with specific responsibilities in the field of occupational safety and health, and of the Occupational Safety and Health Commission (OSHC) is mandatory in the assessment of occupational hazards, as workers know the most dangerous situations and places at workstations and the OSHC is formed and informed about the dangers specific to the company's activities.

As almost all the production sites of this company meet the welding process * (from design to assembly) [5], this paper will assess the chemical risk through inhalation in this technological process [1-2].

Welding assembly is a highly polluting technology process, especially of the atmosphere and soil. The formation of gases in the welding process is the result of electrodes and fluxes burning, forming the molten metal bath, and welded seams developing. Welding operators are exposed to smoke and toxic gases resulting from the welding process, which may be hazardous to health. Many acute intoxications due to excessive exposure or short-term exposure to smoke and gas resulting from the welding process have been studied over time. However, in addition to general effects such as respiratory irritation, long-term effects have been directly attributed. The influence of smoke and toxic gases, for example, the chronic effects due to the presence of chromium, nickel and aluminum. Since in any technological process, the main pollutants emitted in the working environment or the natural environment are chemical substances, the greatest risk that occurs in these jobs is the chemical risk produced by a chemical agent. Regarding chemical risk, the assessment procedure is often difficult because of the many chemical agents and preparations used, as well as the lack of awareness of the dangers they pose.

At work, welding operators may be exposed to hazardous chemical agents, either accidentally (explosions, fires, pipe or tank damage, etc.), or routinely during use, handling or transport. Hazardous chemical agents may cause one or several of the following effects: poisoning, burns, fires, irritations, injuries [3].

2 Chemical risk assessment

The chemical risk assessment method for health, safety and the environment is progressive, calls for simple and easily accessible criteria, being a step-by-step approach to optimizing the collection of information on the main chemical agent (MCA) and workload. The method comprises the following main steps:

- inventory of chemical products and materials used in a location, workshop or workstation;
- potential risk management (PRA);
- risk evaluation.

Product inventory is the most important step because it condition the quality of a risk assessment approach. The inventory of chemicals, raw materials, intermediate products should be as exhaustive as possible. The data collected at this stage are:

- chemical name or trade name of the product,
- the quantity used (in the previous months);
- frequency of use;
- the work area where the product is used,
- hazard information on labels (pictograms, risk phrases);
- the information given in the risk phrases (hazards, physicochemical properties).

In the inventory phase, risk phrases are essential aid in this approach.

2.1 Potential hierarchy of risk (PHR).

Because of the large number of products and raw materials used in a site, it is necessary to prioritize hierarchy and prioritization to begin with the most dangerous products. The hierarchy of the identified products is done using the PHR method [8]. This takes into account the hazards, the potential exposure (health), the potential for ignition (fire, explosion) and the transfer potential (environmental impact). The PHR method takes into account the criteria presented in Table 1.

Table 1. Criteria used to calculate the potential risk of a chemical agent [1]

Effects on health		**Fire explosion**		**Environmental Impact**	
Danger	Potential exposure	Flammability	Potential of ignition	Danger	Transfer potential
Risk phrases	Used quantity	Used quantity	Stored quantity	Risk phrases	Used quantity
	Frequency of use		Sources of ignition	Classification of hazardous waste	Stored quantity
				Aggregation state	

The risk assessment consists of simplifying health, safety and environmental risks in a simplified manner [6-7]. These steps require more information to be collected than in the hierarchy, regarding the handling conditions of various chemical agents.

First of all, a potential hierarchy of potential risks (PHR) is made, based on the following: quantity, frequency, exposure, and hazard. The hazard class and danger class is then determined using Table 2, [9-11].

Table 2. List of chemical agent released and proposals for classification of hazards [1]

Chemical agents emitted from processes		**VLE [mg/m3]**	**The danger class**
Powders from:	Iron	5 (xod)	2
	Stainless steel alloys	0,5 (Cr)	3
	Aluminum	10	1
	Wood and derivatives	1 (for cancer)	4
	Cereals and derivatives	5	2
	Glass fibers	10 (dust without specific effect)	2
	Construction materials (stone, bricks, beton)	10 (dust without specific effect)	2
	Metallic plumb	0,15 (steam)	4
	Talcum	10 (dust without specific effect)	2
	Cement	10 (dust without specific effect)	2
	Ceramic fibers	(R40)	3
	Vegetable fibers	0,5	3
	Asbestos and materials containing asbestos	(R45)	4
	Composite materials (resin + fiberglass, carbon fibers,)	-	2
	Painting with plumb	0,5 (estimate)	3
	Abrazive discs	0,1 (estimate)	3
	Sanding	0,1 (estimate)	3
	Graphite	2	2
Fumes from:	The burning treated wood	-	2
	The welding iron	5	2
	Foundry of metal	-	2
	Stainless welding	0,5 (Cr)	3
	Foundry and process of plumb	0,15	4
	The welding and thermoforming of plastics	-	2
	Vulcanization	-	4
Vapors	Materials of vegetable or animal	-	2

and	origin		
gases	The tar and embers	0,2 (R45)	4
from:	The heating equipment	55 (CO)	2
	Mercury	0,05 (steam)	4
	The processing oils	19 (INRS)	3
	The fuel motors	-	2
	Benzine (fuel)	It contains benzene	4
Aerosols	For cutting oils	1 (INRS)	3
come	The pickling water jet	-	2
from:	Oils of stripping	-	2

3 Risk assessment chemical inhalation

In order to establish objectives of this paper, dangers of each chemical agent that occurs in the process and exposure conditions will be account. Exposure after analyzing workload is estimated by:

- physico-chemical properties (volatility);
- handling conditions (process, temperature, etc.);
- the collective safeguards (ventilation);

Risk assessment involves observing and analyzing real work load for each group of welders analyzed consists in reviewing the different tasks carried out, their duration and their chemical agents.

To be meaningful, the analysis should be carried out together with operators to benefit from their experience. Further risk score can be established. Each chemical agent score is determined, depending on the hazard class (Table 2) who has been assigned in PHR stage and presented in table 3 [4].

Table 3. Determination of score of danger by danger class

The hazard class	Danger score
5	10000
4	1000
3	100
2	10
1	0

Volatility class can be determined. Any chemical agent (CA) may be present in 3 aggregation states: solid (pulverulent, fibrous), liquid or gaseous. Each CA is assigned a volatility class based on the aggregate state, and then a score as follows:

- pulverized (solid) materials can be assigned a volatility class of 1 to 3 (Table 4);
- determining the volatility class of a liquid CA requires knowing the approximate use temperature and its boiling temperature in degrees Celsius.

The volatility class is determined using the diagram presented in Figure 1. Products in paste form and those whose ignition temperature is not indicated are assigned the volatility class 3;

- gasses are assigned the volatility class 1, regardless of the temperature of use.

Table 4. Determination of powdered AC class volatility [1]

Description solid materials	The class volatility
The material is in the form of fine powder, forming the powder remaining in suspension in the air during manipulation	1

The material in the form of a powder consisting of granules (1-2 mm), form powders which are deposited rapidly during manipulation	2
The material is in the form of tablets, granules, scales (from 1 mm up to 1 or 2 cm) slightly friable less emitting particulates during manipulation (sugar cubes, granular plastic materials)	3

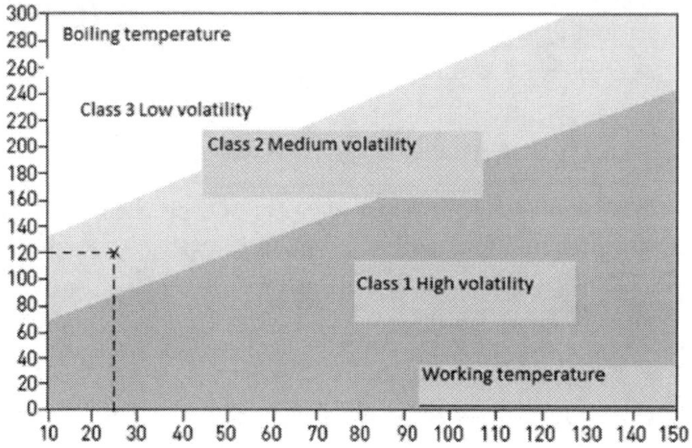

Fig. 1. Determination of liquidity volatility class

The volatility score is then determined. Each volatility class is assigned a score to be used for exposure estimation (Table 5).

Table 5. Score attributed volatility of volatility classes [1]

The class volatility	The volatility score
1	100
2	10
3	1

In the following, the process score is determined. The type of process in which CA is used is determined as indicated in Table 6. A score is assigned to each type of process. In this operation, the most delicate task is to assign a process class by distinguishing between two types of processes, namely:

- processes called "dispersive" which are characterized by an important source of dust, smoke or vapor. By way of example, this type of process corresponds to cutting or welding operations, application of spray products, cleaning of parts in baths with totally discovered solvents;
- the "open" processes, which, by design, are clearly less emissive than dispersive processes. This type can be assimilated to printing machines, presses for metals or plastics, etc

Table 6. Class determination process and the associated score [1]

Dispersion	Open	Closed but opened regularly	Closed permanently
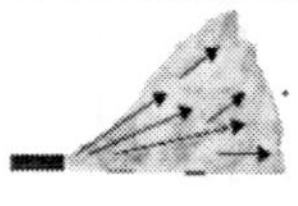	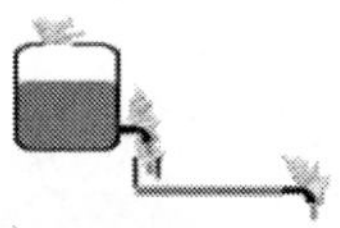		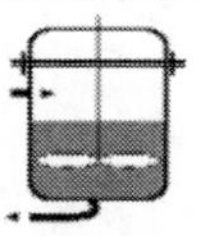

Eexamples: Painting the pistol manually emptying sacks of Arc welding Manual cleansing material, using portable machines (Robots,etc.)	Examples: Reactors, mixers open painting the brush, conditioning stations (Drums,etc.) overseeing printing machines	Examples: Chemical reactors closed to regular AC loads, taking samples, machinery degreasing liquid or vapor phase.	Examples: Chemical reactor
Class 4	Class 3	Class 2	Class 1
Score procedures			
1	0,5	0,05	0.001

The technological process list is not exhaustive

The welding process is analyzed in the paper considers a dispersive process, to fit with in class 4 with a score of process equal to 1.

Taking into account the above elements, the collective protection score is determined. The type of collective protection operating at the station where chemical agent is located is determined by the information shown in Table 7. A score is associated with each class of collective protection means. Then the risk score by inhalation is calculated. For each chemical agent used for a determined load, the inhalation risk score with "S_{inh}" is calculated using the following relationship:

$$S_{inh} = \text{score of danger X score of volatility X score of process X score of collective protection} \quad (1)$$

By substituting in relation (1) the values found for the analyzed case result:

$$S_{inh} = 1000 \times 1 \times 1 \times 1 = 1000 \quad (2)$$

Analyzing the relationship (2) a moderate risk results that probably requires the implementation of corrective measures and an important assessment (emissions determination).

Table 7. Determination of collective protection classes and associated scores [1]

Absence of mechanical ventilation 	**Removing worker from the emission source** 		**Presence of mechanical ventilation** 
Class 4, score = 1	Clasa 3, score = 0,7		
Gas exhaust hood 	**Suction slot** 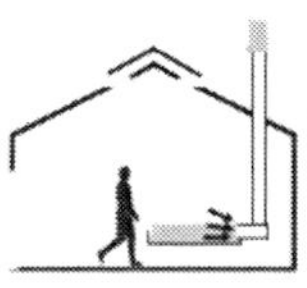	**Aspiration table** 	**Integrated suction equipment** 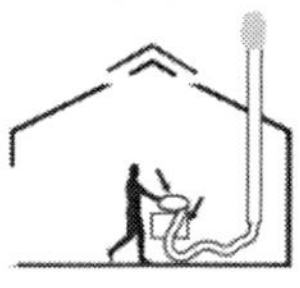
Class 2, score = 0,1			

Small size ventilated cabin	**Horizontal Cabin**	**Vartical cabin**	**Capture envelope: niches of laboratory**
			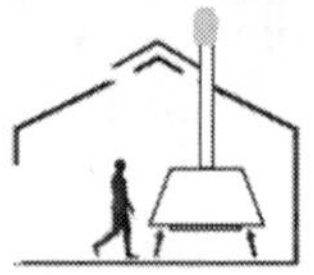
Class 2, score = 0,1			Class1, score = 0,001

Finally, the risk characterization can be done. The risk of each task is characterized by the decision grid shown in Table 8.

Table 8. Risk characterization grid for inhalation and skin

The risk score	The priority action	The risk characterization
≥ 1000	1	Very probable risk that requires immediate corrective action
100-1000	2	Moderate risk requiring probably implementing corrective actions and a thorough evaluation (determination of pollutants)
< 100	3	At low risk does not require modifications

4 Conclusions

- as a result of the determinations made in the steel production hall in different work areas, at the respiratory level it was found that the value of the concentrations of inhalable powders, aerosols and fumes from different metals exceeds the CMA value of the Order 1957/1995, in different places of measurement and in different proportions;
- the determination of the potential risk of welding was done starting from the quantity class, thus determining: the frequency of use class (which is 4); the hazard class (which is 100,000); the priority class (which is high), resulting in a potential risk of large welding;
- inhalation risk assessment was done by following the steps: workload analysis; determining the hazard score according to the hazard class (if analyzed: hazard class = 3 and hazard score = 100); volatility class and volatility grade (here is 100); determining the process score (which in the case being analyzed is 1.00); determining the collective protection score; calculating the risk score by inhalation (which in this case is = 1000), corresponding to a moderate risk;

References

1. Gh. Amza, *Ecotechnology* (Printech Publishing House, 2007)
2. Gh. Amza, *Ecotechnology and sustainable development* (Printech Publishing House, 2009)
3. Gh. Amza, *Treaty of materials technology*, (Academia Romană Publishing House, 2002)
4. Gh.Amza, M.Parvulescu, TQSD, 258 (2008)
5. Gh. Amza, *Materials technology*, **IV**, (BREN Publishing House, 2003)
6. V. Ionescu, *Buletinul Stiintific UPB Seria D*, **2**, 85 (2016)
7. V. Ionescu, Gh.Calea, Gh.Amza, G.Iacobescu, D. Nitoi, A. Dimitrescu, International Scientific Conference SAMRO, 145, (2016)

8. R. E. Knowlton, *An introduction to Hazard and Operability Studies – the Guide Word Approach*, (Chemetics, Vancouver, BC, 1986)
9. Agenţia Europeană pentru Substanţe Chimice (ECHA), P*age dedicated to the use of chemicals safe at work*, http://echa.europa.eu/ro/use-chemicals-safely-at-work
10. Agenţia Europeană pentru Securitate şi Sănătate în Muncă, *theme dangerous substances,* https://osha.europa.eu/ro/topics/ds
11. Inspecţia Muncii, *Page dedicated to IMM, professional chemical risks,* http://www.inspectiamuncii.ro/ssmimm/p1b.html

Contributions to risk injury evaluation of human skin contact caused by welding technology in mechanical engineering a company

Gheorghe Amza[1,*], *Dan Florin* Niţoi[1], *Valentin Dan* Petrescu[2], *Oana Roxana* Chivu[1], and *Zoia* Apostolescu[1]

[1]Politechnica University of Bucharest, Spl. Independentei, 313, sector 6, Bucharest, Romania
[2]Lucian Blaga Universiy of Sibiu, B-dul Victoriei, 10, Sibiu, Romania

Abstract. Assembly by welding is a very important pollutant technological process, mainly of the atmosphere and soil. The formation of gas in welding process is the result of electrodes, fluxes burning, formation of fused bath and welded seam. A large number of human operators feels different adverse effects on health. Most problems appears in respiratory system and include bronchitis, respiratory irritation, fever caused by smoke, changes in lung function, decreased immunity to infection and a possible increase of lung cancer risk. Very little information are available about effects on the organism and skin after exposure to fumes from welding, therefore. In this order, this paper do a risk assessment by skin contact. Starting from the main parameters for the evaluation of a cutaneous risk, it was determined a score of danger, a score of surface and a score of frequency. It results a risk associated with each class with a value S_{cut} = 100.000, that indicates a very high probable risk with immediate proposing corrective measures.

1 Introduction

The technological process of welding is a sum of technological operations that pollutes the atmosphere and soil. Formation of gas and other chemicals in the welding process is the result of burning electrodes, fluxes, formation and developing the welding bath and welded seam. During the welding process, human operators and the auxiliaries personnel are exposed to smoke and toxic gases, arising from the welding process, which can be hazardous to health. Many acute intoxication due to excessive exposure or short exposure to severe smoke and gases that results from the welding process were studied over time [1-6]. However, besides general effects such as respiratory irritation, long-term effects were directly attributable to the influence of smoke and toxic gases, for example, chronic effects due to the presence of chromium, nickel and aluminum. A large number of welders feel some adverse health effects. Respiratory problems encountered in full-time working includes bronchitis, respiratory irritation, fever caused by smoke, changes in lung function.

* Corresponding author: amza@camis.pub.ro

All these problems decreased immunity to infection and a possible increase in lung cancer risk.

Less information are available about effects on the body and skin after exposure to fumes from welding.

Welding is an important technological method that produces high-quality metal components joints. Essentially, all metals and alloys can be welded, some easily, others involving special precautions. Next will be presented a brief consideration of the welder working in the main production shop where the process of welding technology is developing.

Within the department, the welder performs the following:

- making cores and molds fittings, needed for the casting of iron;
- cutting feeders, network hardware and reworking cast ingots with oxyacetylene flame;
- reworking by welding of cast and heat-treated parts: removing defects detected by visual inspection, magnetic control, penetrant liquid control or ultrasonic testing, by arc-air cleaning, filling the areas excavated by welding in accordance with the procedures specific to each type of material;
- creates various welded constructions according to drawings in accordance with welding shop schedule.

In this activity, elements that generates possible health risk are:

- welding equipment and work environment: oxygen-acetylene mixture, welding generator and power transmission cable; welding torches, pliers and accessories (wire brush, hammer welding, protective equipment);
- parts, parts that debits, manufactured and repeated;
- technical equipment in operating area (particularly the cranes);
- the environment in which it operates: its noxious summing with those occurring during welding operations.

The welder executes specific works (autogenous welding and electrical), according its duties.

- organize work, check the technical equipment used in welding;
- regulates the flow of gas (oxyacetylene welding, gas welding protective environment etc.) or connected to power socket welding transformer;
- complies strictly with the obligations incumbent upon it, under the working procedure and working security ruls that has been trained.

Welding work begines after ensuring the safety conditions of labor. As in any process technology, the main pollutants emitted in the working environment or natural surroundings are chemical substances. Highest risk occurring in these jobs is chemical risk produced by a chemical agent.

2 Risk assessment by skin contact

One of the biggest risks to which the welder operator and auxiliary workers are subjected is the cutaneous risk because they come into contact directly or indirectly with a variety of chemicals found in liquids, solids or gaseous [10-12].

The risk assessment derived from direct or indirect manipulation of substances in liquid or solid form (powder) in relation to a dermal exposure is the main objective of this work. Since the process of welding can be classified among dispersive open or closed-open, required parameters for this evaluation are:

- hazard class of the product;
- exposed body surface;
- frequency of the exposure;

In order to make the risk assessment the skin first scoring threat is determined. The hazard classes are determined based on risk classes. Each class will be associated with a score of danger presented in Table 1.

Table 1. Hazard score by hazard class

Hazard class	Hazard score
5	10000
4	1000
3	100
2	10
1	0

Very important is the skin surface exposed to contact with the chemical agent. Therefore, a score of the exposed surface is determined (Table 2).

Table 2. Determination of the score of exposed skin surface

Exposed skin surface	Skin surface score
One hand	1
Two hands A hand and forearm	2
Two hands and forearm A full arm	3
The contact surface comprises upper limbs and torso and/or pelvis and/or feet	10

Welder operator performs numerous operations during the work schedule, hourly, daily, monthly or annually, so a exposure frequency score is determined also (Table 3).

Table 3. Determination of the exposure frequency score

Exposure frequency	Frequency score
Occasional: < 30 min/day	1
Intermittent: 30 min – 2 h/day	2
Frequent: 2 – 6 h/day	5
Permanent: > 6 h/day	10

Taking all these hnown elements a cutaneous risk score is determined by a the relationship:

$$S_{scut} = \text{Danger Score x Skin Surface Score x Frequency Score} \quad (1)$$

In the case presentd in paper, with reference to the welding processes, the following result was obtained:

$$S_{scut} = 1000 \text{ x } 10 \text{ x } 10 = 100.000 \quad (2)$$

This shows a probably high value of skin risk that implies immediately corrective measures.

3 Assessing the risk of fire explosion

Another permanent risk in welding shop is possible risk of fire - explosion. This should be considered in assessing the overall risk to the exposed welder operator. To characterize the risk of fire- explosion the same scale can be used as that used for inhalation risk characterization presented in Table 4. Based on the process parameters and simplified scheme risk of fire explosion, a simplified scheme of the risk of fire-explosion is presented in figure 1.

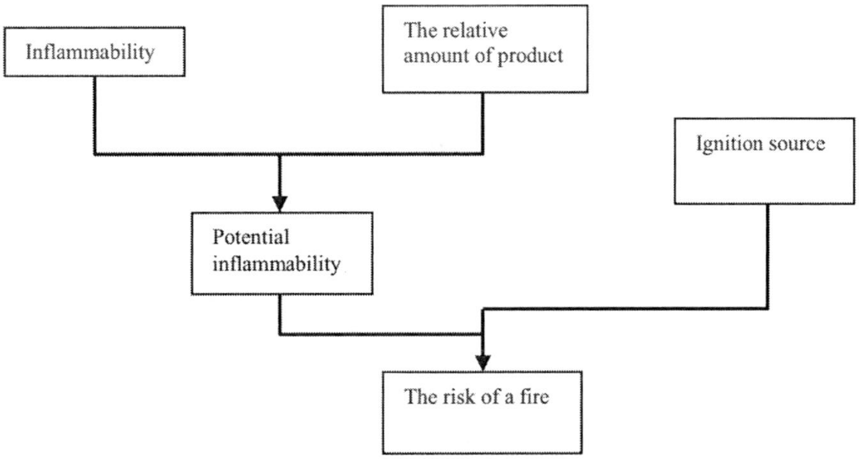

Fig. 1. Simplified scheme of the risk of fire-explosion

For this, paper presents first a classification of potential risk, which aims at classifying chemicals according to their potential flammability and working areas depending on the level of fire risk.

Evaluation data are: the chemical name or trade label, the measured quantity, the ignition source, place of use [13].

Table 4. Class process determination and the associated score

Dispersive	Open	Closed but opened regularly	Closed termanently
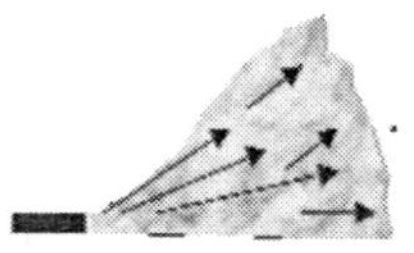	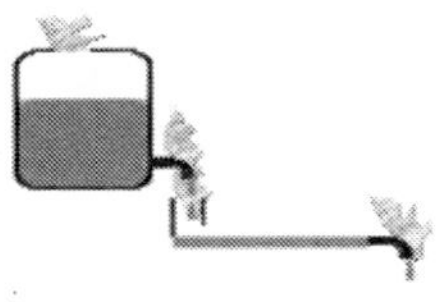	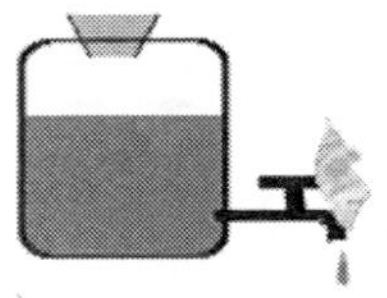	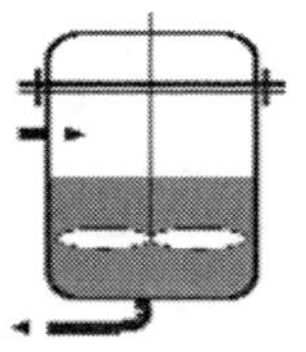
Painting gun, manual emptying of bags, manual cleaning with cloths, the use of portable machines	Reactors, open mixers, paint brush, conditioning shops, surveillance printing machines	Closed chemical reactors, sampling, degreasing machinery in liquid or vapor phase	Chemical Reactor
Class 4	*Class 3*	*Class 2*	*Class 1*
Procedures score			
1	0,5	0,05	0,001

Evaluation data are: chemical name or trade label, the presented quantity, ignition source, place of use. In the case of conditioned products, flammability hazard class is determined from the information mentioned in the FDS or packaging. When a product has more risk phases, the highest hazard class will be selected. Regarding combustible materials, hazard class is determined by the type of material as shown in table 5.

Table 5. Determination of the inflammability classes

Class	Symbol	Risk phases
1	Without	Compact solid materials (wood pieces, blocks of resin, paper charts ...)
2	Without	Divided combustible solids (chips,cloth,trays...) Liquid fuel materials (burning), vegetable oils, lubricating R14, R15, R14/15, R15/329 and the probability of accidental contact with water
3	Without	R10
4	 F - Very flamable	R11, R30
4	 F⁺ - Extremly flamable	R14, R15, R14/15, R15/29 and the probability of accidental contact with water
5	 F ⁺ - Extremly flamable  E - explosive  Oxidant	R14, R15, R14/15, R15/29 and the probability of accidental contact with water R1, R2, R3, R4, R5, R6, R7, R8, R9, R12, R16, R17, R18, R19, R44 Bulk organic matter suspended in the air

In terms of fire expousure, amount class is a function of the quantities. Determination of classes amount shall be based on the quantity of items found in the greatest amount during inventory (Qmax) and the amount of product i (Qi). The grid used is the same with methodology used in ESRS (Tab. 6).

Table 6. Defining the quantity thresholds according to the flamability classes

Flamability classes	*Maximal value*
5	10 g
4	100 g
3	1 kg
2	10 kg
1	100 kg

Depending on the class of product or material, danger thresholds will be applyed for the chemical principal amounts according to Table 6. Products whose quantity is lower this scale are not taken into account when calculating potential flammability.

Risk analysis of explosion- fire continues taking into account existing sources of ignition, storage, detection and protection (security organization and means of prevention etc).
Using Table 7, the grades of fire ignition will be determined, which in the present case is 5. In the case when for a working place there are several sources of ignition, the highest value of the class will be retained.

Table 7. Determination classes of ignition sources

Ignition source class (S_a)	Ignition source example	The frequency of the source of ignition
5	Hot flames in equipment or processes	Permanent presence of ignition source
4	The steps of heating equipment for cleaning, the heat sealing operation, etc,	Ocasional presence of the process
	Fumes	Ocasional presence of the process
3	Work in hot spots	The presence is related to the maintenance
	Withdrawal / loading of organic or flammable products	The ignition source is linked to the emergence of static electricity
2	Electric incident	The ignition source is caused by a malfunction, wear or handling error...
1	Sabotage or natural phenomenon	External or accidental source of natural origin (lightning)

With the above data the potential flammability classes, denoted by Ip can be determined.Potential flammability results by combining hazard classes and quantity. Potential flammability classes are determined using the grid shown in Table 8, which when analyzed section is 4.

Table 8. Determination of potential flammability classes

Flammability classes						
5	3	4	5	5	5	
4	3	3	4	4	5	
3	2	2	3	3	4	
2	1	1	2	2	2	
1	1	1	1	1	1	
	1	2	3	4	5	Quantity class

Starting from potential flammability class Ip of a product used in a location and the presence of ignition sources SA, the risk score potential for starting a fire can be determined. These scores are in addition to prioritize the different entities depending on the level of risk. Score potential risk of starting a fire is determined according to the chart presented in table 9 resulting in the present case the value of 5000.

Table 9. Determination of potentially risk score of starting a fire

Potential flammability class Ip						
5	2000	5000	10000	30000	100000	
4	300	1000	2000	5000	10000	
3	30	100	300	1000	2000	
2	3	10	30	100	300	
1	1	1	3	10	30	
	1	2	3	4	5	Class ignition source

Finally, with all results presented above, the risk characterization of rough fire ignition can be done according to Table 10.

Because, in the present case, the results obtained is 5.000 important potential fire hazard is present.

Table 10. Characterization of the fire potential risk

Scor	*≥ 10000*	*100-10000*	*10-1000*	*<10*
Potentioal risk characterisation	Very important	Important	Moderate	low

This risk estimate does not account for the actual conditions of use of the product and the means of protection. Applying this method of grading allows for products and / or workshops to be analyzed first. Situations characterized by values of lead more than 10,000 correspond most likely non-compliant situations (eg presence of an open fire in a work area where highly flammable products are used).

4 Conclusions

Analyzing the experimental results can draw the following conclusions:

- production of machine-building enterprises is extremely varied and flexible technological processes are extremely complex and using different materials and technological equipment, highly diverse, technical and economic conditions totally different;
- there are many parameters that influence technological process differently emission of pollutants in the workplace based on physical phenomena, chemical, mechanical, electrical, magnetic and nuclear occurring during the technological process;

- in the determinations made in the welding shop, in different areas, on the breath level, it was found that the amount of inhalable particulate concentrations, aerosol and fumes from various metals exceeds the CMA Order 1957/1995, in different places and in different measured proportion;
- risk assessment by skin contact was calculated taking into account: product hazard class; expusp body surface and frequency of exposure, resulting in a very probable risk (S_{cut} = 100,000) that requires immediate corrective action;
- simplified risk assessment fire explosion was done by going through the steps of ranking potential risk; determining the flammability class; risk score to determine potential for starting a fire (which in the present case is equal to 5000), which means that the potential fire hazard is important.

References

1. Gh. Amza, V. Aionesie, *TQSD*, (Bucuresti, 2008)
2. Gh, Amza, *TQSD*, **157**, (Bucuresti, 2008)
3. Gh, Amza, *Ecotehnologie*, **972** (2007)
4. Gh, Amza, *Ecotehnologie si dezvoltare durabila*, 1258 (2009)
5. Gh, Amza, *Tratat de tehnologia materialelor*, 1682 (2002)
6. Gh, Amza, *Tehnolgia materialelor*, **IV** (2003)
7. http://reach.anpm.ro/Helpdesk.aspx, Agenţia Europeană pentru Substanţe Chimice (ECHA), pagina dedicată utilizării produselor chimice în condiţii de siguranţă la locul de muncă
8. http://echa.europa.eu/ro/use-chemicals-safely-at-work, Agenţia Europeană pentru Securitate şi Sănătate în Muncă, tema Substanţe periculoase
9. https://osha.europa.eu/ro/topics/ds, Inspecţia Muncii, pagina dedicată IMM-urilor, riscuri profesionale chimice
10. http://www.inspectiamuncii.ro/ssmimm/p1b.html, Afişul Etichetele produselor chimice se schimbă – cum vă afectează
11. H. Kumamoto, J.E. Henley, IEEE Transactions on Reliability, **R–27**, 242 (1978)
12. E.W. Lawless, M.V. Jones, *Risk Evaluation and Management*, (Plenum Press, 1986)
13. S. Pece, Risc şi securitate în muncă, **34** (1994)

MATEC Web of Conferences **121**, 11003 (2017) DOI: 10.1051/matecconf/201712111003
MSE 2017

Modern principles used in conformity assessment of machinery from forestry sector

Anca Elena Antonov[1,*], *Georgeta* Buica[1], and *Constantin* Beiu[1]

[1]INCDPM „Alexandru Darabont", Certification Body, 35A Blvd, Bucharest, Romania
[2]INCDPM „Alexandru Darabont", Electrical and Mechanical Risks Laboratory, 35A Blvd, Bucharest, Romania

Abstract. The paper is aiming to implement the general principles of risk prevention at employer's level, with respect to occupational risks evaluation, the elimination of risk and accident factors, and information of workers which are using the machinery in the forestry sector. For the use of machinery in the forestry sector in terms of economic performance and a level of maximum safety, it is necessary to ensure the user guides set by the manufacturer in terms of commissioning, use and to provide appropriate safe working operations and interventions and to guarantee the technical and environmental requirements, including appropriate measures and means of protection against accidents and occupational disease. The impact of occupational risks for machinery used in this sector can be reduced through the application of modern principles in conformity assessment and certification and, where appropriate, through technical diagnostics and inspection, taking into account the provisions of the new Machinery Directive 2006/42 / EC which is imposing the obligation of manufacturer to implement conformity assessment procedures in accordance with the methods of assessment and verification of safety at the certification bodies, notified at the European Commission. The paper aims to develop modern technical tools for conformity assessment and verification of this category of machines used in the forestry sector that would provide prerequisite for increasing competitiveness of employers in the market economy. Applying these tools of modern technology for manufacturers and users of this category of machinery provides the necessary conditions for placing on the market of safe products with a appropriate safety level, in the intended using conditions, in order to guarantee the essential requirements for safety and health, technical and environmental conditions, including measures and means of protection. The result of this research is to develop technical tools needed to assess compliance on effective occupational risk management for professional machinery used in the forestry sector.

1 Introduction

A priority objective in the Romania nowadays is the implementation of national legislation harmonized with EU legislation, as an important condition for economic integration, taking into account the provisions of Article 70 of the Law on Ratification of the European Association Agreement, which provides: "employee's safety, health and life of humans are priority areas for harmonization." In this respect, by harmonizing Romanian legislation,

* Corresponding author: aantonov@protectiamuncii.ro

especially for safety and health at work, with Community law aimed, on the one hand creating working conditions at European level by Romanian regulations complying with the provisions of Directives deriving from art. 118A of the Treaty of Rome (article 137 * of the Treaty of Amsterdam, as amended by the article153 from the Treaty on functioning of the European Union) and on the other hand to ensure a maximum level of safety for workers by using work equipment (WE) and appropriate personal protective equipment (PPE) whose safety/protection quality to be ensured from the conception and certified by regulated procedures.

In case of work equipment, the concept of quality should be based primarily on terms of meeting the basic, function in use or safety function, relative to the user and goods /people/ animals from working environment. [1]

2 Statistics of accidents at work in the wood processing industry, manufacture of wood

The information presented are the results of the processing of statistical data on accidents at work and occupational diseases recorded in the wood processing industry, wood and cork products of, excepting furniture, in Romania. These data are based on activity reports for 2011- 2015 of Labour Inspection and data provided by the NSI for 2008-2015, wood processing industry is a sector whose activity has an important share, both economically and workforce involved, widely recognized as high risk, similar to constructions sector. [1, 3-4].

2.1. Statistics of accidents at work in wood processing industry

The analysis of statistics on accidents at work and occupational diseases recorded in the wood processing industry had as a starting point the share of the workforce and workers who worked in 2008-2015. The data recorded shows a decrease in employment from 2009 to 2011 and a small increase in 2014 and 2015.

Figure 1 shows the evolution of employment in the wood processing industry in 2008-2015, according to the NSI in section C-Manufacturing, Code 16 Wood, wood and cork products of, excepting furniture.

Analysing the statistical data, it has found that wood processing industry remains on top as the number of injured produced at work.

The statistics on accidents at work produced at national level, shows that there is a decrease in 2010-2011, followed by an increase of the number of work accidents produced in activities in this sector, in 2014 and 2015.

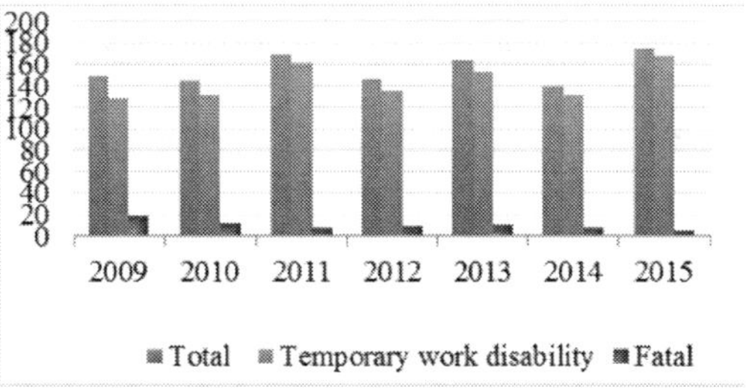

Fig.1 Evolution of workforce in wood processing sector in 2008-2015

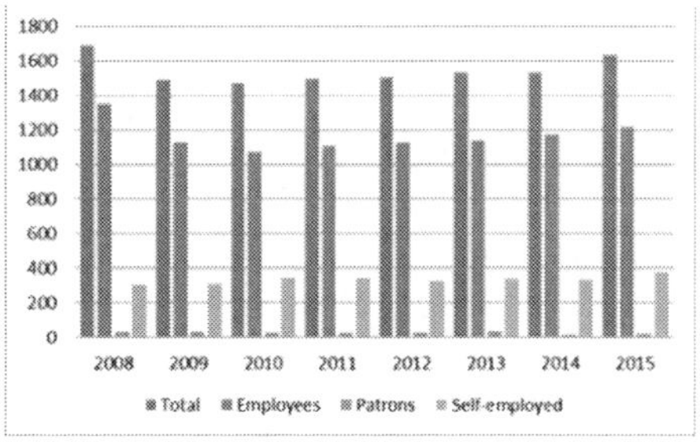

Fig.2 Statistics on accidents at work in the wood processing sector in 2008-2015

There is a very small reduction in the number of workers involved in accidents at work in recent years and also a reduction of the number of collective accidents at work in wood processing industry, mainly due to fluctuating workload in this sector.

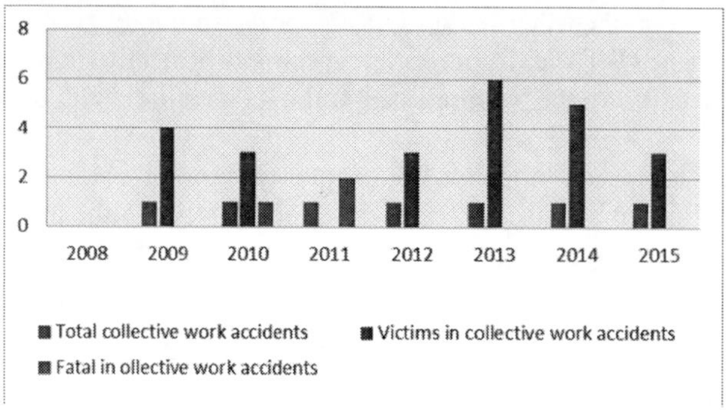

Fig.3 Statistics on collective accident at work in wood processing sector in 2008-2015

2.2 Statistics on occupational diseases recorded in wood processing industry

Statistics show that in the wood processing industry, manufacture of wood, the most common disorders are musculoskeletal system diseases.

3 Principles relating to the conformity assessment of machinery used in the forestry sector

In order to carry out conformity assessment and product certification, it applies different conformity assessment modules, according to New Approach concept, or variants of these modules. European legislation on the quality of work equipment used by employees during work is based on art. 100A of the Treaty of Rome, as amended by art. 94 ÷ 97 of the Amsterdam Treaty, as amended by Article 113 ÷ 117 respectively of the Treaty on functioning of the European Union (Treaty of Lisbon).

European directives arising from Article 100A contains a number of common provisions, regardless of the type of equipment to which it applies and its specific provisions, particularly in the form of essential safety requirements, including technical provisions (characteristics, parameters admissible, test methods) [1-2].

For machinery, main provisions are contained in the European Directive 2006/42/EC, which came into force from 29.12.2009. The Directive 2006/42/EC have set new safety requirements for machinery, the application of some modules for conformity assessment and ensuring total quality of manufacturing, complementary to the provisions of Directive 98/37/EC. Directive 2006/42/EC have set new safety requirements for machinery, the application of modules conformity assessment and ensuring full quality assurance of manufacturing to the provisions of Directive 98/37 / EC. According to the regulation issued by the Machinery European Directive 2006/42/EC does not apply to the lifting of persons by machinery that are not built for this purpose. This Directive allows, however, for Member States to take all measures at national level, according to the European Treaties, for machinery, in application of the Directive 89/655/EEC which refers to the harmonization in Member States of the minimum safety and health requirements for the use of work equipment by workers at work. Regarding the risks related to the use of machinery covered by Directive 2006/42/EC, it has been established procedures for assessing the conformity with the essential health and safety requirements, these procedures are designed according to the importance of the inherent risk that can arise from these machineries.

Consequently, each category of machinery should be harmonized for an appropriate procedure to comply with Decision 93/465 / EEC of the Council from 22 July 1993, concerning the modules for the various phases of conformity assessment and rules on the application and use of CE marking, modules which are intended to be used in the technical harmonization directives and which contains provisions relating to technical verification of these machines.

Directive 2006/42/EC introduces a number of conformity assessment procedures for machinery, according to article 12:

- for machines that are not listed in Annex IV of the Directive, the manufacturer or his representative applies the conformity assessment procedure with internal control of production of machinery, according to Annex VIII of the Directive;
- for machines form Annex IV of the Directive and which are manufactured according to harmonized standards provided in article 7, par. (2), and for all standards which provide all the applicable essential health and safety requirements, the manufacturer or his representative applies one of the following procedures:

a) Conformity assessment procedure with internal checks on the manufacture of machinery, according to Annex VIII of the Directive;

b) The EC type-examination procedure provided by Annex IX and internal checks on the manufacture of machinery according to Annex VIII, section 3 of the Directive;

c) Full quality assurance according to Annex X;

- for machines in Annex IV of the Directive and not manufactured according to harmonized standards provided in article 7, par. (2), or only partially, or if the harmonized standards do not provide all the applicable essential health and safety requirements or no harmonized standards exist for the machine in question, the manufacturer or his representative shall apply one of the following procedures:

a) The EC type-examination procedure provided by Annex IX and internal checks on the manufacture of machinery according to Annex VIII, section 3 of the Directive;

b) In accordance with Annex. VIII, section 3 of the Directive;

c) Full quality assurance procedure according to Annex X;

- for complete parts of the machinery, the manufacturer of such party or its representative, before placing on the market must comply with Article 13 of the Directive and to adopt the following:

a) Specific technical documentation, according to Annex VII, Part B;

b) Instructions for assembly, according to Annex VI;

c) Declaration of incorporation, according to Annex II, Part 1, Section B.

Romanian legal provisions relating to conformity assessment of products, which should cover the machinery, establish the evaluation and certification activities to be carried out by competent technical bodies notified by national authorities, for each technical regulation. Guidelines for safety and health at work of legal and technical regulations to develop the course module are:

- Law no. 319/2006 on safety and health at work;
- Norms for implementing Law No 319/2006 on safety and health, approved by GD no.1425 / 2006, amended (HG nr.955 / 2010);
- GD no.1029 / 2008 on the conditions of putting on market of machinery;
- GD no.457 / 2003 reissued, on concerning the safety of users of low voltage electrical equipment;
- GD no.1146 / 2006 on the minimum safety and health requirements for use of work equipment by workers at work, Romanian standards of product safety, etc.

Law no. 319 / 2006 on occupational safety and health provisions of the Directive Council no.89 / 391 / EEC on the introduction of measures to encourage improvements in the safety and health of workers at work, legislation which establishes that for WE to take, from research stage, design and execution, as well as the development of manufacturing technologies, solutions to comply with legal provisions on health and safety at work, by whose application to removed or diminished the risks of injury and occupational disease for workers. In order to achieve WE without risk to health and safety, it should be ensured in the design and manufacturing, health and safety requirements laid down by technical directives or requirements of national technical regulations. WE not covered by technical regulations provided by EO no.20 / 2010, must ensure the safety requirements of standards, technical and health and safety at work regulations, etc. In order to ensure the use of safe machinery, the designer and manufacturer of such an work equipment (WE) must prove compliance with GD nr.1029 / 2008, which is the legal document that transpose the 2006/42/EC Directive. For major electrical risk (machinery and appliances under Directive machines, except those provided for in Directive LVD 2015/35/EU which replaces Directive 2006/95 / EC) the provisions of Directive 2006/42 / EC are applied and related to emission and immunity to electromagnetic field the provisions of Directive 2014/30 / EU on electromagnetic compatibility are applied. For major electrical risk (with the exception of machinery and appliances under Directive machinery) the provisions of Directive 2006/95 / EC relating to electrical equipment designed for use within certain voltage limits are applied. WE intended for use in potentially explosive atmospheres the provisions of Directive 2014/34/EU concerning equipment and protective systems intended for use in potentially explosive atmospheres, replacing the previous Directive 94/9/EC.

In order to ensure essential health and safety requirements applicable to machines from the earliest stages of design, production and marketing, must be met the GD nr.1029 / 2008 from 29.12.2009.

Work equipment may be new, used, or from occasion, in this category are the work equipment and protective means which have already been used in an EU country and it is, again, placed on the EU market. This new placement on the market means that the occasion machinery passed the first placement on the market state and initial operation in the European Union. Reconditioned machinery that begin with the prefix indicates repeated action or return to the previous state (retro) being assigned to the existing machines that have undergone technical interventions that aimed at changing their state, performance, safety, etc. This work means changing the machine in a greater or smaller way. A "Rebuilt" or "renewed" Machine is a new machinery, built, partially or entirely, from parts from an old machinery. European technical rules that can be applied to a used or renewed machines are:

- "New Approach" Directives were designed exclusively for new products or for products considered as new.
- From technical point of view, implementation of the directives to occasion machinery is generally unrealistic.
- Although there is no legal obstacle to regulating occasion machinery through a European Directive, based on Article 100A of the Treaty of Rome, it is difficult to see how technical harmonization at different generation of machines is applied.

Health and safety requirements applicable to new, used or occasion work equipment used by employers during work activities, are provided in GD no.1146 / 2006 on safety and health at work using WE and application of these requirements must achieve the compliance with WE in use.

In this regard, for WE in use, it must be carried out an initial inspection, as their safety depends on the installation conditions, conditions after installation and before being put into service or inspection after assembly at headquarters or in a new location and / or periodic and special inspections and, where appropriate, testing of WE exposed to conditions that can cause damage that can result in dangerous situations [5].

Conformity assessment of woodworking machinery industry must be associated with a systematic risk assessment of workplaces in the forestry sector. There are some research that promote different methods of risk assessment at workplaces, and studies aimed at "Developing a rational systematization of 31 occupational risk assessment methods. It gives it clear selection criteria for adequate most practical application of tools. The results obtained will support the Decision - making process within the health and safety management process in industrial work environments"[6], which determines the need to apply a specific procedure of risk assessment at workplace, chosen in accordance with WE subject to compliance procedure.

Conformity assessment of woodworking machinery industry, certification and / or diagnosing of technical evaluation is performed under the compliance with applicable essential health and safety requirements and the presumption of conformity with the applicable standards to this category of machines for woodworking – Band sawing machines. The case study was made for woodworking machines which are summarized in Annex 4 of GD nr.1029 / 2004, item 1 (Annex 4 of Directive 2006/42 / EC). This study covers the identification of risk situations and hazardous events that may generate risks for workers in woodworking machines for Band sawing machines (table band saws and band re-saws) [1, 7].

It was identified safety checks applicable to a machinery that is subject to the "EC-type examination" ((Article 12, par.3, letter b) and point 3 / Annex 8 of Directive 2006/42 / EC and Article 13, par (3), b) Annex 9 and item 3 / Annex 8 of GD nr.1029 / 2008, in preparation for the initiation / certification procedure of machine made and marketed by a manufacturer from Alba county. The main relevant European standards for the design and manufacture of band saws and log saws to apply conformity assessment process are: EN ISO 12100: 2010 / EN ISO 12100: 2011 [8], SR EN 1807-1: 2013 / SR EN 1807-1: 2013 [9] SR EN 1807-2: 2013 / SR EN 1807-2: 2013 [10], plus the conformity assessment process and type B safety relevant standards.

Also, during the investigation, it has established the list of essential requirements and characteristics applicable to conformity assessment of woodworking machines - saws and a list of safety checks on electrical risk to be carried out in a process of conformity assessment and certification, list which is established in terms of the essential safety requirements stipulated in Directive 2006/42 / EEC (GD 1092 / 2008), GD 1146 / 2006, SR EN 1807-1: 2013 SR EN 60204-1: 2007 / AC : 2013 [6].

Table 1. Safety checks on electrical risk (excerpt)

Nr. crt.	Name checking	Technical condition				Check method	
		GD 1029/ 2008 (Annex 1)	GD 1146/2006 (Annex 1)	SR EN 1807-2: 2013	SR EN 60204-1: 2007/ AC:2013	SR EN 1807-2: 2013	SR EN 60204-1: 2007/ AC:2013
1	Checking the electrical equipment of machines	1.2.	2.1, 2.2	5.2	4.3	5.2	18.6
2	Checking the control circuits and control functions	1.2	2.2, 2.3 2.4	5.2	9.2 10	5.2	18.6
3	Checking the devices for safe disconnecting and switching off	1.2.4	2.1, 2.4	5.2	9.2 10.7	5.2	18.6
4	Checking the control	1.2.6	2.3	5.2, 5.4	9.4	5.2, 5.4	18.6

Nr. crt.	Name checking	Technical condition				Check method	
		GD 1029/ 2008 (Annex 1)	GD 1146/2006 (Annex 1)	SR EN 1807-2: 2013	SR EN 60204-1: 2007/ AC:2013	SR EN 1807-2: 2013	SR EN 60204-1: 2007/ AC:2013
	system in case of failure of power supply and control system failure						
5	Dielectric test	1.5.1	2.19, 3.3	5.4	6	5.4	18.4
6	Checking correspondence with wiring diagrams	1.7	-	5.4, 6.3	17.6	5.4, 6.3	18.1
7	Verifying the existence, content and sustainability of markings	1.7	2.1	6.2	16	6.2	16

In the research study carried out, were applied principles of conformity assessment of machinery used in the forestry sector, principles implemented by the competent certification / notified body (INCDPM) through a set of technical tools of modern conformity assessment and certification of these machines using 3EV methods: applicable evaluation, testing, certification, and WE in use in terms of principle: inspection-testing-assessment-compliance through a process of compliance.

The research study shows the need to approach all mandatory steps for conformity assessment and certification and / or compliance of the WE, regardless of the sector they are intended to be used.

4 Conclusion

The work aimed to develop modern technical tools for conformity assessment and verification of this category of machines used in the forestry sector, to ensure employers prerequisite for increasing competitiveness in the market economy.

Applying these tools of modern technology for manufacturers and users of this category of machines, provides the necessary conditions for puting on to market of safe products with a level of safety appropriate to the conditions of intended use in order to guarantee the essential requirements for safety and health, technical and environment, including measures and means of protection.

The research results consist of the development of technical tools needed to assess compliance on effective management of professional risks for machines used in the forestry sector.

References

1. INCDPM, Tools for prevention of occupational risks for SMEs to ensure design and manufacturing requirements for work equipment used in the wood and metal processing industries - prerequisite for improving SME productivity through safe and healthy workplaces, research study (2015)
2. INCDPM, *Safety and health at work in the Construction sector. Training coordinators for safety and health at work, course module*, (119-145, 2017)
3. A.E. Antonov, G.Buica, D.C.Darabont, C.Beiu, 7th International Multidisciplinary Scientific Symposium, Conference Proceedings, **1**, 235-240, (published "Calitatea – acces la succes" Journal, **18** (S1) (2017)

4. INCDPM- Guidelines for safety and health at work in the construction sector, developed in the project POSDRU/81/3.2/S55075 funded by European Social Fund - Sectorial Operational Programme Human Resources Development 2007 – 2013, 15-18 (2012)
5 European Commission (2010), Guide for application of the Machinery Directive 2006/42/EC, 126-129 (in Romanian)
6. Moraru, R. I., Băbuţ, G. B., Cioca, L. I.- Rationale and criteria development for risk assessment tool selection in work environments, Environmental Engineering and Management Journal, **13** (6), 1371-1376 (2014)
7. INCDPM - Maintaining the competences ICSPM-CS certification body in accordance with procedures ICSPM-CS and CB accreditation criteria - case studies (2017)
8. EN ISO 12100:2010 Safety of machinery - General principles for design - Risk assessment and risk reduction (ISO 12100:2010)
9. EN 1807-1:2013 Safety of woodworking machines. Band sawing machines. Table band saws and band re-saws
10. EN 1807-2:2013 Safety of woodworking machines. Band sawing machines. Log sawing machines

MATEC Web of Conferences **121**, 11004 (2017) DOI: 10.1051/matecconf/201712111004
MSE 2017

Quantitative and qualitative differences worth considering in approaching critical infrastructures resilience

Dorel Badea[1,*], *Ghiță* Bârsan[1], *Ioan* Virca[1], and *Dumitru* Iancu[2]

[1] "Nicolae Bălcescu" Land Forces Academy, Department of Technical Sciences, 550170 Revoluției 3-5, Sibiu, Romania
[2] "Nicolae Bălcescu" Land Forces Academy, Department of Management and Administrative Sciences, 550170 Revoluției 3-5, Sibiu, Romania

Abstract. Through the implicit connections between risk, safety, protection and, last but not least, resiliency, that need to be managed, critical infrastructures have become a subject of increasing interest both for specialists and for academics. Based on this premise, this article analyzes in terms of operational performance several internationally used solutions for assessing resilience, having in view the desire to choose the best in order to be presented in some specialized disciplines of the postgraduate curricula that are annually presented in the educational offer of the Land Forces Academy. The authors also present their own vision on approaching the resilience of critical infrastructures so that it can harmoniously be integrated into the security plans of the holders or operators of such special assets.

1 Theoretical considerations

For all humanity, especially after the 2000s, it has become clear that the sources of dysfunctionality, seriously affecting the development of society, are no longer of a strictly military nature, others manifesting themselves intensively and extensively, with at least disturbing effects in terms of ensuring continuity in the daily activities. Reconsidering the role and importance of critical infrastructures has thus become more than a necessity, the volume of the works elaborated in the field of applied scientific research being an indicator in this sense. However, one can state that there is a convergence trend for certain points of view concerning issues specific to critical infrastructures, but at the same time complex issues that will be clarified in years ahead. In this context, we bring into discussion concepts such as security, protection, resilience, and lay stress on their complementary character, our objective being to focus our attention on the last concept in the previous enumeration.

National Infrastructure Advisory Council [1] approaches resilience of critical infrastructures as being the capacity to reduce the amplitude and/or duration of certain disturbing events, offering details on the fact that the efficiency of an infrastructure or

* Corresponding author: dorel.badea@yahoo.com

organization resilience depends on its capacity of anticipating, absorbing, adapting to and/or quickly recovering after the emergence of a disturbing event. But the problem consists in approaching resilience on the basis of a SMART determination, the results of the research carried out especially at European level (CIPRNet - Critical Infrastructure Preparedness and Resilience Research Network; CRISADMIN - Critical Infrastructure Simulation of Advanced Models on Interconnected Networks Resilience) or in the United States, still not converging towards a certain standard methodology.

The difficulty lies in providing some series of input data that are simulated based on robust models and allow for *what if* type of analyses. For a national approach, in the work coordinated by G. Bârsan, A.Dinicu and D.Badea [2], an annual questionnaire-based research is proposed, applied to all critical infrastructure operators through which information is to be collected on:

- the number of events recorded in the reference period,
- the average time to restore the functionality of infrastructure, important factors influencing the performance (implementation of risk management standards, SCADA state of equipment readiness, staff participation in specialized training activities, etc.),
- the staffing (by management and execution category) in the reference period (existing/required), number of training exercises conducted to simulate scenarios.

Basically, qualitative information is translated into quantitative information, which, after being compared with some specific data characterizing the existence of that system, becomes subject to processing by software and specialized tools. Acknowledging the fact that through the first research model mentioned we just try to understand some patterns of behavior, motivations, opinions and attitudes - and, subsequently, the results can be interpreted only in a certain context. The component of numerical data is the defining element of the quantitative research, and these data are obtained both in real environments and in laboratory conditions, the quantitative analysis methodology being used very often.

2 Combining qualitative and quantitative components in approaching resilience of critical infrastructures

In this section of the paper we started from the hypothesis absolutely accepted in the literature in the field, according to which the protection of critical infrastructures cannot be ensured 100%. Subsequently, we consider that the solutions that address resilience do not yet have a high level of robustness, regardless of the methods applied and the computer instruments used, simplifying working models being necessary. In this framework, as an applied study, we chose to investigate the possibilities offered by the Analytic hierarchy process (AHP) method. The principle of AHP method consists in converting some subjective assessments made on the relative importance of some criteria into scores and weights. The method, developed by Saaty in 1980 [3], proved to be the most popular form of multi-criteria analysis, being used in many types of applications in different fields. The input data for the specific AHP algorithm are answers to interrogations such as "How important is criterion A relative to criterion B?" So comparisons in pairs are obtained, giving scores or weights, which will determine the final decision. According to the desideratum assumed by the title of the article, one should emphasize that this method does not take into account the alternatives upon which the criteria are applied in order to calculate the importance coefficients (weights) of the criteria. In most of the situations encountered in reality, both qualitative and quantitative criteria appear, the latter being normally expressed in different units of measure, so the question of the non-homogeneity of criteria is being raised. [4]

The created scenario brings into attention an institution holding regionally valuable critical infrastructures, which, given the legislative imperatives at EU and national level, as well as the history of security incidents that took place in the past, sets as its primary goal the carrying out of a reengineering stage based on the implementation of an appropriate strategy.

Table 1. Presentation of taken into account strategies and criteria

<table>
<tr><th rowspan="2">Criteria</th><th colspan="3">Types of strategies</th></tr>
<tr><th>Technology-based</th><th>Human resource-based</th><th>Procedure-based</th></tr>
<tr><td>Robustness</td><td rowspan="4">The extensive use of SCADA systems is considered, with state-of-the-art automatic and sensorization elements, that ensure good interoperability with other systems up and down the interdependency chain and, at the same time, a reduced time for detecting deficiencies. Integration within expert decisional systems is also considered.</td><td rowspan="4">It will focus on an intensive training system of employees within a basic component providing general competences for a systemic, technical and operational understanding of the issues related to critical infrastructures, and a specialty component stressing the operational particularities of the system itself.
It insists on the employees' becoming aware of their roles within the system.</td><td rowspan="4">The focus is on the standard operating procedures for a various range of dysfunctions, approached in an integrative manner within the specific plan of risk management system.
The need of being clearly defined in the form of a viable algorithm is made aware of, no major changes being necessary when social-technical modifications take place.</td></tr>
<tr><td>Necessary initial investments</td></tr>
<tr><td>Operational compatibility with other interdependent critical systems</td></tr>
<tr><td>Reaction capacity</td></tr>
</table>

After carrying out an audit, a firm specializing in consultancy on industrial security concludes that the strategic objective must be to ensure resilience and proposes three technology-based strategies, people and procedures, with common and differentiation elements, as shown in table 1, being the general manager's responsibility to decide (Figure 1) which of the options presented should be implemented.

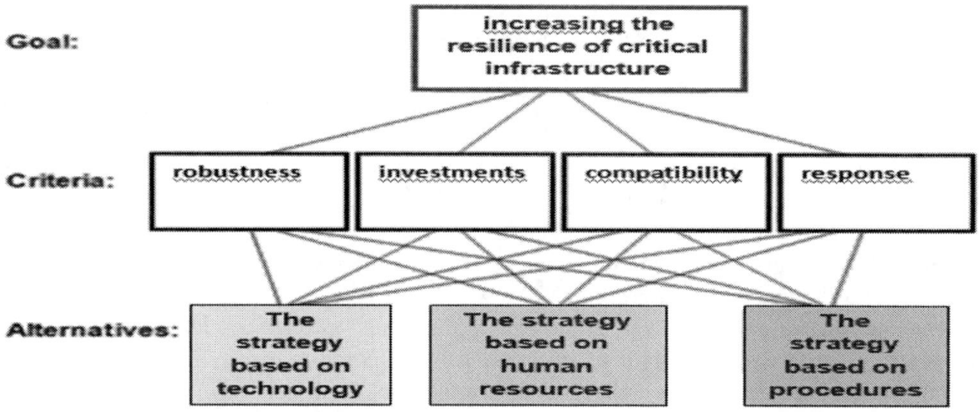

Fig. 1. The created decisional method

According to the AHP methodology and to an often cited and used practical example [5], the three strategies will be compared, each with each, relative to the four criteria considered, and based on the method of assigning weights shown in Table 2.

Table 2. Methodology of assigning [6]

The Fundamental Scale for Pairwise Comparisons		
Intensity of Importance	**Definition**	**Explanation**
1	Equal importance	Two elements contribute equally to the objective
3	Moderate importance	Experience and judgment moderately favor one element over another
5	Strong importance	Experience and judgment strongly favor one element over another
7	Very strong importance	One element is favored very strongly over another; its dominance is demonstrated in practice
9	Extreme importance	The evidence favoring one element over another is the highest possible order of affirmation
Intensities of 2, 4, 6 and 8 can be used to express intermediate values. Intensities of 1.1, 1.2, 1.3, etc. can be used for elements that are very close in importance.		

The final results obtained after going through the steps of the AHP method are shown in Table 3. According to the scores and given the scenario created, the best decision is to apply the human resource-based strategy in order to achieve the objective of increasing the resilience of a critical infrastructure.

Table 3. Centralizing the scores obtained for each strategy

Strategies	Criterion	Criterion Weight	S's weight	Weighted Score
S/Technol.	*Robustness*	0.550	0.163	0.089
	Investments	0.034	0.308	0.010
	Compatibility	0.177	0.151	0.026
	Response	0.236	0.100	0.023
				SECOND BEST - SUM 0.151
S/Hum.Res.	*Robustness*	0.550	0.539	0.297
	Investments	0.034	0.109	0.003
	Compatibility	0.177	0.630	0.112
	Response	0.236	0.673	0.159
				BEST - SUM 0.572
S/Proced.	*Robustness*	0.550	0.296	0.163
	Investments	0.034	0.581	0.020
	Compatibility	0.177	0.218	0.038
	Response	0.236	0.225	0.053
				LAST - SUM 0.275

3 Conclusions

Although useful especially in a situation with a large number of alternatives and criteria, the method involves a level of substantial subjectivity, starting with the first stage when the alternative strategies were compared based on each separate criterion (e.g. comparison relative to robustness, understood as viability over time, the technology-based strategy was assessed with a weight of 6-1 compared with the human resource-based strategy).

Further sensitivity analyses may contribute to increasing the relevance of the analysis. Specifically, the approach of resilience can only be made by using simplifying representations, being the only option to reduce complexity for research purposes, which was comprehensively highlighted by Roland Pulfer and Olga Bucovetchi [7]: "Making complexity transparent and manageable is the key factor to support innovation, to guarantee that innovation can be implemented into practice, to apply the highest professionalism and to recognize and apply possibilities for improvement." As a didactic value, this scientific approach can form the basis for laboratory work at the discipline Protection of critical Infrastructures.

Case studies within the postgraduate program organized in the institution, in which individual analyses can be conducted on the excel format that has been used for this study in order to investigate the range of values within which the optimal solution can vary and, at the same time, some principles of analysis can be established for reducing the bias in comparing in pairs.

References

1. ***, NIAC: *A Framework for Establishing Critical Infrastructure Resilience Goals,* National Infrastructure Advisory Council (2010)
2. G.Bârsan, A. Dinicu, D. Badea, (coordinators), *Analiza şi modelarea conceptuală a situaţiilor complexe – studii interdisciplinare,* ("Nicolae Bălcescu" Land Forces Academy Publishing House, Sibiu, 2016)
3. T.L.Saaty, *The Analytic Hierarchy Process*, (McGraw-Hill, New York, 1980)
4. S.L. Constantin, B.V. Constantin, *Methodology based on AHP for choosing between GPRS, UMTS and WLAN* article available at: http://www.agir.ro/buletine/859.pdf, retrieved on 22.02.2017
5. www.ccunix.ccu.edu.tw/.../AHP/AHP1%20Example%201%, retrieved on 22.02.2017
6. https://en.wikipedia.org/wiki/Analytic_hierarchy_process_%, retrieved on 22.02.2017
7. R. Pulfer, O. Bucovetchi, Sci. Bull., Series B, Vol. 78, Iss. **4**, 140 (2016)

Means to improve underground coal mine safety by automated control of methane drainage systems

Gabriel Bujor Babut[1,*], *Roland Iosif* Moraru[1] , and *Mihai* Popescu-Stelea[1]

[1]University of Petrosani, Faculty of Mines, 332006 Petrosani, Romania

Abstract. Based on the critical analysis of the presently employed management of methane drainage systems operation in Jiu Valley collieries, the paper aims to assess the basic elements required to develop an automated monitoring and control system of these. The results obtained after studies and researches carried out also allowed formulating certain proposals regarding the modification of manual control procedures of methane drainage systems operation, in order to correlate them with the prescriptions of legislation requirements from countries having a well-developed mining industry. Putting in practice the mentioned proposals could have immediate and beneficial effects on increasing the methane drainage process efficiency, leading meanwhile to an improved working environment and, implicitly, to a higher level of occupational safety and health in Jiu Valley collieries.

1 Introduction

In a modern mine a methane drainage system should fulfill the following requirements, sometimes contradictory [1-2]:

- reducing the concentration of methane in ventilation airflows, by draining as much as high possible gas quantities before these are reaching the mine working's atmosphere;
- capturing a gas mixture with a high value calorific power needed for beneficiation purposes.

The management of methane drainage systems requires performing numerous and regular controls which enable programming interventions aimed at limiting the influence of operational anomalies that inevitably have repercussions on the methane concentration in the working environment of coal mines [3-5].

The automated monitoring and control of methane drainage systems, reliable and effective, is therefore currently an imperious necessity.

In this context, the paper aims to identify the elements necessary to develop an automatic management system functioning degassing plants, taking into account the special role they still play, both in terms of output and especially of workplace safety.

The development of procedures for monitoring and control at methane drainage facilities used in Jiu Valley coal basin was this was done through the studies and research carried out at the central methane stations from Lupeni, Paroseni and Livezeni collieries. A major role

[*] Corresponding author: gabriel_babut@yahoo.com

in accomplishing in carrying out this research returned to the expertise gathered from discussions with a large spectrum of experts involved in the coal seams drainage operations (mine technical staff, researchers from the National Institute for Research and Development in Mine Safety and Protection to Explosion - INCD INSEMEX Petrosani, labor inspectors). The main aspect considered was the variation of parameters describing the methane drainage systems operation.

In formulating proposals and conclusions of the studies and research carried out in Jiu Valley collieries were taken into account, where appropriate, the observations made on the length of specialization and documenting stages made in coalfields from Poland, France, Russia and Czech Republic.

2 Monitoring and control of methane drainage systems performed at underground drainage ducting network level. A critical analysis

2.1 Preliminary controls

This kind of control refers to the checking, along the entire length of the drainage pipes, of the following issues:

- valves operation;
- water-gas-detritus separators state;
- location of any aspiration of air within the drainage piping.

While the, occupational safety and health (OS&H) statutory requirements in force in Romania do not require to achieve this type of control, it is only performed sporadically [6]. To increase the efficiency of the coal degassing we propose to implement this type of control on daily basis, by qualified personnel according to a schedule set by the head of ventilation and OS&H department of the mine.

2.2 Methane concentration monitoring

Romanian OS&H regulation in force stipulate to determine the concentration of methane the following intervals: monthly, at the aprons and drainage locations (niches) and at drainage points coupled to the collector pipes; bimonthly at the measuring diaphragms of the drainage pipe route. The low frequency in the determination of the methane concentration does not allow a proper and effective of the methane drainage systems. Experience acquired worldwide recommend the determination of the methane concentration in drainage pipes to be carried out on daily basis, and in some cases the need for several measures during the same day is stipulated [7-8].

The allowed lowest methane concentration value in the main ducting network of central drainage systems is 30 %. In pipes linking the degassing locations to the main or secondary drainage pipes the allowed lowest methane concentration value is 20 %. While the present frequency imposed for methane measurements by the OS&H regulations in force is not allowing to perform adequate adjustments of the system, we are further presenting the control and adjustment procedures involving daily measurements of this parameter.

If in the moment of methane concentration measurement, the value obtained is near the 30% limit, one will pursue to the methane-air mixture flow reduction, through the adjustment of the regulating valve. This operation is aimed at providing a methane concentration increase in the drained methane-air mixture.

After this adjustment is completed, a new control is performed. If the methane concentration did not exceeded 30 % and the oxygen content equals or is higher than 7 %, the adjustment valve will be completely closed. If, contrarily, a significant increase is registered in the methane concentration simultaneously with a lower than 7 % oxygen concentration, then the new parameter values will be recorded and the valve will be maintained in the same position.

The staff members involved in the control process will dispose permanently of the data representing the measurement results from the previous days. If the methane concentration value measured in day N, compared to values measured in days N - 1, N - 2, ... are emphasizing a slow decrease trend, of a few percents per day, then the causes of this phenomenon can be one of those listed below:

- a barometric pressure increase, with a subsequent alteration of the gassy-dynamic regime;
- the depletion of the gas reservoir.

If, although this methane concentration decreases, this parameter's value is placed above the imposed limit (e.g. at values comprised between 45 - 50 %), this value will be recorded, but no adjustment operation will be required. Nevertheless, for that measuring station the frequency of controls will be increased (e.g. one measurement per working shift).

If a sudden methane concentration decrease is occurring from one day to another (having a magnitude of tenths of percents methane), an adjustment operation is required immediately.

The following cases can be identified:

- despite the sudden decrease, methane concentration remains above the regulated limit: in this situation the gas flow will be diminished, the reduction being proportional to the flow quantity value; after adjustment is completed, methane concentration and flow will be monitored on regular basis; if concentration is still decreasing, the flow will be further reduced and the previous procedure will be repeated; otherwise, namely in concentration stabilization conditions, the parameters values will be recorded and no further adjustment will be carried out;
- the sudden decrease of methane concentration brings this parameter in the vicinity of the regulated limit:
  - if the source supplies a high gas quantity (> 500 m^3/h) the flow will be consistently reduced and the methane and oxygen concentrations will be monitored; the previously described procedure will be applied;
  - if the source supplies a high gas quantity (100 - 200 m^3/h), this source will be shut down; a visual control of the drainage pipe will be performed upstream the measuring station for detecting the potential air inleakages; based on the information provided by the staff members involved in the monitoring process, the drainage officer from the mine ventilation and OS&H department will decide if certain adjustments should be done to mitigate the occurred situations.

2.3 Gas flow determination

Gas flow determination in the drainage pipes is done indirectly, employing standardized diaphragms. To simplify the gas flow measurements, the INCD INSEMEX Petrosani has elaborated nomographic charts for each kind of diaphragm.

Romanian OS&H regulations in force are stipulating the compulsoriness of bimonthly determination of gas flows in the drainage locations and at the measuring nozzles on the drainage ducting route. Being that the drained gas flow represents one of the parameters determining the effective operation of the methane drainage system the aforementioned measurement frequency is inadequate [9]. For this reason, we propose the daily determination of gas flow in the drainage locations and at the measuring nozzles on the drainage ducting route.

Next there are presented the proposed control and adjustment procedures for the case of daily frequency gas flow measurement:

- sudden increase of rough and net flow: this increase can be either triggered by a barometric pressure fall, or by the change of operational parameters of the drainage plant; this situation is rather favorable and do not requires any flow adjustment intervention;
- increase in the rough flow value only: if an increase in the rough flow value only is registered simultaneously with pressure decrease, there is a possibility that incidental inleakages occur through the drainage ducting; a visual control is needed upstream the measuring point;
- decrease of rough and net flow: as rule-of-thumb, a gas flow decrease is accompanied by a pressure fall; comparing the measured flow value with the previously determined values it can be emphasized a slow flow decrease trend, not followed by a high pressure drop; this situation is due to the drainage pipe impairment, or else, more probably, to the obstruction or shearing of the drainage borehole; if drainage operations in a production sector are involved then the pressure drop will be risen by opening the adjustment valve (respecting the guard of water-gas-detritus separator); a new measurement series will allow to assess the effects induced on the situation; if the presence of additional drainage possibilities arose, the water-gas-detritus separators will be shut off to apply the maximum available pressure drop; the adjustment valve will be reopened, while the main goal of drainage operation still is the methane concentration reduction in the underground atmosphere.

2.4 Pressure drop determination

According to the OS&H regulations in force in Romania, pressure drop determination should be done bimonthly in the drainage locations and at the measuring nozzles on the drainage ducting route. The considerations mentioned for the case of methane concentration and gas flow measurements are also valid in the case of pressure drop determination in drainage pipes. For this reason we proposed for pressure drop also the measurement on daily basis.

The proposed control and adjustment procedures are the following:

- low variations of pressure drop: in the case of low pressure drop variations the variation of barometric pressure will also be determined, while they generally are in direct interdependence; no adjustment will be required in this situation;
- pronounced pressure drop increase: in this situation, the gas flow and the methane concentration will be also measured; if fluctuating gas velocity is detected, this emphasizes the presence of a water plug in the drainage pipeline; the status of water-gas-detritus separators will be checked, opening them at once if they were found closed; the adjustment valve will be partially shut down, to reduce pressure drop value; once the operational regime is stabilized, a new series of measurements will be performed and the need of other regulation/adjustment operation will be assessed; if the water plug was eliminated, the methane concentration is high and a margin of safety remains on handling the adjustment valve, then the valve can be reopened to obtain a further increase the pressure drop on the pipe;
- pronounced pressure drop decrease: the pronounced pressure drop decrease in the drainage pipe is generally accompanied by rough and net gas flow reduction; procedures to be applied are similar to those described in section 2.3.

2.5 Carbon monoxide concentration determination

Romanian OS&H regulation in force stipulate to determine the concentration of carbon monoxide monthly, only at the aprons and drainage locations (niches). Because the concentration of carbon monoxide can provide information for early detection of spontaneous combustion, we propose its measurement on regular basis. Similarly with the case of the other aforementioned parameters we propose the daily measurement of the carbon monoxide concentration.

For the specific conditions of the Jiu Valley collieries there were not determined the maximum allowable limits of the carbon monoxide concentration. The worldwide experience in the field of coal seam methane drainage indicates that the following limit values can be employed [10]: at drainage locations level - 90 ppm CO; at drainage aprons level - 45 ppm CO.

These values can be also used in a first stage for the methane drainage systems operating in Jiu Valley coal basin, following that they will be further adapted to the specific geological and mining conditions met in this coal basin.

The proposed control and adjustment procedures are the following:

- slow carbon monoxide concentration increase: in this case, we consider a 2-3 ppm/day increase; if the carbon monoxide concentration is not near maximum allowable limit, no adjustment is required;
- fast carbon monoxide concentration increase: can be defined as an increase higher or equal to 10 ppm/day; in this situation, two types of actions can be envisaged:
  - partial closure of the adjustment valve, to decrease the applied pressure drop;
  - in that measuring point the measurement frequency will be increased.

2.6 Oxygen concentration determination

Romanian OS&H regulation in force do not stipulate the need to determine the oxygen concentration in the drainage network. Since the expertise acquired in countries with well-developed mining industry is highlighting that, besides the methane concentration, the oxygen concentration represents a main parameter determining the safe operation of the methane drainage system, we propose the introduction in the national OS&H regulations of the compulsoriness of daily measurement of the oxygen concentration in the drainage locations and at the measuring nozzles on the drainage ducting route.

The maximum value of the oxygen concentration indicated in worldwide literature is 7% [10] and the control and adjustment procedures are synthesized bellow:

- slow oxygen concentration increase: by slow oxygen concentration increase it is understood a modification lower than one percent per day; if such an increase is recorded, the measuring frequency will be higher;
- fast oxygen concentration increase: the meaning of the notion of fast oxygen concentration increase is defined as a change with several percent's of the concentration, above the limit of 7 %, during one day period; in this case, in the first stage a visual inspection of the drainage pipes will be done, to detect a potential incidental air inleakage; furthermore, the adjustment valve will be partially closed to diminish the pressure drop exerted; if the oxygen concentration does not decrease below 7 %, the gas source should be shut down.

2.7 Drained gas temperature determination

Romanian and worldwide OS&H regulation in force do not stipulate the need to determine the temperature of the drained gas. In these conditions, this parameter is measured only exceptionally. Because the temperature of the drained gas can provide information for early detection of spontaneous combustion, we propose its measurement on regular basis.

3 Controls at methane drainage plant level

If the methane concentration in the exhaustion pipe (after the vacuum pump) decreases below 30 %, the operation of the methane drainage system is automatically shut down and an alarm signal is generated.

Romanian OS&H regulation in force do not stipulate the need to determine the oxygen concentration at the drainage system's level. For the reasons mentioned in section 2.6, it is proposed to continuously measure and record the oxygen concentration and introducing the 7 % value for automatic shutdown of the methane drainage system and generation of an alarm signal.

The central methane drainage system should include the following monitoring and control apparatus [6]:

- gas analyzer with continuous indication and recording of the extracted gas concentration;
- flow meter with continuous indication and recording of the extracted gas quantity;
- indicator and recorder vacuum-meter for pressure drop monitoring in the suction pipe;
- indicator and recorder manometer for pressure monitoring in the discharge pipe;
- indicator and recorder thermometers for gas temperature monitoring in the suction and discharge pipes, and for cooling water control;
- gas analyzer for automated control of methane concentration in the chamber containing the suction equipment and in the command room; it should be mounted in the highest location in which methane accumulation is possible.

4 Assessment of elements required for developing automated monitoring and control methane drainage systems

The analysis of the results of studies and research performed, combined with the proposals set in the previous sections allows to assess the main elements required for the development of an automated monitoring and control methane drainage systems [11-12]. These elements can be summarized as it follows:

- parameters to be determined and their variation range:
  - methane concentration of drained gas: 0 - 100 %;
  - carbon monoxide concentration of drained gas: 0 - 300 ppm;
  - oxygen concentration of drained gas: 0 - 21 %;
  - differential pressure (for indirect flow measurement): 0 - 6000 m^3/h;
  - absolute pressure in the drainage pipe: 0 - 10^5 Pa;
  - drained gas temperature: 0 - 120 °C.
- limit values for certain parameters (stipulated in OS&H regulations in force or assessed based on experience acquired):
  - methane concentration: if the methane concentration in the suction pipe (after the vacuum pump) decreases below 35 % an alarm signal will be triggered; if the methane concentration in the suction pipe decreases below 30 % the suction equipment will be automatically shut down; the minimum allowable methane concentration in the underground drainage pipes is 30 %; in the connection pipes between the drainage points and the main or secondary pipes, the minimum allowable methane concentration is 20%;
  - oxygen concentration: maximum allowable oxygen concentration is 7 %;
  - carbon monoxide concentration: maximum allowable carbon monoxide concentrations are: at drainage locations level - 90 ppm; at drainage aprons level - 45 ppm.

Table 1 summarizes the major abnormalities which can affect the normal operation of a methane drainage system and the variation pattern of the main parameters emphasizing these abnormalities.

Table 1. Potential abnormalities possible to occur in a methane drainage system's operation

Specification	CH_4 [%]		O_2 [%]		CO [%]	Gas flow [m^3/s]			Absolute pressure [Pa]		Temperature [°C]
	↓ fast	↓ slow	↑ fast	↑ slow	↑	↑ fast	↓ fast	↓ slow	↑	↓	↑
Drainage pipe rupture or drainage borehole breakage upstream the measuring point	X		X			X			X		
Drainage pipe rupture downstream the measuring point							X		X		
Parallel mounted drainage pipe rupture							X		X		
Drainage pipe impairment							X			X	
Water plug							X			X	
Spontaneous combustion triggering					X						X
Total or partial shutdown of the drainage system							X		X		
Methane source depletion		X		X				X	X		

5 Conclusions and proposals

The critically analysis of control and adjustment procedures presently applied in methane drainage systems within Jiu Valley collieries let us emphasize that they do not allow an adequate and effective enough management of their operation.

The main findings imposing this conclusion are related to the low frequency of measurements (generally on monthly or bimonthly basis) and the lack of controls otherwise well established by regulations in countries with developed mining industry (e.g. oxygen concentration monitoring). Also, it should be mentioned that adjustment procedures are not precisely established, there necessity and ways to perform them being left to the expertise level of the staff members charged with drainage monitoring and management activities.

To set these drawbacks practical proposals concerning the introduction in the IIIrd Chapter - Mine drainage, from the Vth section of the Occupational Safety and Health Regulation in force at the National Hard Coal Company - Petrosani of certain provisions regarding:

- the compulsoriness of carrying out on daily basis the measurement of the methane drainage system's parameters;
- the compulsory oxygen concentration determination at drainage pipes network and methane drainage system's level;
- the compulsoriness of carrying out carbon monoxide concentration and temperature at all methane sources level (drainage locations, aprons etc), considering that monitoring these parameters is a valuable information source for early detection of spontaneous combustion.

At the same time, we developed the procedures of methane drainage system operational adjustment, considering the daily frequency of controls.

Putting in practice the presented proposals is likely to generate a significant improvement of the hand powered control and adjustment procedures presently employed, with immediate beneficial consequence regarding the raise of coal methane drainage systems efficiency. Nonetheless, should be stated that the proposed control and adjustment procedures are based on the reliability of the human factor, raising a potential risk of inadequate assessment of the dynamic situation changes occurring, with direct negative impact on work safety. The uncertainty related to the error range brought by human operator, with all his limits and possible faults, constitutes an important drawback of the actual manually operated methane drainage systems.

Under these circumstances, beyond the identification of improvement possibilities of the present methane drainage systems operation, the analysis carried out allowed also to establish the elements converging to the development of an automated monitoring and control methane drainage systems, representing the initial step to the development of such a system.

References

1. E. Cichowski, Archives of Mining Sciences, **48 (1)**, 81 (2003)
2. E. Krause, K. Krzemień, Journal of Sustainable Mining, **13(2),** 6 (2014)
3. C.Ö. Karacan, F.A. Ruiz, M. Cotè, S. Phipps, International Journal of Coal Geology, **86(2-3),** 121 (2011)
4. F.N. Kissell, *Handbook for methane control in mining*, (U.S. Department of Health and Human Services, Public Health Service, Centers for Disease Control and Prevention, National Institute for Occupational Safety and Health, DHHS (NIOSH) Publication No. 2006-127, Information Circular 9486, Pittsburgh, PA: U.S., Accessed on 11 March 2017 at: https://www.cdc.gov/niosh/mining/UserFiles/works/pdfs/2006-127.pdf, 2006)
5. R.I. Moraru, G.B. Babut, L.I. Cioca, Proceedings of the 13th SGEM GeoConference on Science and Technologies in Geology, Exploration and Mining, **I**, 731 (2013)
6. N.H.C.C., Occupational Safety and Health Regulation of National Hard Coal Company - Petrosani (in Romanian), National Hard Coal Company, Petrosani, Romania (2007)
7. J.M. Mutmansky, *Guidebook on coalbed methane drainage for underground coal mines: White Paper*, (U.S. Environmental Protection Agency, Washington, DC: U.S., 1999)
8. Q. Ye, Y. Pi, Z.Z. Jia, H.Z. Wang, Applied Mechanics and Materials, **295-298**, 3023-3026 (2013).
9. N. Szlązak, M. Borowski, D. Obracaj, J. Swolkień, M. Korzec, Archives of Mining Sciences, **59(3),** 655 (2014)
10. UNECE, *Best practice guidance for effective methane drainage and use in coal mines*, (U.N. Economic Commission for Europe, ECE Energy Series, No. 31, Geneva, Switzerland, Accessed on 11 March 2017 at: http://www.unece.org/fileadmin/DAM/energy/se/pdfs/cmm/pub/BestPractGuide_MethDrain _es31.pdf, 2010)
11. G.B. Babut, R.I. Moraru, Proceedings of the 14th SGEM GeoConference on Science and Technologies in Geology, Exploration and Mining, **III**, 71 (2014)
12. W. Dziurzynski, A. Krach, T. Palka, Archives of Mining Sciences, **54(2),** 159 (2009)

MATEC Web of Conferences **121**, 11006 (2017)
MSE 2017
DOI: 10.1051/matecconf/201712111006

The impact of occupational risks in the operating and maintenance activities in the energy sector

Georgiana Buica[1,*], *Anca Elena* Antonov[2], *Constantin* Beiu[1], *Dragos* Pasculescu[3], *and Remus* Dobra[4]

[1] INCDPM „Alexandru Darabont", Electrical and Mechanical Risks Laboratory, 35A Blvd, Bucharest, Romania
[2] INCDPM „Alexandru Darabont", Certification Body, 35A Blvd, Bucharest, Romania
[3] University of Petrosani, Department of Control Engineering, Computers, Electrical Engineering and Power Engineering, 20 Universitatii St., Petrosani, Romania
[4] 1 Decembrie 1918 University of Alba Iulia, Department of Computer Science and Engineering, 5 Gabriel Bethlen St., Alba Iulia, Romania

Abstract. This paper presents a synthesis of studies in electrical installations from the energy sector, relating to occupational hazards, in terms of history of operations, maintenance, including situations that arose after some events that involved accidents at work and occupational diseases. Over the years, the energy sector has undergone a number of changes to the organizational level, modernization of electrical installations and implementations of new technologies and working procedures. As a result of these changes, has increased, over time, the level of safety of electrical installations, as well as the activities carried out by the operating and maintenance personnel that requires a high degree of training, as determined by the responsibilities assigned by job descriptions and procedures and as a result of physical and psychological stress to which the worker is subject during carrying out the activity. The research study aims to highlight the safety measures and health needed to be taken and their effectiveness to mitigate risks in the energy sector, in order to manage, more effectively, the work carried out by the operating and maintenance personnel of electrical installations, in order to ensure the healthy workplaces and to fulfil the occupational management objectives.

1 Introduction

Investigation on working conditions in the electricity distribution installations, for 1969-2015 and the assessment of risk level in electrical installations of production, transmission, distribution and use of electricity, together with technical and judicial expertise on accidents at work, led to analyse the solutions adopted by employers in order to fulfil the objectives of the management plan for quality of service, the level of safety in electrical installations, the safety provided for personnel, population and environment.

* Corresponding author: gbuica@protectiamuncii.ro

DOI: 10.1051/matecconf/201712111006

The statistics on accidents at work at national level, processed in research studies indicate that there is a low decrease in number of accidents, according to records of National Statistics Institute (NSI) in the *Production and supply of electricity, gas and water section*, presented in Figure no.1 [1, 2].

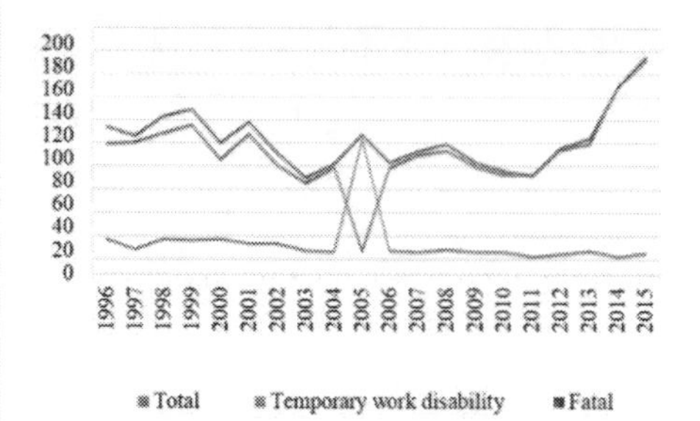

Fig. 1. Total work accidents produced in Romania, during 1996-2015

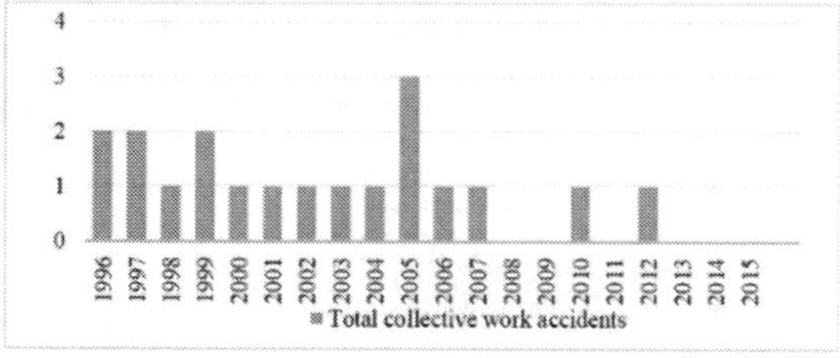

Fig. 2. Total collective accidents at work produced in Romania, during 1996-2015

The analysis allowed the identification of the increasing number of workers involved in work accidents and collective work accidents at is missing in the last three years.

2 Summary of studies on the risks identified in operating - maintenance activities of electrical installations

In order to diagnose the technical condition and safety of electrical installations, research studies consisted in identifying occupational risk factors related to the operating and maintenance of these installations, through the interaction of the four elements that constitute the work system.

In carrying out the operation and maintenance of electrical installations, the principle of division of work, by designating specific, well defined, of workers' participation and by establishing individual responsibility throughout whole activity, from the disposition of work, its organization, to execution and reception.

Analysing the technical, safety documents and job descriptions, it is noted that the work carried out by the personnel of operation and maintenance is a complex one and involves a high degree of training, accountability and require neuropsychological stress, e.g. due to the short time in which the work must be carried out, the need to take quick decisions that can have serious consequence, because of possible errors in carrying out the operations.

The electrical installations, depending on their field and category, have in composition the same types of electrical equipment, based on some standard operating diagrams, schemes that may have certain particularities, depending on the role in the electrical power system.

The degree of protection for the work equipment in electrical installations, confirms a good level of safety. It should be noted, however, that some of the equipment is in the last period of reliability, which can increase the probability of defects and, therefore, can decrease the safety level [3].

In recent years there has been a growing number of upgraded power stations leading to the elimination of these risks, and increasing the number of manoeuvres carried out by the operating personnel, during the upgrading activities.

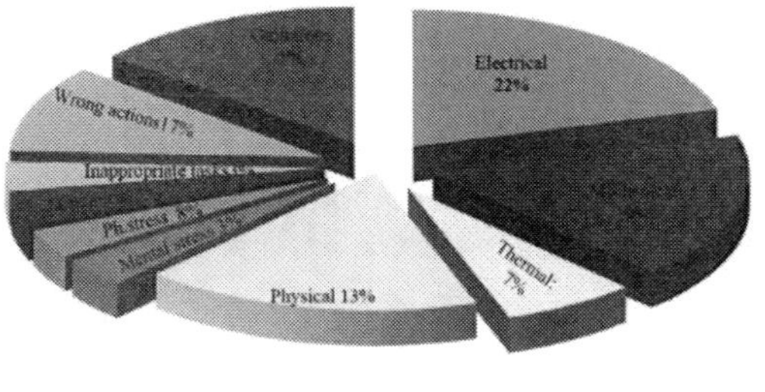

Fig. 3. Risk factors share identified in substations, on operations and maintenance activities

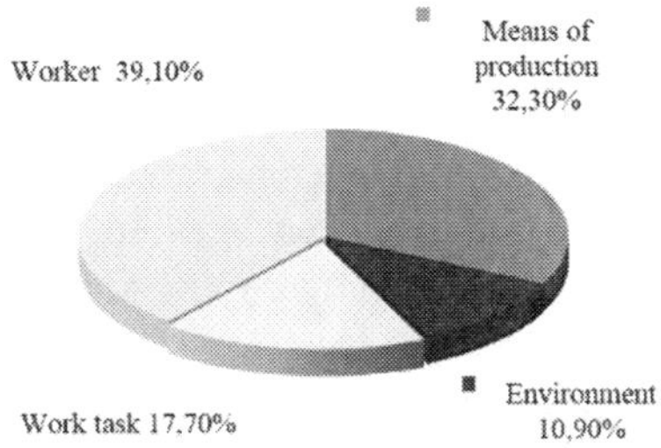

Fig. 4. The distribution of risk factors based on components of work components in substations, on operations and maintenance activities

The largest share of the risk factors identified in electric substations, from all the risk factors identified in the assessment process, are electrical risk factor, with 23% and 19% mechanical risk factors. Figure 3 presents the share risk factors identified in substations, on operations and maintenance activities.

In figure 2 is presented the risk factors share identified in the electrical installations for the overhead power lines, and found that mechanical risk factors and risk factors generated by the workers have the highest percentage 18 %. The electrical risk factors share identified are only 17% [5].

In recent years there has been a decrease in the weighting of electric risk factors, from 28% to 23%, of the factors generated by inappropriate tasks, from 9% to 4%, due to upgrades or modernization and introducing of remote control systems, correlation of work procedures with safety requirements and with causes of accidents at work.

Modern numerical protections with advanced techniques and technologies ensure the efficient operation of the entire coordination and control protection system, fact that is an essential contribution for safe electric energy supply. [4]

Grouping the major risk factors on system work components, shows that the largest share of risk factors is given by risks generated by workers, with a rate of 39.1%, followed by those generates by means of production / work equipment, with a percentage of 32.3%.

Fig. 4 shows the distribution of major risk factors on operations and maintenance activities in power substations, divided by work system components.

From the graphic representation of the distribution of risk factors on work system components shown in Figure 3, fit appears that the most accidentogene situations, with potential risks of injury are those given by worker, with 36% and the means of production with 24%.

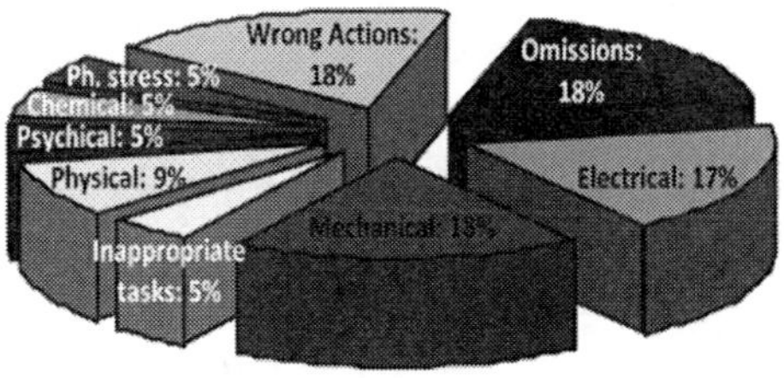

Fig. 5. Risk factors share identified in the electrical installations (overhead power lines)

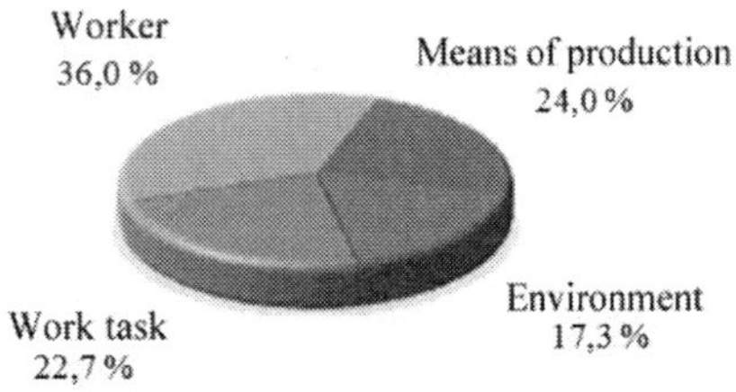

Fig. 6. The distribution of risk factors based on components of work components in the electrical installations (overhead power lines)

In case of the risks identified in the maintenance activity of the HV overhead lines (OHL) it is found that the largest share is held by the mechanical risk factors, risk factors generated by workers, with 18% and electrical risk factors with 17%. Fig. 5 shows the share of the identified risk factors for maintenance activities of HV OHL [5].

Grouping the risk factors on work system components, it shows that the most dangerous situations in maintenance activities of the HV OHL, including potential risks of injury, are given by the worker, with 36% and means production / work equipment, with 24%. The graphical representation of the distribution of risk factors on work system components is shown in Fig. 6 [5].

The weight of these risk factors is in decline due to upgrading programs and National Power System (NPS) interconnection with neighbouring power systems and due to expansion of activities carried out live Working (LW).

The maintenance activity on OHL live working is complex, the maximum foreseeable consequence for accidents at work, death and disability, requiring strict compliance of work technology for live working and of technical and organizational safety requirements in work area, use of work and safety equipment for live working, in accordance with the safety and technical requirements set by the applicable regulations, checked and tested regularly.

Taking into account the risks to which workers are exposed, complexity of the work, its specificity, training and education of workers must be continued and is necessary because during the execution of work, live working workers reaches deliberately the live parts of electrical installation or enter deliberately in live area [6].

The approach to the training an evaluation of Live Working personnel must have a scientifically base and must be a continuous and methodical concern for live working managers [7].

According to the top management of the national energy system, after more than 35 years in the LW, there has been no accident at high voltage.

This cannot be said in the case of live working in low-voltage installations where the risk factors generated by inappropriate technology, procedures or live working methods, risk factors generated by workers, by wrong actions and omissions hold the highest percentage.

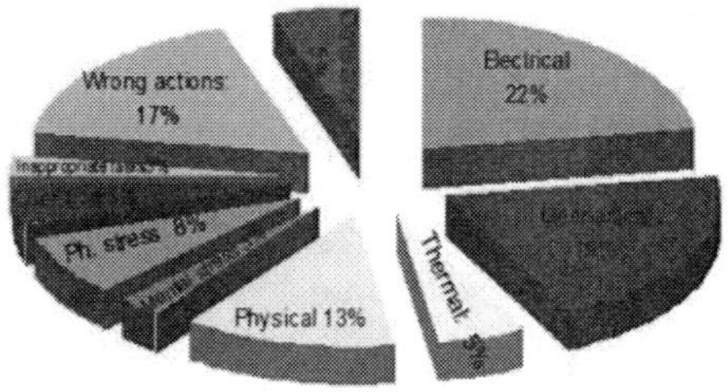

Fig. 7. The weight of the identified risk factors in electrical installations (UGL - 0.4,-110 kV)

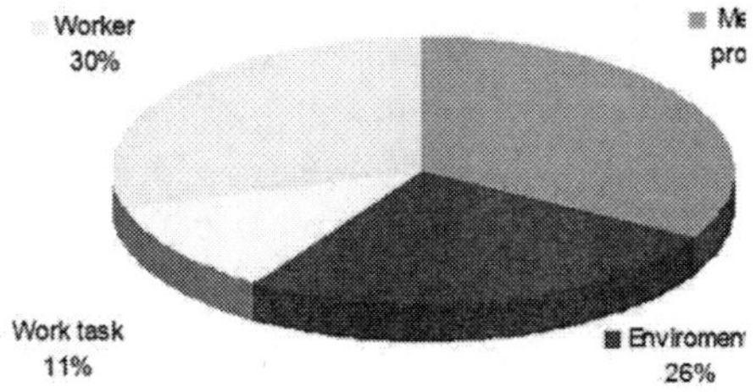

Fig. 8. The distribution of risk factors based on components of work components, in electrical installations (UGL - 0.4,-110 kV)

In the maintenance activities for underground power lines (UPL), the highest share of the identified risk factors it have the electrical risks, with 22% and mechanical risks, 19%. Fig. 7 presents the share of identified risk factors in the maintenance activity of UPL.

Part of the work equipment, especially cables, were replaced, which led to the elimination of risks (risk of electric hazard, heat, fires, explosions, etc.), but many work equipment, especially in the low-voltage distribution transformers joints, is exceeded technical lifetime specified by the manufacturers [8].

From the representation of the distribution of risk factors on work system components, shown in Fig. 8 shows that the most dangerous situations, including potential risks of injury are those given by work equipment, with 33% and by worker with 30%.

Due to the fact that professional risk factors generated by the use of work equipment and by the workers are generally the basis of the technical or human hazards, resulting in the most cases invalidity or death, it is necessary to adopt some technical tools for managing the operation and maintenance activities of electrical installations.

The impact of these occupational hazards in companies require to continue to take technical and safety measures, modernization of work equipment, as well as continuous training of workers who undertake work in this sector

The research made by INCDPM over the years, led to the development of tools for the management of occupational hazards in the national economy in order to reduce the impact of occupational risks both in energy sector and related to operation and maintenance activities of work equipment in use.

For this purpose „Guidance on risk assessment and prevention of electrical risks" and "Risk management within professional equipment maintenance activities to prevent work accidents and occupational diseases - a prerequisite for increasing the competitiveness of employers on the market" developed in the research papers, are instruments that highlights the safety and health measures needed to be adopted in the national economy sectors, in order to reduce the impact of occupational risks and to manage more effectively the work carried out by the operating and maintenance workers in electrical installations and to ensure safe and healthy workplaces and to fulfil the objectives of occupational management.

3 Conclusion

Conducted studies over the years in electrical installations for use, distribution, transmission and production of electricity have allowed finding and drawing conclusions on the role, status and technical deficiencies of work equipment through the interaction of the four elements of work system.

In their activity, workers in electrical installations are subject to a number of factors risks from work equipment (e.g. electrical, mechanical, thermal, chemical risks), work environment (e. g. high / low temperature, air currents, humidity, weather, insufficient light - especially at night - dust, toxic gases, biological risk) and risk factors related to the workload that causes physical strain (e.g. the execution of manual work, forced postures, in narrow spaces, requiring professional gestures which requires strain of musculoskeletal and articular system when handling work equipment, various devices and tools) and mental strain (e.g. the performance of work in a short time, decisions in short time).

It is necessary the continuation of modernization of electrical equipment, the awareness of the importance and compliance of preventive maintenance programs, implementation of new technologies and procedures, correlation of technical and safety and health documents with the causes of accidents at work and occupational diseases, training, annual examination in terms of OHS for workers, all of these being technical and health and safety measures that can be adopted, resulting in reducing the impact of risks on operation and maintenance workers of electrical installations and to fulfil the objectives of occupational management.

References

1. A Antonov, G. Buica, C.Beiu, EEMJ, **13**, (2014)
2. NSI - TEMPO program 2017
3. Dobra R., Buică G., Beiu C, Păsculescu D, WSEAS International Conference, Athena, Greece, **1**, 58 (2013)

4. Păsculescu V, Şuvar M, Păsculescu D.,Transactions on Electronics and Communications, 56(70) Fasc. S1 (2011)
5. Buică G., Contributions of diagnostic methods of safty and health at work in electrical installations, Universitas, Petroşani, 70 (2010)
6. ASRO, SR EN 50110-1:2013 Operation of electrical installations. General requirements (2013)
7. MN Oltean, T Fagarasan, G Cotrigasanu, CIGRE Regional South-Est European Conferince, E111, 54-61 (2012)
8. INCDPM, Technical expertise on the assessment of working conditions of electrical installations for SC Electrica SA, between 1969-2001 required to establish group I work conditions (2009)

 DOI: 10.1051/matecconf/201712111007

Key elements on implementing an occupational health and safety management system using ISO 45001 standard

Doru Costin Darabont[1,*], *Anca Elena* Antonov[1] , and *Costică* Bejinariu[2]

[1]National Research and Development Institute on Occupational Safety – INCDPM "Alexandru Darabont", Ghencea 35A, Bucharest, Romania
[2]"Gheorghe Asachi" Technical University of Iaşi, Dimitrie Mangeron 67, Iaşi, Romania

Abstract. Occupational health and safety (OHS) management system is one of the main elements of the company's general management system. During last decade, Romanian companies gained a valuable experience on implementing this type of management systems, using OHSAS 18001 referential and standard. However, the projected release of the ISO 45001 represents a new approach which requires the companies to take in consideration new key elements for a successful implementation of the OHS management system. The aim of the paper is to identify and analyse these key elements, by integration of the following issues: standard requirements, Romanian OHS legislation and good practice examples, including the general control measures for new and emerging risks such as psycho-social risks, workforce ageing and new technologies. The study results represent an important work instrument for each company interested to implement or upgrade its OHS management system using ISO 45001 standard and could be used regardless the company size or activity domain.

1 Introduction

The expected publication of the new ISO 45001 standard represents an important step in the management of occupational health and safety and, in the same time, is an predictable appearance, taking in account the evolution of other management systems approaching, such as quality and environment. Analysing the history of the quality domain standards, it can be observed that they evolved from the first standards with very narrow applicability, such as MIL-Q-9858 in the military domain in 1942-1952, to the British national standards as BS 9000 and, finally, to international standards in the ISO 9000 series, with the latest version ISO 9001:2015 [1].

A similar evolution could be observed at the environmental management standards, amplified as an effect of the United Nations Conference on Environment and Development (UNCED), held in Rio de Janeiro, in 3-14 June 1992, with the participation of 172 states, having as principal themes environment and sustainable development. Thus, starting from

* Corresponding author: darabont_d@yahoo.com

regional and national standards, the approach has evolved to the ISO 14001 standard, with its latest edition in 2015.

A new approach of the occupational health and safety management is necessary, as the worldwide companies are facing with new and emerging risks as well as with substantial changes on economical environment.

The publication of the ISO 45001 is expected for February 2018 [2], but we consider that the companies should start prepare for its implementation. Thus, the aim of the paper is to identify and analyse the key elements for a successful implementation of the OHS management system, by integration of the following issues: standard requirements, Romanian OHS legislation and good practice examples, including the general control measures for new and emerging risks, such as psycho-social risks, workforce ageing and new technologies.

2 The role of ISO 45001 in organization's management

According to International Labour Office [3], every 15 seconds a worker dies from a work-related accident or disease and 153 workers suffer a work-related accident. This statistic stuck out that more than 2.3 million workers die annually due to work-related accidents or occupational diseases. Moreover, in the last years, the perception of risk has been changed substantially as well as the macro economic context in which organizations have to perform. The main problems raised by the new and emerging risk are related to psychosocial hazards, introduction of new technologies or globalization or ageing of the population. The population across the EU has been ageing significantly over the last decades and this trend will maintains in the next decades. The EU average age was 33 years in 1960 and is expected to grow to 44 years in 2020 and to 46 in 2060 [4].

The main obligation of the employer is to ensure the safety and health of workers in every aspect related to work [5-6]. In order to fulfil this obligation, and faced to the changes in micro and macro economic environment and to the new and emerging risks, the employer needs to adopt a new approach and new instruments to perform an efficient management of occupational health and safety.

In this context, the main role of the ISO 45001 standard is to serves as a useful instrument to enable an organization to proactively improve its occupational health and safety performance, regardless the size, type and nature of the organization [7-8]. But observing the requirements of the standard is not enough. An efficient implementation of the standard requires, in the same time, a very good knowledge and compliance with the legal requirements and other requirements, a complete and up-dated risk assessment and a strong safety culture inside the organization. In implementation of the ISO 45001, the organization should not start from zero, but it should put in value the achievements that it already have.

The structure of the ISO 45001 corresponds to ISO Annex SL, guideline for writing standard for management systems. The main advantage of this approach is that all management standards have the same high-level structure, facilitating the implementation of an integrated management system, especially with ISO 9001:2015 and ISO 14001:2015. The structure of the ISO 45001 standard is shown in Table 1 [9].

Is expected that an ISO 45001 based occupational health and safety management system will enable the organization to improve its performance by [7]:

- developing and implementing an OHS policy and objectives;
- establishing systematic processes which consider the organization context, its risk and opportunities and its legal and other requirements;
- determining the hazards and risks associated with its activities;

- establishing operational control to manage the OHS risks as well as legal and other requirements;
- increasing awareness of its OHS risks;
- evaluating its performance and seeking to improve it;
- ensuring workers to take an active role in OHS issues.

Table 1. The structure of ISO 45001 standard

Chapter no.	Title
1	Scope
2	Normative references
3	Terms and definitions
4	Context of the organization
5	Leadership and worker participation
6	Planning
7	Support
8	Operation
9	Performance evaluation
10	Improvement

3 OHS key elements related to ISO 45001

3.1 Context of the organization

In chapter 4, ISO 45001 requires the organization to address the following issues:
- understanding the organization and its context;
- understanding the needs and expectations of workers and other interested parties;
- determining the scope of the OHS management system.

To determine and understanding the context of the organization, the ISO 45001 standard recommends to address both internal and external context issues, such as: cultural, social, political, legal aspects, introduction of new competitors, contractors, suppliers, partners and providers, new technologies, culture in the organization, form and extent of contractual relationship, working time arrangements etc. [9].

The interested parties, of which expectations should be taken in consideration, include stakeholders, authorities, parent organizations, suppliers, contractors, trade unions and employers' organizations.

Before establishing the scope of the OHS management system, a detailed analysis of the organization's context and expectations of workers and other interested parties should be performed. The main instruments that should be used to perform this analysis are:
- meetings with each interested party in order to document their needs and expectations;
- reviews of the legal requirements which are relevant for organization as well as of other requirements, such as contractual clauses or other interested parties requirements;
- reviewing the good practice examples, in order to adopt the relevant ones;

- internal audits for evaluation of compliance with legal and other requirements, including the requirements of other interested parties;
- consultations with OHS experts and authorities, i.e. National Research and Development Institute on Occupational Safety or Territorial Labour Inspectorate.

The results of applying the above mentioned instruments should be used to define the scope of the OHS management system.

3.2 Leadership and worker participation

Chapter 5, "Leadership and worker participation" address the following issues [9]:
- leadership and commitment;
- OHS policy;
- organizational roles, responsibilities, accountabilities and authorities;
- participation and consultation.

The main idea that all members of an organization, from top-management to each worker, should understand is that they must have an active role in the OHS management system, and the efforts of all members should be synergic.

At this phase, the following instruments should be used, additionally to the specific requirements of the standard:
- training courses for all managing levels of the organization, especially for top-management, taking in account that in many situations, top-managers have a poor knowledge of occupational health and safety principles and legislation;
- training courses for all workers, in order to make them understand their role as a support for management and to strength the organization safety culture;
- activating and involving the actors in OHS domain, such as OHS Committee, OHS worker's representatives, internal/external preventive and protective service(s), occupational medicine service.

The training courses should be tailored for each level and OHS role in organization and should be oriented on practical issues which are relevant for the focus-group.

3.3 Planning

Planning the OHS management system is presented in chapter 6 of ISO 45001 and refers to the followings [9];
- actions to address risks and opportunities, including hazard identification and OHS risks assessment, and determination of applicable legal requirements and other requirements;
- OHS objectives and planning to achieve them.

Hazard identification and OHS risk assessment is one of the most sensitive activities which could decide the success or failure of the OHS management system. The method selected as an instrument for hazard identification and risk assessment should respond to the following main requirements:
- it should enable the identification of hazards and risk assessment for each work system components, respectively, worker, work task, work means/work equipments and work environment for each workplace, as requires the G.D. no.1425/2006 regarding the approval of the Methodological norms for applying the provisions of Law no.319/2006 on occupational health and safety, art.15(1) pt.1 [10];
- it should be fitted to the activity type of the organization;
- it should enable to identify and assess all OHS risks related to the organization's activities.

For example, the risk assessment method developed in 1998 by the National Research and Development Institute on Occupational Safety, and widely used by Romanian companies and external preventive and protective services, responds to requirements of art.15(1) pt.1 of

G.D. no.1425/2006 and is fitted for industrial activities, but is too detailed for more simple activities such as office activities. It is recommended to review studies which analyse multiples assessment methods and provides clear selection criteria for practical application of most adequate tools [11].

It is very important that the risk assessment method should enable to identify and assess all OHS risks related to the organization's activities, including new and emerging risks or specific risk. Also, for specific risks, such as fire and/or explosion risks in coal mines, it is more suitable to develop and use additional specialised methods, as shown in [12].

3.4 Support

Chapter 7 of the ISO 45001 is related to the following issues:
- resources;
- competence;
- awareness;
- information and communication;
- documented information.

In this phase, the main instruments that could be used to fulfil the standard requirements are the following:
- trainings and courses to maintain qualification or to acquire a new qualification;
- formal or informal meetings with different interested parties on OHS aspects, as a support for information and communication process;
- awareness programmes for contractors, temporary workers, visitors etc. in accordance to OHS risks at which they are exposed.

3.5 Operation

Regarding operation of OHS management system, the following issues should be addressed [9]:
- operational planning and control;
- management of change:
- outsourcing;
- procurement;
- contractors;
- emergency preparedness and response.

The main instruments that could be used in this phase, additionally to the requirements of standard, are the following:
- trainings for ensuring the competence of workers, to update their competence as a part of management of change or to rise the response capacity of workers in emergency situations:
- internal audits focused on compliance with preventive regulations and manufacturer's instructions for equipments;
- internal audits focused on compliance with essential requirements of safety for machinery;
- checklists and questionnaires on OHS issues for suppliers and contractors.

3.6 Performance evaluation

Chapter 9 of the ISO 45001 is related to the following clauses:
- monitoring, measurement, analysis and evaluation, including evaluation of compliance with legal requirements and other requirements;

- internal audit;
- management review.

Practice shows that, if the internal audit of the OHS management system is performed on a proper basis, the evaluation of compliance with legal and other requirements is made, in most cases, on a superficial manner. Among the main causes of this poor approach, we could mention:

- lack in identification of legal and other requirements relevant for organization;
- insufficient information and training of the OHS managers and internal/external auditors regarding this requirement.

A proper approach of the evaluation of compliance with legal requirements and other requirements should start with an exhaustive and updated identification of the requirements which is relevant for the organization, including the organization's internal OHS instructions. An efficient instrument for evaluation of compliance should be structured on checklists and should address each legislation article or requirement that is relevant for the organization. An example of method for evaluation of compliance with legal requirements is presented in [1] and is composed by a set of checklists each related to a specific norm, as follows:

- Checklist A "Employer's obligations" and
- Checklist B "Rights and obligations of workers" – both related to the provisions of Law no.319/2006 and G.D. no.1425/2006 – applicable for the organization's management;
- Checklists C "Specific requirements", from C.1 to C.23 each related to a specific norm, applicable selectively according to the type of activities performed in organization;
- Checklists D "Internal OHS instructions and other requirements", containing checklists elaborated on the internal OHS instructions of the organizations and other relevant requirements, i.e. contractual OHS requirements.

Each checklist is composed by a number of items linked to specific articles of the related norms, and each item could be evaluated by a score from 0 to 3 points, reflecting the compliance with the requirements.

The method enables to determine two qualitative indicators which define the compliance with legal requirements:

- compliance level – showing the global level of compliance with requirements;
- safety level – which considers the potential consequences of the risks related to the referred requirements.

3.7 Improvement

Chapter 10, "Improvement", of the ISO 45001 refers to the following issues:

- incident, nonconformity and corrective action;
- continual improvement.

Instruments used in this context should be focused on the identification of opportunities, such as [9]:

- new technologies;
- good practices, both internal and external to the organization;
- suggestions and recommendations from interested parties;
- new knowledge and understanding of OHS issues which is relevant for the organization;
- new or improved materials;
- changes in worker capabilities or competence;
- achieving improved performance with fewer resources.

In each case, attention should be paid to a proper risk assessment prior to implementing the improvement resulting from the above mentioned opportunities.

DOI: 10.1051/matecconf/201712111007

4 Conclusions

Occupational health and safety management system is one of the main elements of the company's general management system and its implementation should be made by integration with other management systems which are relevant for the organization, such as quality, environment or social responsibility.

Romanian companies gained a valuable experience on implementing this type of management systems, using OHSAS 18001, but the projected release of the ISO 45001 (planned for February 2018) represents a new approach which requires the companies to take in consideration new key elements for a successful implementation of the OHS management system.

The presented study was performed on the present draft version of the ISO 45001, considering that the final version of the standard could mark some differences, but not essentially for the relevance of the proposed instruments as support in implementing and maintaining an OHS management system.

The study results represent an important work instrument for each company, regardless its size and activity domain, interested to implement or upgrade its OHS management system using ISO 45001 standard.

References

1. D. Darabont, *Managementul securităţii şi sănătăţii în muncă – Ghid de evaluare a conformării cu cerinţele legale*, (Editura AGIR, Bucureşti 2010)
2. ***- https://www.iso.org/iso-45001-occupational-health-and-safety.html
3. *** - http://www.ilo.org/global/topics/safety-and-health-at-work/lang--en/index.htm
4. European Agency for Safety and Health at Work, E-Guide Healthy Workplaces for All Ages (2015)
5. Directive 89/391/EEC on the introduction of measures to encourage improvements in the safety and health of workers at work
6. *** - Law no.319/2006 on Occupational Health and Safety
7. *** ISO 45001 – Briefing notes - https://www.iso.org/files/live/sites/isoorg/files/standards/docs/en/iso_45001_briefing_note_EN.pdf
8. International Organization for Standardization - https://www.iso.org/iso-45001-occupational-health-and-safety.html
9. International Organization for Standardization – ISO 45001 (Draft) – Occupational Health and Safety Management Systems – Requirements with Guidance for Use (2016)
10. *** - Government Decision no.1425/2006 on approving the Methodological norms for applying the provisions of Law no.319/2006 on occupational health and safety
11. Moraru, R. I., Băbuţ, G. B., Cioca, L. I., Environmental Engineering and Management Journal, **13(6)**, 1371 (2014)
12. Cioca L. I, Moraru R. I., Archives of Mining Sciences, **57(1)**, 53-60 (2012);

MATEC Web of Conferences **121**, 11008 (2017) DOI: 10.1051/matecconf/201712111008
MSE 2017

The ergonomics interventions evaluation. A study based on usability

Anca Draghici[1,*], *Larisa* Ivascu[1], *Alin* Gaureanu[1], and *Anca* Mocan[1]

[1]Politehnica University of Timisoara, Faculty of Management in Production and Transportation, 14 Remus str., 300191 Timisoara, Romania

Abstract. ISO 9241-11 and ISO 13407 are recognized as important standards related to usability. ISO 9241-11 suggests that usability measures should cover: effectiveness (the ability of users to complete tasks using the system, and the quality of the output of those tasks), efficiency (the level of resource consumed in performing tasks) and users' satisfaction (subjective perception when using the product, process or system). Based on these basic knowledge on usability concept the article aims to describe a proposed methodology for the global assessments of the ergonomics interventions in the case of industrial system. The research scenario consists in a survey based on a developed questionnaire that evaluate the ergonomics interventions, in terms of usability. In addition, the results of an experimental research demonstrate the testing and validation of the proposed approach. The method could be seen as an ergonomics continuous improvement one.

1 Introduction

The concept of usability has evolved in the man-machine-environment system studies and was strongly debated for users interface, user-centred design subject of interest [1-2]. Usability is "about learnability, efficiency, memorability, errors, and satisfaction" [2]. The above definition refers in addition to efficiency in normal use and satisfaction with use, learnability in early use, memorability after a period of non-use and aspects related to errors during use, in order to avoid undesirable consequences.

ISO 9241 series and ISO 13407 are two important standards related to usability: the former one (part 11) provides the definition of usability and the latter one guidance for designing usability [3-4]. According ISO 9241-11 usability is seen as "the extent to which a product can be used by specified users to achieve specified goals with effectiveness, efficiency and satisfaction in a specified context of use" (this is the main reference of usability) [3]. The standardize approach of the concept represent both users and business requirements: effectiveness means success in achieving goals, efficiency means not wasting time and satisfaction means willingness to use the system. In addition, Figure 1 shows the main aspects that have to be considered when a usability analysis and evaluation is

* Corresponding author: anca.draghici@upt.ro

 DOI: 10.1051/matecconf/201712111008

developed, together with its benefits. The perspective introduce by the ISO 9241-11 standard has been extended to product, service, processes and systems.

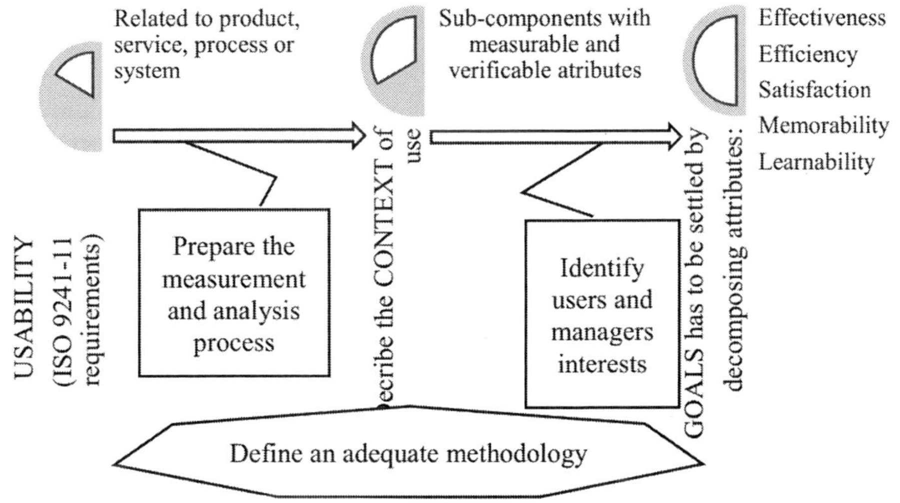

The approach is used to identify the usability measures and context of use when designing, analysis or evaluating usability.	Attributes are used to analysis and measure the extent to which a product, process or system is usable in a particular context.	Measures of usability attributes can provide the basis for comparison and dynamics analysis.

Fig. 1. Relevant aspects related to the usability approach.

In this context, the article describes a proposed methodology for the global assessments of the ergonomics intervention in the case of industrial system. The research scenario consists a survey based on a developed questionnaire that evaluate the interventions impact usability. In addition, the results of the experimental research will validate the proposed approach. The method could be included in the ergonomics continuous improvement programs, in order to collect periodical feed-backs from the affected employees and managers.

2 The proposed methodology

In the case of industrial systems, the continuous improvement dynamics is high. Ergonomics interventions and corrections are correlated with the health and safety strategy, with quality management requirements and finally, to ergonomics improvements that support workplace well-being [5-6].

Usability is seen as an adequate attribute to a purpose (applicable for ergonomics continuous improvement) of any particular product, service, process or system. It means to collect structured feed-back from users that express their opinions regarding a particular context of use, relatively to their work place. This is related to participatory ergonomics.

When selecting usability metrics, it is often desirable to have measures which do not require vast effort and expenses to collect and analyse data. Complex systems or tools of participatory ergonomics were analysed but their main disadvantage is related to additional high costs for external consulting and implementation [1, 7]. Other references have indicated methods or techniques for participatory ergonomics having the primary purpose of: idea generation and concept development, concept evaluation, problem solving or process recording [1, 8].

For practical purposes, there have been adopted the survey approach developed with the support of the work places users, employees of an industrial system. A questionnaire was developed and it was inspire by the System Usability Scale (SUS) described in [9].

Table 1. The proposed questionnaire for the ergonomics interventions evaluation.

#	**Question/usability item evaluated:** **Strongly disagree** **Strongly agree** **1 2 3 4 5**	**Usability capacities or attributes evaluated*:**
1	I think that I would like to use the new working system frequently	S (-1)
2	I found the new working system un-useful un-helpful, too complicated, too complex	M (-5)
3	I suppose that the new working system was easy to use in terms of my work effectiveness including error avoiding	O (-1)
4	I think that I would need guidance from the technical person in order to attend high level of efficiency of my work in the new conditions	E (-5)
5	I found the different handling working movements very easy to be learned within the new working system	L (-1)
6	I think that I can do the different handling working movements more quickly, but after a period of adaptation and conformation	O (-5)
7	I imagine that most of my co-workers would learn to use this new working system quickly	E (-1)
8	I found the new working system very difficult to use: embarrasses me sometime or limited my working movements	S (-5)
9	I am very confident using the new working system	M (-1)
10	I have to learn a lot of things before I could get using the new working system very good	L (-5)

*) E – Efficiency including error avoiding; O – Effectiveness including error avoiding; L – Learnability; S – Satisfaction; M – Memorability

The questionnaire has been developed during three focus groups done with the involvement of the external consultants (the researchers), supervisors and the middle managers of the industrial system. The creative sessions have been take into considerations the objectives followed by the ergonomics intervention.

Based on the final solutions implementation, positive/negative aspects related to usability have been considered for the questions formulation (Table 1). Each item that was evaluate has been measure using a 5 point Likert scale in order to identify aspects or things which lead to extreme expressions of the attitude being captured by users collected opinions.

The questionnaire was distributed after users had a period of time of using (get familiar with) the facilities created by the ergonomics, but before any debriefing or discussion between them or with their supervisors. Users/respondents were asked to record their responses as spontaneous opinions, rather than reflecting about items for a long time. If a user feels that he/she cannot respond to a particular question, this was marked in the centre point of the respond scale (3).

The final score has been calculated similar to the procedure mention in [9]. First there have to be sum the score contributions from each item. Each item's score contribution will range from 0 to 4. For items 1, 3, 5, 7 and 9 the score contribution is the scale position minus 1; for items 2, 4, 6, 8 and 10, the contribution is 5 minus the scale position. The final ergonomics usability interventions evaluation score is the result of multiplying the sum of the scores by 2.5.

3 A practical application of the ergonomics intervention usability evaluation using the proposed methodology

The study was developed in the context of a participatory ergonomics program implementation in an automotive company's practice (an assembly line for automotive cables and wires manufacturing). Participatory ergonomics is defined as "the involvement of people in planning and controlling a significant amount of their own work activities, with sufficient knowledge and power to influence both processes and outcomes in order to achieve desirable goal" [10]. According to previous researches, participatory ergonomics process is believed to encourage workers to be involved in controlling their own work, consequently decreasing work-related risk factors and thereby improve occupational health and safety [11].

In the presented case study, the defined research protocol provide an a priori program logic [12], that describes the pathway from the ergonomics intervention to the reduction of work stress and the usability assessment. There have been anticipated that the employees in the study would be able to identify risks and implement solutions together with the external consultants and their managers, in order to reduce physical work demands (physical exertion and occupational lifting) which in turn would contribute to the improvement of their work comfort (Table 2). The proposed approach, described in section 2 of this article, has been developed, tested and validated, and the ergonomics usability interventions evaluation are shown in Table 3.

Table 2. Inventory of the ergonomics interventions in the studied production system.

Work places or areas of intervention	Improvements operated	Benefits
Control and defect prevention work places (Figure 2a)	Re-organizing work places with respect of succession movements; Re-organization of the local logistics; Re-orientation in an horizontal plan of the control and test panels	Improve employees postures (reduce back pains); Increase work rhythms and efficiency; Use of gravity for handling materials; Improve work place design
Cables assembly work places (manual operation with visual inspection) (Figure 2a)	Correct gripping and grasping of objects; Eliminate repetitive tasks; Tools were suspended when possible; Implement ergonomic chairs	Avoid hands fatigue and adopt good tool-work piece contact; Improve reaching and grasping; Assembly time cycle reduction; Improve employees postures
In-circuit defection work places	Eliminating postures as bending, twisting, eye movements, lifting against gravity by re-organization and re-design	Eliminate back pain, shoulder pain, wrist pain, neck pain etc.; Increase of work efficiency
Packaging work places	Re-design of the work place by intensive using of gravity; Introduce a handling system for big packages (vacuum grip)	Reduce back injuries; Eliminate some human movements because of using gravity; Automate the process of packaging and shipment
In-door training area (Figure 2c)	Design and implementation of an adequate area of training; Define a group of Ergo Coaches to support employees	Reduce time for adopting a new assembly tasks through demonstrations and practical exercises; New employees are taught the proper technique, including safety and quality

a. b. c.

Fig. 2. Details of the ergonomic intervention final solutions: a. re-orientation in a horizontal plan of the control and test panels, b. ergonomic organization of the work places, c. in-door training program.

Table 3. The ergonomics usability interventions evaluation – research results.

Work places or areas of intervention	**Global score gained by each user / questionnaire applied**	**Average**	**Conclusions**
Control and defect prevention work places (10 users investigated)	55, 67.5, 84, 72.5, 70, 77.5, 90, 95, 100, 67.5	77.90	Most of the improvements were made in these areas. Users appreciate well the usability but new developments should positively affect *learnability* and *satisfaction* attributes.
Cables assembly work places (manual operation with visual inspection) (24 users investigated)	75, 80, 77.5, 55, 62.5, 87.5, 84, 90, 90, 55, 80, 80, 65, 67.5, 84, 72.5, 70, 77.5, 90, 95, 100, 82.5, 90, 77	78.65	
In-circuit defection work places (10 users investigated)	65, 77.5, 67.5, 90, 84, 72.5, 70, 77.5, 90, 84	77.80	
Packaging work places (6 users investigated)	75, 80, 77.5, 82.5, 70, 95	80.00	Usability has been very well appreciated. Improvements can be done for *effectiveness* attribute.
In-door training area (10 users investigated)	90, 82.5, 87.5, 84, 90, 95, 100, 95, 82.5, 87.5	89.40	
Total users investigated: 60 subjects		80.75	Very good score gained in comparison with the investments done for the ergonomic interventions.

4 Conclusions and future work

The aim of this study was to analyse the effectiveness (in terms of usability) of the implementation of a participatory ergonomic program taking into account an ergonomic evaluation and workers' work comfort together with their perception about the work stress and the risk factors related to task performance. The ergonomics intervention attempted to improve the fit between the worker possibilities and the work task (processes and associated activities) by decreasing the physical workload.

The presented research approach has underlined the strong link of usability with ergonomics (improvements done exploiting ergonomics knowledge have a positive, direct impact on usability attributes). The perspective of usability is closely aligned with business goals: effectiveness, efficiency and satisfaction have a direct impact on work productivity and finally, on profitability.

The presented approach has demonstrate that the framework defined by ISO 9241-11 and ISO 13407 standards (refers to visual display terminals work places) could be extended to other working areas, from different industries. The developed questionnaire has proved to be a valuable evaluation tool, accepted by researchers and managers, practitioners from the

DOI: 10.1051/matecconf/201712111008

industrial system; the questionnaire is considered robust and reliable. Research results demonstrate a significant improvement regarding task performance indicating a workload decrease, suggesting that workers' satisfaction has considerably increased.

Future applicative researches should extend the data process, as suggested by the SUMI inventory, in order to refine the description of users' behaviour related to the usability attributes.

This article has presented the research results obtained through our implication in the national project: Researches based on Knowledge Management Approach Concerning Industry–University Collaboration in the Open Innovation Context (UNIinOI). This work was undertaken through the Partnerships in Priority Domains Programme-PN II, developed with the support of MEN-UEFISCDI, Project no. 337/2014 in Romania. Any findings, results, or conclusions expressed in this article belong to authors and do not necessarily reflect the views of the national authority UEFISCDI.

References

1. W. Karwowski (Ed.), *International Encyclopedia of Ergonomics and Human Factors*, (second edition, CRC Press, Taylor & Francis Group, New York, 2006)
2. J. Nielsen, *Usability Engineering*. (Academic Press Inc., San Diego, 1993)
3. ISO 9241-11, *Ergonomic requirements for office work with visual display terminals (VDTs)*, Part 11, Guidance on usability (1998)
4. ISO/IEC. 13407, *Human-Centred Design Processes for Interactive Systems* (1999)
5. A. Yazdani, W.P. Neumann, D. Imbeau, P. Bigelow, M. Pagell, N. Theberge, R. Wells, Scandinavian Journal of Work Environment & Health, **41(2),** 111 (2015)
6. J.R. Wilson, Applied ergonomics, **45(1)**, 5 (2014)
7. M. Robertson, R. Henning, N. Warren, S. Nobrega, M. Dove-Steinkamp, L. Tibirica, Journal of Occupational and Environmental Medicine, **55**, S86 (2013)
8. P. Waterson, M.M. Robertson, N.J. Cooke, L. Militello, E. Roth, E., N.A. Stanton, Ergonomics, **58(4),** 565 (2015)
9. J. Brooke, SUS - A quick and dirty usability scale, Usability evaluation in industry, **189(194),** 4 (1996)
10. H. Haines, J.R. Wilson, P. Vink, E. Koningsveld, Ergonomics, **45**, 309 (2002)
11. C.D.N. Rasmussen, N.K. Lindberg, M.H. Ravn, M.B. Jørgensen, K. Søgaard, A. Holtermann, Applied Ergonomics, **58**, 491 (2017)
12. C.D.N. Rasmussen, A. Holtermann, O.S. Mortensen, K. Søgaard, M.B. Jørgensen, Prevention of low back pain and its consequences among nurses' aides in elderly care: a stepped-wedge multi-faceted cluster-randomized controlled trial, BMC Public Health, **13** 1088 (2013)

Safety-at-work competences as a driver of corporate social responsibility

Adam Górny[1,*]

[1]Poznań University of Technology, Faculty of Management Engineering, Chair of Ergonomics and Quality Management, 11 Strzelecka Str., PL-60-965 Poznań, Poland

Abstract. In order to operate effectively in a continuously changing and frequently turbulent markets, companies must account for the needs and expectations of both their management and lower-ranking employees. To that end, it is essential that business organizations identify ways to adopt changes that will guarantee their success. One way to improve the market position of a company is to employ the principles of corporate social responsibility. A key requirements as well as a key area of such responsibility is occupational health and safety, whose guidelines fall within the scope of the overall practices enshrined in labor law. A prerequisite for the effective fulfillment of such requirements is to secure competent contractors who will undertake all measures associated with this field. The article notes the issue and examines it against the standards set forth in ISO 26000. The author demonstrates the need to acquire competences that will enable the concerned company to ensure the safe performance of work and the fulfillment of occupational health and safety requirements in conformity with the principles of corporate social responsibility. Only by embracing the rules of CSR under such an approach will a business be able to achieve the desired outcomes.

1 Introduction

In the course of their activities, today's business organizations increasingly follow the guidelines of corporate social responsibility. To become socially responsible in their day-to-day business, enterprises need to strive to fulfill the expectations of all of their stakeholders. They should therefore examine multiple aspects of any relevant issue [1].
One significant area having a major influence on the effectiveness of an organization's efforts is occupational safety. Safety should be seen as a necessary condition for an enterprise's success. To attain safety, a company has to create worker-friendly working conditions that will enable its workers to function properly in the working environment [2].

Any actions depend heavily on man seen as both their beneficiary and their performer. The job of humans is to take measures aimed at ensuring proper conditions in the working environment [3] and strive to make such measures as effective as possible [4-5]. The consequence of such efforts is a reduction in loads and, ultimately, the lowering of accident

* Corresponding author: adam.gorny@put.poznan.pl

rates and risk levels [6-7], which in turn boosts the bottom line providing the company with a competitive advantage [4].

2 Characteristics of issues in question

2.1 Corporate social responsibility as an approach to doing business

The corporate social responsibility (CSR) can be defined as the voluntary incorporation of the social and ecological dimension into any activities and structures of a business organization relating to their stakeholders [8-9]. A socially responsible enterprise needs to do its business in keeping with the expectations of any concerned parties to ensure multifaceted development of humans [10]. A substantial role in the development of a company that abides by the principles of CSR played by ethical values, respect for both the natural environment, and people (employees).

In addition, an organization's strategy commonly incorporates the pursuit of sustainable growth. Such growth depends heavily on satisfying human needs. Research finds that companies that comply with CSR guidelines tend to experience improvements in the efficiency, quality and effectiveness of their workers [2, 11].

When embracing CSR principles, businesses should recognize the limitations of humans who need to be seen as a critical factor for improvement that is vital for fulfilling customer needs and expectations [8, 12-13]. Commonly referred to as the human factor, people are of great importance for ensuring the effective and efficient achievement of an organization's objectives. The human factor determines the ability to adopt the changes that result from the efficient management of every corporate function. In order for CSR to enable a business to accomplish its intended aims, corporate social responsibility has to be incorporated into its strategy, its elements permanently implanted into the company's key operating areas. One can safely assume that some such criteria pertain to the working environment.

2.2 Occupational safety as a precondition for effective task completion

Occupational safety is traditionally, viewed as the status of working conditions; working environments, work design and employee behaviors that ensures the required level of health and life protection against risks arising in the working environment [4-5].

Occupational safety may be seen as the absence of any unacceptable risks and untoward factors that result in damage. In this sense, companies seeking to safeguard safety and protect worker health must adopt measures and solutions that comply with relevant laws and regulations while accounting for the role and significance of employees for ensuring the companies' effective operation.

The aim behind occupational safety improvement measures is to achieve an equilibrium between man, machine and the working environment that will ensure that all three operate reliably in unison. To achieve that, it is crucial to account for methodological complexity and apply complementary actions for people and technologies covering social, technical, economic and organizational considerations. Success in ensuring safety and reducing strains may depend on the degree of success in the achievement of various objectives associated with work design, technical measures, the use of personal protection equipment, etc. In pursuing such objectives, organizations need to be mindful of the specific working conditions and the conditions in which man operates in the working environment and at the workstation. What is essential is that enterprises recognize all factors that may potentially exert a negative influence on working conditions. In particular, emphasis should be placed on the factors that can be modified to achieve the desired outcomes.

The prime goal in employing improvement measures is to mitigate threats and strains to optimize loads and ultimately boost working effectiveness. Success in protecting and promoting humans naturally, can be, linked with innovation. Many of the criteria that are vital for innovative pursuits tie inextricably to the ability to ensure a safe working environment or reduce untoward strains during the performance of work duties. What is critical is for the process to be continuous and to create conditions for the growth of the concerned organization that conform to the principles of corporate social responsibility.

2.3 Worker competences as a determinant for the adoption of occupational safety solutions

Competences may be viewed as a body of theoretical knowledge and practical skills that support work performance. Competent individuals complete their tasks more easily, performing more efficiently and effectively and are therefore more likely to meet relevant expectations.

Pursuant to ISO 9001 [14], competences may be seen as a demonstrated ability to apply knowledge and employ skills. Defined in this way, competences can be designed as:

- Having knowledge in a given field (I know what to do),
- Skills (I know how to do it),
- Attitude (I want to and am prepared to use my knowledge),
- Personality traits.

A competent person, i.e. one that satisfies the above criteria, is one empowered to act and make decisions and qualified to make judgments and offer opinions and assessments [15]. With respect to the subject matter of this article, the actions of a person who is competent in a specific field should respect the boundaries and satisfy the requirements that are in place in a given community and that make it possible to operate in a specific environment. Workers charged with the performance of tasks associated with specific threats, be it to themselves and/or other workers in their organization as well as any individuals from outside the company, should have pertinent competences documented with educational and training credentials and/or experience that matches specific requirements [15].

According to ILO guidelines [16], an employer should have or ensure occupational health and safety competences that are sufficient to identify threats, eliminate or reduce occupational risks as well as adopt occupational health and safety solutions. Such competences should be viewed as mandatory for persons responsible for safety. Among the required competences, emphasis should be placed on such knowledge and skills as allow one to [17]:

- Perform work in a manner that eliminates threats seen as a source of potential risks,
- Assess the impact of any threats that have not been eliminated,
- Eliminate threats at source and prevent their spreading,
- Adjust working conditions and work processes to the individual needs and capacities of workers,
- Employ state-of-the-art technologies and the latest technical solutions and organizational measures as available and applicable in particular circumstances,
- Replace dangerous technological processes, high-risk equipment, hazardous substances and other dangerous materials with safer / less dangerous equivalents,
- Give priority to the use of collective protective measures before resorting to personal protection equipment that may cause secondary threats and strains,
- Train and instruct workers in occupational health and safety ensuring they are capable of performing their work safely and make proper use of improvement measures.

3 Conditions for adopting social responsibility

3.1 The essence of embracing social responsibility

Under the current paradigm, organizations need to adopt growth strategies aligned with new sources of competitive advantages that will allow them to respond rapidly to changes in the global environment. Although the task may be daunting, a company that is aware of its needs and abilities will find it achievable. The need to ensure an ever-greater level of competitiveness may be considered as helpful for an organization's market credibility. To become increasingly competitive, enterprises should engage on multiple fronts in actions aimed at establishing good relationships with their stakeholders, including their employees. Such actions become a foundation and define guidelines for the performance of tasks in conformity with CSR principles [11-12].

One of the factors that critically affect success in becoming a socially responsible organization is occupational health and safety, which defines the ability to operate effectively in a specific environment. The extent to which a company succeeds in keeping its workers safe translates into the level of its social responsibility defined based on measures taken to achieve the status of a safe organization that cares for the growth of its environment and that pursues tasks in keeping with working environment requirements.

Occupational health and safety is one of the main areas of practice in the general field of work. Its place in the structure of labor law practices is presented in Table 1.

Table 1. Labor-law related practices significant for the assumption of proper social responsibility.

No.	Key areas and issues	Requirement area
1	Employment and employment relationships	Employment is subject to laws, which impose the same obligations on workers and employers. The key responsibility of an organization is to adopt solutions that will eliminate the danger of benefiting from unfair employment practices that involve exploitation and fraud.
2	Terms of employment and employment relationships	Terms of employment including pay, work and rest times, disciplinary practices, redundancy practices, maternity protections, etc. affect the quality of life of both workers and their families. The scope of such terms makes it necessary to approach implementation in a fair and attentive fashion.
3	Social dialogue	Social dialogue applies to issues of popular interest, including economic and social matters. It requires the recognition that both the employer and the employees have shared as well as opposed interests. Such dialogue can be used to design competence development programs that help people improve and become more efficient at work or to reduce adverse impacts of social change that affects organizations.
4	Occupational health and safety	The field of occupational health and safety concerns measures aimed at maintaining workers in maximum physical, mental and social fitness. The occupational health and safety requirements adopted in organizations protect workers against the health risks that arise in the working environment and help adjust that environment to reflect employees' physical and mental capabilities.
5	Human improvement and workplace training	The measures taken should guarantee the improvement of humans that is vital for the growth of society. Initiatives are launched to boost the potential of employees and improve their fitness for work. To achieve this, it is essential to continually increase and improve experience, competences and qualifications.

3.2 Occupational safety as a field for adopting social responsibility

The occupational health and safety of workers can be viewed as being of particular significance for the proper operation of businesses. This applies in particular to the requirements that go beyond the measures mandated by law. The outcomes achieved through social responsibility can be viewed as the internal and external impacts of a company. To ensure a safe environment for its workers, measures that address the internal impacts are particularly significant. These include [4, 7-8, 16, 18]:

- Developing, implementing and maintaining occupational safety rules in the recognition that strict safety standards and financial and organizational concerns within enterprises complement one another,
- Recognizing that the rules for the development of occupational health and safety principles rely on the involvement and commitment of workers, who are entitled to define and voice their own expectations regarding the way their work is to be performed,
- Properly applying occupational health and safety rules, including respect for the hierarchy of methods for eliminating threats inherent in working conditions,
- Assessing and monitoring risks and adopting best practices to eliminate threats or mitigate their adverse impact on workers,
- Ensuring proper communication to guarantee that any concerned persons are aware of the applicable rules and ways to comply with the adopted procedures that are based on legal regulations and good practices,
- Ensuring access to the required personal protective equipment for the prevention of injuries, occupational diseases and accidents at work,
- Proper and effective emergency responses,
- Recording and investigating all incidents and issues relating to safety at work with a view to eliminating them, preventing their recurrence or mitigating their impact,
- Eliminating psychological and social threats occurring at the work place and leading or contributing to stress and any related health conditions,
- Gaining the ability to properly train all workers regardless of the form of and legal basis for their employment,
- Allowing worker involvement in health and safety decisions,
- Taking measures to boost safety while protecting any persons charged with conducting such measures and responsible for their implementation from any negative consequences.

It is equally critical to ensure that the measures taken are comprehensive. Such an approach is particularly crucial in organizations in which irregularities are of a complex nature.

A key consideration in such measures is the organization's internal working environment. Factors that appear in such an environment are essential for assessing the conformity of a working environment with the optimal requirements for a given category of work task performance. Viewed in such a manner, the principles of corporate social responsibility (CSR) allow one to ensure an optimal approach to the loads in the working environment. The principles extend to all significant areas that fall within the category of sustainable growth. If adopted properly, they ensure business growth while promoting a proper quality of life for the workers, in the local community and in the business environment at large.

Every organization has a responsibility for ensuring occupational safety. An organization's size often determines its ability to adopt better or worse OHS practices [19]. The measures taken are intended primarily to ensure the protection of health and the establishment of safety during the performance of work. Proper work design and the adoption of adequate technical solutions help not only to enhance the working environment but also to boost the effectiveness of the labor force. A precondition for the achievement of the desired outcomes is to select the right solutions and implement them in an appropriate manner.

4 Competences as a safety driver within the framework of social responsibility

Changing environments force the organizations that abide by the rules of social responsibility to search for new areas of improvement that will provide them with benefits. Many find it advantageous and critical to examine the specific nature of the human factor in such a search. The human factor proves to be an ever more vital driver of enterprise growth. Organizations that fail to account for the human aspect of their operations find it impossible to adopt modern enterprise growth concepts. However, success in these fields requires specific measures aimed at achieving the desired outcomes and genuine commitment thereto on the part of the organization [3].

Seen as a factor in labor practices, safety depends on a company's competences needed to ensure that work is performed safely and adopt the principles of corporate social responsibility. The need for such competences arises across a wide range of areas that are:

- Necessary for adopting social responsibility,
- Necessary for ensuring occupational safety,
- Instrumental in implementing OHS requirements within the framework of social responsibility principles.

The related interrelationships are summarized in Figure 1.

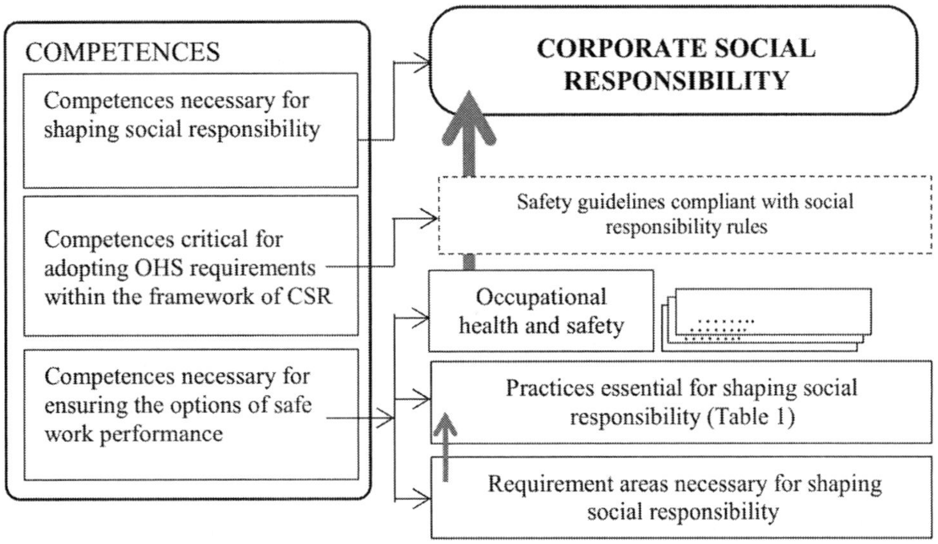

Fig. 1. Interrelationships among competence requirements in the fields of occupational health and safety and corporate social responsibility.

Competences essential for ensuring the effective performance of work among one's employees can be considered as a prerequisite for safety improvement. In actual practice, it is frequently difficult to reconcile business growth with occupational health and safety rules, as laid down in relevant regulations. Despite many advanced safeguards in workplaces, regular inspections and preventive measures, accident rates remain unaffected or even continue to rise [20-21]. Threats to human life and health frequently underrated due to insufficient awareness among both the workers and their employers. It is therefore even more necessary to boost the currently low occupational health and safety levels. Note also that organizations ignorant of the risks of major failures and that underestimate injury risks do not feel motivated to detect threats and undertake preventive measures [2]. Under such circumstances, CSR can be seen as a factor that urges organizations to take action.

Studies in various fields of business activities find that CSR helps improve the quality of life, of which the human working environment is an integral part [1, 22-23]. CSR affects relationships with workers, the building of trust and the implementation of voluntary measures that enable workers to grow personally and obtain the required qualifications. Such efforts produce the positive synergy effect of reducing losses by lowering the number and severity of accidents [24]. These benefits all parties involved in the organization's activities.

4 Conclusions

An effort to comply with OHS requirements with a view to adopting the principles of corporate social responsibility requires adequate solutions. The outcomes of such an approach to the protection of health includes:

- A level of work comfort that guarantees employee health,
- Conformity with the principles of safety in employment resulting from the proper treatment of workers and the protection of human rights,
- Providing employees with a free choice of their preferred form of employment,
- Care for society and the strife to prevent any adverse impact of the organization's business.

To a varying degree and extent, the multiple safety factors include the level of accident hazard. In practice, the likelihood of an accident occurring depends on factors linked to an organization's ability to ensure proper performance of occupational tasks. Such factors include the competences required to conduct work. To incorporate CSR rules in the recognized working environment matrix of safety factors, enterprises need to establish proper relationships that reflect the expectations of all concerned parties.

In practice, the most crucial feature of the new approaches, as adopted by today's businesses, is to remain open to change and flexible, which supports the validity of the above precepts. Therefore, all aspects pertinent to creating a proper working environment appear to be critical and need to be considered in adopting new solutions. It is also vital to identify the requirements of the concerned parties, including a company's employees for whom working conditions are a significant job satisfaction criterion.

References

1. E. Ozdora-Aksak, M. A. Ferguson, S. A. Duman, Public Relations Review, **42(1)**, 79 (2016)
2. A. Kawecka-Endler, B. Mrugalska, Advances in Social and Organizational Factors, Advances in Human Factors and Ergonomics Series, 672, AHFE (2012)
3. A. Górny, Procedia Manufacturing, **3**, 4700 (2015)
4. A. Górny, MATEC Web of Conferences (the 4th International Conference on Computing and Solutions in Manufacturing Engineering 2016, vol. 94, no 04011 (2017)
5. A. Górny, Occupational Safety and Hygiene, SHO 2016, 105, Guimarães (2016)
6. A. Górny, 2015, Occupational Safety and Hygiene, SHO 2015, 143, Guimarães (2015)
7. M. Rembiasz, MATEC Web of Conferences (the 4th International Conference on Computing and Solutions in Manufacturing Engineering 2016, vol. 94, no 07009 (2017)
8. ISO 26000:2010 (PN-ISO 26000:2012), Guidance on social responsibility, PKN, Warsaw (2012)
9. *Green Paper for Promoting a European Framework for Corporate Social Responsibility*, COM(2002), 366 final, Brussels
10. *Communication for Commission to the European Parliament, the Council and the European Economic and Social Committee Implementing the partnership for growth and jobs: Making Europe a pole of excellence on Corporate Social Responsibility*, Commission of the European Communities, COM(2006), 136 final, Brussels

11. Bernal-Conesa, J., Briones-Peñalver, A., De Nieves-Nieto, C., Eur. J. of Management and Business Economics, **25(3)**, 121 (2016)
12. A. Grover, Procedia - Social and Behavioral Sciences, 157, 103 (2014)
13. R. Kudłak, K. Y. J. Low, Long Range Planning, **48(3)**, 215 (2015)
14. EN ISO 9001: 2015 (PN-EN ISO 9001: 2015-10), Quality management systems. Requirements, PKN, Warsaw (2015)
15. PN-N-18001:2004, Systemy zarządzania bezpieczeństwem i higieną pracy. Wymagania, PKN, Warsaw (2004)
16. *Guidelines on occupational safety uidelines on occupational safety and health management systems and health management systems ILO-OSH 2001*, International Labour Office, Geneva (2001)
17. Council Directive 89/391/EEC of 12 June 1989 on the introduction of measures to encourage improvements in the safety and health of workers at work, OJ L 183, 29.6.1989, pp. 1–8, as amended
18. A. Grzelczak, A. Borowiec, A. Górny, *Foresight przedsiębiorstw, vol. 2: Zastosowanie metodyki myślenia sieciowego*, Wydawnictwo Politechniki Poznańskiej, Poznań (2013)
19. H. Nordlöf, B. Wiitavaara, H. Högberg, R. Westerling, Safety Science, **95**, 92 (2017)
20. *Accidents at work and work-related health problems*, Central Statistical Office, Warsaw (2014)
21. *Working condition in 2015*, Central Statistical Office, Warsaw (2016)
22. H. L. Kim, Y. Rhou, M. Uysal, N. Kwon, Int. J. of Hospitality Management, **61**, 26 (2017)
23. Y. Rhou, M. Singal, Y. Koh, Int. J. of Hospitality Management, **57**, 30(2016)
24. K. Čarnogurský, A. Diačiková, A. Ďaňková, M. Lach, Procedia Economics and Finance, **34**, 244 (2015)

Assessment of the compliance of OSH competences with student expectations in the post-graduate programs offered by the Poznań University of Technology

Adam Górny[1,*], and *Małgorzata* Rembiasz[2]

[1] Poznań University of Technology, Faculty of Management Engineering, Chair of Ergonomics and Quality Management, ul. Strzelecka 11, Poland
[2] Poznań University of Technology, Faculty of Management Engineering, Chair of Entrepreneurship and Business Communication, ul. Strzelecka 11, Poland

Abstract. In order to fulfill the duties associated with ensuring occupational safety at work, a broad range of competences are required that are critical to perform work and comply with relevant laws. One way to obtain such competences is to complete a post-graduate program of studies. The Poznań University of Technology offers precisely such an opportunity. Its post-graduate OHS program provides students with the knowledge and skills necessary to work in the field while satisfying their expectations. Student expectations are ascertained in surveys designed to assess the extent to which the knowledge and skills taught meet the needs they have indicated. The findings of such surveys are used to improve course design with a view to enabling the graduates to conduct OHS work in both the manufacturing and service sectors. However, to ensure that the courses designed to satisfy student expectations do not fail to teach the skills necessary in real life, criteria other than the survey results alone need to be adopted in defining program content.

1 Introduction

In order to ensure proper working conditions, including occupational health and safety, for workers, it is critical to either employ competent OHS experts or outsource the function to knowledgeable specialists. The Polish law allows a certain flexibility in fulfilling this obligation [1-2]. The adequate approach depends on the number of workers and the magnitude of risks in the working environment. Regardless of how the obligation is satisfied, persons in charge of occupational health and safety must be properly qualified. OHS officers are expected to have the competences that match their job descriptions and therefore need to have completed such occupational health and safety training as designed for such workers [2-3].

* Corresponding author: adam.gorny@put.poznan.pl

To ensure that the workers responsible for shaping working conditions perform their duties effectively, it is vital that they improve continuously. The issue is seen as complex and challenging [4-8].

There is definitely a need to improve professionally on an ongoing basis to acquire new specialized knowledge that is necessary for the performance of work. Therefore, safety officers must stay abreast of the latest developments in science, technology, work design, and learn to use some of the latest safety instruments and solutions.

2 Aim of research

2.1 Scope of responsibilities of OHS services

The primary responsibility of OHS services is to oversee working conditions and advise the employer on ways to complete tasks associated with ensuring safety in the working environment. In view of the nature and scope of OHS duties, the relevant services require all-around preparation for their work, including the knowledge of the available technical and organizational solutions that help mitigate threats. The basic scope of the responsibilities and obligations of OHS services has been defined in the relevant laws. The job of an OHS officer or any person charged with the related responsibilities is to [2]:

- Identify any threats and untoward factors associated with the performance of work in their organization, assess the related occupational risks and propose measures designed to eliminate them,
- Prepare methods to assess safety including proposed technical and organizations measures aimed at averting any existing threats to worker health and lives and improving working conditions,
- Assess precepts, contribute to the development of workplace modernization and improvement plans and propose technical and organizational solutions aimed at boosting safety and hygiene levels at work,
- Formulate findings from surveys of occupational accident causes and circumstances as well as cases of occupational diseases,
- Have external experts oversee the ways in which work is performed and supervise surveys and measurements of factors that are deleterious to health or any other untoward conditions in the working environment,
- Collaborate with physicians providing preventative health care to workers, the social labor inspectorate and company trade unions regarding the performance of tasks associated with employee safety and health,
- Participate in drafting internal regulations governing terms of employment and work methods, including internal ordinances, rules and general instructions regarding occupational health and safety,
- Follow recommendations on compliance with ergonomic requirements applicable to specific workstations,
- Support the employer in assigning responsibilities to supervisors and managers to the extent of the duties associated with ensuring safe working conditions for their subordinates.

2.2 Opportunities for acquiring the required credentials and competences

In view of the scope and nature of OHS responsibilities, it is essential that OHS officers are provided with opportunities to stay up to date on changes in the field and any new methods and solutions for ensuring occupational safety and protecting worker health [9]. The easiest and simplest way to acquire the necessary knowledge is to enroll in dedicated occupational health and safety training. Such training is offered to enable members of OHS services to [10]:

- Become familiar with any working environment factors found at workstations that may endanger worker health and safety during their work,
- Learn about the current laws and occupational health and safety regulations to the extent necessary to perform their duties and fulfill their OHS-related obligations,
- Acquire practical skills to perform work in ways that are safe for themselves and others, respond to emergencies and aid accident victims.

It is crucial for such measures to benefit the company by generating profit [6, 11]. To ensure that, OHS services must be adequately skilled in performing their professional responsibilities [11]. Considering the specific nature of their duties, members of such services need specialized education in occupational health and safety [4].

The most advanced path to obtaining occupational health and safety-related professional qualifications available in Poland is to enroll in a post-graduate program of studies in the field. Such programs are geared to updating and/or providing such specific knowledge and skills as are not available in undergraduate or graduate programs [12]. Post-graduate studies are designed for holders of lower-level higher education degrees. The laws of Poland restrict the number of institutions of higher learning permitted to offer post-graduate OHS programs to those that teach occupational health and safety, ergonomics and the related subjects to ensure worker protection in working environments with particular emphasis on deploying technical and organizational measures [9].

Market demand as well as the long-standing record and vast experience in teaching methodology and education in the fields of OHS, ergonomics and health protection held by the Poznań University of Technology has prompted the institution to offer a post-graduate program specializing in occupational health and safety.

2.3 Educating OHS workers at Poznań University of Technology

The presumption made by the Poznań University of Technology was that the graduates of its post-graduate program should walk away equipped with the competences needed to perform work in the field of OHS at least in the position of a Senior OHS Inspector. Access to employment in higher-ranking positions depends on the years served in occupational health and safety. Another presumption made was that the knowledge provided to the students must provide them with opportunities for promotion and continuing work in other OHS capacities, up to Chief Specialist, as long as the concerned individual has obtained all other required credentials [2, 4]. Since the job descriptions in such positions often extend beyond the traditional duties of an OHS officer, the University recognized the need to offer additional knowledge and skills. Such knowledge includes occupational health and safety management, the ability to analyze a company is operating expenses in connection with occupational health and safety issues, and business project planning.

Of equal importance, and considered in developing the curriculum, are the abilities to teach specific OHS skills to help others update their knowledge and acquire the necessary abilities. Such educational skills include human-factor-related, technical, legal, organizational and economic issues.

The thematic areas and knowledge content offered to students were, selected to ensure their mutual correlation. Consequently, the graduates should be able to combine e.g. technical knowledge with organizational, legal and economic considerations while not forgetting about the central role of humans in shaping the working environment. The program of studies continually updates to account for legislative changes and, primarily, the expectations of post-graduate students. The thematic areas of the current post-graduate curriculum adopted by the

Poznań University of Technology in the field of occupational health and safety are summarized in Table 1. A large proportion of the classes are practical workshops designed to move beyond theoretical knowledge and help acquire actual working skills.

Table 1. Post-graduate occupational health and safety classes offered by the Poznan University of Technology.

No.	Name of course (thematic area)	Course description, format and number of hours
1	2	3
1	Fundamentals of occupational health and safety	Introduction to the material covered further into the program. lecture = 4 hours
2	Ergonomics in shaping working environment	A broad discussion of the ergonomic aspects of shaping the working environment. lecture = 10 hours
3	Work loads	Issues associated with musculo-skeletal loads at work linked to workstation design and equipment. lecture = 6 hours, laboratory = 6 hours
4	Working environment – preconditions for ensuring occupational health and safety	Issues associated with measurements, result interpretation and improvement measures, primarily for factors in the physical working environment. lecture = 12 hours, workshop = 10 hours
5	Human factor in shaping occupational safety	Issues associated with the impact of the human factor on the ability to ensure safe work performance. lecture = 6 hours
6	Ecology in shaping working environment	Discussions of issues and requirements associated with the environmental impact of business organizations. lecture = 10 hours
7	Comprehensive work strain analyses	Discussion of comprehensive methods of examining working conditions. lecture = 8 hours
8	Organizational safety systems	Issues linked to ensuring safety in organizational units (and consequently in enterprises). lecture = 6 hours, workshop = 8 hours
9	Interpersonal communication in OHS	A discussion of factors affecting vertical and horizontal communication and its effectiveness. lecture = 6 hours
10	Oversight over working conditions – recordkeeping by OHS services	A discussion of rules for overseeing working conditions and documenting work performance and working conditions with a view to ascertaining compliance with relevant laws and regulations. lecture = 6 hours, workshop = 14 hours
11	Occupational risk assessment and management	Focused on occupational risk and employer measures compliant with occupational risk assessment and management guidelines to ensure safety at work. lecture = 6 hours, workshop/project = 12 hours
12	Fire safety at the workplace	Issues associated with ensuring fire safety at the workplace. lecture = 6 hours
13	Economic aspects of improvements in occupational health and safety	A discussion of economic aspects of occupational health and safety, including the financial cost of safety deficiencies (enterprise losses). lecture = 4 hours

1	2	3
14	Occupational health and safety – laws and standards	Labor law issues, as enshrined in international, European and national law, as well as relevant standards. lecture = 10 hours
15	Technical and organizational aspects of shaping working environment	Technical and organizational requirements determining proper working conditions. lecture = 10 hours
16	Safety in the operation of work machinery, equipment and tools	Issues associated with ensuring occupational safety when operating machinery, equipment and tools. lecture = 12 hours
17	Health protection. First aid in the working environment	Issues associated with providing paramedical first aid to victims. lecture = 4 hours, workshop = 8 hours
18	Introduction to occupational health and safety management	Issues associated with the systemic shaping of working conditions. lecture = 6 hours, workshop/ project = 8 hours
19	Teaching methodology for OHS training	Issues whose awareness is critical for proper worker training and instruction. workshop = 10 hours
20	IT for shaping occupational safety	Discussion of the software used to shape occupational safety, including occupational safety management. lecture = 4 hours, workshop = 6 hours

3 Evaluation of the effectiveness of the OHS program

A survey of the effects achieved in developing competences was conducted among the students of two editions of the post-graduate occupational health and safety program, as held in the academic years 2014/2015 and 2015/2016. The survey, which extended to a total of 79 persons, produced 41 complete responses which could then be used for further analysis. Three of the respondents were employed in an OHS service based on their prior credentials. The remaining 38 intended to take such employment immediately upon completing the post-graduate program.

Each of the respondents was given three options for answering questions regarding the usefulness of specific thematic areas (courses) designed to prepare them for actual employment. They could mark a specific area as:

– useful (+),
– moderately useful (+/–),
– useless (–),

Each respondents were familiar with the job descriptions of OHS officers, which they learned in prior lectures. The participants were asked not to assess the way the knowledge was taught and focus solely on the usefulness of a given area, as covered in their course.

Based on their responses and in keeping with the formula below, a usefulness index was developed that, in the respondents' view, reflected the usefulness of the specific areas in their future jobs:

Usefulness index = 2 points × number of assessments (+) + 1 point × number of assessments (+/–) (1)

The resulting indices of the usefulness of specific areas covered in the program enumerated in Table 2 and presented in Figure 1.

Table 2. Usefulness assessments by thematic area.

No.	Name of course (thematic area)	Number of assessments			usefulness index
		+ 2 points	+/– 1 point	– 0 point	
1	Fundamentals of occupational health and safety	36	4	1	76
2	Ergonomics in shaping working environment	30	5	1	65
3	Work loads	24	15	2	63
4	Comprehensive work strain analyses	41	0	0	82
5	Human factor in shaping occupational safety	31	10	0	72
6	Ecology in shaping working environment	11	15	15	37
7	Comprehensive work strain analyses	17	15	9	49
8	Organizational safety systems	10	20	11	40
9	Interpersonal communication in OHS	10	15	16	35
10	Oversight over working conditions – recordkeeping by OHS services	40	1	0	81
11	Occupational risk assessment and management	41	0	0	82
12	Fire safety at the workplace	28	12	1	68
13	Economic aspects of improvements in occupational health and safety	25	4	12	54
14	Occupational health and safety – laws and standards	26	3	12	55
15	Technical and organizational aspects of shaping working environment	15	20	6	50
16	Safety in the operation of work machinery, equipment and tools	39	1	1	79
17	Health protection. First aid in the working environment	20	10	11	50
18	Introduction to occupational health and safety management	15	21	2	51
19	Teaching methodology for OHS training	38	2	1	78
20	IT for shaping occupational safety	36	2	3	74

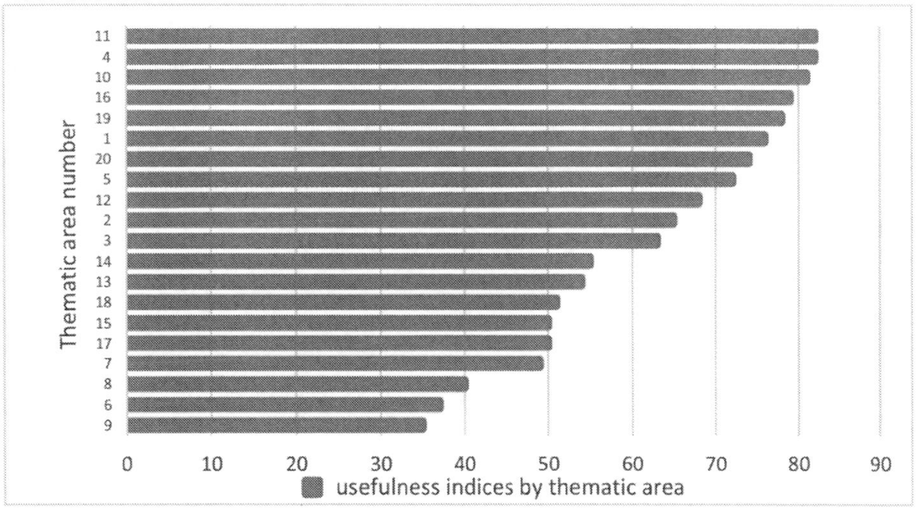

Fig. 1. Distribution of thematic area indices.

The survey helped identify what the student felt were the most useful thematic areas of education as well as the fields they thought were the least valuable for their future jobs. The areas considered the most useful for the proper performance of work were:

- /11/ Occupational risk assessment and management (82 points),
- /4/ Working environment – preconditions for ensuring occupational health and safety (82 points),
- /10/ Oversight over working conditions – recordkeeping by OHS services (81 points),
- /16/ Safety in the operation of work machine, equipment and tools (79 points).

Among the areas evaluated as the least useful, the respondents selected:

- /9/ Interpersonal communication in OHS (35 points),
- /6/ Ecology in shaping working environment (37 points),
- /8/ Organizational safety systems (40 points),
- /7/ Comprehensive work strain analyses (49 points).

The usefulness indices varied considerably across the areas covered in the program. This is due largely to differences in the perception of their importance and, indirectly, to the way the students imagined the nature of their future jobs. Low usefulness indices of specific courses should not be, considered as justifying their removal from the curriculum as future OHS officers have to be fully prepared for fulfilling their OHS-related duties. It is, however, essential to re-examine curricular design and modify curricula to meet student expectations to the extent reasonable.

4 Conclusion

As organizations are bound to comply with relevant laws, OHS officers must have specific competences and credentials, including a diploma certifying completion of a higher education program or specialty in occupational health and safety or at least a certificate of completion of a post-graduate program in OHS. Furthermore, within one year of commencing their work, OHS specialists are required to complete periodic training tailor-made to suit their specific needs. The rules applicable in Poland do not provide detailed descriptions of such programs of studies. However, the scopes of knowledge and skills that are offered should be sufficient to enable the students to carry out their job duties adequately.

A prerequisite for effectively educating occupational health and safety specialists and especially self-employed safety inspectors in places of business is to establish a proper teaching structure and offer instruction covering an adequate scope of material. OHS experts require broad knowledge on the functioning of various fields of industry and services [2]. In addition to meeting formal requirements, occupational health and safety workers need to be properly versed in occupational safety, health protection and ergonomics as well as possess an overall technical expertise. The scope of the knowledge they are offered needs to meet their expectations while fulfilling the formal requirements that must be satisfied to enable them to carry out their jobs.

References

1. Law of 26 June 1974 - Labor Code, consolidated text: Journal of Laws of 2016, Item 1666, as amended (2016)
2. Regulation of the Council of Ministers of 2 September 1997 on occupational health and safety services, Journal of Laws of 1997, No. 109, Item 704, as amended
3. Council Directive 89/391/EEC of 12 June 1989 on the introduction of measures to encourage improvements in the safety and health of workers at work, OJ L 183, 1-8 (1989)

4. J. S. Marcinkowski, L. M. Pacholski et al. (Eds.) The role of education and researches in ergonomics and work-safety in health care of population, Poznan University of Technology, Poznan (2006)
5. P. Pryor, Safety Science, **81**, 5 (2016)
6. G. Paul, W. Pearse, Safety Science, **81**, 13 (2016)
7. O. Broberg, I. Hermund, Int. J. of Industrial Ergonomics, **37**(9-10), 810 (2017)
8. M. Spychała, M. Butlewski, In: P. Arezes et al. (Eds.) Occupational Safety and Hygiene, SHO 2015, pp. 341, Portuguese Society of Occupational Safety and Hygiene (SPOSHO), Guimarães (2016)
9. A. Górny, Proceedings of the 17th World Congress on Ergonomics, Beiging (2009)
10. Regulation of the Minister of Economy and Labor of 27 July 2004 on occupational health and safety training, Journal of Laws of 2004, No. 180, Item 1860, as amended
11. A. Górny, *The use of Ishikawa diagram in occupational accidents analysis*, In: P. Arezes et al. (Eds.) Occupational Safety and Hygiene, 162, Portuguese Society of Occupational Safety and Hygiene (SPOSHO), Guimarães (2013)
12. Law of 27 July 2005 on higher education, consolidated text: Journal of Laws of 2016, Item 1842, as amended

Prospective analysis of ambiental and weather effects on conform displaying of fireworks

Ciprian Jitea[1,*], *Attila* Kovacs[1], *Gabriel* Vasilescu[1], *Edward* Gheorghiosu[1], Daniela Rus[1], and *Sorin* Bordoş[1]

[1]National Institute for Research and Development in Mine Safety and Protection to Explosion – INSEMEX Petroşani, 32-34 G-ral Vasile Milea Street, Postcode: 332047, Petroşani, Hunedoara County, Romania

Abstract. The paper presents the results of research on theoretical and practical relating to establishing and analysis of effects of ambiental and weather effects on conform displaying of fireworks used to organizing fireworks. Scientific research highlighted in this article was conducted in the project no. PN 16 43 03 04, study conducted through Nucleu program entitled -„ Research on the influence of microclimate factors on parameters of the products, while testing the functioning of pyrotechnic articles ".

1 Introduction

Pyrotechnic articles are under classification of dangerous products, risk Class 1 - explosives. At European and national level is in force the Directive 2013/29 / EU of the European Parliament and of the Council of June 12, 2013 respectively Decision of Romanian Government no. 1102 of 10 December 2014 laying down conditions making available on the market of pyrotechnic articles. In these documents are established the essential safety requirements to which these products must meet. Essential safety requirements be taken into account on the one hand in the design and manufacturing of products and tests and assessment by third party entities (products certification body and accredited laboratory). In case of complies and suspicions on conformity the market surveillance authorities may require additional repetition of such tests. Depending on the type and category of pyrotechnic articles usually should check their conformity by application of harmonized European standards which specific test methods with all the technical and organizational measures.

The paper highlights the degree of improvement the infrastructure on test of pyrotechnic articles (fireworks) whose attempt to apply the SR EN 15947-4: 2016 - Pyrotechnic articles. Fireworks, Categories F1, F2 and F3. Part 4: Test methods SR EN 16261-3: 2013 - Pyrotechnic articles. Fireworks, Category 4. Part 3: Test methods. The Directive states that attempts to use products under conditions that reproduce as closely is possible as, the real ones that can meet in their use. The project has in view products whose intended use is specified in the safety data sheet, that instructions are known within the parameters of micro-climate that can meet (temperature, humidity, wind - for those used outside, respectively that with inside use temperature, humidity, generating toxic fumes).

* Corresponding author: ciprian.jitea@insemex.ro

In this respect a modern "Equipment for professional monitoring and recording micro-climate parameters,, allows simultaneous management of all micro - climate factors which are variable and their unusual effects can affect health and safety requirements, comparative to the normal expected values [1-2]. Also, compliance with regulatory technical requirements and test procedures led to improved technical and qualitative level to perform the functioning test of the pyrotechnic articles, regarding the influence of micro climate and the possibilities to conduct tests operation with automated data processing results.

2 Material and Method

Establishing the specific functional parameters of fireworks that may be affected by ambiental and weather effects. European Directive provides that tests of products to be done under conditions that replicate as closely as real ones which can meet on their use. Being products whose intended use it is specified in the safety data sheet respectively instructions for use are known the limits of micro-climate that may be encountered (temperature, humidity, wind - for the outdoor respectively temperature, humidity, generation of toxic compounds, to the indoor use).
Products intended for interior can be considered as being banal, whereas usually factors contained are not exceptional in use (20 ± 40 ° C, relative humidity 60 to 80% fairly ordinary, unnoticeable air currents, excluding compounds from burning pyrotechnic composition which generates toxic agents). The situation of fireworks from category F2, F3 respectively F4 it is very variable which means micro-climate conditions in which is used respectively product is tested for assessments. [3-4]

The purchase of a performant multifunction device for monitoring and recording parameters of microclimate with incorporated software according to European applicable harmonized standards.

In order to implement an integrated and comprehensive solutions to ensure the measurements and determinations of fireworks displaying (F1, F2, F3 and F4) under conditions of high accuracy in accordance with standards methods applied, a multifunctional equipment was purchased, as can be seen from figure 1.

Fig. 1. Performant multifunction device for monitoring and recording parameters of microclimate

2.1 Case Study

In order to assess how is the compliance of the fireworks functioning in agreement with the legislative requirements and harmonized applicable method standards, following is an example of applying the procedure INSEMEX GLI-PI-98 on a Battery of shot tubes F2, as can be seen from figure 2.

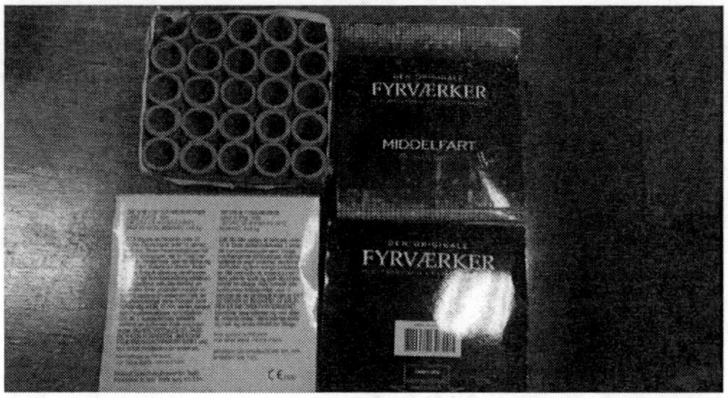

Fig. 2. Battery of shot tubes F2

The test methods have been conducted according to INSEMEX testing procedure GLI-PI-98 which is based on EN 15947–4, as can be seen from table 1.

Table 1. Test Requirements - Verification

No	**Points required by client from EN 15947-5:2010**	**Tests which will be performed by INSEMEX from EN 15947-4:2010**
1.	4.1.1 - Construction materials - General requirements - Critical	Visual examination
2.	4.1.2 - Construction materials - Specific requirements - Critical	Visual examination
3.	4.3 - Permitted elements in batteries and combinations - Major	Visual examination
4.	6.2 - Protection of initial fuse – Major	Visual examination
5.	6.3 - Attachment of means of ignition – Major	6.1.6 - Fixing the ignition fuse
6.	6.4.1 - Start time and duration of burning of initial fuse – Major	6.6.2.1 - Ignition time 6.6.2.2 - Initial fuse time
7	6.4.2. Invisible burning of Roman Candles; Ignition time of sparkles - Major	6.6.2.3 - Invisible burning time
8.	7.1.2.1 - Integrity – General requirements – Major	Visual examination
9.	7.1.2.2 - Integrity – Specific requirements - Major	Visual examination
10.	7.1.3 – Stabilization of flight – Critical	Visual examination
11.	7.2.2 - Functioning – Major	6.17 - Function test
12.	7.2.3 – Angle of ascent or flight	6.4 – Angle of ascent and height of effects
13.	7.2.4 – Motion – Applies to jumping ground spinners only - Major	6.4.3 Procedure for items other than Spinners
14.	7.2.5 - Stability during functioning – Critical	6.17 - Function test

No	Points required by client from EN 15947-5:2010	Tests which will be performed by INSEMEX from EN 15947-4:2010
15.	7.2.6 - Height of explosion - Major	6.4 - Angle of ascent and height of effects
16.	7.2.7 - Sound pressure level – Major	6.5 - Measurement of sound pressure level
17.	7.2.8 - Explosions and other failures – Critical	6.17 - Function test
18	7.2.9.-Burning or incandescent matter-Major	6.12 - Incandescent matter
19.	7.2.11 – Projected debris - Major	6.11 – Projected debris (outdoor)
20	7.3.2. – Plastics body – Major	Visual tracking in operation

3 Results

The results after applying testing procedure GLI-PI-98, as can be seen from table 2.

Table 2. Test results

No	Description	Sample										Conclusions
		1	2	3	4	5	6	7	8	9	10	
1	The body of the firework case shall be made of paper, cardboard or plastics; The base (end closures) or means of fixing shall be made of non-metallic material (wood, staples, nails, aluminum coated foil or binding wires).	√	√	√	√	√	√	√	√	√	√	Conformity 10/10
2	For batteries of mines, Roman candles or shot tubes: the tubes shall have a maximum angle of 30° to the vertical	√	√	√	√	√	√	√	√	√	√	Conformity 10/10
3	The following elements can be used in batteries: bangers and flash bangers, Bengal flames, crackling granules, fountains, ground spinners, mines, rockets (assembled in a launcher),Roman candles, spinners, shot tubes and wheels;	√	√	√	√	√	√	√	√	√	√	Conformity 10/10
4	The initial fuse and reserve fuse, shall be protected by an orange fuse cover, a primary pack or a selection pack;There shall be no exposed instantaneous fuse.	√	√	√	√	√	√	√	√	√	√	Conformity 10/10
5	Using the timing device, determine and record the time until the initial fuse becomes detached. If this time is longer than 10 s stop the test and record as "longer than 10 s.	√	√	√	√	√	√	√	√	√	√	Conformity 10 /10

No	Description	Sample										Conclusions
		1	2	3	4	5	6	7	8	9	10	
6	For indoor fountains of category 1: the fountain shall ignite within 5,0 s. Ignition time max. 10 s; Initial fuse must be visible and must be indicated by labeling / instructions;	1s	1s	2s	1s	1s	1s	1s	2s	1s	1s	Conformity 10/10
	Duration of burning of initial fuse 3-8 s cat 1 and 2; Duration of burning of initial fuse 5-13 s cat 3.	6,08 s	6,60 s	6,02 s	6,32 s	6,12 s	6,25 s	6,74 s	6,02 s	6,35 s	6,68 s	Conformity 10 /10
7.	The period of invisibleburning occurring after the preliminary effect shall not exceed 5,0 s (cat. 2) or 10,0 s (cat. 3).	NA	NA	NA	NA	NA	NA	NA	NA	NA	NA	NA
8	There shall be no holes, splits, dents or bulges in the body of the firework case, except those technically necessary for the correct functioning of the firework.	√	√	√	√	√	√	√	√	√	√	Conformity 10/10
9	Each individual element shall be securely attached, other than by the transmitting fuse(s) alone, to the other elements or to the framework. For a battery of rockets this requirement shall apply to the rocket launchers, not to the rockets; Elements in banger batteries and flash banger batteries may be joined entirely by the transmitting fuse only if it is sufficient to keep the elements joined together during normal handling.	√	√	√	√	√	√	√	√	√	√	Conformity 10/10
10	The article shall be fitted with a stick or sticks as a means of stabilisation	NA	NA	NA	NA	NA	NA	NA	NA	NA	NA	NA
11	All pyrotechnic units functioned completely; The related principal effect; No explosion or rupture occurs during function; The elements of the batteries and combinations are attached securely.	√	√	√	√	√	√	√	√	√	√	Conformity 10/10
12	For rockets: the angle of flight of the rockets shall not exceed 15° to the vertical up to a height of 20 m above ground.	NA	NA	NA	NA	NA	NA	NA	NA	NA	NA	NA
13	The article shall not move more than 8,0 m away from the testing point; the height of ascent shall be not greater than 3 m.	NA	NA	NA	NA	NA	NA	NA	NA	NA	NA	NA
14	Shall remain upright whilst functioning:- batteries	√	x	√	√	√	x	√	√	√	√	Conformity **8/10**
15	No explosion shall occur below a height of 8 m.	NA	NA	NA	NA	NA	NA	NA	NA	NA	NA	NA

No	Description	Sample										Conclusions
		1	2	3	4	5	6	7	8	9	10	
16	Max. 120 dB for category 2, the microphone is set up at a horizontal distance of 8,0 m from the testing point;	114,5 dB	**120,6 dB**	113,4 dB	115,5 dB	113,8 dB	**124,6 dB**	115,1 dB	114,6 dB	115,4 dB	114,3 dB	Conformity **8/10**
17	For batteries and combinations: the article shall not produce an explosion damaging the integrity of the firework. No explosion or rupture occurs during function.	√	√	√	√	√	√	√	√	√	√	Conformity 10/10
18	No burning or incandescent matter shall fall to the ground more than 1,0 m (category 1), 8,0 m (category 2) or 15,0 m (category 3) from the testing point	√	x	√	√	√	x	√	√	√	√	Conformity **8/10**
19	Record whether any debris is projected laterally more than the relevant safety distance (1m cat.1, 8 m cat.2 and 15m cat.3)	√	√	√	√	√	√	√	√	√	√	Conformity 10 /10
20	If the article has a plastics body, the body shall not splinter.	NA	NA	NA	NA	NA	NA	NA	NA	NA	NA	NA

NOTE: √- Conformity; X - Non conformity; NA - Not applicable; Wind speed under 5 m/s at function test period.

Analysis of ambiental and weather effects on the functioning of fireworks (F1, F2, F3, F4) Situation on functioning behaviour of fireworks F2, F3 and F4 it is highly variable, in what it means micro-climate conditions of the use of these products in question is tested for evaluation.

Thus, the rate of combustion and the intensity of emission of heat energy is directly proportional to the temperature variation, or kinetic energy, high humidity also adversely affect the functioning of the product slowing down the speed of combustion or misfire situation (stop working); but pressure is also an influencing factor, in the sense that its growth also leads to higher burning velocity. One major factor is relatively quite difficult to monitor at operation, the influence of wind that occurs mainly on deviation of the upward trajectory of pyrotechnic articles, the noise, and the spread of toxic fumes downwind outside the safety zone. During operation, the wind speed vector changes randomly and assess the influence of global trajectory and wind speed is difficult to provide and quantify.

By analysing the specific operating parameters of these types of fireworks, could be established on scientific bases for each type / category of fireworks tested the possible influence of external factors specific of environment test and work which can affect their smooth functioning, among which may be mentioned in particular: the wind (the intensity and direction), ambient temperature, humidity, pressure, nature and consistency / quality of soil / platform on which are placed the fireworks before the test of its functioning etc.

4 Conclusions

For a better understanding of how to evaluate the ambiental effects and weather on the functioning of these types of fireworks, it was considered the knowledge / documentation of alternative solutions given in the informative annexes contained in the applicable standards method, in view of solving the whole range of issues related to the testing of these types of fireworks.

In order to implement an integrated and comprehensive solutions to ensure the measurements and determinations of fireworks displaying it was purchase a performant multifunction device for monitoring and recording parameters of microclimate with incorporated software according to European applicable harmonized standards, type RainWise CC-3000, produced in the USA.

Following the test operation of the fireworks it was found that microclimate factors have a direct influence on their proper functioning, affecting the accuracy of outcome indicators both in terms of absolute value and is required a correction given by statistical uncertainty and in terms of the appreciation of their quality, influencing how perception often in situations of unfavourable ambiental and weather effects, is one eronated.

References

1. R. I. Moraru, G. B. Băbuţ, L. I. Cioca, Environmental Engineering and Management Journal, **13**, (6), 1371-1376 (2014)
2. L. Ivaşcu, L. I. Cioca, The 2nd International Conference on Psychology, Management and Social Science, Psychology, Management and Social Science, Shenzhen, China, Book Series: Advances in Education Research, **49**, 77-80 (2014)
3. C. Jitea, Research on aspects related to structuring exercise and psychosocial training to optimize the process of intervention and rescue personnel in toxic / flammable / explosive environments PN 07 45 02 30, NRDI INSEMEX
4. *** Directive 2013/29/EU of the European Parliament and of the Council of 12 June 2013 on the harmonisation of the laws of the Member States relating to the making available on the market of pyrotechnic articles.

Study on cognitive behavioural coping of intervention and rescue personnel in toxic / flammable / explosive environments

Izabella Kovacs[1,*], *Artur George* Găman[1], *Cosmin* Ilie[1], *Alin* Irimia[1], and *Andrei* Gireadă[1]

[1]National Institute for Research and Development in Mine Safety and Protection to Explosion – INSEMEX Petroşani, 32-34 G-ral Vasile Milea Street, Postcode: 332047, Petroşani, Hunedoara County, Romania

Abstract. In any given field, the psychological examination represents a prerequisite for ensuring that the work process is properly and appropriately directed towards increasing its efficiency. An important aspect of the psychological examination is to identify risk and protective factors associated with developing and maintaining emotional and behavioural problems. Special conditions resulting from emergency situations are likely to lead to physical and emotional tensions. In some intervention and rescue personnel these are accompanied by mobilization of internal resource, while in others they can generate inadequacy phenomena and symptoms of mental distress. From this perspective, stress is regarded as a result of the marked disparity between environmental requirements and the individual's response possibilities. To highlight both cognitive and behavioural coping strategies most often used by rescuers trained in NRDI INSEMEX we used two instruments, namely Strategic Approach to Coping Scale SACS and Cognitive Emotion Regulation Questionnaire CERQ. This paper displays the results of the project no. PN 16 43 01 12, study conducted through Nucleu program, implemented with the support of NASR.

1 Stress in the context of intervention and rescue activities

The work of rescuers falls into the category of "hard work" in terms of effort deployed, far exceeding the average calories global consumption of workers from active work fronts. The character of hard work also comes from the fact that the occupational effort contains many high amplitude oscillations concerning the efforts deployed. The greater their duration is, the harder the work is. Rescue activities have an occasional character (they are not permanent) involving the change of daily habits from the basic profession. Also, rescuers have to perform training in conditions similar to those of the damaged area, meant to make them able to act efficiently and safely. [1-2]

The way rescue personnel cope with stressful situations depends on several variables: the circumstances of the event, personal training, pre-existing organizational stressors and

* Corresponding author: izabella.eisler@insemex.ro

pre-existing personal stressors. In general, most of intervention and rescue personnel seem to favour coping responses that involve taking steps to solving problems or using logical analysis to understand the stressors related to work. Some rescuers capitalize and benefit from retreat in solitude, while others seek the company of others.

Given the short timeframe in which mental health professionals have contact with intervention and rescue personnel it is difficult to assess the effectiveness of these individual processes of adaptation. It is well known that assistance and psychological support programs are not addressed only to persons who are victims of various types of adverse events and disasters. Intervention and rescue personnel, exposed to emotional stress, may experience state of crisis as a result of professional activity, situation often found in young people with less experience. The traditional heroic role of rescuers includes the expectations that they are selfless, tireless and to some extent "superheroes". Of course, these expectations are largely unrealistic and rescue and intervention personnel are also affected by their professional activity.

Stress is inherent in many of intervention and rescue activities. Apart from the usual work-related stress in the intervention and rescue activities the following situations represent specific challenges [3]:

- Being part of a collective crisis;
- Repeated exposure to grim experiences such as handling dead bodies, dealing with multiple victims, ravishing stories of affected people;
- Performing physically difficult, strenuous or hazardous tasks;
- Lack of sleep and chronic fatigue feeling, due to the demanding tasks;
- Facing the perceived inability to ever do enough;
- Facing moral and ethical dilemmas;
- Exposure to anger and apparent lack of gratitude of the people affected;
- Exiting the home support system;
- The feeling of frustration given by the policies and decisions of superiors within the organization.

The rescuers are exposed to unusual personal demands in their desire to meet the needs of survivors from a disastrous event. The multitude of feelings associated with the intervention, such as giving first, closeness to centre of the disturbing event and not least dealing with emotionally troubled and physically hurt people, must be addressed and confronted. Rescuers, who are often poorly prepared for their own emotional reactions during rescue intervention, can be supported in several key moments in order to reduce the likelihood of developing stress-related problems. Primary intervention consists of solid information about stress and how to cope with emotional reactions in difficult situations. [4] Such information educates and prepares rescuers for detection of their own reactions and provides active options for self-help and peer support. If left unaddressed, these stressors are likely to affect the welfare and quality of rescuer's work. Thus, self-help is both an essential pre-requisite to effectively help others and a means of maintaining the capacity to continue the work of rescue. [5]

Some people have a much higher risk than normal, for developing severe symptoms of stress and lasting posttraumatic stress disorder if they have a history of:

- Exposure to trauma (serious accidents, abuse, aggression, war, business rescue);
- Medical illness or chronic psychological disorder;
- Chronic poverty, homelessness, unemployment or discrimination;
- Major stressors or recent emotional strain (such as the condition of being a single parent).

2 Coping strategies used by rescuers trained within NRDI INSEMEX Petrosani

Disaster related stress can bring back to the fore memories of earlier trauma and may enhance social, economic, spiritual, psychological or medical background. In order to highlight both cognitive and behavioural ways of coping most often used by rescuers trained within NRDI INSEMEX, we have used two instruments, namely Strategic approach to coping scale SACS and Cognitive Emotion Regulation Questionnaire CERQ. The results obtained by using these instruments on a number of 38, respectively 70 people are discussed below.

Strategic approach to coping scale SACS is a tool that assesses behavioural coping, taking into accounts the social aspects of the strategies by which one faces stressful situations. SACS highlights a person's coping strategies where they particularly manifest, respectively in a social context. This scale assesses behavioural coping strategies on the following dimensions: active-passive pro-social-antisocial, direct-indirect, including nine scales (behavioural coping strategies):
1. Active-passive dimension
- Assertive action: addressing the situation firmly, spontaneously, honestly and directly, without retreat from problems, pursuing personal interests without harming others.
- A high score for this strategy is generally associated with low emotional distress and low scores may be associated with emotional problems or psychopathological symptoms. The observation that can be made about this result is that this way of coping is little used by respondents although in the context of intervention and rescue activities, addressing stressful issues firmly, honestly and directly is a desirable coping strategy, as shown in the results achieved by applying SACS on part of the rescuers trained within INCD INSEMEX and illustrated in table 1.

Table 1. Results for assertive action scale

	Frequency	Percent	Cumulative Percent
Under population average	3	7.9	7.9
Population average	30	78.9	86.8
Above population average	5	13.2	100.0

2. Pro-social-antisocial dimension
- Social networking: through which the individual joins others to face the situation together or by which the individual helps others, taking into account their needs;
- Seeking social support: the individual turn to others for help and emotional support and he consults with loved ones (family, friends) about what to do;
- These strategies to cope with stressful situations are frequently used by respondents, as can be seen from table 2, studies indicating that frequent use of these coping strategies is associated with increased levels of positive emotions.

Table 2. Results for social networking and seeking social support scales

	Social networking		Seeking social support	
	Frequency	Percent	Frequency	Percent
Population average	24	63.2	26	68.4
Above population average	14	36.8	12	31.6

- Prudent action: by which the individuals take all precautions before action and carefully evaluate their options to protect themselves from any danger;

- Instinctive action: by which the individual rests upon his own intuition and he reacts by momentary impulses to solve problems;

Table 3. Results for prudent action and instinctive action scales

	Prudent action		Instinctive action	
	Frequency	Percent	Frequency	Percent
Under population average	2	5.3	2	5.3
Population average	25	65.8	18	47.4
Above population average	11	28.9	18	47.4

- Prudent action and instinctive action are in opposition: while the first strategy refers to careful and rational evaluation of the situation and involves consideration for others, instinctive action, which is predominant compared to prudent action among respondents in the research, situation shown in table 3, involves impulsive action without much time to think and less consideration for others. Of course, in the context of intervention and rescue activities the prudent action strategy for coping with stressful situations seems to be preferable to instinctive action. This does not mean that this second coping strategy may not prove to be beneficial in some particular contexts.

3. Direct-indirect dimension

- Avoidance: by which the individual engages in other activities so that he no longer has to deal with the problem, by which he withdraws or expects things to solve themselves.
- According to studies, this strategy is maladaptive and is associated with high emotional distress and psychopathological symptoms. Although in lower number, as shown in table 4, this strategy is nevertheless used by rescuers who are part of the population of present research respondents.

Table 4. Results for avoidance scale

	Frequency	Percent	Cumulative Percent
Under population average	6	15.8	15.8
Population average	23	60.5	76.3
Above population average	9	23.7	100.0

- Indirect action: by which the individual manipulates the situation to make others believe that they are in control, while he solves the problem, opaquely, behind the scenes;
- Anti-social action: by which the individual pursues his interests even if they hurt others;
- Aggressive action: by which the individual act decisively, quickly, to take others by surprise in order to take control.
- As shown in table 5, the responders in the current research use these strategies just as much as the average population.

Table 5. Results for indirect action, antisocial action and aggressive action scales

	Indirect action	Antisocial action	Aggressive action
	Percent	Percent	Percent
Under population average	21.1	34.2	13.2
Population average	73.7	63.2	86.8
Above population average	5.3	2.6	0.00

Active and pro-social coping strategies tend to be associated more often with positive emotional consequences, with low levels of anxiety and depression, while antisocial coping is associated with increased levels of anger and the avoidance is associated with increased levels of anxiety and depression.

Cognitive Emotion Regulation Questionnaire CERQ is designed to identify cognitive coping strategies that someone uses after experiencing negative events (thus only referring to a person's thoughts). CERQ distinguished nine different types of cognitive coping strategies:

1. Self-blaming, referring to the thoughts by which an individual blames himself for what happened. A high degree of concern regarding self-blame may be associated with psychopathological symptoms. As shown in table 6 this coping strategy is used by a significant proportion of respondents (more than half of them), situation that s worrisome.

Table 6. Results for self-blaming scale

	Frequency	Percent	Cumulative Percent
Very low/low	20	28.6	28.6
Under average	1	1.4	30.0
Average	10	14.3	44.3
Above average	20	28.6	72.9
High/very high	19	27.1	100.0

2. Acceptance, which refers to the thoughts by which an individual resigns himself to what happened. Although self-acceptance itself is a beneficial process for most events, a very high level of acceptance, as occurs in almost a quarter of the respondents of the present research and is shown in table 7, can result in a negative form of resignation, meaning the inability to be able to influence events that can be associated with psychopathological symptoms.

Table 7. Results for acceptance scale

	Frequency	Percent	Cumulative Percent
Very low/low	14	20.0	20.0
Under average	10	14.3	34.3
Average	18	25.7	60.0
Above average	11	15.7	75.7
High/very high	17	24.3	100.0

3. Rumination: the individual thinks continuously of feelings and ideas associated with the event. High and very high score obtained by almost a quarter of respondents, as seen in table 8, indicates that rumination its frequently used as a cognitive coping strategy. A high score is certainly associated with emotional problems or psychopathological symptoms; therefore the case of the responders in the current research, as it is shown in the table below, is at least worrying.

Table 8. Results for rumination scale

	Frequency	Percent	Cumulative Percent
Very low/low	14	20.0	20.0
Under average	16	22.9	42.9
Average	8	11.4	54.3
Above average	15	21.4	75.7
High/very high	17	24.3	100.0

4. Positive refocusing: the individual is thinking about other pleasant things, not the event itself.

Table 9. Result for positive refocusing scale

	Frequency	Percent	Cumulative Percent
Very low/low	9	12.9	12.9
Under average	3	4.3	17.1
Average	14	20.0	37.1
Above average	9	12.9	50.0
High/very high	35	50.0	100.0

Studies show that this strategy may have positive effects on the wellbeing of the individual, therefore the higher the scores the more we can talk about high levels of positive emotionality. Very few of our responders reported of not using this coping strategy, as can be seen in table 9, with half of them reporting to use it often or very often.

5. Refocusing on planning, by which an individual thinks about the steps they will take to face the event.

6. Positive reappraisal by which an individual thinks to assign a positive meaning to the event, in terms of personal development.

High scores on refocusing on planning and positive revaluation scales, as can be seen in table 10, show the frequent use of these coping strategies. These are functional cognitive coping strategies, but this is true only if, at the same time, something is done to solve the problem, if the individual addresses the problematic situation. If else, the score is associated with emotional problems.

Table 10. Results for refocusing on planning and positive reappraisal scales

	Refocusing on planning		Positive reappraisal	
	Frequency	Percent	Frequency	Percent
Very low/low	8	11.4	6	8.6
Under average	4	5.7	11	15.7
Average	10	14.3	12	17.1
Above average	18	25.7	12	17.1
High/very high	30	42.8	29	41.4

7. Putting things into perspective or thoughts that minimize the seriousness of the event when compared to other events.

As shown in table 11, the responders in the current research use these strategies above the average population.

Table 11. Results for putting things into perspective scale

	Frequency	Percent	Cumulative Percent
Very low/low	14	20.0	20.0
Under average	9	12.9	32.9
Average	14	20.0	52.9
Above average	11	15.7	68.6
High/very high	22	31.5	100.0

8. Catastrophizing by which the individual emphasizes the terror caused by the event.

High scores obtained for catastrophizing, as obtained by a quarter of the respondents in our research and shown in table 12 are almost certainly associated with emotional problems or psychopathological symptoms.

Table 12. Results for catastrophizing scale

	Frequency	Percent	Cumulative Percent
Very low/low	12	17.1	17.1
Under average	15	21.4	38.6
Average	10	14.3	52.9
Above average	15	21.4	74.3
High/very high	18	25.7	100.0

9. Blaming others by which one thinks to blame others for what happened. As shown in table 13, the responders in the current research use these strategies just as much as the average population.

Table 13. Results for blaming others scale

	Frequency	Percent	Cumulative Percent
Very low/low	20	28.6	28.6
Under average	3	4.3	32.9
Average	24	34.3	67.1
Above average	14	20.0	87.1
High/very high	9	12.9	100.0

Cognitive coping strategies are defined as cognitive strategies for emotion regulation which involve cognitive regulation of emotional responses to events resulting in worsening individual emotions. These coping strategies refers to rather constant styles of coping cope with negative events but at the same time they can be influenced, e.g. changed through learning or forgotten, for example through various intervention programs or personal experiences.

As stress as a whole cannot be removed, the main objective in coping with stress is to feel particularly eustress in the detriment of distress and to prevent distress to degenerate into dysfunctional behaviours. In the context of intervention and rescue activities, dysfunctional behaviours such as alteration of decision-making capacity, mistakes, impulsive and disorganized behaviour, behaviour paralysis or panic can lead to further damage and loss of human lives.

Therefore, we find that respondents, intervention and rescue personnel trained within NRDI INSEMEX, use both adaptive strategies and strategies whose use is associated with emotional problems and even psychopathology. On one hand being aware that they use a strategy or another (with benefits and drawbacks involved) and, on the other hand psychological training interventions and personal development can help individuals learn the use of adaptive strategies to the detriment of those associated with emotional distress.

3 Conclusions

The work of rescuers falls into the category of "hard work" in terms of effort deployed, the plurality, diversity and especially the nature of the tasks involved in the intervention and rescue activities. Rescue activities have an occasional character; they are not permanent, involving the change of daily habits from the basic profession. As well as performing training in conditions similar to those of the damaged area, which make the rescuers able to act effectively and safely, intervention and rescue personnel, that during or after intervention, experiences or is direct witness of violence, injuries, extreme destruction, prolonged exposure to fatigue, bad weather, starvation, toxic contamination is at high risk for developing severe symptoms of stress and lasting problems regarding recovery.

The results attained by applying SACS and CERQ instruments on a total of 38, respectively 70 individuals indicate that intervention and rescue personnel trained within NRDI INSEMEX uses both adaptive strategies and strategies whose use is associated with emotional problems and even psychopathology. On one hand being aware that they use a strategy or another (with benefits and drawbacks involved) and, on the other hand psychological training interventions and personal development can help individuals learn the use of adaptive strategies to the detriment of those associated with emotional distress.

The training process of intervention and rescue personnel within NRDI INSEMEX would capitalize on a psychological training programme, to help individuals learn the use of adaptive coping strategies.

References

1. C. Ilie, *Research on aspects related to structuring exercise and psychosocial training to optimize the process of intervention and rescue personnel in toxic / flammable / explosive environments* (NRDI INSEMEX Research study, 2012)
2. D. Pupăzan, C. Ilie, *Intervention and rescue in toxic / flammable / explosive environments* , (NRDI INSEMEX Training material for rescuers, 2014)
3. https://mhpss.net/?get=58/1354772532-Mod06CommBasedPSSSupport-HelpingtheHelpers.pdf
4. http://www.ceep.ca/resources/Guterman2005_Psychological_Preparedness_for_Disaster.pdf
5. http://www.ptsd.va.gov/professional/treatment/early/disaster-rescue-response.asp

The application of event tree analysis in a work accident at maintenance operations

Radu Mareş[1,*], and *Mihai Popescu - Stelea* [1]

[1]University of Petroşani, 20 Universităţii st., Petroşani 332006, Romania

Abstract. Event tree analysis is an inductive method, which involves determining the events flowing from the failure of a component or part of the system. Starting from an initiator event or an origin failure, the event tree analysis allows the estimation of system deviation, taking into account in a systematic manner the functioning or failure of detection devices, alarm, prevention, protection or intervention. This paper follows the steps in determining the causes that led to a work accident through the Event Tree Analysis method. The construction of event tree analysis consists of graphical representation of operations which through a succession of decisions lead to an accident during maintenance of work equipment.

1 Introduction

For understanding any risk is necessary to know its structure and its mode of expression [1, 4-5]. It must therefore be addressed complexity and all its required knowledge fields. Extensive efforts are needed since life is a continuous risk and there can be progress to increase our comfort without taking some risks.
This progress involves becoming more a rejection of imposed risks resulting from human activities conducted within work processes. But there are risks which are not imposed (related to road traffic, from domestic activities or the leisure) which shall neither they disregard [1].

2 Theoretical fundamentals

Event tree analysis is a method that allows determining, from an initiating event, the chaining of events that may lead to a potential accident [2, 4].

The method finds particular applicability to study architecture safety means (prevention, protection, intervention) present or applicable in a system. On the other hand, if used indiscriminately in complex systems, the method becomes very difficult to practice. The method is commonly used in the analysis and retrospective explanation for the observed consequences arising from a failure of the system.

The procedure for applying the method - generally applicable approach to achieving risk analysis trough events tree involves the following steps:

* Corresponding author: radumares50@gmail.com

I. Defining initiator event;
II. Identification the functions safety - preventive;
III. Building the tree;
IV. Analysis and valorization of tree [2].

3 Case study: Work accident - Mechanical department

Occupational accident investigation file of worker's temporary work incapacity must contain:
A) identification of work accident
B) accident investigation report
C) spot observation note
D) technical report of bridge crane inspection
E) official report on the technical verification of bridge crane
F) sketch of the accident site
G) pictures from the scene
H) injured declaration
I) witness declaration
J) leader job declaration

We try to identify membership in the four elements of tree events for all elements of the research dossier of work accident.

I) In the mechanical department during the operation of compressed air tube has been a work accident by cutting off the thumb and index the left hand of the worker X by crane operated by the worker Y. The event was produced at the level of travel path of the crane located at a height of about 4, 5 m around the time 10,30, in 15 01 2011.

II) Accident investigation report:

a - date of the conclusion of official report: 20 01 2011

b – names of persons performing accident investigation report: AB undersigned chief internal service desk OSH inspector MR SSM domestic service, CI Eng. HR, the members of the investigation committee under Decision 1/2011, I researched the event whose victim was working X

c – time and place where the investigation was conducted, the circumstances and causes which led to the accident occurred during 15 01 2011 - 20 01 2011 and have been carried out at the place of production of the event at the mechanical processing

d – the object of investigation: was done in order to establish the circumstances and causes that led to the accident, legal regulations violated, responsibilities and necessary measures to be taken to prevent the occurrence of other similar cases and to track the accident

e – date- event time: the event was produced in 15 01 2011 around 10,30

f – place of the event: event occurred in the mechanical processing department at the rolling track of bridge crane

g – employer identification data: SC ABC SRL, mechanical processing department

h – identification data of injured worker: name X, 40 years, romanian citizen, mechanical maintenance, 22 years work experience, seniority at work 20 years

i – detailed description of the place, equipment, circumstances and manner in which the event occurred: On 15.01.2011, 10:30, Mr. Y - supervisor in the mechanical department announced by phone the workers from prevention and protection service of society, about an event in mechanical department. The event consists in worker X injury at mechanical maintenance department, respectively worker injury by trapping the worker hand between gears from the translational movement of the bridge crane. Investigation commission appointed by decision of the company manager has moved onsite for research in this event by compiling a note on the findings.

i.1 – detailed description of the place of event: From the statements on the record event occurred on the premises, mechanical department, on the rails of the bridge crane located at a height of about 4.5 meters from the ground. Machined parts within the section is handled from one work station to another by overhead crane operated from the ground by a crane authorized I.S.C.I.R. A team of three workers of the maintenance department works for compressed air network expansion in order to connect to a new chamfered machine. The work is performed near the path of the bridge crane. Within the department the main business is sanding, cutting.
i.2 – In the mechanical department the workers are using the following work equipment: oxy-fuel cutting torch, keys and specific devices plumber profession compressed air installations. In cut marks section is handled from one work station to another by bridge crane operated from the ground by crane authorized I.S.C.I.R. The bridge crane having last check parameters: nominal load t Sn = 5 x 16.5 m wide. Last check carried out in accordance with inspection report no. 23-37 / 2011 consisting of technical inspection in use with a maturity of verification until October 31, 2011.
i.3 – detailed description of the circumstances: On 15.01.2011, the mechanical department workers led by RD, prepared the ground pipes were to be fitted to measure welded thread whereon measure valve and pipe insulation necessary for connection to the compressed air network of chamfered machine. Once executed this operation after X and G to climb on the rails to take measure pipe that was to be mounted on top of the rails. At around 10.30 X on its own initiative asked G to climb to access path next to the bridge crane. Arriving at the place for them to connect the chamfered machine measures they have raised the bulb with welding hoses on the crane platform. According to RD supervisor after he left gate valve and fitting to be mounted on pipes attached to the undercarriage of the crane, work that would be carried out during a lunch break at 11:00 when all work equipment in department were to be stopped.
i.4 – detailed description of how the event occurred: At around 10.30 X and G were on the platform next to the crane runway, they took measures of pipe that was to be mounted on top of the path of the crane. Meanwhile in the mechanical department the bridge crane operated by FI (authorized ISCIR, license no. C, carry parts to chamfered. When movement of bridge crane the two workers from the maintenance department were near the rolling track of the crane. From G's statement shows that it was on the 1st line support pole near the runway crane and X on line 2. X, according to the statement, the board decided to cut with oxyacetylene burner gas welding being facing the pole on line 2, at which the bridge crane it was hanging on the padded coat, it twisted and injured by pressure from the bridge crane and the metal structure of the access on his legs and left hand by gearing gears from mechanism translation. The root cause: Making improper commands, maneuver.
j. – Consequences of the event and / or consequences suffered by injured: According to medical conclusions issued by the Arad County Emergency Service, in address nr./.2011 X suffered multiple trauma; crushing injuries left hand with: amputation inter phalangeal, metacarpal joint-phalangeal dg. 2; dorsal contusion wound finger 3; volar soft tissue avulsion f2 - f3 finger 4; 3 finger amputation phalanx 5; right ankle fractures; malleolus left fractures
k. – cause of the event: From the above analysis and documents on file, based on the circumstances in which the event occurred as a result the main reason that the victim X is due to improper command-maneuvers. Improper handling of the bridge crane of a worker's injury X F and I was in the area of the crane runway led to violation of art 22 of Law no 319/2006 occupational safety and health updated: "Every worker must operate in accordance with his training and instruction and instructions of the employer, so as not to expose themselves to risk of injury or occupational disease both self and other persons who may be affected by the actions or omissions during the work "of the provisions of article 2,

section. 2.2, chapter II of internal regulation: „ ... in addition, all employees have a legal obligation to take care of them themselves and any other person who may be affected by their acts or omissions at work, cooperate with the company and all other persons who are directly or indirectly involved in the activities of the company. All working methods should be regularly checked to ensure that it applies the most secure methods possible. ". Also not been complied with regulations PT R1 / 2010 ISCIR, edition 1„ responsibilities crane lifting machines during work of art. 121 lit. a: „ - to maneuver only on the orders in accordance with the code signaling load slinger designated; The stop signal them to consider the other person " and lit. s, - to not allow anyone to climb the crane when it is connected to the power supply circuit.". According to official report no 6C 2011, I.S.C.I.R., Oradea city. Code 23 -03 carrying out orders improper maneuvers.
l. – Other causes which competed at the event: 1. bullish X's own initiative and requested that the G to climb the access path to tread the crane were exposed and led to violation of art 22 of law no 319/2006 occupational safety and health. updated: "Every worker must operate in accordance with his training and instruction and instructions of the employer, so as not to expose themselves to risk of injury or occupational disease both self and other persons who may be affected by the actions or omissions during the work "; the provisions of article 2, section. 2.2, chapter II of internal regulation: „,... In addition, all employees have a legal obligation to care for themselves and any other person who may be affected by their acts and omissions during work, cooperate with the company and all other persons who are directly or indirectly involved in the company's activities and art. 17 of Instruction own health and safety at work for work at a plumbing and gas, water, heating „ staff worker will perform only works entrusted by the leader of the band for work and only those who are qualified and authorized if it have deactivated all technical equipment that workers can get in touch. "2. When carrying out reconstruction work at the event was found to lack defenders protective gear gears from translational movement of the crane-crane. Were violated GD 1146/2006 Annex 1 art. 2.8.1„ where mobile devices of work equipment risk of accident by mechanical contact, they must be provided with guards and protective devices to prevent access to dangerous areas ... "
m. – other observations during the event investigation: 1. Employer has ensured the organization of prevention and protection unit by setting up a national prevention and protection services in the ministry's coordination. 2. The employer is in possession of a risk assessment of injury for workplace where the event occurred, based on risk assessment being drafted injury prevention and protective plan. 3. Victims event was instructed in terms of safety and health at work last regular training being carried on 03.01.2011. 4. There is OSH instructions in relation to work. 5. The job descriptions were established duties and responsibilities regard to occupational safety and health of the worker X, corresponding to the performed position.
n. – the persons responsible for violation of legal regulations of chapters from point K
1. – F.I. locksmith for those established in chapter k
2. – X maintenance mechanic and G plumber for those established in chapter L.1
o.- proposals for administrative sanctions and disciplinary: FI is punishable by suspension of crane authorization over a period of three months, X and G is punished with disciplinary warning according to Labor Code 2003 (Law 53/2003) art. 248 lit a.
p. – proposals for penal investigation: this is not the case
q. – character of the accident: according to article 32 paragraph 1 of the law 319/2006 updated with regard to art. 135, paragraph 1 of H.G. 1425/2006 accident suffered by X is recorded as accident with temporary work incapacity
r. – ordered measures to prevent further similar events and those responsible for their implementation:

1. Processing causes and circumstances of the incident with all working personnel within the company. Deadline: 10 days from the receipt of the investigation report
2. Mounting the protective elements of the gearing gears from translational movement of the bridge crane. Deadline: Immediately.
3. It will review the action in prevention and protection plan and taking into account the causes of the work accident and other circumstances occurring after its preparation as required by art. 46 paragraph 1in methodological norms of Law no 319/2006 occupational safety and health updated, approved by H. G. 1425/2006 and amended by H. G. 955 / 2010.Deadline: 10 days of receipt of the investigation
4. Occupational risk assessment will be updated as required by art 15 pt 1 of the methodological norms for the application of Law No 319/2006 occupational safety and health updated, approved by H. G. 1425/2006 and amended by H. G. 955/2010. Deadline: 10 days from the receipt of the investigation report..
s. – Reporting deadline to the Territorial Labor Inspectorate regard the implementation of the measures referred to in subparagraph s: Territorial Labor Inspectorate will be informed in writing regard the implementation of measures ordered within 15 days of the approval of investigation dossier.
t. – The number of copies that ended official report and their distribution: This official report ended in quadruplicate and submitted as follows:
- a copy for the company
- a copy to the Territorial Labor Inspectorate
- a copy to the County Pension
- a copy to the victim X
u. – employer stamp

III) Representation of events tree for maintenance operation:

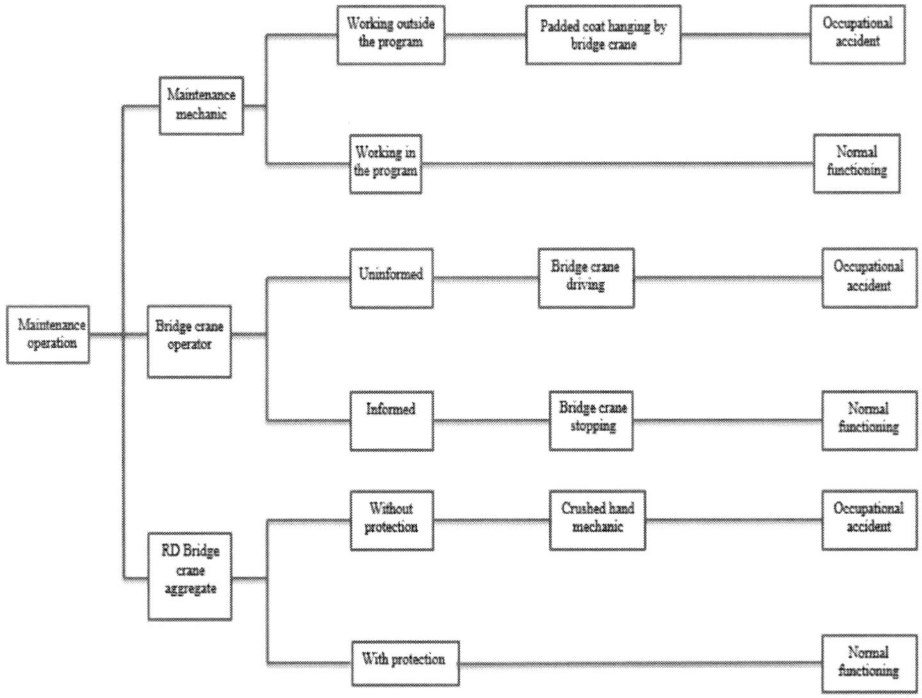

Fig.1. Events tree

IV) Analysis and exploitation of tree:

- If during the maintenance operation worker X will carry out activity during the program and will inform the crane bridge operator on the fact that he operates on the runway, then the whole system will operate normally;
- If during the maintenance operation worker X will carry out the activity outside the program and will not inform the bridge crane operator that he is at work will result a work accident;
- If during the execution of maintenance operation the worker X will be hanging from the bridge crane and it will have R D covered with a shell, then the probability of having an work accident and catching worker X hand between R D and the system will operate normally;
- If during the execution of maintenance operation the worker X will be hanging from the bridge crane and it will not have R D covered with a shell, then the probability of having a work accident and catching workers hand between R D and will result in a work accident.

4 Conclusions

Event Tree Analysis method of a work accident is a quick and efficient method that allows starting from an initiator event and browsing systemic causes of the accident to establish the sequence of events. What articles of the legislation, the specific regulations and internal rules, were not complied. To establish technical and organizational measures necessary to prevent further accidents. The method also allows establishing relations with enough finesse to determine: cause - effect.

References

1. Moraru R.I., *Securitate si sănătate în muncă: tratat univeristar,* (Editura Focus, Petrosani, 2013)
2. Moraru, R. I., Băbuţ, G. B., Cioca, L. I., Environmental Engineering and Management Journal, **13** (6), 1371-1376 (2014)
3. Torretta, V., Rada, E. C., Ragazzi, M., Trulli, E., Istrate, I. A., Cioca, L. I., Waste Management, **45** (SI), 152–160 (2015)
4. Cioca L. I, Moraru R. I., Archives of Mining Sciences, **57** (1), 53-60 (2012)
5. Ivaşcu, L., Cioca, L. I., The 2nd International Conference on Psychology, Management and Social Science, Psychology, Management and Social Science, Shenzhen, China, Book Series: Advances in Education Research, **49**, 77-80 (2014)
6. Cioca, L.I., Moraru, R.I., Băbuţ, G.B., Proceedings of the 15th International Scientific Conference „The Knowledge Based Organization", **2**, section: Management, 43-48, Land Forces Academy Sibiu (2009)
7. Legea nr. 319/2006 a securităţii şi sănătăţii în muncă
8. Hotararea de Guvern 1425 /2006. HG pentru aprobarea Normelor metodologice de aplicare a prevederilor Legii securitatii si sanatatii in munca nr. 319 din 2006

Workforce active ageing case study in a Romanian manufacturing company

Roland Iosif Moraru[1,*], *Lucian-Ionel* Cioca[2], and *Gabriel Bujor* Babut[1]

[1]University of Petroşani, Faculty of Mines, 332006 Petroşani, Romania
[2]University "Lucian Blaga" of Sibiu, Faculty of Engineering, 550025 Sibiu, Romania

Abstract. Ageing workforce is a major trend that characterizes the demographic evolution of Europe, which generally affects the society, but also the organizations from economic and social point of view. Employers should change their negative perceptions of ageing workforce, whereas the changes that occur along with ageing are unable to influence the efficiency of work, assuming that a suitable strategy implemented to prevent and manage potential problems. The paper includes European and national statistics on the issue, a case study analysing the workforce ageing management in a Romanian manufacturing company, which has succeeded in achieving the active ageing target, and - on this basis - several good practice recommendations given on the approach employers should have on the workforce with regard to age.

1 Introduction

Worldwide, lately was observed a trend of an aging workforce, on the one hand due to increasing life expectancy, but also due to the increase in retirement age [1-2]. Regardless of workers age, the employer has the same responsibilities when it comes to health and safety [3-4]. Meanwhile, employers when hiring in companies cannot invoke the age issue. Each age has advantages regarding the conduct of professional activity. Thus, elder employees have a wide range of skills; have rich experience gained during their careers and extensive knowledge about the operations, which they perform. Younger workers are bringing new life and a fresh spirit in the activities they are involved in; they are very often innovators, have a strong desire for affirmation and have a greater capacity to work. From the proper mixture of the two types of employees can benefit the employer, both in terms of workload and of quality of provided products or services.

2 Statistics at European and national level, by age

In 2014, the employment rate of labor force in the European Union for people aged between 15 and 64 years, as measured under the EU labor force survey, was 64.9% (figure 1). The employment rate of older workers (aged 55 to 64 years) continued to grow fast despite the

* Corresponding author: roland_moraru@yahoo.com

financial and economic crisis that manifested itself in the Member States of the European Union. In the European Union, the employment rate of older workers reached 51.8% in 2014 (figure 2); this rate has increased every year since 2002 to the present [5].

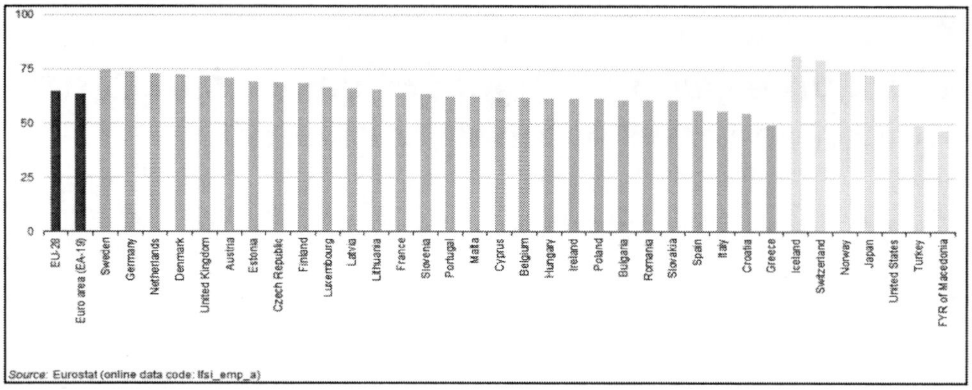

Fig. 1. Employment rate in EU, age group 15 - 64, 2014 [5]

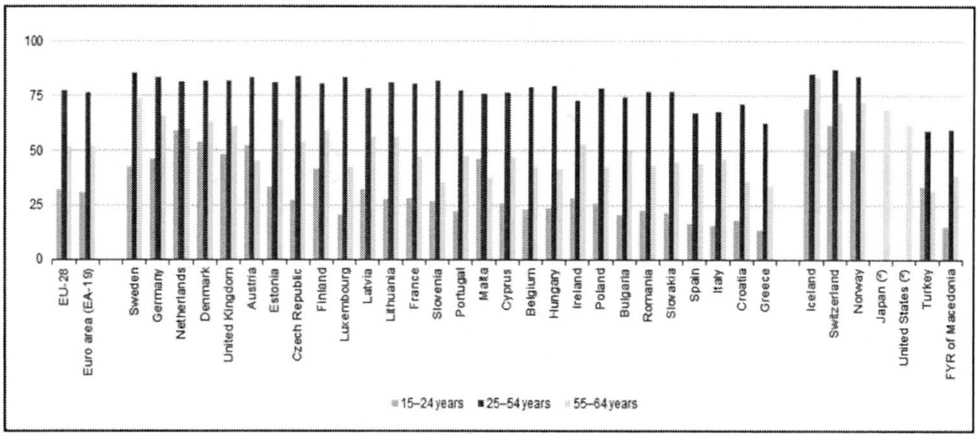

Fig. 2. Employment rate in EU, by age groups, 2014 [5]

In 2015, in eleven EU Member States the employment rate for senior workers was between 50% and 66%, while the highest rate recorded in Sweden (74%) [6, 7, 9]. Employment statistics are at the heart of many EU policies. In March 2010, the European Commission launched the Europe 2020 strategy for smart, sustainable and inclusive growth; the European Council formally adopted it in June 2010 [9]. The European Council agreed on five main objectives, the first being to increase the employment rate for women and men aged from 20 to 64 years up to 75% in 2020 [10].

Population aging effects were also felt in Romania, being reported the phenomena of decrease of young workers employment, continuing the high levels of the active adult population and increase their number of elderly in the labor market. According to data from the National Institute of Statistics, in 2015 the Romania's active population was 9.159 thousand people, decreasing by 84 000 people related to 2014 [11].

Those with medium level of education (57.6%), by males (57.2%) and those living in rural areas (54.7%) held the largest share of total active population. Among the active population, 96.7% were in the age group (15-64 years). Young people accounted for 7.5% of the active population, more than half (62.3%) residing in rural areas. The activity rate of the population aged 20-64 years was 70.8%. In 2015, the employed population comprised 8.535 thousand people; of these, the people able to work (15-64 years) represented 96.5% [12].

As highlighted in figure 3, the employment rate of working age population (employed people of working age in the total population of working age) was 61.4%, with higher values for males employed (69.5% versus 53.2% for women) and close by residential (61.7% in rural areas compared to 61.3% in urban areas).

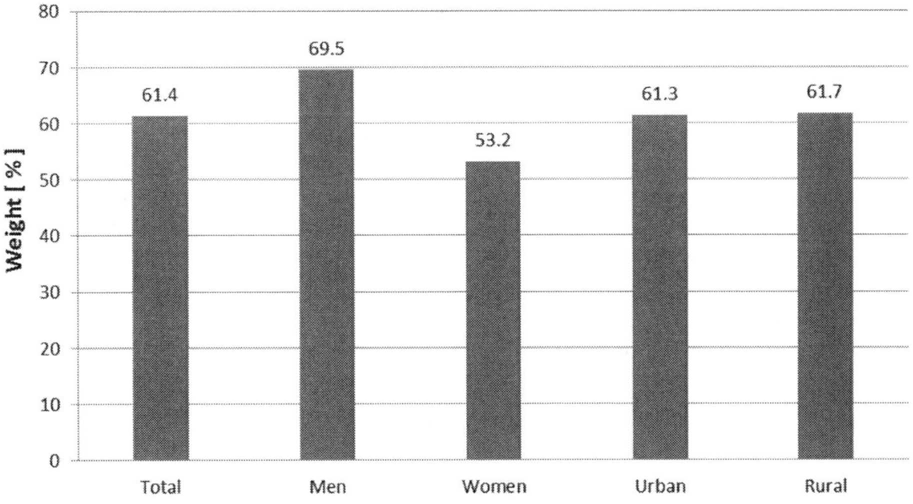

Fig. 3. The employment rate for working age people, by gender and area, 2015 [12]

Comparatively, the data from the National Institute of Statistics in the period 2014-2015 shows a decrease in the employment rate of the population aged over 15 years.

Table 1. The employment rate of the population over 15 years, on age groups in 2014-2015 [12]

Year	Total population aged over 15	Total 15-64 years	Age groups					
			15-24 years	25-34 years	35-44 years	45-54 years	55-64 years	Over 65
2014	51.1	61.0	22.5	75.1	80.1	75.6	43.1	10.8
2015	50.8	61.4	24.5	74.9	81.1	75.7	41.1	8.9

The evolution of staff by age provides information on the existence of a psychological balance of the enterprise (in case of uniform distribution of staff between different categories of age) or, conversely, a pronounced imbalance in the age structure that can harm psychological balance of the company (massive employment of young people hampers their opportunities for promotion).

In this context Romania has adopted the National Strategy for Employment 2014-2020 which aims to boost efforts to achieve the employment target set by Romania for 2020, namely a 70% employment rate for the population aged 20 -64 years [13]. Fulfilling this target will contribute to achieving the goal of Romania in the European employment strategy established by 2020, a strategy that proposes that Europe grow smart, sustainable and inclusive and to find ways to create new jobs. The strategy developed consider both national issues and challenges identified through thorough analysis of economic and social developments in the first decade of the third millennium and the challenges facing Romania in the economic crisis and recovery of the gaps compared to the EU average. In addition, the development strategy took into account strategic reference framework at European level - Europe 2020 strategy and the European Employment Strategy.

3 Case study

3.1 Manufacturing company brief description

Next, an analyzes is carried out, based on data from the past 25 years, regarding the factors that led to active aging workers who have "aged" at their jobs. The studied company has the average age of workers of 54 years and is active in the machinery industry, manufacturing, sales and repair in the field of construction machines (cranes), execution of complex steel structures. Most of the workers are older than 30 years in the company.

A considerable share relative to other companies is working after retirement in the same job. We analyzed a series of staffing resources, labor conditions, medical monitoring procedures from company archives to find good practice guide, which competed in achieving the goal of active aging.

Currently the company performs complex projects of steel structures and operates in a hall of 10,000 square meters equipped with cranes that can lift loads of up to 20 ft. and a depot equipped with gantry crane 12 ft.

The company has adapted to economic and technological change preserving human resources in the company. They installed automated welding lines; they have adopted new technologies (MIG MAG welding system) with electrode wrapped in protective gas atmosphere, it was replaced the blasting and cutting technology. Activities in the production areas held over time at temperatures of 7-8 degrees Celsius in winter and summer temperatures of 25-28 degrees Celsius. During hot periods soda was granted to workers from the own station. In the company has operated a labor union until 2008, who participated in decisions on health and safety at work.

The analysis of pollutants measurements records in 1989, 1991, 1994, 2015 show that the pollutants did not exceeded the maximum level at the workplace. Certain pollutants level decreased, other pollutants vanished through the reorganization of the company and its refurbishment [14]. There were no occupational illnesses recorded. The company recorded eight serious accidents, including one event with five victims due to human error.

3.2 Workforce structure analysis

The company has in his personnel structure over 70% of persons older than 30 years at the same job. Currently the company employees 72 people with an average age of 54, working as: welders (16), mechanics (22), welders (2), crane operators (9), engineers and supervisors (line management -12).

The analysis done for a staff number of 67 workers from the productive activity at company level (Figure 4) and for each job. Workers older than 50 are at the rate of 74%. The percentage of 74% includes 14 retirees, former employees of the company, working as welders and mechanics. Within the company, 46% of the workers have higher seniority than 30 years.

They are specialists who trained in the vocational, secondary and postsecondary education system, existent until the 1990s. A percentage of 45% workers between 10-20 years seniority, are over 50 years, with experience gained in other companies. They employed mostly in 2004-2006. Some of them were trained in occupations they practice working with their more experienced colleagues. Only four workers are under the age of 30 years.

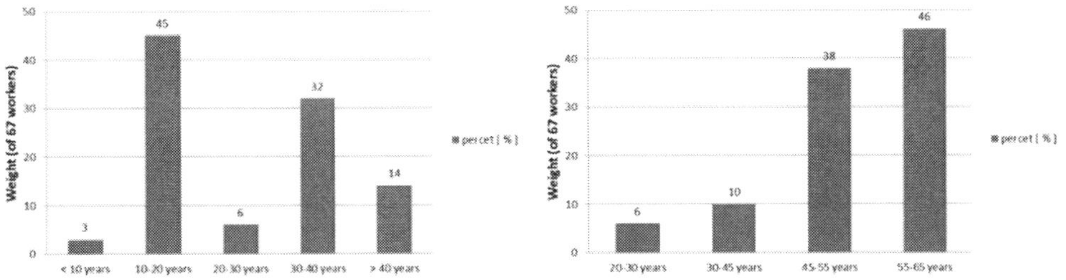

Fig. 4. Global structure of workforce: seniority category (left); age category (right)

Analysis of welders by seniority and age (figure 5) within the company shows that 88% of welders are aged over 45 years, 87% were over 20 years in seniority, of which 37% are between 30 and 40 years in seniority. The share of 50% for the 10-20 year seniority in the company is due to refurbishment and a wave of hiring in the years 2004- 2006.

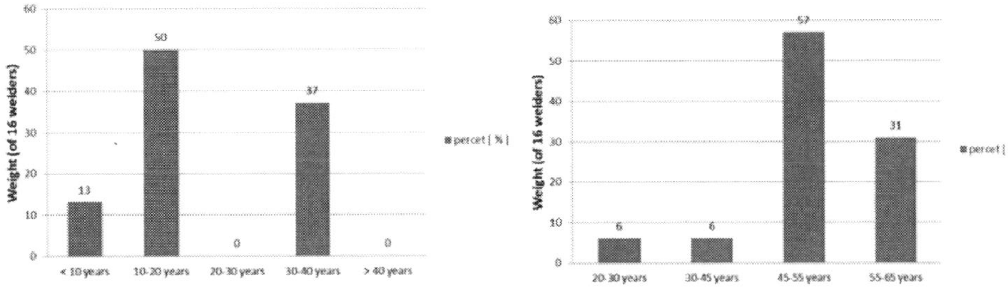

Fig. 5. Structure of welders by: seniority category (left); age category (right)

Mechanics are another category of basic productive workforce category. It observes (figure 6) the same peak at 10-20 years seniority group, related to period 2004-2006 related to the refurbishments made. Over 90% of the mechanics are older than 45 years.

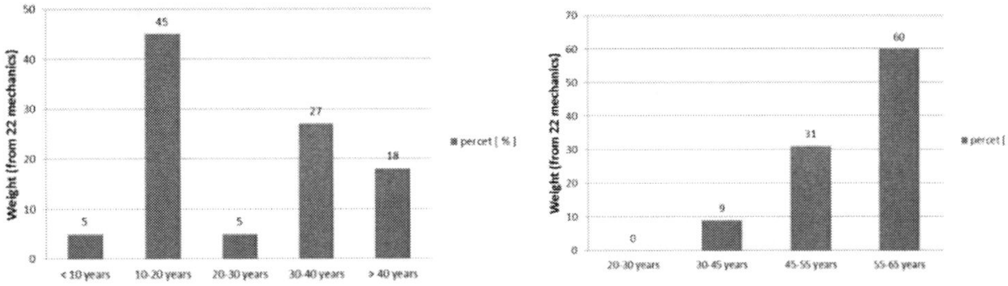

Fig. 6. Structure of mechanics by: seniority category (left); age category (right)

The crane operator is a special profession in professional careers within this company because of the special skills to acquire and keep as long as possible. Most crane operators are located in a range with 20-30 years of seniority. They participate in activities including repair, installation and operation of cranes. Teams are mixed in terms of age, since in addition to experience and rigor needed there is also required a significant level of physical activity. It is the only profession where workers distributed relatively evenly over a wide age range, comprised between 20-65 years (figure 7).

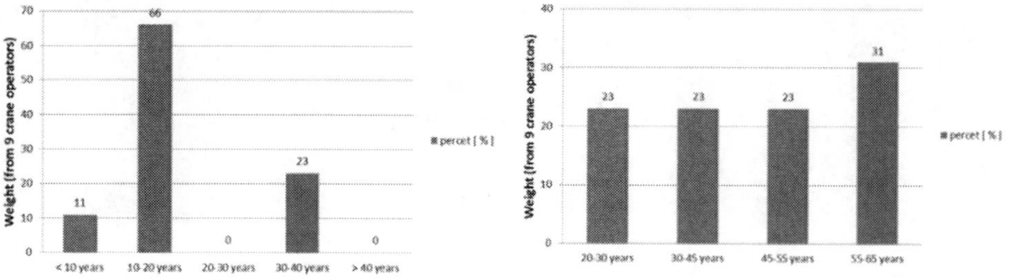

Fig. 7. Structure of crane operators by: seniority category (left); age category (right)

The company's management, on all segments of decision made mainly by people with experience of over 30 years in the company. An overwhelming majority of engineers and supervisors, over 76% of them are people who have an age over 55 years. A percentage of 34% engineers and supervisors, heads of department and directors have over 40 years in the company (figure 8).

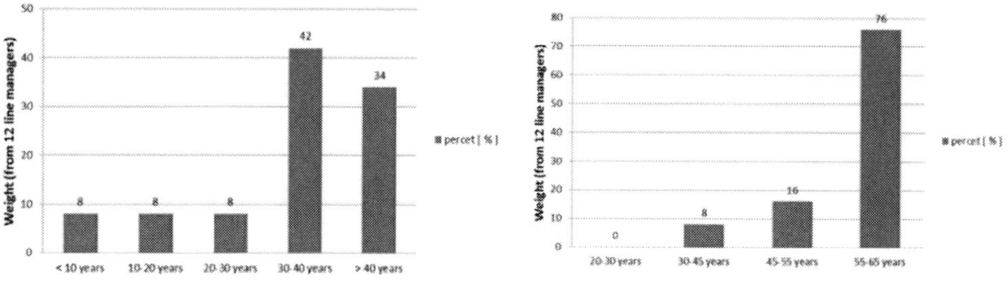

Fig. 8. Structure of line management (engineers, supervisors) by: seniority category (left); age category (right)

Occupations in which workers have in a percentage of over 80% ages over 50 years are welders, mechanics, engineers and supervisors. Gaining experience is an asset both as individual career development and as company growth by adapting legally and economically the company to the requirements of an increasingly competitive market in four occupations.

3.3 Discussion

The activity of people aged over 50 and those retired promoted by analyzed company policy in a complex context. The company management achieved through a core of people who have over 40 years of seniority in the company. The process of leadership ageing in this company along with the largest part of the employees has led to a relationship of loyalty unmatched by other strategies. Career development of leadership occurred in parallel with that of workers and linked to development of the company and moving together through a series of reorganizations and periods, better or worse. The general policy of the company to create premises for active aging applied throughout the years and has been beneficial for both management and workers.

Longevity of management is an important and positive factor in providing continuity of the policies and decisions taken over the years that lead to achieving long-term goals. The specific character of the company's activities required workers with high specialization and expertise. Although it is near the city of Bucharest, the labor market demand for trades exercised in the company is relatively small and casual.

Recruitment policies of specialized personnel aged over 50 years and policies to maintain elderly and experienced personnel have adapted to the realities of the Romanian society. Concerning Romanian education system, which no longer produce the necessary specialists in all regions of the country, for all activities in the economy also adapted to the phenomenon of migration in European space of skilled workforce.

The analyzed company has developed own mentoring activities and processes. Experienced workers transmitted to their younger coworkers an information flow that helped the last ones to evolve careers in the trades they practice. The educational gap between generations can still totally complemented by mentoring in a company. Elder colleagues can submit information for career development of younger colleagues to a certain level, according their own the level of training in school. If practicing a job can increase, young workers skills, the lack of school in a particular area cannot fully compensated. An example of this would be in the company that experienced workers and having technical studies may lead a project from the execution to mounting, respectively to read and execute from the drawings and sketches, which a worker without technical learning pathways cannot do. On the other hand, younger workers have taken over the activities that required physical demands and were performing more jobs in differing workplaces.

Although in the company there were working environments with chemical pollutants, most workers have not accused major influences of the working environment on their health during their working life. The monitoring of their health state through annual checks with physicians specialists in occupational medicine revealed that just in a small percentage it was recommended the admission to work conditioned by disorders of the ear, by musculoskeletal illnesses or restrictions of work at heights. Solutions applied to compensate dysfunctions installed over time by developing other skills useful in the production process. The company not employed unskilled workers. The company has shown openness to the needs and expectations of older people. A relaxing policy installed regarding the holiday periods, rest time, recovery holidays for physical dysfunctions. Workers received days off when social obligations inherent to their age were requiring that.

This combined with the specificity of workers of this company. Most are from the suburban areas and are working off-job in farmland. Outside their working time, they are most active, working their land with their families, feeding with their own products and have a healthy lifestyle. Outdoor activity, physical labor in agriculture, a healthy lifestyle is a key factor in maintaining a healthy and active body into old age. The fact that a total of 14 pensioners are still working shows that in addition to a healthy private life, which has a main role in aging, were created in this company premises of a prolonged professional route and extensively was maintained the state of a good physical and mental health at a level that allows an active life. All management measures taken have led to the development of a competitive company, which conducted the changes imposed by economic and legislative conditions using in its core business development the human resources. Around the human resource, strategies thought and implemented to value the opportunities for change and development. Although employees get older, today the average age being 54 years, the company uses its resources effectively, which places it in the top of complex metal construction and repair and maintenance of tower cranes activities.

4 Conclusion

Diminishment of physical and cognitive abilities related to aging in general, does not necessarily affect performance and productivity and there is no need to treat older workers differently than young workers, as long as employers are aware that with age the physical and mental capabilities are slightly reduce. Any obstacles identified and compensated through a series of preventive measures, through education and training, and remedial action such as training for older workers lacking specific skills to use new technologies. Through proper management, valuable experience and skills of older workers can retained within organizations, ensuring a balance between the energy and dynamism of youth and experience of the elderly.

The manufacturing company analyzed in the case study is a perfect example of an extended professional development, having the employees' age average of 54 and half of them working for the company for more than 30 years. The company has succeeded in achieving the workforce active ageing target.

References

1. B. Rechel, E. Grundy, J.M. Robine, J. Cylus, J.P. Mackenbach, K. Knai, M. McKee, Lancet Series on Europe, **381**, 9874, 1312–1322 (2013)
2. Eurostat, Archive: European population compared with world population, http://ec.europa.eu/eurostat/statistics-explained/index.php/Archive:European_population_compared_with_world_population (2012)
3. L.I. Cioca, R.I. Moraru, G.B.Băbuţ, Proceedings of the 15th International Scientific Conference „The Knowledge Based Organization", section: Management, **2**, 43-48, Land Forces Academy Sibiu, Romania (2009)
4. L.I. Cioca, R.I. Moraru, Explosion and/or fire risk assessment methodology: a common approach, structured for underground coalmine environments, **57** (1), 53 (2012)
5. Eurostat, Statistics Explained" European Union Labor force survey - Annual results" http://ec.europa.eu/eurostat/web/microdata/european-union-labour-force-survey, (2014)
6. Eurostat, Employment rate of people ages 20 to 64 in the EU above to 70% in 2015, http://ec.europa.eu/eurostat/documents/2995521/7240293/3-26042016-AP-EN.pdf/2be72dc1-dcbd-4beb-bad2-42c3ab792e62 (2015)
7. Eurostat, Population structure and ageing, Eurostat website, http://ec.europa.eu/eurostat/statistics-explained/index.php/Population_structure_and_ageing (2015)
8. Health and Safety Laboratory, An update of the literature on age an employment, RR832 Research Report (2011)
9. European Parliament, Pilot project on health and safety of older workers, (2016) Accessed on 20 December 2016 at https://osha.europa.eu/en/themes/osh-management-context-ageing-workforce/ep-osh-project (2016)
10. EU-OSHA, Rehabilitation and return to work: an analysis of EU and Member State systems and programs, https://osha.europa.eu/en/tools-and-publications/publications/rehabilitation-and-return-work-analysis-eu-and-member-state/view (2016)
11. L. Ivaşcu, L.I. Cioca, L., The 2nd International Conference on Psychology, Management and Social Science, Psychology, Management and Social Science, Shenzhen, China, Book Series: Advances in Education Research, **49**, 77-80 (2014)
12. ***, National Institute of Statistics of Romania, Press release, http://www.insse.ro/cms/sites/default/files/com_presa/com_pdf/somaj_2015r.pdf (2015)
13. Romanian Government, Government Decision no. 1071 of December 11, 2013 approving the National Strategy for Employment 2014-2020 and 2014-2020 Action Plan for implementing the National Strategy (2013)
14. V. Torretta, E.C. Rada, M. Ragazzi, Marco, et al. Waste Management 45 Special Issue: SI, pp: 152 (2015)

Advances in impact resistance testing for explosion-proof electrical equipment

Vlad Mihai Pasculescu[1,*], *Nicolae Ioan* Vlasin[1], *Gheorghe Daniel* Florea[1], *Marius Cornel* Suvar[1], and *Cosmin Ioan* Colda[1]

[1]National Institute for Research and Development in Mine Safety and Protection to Explosion – INSEMEX, 32-34 G-ral Vasile Milea Street, 332047, Petroşani, Hunedoara County, Romania

Abstract. The design, construction and exploitation of electrical equipment intended to be used in potentially explosive atmospheres presents a series of difficulties. Therefore, the approach of these phases requires special attention concerning technical, financial and occupational health and safety aspects. In order for them not to generate an ignition source for the explosive atmosphere, such equipment have to be subjected to a series of type tests aiming to decrease the explosion risk in technological installations which operate in potentially explosive atmospheres. Explosion protection being a concern of researchers and authorities worldwide, testing and certification of explosion-proof electrical equipment, required for their conformity assessment, are extremely important, taking into account the unexpected explosion hazard due to potentially explosive atmospheres, risk which has to be minimized in order to ensure the occupational health and safety of workers, for preventing material losses and for decreasing the environmental pollution. Besides others, one of the type tests, which shall be applied, for explosion-proof electrical equipment is the impact resistance test, described in detail in EN 60079 which specifies the general requirements for construction, testing and marking of electrical equipment and Ex components intended for use in explosive atmospheres. This paper presents an analysis on the requirements of the impact resistance test for explosion-proof electrical equipment and on the possibilities to improve this type of test, by making use of modern computer simulation tools based on finite element analysis, techniques which are widely used nowadays in the industry and for research purposes.

1 Introduction

The explosion risk is one of the main industrial hazards from the economy, this risk being present in all units which produce, process, manipulate, store or transport hazardous substances, respectively products with flammable and explosive characteristics. Explosion protection is of high importance for the occupational health and safety field, explosions endangering the life and health of workers due to their resulting flames and gas pressures,

* Corresponding author: vlad.pasculescu@insemex.ro

to the presence of hazardous products and to the consumption of oxygen from the atmospheric air. [1-2]

In the protection to explosion field, within the Romanian legislation have been transposed the two European Directives, the so called "ATEX" Directives, which regulate the placing on the European market of products intended to be used in potentially explosive atmospheres, respectively their usage in safety conditions.

Starting from April 20th 2016, the new ATEX Directive 2014/34/EU of the European Parliament and of the Council of 26 February 2014 on the harmonisation of the laws of the Member States relating to equipment and protective systems intended for use in potentially explosive atmospheres (recast) has entered in force, substantially amending and replacing Directive 94/9/EC.

The ATEX "Workplace" Directive 1999/92/EC which deals with the minimum requirements for improving the level of health and safety protection of workers potentially at risk from explosive atmospheres complements the responsibilities of manufacturers under Directive 2014/34/EU by setting out the responsibilities of employers.

For the authorities of the Member States to trust in the products which are placed on the community market are required guarantees on the products' conformity within a system of mutual recognition of normative requirements. A high safety level in areas with explosion hazard may be guaranteed only through an efficient and close cooperation between all parties involved: manufacturers, certification bodies, authorities, end-users of equipment, as well as persons who install equipment within the installations. [3-4]

Placing on the market and/or into operation of Ex products (Ex equipment, protective systems and devices) is allowed only if they do not harm the health and safety of persons, or, if the case, of domestic animals or goods, when they are properly installed, maintained and used in accordance with their destination. For these conditions to be fulfilled, Ex equipment, protective systems and safety devices have to comply with the essential health and safety requirements (EHSR) which are applicable for their intended destination. [5-8]

2 Impact resistance testing for explosion-proof electrical equipment

Electrical equipment used in potentially explosive atmospheres may represent an ignition source for such atmospheres. In order to decrease this risk, it has been established that they have to be subject to type tests in order to establish if they are compliant to be used in explosive atmospheres [9].

One of these tests is the impact resistance test. Therefore, in accordance with Standard EN 60079-0, Explosive atmospheres - Part 0: Equipment - General requirements, the electrical equipment gas to be subjected to the effect of a 1kg weight test mass falling (Fig. 2) from a specific height h. The height h is specified in the standard in relation with the destination of the electrical equipment and is presented in Table 1.

Table 1. Tests for resistance to impact

	Drop height *h* with 1 kg mass m			
Equipment grouping	**Group I**		**Group II**	
Risk of mechanical danger	**High**	**Low**	**High**	**Low**
a) Enclosures and external accessible parts of enclosures (other than light-transmitting parts)	2	0,7	0,7	0,4
b) Guards, protective covers, fan hoods, cable glands	2	0,7	0,7	0,4
c) Light-transmitting parts without guards	0,7	0,4	0,4	0,2
d) Light-transmitting parts with guard having individual	0,4	0,2	0,2	0,1

	Drop height *h* with 1 kg mass m			
openings from 625 mm^2 to 2 500 mm^2				
NOTE A guard for light-transmitting parts having individual openings from 625 mm^2 to 2500 mm^2 reduces the risk of impact, but does not prevent impact				

The test mass has to be fitted with an impact head made of steel, of semi-spherical shape, having 25 mm in diameter.
Before each test, there shall be verified of the surface of the impact head is in good shape.

The impact resistance test has to be performed on the completely assembled and ready for use electrical equipment; however, if this is not possible (i.e. light-transmitting parts), the test has to be performed on the un-mounted relevant parts, but which are fixed in their mounting support or in an equivalent one. Tests upon an empty enclosure are allowed only with a proper justification provided in the documentation.

The impact points have to be the places which are considered to be the weakest and they have to be external parts which might be exposed to impact. If the enclosure is protected by another enclosure, only the external parts of the assembly have to be subjected to impact resistance tests.

The electrical equipment must be mounted on a steel base, in such a way that the impact direction to be perpendicular to the surface to be tested if this is a flat one, or perpendicular to the tangent to the surface at the point of impact, if it is not a flat one. The pedestal must have a mass of at least 20 kg or to be rigidly fixed to or recessed in the floor, for example, set in concrete. Figure 1 shows an example of a stand used for performing the impact resistance tests for electrical equipment, being schematically presented in Figure 3.

The impact resistance test has to be performed at least for two specimens. For light-transmitting parts made of glass, the test has to be performed only once for each specimen. In any other case, the impact resistance test has to be performed in two separate places on each specimen. [10]

Fig. 1. Stand for impact resistance testing

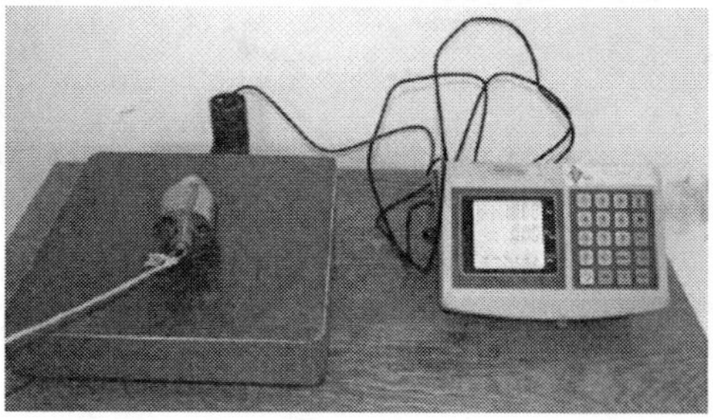

Fig. 2. 1 kg steel mass.

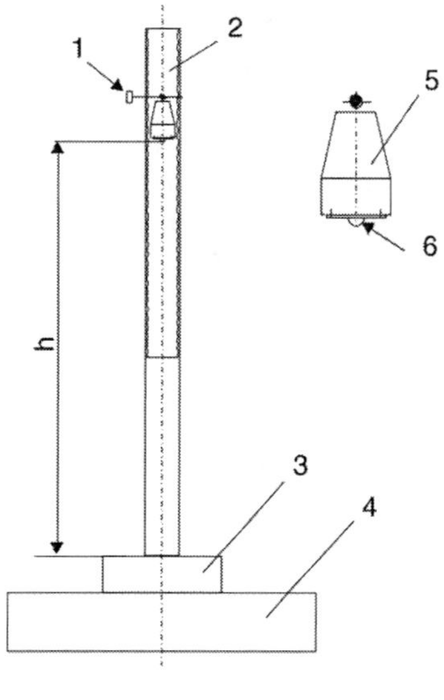

Legend:

1 Adjustment pin
2 Guidance tube made of plastic
3 Equipment to be tested
4 Steel base (mass ≥ 20kg)
5 Steel mass - 1 kg
6 Impact head, 25 mm in diameter, made of hardened steel
h Falling height

Fig. 3. Representation of the impact resistance-testing stand

3 Quantitative tests for determining the impact resistance of explosion-proof electrical equipment

Normally, the impact resistance test does not provide a value as result, the result being only qualitative, namely the equipment "passed" or "did not pass" the test.

In order to be able to record the load on the equipment generated from the falling test mass, the impact resistance test stand has been equipped with a 50kN tension load cell force transducer (Fig. 4) and the values has been recorded using MaxTest software.

Fig. 4. 50 kN tension load cell force transducer

There have been subjected to the impact resistance test several equipment in Group II (Fig. 5-9), with different types of protection to explosion, for different falling heights of the test mass, in accordance with the type of equipment and the mechanical danger risk. MaxTest software generated the load-time graphs for each test, these being presented in Fig. 10.

Fig. 5. Impact resistance test carried out for equipment no. 1

Fig. 6. Impact resistance test carried out for equipment no. 2

Fig. 7. Impact resistance test carried out for equipment no. 3

Fig. 8. Impact resistance test carried out for equipment no. 4

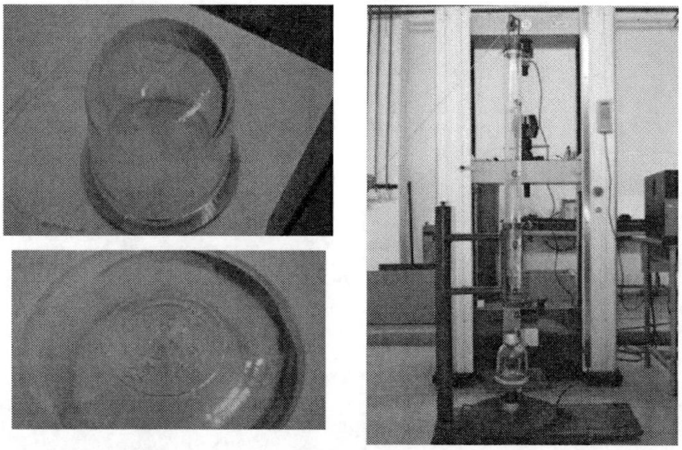

Fig. 9. Impact resistance test carried out for equipment no. 5.

DOI: 10.1051/matecconf/201712111015

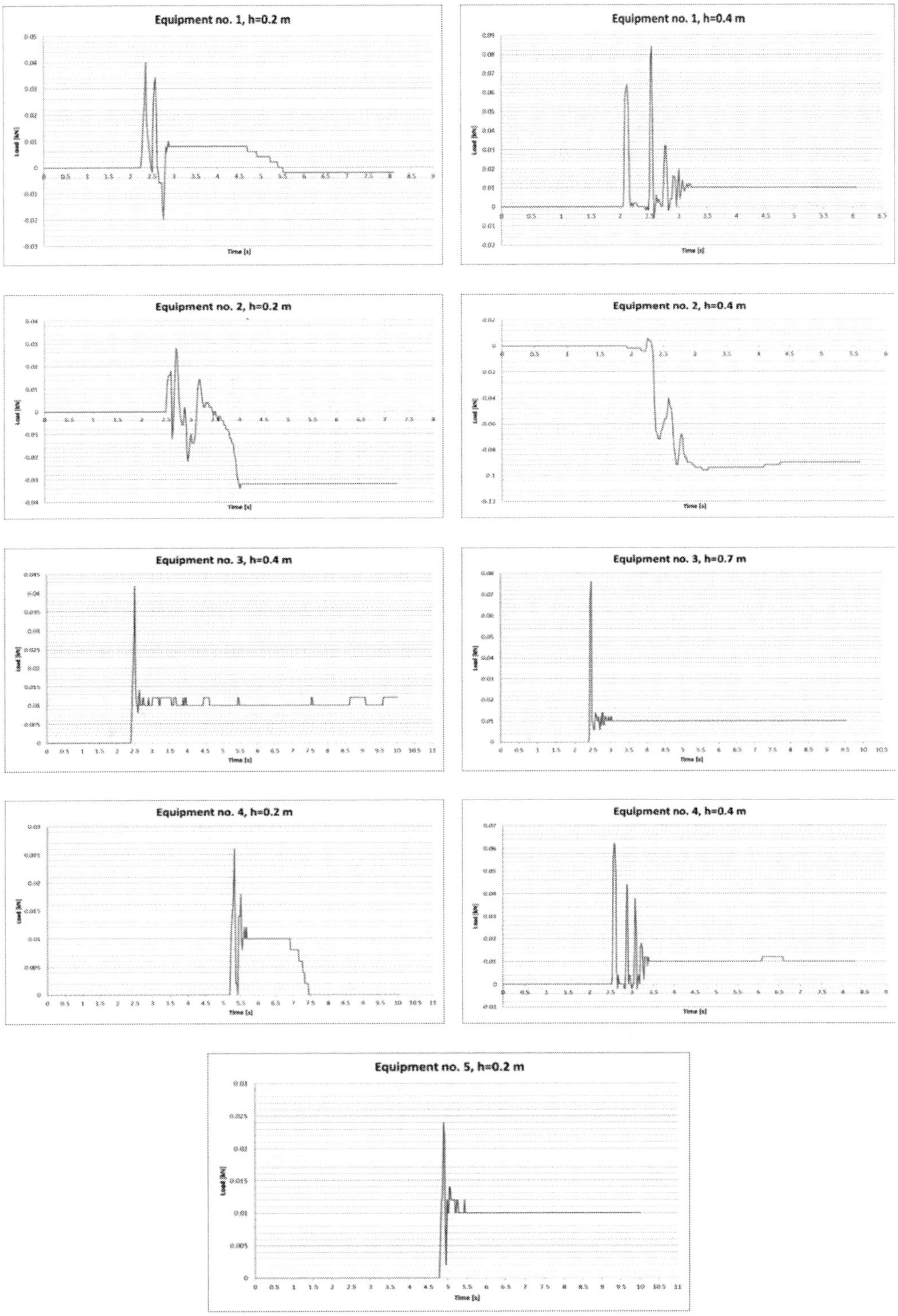

Fig. 10. Load-time graphs for the equipment subjected to impact tests

4 Conclusions

Results obtained and presented in the paper are going to be analysed in order to calibrate the computational simulation tools of which INSEMEX disposes on the tests for impact resistance of explosion-proof electrical equipment. It has been established in previous studies that ANSYS Structural represents the best solution for performing the computational simulations for impact resistance tests, the results obtained are going to be used as input data for these simulations, using geometries which describe as close as possible the types of equipment tested. [10-12]

Therefore, the currently known methods for impact resistance tests for explosion-proof electrical equipment are going to be improved by using modern computational simulation techniques, resulting in an improved occupational health and safety level for industries which process, transport, store etc. flammable substances and in which there is the possibility for an explosive atmosphere to occur.

This paper was developed within the Nucleu-Programme, carried out with the support of ANCSI, project no. PN-16-43-03-06

References

1. R.I. Moraru, G.B. Babut, L.I. Cioca, Environ Eng Manag J, **13** (6), 1371-1376 (2014)
2. V. Torretta, E. Rada, M. Ragazzi, E. Trulli, I.A. Istrate, L.I. Cioca, Waste Manage, 45, 152-160 (2015)
3. L.I. Cioca, R.I. Moraru, Arch Min Sci, 57 (1), 53-60 (2012)
4. I. Ivascu, L.I. Cioca, *Advances in Education Research*, **49**, 77-90 (2014)
5. D. Pasculescu, L. Lupu, V. Pasculescu, I. Inisconi, M. Suvar, Environ Eng Manag J, **11** (7), 1267-1271 (2012)
6. D.C. Petrilean, S.I. Irimie, V. Baleanu, S. Stanila, Environ Eng Manag J, **13** (6), 1383-1388 (2014)
7. G. Buica, A.E. Antonov, C. Beiu, I. Iorga, Environ Eng Manag J, **11** (7), 1247-1255 (2012)
8. M. Prodan, E. Ghicioi, D. Oancea, Environ Eng Manag J, **13** (6), 1409-1414 (2014)
9. M. Magyari, S. Burian, L. Moldovan, D. Fotau, C. Colda, Proceedings of the 15th International Multidisciplinary Scientific GeoConference, **1** (1), 835-842 (2015)
10. V.M. Pasculescu, G.D. Pricop, M.S. Morar, V. Florea, Proceedings of the 15th International Multidisciplinary Scientific GeoConference SGEM, **1** (2), 291-298 (2015)
11. D.G. Pricop, V.A. Florea, Proceedings of the 15th International Multidisciplinary Scientific GeoConference SGEM, **1** (1), 697-704 (2015)
12. D. Pasculescu, T. Niculescu, Proceedings of the 15th International Multidisciplinary Scientific GeoConference SGEM, **1** (2), 323-330, (2015)

MATEC Web of Conferences **121**, 11016 (2017) DOI: 10.1051/matecconf/201712111016
MSE 2017

Aspects regarding explosion risk assessment

Mihaela Părăian[1,*], *Sorin* Burian [1], *Mihai* Magyari[1], and *Lucian* Moldovan[1]

[1]National Institute for Research and Development in Mine Safety and Protection to Explosion – INSEMEX Petroşani, 32-34 G-ral Vasile Milea Street, 332047, Petroşani, Romania

Abstract. Explosive risk occurs in all activities involving flammable substances in the form of gases, vapors, mists or dusts which, in mixture with air, can generate an explosive atmosphere. As explosions can cause human losses and huge material damage, the assessment of the explosion risk and the establishment of appropriate measures to reduce it to acceptable levels according to the standards and standards in force is of particular importance for the safety and health of people and goods.There is no yet a recognized method of assessing the explosion risk, but regardless of the applied method, the likelihood of an explosive atmosphere occurrence has to be determined, together with the occurrence of an efficient ignition source and the magnitude of foreseeable consequences. In assessment processes, consequences analysis has a secondary importance since it's likely that explosions would always involve considerable damage, starting from important material damages and up to human damages that could lead to death.The purpose of the work is to highlight the important principles and elements to be taken into account for a specific risk assessment. An essential element in assessing the risk of explosion in workplaces where explosive atmospheres may occur is technical installations and personal protective equipment (PPE) that must be designed, manufactured, installed and maintained so that they cannot generate a source of ignition. Explosion prevention and protection requirements are governed by specific norms and standards, and a main part of the explosion risk assessment is related to the assessment of the compliance of the equipment / installation with these requirements.

1 Introduction

An essential element in assessing the risk of explosion in workplaces where explosive atmospheres may occur are facilities which must be designed, installed and maintained so as to keep flammable substances in the enclosed site to prevent the formation of explosive atmospheres around them and not to generate sources of ignition.
SR EN 15198 defines the *ignition risk* as the probability of occurrence of an ignition source that is capable of igniting an explosive atmosphere [1].

Explosion risk may be defined as probability of occurrence of an explosive atmosphere simultaneously with occurrence of an efficient potential ignition source and the effects induced by it [2].

* Corresponding author: mihaela.paraian@insemex.ro

Explosion risk identification and assessment

In conformity with the requirements of the European Directive 1999/92/CE, harmonized in Romanian legislation by Government Decision no. 1058 of 9th August 2006 [3], regarding the minimum requirements for improving health safety and protection of workers which can be exposed to risks due to explosive atmospheres, the employer responsible for the workplace, according to legislation, shall coordinate application of all precautions related to workers health and safety and he has to declare in the document regarding explosion protection, the purpose of coordination. the implemented measures and procedures that had bed adopted.

When setting out the appropriate technical and organizing measures to ensure workers safety and health by the employer, several fundamental principles have to be taken into consideration, as the following ones: preventing formation of explosive atmospheres, avoiding ignition of explosive atmospheres and limiting the damaging effects of explosions.

In order to fulfil the obligations provided by the European Directive 1999/92/EC, the employer has to ensure drawing up and update of a document called *Explosion protection document.*

According to the guide of good practice for implementing the 1999/92/CE directive, the structure of the explosion protection document comprises a description of the workplace and work areas with explosion hazards, as well as a description of technological process stages or activities, a description of safety substances / parameters used, the risk analysis results, the explosion protective measures adopted and the implementation of explosion protection measures.

In regard to explosion risk assessment, this stage focuses initially on formation of dangerous environments, and farther on ignition sources presence and activation.

In assessment processes, consequences analysis has a secondary importance since it's likely that explosions would always involve considerable damage, starting from important material damages and up to human damages that could lead to death. The quantitative risk approach is also secondary, in explosion protection, in relation to preventing formation of dangerous explosive environments.

Each work process as well as each operating condition of an installation and each alteration of such conditions shall be subject to assessment. During new or existing installation assessment the following operating conditions shall be taken into attention: normal operation including maintenance, starting and shutting down operations, operation accidents and foreseeable malfunctions, inappropriate use that can reasonably be foreseen.

Methods for assessment of explosion risks

The concept of explosion risk consists of two elements: firstly, the probability that an explosion will occur (explosion sensitivity); and secondly, the adverse effects resulting from the occurrence of an explosion (explosion severity).

Explosion Sensitivity

Given the presence of a combustible or explosive mixture in a certain space, the probability relates to the ease of ignition.

Explosion Severity

The adverse effects of explosions relate to the amount of energy released and the time required to release this amount of energy (power), as well as to its volume density (energy/volume) and the maximum explosion temperature reached. Example of estimation and assessment of explosion risk [4]

For estimation and assessment of explosion risk must be determined two elements: severity of possible negative events and likelihood of occurrence of such events (sensitivity). Severity or consequence of explosions can many times be correctly assessed, but likelihood of occurrence is usually more difficult to quantify.

DOI: 10.1051/matecconf/201712111016

Severity can be expressed as a pre-defined level, one or more levels that may result from each dangerous event. Thus, in terms of person or material damages in the system, severity may be expressed as follows: catastrophic, major, minor, negligible. In order to assess frequency of each severity level, as a first step the screening technique can be applied for determination of each event separate likelihood. The occurrence frequency can be expressed quantitatively as: frequent, probable, occasional, less probable, improbable.

Definitions of severity and frequency levels are given below:

- risk level A: High risk level
- risk level B:
- risk level C:
- risk level D: Low risk

Table 1. Levels of severity of the identified risks

SEVERITY	The definition of adverse event
Catastrophic	Death and destruction system
Major	Severe injury, severe illness or major system damage
Minor	Minor injury, minor illness or minor system damage
Negligible	Less than minor injury, illness or damage to system

Table 2. The frequency of risk

Frequency of occurrence	Individual specific element	Particularization
Frequent	Likely to occur frequently	It happens continuously
Probably	Will appear several times over the life of an item	Will appear frequently
Occasionally	Likely to occur sometime in the life of an item	Will appear several times
Little probably	Unlikely but possible to occur over the life of an item	Unlikely but it is reasonable to expect that there
Unlikely	So unlikely that it can be estimated that there will appear	Unlikely to occur but possible

Table 3. Matrix frequency /severity of risk levels

FREQUENCY OF OCCURRENCE	SEVERITY			
	Catastrophic	Major	Minor	Negligible
Frequent	A	A	A	C
Probably	A	A	B	C
Occasionally	A	B	B	D
Less probably	A	B	C	D
Unlikely	B	C	C	D

After risk estimation, the risk assessment should be performed to determine if it is necessary to reduce the risk or has reached the level of security.

It is obvious that if the risk estimate is obtained a level of risk A, then the risk is so high that it is intolerable and additional measures are required to reduce further risks. Similarly, the risk level D can be considered acceptable and no further action is needed to reduce risk.

Thus, the risk can be described as:

Intolerable If the risk falls into this category, then you have taken appropriate security measures to reduce the risk; or:

Acceptable If the risk falls into this category then it is necessary to reduce the risk and the risk assessment is complete.

Risk levels B and C are intermediate and normally will require risk mitigation measures to make acceptable level of risk. However, the magnitude of these measures will be lower if the risk level C usually organizational risk reduction measures will suffice.

Table 4. Matrix risk levels/actions

A	High risk (intolerable)	Adequate actions, organizing and/or technical, with priority
B	Medium risk	Adequate actions, organizing and/or technical, on short or medium term
C	Low risk	Maintains the present safety state
D	Null risk (acceptable)	No action

2 Risk assessment

An essential element in assessing the risk of explosion in workplaces where explosive atmospheres may occur are facilities which must be designed, installed and maintained so as to keep flammable substances in the enclosed site to prevent the formation of explosive atmospheres around them and not to generate sources of ignition. They must be designed, manufactured, installed and maintained in accordance with explosion protection requirements and standards established by specific normes and standards, and a main part of the explosion risk assessment relates to the assessment of the conformity of the equipment / installation with those requirements.

The assessment has to determine if a potential ignition source may be present and to identify if an explosive atmosphere is present. The assessment begins usually by analysing the equipment during normal operation, and then is extended to take into consideration the foreseeable and rare malfunctions, in order to establish the level of explosion safety. An assessment of the likelihood of occurrence of an ignition source and its efficiency in igniting the explosive atmosphere; this requires detailed information on the flammability and explosion characteristics of the explosive atmosphere. Practically, first the areas where the explosive atmosphere may occur shall be defined, as well as the likelihood of its occurrence by applying a "Zoning" method; then the possible ignition sources shall be identified, taking into account normal operation conditions and the foreseeable and rare malfunctions [2].

In regard to equipment, protective systems and components in installation and all safety devices placed outside the explosive atmosphere but having a role of controlling the equipment safety within the Ex zones under the ATEX Directive's scope[5], the problem is sort of solved already, since these, before being placed on the market, had already been assessed by the manufacturer and are guaranteed for certain fields of use (the intended use).

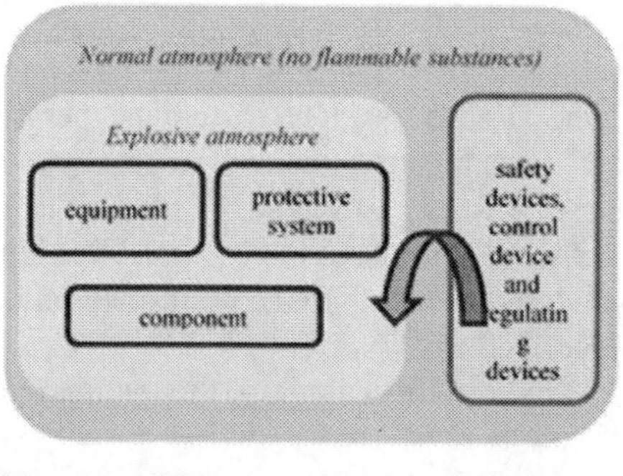

Fig. 1. Aplication domain of European Directive ATEX 2014/34/UE

The users have only to classify their dangerous areas and then adequately choose the equipment, according to the table 5[3,5-6].

Table 5. Level of protection required, in function of the explosive atmosphere

ZONE	Presence of an explosive atmosphere	Ignition sources avoidance	Level of protection required	Group II category	EPL
2	Infrequent or only on a short period of time	during normal operation	NORMAL	3G	Gc
1	Likely to occur	also during foreseeable malfunctions (one defect)	HIGH	2G	Gb
0	Continuously, for long periods of time or frequently	also during rare malfunctions (two defects independent)	VERY HIGH	1G	Ga
USERS Directive1999/92/EC (HG 1058/2006)		MANUFACTURERS European Directive 2014/34/UE (HG 245/2016)			

Following, the installation as a whole shall be assessed in regard to possible ignition sources.

All electric and non electric equipment, as well as the associated connected devices shall be taken into consideration.

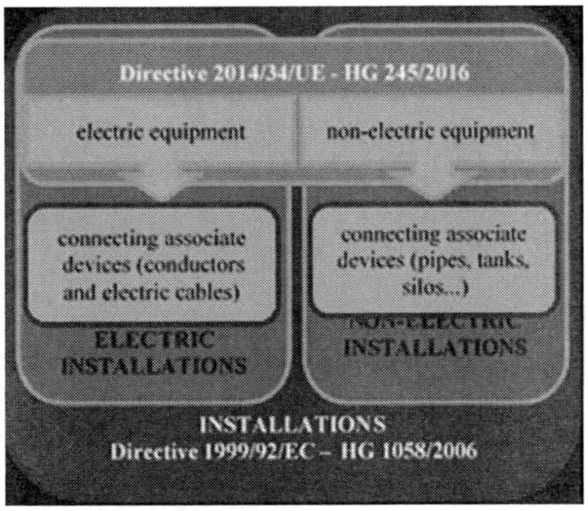

Fig. 2. Analysis of installation as a whole

Each ignition source according to EN 1127-1:2011[7] must be analyzed.

Table 6. Ignition sources according to

Ignition sources	
Possible	**Reason**
Flames and hot gases (including hot particles)	
Mechanically generated sparks	
Electrical apparatus	
Stray electric currents, cathodic corrosion protection	
Static electricity: Spark discharges; Corona discharges; Brush discharges; Propagating brush discharges; Cone discharges	
Lightning	
Electromagnetic fields in the frequency range from 10^4Hz to 3×10^{12}Hz	
Electromagnetic radiation, frequency range from 3×10^{11}Hz to 3×10^{15}Hz	
Ionizing radiation	

Ignition sources	
Possible	**Reason**
Ultrasonics	
Adiabatic compression and shock waves, gas flows	
Chemical reactions - Exothermic reactions, including self-ignition of dusts	
Ex Classification: Zone1 and Zone 2	
The equivalent level of protection corresponds to the zone where the potential ignition source is identified.	

If is not possible to avoid explosive atmospheres and sources of ignition, protective systems should be used to limit the effects of an explosion to an acceptable extent.

When using PCE (process control engineering – PCE) to prevent the occurrence of hazardous explosive atmospheres, ignition sources or to mitigate the harmful effects of an explosion, the reliability of the safety function of PCE devices and their components must be assessed.

The risk assessment shall take into consideration also the personal protective equipment and the floorings that have an important role in prevention of dangerous static discharges from persons, which could lead to ignition of explosive atmospheres [8]. Persons insulated towards earth may easily acquire and build up electrostatic charges.

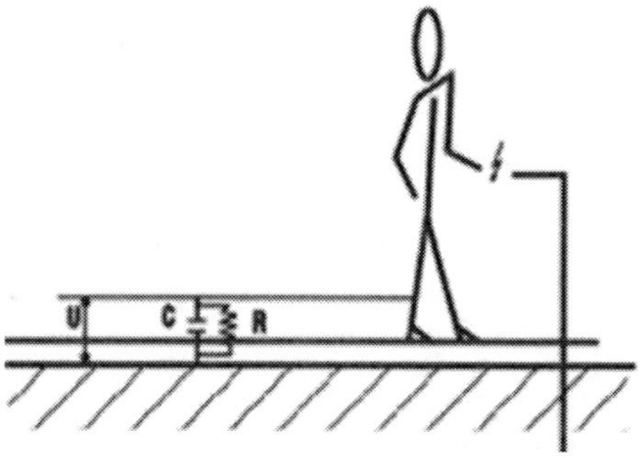

Fig. 3. Electrostatic charging of persons

In order to mitigate explosion risks, installations inspection and maintenance have also an important role.

3 Explosion protection - the actors

The personnel with attributions and responsibilities in the field of explosion protection and prevention: not only the competency of personnel involved in explosion risk assessment is of importance in order to mitigate explosion risk but also all the persons having attributions and responsibilities in the field of explosion protection and prevention: equipment manufacturers and users, personnel implied in repairing, service and inspection activities, as well as the control and assessment third-party bodies, named in the regulating legislation (INCD-INSEMEX, Territorial Labor Inspection bodies).

According the normative NEx 01-06/2007 at commissioning of a technical installation operating in potentially explosive atmospheres a verification of the technical documentation and an on-site assessment must be performed by INSEMEX Petrosani [9]. This is to assess the technical installation in order to check if all the measures required to ensure the protection to explosion are adopted, especially those regarding the equipment/protective systems used in Ex classified areas. Also, the legal entities engaged in the design, installation, maintenance and / or repair of technical installations operating in potentially explosive atmospheres, for carrying out specific activities must possess trained personnel (by INSEMEX) and have procedures / work instructions for specific activities for technical installations that operate in potentially explosive atmospheres . Also, INSEMEX can assess the capability of the organizations to perform specific activities related to technical installations operating in potentially explosive atmospheres.

Fig. 4. Correlation between the actors having a role in explosion protection

4 Conclusion

Assessment of explosion risks presumes determination of the likelihood of an explosive atmosphere occurrence, together with the occurrence of an efficient ignition source and the magnitude of foreseeable consequences. In assessment processes, consequences analysis has a secondary importance since it's likely that explosions would always involve considerable damage, starting from important material damages and up to human damages that could lead to death.

The explosion risk assessment methods have to start by classifying the Ex dangerous areas, then verifying the appropriate selection of equipment and protective systems, followed by an analysis of the installation as a whole, from the possible ignition sources standpoint. Risk / safety level assessment shall relate to the safety requirements provided in the norms and standards in the field, whilst there is a presumption of ensuring a low, acceptable level of risk level, by conforming to norms.

The risk assessment shall take into consideration also the personal protective equipment and the floorings that have an important role in prevention of dangerous static discharges from persons, which could lead to ignition of explosive atmospheres.

Setting out competencies and responsibilities for those involved in installing, maintaining and inspecting installations has a particular importance for the purpose of ensuring safety against explosions.

References

1. 1.SR EN 15198: 2008 Methodology for the risk assessment of non-electrical equipment and components for intended use in potentially explosive atmospheres (2008)
2. M. Părăian, G.A.Găman, E. Ghicioi, S. Burian, F. Paun, Buletinul Institutului Politehnic din Iaşi, Tomul **LV (LIX)**, Fasc. 1, 138-144 (2009)

3. HG. 1058 / 2006 sets out the minimum requirements for improving the health and safety protection of workers potentially at risk from explosive atmospheres (Directive 1999/92/CE) /Non-binding guide to good practice for implementing Directive 1999/92/EC "ATEX" (explosive atmospheres)
4. Methodology for the Risk Assessment of Unit Operations and Equipment for Use in Potentially Explosive Atmospheres, EU Project No: SMT4-CT97-2169
5. *** Directive 2014/34/EU on the harmonisation of the laws of the Member States relating to equipment and protective systems intended for use in potentially explosive atmospheres (recast), Official Journal of the European Union, **96** (2014)
6. SR EN 60079-14: 2014 Explosive atmospheres – Part 14: Electrical installations design, selection and erection, ASRO (2014)
7. SR EN 1127-1: 2008 Explosive atmospheres - Explosion prevention and protection - Part 1: Basic concepts and methodology (2008)
8. CEN/CLC/TR 16832:2015 Selection, use, care and maintenance of personal protective equipment for preventing electrostatic risks in hazardous areas (explosion risks) (2015)
9. *** Normative NEx 01-06/2007, Normative of 2 may 2007 regarding explosion prevention for designing, installing, commissioning, using, repairing and maintenance of technical installations operating in potentially explosive atmospheres

DOI: 10.1051/matecconf/201712111017

Seasonal monitoring of microbiological air contamination from a museum - a case study

Iuliana Pamela Scarlat[1,2*], *Veronica* Lazăr[2], *Raluca* Stepa[1], *Mariana Carmen* Chifiriuc[2] *Lia-Mara* Dițu[2], and *Maria* Haiducu[1]

[1]The National Institute of Research and Development on Occupational Safety "Alexandru Darabont" Bucharest, Laboratory Chemical and Biological Risks Bucharest, Romania,
[2]University of Bucharest, Department of Microbiology, Faculty of Biology, Bucharest, Romania

Abstract. The aim of this study was to monitor microbial contamination in the work place of a museum, for estimating the influence of seasons, microclimatic parameters and tourists. The analysis was carried out in artefacts deposits, restoration laboratories and exhibition spaces. The media used to demonstrate the presence of biological agents, non-pathogenic and with pathogenic potential (opportunistic bacteria and filamentous fungi with haemolytic properties) were: simple agar, blood agar and Sabouraud with chloramphenicol. The isolated bacterial strains were identified by API tests and the fungal strains by examination of features with taxonomic value, aspects based on macroscopic observations of colony/cultures and microscopic assessment of mycelia. The counts of the microorganisms in the air were higher especially in the textiles deposits (the mesophilic microorganisms ranged from 200 to 1.4×10^4 CFU/m^3). The filamentous fungi most frequently isolated species were: *Aspergillus* spp.; *Penicillium* spp., *Cladosporium* spp. The most frequently isolated bacteria swere: *Staphylococcus* (*S. epidermidis, S. xylosus, S. capitis*); *Bacillus (B. cereus, Bacillus* spp). The results of this study indicate a microbial air load containing both filamentous fungi and bacteria strains, including some opportunistic pathogens which can represent a health risk for the museums workers, especially in the seasons with higher exposure.

1 Introduction

Museums and the heritage deposits store, exhibit and sometimes preserve and restore old and valuable objects which can have a high degree of microbial contamination and degradation. The personnel involved in the surveillance, manipulation and restoration are exposed to microbiological health risk factors, which have to be monitored and reduced as much as possible.

According to the National Statistics Institute (INS), in Romania, in 2014 there were 739 museums and public collections with 10, 823,706 visitors [1]. In these institutions 6,554 employees work, of which: 1178 museum curators; 501 conservation workers; 390

* Corresponding author: scarlatpamela@yahoo.com

restorers; 258 researchers; the rest is accounted for by administrative staff/ supervisors (INS/ 2015).

The museum selected for monitoring is situated in a historical building, constructed at the end of the XIX century and is one of the most visited Romanian museums all year round, the highest number being during summer vacation. This museum has its own storage spaces for exhibits, not all of them in the main building of the museum, but in the same area.The weather in the museum's area is cold and moist: multi-annual average temperature is 8^0C and the average humidity is between 71% and 76% [2]. The building is surrounded by a lot of vegetation.

The number and type of microorganisms are influenced by the presence of nutrients in the contaminated objects and by temperature, humidity and seasons [3]. The present study investigated these influences on the microbiological contamination of air. These aspects are important because they may also impact the heath of museum employees, besides affecting patrimonial artefacts.

2 Materials and methods

Sampling was done in seasonal dynamics during one year, collecting a total of 420 samples, 105 samples per season. Each season the sampling was done in the same places: 10 heritage deposits (for decorative art objects, archives, and warehouses of furniture, paintings or other heritage objects) and 25 museum exhibition rooms; of these 25 rooms, in ten of them the access of visitors is forbidden, only watching from the doorway being allowed.

Microbiological sampling was performed according to EN 14042:2003, with the device SAS Super Sampler 100, for one minute per sample, with a flow rate of 100 l/min [4].

The culture media used for sampling and culturing were specific for development of mesophilic biological agents, i.e., haemolytic bacteria and microfungi: simple agar, blood agar and Sabouraud with chloramphenicol, respectively.

Microbial counting was conducted using the automated colony counter 300 SCAN, the correction being performed according to the chart sampler with 219 holes in 90 mm Petri dishes.

The calculation formula:

$$X = \frac{Pr \times 1000}{V} \tag{1}$$

V = volume of air sampled;
r = number of the counted colony forming units (C.F.U.);
Pr = probable number of holes obtained by positive correction;
X = number of C.F.U / 1000 litres of air.

The measured values of total mesophilic bacteria and filamentous fungi have been compared to those specified in the European Collaborative Action - Indoor Air Quality & ITS Impact on Man (ECA-IAQ) [5], as recommended limit values according to tables 1 and 2 respectively.

Table 1. The recommended values for mesophilic bacteria according to ECA IAQ

Category	Homes	Non industrial indoor environments
Very low	<100	< 50
Low	<500	< 100
Intermediate	<2500	< 500
High	<10 000	< 2 000
Very high	> 10 000	> 2 000

Table 2. The recommended values for fungi according to ECA IAQ

Category	Homes	Non industrial indoor environments
Very low	<50	< 25
Low	<200	< 100
Intermediate	<1 000	< 500
High	<10 000	< 2 000
Very high	> 10 000	> 2 000

API biochemical tests (BioMerieux, France) were used to identify the isolated bacterial strains. Fungal taxonomic units/at genus level were identified based on macro- and microscopic mycelia assessment.

Measurement and calculation of the parameters of the microclimate were performed according to EN 27243:1996 - "Warm ambience. Estimation of the thermal stress on working of human in WBGT index (wet temperature and globthermometer)" [6]. For the determination of the physical parameters were used: for dry air temperature (^{0}C) - measuring the digital camera model HYGRO-THERMOMETER HTA 4200; for air speed (m/s) digital camera sensor SUPER SCIENTIFIC DIGITAL hand TERMOANEMOMETER 840 030.

3 Results and Discussion

3.1 Microbiological assay

3.1.1 Quantitative analysis

The analysed samples revealed a high level of the air contamination in almost all the rooms where measurements were performed.

The minimum and maximum values as well as standard deviation of air microbial contamination are shown in Table 3.

Table 3. The range of microbiological contamination levels in the museum.

Microorganisms	No of samples	Minimum values [CFU/m^3]	Maximum values [CFU/m^3]	Standard deviation	Percentile (90%)
mesophilic	415	200	$1.4x10^4$	$3.2x10^3$	$4.6x10^2$
haemolytic bacteria	415	20	350	83	273
fungi	415	170	$3 x10^3$	$9x10^2$	$2.6x10^3$

The standard deviation for mesophilic bacteria and haemolitic bacteria is quite high, showing a large variation in the number of CFU/m^3 for these type of microorganisms. In the case of fungi there is a smaller variation.

The variation of air microbial contamination level values in each season and type of room (with full access/ doorway watching/deposit) are presented in figures 1, 2 and 3 for each type of monitored microbial groups.

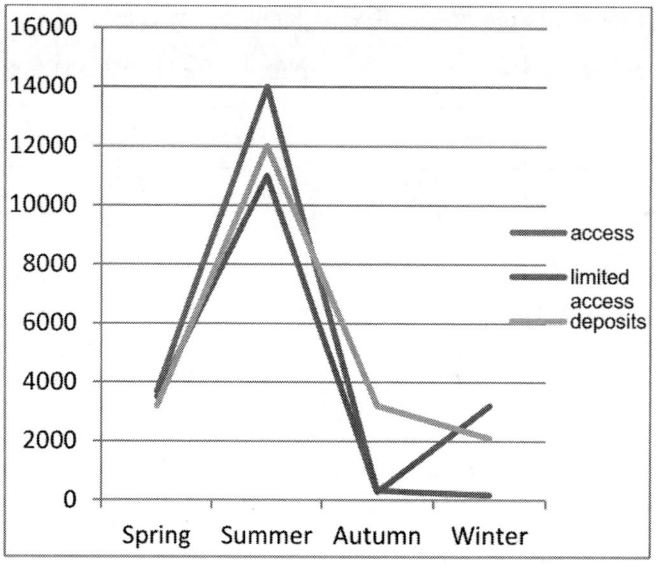

Fig. 1. The seasonal variation of airborne mesophilic bacteria (number of C.F.U /1000 l of air) in different museum spaces.

Concerning the seasonal variation of mesophilic bacteria, the highest values of indoor air contamination (correlated with the outdoor one) have been registered during the summer, in all analysed spaces. This may be explained by their resistance to high temperatures, especially Gram positive ones (see the qualitative analysis): the bacilli and clostridia which produce endospores; the staphylococci, non-spore forming bacteria, are also present but with a natural resistance in environment.

In general, the concentration of mesophilic bacteria was higher in the rooms with full visitor access than in those with limited access.This may show a relation between the presence of visitors (particularly numerous in summer) and the level of air contamination.

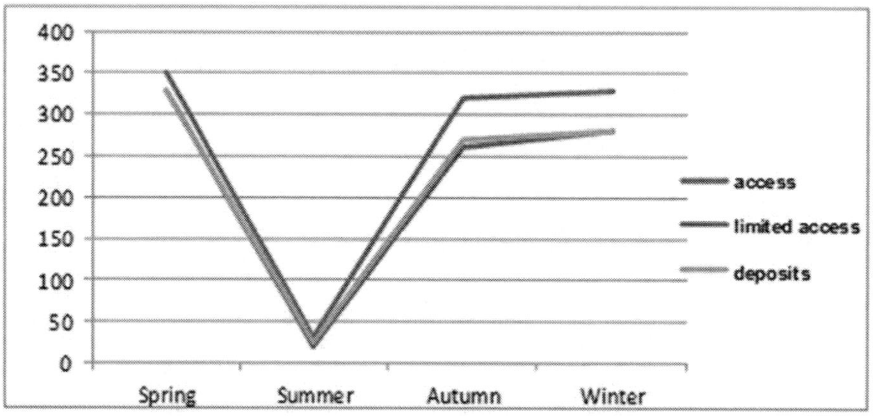

Fig. 2. The seasonal variation of haemolytic microorganisms counts (number of C.F.U / 1000 l of air) in different museum's spaces.

Contrary to mesophilic bacteia, the seasonal variation of haemolytic microorganisms shows that the values of indoor air contamination (mainly correlated with the upper respiratory tract of human carriers) have been the lowest during summer, in all analysed spaces, which may explained by their susceptibility at high and dried temperatures and by the decreased cases of visitors with seasonal respiratory problems during summer.

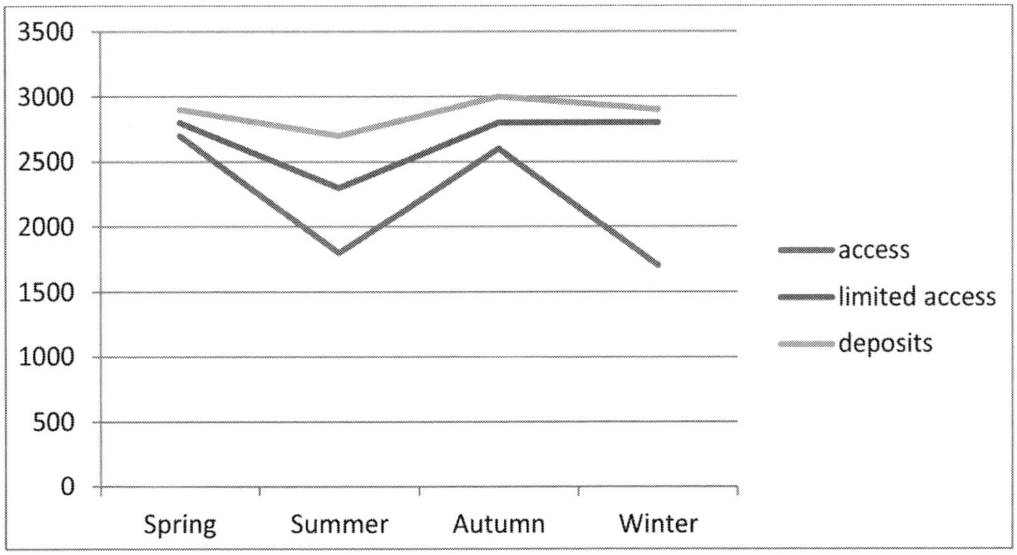

Fig. 3. The seasonal variation of filamentous fungi counts (number of C.F.U / 1000 l of air) in different museum spaces.

Concerning the seasonal variation of microfungi, the highest values of indoor air contamination (correlated with the outdoor one, but also with the contaminated exhibits) have been observed during spring and autumn, in all analysed spaces, values correlated with the air humidity during these seasons; the microfungi are resistant at hot and dried temperatures, but their growth is favoured by humidity too, and this condition seems to be accomplished in deposits and rooms with limited access of visitors, probably due to a bad ventilation.

3.1.2 Qualitative analysis

Concerning the bacterial strains isolated from the analysed air samples, it was found that 80% of them are Gram – positive bacteria, the most common isolates belonging to some species of the genera Bacillus (B. cereus and other species); Clostridium, Staphylococcus (S. epidermidis, S. xylosus, S. capitis) and Cellulomonas spp. Among the Gram – negative bacteria, the most frequently identified belong to the genus Pseudomonas (P. fluorescens and other species). The Gram -positive bacteria are more specific to indoor environment, while the Gram- negative ones are specific to external environment, according to some authors [7].

For the filamentous fungi, the most frequently isolated strains belong to: Aspergillus spp.; Penicillium spp., Cladosporium spp., Alternaria spp. and Rhizopus spp.

3.2 Variation of the microclimate parameters on the air museum microbiological quality

Microclimate parameters were monitored in the same rooms and the same period in which microbiological sampling were carried out. Average values of microclimate parameters depending on room type are presented in table 4.

Table 4. The range of the microclimate parameters from museum

Season	Type of room	Microclimate parameters			
		Temperature (0C)		Relative humidity (%)	
		Minimum values	Maximum values	Minimum values	Maximum values
Spring	access	17.8	18.2	55	55
	limit access	17.8	18	55	55
	deposits	9.3	10.5	73.7	74.3
Summer	access	20.8	21.4	52.7	53.8
	limit access	20.7	21.3	51.7	52.2
	deposits	29	30.7	73.7	74.3
Autumn	access	20.8	21.2	53.8	54.2
	limit access	21	21	54	54
	deposits	26.9	27.2	72.8	73.1
Winter	access	16.8	17.3	55	55
	limit access	17	17	55	55
	deposits	6.8	7.2	75.8	76.3

The micro-climate parameters do not have large variations in the museum rooms (with full or limited access) or in the deposits. This may indicate that they have a small influence on the variation of the microbiological air load. However, when comparing room with deposits, it can be seen there are difference of 10 ^{0}C for temperature and 20% for humidity, which may explain why in deposits the number of fungi colonies is higher.

The high number of CFU/m^3 for all species in the present study confirms the findings of other researchers that the museum environment is favourable for the growth of microorganisms, ensuring conditions for survival and propagation an environment with high humidity and a nutrients [8], particularly the patrimony objects of textile and leather being a growth supportive substratum for developing moulds and bacteria [9, 10]. According to the results of the quantitative analysis the recommended limits have been exceeded in almost all deposits but also in many of the rooms with full visitors access. This can represent a serious risk to the health of the museum and deposit personnel.

The species identified were Bacillus spp. (B. subtilis, B. cereus), Staphylococcus spp., Streptococcus spp., Enterobacter cloacae, Pseudomonas fluorescens, and Clostridium beijerinckii.

Genera as Bacillus, Staphylococcus, Streptococcus were isolated by other authors in museums, libbers and archive indoor environment [1, 9, 11-12].

Genera as Aspergillus spp.; Penicillium spp., Cladosporium spp., Alternaria spp, and Rhizopus spp. were isolated also by other authors in museums or similar spaces [12-13].

The species accounting for mycological contamination of the air from museums and heritage deposits are Aspergillus spp.; Penicillium spp., Cladosporium spp., Alternaria spp, Rhizopus nigricans; these are also known as an etiological factor of several allergic diseases [14-15].

There are many studies that show the link between exposure of workers to bioaerosols and occupational diseases [16-17]. However there are very few studies that have been done a seasonal monitoring. Such studies are however important since they may contribute to a better understanding of the phenomena that cause patrimonial degradation and may also cause occupational diseases; they will also help to establish occupational exposure limit values of workers.

4 Conclusions

The study shows there are seasonal variations in the microbiologic air contamination, especially for mesophilic bacteria and hemolytic bacteria and to a lesser extent for fungi. Since the micro-climate parameters are quite stable their influence may be assumed to be limited. The influence of the seasons themselves (weather, life cycle) and the seasonsal variation of tourists (more numerous in summer) may have a more important influence on microbiological contamination of the air in teh museum.

The present study highlighted a high level of mesophilic microorganisms recorded in the summer in the monitored museum rooms which can be explained by the large number of tourists and the rich vegetation of the close surrounding area.

The highest level of haemolytic microorganisms was recorded in spring, in areas with tourists access, their strong influence on the results being explained by the high rate of respiratory opportunistic infections and carriers too of such infectious agents during this season.

The maximum levels of air pollution were found in the heritage deposits (those outside the museum building). It can be seen that in these deposits the relative humidity is quite high which favours the growth and multiplication of microorganisms, especially microfungi. Aspergillus and Penicillium are the most frequently identified microfungi. Aspergillus niger is associated with onychomycosis [18], skin infections [19] but there are also some reports on lung diseases.

The variation of the amount of aeromicroflora may be very fast, depending on environmental conditions and the life cycle of organisms, making it difficult to establish a medium level of exposure and thus a limit value of occupational exposure.

This work was funded by the ANCS, PN-II-PT-PCCA-2011-3-1742 project, "Improvement of occupational environment quality in cultural heritage deposits. Validation of gamma radiations treatment of textile and leather cultural goods (TEXLECONS)" contr. no. 213/2012.

References

1. National Statistical Institute, Units cultural and artistic activity in 2014 (2015)
2. https://romanian.wunderground.com
3. B. Gutarowska, J. Skóra, K. Pielech-Przybylska, Aerobiologia, 33–44 (2015)
4. EN 14042:2003, Workplace atmosphere – Guide for the application and use of procedures for the assessment of exposure to chemical and biological agents
5. EUR 14988 EN, Commission of the European Communities, Biol. Particles in indoor environments, **12** (1993)

6. EN 27243:1996, Warm ambience. Estimation of the thermal stress on working of human in WBGT index (wet temperature and globthermometer)
7. H. Aydogdu, A. Asan, M. T. Oktun, Environ. Monit. Assess., 53-66 (2010)
8. J.P. Brown, B.W. Rose, APT Bulletin, **27** (1996)
9. S. Fekadu Hayleeyesues, A. M. Manaye, Asian Pac J Trop Biomed, S312 – S317 (2014)
10. M. Lazaridis,E. KatsivelaI, I. Kopanakis, L. Raisi, G. Panagiaris, Heritage Science, 3-34 (2015)
11. J. Skóra, B. Gutarowska, K. Pielech-Przybylska, Ł. Stępień, K. Pietrzak, M. Piotrowska, P. Pietrowski, Aerobiologia, 389–401 (2015)
12. J. Mandal, H. Brandl, The Open Environm & Biol. Monitoring Jour., 83-96 (2011)
13. M. L. Grbic, M. Stupar, J. Vukojevic, I.Maricic, N. Bungur, Arch. Biol. Sci., 955-962 (2013)
14. M. Z. R. Gomes, R.E. Lewis, D.P. Kontoyiannis, Clin. Microbiol Rev. 411-445 (2011)
15. H. Bouziane, J.P. Latgé, C. Fitting, S. Mecheri, M. Lelong, B. David, Allergol. Imunopathol., **33** (2005)
16. J. Skóra, K. Zduniak, B. Gutarowska, D. Rembisz, Med Pr., 153-65 (2012)
17. P. Eggleston, K. Bush, J. Allergy Clin Immunol, 403-405 (2001)
18. J. Araiza, P. Cansesco, A. Bonfaz, PubMed,251-254 (2006)
19. KW. London, A.P. Coke, J.P. Burnie, AJ Oppenheim, C.Q. Morris, PubMed, 191-198 (1996)

DOI: 10.1051/matecconf/201712111018

Liquid hydrocarbons from coal beds – risk factor for the underground work environment - Case study

Cristian Tomescu[1,*], *Constantin* Lupu[1], *Maria* Prodan[1], *Ion* Gherghe[1], and *Florin* Radoi[1]

[1]National Institute for Research and Development in Mine Safety and Protection to Explosion – INSEMEX, 32-34 G-ral Vasile Milea Street, 332047, Petroşani, Hunedoara County, Romania

Abstract. Liquid hydrocarbons from the coal bed and surrounding rocks, besides the stored gases, methane, carbon dioxide, carbon oxide, generate the increase of the risk factor from the occupational health and safety point of view. If for reducing the gas concentrations level and the methane emissions in order to increase the safety in exploitation exist well-known solutions and methods, the oxidation or self-oxidation of the hydrocarbons from the coal bed generate a series of compounds, reaction products over maximum admitted concentrations which give birth to a toxic atmosphere and which is hazardous for workers, at the same time inducing an error in noting the occurrence of a spontaneous combustion phenomena, a major risk for the workers and for the mineral resource. This paper represents a case study performed in one underground mine unit from Jiu Valley and presents the analysis for underground environment factors monitoring and for solutions for diminishing the OHS risk factors.

1 Introduction

Located in the west area of the Jiu Valley coal, the Uricani mine is a unit that operates coal mining for about 70 years. Being in a closure plan, the mine unit conducts the extraction process unit in parallel with the closure plan according to the Project schedule.

The production capacity - 8,000 tonnes / month of operation is given by the exploitation of the 3rd layer form the horizon 300 in an undermined front bench stope. The good progress of the extraction process is disrupted by the risk factors of the working underground environment: the development of spontaneous combustion phenomenon with toxic gases emissions and in the most undesirable case, the forming of methane-air mixtures that can generate an explosion for the methane concentrations between 5-15 % vol. An additional danger coefficient is given by the presence in the coal layer of the liquid hydrocarbons specifics to the mining field of Uricani Mine and whose oxidation products lead to the mislead interpretation of some phenomena and implicitly in applying the proper preventing measures.

* Corresponding author: cristian.tomescu@insemex.ro

The case study is focused in two vectors: the analysis of direct and indirect measurements and gas concentrations, determinations and interpretations, which may highlight an evolution/ involution of a phenomenon of spontaneous combustion and the analysis of a mixture of explosive gases that decrease the lower explosion limit of methane.

The measurement locations were located in different points of the related mining works of the undermined bench stope panel 7, 3rd layer, block IV, horizon 300, in the base gallery, the base profile, the stair profile and in the coal bench drillings.

2 Liquid hydrocarbons and their oxidations

In the Uricani minefield, the liquid hydrocarbons are frequently scattered in the coal layer 5 and sporadic in layer 3. The structure of liquid hydrocarbons from carbonaceous mass same as the oily ones, is expected not to suffer chemically changes under certain conditions. However, occurs an uncontrollable slow oxidation, also called autoxidation, which can occur even in a stable storage medium [1].

The conditions needed for the autoxidation to takes place in a system are: suitable pressure of oxygen 0.07 bar, the temperature between 30-120 ° C, the presence of a certain quantity of hydroxides to initiate the process or other factors that favours the formation of free radicals such as for example, the active metal ions, water, microorganisms [2]. Since the reactions that occur during the phenomenon of autoxidation results carbonyl compounds as the case of carbon monoxide, forming the final stable compounds such as alcohols, ketones, acids which can react further for the formation of macromolecular substances responsible for the high viscosity of hydrocarbon subjected to oxidation [3,5].

From the Uricani Mine samples it was observed the liquid hydrocarbons have a high viscosity, similarly with the hydrocarbons subjected to oxidation process. The presence of this kind of hydrocarbons, their high viscosity and the presence of the factors that favours the initiation of the oxidation process such as metallic ions from coal suggest the fact that the liquid hydrocarbon that are in the mining works can be subjected to this phenomenon that can cause as reaction product carbon monoxide.

3 The exploiting of the panel 7, gas emanations and measurements

The front stope panel 7, 3rd layer, block S IV, hor. 300 was put into operation using the operating undermined bench under natural ceiling, which involves the exploitation of thick layers of coal and reduced inclination (the inclination of the layer α layer ≤ 22^0), method of operation which falls in those who provide descending extraction of the deposit sliced / undergrowth, with long directional slope, with the directing the mining pressure by total collapse of coal and the surrounding rocks behind the front. This panel sub traverse the individual stope front panel 1, 3rd layer, block S IV, hor. 300, which has been exploited in the previous years.

The coal exploited in the undermined bench panel front 7, 3rd layer, block S IV, hor. 300 is characterized as a low self-igniting risk coal and is classified in Group II in terms of risk (tendency) to self-ignition. The needed air flow for stope ventilation is Qr = 335 m3 / min.

The production capacity was monitored by direct measurement with portable gas-analysers (O_2, CO_2, CH_4 and CO), and indirectly through gas chromatographic analysis of samples of air taken from the underground atmosphere to determine the fire indices (index Graham-R_1, breathing R_2 index, ethylene and acetylene C_2H_4, C_2H_2) through which is evaluated the security status of the coal face in terms of the occurrence of a spontaneous combustion (self-heating to self-ignition) and determinations the working environment temperature.

During the monitoring period were identified variable concentrations of carbon monoxide, measured directly with portable gas analysers, highlighted in mining nest from the exploited space and from research drillings with an increasing trend over 100 times the maximum permitted concentration CO≥2000 ppm (maximum permissible concentration of SSM Regulation, CO = 20 ppm), concentrations $CO_2 \geq 5\%$ vol., methane $CH_4 = 5.0\%$ vol., with the detection of specific odours of hydrocarbons. The evolution of the temperature of the coal face including the undermined dislodged coal bench did not exceed 22 ° C [6].

The measurements and analysis on gas chromatograph air samples taken from the monitored area revealed that the concentrations of carbon monoxide was not the same as that measured directly, but 3-5 times lower, and ethylene exceeded 50 times (over 500 ppm) concentration of self-heating phase (10 ppm) with to 0,8 ppm acetylene (trace). They were found high concentrations of heavy hydrocarbons (methane, ethane, propane, propene, iso-butane, n-butane) and high concentrations of hydrogen until $H_2 = 0.43\%$ vol.

4 The interpretations of results

4.1 The risk of evolution and the development of spontaneous combustion

It is known from literature and from the good engineering practice undertaken at INCD-INSEMEX Petrosani, that the key indicator element for assessing the emergence and development of a spontaneous combustion phenomenon consists of the concentration of carbon monoxide, correlated with certain temperature thresholds from the self-heated coal. At high detected concentrations the interpretation of the phenomenon would have a diagnosis for auto-ignition of the coal mass, with the measured values and compared to the model of evolution and involution of the phenomenon of spontaneous combustion based on the correlation between the variation of carbon monoxide in time (τ) and developed temperature (t) during the process - Bystron diagram (Figure 1).

The research, observations and findings proved otherwise, because, the absence of the factors specific to the self-heating self-igniting phenomenon: smoke, fog, dew, heated coal mass (coke), a burning smell and a rise in temperatures air from the underground. The big difference between the concentrations of carbon monoxide measured with portable equipment and the ones determined by Gas Chromatography, represents the result of interference from the presence of high concentrations of heavy liquid hydrocarbons, including those aromatics.

The big difference between the concentrations of carbon monoxide determined by the two methods are reflected in the variation of the fire indices, thus, taking into account the most critical periods, with the highest values recorded (September and October from monitoring times, May-December 2016).

To have a clear view and an element of comparison, from the literature is given by the value interpretation of the fire indices used in Romania (Graham - R_1, respiration index R_2, ethylene and acetylene C_2H_4, C_2H_2):

DOI: 10.1051/matecconf/201712111018

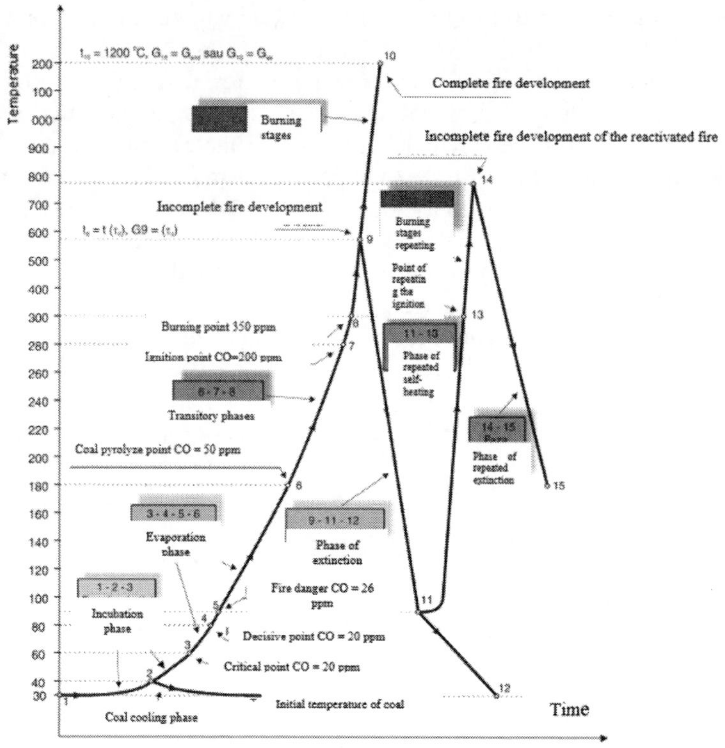

Fig.1. Bystron Diagram

➢ Graham Index (R_1),

$$R_1 = \frac{+\Delta CO}{-\Delta O_2} x100 \tag{1}$$

where:

$+\Delta$ CO – represent the CO concentration growth towards the reference level

$-\Delta O_2$ – represents the O2 concentration decrease towards the initial concentration of O_2 from the fresh gas

Interpretation:

- R_1<0,4 – normal situation
- R_1≥0,4 – start of self-ignition
- R_1=2-3 – lower limit of self-ignition
- R_1=25 – endogenous fire

➢ The breathing index (R_2)

$$R_2 = \frac{+\Delta CO_2}{-\Delta O_2} x100 \tag{2}$$

where:

$+\Delta$ CO2 –represent the CO_2 concentration growth towards the reference level,

$-\Delta$ 02 – represent the O_2 concentration decrease towards the initial concentration of O_2 from the fresh gas. (20,9% vol.)

Interpretation:

- R_2≤60 normal situation

• R_2>60 shows the existence of a spontaneous combustion phenomena, where CO, CO_2 and O_2 gas concentrations are in % volume, measured at the outlet downstream of the oxidation furnace

➢ Ethylene index %C_2H_4
• $C_2H_4 \geq 0{,}001\%$ vol. Existence of a self-ignition process
• $C_2H_4 \geq 0{,}06\%$ vol. endogenous fire

➢ Acetylene index %C_2H_2
• C_2H_2 traces - self heating process
• $C_2H_2 \geq 0{,}0001\%$ self-ignition process

For the characterisation of the phenomenon are taken into account these indices, and for the interpretation of the phenomenon are considered the values that indicates the most dangerous situation [6].

Thus, it was determined the values of the indices from the critical period:
• The Graham Index R1, calculated in the case of portable equipment for CO, varies between R1=0,22-5,10;
• The Graham Index R1, calculated in the case of gas-chromatographic determination of CO varies between R1=0,08-1,31;
• The breathing index R2, calculated in the case of portable equipment for CO2, varies between R2=20,8-147,3;
• The breathing index R2, calculated in the case of gas-chromatographic determination of CO2 varies between R2=43,1-66,8;
• Ethylene, C2H4 = 0,0001-0,0542 %;
• Acetylene, C2H2=0,8 ppm (traces);
• The hydrogen concentration determined in the laboratory, H2= 0,0014-0,315%

After analysing the results of measurements and determinations, it appears as logical consequence of interpretation that across exploited space of the undermined bench stope, panel 7, 3rd layer, block IV, horizon 300, found an atypical phenomenon similar to the self-heating phenomenon at auto-ignition limit, specific to the oxidation of coal, but generated by the oxidation of liquid hydrocarbon, thus a pseudo-phenomenon of spontaneous combustion

4.2 The estimation of the gas mixtures explosivity

Estimation of the degree of explosiveness of the mixture of gases in the samples analyzed was conducted solely on the basis of the concentrations determined in the samples. Flammable gas analysed were the following: hydrogen, carbon monoxide, methane, ethane, ethylene, propane, propene, acetylene, iso-butane and n-butane.

The lower explosive limit of a combustible gas mixture can be estimated using the derivation of the Le Chatelier's. The derivation uses only the flammable components of the mixture with the molar fractions corresponding to the adjusted relative percentage as follows:

$$\%LIE_{mix} = \frac{1}{\sum \frac{x_{i\,inflam}}{\%LIE_i}} x100 \tag{3}$$

where:
• % LIE_{mix} = lower explosion limit (LIE) of the mixture;
• Xi_{inflam} = the molar fraction correspondent to the fuel gas;
• LIE = lower explosion limit (LIE) of the individual component.

Explosivity of a gas mixture can be determined by estimating the lower explosion limit of the mixture applying Le Chatelier's principle and then comparing the result with the overall percentage of flammable substances in the mixture. If the total percentage of flammable substances in the mixture is higher than the lower calculated limit of the mixture, the mixture is considered explosive.

Explosivity estimation of a gas mixture using the enounced principle only take into account the fuel components determined in the mixture. Because an explosive mixture maintains and propagate the flame, the oxygen concentration must be greater than the limiting oxygen concentration [1].

In the samples analysed, the major component is methane and oxygen concentration value limit varies between 10.7% and 12% depending on the explosion vessel used for determining explosion [4].

For sample no. 1, taken from exploited space the date are given in Table 1.

Table 1. Sample no. 1 results analysis

Fuel gas component	Concentration [$\%_{vol.}$]	LIE [%]
Hydrogen	0,430	4
Methane	9,260	5
Ethane	0,029	3
Ethylene	0,015	2,70
Propane	0,170	2,1
Propene	0,004	2
Iso-Butane	0,022	1,8
N-Butane	0,017	1,6
Carbon Monoxide	0,050	12

$\mathbf{LIE_{mix1}}$ = 4,79 %

The total percentage of flammable substances in the mixture = 9.99%. According to the principle stated above, because the percentage of flammable substances in the mixture is greater than the LIE calculated, analyzed gas mixture is considered explosive.

No. 2 sample, taken from exploited space the date are given in Table 2.

Table 2. Sample no. 2 results analysis

Fuel gas component	Concentration [$\%_{vol.}$]	LIE [%]
Hydrogen	0,27	4
Methane	5,17	5
Ethane	0,009	3
Ethylene	0,008	2,70
Propane	0,089	2,1
Propene	0,002	2
Iso-Butane	0,146	1,8
N-Butane	0,012	1,6
Carbon Monoxide	0,019	12

$\mathbf{LIE_{mix2}}$ = 4,61 %

The total percentage of flammable substances in the mixture = 5,72 %

Because the percentage of flammable substances in the mixture is greater than the lower explosion limit calculated, analysed gas mixture is considered explosive.

No.3 sample, taken from the gallery profile headaches have input in Table 3.

Table 3. Sample no. 3 results analysis

Fuel gas component	Concentration [$\%_{vol.}$]	LIE [%]
Hydrogen	0,009	4
Methane	0,740	5
Etane	0,003	3
Etylene	0,0008	2,70
Propane	0,0155	2,1
Propene	0,0001	2
Iso-Butane	0,0033	1,8
N-Butane	0,0004	1,6
Carbon Monoxide	0,0079	12

$\mathbf{LIE_{mix3}}$= 4,82 %

The total percentage of flammable substances in the mixture = 0.78%. The percentage of flammable substances in the mixture is lower than the lower explosion limit calculated that, analysed the gas mixture is not considered explosive.

5 Action measures for the risk minimisations

To reduce the risk factors were taken a series of preventive measures to prevent / combat the emerging phenomenon of spontaneous combustion:

- Setting up a pillar protection and enforcement of a new alignment of the stope panel 7;
- Monitoring the atmosphere of the front stope front in terms of gas concentrations (per shift);
- Monitoring of air flow rates in sub-circuit bench stope front panel undermined 7, 3rd, block S IV, hor. 300 (weekly);
- The intensively treatment of the exploited space with aerosols (trisodium phosphate - 2400l/day) and sodium bicarbonate in an (8-10 kg / m2);
- Instrumentation of fly ash from exploited space through the liner or by means of boreholes siltation (intensively);
- Monitoring of upper areas of construction insulation layer (5) respectively in adjacent areas (layer 3).

6 Conclusions

The research paper is represented by a case study at Uricani mine from the Jiu Valley coal basin, the deposit that has liquid hydrocarbons and directed on two analyses: monitoring gas concentration, determinations and interpretations that may highlight a trend of evolution / involution of a combustion phenomenon and forecasting the explosion degree of the gas mixtures.

- Assessment of the risk and developing a spontaneous combustion phenomenon was performed using fire indices Graham-R_1, breathing R_2, Ethylene - C_2H_4, acetylene - C_2H_2 determined from direct measurements of gas concentrations O_2, CO_2 and CO and indirectly by analysis by gas - chromatograph of air samples taken from the underground atmosphere and temperature measurements;

• Direct measurements in places established revealed an overrun of about 100 times the maximum permitted concentration $CO \geq 2000$ ppm, compared to the maximum allowed by Regulation SSM, concentrations $CO_2 \geq 5\%$ vol., Methane $CH_4 = 5.0$ % vol., with the detection of specific odours of liquid hydrocarbons. The evolution of the temperature of the coal face including the undermined coal bank dislodged from the temperature did not exceed 22 ° C;

• Gas chromatograph analyses revealed heavier hydrocarbons presence (methane, ethane, propane, propylene, iso-butane, n-butane) and a difference in the concentration of carbon monoxide than those measured directly by portable analysers, as a result interference measurements with their influence on components of appliances and sensors default difference value indices of fire;

• After analysing the results of measurements and determinations, it was found that in the bench stope front panel undermined 7, 3rd, block S IV, hor. 300 appeared an atypical phenomenon similar to the self-heating to autoignition limit specific to coal oxidation, but generated by the oxidation of liquid hydrocarbons, ie a phenomenon of pseudo-spontaneous combustion, because of the absence of the other specific elements of a self-heating to self-igniting phenomenon: smoke, fog, dew, heated coal mass, burning smell and a rise in temperature of the working environment;

• Estimating the degree of explosiveness of the mixture-gas samples analysed was made solely on the basis of concentrations determined in the samples. Inflammable gas samples were pursued: hydrogen, carbon monoxide, methane, ethane, ethylene, propane, propene, acetylene, Iso-butane and n-butane. The samples were analysed to estimate the lower explosion limit, respectively explosivity, samples that were found the highest or lowest gas concentrations and average sample. In the samples analysed, the major component is methane and oxygen concentration value limit varies between 10.7% and 12%;

• To minimize the risks and to increase the safety of mineral resources, health and safety of workers in underground, were taken a series of technical and organizational measures, prophylactic prevention / combating the emerging phenomenon of spontaneous combustion, which gave great results by eliminating syncope in extraction process.

References

1. A., Isaac, G., Zlochower, M. Green, Journal of Loss Prevention in the Process Industries, **22** (4), 499-505 (2009)
2. D., Crowl, *American Institute of Chemical Engineers*, (New York, 2003)
3. D., Cioclea, I., Toth, *Prevenirea combustiilor spontane la extragerea carbunilor prin metoda de exploatare cu banc subminat*, (Editura Agora, 2004)
4. C., Tomescu, *Masuratorile efectuate in teren la Mina Uricani, Rapoarte tehnice lunare*, (INCD-INSEMEX Petrosani, 2016)
5. I., Dunn, Biotechnology and Bioengineering, **10** (6), 891–894 (1968)
6. X., Ma, A., Shou, A., Song, Catalysis Today, **123** (1–4), 276–284 (2007)

MATEC Web of Conferences **121**, 12001 (2017) DOI: 10.1051/matecconf/201712112001
MSE 2017

Replicating enterprise environment using Office 365 to enhance graduates' employability

Nicoleta Luminita Carutasu[1,*], and *George* Carutasu[2]

[1]University POLITEHNICA of Bucharest, Department of Machines and Manufacturing Systems, 060042 Bucharest, Romania
[2]Romanian-American University, Department of Informatics, Statistics, Mathematics, 012101, Bucharest, Romania

Abstract. The need of faster insertion of graduates into labor market and enhancing professional and soft skills of graduates required by employees, conduct to new learning method necessity. Starting from stated foreseen of Europe 2020 strategy, creativity, entrepreneurship and intensive use of ICT should be enhanced soon to all academic levels. Also, the entrepreneurs require that graduates to have strong organizational knowledge, to quickly integrate into company' business processes. The traditional assessment of students implies an individual form, team assessment being avoided to be sure of each individual contribution. Also, the future Industry 4.0 implementations will ask for interdisciplinary skills regarding ICT use and specific digital workflows. The proposed enterprise environment replication uses ERP as backbone of IT infrastructure and Office 365 as business workflow tool management. The experience in using ERP as laboratory IT infrastructure for multiple subjects of academic curriculum of the same academic program showed that graduates could focus on internal business process and documents flow rather than learning how to use the software. The Office 365 is used to experience internal workflow of companies, implemented on existent tenants. To enhance the entrepreneurship and innovation, the learning method is completed with simulated enterprise specific activities.

1 Introduction

Recent studies [1] highlight the importance of graduates' employability readiness and soft skill awareness. In this context, one response could be foreseen by developing a clear roadmap of skills and competencies [2] on national scale. In Europe, a document signed [3] by 47 Ministries of European Higher Education Area reveals the significance of transversal, multidisciplinary and innovation skills. The document states also the aim of signatory parties to enhance the employability and personal and professional development of graduates. The Europe 2020 strategy [4] was conceived as a response to the economic crisis. It covers also some aspects regarding labor market and skills development. The strategy is built around

* Corresponding author: nicoletacarutasu@yahoo.com

seven flagship initiatives. From this paper perspective, the most important aims of this strategy are:
- Increasing the employability rate of 20-64 aged population, from 69% in 2010, to 75% in 2020,
- Targeting of investing 3% of GDP in R&D,
- Reduce the early school leavers drop rate from 15% in 2010, to 10% in 2020 and rise of share of population aged 30-34 graduating tertiary education, from 31% to at least 40% in 2020.

Furthermore, the Europe 2020 strategy, through flagship initiatives, the European Commission envisages various aspects, like agenda modernization of higher education and promotion of entrepreneurship, in "Youth on the move", enlarging access to ICT, in "A Digital Agenda for Europe" and modernizing labor markets, in "An Agenda for new skills and jobs".

Having the view defined from above statements and studies, it could be emphasized a series of requirements for graduates, soon:
- Demonstrate strong professional skills, adapted to a continuous changing structure of jobs,
- Have an enhanced entrepreneurial thinking, conducting to new businesses establishment,
- Use extensively ICT in working environment, as communication tools or problems-solving situations,
- Develop innovation and innovative ways of processes approach, regarding new forms of businesses, technological aspects etc.

Regarding Higher Education, also using the same premises, a set of needed measures might be enlisted:
- Sustaining the creativity and creative thinking through learning process,
- Boost the entrepreneurial skills, by opposing students to various forms of learning activities,
- Develop learning methods of enhancing student soft skills.

Regarding these issues, several proposals were made, concerning the learning process [5] and learning environment and teaching materials [6]. The first study proposes the concept of creative classroom. This concept is defined by authors as "innovative learning environments that fully embed the potential ICT to innovate and modernize learning and teaching practices". Furthermore, creativity is conceived as "innovation and modernization of learning and teaching practices through technologies (collaboration, personalization, active learning and entrepreneurship)". The second one debates the creative problem solving fostered by creative learning environments supported by ICT. Another interesting approach [7], proposes virtual internships as an innovative learning technique. The paper emphasizes also the concept and implications of this form of internship and, more important, a communication guideline between the internship hosting companies' representatives and "virtual interns". Regarding entrepreneurship enhanced in Higher Education programs, another recent study [8] reveals that "aim of entrepreneurship education is to teach students the attitude, skills and information that are needed later in working life regardless of whether one works for someone else or as an independent entrepreneur". Another view of entrepreneurship into learning environment [9], examines the self-regulated learning and team learning using lean start-up. The author shows that self-regulated learning has positive impact to individual exam score and no impact to group level performance, measured using group scores and entrepreneur's assessment. The group learning affects positively individual skills acquired and group level performance.

To summarize the above-mentioned requirements for graduates and possible teaching methods used to meet those requirements, therefore are underlined some preliminary conditions:
- Learning environment and teaching method should enhance, both individual and group performance, conducting to positive impact of professional and personal skills acquirement,

- Creativity and innovation must be the primary key of learning, assessed through disruptive ways of problem-solving and technological and business processes approach,
- Entrepreneurship, highly requested by employers, should be enforced, through various teaching activities, seeking for entrepreneurial attitude, knowledge of entrepreneurship and entrepreneur skills enhancement.

The competency term it seen in the literature "as dispositions to self-organization, comprising different psycho-social components, existing in a context-overlapping manner, and realizing themselves context-specifically" [10] or as a compound of talents, skills and capabilities of HE graduates that contribute to multi-factor productivity gains [11]. A full description of competency concept, etymology and use in Higher Education as competencies-oriented learning is also debated into the literature [12]. The professional competencies target more the ability of the graduate to solve a specific field problem, being technical knowledge. Regarding transversal competencies are more generic, available to major fields of study and more related to group competencies (e.g. communication, teamwork, creative thinking etc.).

2 Academic programs and cloud collaborative tools

Starting from enterprise environment replication, as a base for the teaching method, there are inquired the actual enterprise information system. The backbone of the simulated enterprise system is represented by the ERP (Enterprise Resource Planning). In this case, ERP is a software platform which integrates all business processes and optimize the usage of available resources at enterprise level [13]. The ERP system manages the information flow (products data, customers, suppliers etc.), material flow (inputs and outputs of raw materials, goods, assets etc.) and financial flow (between company and suppliers, customers, third parties etc.).

Previous efforts to replicate the company environment into learning purposes were made using simplified solutions. For instance, Central Training Firms Network / Simulate Enterprises in Romania (ROCT) offers a dedicated software solution regarding company environment replication, being used largely by high school students. The solution is failing to reveal details regarding transactions, internal resource planning and consumption (only wages are registered) and a student activity history. Other initiatives [14] solve partially the student activity history issue but having reduced features in terms of business activity. What is relevant regarding the initiatives presented above, learning methodology development using a company as center of case studies.

Regarding collaborative tools/environment, virtualization comes under various forms. In literature, the collaborative environment is mentioned in terms of virtual teams [15], collaborative tools or collaborative environment [16]. Charting those concepts, in this paper perspective, the collaborative environment it is a set of software components used to facilitate sharing information regarding business and technical processes inside the enterprise.

In the following are presented the academic programs used by both of universities and that might be used in the proposed learning method. The first academic program, which is Microsoft Dynamics Academic Alliance [17] enables student and teachers access to ERP and CRM software license, service and support, access to Microsoft Dynamics Curriculum and training materials and student certificate programs. Both universities deliver subjects regarding ERP, as a system, or, furthermore, is used as infrastructure for various others (accounting, logistics, SCM, retail etc.). The used Dynamics 365 infrastructure is part of Office 365 suite, being delivered as Software as a Service. The second program, oriented to collaboration tool delivery is Microsoft Imagine [18], that offers free software access, training resources and certifications. It is important for this framework, by offering a public cloud infrastructure for Office 365 and SharePoint, which integrates also Dynamics 365. Office 365 deployment has an Active Directory service that might be used for users' identification and group creation.

3 Enterprise-centric learning method

In the beginning, it should be stated the objective of the learning method. Having in mind facts and actions revealed in the first section, the enterprise-centric learning method proposes to enhance individual and group skills, creativity and entrepreneurship using as central case the enterprise, replicated using ERP and collaborative tools. To achieve those objectives, it is very important to train students not only as individuals but also as part of groups. Other challenge is to focus the student training towards creativity and entrepreneurship, enhancing competition as positive context.

Furthermore, the learning method is described, using four perspectives: learning scenario, learning environment, and software infrastructure and method integration in the actual curriculum.

3.1 Learning scenario – adoption of cloud technologies

The learning scenario starts with an initial evaluation of students. The objective of this evaluation is to determine if the student is more fitted to employee or entrepreneur. This evaluation is made using vocational and personality test and is completed an interview with the university counsellor. After this initial evaluation, the student has the first image of his psychological profile, what kind of jobs are more suitable for him and what skills should enhance during his university activity. After the initial evaluation, students might choose between employee and entrepreneur position. For the entrepreneurs, the learning scenario continues with the first phase of the enterprise lifecycle: the establishment. This procedure replicates entirely the real enterprise registration, comprising activity field definition using CAEN codes, initial association contract signed by entrepreneurs, initial capital and for didactical reasons, all enterprises act in the beginning as Limited Liability Company. If, in the future, the enterprise results allow, the organization form of the enterprise might evolve to joint-stock company. Students-entrepreneurs having teachers support and forwarded to Enterprise Central Register for approval and registration fill in the documents. This phase enhances the creativity of business approach, giving the possibility to students to test their business ideas'. Part of registration documents and procedures are developed in several forms [19]. The enterprise used for teaching reasons it is named simulated enterprise, training firm or virtual enterprise.

The second phase of the enterprise lifecycle is enterprise setup. After the core business coordinates are established, the entrepreneurs decide the organizational chart, the major business and technological flows. With those details uncovered, they start to hire personnel for key positions in the enterprise: CEO, CEF and HR Chief of Department. Having those positions occupied, further candidates are seeking to fill the rest of position described in the organizational chart. In the same time the Steering Committee (key management positions) redacts the Internal Regulation Policy, Job Description for all position, employment and all documents which further define the internal environment of the enterprise. The next phase is given by enterprise operation, corresponding to ordinary activity of the enterprise. The enterprise setup and operation is registered into ERP systems, in terms of goods offered, product manufactured, production system etc. The role of the teacher in all enterprise phases is limited to mentoring, all decision being made by students, giving them the possibility to encounter success or failing, depending on their decision made. If the employee path is chosen, then the student follows the carrier guidance actions, where they are guided on how to write a Letter of intent, CV and how to apply for job. The end of this path is the employing stage, when the students are employed, by already established enterprises. The object of the employing contract that occur between enterprise and student-employee might be the outcome of teaching projects delivered to various subjects. Is very important to keep the enterprise close to traditional learning path. In terms of assessment, the enterprise-centric learning method combines individual scoring, for each student activity at the end of the academic semester, judging its activity as entrepreneur or employee, with peer assessment from "employers" part. Therefore, the rest of enterprise staff should manifest their appreciation regarding a student activity. This assessment should measure specific activity features, like creativity, entrepreneurship attitude, teamwork, leadership etc., all being expressed during group activity inside the enterprise.

After acting as entrepreneur or employees inside of an enterprise, students could change sides, experiencing as many positions as they can during their university activity. In fact, job rotation is highly recommended, giving the opportunity to students to experiment various position, to see how it fits to its personal skills and competencies.

At the end of academic cycle, the student activity is re-evaluated, taking into consideration the performances obtained and positions occupied. Together with professional individual scores registered at regular exams, the guidance counsellor could advice the student what kind of job is more suitable for him or what internship is closer to his profile.

3.2 Enterprise-centric learning environment

The learning environment has the enterprise as center. However, the enterprise cannot work as an individual and it needs a higher and more complex framework. In this case, the market is recreated by the totality of enterprises established at university level. Replicating the real world, the enterprises could form alliances under the form of clusters, in their pursuit of conquering new markets, to establish monopole positions or to share the operational risk. The missing link into the supply chain is replicated using external environmental, as express business opportunities, given by the tutor environment, in such way that each enterprise to have a least a weekly registered commercial act.

The learning environment replicates the external business environment by guidance teachers, which can impose temporal rules for conducting businesses. In addition, the general framework could imply the application of law or business national frameworks changes (changing of VAT, rules for SME etc.). The learning environment supposes the following actors:

- Creativity and innovation must be the primary key of learning, assessed through disruptive ways of problem-solving and technological and business processes approach,
- Students, acting as entrepreneurs or employees,
- Vocational counsellors, to establish the psychological profile and initial skills pulls,
- Tutor of each enterprise that oversees the overall activity,
- Guidance counsellor, for student final evaluation based on entire student academic activity.

4 Office 365 deployment

The software infrastructure needed to operate the enterprise-centric learning method is emphasized in Fig. 1. The internal enterprise environment is replicated using ERP system. In this case, for each enterprise established one instance is created using the same initial setups as in real implementations, meaning importing chart of accounts, creating roles for users, defining production lines, warehouses, locations etc. If cluster organization is chosen (e.g. holding), the Intra-Company feature could be used to consolidate the financial results, stocks of goods, available resources. The students are identified using Microsoft Imagine accounts, created after admission. The tested implementation was done using tenant student.rau.ro of Romanian-American University Office 365 implementation. The deployed tenant has already defined all students in Active Directory, learning facilities are defined as resources and

Calendar function is already implemented for learning activities. Using this experience, a complete company lifecycle was tested in Office 365 environment.

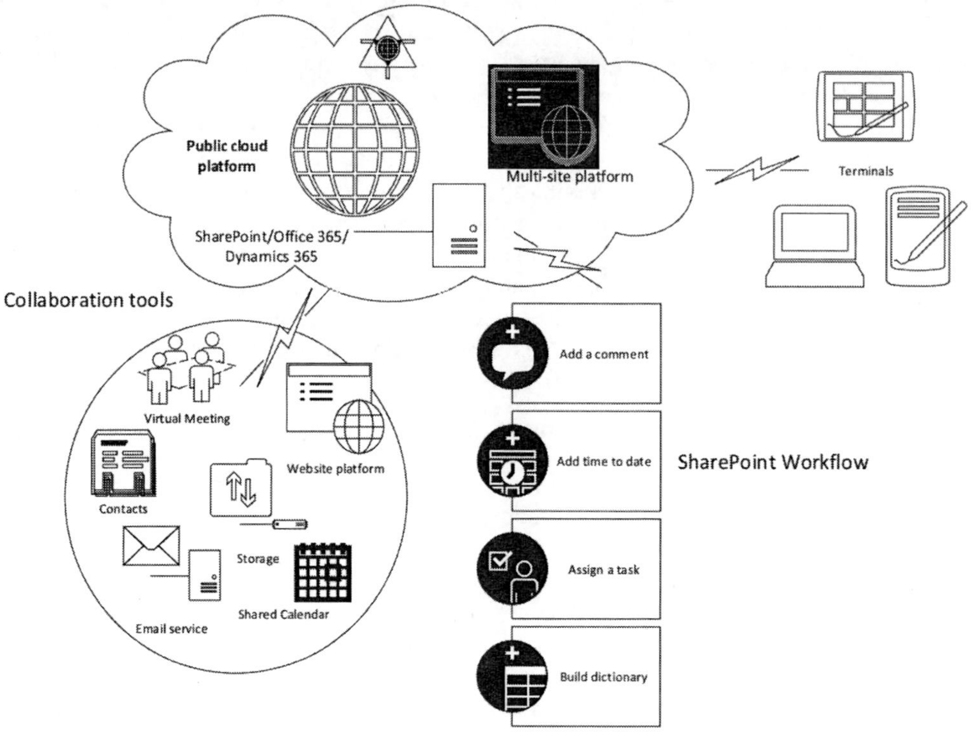

Fig. 1. Office 365 platform.

The company lifecycle, presented in figure 2, shows the correspondence between company life stages, technologies used and software settings needed.

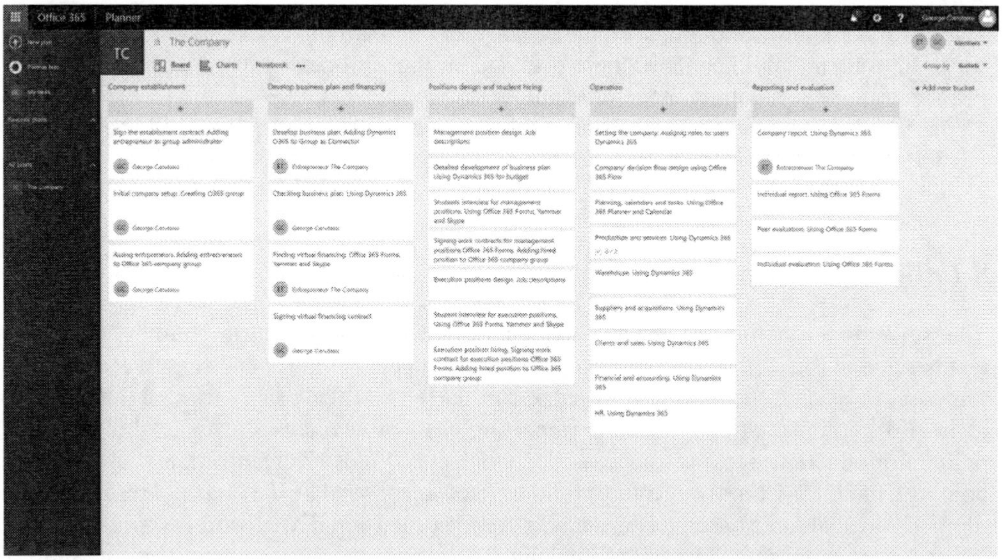

Fig. 2. Company lifecycle captured from Office 365 Planner for The Company group.

For the first stage, company establishment, the tutor sets a new Office 365 Group, in the above-mentioned example is The Company. After initial selection of the student, that will act as entrepreneur, the tutor sends an invitation using Office 365 Form to the entrepreneur. The entrepreneur fills the form about company establishment, and tutor verify the completeness.

Further, the tutor sets the entrepreneur account as group administrator. All the establishment documents filled using Office 365 Form, are stored into company group OneDrive. If OneNote Class Notebook is preferred, the documents are posted also in Content Library as reference for future company employees.

The second stage has as objective to develop a business plan and financing. The tested implementation used as Connector in group page, choosing what application is fitted to the company, related to company activity field. A complete list of Dynamics 365 apps is given [20]. If Dynamics 365 isn't available for the region, a client-server instance of Microsoft NAV 2017 can be used. After budget setting in Dynamics by entrepreneur, it is validated by the tutor. The entrepreneur also fills several predefined forms to elaborate the business plan, using Office 365 Form. All documents are revised by a tutors' committee and the virtual funding is approved.

The third stage envisages position design and students hiring. There are two sequential flows, management positions employment, students assigned to these positions will become also group administrators for Office 365 group and execution positions employment, where students assigned will become members only of this group. Job application can be done using Office 365 Form and the interview may have the support of Yammer or Skype. The assigned students for CEO, CEF and HR are participating to employment of execution positions.

Each settled department of the company uses the Office 365 Planner to plan future activities and assign tasks for execution employed position, monitoring the task accomplishment. Also, the decision flow is decided and implemented using Office 365 Flow The execution positions receive via email notification and tasks are posted in Office 365 Planner personal section. The assignments are also posted into OneDrive as documents and in OneNote Class Student Notebook section or Collaboration Space if it is a shared task. Posting assignments in Student Notebook which is a private space accessed only by tutor and each individual student allows tutor to properly asses the student activity. The activity impact, depending on subject, is registered by each student into Dynamics 365, having a live status of company operations. Furthermore, all students may use Calendar to make appointments, being integrated with education activities calendar and resources, so if an activity is planned to a specific room, the organizer can check the resource availability in real time.

At the end of semester, the entrepreneur and management positions report the state of the company, using Dynamics features. This is a global perspective of company efficiency, being enhanced with peer assessment (made by all employees) and individual assessment (made by tutor). The assessment is deployed using Office 365 Form.

4 Conclusions and proposals

The enterprise-centric learning method enhance the student individual and group skills and competencies. It allows that students to experience the personal success and failure without the fear of direct negative results. The inquired method conducts to a better enhancement of creativity, entrepreneurship and using collaborative tools, objective stated by European Commission.

The method does not replace the traditional learning method, being complementary. It is suited for either engineering or business major field of study. By acting in the role of an entrepreneur or employee, students learn faster the fact that any decision made reflects on enterprise level, affecting the entire activity. By using this feeling of ownership, the students are stimulated to obtained better results, by solving theoretical and practical issues, not for a certain enterprise as a case study, they ae finding solution for their enterprise. Job rotation assure a broad experience, very useful when the real employment time comes. The method implies the activity of various categories of actors: students, vocational and guidance counsellors, enterprise tutors, and teachers, combining individual scores with peer assessment. The software infrastructure, even it seems to be very complicated at the first sight, replicate the real enterprise environment. University POLITEHNICA of Bucharest uses Dynamics NAV intensively as laboratory IT support, for Industrial Logistics bachelor program. The Romanian-American University is on the third-year experience using enterprise-centric learning method, organizing various activities (including Virtual Enterprise Fair in 2017 with ten exhibitors, with 100 students as audience). Enlarging this activity requires process automation. After successful implementation of Office 365 platform, it is expected that starting from next academic year the enterprise-centered learning method to be fully automated using presented sections. The process automation will reduce the time of filling various documents and provides a collaboration environment. Binding these initiatives may conduct to common enterprise-centric activities where engineers collaborates with IT specialist and economist, as in real companies.

References

1. M. Stevens, R. Norman, Proceedings of the Australasian Computer Science Week Multiconference, 13 (2016)
2. N.M. Agrawal, M.R. Rao, S. Venkatesh, India: Preparation for the World of Work, 311 (2016)
3. Communiqué, Conference of Ministers responsible for higher education, Bucharest, 26 (2012)
4. European Commission. Europe 2020: A Strategy for Smart, Sustainable and Inclusive Growth (2010)
5. S. Bocconi, P.G. Kampylis, Y. Punie, Innovating Learning: Key Elements for Developing Creative Classrooms in Europe (2012)
6. L.H. Horn, M. S. Khalid, *Handbook of Research on Creative Problem-Solving Skill Development in Higher Education*, 407 (2016)
7. A.R. Medeiros, D. Icen, E. A. Morciano, M. Cortesao, Global Engineering Education Conference *(EDUCON)*, 262 (2015)
8. K. Ernest, S.K. Matthew, A.K. Samuel, Hi. Ed. St., **5**, 20 (2015)
9. R. Harms, Tech. For. and Soc. Ch., **100**, 21 (2015)
10. M. Barth, J. Godemann, M. Rieckmann, U. Stoltenberg, Int. J. of Sust. in Hi. Ed., **8**, 416 (2007)
11. A.García-Aracil, R. Van der Velden, 2008. Hi. Ed., **55**, 219 (2008)
12. M. Mulder, J. Gulikers, H. Biemans, R. Wesselink, J. of Eur. Ind. Train., **8/9**, 755 (2009)
13. N. Carutasu, G. Carutasu The 7th International Conference of Management and Industrial Engineering "Management – The Key Driver for Creating Value ", 15 (2015)
14. A. Vasilateanu, M. I. Dascalu, I. B. Pavaloiu, G. Dragoi, G. Carutasu, *eLearning and Software for Education*, 210 (2015)
15. C. Aldea, A. Draghici, G. Dragoi, New Perspectives of Virtual Teams' Collaboration *Emergent Technologies and Tools*, 176, (2012)
16. K. Schuster, K. Groß, R. Vossen, A. Richert, S. Jeschke, Automation, Communication and Cybernetics in Science and Engineering 2015/2016, 417 (2015)
17. Microsoft Dynamics Academic Alliance – official website, https://www.microsoft.com/en-us/dynamics/academic.aspx retrieved at 3/03/2017
18. Microsoft Imagine – official website https://imagine.microsoft.com/en-us / retrieved at 3/03/2017
19. E. Moise, C. Coculescu, G. Carutasu, J. Inf. Sys. & Op. Man., **2**, 262 (2014)
20. Microsoft Dynamics – official website https://www.microsoft.com/en-us/dynamics365/home retrieved at 3/03/2017

Hopes and fears of teacher candidates concerning the teaching profession

Daniela Crețu[1,*]

[1]Lucian Blaga University of Sibiu, Teacher Training Department, 34 Calea Dumbrăvii 550324 Sibiu, Romania

Abstract. Romanian university students have the opportunity to enroll in the education program for the teaching profession starting with their first year at the university. Most undergraduates choose the education program for the teaching profession, although some of them do not see themselves as teachers. The schooling experiences, the motivation for entering the teacher education programs, the initial teacher education experiences and their future plans can be important factors in considering the perspective of becoming a teacher or not. Focused on a future orientation, the goal of this study was to investigate teacher education candidates' hopes and fears concerning a possible career as teacher. Using an open-ended questionnaire as a research tool, a qualitative analysis was made on the hopes and fears expressed by the teacher candidates. Our findings showed different categories of hopes and fears and generated an enriched understanding of teacher candidates' views on this subject. The study can be a valuable support for teacher educators preoccupied to understand teacher candidates' perceptions and to assist them in getting closer to their ideals by improving the education program.

1 Introduction

Teacher education programs have an essential role in the professional development of future teachers- who are confronted with complex requests in a world with an accelerating dynamic. Romanian university students have the opportunity to enroll in the education program for the teaching profession beginning with their first year at the university. Most students enrolled in various study programs also choose the training program for the teaching profession, although not all of them are convinced that they want to become teachers. Becoming a teacher is not an easy decision for those interested in the profession. The schooling experiences, the motivation for entering the teacher education programs, the initial teacher education experiences and the future plans can be important factors in considering the perspective of becoming a teacher or not.

The perceptions and beliefs of teacher candidates about the work of teachers and the specifics of this profession are influenced by past and present experiences. When we say past experiences, we talk about the learning situations generated by the interactions with

* Corresponding author: daniela.cretu@ulbsibiu.ro

dozens of teachers from different fields throughout the 12 years of school, as former students. Present experiences are related to their initial teacher education, which begins in their first year at the university. In the university where this work was conducted, teacher education involves a three to five-year undergraduate course in the specific field chosen by the student, along with a three-year pedagogy program. Within this teacher education program, students can study disciplines, such as: Educational Psychology, Pedagogy I (Education and Curriculum), Pedagogy II (Instruction and Assessment), Specialty Didactic - the specialty (major course), Educational Management, Pedagogical practicum and Computer Assisted Training. Such a program has the purpose to support the development of professional competencies that a teacher needs. The fact that most students choose this program is proof that their future plans include - as a priority or, in some cases, as a safety net- the possibility of working as teachers.

2 Theoretical background

Based on the past and the present experience, thinking about the future can provide the context for analyzing options, setting goals, for commitment, for identifying hopes and fears regarding the teaching profession. According to Malmberg & Norrgard, hopes include aspirations, desires, and goals to achieve, while fears include things which we want to avoid [1]. Both hopes and fears shape the behavior to fulfill goals or avoid situations that are not perceived as pleasing. Hope expresses a desire for a better condition in the future, commonly associated with a positive, successful, desired image of the self, while fear is related to anxiety about a worse or undesirable condition in the future.

There are studies that investigate the problem of hopes and fears of future teachers regarding the teaching profession. We are presenting some of the studies which constituted a real support in the research stage of this investigation.

Markus and Nurius theory of possible-selves [2] provides a conceptual link between cognition and motivation, by taking into consideration individuals' ideas of what they might become, what they would like to become, and what they are afraid of becoming. Possible-selves are the cognitive components of hopes, fears, goals, and threats. According to Markus and Nurius possible selves are the ideal selves that we would very much like to become. They are also the selves we could become, and the selves we are afraid of becoming [2]. This theory helps to focus on the students'views of themselves in the future.

Hamman et al. [3] suggest that examining teachers' views in terms of possible-selves provides a better understanding of their present identity, and also of their identity in the future. In their study, Hamman et al. explored expected and feared possible selves of beginning teachers and identified four main categories: interpersonal relationships, classroom management, instructional strategies, professional qualities [3].

Johnson [4] considers that a teacher's knowledge and beliefs have a tremendous impact on how they view themselves as teachers, what they learn from their professional course work, how they reason about their teaching, and how and why they teach the way they do. Hamilton and Pinnegar [5] suggested that "who our students are, what they know and believe, and what experiences and talents they bring with them is an important but often overlooked component of teacher education" (p. 237).

In their study about the intern teacher development during an internship program in teaching, Conway & Clark [6] consider that"attention to both hopes and fears together provides the opportunity to present a more balanced and expansive view of prospective teachers' actual experience and anticipations about learning to teach, than a concerns-only focus" (p.468). Recently, Shoyer & Leshem [7] investigated teacher education candidates' hopes and fears concerning their future career as teachers. Two main categories related to hopes were identified: who the hopes are addressed to (self, the other pupil, or society) and the content of hopes (emotional, educational, or social). The analysis of fears revealed three main categories: emotional, cognitive and school-related.

In an investigation about teacher candidates' expectations and opinions related to practice school [8], Şimşek found that the anxieties and concerns were gathered under the following categories: personal, professional colaboration, school management, classroom management, student status, school status and parent colaboration.

Subscribing to the same field of interest, we believe that investigating students' hopes and fears about the teaching profession is a beneficial approach, useful for the following reasons: it can offer student teachers the opportunity to reflect on the teaching profession in a structured context; the teacher educators involved in the teacher education programs can discover and take into account the candidates' hopes and fears when building the foundations of the professional practice that students need to develop for their future teaching career, in terms of assisting them in moving closer to their aspirations and supporting their development in order to overcome their fears.

3 Research design

Based on a qualitative research design, this study had as a goal to explore teachers candidates' hopes and fears regarding the teaching profession. Two research questions guided the investigation: What are the teacher candidates' hopes about the teaching profession? and What are the teacher candidates' fears about the teaching profession?

An open-ended questionnaire was used as an investigation tool. The first part of the questionnaire was related to personal details such as: age, gender, study major and the intention of becoming or not a teacher. The second part contained two open-ended questions, as follows:

1. What are your hopes about the teaching profession? Identify them and comment on them.
2. What are your fears about the teaching profession? Identify them and comment on them.

The questions were simple and wide, in order to allow the participants to express themselves as freely and sincerely as possible. Student teacher participants were recruited at the end of an exam, during the assessment period (January- February 2017). They were informed about the goal of the research and volunteers were given the questionnaire to complete. With no time limit, the participants were invited to write as many hopes and fears as they wished.

A total of 223 teacher candidates enrolled in a three years teaching program at *Lucian Blaga* University participated in this study. At the time of this study, 35.53% of participants were at the end of the first semester of their second year in the teaching education program and 64.46% were at the end of the first semester of their third year in the teaching education program. The latest category also had the experience of didactic practice for a semester. The study had students from the following majors: Letters - Literature and Foreign Language (n=86), Engineering (n=42), Sciences (n=30), Social Sciences (n=23), Theology (n=23), Medicine (n=21). 70.85% of participants were female and 29.14% were male, aged twenty to forty-six.

4 Results and discussion

The content analysis technique was used for the undergraduates' responses to the two questions. Specific codes emerged from the analysis. Thus, we identified ten categories for hopes and for fears, each of them with more subcategories. We begin with the analysis of the hopes. Their frequency is presented in Table 1. Some of these categories were similar to some already identified in previous investigations and were defined accordingly. Other categories that resulted from the students' answers analysis were added to them.

Table 1. Teachers candidates' hopes concerning the teaching profession

Hopes (H) Category/ Subcategories	**Frequency**
H1: Generic motives/ Teaching mission - help children to learn/to succeed, support child development, inspiring/motivating the new generation, be a model	104
H2: Interpersonal relationships - with pupils (communication, receive respect); with parents (communication, receive respect); with other teachers/colleagues (communication, be accepted, receive support)	73
H3: Perception of the teaching profession - increasing social status, increasing financial status, employment (finding a job)	69
H4: Classroom management - maintain discipline, handle difficult behavior or situations, create positive learning environment	54
H5: Identity/personal issues - vocation, suited for teaching, personality traits (friendly, empathetic, energetic, creative, patient)	39
H6: Instruction - instructional strategies (variety, adequacy), students centered instruction	33
H7: Professional development - life-long learning, succeeding in professional exams	32
H8: School conditions - improving school conditions, teaching - learning equipment, school climate/ prestige	26
H9: Curriculum - less theoretical, more practical, less congestion	18
H10: School system - less bureaucracy, more stability	15
Total	**463**

Out of 779 statements, 463 (59.43%) were related to hopes and 316 (40.56%) were related to fears. The analysis of hopes (H) revealed ten major categories. Next, these categories are presented with a brief description and some examples were extracted from the questionnaires completed by the undergraduates.

H1: Generic motives - generic reasons related to pre-service teachers' goals of becoming a teacher and to the teaching career choice [9]: *"I want to work with children, help them learn, understand the world, to support them in forming character and choosing values, also to prepare them for the future."*

H2: Interpersonal relationships - interactions with persons or groups found within a school [3]: *"I want to have a good communication with students, based on trust and respect."*

H3: Perception of the teaching profession - how the teaching profession is perceived [9]: *"I wish the teaching profession would get more respect in our society."*

H4: Classroom management - management of the classroom, discipline issues [3]: *"I hope I will be able to manage difficult situations and I'll have patience with disruptive students."*

H5: Identity/personal issue - self-perceptions about teaching abilities and identity related motives [9]: *"I am an energetic and communicative person and I hope this will help me in my teaching career."*

H6: Instruction - the use of strategies, the focus on instructional process [3]: *"I hope I can use the methods that are appropriate for the children I'll work with. They are so different."*

H7: Professional development - continuing learning process: *"If I become a teacher I expect it to be a continuous learning experience."*

H8: School conditions - characteristic of schools in terms of instructional facilities, climate, prestige: *"I wish I'd work in a school where I'll have optimal work conditions and the necessary means to teach"*
H9: Curriculum - characteristic of curriculum: *"I would like a less congested school curriculum that leaves more freedom to the teacher."*
H10: School system - characteristic of school system: *"I would like a more stable teaching system, that doesn't change so much each year."*

The results of research demonstrate that teacher candidates have a lot of diverse hopes concerning the teaching profession. Most of these (22.46%) are connected to the desire to support pupils in their development and to be a role model for them, which demonstrates their good understanding of the teacher's mission. 15.76% of their hopes are linked to the possibility of developing positive relationships- with the students but also with the student's parents and with their fellow teachers. Teacher candidates hope (14.90%) in an improvement of the social and financial status of the teaching profession and the possibility to work in a good school (or one that is closer to home). Teacher candidates hope to manage the classroom appropriately (11.66%), to capitalize on their abilities/personal strengths when teaching (8.42%), to offer quality teaching- student centered- by choosing varied and optimal methods (7.12%), to continuously grow professionally (6.91%). Also, teacher candidates wish for better working conditions in schools (5.61%), a less congested and less theoretical curriculum, with more focus on practical aspects (3.28%), less bureaucracy and more stability in the educational system (3.23%).

Out of 463 hopes, 262 (56.58%) are related to the self and 201 (43.41%) related to the other. The students have demonstrated understanding that they need to personally resonate with the teaching profession, but also consider the external factors that influence it.

Regarding the fears (F) and the problems they believe they will have to face, the analysis revealed the same categories as in the case of hope, with the addition of another one- having no fear. The frequency of each category and their sub-categories are presented in Table 2. Next, the fears categories are presented with a brief description and some examples extracted from the questionnaires completed by the undergraduates.
F1: Perception of the teaching profession – negative or unsatisfactory perception of the teaching profession: *"Unfortunately, the teaching profession is not appreciated in our country and I am afraid I can't support even myself with a beginner teacher's payment".*
F2: Interpersonal relationships - negative or a poor quality of interaction with persons or groups found within a school [3]: *"I am afraid the new generations of students are disrespectful to teachers."*
F3: Classroom management - being ineffective at orchestrating the matters of discipline, student behavior or classroom procedures [3]: *"I fear I won't be able to handle more delicate situations / problematic student behavior (violence, behavioral disorders)."*
F4: Identity/personal issues - negative self-perception about teaching abilities and behavior, undesirable characteristics: *"I don't think I'd be a good teacher. I often lose my patience."*
F5: Generic motives/ Mission of teaching - fears about failing the teacher' mission: *"I am afraid I will not be able to support my pupils in the learning process."*
F6: Instruction - being ineffective with instruction, using ineffective strategies, or not being successful in some aspect of the instructional process [3]: *"I don't know if I'll be able to keep the students' attention for an entire class."*
F7: Professional development – obstacles, difficulties in the professional development: *"I know a teacher has to take a lot of exams. I am kind of scared of this perspective".*
F8: School conditions and management - unsatisfactory school characteristics, in terms of instructional facilities, climate or school management: *"I am afraid I am going to end up in a school that is lacking even the minimal teaching conditions."*

F9: School system – negative aspects of the school system: *"I am rather scared of the many changes in the system, confusing for both teachers and students."*
F10: Curriculum – negative/unsatisfactory characteristic of curriculum: *"The curriculum has too much content, I am afraid I won't be able to get through all of it."*
No fear: *"I have no fears regarding the teaching profession."*

Table 2. Teachers candidates' fears concerning the teaching profession

Fears Category/ Subcategories	Frequency
F1: Perception of the teaching profession - low financial status, low social status, employment (difficult to find a job)	68
F2: Interpersonal relationships - with pupils (being disrespected, students' lack of interest, not being understood, not being accepted or liked); with parents (barriers in communication, not being respected, lack of interest and involvement from parents); with other teachers/colleagues (not being accepted or appreciated, not receiving support)	59
F3: Classroom management - difficulties in maintaining class discipline, manage difficult behavior or situations, classroom size (too many students), being too authoritarian or too permissive	47
F4: Identity/personal issue - not good enough, losing patience, tiredness, stress, loss of health, negative emotions (scared, fear of public speaking)	30
F5: Generic motives/Teaching mission - fear of failure to help the students to learn and develop, not inspiring/motivating new generation	25
F6: Instruction - difficulties in instructional design, difficulties in capturing and maintaining attention, failure to use effective strategies, working with pupils with special educational needs	24
F7: Professional development - routine, self-sufficient, not keeping pace with changes, failing in professional exams	19
F8: School conditions and management - lack of school facilities, negative school culture/ prestige, ineffective school management	18
F9: School system - to much bureaucracy, to many changes	15
F10: Curriculum - to congested, too theoretical	11
Total	**316**
No fear	7

The findings of the research demonstrate that the majority of teacher candidates have many, diverse fears concerning the teaching profession. Only seven of them reported no fears. Most of the fears (21.51%) are connected to how the teaching profession is perceived: low financial and social status, hard to find a position in the system. 18.67% of fears are linked to interacting with future pupils, parents or colleagues. Students consider that the new generations of students are more disrespectful and lack an interest for learning; also, they fear being disliked or misunderstood. Regarding the teacher - parent relationship, candidates are worried that they won't be able to properly communicate with parents due to their lack of respect for the teaching profession. Some talk about the fear of not being accepted or appreciated by their fellow teachers. Students fear the responsibility of managing student classrooms, maintaining discipline despite having too many students in a class, being too authoritarian or too permissive (14.87%). Some students don't feel good enough/ prepared enough for being teachers, others lack patience, are afraid of the stress resulting from the job demands, or are afraid of public speaking and failing due to nerves. They have the fears of not being able to help the students to learn and develop or not being a good role model (7.91%). 7.59% of the fears expressed by teacher candidates are related to specific instruction issues, such as: planning classwork instruction, capturing and maintaining student's attention, using effective teaching strategies or working with special educational needs pupils). Some fears appear in smaller percentages, such as those relating to self-sufficiency, routine, not being able to pass the multitude of exams teachers have to take during their careers (6%), fears about the bad working conditions in some schools, negative climate or bad management (5.69%), fears concerning the bureaucracy and the many changes in the system (4.74%), fears about the congested, over-loaded, mostly theoretical curriculum (3.48%).

Out of 316 fears, 145 (45.88%) were related to the self and 171 (54.11%) related to the other. Students are more afraid of external factors, such as: professional status, students, parents, colleagues, system, curriculum, school conditions and management, than they are of internal factors. In the end, only 56.50% of the participants in this study have expressed a firm option for becoming teachers, 38.56% are not decided, and 4.94% don't want to become teachers, even though they are all taking the same teacher - training program. Some aspects have indeed contributed to making the teaching profession less attractive: the continuous degradation of the teacher's social status for the last 25 years in Romanian society, the growth of violent acts in schools, the worrisome lack of motivation recorded in students, the abandonment of the educative attributions of parenthood in some parents, the lack of continuity and coherence in school policies, the poverty of some schools, excessive bureaucracy.

It is necessary that teacher candidates develop a realistic understanding of the teaching career. Thinking about the future, realizing the pleasant aspects, considering the challenges and the difficulties associated with the didactic profession is an useful exercise for undergraduates that can prepare them for the transition from the university to the outside world. The reflection exercise allowed the students to explore- both cognitively and emotionally- their own universe (hoped-for and feared possible selves), to become aware of the complexities of the teaching profession.

We consider that capturing the teacher candidates' hopes and fears concerning the teaching profession is a great resource for teacher - educators preoccupied to tailor the teacher programs to meet the individual needs of future teachers. Who our students are, what they know and believe, what they hope or fear are important issues which should be taken into consideration in teacher education program. These hopes and fears may be an indication of the effect the educator preparation programs, working within a specific context, have had and also of the influence past interactions with the school system and their teachers have had on them. The experiences they have had may have influenced their ability to consider what they want to become or find and what they want to avoid.

Since the past can't be recovered, we can focus on modifying the present for a better future. Conway & Clark [6] consider that "hopes and fears alone or together can be the basis for both a supportive and challenging conversation with student teachers about their beliefs and motives, about classrooms, schools, students, learning and teaching, and relationships... a focus on both aspirations and concerns, within an inquiry approach to teaching, can provide a helpful lens with which to understand and support the development of prospective teachers" (p. 468).

5 Conclusions

The study of hopes and fears of teacher candidates regarding their future as teachers revealed their perceptions about themselves and about multiple aspects of the profession. More hopes than fears were expressed by participants. Ten categories for both hopes and fears, each of them with more subcategories, were extracted from the participants' responses. In their reflections, the candidates invoked both self-related and others-related aspects. When expressing hopes, self-related factors prevailed and when expressing fears there were more external-related factors to count in their answers.

The current investigation has limitations because is based on a one-time look at teachers candidates' hopes and fears. A longitudinal approach can be used for future research.

It is needed that teacher training programs tailor curricula and professional practices to meet the individual needs of future teachers. As teacher educators, we may value the candidates' hopes by building positive professional practices that can than be transferred towards building their early career. The enriched understanding of the undergraduates' fears can help the teacher educators to focus on giving specific support for overcoming them. It is true that some fears can't be checked by the intervention of teacher educators (such are those relating to the social and financial status of the profession, those concerning working conditions in schools, finding a good position, curriculum, the way the school system works), but others are under their direct influence (classroom management, communication in relationships, instruction, professional development).

Specific measures can be taken, for instance: enhancing class management exercises, more examples of dealing with difficult students exhibiting deviant behaviors (by case studies), creating more opportunities for students to practice public speaking (during seminars or student meetings, conferences), more micro-teaching practice for stimulating their presentation skills and for using various instructional strategies, more communication exercises in order to improve parent/ student communication (role- playing or scenario analysis), meeting with students, parents and other teachers before school practice or in other contexts, giving more attention to the problems inclusive education and teaching special needs pupils poses. Measures like these can support teacher candidates in overcoming their fears regarding a teaching career, reaching for their goals and preparing them for the school that they will possibly serve in the future.

References

1. L.E. Malmberg, S. Norrgard, J.of Adolescence **22**, 33-47 (1999)
2. H. Markus, P. Nurius, Am. Psy. **41**, 954-969 (1986)
3. D. Hamman, K. Gosselin, J. Romano, R. Bunuan, Teaching and Teacher Ed **26** (7), 1349-1361 (2010)
4. K. Johnson, *Understanding Language Teaching: Reasoning in Action*. (Boston: Heinle and Heinle Publishing Company ,1999)
5. M. L., Hamilton, S. Pinnegar, J., Teacher Ed **51** (3), 234-240 (2000)
6. P. F. Conway, C. M. Clark, Teaching and Teacher Ed., **19**, 465–482, (2003)
7. S. Shoyer, S. Leshem, Cogent Education **3**: 1139438 (2016)
8. C. Şimşek, Inter.J. Edu. Res.Technol **5** (2), 55-63 (2014)
9. M. M. Thomson, C. Palermo, Teaching and Teacher Ed. **44,** 56-68 (2014)

Some considerations on adapting academic qualifications to the Romanian labor market

Calin Denes[1,*], and *Valentin* Grecu[1]

[1]Lucian Blaga University of Sibiu, Department of Industrial Engineering and Management, 550024, 10 Victoriei Bd, Sibiu, Romania

Abstract. The paper presents the results of some research carried out by the authors during the activities of the "Academic Entrepreneurial Laboratory – Progress through Innovation and Practice – (Univers Practic)" and seeks to evince certain key aspects pertaining to the adaptation of university qualifications to the needs of the Romanian workforce. First, the general context of higher education and the current state of affairs, as reflected by the analysis of relevant works in the field are presented. This is followed by a presentation of the main results of a questionnaire-based study that evinces the opinions of students, graduates, and company managers with regard to the necessity of adapting academic qualifications to the Romanian labor market. Lastly, the conclusions of the paper are delineated and several recommendations concerning the relationship of universities with the business environment and other stakeholders are elaborated, with the aim of improving the interactions of the university with the entire society.

1 Introduction

The contemporary university has more functions than we might be tempted to believe at a first glance. The expectations of modern society, based on the principle of democratic freedoms, the market economy, and the competition between operators activating on the same segment of socio-economic life cannot be satisfied except through the concomitant exercising of functions that any higher education institution should endeavor. Of all these functions, the formative and the one of ensuring the occupational mobility of the workforce are often understood, nuanced, and exercised differently depending on the context in which each university operates.

The extremely dynamic social, political, economic, and cultural context of the last 50 – 60 years has had a powerful impact on the quality of higher education. In their race to acquire students and resources, the providers of academic studies have adopted various attitudes and strategies. The consequences are easily noticeable in many countries with diversified higher education systems. While traditional universities have striven to maintain their conduct in assuring the quality of the academic education, many recent providers of study programs have relegated the quality of the education offered to a secondary position. Commercial, financial, or personal interests have prevailed and continue to do so in many of the more recent entities that compete against the traditional academic offer.

* Corresponding author: calin.denes@ulbsibiu.ro

Given this fluid context of higher education, during the latter part of the 1970s, the first quality accreditation authorities have appeared in various European countries and beyond the Old Continent. This way, *the quality of higher education is defined and ensured not by the university, but by entities outside the university.* Some countries initially chose to relegate this public responsibility to ministries or departments of higher education. Subsequently it was proven that too often the license to operate and / or the accreditation resulted in the atrocious encroachment on university autonomy.

In many of the academic systems of developed, but also emerging countries, to constitute an authority – either independent or under the patronage of the authorities – has been deemed to be a much more suitable formula of ensuring the quality of the recent providers of academic education and study programs, especially given that these authorities were not operating with public servants, but rather invited teaching staff from traditional universities, and sometimes academic experts from abroad, in order to carry out peer-review academic evaluations.

2 The current status of higher education within the context of the European Union

In order to formulate the study entitled "Analysis of the Need to Adapt the Academic Curriculum to the Competences Demanded by the Labor Market and the Improvement of Relationships with Internship Partners", the authors of the paper have analyzed numerous papers relevant to the study matter. As a result of the analysis carried out in [1], corroborated with the information presented in [2] and [3], the authors' opinion was shaped with regard to the current status of the Romanian learning system from the perspective of adapting it to the current needs of society, which is succinctly presented hereafter.

The Romanian Agency for Quality Assurance in Higher Education (ARACIS) operating in Romania has a vital role in ensuring and improving the quality of the education provided by Romanian institutions of higher education. Also, at European level, The European Association for Quality Assurance in Higher Education (ENQA) disseminates information, experiences, and good practices in the field of quality assurance in higher education, to European quality assurance agencies, public authorities, and higher education institutions.

In Europe, the reference instrument required to compare the levels of competence of the various national qualification systems has been constituted by EQF - The European Qualifications Framework. The latter promotes both life-long learning, and the equality of chances in the knowledge-based society, as well as the continuous integration with the European labor market, honoring at the same time the great diversity of national education systems. In our country, the National Framework of Qualifications (CNC), together with the National Registry of Qualifications (RNC) are in operation, having been elaborated, updated, and implemented by the National Authority of Qualifications (ANC). A special space within it is occupied by higher education, for the purpose of which the National Framework of Qualifications in Romania (CNCIS) and the National Registry of Qualifications in Higher Education (RNCIS) are distinguished. CNCIS is the instrument used in establishing the structure of qualifications in higher education, and is intended to ensure the coherency of qualifications and titles obtained from higher education. At the same time, CNCIS ensures national recognition, as well as international compatibility and comparability of higher education qualifications.

In the years following the signing of the 1999 Bologna Declarations, universities and their academic communities have understood better and better that it is in their interest to consolidate the internal assurance of quality; they need to promote the transparency of their own measures of increasing quality expressed in the form of learning results and they need to develop a new communication strategy with their circles of stakeholders outside the academic environment, involving them in the development of institutional strategies, in the design of new study programs that best and most visibly answer their demands and expectations.

The reports of certain pieces of qualitative research from various fields of the business environment illustrate a variety of practices coming from both employers and providers of higher education [2]. There are certain examples of practices that perpetuate the specific discrepancies between academic qualifications and workplace demands, just as there are good practices based on the dialogue between universities and the business environment through which discrepancies are eliminated. Discrepancies are materialized either through the over-qualification or overspecialization of the graduate (given that the market does not absorb such specialists, it needing execution specialists of certain decisional options), or through the insufficient training of the graduate (in the context of the need that specialists from different fields be able to interact in order to carry out the tasks in their workplace).

Changes in the nature of the qualification demand are often dictated by labor market constrains or market pressures. In general, it is a known fact that workforce planning is too loosely actively connected to long- or medium-term business planning [2]. On the other hand, educational providers, especially universities, ought to be more sensitive to the needs of a permanently changing labor market. This requires on the one hand, to strengthen the dialogue with employers and other stakeholders in the process of elaborating a new curriculum and in the delivery of study programs, as well as to improve communication with the rest of society, in order to make higher education reforms better understood. On the other hand, to alter the institutional culture of quality in universities, so as the employability of graduates may become a key problem of education / training to each member of the academic community (instructor, researcher, student, alumnus, and member of the administrative staff).

Within the European Higher Education Area (EHEA) there is raised awareness of the fact that the concern for quality must lie at the center of the system. On the whole it is a known fact that quality assurance in higher education consists of a set of measures and actions meant to augment the institutional capability of elaborating, planning, and putting into practice study programs, all of which should convince the beneficiary that the learning provider fulfills the appropriate quality standards. Ensuring quality reflects the provider's capability of offering study programs which are in accordance with the standards that they have mentioned beforehand. In order to manage the issue of quality, the academic learning provider must run a continuous quality-monitoring internal process which is to be extended to every component of the teaching – learning process.

The current state of higher education in our country has always been – and still is – determined by a series of limitations, but also by the general context of Romanian society. Concerning the general strategy of higher education, the overall perspectives of the educational system, together with the salient transformations that have occurred in Romanian society over the past few decades have been left out.

It is necessary that the shift occur from the traditional, inertial learning system that has been in force hitherto, to that learning system marked by radical changes, by the synchrony with the latest and future developments, the compliance with the 1999 Bologna and 2003 Berlin Declarations, as well as with the Law of National Education. The process of radical renewal of the higher education is vital; otherwise, to neglect to achieve this process is to risk generating long-term effects that will be hard to correct. This process is generated mainly by three factors: increasing the individual request of higher education – the shift from selective to mass higher education occurs; the labor market – it is noticeable that there exist significant chasms between the qualifications demanded by the labor market and the ones provided by universities; the information and communication technologies – which ensure the fundamental resources for the intensification of the socio-economic development.

In this context, the concern with aligning the workforce with the new socio-economic specific conditions must become a primary concern. The qualifications obtained by higher education graduates have to match the demands of the labor market – which undergoes increasingly dynamic changes – as faithfully as possible.

3 Adapting the university curriculum to the competences demanded by the Romanian labor market

The conceptual – methodological model of describing qualifications in higher education (the CNCIS model) represents a reference framework which was developed to analyze, describe, and interpret higher education qualifications (according to the OMECTS 5703/2011 order of the Ministry of Education. The CNCIS model is generally compliant with the vision put forth by the European Qualifications Framework (EQF), and with the learning results specified by the latter for qualification levels 6, 7, and 8. The structure and content of the model valorize the descriptors of the general qualifications framework regulating the European space of higher education, as well as the containing elements of models that already enjoy the appreciation of European experts. At the same time, the CNCIS model has its own identity; it integrates types and categories of competences, levels of qualifications, and specific descriptors.

The essential elements of the model resulted in the CNCIS Matrix and the complementary grids: Grid 1 (the description of the field / higher education study program through professional and transversal competences), Grid 1s (specific descriptive elements of the field / study program), and Grid 2 (establishing the correlations between competences and transversal competences and content areas; study disciplines and credits assigned). The CNCIS Matrix comprises: qualification levels, learning results expressed through knowledge, abilities, and competences, as well as the level descriptors of higher education qualifications.

The procedure to be followed by higher education institutions in order to gain the right to certify a qualification achieved by their graduates is schematized in Figure 1. Universities need to initiate the procedure by putting together a validation file for every qualification achieved upon graduating from an academic study program; the file will then be forwarded to the National Authority of Qualifications (ANC). ANC, in consultation with ARACIS and the Ministry of Higher Education (MEN), and by taking into account the demands of the labor market (the way in which this occurs is not explained), evaluates the documentation received and certifies (or refuses to do so) the university qualification in question, by registering it (or not, as the case may be) with CNCIS/RNCIS and notifying the university with regard to the resolution reached.

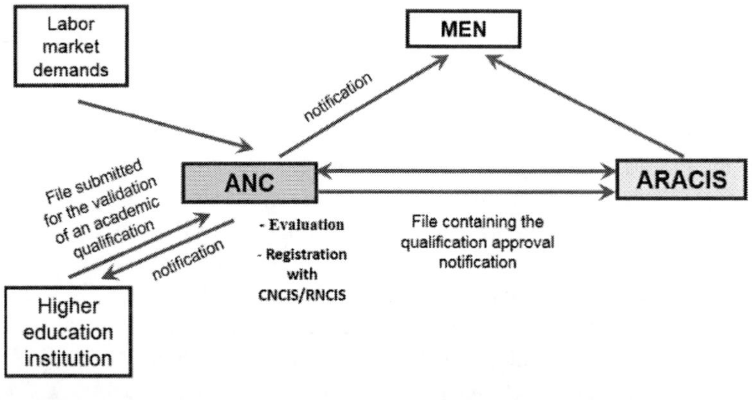

Fig. 1. The certification procedure of a professional academic qualification

DOI: 10.1051/matecconf/201712112003

The previously described current procedure presents certain obvious shortcomings that directly affect the labor market. On the one hand, the demands of the labor market are not fully addressed, whereas the information that the ANC possesses is debatable with regard to its degree of actuality. On the other hand, no attention is paid to the opinion of the future students or alumni. These categories of people could make useful contributions to the more rigorous and more complete definition of a professional academic qualification. Furthermore, the ANC is constituted along the lines of a public institution having a juridical personality and being in a position of coordination with MEN, which leads to subjectivity, since the ANC can no longer be considered as a third party in the relationship of universities with MEN.

In order to forward the validation file of a professional academic qualification, universities must first clarify the defining elements of the respective qualifications, by explaining professional roles, occupations, competences, competence units and study disciplines.

Detailing a professional academic qualification must start from the identification of obvious needs that appear due to the evolution of the labor force market. Once the necessity and opportunity of a new qualification has been clarified, one must delineate the possible professional roles and occupations of the people attending the respective qualification, which must comply with the Classification of Occupations in Romania and with the International Standard Classification of Occupations (ISCO-88). Afterwards, the (general and specific) professional and transversal competences, the competence units and study disciplines, together with their content and share are clarified (refer to the methodology described in Annex OMECTS 5703/2011).

Not until an academic qualification has been certified can one proceed to organize the educational student training process in accordance with the exigencies of the respective qualification. Interested universities will have to undergo the internal quality evaluation concerning the institutional capability, educational efficiency, and quality management, according to the ARACIS evaluation methodology. Following this is the external evaluation carried out by ARACIS at the request of universities, and based on the current legislation, methodology, guides, and procedures. Subsequent to the external evaluation, ARACIS recommends the provisional authorization or accreditation of higher education and study programs providers.

Harmonizing the demand and supply in the field of the labor market cannot be left up to the state, the university, or employers. This is a social problem, on which all of our lives depend. Since it is everyone's problem, it is only natural that everyone should contribute to it and collaborate in a productive way in order to solve it. The opinions of everyone must be considered valuable and, by integrating them, together with the collected accountability, the best decisions have to be made and consistent work must be done to implement them, in order to ensure the durable development of the entire society.

The continuous development of the labor market will undoubtedly prompt the periodic review of the professional academic qualifications. At the same time, all stakeholders will have to work together to carry out the periodic review of professional academic qualifications, as mentioned above.

Compared to the certification procedure of a professional academic qualification, shown in Figure 1, we recognize that the definition and review of a qualification can be done in a much better way by consulting the main stakeholders concerned with the activity of universities: employers and the students / alumni. This aspect can be schematically presented as shown in Figure 2.

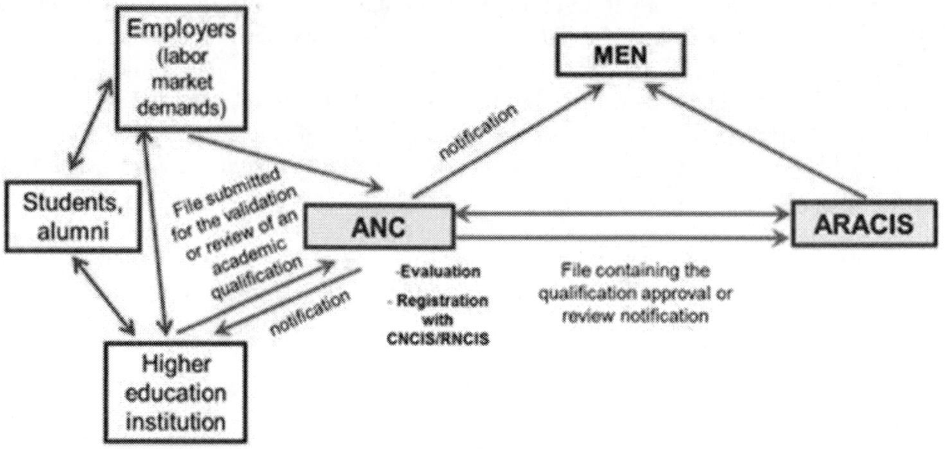

Fig. 2. Defining and reviewing a professional academic qualification – proposal

In order to further illustrate this point of view, a research has been carried out targeting the harmonization of specialty structures (program studies) from Romanian universities with the labor market, by taking into account the employers' opinions – expressed by the managers of the employing organizations – and the opinions of Romanian students and alumni. For the purpose of the research, the subjects were questioned in connection with the necessary knowledge and competences, by avoiding (so as not to generate misunderstandings) the usage of the concept of competence with the meaning employed by CNCIS.

4 A few results of a study concerning the need of adapting academic qualifications to the needs of the Romanian labour market

What actually happened within the "Academic Entrepreneurial Laboratory – Progress through Innovation and Practice – (UniversPractic)" project that took place in 2014 – 2015 at Lucian Blaga University of Sibiu was a series of opinion polls resulting in three strategic documents relating to the theme of the project, one of them being none other than paper [1]. Two of the polls sought to determine the opinions of students and alumni (on the one hand) and of the employers (on the other hand) regarding the adaptation of academic qualifications to the needs of the Romanian labor market. The polls, exhaustively presented in [1], were conceived so that the probability of guaranteeing the results of the research (the level of trust) be 95%, and the margin of error ±5%. To collect the information, two questionnaires were utilized, one addressed to students and alumni, and the other to firm managers, the latter being deemed as relevant representatives of the employers. The researched has brought about the collection of some very interesting information, some of which will be presented over the next lines.

Both categories of subjects reveal similar opinions concerning the training provided by the universities. What became evident was that the professional training was prevailingly theoretical, whereas the practical and applicative training was deficient. As concerns a good theoretical training, both students and managers shared similar opinions, whereas, with regard to the practical and applicative training, managers were more critical (given that their answer encapsulates the actual situation in their companies), whereas the students were less aware of this aspect.

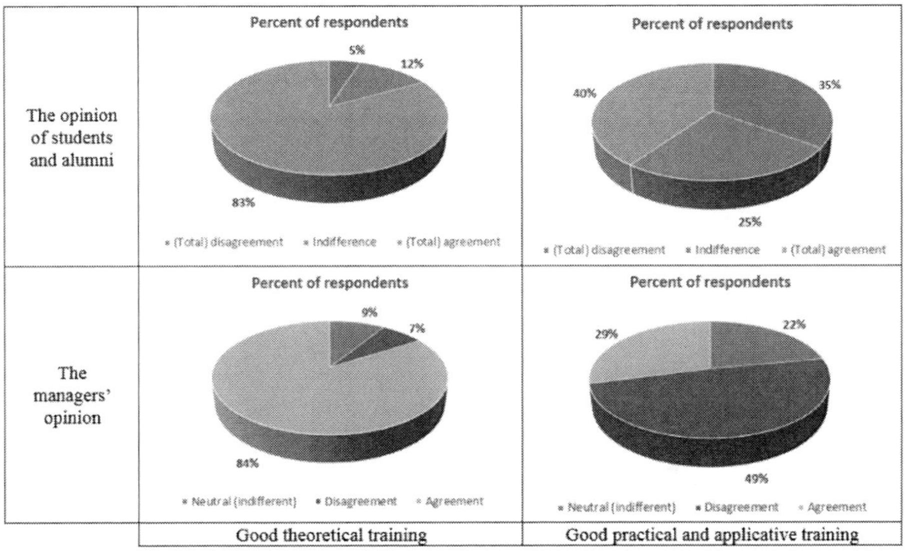

Fig. 3. The opinions of respondents regarding the professional training offered by universities

The main skills needed in practicing one's profession, according to students and alumni, are the ones shown in Figure 4. These do not exactly coincide with the opinions of the managers about the same issue (Figure 5), the latter being aware of the fact that college graduates successfully pursue their professional training after being hired by the company, so long as they possess communication and teamwork skills, corroborated with solid computer-operating skills and with the proficiency in speaking some international languages.

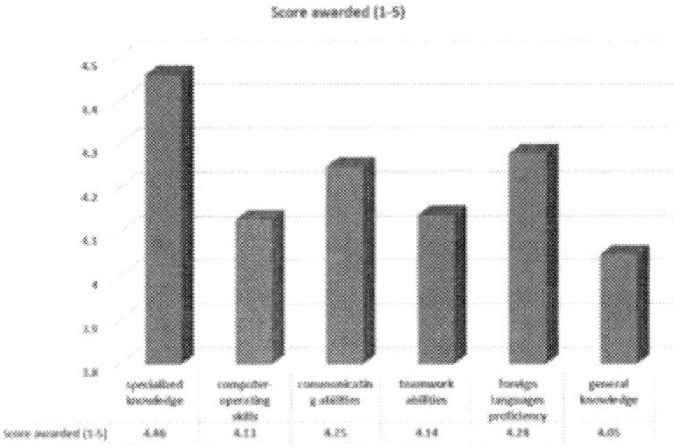

Fig. 4. The opinion of students and alumni concerning the main skill needed to practice one's profession

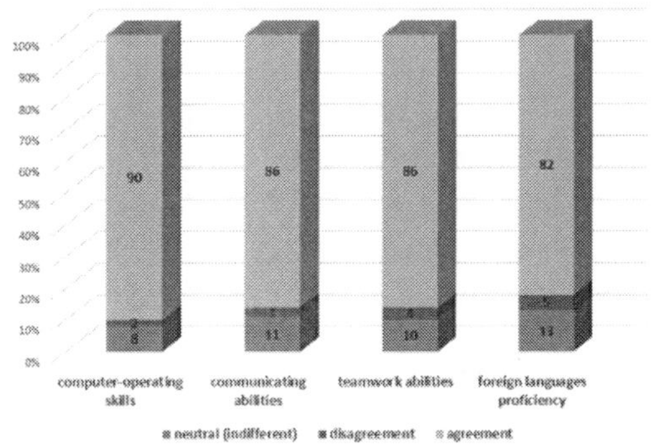

Fig. 5. The opinion of managers concerning the main skills needed to practice one's profession

A considerable number of students and alumni questioned (58%) consider that the professional training accrued in university, although exceptionable, enhances their chances of immigration, especially within the European Union. At the same time, the questioned subjects argue that certain undertakings are necessary to increase the quality of higher education (72%) and to adapt it to the needs of society (86%).

Managers estimate that there is a 73% chance that over the next few years universities will adapt their educational offers to the demands of the labor market and a 74% chance that the Romanian higher education system become more flexible. 71% of the managers assume that the Romanian higher education system will significantly increase its quality over the next 5 – 10 years, especially thanks to Romania's integration in the European Community (78% of the respondents indicate this as a main reason).

5 Conclusions

The generalizing conclusion is that post-1990 society has undergone a rapid evolution in various respects – economic, social, political, informational, internationalizing, etc. – and that universities – Romanian universities, in particular – failed to keep up with the social dynamic, the rapid change in the context of highly qualified labor market, social and individual expectations of university education beneficiaries. It is no less true that, in spite of isolated warnings coming from the very bosom of the academic world [2], the contemporary university has opened itself up very little, showing a vacillating professionalism and a diminished enthusiasm for the world around it, and engaging in feeble dialogue with the society it serves, often falling behind in the competition against other providers of tertiary training [2].

The main conclusions concerning the adaptation of academic qualifications to the Romanian labor market, resulting both from this paper, as well as from papers [1-2] are the following: the university qualifications achieved upon graduation from current study programs are not completely underpinned by the demands of the labor market; there is no objective underpinning at study program / field level of certain appropriate measures targeting the adaptation of the current study offer to the demands of the labor market; the universities' dialogue with business associations and employers, students and the teaching body, as well as other stakeholders is deficient.

There are two strategic directions to be followed by Romanian universities. The first direction requires focusing on the actual and efficient implementation of a continuous internal operative monitoring mechanism (semester by semester, lecture by lecture), aimed at ensuring the quality of academic training. Similar to other European higher education systems, the Romanian one is in dire need of a new quality culture in each university and throughout the whole system. The new quality culture must be founded on the strategic option of real and objective openness towards the society, actively promoted from within the university.

The second strategic direction entails the development of an active and continuous dialogue (initiated by the university) that involves all stakeholders and demands the honest and continuous collaboration with the latter. The consultation of representative actors from the ranks of the employers, as well as the maintenance of contact with alumni associations, are now perceived as efficient means of learning the needs of the labor market, and also of educating it, so that the proper understanding of the complex system of academic qualifications, subsequent to the three Bologna academic study cycles, may occur. The presence of partners from the business environment, public administration institutions, business associations, as well as the ranks of former students, is likely to stimulate the debate and identification of means of adapting the study offer to the needs and expectations specific to the highly-qualified labor market.

References

1. C. Deneş, s. a., *Analiză privind nevoia de adaptare a curriculei universitare la competenţele cerute de piaţa muncii şi îmbunătăţirea relaţiilor cu partenerii de practică* (Universitatea Lucian Blaga din Sibiu, 2015)
2. M. Korka, s. a., *Educaţie de calitate pentru piaţa muncii*. Quality Education for Labor Market (Editura Universitară, Bucureşti, 2009)
3. Multi Consulting Group SRL, *Studiu de perspectivă privind necesarul de forţă de muncă calificată* (2011)

Reflective practice – an objective of the career counselling and orientation process

Cristian Doicin[1,*], *Sorina* Chircu[2], and *Mihaela* Ulmeanu[1]

[1] Faculty of Engineering and Management of Technological Systems, Politehnica University of Bucharest, 313 Splaiul Independentei, District 6, Bucharest, Romania
[2] Department of Teacher Training and Social Sciences, Politehnica University of Bucharest, 313 Splaiul Independentei, District 6, Bucharest, Romania

Abstract. The correlation of theory with practice in the educational process, materialized in various forms, one of which being the professional practice, brings with it a series of benefits for the students: from verifying and validating the knowledge to covering all the learning styles of students. It is also an important milestone in guiding student careers by facilitating reflexive approaches, with a significant contribution to the crystallization of the professional route. Having the study case as a central research method, this paper highlights the mentioned benefits of the professional practice. The sample consist of 100 respondents (students in the 3rd year of study), data processing being done in SPSS 20.

1 Introduction

Formal education proves its effectiveness to the extent it enables a fast and successful integration on the job market. This means, on the one hand, that a young graduate might find a job related to his/her expertise that implied time, effort and hope, and on the other hand, to prove that he/she has such social skills that enables his/her integration in the organizational culture of that job (a critical condition for a good work efficiency and maintaining stress and work anxiety under control). There are several factors whose implementation may facilitate the coagulation around this standard of effectiveness of formal-type education, among which, one of maximum impact, the impact of correlating theory with practice. Based on the cognitive structuralism theory [1] which promotes the proactive-practical way of learning by discovering, the principle of relating theory with practice (actually, one of the major principles governing academic activity) reconsiders the sense and purpose of learning, i.e. knowledge by understanding, for application, adaptation and progress. The orientation towards combining theory with practice in the teaching/learning process allows the valorisation and coverage of all the learning styles of the beneficiaries of education acquiring besides the theoretical component, that of the proactive, reflective-pragmatic one. One of the variants of implementation of this principle at the level of higher education involves the correlation of the educational offer of the university with the job offer on the market, which implies a dialogue between universities and companies-institutions (e.g.: Ministry of Labour,

* Corresponding author: cristian.doicin@cont-edu.pub.ro

Ministry of Education, ANOFM- The National Employment Agency etc.) responsible for welfare and quality of life. Economic growth and employment measures within the knowledge society, as set out in the Lisbon Strategy (adopted by the European Council in 2000 and implemented in three cycles until 2013) [2] which proved to be successful but also poorly implemented, are pursued in Europe 2020 strategy, whose European objectives are transposed into national objectives. The Sectorial Operational Program for Human Resources Development 2007-2013 (POSDRU) financed by the European Union through the Structural Instruments, namely the European Social Fund [3], has as a general objective "to develop human capital and increase its competitiveness by connecting education and learning throughout life with the labour market and ensuring increased participation in a modern, flexible and inclusive labour market for 1.650.000 people" [4].

University POLITEHNICA of Bucharest (UPB) is one of the oldest and reputable schools for engineers in Romania, with a considerable experience in technical training. UPB mission is to train engineers according to the modern principle of direct participation in choosing his/her formative course, including in a learning process which might assure him/her real chances on the free job market, such engineers who might have economic and managerial knowledge and be capable of adapting to the requirements of market economy and to the new technologies. Moreover, UPB has a wide experience in project management, having participated as a lead partner and a partner, in over 60 POSDRU and OS CCE projects [5] POSDRU/161/2.1/G/136211 project, „A successful career in engineering sciences", conducted by UPB, the Faculty of Engineering and Management of Technological Systems (IMST), was carried out under the priority axis 2 concerning „Correlation of long-life learning with the job market", major domain of intervention 2.1 „Transition from school to active life", with the purpose of facilitating the shift from the student to employee status. The above-mentioned project aimed at the achievement of the European Social Fund objectives (while also aligned to the priorities of 2020 Strategy) [6] i.e.: getting aware of the real competences needs of the graduates reducing the differences between the knowledge acquired in the academic system and the employers' requirements, enhancing sustainable economic growth using smart networks and digital economy, a better training of the young people [7] for the job market and equality of chances. The overall objective of the above-mentioned project, of assuring the transition of 400 students (per project) from Bucharest-Ilfov area to the job market, was attained by a few specific objectives materialized in two major activities: (1) professional counselling and orientation of the undergraduates in order to develop a career and to be aware of their rights and obligations as employees; (2) professional internships for the correlation of the skills and competences acquired by the undergraduates by means of higher education with the employers' expectations. The project benefitted at the same time from an online platform, for the dissemination of the project results and providing online support to undergraduates.

2 Methodology of research

This paper aims at highlighting the fact that professional counselling and orientation services provided for the undergraduates enable their development and manifestation of a reflexive attitude on practical academic education, beneficial for facilitating the transition to the job market.

The central research method selected for the verification of our theory mentioned above as the purpose of our research, is the case study, because „it develops a situation where we can notice the game of a large number of factors interacting, allowing to recognize the complexity and richness of social situations and explanation of the relationships which are too complex for survey or experimental strategies" [8]. The case is understood as an integrated system, holistically approached, as the researcher is interested more in 'how' and not 'why' the phenomenon under study (a person, a community, an institution) presents itself in a certain way. In our paper, we use a case study of instrumental interest (it aims at verifying a theory, this particular case being of secondary interest), of qualitative type (because we use qualitative methods – the questionnaire and social document-based interview). On the other hand, when drafting this case study, we used the question-answer type format (consisting in the presentation of questions that the researcher asked himself/herself and the answers he/she got) with a compositional structure in suspense, which means that „the laniary-analytical structure is inverted: conclusion – the result of the case study – is presented at the beginning of the research report, the following chapters developing the reasoning, the explanations" [9].

In order to get answers to our questions raised during the case study and which contributed to the validation of our theory as a final result and conclusion of this investigation, we used the study of social documents and the structured interview. The study of the social documents refers to those documents or official deeds which provide information and based on which a fact can be proved. The method is often used to verify and supplement the data obtained through other methods. The documents studied to reveal certain information which add to our investigation are both digital and non-digital, of public nature: the register of the project target group, the assessment reports of the beneficiaries related to the internships, the web page of the project. The structured interview – based on a questionnaire, also called standardized interview, widely used in sociological and psychological research, enables the comparison of the points of view of the interviewed persons. The mini-questionnaire underlying our interview, aimed at the comparative investigation (before and after the internship) of the opinions of the beneficiaries related to the utility of the internship for their transition to the job market made out of 3 pre-codified items (because they allow the best statistical processing), but the lists of potential answers necessarily include a variant of an 'other mentions'-type of multiple choice test. Out of a total of 117 students participating in the internship, 100 responded to our investigation. We used a multistage stratified sampling. The multistage sampling: the first phase when the sampling unit was the faculty, then the second step, the sampling unit was specialization. Stratified sampling participants' genders are presented in tables 1 and 2.

Table 2. Participants' demographics

Age range	P1-Gender
M_{age} = 21,7 years	75% male 75% male

2.1 Case study - results

The result of this case study is that the professional counselling and orientation services (provided by means of the POSDRU/161/2.1/G/136211 project, „A successful career in engineering sciences") enabled the students to develop and manifest a reflexive stance on the practical academic-type learning process, beneficial for the awareness and realistic self-assessment of the competences and necessary for the transition to the job market and practicing the engineer profession.

We reached this conclusion by turning to good account the opinions and experiences of 100 students, as beneficiaries of the two above-mentioned projects. A total of 400 students were registered in the target group selected from the four years of studies according to a procedure which clearly set the criteria, aiming at the respect for the equality of chances and relying on the students' interest. The professional counselling and orientation activity was performed throughout the whole project, all the students belonging to the target group benefitted from the counselling support. The internships covered a three-month period out of

17 months allocated to each project, and only 3rd Year students were selected, out of 117. The recommendations of the professors and experts in professional counselling were considered in the selection and distribution of students for internships.

The answers obtained by the authors to the questions asked and which confirm the theory of this investigation, are given below.

What are the specializations in engineering of the faculty which run the project?

The webpage of the faculty shows that the three departments of the Faculty of Engineering and Management of Technological Systems of POLITEHNICA University in Bucharest provides for the following specializations of bachelor and master in the line of engineering: the Department of Machine Manufacturing Technology (specializations: Machine Manufacturing Technology, Industrial Economic Engineering, Nanotechnologies and Unconventional Systems), Department of Machines And Manufacturing Systems (specializations: Machine-tools and manufacturing systems, Robotics, Industrial Logistics), Department of Materials Technology and Welding (specializations: Engineering and quality management, Welding engineering, Engineering of industry security) [10].

How was internship designed in projects?

According to the studied documents (mentioned above), during the academic years 2014-2015, internship was completed within each mentioned project at IMST – Faculty of Engineering and Management of Technological Systems, where 3rd Year students selected from among the counselled students participated. The internship summed up 360 hours, of which 120 hours – allotted to the CAD design stage under the guidance of the internship partner 1, and 240 hours – allotted to technological internship with internship partner 2. The CAD design stage was completed under the coordination of Partner 1 - S.C. VEGRA INFO S.R.L., and the technological internship was organized under the coordination and at the head office of 27 companies in Bucharest and surrounding areas. For the completion of these stages, tripartite Internship Agreements were signed in compliance with the provisions of the project between University POLITEHNICA of Bucharest, by the IMST – Faculty of Engineering and Management of Technological Systems, the above-mentioned partners and every intern student, with the strict observance of the legal provisions mentioned above. They included also, as an appendix, internship portfolios which set the rights and assignments of the signatory parties and the competences to be acquired during the technological internship and were signed by the Dean of the Faculty of Engineering and Management of Technological Systems, the Directors of the internship partners, by every intern student and by the Internship Supervisor on behalf of University POLITEHNICA of Bucharest and the tutors designated by the internship partners, with the strict observance of the Order of the minister of education, research and youth no. 3955/2008 regarding the internship stage during the academic programs, of the specific internship rules of University POLITEHNICA of Bucharest and the laws related to labour protection.

How was the CAD internship stage carried out in the academic year 2014-2015?

The CAD design internship was carried out at the internship organiser's place - POLITEHNICA University in Bucharest, at the Faculty of Engineering and Management of Technological Systems, under the guidance of the internship partner, S.C. VEGRA INFO S.R.L. a supplier of professional services in the country and abroad (working mainly for clients in France, Germany, Holland, Belgium, UK and USA) in areas such as CAD (Computer Aided Design), 3D engineering, GIS (Geographic Information System) and CAFM (Computer Aided Facility Management). All the students selected for the internship at SC Vegra Info SRL took part in various production processes under the guidance of the company tutors, and had all the expertise and support provided for the UPB students' internship, in a key-field extremely useful to any company they will be employed with: automatic mechanical design, to build parameterized models of the components and subsets, thus covering the whole production process – from design to the finite product, according to the internship report book. Vegra Info has a wide experience perfectly adapted to the scope of the project. Since its establishment, the company management was aware of the importance of the development of the young people (students and graduates) placing special emphasis both on the development of practical skills and on specific competences requiring high qualification. The training of these young people and co-opting them in the team provided for the company growth resources, which became a market leader in its line of business. To this day, approximately 30% of the company staff has been employed directly from the faculty.

Concerning the proceedings of internship, the students were divided into sixteen groups of 4 or 5 students each, every group was guided by a tutor designated by the internship partner. The CAD design internship, according to the internship portfolio, had to develop in students professional competences by specific activities (table 3).

Table 3. Competences resulted from the design internship

<table>
<tr><th>No.</th><th>Competency</th><th>Planned and carried out activities</th></tr>
<tr><td>1.</td><td>Associating knowledge, principles and methods from technical sciences of the field to graphic representations to solve specific tasks.</td><td rowspan="2">Initiating the use of SolidWorks and CAM Works specialized software packages, for the constructive and technological design of the products, management and/or economic analysis of the processes/ systems of production etc.</td></tr>
<tr><td>2.</td><td>Using software applications for the assisted design of products and/or the solution of tasks typical for engineering and management.</td></tr>
</table>

To this purpose, throughout the 120 hours of design internship, the students were initiated and instructed by the internship tutors for the successful use, in design, the SolidWorks software. In accordance with the provisions of both projects, Partner 1 used for instruction an assisted design guide elaborated in the first part of the project, when the software applications the students worked with were defined (e.g.: automated mechanical design – parameterised models of the components and subsets, views in drawing, manipulation of dimensions, study of the fluid flows etc.). In the last week of internship, the students used the knowledge accumulated with the purpose of designing a part.

The internship supervisor and the internship tutors assured the planning, organization and supervision of the internship, participated in setting themes and competences and made efforts for the internship to be in compliance with the agreements and the students observe the conditions and requirements of the internship stage. At the end of the CAD design internship, the students filled in the observation sheets, which the tutors endorsed later. At the same time, the tutors filled in the Activity Reports, and then they filled in the student assessment sheets by the tutors, according to the provisions of the Procedure of assessment of the practitioner students.

How was the technological internship carried out in the academic year 2014-2015?

For 7 weeks (240 hours), the students participated in the technological internship. They watched the way in which the production activities were carried out at the internship partners 2 (27 technical companies), with which partnership agreements were signed.

The students' allotment to internship partners 2 considered the overall grade of the 2nd year of studies, the score of the general technical knowledge, the recommendations of the specialists in professional counselling activities, but also every one's options.

The professional competences to be acquired by the students after completion of the technological internship stage by carrying out specific activities, according to the internship portfolio (table 4).

Table 4. Competences resulted from the technological internship

<table>
<tr><th>No.</th><th>Competence</th><th>Planned and carried out activities</th></tr>
<tr><td>1.</td><td>Elaborating and interpreting the technical documentation corresponding to the technological manufacturing processes.</td><td rowspan="4">- Preparing the documentation for the programming, planning and monitoring the production in certain organizational structures of the company / the research institute where the internship was carried out.
- Knowing, using and preparing the information necessary for the design and implementation of the technological processes by means of real and virtual processes (drawings, forms, technology sheets, operation plans, specifications, 3D models etc.)
- Using certain software instruments of the internship partner for the technological design of some products or for production management
- Knowing and analysing certain process projects and production system of the company/research institute where the internship is carried out: flows, locations, programs, load tracking, quality control or other specific examples.</td></tr>
<tr><td>2.</td><td>Using the software applications and the information technologies to solve tasks specific for engineering and management.</td></tr>
<tr><td>3.</td><td>Technical and economic design and product and industrial processes improvement.</td></tr>
<tr><td>4.</td><td>Economic evaluation, planning and conducting processes and logistic and production systems, assuring the quality of the manufactures.</td></tr>
</table>

The internship program generally observed the partners' working hours, as provided by the internship agreements, within the limit of 7 hours / day. Internship was carried out under the guidance of the tutors, and the students acquired abilities for the realisation of parameterised models, of components and sets and covering the whole production process – from design to finite product, according to the internship notebooks. The internship notebooks were filled in under the guidance of the tutors, of the expert in the elaboration of the internship documentation and of the internship supervisor.

The internship supervisor, along with the internship tutors, provided for the planning, organization and supervision of the internship (by periodical visits to the internship partners) and supervised the internship to be in accordance with the commitments set out and the practitioner students observe the requirements of the internship and internship partners. At the end of the technological internship, the students completed the Observation Sheets that we forwarded, and the tutors endorsed. The tutors filled in the Reports on the activities carried out, and then they filled in the Student assessment sheets by the tutor, according to the provisions of the Procedure of practitioner students' assessment. We endorsed these assessment sheets and the students' attendance. According to the same procedure the internship was completed by a colloquium where the students presented their achievements and answered the questions of the examining board consisting of the Project Manager, the expert with the selection of the target group and the internship supervisor. The students' assessment at the colloquium was done by filling in the Internship Final Assessment Sheets by the members of the assessment board was done by filling in the Internship Final Assessment Sheets by the members of the assessment board following the consultations and was completed by drafting the Summary table of the students' results and the Ranking of the award winning students.

What is the real satisfaction of the internship beneficiaries?

The final assessment sheet aimed at observing the six criteria of assessment with different percentages of the final grade, as below (table 5):

Table 5. The internship final assessment sheet

No.	Assessment criteria	Marking points	Marked - obtained* points
1	The tutors assessment (in the CAD design internship and the technological internship)	15	14.55
2	The quality, volume and originality of the internship notebook on the description of the daily routine of their internship (of CAD and technological design)	30	28.48
3	The quality, consistency and originality of the presentation the day of the assessment	10	9.20
4	The degree of participation in internship (of CAD and technological design)	10	9.63
5	The accuracy and consistency of the project for the part received as an internship theme	15	14.73
6	The degree of acquiring professional technical and managerial skills aimed by the CAD design and technological internship stage (answers to the questions of the board)	20	18.89
TOTAL SCORE		100	100

*The arithmetic mean of the points obtained per each criterion of assessment was calculated and the answers of the 100 beneficiaries of the internship stage were centralized.

In accordance with the provisions of the project, 50% of the students who participated in the internships organized in the academic year 2014-2015 were rewarded, respectively the 2nd Year of the project, i.e. 40 students, who had special results and demonstrated the acquisition of skills expected from the internship stage.

What are the opinions of the beneficiary students in relation with the importance of internship in facilitating the transition to the job market?

Any career counselling service for the beneficiaries of education – in this case, students – must start in the design phase from the specifics of the area of study. Technical thinking and skills are dominants of the technical higher education. We find the principle of the relationship between theory and practice very well and pungently illustrated in technical education. In technology, technical thinking has specific elements: pragmatism (design for solving practical problems), intuitive-action thinking (which determines the concrete orientation into an intuitive situation and the rapid passing to action, for the quick exit from the unforeseen functional-technical difficulty), constructive-functional thinking, economic and organizational thinking [11]. The UPB students, irrespective of their major, have a pragmatic style of approaching and solving problems – it is a defining note and at the same time a particularity in choosing and browsing these studies. Yet, pragmatic thinking has efficiency and qualitative results only when accompanied by an important reflexive component (because it allows the anticipation of the problems that may occur, the identification of certain dysfunction of the technological process, it enables quality control etc.). Reflexive thinking must be supported during the academic years by all the didactic approaches, so that the main objective of the career counselling process intended for the students in the project who benefitted from professional internship, was to encourage this component in the group and individual counselling sessions. The mini-questionnaire on the importance of internship in facilitating the transition to the job market was filled in first (before starting the internship stage of the project) during the individual counselling session. This questionnaire was followed by a student-counselling expert discussion, both to debate and to clarify the contribution of the internship stage in the formation of professional competences, and to clarify other aspects related to the profession-personality concordance (and materialized in a career plan). To capture the impact of internship, we applied this mini-questionnaire on completion of the internship stage, and the results were as below (tables 6, 7, 8).

Table 6. Importance of participation in the internship stage to get a job

How important do you think that the participation in these internship stages of the project to get a job?	Very little importance		Little importance		Average importance		Important enough		Very important	
	A*	B*	A	B	A	B	A	B	A	B
	0%	0%	8.4%	0 %	28.2	2.2 %	20%	21.3 %	43.4 %	76.5%

*A= Before starting the internship stage; B= After completion of the internship.

The students were aware of the benefits of professional internship as very important (43.4% before internship – 76.5% after internship) during the studies: overcoming the impediment of the lack of work experience on employment of the debutant, verifying the type of work and appropriate work environment and an objective feedback related to the level of acquired competences.

Table 7. Considerations of the theoretical training received during the academic years for the specialized field

How do you assess the theoretical training you get during academic years, for your area of expertise?	Very little importance		Little importance		Average importance		Important enough		Very important	
	A*	B*	A	B	A	B	A	B	A	B
	0%	0%	26.6%	4%	33%	5.3%	31.9%	31.4%	8.5%	58.8%

*A= Before starting the internship stage; B= After completion of the internship.

The opportunity of performing practical activities in the specific field enabled a better valuing of the theoretical component taught in the faculty. Thus, the students found a lot of theoretical information acquired during the courses as essential conditions for the application of the technological process. The re-signification of the theoretical component at the level of the meaning and utility is considerable in this context (from 8.5% very important before starting internship to 58.8% very important after completion of professional internship).

Table 8. Relevance of practical training obtained during academic years for the specialized field

How do you assess the practical training you received during academic years for your area of expertise?	Very little importanc e		Little importance		Average importance		Important enough		Very important	
	A*	B*	A	B	A	B	A	B	A	B
	0%	0%	8%	2.3 %	66 %	73%	21 %	8%	5%	3.7%

*A= Before starting the internship stage; B= After completion of the internship.

The question aimed at the students' assessment of the quality of practical training through academic studies, except for the internship opportunity in the project. Turning to good account this essential component to train the engineer as average importance refers less to the quality of the activity itself but rather to the fact that, in the students' opinion, it is achieved insufficiently in point of frequency.

3 Interpretations

Carrying out the project internship enabled from the didactic point of view the correlation of theory with practice, and from the point of view of the professional counselling process, encouraging reflexive thinking – considered as the second nature of the engineer's pragmatic thinking. The cognitive-reflexive approach of the investigated students facilitated by the professional counselling sessions resulted in the following remarks:

Orientation to jobs in the line of design engineering – the students find this field more appropriate, the production work comes second.

Need to cover several disciplines of the academic curriculum design type, in their opinion, the competences acquired in this area of expertise provide higher chances of employment and integration on the current labour market.

Extension of internship opportunities similar to those in the project, at the level of all the academic years (not only on the 3rd Year).

Interest in revising certain theoretical knowledge taught in the previous years with major disciplines for the formation of engineering bases, through the practical activities carried out.

4 Conclusions

The data obtained and the students' remarks as a part of supporting the cognitive-reflexive approach during the counselling session reveal the various hypostases in the learning process provided by turning to good account the relationship between theory and practice in the teaching/learning process, i.e. the application of certain theoretical knowledge to solve certain tasks – a preparatory exercise to solve the future tasks specific for the work place – and internship – valued as starting point in learning by discovering, collaboration and research.

The work has been funded by the Sectorial Operational Programme Human Resources Development 2007-2013 of the Ministry of European Funds through the Financial Agreement POSDRU/161/2.1/G/136211.

References

1. J.S., Bruner, *Towards theory of instruction*, (Editura Didactică şi Pedagogică Publishing House, 1970)
2. Europe 2020 Strategy - The Lisbon Strategy in short Last accessed on 15.03.2017, Available at: https://portal.cor.europa.eu/europe2020/Profiles/Pages/ TheLisbonStrategy inshort.aspx
3. European Paralament and Council Regulations no. 1081/2006 regarding European Social Fund, Last accessed on 07.03.2017, Available at: http://www.consilium.europa.eu/en/ european-council/
4. European Social Fund – Officilal web portal, Last accest on 02.03.2017, Available at: www.fseromania.ro
5. University Politehnica of Bucharest, Official web site, Last accessed on 18.03.2017, Available at: http://www.upb.ro/en/
6. Europe 2020 Strategy, Last accest on 22.02.2017, Available at: http://ec.europa.eu/europe2020/index_en.htm

7. POSDRU/161/2.1/G/136211 project – Official web site, Last accessed on 25.02.2017, Available at: http://cariera ininginerie.ro/consiliere-online/
8. P., Collerette, *Etudes de cas au service de la recherche. Dans A. Mucchielli* (coord). Dictionnaire des méthodes qualitatives dans les sciences humaines et sociales. Iasi, (Editions Polirom 1996)
9. S., Chelcea, *Metodologia cercetării sociologice. Metode cantitative şi calitative (Méthodologie de la recherche sociologique. Méthodes qualitatives et quantitatives), éditions II-eme.* (Editions Comunicare.ro, Bucarest, 2004)
10. Faculty of Engineering and Management of Technological System – Official web site, Last accest on 25.02.2017, Available at: http://imst.pub.ro/index.php/ro/licenta/examen-diploma
11. M., Mircescu, *Pedagogie (Pédagogie)*, Bucarest, (Printech Publishing House, 2004)

How important is mentoring in education?

Carmen Sonia Duse[1,*], *Dan Maniu* Duse[2], and *Maria* Karkowska[3]

[1] Lucian Blaga University of Sibiu, Department for Teacher Training, 34 Calea Dumbravii, Sibiu, Romania
[2] Lucian Blaga University of Sibiu, Department of Industrial Machines and Equipment, 4 Emil Cioran, Sibiu, Romania
[3]Krakowskie Centrum Zarzadzania i Administracji Sp. z o.o. 31-157 Krakow, Plac Matejki 10/3, Poland

Abstract. The paper aims to provide the mentoring experience made in Romanian schools, developed as a result of the Erasmus + project entitled "Mentoring Between teachers from secondary and high schools" project No. 2014-1-PL01-KA200-003335. Thus, we seek to present the work and the general considerations in connection with the mentoring theme, how we decided on the curriculum in order to form mentors, how mentors were selected and what was the result of their mentoring work. It is important to observe the main obstacles in the mentoring process were and how these problems could be solved. We will also highlight the impact of the mentoring activity on mentees and their perceptions about the usefulness of such activities. The project design and the activities undertaken are the results of research and brainstorming made by teams of specialists who participated in the project and are conceived by university professors and researchers from Poland (project coordinator), Portugal, Spain, Greece and Turkey and also Romania. The teachers involved in the project were from the secondary and high schools from the Counties of Sibiu, Brasov, Cluj-Napoca and Iasi. We thought the geographical dispersion will be very useful to establish comparative elements between the involved schools.

1 About Mentoring

Formally, mentoring as a problem in Romania would be described nowadays as a new one. Informally, there have always been people who programmatically or not, behaved as mentors.

But we have to ask ourselves: what is a mentor and who can play this role? Mentor can be considered "a trusted counsellor" or "a guide".

The term's origin is attributed to Homer, and binds to ancient Greek mythology, when talking about the character reminded, the Greek epic Trojan War, Odysseus Mentor was a friend to whom he entrusted his son Telemachus for support and guidance. Athena, the goddess of wisdom, got sometimes into the role of a Mentor and offered encouragement and support to Telemachus, for learning and enlightenment. From this story, the word "mentor" is derived and has come to mean a "father" or even a "mother figure" (according

*Corresponding author: carmen.duse@ulbsibiu.ro

to the wisdom and advice of Athens) for youngsters. Later, being associated with the terms of a teacher or a counsellor, the term "mentor" will be seen and used in the schools of Socrates and Plato, and in Aristotle's academy.

In other social and literary contexts the term "mentor" can be seen as "advisor", "instructor", "teacher", "guru", "Sensei", "coach". In some specific situation, the word mentor can be used for anyone who has a positive or a guiding influence on another person in his life.

The definitions of the term "mentor" are manifold. Thus, as recorded by Wai-Packard "Mentoring is a term generally used to describe a relationship between a less experienced individual, called a mentee or protégé, and a more experienced individual known as a mentor" [1].

In [2] it is stated that "mentoring is most often defined as a professional relationship in which an experienced person (the mentor) assists another (the mentee) in developing specific skills and knowledge that will enhance the less-experienced person's professional and personal growth."

An interesting part is that the parent of the management - Frederick Taylor (1911) stated in "Principles of Scientific Management" that: "what we all want is a competent person, already prepared; someone, who has been trained by someone else. But we will not find ourselves on the path to national efficiency as long as we will not realize that it is not only our duty, but also our chance consists in the systematic collaboration for education and training of this competent person, instead of seeking a man who already prepared others."

Mentoring the next generation of teachers is a huge responsibility project for current teachers and school managers. The need to maintain mentoring in secondary and high schools, but not only, is pressing at the moment, mainly because of the lack of attractiveness of the job so that young people tend to abandon the teachers profession mainly because of economic reasons. As John Holloway [3], director of the Educational Testing Service's Teacher Quality Initiative stated "One of the greatest reasons new teachers leave the profession is lack of support and assistance in dealing with the many frustrations they face. Mentoring plays a nurturing role in guiding those new teachers through the first critical years and in providing them with tools and skills needed in actual practice."

Therefore the Erasmus + Project entitled "MENTOR - Between Mentoring teachers in secondary and high schools" No. 2014-1-PL01-KA200-003335 has come to meet this need [4].

2 The MENTOR Project

The Mentor project included six partners from six different countries (Poland, Romania, Portugal, Spain, Greece and Turkey), being coordinated by the Krakowskie Centrum Zarzadzania i Administracji Sp. z o.o. and was conducted over a period of 3 years (2014-2017). The MENTOR project's focus was to support the growth and professional development of new teachers in secondary and high schools by means of mentoring. The proposed specific objectives of the MENTOR project were:

1. To organize the desk research about the methods (models) of mentoring, especially between teachers in secondary and high schools in EU countries.
2. To elaborate the method (model) of mentoring between teachers in secondary and high schools.
3. To design and develop the training modules for secondary and high school teachers mentors.
4. To organize a set of Training for Teachers Mentors (TTM) aimed in familiarizing teachers in the mentoring skills using an on-line/blended approach.

5. To establish a joint infrastructure in order to facilitate sustainable accessibility to the mentoring approach for schools as well as the exchange of scientific information.
6. To test delivered mentoring method in the secondary and high schools.
7. To monitor the dissemination, impact measurements and the quality assessment of the project.

2.1 Desk research about the methods (models) of mentoring in EU countries

The bibliographical research undertaken by the participating teams has gathered a large number of bibliographic materials from which later started the conceiving of the training method (model) [5-7].

We opened the desk research using key - words like: mentor, mentoring process, beginner teacher, mentees, new entrants. It was useful, for the beginning to understand more about who can be a mentor, what qualities are needed, if he needs previous experience as mentor and what kind of experience, some about the ethic of mentoring process and other more.

More than 50 resources, blogs, sites regarding the Mentor and Mentoring process was analysed in order to understand the way of thinking of these specialists in this field and about 38 of them were chosen as a model and for a deeper analyse. In the first step, the research was focused on the mentoring process in general, and in the second step on the educational field. It was interesting for us to understand how the problem of beginner teacher was solved in other countries or continents, as we collect information from all around the world.

In this first step we found differences between the European countries from the mentoring point of view. As an example, there were countries in which there is no idea about the mentoring process, as well as there are European countries which have a specific legislation for that. All these situations were examined and judged in order to find the best way of mentoring proposal.

2.2 The development of the training method

Bibliographical research helps us have a clear image over the best option in training the junior teachers. In order to achieve the training curriculum several lots of teachers from partner countries were investigated over the usefulness and importance of certain skills and essential detailed knowledge of beginning teachers. Initially we proposed that this part of formal curriculum to be one small stretch, not to overburden the novice teachers.

To test the importance of training curriculum, each participant has worked with a number of minim 20 mentors. These over 100 mentors have been trained in a special session, conducted during one day, in which the produced materials were presented and the teachers were asked about their opinion on the appropriateness/suitability of these materials in order to work with beginning teachers.

The offered feed-back by these mentors of the 6 participating countries was extremely useful to complete the first theoretical part. Thus, besides the short version of the handbook's work a supplement to the handbook has been done as well. The logic of its occurrence related discussions resulted in the initial phase of the mentors training, where the idea of existence of more materials dedicated to the individual study for either the mentors and teachers juniors, or even other teachers was present.

Secondly, after the feed-back received from the mentor training, a training kit has been created, a collection of worksheets, PowerPoint presentations and case studies for the training of the junior teachers.

Once these instruments were operationalized, the testing phase of these instruments went on directly in schools.

Thus, the process of training of junior teachers began in every participating country using the tools designed for the first two years of the project.

3 The Training of beginners

3.1 Teachers through Mentoring

The training step for novice teachers was conducted over a period of nine months, during which the novice teachers and the mentors met weekly at first, then once every two weeks. Practical training was conducted between September 2016 and the end of May, 2017. Questionnaires to the mentors and to the mentored were applied at the end of the first half and at the end of the training stage in order to collect the feed-back of that activity.

For starters, in August 2016, the most difficult problem that had to be urgently solved was the establishment of the included schools in the testing phase, because the number of the junior teachers in Romanian schools was found to be becoming smaller. But the even bigger surprise was when they found the same situation in other participating countries. This can be blamed either on the negative demography, but also on lack of attractiveness of the teaching profession for young graduates. As a result, for Romania, the test group included six mentors and six mentees form four schools. There were involved one secondary school and three vocational high schools, one of them from a rural area.

The project directors from the involved schools and high schools have expressed their agreement by signing a cooperation protocol. They also assured their mentoring support for the first 4 months.

Out of the six mentors, five of them were women and only one was male. Their age was between 30 and 50 and four of them were aged between 30 and 40 years (66.6%), while the other two were between 40 and 50 years old (33.3%). Among the surveyed mentors, only one had previous experience of mentoring.

The reason they accepted the mentoring activity was because they considered this activity as a challenging and interesting one.

Also, the middle age of the involved teachers as a mentor is one that demands new challenges. Most of them are teachers who have achieved professional degrees, academic degrees (grade II and I, which in the Romanian educational system scale is the maximum for professional development in pre-university schools) and now they are looking for new ways of self-realization.

The mentees group comprises 6 people, 5 women and one man. Their age is as follows:
- one person (female) under 25 years (16.6%);
- three persons between 25-30 years (50%);
- two persons between 31-35 years (33.3%).

None of the mentors had previous experiences of mentoring. The reasons they accepted mentoring activity was because:
- It appeared as an exciting and challenging activity;
- They were convinced by mentors involved in the program;
- They felt the need for personal development.

The meetings between the mentors and the mentees were held mostly face-to-face, for a period of more than an hour, but also by communicating through e-mail or chat. B means of the face-to-face meetings both mentors and mentees were easier to clarify certain aspects of mentoring.

What made it more remarkable was that the meetings were held in their free time, usually in the school. Opinions have shown that the most difficult problem was to find their free time.

Even if a lot of the factors were considered very important in the mentoring relationship, the first place was taken by mutual understanding, acceptance and regular feedback from both sides. This may betray the mentees uncertainty and their need to receive confirmation that they are on the right way.

As shown above, questionnaires to observe the conduct and effectiveness of the mentoring program were applied. Thus, both mentors and mentees were questioned as on the question "what were the most important / valuable benefits mentors and mentees individually received so far from this cooperation?"

The mentors came up with the following answers:
- The mentees involvement.
- The decision of a mentee to consult the mentor even outside the scheduled meetings.
- One mentee hugged the mentor one day and thanked him for everything they have shared together, saying that their relationship there is still work and practice to do for the methods and teaching techniques.
- They consider everything could be solved through trust.
- They learned to manage the personal resources.
- The self-confidence, the mentee earned, which was visible in the accurately drawing of documents for the school and the grade she received in the completed inspection.
- They learned to be more confident.
- They learned to discover their own resources needed for conducting an effective teaching process.

From the mentors' perspective, one mentor said: "This cooperation really helped me to evaluate myself both as a teacher and a friend / colleague, open to share experiences and examples of good practice to stimulate a person willing to apply their knowledge in a personal activity. I found, that it's wonderful to give, to stay close to others with love and to communicate with the other, who is interested in the knowledge." Another mentor considered "I'm available, more than I thought".

The answers of others conclude, that: was a real help for his work in the classroom; the mentoring trust that she seems to receive, as she advances both in her emotionally and professionally life.

For the mentees, most important / valuable benefits were:
-The job as a teacher is made with much, much more love and many sacrifices for the teacher.
- They learned to work more effectively with students.
- The professional behaviour, which is required for a teacher.
- The richness of the information which they were provided with.
- Application related design of the lesson plans, achieving the initial tests.
- Facilitating best practice by sharing experience.

Mentors and mentees were asked "what they have learned about themselves, their abilities, knowledge and attitudes so far from this cooperation?" as well. There are different answers, from mentors and mentees. For example, the mentees stated, that:
- They consider it to be very important to know how to control emotions, not to panic, to be patient and always act calmly.
- They understand they can give more than they thought.
- There is always something new to learn.
- Through the careers it can be possible to improve methodical knowledge and skills. This is visible when someone is trying to share their knowledge with other teachers.

- It is good sometimes to see again the teaching profession from the perspective of a newcomer in education.

The last question was related to the additional comments from mentors and mentees (thoughts, feelings, suggestions) about the Testing Phase progress. You can see here examples such as:

"The mentoring activity held during this period encouraged me to keep being available to those who needed support to start their teaching career. The mentee opening her desire to do everything better and more efficiently, foster cooperation and to share joy, to determine experiences, knowledge, materials and much love. Through our work with mentee, we noticed visible progress in the quality of her educational activity."

Other opinions mentors gave, were:
- There were moments of fatigue, but they overcame them.
- With mentoring activity there were only positive feelings connected.
- Mentoring is an exercise for success. This program implemented in schools would only bring benefits for novice teacher, as well as for the school. Through this program the relationships between colleagues could improve a lot.
- It's an enjoyable and exciting experience.

The 6th mentor had a reflection moment: "In this test phase I realized that mentoring is extremely important because it makes you see with new eyes things already tested and learned and to understand the approach of beginning of a teaching career."

Mentees gave important feed-back as well:

"Activities carried out with mentor in this period were very useful for me. Due to the opportunity I had, I learned many practical and beneficial things for my teaching career. The warmth and love that we received during this period from the mentor and my colleagues made me want to continue this path, which I have chosen in education."

Another mentee said: "It was a very interesting and formative experience and I would have preferred to have this experience during my university time."

Other feed-back from the novice teachers included:
- Mentoring is a very useful experience for any newcomer in school.
- It is a very useful experience for the novice teachers. Everyone should benefit from it.
- The introduction of this program in all Romanian should be mandatory.

4 Conclusion

Despite the problems in finding the needed time, which is always too short because of their duties and responsibilities, mentors and mentees found a special time for them and their relationship. For mentors it was a good moment to understand the novice teachers' needs and to remember their first steps in teaching career. For mentees it was encouraging to have a person who can help them in different situation. From the point of view of subject discussions, the focus was in didactics and pedagogical design as well as in organizational culture and classroom management.

A good and encouraging point was for the mentors the desire of progress and development of the mentees.

From the mentees perspective, the most appreciated feature of the mentors was the desire for help and the open relation which was built.

A strong professional development of the mentors facilitates the relationship, but in the same way, it was important for the development of these good relationship concerning mentors human qualities.

And on the first place here is the Emotional Intelligence of the mentors which was proved.

Looking of the geographical dispersion of the mentors, an important conclusion can be seen. The relationship between mentors and mentees in the Counties of Iasi and in Cluj are warmer, with more positive emotions. In the high-schools and secondary schools from Sibiu and Brasov County, which are in the heart of Romania, the relationships are more formal, more distant. This could be in connection with the specific culture of these parts of the country. The people from Transylvania are seen colder and more distant in relationships then the habitants from the other parts of Romania.

Following the experience of mentoring achieved in the six partner countries, a conclusion can be drawn: Mentoring is a must that has to exist, no matter what we are talking about: young teachers starting their career or teachers in a new school.

Each of us needs some time for integration and learning / suitability in the organizational behaviour and as short as we can make it, by keeping a high standard of quality, the more productive we are as professionals, regardless of the area in which we work and the more in the educational environment.

The specialty literature presents the idea of the unconscious incompetence, when we are not aware of what we are wrong, and persist in these errors, conscious incompetence when we realize that we do something new than we would like, but we do not know how to get it right, conscious competence, that we think we could do a certain thing better, if we focus on it, and unconscious competence when everything becomes normal and we can turn our attention to other things. The presence of the mentor is to ensure the implementation and strengthening awareness, which is the key element in achieving individual progress.

Shaw presents us: "As a form of professional development, mentoring affects both the new teachers who are being mentored, and the experienced teachers, who will serve as mentors" [6]. Through this, we can understand that the winners of the mentoring process would be everyone: mentees, mentors, school members, students and parents, community.

And as the writer Kamand Kajouri once said: "Do not give them a candle to light the way, teach them how to make fire instead. That is the meaning of enlightenment."

References

1. B. Wai-Packard, Science Mentoring Research, *Definition of mentoring*, http://ehrweb.aaas.org/sciMentoring/Mentor_Definitions_Packard.pdf (American Association for the Advancement of Science, 2003)
2. D. Cutler, *Why New Teachers Need Mentors*, https://www.edutopia.org/blog/why-new-teachers-need-mentors-david-cutler (Edutopia, George Lucas Educational Foundation, 2014)
3. Management Mentors, http://www.management-mentors.com/resources/corporate-mentoring-programs-resources-faqs (North America/Europe/Africa, 2015)
4. The Erasmus + Project *Mentor - Mentoring between teachers from secondary and high schools*, Project No. 2014-1-PL01-KA200-003335, http://edu-mentoring.eu/ (2014)
5. Education World Inc., *Guidance from the Get-Go: Mentoring New Teachers*, http://www.educationworld.com/a_admin/admin/admin260.shtml (2016)
6. R. Shaw, E. Villegas-Reimers, *Teacher professional development: an International review of the literature* (UNSECO: International Institute for Educational Planning, 2003)
7. G. Waugh, J. Lovatt, *Guide for mentors 2014-2015*, http://www.ianphi.org/ (2014)

The importance of OER in engineering education

Carmen Sonia Duse[1,*], *Dan Maniu* Duse[2], and *Philip* Bonnano[3]

[1]Lucian Blaga University of Sibiu, Department for Teacher Training, Calea Dumbravii nr. 34, Sibiu, Romania
[2]Lucian Blaga University of Sibiu, Engineering Faculty, Department of Industrial Machines and Equipment, Emil Cioran 4, Sibiu, Romania
[3]University of Malta, Faculty of Education, Msida, Malta

Abstract. The problem of education is a very important one and occupies an increasingly wider register among educational specialists, who see these changes directly related to current generations of students. It speaks increasingly more and more about the desire of young people to learn when they want, how they want and how much they want, in their own way, without any constraints imposed by fixed educational program, established by teachers, according to their idea how to go on an educational route. In this context, open education provides crucial support of any person who is interested, and covered a complete and appropriate educational needs, in connection with time and availability of the person. The paper aims to bring attention to the problem of open education and open learning resources, in the context of aligning the educational system at European and world requirements regarding accessibility of education. Also, we use work experience resulting from Erasmus + project entitled "EU-store" European Standards for Open Educational Resources, project no. 2014-1-RO01-KA202-002985, project leaded by our university.

1 What are OER

1.1 OER Definitions

The concept of Open Educational Resources was defined for the first time in 2002, within a UNESCO meeting in Paris. Open Educational Resources (OER) are materials for learning, teaching, research or other educational purposes that can be used, adapted and redistributed freely, without constraints - or with very few constraints - related to copyrights. Such materials can be courses, lesson plans, presentations, books, handbooks, homework, questionnaires, activities for the classroom or for the laboratory, games, simulations, tests, audio or video resources and many more, available in digital format or on a physical support, to which the access is free.

*Corresponding author: carmen.duse@ulbsibiu.ro

Open Resources are resources to which the access is free or for a moderate fee, from anywhere in the world and from any type of platform – from the computer, from the table or from the phone. However, OER are not only digital resources or those associated with e-Learning. OER can also be resources in physical, printed format.

In order to be called open educational resources, they must be free or have a reduced cost, to not have a commercial purpose, to be accessible from anywhere in the world and with any type of technology and, moreover, to have a free license. However, this is not an universally accepted point of view and there exist several definitions for Open Educational Resources. For example, in the definition adopted by UNESCO there is comprised also the part of free licensing, whereas OER Commons does not mention the type of license that resources need to have.

The first question which can arrive is where to find these OER and the second one is how free these are? A complete and certain list of all OER initiatives and databases is impossible, given their dynamic nature. Thus, even if the number of freely available documents increases daily, the finding of documents of interest remains a task for the user. On the other hand, there is very much need to be checked the license under they are published.

The most important universities are developing OER concerned on a variety of subjects for learning and development. There are a lot of other OER websites for people with small budgets and no formal training in education, especially at higher levels. The aim is to make these courses accessible, to share knowledge and therefore to make society more performant and tolerant through learning and development.

In point A of the Paris OER Declaration are presented Open Educational Resources as an important step in education, this Declaration recommends to: "Promote and use OER to widen access to education at all levels, both formal and non-formal, in a perspective of lifelong learning, thus contributing to social inclusion, gender equity and special needs education. Improve both cost-efficiency and quality of teaching and learning outcomes through greater use of OER" (UNESCO 2012).

A resource is considered educational if, as Camilleri, Ehlers, & Pawlowski [1] said:
"- it has been produced / released specifically for use in formal or non-formal education;
- it is used as course-material in an educational course or program;
- it is included in a repository of materials intended to support formal/non-formal education".

Speaking about what can be considered an Open Educational Resources, Ischinger [2] include:
1. Learning content: full courses, courseware, content modules, learning objects, collections and journals.
2. Tools: software to support the development, use, reuse and delivery of learning content, including searching and organization of content, content and learning management systems, content development tools and learning communities, and
3. Implementation resources: intellectual property licenses to promote open publishing of materials, design principles of best practice and localize content.

Johnstone made a classification of three types of OERs [3], as:
- Learning resources and courseware, content modules, learning objects, learner-support and assessment tools, online learning communities;
- Resources to support teachers – tools for teachers and support materials to enable them to create, adapt and use OER;
- Resources to assure the quality of education and educational practices.

1.2 What are Open Licenses

There is existing a non-profit organization named Creative Commons „which has released several copyright-licenses known as Creative Commons licenses free of charge to the public. These licenses allow creators to communicate which rights them reserve, and which rights they waive for the benefit of recipients or other creator" [4]. For education, the idea that content should be free and openly available is one of the most important, because that means anyone will have access from all over the world at the educational content. The rights of using of different open content are expressed in the "5Rs Framework": retain, reuse, revise, remix, and redistribute. These permissions are granted to users free of charge, or with just a small amount of money.

In January 2016 there were an estimated 1.1 billion works licensed under the various Creative Commons licenses and that is speaking about the richness of the educational (not only) materials, which can be use in educational purposes.

2 Why use OER

The characteristics of the Millennials, which have been described by Howe and Strauss [5] and more recently of the Generation Z is speaking about individuals who [6]:
- have a strong relationship with technology as social environment;
- take into account change, the innovation being the pillars of their existence;
- believe it's cool to be smart;
- are fascinated by new technologies;
- are characterized by the idea that "if it has not been created, I can do it ".

The Children of these generations were the ones, who threw out the toys in favor of computers and devices and learned intuitively by themselves how to use them. So, we will not be surprised that, as highlighted by the current research Sparks & Honey [7], children of the generation Z spend about 41% of their time outside school on the computer or on other devices, suffering from FOMO (fear of missing out).

Especially for Z Generation, the Internet and technologies have had a major influence on their culture and development. Millennials and Z Generation students have never known a world without computers, the World Wide Web, video games, cellular phones and other devices.

As researchers said [8], the brains of Generation Z are structurally different than those of earlier generations and this is not because of the genetics, but in connection we are using our brains to respond to things in our environment. Darla Rothman says the part of the brain which is responsible for visual ability is far more developed, making visual forms of learning more effective, while auditory learning (lecture and discussion) is very strongly disliked in Generation Z case. The result is, said the researcher „interactive games, collaborative projects, advance organizers, challenges, and anything that they can try and see are appreciated".

Other differences connected with this generation are:
- a very developed visual intelligence – they are able to understand, to read the images and to use the images for sending a message;
- a very developed spatial intelligence as a consequences of their playing video games from the earlier ages;
- the distributive attention which sometimes is called by teachers superficial;
- the ability to have fast responses in a short time and they need a quick response as well;
- the pleasure and the capacity to learn by doing, to learn from their discovery and not from the books or brochure.

The conclusion about generation Z was they need to use collaborative work, to work in teams, to use experiential learning as part of their academic activities.

But the main focus for an effective and relevant educational process, in connection with Generation Z is the development of Knowledge society skills, with digital competence having both a leading and an enabling role. International reports such as 'Digital Education – Making the change happen' MCEETYA (2008); UNESCO ICT competence for teachers Hine, P. (2011), the on-line document from the 'Partnership for 21st century skills', European Commission Policy Report (2013): 'DIGCOMP: A Framework for Developing and Understanding Digital Competence in Europe' emphasize a developmental approach to bring about renovation of Education. In this process digital technologies and open educational resources serve as major catalysts in bringing about pedagogical innovation. Universities need to be changed from teaching institutions, driven by content-centered approaches, to learning institutions that use the context of curricular subjects to nurture knowledge society competences. The curricular content platform should be used to develop identified competences.

These reports recommend that Education should evolve along a three stage developmental trajectory. The initial Technology Literacy stage enables learners to use ICT to learn more efficiently. In the second Knowledge Deepening level learners acquire in-depth knowledge of their school subjects and apply it to complex, real-world problems. The third level focuses on Knowledge Creation and serves to develop competences to create the new knowledge required for more harmonious, fulfilling and prosperous societies.

In this evolving process, the role of Digital Technologies and Open Educational Resources is to catalyze the much deserved pedagogical change from one focused on knowledge acquisition to one based on knowledge application and knowledge creation [9]. The current predominant teaching-centered approach should be complemented with truly student-centered approaches that employ digital technologies to mediate different modes of learning and assessment. Consequently the ability to identify, use, evaluate and create OER becomes a mandatory competence to be developed and refined throughout the educational process, and in particular with the professional development of teachers.

The perennial need for innovating teaching and learning practices brought about by the constant change in the nature and organization of knowledge, the continual emergence of new digital technologies and accompanying nascent behaviors, and the affinity of the younger generations with technology can only be mitigated by adopting a learning design attitude and approach that exploits the full potential of OER.

3 OER in Engineering Education

In Engineering Education and beyond the basic use for content acquisition, in the context of 21st century education, OER are used to promote identified processes through which communication, different forms of learning, socialization, knowledge creation and sharing are nurtured. For example the ePortfolio open educational resource Mahara could be used, not just to showcase knowledge acquisition, but more important to promote autonomous learning, collaborative learning, individual and collective reflections and the nurturing of interactions within social networks. The different processes are the main outcomes of using the educational resource. The same principle can be applied to MOOCs in which interactive experiences beyond the acquisition of domain knowledge characterized the learning experience. Process-oriented pedagogical models, such as those proposed by Bonanno 2011 & 2015, that organize learning and its assessment along different levels and dimensions of interactions, are more appropriate for using and evaluating OER. Consequently, OER can be categorized according to the pedagogical strategies they can mediate and according to the competences they can develop.

The Framework for 21st Century skills describes how learning in core subjects and 21st century themes should not be the ultimate goal of the educational process, but more to provide the pedagogical context for developing 21st century competences. Three major competence categories are identified: 'Life and Career skills', 'Learning and Innovation Skills' and 'Information, Media and Technology Skills'.

Life and Career skills are manifested through one's ability to navigate the complex life and work environments in the globally competitive information age requiring flexibility and adaptability, initiative and self-direction, social and cross-cultural skills, productivity and accountability, together with leadership and responsibility. Flexibility is manifested through one's ability to understand, negotiate and balance diverse views and beliefs to reach workable solutions, particularly in multi-cultural environments. One's ability to adapt to change is shown through one's adjustment to varied roles, jobs responsibilities, schedules and contexts.

Initiative and self-direction is shown through one's ability to manage goals, work independently and manage own learning. Social and cross-cultural skills are manifested by one's ability to interact effectively with others and work effectively in diverse teams. Productivity and accountability is shown by skills in managing projects and produce results. Leadership and responsibility involve skills for guiding and leading others and for being responsible to others.

The competence for Creativity and Innovation is manifested through critical thinking and problem solving and through communication and collaboration. One's creative and innovative potential is shown through creative thinking, creative collaboration and the ability to implement innovations. Critical thinking and problem solving involves one's ability to reason effectively, use systems thinking, make judgement, take decisions and solve problems. Communication and collaboration involves the ability to communicate clearly and to collaborate with others.

Information, Media and Technology competences are also described by this model through sub-categories of skills. Information literacy includes one's ability to 'access and evaluate information' and 'the ability to use and manage information'. Media literacy involves one's ability to analyze media and to create media products. ICT literacy is manifested through one's skill in applying digital technologies effectively as a tool to research, organize, evaluate and communicate information and to communicate and network with others. The Digicomp framework details the various aspects of digital competence by categorizing 21 competences into the following five major categories: Information, Communication, Content-creation, safety and problem-solving. Each of the competences is described in terms of knowledge, skills, and attitudes.

This competence framework offers the possibility to reflect about OER from a learning outcomes perspective. A resource is evaluated and defined by the type of competence or competences a user develops as a result of interacting with it. This user-centric approach is much more valid and relevant to designing and evaluating learning activities being driven by pedagogical principles and experience. Thus one way to evaluate OER is to consider the pedagogical principles and learning experiences they are capable to promote and support

There are many studies which present the importance of OER using in engineering education. As examples, there are Engineering textbooks along with additional material and a few lecture videos. Offered by professors from prestigious universities from America, Asia, Australia and, of course, Europe. As example, Dr. Kurt Gramoll from University of Oklahoma is offering an eCourse management system features content about Statics, Dynamic, Fluid Mechanics, Thermodynamics, Calculus, and Solid Mechanics [10].

In the same way, a huge number of universities are offering Open Educational Programs, in order to prepare future specialists in different fields. In a recent research about OER, realized in Erasmus + project called "EU-StORE- Creating European Standards for

Open Educational Resources and Open Learning Resources" [11] in 2015, we found different universities from Europa, Asia and America, with very strong educational programs in Engineering education with a very high demand from the target group.

4 Conclusion

We can affirm, the number of OER applications in engineering education will continue to increase and to cover all the engineering domains.

An increasing number of the universities developed programs, libraries with textbooks collections, other supplementary materials, in order to help professors and students, or any potentials to conduct a course and to learn different subjects without costs, with a good connection and communication between students and instructor/professor. For students there will be a very easy way to learn from across multiple locations, platforms, or formats.

The development of OER and OER initiatives will facilitate the teaching process and will help teachers to be more adequate to the students' needs. And finally this will increase the learning motivation for the students.

References

1. A. Camilleri, U. Ehlers, J. Pawlowski, State of the Art Review of Quality Issues related to Open Educational Resources, retrieved from http://www.pedocs.de/volltexte/2014/9101/ (2014)
2. B. Ischinger, *Giving Knowledge for Free: The Emergence of Open Educational Resources*. (OECD, 2007)
3. S. M. Johnstone, Open Educational Resources Serve the World, Educause Quarterly **3** (2005).
4. https://creativecommons.org/, accessed in 09.02.2017
5. W. Strauss, N. Howe, *Generations* (Quill, New York, 1991)
6. C. S. Duse, D. M. Duse, The European Proceedings of Social & Behavioural Sciences EpSBS (2016)
7. Sparks & Honey at http://www.slideshare.net/sparksandhoney/generation-z-final-june-17/8-They _ are _ mature_and_in, accessed in 24.10.2015
8. D. A. Rothman, Tsunami of Learners called Generation Z, available at http://www.mdle.net/Journal/A_Tsunami_of_Learners_Called_Generation_Z.pdf, accessed 6.06.2016
9. P. Bonanno, Ubiquitous Learning: A Survey of Applications, Research and Trends (2011)
10. http://guides.ou.edu/oer/engineering, accessed in 13.03.2016
11. http://eustore.eduproject.eu/, accessed in 16.09.2016

Benefits of entrepreneurship education and training for engineering students

Valentin Grecu[1,*], and *Calin* Denes[1]

[1]Lucian Blaga University of Sibiu, Department of Industrial Engineering and Management, 550024, 10 Victoriei Bd, Sibiu, Romania

Abstract. In the present economic situation, having knowledge of an academic subject is no longer sufficient for a new graduate. Students are increasingly required to have skills and abilities which will increase their employability, such as: the retrieval and handling of information; communication and presentation; planning and problem solving; and social development and interaction. Entrepreneurial education and training provides individuals with the ability to recognize commercial opportunities, self- esteem, knowledge and skills to act on them. It includes instruction in opportunity recognition, commercializing a concept, managing resources, and initiating a business venture. It also includes instruction in traditional business disciplines such as management, marketing, information systems and finance. Entrepreneurs or the move towards self- employment is, and will continue to become, an increasingly important element of economic growth and development. It is essential to have the infrastructure required to facilitate entrepreneurial mind-set and encourage self-employment. Having a culture of the creation of a new enterprise is a critical aspect of this infrastructure, as it will encourage students to take the risk of starting a business. The purpose of this paper is to describe the design and introduction of the entrepreneurial mindset for engineering students.

1 The Challenges of the Global Economy

Discoveries and inventions, revolutions and social movements have been the triggers of progress throughout history. Mankind has to face new challenges in the 21st century such as globalization, the rapid pace of innovation, the fast spread of technology and its high speed adoption in our lives [1]. These factors and many others are changing not only how businesses and economies are functioning, but also the job market landscape. The knowledge and skills required by the present and future jobs are changing and consequently the education system at all levels has to respond and adapt to the new challenges [1].

Cheaper transportation and communication facilitate greater mobility and digital technologies support long distance exchange of large amounts of information, thus transforming the global labour markets. International migration used to be a one-way process, but especially for the ones with technical skills, it has become a reversible choice.

* Corresponding author: valentin.grecu@ulbsibiu.ro

Scientists and engineers from developing countries are now able to work on increasingly complex tasks and collaborate with counterparts located at great distances and therefore they can contribute to their home economies while having social and professional connections in more advanced economies, without being forced to choose between settling abroad and returning home to far less attractive professional opportunities. Some even become 'transnational' as they work, and even maintain residences and citizenship, in more than one nation [2].

Given the dynamic economic context marked by profound and rapid changes, higher education institutions can no longer equip their graduates with all the knowledge and skills needed to perform in their professional life. Knowledge and skills needed for effectively functioning in a complex global world may be changing. In the information age, industries will have dynamic needs. According to Tony Dolphin [3], these drivers for change will continue "to cause dislocations and disruptions in the labour market", so that there will continue to be "fundamental shifts in the types of jobs that will be available and skills demanded by employers". The new jobs that will be created in the future will require individuals with "entrepreneurial, scientific, creative and emotional skills" [1,3].

2 Understanding Entrepreneurship

The link between economic growth and entrepreneurship has been highlighted by many studies and facts and this connection can easily be demonstrated also by common sense, economic observation or just by simple intuition: entrepreneurship is based on activities that convert ideas into economic opportunities. As a source of change and innovation, entrepreneurship boosts economic competitiveness and increases productivity [4].

The increasingly globalized world economy challenges organizations to increase competitiveness, productivity, flexibility and knowledge, factors that are closely associated with entrepreneurship. The OECD countries have experienced a shift in the structure of the industry between the late '70s and early '90s, moving from concentration towards decentralization, indicating the development of entrepreneurship and the entrepreneurial mind-set. The assumption that supporting entrepreneurship is closely related with fostering a country's competitiveness appears today more valid than ever, given the technological change and the intensified global competition brought about by globalization and economic liberalization.

The fact that entrepreneurship is not a static phenomenon, but rather a dynamic process is demonstrated by most economic, psychological and sociological research on this topic [5]. Entrepreneurship is usually associated with issues related to decision making and represents more than just a mechanical economic factor.

The functional role of entrepreneurs is often highlighted in the existing definitions of entrepreneurship [6] and refer to decision making, allocating resources, coordination, supplying capital, bearing uncertainty and innovation [7-8]. Major schools of thought frequently associate these functional roles of the entrepreneurs with entrepreneurship:

- The Cantillon or Knightian entrepreneur is associated with risk seeking and is willing to take the risk associated with uncertainty
- The Schumpeterian entrepreneur is associated with innovation because he/she accelerates the generation, dissemination and application of innovative ideas
- The Kiznerian entrepreneur is associated with opportunity seeking, as he perceives end seizes new profit opportunities [9-10]

Entrepreneurship has undeniably impacted the world of business forever and has taken hold across the globe [11]. The powerful emergence of entrepreneurial activity in the U.S. in the past decade is just one example that supports this statement. Hundreds of thousands of small firms have been established each year. According to Kuratko [11], millions of new jobs have been created and for 94% of these new jobs are responsible 15% of the fastest-growing new firms (i.e., "gazelles") and less than one third of these gazelles was involved in high technology [11]. Sixteen percent of all U.S. firms have been in existence for less than 1 year. Sixty-seven percent of all new inventions are created by smaller firms [12].

These facts prove that entrepreneurial firms are part of the process that re-defines the market economies. These small and medium sized enterprises play a crucial role in the innovations that lead to the growth of productivity and the changes in technology that redefined the market structure and therefore the competition. The market economies are always in the process of "becoming" as they are dynamic organic entities, being oriented towards the future, not focusing on the inheritance of the past [13].

Millions of people, including minorities, women and immigrants enter the economic mainstream through the entrepreneurial firms that enable them to access the pursuit of economic success. Entrepreneurship is crucial in this evolutionary process, playing an indispensable role of providing the "social glue" that connects "Main Street" activities and high-tech innovation [14].

3 The Role of Education

Education is the most effective means available to society to challenge the future [15]. Progress depends increasingly more on the capacities to research, innovate and adapt of the new generations. Without education, youth participation in the cultural and socio-economic life is impossible. Education will obviously not solve all the problems that humanity faces today, but it is essential in the effort to connect the members of the society, generate new relationships and respect to environmental needs [16].

School instruction, or formal education is not everything. Education also values the role of family and community and includes non-formal and informal sides. The immense community of educators is not being efficiently used to develop the entrepreneurial mind-set for the younger generation, as they are a crucial human resource whose contribution can be useful in all local communities.

Entrepreneurship education has become popular for many reasons [17]. Learning about developing business plans and creating a company allows students to better understand and integrate finance, economics, accounting, marketing and other business disciplines, offering them an integrative and enriching educational experience. Entrepreneurship education encourages the founding of new businesses by students and alumni and equips them with critical decision-making skills that enhance the success of graduates in the job market. Furthermore, the entrepreneurial mind-set increases the transfer of technology to the market, from the university, through the development of technology-based business plans and student involvement with technology licensing. Finally, entrepreneurship education creates links between the academic and business communities. Education for entrepreneurship is considered a useful, applied approach to the study of business and the economy.

All these reasons are solid arguments for established entrepreneurs to support and fund entrepreneurship programs in universities and to get involved in developing the entrepreneurial eco-system in the community where they activate. Entrepreneurship offers also the possibility to experiment with pedagogy and curricula, as it is generally outside traditional discipline boundaries, allowing guest speakers and more practical and applied tasks. These experiments have also impacted and enhanced other business related courses, thus improving the experience of students.

4 The Entrepreneurial University

As argued by Etzkowitz, Webster, Gebhart, and Terra [18], there is an increasing trend observed in universities around the world that aim to become entrepreneurial universities, incorporating the role of active contributors to the development of new companies in their local communities, in addition to their traditional role of scientific knowledge creators and educators Therefore, through their closer and more intense interaction with private industries and government institutions, universities become an increasingly important component of the national innovation system [19].

Interest for entrepreneurship education is more and more present among individuals outside the fields of business and engineering, although at first entrepreneurship education was firmly connected with these two fields [20]. Entrepreneurship education should be seen from a university-wide perspective and the increasing interest for this topic in a broader set of fields, combined with the enthusiasm of alumni from all fields to introduce a real-world dimension to their home schools encourages the implementation of the entrepreneurial mind-set throughout the whole campus. The gap between real-life situations and the classroom theories and concepts can be bridged with the help of entrepreneurship education for students with different specialization fields. Bringing students from non-business fields closer to the business school is the approach of some institutions that aim to educate students in an entrepreneurship classroom characterized by diversity. A different approach of university-wide entrepreneurship education consists of presenting lessons of entrepreneurship from within a specific field, thus giving a message relevant to the field itself.

As Streeter, Jaquette and Hovis [20] describe, there are more approaches of entrepreneurship and entrepreneurial education in universities. Streeter et al [20] recommend a radiant model of an entrepreneurial university, based on an extensive research. Such a model involves all stakeholders, being a more complex body of politics, academics, and finances. In order to move forward such a complex organism, the program leaders must coordinate and adjust the self-interests of individual stakeholders, thus creating a radiant program is a process that takes time. Even if the effort of each academic unit is only local, they must perceive that they can lay claim to the larger university-wide program.

However, the radiant model is rather difficult from an academic perspective. Finding justification in terms of curriculum and availability of the faculty to have entrepreneurship classes in non-business fields is challenging. The importance given to the bridge between the work world and the education and preparation phase is crucial for this justification. Not all majors accept that business education is useful and challenging for the graduates of those study fields. Therefore, the magnet model presented by Streeter et al [20] will be more likely to be maintained in higher education institutions that don't see entrepreneurship as rigorous outside business and/or engineering majors.

The model proposed in figure 1 focuses on a triple approach of the entrepreneurial university: an internal approach, an external approach and an operational one. The model highlights the importance of each of these three dimensions, acknowledging the need of having the appropriate infrastructure and creating a space for students to meet, interact and foster creativity and innovation. The external approach recognizes the role of the community in building an entrepreneurial ecosystem and the symbiotic relationship that the university should have with the business environment and the local community. The third approach described in the model, namely the operational one, is itself divided into three components that summarize the activities of a university: research, formal and informal teaching and learning.

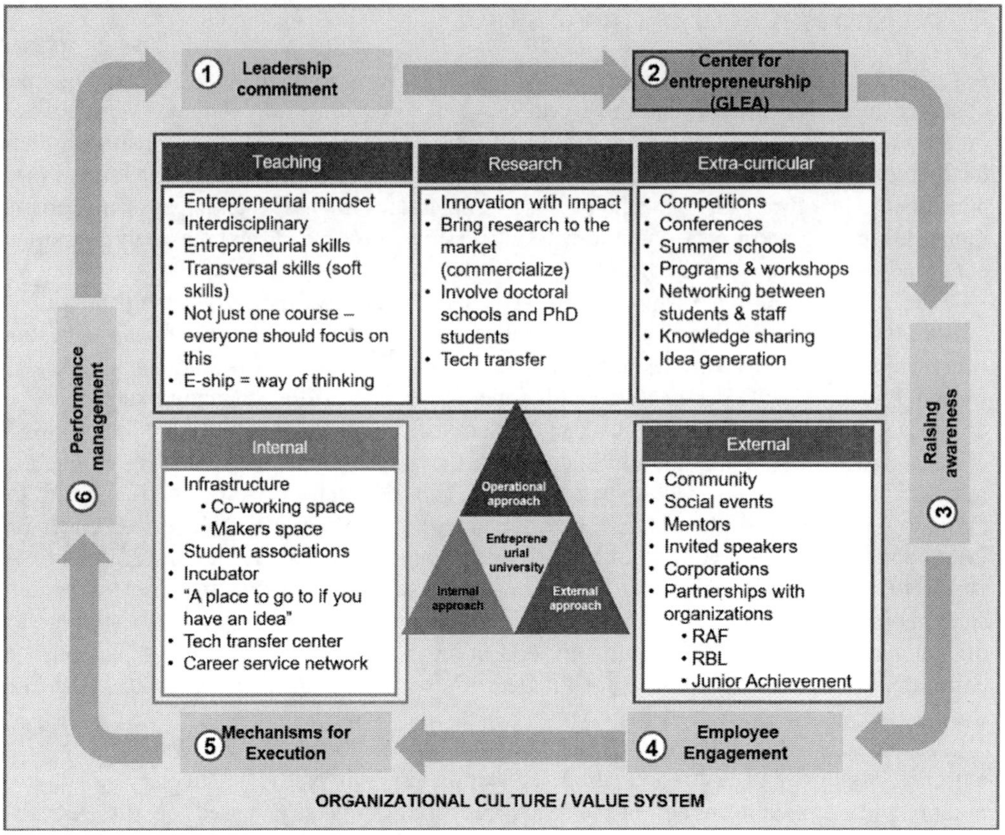

Fig. 1. Model of the entrepreneurial university.

The model presents also the steps that need to be followed in order to implement the entrepreneurial mind-set across the campus for achieving the radiant model of an entrepreneurial university. It all starts with the commitment of the leadership (university administrators). The following step is creating a structure that will coordinate and monitor the implementation of the measures needed to transform the university into an entrepreneurial one. Raising awareness of the importance of entrepreneurship, both for the economy and for the future of the university is an ongoing process and it should target students, alumni, faculty, administrative staff and the whole community and business environment. Engaging the employees in this process is a difficult task, but it is essential that everyone understands the short and long-term necessity of commitment towards entrepreneurship education. The mechanism of execution refers to offering all the stakeholders the necessary means to implement the plan to move forward towards a university-wide approach of entrepreneurship.

5 Conclusions

Regardless the approach, it is widely recognized that it is increasingly necessary for students to have transversal skills which will increase their employability. Entrepreneurial education equips students with abilities that increase their employment potential and include: the abilities to solve problems, to develop social interaction, abilities to find information and to handle it for decision making, planning, communication and presentation skills, etc. Entrepreneurial education and training provides individuals with the ability to recognize commercial opportunities, self- esteem, knowledge and skills to act on them. It includes instruction in opportunity recognition, commercializing a concept, managing resources, and initiating a business venture. Traditional business disciplines such as marketing, management, finance and information systems are also taught. Thus, the necessity of entrepreneurship education is undeniable.

Individuals must continuously improve their knowledge and skills as a response to today's rapid pace of change, otherwise economic stagnation will be in place. Training organizations, such as universities, find it difficult if not almost impossible to correlate their capacity to adapt the education and curricular offering with the pace of change in labour market needs. In a global, dynamic world it becomes more important that skills taught in schools are relevant for the workplace and are maintained and improved during the working life. Individuals can't afford anymore to stop their formal education once entering the labour market. Building a solid foundation of skills during school years and providing lifelong learning is probably the viable solution to reduce the skill gap [1].

A long-term strategy that involves governments, employers and educational institutions (like universities) is needed in order to reduce skill mismatch effectively. Only bringing together education and the working world can guarantee success. Individuals have to enter the job market with an appropriate, balanced mix of knowledge and skills that will allow over the professional life further education in order to cope with the new demands. There should be coherent guidance in choosing a career path, for a smooth transition from school to work and for skill improvement over the working life, to better exploit the skills of individuals and meet the needs of employers.

References

1. L. Ilie, I. Bondrea, Changing Labour Market Needs and the Challenges for Academic Leadership. In: ECMLG 2016-Proceedings of the 12th European Conference on Management, Leadership and Governance, **80** (2016)
2. A. Saxenian, International Mobility of Engineers and the Rise of Entrepreneurship in the Periphery. Research Paper, UNU-WIDER, United Nations University (UNU) 2006
3. T. Dolphin, (ed), Technology, globalization and the future of work in Europe: Essays on employment in a digitised economy, IPPR. http://www.ippr.org/publications/technology-globalization-and-the-future-of-work-in-europe (2015)
4. S.E. Sanyang, H. Wen-Chi, International Entrepreneurship and Management Journal **6.3**, 317-329 (2010)
5. A. Pirich, An interface between entrepreneurship and innovation: New Zealand SMEs perspective. Paper prepared for the 2001 DRUID Conference, Aalborg, Denmark (2001)
6. H. Barreto, (1989). The Entrepreneur in Microeconomic Theory: Disappearance and Explanation. London: Routledge.
7. C. Friijs, T. Paulsson and C. Karlsson, Entrepreneurship and Economic Growth: A Critical Review of Empirical and Theoretical Research. Östersund, Sweden: Institutet för tillväxtpolitiska studier (2002)
8. M. Jääskeläinen, Entrepreneurship and Economic Growth. Helsinki: Institute of Strategy and International Business (2000)
9. Organisation for Economic Co-operation and Development (OECD), Fostering Entrepreneurship. Paris: OECD (1998)
10. M. Carree, and A.R. Thurik,, International Handbook of Entrepreneurship Research, Boston/Dordrecht: Kluwer (Academic Publishers, 2003)
11. D.F. Kuratko, *The emergence of entrepreneurship education: Development, trends, and challenges. Entrepreneurship theory and practice* 29.5, 577-598 (2005)
12. P.D. Reynolds, M. Hay, & S.M. Camp, Global entrepreneurship monitor. Kansas City, MO: Kauffman Center for Entrepreneurial Leadership (1999)
13. D.F. Kuratko, & R.M. Hodgetts, Entrepreneurship: *Theory, process, practice*, (Mason, OH: South-Western College Publishers, 2004)

14. Small Business, Administration,The new American revolution: The role and impact of small firms. Washington, DC: Small Business Administration, Office of Economic Research (1998).
15. UNESCO, Higher Education in the Twenty-first Century Vision and Action, World Conference on Higher Education, Paris (1998), retrieved 20.09.2010 from: http://unesdoc.unesco.org/images/0011/001163/116345e.pdf
16. C. Bucovală, Dezvoltarea durabilă a Mării Negre, (2009) retrieved 3.12.2010 from http://lefo.ro/iwlearn/curriculum.html
17. A. Charney, and G.D. Libecap. Impact of entrepreneurship education. Kansas City, MO: Kauffman Center for Entrepreneurial Leadership (2000)
18. H. Etzkowitz, A. Webster, C. Gebhart, B.R.C. Terra, Research Policy, **29** (2), 313–330 (2000)
19. W. Poh-Kam, Y.P. Ho, and A. Singh, World Development 35.6 (2007), 941-958 (2007)
20. D.H. Streeter, J.P. Jaquette Jr, and K. Hovis, University-wide entrepreneurship education: Alternative models and current trends. No. 127271 (2002)

Teaching issues of contemporary history using historical sources and modern teaching methods

Gabriela Gruber[1,*]

[1]Lucian Blaga University of Sibiu, Sibiu, Romania

Abstract. The study of history is becoming increasingly less interesting to students, despite the fact that the history teaching process has been continuously modernized during recent years. It is an observation which can be perceived even if we don't make an elaborated research in the field. Some empirical data show us that students in secondary and High Schools are less interested in studying History than in studying Geography or other social sciences. The number of students who are determined to study History in universities has significantly dropped in recent years [1]. Of course, there are multiple causes and the factors behind this change are numerous and varied. In this paper we handle only some changes in teaching History in High Schools, as they are designed in History Curricula and in History textbooks. Therefore during the first sequence of this paper we shall analyze the History Curricula for High School, 11th and 12th grades, regarding their finalities (competencies), some relevant contents and the recommended pedagogical approaches about the teaching methods and the auxiliary material. In the second part of the paper we propose some teaching activities through which students would practice the specific competencies from their Curriculum for History. We aim at presenting attractive teaching material and learning methods and applying the methodological recommendations from the High school Curricula for History, 11th and 12th grades.

1 A short analysis on the 11-th and the 12-th grades Curricula for History

The Romanian education reform has led to major changes in teaching history. School programs have received a curricular structure for all school subjects. Thus, they include an introductory note, general and specific competencies, contents, values and attitudes and methodological suggestions. Competencies are defined as knowledge, abilities and attitudes that are assimilated by students during the years of studying a specific school subject. The general competencies listed in the History Curricula for High School are: the efficient use of specific historical terminology in communication; the practice of democratic approaches and civic actions; the application of the principles and appropriate methods to address

*Corresponding author: gabriela.gruber@ulbsibiu.ro

historical sources and the use of resources that support lifelong learning. [2] The specific competencies are those finalities in studying History to which every student has to arrive from one year of study to another. Thus, the specific competencies are different from one class to another, becoming more complex and more diversified. A comparative presentation of these specific competencies in History study, from the 11th to the 12th grades is presented in the table below:

Table 1. General and Specific competencies from Curricula of History, High School, 11th and 12th grades (High School Curricula for History, 11th and 12th grades, Bucharest, 2009)

General competencies	Specific competencies	
	11th grade	**12th grade**
1.Efficient use of specific historical terminology in communication; practicing democratic approaches and civic actions	1.1 Formulating written and oral opinions on a topic of history; 1.2 Developing an oral or written argumentation ; 1.3 Comparison on different opinions and arguments regarding an historical issue. *1.4 Supporting an argued point of view in a discussion/ referee on topics of history	1.1 Building explanations, intra and multidisciplinary arguments about historical events and processes; 1.2Use of terms /concepts specific to History in contexts involving interdisciplinary interpretations and explanations
2.Practicing democratic approaches and civic actions	2.1 Knowing and practicing the values of democratic citizenship; 2.2Analyzing the institutions, norms and procedures of governance 2.3 The use of strategies of negotiation and civic cooperation	2.1Building analytic approaches involving economic, social, political or cultural contexts; 2.2 Designing cooperative approach to identify and develop common goals; 2.3Discovering the permanency in conducting historical phenomena 2.4Comparison and assessment of different reasons to formulate their own judgments.
3.Applying the appropriate principles and methods to address historical sources	3.1 Selecting and commenting historical sources to support/combat a point of view 3.2Comparing the relevance of historical sources in approaching a subject 3.3 The discovery of different perspectives on historical events and processes in sources of information	3.1 Comparison of historical sources to establish their credibility and validity; 3.2 Analyzing the messages sent by different historical sources through comparison the terminology they are using
4.Using resources that support lifelong learning	4.1 Designing a research with an historical topic 4.2 Using resources and technologies of information and communication to investigate an historical event or process	4.1Discovering opportunities in history research as a source of permanent learning; 4.2 The integration of knowledge obtained from non-formal learning environment in analyzing historical phenomena; 4.3 Analyzing similar, opposite and complementary points of view related with the historical phenomena that are studied; 4.4 Making connections between information provided by historical sources and the context of their daily life.

These competencies are related with complex History contents during the 11th and the 12th grades. The Curricula organizes History contents differently from the inferior to the superior curricular cycles of the High School. Thus, the contents are chronologically organized during the 9th and the 10th grades, but thematically during the 11th and the 12th grades. That means that during the 11th and the 12th classes students have to synthesize their previous knowledge of History and have to approach it from multiple points of view. In order to develop the competencies above, students have to express their opinions sustained with facts, logics and evidences, they have to argument and to compare events and processes. They also have to prove competencies in using technology for information.

The Curricula for History organizes the contents in greater areas, such as "Peoples and historical places", "People, Society and the World of Ideas", "The State and the Politics", "International Relationships" or "Religion and Religious Life". Therefore, the themes are complex, aimed at integrated approaches of History of the world, both that of Europe or the national one. Contents often provide controversial and sensitive topics [3-4]. Some examples for the 11th class curriculum might be "Europe and the World during the 20th century", "The Economy and Society during the Postwar World", "Cooperation and Conflict", or "Religion in the Contemporary World". The 12th grade curriculum for History proposes complex subjects, but this time the focus is on Romanian history, integrated in the European one. "Discoverers of new places and cultures during the 15th and 19th centuries", "Romanian travelers at home and in the World", "Cultural Policies and Romanians abroad", "Redefining the roles of the State from the First World War to Schumann Plan" are some relevant examples. Such organization of the content in studying History in High school has been recently criticized by important representatives of The Romanian scientists, such as Ion Aurel Pop, who appreciated that the curricula for History does not comply with "the specificity (essence) of the discipline, meaning its diachronic and the placement of historical events in space" [5].

Methodological Suggestions end the Curriculum for History for every year of study. They focus on the new role of the student, that of learner through cooperation, or learner in formal and non-formal contexts. A nuanced methodology of using and interpreting historical sources, depending on the formal type of the source, the significance and techniques of applying the "multi-perspective" concept are some of the most important methodological aspects that must be in the attention of a modern teacher of History.

The textbooks for History in High school are not very well adapted to the newest requirements of Curricula for History. The textbooks provide the texts of the lessons and the teaching material that allow modern teaching approaches. But the applying exercises do not explicitly target the competencies provided by the Curricula. No matter what class they were created for, the text books include images, fragments of historical texts/documents, historical maps, statistical tables or charts, chronological landmarks or vocabulary explanations. Since there are alternative textbooks, the contents of the lessons are very different. However, a textbook is an auxiliary material for both teacher and student. The teacher may be creative with his teaching design, by using his own teaching material and methods that enable his students to develop the competencies required by the Curriculum. Students, on the other side, have to "create their own new understandings on the basis of an interaction between what they already know and believe and ideas and knowledge with which they come into contact" [6]. Thus, the constructivist pedagogical paradigm could complete the traditional teaching process and make the learning process more attractive and interactive. Some trends that are common to such a way of teaching, as synthesized by research, could consist of: attention to the individual and respect for students' background, facilitation of group dialogue that explores an element of the domain; planned or unplanned introduction of formal domain knowledge into the conversation through direct instruction, reference to text, opportunities for students to change or add to existing beliefs and understandings through engagement in tasks that are structured for this purpose [6].

2 Constructivist Teaching Activities for Exercising Specific Competencies through Topics of World or of Romanian History

The considerations above are the basis from which we begin the teaching design of some learning activities that are aimed to be attractive and efficient for students. Our target is that the designed teaching activities involve the students so that they are able to build their own knowledge and achieve the competencies required by their Curriculum for History. The teaching and learning activities are designed for the lesson "The World after 1989", 11th class. The topics which interest us from the lesson concern the most important events during the revolutionary year 1989, the main causes of the revolutions and the principal changes in the world after 1989. The designed teaching activities are aimed to achieve the specific competencies 1.1; 2.1; 3.2; 3.3 and 4.2 from the table above. As an introduction, the teacher will project for 6 minutes two fragments of the films entitled "The Revolutions of 1989" [7] and "Romanian Revolution 1989. What really happened" [8]. Before starting the projection, the teacher distributes working individual sheets to students, explaining them which are the requirements. While watching the videos students have to note: 1) the events that are described in the clips; 2) the immediate consequences of the revolutions, which they can observe in each situation. When the projection is ended the students are asked to present their answers to the tasks on the cards. They can complete each other if necessary. Through this activity, students prove the specific competence 1.1 (Formulating oral opinions on a topic of history), using discovery as a teaching-learning method.

For the second activity, students are working frontally. They are asked to read silently the document 2 from their History textbook, "About dignity in dictatorial regime" (Michnik, 1997). Then they have to find the link between the content of the document and the topic of the lesson. The expected answer should reveal that the violation of individuals` dignity, violence, torture and terror were the most important causes of the 1989 uprisings against dictatorial regimes in Eastern Europe. The teaching methods that are used in this case are conversation and debate and by interpreting this text students are practicing the competence 2.1 from their Curriculum for History, (knowing and practicing the values of democratic citizenship).

To discuss the main changes that have occurred into the world after 1989, the class will be divided into four groups. Each group will receive a working sheet with different images and information. Students from each group have to answer to the following tasks: Group 1 receives a sheet that contains a map with the former Soviet states and information about the Baltic States and about the Eastern European ones; Group 2 receives the same map, but information about Transcaucasian Republics and about The Central Asian States. The two groups have to identify and to locate the groups of states that they hold information about on the map and exchange information about their population and their about religions. Group 3 receives the map of the former Yugoslavian states and a text entitled "Sloboda Milosevici's Radical Opportunism" (Tismaneanu, Daily, March, 31, 2006). The students in that group have to identify factors that led to the disintegration of Yugoslavia and locate the new states formed after the war in these territories. During discussions students have to understand and use specific terms such as "mafia group", "extreme nationalism", "hegemonic party", which they come into contact with during their lecture. Group 4 receives a fragment of the text "The Century of Extremes" (E. Hobsbawn). The working tasks for the students in this group are: to explain the significance of the apartheid policy in South Africa, as they can understand it from the analyzed text (unlimited power of a minority over a majority); to argument the failure of the apartheid policy and use the ideas from the text in their argumentation (the brutality of the governors, ignorance of the political levers of Government, the lack of legitimacy of the political regime). After the groups have presented their answers, students are asked to compare the sources that they have analyzed and appreciate which one is the most relevant in studying the changes in the world after 1989. They have also to argument their selection. By fulfilling these tasks students are practicing the specific competencies 3.1 and 3.2 from their Curriculum for History. Thus, they comment historical sources to support/combat a point of view (3.1) and compare the relevance of historical sources in approaching a subject (3.2). Through these activities students use discovery, group-working and debate as teaching and learning methods and contribute at constructing their knowledge.

The learning activities for the 12th class are centered around the lesson titled "Building the Post revolutionary (Romanian) Democracy". The issues of the lesson which we design teaching activities for refer to: political parties and political groups appeared in Romania during the year 1990; new principles in organizing the State; conflict relationships between political parties and groups in Romania after 1989 and features of the Romanian democracy during transition. As an introduction, students are asked to read the chronological landmarks from their textbooks. Then the teacher distributes working sheets extracted from the Decree-Law No 8/31.12.1989, concerning the legalization of a multiparty political system. The tasks written on these cards are: to identify the political group that has signed the analyzed Decree-Law; to name the principle that has been set at the basis of this document. During this activity students are working individually and aim to achieve the competence 2.1 from the 12th class Curriculum for History, (building analytic approaches involving economic, social, political or cultural contexts). Thus, students discover that the Decree- Law was signed by The Front of National Salvation (FSN) which was the political group that had taken power. They also find out that the document legalized the principle of political pluralism.

The next activity is meant to help students to find out or renew their knowledge about the relationships between political parties born in Romania during 1990. The teacher organizes the class into four groups and gives them working sheets containing information about the result of the Romanian elections in May 1990. Then, a short fragment from the film "23 years from the First Elections Held after 1989 Revolution" [9] (in Romania) is projected by the teacher. Students are asked to establish which were the political parties and groups participating in the 1990 elections. Watching the film and using the information on their sheets, students are expected to show that Ion Iliescu, the candidate of the Front of National Salvation won the elections, while Radu Campeanu (NLP) and Ion Ratiu (NPPCD) lost them. The main political parties that students could identify from their working sheets are FSN (The Front of National Salvation), UDMR (Democratic Union of Hungarians in Romania), PNȚCD (The National Peasant Party Christian and Democratic), PSDR (The Social Democratic Party in Romania), etc. Students are also asked to discuss in their groups and argument economically, politically or culturally to explain the outcomes. To do this, students have to use specific historical terms in contexts involving interdisciplinary approaches. Thus, they achieve the specific competencies 1.2 and 2.1 from their Curriculum for History.

The characteristics of the Romanian democracy during the years of transition are discussed during the next teaching activity. Students receive working sheets with two texts to be analyzed. The first text contains a fragment from the "Proclamation in Timisoara", point eight, calling for prohibition of the former communist leaders or former Security officers to participate in public life. The second text represents a short fragment written by Timothy Garthon Ash in his "Truth about the revolutions in 1989", published in 1999. The text on students` sheet defines the "civil society" as "organizations which aren't controlled by parties or by a State party". The class is divided into groups of four that have to choose the text which they will analyze during the next minutes. Organized in pairs, students in each group have to write 4 minutes for and against depositions about the justice of the eight point of Proclamation in Timişoara, or about the necessity of the civil society in Romania after 1989. After 4 minutes, pairs split up, seeking for a fellow from another pair in the classroom with the same type of arguments and the same analyzed document. The new pairs work together for another 5 minutes and share their arguments. Then, the initial pairs reunite and bring their enriched arguments into the initial groups. For six minutes the two pairs in the group debate their problem based on the opposite arguments which they have produced by cooperation. Thus, students practice the 4.3 specific competence from their Curriculum for History (Analyzing similar, opposite and complementary points of view related with the historical phenomena that are studied), learning by cooperation and using the so called academic controversy as a teaching and learning method.

The last teaching activity proposed by this paper occurs with the so called phenomenon of "Miners", which took part in Romania during 1990. The Romanian society was dividing into two groups: those who wanted a major change in the governance policy towards a real democracy and those who wanted to maintain the power of the survivors of the communist regime. The miners, a group of professionals from the mining Romanian counties came to Bucharest to "restore order" or to spread the ones who protested peacefully in the University Square. Violence resulted in deaths and injuries. The teaching activity which we are proposing is aimed at making students practice the specific competences 2.4 and 3.1 from the 12th grade Curriculum for History (Comparison and assessment of different reasons to formulate their own judgment and Comparison of historical sources to establish their credibility and validity). For this activity, the teacher prepares four texts with different opinions of the participants at the events in Bucharest in the June 1990. The first text represents the views of a man in the street who was aggressed by two miners, was hit in the face and fell down, losing his consciousness. The second text represents the opinion of a miner who participated in the events. He declares that there were drugs on the street, without knowing for sure how the drugs looks like. The third text represents a declaration of Miron Cozma, the miners` leader, who said that he was called "by someone important at Cotroceni" and was asked to go to Bucharest. Miron Cozma also said that he was expected by Gelu Voican-Voiculescu, the vice-president of the Government of that time, whom he had a discussion with. The fourth text presents a fragment from an article in the Romanian journal, "The Truth". The author describes a violent scene in the Police headquarter in Bucharest. Into a fire burned decor, the author presents some miners` victims, describing them as individuals full of blood, drugged and hysterical. Each of the four groups of students receives a sheet that contains two of the four texts, differently combined. The different combination of the texts is meant to allow groups to make various comparisons between the perspectives of the stakeholders, but also between the credibility of the sources. The working tasks are the same for all the groups:1) to compare the opinions related by the documents and appreciate the credibility of the sources; 2) to sustain their points of view, based on the ideas expressed by sources but also on the author's role in the events that they are describing. These activities cover only parts of the contents of the cited lessons. We chose to design these teaching activities as examples of possible good practices in achieving the requirements of the Curricula for History by valorizing interesting teaching material and modern teaching methods.

3 Conclusion

The reform in the Romanian education determined important changes in the Didactics of History. The disciplinary competencies, the new approaches of the contents assimilated by students, the new methods and teaching material allow the teacher of History to be creative in designing his teaching activities. Some aspects of teaching History in High schools supported major criticism from specialists in History. The value of some competencies from the Curricula for History is criticized, contents are sometimes poorly organized, the textbooks are old or too crowded with information. The textbooks may have few teaching materials, which are not always the most compatible with the competencies required by the Curricula for History. However, a careful teaching design, a proper interpretation of the different aspects of the contents and a judicious correlation between the contents and the competencies lead to attractive teaching activities, correctly oriented towards the competencies in History and towards the correct historical contents. Students learn by cooperation, they can use information assimilated from non-formal learning and build their knowledge together. But they need well selected teaching material, modern and active learning scenarios which can be designed and realized by modern and interested teachers, with effort and creativity.

References

1. Ministry of National Education and Scientific Research, *Report on the status of Higher Education in Romania*, accessed 21.02.2017 https://www.edu.ro/sites/default/files/Raport%20Stare%20invatamant%20superior%202015.pdf (Bucharest, 2015)
2. Ministry of Education, Research and Innovation, School Curricula. History, 12th grade, http://www.isjcta.ro/wp-content/uploads/2013/06/istorie_12.pdf., accessed 21.02. 2017
3. R. Stradling, *Teaching the XXth Century History,* (Corint, 2003, 87)
4. S.P. Bolovan, *Teaching History* (Presa Universitara Clujeana, 2007, 85)
5. F. Pop, *Do we need History in Schools*? The Truth, accessed 20.02.2017 from http://adevarul.ro/locale/cluj-napoca/de-nevoie-istorie-scolil-rectorul-ubb-la-tara-matematicafizica-chimie-nu-inteleg-profesorii-ceea-predau-daramite-elevii-1_56c741bf5ab6550cb8fc302f/index.html (2016)
6. V. Richardson, *Constructivist Pedagogy*, in TCR **105(9),** 1623-1626, (2003)
7. https://www.youtube.com/watch?v=hc2XGHA7NK4, accessed 24.02.2017
8. https://www.youtube.com/watch?v=VeXJevm1cQc, accessed 24.02.2017
9. http://stiri.tvr.ro/23-de-ani-la-primele-alegeri-libere-desfasurate-dupa-revolutia-din-decembrie-1989_30707.html, accessed 25.02.2017

The interaction of Universities with high-tech enterprises: commercialization of innovations

Irina Kalinina[1,*], and *Karina* Chernitsova[2]

[1]Department of Theory of Management and Business Technologies, Plekhanov Russian University of Economics in Moscow, Moscow, Russia
[2]Department of Theory of Management and Business Technologies, Plekhanov Russian University of Economics in Moscow, Moscow, Russia

Abstract. The dynamics of innovative development greatly depend on the efficiency of the processes employed in creating and organizing active small innovative businesses. In this area, universities, with their innovative infrastructure and collaborative arrangements with businesses that provide a synergistic effect, play an important role. The goal is to substantiate the recommendations for commercializing the results of intellectual activities using networking mechanisms of interaction between universities and high-tech enterprises. The methods employed include process modeling and ABC-analysis, as well as a systematic approach to the networking of universities and enterprises producing high-tech products. The model for the processes of commercializing innovations with the participation of universities was based on the business design methodology. The article contains an analysis of the development state of small innovative enterprises, as well as prospects for the development of interactive network mechanisms between universities and high-tech enterprises. The article defines the special role played by universities in the process of commercializing innovations; proposes models of classification, and reveals the methods of approach to an organizational and economic model of collaboration between universities and enterprises producing high-tech products. The results of the study are intended for use by heads of universities, small enterprises in the field of innovation, as well as high-tech enterprises.

1 Introduction

The interaction of universities with high-tech enterprises is the primary focus of attention of European researchers who are studying the participation of universities in the dissemination of innovative solutions from different perspectives. First, cooperation between universities and industry is increasingly being viewed as a means to enhance innovation through the exchange of knowledge [1]. In this case, structured relationships are effective in developing the concept of the university-industry-government triple helix [2-4].

In light of this, universities are relevant for high-tech enterprises as a source of

* Corresponding author: karinaaa2004@mail.ru

knowledge and the ability to use scientific research and receive innovation support. In this regard, the work done by Thomas Clauss and Tobias Kesting show a significant effect of the interaction of management mechanisms with the achievement of common goals of universities and enterprises producing high-tech products [5].

The issue of finding solutions to the problem of having organizational borders that may limit the ability of companies to obtain the necessary access to the university R & Ds is also an important aspect of this research. The transmission of the information on the results of the research acts as a demotivating factor for cooperation, especially with respect to networking [6-7]. In this case, public-private partnerships play an important role.

The interaction between universities and high-tech enterprises is economically beneficial to both parties, as long as they are able to obtain reliable information about changes in the university R & D portfolios and identify on this basis the potential for cooperation. In this case, it is important not only to find common interests, but also to organize the monitoring and evaluation of joint research programs [8, 9].

The dynamics of the interaction can be assessed through quantitative analyses based on wavelet transforms [10].

The most promising form of interaction between universities and high-tech enterprises is networking in which researchers act as knowledge experts, carrying out task assignments as consultants and innovation intermediaries. Universities scholars working for practice communities are particularly involved in the development of multilateral initiatives, with the aim of choosing ways and mechanisms of cooperation between different players (universities, research organizations, industrial enterprises, innovation agents and the like) [11].

An example of the role of a modern university in the field of intellectual property management in the form of the entrepreneurial activity of researchers is presented in the form of a synergy model between researches, entrepreneurship and technology commercialization [12].

2 Relevance of the research

The pace of the implementation of innovation depends on the activity of the interaction of economic agents at different stages of the innovation cycle. Higher education institutions that ensure the distribution of their innovations in high-tech enterprises by organizing small innovative enterprises (SIE) play an important role in this process.

SIE activity is an important part of the policy of the Russian Government, which is taking drastic measures to establish and implement the scientific potential of universities, as well as increase the level of scientific and technical development.

An analysis of the distribution of university developments in critical technologies, shown in Table 1, showed that sectors like digital manufacturing, mechanical engineering, nano - and microtechnology account for the major share (95 per cent). This indicates the orientation of university activities to the priority areas of high-tech development (Figure 1).

DOI: 10.1051/matecconf/201712112009

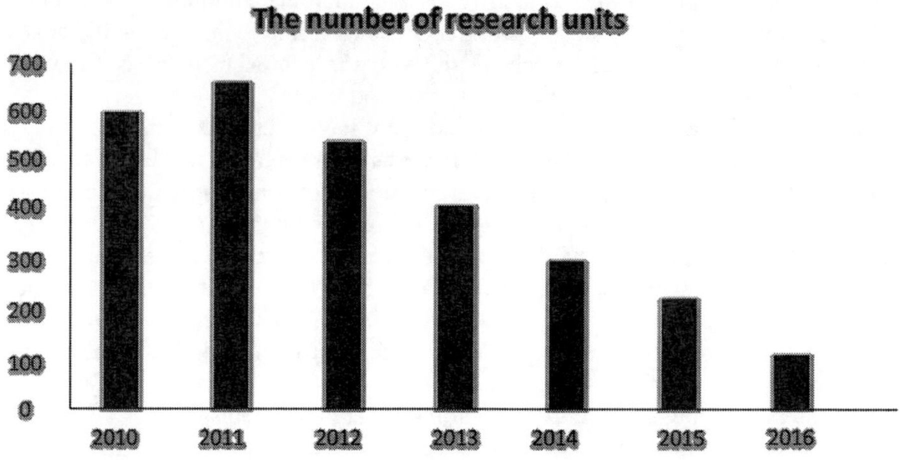

Fig. 1. Distribution of university developments in critical technologies (According to 2015 data)

The indicators for the distribution of university research developments based on stages of readiness for production show that industrial production contributed only 5 percent.

The data indicate the need to strengthen the role of the distribution of innovations in the work of university scientists. (Table 1, Figure 2).

Table 1. Distribution of university developments on critical technologies (according to 2015 data)

Group	Description	Number of developments, units	Percentage of total
1	Digital manufacturing	507	31%
2	Mechanical engineering	363	22%
3	Nano- and microtechnology	283	17%
4	Bioengineering	236	14%
5	Human safety and medicine	141	9%
6	Mineral resource exploration and extraction	80	5%
7	Nuclear power	24	1%
	TOTAL	1634	100%

Calculated by the authors based on the data [13].

The main growth in the creation of SIEs by Russian universities took place in the period between 2010 and 2013. Presently, the growth potential has been exhausted. Moreover, many SIEs were abandoned without having been fully developed, while part of the registered SIEs either failed to proceed to launching into practice, or are currently demonstrating low performances.

In these circumstances, it becomes relevant to analyze the opportunities and find ways to improve the efficiency of higher educational institutions for the dissemination of innovations in cooperation with enterprises producing high-tech products.

DOI: 10.1051/matecconf/201712112009

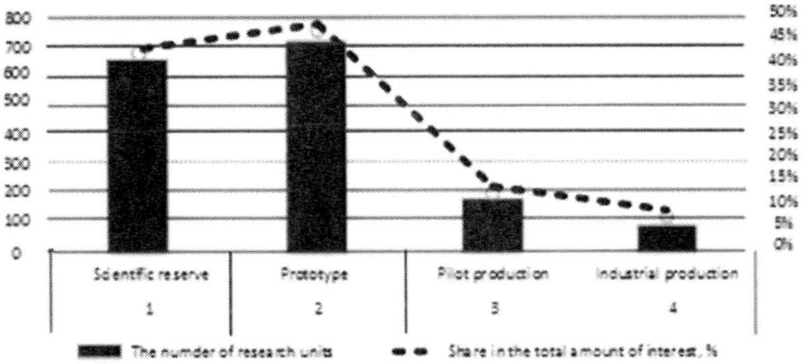

Calculated by the authors based on the data [13].

Fig. 2. Distribution of research developments of universities in stages of readiness for production (according to 2015 data)

The mechanism of interaction between universities and high-tech enterprises is based on a license agreement on the use of the intellectual deliverables (results of intellectual activity), which include the following results: computer programs; databases; inventions; utility models; – industrial designs; selective breeding results; topographies of integrated microcircuits; trade secrets (know-how). Thus, the exclusive rights to the intellectual deliverables (results of intellectual activity) are stored for further analysis by higher education institutions.

Research hypothesis consists of the following: the main problem in the interaction of university SIEs with high-tech enterprises is the insufficient demand for the SIEs' goods or services by industrial enterprises. A new networking model is required.

Research methodology. The methodological basis of the study served as a statistical analysis of the SIE data created within the universities, the systematic and structural analysis of the elements of the mechanism of interaction between universities, SIEs and companies, as well as a systematic analysis of the activities of universities in organizing research and developing innovative infrastructure.

Therefore, recommendations on improving the mechanism of interaction should be directed towards solving this problem, based on the development of university innovation infrastructure, that is, the creation of new units designed to be used in the commercialization of intellectual deliverables:

- technology transfer centers
- R & D commercialization departments
- intellectual property departments
- marketing centers

Such a change in the interaction of universities with high-tech enterprises requires changes in the organization of scientific research itself:

- focusing on purchasing power in the research topics;
- professional development of the levels of intellectual deliverable substantiation;
- increase of the level of readiness of scientific and research developments for practical use.

This contributes to the formation of new institutional models in the Russian scientific and educational sector. First, we note the allocation of national research centers and national research universities that are designed to become key centers of excellence on the global scientific and technological platform.

However, the following range of issues related to the establishment and functioning of SIEs at universities have to be resolved currently:

- the development on the part of the university of special support measures compensating for the significant financial costs at the initial stage of development – lease payments, tax burden, purchase of equipment, materials and so on;
- the development in the university SIE of commercially-viable structural units of innovation infrastructure;
- the formation of a network model of interaction between the university SIE and enterprises producing high-tech products.

The networking model could primarily be focused on the scientific, technical and managerial capacity of the university SIE, which forms the organizational basis as an innovative link for scientific research organizations and companies. Within this context, the SIE shall act as the organizational basis for creating a separate institution (Single Institute Model), or in conjunction with the enterprises (Multi-Institute Model).

Network interaction is aimed towards full-scale implementation by participants of research and innovation projects as part of a long-term joint program to accelerate the commercialization of research developments by universities, with the implementation of the experience of their industrial partners.

The initial principles of effective interaction between universities and a high-tech enterprise should recognize the high commercial activity of the SIE, as well as the self-financing of its activities.

The "elements of success" essential for a networking model are:

- availability of the scientific, technical, technological and resource potential of the networking participants, providing a competitive advantage;
- providing support to the management of scientific organizations;
- gaining access to various sources of financial support at different stages of the distribution of innovations;
- availability of a university leader of the commercialization process, as well as a team of professionals (well-rounded human resources);
- flexibility in the adoption and implementation of solutions, including issues on intellectual property, including its defense within the country and abroad, as well as the development of mechanisms for the commercialization of intellectual property;
- sufficient funding;
- practical experience in commercialization of the participants of the networking.

3 Conclusion

The study produced the following results:

- an analysis of the activities of Small Innovative Enterprises of Russian universities was conducted using an information and analytical system of monitoring the Small Innovative Enterprises of universities;
- an analysis of the interaction of Small Innovative Enterprises of universities and enterprises producing high-tech products was conducted;
- approaches to developing a model based on an example of the interaction of business entities with mechanical-engineering enterprises producing high-tech products were formulated and substantiated.

The research results make it possible to say that the following aspects need to be improved in the interaction process between universities and high-tech enterprises:

a. To stimulate the demand for developments offered by universities to commercialize them by enterprises. For this, first, it is possible to transfer to universities the right to independent disposition of dividends and incomes from participation in joint enterprises. This also applies to the possibility of the high-tech equipment transferring to joint use with universities.

b. To support the innovation infrastructure of universities when preparing applications for various competition participations to attract the public and private investments. This aspect can, by means of partnership with universities, help high-tech enterprises to promote the results of their activities to the market, and promote the realization of their competitive advantages.

c. To develop technologies for the successful technology transfer from universities to high-tech enterprises through spin-off university companies, as well as to improve the transferring technology procedure from universities to industrial enterprises and companies, based on the realization of the right to sell licenses and assign patent rights.

References

1. S. Ankrah, O. AL-Tabbaa, Scandinavian Journal of Management, **31**, Issue 3, 387-408 (2015)
2. V.V. Maslennikov, I.A. Kalinina, Mediterranean Journal of Social Sciences, **6**. № 5 S4. 76-83 (2015)
3. M. Heitor, Original Research Article, Technological Forecasting and Social Change, **95**, June 2015, 276-293 (2016)
4. V.V. Maslennikov, I.A. Kalinina, International Business Management, **9**, 6. 1141-1146 (2015)
5. T. Clauss, T. Kesting, Industrial Marketing Management, In Press, Corrected Proof, Available online 20 September (2016)
6. M. Steinmo, E. Rasmussen, Journal of Business Research, **69**, Issue 3, 1250-1259 (2012)
7. M. Perkmann, H. Schildt, Research Policy, **44**, Issue 5, 1133-1143 (2015)
8. J. Ekboir, G. Blundo Canto, C. Sette, Evaluation and Program Planning, **61**, 64-75 (2017)
9. M. Simeth, J.D. Raffo, Research Policy 42, 1531–1543 (2013)
10. F. de Melo, V.V. Maslennikov, E.V.Popova, T.L. Bezrukova, I.V. Kyksova, Asian Social Science, **11**. 20, 66-73 (2015)
11. O. Alexy, G. George, A.J. Salter, Academy of Management Review, **38**, 270–29 (2013)
12. J. G. Payumo, P. Arasu, A. M. Fauzi, I. Zulkarnaen Siregar, D. Noviana, World Patent Information, **36**, 22-31 (2014)
13. http://inno-exp.ru/archive/16/innov_2016-1_145-151.pdf

Models of e-testing

Sara Konnerth[1,*], and *Gunter* Lobin[2]

[1]Lucian Blaga University of Sibiu, Department for Teacher Training, Calea Dumbrăvii 35, Sibiu, Romania,
[2] University Paderborn Germany, Fakultät für Kulturwissenschaften, Warburger Straße 100, Paderborn, Germany

Abstract. Transdisciplinary interest and preoccupations regarding computer-assisted instruction have stirred our urge to perform a thorough study of the field and trends as well as to uphold certain approaches that have emerged in specialized scientific research. We have decided to focus our analysis primarily on the assessment process as a part of computer-assisted instruction; this in fact represents the major objective of our research, whereas the results of the experiments, performed in the context of the research projects undertaken, enabled us to set forth relevant conclusions about e-testing. Our computer-assisted assessment primarily includes tests with objective or semi-objective items, which inherently evince certain disadvantages specific to these items. Beginning with the year 2000 we succeeded to create such e-tests for different examinations for basic mathematics; these e-tests were used continuously in the last 17 years in different schools, especially for children with learning-problems. We also created e-tests for vocabulary-testing in language-instruction, which we applied at the University of Sibiu by student-evaluations but also in different activities with children of different ages. Some of these products were also published on different web-sites. We present these applications here, considering that they can be models for anyone, who wants to use the advantages of e-learning.

1 Introduction

In 1990 Prof.dr.h.c.mult. Helmar Gunter Frank from the University of Paderborn realized the first contract of collaboration between the university of Sibiu and the "Institut fur Kyberbnetik Paderborn" in order to promote research and instruction in the field of cybernetic-instruction (Bildungskybernetik),[2]. During the following years several teachers and scientists from the university of Paderborn (Prof. on. Dr. Brigitte Meder, Prof. on. Dr. Gunter Lobin, Prof. on. Dr. Vera Barbadovska-Frank) but also from other universities, members of the "Akademio Internacia de la Sciencoj San Marino" (Prof. Dr. Fabricio Pennachietti from Torino, Prof. Dr. Carlo Minnaja from Padua, Prof. Dr. Hans-Dietrich Quednau from Munchen-Freising, Prof. on. Dr. Reinhard Fossmeier) visited the University

*Corresponding author: sara.konnerth@ulbsibiu.ro

of Sibiu in order to teach students or to present their research results to the stuff at different international conferences organized here.

These contacts permitted us to progress in the field of objective instruction realized on computers by using specific software. But only after ten years of theoretical research in the year 2000 we succeeded to finalize the first practical application.

2 Basic mathematical-knowledge e-tests

Our first applications of e-tests were destined to the pupils from the „Centrul Scolar pentru Educatie Incluziva nr. 2 din Sibiu" for pupils with hearing-problems. This school received in 1996 the version Authorware3 from their supporting association "Freunde des Taubstummenheims Hermannstadt" from Siegen/Germany, in order to create applications for children with hearing difficulties, [1]. A collaboration-project of the „Centrul Scolar pentru Educatie Incluziva nr. 2 din Sibiu" with our research team from the University of Sibiu was started. So, we began to use the offered software in order to create useful applications for the children of the school.

The school-programs for the children with special educational needs (SEN) are centered on targets and competences specific for these institutions, with targets and references established for each branch of study. Important for every child in the primary school are the basic operations of addition and multiplication. Here the rapid oral exercises at the beginning of every instructional hour are important for fixing the knowledge in basic calculations. For pupils with hearing-difficulties such exercises can't be performed. So we produced for them suitable e-tests on the computer, in order to be used for rapid exercising of the basic mathematical operations: addition and multiplication. These exercises were also used in different user schools for children with SEN. We want to describe here one of our experiments.

The instruction and testing units, designed and used throughout the e-test-application for basic mathematics performed at the „Lucian Blaga" University of Sibiu (LBUS), have been developed in the authoring language Authorware, [3].

Programming of the e-test Addition till 10: The 45 items with exercises for adding the numbers from 1 to 9, without passing over the value 10, were distributed in 5 test-modules A_i $(i = \overline{1,5})$ with the same level of difficulty with 9 operations each.[4]

When the children have finished the exercise all answers remain on the monitor, so that they can see the wrong answers and so the testing can help them to improve their knowledge and next time have a better result.

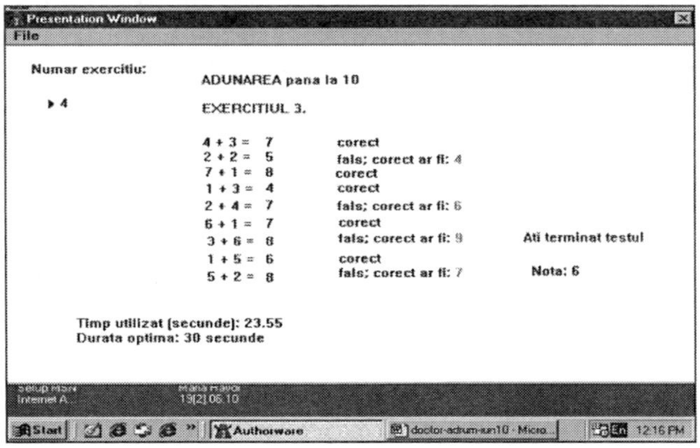

Fig. 1. Aspect of the desktop after finishing an exercise

In the figure above one can observe that the software-product fulfills the following operations: appreciates the results of the assessment delivering a *mark* for the number of correct answers and measures the *pedagogical time* (time for thinking), [5] necessary for finding the respond. We mention that the authoring language „Authorware", used for the realization of the didactical projects, has an option „time in section" (figure 2) which measures only the time sequence while the user is in the interrogation-cycle – the time-sequence necessary for creating the response – but not the time necessary for going from one exercise to the following; so the system permits to pause between the exercises.

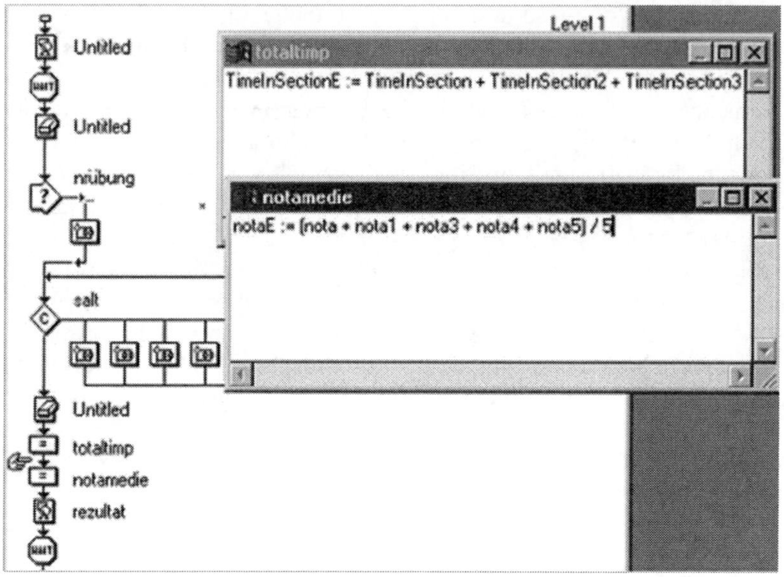

Fig. 2. Creation of the mark (nota) and the measuring of the pedagogical time

We want to mention here that, relating to the pedagogical time, the software permits a limitation of the time until one can respond, but we didn't use this option, considering that this would have negative psychological influence on the children. We wanted to be sure that first of all they know the correct answer and the rapidity of answering will be exercised only in the moment, when the knowledge-content is reached. For each exercise we indicated a pedagogical-time that should be reached (in this case for solving 9 additions we indicated 30 seconds) so that the learner has an indication how good his time was. We did not include the time-factor in the mark, because the rapidity of the answer might also depend on competences in typing on the computer, a factor we don't want to measure here.

This arithmetical e-test with five assessment-models A_i was used in the school-year 2009-2010, for the final evaluation in mathematics for four groups of subjects:

• G1: Centrul Scolar pentru Educatie Incluziva nr. 2 din Sibiu (Scholl center for included education nr. 2 from Sibiu) - form I; children with hearing problems;

• G2: Scoala cu clasele I-VIII nr.10 Sibiu (School with the forms I-VIII nr. 10 Sibiu) - form I;

• G3: Centrul Scolar pentru Educatie Incluziva „Primavara", Caras-Severin (School - center for included education for children with SEN from the school with the forms I-VIII nr. 9 Resita, Caras-Severin) - form V;

• G4: Centrul Scolar pentru Educatie Incluziva „Primavara", Caras-Severin (School - center for included education for children with SEN from the school with the forms I-VIII nr. 9 Resita, Caras-Severin) - form VI.

The results obtained after the application of the e-test at 158 subjects (41 - G1, 66 - G2, 24 – G3 and 27 - G4) were registered in tables and the following graphics present a sample of the analyze performed in this experiment.

At the group G1, consisting of children with hearing problems, one can observe that the marks are not increasing from one exercise to the other – not even when the same exercise is repeated.

Table 1. **E**-test results

	A1	A2	A3	A4	A5	A1	A2	A3
S1	10	10	9	10				
S2	10	10	10	10				
S3	9	10	10	9	10	9		
S4	6	3	3	3	3	4	3	2
S5	6	2	3	3				
S6	9	7	7	6	4	7		
S7	10	9	10	9	10	10	10	

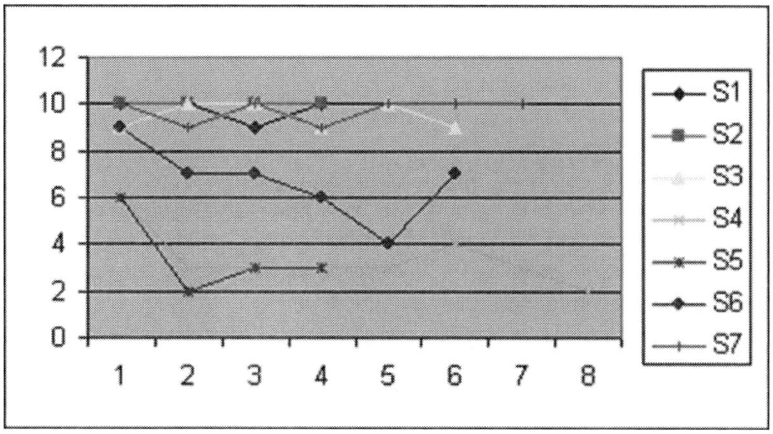

Fig.3. Marks obtained with the modules A_i at the group G1 (7 subjects registered as S_i)

But the pedagogical time necessary for solving the exercises was decreasing almost at every pupil. This decreasing of the pedagogical time made us conclude that, during performing the e-test, the pupils were adapted to the new difficulties caused by the situation of being tested by a computer-test, [6].

Although the "Proiectul aritmetic: Adunarea pana la 10" (Arithmetic-project: Addition until 10) composed from the modules A_i was based on simple items, the capacity of attention of the subjects decreased rapidly after solving only 9 of the 45 possible operations, so that we could conclude, that the intellectual effort made by these pupils with a learning-deficit, could not be more than to solve two modules in the same day – with a pause between these modules. This shows what efforts must be made by the didactical stuff in order to obtain at least some favorite results by teaching children with special educational needs, even if one takes only the basic mathematical elements.

Analogue we also created ten e-test-modules M1-M10 - with 10 items each - for testing the knowledge of the multiplication-table, to complete our Proiectul aritmetic (Arithmetical project). The application realized with these e-tests for testing the basic knowledge of the mathematical operations – addition or multiplication -, enabled us to get assessment results with a great level of objectivity, fidelity and validity. The testing of basic mathematical knowledge is important for pupils at the end of a school-year, when one has to decide if the assessed pupil can pass the finished form, has to repeat the form or has no chance to reach the purpose and must be declared to be a child with SEN, which must be instructed with specific methods and the instruction-plan for this pupil must be reduced.

The experiences gained with these applications encouraged us to create also larger tests. So we developed e-tests for measuring the competences in using the articles „der, die, das" in the German language. This experiment was described in several publications, [4,7].

The most complex e-test we created is made of 1008 items. We will describe it summery in the following:

3 Vocabulary-e-test

First we created an instruction-e-test by using a large Instruction Matrix consisting of 42 instruction-vectors. Each instruction module is laid on a separate line of the instruction matrix in order to obtain the vector structures:

$MI_1 = [\ a_{1\,1};\ a_{1\,2};\ a_{1\,3};\ ...;\ a_{1\,q}\ ]$ Where a_{ij} represent instruction elements

$MI_2 = [\ a_{2\,1};\ a_{2\,2};\ a_{2\,3};\ ...;\ a_{2\,q}\ ]$ (the index *i* indicates the instruction module,

$MI_3 = [\ a_{3\,1};\ a_{3\,2};\ a_{3\,3};\ ...;\ a_{3\,q}\ ]$ and *j* shows the element ranking

... in the instruction module;

$MI_m = [\ a_{m\,1};\ a_{m\,2};\ a_{m\,3};\ ...;\ a_{m\,q}\ ]$ i,j,m,q are non-zero natural numbers; $i \le m,\ j \le q$)

The Assessment Item Bank is made up of the totality of the elements in the instruction matrix A, while an assessment cycle includes at least one element from each MI_i instruction module ($i = \overline{1,m}$) randomly selected by the computer-assisted assessment program. Therefore the structure of an Assessment Module ME is:

$a_{1\,k(1)}$

$a_{2\,k(2)}$

$a_{3\,k(3)}$

$ME =$ ... The elements $a_{i\,k(i)} \in MI_i \mid i \in \{1,2,\ ...,\ m\}$ are randomly selected

from the corresponding modules in keeping with

$a_{m\,k(m)}$ $\forall k(i),\ i \in \{1,2,\ ...,\ q\}$

This way of choosing the assessment-items represents a hindrance in running through all the elements of a comprehensive item bank during the final assessment of a candidate and requires that the assessment modules applied in any examination process should differ from the instruction modules.

The random selection of assessment elements from the multitude of the assessment item bank performed by a software product is meant to assure the objectivity of the assessment process. Furthermore, the assessment validity is substantiated by the fact that no instruction module is left untested.

The informational content selected for this project focused on the identification of Romanian equivalents to ILo (Internacia Lingvo de doktoro Esperanto) vocabulary elements, from a fundamental vocabulary of 1008 words (42 groups each with 24 words).

The design of the assessment program prevented the random selection of elements from the entire item bank of 1008 elements, so that the assessment occurs cyclically, and a word is selected from each instruction module in order to accomplish an assessment module and thus assure a standardized assessment of all instruction modules.

The 42 instruction modules are placed sequentially in the program diagram performed in Auhorware (as shown in the two icons in the left-hand side of figure 4 given below), whereas the right-hand icon shows the representation of a_{ij} items in the loop attached to the decision icon referring to the MI_1 instruction module:

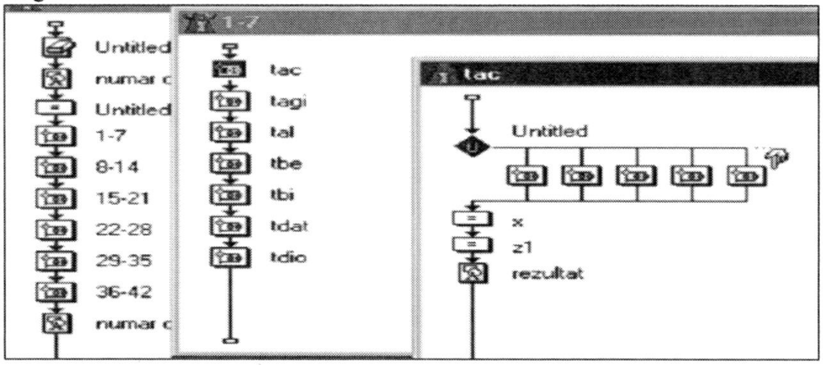

Fig.4. Scheme of the program for the vocabulary-e-test in Authorware

In the authoring language Authorware the assessment items are randomly selected from the content of the instruction module by means of the „decision icon" whose purpose – by means of the option „Randomly to Unused Path" – is to assure, for each assessment cycle the selection of a different item from the instruction module, thus avoiding item repetition in an assessment module since it might trigger redundant elements in the assessment program.
The web-publishing of computer-assisted instruction and assessment programs, was made available on the website of the International Academy of Sciences San Marino at www.ais-sanmarino.org/kursoj; the image captured in figure 5 is available on this site.

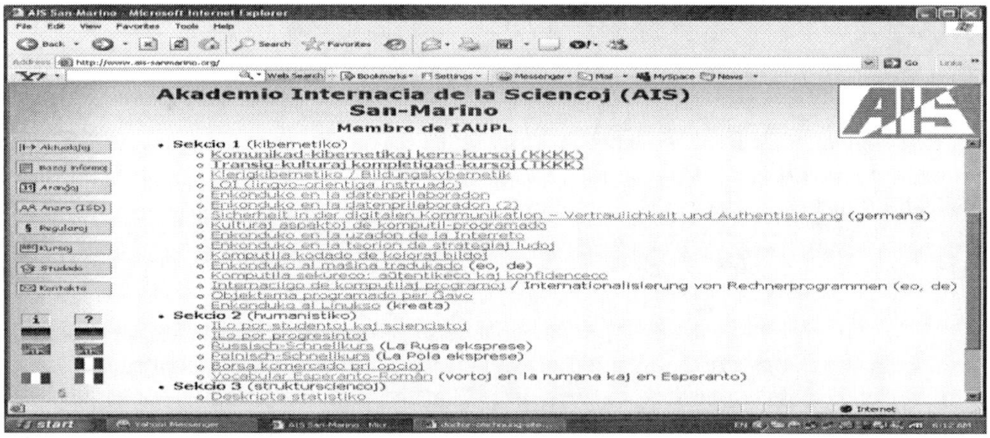

Fig.5. Image of the vocabulary-e-test published in the web, [8]

The instruction modules can be randomly accessed in order to select an assessment element provided that the instruction modules are not placed sequentially; instead they may be accessed by random selection for developing the *ME* assessment module.

The analysis of results obtained by these e-tests is meant to testify the efficiency of computer-assisted assessment programs, thus accomplishing computerized assessment models.

The term e-learning, or electronic learning, is very general and defines the teaching and learning with or from a technical device, which presents the contents of the teaching (also called teaching programs). The authoring systems, such as Authorware and Toolbook, use multi-medial systems of finished programs, which can be used to present the most diverse subject contents in a multi-medial manner and to control the learning progress through interaction with the system. They are teaching systems, which take over the tasks of the teacher in the context of their technical possibilities. The learner interacts or communicates with the teaching system independently according to previously implemented rules and contents. Regardless of the course offer, these systems are also good for testing teaching content as described in this article.

We already mentioned that the software used in the presented application was created in the authoring language Authorware. At the University of Paderborn the authoring language used was Toolbook. We always hoped that there will appear a software which will be used for didactical purposes, that will be accepted generally in schools – such as Word is used for most people as a text editor. But this didn't happen. Most schools, which use e-learning-methods, use ready delivered instructional products, produced by specialized firms. The teachers from these schools do not produce their e-test themselves. So we considered that there is no need to instruct to future teachers a software-program they will never use and we produced our own application (a program created in Java) which we offered - free on charge – to our students. This software-product we called QUICK E-TEST.

We will not present here the information-content of the QUICK E-TEST created by a computer-engineer, because we are not interested in technical details of software-programs. We will analyze the way the common users can use the application QUICK E-TEST to create their Web-Based-Training-application. This application has four „bricks"[9], as it is shown by the symbols from the monitor-representation of the folder presented in the following figure 6:

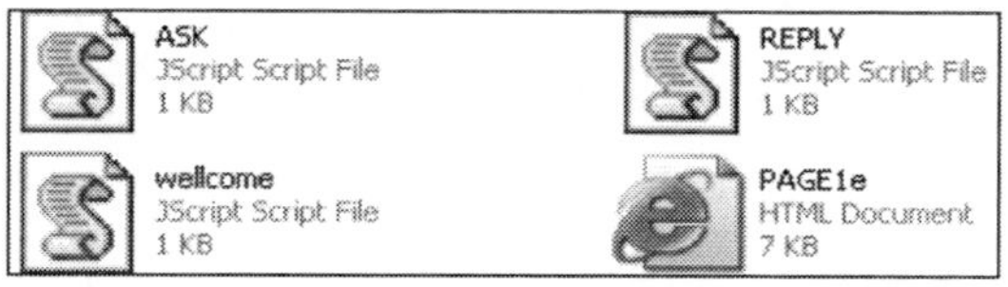

Fig. 6. The QUICK E-TEST files

The content of the files „wellcome", „ASK" and „REPLY" can be modified by the student using a simple text-editor (for example Notepad). In the file „wellcome" one will publish the indication for the user, in „ASK" are introduced the sentences with a missing word and in „REPLY" the correct answer-possibilities. The software-program allows us to indicate more than one – so synonymous – correct answer; we underline here, that even if there are more than one possible correct answer, these answers must be precisely. In an e-test we can't use items with an uncertain answer. The „objectiveness" of our QUICK E-TEST consists in the establishment of the set of possible „correct" answers. We can't accept sentences with an infinite set of possible correct answers, like: „On the table there is a ..." in order to be programmed, because in the file „REPLY" the programmer has to indicate all possible correct answers.

DOI: 10.1051/matecconf/201712112010

The final product has on the monitor of the user the following aspect, Fig.7:

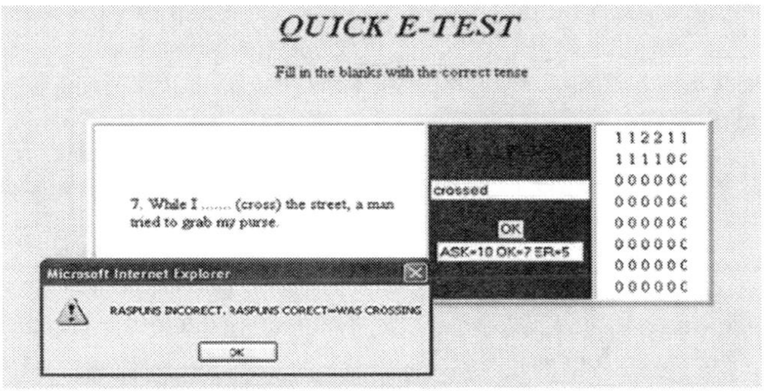

Fig. 7. Monitor-aspect of a sequence during the training with the QUICK E-TEST

The exercises published in our QUICK E-TEST can be used for training in the following way:

The user is indicated to write the missing word in the pattern „RASPUNS". If he doesn't find the proper word, the program will indicate the right answer. The test is continuing linear till the last answer of the exercise is reached. Than the program will turn back and present again the sentences, where the user didn't indicate the right answer. So we have a structure-model of linear training with cyclic repeating of unsolved items, [10]: „Classification of the teaching-algorithms in function of their macrostructure". The training is finished, when there are no unsolved items anymore left – the indication of the as „right" considered answer (by the author) guarantees the possibility of reaching the end of the training. The risk, that the author of the training-vector indicates a false answer in the file „REPLY", must be reduced by the corrections made by the scientific coordinator of the training-module.

The table in the right side of the application indicates the number of necessary repeats, until the user found the right answer in the current training-stage. This table can be exported to a spreadsheet and can be used for assessment, [9].

The feedback of assessment procedures on instruction methods is very important, because this reaction can not only correct the learning procedures but also the instruction-methods.

The file „PAGE1e" allows the pretesting of the created training-vector. By activating this symbol the student can try the test he created. If the application doesn't work, there can be a lot of reasons. One of the reasons may be the use of unaccepted symbols – for example the use of the symbol „/" during a sentence is not allowed. If the number of sentences from the file ASK is not equal with that of the file REPLY the application will not work. These problems are discussed between the university-teacher (or the tutor) and the particular student.

The software-application created after the logical scheme of the university-teacher was created as simple as possible, so that students, who have only a common level of digital literacy can use it properly, [10].

During the last 24 years, since I work at the University of Sibiu, we created different applications of e-learning and the experience proved, that the more complicated an application is, the more students avoid using it. If e-learning is going to be an instrument for each school, than we must find a simple way of approach for every teacher.

4 Conclusions

In the past it proved, that creating products of e-learning needs a lot of financial efforts for the educational institution, which wants to apply such methods. Even if the software is free, the efforts made by the creators of such products still cost a lot of time. But if the results are published in the web and a lot of people – including the whole group of producers – can use them often, then it can also financially worth to create and publish such applications.

The scientific usefulness of such applications is always guaranteed. That's why we described these e-tests, considering that they can be *models* for those, who want to apply methods of e-testing.

References

1. *Dokumentoj pri la "denaska internacieco" de ULBS*, in *Carte omagiala 10 ani de Pedagogie cibernetica la Universitatea "Lucian Blaga" din Sibiu,* (Psihomedia, Sibiu 2003)
2. G. Lobin, *Bildungsokonomische Analyse zu verschiedenen Modellen des Fruh-Fremdsprachenunterrichts,* GrKG , H **4**, 126-132. (1978)
3. http://en.wikipedia.org/wiki/Macromedia_Authorware#History, accessed in 08.02.2017
4. S. Konnerth, *Modele şi tehnici de evaluare in instruirea asistata de calculator* (Editura Universitatii „Lucian Blaga" din Sibiu, 2012)
5. M. Stefan, *Lexicon pedagogic* (Editura Aramis, Bucuresti, 2006)
6. S. Konnerth, *Instruire asistata de calculator. Evolutia instruirii asistate de calculator: curs universitar* (Editura Universitatii „Lucian Blaga" din Sibiu 2009)
7. S. Konnerth, *O formula de calcul pentru stabilirea unei limite a rentabilitatii* (Neutrino, Resita, 2009)
8. http: //www.ais-sanmarino.org/kursoj/, accessed in 10.02.2017
9. C. G. Apostol, S. Konnerth, The Tenth Conference on Informatics in Economy: Education, Research & Business Technologies, Bucuresti (2011)
10. H. G. Frank, Helmar, B. Meder, *Einfuhrung in die kybernetische Padagogik.* In: Brigitte S. Meder, Wolfgang Schmid (Ed), *Kybernetische Padagogik.* Schriften 1958-1972. Vol. 5 (Stuttgart Berlin Koln Mainz, Germany: Publishing House W. Kohlhammer, 1974)
11. A. Stoica, R. Mihail, N. Litoiu, M. Mandrut, *Evaluarea curenta si examenele.* Ghid pentru profesori (Serviciul National de Evaluare si Examinare, Bucuresti, 2001)

The benefits of inclusive education: new challenges for university teachers

Alina Georgeta Mag[1,*], *Sandra* Sinfield[2], and *Tom* Burns[2]

[1] Lucian Blaga University of Sibiu, Faculty of Social and Human Sciences, Department for Teacher Training, Calea Dumbravii Nr. 34, 550324, Sibiu, Romania
[2]London Metropolitan University, London, N7 6PP, United Kingdom

Abstract. Inclusive education is a child's right, not a privilege. The amount of current attention given to children and to childhood is unprecedented. In a world affected by financial crises, attitudes about children's needs are very important. Recent years have seen innovations in the focus of research, as political moves that challenge the ways different children have been excluded and silenced. Inclusive education is a real and urgent need for all educational systems around the world. Inside our university, the Teachers Training Department needs to take serious measures to create a professional environment for all the students involved in the initial training for becoming teachers at all levels. This study reveals the interest and the planning of our trainers to develop their competencies for inclusive education, supported by new ways of international support with the professional involvement of two experts in university education from London Metropolitan University. We discovered that inclusive education is a serious priority, most of their courses and seminars need continue improvement in this area and their interest to develop new skills is high. We need to increase attention to our professional development as a factor that can contribute to a healthy society.

1 Introduction

The education system in the entire world is changing and evolving constantly. Experts in Educational Sciences are directly involved in developing their own skills to be competent and able to meet the challenges of a socio-economic environment in a continuous process of change.

Inclusive education is one of the most sensitive ongoing issues, because not all children enjoy an education experience that enables them to develop to their full potential [1-2]. Inclusion in the UK context means primarily inclusion of those with physical or learning disadvantage – whilst the thrust here is also about equality of educational opportunity in re the socio-economic circumstances of the child. Inclusion in education means if it means anything is respect for the diversity of all our children, and if this is to be realised must include several changes of substance in terms of educational content, pedagogical and

* Corresponding author: alina.mag@ulbsibiu.ro

didactical approach, the structure of the education system and educational strategies. Although many programs and projects are implemented to enable inclusion of children with different special needs in mainstream schools, in many countries, the results are often positive, but dissemination of results and best practices is not systematically undertaken.

2 Problem statement

Children can change the world and education can change the way children perceive the world. Inclusion refers to all children, because inclusive education is one in which all children have access to quality education in an inclusive environment that provides three components:

• all children mean "children with disabilities, gifted children, homeless children, children belonging to nomadic groups, children belonging to linguistic minorities, ethnic or cultural, children with HIV or belonging to any other vulnerable groups" per Salamanca's Declaration (1994);

• quality education - means creating real opportunities to achieve success in the learning experience; in other words, the quality of education is not measured in terms of performance, graduation, infrastructure, technology and comfort, but the system capacity, the school and the teacher to create learning opportunities for every child;

• inclusive environment - which means that environment is welcome, protect and educate all children, regardless of gender, physical, intellectual, economic, linguistic or other characteristics (UNESCO).

Inclusive education is one of the top challenges in today's world, and whilst educational systems make efforts to become more inclusive, new teachers must be developed to be more inclusive in their future practice. University teachers involved in Education Studies arguably must address this in the theory and practice that they deliver on Teacher Training courses.

Nowadays Romania is facing huge economic and political crises. The level of the economic stress is high, affecting both adults and children. It is therefore appropriate to explore the conditions of inclusive education in Romania today and ask how far the current anxiety about children's rights is justified – and to ask ourselves: How can university teachers develop their own competencies to become truly inclusive? Is this theme a current pressing priority for their own training? These important questions were investigated in a university from Romania. The attitudes, opinions and lived experiences of inclusive education of future teachers were investigated and this research is focused on the opportunities created by listening to their opinions and the challenge is to harness their richness to produce better solutions. The vision of an expert in education from London Metropolitan University, United Kingdom was a real support during the study.

3 Theoretical foundation and related literature

Recently we have seen innovations in the focus of research, as political moves that challenge the ways different children have been excluded and silenced [3-4]. Inclusive education is a child's right, not a privilege. The amount of current attention given to children and to childhood is unprecedented [5]. Inclusion promotes quality and equitable education for all, without exclusion, including those who may be potentially marginalized by learning need or social position. In a world affected by financial crises, attitudes about children's needs are very important.

Inclusive education is now firmly established as the main policy imperative with respect to children who have special education needs or disabilities. It is championed to remove barriers, improve outcomes and remove discrimination. Inclusion is however, a complex and contested concept and its manifestations in practice are many and various [6].

Recent studies demonstrated that if children's educational needs are not fulfilled in the first years of life, serious gaps will appear in their development [7-8]. The attitudes of adults, in all their different roles and relationships, affect children deeply. Children have been silenced and excluded in many ways, because they were underestimated. They have different needs or special needs, and they need inclusive education for a proper development. In fact, all children are born with sophisticated capacities and adults should not restrict or limit their potential. Educators should work with children in new ways driven by a rights perspective [5].

Politicians and the media debate today's children's educational needs. According to recent studies children's mental and emotional health depends on the efficiency of the relationships that adults construct with them from conception [9]. Parents and teachers are responsible for children's health and development and they should practice inclusive education. New studies into children's competencies from birth changed the vision on children's potential and needs [10]. In the century of change, adult's abilities are vital, such as supporting children's emotions, understanding stressful situations, developing relationships and communication skills, promoting critical and creative thinking and demonstrating empathy.

Inclusive education in Romania is based on several national documents like: the Constitution of Romania 1991; Law no. 53/1992 on special protection of disabled persons and Law no. 57 on the emplyoment of disabled persons (amended by Ordinance no. 47/1997, subsequently repealed by Ordinance no. 102/1999 on special protection and employment of disabled persons) and Education Law no. 84/1995. The main international documents for inclusive education are: World Declaration on Education for All (1990); CoE recommendations on coherent policy for the rehabilitation of persons with disabilities (1992); Standard rules on equalization of opportunities for persons with disabilities (1993); Declaration of World Conference on Special Education, Salamanca (UNESCO, 1994). All efforts inside the educational departments in different universities all over the world relay on these important documents.

Educational partnership is one of the key words of the contemporary pedagogy [11]. Partnerships are indispensable for a qualitative inclusive education. Anxieties among teachers because of the financial crisis, too many changes brought by educational reform, anxieties about parents' involvement in education, the growing number of children with behavioural and emotional disorders are important issues in today's education in our country. The efforts to create an inclusive environment for Romanian children are still a challenge [1].

4 Purpose of the study

In this paper, we intend to assess what university teachers from the Teacher Training Department of "Lucian Blaga" University from Sibiu, Romania, consider to be their key efforts to support inclusive education, at the beginning of this year. The key argument of this study is that inclusive education is a real and urgent need for our educational system, starting with early childhood and ending with university's education. The main factors of inclusion are: the teachers, the colleagues in the classroom or school and the school environment. It's obvious that inclusion depends largely on the teachers' training, on their attitude and on their way to report to the children with SEN, as well as on the teaching strategies used.

The aim of this investigation is to find new ways of international support, of collaboration and good practice, with the professional involvement of experts in university education from London Metropolitan University, so that we can take new measures to create a professional environment for all our students that will become the next generation of teachers at all levels.

5 Methodology

The study started with a systematic review of current guidelines in teacher training from Inclusive Education frameworks available nationally and internationally and of the competencies required in this area for university teachers today. The next step was an analyse of the current situation and needs for training in inclusive education in the Romanian Teacher Training Department. In recognition of the importance of professional learning we conducted a survey with open questions to identify teachers' ongoing professional learning needs about inclusive education. This is a big challenge for all but it is an opportunity to advance the Teacher Training Department as a change factor that promotes dialogue and participation, making possible well-being through an education of quality for all without exception, for the commitment of the community. We also used focus group as a technique to collect data from 20 university teachers in our Department in Sibiu, Romania. They were voluntarily involved in the study. The survey was built to reveal the next professional concerns:

• What is inclusive education in the vision of the teachers from the Teacher Training Department?

• How do university teachers specialised in education made the distinction between inclusive education as a right and inclusive education as efficacy in their own practice?

• How do university teachers develop their own competencies in the field of inclusive education?

• What are the benefits of inclusive education for their perspectives?

We examined the personal arguments and the evidence presented for their personal professional development in this area, having the support and expertise for analysing the findings from London Metropolitan University, United Kingdom.

6 Findings

Evidence shows that the quality of teaching has one of the largest impacts on students learning. Building teaching excellence through pre-service and in-service training is central to making educational settings more inclusive. The teachers involved in the study started with a swot analysis of the inclusive education in Romania revealed that this is a real challenge for our staff. Findings revealed that the main strong points are: children with soft and medium disabilities are included; intervention team; intervention instruments; efficient learning strategies: individualised learning, cooperative learning; partnership with family and local community; initial and continuous development of teachers. The low points are: discontinuities in educational policies and incoherence in reforms application; having a certificate of SEN; insufficient number of support teachers and hours for activities with children with SEN; a great number of children in the class; different assessment in primary school and gymnasium; national exams are not different. We also evaluated the opportunities - Romania Integration in European Union, accessing European money through national and international projects, teacher continuous development - and the main barriers: stereotypes and misjudgement; financial resources crises; infrastructure and inadequate material resources.

This radiography of the Romanian inclusion was analysed by the 20 university teachers from the Teacher Training Department. They consider that inclusive education is a serious priority and they admit that most of their courses and seminars need continue improvement in this area and their interest to develop new skills is high. Their motivation to develop the opportunities, to reduce the low points and to fight the barriers starts with their own preparation to promote inclusive education. During the focus group, the university teachers were interviewed with all the questions mentioned, so they offered an objective overview of their perspectives, efforts, actions and plans to develop their competencies in this area. We will present the findings in a synthetic perspective, to each question.

At the first question, they admit that inclusion requires a large vision and specific competencies for all teachers. Now teachers need to know that diversity is present in the classroom, and that they should attend to learners with a range of diverse needs. In this frame, it is imperative to develop the following professional aptitudes: researcher, strategic and resilient. They identified important educational aspects that every teacher needs to be inclusive: equality; promoting the same opportunities for all, quality; offering functional and meaningful learning and equity; responding to special educational needs.

At the second question part of the teachers report that a lack of teacher training in facilitating the learning of students with disabilities is a barrier to individuals being able to succeed in education. Knowing students' rights, it's not sufficient. Teachers' educational efforts are valued by the student success in time, not just in school. As with all professional learning activities, it is important to stay up-to-date with current thinking and keep building professional expertise as new research and knowledge emerges.

At the third question, the teachers presented all their actions and plans to develop the skills to provide an inclusive environment, where every student can learn and achieve their full potential. The last years' most of them were engaged in relevant professional learning activities to prepare their own development in this area: research, observation and practice, training courses and seminars, conferences, workshops, an important project about mentoring in inclusive education managed by one of our professor in which four colleagues were trainers in inclusive education. Some of them already published relevant studies about inclusive education.

Professional learning is most effective when it is relevant, collaborative and future focused, and when it supports teachers to reflect on, question and consciously improve their skills. According to the teachers interviewed, they also need to have high expectations for all (inclusive vision), develop inclusive projects including diverse teaching strategies and support systems (inclusive practices) and participate in a collective work. The group investigated from the Teacher Training Department proposed the development of a new master program about Interventions in inclusive school.

The last question proved the benefits of inclusive education per the university teachers involved in the study. All children benefit from inclusive education, because it allows them to: develop individual strengths and gifts, with high and appropriate expectations for each child, involve their parents in their education, foster a school culture of respect and belonging, inclusive education provides opportunities to learn about and accept individual differences, lessening the impact of harassment and bullying, develop friendships with a wide variety of other children, each with their own individual needs and abilities and positively affect both their school and community to appreciate diversity and inclusion on a broader level. The inclusive teacher has a holistic educational view with strong skills and experience to participate in diverse contexts.

Findings were analysed together with two London experts to find new ways of international support, of collaboration and good practice that can help us improve our perspectives. A creative way that we approached this in one UKHE setting was to develop a first-year undergraduate module for Education Studies students that modelled inclusive practice in learning, teaching and assessment.

The module was creative and emancipatory, designed to challenge and engage ALL the participants, including those with their own Specific Learning Needs (typically this included students with Dyslexia) and those who had experienced previous educational rebuff because of socio-economic status (this included what we term 'non-traditional' students and those who had been 'looked after'/in care).

The module "Becoming an Educationist" was delivered mainly in 'workshop' mode, with role play and simulations, draw-to-learn and music workshops, real qualitative research, student-led 'performances', active-, question- and project based learning - and choice – for example, choice about which projects to submit for the final assessment. Students were required to 'blog' their learning as part of reflective practice – and to develop a writing habit that thus also improved their formal academic writing.

The goal was to develop student self-efficacy at the same time as expanding their notion of what education was, what it could be and how it could promote inclusion and equality for all. The graduates from this module reported increased self-confidence and increased academic success. In a succeeding Becoming a Teacher module, they all cited "Becoming an Educationist" as a role model of good practice to harness in their own future teaching practice. Ideally, we would have such a module at every level of an Education Studies student's under- and postgraduate programmes. Inclusivity has to part of a trainee teacher's own tuition if it is to be realised in a future teacher's practice.

7 Conclusions

Children need to be listened to, valued and re-evaluated. Their rights should be respected, no matter what economic issues affects adult's lives. Children need to be respected by parents, carers and professionals, but this respect is shaped by their attitudes towards children – and their own experiences of teaching, learning and assessment. Certain attitudes and practices affect many children in ways that are not beneficial, and which can harm or restrict them. The language and actions of power and regulation held by adults in making and overseeing children's lives has been explored through the ways in which research, thinking and practice around inclusive education is reconstructing education and children in new ways.

It is a fact, every educational system in the world needs to improve and work towards the best results for all. By respecting children's educational needs, and developing empowering curricula and pedagogy, we have the chance to regain essential universal values for peace, justice and welfare. We must reach out with empathy to those next to us, listen to children, and fight for justice, for principles and for values. The study had a significant impact in our teachers' motivation to practice inclusive education. We need to increase attention to our professional development as a factor that can contribute to a healthy society and to a better live for children in Romania.

References

1. T. Vrasmas, *Invatamantul integrat si/sau incluziv* (Aramis Pub., Bucharest, 2001)
2. D. Mara, *Strategii didactice in educatia incluziva*, 2nd Ed. (Didactic and Pedagogical Pub, Bucharest, 2009)
3. C. Smith, L. Strick, *Dizabilitati legate de invatare, explicate de la A la Z,* (Aramis Pub., Bucharest, 2011)
4. A. Ghergut, *Psihopedagogia persoanelor cu cerinte speciale. Strategii de educatie integrata* (Polirom Pub., Iasi, 2001)
5. P. Jones, *Rethinking childhood. Attitudes in Contemporary Society*, New childhoods (Continuum, New York, 2009)

6. G. Lindsay, *Inclusive education: a critical perspective*, British Journal of Special Education **30(1)**, 3-12 (2003)
7. American Psychiatric Association, *Diagnostic and Statistical Manual of Mental Disorders,* 5th Ed., DSM-5 (APA, Washington DC, 2013)
8. A. Rosan, (coord.), *Psihopedagogie speciala Metode de evaluare si interventie* (Polirom Pub., Iasi, 2015)
9. C. Cable, L. Miller, *Working with Children in the Early Years.* 2nd Ed. (Routledge Taylor Francis Group, London and New York, 2010)
10. J. Johnson, *Positive and trusting relationships with children in Early Years Settings* (Learning Matters, Great Britain, 2010)
11. D. Mara, *Partnership in Inclusive Education*, Int. Journal of University Teaching and Faculty Development, **5(1)**, (Nova Science Publishers, 2014)

The function of mirror neurons in the learning process

Daniel Mara[1,*]

[1]"Lucian Blaga" University of Sibiu, Department for Teacher Training, 34 Calea Dumbravii, Sibiu, Romania

Abstract. In the last years, Neurosciences have developed very much, being elaborated many important theories scientific research in the field. The main goal of neuroscience is to understand how groups of neurons interact to create the behavior. Neuroscientists studying the action of molecules, genes and cells. It also explores the complex interactions involved in motion perception, thoughts, emotions and learning. Brick fundamental nervous system is the nerve cell, neuron. Neurons exchange information by sending electrical signals and chemical through connections called synapses. Discovered by a group of Italian researchers from the University of Parma, neurons - mirror are a special class of nerve cells played an important role in the direct knowledge, automatic and unconscious environment. These cortical neurons are activated not only when an action is fulfilled, but when we see how the same action is performed by someone else, they represent neural mechanism by which the actions, intentions and emotions of others can be understood automatically. In childhood neurons - mirror are extremely important. Thanks to them we learned a lot in the early years: smile, to ask for help and, in fact, all the behaviors and family and group norms. People learn by what they see and sense the others. Neurons - mirror are important to understanding the actions and intentions of other people and learn new skills through mirror image. They are involved in planning and controlling actions, abstract thinking and memory. If a child observes an action, neurons - mirror is activated and forming new neural pathways as if even he takes that action. Efficient activity of mirror neurons leads to good development in all areas at a higher emotional intelligence and the ability to empathize with others.

1 Neurosciences

The term of neurostiinte comes from the English neurosciences, a neoligism coined in 1962 by the American neurophysicist Francis O. Schmitt. He realised that, in order to fully understand the complexity of the brain functions, one must eliminate any existant obstacle among disciplines, by means of merging available resources and attempts.

Therefore, he used the word neurosciences (Neurosciences Research Program) to name his research team, which incorporated scientists belonging to various fields of study. The assemblage, bearing the name of neurosciences, represents an interdisciplinary branch that

* Corresponding author: daniel.mara@ulbsibiu.ro

combines pieces of information from study domains such as mathematics, physics, chemistry, nanotechnology, engineering, computer science, psychology, medicine, biology and philosophy. It follows an antipodal direction of research, when it comes to brain study and technique, than the one utilised previously.

Neurosciences cover a broad variety of aspects, such as the development, the aging process and the maintenance of the nervous system or its functional and anatomical structure, observing carefully the brain itself, along with the role it plays regarding behaviour and knowledge. Neurosciences seek to explain not only the normal mechanism of the nervous system, but the breakdowns, the malfunctions that occur in development disorders, both psychiatric and neurologic, for the purpose of finding new treatments or preventive methods.

Eric Kandel, a Nobel Prize laureate, used to say: "the main purpose of neurosciences is to explain the behaviour of the brain. How is it that the brain can direct millions of individual cells, in order to trigger a certain behaviour, and how is it that these cells can be influenced by the environment? The last barrier of neurosciences, their ultimate challenge, is to understand the biological essential features of consciousness, as well as the mental devices that allow us to perceive, react, learn and remember" [1].

2 Mirror Neurons

Neurosciences have encountered a great development in the last recent years and many important theories have come into existance. The main purpose of neurosciences is to understand how certain neuronal groups interact in order to create behaviour. Neurologists study the action of molecules, genes and cells, as well as the complex interplay involved in movement, perception, thought, emotions and learning. The main component of the nervous system is the brain cell, the neuron. Neurons exchange pieces of information, by means of sending electrical signals and chemical substances via the connections named synapses.

The mirror-neurons were discovered at the beginning of the 90's by the neurophysicist Giacomo Rizzolati while studying primates and they bear this specific name because they activate in order to imitate other persons' actions and gestures. Scientists consider that there is a myriad of elements yet to be discovered, besides the physical and functional extents of mirror neurons, which require further investigations.

Giacomo Rizzolatti and his researchers have implanted electrods in the premotor cortex of several monkeys. The neurologists have noticed that the mirror neurons actuated electrical signal at the very moment when the monkeys performed movement in order to feed themselves. It was only when they noticed that the same neurons were activated when the monkeys watched other peers feeding themselves, that the scientist realised the importance of this discovery. Hence the term of mirror designated.

There are mirror-neurons that activate electrical signals in our brain when we see someone being touched, when we watch other people's movements or, only as a sign of empathy (as explained by neurologist V.S. Ramachandran and shown by the tests realised by Christian Keysers). If you watch a wounded, bleeding person, you will have a certain facial expression and experience certain certain feelings related to this situation (this is not the case for psychopaths, whose brains function differently).

The magnetic resonance imaging has shown that the same neuronal structures activate when we observe the emotional state of a human being or when we experience them ourselves, too. We can perceive mirror neurons as a way people are given the ability to connect with others (but not at a conscious level) and having as an outcome the learning process itself (learning by imitating), understanding and offering help (empathy). This connection is not realised directly, but mediated by one's personal experience. Mirror neurons help us put ourselves in other people's shoes. Yet, this neuronal system does not stop us of from ignoring the ones suffering, or from causing sufferance. Significant changes occur rarely and slowly, but this mechanism embedded in our brain (and therefore, in our psychology) did not appear out of the blue. There is more to this neuronal system than the evolutional function, their discovery and roles- the mirror neurons can be perceived as an indicator of the fact that we become more and more connected, at an individual or group level, or at a broader dimension, focusing on organizations and societies [2].

Mirror neurons represent a special category of brain cells, and play a very important role in the direct, automatic and instinctive understanding of the environment. These cortical neurons are activated when we observe how a specific action is performed, not only when we actually carry it out ourselves. Therefore, they become the exponents of a neuronal mechanism which helps us automatically comprehend other people's actions, intentions and feelings [3].

Mirror neurons are very important during childhood. Thanks to them, we learn a great amount of things in our early ears of life: to smile, to eat, to learn and to embrace the customs, rules and traditions of our families and groups. This is a means of learning by seeing and feeling what others see and feel. Mirror neurons are highly important when it comes to discern and assimilate other human beings' actions and intentions, and to grasp new skills by mirroring (imitating). They are involved in thought planning and controlling, in abstract thinking and memory. If a child observes an action, the mirror-neurons will activate and form new neural connections, as if he or she undertook the action herself/himself. An efficient functioning of the mirror neurons leads to an excellent development in any field, to a greater emotional intelligence and to a higher level of empathy.

It is considered that the malfunctions of these cells are linked to several ilnesses or disorders, such as autism. This fact turns the study of mirror-neurons into a very important sequence of research that could lead to finding new ways of treatment. These neurons were, initially, discovered in the premotor frontal cortex of the monkeys, and, in later studies, it was found that they were also present in the inferior parietal lobe. Their main characteristic is given by the tight connection between the motor pieces of information they codify and the visual data they respond to, bridgeing the areas of the brain in charge with both motor and sensuous processing.

According to the latest studies, the human brain also contains mirror neurons, and a large number of hypotheses were created in order to explain their functional purpose, such as understaning actions and intentions, imitating and empathizing.

The first hypothesis regarding the function of mirror neurons was the one of actions distinguishing, as peculiar as it may sound the fact that, in order to observe them, the motor system has to be activated. Actually, a simple visual perception, without the involvement of the motor system, would only offer pieces of information regarding the discernible aspects of the movement, and would not give data about its inner components or sense of direction. Approaching the issue from this point of view, mirror-neurons might be the core mechanism for a myriad of functions. Among them is the imitation, defind as the ability to reproduce or learn new actions. There are many obvius proofs that the mirror-neurons system is involved both in the directly imitation of an action performed by other people, as well as in learning behaviours by reproductions. Observing an action offers two types of pieces of information related to what and why happens, and the mirror-neurons are able not only to react to a simple stimulus, but to even understand the sense and purpose of the input, therefore being in charge with the comprehension of intentions. It was also suggested that there is a certain connection between autism and the mirror-neuronal system, the inability of autistic children to adjust to other human beings and life situations, and that this disorder might be also caused by a mirror-neurons malfunction.

The term of Autistic Spectrum Disorder (ASD) describes a heterogenous group of neurodevelopment disorders, including the Asperger syndrome, the Rett syndrome, the Heller syndrome or the childhood disintegrative disorder, and the prervasive childhood disorder- which is not defined otherwise (typical autism). Children suffering from ASDs encounter difficulties in social interactions, verbal and non-verbal communication and show a limited range of interests and behaviours. They are often linked with repetitive, stereotypical, obsessive gestures as well as operating deficiencies. These children can also have a high level of sensitivity when it comes to sounds, smells, tastes and textures. The main three features specific to autism are: malfunctional social interactions, inability to develop relationships with peers and lack of initiative when it comes to spontaneous activities. They can also show belated speaking skills, the inability to initiate or maintain conversations, stereotypical behaviour and limited concernments, inflexible routines and repetitive motor sequences (palms and fingers movements).

ASD affect 1 in every 166 children and are four times more frequent in boys than girls. The diagnosis is usually given around the age of 2 and 3, but symptoms can be observed earlier, after 6 months. The disorder has a liniar development, without any periods of improvement, and it continues to manifest itself even at manhood, yet in less severe forms.

The ability of an individual to understand another's emotions, to perceive what the others perceives, or, in other words, to empathize, it is also closely related to mirror-neurons. The magnetic resonance imaging has shown that, while observing someone else's emotional state, the same neuronal structures are being activated as when we experience the emotions ourselves.

Some researchers connect mirror neurons with speech development, because it seems that they are located near the Broca area, region considered in charge with this function.

Representing one of the most interesting discoveries in the field of neurosciences, mirror-neurons radically change the old conceptions about human brain functioning which states that the brain works in a sequentially manner. This means that signal perception is followed by a cognitive reprocessing, a decisional one and, then, by the enforcement of the motor task.

It is quite obvious that, the ability of the specimen to react to other person's emotions and actions has a significant importance in behaviour adjustment. One could consider this statement valid because people would not efficiently work together if they did not accomodate to the emotional state of the people around them. And it is the role of the mirror-neurons to interlock our personal experiences with the ones perceived by our peers [4].

3 The Connection Between Feuerstein's Theory and the Breakthroughs in the Field of Neurosciences

Neurosciences can consider the biological aspects of the structural cognitive modifiability conceived by Reuven Feuerstein, by describing the cerebral mechanisms on which the process is based. The cognitive neurosciences haven't produced a set of fixed axioms, therefore, their results are seldom taken with a grain of salt. It is highly important to interpret and analyze the main results of neurosciences, which can be easily put in conjunction with other fields of study. Building on studies regarding structural cognitive modifiability, there must be endorsed a reciprocity among neurosciences and education, mainly via instrumental enrichment. There are a number of experimental and non-experiemental researches that support instrumental enrichment, showing that, along with mediation, it promotes cognitive modificability.

Neurosciences also offer similar proof to support The Intrumental Enrichment Program. Theories provided by cognitive neurosciences about the way brain activity influences the brain itself, and brain activity is influenced by cerebral modification, as well as those supporting the theory of cognitive modifiability, all of them emerge from studies on brain mouldability. This is the feature of the brain that allows cognitive modifiability, and is integrated in all the stages of the learning process. The embedded mechanisms create two different functions of adjustement: the capacity to adapt to any type of environment, the ability to improve the acclimation to specific surroundings. Such characteristics of the brain are influenced by many conditional factors which are mainly of a socializing nature, and can be split into two main categories: emotions and beliefs.

The first type of modifiability, which focuses on the capacity to adapt to any type of environment, depends on the working memory and its functioning. The second one, regarding the ability to improve the acclimation to specific surroundings, relies on the formation of new models of neural systems, the core of competences. These two processes are linked together but have different roles: the working memory is required in order to create new models of neural systems, which, by their sheer nature are built via synoptic connections. Therefore, they are also created as a result of synoptic usage, requiring a cloister of the contiguous nervous tissues. Consequent to this point of view is the fact that, due to synoptic consolidation, the tasks of the working memory which involve a specific neural-map-dependent domain are being facilitated. The performance of the working memory relies on the synopic power. Even more, the entire mental activity coincides with the working memory, and the mental activities continuously blossom the brain, by means of new models of neural systems.

In spite of all these, when an organism constantly relates to the same environment, or does not learn anything new, the mental activity has the tendency to maintain the adaptive structures, avoiding the aggradizement of significant changes. When the organism confronts new situations, the mental activity immediately starts to alter the brain structures, and its ability of efficient adjustement will depend on the efficiency of synoptic intercommunication.

The working memory is an important function of such intertrades, and constitutes the cognitive component. Consequently, on one hand, the activation of the working memory has the bent for creating new models of neural systems, while, on the other hand, new models of neural systems enlarge synoptic intercommunication, creating a greater local efficienty for tge working memory.

4 The Concept of Mediated Learning Experience

Mediation represents the deliberate, active, intervention provided by an educator to the person he/she interracts with, aiming for a better capaciousness which also involves a gradual autonomy level.

According to the theory of modificability and mediation, at the core of the higher mental functions stays the learning process, and the role of the educator is fundamental. The forms of interaction among environment, educator and child underlie the majority of the structural changes that occur in the human cognitive apparatus. The mediator must be able to identify, choose and encompass some of the stimuli reaching the target, in order to make them more approachable. The educator will not intercede between the human being and the outside world, because his main role will be to identify the stimuli needed to be valorized in an optimal manner, to focus his/her attention on relevant data, encompassing interest, attentiveness and awareness [5].

The mediator represents a strainer for the environmental stimuli. When the learning process occurs due to alterations of the ability to adjust to the environment, mediation is supported by the capacity to co-imitate the adaptative modifications. Intentionality and reciprocity, two concepts considered by Feuerstein as the fundamental parts of mediation, are inherent to human relationships. The lack of mediation, or the lack of tendance, is a form of repudiating interpersonal embodiment [6].

It is important for us to emphasize the fact that, the aim of the mediated learning experience is to select, mark out and offer stimuli by manipulating the environment in which the mediated person tries to accomodate himself/herself. In other words, as long as humans adapt and then learn, not only by themselves, but also by means of social interactions, they try to connect with an exterior world. They have to explore it independently in order to embrace behaviours and mental performances which are co-imitated. The cerebral activity between the mediator and the mediated person entails changes for both of the involved parts, through interplay. Each human being contiguously changes its mechanisms due to environmental interactions, by means of variation and selection, so that the brain structures configurate accordingly. Their mediation is an expression of interpersonal affiliation and, it always entails the learning process. Mediation involves inclusion and cooperation. Given this fact, mediation is always considered to be a good moment for adjustments and accomodation. Besides the individual issues of the mediator and of the mediated person, this fact also confronts the problem of environmental acclimatization.

The lack of mediation can encompass cultural poverty, yet the purport of the mediation may not pe environment-appropriate, meaning that is is not accordant to subjective learning. This is the reason why Feuerstein states that mediation does not have a specific content, because its main purpose is to enhance the behaviouristic flexibility, not to adapt only to a certain environment. This fact shows the difference between mediation and the plain cultural conduction. Throughout the process of mediation, the environment is being manipulated and harnessed according to the needs of the mediated person, and more than just the mental activities of the mediator and mediated person are being involved. The mediator also moulds the learning environment of the other human being, in order to offer as many adjustment opportunities as possible. It is quite obvious that this particular aspect grants mediation the status of a simple and educational human interaction.

The mediator percolates the environment, in order to facilitate autoplasticity, defined by Feuerstein as the way an organism changes itself, as a result of occuring environmental changes. At the other end of the spectrum is the haloplasticity, described as the set of changes meant to adjust the external reality of the organism.

Feuerstein uses the concept of mediation, or mediated learning experience, as a means of transforming the mediated person. In other words, mediation regards the manipulation of the learning environment. Its main purpose is to offer the possibility of an adaptative and educative praxis, focused on co-imitative behaviours related to the mirror-neuronal system.

5 Conclusions

Thanks to the discovery of the mirror-neurons a new and revolutionary research perspective has been commenced. The mirror-neurons have been, initially, detected, approximately twenty years ago and have been immediately connected to action and object recognition.

A special emphasis was put on the fact that, while observing an action conducted by another human being, the neuronal system activates as if the person itself would carry out the activity. It is considered that mirror-neurons facilitate the imitative learning process, as well as the verbal communication, and their malfunction entails disorders such as the autism [7].

References

1. *Cosa sono le neuroscienze*, Accessed 31.01. 2017 http://www.neuroscienzedipendenze.it/neuroscienze.html, (2008)
2. C. Jelea, *Sistemul neuronilor oglinda*, https://costinjelea.wordpress. com/2012/01/06/sistemul-neuronilor-oglinda/ (2012)
3. G. Rizzolatti, C. Sinigaglia, *So quel che fai. Il cervello che agisce e i neuroni specchio,* (Raffaello Cortina Editore, Milano, 2008)
4. Spitalul Clinic Nicolae Malaxa, *Neuronii oglinda - traducerea neurologica a empatiei si evolutiei umane*, http://www.spitalmalaxa.ro/index.php?option=com_content&view=article&id=112:neuronii-oglinda-traducerea-neurologica-a-empatiei-si-evolutiei-umane&catid=45:articole-medicale&Itemid=222, accessed 26 .01. 2017
5. J. Kopciowski Camerini, *L'apprendimento mediato. Orientamenti teorici ed esperienze pratiche del metodo Feuerstein,* (La Scuola, Brescia, 2002)
6. R. Feuerstein, J. Rand, J. Rynders, *Non accettarmi come sono,* (Sansoni, Milano, 1995)
7. M. Iacoboni, *I neuroni specchio. Come capiamo ciò che fanno gli altri,* (Bollati Boringhieri, Torino, 2008)

New perspective of learner-centered education in nowadays didactics

Elena Lucia Mara[1,*]

[1]Lucian Blaga University of Sibiu, Teacher's Training Department, 34 Calea Dumbravii, Sibiu, Romania

Abstract. An approach to education that is based on the educational context in which the student comes and track progress in achieving the learning objectives. The learner-centred education learning places the responsibility on the shoulders of students, while the teacher takes responsibility for facilitating the education process. The approach tends to be unique, flexible, competency-based methodology and not always varied in time and space constrained. Teaching will have complex changes, so if, traditionally, it is just a business communication / transmission of knowledge, actual perspectives follows specifically aims in organizing and managing learning processes. According to the researchers in the field, developing paradigm / meanings in teaching practice in teaching, learning and assessment, respectively constructivist paradigm and the paradigm of interactive, make an important contribution. A student-centred educational activity is if it satisfies the following requirements: it is based on personal characteristics of the students; it aims at developing skills (main objective) and transmission of specific content (secondary); contribute actively student in planning, execution and evaluation of them; it is a positive learning experience; and allows transfer of experience to another educational situation. As a general matter, the way of learning is the way of influence how students learn how to teach; also, the content being taught (notions, concepts, procedures, problem solving, attitudes etc.) must determine how to present the information, and the logical sequence and structure, while achieving effective transposition of internal staff. The teacher is no longer just providing information to the student (what to teach but also to help students how to learn).

1 Modern Didactics

The concept of Modern Didactics, develops under the impact of developments in psychology and sociology, with emphasis on the action of learning, with the two protagonists, the teacher and the student. The focus is on the interaction teacher - student in the action of teaching and learning. Miron Ionescu believes that "modern/current didactics incorporates the entire scope of traditional didactics and extends its content through:
- Studying the formal and even informal educational influences;

* Corresponding author: lucia.mara@ulbsibiu.ro

- Valuing the elements that ensure self - instruction (and self-education);
- Studying the issue of instruction and education for all age groups" [1].

Easy to understand why modern didactics concerns include new challenging themes that also aim interdisciplinary approaches, such as: learning by machines, instruction and self-instruction computer-assisted; multimedia learning; using technical means training and self - training; pedagogical programming; studying the effectiveness of different ways of organizing the instructive - educational activities. Comparative analysis between traditional didactics and modern didactics also changes the general concept on the teaching-learning process. It goes from the conception of magistro-centrist type at that of psycho-centrist or socio-centrist. Modern didactics becomes a didactic of active participative methods, where the student becomes an active subject of his own training, trained in all phases of the educational process. In formal and informal contexts; exploiting certain forms of work that are complementary to the lesson, training the rational style of intellectual work; leisure pedagogy; self-training, etc.

2 Nowadays Didactics

The post modernization effect of the school is seen as a current solution because focuses as a centre of all concerns the student/subject of research, but not an abstract, ideal student that is out of the reality in which he lives, but a specific student with his difficulties and achievements. In agreement with these the school must also meet the requirements of the individual. The idea is captured very well by E. Paun, "we can speak of a return of the individual as an actor in the social space and a resurrection of the student as a person with his specific - differentiating features that have to be maximal valued, constitutes the dominant dimension of postmodern pedagogy" [2]. This remembrance of the beginnings of traditional didactics to the postmodern one, emphasizes the evolution of didactics concept, evolution that captures the ample openings and meanings that were captured over time by the fundamental processes (teaching, learning, assessment) of education, main actors (student and teacher) and the very concept of educational process.

General school didactics focuses on school practice experience, illustrated in methodical guides, rules that apply to the learning process as a whole. It also, designates necessary principles for this process, ensuring the basis for developing special didactics of specialty, also offering the required orientation to solve specific issues of any special didactics. Thus, general didactics offers to special didactics the necessary theoretical basis, guiding framework: knowledge, guidelines, trends, norms, rules, principles, etc. Special didactics of specialty, target specific issues arising from the application of general school didactics in different subjects, namely: educational objectives of their teaching and learning, specific content, specific teaching principles, appropriate methodology and assessment. Special didactics of specialty do not support the takeover, rigid and uniform application of general didactics recommendations, but focus on their application. Make available practical material for theoretical developments and generalizations of general school didactics. Between general school didactics and special didactics of specialty are determined interdependence relationships. Pedagogic training of a teacher can not be restricted to special teaching. It must include the theory of objectives, forms and teaching technologies, knowledge of human development factors, of its bio-psycho-socio-educational condition.

Moving to university didactics, we note that it "analyzes the processes of scientific teaching and learning, taking account of the organizational conditions of personnel, content, methods and teaching materials, recommending, based on research, effective means of teaching, learning and assessment" [3]. It became increasingly apparent the awareness of the need for pedagogical culture, at university level, overcoming the prejudgments according to that the training of necessary pedagogical skills at university level would require only knowing the taught specialty. Miron Ionescu shows that macro pedagogy/pedagogy of systems, "perceives the school as a whole comprehensive integrated subsystem of the society"; micro pedagogy/learning pedagogy and self-learning "addresses in mono - or interdisciplinary manner the learning, fundamental human activity, decisive in solving all social problems" [1].

The trends of modern didactics have known an explosive growth also in today's society, so we enumerate the premises of these developments by identifying in the literature, the following directions:

- Action of prospecting, as a way of anticipative studying of training and education of future society. G. Vaideanu, characterizes training and education "are ordered more and more by the future" [4], which requires, obviously, the need of their exploring their future society. "Until now", M. Ionescu said, "pedagogical research studied more the state than the change, adaptation than creation, the present than the future, the state of things that cannot be accepted" [5].
- The comprehensive, inclusive and global approach of the training and education process, namely, its components.
- Focusing the educational research on learning, in structural and functional relationship with teaching and assessment. It speaks more and more now that learning causes teaching (teaching should meet the needs of learning), and that assessment should assist, regulate, direct and optimizes the process throughout its duration and through this, and trainee's performances.
- In agreement with the theory of curriculum, redesigning the existing relations in the educational process between the "subject" and "object" of education, between the scheduled studied discipline and its formative effects on short, medium and long term.
- Re-assessment and selection of content education both in the sense of its openness to the "new educations", as well as its selection after the formative value and, especially the restriction by deepening the reference sphere of education content to specific pedagogical values.
- Defining the content of education by its reporting to all types of curriculum specific to an efficient education: common curriculum - differentiated curriculum; compulsory curriculum - school curriculum; formal curriculum - curriculum informal etc.
- Identifying certain working strategies and some forms of organization of teaching activity that responsibly engage students, while being operant and productive. Learning by cooperation asks for relational components of activity and better values the intellectual potential, of initiative and creativity of students, forming at the same time, availability of communication and cooperation.

3 Educational System

Also in the Romanian education system, the framework of curricular and extracurricular educational activity constitutes the space that is able to meet the challenges of contemporary society, in the extent in which its flexible conception allows a continuous update of content learning and student-centred teaching methods and also a monitoring and assessment of learning outcomes. Also, specific features facilitate the implementation of new teaching approaches by which the student becomes a resource, producer, opinion leader, so active participant. To stimulate the cognitive, spiritual, interpersonal and social development, the educational and extracurricular educational activity has always the need to adapt to individual requirements, different of all children, to the interests of knowledge and to their potential.

The strategy emphasizes the importance of multiplying the positive experiences registered in school and extracurricular educational activities and requires the extension of the intervention space in the educational curriculum process, in order to capitalize all educative strengths of educational learning content in the best interests of the child. Comparing the traditional view on the instructive educative activity and that on the differentiated activity, including that centred-on student/learning subject, in order to highlight the impact of the latter on the development of the student in relation to own potential, it signalled the difference between them [6].

Table 1. Comparative presentation, traditional educational – instructional activity and differentiated activity

Traditional educational – instructional activity	Differentiated activity
• The differences between students are masked; Is acted upon them only when they become problematic; • The interests of the child are rarely called; • Dominates the work with the entire class	• Differences between students are studied and accepted as a start in the instructive – educational process; • Students are guided to their own interests and purposes of the elections guided by them; • Are used different forms of organization of work groups, teams, frontal or individual.
• Time is inflexible.	• Time is flexible, according to the students' needs.
• Are allowed unique interpretations of ideas and events	• Are allowed, even are searched multiple perspectives on events and ideas.
• The teacher directs the student's behaviour.	• The teacher facilitates the independent activity of students.
• The teacher solves his class issues.	• The students help each other, assisted in solving issues.
• The teacher plans sequences, training objectives, provides standards for assessing the work.	• Students work together with teachers to establish goals and assessment criteria.
• Observe what should be learned	• Observe for whom they learn.

Also from a systemic perspective, any educational situation is a system of transactions and simultaneous interactions between the subject of education, the object of education and environment. Such an approach is useful in understanding that any particular situation has several features [7]:
- It is determined by the purposes of education;
- It is dependent on the personal characteristics of the student and teacher;
- Is oriented to achieve certain specific goals;
- It is limited by strategic options, methods, instruments and forms of organization that are available and used;
- It is possible within the limits of the socio-cultural-educational context.

Table 2. The object of education and environment

Purposes	Personal characteristics	Needed cognitive processes	Type of learning
Learn to learn	I know that I know	Metacomponents	Metalearning
Learn what is	What I already learn	Acquisition components	Declarative learning
Learn how to do	What can I do	Performance components	Procedural learning

The teaching activity acquires more complex changes, so if, traditionally, it is just a communication/transmission activity of knowledge, now mostly aims the organization and management of learning processes. According to researchers in the field, developing paradigm/meanings in teaching practice, in teaching activity, in learning and assessment, namely the constructivist paradigm and the interactive paradigm, have an important contribution. In the didactics of science, specifically, it is found that in not emphasized enough the process of building knowledge, but is especially focused on the transmission and acquisition of content of school subjects through mere accumulation of data, facts, information, by use of expository strategy of introducing and explaining concepts. According to researchers in the field, developing certain paradigms/meanings in teaching practice, in teaching, learning and assessment activity, namely constructivist paradigm and interactive paradigm, make an important contribution. In teaching science, it is found that is not insisted enough on the process of building knowledge, but is especially focused the process of transmission and acquisition of content of school subjects through the mere accumulation of data, facts, information, by the use of expository strategies of introduction and explanation of concepts.

4 Learner Centred Perspective

Thus, the researches of cognitive psychology, have an important contribution in increasing the quality of teaching and learning activities. The major characteristic of constructivism is the centred approach on the one that is educated, namely on its activity of building the knowledge, the focus is on individual construction of personal knowledge, on the way that the person acquires knowledge. In other words, the subject customizes the working methods and builds strategies of action tailored to its own training and personality. Musata Bocos, 2002, shows the design of constructivist concept through a set of principles [8]:

• Supporting the subject of knowledge to acquire experience in the process of building knowledge.
• Supporting self-awareness for the building process of knowledge.
• Promotion the use of multiple modes of representation.
• Inclusion of learning in social experience.
• Encourage the subject of knowledge to organize its own learning process and to be involved in its development.
• Contextualization of learning in real and relevant frameworks for the knowing subject.
• Supporting the subject of knowledge to gain experience across the multiple perspectives of addressing topics and appreciation/assessment thereof.

Also, constructivist teaching highlights the focus on student - centred and the teacher pays special attention to pupils' motivation, skills training for autonomous learning and even a certain freedom to choose beyond the core compulsory curriculum, certain disciplines of learning, namely specific contents that they want to learn or other auxiliary that they use. In essence, the entire teaching process becomes one centred on person which means: on the one hand, investigating the knowledge base of the student, knowledge of his cognitive peculiarities; on the other hand, in relation with the identified needs, contents are adapted and the strategies are taught, learning and assessment are individualized.

As a general matter, the way the students learn influences the way in which it should be taught; also, the content that should be taught (notions, concepts, procedures, problem solving, attitudes etc.) determine the way information should be described, logical sequence and structure, and in the same time, effective achievement on internal didactics transposition. The teacher does not only provide information to the student (what to learn, but also helps students how to learn). In this context, contemporary pedagogy in connection with cognitive psychology and to modalities of activation of students proposes the

reconsideration of the learning activity, as a fundamental activity. Learning is described as an individual process, it is done through direct experience with environmental situations; It is strongly determined by previous knowledge base and the teaching process becomes one centred on the learner. Ioana Todor quoting Jonassen, 1994, presents the learning environment characteristics in terms of constructivist theories:
- to offer multiple representations of reality (contributing to the increase of the cognitive flexibility);
- these multiple representations should be adaptive, preventing over - simplification and showing real environmental complexity;
- to facilitate the construction of knowledge in instead of their reproduction;
- to facilitate the solving of genuine tasks in relevant contexts, instead of abstract decontextualized instruction, to involve the application of opened and authentic teaching methods, like laboratory investigation or case study in lieu of rigidly structured methods;
- to encourage the personal reflection on experience;
- to allow the construction of knowledge associated with different contexts of learning;
- to facilitate the collaborative construction of knowledge by the social negotiation of significations [9].

Learning, is achieved in a direct interaction with the environment, proposes a direct experience where the subject is actively involved, these are conditions which ensure success in the learning activity. The constructivist teachers emphasize the need for the process of building schemes/structures of knowledge, knowledge personalization, and student role is to reflect to the new material by using cognitive strategies and metacognitive strategies in their assimilation. Essentially, the constructivist dimension makes two fundamental issues with a focus on which that one that represents the focus on the student, namely: the subject builds knowledge within its own activities; and the activity aims its own knowledge through the elaboration of knowledge by the student himself. As a complementary to the constructivist pedagogy we also enunciate the role of active and interactive learning promoted by the interactive instruction. As Serbanescu Laura highlights, "modern didactic recommends that learning performed by students to be active and interactive, to assume the full involvement of students - intellectual, motor and affective-volitional, to be based on practical activities and experiments, to be achieved by intra - and interdisciplinary connections on an heuristic and problematical context, to stimulate active, critical and creative thinking of students" [10].

5 Conclusions

Thus, the paradigm of student-centred learning determines the need for decentralization and autonomy of education, namely the need for an educational system organized after the model of the network that provides the opportunity for a real adaptation to the needs and possibilities of the student, promoting inclusive school, to social changings through the high level of adaptability of graduates or by focusing on skills and social requirements through the wide range of diversity of "inputs" and "outputs" in the system.

The paradigm of school adaptation to the requirements and possibilities of training of student represents the product of flexible structure of the system, but also the result of multiplying the types of curriculum.

The programs that reflect the flexibility assumed by the network organized educational systems are the open school, school/education for all, inclusive school, school centred on resources, integrated teaching and especially the constructivist paradigm, the paradigm of interactive training etc.

References

1. M. Ionescu, M. Bocos, *Tratat de didactica moderna* (Parallel 45 Pub., Pitesti, Romania, 2004)
2. E. Paun and D. Potolea, *O lectura a educatiei prin grila postmodernitatii. Pedagogie. Fundamentari teoretice si demersuri aplicative* (Polirom Pub., Iasi, 2002)
3. *Dictionary of Pedagogy* (2001)
4. G. Vaideanu, *Educatia la frontiera dintre milenii* (Political Pub., Bucharest, 1988)
5. M. Ionescu, *Didactica moderna* (Dacia Pub., Bucharest, 2003)
6. A. Rotaru, L. Soitu, *Psihologia elevului* (Timisoara, 2002)
7. M. Predescu, L. Soitu, *Psihologie*, (Ex Ponto Pub., Timisoara, 2006)
8. M. Bocos, Instruirea interactiva. *Repere pentru reflectie si actiune,* 2nd Ed. (P.U.C., Cluj Napoca, 2002)
9. I. Todor, *Invatarea de la adulti. Aplicatii in didactica universala*, In Ghid metodologic de pedagogie universitara (Aeternitas Pub., Alba Iulia, 2010)
10. L. Serbanescu, Conferinta Nationala de educatie a adultilor, Timisoara, 2010

University and non-formal education

Liliana Georgeta Popescu[1,*]

[1]Lucian Blaga University of Sibiu, Industrial Engineering and Management Department, 550025, Victoriei Bld. no.10, Sibiu, Romania.

Abstract. Young students place great importance on their personal, professional and educational development alike but in the same time are actively involved in leisure activities. Through non-formal and informal activities the university can help students to develop new skills, can change or increase certain preferences regarding cultural consumption, sports and recreational activities. This paper presents the results of a study based on students attending universities across three cities. It aims to demonstrate that during the years spent at university, students are significantly less influenced by their parents in terms of behaviour and cultural preferences; instead these aspects as well as recreational activities are undertaken by universities and their group of friends and colleagues. For a meaningful analysis and correct interpretation of data, specific tools of quality management were used.

1 Introduction

The current European cultural situation of youth targets mechanisms such as: juvenility (extension of lifestyles of young people throughout the entire life cycle), acculturation (opening young people to the diversity of cultural expressions at a higher level than in adults), marketing (the development of goods and cultural products in the global market), consumerism (simultaneity of the quality of the producer and consumer culture), individualization (expanding the social category of youth in terms of identity), fragmentation (diversification trends in youth culture in heterogeneous groups), globalization versus localization (interdependence and reduced distances combined with a focus on immediate interactions with the environment), transculturalism (increased identification of the marginalized and cultural hybridization), social activism (using culture against forms of discrimination), engineering (development of digital practices as everyday practice) [1].

Young students have various interactive ways and practices of cultural consumption, influenced to a large extent by the use of new technology that allows them to access culture channels that are "dynamic and often informal" [2]. Worldwide there are studies that indicate that there isn't a complete replacement of old practices with new ones, but rather there is a change in the way in which the cultural consumption is done and that new technologies carry out a democratization of access to cultural content [3]. In this regard, students use new technologies in order to obtain information mainly about leisure and

* Corresponding author: liliana.popescu@ulbsibiu.ro

opportunities of school education, jobs and occasional study trips and work but are also interested in social issues (globalization, human rights, social commitment and voluntary) [4].

2 Leisure activities of students

According to the Eurobarometer (published in 2015) carried out at EU level, on a sample of 13,000 young people aged between 15 and 30 years, who were questioned about their participation in society, focusing on their involvement in cultural activities 80% of the respondents to the survey answered that they went to the cinema or a concert in the last 12 months [5].

At a national level, in 2015 young respondents to a survey by the National Institute for Cultural Research and Training [6] chose to consume dynamic, varied, interactive types of culture: 70.7% used culture in the public space through going to the cinema, while 61.6% of young people went to music or entertainment shows, and almost 50% went to the theatre.

Festivals are the main types of cultural activities where young viewers are passive consumers of culture. According to the latest study by the National Institute for Cultural Research and Training in 2015 on a sample of 375 young people age between 14 and 30, the factors that lead young people to attend festivals to the greatest extent were: the possibility to share this experience with other family members - 52.7%; the opportunity to relax and stand out - 46.1%; opportunity to meet people with similar preferences and cultural habits - 42.7% (Table 1). Young people want to experience the festivals in a personal way as much as possible and view them as a leisure activity, their very nature being one of celebration, festivity and are something attractive because it is perceived as "special occasion", as something which brings several "opportunities" together [6].

Table 1. Determining factors of youth participation in festivals [6]

How important are the following aspects when you decide to go to a festival?	not important	Not very important	Important enough	Very important
The opportunity to go with other family members	1,0 %	9,2%	37,1 %	52,7 %
The ability to relax / rest / detach	1,5 %	8,6 %	43,8 %	46,1 %
Opportunity to meet people like you with the same cultural preferences, habits etc.	2,4 %	14,5 %	40,4 %	42,7 %
Famous artists are performing	3,3 %	12,4 %	41,8 %	42,5 %
Ability to learn new things, to enrich your knowledge	0,7 %	14,1 %	43,9 %	41,3 %
The opportunity to experience more cultural products in one place (music, film, dance, theater, etc.)	3,4 %	11,7 %	45,2 %	39,8 %
To have free access	4,6 %	18,4 %	38,1 %	39 %
The opportunity to experience new things	4,5 %	8,6 %	48,2 %	38,7 %
To take place outdoors	4,5 %	22,6 %	37,5 %	35,4 %
To be recognized as a festival with tradition	6,7 %	20,9 %	37,4 %	35 %

The opportunity to experience different cultures and habits other than your own/where you live	3,4 %	13,3 %	48,9 %	34,4%
The ability to see / help / get involved in events and activities that typically you would not have access to	4,7 %	8,9 %	56,4 %	28,9 %

3 Case Study

This case study was conducted by the network of University Cultural Centers from Romania, institutions, which, in association with young people, students and student organizations implement movements that allow young people to become cultural actors, with access to knowledge and to the practice of culture and creation. The activity of University Cultural Centres is to ensure conditions for students to be able to express their initiatives in a complex way, and to show the depth of the student energy and thought. As a culture provider, the University Cultural Centres have to always be innovative, cultural offerings must be always attractive, creative and interactive.

This study follows the relationship between young people and culture, focusing on issues such as the access of students to culture, the level of public cultural consumption, the role of cultural education in access and increasing youth interest in culture.

University Cultural Centres annually organize festivals for national or international students on different areas: music, dance, folklore, fashion and fashion design, literary, visual arts, etc. These types of events are composed of creator-students, performers, producers of culture but also of those who passively consume culture (spectators). For both groups the events deliver cultural education. It should be emphasized that the University Cultural Centre includes all cultural areas, therefore the only cultural institution where the resulting cultural product is as much a concert as show of folklore, as theatre, as dance, as music, etc.

Students represent the category of youngsters who place great importance on their personal, professional and educational development, but are also actively involved in leisure activities. Practical reasons why young people pursue artistic or creative activities, according to a study undertaken in universities from Sibiu, Cluj-Napoca and Ploiesti on a sample of 500 students, are shown in Figure 1.

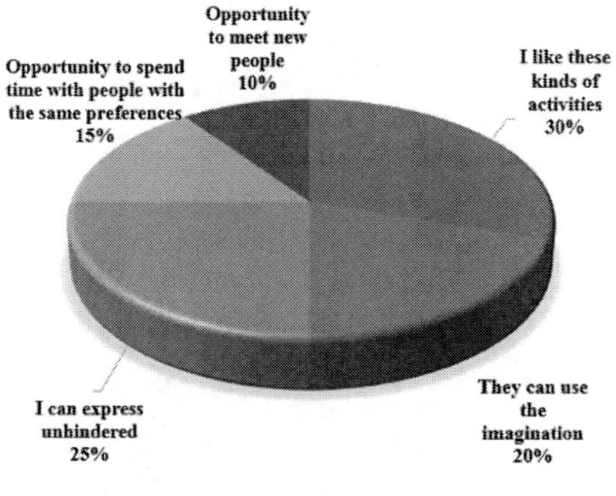

Fig. 1. Reasons why young people practice artistic activities.

According to the questionnaires drawn up annually by the University Cultural Centre of Sibiu the members of the bands and groups working in the institution, 350 respondents said they are interested in learning something new (35%), the institution supports their launch /

promotion at national level (24%), the institution offers them the chance to visit new places by participating in national and international festivals (20%), they have the opportunity to spend time pleasantly in a group of friends (10%) that want to develop ideas outside the university (10%), others (1%). (Figure 2.)

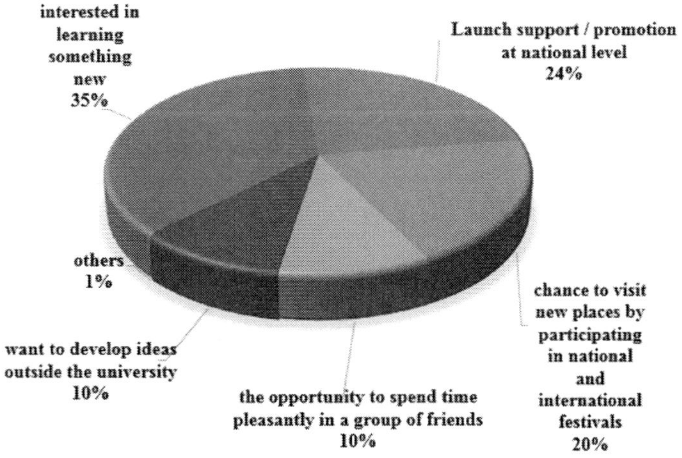

Fig. 2. Reasons for students working in artistic societies.

Nationally the most common leisure activities of young people, regardless of the age or gender are photographing and filming. (Figure 3, Figure 4).

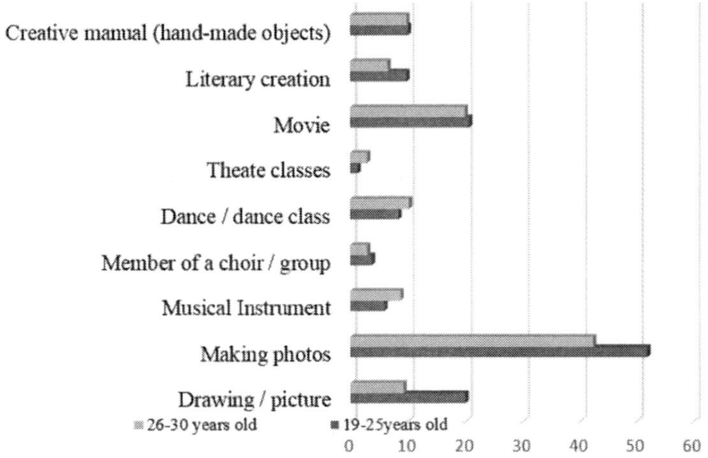

Fig. 3. Youth arts activities, analysis by age.

The graph in Figure 3 contains the values for the answers "YES, I have done..." and the distribution by gender (female / male) responses to the questions asked is shown in the graph in Figure 4.

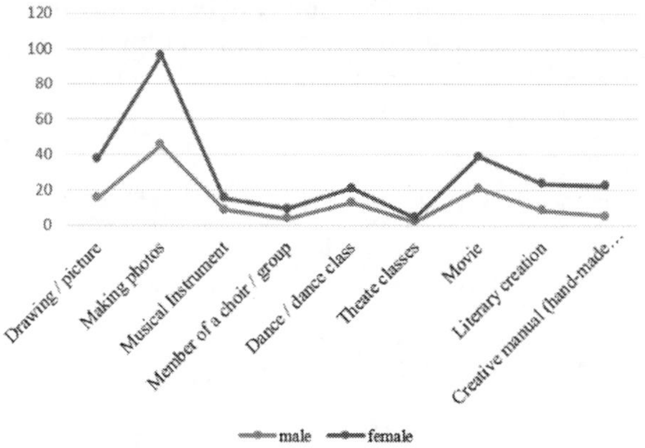

Fig. 4. Youth arts activities, analysis of gender categories.

Having in view these results, and the results from other studies [7, 8] the universities or the universities cultural centres could develop new extra-curricular activities, non-formal activities for students in these fields. Photography is the most appreciated activity whatever the field of study in which students are enrolled. Also, they are interested in learning something new. Students and students' associations have to be stimulated by the universities or by the university cultural centres to develop extra-academic activities by:

- providing support for the operation of the educational process;
- development and support of the creative activity;
- support the practical training of students;
- involvement of students in the formative process modernization;
- increasing artistic competence of students.

The problem linked to the students' involvement in projects remain open for the future. A quality management system, an essential component of the overall management for any organization, must allow actions and procedures for continuous quality improvement [9]. The ways for involving students in the formative process modernization is achieved by:

- defining the priority directions of development consistent with existing human resources and infrastructure and tightly correlated with the students' demands;
- identifying and organizing new extra-curricular activities taking into account the space used;
- boosting enrolment in such activities through their participation in various national or local events;
- identifying resources for financing to increase participation in domestic and international competitions fundraising.

Art and culture forms multiple and constantly changing, according to tastes, places and period. They are, in a way, part of the heritage of the past, present and future, personal and collective, to which successive generations contribute. There is a kind of reflection of each society. Students through their practice of culture and their capacity for initiative, exploration and innovation, build and play a role in these cultural developments. It is therefore important to allow access to culture in all its forms and to promote their possibilities for creative activity including in new areas [1].

4 Conclusions

Knowing how young people spend their time outside the university is very important in terms of offers and services provided by cultural or other organizations. Considering the lack of extensive research on the relationship between young peple and cultural activities, as well as the constant development of new technologies and it's implications in leisure practices, there should be an increased focused on the offer that cultural centers put forward and they should adapted based directly on the requirements of young consumers.

Currently there is a change in the way in which the cultural consumption and new technologies is accomplishing a democratization of access to cultural content. This study followed the relationship between young people and culture centered on issues such as the access of young people to culture, the level of public cultural consumption, the role of cultural education in access and increasing youth interest in culture.

References

1. https://www.coe.int/t/dg4/youth/Source/Coe_youth/Participation/COE_charter_participation_ro.pdf, Carta Europeana revizuita privind Participarea tinerilor în viata locala si regionala, Congresul Autoritatilor Locale şi Regionale din Europa
2. ***, Interarts Foundation, *Access of Young People to Culture.* Final Report, European Commission, Directorate-General for Education and Culture (2010)
3. ***, Observatorul Cultural Basc, *Young people, culture and new technologies*, 11
4. D. Pozzolo, *Youth and cultural consumption in Piemonte – a first glance,* in Osservatorio Culturale del Piemonte, (www.stat.gouv.qc.ca/statistiques/ culture/pratiques- culturelles/consommation-piedmont.pdf)
5. ***, Comisia Europeană, Flash Eurobarometer 408, Tinerii din Europa, http://ec.europa.eu/youth/news/2015/docs/ youth- eurobarometer_en.pdf (2015)
6. C. Croitoru, *Barometrul de consum cultural 2015 Preferinţe, practici şi tendinţe, (*Editura Pro Universitaria, Bucureşti, 223 2016)
7. G.M. Moraru, L. G. Popescu, M.V. Zerbes, Proceedings of International Multidisciplinary Scientific Conferences on Social Sciences and Arts, Bulgaria (2015)
8. L.G. Popescu, G.M. Moraru, Proceedings of 10th International Technology, Education and Development Conference, Spain, 7675 (2016)
9. L.G. Popescu, Proceedings of EDULearn - 7th International Conference on Education and New Learning Technologies, Spain, 0681 (2015)

Student entrepreneurship – research on development

Małgorzata Rembiasz[1,*]

[1]Poznań University of Technology, Faculty of Engineering Management, Strzelecka 11, Poznań, Poland

Abstract. The article discusses selected theoretical aspects of student entrepreneurship and the findings of the author's research in this field. Academic entrepreneurship programmes constitute a significant component of higher education. In the empirical part of the study, which relied on auditorium surveys, the author found that students are strongly motivated to engage in business pursuits and identified barriers to their entrepreneurial engagement. The respondents varied widely in their opinions on the nature of entrepreneurship and enterprises. Such a broad range of opinions underscores the complexity of entrepreneurship in today's economies. Programmes of entrepreneurship education in higher education systems should place greater emphasis on the formalities required for starting a business and raising funds for this purpose. The development of entrepreneurial skills from a young age should be recognised as critical for creating a community capable of functioning in a knowledge-based economy.

1 Introduction

In the classical view represented by P. Drucker, entrepreneurship is seen as a specific approach to management, which involves "adopting innovations that constitute the basis of future business" while "making the best possible use of the available resources" [1].This suggests a strong link between innovation and entrepreneurship. The existence of the link was posited earlier by J. Schumpeter who proposed that one only becomes an entrepreneur provided one engages in innovation and that one stops being one as soon as one settles into an entrenched market position and routine ways of operating one's business. Furthermore, according to Schumpeter, entrepreneurship is a driver of all rapid changes in economies [2].

A contemporary approach consistent with the definition proposed by the Global Entrepreneurship Monitor views entrepreneurship as "any attempt at new business or new venture creation, such as self-employment, a new business organisation, or the expansion of an existing business, by an individual, a team of individuals, or an established business" [3].

The term *entrepreneurship* has multiple meanings. Due to this variety, both the theoretical and practical aspects of entrepreneurship provide fertile ground for the

* Corresponding author: malgorzata.rembiasz@put.poznan.pl

development of research in a wide range of fields, predominantly management, economics, sociology, psychology and education. Hence, entrepreneurship literature also abounds with interdisciplinary publications. As a consequence, a number of definitions and classifications of entrepreneurship have been proposed ranging from simple to highly complex [4].

The central role of education in this context across nearly all fields and levels is not only to offer knowledge that matches the needs of existing businesses but also to encourage and prepare students to be open to self-employment and job creation. The article presents selected theoretical aspects of student entrepreneurship along with findings of own research in the field carried out in two Poznań institutions of higher learning.

2 Academic entrepreneurship – selected theoretical aspects

2.1 Academic entrepreneurship as a driver of growth of national economies

Entrepreneurship propels the growth of Poland's economy in a number of ways. Small and medium-sized enterprises exert a significant impact on numerous macroeconomic parameters. SMEs account for a growing share of international trade and for 65-70% of the total employment in business organisations across Poland and many other countries [3].

The performance of SMEs also reflects engagement in business and contributes to the rise of the middle class in society at large.

Against the background of Poland's overall economic growth, the advancement of the country's small and medium-sized enterprises is contingent upon the successful adoption of various forms of innovation on a larger scale [5].

Defined broadly as the integration of science and business, student entrepreneurship may take any of a number of various forms, the most prominent of which being educational cooperation, business counselling, the organisation of joint research programmes and the creation of enterprises by researchers and, above all, by university students and graduates. Academic entrepreneurship relies chiefly on the creativity of student communities and the extent to which their outcomes are applied in business practice. Efficient cooperation between the academic and economic sectors should be a cornerstone of the national economic policy [6].

A major role in the economic development of any country is played by family firms. This is because the values cultivated by generations of family entrepreneurs tend to inspire young people and family business successors [7]. Note also the strong link that exists between family and student entrepreneurship. As demonstrated by K. Wach by means of empirical research, the family business exposure has the effect of encouraging students to engage in entrepreneurial activities. Such exposure drives the young generation to embrace entrepreneurial values and appreciate the prominent role of entrepreneurs in society. As a consequence, young people show a greater propensity to taking business risks [8].

Note also that the entrepreneurial spirit can be instilled in young people even in childhood. As the bulk of child rearing and education in Poland and many other countries is in the hands of mothers and grandmothers, it is essential for the state to support women's entrepreneurship across all age brackets and family backgrounds [9]. This applies in particular to single mothers. Since more than 20% of children in Poland are raised by single parents, it is crucial to enable single mothers to contribute actively to the social and economic growth of the country also by starting and running their own businesses. This will enable them to become entrepreneurial role models for their children [10]. Instilling entrepreneurial attitudes from an early age should be recognised as a critical part of child development and, consequently, as a prerequisite for creating a society capable of operating successfully in modern knowledge-based economies.

Just as any form of engagement in business, student entrepreneurship ties closely to the political and public approval for self-employment. It appears that not enough attention is paid to the role of entrepreneurship in today's market economy, including that of Poland, the role of small and medium-sized businesses in national economies, entrepreneurship education, the popularization of self-employment as a viable career option and the encouragement of self-employment. To remedy these shortcomings, academia need to work closely together with business practitioners [6].

2.2 Higher entrepreneurship education – selected theoretical considerations

Entrepreneurship ranks among the so-called key competences in the European educational system. The idea of including entrepreneurship education at every level of learning holds a prominent place in European Union policies, especially since the start of the 21st century [11]. Entrepreneurship education is a priority in the EU's education policy. The Bruges Communiqué on enhanced European Cooperation in Vocational Education and Training for the period 2011-2020 called upon the participating countries to "support newly-established and future entrepreneurs by encouraging start-ups for VET graduates and by promoting learning mobility for young entrepreneurs" [12].

A particular role in encouraging student entrepreneurship should be played by education in the use of modern IT tools in business. For instance, companies are increasingly promoted not only with their own websites but also through a range of social media [13, 14].

According to Okoń-Horodyńska, education today should rely primarily on teaching students to act independently, self-organise quickly and adapt their business to changing micro- and macroeconomic environments. In the process of education, pro-entrepreneurial behaviours should be strengthened by providing multidimensional knowledge in economics, psychology, cultural sociology, anthropology, information technology, social communications and law [15].

3 Student entrepreneurship – empirical research findings

3.1. Description of survey

In January 2017, a survey was conducted of the students of three institutions of higher education: the Poznań University of Technology (students majoring in Technical/IT Education), the WSB of Poznań (students majoring in Business Law, Logistics and Tourism and Recreation) and the State School of Higher Vocational Education in Leszno (students majoring in Economics). The study sample comprised of the students of economics programmes as well as those of technical studies leading to engineering degrees.

The study relied on quantitative methods and specifically the auditorium survey. The survey form was completed by a total of 166 respondents. The majority of them (68%) were women. 89% belonged to the 18-25 age group. 42% of the respondents resided in rural areas, 19% were large city dwellers while the remaining participants indicated they had come from smaller towns (with populations of less than 200,000). The majority of the respondents (57%) were enrolled in the second year, another 31% in the first year while the remaining ones attended the third, fourth and fifth years of their programmes. Only 10% of the surveyed had already completed a higher education programme, meaning they were pursuing a second degree.

3.2. Attitudes towards entrepreneurship by surveyed students

Attitudes towards entrepreneurship correlated closely with the respondents' intent to set up a business. The results are shown in Table 1.

Table 1. Respondent declarations of intent to start and run own business

Current employment and plans for the future	Percentage of responses [%]
I operate my own business of which I am either an owner or co-owner	7.2
I am employed but want to start my own business	10.2
I am not employed but want to start my own business	21.7
I am employed and do not know if I want to start my own business	22.3
I am not employed and do not know if I want to start my own business	28.3
I am employed and do not want to start my own business	3.6
I am not employed and do not want to start my own business	6.6

Source: own research

The above summary of responses shows that 39.1% of the respondents displayed an active entrepreneurial mind-set: 31.9% planned to start their own business while 7.2% already owned one. Meanwhile, only 10.2% of the students indicated having no entrepreneurial intent. The largest group (of 50.6%) were the undecideds, who were unable to define their intents concerning their future self-employment.

A significant aspects of the research on the respondents' attitudes towards entrepreneurship was their self-perception as enterprising people. However, a survey held in 2010 and repeated in 2016 in the institutions of higher education in the Polish region of Wielkopolska showed that a person's self-perception as being an enterprising individual did correlated poorly with plans to set up one's own firm. The 2016 phase of the study demonstrated that students tended to view themselves as enterprising. Such a self-perception was indicated by 68.8% of female and 59% of male first-year students as well as 64.9% female and 70% of male third-year students. However, only a certain proportion of such respondents were prepared to use such confidence to actually start a business. Studies have found that 50.1% of female and 38.5% of male first-year students as well as 37.9% of female and 47% of male third-year students intended to start up their own company in either near or more remote future. Note that a comparison of the 2010 and 2016 findings shows that although the respondents to the 2016 study thought of themselves as being more enterprising, the number of students planning to go into business for themselves declined whereas the number of undecideds and those determined not to engage in a business venture went up [4].

Note also the findings of a study comparing the attitudes towards entrepreneurship displayed by the students of programmes in economics, social studies and the humanities. The study found that the majority of the students consider themselves to be enterprising, which they understand as being able to take risks, be ingenious and creative and show initiative. However, some 40% of the student sample did not, during their studies, holds jobs associated with their majors. According to the authors, the main barriers to such engagement were the students' reluctance to build their careers shortly after enrolment with their universities and lacking a business idea for starting and operating their enterprise [16].

One other factor of possible significance in this context is the way entrepreneurship is viewed by students. The above-mentioned 2016 study requested that students define the notion of "entrepreneurship". The respondents offered a wide range of interpretations. Some perceived entrepreneurship mainly as running one's own business and generating an income. Others focused on the other meaning of the Polish equivalent of the term focused on "enterprise and resourcefulness". Such students saw entrepreneurship as a set of personality traits. All in all, the study provided further evidence for the great complexity of entrepreneurship in today's economy [4].

3.3 Barriers to the advancement of student entrepreneurship

Factors for the advancement of student entrepreneurship should be identified against the background of barriers to entrepreneurship found in the economy. Literature frequently refers to the main five categories of barriers identified by the Polish Agency for Enterprise Development, which are legal, economic, managerial, educational and social.

In the study in question, the respondents were requested to select what they believed were the top three factors that tend to frustrate the intents to start and run a business. Their opinions are summarised in Table 2.

Table 2. Key barriers to starting and operating own business

Factors hampering the development of entrepreneurship	Percentage of responses [%]
Tax system	59.6
Difficulties with securing funding	59.6
Absence of a "good business idea"	54.2
Little time left for personal life	30.7
No knowledge on how to operate a business	42.8
No support from economic policies	34.9
No support from family	4.2

Source: own research

According to the respondents, the key barriers to starting and running one's own business in Poland were the tax system and difficulties with securing funds. The very same barriers have been identified in a variety of reports published by the Polish Agency for Enterprise Development and in a number of scientific studies. Small and medium-sized business owners frequently point to difficulties accessing third party capital as a major inhibitor of entrepreneurship. Precisely this conclusion has been proposed by T. Łuczka [17, 18]. Note that the research described in this article suggests that only a mere 27% of the respondents searched the Internet for ways to finance a business venture over the preceding year.

Note also that 42.8% of the respondents pointed to the lack of knowledge on how to operate a business as one of the top three obstacles. This particular barrier can be dismantled by offering entrepreneurship education at all levels of schooling, particularly in higher learning.

A study by K. Dąbrowska and M. Skowron found that young people rarely notice good opportunities for starting a business in their environments. Therefore, one of the main responsibilities of institutions of higher education should be to promote entrepreneurship by encouraging young people to take advantage of the opportunities by starting their own businesses. To make that possible, it is vital that an entrepreneurship-friendly atmosphere be created in the market environment not only at the regional but also the local level, thereby strengthening the resolve of potential entrepreneurs [16].

3.4. Student motivations to start/run a business

A number of systems are available for classifying drivers for the establishment and advancement of entrepreneurship, including student entrepreneurship. One of them, designed by D. Piróg, resulted from an overview of multiple perspectives on the issue. Piróg's classification refers to both external factors: (1) basic macroeconomic indicators, and (2) social influences – the social capital, as well as external variables: (3) competences and qualifications – education and experience, and (4) the personal traits of individuals [19].

Research conducted by the students of the Pedagogical University of Kraków has identified a plethora of factors that affect young people's entrepreneurial intent. In keeping with planned behaviour theory, some of the most important of them are views on private enterprise, the propensity to take risks, subjective public perceptions of private enterprise, the self-perceived effectiveness in business ventures and early exposure to business within one's family [20].

The participants in the research carried out by the article's author pointed not only to barriers to but also to factors that foster the starting and running of one's own enterprise (they were asked to select the top three factors). The students' choices of the three most influential drivers of student entrepreneurship are provided in Table 3.

Table 3. Key incentives to start and run a business

Entrepreneurship driver	Percentage of responses [%]
Opportunity to earn higher income	73.5
Opportunity to pursue one's passions	65.7
Independence ("no boss")	60.8
Opportunity to balance work and personal life	39.2
Necessity (no jobs available)	4.8
Access to state support available under economic policies	4.2

Source: own research

The bulk of research on entrepreneurship points to the opportunity to earn a higher income as the top driver of entrepreneurship. The income factor was also the incentive most frequently selected by the student respondents to this study. The second most commonly chosen factor behind business start-up was the chance to pursue one's passion, the third: independence. Such findings are consistent with the overall profile of the so called generation Z, which is currently entering the labour market and to which the majority of the respondents belong. This demographic cohort is highly mobile, individually-minded, creative and valuing self-improvement and independence. Note also that a substantial proportion of the respondents (39.2%) selected the opportunity to balance one's work and personal life. This emphasis can be associated with the fact that the study sample was comprised of predominantly female respondents. For the sake of comparison, an impressive 30.7% chose limited time for personal life as a factor discouraging them from starting a business.

As noted earlier, students' attitudes towards entrepreneurship are largely affected by their social backgrounds, including exposure to enterprise pursued by either their family members and close friends. The study conducted for the purposes of this publication showed that 24% of the respondents did not know closely anyone who ran their own company. 45% of the students came from families whose close members (parents, grandparents, siblings, aunts or uncles) owned their own businesses whereas 31% indicated that an extended family member or a close friend was a business proprietor. The conclusion therefore is that exposure to the business experience in one's family had little effect on encouraging students to set up their own ventures. Perhaps being witness to the hard work and constant strife against many odds (including a rather unfriendly tax system) as well as the recent decline in unemployment in Poland discourages young people from engaging in their own business pursuits.

4 Conclusions

Enterprise ranks among the key factors for improving the competitiveness of businesses and promoting the social and economic development of countries. Whereas student entrepreneurship, as examined in terms of selected aspects, appears to be a highly complex issue, its advancement can certainly be ensured by promoting self-reliance and encouraging entrepreneurial intent, mainly in universities. Today's free-market economy clearly requires an entrepreneurial spirit and managerial skills on the part of graduates in nearly every field, not just economics and business management.

Particular importance can be ascribed to, on the one hand, the incorporation into university curricula of legal, economic and institutional aspects of entrepreneurship, the commercialization of science and technology and intellectual property rights and, on the other, collaboration between institutions of education and business practitioners, particularly enterprises and their inclusion in teaching practice, as well as the development of entrepreneurship infrastructure institutions and programmes (Poznańska 2014).

Research has shown that nearly 40% of the surveyed students demonstrate an active entrepreneurial mind-set while only 10% categorically rule out the option of running their own company. Entrepreneurship education should focus on reducing the number of the undecideds (who account for approximately 50% of the surveyed) who are unable to define their intent to start up their own business. Specifically, such education should focus on identifying the benefits of actively committing to a business venture and on providing knowledge that can be transformed into entrepreneurial competences and lead to starting a business. This is particularly critical in view of the fact that a staggering 43% of the respondents named the lack of business knowledge as one of the top three barriers to starting and running an enterprise. Furthermore, according to the respondents, the prime reasons for choosing to start one's own business are opportunities to earn a higher income and pursue one's passions as well as independence. Such an outcome stems from the fact that the majority of the respondents belonged to the so-called generation Z, which is characterised by great self-reliance, creativity and emphasis on self-improvement.

Note that according to J. Wasilczuk, who is an experienced researcher of entrepreneurship in Poland, one cannot escape subjectivism in either qualitative or quantitative research on entrepreneurship. By adopting a subjective view, scholars are likely to propose new research methods and new approaches. Subjectivism also tends to promote arbitrariness in research making it difficult to provide definitive answers to research questions. Two potential solutions to this problem are to consult research methodology design decisions within larger teams of researchers concerned with a specific area and to abandon large-scale studies on heterogeneous samples in favour of smaller-scale projects using homogenous samples, which are better positioned to offer more profound insights [21].

The author realises that the discussion provided in her article is fairly general. She is therefore planning to conduct more research at a greater level of detail on student entrepreneurship and its drivers and, in particular, the needs of entrepreneurship education programmes seen as providing a fundamental social competence. Continued research in the field may help strengthen entrepreneurial attitudes among young Poles.

References

1. P. Drucker, *Innovation and entrepreneurship. Practice and principles*, (Heinemann, 1992)
2. J. Schumpeter, *The Theory of Economic Development: An Inquiry into Profits, Capital, Credit, Interest and the Business Cycle*, New Brunswick and London (2008)
3. *Report on Global Entrepreneurship Monitor – Poland 2015*, PARP (2016)

4. T. Łuczka, M. Rembiasz,, Horyzonty Wychowania, **15 (34),** 27-47 (2016)
5. A. Borowiec, *The Use of Electronic Tools in Public Procurement as a Factor Spurring Demand for Innovative Solutions in the Economy, according to Contractors, IMECS 2016,* 418-431 (2016)
6. K., Poznańska, *Przedsiębiorczość akademicka – cechy i znaczenie w gospodarce światowej i polskiej.* in: T. Kraśnicka (ed.), *Innowacyjność współczesnych organizacji. Kierunki i wyniki badań [II].* Katowice: UE, 164-172 (2014)
7. B. Roszkowska-Mądra, A. Parfieniuk, M. Studnicki, Studia Ekonomiczne **6 (72),** 217-235 (2014)
8. K. Wach, L. Wojciechowski, EBER **4 (1),** 83-94 (2016)
9. P. Siemieniak, T. Łuczka, *Przedsiębiorczość kobiet. Wybrane aspekty ekonomiczne i psychokulturowe*, Poznan (2016)
10. M. Rembiasz, Zeszyty Naukowe Politechniki Poznańskiej. Organizacja i Zarządzanie **68,** 167-179 (2016)
11. K. Wach , Horyzonty Wychowania **13(26),** 11–32 (2014)
12. http://ec.europa.eu/education/policy/vocational-policy/doc/brugescom_pl.pdf (as accessed on 15.01.2016)
13. J. Pawlak, P. Siemieniak, Zeszyty Naukowe Politechniki Poznańskiej. Organizacja i Zarządzanie **68**, 179-190 (2016)
14. P. Siemieniak, M. Rembiasz, J. Pawlak, *Internet as a tool to promote women's entrepreneurship. Selected aspects,* Studia i Materiały (to be published)
15. E. Okoń-Horodyńska,. Nauka i Szkolnictwo Wyższe **1(31),** 34-54 (2008)
16. K. Dąbrowska, M. Skowron, Annales. Etyka w życiu gospodarczym, **18** (**3**), 121-131, (2015)
17. T. Łuczka T., *Makro- i mikroekonomiaczne determinanty struktury kapitału w małych i średnich przedsiębiorstwach*, Poznan (2013)
18. J. Małecka, T. Łuczka, *The Structure of Venture Capital Raising by Companies in Poland and Central and Eastern Europe: Selected Aspects, IMECS 2016,* 418-431 (2016)
19. D. Piróg,. Przedsiębiorczość – Edukacja **10**, 306-315 (2014)
20. T. Rachwał, K. Wach., Przedsiębiorczość – Edukacja **12,** 405-415 (2016)
21. J. Wasilczuk, Problemy Zarządzania **1** (**51**), 13-25 (2015)

Improvement of learning quality in manufacturing technology: a case study

Ionela Magdalena Rotaru[1,*]

[1]Lucian Blaga University of Sibiu, Industrial Engineering and Management Department, Victoriei, 10, 550025, Sibiu, Romania

Abstract. The paper presents the initial part of a study designed to understand the place of the present generation of engineering students in the global picture of their working life. The education represents an inestimable good that accompanies peoples through their existence. Higher education is the place were teenagers with different profiles modelled by their families, studies; experiences and so on come to meet, in most cases, the final education platform before employment. In light of this the higher education has the major role of prepare the students with the knowledge and skills appropriate to the labour market. That means with the knowledge and skills they need and their future employers are looking for. It involves a huge effort, dedicated to the transfer of knowledge and skills with a success rate as high as possible. To achieve this the study programs and the teaching methods must embody both objectives: to be tailored for the students and to meet the employers needs. That places the higher education on the border of the two parties, its role in achieving the outlined objectives being major.

1 Introduction

The role of higher education to ensure the transition from student to employee requires the fulfillment of major demands such as:

- Provide specific knowledge aligned with current requirements. In automotive area, where the boundaries are in continuous movement, the alignment to the new technologies is a necessity in order to survive in the competition "Today, the practice of engineering needs to change further because of the demands for technologies and products that exceed existing knowledge bases and because the changed professional environment in which engineers need to operate. That change must be encouraged and facilitated by change in engineering education" [1].
- Engineer's role is becoming more complex and this represents one of the challenges the education faces "We have to change what we expect from engineers, and we have to turn out our graduates with broader skills, interests, and abilities...I need graduates who can speak before an audience to make a point...Engineers must be able to write reports, studies...." [2].
- The present generation, known also as Millennial generation, differentiates from the previous generations through their preferences for collaborating, connecting and creating

* Corresponding author: ionela.rotaru@ulbsibiu.ro

social change. [3] Connecting represents the major characteristics. Their unlimited connectivity to the digital technologies is underlined also by the fact that they are known also as "digital natives" [4]. That means not only unfettered electronic access to the information's, friends, instructors, classmates [3] through their own devices and social networks but also a current personal status highly influenced by this state of facts.

- Their experiences, perceptions, skills and needs are facts that must be taken in considerations long as all of these are reflecting in their expectations regards the educational process.

2 Objective

The major objective of this study is to model an algorithm that may be used in the practice classes of manufacturing technology. This objective must embody all the major needs presented above. It should be created a learning environment that enables learning by appealing to the specific characteristics of this generation. This are integrated into the teaching activity by the teaching methods used and it refers at issues like the ones presented below:

- collaborating, team working - the teacher role is to lad activities designed in order to facilitate the creation, structuring and retention of the required knowledge; the way to integrate it into the class activity is trough group discussions, Concept Map elaboration, results: this activity may eliminate the lack of attention, uses their natural abilities and in this way, engage the students in the proposed activities in and outside the classrooms;
- support active engagement in- and outside the class room - the teacher role is to prepare and sustain company visit with a specific topic, group discussions, design of concept maps; results: this activities helps the students to understand in profound the phenomena, eliminate the situations in which the participants doesn't know what should follow, has no questions, and the retained information is superficial.

3 Methodology

The paper presents a case study that started from the necessity of transforming aspects relating with: engaging of all students in the classroom, maintaining interest, a better understanding of issues and propose a combination of activities to ensure a better knowledge retention.

The participants were the students from the final year of Manufacturing Technology study program. Their profile is outlined in the following:

- age average: 22 years;
- female 19%, male 81%;
- foreign languages known: english 85,71%, two foreign languages known 38,1%, from wich 23,8% english and german, 19% english and french;
- activity on social networks: Facebook - 100%, Instagram 28,9%, WhatsApp 85,71%, SnapChat 19%;
- specific engineering software known: CATIA 85,71%, Autocad 66,66%, ProE 23,80%;
- employed 61,90% from wich just one in a field other than engineering;
- hired by 10 months (the average);
- reasons behind the start of the activity: 54,4% experience, 33,33% experience and gain.

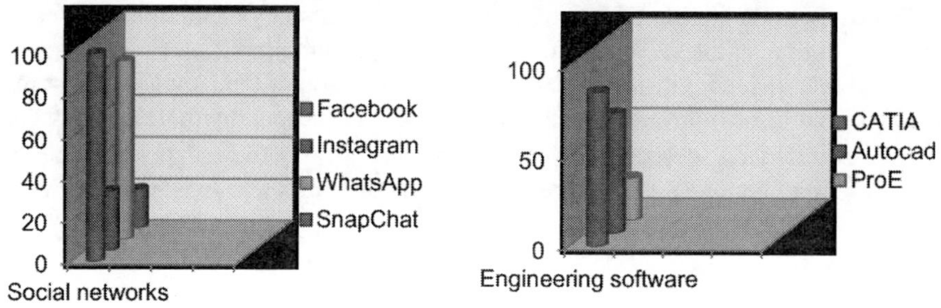

Fig. 1. Computer abilities: social and specific

All this underscores the unlimited access to information, the area of interest and the concerns that define them (closely related to digital technologies), the desire to learn and confirm the necessity to take into account this status that characterizes them.

In order to do this the first step consists in a systematic literature review on current teaching strategies, knowledge management and research concerning the profile of the present generation.

In the light of all these facts a solution is to treat the practical classes based on knowledge management principles. This means the transmission of knowledge in the most appropriate manner, taking into account the type of knowledge to be transmitted (charts, diagrams, tables, theorems, equation, videos, sketches, photos, effective machining - Figure 1) and capturing, structuring, presenting and storage the knowledge using Concept Maps. Their completion will do with a working visit designed so, that by cooperating of the teacher with the company representative the participants to be guided so that they can find, observe, understand and settle as many of the initial knowledge acquired over several practical classes. The first phase has already been checked over the years and the findings indicates its sustainability [6-7]. However the need for a qualitative improvement of the educational process led to the introduction of the second phase. Its stages consists of the steps presented below:

- preparing a file for each participant, the file contain a questionnaire and the technological process for the part to be followed along the process flow - an axe from a shock absorber - Figure 2;
- elaborating a questionnaire that highlights, grouped by series of questions, the knowledge gained over several practical classes - knowledge regarding the manufacturing process, component elements of a technological system, the structure of a technological process, the optimum sequence of operations, the specific technology for axes family. The questionnaire was distributed to the participants in the previous week of the visit and its role is to set a clear objective for the visit and to prepare the participants for it, allowing them to review the proposed subjects in order to identify the topics for wich them have questions. At the base of the questionnaire preparation there were the personal experience and reference works in the field, such as: Mechanical Engineering Design [8], Manufacturing Engineering and Technology [9], Standard Technologies [10], Workbook for Technology of Machine Tools [11], Manufacturing Technology [12], Engineering Design [13], Precision Machining Technology [14];
- at the beginning of the visit, the company representative highlight the functional role of the axe in the assembly shock absorber. In these way the necessity of each stage of the
- manufacturing process was underlined and the manufacturing process as a hole were better understood. Also, the hole technological flow was presented and explained from the warehouse of the semifinished part to the final product;

- then followed the visit itself. At each workplace explanations have been given, the employers shared knowledge also, and the raised questions received answers;
- the final stage consists of discussion session in wich questionnaires have been brought and there were discussions sessions in order to evaluate the activity as completely as possible, there were bothe individual and group interviews.

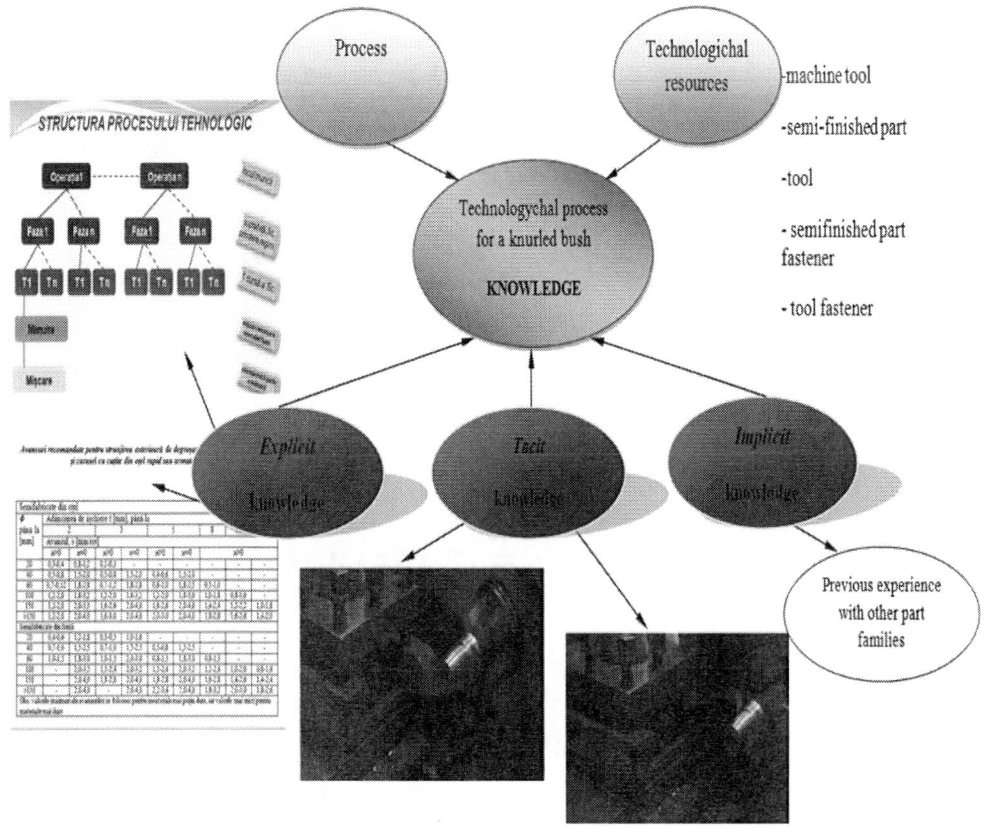

Fig. 2. Knowledge types and their transmission channels [6]

4 Results and discussions

In conclusion, it was proposed a learning improvement cycle wich brings together the practical and theoretical knowledge, the academical and companies environment and demands through sustainable principles dedicated to a group of people wich are characterized by the desire for active learning, connectivity, working in groups and, very important, with a major interest for knowledge acquisition when they understand their place in a current activity, when they see their finality.

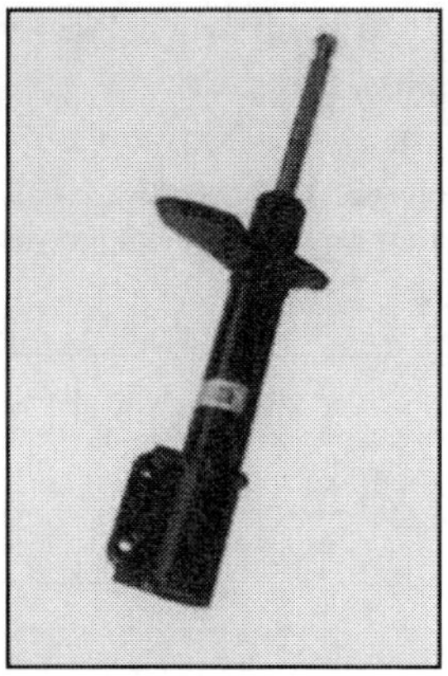

Fig. 3. Shock absorber http://www.thyssenkrupp-bilstein.de/en/produkte/2-rohr-daempfer.php

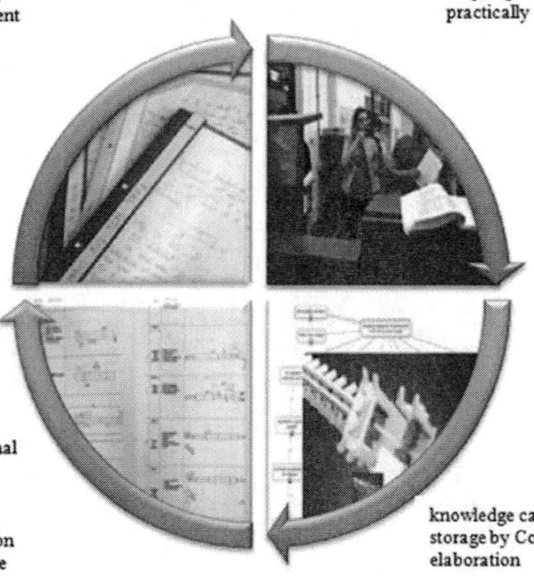

Fig. 4. Learning improvement cycle

The evaluation of the knowledge transfer was made by analyzing the questionnaires and by individual and group interviews. The questionnaires responses were scored on four points on Likert scale - 1 - insufficient, 2 - satisfactory, 3 - good and 4 very good. All the questions proposed in the questionnaire received responses. The quality of the responses is indicated by the test values 21,11% very good, 50,12% good, 19,16% satisfactory and 9,61% insufficient.

The questionnaire has a group of 32 questions grouped in six sections regarding: manufacturing process, component elements of a technological system, semifinished part, the structure of a technological process, the optimum sequence of operations, the specific technology for axes family.

At the first questions, regarding the part group/family from 36 forms only 2 was wrong.

From the group of question concerning the semifinised part everyones answers were good. Their responses to the 4 questions from the group (1. Which was the kind of the semifinished part elected? 2. Why. Justify the answer. 3. Which was the first operation of the technological process? 4. Name another semifinished part for which the first operation would not be needed.).

The structure of the technological process had a group of 10 questions like: which is the first operation of the technical process? Which is the role of the first operation, which kind of operations (outside the machining operations) do we find? It would be a different process structure if we have just 8 parts to machine? Justify your answer. Each of them responded to the questions and even there were some wrong answers, it indicates a better understanding of the problem, the fact that the knowledge reached to them.

The student's perception was positive about the entire practice classes and regarding the visit, at the question "What do you liked concerning the visit?" they were having remarks as: "I find the visit interesting", "I like the idea and the initiative", "the fact that I saw the entire technological manufacturing process for the part", " the collaboration with the company representative and the professionalism with which the visit was treated". At the question "What didn't you like during the visit?" the responses was concerning the short time (two hours for a group), the number of participants from a group (around 12 participants), the fact that near the machines they could not hear the explanations very well.

In conclusion, active participation, the acquired knowledge, find their place and meaning in the complex mechanism of a producing company. The integration of knowledge in the reality indicates the finality of the hard work done and encourages the interest and sustained effort.

References

1. National Academy of Engineering, *Adapting Engineering Education to the New Century,* (Aviable at: http://c.ymcdn.com/sites/www.acectx.org/resource/resmgr/Learning_Center/ Educating_The_Engineer_of_20.pdf, 2005)
2. C. T., Kennedy, The Bridge, **36** (2006)
3. T., Grover, C., Groscurth, Occasional Papers, Center for Research on Learning and Teaching, University of Michigan, **26** (2009)
4. M., Prensky, On the Horizon, **9 (5)** (2001)
5. I. M. Rotaru, S. Borza, The 6th Balkan Region Conference on Engineering and Business Education & The 5th International Conference on Engineering and Business Education & The 4th International Conference on Innovation and Entrepreneurship, Sibiu, Romania, **1** (2011)
6. R. Budynas, K. Nisbett, *Mechanichal engineering design,* (Publisher McGraw-Hill Education, New-York, 2015)
7. S.Krar, A. Gill, P. Smid, *Workbook for Technology of Machine Tools,* (Publisher McGraw-Hill Education, New-York, 2011)
8. S. Kalpakian, S. Schmid, *Manufacturing Engineering and Technology,* (Publisher Prentice Hall, New-York, 2010)
9. D.M. Duse, O. Bologa, *Tehnologii de prelucrare tipizate,* (Publisher, Editura Universitatii Lucian Blaga din Sibiu, Sibiu, 1995)
10. P. N. Rao, *Manufacturing Technology,* (Publisher McGraw-Hill Education,India, 2013)
11. G. E. Dieter, L.C. Schmidt, *Engineering Design,* (Publisher McGraw-Hill Education, New-York, 2009)

Knowledge and usage of methods of online promotion by students as potential entrepreneurs

Paulina Siemieniak[1,*], and *Jakub* Pawlak[1]

[1]Poznań University of Technology, Faculty of Engineering Management, ul. Strzelecka 11, 60-965 Poznań, Poland

Abstract. According to popular belief, it has never been easier to raise funds to run and manage one's own business. In an era of startups popping up like mushrooms, crowdfunding, business angels, venture capital and mobile Internet, investing in a business is a much easier undertaking than it was two decades ago. Furthermore, Augmented Reality available through mobile devices like smartphones or tablets, the Internet of Things, which makes the consumer feel like a science fiction character in their own home, or Virtual Reality shopping, a Matrix movie-like experience in a testing phase at the moment, provide increasingly more innovative ways to take one's first steps in business, gather momentum, and win loyal customers. According to the results, running one's own business is not a popular professional plan among young people, which is especially surprising given the fact that the majority of the respondents were students of specialisations that equipped them with knowledge facilitating entrepreneurial activities. Running one's own business seemed to be an attractive career path for about 30% of the respondents. Although they realise the significance of the Internet in running their own business, they do not believe that this is always a necessary tool. The students declared knowledge of multiple methods of communication and promotion and often use social media channels, but rarely expressed the desire to use these measures. A particular way of using the Internet to foster entrepreneurship are methods of online promotion like Social Media Marketing or Search Engine Optimization. Those tools enable reaching potential customers and engaging in an active dialogue with them about their favourite brands, products and services. They also allow exchange of information with the entrepreneur and are currently one of the fundamental tools for creating a company's success. The article presents the results of research on the knowledge of methods of online promotion and networking by students as potential entrepreneurs.

* Corresponding author: paulina.siemieniak@put.poznan.pl

1 Introduction

According to popular belief, it has never been easier to raise funds to run and manage one's own business. In an era of startups popping up like mushrooms, crowdfunding, business angels, venture capital and mobile Internet, running a business seems to be a much easier undertaking than it was two decades ago [1-2]. Furthermore, Augmented Reality available through mobile devices like smartphones or tablets, the Internet of Things, which makes the consumer feel like a science fiction character in their own home, or Virtual Reality shopping, a Matrix movie-like experience in a testing phase at the moment, provide increasingly more innovative ways to take one's first steps in business, gather momentum and win loyal customers [3]. The usage of networking apps, such as LinkedIn, and all the various ways of promotion through Online Marketing should make it easy for beginning entrepreneurs – at least in theory.

2 The study

A study was conducted in February 2017 among 166 students from three universities in Wielkopolska. In that group, 65 people studied Economics, 17 – Education in Technology and Informatics, 30 – Logistics, 16 – Tourism and Recreation, 34 – Law in Business, and four – Management. Most of the respondents were women, representing 67.5% of the total. Male students accounted for 32.5% of all respondents. The students usually lived in the countryside or in large cities (cf. Table. 1). The vast majority (88.5%) were people aged between 18 and 25, and only 11% of the respondents were over 26 years old.

Table 1. Place of residence of the students surveyed

Place of residence:	**Quantity:**	**Percent:**
Village	70	42.17
City up to 10K inhabitants	19	11.45
City from 10K to 50K	24	14.46
City from 50K to 200K	22	13.25
City above 200K	31	18.67
Total:	166	100

Source: Own research.

Respondents were also asked about their current job situation and plans for the future. The largest group were people who were not sure yet if they wanted to set up their own business. It is worth noting that they were mainly students who were not employed at the time and, therefore, did not have the opportunity to gain any work experience, which may somewhat restrict the choices related to their career [4-6].

Respondents who intended to start their own business accounted for about 32%, and those who were entrepreneurs at the time represented about 7% (cf. Table 2).

Table 2. Current employment situation and future plans of the students surveyed

Professional situation:	Quantity:	Percent:
Currently, I am not working professionally and I do not know if I want to start my own business.	48	29.8
I am not working professionally, but I want to start my own company	35	21.73
I am not working professionally and do not want to start my own business.	10	6.21
I am working professionally, but I want to start my own company.	18	11.2

I am currently working and I do not know if I want to start a business.	38	23.60
I am running my own company and I own it.	10	6.21
I am running my own company and I am a co-owner.	2	1.24
Total:	161	100

Source: Own research.

The vast majority of the respondents could not imagine running their own business without a website (89.87%), but there were also those who did not exclude this possibility (10.13%). The latter supported their conviction that they "could imagine conducting a business without a website" with the following arguments:

- "Not every business needs a website."
- "Cooperation with only one entity, no need to acquire new contacts."
- "This industry does not require me to provide a website."
- "The industry in which I plan to be active is based on direct contact with the customer. A website would have a negligible impact on work performance."
- "The industry (trade – a grocery store) does not need online advertising, in my opinion."
- "There are other forms of advertising."
- "Because I'd like to acquire customers only from the immediate neighbourhood."
- "Certain companies do not need one."
- "I have a fixed number of customers, and I cannot take more at the moment, so I don't need additional forms of promotion."
- "It is possible without a website if you have a good location and well-developed marketing."

3 Knowledge on gaining professional contacts and online promotion

Respondents were asked about their knowledge of crowdfunding internet portals that could potentially help them to raise money for their starting capital. More than half of the students knew the Polish digital platform *polakpotrafi.pl* only. Other, especially foreign ones did not seem to be popular among them (cf. Table 3).

Table 3. Knowledge of internet portals according to students' declarations

Crowdfunding portal	**kickstarter.com**	**indiegogo.com**	**polakpotrafi.pl**	**wspieram.to**
Known (%)	20.48	14.46	55.42	28.31
Not known (%)	79.52	85.54	44.58	71.69
Total	166	166	166	166

Source: Own research.

Networking services like LinkedIn and its Polish counterpart, GoldenLine, are popular tools for networking – gaining new business contacts, maintain relationships with people from one's industry, and taking part in discussions in topic-related groups.

The study showed that networking websites did not facilitate the respondents in obtaining professional contacts. It should be noted, however, that perhaps only a small number of them (4 to 5%) were actively looking for such contacts at the time, while at university (Table 4). This presumption is supported by the fact that only about 30% of the respondents were expecting to start their own business. Nevertheless, it was interesting that all students who had a GoldenLine account had a LinkedIn account too.

Table 4. Knowledge of internet portals according to students' declarations

Question:	Are you aware of the existence of the Goldenline.pl social network?	Do you have an account on the Goldenline.pl social network?	Have you acquired at least one contact on Goldenline.pl	Are you aware of the existence of the LinkedIn.com social network?	Do you have an account on the LinkedIn.com social network?	Have you acquired at least one contact on LinkedIn.com	Are you aware of the existence of the Xing social network?	Do you have an account on the Xing social network?	Have you acquired at least one contact on Xing?
Yes (%):	76.51	11.44	4.22	64.46	11.44	5.42	27.71	01.81	0
No (%):	23.49	88.55	95.78	35.54	88.55	94.58	72.29	98.192	100
Total:	166	166	166	166	166	166	166	166	166

Source: own research

A wide variety of web promotion methods and tools has evolved along with technological advances over the last three decades. Their selection and scope of use in shaping one's promotion on the Internet depends on the knowledge of the market and the target consumer group, the activities of competitors, and the marketing strategy developed.

Knowledge of the methods of communication and promotion on the Internet among the respondents is shown in the table below (Table 5).

Table 5. Students' knowledge of methods of communication and promotion on the Internet

Method	**Yes (in %)**	**No (in %)**	**Total**
Social Media Marketing (published content)	59.15	40.85	164
Social Media Marketing (paid advertising)	57.66	42.33	163
Blogs	65.85	34.15	164
Display advertising (no PPC such as AdWords)	39.27	60.73	163
PPC Advertising (e.g. AdWords)	30.06	69.94	163
SEO	27.00	73.00	163
Affiliate Marketing	29.27	70.73	164
Couponing (e.g. Groupon)	40.50	59.50	163

Source: Own research.

The respondents were mostly familiar with Social Media Marketing (59.15%), including Blogs (65.85%), while Search Engine Optimization was known to the smallest group (27%) [7-8]. It is worth noting that the respondents had generally good knowledge of those methods of online promotion. Even those scoring the lowest were still known to approximately 30% of the students. On the other hand, it was surprising that those young people did not realise the advantages of using Social Media for business promotion (40.85%). Only six respondents used Social Media for business purposes in the past. It is worth remembering, however, that only 12 people among the respondents had their own company, and more than half of them (53%) was professionally inactive.

Table 6. Usage of social media by respondents

Social Media channel	Currently using	Will use for business
Facebook	74.70	29.52
Twitter	10.84	5.42
YouTube	53.61	10.84
Google+	39.76	10.84
Instagram	38.55	16.87
nk.pl	4.81	1.81
Snapchat	30.72	6.02
Hangout	3.61	3.61

Source: Own research.

4 Summary

In summary, running one's own business is not a popular professional plan among young people, which is especially surprising given the fact that the majority of the respondents were students of specialisations that equipped them with knowledge facilitating entrepreneurial activities.

Running one's own business seemed to be an attractive career path for about 30% of the respondents. Although they realise the significance of the Internet in running their own business, they do not believe that this is always a necessary tool.

The students declared knowledge of multiple methods of communication and promotion and often use social media channels, but rarely expressed the desire to use these measures. It can, therefore, be assumed that they will not make full use of the opportunities offered by that medium, which will most probably limit the development of their professional activities in the future.

References

1. M. Rembiasz, P. Siemienak, J. Pawlak, *Internet as a tool to promote women's entrepreneurship. Selected aspects, Research Reports* (to be published) (2016)
2. M. Rembiasz, Zeszyty Naukowe Politechniki Poznańskiej. Organizacja i Zarządzanie 68, 167-179 (2016)
3. J. Pawlak, J., Przybysz, *Technologie ICT w kreowaniu innowacyjności MSP*, in: Wybrane problemy zarządzania. Teraźniejszość i przyszłość, S. Trzcieliński (ed.), Wydawnictwo Politechniki Poznańskiej (2013)
4. P. Siemienak, *Metody wspierania przedsiębiorczości kobiet*, Praca doktorska. Politechnika Poznańska (2014)
5. P. Siemieniak, T. Łuczka, *Przedsiębiorczość kobiet. Wybrane aspekty ekonomiczne i psycho-kulturowe,* Poznań, Wydawnictwo Politechniki Poznańskiej (2016)
6. T. Łuczka, M. Rembiasz, Horyzonty Wychowania, **15** (34), 27-47 (2016)
7. J. Pawlak, *Promocja internetowa mikro i małego przedsiębiorstwa w portalu społecznościowym Facebook,* (in:) Uwarunkowania rynkowe rozwoju mikro, małych i średnich przedsiębiorstw. Mikrofirma 2013, A. Bielawska (ed.), Uniwersytet Szczeciński (2013)
8. B. A. Nardi, D. J. Schiano, M. Gumbrecht, Proceeding CSCW '04 Proceedings of the 2004 ACM conference on Computer supported cooperative work, ACM, New York (2004)

AUTHOR INDEX

AUTHOR INDEX

AUTHOR INDEX

AUTHOR INDEX

EDP Sciences – Web of Conferences
17, Avenue du Hoggar
Parc d'Activité de Courtabouf
BP 112
F-91944 Les Ulis Cedex A
France

ISBN 978-1-5108-4705-7

9 781510 847057

CURRAN ASSOCIATES INC.
proceedings.com